Regulation of Organelle and Cell Compartment Signaling

Regulation of Organelle and Cell Compartment Signaling

Editors-in-Chief

Ralph A. Bradshaw
Department of Pharmaceutical Chemistry,
University of California, San Francisco,
San Francisco, California

Edward A. Dennis
Department of Chemistry and Biochemistry,
Department of Pharmacology, School of Medicine,
University of California, San Diego,
La Jolla, California

AMSTERDAM • BOSTON • HEIDELBERG • LONDON • NEW YORK • OXFORD
PARIS • SAN DIEGO • SAN FRANCISCO • SINGAPORE • SYDNEY • TOKYO

Academic Press is an imprint of Elsevier

Academic Press is an imprint of Elsevier
525 B Street, Suite 1900, San Diego, CA 92101-4495, USA
30 Corporate Drive, Suite 400, Burlington, MA 01803, USA
32 Jamestown Road, London NW1 7BY, UK
360 Park Avenue South, New York, NY 10010-1710, USA

First edition 2011

Material in the work originally appeared in Handbook of Cell Signaling, Second Edition, edited by Ralph A. Bradshaw and Edward A. Dennis (Elsevier, Inc. 2010)

Library of Congress Cataloging in Publication Data
Regulation of organelle and cell compartment / editors-in-chief, Ralph A. Bradshaw, Edward A. Dennis. – 1st ed.
p. ; cm.
Summary: "Cell signaling, which is also often referred to as signal transduction or, in more specialized cases, transmembrane signaling, is the process by which cells communicate with their environment and respond temporally to external cues that they sense there. All cells have the capacity to achieve this to some degree, albeit with a wide variation in purpose, mechanism, and response. At the same time, there is a remarkable degree of similarity over quite a range of species, particularly in the eukaryotic kingdom, and comparative physiology has been a useful tool in the development of this field. The central importance of this general phenomenon (sensing of external stimuli by cells) has been appreciated for a long time, but it has truly become a dominant part of cell and molecular biology research in the past three decades, in part because a description of the dynamic responses of cells to external stimuli is, in essence, a description of the life process itself. This approach lies at the core of the developing fields of proteomics and metabolomics, and its importance to human and animal health is already plainly evident"–Provided by publisher.
Includes bibliographical references and index.
ISBN 978-0-12-382213-0 (alk. paper)
1. Cellular signal transduction. 2. Cell organelles. 3. Transcription factors. I. Bradshaw, Ralph A., 1941- II. Dennis, Edward A.
[DNLM: 1. Signal Transduction. 2. Cell Cycle Proteins. 3. Gene Expression Regulation. 4. Organelles. 5. Transcription Factors. QU 375]
QP517.C45R45 2011
571.6'5–dc22
2011001762

British Library Cataloging in Publication Data
A catalog record for this book is available from the British Library

ISBN : 978-0-12-382213-0

For information on all Academic Press publications
visit our website at www.elsevierdirect.com

Printed and bound by CPI Group (UK) Ltd, Croydon, CR0 4YY

Editorial Advisory Board

Contents

Preface

Since cell signaling is a major area of biomedical/biological research and continues to advance at a very rapid pace, scientists at all levels, including researchers, teachers, and advanced students, need to stay current with the latest findings, yet maintain a solid foundation and knowledge of the important developments that underpin the field. Carefully selected articles from the 2nd edition of the *Handbook of Cell Signaling* offer the reader numerous, up-to-date views of intracellular signal processing, including membrane receptors, signal transduction mechanisms, the modulation of gene expression/translation, and cellular/organotypic signal responses in both normal and disease states. In addition to material focusing on recent advances, hallmark papers from historical to cutting-edge publications are cited. These references, included in each article, allow the reader a quick navigation route to the major papers in virtually all areas of cell signaling to further enhance his/her expertise.

The Cell Signaling Collection consists of four independent volumes that focus on *Functioning of Transmembrane Receptors in Cell Signaling, Transduction Mechanisms in Cellular Signaling, Regulation of Organelle and Cell Compartment Signaling,* and *Intercellular Signaling in Development and Disease*. They can be used alone, in various combinations or as a set. In each case, an overview article, adapted from our introductory chapter for the Handbook, has been included. These articles, as they appear in each volume, are deliberately overlapping and provide both historical perspectives and brief summaries of the material in the volume in which they are found. These summary sections are not exhaustively referenced since the material to which they refer is.

The individual volumes should appeal to a wide array of researchers interested in the structural biology, biochemistry, molecular biology, pharmacology, and pathophysiology of cellular effectors. This is the ideal go-to books for individuals at every level looking for a quick reference on key aspects of cell signaling or a means for initiating a more in-depth search. Written by authoritative experts in the field, these papers were chosen by the editors as the most important articles for making the Cell Signaling Collection an easy-to-use reference and teaching tool. It should be noted that these volumes focus mainly on higher organisms, a compromise engendered by space limitations.

We wish to thank our Editorial Advisory Committee consisting of the editors of the Handbook of Cell Signaling, 2nd edition, including Marilyn Farquhar, Tony Hunter, Michael Karin, Murray Korc, Suresh Subramani, Brad Thompson, and Jim Wells, for their advice and consultation on the composition of these volumes. Most importantly, we gratefully acknowledge all of the individual authors of the articles taken from the Handbook of Cell Signaling, who are the 'experts' upon which the credibility of this more focused book rests.

Ralph A. Bradshaw, San Francisco, California
Edward A. Dennis, La Jolla, California
January, 2011

Contributors

Géza Ambrus (2), Department of Cell Biology, Scripps Research Institute, La Jolla, California

Sally A. Amundson (20), Center for Radiological Research, Columbia University Medical Center, New York

Carl W. Anderson (24), Biology Department, Brookhaven National Laboratory, Upton, New York

Peter Angel (7), Deutsches Krebsforschungszentrum, Division of Signal Transduction and Growth Control, Heidelberg

Ettore Appella (24), Laboratory of Cell Biology, National Cancer Institute, National Institutes of Health, Bethesda, Maryland

Juan Ausió (17), Department of Biochemistry and Microbiology, University of Victoria, Victoria, British Columbia, Canada

Brandon J. Baird (31), Laboratory of Molecular Pharmacology, Center for Cancer Research, National Cancer Institute, National Institutes of Health, Bethesda, Maryland

Dafna Bar-Sagi (50), Department of Molecular Genetics and Microbiology, State University of New York at Stony Brook, Stony Brook, New York

Alicia A. Bicknell (39), University of California San Diego, Division of Biological Sciences, La Jolla, California

William M. Bonner (31), Laboratory of Molecular Pharmacology, Center for Cancer Research, National Cancer Institute, National Institutes of Health, Bethesda, Maryland

Ralph A. Bradshaw (1), Department of Pharmaceutical Chemistry, University of California, San Francisco, CA

Paul K. Brindle (10), Department of Biochemistry, St. Jude Children's Research Hospital, Memphis, Tennessee

Michael S. Brown (37), Department of Molecular Genetics, University of Texas Southwestern Medical Center at Dallas, Dallas, Texas

Anne Brunet (6), Division of Neuroscience, Children's Hospital and Department of Neurobiology, Harvard Medical School, Boston, Massachusetts

Dmitry V. Bulavin (29), Institute of Molecular and Cell Biology, Proteos, Singapore

Tara L. Burke (13), Department of Biochemistry and Molecular Genetics, University of Virginia School of Medicine, Charlottesville, Virginia

Denise A. Chan (27), Division of Radiation Biology, Department of Radiation Oncology, Stanford University School of Medicine, Stanford, California

Harry Charbonneau (52), Department of Biochemistry, Purdue University, West Lafayette, Indiana

Zhijian J. Chen (9), Department of Molecular Biology, Howard Hughes Medical Institute, University of Texas Southwestern Medical Center, Dallas, Texas

Philip Chen (23), Queensland Cancer Fund Research Laboratory, Queensland Institute of Medical Research, Brisbane, Australia

Chyi-Ying A. Chen (32), Department of Biochemistry and Molecular Biology, University of Texas, Medical School at Houston, Houston, Texas

Peter Cheung (16), Department of Medical Biophysics, University of Toronto, and Division of Signaling Biology, Ontario Cancer Institute, Toronto, Ontario, Canada

Aaron Ciechanover (38), Vascular and Tumor Biology Research Center, Bruce Rappaport Faculty of Medicine, Technion-Israel Institute of Technology, Haifa, Israel

Patrice Codogno (48), INSERM U756, Faculté de Pharmacie, Université Paris-Sud 11, Châtenay-Malabry, France

Peter J. Cook (4), Howard Hughes Medical Institute, Department and School of Medicine, University of California, San Diego, La Jolla, California

Fernando Cruz-Guilloty (11), Department of Pathology, Harvard Medical School and the Immune Disease Institute, Boston, Massachusetts

Xiaoping Cui (22), University of Southern California Keck School of Medicine, USC Norris Comprehensive Cancer Center, Los Angeles, California

Bruce Demple (21), Department of Pharmacological Sciences, Stony Brook University Medical Center Stony Brook, NY 11794, Department of Genetics and Complex Diseases, Harvard School of Public Health, Boston, Massachusetts

Edward A. Dennis (1), Department of Chemistry and Biochemistry and Department of Pharmacology, School of Medicine, University of California, San Diego, La Jolla, CA, USA

Jennifer S. Dickey (31), Laboratory of Molecular Pharmacology, Center for Cancer Research, National Cancer Institute, National Institutes of Health, Bethesda, Maryland

Ben Distel (41), Department of Medical Biochemistry, Academic Medical Center, University of Amsterdam, Amsterdam, The Netherlands

Ryan J.O. Dowling (35), Department of Biochemistry, Rosalind and Morris Goodman Cancer Centre, McGill University, Montreal, Quebec, Canada

Michael R. Duchen (45), Department of Physiology and UCL Mitochondrial Biology Group, University College London, England, UK

Peter J. Espenshade (37), Department of Cell Biology, Johns Hopkins University School of Medicine, Baltimore, Maryland

Andrei D. Fagarasanu (46), Department of Cell Biology, University of Alberta, Edmonton, Alberta, Canada

Pier Paolo Di Fiore (47), IFOM, Fondazione Istituto FIRC di Oncologia Molecolare, Milan, Italy, Dipartimento di Medicina, Chirurgia ed Odontoiatria, Universita' degli Studi di Milano, Milan, Italy, Dispartimento di Oncologia Sperimentale, Istituto Europeo di Oncologia, Milan, Italy

Albert J. Fornace Jr. (20), John B. Little Center for the Radiation Sciences and Environmental Health, Harvard School of Public Health, Boston, Massachusetts, Lombardi Comprehensive Cancer Center, and Department of Biochemistry and Molecular and Cellular Biology, Georgetown University, Washington, DC

Magtouf Gatei (23), Queensland Cancer Fund Research Laboratory, Queensland Institute of Medical Research, Brisbane, Australia

Larry Gerace (2), Department of Cell Biology, Scripps Research Institute, La Jolla, California

Amato J. Giaccia (27), Division of Radiation Biology, Department of Radiation Oncology, Stanford University School of Medicine, Stanford, California

Vincent Giguère (5), The Rosalind and Morris Goodman Cancer Centre, Faculty of Medicine, McGill University, Montréal, Québec, Canada

Christopher K. Glass (3), Department of Cellular and Molecular Medicine, School of Medicine, University of California, San Diego, La Jolla, California

Joseph L. Goldstein (37), Department of Molecular Genetics, University of Texas Southwestern Medical Center at Dallas, Dallas, Texas

Myriam Gorospe (28), Laboratory of Cellular and Molecular Biology, National Institute on Aging-IRP, Baltimore, Maryland

Patrick A. Grant (13), Department of Biochemistry and Molecular Genetics, University of Virginia School of Medicine, Charlottesville, Virginia

Douglas R. Green (54), Division of Cellular Immunology, La Jolla Institute for Allergy and Immunology, San Diego, California

Michael E. Greenberg (6), Division of Neuroscience, Children's Hospital and Department of Neurobiology, Harvard Medical School, Boston, Massachusetts

Linda Hendershot (40), Department of Genetics and Tumor Cell Biology, St. Jude Children's Research Hospital, Memphis, Tennessee

Jochen Hess (7), Deutsches Krebsforschungszentrum, Division of Signal Transduction and Growth Control, Heidelberg

Daniel R. Hyduke (20), Department of Bioengineering, University of California-San Diego, La Jolla, California, John B. Little Center for the Radiation Sciences and Environmental Health, Harvard School of Public Health, Boston, Massachusetts

Miho Iijima (42), Department of Cell Biology, Johns Hopkins University School of Medicine, Baltimore, Maryland

Alberto Inga (25), Unit of Molecular Mutagenesis and DNA repair, Department of Epidemiology and Prevention, National Institute for Cancer Research, IST, Genoa, Italy

Toyotaka Ishibashi (17), Department of Biochemistry and Microbiology, University of Victoria, Victoria, British Columbia, Canada

Jennifer J. Jordan (25), Laboratory of Molecular Genetics, National Institute of Environmental Health Sciences, NIH, Research Triangle Park, North Carolina, Curriculum in Genetics and Molecular Biology, University of North Carolina, Chapel Hill, North Carolina

Amanda Kijas (23), Queensland Cancer Fund Research Laboratory, Queensland Institute of Medical Research, Brisbane, Australia

Jong Heon Kim (33), Program in Molecular Medicine, University of Massachusetts Medical School, Worcester, Massachusetts, Research Institute, National Cancer Center, Goyang, Gyeonggi, Korea

Albert C. Koong (27), Division of Radiation Biology, Department of Radiation Oncology, Stanford University School of Medicine, Stanford, California

Daniel Kornitzer (38), Department of Molecular Microbiology, Bruce Rappaport Faculty of Medicine, Technion-Israel Institute of Technology, Haifa, Israel

Sergei Kozlov (23), Queensland Cancer Fund Research Laboratory, Queensland Institute of Medical Research, Brisbane, Australia

Gary M. Kupfer (30), Departments of Pediatrics and Pathology, Yale University School of Medicine, New Haven, Connecticut

Priscilla Nga Ieng Lau (16), Department of Medical Biophysics, University of Toronto, and Division of Signaling Biology, Ontario Cancer Institute, Toronto, Ontario, Canada

Martin F. Lavin (23), Queensland Cancer Fund Research Laboratory, Queensland Institute of Medical Research, Brisbane, Australia, Department of Surgery, University of Queensland, Brisbane, Australia

Andra Li (17), Department of Biochemistry and Microbiology, University of Victoria, Victoria, British Columbia, Canada

Xialu Li (51), National Institute of Biological Sciences, Beijing, China

Michael R. Lieber (22), University of Southern California Keck School of Medicine, USC Norris Comprehensive Cancer Center, Los Angeles, California

Fernando Macian (11), Department of Pathology, Albert Einstein College of Medicine, Bronx, New York

James L. Manley (51), Department of Biological Sciences, Columbia University, New York

Clare H. McGowan (49), Department of Molecular Biology, Department of Cell Biology, Scripps Research Institute, La Jolla, California

Alfred J. Meijer (48), Department of Medical Biochemistry, Academic Medical Center, Amsterdam, The Netherlands

Daniel Menendez (25), Laboratory of Molecular Genetics, National Institute of Environmental Health Sciences, NIH, Research Triangle Park, North Carolina

Frank Mercurio (8), Signal Research Division, Celgene, San Diego, California

Richard I. Morimoto (26), Department of Biochemistry, Molecular Biology and Cell Biology, Rice Institute for Biomedical Research, Northwestern University, Evanston, Illinois

Thomas D. Mullen (55), Department of Medicine, Division of General Internal Medicine and Geriatrics, Medical University of South Carolina, Charleston, South Carolina

Asako J. Nakamura (31), Laboratory of Molecular Pharmacology, Center for Cancer Research, National Cancer Institute, National Institutes of Health, Bethesda, Maryland

Gioacchino Natoli (19), Department of Experimental Oncology, European Institute of Oncology (IEO), Milan, Italy

Maho Niwa (39), University of California San Diego, Division of Biological Sciences, La Jolla, California

Lina M. Obeid (55), Ralph H. Johnson Veterans Administration Hospital, Charleston, South Carolina, Department of Medicine, Division of General Internal Medicine and Geriatrics, Medical University of South Carolina, Charleston, South Carolina

Yuki Okuda-Shimizu (40), Department of Genetics and Tumor Cell Biology, St. Jude Children's Research Hospital, Memphis, Tennessee

Lisa J. Pagliari (54), Division of Cellular Immunology, La Jolla Institute for Allergy and Immunology, San Diego, California

Michael J. Pinkoski (54), Division of Cellular Immunology, La Jolla Institute for Allergy and Immunology, San Diego, California

Gratien G. Prefontaine (4), Howard Hughes Medical Institute, Department and School of Medicine, University of California, San Diego, La Jolla, California

Richard A. Rachubinski (46), Department of Cell Biology, University of Alberta, Edmonton, Alberta, Canada

Arun Radhakrishnan (37), Department of Molecular Genetics, University of Texas Southwestern Medical Center at Dallas, Dallas, Texas

Anjana Rao (11), Department of Pathology, Harvard Medical School and the Immune Disease Institute, Boston, Massachusetts

Christophe E. Redon (31), Laboratory of Molecular Pharmacology, Center for Cancer Research, National Cancer Institute, National Institutes of Health, Bethesda, Maryland

Michael A. Resnick (25), Laboratory of Molecular Genetics, National Institute of Environmental Health Sciences, NIH, Research Triangle Park, North Carolina

Joel D. Richter (33), Program in Molecular Medicine, University of Massachusetts Medical School, Worcester, Massachusetts

Michael G. Rosenfeld (4), Howard Hughes Medical Institute, Department and School of Medicine, University of California, San Diego, La Jolla, California

Guy S. Salvesen (53), Program in Apoptosis and Cell Death Research, Burnham Institute, San Diego, California

Veronica De Sanctis (25), Centre for Integrative Biology, CIBIO, University of Trento, Italy

Immo E. Scheffler (43), Division of Biology (Molecular Biology Section), University of California, San Diego, La Jolla, California

Christian Schindler (12), Department of Microbiology and Medicine, College of Physicians and Surgeons, Columbia University, New York

Klaus Schwamborn (8), Signal Research Division, Celgene, San Diego, California

Giorgio Scita (47), IFOM, Fondazione Istituto FIRC di Oncologia Molecolare, Milan, Italy, Dipartimento di Medicina, Chirurgia ed Odontoiatria, Universita' degli Studi di Milano, Milan, Italy

Jennifer Scorah (49), Department of Molecular Biology, Scripps Research Institute, La Jolla, California

Olga A. Sedelnikova (31), Laboratory of Molecular Pharmacology, Center for Cancer Research, National Cancer Institute, National Institutes of Health, Bethesda, Maryland

Anand Selvaraj (34), Department of Cancer and Cell Biology, Genome Research Institute, University of Cincinnati, College of Medicine, Cincinnati, Ohio

Hiromi Sesaki (42), Department of Cell Biology, Johns Hopkins University School of Medicine, Baltimore, Maryland

Edward Seto (14), Molecular Oncology Program, H. Lee Moffitt Cancer Center and Research Institute, Tampa, Florida

Sonia Sharma (11), Department of Pathology, Harvard Medical School and the Immune Disease Institute, Boston, Massachusetts

Ying Shen (40), Department of Genetics and Tumor Cell Biology, St. Jude Children's Research Hospital, Memphis, Tennessee

Ann-Bin Shyu (32), Department of Biochemistry and Molecular Biology, University of Texas, Medical School at Houston, Houston, Texas

Leah J. Siskind (55), Ralph H. Johnson Veterans Administration Hospital, Charleston, South Carolina, Department of Medicine, Division of General Internal Medicine and Geriatrics, Medical University of South Carolina, Charleston, South Carolina

Nahum Sonenberg (35), Department of Biochemistry, Rosalind and Morris Goodman Cancer Centre, McGill University, Montreal, Quebec, Canada

Li Song (12), Department of Microbiology, College of Physicians and Surgeons, Columbia University, New York

Subramanya Srikantan (28), Laboratory of Cellular and Molecular Biology, National Institute on Aging-IRP, Baltimore, Maryland

Joshua D. Stender (3), Department of Cellular and Molecular Medicine, School of Medicine, University of California, San Diego, La Jolla, California

Li-Ping Sun (37), Department of Molecular Genetics, University of Texas Southwestern Medical Center at Dallas, Dallas, Texas

Carolyn K. Suzuki (44), Department of Biochemistry and Molecular Biology, University of Medicine and Dentistry of New Jersey, Newark, New Jersey

György Szabadkai (45), Department of Physiology and UCL Mitochondrial Biology Group, University College London, England, UK

Yasushi Tamura (42), Department of Cell Biology, Johns Hopkins University School of Medicine, Baltimore, Maryland

Laura J. Taylor (50), Department of Molecular Genetics and Microbiology, State University of New York at Stony Brook, Stony Brook, New York

George Thomas (34), Department of Cancer and Cell Biology, Genome Research Institute, University of Cincinnati, College of Medicine, Cincinnati, Ohio

Hien Tran (6), Division of Neuroscience, Children's Hospital and Department of Neurobiology, Harvard Medical School, Boston, Massachusetts

Jean Y. J. Wang (36), Moores Cancer Center, Division of Hematology-Oncology, Department of Medicine, University of California at San Diego, La Jolla, California

Vikki M. Weake (18), Stowers Institute for Medical Research, Kansas City, Missouri

Sandy D. Westerheide (26), Department of Biochemistry, Molecular Biology and Cell Biology, Rice Institute for Biomedical Research, Northwestern University, Evanston, Illinois

John K. Westwick (8), Signal Research Division, Celgene, San Diego, California

Johnathan R. Whetstine (15), Harvard Medical School and Massachusetts General Hospital Cancer Center, Charlestown, Massachusetts

Chris Williams (41), Department of Medical Biochemistry, Academic Medical Center, University of Amsterdam, Amsterdam, The Netherlands

Stacy A. Williams (30), Departments of Pediatrics and Pathology, Yale University School of Medicine, New Haven, Connecticut

Jerry L. Workman (18), Stowers Institute for Medical Research, Kansas City, Missouri

Ming Xu (9), Department of Molecular Biology, University of Texas, Southwestern Medical Center, Dallas, Texas

Xiang-Jiao Yang (14), Molecular Oncology Group, Department of Medicine, McGill University Health Center and McGill Cancer Center, McGill University, Montréal, Québec, Canada

Section A

Overview

Chapter 1

Organelle Signaling*

Ralph A. Bradshaw[1] and Edward A. Dennis[2]

[1]Department of Pharmaceutical Chemistry, University of California, San Francisco, CA

[2]Department of Chemistry and Biochemistry and Department of Pharmacology, School of Medicine, University of California, San Diego, La Jolla, CA

Cell signaling, which is also often referred to as signal transduction or, in more specialized cases, transmembrane signaling, is the process by which cells communicate with their environment and respond temporally to external cues that they sense there. All cells have the capacity to achieve this to some degree, albeit with a wide variation in purpose, mechanism, and response. At the same time, there is a remarkable degree of similarity over quite a range of species, particularly in the eukaryotic kingdom, and comparative physiology has been a useful tool in the development of this field. The central importance of this general phenomenon (sensing of external stimuli by cells) has been appreciated for a long time, but it has truly become a dominant part of cell and molecular biology research in the past three decades, in part because a description of the dynamic responses of cells to external stimuli is, in essence, a description of the life process itself. This approach lies at the core of the developing fields of proteomics and metabolomics, and its importance to human and animal health is already plainly evident.

ORIGINS OF CELL SIGNALING RESEARCH

Although cells from polycellular organisms derive substantial information from interactions with other cells and extracellular structural components, it was humoral components that first were appreciated to be intercellular messengers. This idea was certainly inherent in the 'internal secretions' initially described by Claude Bernard in 1855 and thereafter, as it became understood that ductless glands, such as the spleen, thyroid, and adrenals, secreted material into the bloodstream. However, Bernard did not directly identify hormones as such. This was left to Bayliss and Starling and their description of secretin in 1902 [1].

Recognizing that it was likely representative of a larger group of chemical messengers, the term *hormone* was introduced by Starling in a Croonian Lecture presented in 1905. The word, derived from the Greek word meaning 'to excite or arouse,' was apparently proposed by a colleague, W. B. Hardy, and was adopted, even though it did not particularly connote the messenger role but rather emphasized the positive effects exerted on target organs via cell signaling (see Wright [2] for a general description of these events). The realization that these substances could also produce inhibitory effects, gave rise to a second designation, 'chalones,' introduced by Schaefer in 1913 (see Schaefer [3]), for the inhibitory elements of these glandular secretions. The word 'autocoid' was similarly coined for the group as a whole (hormones and chalones). Although the designation chalone has occasionally been applied to some growth factors with respect to certain of their activities (e.g., transforming growth factor β), autocoid has essentially disappeared. Thus, if the description of secretin and the introduction of the term hormone are taken to mark the beginnings of molecular endocrinology and the eventual development of cell signaling, then we have passed the hundredth anniversary of this field.

The origins of endocrinology, as the study of the glands that elaborate hormones and the effect of these entities on target cells, naturally gave rise to a definition of hormones as substances produced in one tissue type that traveled systemically to another tissue type to exert a characteristic response. Of course, initially these responses were couched

* *Portions of this article were adapted from Bradshaw RA, Dennis EA: Cell signaling: yesterday, today, and tomorrow. In Bradshaw RA, Dennis EA, editors. Handbook of cell signaling. 2nd ed., New York: Academic Press; 2008; pp 1–4; Karin M. Introduction. In Bradshaw RA, Dennis EA, editors. Handbook of cell signaling. 1st ed., vol. 3. New York: Academic Press; 2003; pp 3–4.*

in organ and whole animal responses, although they increasingly were defined in terms of metabolic and other chemical changes at the cellular level. The early days of endocrinology were marked by many important discoveries, such as the discovery of insulin [4], to name one, that solidified the definition, and a well-established list of hormones, composed primarily of three chemical classes (polypeptides, steroids, and amino acid derivatives), was eventually developed. Of course, it was appreciated even early on that the responses in the different targets were not the same, particularly with respect to time. For example, adrenalin was known to act very rapidly, while growth hormone required a much longer time frame to exert its full range of effects. However, in the absence of any molecular details of mechanism, the emphasis remained on the distinct nature of the cells of origin versus those responding and on the systemic nature of transport, and this remained the case well into the 1970s. An important shift in endocrinological thinking had its seeds well before that, however, even though it took about 25 years for these 'new' ideas that greatly expanded endocrinology to be enunciated clearly.

Although the discovery of polypeptide growth factors as a new group of biological regulators is generally associated with nerve growth factor (NGF), it can certainly be argued that other members of this broad category were known before NGF. However, NGF was the source of the designation *growth factor* and has been, in many important respects, a Rosetta stone for establishing principles that are now known to underpin much of signal transduction. Thus, its role as the progenitor of the field and the entity that keyed the expansion of endocrinology, and with it the field of cell signaling, is quite appropriate. The discovery of NGF is well documented [5] and how this led directly to identification of epidermal growth factor (EGF) [6], another regulator that has been equally important in providing novel insights into cellular endocrinology, signal transduction and, more recently, molecular oncology. However, it was not till the sequences of NGF and EGF were determined [7, 8] that the molecular phase of growth factor research began in earnest. Of particular importance was the postulate that NGF and insulin were evolutionarily related entities [9], which suggested a similar molecular action (which, indeed, turned out to be remarkably clairvoyant), and was the first indication that the identified growth factors, which at that time were quite limited in number, were like hormones. This hypothesis led quickly to the identification of receptors for NGF on target neurons, using the tracer binding technology of the time (see Raffioni *et al.* [10] for a summary of these contributions), which further confirmed their hormonal status. Over the next several years, similar observations were recorded for a number of other growth factors, which in turn led to the redefinition of endocrine mechanisms to include paracrine, autocrine, and juxtacrine interactions [11]. These studies were followed by first isolation and molecular characterization using various biophysical methods and then cloning of their cDNAs, initially for the insulin and EGFR receptors [12–14] and then many others. Ultimately, the powerful techniques of molecular biology were applied to all aspects of cell signaling and are largely responsible for the detailed depictions we have today. They have allowed the broad understanding of the myriad of mechanisms and responses employed by cells to assess changes in their environment and to coordinate their functions to be compatible with the other parts of the organism of which they are a part.

RECEPTORS AND INTRACELLULAR SIGNALING

At the same time that the growth factor field was undergoing rapid development, major advances were also occurring in studies on hormonal mechanisms. In particular, Sutherland and colleagues [15] were redefining hormones as messengers and their ability to produce second messengers. This was, of course, based primarily on the identification of cyclic AMP (cAMP) and its production by a number of classical hormones. However, it also became clear that not all hormones produce this second messenger nor was it stimulated by any of the growth factors known at that time. This enigma remained unresolved for quite a long time until tyrosine kinases were identified [16, 17] and it was shown, first with the EGF receptor [18], that these modifications were responsible for initiating the signal transduction for many of those hormones and growth factors that did not stimulate the production of cAMP.

Aided by the tools of molecular biology, it was a fairly rapid transition to the cloning of most of the receptors for hormones and growth factors and the subsequent development of the main classes of signaling mechanisms. These data allowed the six major classes of cell surface receptors for hormones and growth factors to be defined, which included, in addition to the receptor tyrosine kinases (RTKs) described previously, the G-protein coupled receptors (GPCRs) (including the receptors that produce cAMP) that constitute the largest class of cell surface receptors; the cytokine receptors, which recruit the soluble JAK tyrosine kinases and directly activate the STAT family of transcription factors; serine/threonine kinase receptors of the TGFβ superfamily; the tumor necrosis factor (TNF) receptors that activate nuclear factor kappa B (NFκB) via TRAF molecules, among other pathways; and the guanylyl cyclase receptors. Structural biology has not maintained the same pace, and there are still both ligands and receptors for which we do not have three-dimensional information as yet.

In parallel with the development of our understanding of ligand/receptor organization at the plasma membrane, a variety of experimental approaches have also revealed the general mechanisms of transmembrane signal transduction in terms of the major intracellular events that are induced by these various receptor classes. There are three principal means by which intracellular signals are propagated: protein posttranslational modifications (PTMs), lipid messengers, and ion fluxes. There are also additional moieties that play significant roles, such as cyclic nucleotides, but their effects are generally manifested in downstream PTMs. There is considerable interplay between the three, particularly in the more complex pathways.

By far the most significant of the PTMs is phosphorylation of serine, threonine, and tyrosine residues. Indeed, there are over 500 protein kinases in the human genome with more than 100 phosphatases. Many of these modifications activate various enzymes, which are designated effectors, but it also has become increasingly clear that many PTM additions were inducing new, specific sites for protein–protein interactions. These 'docking sites' introduced the concept of both adaptors, such as Grb or Shc proteins, and the larger, multisite scaffolds, such as insulin receptor substrate (IRS) that bound to the sites introduced by the PTMs through specific motifs and as the process is repeated, successively built up multicomponent signaling structures [19]. There has now emerged a significant number of binding motifs, recognizing, in addition to PTMs, phospholipids and proline-rich peptide segments to name a few, that are quite widely scattered through the large repertoire of signaling molecules and that are activated by different types of receptors in a variety of cell types.

TRANSCRIPTIONAL RESPONSES

Although the intracellular signaling pathways are characterized by a plethora of modifications and interactions that alter existing proteomic and metabolomic landscapes, the major biological responses, such as mitosis, differentiation, and apoptosis, require alterations in the phenotypic profile of the cell and these need be directed by changes in transcription and translation (see Figure 1.1). Indeed, signaling can be thought of at two levels: responses (events) that affect (or require) preexisting structures (proteins) and those that depend on generating new proteins. Temporally, rapid responses are perforce of the first type, while longer-term responses generally are of the second. Thus, it may be viewed that the importance of the complex largely cytoplasmic machinery, involving receptors, effectors, adaptors and scaffolds, has two purposes: to generate immediate changes and then to ultimately reprogram the transcriptional activities for more permanent responses.

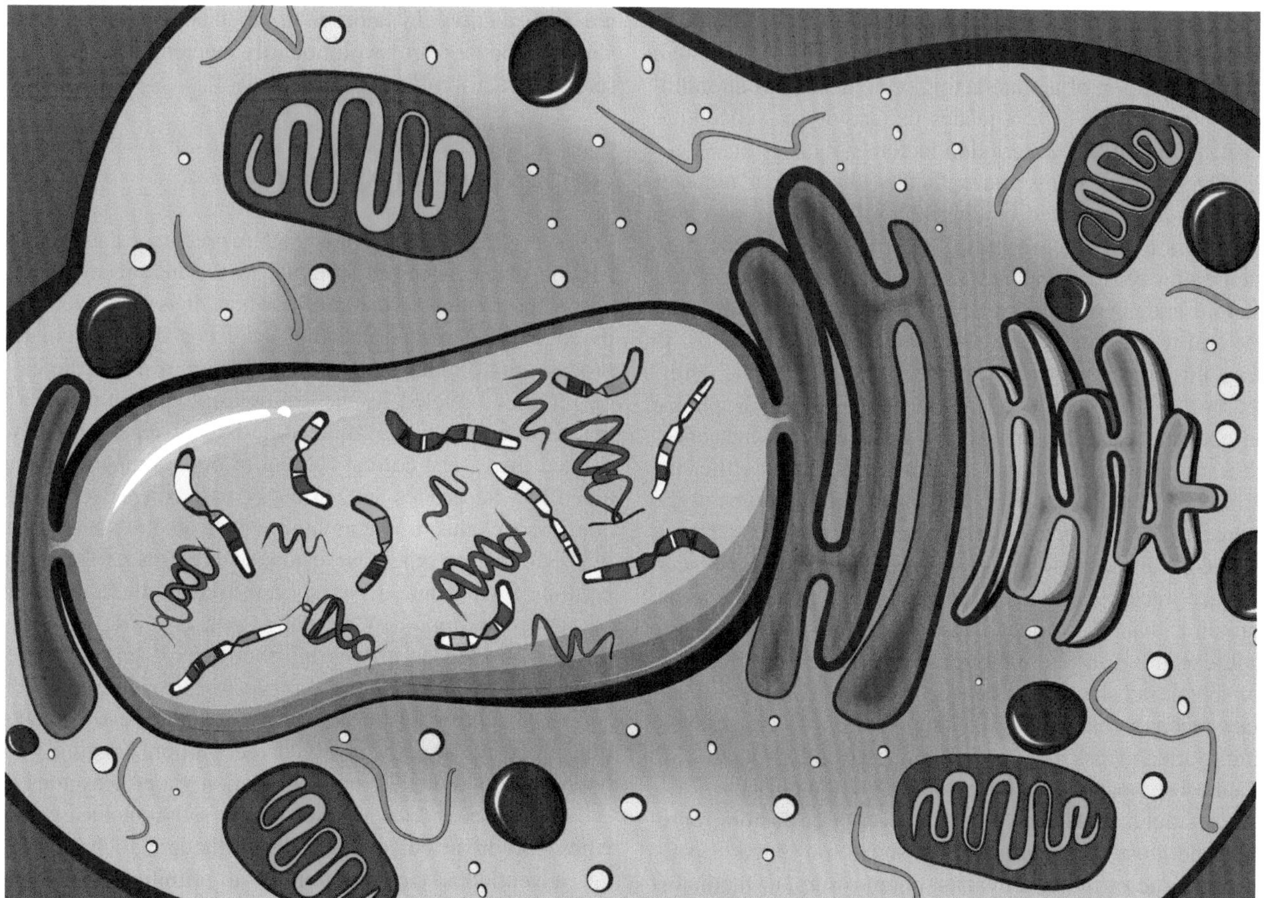

FIGURE 1.1 Subcellular organelles play critical roles in compartmentalizing signaling events. Of central importance are the numerous nuclear receptors/effectors and the subsequent regulation of transcriptional and translational processes. Signaling in various compartments and organelles, such as mitochondria, the Golgi, and the endoplasmic reticulum, as well as peroxisomes, lysosomes, and other vesicles, play critical roles in converting extracellular signals to meet specialized cellular requirements.

The process of gene expression in eukaryotes can be considered at several levels: the generation of the primary RNA transcript, its processing and transport, translation of the mRNA into protein, and finally its turnover. Since the amount of the potential activity associated with a given protein is fundamentally dependent on both its rate of synthesis and its rate of degradation, the turnover of the protein itself is also critical to signaling processes and is certainly largely, if not completely, affected by signaling events, too. In eukaryotes, transcription and mRNA processing take place in the nucleus; translation and mRNA turnover are cytoplasmic events. All of these processes are controlled or affected by signal transduction pathways.

The most common form of regulation is based on the phosphorylation of either sequence-specific transcription factors or proteins that directly interact with such transcription factors. These events can occur in the cytoplasm by kinases activated during signal transduction or by activated kinases that are transferred to the nuclear compartment. Thus, the phosphorylation event(s) can affect the subcellular distribution of the transcription factor (e.g., NFAT, NF-κB), that is, it is present in the cytoplasm and modification directs its nuclear transport, its ability to bind DNA, or its ability to activate or repress transcription (e.g., CREB, c-Jun). The regulation can be achieved through phosphorylation of the transcription factor itself or through phosphorylation of an interacting protein, such as an inhibitor (e.g., IκB), which regulates the activity or subcellular distribution of the transcription factor.

One class of transcription factors, the nuclear receptor family, requires ligand binding before they are functional. Members of this family form the core of signal transduction pathways that regulate gene expression in response to steroid and thyroid hormones, fatty acids, bile acids, cholesterol metabolites, and certain xenobiotic compounds. In fact, this can be viewed as an extension of lipid signaling, as most of the ligands for these receptors are hydrophobic in character. The ligands exert their affects through allosteric regulation, which has a dramatic effect on either the DNA binding or transcriptional activation properties of the transcription factor. Unlike the multicomponent pathways that control transcription in response to activation of cell surface receptors, nuclear receptors are multifunctional proteins that incorporate signal detection, amplification, and execution in one molecule. This branch of the family of signal transduction mechanisms does not utilize cell surface receptors but are activated by ligands that are passively transported across the plasma membrane and associate with their receptors either in the cytoplasm or the nucleus.

Although sequence-specific transcription factors represent the most common target for signal transduction pathways, some of the coactivators, corepressors, or mediators with which these factors interact, may also be subject to regulation. Coactivators, co-repressors, and mediators are often large multicomponent protein complexes that are recruited to promoters or enhancers through interactions with sequence-specific transcription factors. These protein complexes may act either through chromatin modifications or direct interactions with the RNA polymerase holoenzyme. In addition to modulation of chromatin structure via recruitment of chromatin modifiers to sequence-specific transcription factors, signal-responsive protein kinases may directly phosphorylate histones and regulate chromatin structure via a more direct route. Additional posttranslational modifications, such as the acetylation, methylation, and ubiquitinylation, that modify the N-terminal region of these nucleosome components and contribute to the 'histone code', are an essential part of the epigenetic mechanisms that also regulate gene expression, although the connection of these events, in terms of both modification and demodification, to transmembrane signaling has not yet been well defined.

The importance of transcriptional and posttranscriptional control of gene expression in adapting to adverse environmental conditions is underscored by the various stress responses that cells can undergo. Heat shock and UV and the different responses that are elicited by DNA damage, provide valuable insight relevant to transcriptional responses to many other aspects of cell regulation and signal transduction. In addition to metabolic control, these stress responses are evolutionarily ancient and are conserved in many eukaryotic orders.

ORGANELLE SIGNALING

Following the synthesis of a mRNA precursor and its conversion, by exon-intron splicing, to the mature mRNA, it is transported to the cytoplasm where it is translated into its cognate protein. Translation itself is a tightly regulated process, taking place on soluble ribosomes, or in the case of proteins targeted to the endoplasmic reticulum (ER), extruded across the ER membrane by ribosomes that have docked there. The correct folding in both compartments is aided by chaperones and, in either case, there are quality control mechanisms and pathways dedicated to the removal of misfolded or otherwise damaged proteins as these can be quite toxic if not efficiently removed. In the ER, this is known as the unfolded protein response (UPR) and is of marked importance in insuring that the ER protein secretion pathway, which is responsible for providing new cell surface receptors, is functioning properly. These degradation processes usually involved recognition, tagging with polyubiquitin moieties, and degradation via proteasomes.

The mitochondrion is a seemingly self-contained entity, whose origin in eukaryotic cells is thought to have been via adventitious incorporation of a primitive prokaryote into an early precursor to form a symbiotic relationship. Its principal role appeared for a long time to be the major organelle responsible for generating cellular energy currency, particularly nucleotide triphosphates. As such, it was not generally thought of as being important in signaling activities. However, its critical role in apoptosis (by releasing cytochrome c and other programmed cell death participants) dramatically altered this view. Mitochondria do not, as a rule, actively export macromolecules – rather they import the majority of their constituent proteins, whose synthesis is directed by nuclear chromosomes and occurs in the cytoplasm, via a mechanism, related to but distinct from, the ER transport system – but they do release a variety of ions and metabolites that act as small molecule messengers. These are controlled by a number of inner membrane-bound channels and transporters (the best known of which is the ADP/ATP transporter, putatively the most abundant eukaryotic protein). These can variously affect metabolism, largely as allosteric effectors, and gene expression. Thus, they are important contributors to the overall signaling capacity of the cell.

Two biological phenomena of critical importance in all organisms are cell generation (cell division or mitosis/meiosis) and cell death (apoptosis and necrosis). Both are extensively regulated and not surprisingly, much of this control is under the aegis of cell signaling events. The progression through the cell cycle and its various checkpoints is a symphony of protein modifications coupled to programmed protein turnover. The key players are a complement of kinases, known as cyclin-dependent kinases (Cdks), whose activation and deactivation are involved in every stage of the cycle. Interaction with cyclins, required for their activity, allows them to cycle in an on-off manner, and the ubiquitin-dependent degradation of the cyclins controls the vectoral nature of the cycle. The cyclin–Cdk complexes can be further regulated by phosphorylation or complexation with other proteins, which also allows for pausing at checkpoints if the cell senses it should not continue with the division process. There are also feedforward mechanisms that allow early steps to regulate successive ones. Apoptosis is equally tightly regulated and its progression easily recognized by distinct phenotypic responses (membrane blebbing, cell shrinking, and chromosomal condensation) as the cell progresses to its end. It is predicated on a family of cysteine proteases, called caspases (because they cleave their substrates to the C-terminal side of aspartic acid residues) that are activated in either an extrinsic or intrinsic pathway. The ten caspases generally exist as inactive precursors (zymogens) and can be subclassified into executioner, initiator, and inflammatory types. These have different structural features and different roles in apoptosis. One apoptotic pathway is directly related to the TNF superfamily, transmembrane receptors that contain a death domain. When activated, these lead to the activation of caspase 8, which in turn, activates the executioner caspase 3. Apoptosis is also triggered by cellular stress, and this leads to the involvement of the mitochondria (as noted previously). In a complex pathway involving many proteins, an apoptosome is formed which also leads to the eventual activation of the executioner caspases. Clearly, the connections between these two fundamental processes are of great importance and are closely related to a number of human diseases, notably cancer and neural degeneration.

FOCUS AND SCOPE OF THIS VOLUME

The chapters of this volume have been selected from a larger collection [19] and have been organized to emphasize transcriptional regulation and the function of nuclei and other subcellular organelles in signaling activities. They have been contributed by recognized experts and they are authoritative to the extent that size limitations allow. It is our intention that this survey will be useful in teaching, particularly in introductory courses, and to more seasoned investigators new to this area.

It is not possible to develop any of the areas covered in this volume in great detail, and expansion of any topic is left to the reader. The references in each chapter provide an excellent starting point, and greater coverage can also be found in the parent work [19]. It is important to realize that this volume does not cover other aspects of cell signaling such as receptor organization and function, transduction mechanisms, and organ-level manifestations, including disease correlates. These can be found in other volumes in this series [20–22].

REFERENCES

1. Bayliss WM, Starling EH. The mechanism of pancreatic secretion. *J Physiol* 1902;**28**:325–53.
2. Wright RD. The origin of the term "hormone". *Trends Biochem Sci* 1978;**3**:275.
3. Schaefer EA. *The endocrine organs*. London: Longman & Green; 1916; p. 6.
4. Banting FG, Best CH. The internal secretion of the pancreas. *J Lab Clin Med* 1922;**7**:251–66.
5. Levi-Montalcini R. The nerve growth factor 35 years later. *Science* 1987;**237**:1154–62.
6. Cohen S. Origins of Growth Factors: NGF and EGF. *J Biol Chem* 2008;**283**:33793–7.
7. Angeletti RH, Bradshaw RA. Nerve growth factor from mouse submaxillary gland: amino acid sequence. *Proc Natl Acad Sci USA* 1971;**68**:2417–20.
8. Savage CR, Inagami T, Cohen S. The primary structure of epidermal growth factor. *J Biol Chem* 1972;**247**:7612–21.
9. Frazier WA, Angeletti RH, Bradshaw RA. Nerve growth factor and insulin. *Science* 1972;**176**:482–8.
10. Raffioni S, Buxser SE, Bradshaw RA. The receptors for nerve growth factor and other neurotrophins. *Annu Rev Biochem* 1993;**62**:823–50.
11. Bradshaw RA, Sporn MB. Polypeptide growth factors and the regulation of cell growth and differentiation: introduction. *Fed. Proc* 1983;**42**:2590–1.
12. Ullrich A, Bell JR, Chen EY, Herrera R, Petruzzelli LM, Dull TJ, *et al.* Human insulin receptor and its relationship to the tyrosine kinase family of oncogenes. *Nature* 1985;**313**:756–61.

13. Ullrich A, Coussens L, Hayflick JS, Dull TJ, Gray A, Tam AW, *et al.* Human epidermal growth factor receptor cDNA sequence and aberrant expression of the amplified gene in A431 epidermoid carcinoma cells. *Nature* 1985;**309**:418–25.
14. Ebina Y, Ellis L, Jarnagin K, Edery M, Graf L, Clauser E, *et al.* The human insulin receptor cDNA: the structural basis for hormone transmembrane signalling. *Cell* 1985;**40**:747–58.
15. Robison GA, Butcher RW, Sutherland EW. *Cyclic AMP*. San Diego: Academic Press; 1971.
16. Eckert W, Hutchinson MA, Hunter T. An activity phosphorylating tyrosine in polyoma T antigen immunoprecipitates. *Cell* 1979;**18**:925–33.
17. Hunter T, Sefton BM. Transforming gene product of Rous sarcoma virus phosphorylates tyrosine. *Proc Natl Acad Sci USA* 1980;**77**:1311–5.
18. Ushiro H, Cohen S. Identification of phosphotyrosine as a product of epidermal growth factor-activated protein kinase in A-431 cell membranes. *J Biol Chem* 1980;**255**:8363–5.
19. Bradshaw RA, Dennis EA, editors. *Handbook of cell signaling*. 2nd ed. San Diego, CA: Academic Press; 2008.
20. Bradshaw RA, Dennis EA, editors. *Functioning of transmembrane receptors in cell signaling mechanisms*. San Diego, CA: Academic Press; 2011.
21. Dennis EA, Bradshaw RA, editors. *Transduction mechanisms in cellular signaling*. San Diego, CA: Academic Press; 2011.
22. Dennis EA, Bradshaw RA, editors. *Intercellular signaling in development and disease*. San Diego, CA: Academic Press; 2011.

Section B

Nuclear Signaling

Part 1

Transcription

Chapter 2

Signaling at the Nuclear Envelope

Géza Ambrus and Larry Gerace
Department of Cell Biology, Scripps Research Institute, La Jolla, California

INTRODUCTION

In eukaryotic cells the nuclear envelope (NE) separates the cytoplasmic and nuclear compartments [1]. It controls a diverse range of cellular functions and properties, including nucleocytoplasmic transport, chromatin organization, gene expression, nuclear architecture, and signal transduction. The NE consists of the outer nuclear membrane (ONM), the inner nuclear membrane (INM), and nuclear pore complexes (NPCs) (Figure 2.1). In higher eukaryotic cells, the NE also contains the nuclear lamina, a protein meshwork that lines its inner surface. The ONM and INM, which are separated by the perinuclear luminal space, are connected at the NPCs via the "pore membrane." NPCs are giant proteinaceous assemblies that provide channels across the NE for nucleocytoplasmic transport. The ONM is continuous with the more peripheral endoplasmic reticulum (ER) and is functionally similar to rough and smooth ER membranes, whereas the INM has distinctive functional properties, in large part related to the nuclear lamina. This review focuses on the role that INM proteins and lamins play in cellular signaling pathways. Cellular regulatory functions of NPC proteins separate from their role in nucleocytoplasmic trafficking also are briefly discussed.

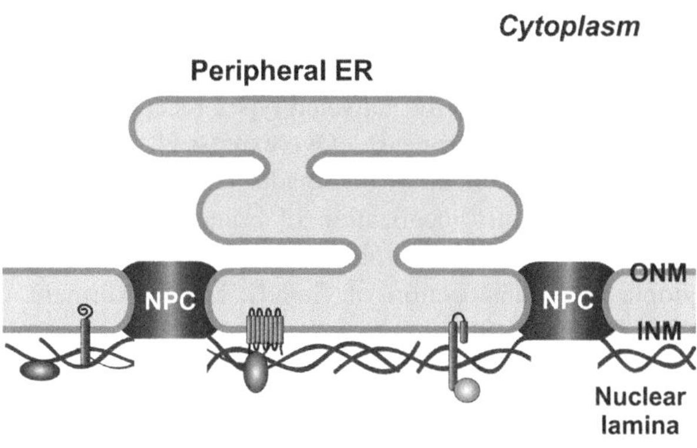

FIGURE 2.1 In eukaryotic cells the nuclear envelope (NE) separates the cytoplasmic and nuclear subcellular compartments.
The NE consists of the outer nuclear membrane (ONM), the inner nuclear membrane (INM), an underlying lamina network scaffold, and the nuclear pore complexes (NPCs). The ONM and INM are lipid bi-layers separated by a lumenal space. They are connected at the NPCs, giant proteinaceous assemblies of nucleoporins and the sites for nucleocytoplasmic transport. The ONM is continuous with the ER whereas the INM is connected to the nuclear lamina through INM proteins that bind lamins.

The nuclear lamina provides mechanical support for nuclear membranes and serves as an anchoring site for chromatin and NPCs [2, 3]. The backbone of the lamina is formed by a polymeric assembly of nuclear lamins, type V intermediate filament proteins. In vertebrates, the major lamin isotypes are lamins A and C (A-type lamins), which are alternatively spliced products of the same gene that usually are expressed near the time of cell differentiation, and lamins B1 and B2 (B-type lamins), which are products of separate genes and are expressed in most cells throughout development. Like other intermediate filament proteins, lamins contain a central α-helical "rod" domain flanked by non-α-helical "head" and "tail" domains. *In vitro*, lamins form parallel, unstaggered dimers through their coiled-coil domains, and further associate by head-to-tail interactions, and by antiparallel partially staggered associations [4]. Whereas most lamins are concentrated at the INM, a second population of lamins (particularly A-type lamins) has been described in the nucleoplasm of cultured cells, and has been suggested to function in cell cycle control [5].

About 20 transmembrane proteins that are highly enriched at the NE have been characterized in detail in vertebrates, and over 50 additional candidate nuclear membrane proteins have been identified [6]. Of the former group, most are localized at the INM, three are found at the nuclear pore membrane (gp210, POM121, and NDC1), and several others appear to be restricted to the ONM (certain splice isoforms of the nesprin family, which connect the nuclear lamina to the cytoplasmic cytoskeleton).

Most INM proteins interact with lamins and/or chromatin, making them relatively resistant to Triton extraction and immobile in fluorescence recovery after photobleaching (FRAP) analysis. Several of the best characterized transmembrane proteins of the INM contain the so-called LEM domain, a 43 amino acid motif originally described in lamina associated polypeptide 2 (*L*AP2), *e*merin, and *M*AN1. The LEM motif binds to barrier-to-autointegration factor (BAF), an essential DNA binding protein. The gene for LAP2 encodes multiple splice isoforms, all with the LEM domain. One of these, LAP2α lacks a transmembrane segment and is peripherally associated with the INM and is also present in the nuclear interior, whereas the remaining isoforms all contain a transmembrane segment and localize to the INM. Emerin and MAN1 are integral membrane proteins of the INM and associate with lamins [7].

Certain mutations in lamins and lamina associated transmembrane proteins give rise to tissue specific diseases collectively called "laminopathies" [8]. Laminopathies are associated with a wide spectrum of pathologies including muscular dystrophies, lipodystrophies, premature aging, and disorders affecting bone/connective tissue. Most disease causing laminopathy mutations are mapped to the gene for lamin A, and a smaller number are linked to B-type lamins and to certain INM proteins. Two models that are not mutually exclusive have been proposed to explain the molecular basis of these diseases. The "mechanical stress model" argues that mutations in lamina components weaken the NE and make it more susceptible to mechanical damage. The "gene expression model" on the other hand proposes that mutated NE proteins are directly responsible for alterations in gene expression patterns and/or cell cycle progression, which in turn lead to disease. Recently, significant evidence has been provided in support of both models.

LAMINS AND LAMIN ASSOCIATED PROTEINS IN CELL SIGNALING

Transforming Growth Factor β and Bone Morphogenic Protein Signaling

Transforming growth factor β (TGFβ)/bone morphogenic protein (BMP) signaling directs tissue formation through regulating cell proliferation, differentiation, and cell migration [9]. TGFβ and related factors bind to type II and type I TGFβ receptors that activate receptor regulated Smads (R-Smads). R-Smads form heteromers with co-Smad and translocate into the nucleus to induce transcriptional activation or repression. The NE transmembrane proteins MAN1 and Dullard/NET56 have been implicated as regulators of TGFβ/BMP signaling.

The role of MAN1 in the TGFβ/BMP signaling pathway is the best documented example of signaling regulation by an INM protein [10, 11]. The first lines of evidence came from studies with XMAN1, the *Xenopus* ortholog of human MAN1. XMAN1 was identified in a functional screen for neuralizing factors antagonizing BMP signaling in *Xenopus* development [12]. A two-hybrid screen in *Xenopus* with Smad1 as the bait also identified XMAN1 ("SANE") as a binding partner [13].

MAN1 appears to fulfill similar functions in TGFβ/BMP signaling in mammals. A loss-of-function mutation of MAN1 (also called LEMD3) in humans causes osteopoikilosis, a bone disorder caused in part by misregulated TGFβ/BMP signaling [14]. Additional knockout, overexpression and silencing studies confirm that MAN1 antagonizes TGFβ/BMP signaling in mammals. In mice with a homozygous functionally null allele of *MAN1,* the Smad2/3 pathway is abnormally activated, which is accompanied by abnormal yolk sac vascularization suggesting a regulatory role for MAN1 in yolk sac angiogenesis by TGFβ signaling [15, 16]. In cultured cells, overexpression of MAN1, but not a Smad binding incompetent point mutant, reduces transcription from TGFβ, activin, and BMP responsive promoters, whereas silencing MAN1 increases expression from these reporters [17]. Mammalian MAN1 was shown to bind the R-Smads. Smad1, Smad2, Smad3. and Smad5. through its C-terminal RNA recognition motif (RRM) in pull-down assays, co-immunoprecipitation studies and yeast two-hybrid screens [14, 17, 18], and a portion of Smad1 and Smad3 co-localizes with MAN1 at the NE [17]. MAN1, however, does not bind Smad4 [17, 18].

The most attractive explanation to MAN1's antagonistic effect on TGFβ/BMP signaling is that it affects the phosphorylation of R-Smads, which results in their cytoplasmic localization through nuclear export, thereby attenuating TGFβ/BMP signaling [10, 12]. This idea is supported by studies in mammalian cells with MAN1 and XMAN1. Overexpression of MAN1 does not affect the half-life of R-Smads but reduces the activated type I receptor mediated phosphorylation of Smad1. Overexpressed MAN1 also blocks heterodimerization of Smad3 and Smad4 resulting in the cytoplasmic localization of Smad3 [17]. XMAN1 appears to inhibit Smad1 phosphorylation resulting in the cytoplasmic translocation of Smad1 and impairment of BMP signaling [13].

Another NE protein that participates in BMP signaling is Dullard/NET56, first identified as an essential factor for neural development in *Xenopus* [19]. In a proteomics screen Dullard/NET56 was found to be enriched in an NE fraction and was shown to target to the NE [6]. Studies in *Xenopus* involving overexpressed Dullard/NET56 suggest that Dullard/NET56 is a phosphatase that antagonizes BMP signaling [20]. It has been proposed that Dullard/NET56 impairs BMP signaling through binding to BMP receptors and promoting their proteosomal degradation via the lipid raft-caveolar pathway [20]. However, in its simplest form this model is inconsistent with the NE/ER localization of endogenous Dullard [21].

Cell Cycle Control at the G_1/S Checkpoint

A classical tumor suppressor is the retinoblastoma protein (pRB) [22]. Proteins of the pRB family repress gene expression from E2F responsive promoters. Hyperphosphorylation of pRB by cyclin dependent kinases (CDKs) releases suppression and allows the cells to transition from G_1 phase into S-phase. In addition to cell cycle progression, pRB has been suggested to control numerous cellular events such as DNA repair, apoptosis, and differentiation. There is strong evidence that pRB is required for switching from muscle myoblast proliferation to differentiation state during muscle cell development [23, 24]. It is also believed to activate MyoD, a transcriptional regulator of skeletal muscle differentiation [25]. The nuclear envelope proteins lamins A/C and emerin also have been reported to be required for muscle regeneration and pRB appears to be involved in this process [26–29]. Myoblasts lacking A type lamins or emerin have lower levels of pRB, MyoD, desmin, and M-cadherin and have impaired differentiation potential [26]. Expression of MyoD or desmin can rescue differentiation in these cells. A role for emerin in pRB mediated muscle cell differentiation is underscored by the observed delays in transcriptional activation of MyoD targets and impaired repression of pRB targets in an emerin deficient mouse [27].

There is mounting evidence that the NE proteins lamin A/C and LAP2α bind pRB and play a role in pRB signaling [5]. These regulatory interactions, however, may not occur at the NE, and are proposed to be mediated by nucleoplasmic pools of lamin A/C and LAP2α complexes [28,30]. There are indications that lamin A/C and LAP2α regulate pRB signaling through influencing the localization of hypophosphorylated pRB [28, 31–34], but it is also possible that they indirectly control the phosphorylation state of pRB [28, 33, 35] or prevent it from proteosomal degradation [36, 37]. Further studies will be needed to elucidate the precise control mechanism [38].

Ca^{2+} Signaling

Ca^{2+} as an intracellular signal is central to many signaling pathways [39], including those that drive proliferation, development, and secretion, as well as those active in the physiological processes of muscle contraction, learning, and memory. In spite of the high permeability of nuclear pore complexes for small ions like Ca^{2+}, nuclear and cytosolic Ca^{2+} pools appear to be independently regulated [40]. The cytosolic Ca^{2+} flux from intracellular stores is initiated by either Ca^{2+} itself or by messengers, such as inositol-1,4,5,-triphosphate (IP_3). IP_3 regulated Ca^{2+} release has been suggested to regulate early gene expression in myotubes [41–43] through the phosphorylation of cAMP response element binding protein (CREB) [44]. IP_3 could be generated *in situ* in the nucleoplasm from its precursor phosphatidylinositol-4,5-bisphosphate (PIP_2) as the β_1 isoform of phosphoinositide specific phospholipase C (PI-PLCβ_1), which, similar to other PI-PLCs catalyzes IP_3 formation from PIP_2, can be found in the nucleus [45]. Signals from IP_3 could be captured and transmitted through type-1 IP_3 receptors, which have been shown to preferentially localize to the inner nuclear membrane of cultured rat skeletal myotubes [44] and in the myotubes of the mouse 1B5 myotubes [46]. IP_3 mediated signaling can in turn lead to protein kinase C translocation to the nuclear envelope [47] potentially resulting in the phosphorylation of CREB [48]. Although this field is still controversial, there is growing evidence that the nuclear envelope harbors elements of Ca^{2+} signaling pathways that influence transcriptional regulation.

Insulin Signaling

Lipodystrophies are characterized by the loss of subcutaneous fat from certain tissues, frequently resulting in insulin resistance and the metabolic syndrome [49]. Lamin A/C mutations, as well as mutations in perixosome-proliferator activated receptor γ (PPARγ) are known to be associated with familial partial lipodystrophies (FPLDs) [50, 51]. Insulin has long been known to induce preadipocyte differentiation [52] and recent studies linking lamin A/C functions to insulin signaling through PPARγ have emerged. Overexpression of lamin A or a lamin A mutant that causes Dunningan-type FPLD were found to inhibit adipocyte differentiation in 3T3-L1 preadipocytes, manifested by a deficiency in lipid accumulation, triglyceride synthesis, and the expression of the adipogenic markers PPARγ2 and Glut4 [53]. By contrast, others did not find defects in 3T3-L1 adipocyte differentiation when lamin A was overexpressed [54]. Nevertheless, knockout experiments suggest a regulatory role for A type lamins in insulin signaling [53]. In *LMNA*$^{-/-}$ MEFs phosphorylated AKT1 levels are elevated [53], which, in turn, could induce the expression of the adipocyte transcription factor sterol regulatory element binding protein 1 (SREBP1) a transcriptional activator of PPARγ2 [55, 56]. Lamin A carrying the FPLD associated mutation R482L has been shown to sequester SREBP1 *in vivo* [57, 58]. This gives rise to the speculation that in Dunnigan-type FPLD the mutations in lamin A/C may inhibit SREBP1 binding to DNA, which, in turn, reduces PPARγ expression causing impaired adipocyte differentiation [59]. More experiments will be needed to confirm this interesting hypothesis.

Dullard/NET56 is also implicated in insulin signaling through the dephosphorylation of lipin, a mammalian phosphatidic acid phosphatase [21]. Lipin is required for normal adipose tissue development and is central to nuclear membrane biogenesis and lipid signaling [60]. Phosphorylation and localization of lipin is in part regulated by insulin [61].

This suggests the involvement of Dullard in insulin signaling but further research is needed to elucidate its exact role.

Mitogen Activated Protein Kinase Signaling

The mitogen activated protein kinase (MAPK) pathway is involved in many cellular processes including differentiation, proliferation, apoptosis, motility, and metabolism [62]. Recent studies have linked lamin A/C and the INM protein emerin to the MAPK signaling pathway. Certain mutations in the genes encoding emerin and lamin A/C are known to cause X-linked and autosomal dominant Emery-Dreifuss muscular dystrophy (EDMD), respectively. However, the molecular mechanism of the pathological processes is poorly understood [63–65]. A genome-wide analysis of gene expression in *Emd* knockout mice (a model for X-linked EDMD) indicated the activation of the ERK1/2 branch of the MAPK signaling pathway [66]. Similarly, the MAPK signaling pathway and its downstream targets were activated in the heart muscle and in isolated cardiomyocytes of an autosomal dominant EDMD mouse model with mutant knockedin lamin A/C [67]. These activated downstream targets were implicated in the pathogenesis of cardiomyopathy by impacting sarcomere structure, cardiomyofiber organization, and other aspects of heart function. This suggests that the activation of MAPK signaling is the basis for the development of heart disease in both autosomal dominant and in X-linked EDMD. The molecular basis by which emerin controls MAPK signaling, however, is not established.

Wnt/β-Catenin Signaling

Canonical Wnt/β-catenin signaling plays a crucial role during developmental stages as well as in adult tissue self-renewal [68]. Upon pathway activation β-catenin localizes to the nucleus and initiates transcription of downstream targets. Emerin was suggested to bind to β-catenin as early as 1997 [69], but clear experimental evidence for this surfaced only recently [70]. *In vivo* and *in vitro* data show that emerin binds to β-catenin through the C-terminal nucleoplasmic adenomatous polyposis coli (APC)-like domain of the former, and inhibits β-catenin transcriptional activity by promoting its nuclear export by the karyopherin CRM1. In emerin null human fibroblasts derived from X-linked EDMD patients and lacking detectable emerin protein, β-catenin accumulates in the nucleus, whereas pRB localization is not affected [70, 71]. Upon withdrawal of growth factors emerin null fibroblasts keep proliferating and c-myc, an immediate downstream target of β-catenin, is significantly upregulated [72]. The expression of GFP-emerin in emerin null fibroblasts renders β-catenin cytoplasmic thereby inhibiting its activity [70]. These observations hint at the possibility that mutations in the emerin gene, in addition to affecting muscle cells, might also promote pathological disorders in heart and skeletal muscle through the adverse growth of fibroblasts and subsequent fibrosis.

NFκB Signaling

Nuclear factor κB (NFκB) is a key regulator of both adaptive and innate immune responses [73]. It acts as an antiapoptotic transcription factor contributing to cell survival and metastatic potential. Therefore, it is plausible that apoptosis is more pronounced in mechanically strained tissue with inhibited NFκB signaling. To investigate the potential role of lamin A/C in biomechanically activated NFκB signaling Lammerding and co-workers subjected lamin A/C deficient mouse embryo fibroblasts to mechanical stress [74]. Interestingly, expression from an NFκB activated reporter gene was significantly impaired in $LMNA^{-/-}$ cells under mechanical stress and there was an increase in the proportion of both apoptotic and necrotic cells. This suggests that $LMNA^{-/-}$ cells have both defective nuclear mechanics and an impaired mechanically activated gene transcription. In contrast, NFκB signaling is not altered in emerin deficient mouse embryo fibroblasts although they exhibit impaired mechanosensitive gene regulation and have higher apoptotic rates under mechanical stress than wild-type cells [75]. The molecular mechanism responsible for impaired mechanotransduction in lamin A/C and emerin null cells is yet to be determined.

Gene Silencing

In differentiated nuclei of metazoan cells, most transcriptionally inactive heterochromatin is localized to the nuclear periphery [76]. Lamins are important in tethering the chromatin to the NE, whereas integral INM proteins such as LAP2β, emerin, and lamin-B receptor (LBR) are suggested to participate in gene silencing [77, 78]. LAP2β, a B-type lamin binding LEM domain protein, promotes gene repression of transfected plasmids through binding to the transcriptional regulator germ-cell-less (GCL), an inhibitor of the heterodimeric transcriptional activator E2F-DP3 [79]. LAP2β also abrogates transcriptional activation of plasmid reporters by p53 and NFκB. In addition, LAP2β was proposed to induce histone H4 deacetylation via its interaction with histone deacetylase 3 (HDAC3) and thus participate in a more general mechanism for gene silencing [80]. Emerin, another LEM domain protein also binds GCL and might modulate gene expression through this interaction, but the implications of this binding are not yet clear [81]. Emerin was also proposed to repress transcriptional activation by sequestering the putative transcription factor Lmo7 [82]. LBR binds lamins but it was also shown to be involved in chromatin binding. In fact, mutations in LBR can cause Pelger-Huet anomaly, which is characterized by abnormal chromatin organization in white blood cells. LBR, in addition to binding the heterochromatin protein HP1, has a sterol reductase activity that makes it a potential candidate to participate in nuclear sterol dependent signaling and transcriptional regulation. A number of additional gene regulators were shown to bind INM proteins but their functional implications remain to be determined [1, 83]. Interestingly, lamin A/C has also been directly linked to gene repression. It was shown to suppress activating protein 1 (AP-1) function by sequestering c-Fos at the nuclear periphery, providing a potentially new mechanism for cell cycle and transcriptional control [84].

THE NPC IN CELL SIGNALING

Recent studies indicate that certain active genes are stably localized near NPCs in yeast, suggesting a role for nuclear pores in transcriptional control in this organism [85–87]. A variety of mechanisms have been described to account for gene localization near yeast NPCs. Co-transcriptional recruitment of the nuclear export machinery to highly transcribed genes may drive their localization to the NPC. *GAL10* and *HSP104* require Mex67, the yeast mRNA export factor and myosin-like protein 1 (Mlp1), an NPC protein, for stable NPC association [88]. Nascent transcripts of pheromone induced genes also associate with the NPC in an Mlp1 dependent manner [89]. Transcription independent mechanisms for gene localization to the NPC have also been described. In an elegant demonstration of epigenetic memory of inducible genes, Nup2 and histone H2A.Z were shown to be required for gene retention at the NPC and this interaction persisted even when RNA-polymerase II mediated transcription was blocked [90].

In a handful of studies, anchorage of specific genes to nuclear pores was required for their optimal expression. Optimal gene expression from a subtelomeric gene, HXK1 (hexokinase isoenzyme 1), that associates with the NPC is lost when positioned away from the nuclear periphery [91]. In another study the recruitment of the *INO1* gene to the nuclear periphery correlated with its increased expression [92, 93]. Nevertheless, not all NPC anchored genes have a requirement for NPC localization for efficient gene expression in yeast. For instance, gene positioning did not have a significant effect on accumulating transcript levels in the case of either the *GAL10* or the *HSP104* genes [88].

In contrast to yeast, transcription is not concentrated at the NPC in higher eukaryotes, but, rather, is distributed throughout the nuclear interior [94]. Indeed, the periphery of the vertebrate nucleus is an area where heterochromatin is concentrated, and has been typically linked to gene silencing. Consistent with this, certain vertebrate genes are active only when they move from the nuclear periphery to the nuclear interior [94]. Similarly, intranuclear to perinuclear positioning of genes can switch between active and repressed states in yeast [94, 95]. The differences between the spatial distribution of actively transcribed genes in higher eukaryotes and yeast may also be explained by the much smaller size of the genome and nucleus of yeast as compared to vertebrates, allowing yeast genes more facile topological access to the NPC. Nonetheless the mammalian NPC has an established role in transcriptional control through the regulated nucleocytoplasmic trafficking of transcription factors and their binding partners. This NPC mediated transcriptional regulation is indirect, since it is achieved through the interaction of nucleoporins with karyopherins, nuclear transport receptors that ferry cargo molecules such as protein components of the cell signaling pathways through the NE. Transport competence can be established by posttranslational modification of signaling molecules: phosphorylation, ubiquitination, sumoylation, methylation, and ribosylation have all been shown to direct nucleocytoplasmic trafficking [96–98]. In addition to their crucial role in nucleocytoplasmic trafficking during interphase, nucleoporins and mobile components of the nuclear transport machinery have been found to be involved in the signaling cascades leading to NE breakdown during mitosis, assembly of the mitotic spindle, and subsequent NE assembly [45, 99].

CONCLUSIONS

The notion that the NE participates in cell signaling has gained a strong foothold in recent years. Substantial evidence for specific contributions of INM proteins and lamins to signal transduction pathways has already been demonstrated. The role of the NE in gene silencing and activation is now also starting to become elucidated. With the majority of the predicted inner nuclear membrane proteins remaining to be characterized, it seems likely that additional functions for NE proteins in signal transduction will surface. Elucidating the molecular functions of lamins and lamin associated proteins in signaling and transcriptional regulation will help us understand how mutations in these factors can result in human disease, such as muscular dystrophies and other laminopathies.

ACKNOWLEDGEMENTS

We are grateful to Brandon Chen for preparing the figure and would like to thank Michael Huber and Kaustuv Datta for their helpful comments on the manuscript. Our research in this field is supported by National Institutes of Health grant GM28521. GA is supported by a Postdoctoral Fellowship Award from the California HIV/AIDS Research Program.

REFERENCES

1. Gruenbaum Y, Margalit A, Goldman RD, et al. The nuclear lamina comes of age. *Nat Rev Mol Cell Biol* 2005;**6**:21–31.
2. Burke B, Stewart CL. The laminopathies: the functional architecture of the nucleus and its contribution to disease. *Annu Rev Genomics Hum Genet* 2006;**7**:369–405.
3. Aebi U, Cohn J, Buhle L, Gerace L. The nuclear lamina is a meshwork of intermediate-type filaments. *Nature* 1986;**323**:560–4.
4. Worman HJ, Courvalin JC. Nuclear envelope, nuclear lamina, and inherited disease. *Int Rev Cytol* 2005;**246**:231–79.
5. Dorner D, Gotzmann J, Foisner R. Nucleoplasmic lamins and their interaction partners, LAP2alpha, Rb, and BAF, in transcriptional regulation. *FEBS J* 2007;**274**:1362–73.
6. Schirmer EC, Florens L, Guan T, et al. Nuclear membrane proteins with potential disease links found by subtractive proteomics. *Science* 2003;**301**:1380–2.
7. Wagner N, Krohne G. LEM-Domain proteins: new insights into lamin-interacting proteins. *Int Rev Cytol* 2007;**261**:1–46.
8. Worman HJ, Bonne G. "Laminopathies": a wide spectrum of human diseases. *Exp Cell Res* 2007;**313**:2121–33.
9. Derynck R, Zhang YE. Smad-dependent and Smad-independent pathways in TGF-beta family signalling. *Nature* 2003;**425**:577–84.
10. Bengtsson L. What MAN1 does to the Smads. TGFbeta/BMP signaling and the nuclear envelope. *FEBS J* 2007;**274**:1374–82.
11. Worman HJ. Inner nuclear membrane and signal transduction. *J Cell Biochem* 2005;**96**:1185–92.
12. Osada S, Ohmori SY, Taira M. XMAN1, an inner nuclear membrane protein, antagonizes BMP signaling by interacting with Smad1 in Xenopus embryos. *Dev* 2003;**130**:1783–94.
13. Raju GP, Dimova N, Klein PS, Huang HC. SANE, a novel LEM domain protein, regulates bone morphogenetic protein signaling through interaction with Smad1. *J Biol Chem* 2003;**278**:428–37.
14. Hellemans J, Preobrazhenska O, Willaert A, et al. Loss-of-function mutations in LEMD3 result in osteopoikilosis, Buschke-Ollendorff syndrome and melorheostosis. *Nat Genet* 2004;**36**:1213–8.
15. Ishimura A, Ng JK, Taira M, et al. Man1, an inner nuclear membrane protein, regulates vascular remodeling by modulating transforming growth factor beta signaling. *Dev* 2006;**133**:3919–28.
16. Cohen TV, Kosti O, Stewart CL. The nuclear envelope protein MAN1 regulates TGFbeta signaling and vasculogenesis in the embryonic yolk sac. *Dev* 2007;**134**:1385–95.
17. Pan D, Estevez-Salmeron LD, Stroschein SL, et al. The integral inner nuclear membrane protein MAN1 physically interacts with the R-Smad proteins to repress signaling by the transforming growth factor-beta superfamily of cytokines. *J Biol Chem* 2005;**280**:15,992–16,001.
18. Lin F, Morrison JM, Wu W, Worman HJ. MAN1, an integral protein of the inner nuclear membrane, binds Smad2 and Smad3 and antagonizes transforming growth factor-beta signaling. *Hum Mol Genet* 2005;**14**:437–45.
19. Satow R, Chan TC, Asashima M. Molecular cloning and characterization of dullard: a novel gene required for neural development. *Biochem Biophys Res Commun* 2002;**295**:85–91.
20. Satow R, Kurisaki A, Chan TC, et al. Dullard promotes degradation and dephosphorylation of BMP receptors and is required for neural induction. *Dev Cell* 2006;**11**:763–74.
21. Kim Y, Gentry MS, Harris TE, et al. A conserved phosphatase cascade that regulates nuclear membrane biogenesis. *Proc Natl Acad Sci U S A* 2007;**104**:6596–601.
22. Sherr CJ. Principles of tumor suppression. *Cell* 2004;**116**:235–46.
23. Novitch BG, Spicer DB, Kim PS, et al. pRb is required for MEF2-dependent gene expression as well as cell-cycle arrest during skeletal muscle differentiation. *Curr Biol* 1999;**9**:449–59.
24. Huh MS, Parker MH, Scime A, et al. Rb is required for progression through myogenic differentiation but not maintenance of terminal differentiation. *J Cell Biol* 2004;**166**:865–76.
25. Puri PL, Iezzi S, Stiegler P, et al. Class I histone deacetylases sequentially interact with MyoD and pRb during skeletal myogenesis. *Mol Cell* 2001;**8**:885–97.
26. Frock RL, Kudlow BA, Evans AM, et al. Lamin A/C and emerin are critical for skeletal muscle satellite cell differentiation. *Genes Dev* 2006;**20**:486–500.
27. Melcon G, Kozlov S, Cutler DA, et al. Loss of emerin at the nuclear envelope disrupts the Rb1/E2F and MyoD pathways during muscle regeneration. *Hum Mol Genet* 2006;**15**:637–51.
28. Mariappan I, Parnaik VK. Sequestration of pRb by cyclin D3 causes intranuclear reorganization of lamin A/C during muscle cell differentiation. *Mol Biol Cell* 2005;**16**:1948–60.
29. Bakay M, Wang Z, Melcon G, et al. Nuclear envelope dystrophies show a transcriptional fingerprint suggesting disruption of Rb-MyoD pathways in muscle regeneration. *Brain* 2006;**129**:996–1013.
30. Kennedy BK, Barbie DA, Classon M, et al. Nuclear organization of DNA replication in primary mammalian cells. *Genes Dev* 2000;**14**:2855–68.
31. Dorner D, Vlcek S, Foeger N, et al. Lamina-associated polypeptide 2alpha regulates cell cycle progression and differentiation via the retinoblastoma-E2F pathway. *J Cell Biol* 2006;**173**:83–93.
32. Mancini MA, Shan B, Nickerson JA, et al. The retinoblastoma gene product is a cell cycle-dependent, nuclear matrix-associated protein. *Proc Natl Acad Sci U S A* 1994;**91**:418–22.
33. Pekovic V, Harborth J, Broers JL, et al. Nucleoplasmic LAP2alpha-lamin A complexes are required to maintain a proliferative state in human fibroblasts. *J Cell Biol* 2007;**176**:163–72.
34. Markiewicz E, Dechat T, Foisner R, et al. Lamin A/C binding protein LAP2alpha is required for nuclear anchorage of retinoblastoma protein. *Mol Biol Cell* 2002;**13**:4401–13.
35. Mariappan I, Gurung R, Thanumalayan S, Parnaik VK. Identification of cyclin D3 as a new interaction partner of lamin A/C. *Biochem Biophys Res Commun* 2007;**355**:981–5.
36. Johnson BR, Nitta RT, Frock RL, et al. A-type lamins regulate retinoblastoma protein function by promoting subnuclear localization and preventing proteasomal degradation. *Proc Natl Acad Sci U S A* 2004;**101**:9677–82.
37. Nitta RT, Jameson SA, Kudlow BA, et al. Stabilization of the retinoblastoma protein by A-type nuclear lamins is required for INK4A-mediated cell cycle arrest. *Mol Cell Biol* 2006;**26**:5360–72.
38. Nitta RT, Smith CL, Kennedy BK. Evidence that proteasome-dependent degradation of the retinoblastoma protein in cells lacking a-type lamins occurs independently of gankyrin and MDM2. *PLoS ONE* 2007;**2**:e963.
39. Berridge MJ, Bootman MD, Roderick HL. Calcium signalling: dynamics, homeostasis and remodelling. *Nat Rev Mol Cell Biol* 2003;**4**:517–29.
40. Leite MF, Thrower EC, Echevarria W, et al. Nuclear and cytosolic calcium are regulated independently. *Proc Natl Acad Sci U S A* 2003;**100**:2975–80.
41. Araya R, Liberona JL, Cardenas JC, et al. Dihydropyridine receptors as voltage sensors for a depolarization-evoked, IP3R-mediated, slow calcium signal in skeletal muscle cells. *J Gen Physiol* 2003;**121**:3–16.
42. Carrasco MA, Riveros N, Rios J, et al. Depolarization-induced slow calcium transients activate early genes in skeletal muscle cells. *Am J Physiol Cell Physiol* 2003;**284**:C1438–47.

43. Powell JA, Carrasco MA, Adams DS, et al. IP(3) receptor function and localization in myotubes: an unexplored Ca(2+) signaling pathway in skeletal muscle. *J Cell Sci Suppl* 2001;**114**:3673–83.
44. Cardenas C, Liberona JL, Molgo J, et al. Nuclear inositol 1,4,5-trisphosphate receptors regulate local Ca2+ transients and modulate cAMP response element binding protein phosphorylation. *J Cell Sci* 2005;**118**:3131–40.
45. Irvine RF. Nuclear lipid signalling. *Nat Rev Mol Cell Biol* 2003;**4**:349–60.
46. Kusnier C, Cardenas C, Hidalgo J, Jaimovich E. Single-channel recording of inositol trisphosphate receptor in the isolated nucleus of a muscle cell line. *Biol Res* 2006;**39**:541–53.
47. Echevarria W, Leite MF, Guerra MT, et al. Regulation of calcium signals in the nucleus by a nucleoplasmic reticulum. *Nat Cell Biol* 2003;**5**:440–6.
48. Mao LM, Tang Q, Wang JQ. Protein kinase C-regulated cAMP response element-binding protein phosphorylation in cultured rat striatal neurons. *Brain Res Bull* 2007;**72**:302–8.
49. Garg A. Acquired and inherited lipodystrophies. *N Engl J Med* 2004;**350**:1220–34.
50. Agarwal AK, Garg A. Genetic basis of lipodystrophies and management of metabolic complications. *Annu Rev Med* 2006;**57**:297–311.
51. Capeau J, Magre J, Lascols O, et al. Diseases of adipose tissue: genetic and acquired lipodystrophies. *Biochem Soc Trans* 2005;**33**:1073–7.
52. MacDougald OA, Lane MD. Transcriptional regulation of gene expression during adipocyte differentiation. *Annu Rev Biochem* 1995;**64**:345–73.
53. Boguslavsky RL, Stewart CL, Worman HJ. Nuclear lamin A inhibits adipocyte differentiation: implications for Dunnigan-type familial partial lipodystrophy. *Hum Mol Genet* 2006;**15**:653–63.
54. Kudlow BA, Jameson SA, Kennedy BK. HIV protease inhibitors block adipocyte differentiation independently of lamin A/C. *AIDS* 2005;**19**:1565–73.
55. Porstmann T, Griffiths B, Chung YL, et al. PKB/Akt induces transcription of enzymes involved in cholesterol and fatty acid biosynthesis via activation of SREBP. *Oncogene* 2005;**24**:6465–81.
56. Fajas L, Schoonjans K, Gelman L, et al. Regulation of peroxisome proliferator-activated receptor gamma expression by adipocyte differentiation and determination factor 1/sterol regulatory element binding protein 1: implications for adipocyte differentiation and metabolism. *Mol Cell Biol* 1999;**19**:5495–503.
57. Capanni C, Mattioli E, Columbaro M, et al. Altered pre-lamin A processing is a common mechanism leading to lipodystrophy. *Hum Mol Genet* 2005;**14**:1489–502.
58. Lloyd DJ, Trembath RC, Shackleton S. A novel interaction between lamin A and SREBP1: implications for partial lipodystrophy and other laminopathies. *Hum Mol Genet* 2002;**11**:769–77.
59. Maraldi NM, Capanni C, Mattioli E, et al. A pathogenic mechanism leading to partial lipodistrophy and prospects for pharmacological treatment of insulin resistance syndrome. *Acta Biomed* 2007;**78**:207–15.
60. Carman GM, Han GS. Roles of phosphatidate phosphatase enzymes in lipid metabolism. *Trends Biochem Sci* 2006;**31**:694–9.
61. Harris TE, Huffman TA, Chi A, et al. Insulin controls subcellular localization and multisite phosphorylation of the phosphatidic acid phosphatase, lipin 1. *J Biol Chem* 2007;**282**:277–86.
62. Kolch W. Coordinating ERK/MAPK signalling through scaffolds and inhibitors. *Nat Rev Mol Cell Biol* 2005;**6**:827–37.
63. Bione S, Maestrini E, Rivella S, et al. Identification of a novel X-linked gene responsible for Emery-Dreifuss muscular dystrophy. *Nat Genet* 1994;**8**:323–7.
64. Bonne G, Di Barletta MR, Varnous S, et al. Mutations in the gene encoding lamin A/C cause autosomal dominant Emery-Dreifuss muscular dystrophy. *Nat Genet* 1999;**21**:285–8.
65. Raffaele Di Barletta M, Ricci E, Galluzzi G, et al. Different mutations in the LMNA gene cause autosomal dominant and autosomal recessive Emery-Dreifuss muscular dystrophy. *Am J Hum Genet* 2000;**66**:1407–12.
66. Muchir A, Pavlidis P, Bonne G, et al. Activation of MAPK in hearts of EMD null mice: similarities between mouse models of X-linked and autosomal dominant Emery Dreifuss muscular dystrophy. *Hum Mol Genet* 2007;**16**:1884–95.
67. Muchir A, Pavlidis P, Decostre V, et al. Activation of MAPK pathways links LMNA mutations to cardiomyopathy in Emery-Dreifuss muscular dystrophy. *Eur J Clin Invest* 2007;**117**:1282–93.
68. Clevers H. Wnt/beta-catenin signaling in development and disease. *Cell* 2006;**127**:469–80.
69. Cartegni L, di Barletta MR, Barresi R, et al. Heart-specific localization of emerin: new insights into Emery-Dreifuss muscular dystrophy. *Hum Mol Genet* 1997;**6**:2257–64.
70. Markiewicz E, Tilgner K, Barker N, et al. The inner nuclear membrane protein emerin regulates beta-catenin activity by restricting its accumulation in the nucleus. *EMBO J* 2006;**25**:3275–85.
71. Markiewicz E, Venables R, Mauricio Alvarez R, et al. Increased solubility of lamins and redistribution of lamin C in X-linked Emery-Dreifuss muscular dystrophy fibroblasts. *J Struct Biol* 2002;**140**:241–53.
72. van de Wetering M, Sancho E, Verweij C, et al. The beta-catenin/TCF-4 complex imposes a crypt progenitor phenotype on colorectal cancer cells. *Cell* 2002;**111**:241–50.
73. Karin M. Nuclear factor-kappaB in cancer development and progression. *Nature* 2006;**441**:431–6.
74. Lammerding J, Schulze PC, Takahashi T, et al. Lamin A/C deficiency causes defective nuclear mechanics and mechanotransduction. *J Clin Invest* 2004;**113**:370–8.
75. Lammerding J, Hsiao J, Schulze PC, et al. Abnormal nuclear shape and impaired mechanotransduction in emerin-deficient cells. *J Cell Biol* 2005;**170**:781–91.
76. Akhtar A, Gasser SM. The nuclear envelope and transcriptional control. *Nat Rev* 2007;**8**:507–17.
77. Shaklai S, Amariglio N, Rechavi G, Simon AJ. Gene silencing at the nuclear periphery. *FEBS J* 2007;**274**:1383–92.
78. Taniura H, Glass C, Gerace L. A chromatin binding site in the tail domain of nuclear lamins that interacts with core histones. *J Cell Biol* 1995;**131**:33–44.
79. Nili E, Cojocaru GS, Kalma Y, et al. Nuclear membrane protein LAP2beta mediates transcriptional repression alone and together with its binding partner GCL (germ-cell-less). *J Cell Sci* 2001;**114**:3297–307.
80. Somech R, Shaklai S, Geller O, et al. The nuclear-envelope protein and transcriptional repressor LAP2beta interacts with HDAC3 at the nuclear periphery, and induces histone H4 deacetylation. *J Cell Sci* 2005;**118**:4017–25.
81. Holaska JM, Lee KK, Kowalski AK, Wilson KL. Transcriptional repressor germ cell-less (GCL) and barrier to autointegration factor (BAF) compete for binding to emerin in vitro. *J Biol Chem* 2003;**278**:6969–75.
82. Holaska JM, Rais-Bahrami S, Wilson KL. Lmo7 is an emerin-binding protein that regulates the transcription of emerin and many other muscle-relevant genes. *Hum Mol Genet* 2006;**15**:3459–72.
83. Stewart CL, Roux KJ, Burke B. Blurring the boundary: the nuclear envelope extends its reach. *Science* 2007;**318**:1408–12.

84. Ivorra C, Kubicek M, Gonzalez JM, et al. A mechanism of AP-1 suppression through interaction of c-Fos with lamin A/C. *Gen Dev* 2006;**20**:307–20.
85. Ahmed S, Brickner JH. Regulation and epigenetic control of transcription at the nuclear periphery. *Trends Genet* 2007;**23**:396–402.
86. Taddei A. Active genes at the nuclear pore complex. *Curr Opin Cell Biol* 2007;**19**:305–10.
87. Sexton T, Schober H, Fraser P, Gasser SM. Gene regulation through nuclear organization. *Nat Struct Mol Biol* 2007;**14**:1049–55.
88. Dieppois G, Iglesias N, Stutz F. Cotranscriptional recruitment to the mRNA export receptor Mex67p contributes to nuclear pore anchoring of activated genes. *Mol Cell Biol* 2006;**26**:7858–70.
89. Casolari JM, Brown CR, Drubin DA, et al. Developmentally induced changes in transcriptional program alter spatial organization across chromosomes. *Genes Dev* 2005;**19**:1188–98.
90. Brickner DG, Cajigas I, Fondufe-Mittendorf Y, et al. H2A.Z-mediated localization of genes at the nuclear periphery confers epigenetic memory of previous transcriptional state. *PLoS biology* 2007;**5**:e81.
91. Taddei A, Van Houwe G, Hediger F, et al. Nuclear pore association confers optimal expression levels for an inducible yeast gene. *Nature* 2006;**441**:774–8.
92. Brickner JH, Walter P. Gene recruitment of the activated INO1 locus to the nuclear membrane. *PLoS Biol* 2004;**2**:e342.
93. Casolari JM, Brown CR, Komili S, et al. Genome-wide localization of the nuclear transport machinery couples transcriptional status and nuclear organization. *Cell* 2004;**117**:427–39.
94. Misteli T. Beyond the sequence: cellular organization of genome function. *Cell* 2007;**128**:787–800.
95. Andrulis ED, Neiman AM, Zappulla DC, Sternglanz R. Perinuclear localization of chromatin facilitates transcriptional silencing. *Nature* 1998;**394**:592–5.
96. Miyauchi Y, Yogosawa S, Honda R, et al. Sumoylation of Mdm2 by protein inhibitor of activated STAT (PIAS) and RanBP2 enzymes. *J Biol Chem* 2002;**277**:50,131–6.
97. Salmena L, Pandolfi PP. Changing venues for tumour suppression: balancing destruction and localization by monoubiquitylation. *Nat Rev* 2007;**7**:409–13.
98. Terry LJ, Shows EB, Wente SR. Crossing the nuclear envelope: hierarchical regulation of nucleocytoplasmic transport. *Science* 2007;**318**:1412–6.
99. Prunuske AJ, Ullman KS. The nuclear envelope: form and reformation. *Curr Opin Cell Biol* 2006;**18**:108–16.

Chapter 3

Nuclear Receptor Coactivators

Joshua D. Stender and Christopher K. Glass

Department of Cellular and Molecular Medicine, School of Medicine, University of California, San Diego, La Jolla, California

INTRODUCTION

The nuclear receptor (NR) superfamily consists of 48 homologous transcription factors that function as molecular sensors for a diverse set of lipophilic hormones, vitamins, and dietary lipids [1]. Members of the NR superfamily share overlapping and distinct tissue expression patterns resulting in coordinated regulation of transcriptional programs that control the reproductive, developmental, immunological, and metabolic requirements for an organism [2, 3]. Included in this family are proteins that recognize steroid hormones (e.g., estrogens, progestins, and androgens), fatty acids, bile acids, oxysterols, vitamins A and D, and thyroid hormones. In addition, ligands for several members of the NR superfamily have not been identified, and these NRs are collectively referred to as orphan receptors [4]. The NRs share a homologous modular structure consisting of an amino-terminal transcription activation function, which can activate transcription independent of ligand binding, a central DNA-binding domain that allows the NR to interact with sequence-specific DNA elements, and a COOH-terminal ligand binding domain [1]. Activation of a steroid hormone receptor by ligand binding or through posttranslational modifications results in a conformational change that allows it to bind to specific DNA sequences in the regulatory regions of target genes. Many receptors for non-steroidal ligands bind to DNA in the presence or absence of ligand as heterodimers with Retinoid X receptors. Once docked to these regulatory sites, the NRs function as a scaffold to recruit multiple coactivator complexes that modify local chromatin structure that ultimately engage the RNA polymerase II complex on the promoter of the target gene resulting in initiation of transcription [5]. These coactivators fine-tune NR gene regulation, direct the cell and promoter specificity of their ligands, form bridges with the general transcriptional machinery, and catalyze several enzymatic activities that drive the transcription process [6]. The biochemical purification and characterization of several hundred coactivator proteins suggests that they function in multiprotein complexes either sequentially or in a combinatorial fashion. Analysis of coactivator complex recruitment to an estrogen-dependent target gene occurs in a sequential and cyclical fashion, which implies that these complexes perform specific roles in transcription initiation, elongation, and disassembly [5]. Extrapolation of these findings to genome-wide and organ-specific functions of nuclear receptors suggest that different coactivator complexes are utilized in a combinatorial manner to enable the distinct actions of nuclear receptors in development and homeostasis. In this chapter, the general mechanisms for coactivator complexes in enhancing NR gene activation will be addressed with emphasis on the ligand-dependent interaction between nuclear receptors and coactivator proteins and the posttranslational modifications performed by these complexes.

LIGAND-DEPENDENT INTERACTION BETWEEN NUCLEAR RECEPTORS AND COACTIVATORS

For proper control of gene expression by NRs it has become increasing apparent that NRs require ligand-dependent interactions with specific components of coactivator complexes. These interactions between NRs and coactivators are dependent on the three-dimensional structure of the NR, which is influenced by the nature of the ligand and the sequence of the DNA binding site. Crystallographic analysis of several nuclear receptor ligand binding domains (LBDs) revealed a conserved three-dimensional structure in which a three-layered anti-parallel alpha-helical sandwich encloses a central, hydrophobic ligand binding pocket [7, 8]. In unliganded RXR structure, the helix 12 of the LBD extends away from the LBD [9]. In contrast, in the ligand-bound RARγ, TRα, PPARγ, and ER LBD structures, helix 12 is tightly packed against the body of the LBD and

Handbook of Cell Signaling, Three-Volume Set 2 ed.

makes direct contacts with the ligand [10–13]. Therefore, ligand binding induces a conformation change of the LBD of NRs that rearranges helices 3, 4, 5, and 12 that uncovers a hydrophobic groove flanked by two conserved charge clamp residues. This hydrophobic groove is bounded by helices 3 and 12 of the NR and is the site for ligand-dependent coactivator interaction [14, 15]. Extensive analysis of the amino acid sequences within the nuclear receptor coactivator revealed a conserved helical sequence LXXLL (L, leucine; X, any amino acid) that is necessary and sufficient for the ligand-dependent interaction between the NR hydrophobic groove and coactivator proteins [16–20]. Nearly all coactivators that directly interact with NRs in a ligand-dependent manner contain an LXXLL motif, including members of the p160/SRC family, TRAP220/DRIP205, and PGC-1. Listed in Table 3.1 are coactivators whose LXXLL motifs have been experimentally demonstrated to mediate interaction with nuclear receptors. Structural studies of the PPARγ ligand binding domain complexed to a fragment of the SRC1 NR interaction domain revealed that the LXXLL motifs form alpha helices that are gripped at each end of the interaction surface by a "charge-clamp" consisting of a conserved lysine in helix 3 of the NR ligand binding domain and a conserved glutamate in the AF-2 helix [21]. Additional structures of nuclear receptors LBDs complexed to LXXLL peptides from other p160 family members exhibit the same structural basis for interaction [12, 14, 15]. For example, the androgen receptor (AR) LBD bound to the LXXLL motif from SRC2 demonstrates hydrophobic interactions between the LXXLL helical motif and the hydrophobic binding surface on AR [22]. In addition, hydrogen bonds between the LXXLL peptide backbone and the conserved charge clamp residues are also observed. Therefore, the

TABLE 3.1 Coactivators that have experimentally validated LXXLL motifs[1]

Coactivator	Human gene ID	Mouse gene ID	LXXLL motif	References
CIA/NCOA5	57727	228869	QS**LINLL**AD	[36]
HMGCS2/mHMG-CoAS	3158	15360	GC**LASLL**SH	[37]
MNAR/PELP1	27043	75273	AV**LRDLL**RY	[38]
			NH**LPGLL**TS	
NRIF3/ITGB3BP	23421	67733	LK**LDGLL**EE	[39]
PERC	133522	170826	S**LLQKLL**A	[40]
			SI**LRELL**AQ	
PGC1	10891	19017	SL**LKKLL**LA	[41]
PROX1	5629	19130	NV**LRKLL**KR	[42]
SRC1/NCOA1	8648	17977	HK**LVQLL**TT	[17]
			KI**LHRLL**QE	
			QL**LRYLL**DK	
			SL**LQQLL**TE	
SRC2/GRIP1/TIF2	10499	17978	TK**LLGLL**TT	[16, 20]
			KI**LHRLL**QD	
			AL**LRYLL**DK	
SRC3/AIB1/ACTR/RAC3	8202	17979	KK**LLQLL**TC	[43]
			RI**LHKLL**QN	
			AL**LRYLL**DR	
TIP60	10524	81601	AM**LKRLL**RI	[44]
TRAP220/MED1/DRIP205/PIBP	5469	19014	PM**LMNLL**KD	[45]
TRBP/RAP250/NCOA6/PRIP/ASC-2	23054	56406	PL**LVNLL**QS	[46]
VPR	9730	321006	AE**LTTLL**EQ	[47]

[1]The coactivators are listed alphabetically according to their common name along with their corresponding Entrez Gene Ids. In addition, the amino acid sequence corresponding to their LXXLL motifs are noted in bold text.

results of these structure studies identified a conserved conformational change in helix 12 of NRs upon ligand binding that configures a binding surface that specifically interacts with the LXXLL motif of coactivator proteins. Although the LXXLL motif is sufficient for interaction between coactivators and NRs, additional amino acids on the amino and carboxy ends of the LXXLL motif make contacts with the LBD of NRs [22, 23]. These residues are not conserved among coactivators and potentially mediate the specificity of interaction between different NRs.

POSTTRANSLATIONAL MODIFICATIONS PERFORMED BY COACTIVATOR COMPLEXES

Coactivator proteins are generally considered to function by modifying chromatin architecture and/or by serving as adapters that recruit core transcription factors to target promoters. Table 3.2 contains representative coactivators that have enzymatic activities that remodel chromatin structure or modify histone tails. Since chromatin assembly results in the compaction of DNA preventing access and the binding of transcription factors, it exerts a severe repressive effect on gene transcription. Therefore chromatin remodeling needs to be one of the first steps in initiating gene transcription. Consistent with this hypothesis are experiments that examined the dynamic recruitment of transcription factors to the estrogen receptor regulated gene Trefoil Factor 1 (TFF1), which revealed that chromatin remodeling complexes are the first proteins recruited to the TFF1 promoter upon ligand treatment of breast cancer cells [5]. The BRG/BAF complexes, which are highly related to SWI/SNF complexes originally identified in yeast and Drosophila, utilize ATP to reposition nucleasomes, presumably facilitating access of NRs to their sequence-specific DNA binding sites [24].

The recent identification of several modifications of histones has established a posttranslational code that specifies the activation or silencing of gene expression [25, 26].

TABLE 3.2 Representative nuclear receptor coactivators with chromatin remodeling or histone modifying activities

Coactivator	Human gene ID	Mouse gene ID	Enzyme activity	Comment	References
CBP	1387	12914	Histone acetylation	Interacts with SRC family	[48]
p300	2033	328572	Histone acetylation	Interacts with SRC family	[48]
pCAF	8850	18519	Histone acetylation	Interacts with NR DBDs	[49, 50]
GCN5	2648	14534	Histone acetylation	Interacts with the AF2 region of NRs	[51, 52]
BRG1	6597	20586	Chromatin remodeling	Interacts with several NR LBDs	[53, 54]
LSD1	23028	99982	Histone demethylation	Interacts with AR and ER and demethylates H3K4 and H3K9	[32, 33]
CARM1	10498	59035	Arginine methylation	Interacts with SRC family	[30]
PRMT1	3276	15469	Arginine methylation	Interacts with SRC family	[31]
NSD1	64324	18193	Lysine methylation	Interacts with several NR LBDs and methylates H3K36 and H4K20	[55, 56]
G9a	10919	110147	Lysine methylation	Interacts with SRC family and methylates H3K9	[57]
RIZ1	7799	110593	Lysine methylation	Interacts with SRC family and methylates H3K9	[58]
MEN1	4221	30130	Lysine methylation	Component of complex that methylates H3K4	[59]

For example, acetylation of lysine residues on histone H3 and H4 tails results in chromatin opening and subsequent gene activation [26]. Several proteins that function as NR coactivators harbor histone acetylatransferase activity including CBP, p300, and members of the p160/SRC family [27, 28]. In addition, histone methylation, once thought to be an irreversible process, has recently emerged as a modification that is associated with either transcriptional activation or gene silencing, depending on the specific histone and amino acid that is modified [26, 29]. Interestingly both histone methylation and demethylation are critical processes necessary for nuclear receptor gene activation. For example, the nuclear receptor coactivators CARM1 and PRMT1, which methylate arginines on histone H3 and H4, respectively, enhance NR dependent transcriptional activation [30, 31]. More recently, Lysine Specific Demethylase 1 (LSD1), which functions as a H3K4 and H3K9 demethylase, has been shown to be critical for AR, ER, and TR gene activation [32, 33]. Although LSD1 is the only demethylase currently recognized to function as an NR coactivator, it is highly likely that additional histone demethylases also participate in nuclear receptor-dependent gene regulation.

In addition to modifying local chromatin architecture, coactivators also serve as adapter proteins for recruitment of secondary coactivator proteins. For example, the p160 family functions as adapters between NR and the protein methyltransferases CARM1 and PRMT1, whose known substrates include histone tails, CBP/p300, and SRC3 [34, 35]. Methylation of SRC3 by CARM1 results in disassembly of the coactivator complex [34]. Thus these protein methylation events most likely serve as molecular switches that allow the transcriptional complexes to disassemble and recycle to permit successive rounds of transcription [6]. These protein modifications thus function to control the differential recruitment of coactivator complexes, allowing nuclear receptors to fine-tune gene expression in a signal and promoter specific manner.

CONCLUSIONS

The identification and functional characterization of nuclear receptor coactivator complexes represents a major current focus in the nuclear receptor field. The multitude of coactivator complexes allows for integration of multiple signals required coordinately to regulate gene expression on an organismal level. The study of these proteins may lead to therapeutic benefits due to the possible selectivity of interactions between NRs and coactivator complex achieved by synthetic NR ligands. Although synthetic ligands have been developed only for a few NRs, it is likely that ligands with similar properties can be designed for other members of the NR family. Such ligands would have potential applications in a variety of human diseases, including hormone-dependent cancers, atherosclerosis, diabetes, and chronic inflammatory diseases.

REFERENCES

1. McKenna NJ, O'Malley BW. Combinatorial control of gene expression by nuclear receptors and coregulators. *Cell* 2002;**108**:465–74.
2. Bookout AL, Jeong Y, Downes M, Yu RT, Evans RM, Mangelsdorf DJ. Anatomical profiling of nuclear receptor expression reveals a hierarchical transcriptional network. *Cell* 2006;**126**:789–99.
3. Yang X, Downes M, Yu RT, Bookout AL, He W, Straume M, Mangelsdorf DJ, Evans RM. Nuclear receptor expression links the circadian clock to metabolism. *Cell* 2006;**126**:801–10.
4. Benoit G, Cooney A, Giguere V, Ingraham H, Lazar M, Muscat G, Perlmann T, Renaud JP, Schwabe J, Sladek F, Tsai MJ, Laudet V. International Union of Pharmacology. LXVI. Orphan nuclear receptors. *Pharmacol Rev* 2006;**58**:798–836.
5. Metivier R, Penot G, Hubner MR, Reid G, Brand H, Kos M, Gannon F. Estrogen receptor-alpha directs ordered, cyclical, and combinatorial recruitment of cofactors on a natural target promoter. *Cell* 2003;**115**:751–63.
6. Rosenfeld MG, Lunyak VV, Glass CK. Sensors and signals: a coactivator/corepressor/epigenetic code for integrating signal-dependent programs of transcriptional response. *Genes Dev* 2006;**20**:1405–28.
7. Bourguet W, Germain P, Gronemeyer H. Nuclear receptor ligand-binding domains: three-dimensional structures, molecular interactions and pharmacological implications. *Trends Pharmacol Sci* 2000;**21**:381–8.
8. Moras D, Gronemeyer H. The nuclear receptor ligand-binding domain: structure and function. *Curr Opin Cell Biol* 1998;**10**:384–91.
9. Bourguet W, Ruff M, Chambon P, Gronemeyer H, Moras D. Crystal structure of the ligand-binding domain of the human nuclear receptor RXR-alpha. *Nature* 1995;**375**:377–82.
10. Brzozowski AM, Pike AC, Dauter Z, Hubbard RE, Bonn T, Engstrom O, Ohman L, Greene GL, Gustafsson JA, Carlquist M. Molecular basis of agonism and antagonism in the oestrogen receptor. *Nature* 1997;**389**:753–8.
11. Renaud JP, Rochel N, Ruff M, Vivat V, Chambon P, Gronemeyer H, Moras D. Crystal structure of the RAR-gamma ligand-binding domain bound to all-trans retinoic acid. *Nature* 1995;**378**:681–9.
12. Shiau AK, Barstad D, Loria PM, Cheng L, Kushner PJ, Agard DA, Greene GL. The structural basis of estrogen receptor/coactivator recognition and the antagonism of this interaction by tamoxifen. *Cell* 1998;**95**:927–37.
13. Wagner RL, Apriletti JW, McGrath ME, West BL, Baxter JD, Fletterick RJ. A structural role for hormone in the thyroid hormone receptor. *Nature* 1995;**378**:690–7.
14. Darimont BD, Wagner RL, Apriletti JW, Stallcup MR, Kushner PJ, Baxter JD, Fletterick RJ, Yamamoto KR. Structure and specificity of nuclear receptor–coactivator interactions. *Genes Dev* 1998;**12**:3343–56.
15. Feng W, Ribeiro RC, Wagner RL, Nguyen H, Apriletti JW, Fletterick RJ, Baxter JD, Kushner PJ, West BL. Hormone-dependent coactivator binding to a hydrophobic cleft on nuclear receptors. *Science* 1998;**280**:1747–9.
16. Ding XF, Anderson CM, Ma H, Hong H, Uht RM, Kushner PJ, Stallcup MR. Nuclear receptor-binding sites of coactivators glucocorticoid receptor interacting protein 1 (GRIP1) and steroid receptor coactivator 1 (SRC-1): multiple motifs with different binding specificities. *Mol Endocrinol* 1998;**12**:302–13.
17. Heery DM, Kalkhoven E, Hoare S, Parker MG. A signature motif in transcriptional co-activators mediates binding to nuclear receptors. *Nature* 1997;**387**:733–6.
18. Le Douarin B, Zechel C, Garnier JM, Lutz Y, Tora L, Pierrat P, Heery D, Gronemeyer H, Chambon P, Losson R. The N-terminal part of TIF1, a putative mediator of the ligand-dependent activation function

(AF-2) of nuclear receptors, is fused to B-raf in the oncogenic protein T18. *Embo J* 1995;**14**:2020–33.

19. Torchia J, Rose DW, Inostroza J, Kamei Y, Westin S, Glass CK, Rosenfeld MG. The transcriptional co-activator p/CIP binds CBP and mediates nuclear-receptor function. *Nature* 1997;**387**:677–84.
20. Voegel JJ, Heine MJ, Tini M, Vivat V, Chambon P, Gronemeyer H. The coactivator TIF2 contains three nuclear receptor-binding motifs and mediates transactivation through CBP binding-dependent and -independent pathways. *Embo J* 1998;**17**:507–19.
21. Nolte RT, Wisely GB, Westin S, Cobb JE, Lambert MH, Kurokawa R, Rosenfeld MG, Willson TM, Glass CK, Milburn MV. Ligand binding and co-activator assembly of the peroxisome proliferator-activated receptor-gamma. *Nature* 1998;**395**:137–43.
22. Hur E, Pfaff SJ, Payne ES, Gron H, Buehrer BM, Fletterick RJ. Recognition and accommodation at the androgen receptor coactivator binding interface. *PLoS Biol* 2004;**2**:E274.
23. McInerney EM, Rose DW, Flynn SE, Westin S, Mullen TM, Krones A, Inostroza J, Torchia J, Nolte RT, Assa-Munt N, Milburn MV, Glass CK, Rosenfeld MG. Determinants of coactivator LXXLL motif specificity in nuclear receptor transcriptional activation. *Genes Dev* 1998;**12**:3357–68.
24. DiRenzo J, Shang Y, Phelan M, Sif S, Myers M, Kingston R, Brown M. BRG-1 is recruited to estrogen-responsive promoters and cooperates with factors involved in histone acetylation. *Mol Cell Biol* 2000;**20**:7541–9.
25. Jenuwein T, Allis CD. Translating the histone code. *Science* 2001;**293**:1074–80.
26. Kouzarides T. Chromatin modifications and their function. *Cell* 2007;**128**:693–705.
27. Bannister AJ, Kouzarides T. The CBP co-activator is a histone acetyltransferase. *Nature* 1996;**384**:641–3.
28. Chen H, Lin RJ, Schiltz RL, Chakravarti D, Nash A, Nagy L, Privalsky ML, Nakatani Y, Evans RM. Nuclear receptor coactivator ACTR is a novel histone acetyltransferase and forms a multimeric activation complex with P/CAF and CBP/p300. *Cell* 1997;**90**:569–80.
29. Barski A, Cuddapah S, Cui K, Roh TY, Schones DE, Wang Z, Wei G, Chepelev I, Zhao K. High-resolution profiling of histone methylations in the human genome. *Cell* 2007;**129**:823–37.
30. Chen D, Ma H, Hong H, Koh SS, Huang SM, Schurter BT, Aswad DW, Stallcup MR. Regulation of transcription by a protein methyltransferase. *Science* 1999;**284**:2174–7.
31. Wang H, Huang ZQ, Xia L, Feng Q, Erdjument-Bromage H, Strahl BD, Briggs SD, Allis CD, Wong J, Tempst P, Zhang Y. Methylation of histone H4 at arginine 3 facilitating transcriptional activation by nuclear hormone receptor. *Science* 2001;**293**:853–7.
32. Garcia-Bassets I, Kwon YS, Telese F, Prefontaine GG, Hutt KR, Cheng CS, Ju BG, Ohgi KA, Wang J, Escoubet-Lozach L, Rose DW, Glass CK, Fu XD, Rosenfeld MG. Histone methylation-dependent mechanisms impose ligand dependency for gene activation by nuclear receptors. *Cell* 2007;**128**:505–18.
33. Metzger E, Wissmann M, Yin N, Muller JM, Schneider R, Peters AH, Gunther T, Buettner R, Schule R. LSD1 demethylates repressive histone marks to promote androgen-receptor-dependent transcription. *Nature* 2005;**437**:436–9.
34. Feng Q, Yi P, Wong J, O'Malley BW. Signaling within a coactivator complex: methylation of SRC-3/AIB1 is a molecular switch for complex disassembly. *Mol Cell Biol* 2006;**26**:7846–57.
35. Xu W, Chen H, Du K, Asahara H, Tini M, Emerson BM, Montminy M, Evans RM. A transcriptional switch mediated by cofactor methylation. *Science* 2001;**294**:2507–11.
36. Sauve F, McBroom LD, Gallant J, Moraitis AN, Labrie F, Giguere V. CIA, a novel estrogen receptor coactivator with a bifunctional nuclear receptor interacting determinant. *Mol Cell Biol* 2001;**21**:343–53.
37. Meertens LM, Miyata KS, Cechetto JD, Rachubinski RA, Capone JP. A mitochondrial ketogenic enzyme regulates its gene expression by association with the nuclear hormone receptor PPARalpha. *Embo J* 1998;**17**:6972–8.
38. Barletta F, Wong CW, McNally C, Komm BS, Katzenellenbogen B, Cheskis BJ. Characterization of the interactions of estrogen receptor and MNAR in the activation of cSrc. *Mol Endocrinol* 2004;**18**:1096–108.
39. Li D, Desai-Yajnik V, Lo E, Schapira M, Abagyan R, Samuels HH. NRIF3 is a novel coactivator mediating functional specificity of nuclear hormone receptors. *Mol Cell Biol* 1999;**19**:7191–202.
40. Kressler D, Schreiber SN, Knutti D, Kralli A. The PGC-1-related protein PERC is a selective coactivator of estrogen receptor alpha. *J Biol Chem* 2002;**277**:13,918–13,925.
41. Vega RB, Huss JM, Kelly DP. The coactivator PGC-1 cooperates with peroxisome proliferator-activated receptor alpha in transcriptional control of nuclear genes encoding mitochondrial fatty acid oxidation enzymes. *Mol Cell Biol* 2000;**20**:1868–76.
42. Song KH, Li T, Chiang JY. A Prospero-related homeodomain protein is a novel co-regulator of hepatocyte nuclear factor 4alpha that regulates the cholesterol 7alpha-hydroxylase gene. *J Biol Chem* 2006;**281**:10,081–10,088.
43. Li H, Chen JD. The receptor-associated coactivator 3 activates transcription through CREB-binding protein recruitment and autoregulation. *J Biol Chem* 1998;**273**:5948–54.
44. Gaughan L, Brady ME, Cook S, Neal DE, Robson CN. Tip60 is a coactivator specific for class I nuclear hormone receptors. *J Biol Chem* 2001;**276**:46,841–46,848.
45. Yuan CX, Ito M, Fondell JD, Fu ZY, Roeder RG. The TRAP220 component of a thyroid hormone receptor- associated protein (TRAP) coactivator complex interacts directly with nuclear receptors in a ligand-dependent fashion. *Proc Natl Acad Sci U S A* 1998;**95**:7939–44.
46. Zhu Y, Kan L, Qi C, Kanwar YS, Yeldandi AV, Rao MS, Reddy JK. Isolation and characterization of peroxisome proliferator-activated receptor (PPAR) interacting protein (PRIP) as a coactivator for PPAR. *J Biol Chem* 2000;**275**:13,510–13,516.
47. Kino T, Gragerov A, Kopp JB, Stauber RH, Pavlakis GN, Chrousos GP. The HIV-1 virion-associated protein vpr is a coactivator of the human glucocorticoid receptor. *J Exp Med* 1999;**189**:51–62.
48. Ogryzko VV, Schiltz RL, Russanova V, Howard BH, Nakatani Y. The transcriptional coactivators p300 and CBP are histone acetyltransferases. *Cell* 1996;**87**:953–9.
49. Blanco JC, Minucci S, Lu J, Yang XJ, Walker KK, Chen H, Evans RM, Nakatani Y, Ozato K. The histone acetylase PCAF is a nuclear receptor coactivator. *Genes Dev* 1998;**12**:1638–51.
50. Spencer TE, Jenster G, Burcin MM, Allis CD, Zhou J, Mizzen CA, McKenna NJ, Onate SA, Tsai SY, Tsai MJ, O'Malley BW. Steroid receptor coactivator-1 is a histone acetyltransferase. *Nature* 1997;**389**:194–8.
51. Candau R, Zhou JX, Allis CD, Berger SL. Histone acetyltransferase activity and interaction with ADA2 are critical for GCN5 function in vivo. *Embo J* 1997;**16**:555–65.
52. Wang L, Mizzen C, Ying C, Candau R, Barlev N, Brownell J, Allis CD, Berger SL. Histone acetyltransferase activity is conserved between yeast and human GCN5 and is required for complementation of growth and transcriptional activation. *Mol Cell Biol* 1997;**17**:519–27.

53. Fryer CJ, Archer TK. Chromatin remodelling by the glucocorticoid receptor requires the BRG1 complex. *Nature* 1998;**393**:88–91.
54. Ichinose H, Garnier JM, Chambon P, Losson R. Ligand-dependent interaction between the estrogen receptor and the human homologues of SWI2/SNF2. *Gene* 1997;**188**:95–100.
55. Huang N, vom Baur E, Garnier JM, Lerouge T, Vonesch JL, Lutz Y, Chambon P, Losson R. Two distinct nuclear receptor interaction domains in NSD1, a novel SET protein that exhibits characteristics of both corepressors and coactivators. *Embo J* 1998;**17**:3398–412.
56. Rayasam GV, Wendling O, Angrand PO, Mark M, Niederreither K, Song L, Lerouge T, Hager GL, Chambon P, Losson R. NSD1 is essential for early post-implantation development and has a catalytically active SET domain. *Embo J* 2003;**22**:3153–63.
57. Lee DY, Northrop JP, Kuo MH, Stallcup MR. Histone H3 lysine 9 methyltransferase G9a is a transcriptional coactivator for nuclear receptors. *J Biol Chem* 2006;**281**:8476–85.
58. Carling T, Kim KC, Yang XH, Gu J, Zhang XK, Huang S. A histone methyltransferase is required for maximal response to female sex hormones. *Mol Cell Biol* 2004;**24**:7032–42.
59. Dreijerink KM, Mulder KW, Winkler GS, Hoppener JW, Lips CJ, Timmers HT. Menin links estrogen receptor activation to histone H3K4 trimethylation. *Cancer Res* 2006;**66**:4929–35.

Chapter 4

Corepressors in Mediating Repression by Nuclear Receptors

Gratien G. Prefontaine, Peter J. Cook and Michael G. Rosenfeld
Howard Hughes Medical Institute, Department and School of Medicine, University of California, San Diego, La Jolla, California

INTRODUCTION

A critical aspect of regulated gene expression is the ability to effectively request basal gene expression, essentially imposing a "checkpoint" to permit the highest possible amplitude of activation by diverse signals and ligands for nuclear receptors. This ability to repress gene expression is imposed by corepressor complexes, each consisting of a large number of protein components that together modulate gene transcription upon their recruitment by DNA binding transcription factors, such as nuclear receptors. Nuclear receptors, including those activated by lipophilic ligands can exert both activation and repression functions on distinct cohorts of gene targets. Repression by nuclear receptors can occur alternatively either as a result of direct binding to DNA present in regulated gene targets or by acting as corepression, based on recruitment to coregulatory machinery or histone marks on effective transcription units. Corepressor complexes consist of both platform proteins and enzymatic components responsible for a variety of posttranscriptional modifications of histones and other transcriptional regulatory proteins. Covalent modifications on histones serve as molecular beacons for protein complexes, including histone-modifying enzymes that modulate the serial complexes exchanged to impose repression. These complexes exert distinct functions, including formation of repressive heterochromatin or local inhibition at the level of promoter or enhancer. One model of gene regulation suggest that ATP dependent remodeling factors are DNA translocases that move DNA around histones in effect sliding, ejecting, and restructuring nucleosomes [1]. Together the concerted action of these corepressor complexes leads to disruption of gene transcription by decreasing the recruitment of RNA polymerase II to gene promoters, relocation to specific nuclear organelles, or by inducing condensed or repressive forms of chromatin.

The nuclear receptor superfamily consists of 48 orthologs and can be subdivided into three categories [2–6]. Type I or classical steroid receptors include the glucocorticoid receptor (GR), the androgen receptor (AR), the mineralocorticoid receptor (MR), progesterone receptor (PR), and the estrogen receptor (ER) and are distinguished from others by their association with chaperone and heat-shock proteins in the unliganded state and generally localized to the cytoplasm [7]. Upon ligand binding, the receptors dissociate from the chaperone and heat-shock protein complex, homodimerize and relocate to the nucleus where they regulate gene expression. In contrast, type II receptors interact with their DNA response elements in the absence of ligand, usually as a heterodimer with retinoid X-related receptors (RXRs) and are often associated with corepressor proteins [4, 8]. Upon ligand binding, corepressor proteins are exchanged for protein complexes that activate gene transcription. Type III nuclear receptors or orphan receptors were identified based on their homology to ligand-activated receptors [5, 8]. Aptly named, these receptors do not have known ligands or their ligands remain to be determined. In cases where the ligands have been identified often they are reclassified as adopted receptors and relegated to the type II category. Here we briefly recount alternative mechanisms by which nuclear receptor action is inhibited by distinct corepressor complexes.

Nuclear receptors are modular proteins characterized by a central zinc finger coordinated DNA binding domain, a C-terminal ligand binding domain (LBD), and a variable N-terminal domain. Structurally, the LBD consists of a number of layered anti-parallel α-helices assembled to form a ligand binding cavity with a protruding C-terminal α-helix (AF-2) that changes position upon ligand binding [9–16]. In the absence of hormone, corepressors including NCoR and SMRT interact with the receptor LBD through an elongated helix containing variations of the sequence

Handbook of Cell Signaling, Three-Volume Set 2 ed.

LXXI/HIXXXI/L, also referred to as the CoRNR-box [17–20]. NCoR and SMRT are huge adaptor proteins that interact with the same NR hydrophobic pocket contacted by alternate coregulatory factors characterized by LXXLL motifs. Upon agonist binding, the AF-2 helix is reconfigured to form a "charged clamp" where one of its conserved glutamate residues interacts with a lysine residue in helix 3 of the LBD and grips the ends of the helical LXXLL motif. The interaction is stabilized by hydrophobic contributions from both the leucine residues of the LXXLL motif and the nature of the pocket [11–13, 15]. Displacement of the AF-2 helix in the new charged clamp configuration physically limits the ability of the larger LXXI/HIXXXI/L motif to interact with the hydrophic core. This "charge clamp" configuration destabilizes interactions between nuclear receptors and corepressors NCoR/SMRT, resulting in release of these corepressors and associated protein complexes.

COREPRESSORS BOUND TO UNLIGANDED RECEPTOR

Lysine residues of histones can be acetylated usually on protruding tails emanating from the histones. Generally, acetylated lysine is a mark for euchromatin while deacetylated histones can adopt a more closed confirmation conducive to formation of heterochromatin. Histone deacetylases (HDACs) can be divided into two families based on cofactor utilization with the classic HDACs using Zn2+ and Sirtuins using NAD^+ [21–24]. These two families can then be further subdivided into three classes based on sequence homology. Class I includes HDAC 1, 2, 3, 8, and 11; class II includes HDAC 5, 6, 7, 9, and 10; class III consists of silent information regulator 2 (Sir2) related protein families, SIRT 1–7, or Sirtuins.

HDAC3 appears to play a major role in nuclear receptor corepressor activity for many transcriptional targets. Purification of NCoR/SMRT complexes has revealed a number of stably copurifying components including HDAC3, transducin β-related protein-1 (TBL1), TBL-related protein-1 (TBLR1), and G-protein pathway suppressor-2 (GPS2) [25–28]. These components are believed to be the core enzymatic components in nuclear receptor corepressor mediated HDAC activity. HDAC3 activity is mediated by direct association with NCoR/SMRT through interaction with one of the two highly conserved domains originally identified in SWI3, ADA2 and TFIIIB, termed the SANT domain [29]. This interaction maintains HDAC3 in a properly folded configuration for optimal histone deacetylase activity. HDAC3 activity is widely believed to repress transcription primarily by deacetylating histones, but could actually be a required coactivator under certain circumstances by using other non-histone substrates [30–34]. Thus, inactivation of HDAC3 results in a genome-wide increase in histone acetylation [35]. Other, lower affinity components of NCoR/SMRT include Sin3A and HDAC 1, 2, 4, 5, 7, and 9, which form a distinct corepressor complex and are implemented in repressive functions on specific subsets of genes [18, 36–39].

In contrast to the predominant nuclear localization of class I HDACs, HDAC3 can also be localized to the inner cytoplasmic membrane and the cytoplasm, in addition to the nucleus [40, 41]. The primary amino acid sequence reveals two nuclear export signals on HDAC3 and a C-terminal nuclear localization signal (NLS). Subcellular localization of HDAC3 has been attributed to a number of posttranslational modifications [42, 43]. During apoptosis, caspase dependent cleavage of HDAC3 separates the C-terminal NLS from the catalytic domain and HDAC3 enzymatic activity is relegated to the cytoplasm and can associate with IκB [44]. Protein–protein interactions with factors like TAB-2 also relegates HDAC3 to the cytoplasm following IL-1β treatment [45].

Enzymatic activity of HDAC3 can also be regulated by posttranslation modification. When localized to the cellular membrane, HDAC3 is a substrate for a membrane associated tyrosine kinase, Src [40]. Both DNA protein kinase (DNA PK) and casein kinase 2 (CK2) have been shown to phosphorylate HDAC3 at S424, stimulating enzymatic activity while reciprocal removal of the phosphate can be accomplished by serine/threonine phosphatase 4, reducing enzyme function [43]. Hence, it is also a "sensor" for other signaling pathways, integrating transcriptional programs.

CONTROL OF COREGULATOR EXCHANGE

Upon stimulation with ligand, a careful sequential temporal exchange of coregulators is orchestrated by nuclear receptors. Corepressor complexes bound to the receptor in the absence of ligand are rapidly cleared away and are substituted with coactivator complexes [46]. The intricacies of these dynamic events were illustrated by studies with the estrogen and androgen receptors [47, 48]. Activity of the 26S proteosome is a critical component for activation of target genes in response to ligand stimulation. Treatment with the proteosome inhibitor MG132 severely curtails transcriptional activation of estrogen receptor target genes by ER-α after treatment with ligand, demonstrating protein degradation was a critical component of transcriptional activation and most other nuclear receptors [49]. The receptor itself seems to be a target for degradation, as the steady-state level of ER-α protein drops after ligand treatment [50]. This may reflect the increased degradation of ER-α that occurs during the cycling of transcription factors on the promoters of activated target genes. A detailed kinetic analysis of the *PS2* promoter, a classic ER-α target gene, revealed that ER-α and associated coregulators bind in successive waves, with ER-α being cleared off the promoter by proteosomal mediated degradation with 30–40 minute intervals [47].

Rapid dismissal degradation is a crucial aspect of the clearance of corepressor complexes bound to the nuclear receptor in the absence of ligand or in the presence of "antagonists" and maintains silencing of target genes. Two highly homologous F-Box/WD40-domain containing factors, TBL1 and TBLR1 have been identified as components of the NCoR/SMRT corepressor complex (for review see [51]). Single cell nuclear microinjection experiments using TBL1 and TBLR1 neutralizing antibodies and short inhibitory RNAs (siRNAs) demonstrated a requirement for TBL1 and TBLR1 in transcriptional activation by most nuclear receptors upon ligand stimulation [51]. F-Box containing factors often function as specificity subunits for the Skp, Cullin, F-box (SCF) containing complexes, which mediate poly-ubiquitylation of proteins destined for proteosomal degradation (reviewed in [52]). Indeed, TBL1 and TBLR1 were shown to directly bind the E3 ubiquitin-conjugating enzyme UbcH5 via the F-Box domain, and UbcH5 also shown to be crucial for transcriptional activity of ligand-bound receptors (Figure 4.1) [51].

TBL1 and TBLR1 are present on DNA bound by nuclear receptors constitutively independent of ligand, thus providing a connection between ubiquitin mediated proteosomal degradation to nuclear receptor complexes at the promoter level. While TBLR1 was required for activation by all nuclear receptors tested, TBL1 was dispensable for activation by ligand-bound RAR, but required for activation by estrogen receptor (ERα), thyroid hormone receptor (T_3R), and peroxisome proliferator-activated receptor (PPAR)-γ (Figure 4.2) [51].

Therefore, the activities of TBL1 and TBLR1 were not fully redundant suggesting the two factors may have different targets for the degradation. Deletion of both NCoR and SMRT permitted TBLR1 independent activation of nuclear receptor target genes, supporting the main role of receptor ligand binding, is TBLR1 mediated clearance of the NCoR/SMRT corepressor complex. However, NCoR/SMRT deletion did not abrogate the requirement for TBL1, suggesting that it may be involved in clearance of a different corepressor complex. Recent work has shown that TBL1 interacts specifically with the CtBP/RIP140 corepressor complex and functions to mediate its dismissal from target promoters and ultimate degradation [53]. Thus, TBLR1 and TBL1 appear to have non-redundant roles in corepressor clearance. Each component dismisses distinct corepressor complexes, explaining the requirement of both factors for nuclear receptor action.

The related TBL1 and TBLR1 factors that share a high degree of sequence identity, yet function with distinct corepressors, provide an example of the precise regulation of corepressor complexes. Differences lie in the phosphorylation sites for dismissal of corepressor complexes and activation of gene expression. TBLR1 has two phosphorylation sites, S123 (protein kinase C [PKC]-δ) and S199 (CK1/GSK3), while TBL1 has three phosphorylation sites S173 (CK1), T334 (CK1), S420 (CK1/GSK3) [53]. Surprisingly, the key to specifying the targets of the two factors lies solely in a set of differential phosphorylation sites. For example, TBLR1 contains a consensus phosphorylation site for PKC-δ, S123, which is not present in TBL1, following PKC-δ mediated phosphorylation, poly-ubiquitylation ensues with promoter dismissal and targeted degradation of NCoR/SMRT [53]. In phosphorylation site swapping experiments with mutation of TBL1 phosphorylation sites to match those of TBLR1, the mutant TBL1

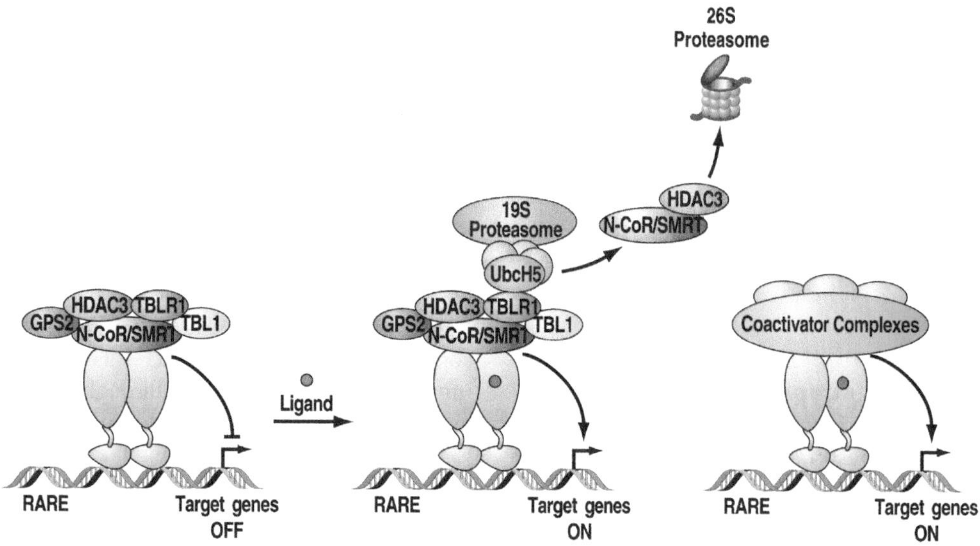

FIGURE 4.1 Coregulator exchange.
In the absence of ligand, NRs are associated with a series of coregulatory proteins that act to repress transcription of target genes. Ligand binding of NRs induces the TBL1/TBL1 dependent recruitment of the E3 ubiquitin-conjugating enzyme UbcH5 and displaces corepressor molecules like NCoR and HDAC3 in exchange for coactivator complexes that promote gene transcription.

TBLR1 and TBL1 Determine Nuclear Receptor Specificity

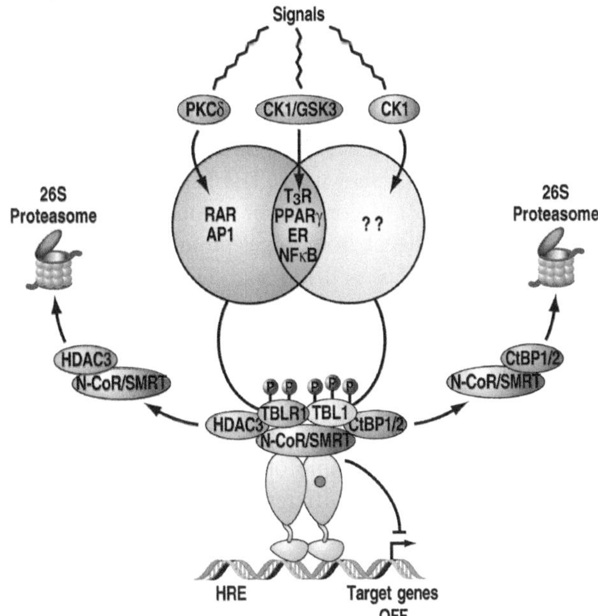

FIGURE 4.2 TBL1 and TBLR1 determine nuclear receptor specificity.
External signals activating protein kinase pathways target distinct phosphorylation sites on TBL1 and TBLR1 and dismiss individual corepressor components determined by NR repression programs. TBLR1 is specifically phosphorylated by PKC-δ at S123 and leads to dismissal and degradation of NCoR/SMRT/HDAC3. Conversely sites on TBL1 phosphorylated by CK1/GSK appear to target NCoR/SMRT/CtBP1,2. RAR is distinct from other NRs by relying solely on TBLR1 mediated dismissal of NCoR/SMRT/HDAC3 for transcriptional activation.

protein switched its substrate specificity from CtBP to NCoR/SMRT imitating TBLR1 specificity toward RAR mediated activation, strongly supporting the hypothesis that these phosphorylation sites are the key determinates for TBL1/TBLR1 specificity. Thus factors serving to control the exchange of coregulatory complexes, with the same corepressor complex, are under the control of different exchange factors. The phospho-regulation of the exchange factors TBL1 and TBLR1 exemplified the posttranslational regulation of components of the nuclear receptor transcriptional repressive machinery with many regulating sites in each protein.

TRANSREPRESSION STRATEGIES

The NCoR/SMRT corepressor complex for gene repression is not limited to nuclear receptors, but extends to many transcription factors, including NFκB, AP1, and the Notch effector recombination signal binding protein for immunoglobulin kappa J region (RBP/J) [54, 55]. Nuclear receptors can act to repress transcription of a number of genes related to inflammation primarily acting through tethering mechanisms with AP-1 and NFκB DNA binding transcription factors. The receptor is tethered to DNA indirectly through protein–protein interactions with DNA binding components of AP-1 and NFκB [56–58]. Sumoylation, or covalent attachment of a small ubiquitin-related modifier appears to a plays central role in governing nuclear receptor mediated transrepression of genes regulated by NFκB and AP-1 in development, inflammation, adipogenesis, and glucose homeostasis [59].

Transrepression by nuclear receptors is essential for biological function. Gene deleted GR mice die very soon after birth presumably due to a lung defect [60]. This lung defect was corrected by GR DIM knock-in mice allowing them to survive past birth [61]. These mice express a GR containing a point mutation that compromises DNA dependent dimerization function, recruitment of coactivators, and gene activation through traditional GREs. Receptor dimerization is not required for transrepression because the GR DIM mutation maintains transrepression function with both AP-1 and NFκB mediated gene regulation demonstrating that nuclear receptor mediated transrepression roles are important for biology [62].

NR can antagonize signal transduction pathways involved in inflammatory responses. Microarray studies of genes involved in inflammatory responses demonstrated that GR, PPAR-γ and LXR can differentially repress specific genes activated by l ipopolysaccharide (LPS) through Toll-like receptors in macrophages [63]. Although GR, PPAR, and LXR specific ligands could inhibit less than one half of all of the LPS induced genes, distinct subsets of genes were not sensitive to all nuclear receptor ligands equally. Combinations of nuclear receptor mediated transrepression is additive on some genes suggesting that for certain genes individual nuclear receptors can act simultaneously on the same target genes. While other combinations are synergistic suggesting individual receptors are acting through separate pathways for transrepression, LPS activation and GR agonist mediated transrepression, could be reversed by TLR-3 signaling following treatment with poly I:C. PPAR-γ and LXR transrepression was not affected by TLR-3 signaling. GR appears to have additional layers of regulation in which PPAR-γ and LXR are immune.

PPAR-γ plays essential roles in lipid metabolism, glucose homeostasis, and immune and inflammatory responses (reviewed in [64, 65]). The same NCoR/SMRT corepressor complex including HDAC3, TBL1, and TBLR1 acting to repress transcription through nuclear receptors is also used by NFκB and AP-1 to repress basal transcription [54, 55]. Using LPS inducible, NFκB responsive inducible nitric oxide synthase (iNOS) gene to study transrepression by ligand activated PPAR-γ, a ligand inducible sumoylation event of PPAR-γ at K365 in the LBD that stabilizes the NCoR–HDAC3 corepressor complex and prevents the recruitment of the E2 conjugating enzyme UbcU5 and exchange for coactivator complexes (Figure 4.3) [59].

This sumo ligation event on PPAR-γ was mediated by unliganded receptor coupled to the E3 ligase protein inhibitor of STAT1 (PIAS) and the E2 conjugating enzyme Ubc9.

Nuclear receptors can regulate ligand dependent transcription through protein–protein interactions by tethering to promoter elements through other transcription factors or by acting through composite DNA response elements. The negative element in the proliferin gene consisted of composite hormone response element (HRE)/AP-1 site and upon stimulation by phorbol ester the AP-1 site becomes occupied by c-jun/c-fos heterodimers and transcription is induced [66]. Upon dual stimulation with glucocorticoids, the GR is activated and joins the c-jun/c-fos heterodimer resulting in transcriptional repression with c-jun as an obligate dimerization partner for the transcriptional repression event. More recent experiments have demonstrated that transrepression occurs because of failure to remove corepressor proteins through an ubiquitin mediated pathway [59]. In a similar example of transrepression, a nuclear receptor (PPAR-γ) is recruited to the gene promoter of iNOS through a tethering mechanism by the p50 component of NFκB dimer [59]. Transrepression results because of the failure to exchange NCoR/HDAC components for activator complexes presumably because of stabilization of the repressor complex by direct sumoylation events on PPAR-γ and failure to recruit the ubiquitylation machinery required for protein degradation pathways.

Nuclear receptors interact with DNA as homo- or hetero-dimers [6]. The spacing and orientation alters the configuration of each bound monomer and allosterically affects the exposure of activation and repression motifs. In the case of RAR/RXR heterodimers, direct repeat (DR) elements with spacing of 5 bps respond to ligand and demonstrate typical exchange of corepressor with coactivator complexes. However, DR elements with 1 bp spacing do not make this exchange resulting in a constitutively repressed outcome.

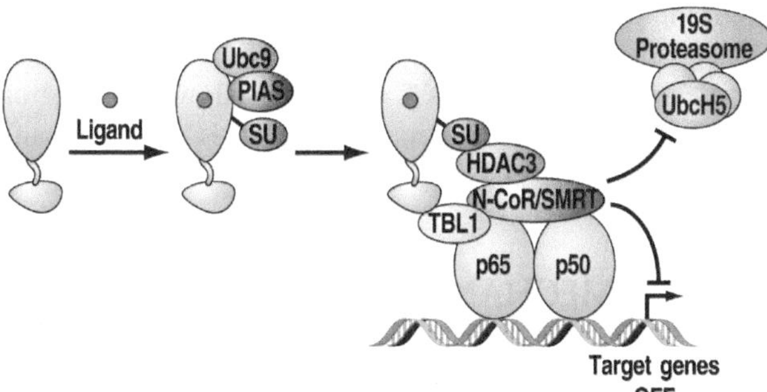

FIGURE 4.3 Transrepression.
Ligand dependent covalent sumolation of NRs stabilize corepressor association with NFκB (p50/p65) and is refractory to corepressor dismissal and degradation and is a general model for transrepression. The sumo ligation event is directed by Ubc9 and PIAS.

COREPRESSORS AS METABOLIC SENSORS

C-terminal Binding Protein (CtBP) was identified by its interaction with the C-terminus of the viral E1A protein, and has emerged as a key corepressor for multiple nuclear receptors. CtBP's function as a repressor was clearly demonstrated in genetic experiments examining its function in Drosophila embryonic development, but the mechanism by which CtBP promotes transcriptional repression has proved elusive [67, 68]. Analysis of the primary amino acid sequence revealed a striking similarity between CtBP and NAD dependent 2-hydroxy acid dehydrogenases. Purified CtBP displayed weak to nonexistent enzymatic activity toward generic dehydrogenase substrates [69]. However, while catalysis may not be involved in CtBP mediated repression, it was clearly demonstrated that CtBP can directly bind to the nicotinamide adenine dinucleotide NAD^+ and reduction product NADH via a conserved binding pocket [69, 70]. Resolution of the crystal structure of CtBP revealed that in the unbound state CtBP adopts a closed conformation, while binding of NAD^+ results in a conformational shift exposing the binding surface for interaction partners such as E1A. Mutation of the NAD^+/NADH binding pocket abolished interaction with E1A and blocked CtBP mediated repression of E1A target genes, as well as repression by unliganded nuclear receptors like RAR [69]. Thus, binding to a metabolic by-product seems to be a key activating event for CtBP. While NAD^+ and NADH are involved in cellular respiration occurring in the cytoplasm, they can freely diffuse into the nucleus where they could influence CtBP function. The binding pocket is potentially able to bind either NAD^+ or NADH based on the structural analysis; however, more recent studies show 100-fold preference for NADH binding with a lower dissociation constant [69, 71]. The requirement of CtBP for NADH binding for effective execution of its repressive program indicates that CtBP may act as a metabolic sensor, testing the overall redox state of the cell based on the ratio of NADH to NAD^+ and regulating its binding partner. In accordance with this idea, exposing cells to hypoxic conditions, a treatment known to increase the NADH/NAD^+ ratio was shown to promote the repression of target genes for the CtBP to ZEB [72].

CtBP is not the only corepressor that could potentially act as a metabolic sensor. The class-III histone deacetylases are characterized by NAD^+ dependent catalytic activity [21]. The founding member of this class is the yeast protein Sir2, which functions as the key mediator of heterochromatic silencing of telomeric regions, silent mating-type loci, and ribosomal DNA loci in yeast [73]. Sir2 is a key regulating enzyme increasing longevity in animals from *C. elegans* to mammals and mediates caloric restriction (CR) [74]. CR in mammals activates white adipose tissue in mammals through PPAR-γ nuclear receptor [75]. The dependence of Sir2 function on NAD^+ binding potentially links it to changes in the cellular metabolic state that could be related to cellular energy level. In *C. elegans*, CR and Sir2 function are linked to insulin signaling. This also may be the case in mammals [76]. The mouse 3T3-L1 adipocyte cells, the Sir2 homolog Sirt1, was shown to block adipogenesis, a process controlled in part by insulin signaling [75]. Sirt1 is recruited to PPAR-γ as part of the NCoR/SMRT corepressor complex and silences pro-adipogenic PPAR-γ target genes. This promotes fat mobilization over fat storage, which could have profound consequences for total lifespan of the mouse. According to one current model for CR, the Sir2 family of NAD^+ dependent HDACs responds to cellular energy level by monitoring free NAD^+ levels and promoting a transcriptional program mediated at least in part by nuclear receptors such as PPAR-γ.

The potential for Sir2 and CtBP to act as metabolic sensors adds another level of complexity to the regulation of corepressor activity on nuclear receptors. While CtBP is responsive to NADH levels, Sir2 is specific to NAD^+, raising the possibility that these factors represent reciprocal functions based on NADH/NAD^+ ratio. A greater understanding of the transcriptional programs regulated by these factors will help us to understand their role in the transcriptional response to cellular metabolic changes.

Allosteric effects could induce a number of different coregulator configurations modulating the efficacy of coregulator function. Initially, it was believed that the majority of coregulator molecules containing LXXLL motifs were coactivators because their interaction with nuclear receptors were induced by agonist. Since then, a number of corepressor molecules recruited through LXXLL motifs to nuclear receptors in a ligand dependent manner have been described including LCoR (ligand dependent corepressor) [77], RIP140 (receptor interaction protein 1) [78], PRAME (preferentially expressed antigen in melanoma) [79], REA (repressor of estrogen receptor activity) [80], MTA1 (metastasis associated factor-1) [81], NSD1 (NR-binding SET domain containing protein 1) [82], and COPR (comodulator of PPAR-γ and RXR-α) [83]. Often coregulatory molecules contain multiple LXXLL motifs, suggesting that individual motifs may play a role in receptor specificity. Recent reviews have described in detail the roles of agonist bound corepressors [84, 85].

The access of sequence-specific transcription factors, including nuclear receptors, to their DNA recognition elements is regulated by chromatin. In the cell, DNA is assembled into nucleosomes, the most basic repeating unit of chromatin, consisting of two copies of histones H2A, H2B, H3, and H4. Each nucleosome has 147 base pairs (bp) of DNA wrapped approximately 1.7 times around the histone octamer. The DNA topology is such that the side of the DNA facing toward the nucleosome is masked or occluded from recognition by DNA binding proteins while the other that faces away from the nucleosome is more accessible. The most common form of DNA makes one helical turn every 10.5 bp of DNA, making accessible DNA surfaces at a periodicity of 10 bp around a nucleosome. Recognition of DNA binding factors like nuclear receptors can be altered by the chromatin state through chromatin remodeling factors that act to slide, eject, or restructure nucleosomes [1].

Chromatin remodeling factors have been separated into distinct families based on homology and activity including SWI/SNF, imitation SWI (ISWI), NuRD/Mi-2/CHD, and split ATPases INO80 and SWR1. ISWI is distinguished from SWI/SNF ATPases by the presence of a C-terminal SANT domain. The role of many of these chromatin remodeling factors in nuclear receptor gene activation programs have been well described; however, the precise link of their actions with respect to transcriptional repression is poorly defined for nuclear receptors. The precise role of chromatin remodeling factors as corepressors including SWI/SNF and ISWI relatively obscure however the prevailing view that they act to move DNA around nucleosomes to an unfavorable position for gene transcription. *In vitro* SWI/SNF factors have been shown to remodel nucleosomes to modulate spacing between nucleosomes that is random [86] or they may simply eject them [87, 88], and histone H1 interactions with mono- and di-H4K20 are suggested to compact interchromosomal distance [89]. ISWI acts to distribute arrays of histones so that for a segment of DNA they are distributed with even spacing [90, 91].

These factors require unacetylated histone tails for recruitment. Unliganded TR recruitment of NCoR/HDAC3 is dependent on the ISWI ATPase remodeling activity of ISWI family member, SNFH2, for repression of the 5' deiodinase type I (Dio1) gene in liver cells [92]. The mechanism of repression is dependent on the histone deacetylase activity of HDAC3 on the histone H4 tail. Deacetylated H4 tail is a binding substrate for SNF2H, an ortholog of ISWI. ISWI can act to space nucleosomes evenly on euchromatin. Negative effects of GR on POMC target genes in pituitary corticotropes depends on HDAC2 and the brg-1 component of SWI/SNF [93]. During POMC activation by CRH acts through PKA and MAPK pathways to hyperphosphorylate NGFI-B to enhance constitutive transcription. During this event the constitutively acetylated tails remain unchanged. GR agonists cause recruitment to NGFI-B based on tethering mechanism with HDAC2 and SWI/SNF with one action of HDAC2 being acetylation of histone H4 tail blocking elongation by RNA pol II.

DISEASE MECHANISMS OF NUCLEAR RECEPTOR DEPENDENT TRANSREPRESSION

In macrophages unliganded PPAR-γ is associated with B-cell CLL/lymphoma 6 (BCL-6) corepressor sequestering it away from NFκB regulated genes resulting in high levels of inflammatory gene expression [94]. Upon ligand binding, PPAR-γ dissociate from BCL-6 allowing it relocate to

TABLE 4.1 Example of actions of methyltransferase/demethylase

Human gene name	Mode of action	Substrate	Function	Reference
JMJD2A/JHDM3a	K demethylase	H3K9/K36Me2/3	Transcriptional repression	[97, 98]
JHDM1	K demethylase	K36Me2	Heterochromatin formation	[99, 100]
JARID1B/PLU-1	K demethylase	H3K4Me2/H3K4Me3	Transcriptional repression	[101]
SUV39H1	K methyltransferase	H3K9	Heterochromatin formation	[102]
SUV39H2	K methyltransferase	H3K9	Heterochromatin formation	[103]
G9a	K methyltransferase	H3K9	Heterochromatin formation	[104]
EuHMTase/GLP	K methyltransferase	H3K9	Heterochromatin formation	[105]
ESET/SETDB1	K methyltransferase	H3K9	Heterochromatin formation	[106]
PRC2 complex	K methyltransferase	H3K27	Polycomb silencing	[107]
RIZ1	K methyltransferase	H3K9	Heterochromatin formation	[108]

and repress NFκB regulated genes (e.g., MCP1) that exert anti-inflammatory effects. Other mechanisms involving nuclear receptor mediated transrepression using corepressor molecules include inhibition of p-TEFβ activity for hyper-phosphorylation of the C-terminal domain of RNA pol II transcription essential for transcriptional elongation [95, 96].

FUTURE DIRECTIONS

A large number of covalent histone modifications have been described. In Table 4.1, we list methylation sites related to transcriptional repression and the enzymes that add or reciprocally erase the marks that will be the next frontier in understanding the players and mechanisms in the next round of studies.

REFERENCES

1. Cairns BR. Chromatin remodeling: insights and intrigue from single-molecule studies. *Nat Struct Mol Biol* 2007;**14**:989–96.
2. Beato M, Herrlich P, Schutz G. Steroid hormone receptors: many actors in search of a plot. *Cell* 1995;**83**:851–7.
3. Bertrand S, Brunet FG, Escriva H, Parmentier G, Laudet V, Robinson-Rechavi M. Evolutionary genomics of nuclear receptors: from twenty-five ancestral genes to derived endocrine systems. *Mol Biol Evol* 2004;**21**:1923–37.
4. Evans RM. The steroid and thyroid hormone receptor superfamily. *Science* 1988;**240**:889–95.
5. Giguere V. Orphan nuclear receptors: from gene to function. *Endocr Rev* 1999;**20**:689–725.
6. Mangelsdorf DJ, Thummel C, Beato M, Herrlich P, Schutz G, Umesono K, Blumberg B, Kastner P, Mark M, Chambon P, Evans RM. The nuclear receptor superfamily: the second decade. *Cell* 1995;**83**:835–9.
7. Picard D. Chaperoning steroid hormone action. *Trends Endocrinol Metab* 2006;**17**:229–35.
8. Mangelsdorf DJ, Evans RM. The RXR heterodimers and orphan receptors. *Cell* 1995;**83**:841–50.
9. Bourguet W, Ruff M, Chambon P, Gronemeyer H, Moras D. Crystal structure of the ligand-binding domain of the human nuclear receptor RXR-alpha. *Nature* 1995;**375**:377–82.
10. Brzozowski AM, Pike AC, Dauter Z, Hubbard RE, Bonn T, Engstrom O, Ohman L, Greene GL, Gustafsson JA, Carlquist M. Molecular basis of agonism and antagonism in the oestrogen receptor. *Nature* 1997;**389**:753–8.
11. Darimont BD, Wagner RL, Apriletti JW, Stallcup MR, Kushner PJ, Baxter JD, Fletterick RJ, Yamamoto KR. Structure and specificity of nuclear receptor-coactivator interactions. *Genes Dev* 1998;**12**:3343–56.
12. Moras D, Gronemeyer H. The nuclear receptor ligand-binding domain: structure and function. *Curr Opin Cell Biol* 1998;**10**:384–91.
13. Nolte RT, Wisely GB, Westin S, Cobb JE, Lambert MH, Kurokawa R, Rosenfeld MG, Willson TM, Glass CK, Milburn MV. Ligand binding and co-activator assembly of the peroxisome proliferator-activated receptor-gamma. *Nature* 1998;**395**:137–43.
14. Renaud JP, Rochel N, Ruff M, Vivat V, Chambon P, Gronemeyer H, Moras D. Crystal structure of the RAR-gamma ligand-binding domain bound to all-trans retinoic acid. *Nature* 1995;**378**:681–9.
15. Shiau AK, Barstad D, Loria PM, Cheng L, Kushner PJ, Agard DA, Greene GL. The structural basis of estrogen receptor/coactivator recognition and the antagonism of this interaction by tamoxifen. *Cell* 1998;**95**:927–37.
16. Wagner RL, Apriletti JW, McGrath ME, West BL, Baxter JD, Fletterick RJ. A structural role for hormone in the thyroid hormone receptor. *Nature* 1995;**378**:690–7.

17. Hu X, Lazar MA. The CoRNR motif controls the recruitment of corepressors by nuclear hormone receptors. *Nature* 1999;**402**:93–6.
18. Nagy L, Kao HY, Chakravarti D, Lin RJ, Hassig CA, Ayer DE, Schreiber SL, Evans RM. Nuclear receptor repression mediated by a complex containing SMRT, mSin3A, and histone deacetylase. *Cell* 1997;**89**:373–80.
19. Perissi V, Dasen JS, Kurokawa R, Wang Z, Korzus E, Rose DW, Glass CK, Rosenfeld MG. Factor-specific modulation of CREB-binding protein acetyltransferase activity. *Proc Natl Acad Sci U S A* 1999;**96**:3652–7.
20. Webb P, Anderson CM, Valentine C, Nguyen P, Marimuthu A, West BL, Baxter JD, Kushner PJ. The nuclear receptor corepressor (N-CoR) contains three isoleucine motifs (I/LXXII) that serve as receptor interaction domains (IDs). *Mol Endocrinol* 2000;**14**:1976–85.
21. Blander G, Guarente L. The Sir2 family of protein deacetylases. *Annu Rev Biochem* 2004;**73**:417–35.
22. Cress WD, Seto E. Histone deacetylases, transcriptional control, and cancer. *J Cell Physiol* 2000;**184**:1–16.
23. Gregoretti IV, Lee YM, Goodson HV. Molecular evolution of the histone deacetylase family: functional implications of phylogenetic analysis. *J Mol Biol* 2004;**338**:17–31.
24. Grozinger CM, Schreiber SL. Deacetylase enzymes: biological functions and the use of small-molecule inhibitors. *Chem Biol* 2002;**9**:3–16.
25. Guenther MG, Lane WS, Fischle W, Verdin E, Lazar MA, Shiekhattar R. A core SMRT corepressor complex containing HDAC3 and TBL1, a WD40-repeat protein linked to deafness. *Genes Dev* 2000; **14**:1048–57.
26. Li J, Wang J, Wang J, Nawaz Z, Liu JM, Qin J, Wong J. Both corepressor proteins SMRT and N-CoR exist in large protein complexes containing HDAC3. *Embo J* 2000;**19**:4342–50.
27. Yoon HG, Chan DW, Huang ZQ, Li J, Fondell JD, Qin J, Wong J. Purification and functional characterization of the human N-CoR complex: the roles of HDAC3, TBL1 and TBLR1. *Embo J* 2003; **22**:1336–46.
28. Zhang J, Kalkum M, Chait BT, Roeder RG. The N-CoR–HDAC3 nuclear receptor corepressor complex inhibits the JNK pathway through the integral subunit GPS2. *Mol Cell* 2002;**9**:611–23.
29. Guenther MG, Barak O, Lazar MA. The SMRT and N-CoR corepressors are activating cofactors for histone deacetylase 3. *Mol Cell Biol* 2001;**21**:6091–101.
30. Chen L, Fischle W, Verdin E, Greene WC. Duration of nuclear NF-kappaB action regulated by reversible acetylation. *Science* 2001;**293**:1653–7.
31. Fu J, Yoon HG, Qin J, Wong J. Regulation of P-TEFb elongation complex activity by CDK9 acetylation. *Mol Cell Biol* 2007;**27**:4641–51.
32. Gregoire S, Xiao L, Nie J, Zhang X, Xu M, Li J, Wong J, Seto E, Yang XJ. Histone deacetylase 3 interacts with and deacetylates myocyte enhancer factor 2. *Mol Cell Biol* 2007;**27**:1280–95.
33. Thevenet L, Mejean C, Moniot B, Bonneaud N, Galeotti N, Aldrian-Herrada G, Poulat F, Berta P, Benkirane M, Boizet-Bonhoure B. Regulation of human SRY subcellular distribution by its acetylation/deacetylation. *Embo J* 2004;**23**:3336–45.
34. Zeng L, Xiao Q, Margariti A, Zhang Z, Zampetaki A, Patel S, Capogrossi MC, Hu Y, Xu Q. HDAC3 is crucial in shear- and VEGF-induced stem cell differentiation toward endothelial cells. *J Cell Biol* 2006;**174**:1059–69.
35. Glaser KB, Li J, Staver MJ, Wei RQ, Albert DH, Davidsen SK. Role of class I and class II histone deacetylases in carcinoma cells using siRNA. *Biochem Biophys Res Commun* 2003;**310**:529–36.
36. Alland Jr. L, Muhle R, Hou H, Potes J, Chin L, Schreiber-Agus N, DePinho RA Role for N-CoR and histone deacetylase in Sin3-mediated transcriptional repression. *Nature* 1997;**387**:49–55.
37. Heinzel T, Lavinsky RM, Mullen TM, Soderstrom M, Laherty CD, Torchia J, Yang WM, Brard G, Ngo SD, Davie JR, Seto E, Eisenman RN, Rose DW, Glass CK, Rosenfeld MG. A complex containing N-CoR, mSin3 and histone deacetylase mediates transcriptional repression. *Nature* 1997;**387**:43–8.
38. Huang EY, Zhang J, Miska EA, Guenther MG, Kouzarides T, Lazar MA. Nuclear receptor corepressors partner with class II histone deacetylases in a Sin3-independent repression pathway. *Genes Dev* 2000;**14**:45–54.
39. Kao HY, Downes M, Ordentlich P, Evans RM. Isolation of a novel histone deacetylase reveals that class I and class II deacetylases promote SMRT-mediated repression. *Genes Dev* 2000;**14**:55–66.
40. Longworth MS, Laimins LA. Histone deacetylase 3 localizes to the plasma membrane and is a substrate of Src. *Oncogene* 2006;**25**:4495–500.
41. Takami Y, Nakayama T. N-terminal region, C-terminal region, nuclear export signal, and deacetylation activity of histone deacetylase-3 are essential for the viability of the DT40 chicken B cell line. *J Biol Chem* 2000;**275**:16,191–201.
42. Jeyakumar M, Liu XF, Erdjument-Bromage H, Tempst P, Bagchi MK. Phosphorylation of thyroid hormone receptor-associated nuclear receptor corepressor holocomplex by the DNA-dependent protein kinase enhances its histone deacetylase activity. *J Biol Chem* 2007;**282**:9312–22.
43. Zhang X, Ozawa Y, Lee H, Wen YD, Tan TH, Wadzinski BE, Seto E. Histone deacetylase 3 (HDAC3) activity is regulated by interaction with protein serine/threonine phosphatase 4. *Genes Dev* 2005;**19**:827–39.
44. Escaffit F, Vaute O, Chevillard-Briet M, Segui B, Takami Y, Nakayama T, Trouche D. Cleavage and cytoplasmic relocalization of histone deacetylase 3 are important for apoptosis progression. *Mol Cell Biol* 2007;**27**:554–67.
45. Baek SH, Ohgi KA, Rose DW, Koo EH, Glass CK, Rosenfeld MG. Exchange of N-CoR corepressor and Tip60 coactivator complexes links gene expression by NF-kappaB and beta-amyloid precursor protein. *Cell* 2002;**110**:55–67.
46. Glass CK, Rosenfeld MG. The coregulator exchange in transcriptional functions of nuclear receptors. *Genes Dev* 2000;**14**:121–41.
47. Metivier R, Penot G, Hubner MR, Reid G, Brand H, Kos M, Gannon F. Estrogen receptor-alpha directs ordered, cyclical, and combinatorial recruitment of cofactors on a natural target promoter. *Cell* 2003;**115**:751–63.
48. Kang Z, Pirskanen A, Janne OA, Palvimo JJ. Involvement of proteasome in the dynamic assembly of the androgen receptor transcription complex. *J Biol Chem* 2002;**277**:48,366–71.
49. Lonard DM, Nawaz Z, Smith CL, O'Malley BW. The 26S proteasome is required for estrogen receptor-alpha and coactivator turnover and for efficient estrogen receptor-alpha transactivation. *Mol Cell* 2000;**5**:939–48.
50. Nawaz Z, Lonard DM, Dennis AP, Smith CL, O'Malley BW. Proteasome-dependent degradation of the human estrogen receptor. *Proc Natl Acad Sci U S A* 1999;**96**:1858–62.
51. Perissi V, Aggarwal A, Glass CK, Rose DW, Rosenfeld MG. A corepressor/coactivator exchange complex required for transcriptional activation by nuclear receptors and other regulated transcription factors. *Cell* 2004;**116**:511–26.
52. Patton EE, Willems AR, Tyers M. Combinatorial control in ubiquitin-dependent proteolysis: don't Skp the F-box hypothesis. *Trends Genet* 1998;**14**:236–43.

53. Perissi V, Scafoglio C, Zhang J, Ohgi KA, Rose DW, Glass CK, Rosenfeld MG. TBL1 and TBLR1 phosphorylation on regulated gene promoters overcomes dual CtBP and NCoR/SMRT transcriptional repression checkpoints. *Mol Cell* 2008;**29**:755–66.
54. Ogawa S, Lozach J, Jepsen K, Sawka-Verhelle D, Perissi V, Sasik R, Rose DW, Johnson RS, Rosenfeld MG, Glass CK. A nuclear receptor corepressor transcriptional checkpoint controlling activator protein 1-dependent gene networks required for macrophage activation. *Proc Natl Acad Sci USA* 2004;**101**:14,461–14,466.
55. Hoberg JE, Yeung F, Mayo MW. SMRT derepression by the IkappaB kinase alpha: a prerequisite to NF-kappaB transcription and survival. *Mol Cell* 2004;**16**:245–55.
56. Jonat C, Rahmsdorf HJ, Park KK, Cato AC, Gebel S, Ponta H, Herrlich P. Antitumor promotion and antiinflammation: down-modulation of AP-1 (Fos/Jun) activity by glucocorticoid hormone. *Cell* 1990;**62**:1189–204.
57. Ray A, Prefontaine KE. Physical association and functional antagonism between the p65 subunit of transcription factor NF-kappa B and the glucocorticoid receptor. *Proc Natl Acad Sci U S A* 1994;**91**:752–6.
58. Schule R, Rangarajan P, Kliewer S, Ransone LJ, Bolado J, Yang N, Verma IM, Evans RM. Functional antagonism between oncoprotein c-Jun and the glucocorticoid receptor. *Cell* 1990;**62**:1217–26.
59. Pascual G, Fong AL, Ogawa S, Gamliel A, Li AC, Perissi V, Rose DW, Willson TM, Rosenfeld MG, Glass CK. A SUMOylation-dependent pathway mediates transrepression of inflammatory response genes by PPAR–gamma. *Nature* 2005;**437**:759–63.
60. Cole TJ, Blendy JA, Monaghan AP, Krieglstein K, Schmid W, Aguzzi A, Fantuzzi G, Hummler E, Unsicker K, Schutz G. Targeted disruption of the glucocorticoid receptor gene blocks adrenergic chromaffin cell development and severely retards lung maturation. *Genes Dev* 1995;**9**:1608–21.
61. Reichardt HM, Kaestner KH, Tuckermann J, Kretz O, Wessely O, Bock R, Gass P, Schmid W, Herrlich P, Angel P, Schutz G. DNA binding of the glucocorticoid receptor is not essential for survival. *Cell* 1998;**93**:531–41.
62. Reichardt HM, Tuckermann JP, Gottlicher M, Vujic M, Weih F, Angel P, Herrlich P, Schutz G. Repression of inflammatory responses in the absence of DNA binding by the glucocorticoid receptor. *Embo J* 2001;**20**:7168–73.
63. Ogawa S, Lozach J, Benner C, Pascual G, Tangirala RK, Westin S, Hoffmann A, Subramaniam S, David M, Rosenfeld MG, Glass CK. Molecular determinants of crosstalk between nuclear receptors and toll-like receptors. *Cell* 2005;**122**:707–21.
64. Berger JP, Akiyama TE, Meinke PT. PPARs: therapeutic targets for metabolic disease. *Trends Pharmacol Sci* 2005;**26**:244–51.
65. Evans RM, Barish GD, Wang YX. PPARs and the complex journey to obesity. *Nat Med* 2004;**10**:355–61.
66. Pearce D, Yamamoto KR. Mineralocorticoid and glucocorticoid receptor activities distinguished by nonreceptor factors at a composite response element. *Science* 1993;**259**:1161–5.
67. Nibu Y, Zhang H, Levine M. Interaction of short-range repressors with Drosophila CtBP in the embryo. *Science* 1998;**280**:101–4.
68. Chinnadurai G. CtBP, an unconventional transcriptional corepressor in development and oncogenesis. *Mol Cell* 2002;**9**:213–24.
69. Kumar V, Carlson JE, Ohgi KA, Edwards TA, Rose DW, Escalante CR, Rosenfeld MG, Aggarwal AK. Transcription corepressor CtBP is an NAD(+)-regulated dehydrogenase. *Mol Cell* 2002;**10**:857–69.
70. Zhang Q, Piston DW, Goodman RH. Regulation of corepressor function by nuclear NADH. *Science* 2002;**295**:1895–7.
71. Fjeld CC, Birdsong WT, Goodman RH. Differential binding of NAD+ and NADH allows the transcriptional corepressor carboxyl-terminal binding protein to serve as a metabolic sensor. *Proc Natl Acad Sci U S A* 2003;**100**:9202–7.
72. Zhang Q, Wang SY, Nottke AC, Rocheleau JV, Piston DW, Goodman RH. Redox sensor CtBP mediates hypoxia-induced tumor cell migration. *Proc Natl Acad Sci U S A* 2006;**103**:9029–33.
73. Lin SJ, Kaeberlein M, Andalis AA, Sturtz LA, Defossez PA, Culotta VC, Fink GR, Guarente L. Calorie restriction extends Saccharomyces cerevisiae lifespan by increasing respiration. *Nature* 2002;**418**:344–8.
74. Tissenbaum HA, Guarente L. Increased dosage of a sir-2 gene extends lifespan in Caenorhabditis elegans. *Nature* 2001;**410**:227–30.
75. Picard F, Kurtev M, Chung N, Topark-Ngarm A, Senawong T, Machado De Oliveira R, Leid M, McBurney MW, Guarente L. Sirt1 promotes fat mobilization in white adipocytes by repressing PPAR-gamma. *Nature* 2004;**429**:771–6.
76. Kenyon C. A conserved regulatory system for aging. *Cell* 2001;**105**:165–8.
77. Fernandes I, Bastien Y, Wai T, Nygard K, Lin R, Cormier O, Lee HS, Eng F, Bertos NR, Pelletier N, Mader S, Han VK, Yang XJ, White JH. Ligand-dependent nuclear receptor corepressor LCoR functions by histone deacetylase-dependent and -independent mechanisms. *Mol Cell* 2003;**11**:139–50.
78. Cavailles V, Dauvois S, L'Horset F, Lopez G, Hoare S, Kushner PJ, Parker MG. Nuclear factor RIP140 modulates transcriptional activation by the estrogen receptor. *Embo J* 1995;**14**:3741–51.
79. Epping MT, Wang L, Edel MJ, Carlee L, Hernandez M, Bernards R. The human tumor antigen PRAME is a dominant repressor of retinoic acid receptor signaling. *Cell* 2005;**122**:835–47.
80. Delage-Mourroux R, Martini PG, Choi I, Kraichely DM, Hoeksema J, Katzenellenbogen BS. Analysis of estrogen receptor interaction with a repressor of estrogen receptor activity (REA) and the regulation of estrogen receptor transcriptional activity by REA. *J Biol Chem* 2000;**275**:35,848–56.
81. Mazumdar A, Wang RA, Mishra SK, Adam L, Bagheri-Yarmand R, Mandal M, Vadlamudi RK, Kumar R. Transcriptional repression of oestrogen receptor by metastasis-associated protein 1 corepressor. *Nat Cell Biol* 2001;**3**:30–7.
82. Huang N, vom Baur E, Garnier JM, Lerouge T, Vonesch JL, Lutz Y, Chambon P, Losson R. Two distinct nuclear receptor interaction domains in NSD1, a novel SET protein that exhibits characteristics of both corepressors and coactivators. *Embo J* 1998;**17**:3398–412.
83. Flores AM, Li L, Aneskievich BJ. Isolation and functional analysis of a keratinocyte-derived, ligand-regulated nuclear receptor comodulator. *J Invest Dermatol* 2004;**123**:1092–101.
84. Christian M, White R, Parker MG. Metabolic regulation by the nuclear receptor corepressor RIP140. *Trends Endocrinol Metab* 2006;**17**:243–50.
85. Gurevich I, Flores AM, Aneskievich BJ. Corepressors of agonist-bound nuclear receptors. *Toxicol Appl Pharmacol* 2007;**223**:288–98.
86. Owen-Hughes T, Utley RT, Cote J, Peterson CL, Workman JL. Persistent site-specific remodeling of a nucleosome array by transient action of the SWI/SNF complex. *Science* 1996;**273**:513–16.
87. Lorch Y, Zhang M, Kornberg RD. Histone octamer transfer by a chromatin–remodeling complex. *Cell* 1999;**96**:389–92.
88. Lorch Y, Maier-Davis B, Kornberg RD. Chromatin remodeling by nucleosome disassembly in vitro. *Proc Natl Acad Sci U S A* 2006;**103**:3090–3.
89. Trojer III P, Li G, Sims RJ, Vaquero A, Kalakonda N, Boccuni P, Lee D, Erdjument-Bromage H, Tempst P, Nimer SD, Wang YH, Reinberg D. L3MBTL1, a histone-methylation-dependent chromatin lock. *Cell* 2007;**129**:915–28.

90. Ito T, Bulger M, Pazin MJ, Kobayashi R, Kadonaga JT. ACF, an ISWI-containing and ATP-utilizing chromatin assembly and remodeling factor. *Cell* 1997;**90**:145–55.
91. Varga-Weisz PD, Wilm M, Bonte E, Dumas K, Mann M, Becker PB. Chromatin-remodelling factor CHRAC contains the ATPases ISWI and topoisomerase II. *Nature* 1997;**388**:598–602.
92. Alenghat T, Yu J, Lazar MA. The N-CoR complex enables chromatin remodeler SNF2H to enhance repression by thyroid hormone receptor. *Embo J* 2006;**25**:3966–74.
93. Bilodeau S, Vallette-Kasic S, Gauthier Y, Figarella-Branger D, Brue T, Berthelet F, Lacroix A, Batista D, Stratakis C, Hanson J, Meij B, Drouin J. Role of Brg1 and HDAC2 in GR trans-repression of the pituitary POMC gene and misexpression in Cushing disease. *Genes Dev* 2006;**20**:2871–86.
94. Lee CH, Chawla A, Urbiztondo N, Liao D, Boisvert WA, Evans RM, Curtiss LK. Transcriptional repression of atherogenic inflammation: modulation by PPARdelta. *Science* 2003;**302**:453–7.
95. Luecke HF, Yamamoto KR. The glucocorticoid receptor blocks P-TEFb recruitment by NFkappaB to effect promoter-specific transcriptional repression. *Genes Dev* 2005;**19**:1116–27.
96. Nissen RM, Yamamoto KR. The glucocorticoid receptor inhibits NFkappaB by interfering with serine-2 phosphorylation of the RNA polymerase II carboxy-terminal domain. *Genes Dev* 2000;**14**:2314–29.
97. Klose RJ, Yamane K, Bae Y, Zhang D, Erdjument-Bromage H, Tempst P, Wong J, Zhang Y. The transcriptional repressor JHDM3A demethylates trimethyl histone H3 lysine 9 and lysine 36. *Nature* 2006;**442**:312–16.
98. Yamane K, Toumazou C, Tsukada Y, Erdjument-Bromage H, Tempst P, Wong J, Zhang Y. JHDM2A, a JmjC-containing H3K9 demethylase, facilitates transcription activation by androgen receptor. *Cell* 2006;**125**:483–95.
99. Tsukada Y, Fang J, Erdjument-Bromage H, Warren ME, Borchers CH, Tempst P, Zhang Y. Histone demethylation by a family of JmjC domain-containing proteins. *Nature* 2006;**439**:811–16.
100. Whetstine JR, Nottke A, Lan F, Huarte M, Smolikov S, Chen Z, Spooner E, Li E, Zhang G, Colaiacovo M, Shi Y. Reversal of histone lysine trimethylation by the JMJD2 family of histone demethylases. *Cell* 2006;**125**:467–81.
101. Christensen J, Agger K, Cloos PA, Pasini D, Rose S, Sennels L, Rappsilber J, Hansen KH, Salcini AE, Helin K. RBP2 belongs to a family of demethylases, specific for tri-and dimethylated lysine 4 on histone 3. *Cell* 2007;**128**:1063–76.
102. Rea S, Eisenhaber F, O'Carroll D, Strahl BD, Sun ZW, Schmid M, Opravil S, Mechtler K, Ponting CP, Allis CD, Jenuwein T. Regulation of chromatin structure by site-specific histone H3 methyltransferases. *Nature* 2000;**406**:593–9.
103. O'Carroll D, Scherthan H, Peters AH, Opravil S, Haynes AR, Laible G, Rea S, Schmid M, Lebersorger A, Jerratsch M, Sattler L, Mattei MG, Denny P, Brown SD, Schweizer D, Jenuwein T. Isolation and characterization of Suv39h2, a second histone H3 methyltransferase gene that displays testis-specific expression. *Mol Cell Biol* 2000;**20**:9423–33.
104. Tachibana M, Sugimoto K, Fukushima T, Shinkai Y. Set domain-containing protein, G9a, is a novel lysine-preferring mammalian histone methyltransferase with hyperactivity and specific selectivity to lysines 9 and 27 of histone H3. *J Biol Chem* 2001;**276**:25,309–25,317.
105. Tachibana M, Ueda J, Fukuda M, Takeda N, Ohta T, Iwanari H, Sakihama T, Kodama T, Hamakubo T, Shinkai Y. Histone methyltransferases G9a and GLP form heteromeric complexes and are both crucial for methylation of euchromatin at H3–K9. *Genes Dev* 2005;**19**:815–26.
106. Wang H, An W, Cao R, Xia L, Erdjument-Bromage H, Chatton B, Tempst P, Roeder RG, Zhang Y. mAM facilitates conversion by ESET of dimethyl to trimethyl lysine 9 of histone H3 to cause transcriptional repression. *Mol Cell* 2003;**12**:475–87.
107. Kuzmichev A, Nishioka K, Erdjument-Bromage H, Tempst P, Reinberg D. Histone methyltransferase activity associated with a human multiprotein complex containing the Enhancer of Zeste protein. *Genes Dev* 2002;**16**:2893–905.
108. Kim KC, Geng L, Huang S. Inactivation of a histone methyltransferase by mutations in human cancers. *Cancer Res* 2003;**63**:7619–23.

Chapter 5

Steroid Hormone Receptor Signaling

Vincent Giguère

The Rosalind and Morris Goodman Cancer Centre, Faculty of Medicine, McGill University, Montréal, Québec, Canada

INTRODUCTION

Steroid hormones are essential regulators of key physiological processes such as reproduction, glucose metabolism, the response to stress, and salt balance. The biological effects of steroid hormones are transduced by intracellular receptors that directly mediate the action of their cognate hormone [1]. Steroid hormone receptors (SHRs) were the first recognized members of the steroid/thyroid/retinoid nuclear receptor superfamily, a class of transcription factors whose activity is regulated by small lipid-soluble molecules. The SHR subgroup includes receptors for estradiol (estrogen receptor α and β – ERα (NR3A1) and ERβ (NR3A2)), cortisol (glucocorticoid receptor – GR (NR3C1)), aldosterone (mineralocorticoid receptor – MR (NR3C2)), progesterone (PR (NR3C3)), and dihydrotestosterone (androgen receptor – AR (NR3C4)). In addition, the SHR subgroup contains three orphan nuclear receptors closely related to the ERs (estrogen-related receptor α, β, and γ – ERRα (NR3B1), β (NR3B2), and γ (NR3B3)) for which a natural ligand remains to be identified [2]. SHRs share a common modular structure composed of independent functional domains [3]. The DNA binding domain (DBD) is centrally located, well-conserved among SHRs, and comprised of two zinc finger motifs involved in both protein:DNA and protein:protein contacts. The ligand binding domain (LBD), located at the carboxy-terminus of the receptor, is moderately conserved and folds into a canonical α-helical sandwich generally consisting of 12 α-helices (H1 to H12) [4]. The LBD contains a ligand dependent nuclear translocation signal, determinants to bind chaperone proteins, dimerization interfaces and a potent ligand dependent activation domain (AF) referred to as the AF-2. A ligand independent activation domain (AF-1) is encoded within the nonconserved amino-terminal region of the receptors.

ACTIVATION BY THE HORMONE

The classic model of SHR action dictates that SHRs interact with chaperones in the cytoplasm, and are dissociated in a ligand dependent fashion, leading to the reorganization of the receptor and exposure of nuclear localization signal(s) and translocation to the nucleus. Although this model is widely accepted for SHRs, there are important exceptions [5]. The ERs are clearly localized to the nucleus despite being part of a complex with chaperone proteins. Perhaps more striking is the finding that the two forms of the PR, PR-A and PR-B, which differ only in the length of their amino-terminal domains, have distinct cellular localization. The PR-A is found predominantly in the nucleus while PR-B is mainly located in the cytoplasm in the absence of hormone [6]. Given that the nuclear localization and chaperone binding functions are identical for both forms of the PR, this data strongly suggests that the interactions of SHRs with complexes containing coregulatory proteins might influence the intracellular distribution of SHRs.

Intracellular redistribution of ligand-bound SHRs is accompanied by the recognition of specific sites on chromatin, referred to as hormone response elements (HREs) [7]. The HREs are short cis-acting sequences located within the promoters and/or enhancers of target genes. SHRs bind DNA as homodimers to HREs composed of inverted repeats of AGGTCA (for the ERs and ERRs) or AGAACA (for the GR, MR, PR, and AR) motifs spaced by three nucleotides. SHR isoforms such as ERα and ERβ or PR-A and PR-B can also form functional heterodimeric complexes on DNA [8–10]. The functional properties of each isoform of the ERs are retained in the heterodimeric complex [11], whereas PR-A is dominant over PR-B [12].

The transcriptional activity of the SHRs is mediated by the two independent domains referred to as AF-1 and AF-2 [13]. The AF-1 is ligand independent and

Handbook of Cell Signaling, Three-Volume Set 2 ed.

constitutive since it can activate transcription in the absence of the ligand when fused to a heterologous DNA binding domain. The structural determinants and mode of action of the distinct AF-1 domains found in each SHR have yet to be characterized. In contrast, the ligand dependent AF-2 is well defined: a short amphipathic α-helix (H12) located at the carboxy-terminal end of the LBD is repositioned upon hormone binding into a hydrophobic cleft formed mainly by residues from helices 3, 4, and 12. The resulting structural change provides a functional interface for coactivator recruitment by the receptor [14]. The AF-1 and AF-2 recruit both common and specific cofactors. These regulatory proteins possess various enzymatic activities such as acetylase, deacetylase, methylase, kinase, and ubiquitinase functions [15]. An RNA molecule known as SRA has also been characterized as an SHR coactivator [16]. Cofactors participate in the remodeling of chromatin, the formation of a stable transcription initiation complex, the association or dissociation of other cofactors within the SHR–cofactor complex as well as recycling and degradation of the receptor [17].

HORMONE INDEPENDENT ACTIVATION

SHRs are phosphoproteins and targets of kinase cascades involved in the response to growth factors and cytokines [18]. Many aspects of SHR function can be modulated in this way, including dimerization, DNA-binding, and both ligand independent and dependent activation [19]. SHRs are the targets of protein kinase A, mitogen-activated protein kinase (MAPK), cyclin dependent kinases, casein kinase, and glycogen-synthase kinase. The molecular mechanisms underlying modulation of SHR activity upon phosphorylation have yet to be elucidated in most instances. However, in the case of ERβ, phosphorylation of two serine residues within the AF-1 promotes the recruitment of steroid receptor coactivator-1 (SRC-1) both *in vivo* and *in vitro* [20]. The physiological relevance of hormone independent activation pathways was clearly demonstrated using the uterus of the ERα mouse knockout as a model [21]. In wild-type animals, the uterus displays growth responses to epidermal growth factor (EGF) and insulin-like growth factor 1 (IGF-1). In the ERα knockout animals, the uterus is unresponsive to the mitogenic actions of EGF and IGF-1, demonstrating an essential requirement for ERα in these biological responses [22, 23]. In cellular and *in vitro* models, EGF, IGF-1, and other agents can induce the phosphorylation of serine 118 in the ERα amino-terminal region [24]. Mutation of this residue abolishes the response of ERα to EGF and considerably reduces the ability of the receptor to respond to its cognate ligand.

CROSS-TALK WITH OTHER TRANSCRIPTION FACTORS

Steroid hormone regulation of a number of target genes does not require the presence of an HRE within the transcriptional unit of those genes. Instead, SHRs can tether to a transcription factor and thus modify its activity. The formation of a SHR/transcription factor complex can lead to changes in cellular localization, DNA binding activity, enzymatic function within the transcription initiation complex, or conversely can provide a hormone dependent transcriptional activation function to a non-HRE site [25, 26]. The biological actions of SHRs have been linked to binding sites for more than a dozen transcription factors, as well as to non-HRE sites for unidentified factors. Recent studies using chromatin immunoprecipitation (ChIP) in combination with genomic DNA microarrays (chip) to identify SHR binding sites in whole genomes have shown that approximately one-third of binding events involved non-HRE sites [27–29]. In addition, these functional genomics studies revealed that SHR action often requires the presence of other transcription factors at their target promoters or enhancers [27, 29–32]. The *in vivo* relevance of the HRE independent pathways was highlighted by the observation that, in contrast to GR deficient mice, genetically engineered mice expressing a non-DNA binding form of the GR are viable [33].

NON-GENOMIC ACTION OF STEROID HORMONES

Steroid hormones have been shown to elicit biological responses too rapid to involve gene transcription and subsequent synthesis of new proteins. Rapid effects of steroid hormones have been reported in blood vessels, bone, breast cancer cells, nervous system, sperm, and maturating oocytes [34–38]. These effects involve changes in the activities of enzymes such as phospholipase C, PI(3) kinase, and adenylate cyclase, leading to increases in intracellular calcium levels, second messengers, and activation of kinase cascades [39, 40]. Distinct pathways are currently known to be responsible for the non-genomic action of estrogens [41]. One involves targeting the classic ER to the plasma membrane via posttranslational modification such as phosphorylation or palmitoylation and subsequent association with signaling proteins such as Csk, Shc, Src, GαI, and MNAR/PELP1 [42]. Another implicates a G protein coupled receptor for estrogen known as GPR30 [43]. GPR30 can be activated by both estrogen agonists and antagonists and initiate multiple intracellular signaling cascades. The PR has also been shown to possess a proline-rich domain within its amino-terminal region that mediates direct hormone dependent interaction with the SH3 domain of a variety of cytoplasmic signaling proteins, including c-SRC [44, 45].

THE ERRS

The ERRs were the first orphan nuclear receptors identified through a search for genes related to ERα [46]. Initial studies of the ERRs showed that they did not bind estradiol or other physiologically relevant steroid hormones and indicated that their biological roles could be quite distinct from those of the classic ERs [47–50]. However, the observation that the ERs and ERRs share target genes and coactivators [51], coupled with the discovery that diethylstilbestrol, a potent synthetic estrogen, and 4-hydroxytamoxifen, a mixed estrogen agonist/antagonist, are ERR ligands [52–54] suggests that the ERRs are bona fide SHRs [51]. However, the physiological roles of the ERRs appear to be quite distinct from those of the classic ERs. The main function of the three ERR isoforms is to cooperate with the coactivators PGC-1α and β to control mitochondrial biogenesis and energy metabolism in different tissues [27, 55–61].

SELECTIVE STEROID HORMONE RECEPTOR MODULATORS

Pharmacological and toxicological studies of synthetic SHR ligands have led to the realization that certain drugs can have distinct effects on the same SHR depending on the target tissue. This concept has mainly emerged from the study of tamoxifen, a molecule that acts as an estrogen antagonist in the breast but as an estrogen agonist in bone and uterus [62]. The term selective estrogen receptor modulator (SERM) is now being used to describe estrogenic drugs that display cell type dependent and context dependent actions, and this concept has now been extended to all SHRs. The mechanisms by which selective SHR modulators exert tissue-specific effects are not yet understood and are likely to be distinct for each class of compounds, targeted receptors, and site of action. However, our better understanding of the modes of action of SHR at the molecular level indicate that these mechanisms may include selective activation of receptor isoforms (e.g., ERα versus ERβ), distinct behavior of the drug-receptor complex in the presence of different corepressor/coactivator ratios, preferential recruitment of specific coregulatory proteins (including other transcription factors involved in cross-talk with SHRs such as AP-1 and NF-κB), and possibly selective activation of non-genomic pathways [63–66]. The future development of selective SHR modulators with improved therapeutic indexes and lack of undesirable side effects will likely constitute the most significant outcome of the vast effort dedicated to understand how SHRs work, and basic molecular and genetic studies of SHR action will continue to be the main driving force behind this process.

ACKNOWLEDGEMENTS

The literature on steroid hormone receptor signaling is extensive and only selected studies and reviews were cited in this chapter due to strict space limitation. I apologize to all investigators whose work was not included. Many thanks to members of my group for revising the manuscript. My laboratory is funded by the Canadian Institutes of Health Research and the Terry Fox Foundation via the National Cancer Institute of Canada.

REFERENCES

1. Evans RM. The steroid and thyroid hormone receptor superfamily. *Science* 1988;**240**:889–95.
2. Tremblay AM, Giguère V. The NR3B subgroup: an ovERRview. *Nucl Recept Signal* 2007;**5**:e009.
3. Giguère V, Hollenberg SH, Rosenfeld MG, Evans RM. Functional domains of the human glucocorticoid receptor. *Cell* 1986;**46**:645–52.
4. Wurtz JM, Bourguet W, Renaud JP, Vivat V, Chambon P, Moras D, Gronemeyer H. A canonical structure for the ligand-binding domain of nuclear receptors. *Nat Struct Biol* 1996;**3**:87–94.
5. Hager GL, Lim CS, Elbi C, Baumann CT. Trafficking of nuclear receptors in living cells. *J Steroid Biochem Mol Biol* 2000;**74**:249–54.
6. Lim CS, Baumann CT, Htun H, Xian W, Irie M, Smith CL, Hager GL. Differential localization and activity of the A- and B-forms of the human progesterone receptor using green fluorescent protein chimeras. *Mol Endocrinol* 1999;**13**:366–75.
7. Glass CK. Differential recognition of target genes by nuclear receptors monomers, dimers, and heterodimers. *Endocr Rev* 1994;**15**:391–407.
8. Cowley SM, Hoare S, Mosselman S, Parker MG. Estrogen receptors a and b form heterodimers on DNA. *J Biol Chem* 1997;**272**:19,858–19,862.
9. Pace P, Taylor J, Suntharalingam S, Coombes RC, Ali S. Human estrogen receptor b binds DNA in a manner similar to and dimerizes with estrogen receptor a. *J Biol Chem* 1997;**272**:25,832–25,838.
10. Pettersson K, Grandien K, Kuiper GGJM, Gustafsson J-Å. Mouse estrogen receptor b forms estrogen response element-binding heterodimers with estrogen receptor a. *Mol Endocrinol* 1997;**11**:1486–96.
11. Tremblay GB, Tremblay A, Labrie F, Giguère V. Dominant activity of activation function 1 (AF-1) and differential stoichiometric requirements for AF-1 and -2 in the estrogen receptor a-b heterodimer complex. *Mol Cell Biol* 1999;**19**:1919–27.
12. Mohamed MK, Tung L, Takimoto GS, Horwitz KB. The leucine zippers of c-fos and c-jun for progesterone receptor dimerization: A-dominance in the A/B heterodimer. *J Steroid Biochem Mol Biol* 1994;**51**:241–50.
13. Aranda A, Pascual A. Nuclear hormone receptors and gene expression. *Physiological Review* 2001;**81**:1269–304.
14. Feng W, Ribeiro RCJ, Wagner RL, Nguyen H, Apriletti JW, Fletterick RJ, Baxter JD, Kushner PJ, West BL. Hormone-dependent coactivator binding to a hydrophobic cleft on nuclear receptors. *Science* 1998;**280**:1747–9.
15. McKenna NJ, Lanz RB, O'Malley BW. Nuclear receptor coregulators: cellular and molecular biology. *Endocr Rev* 1999;**20**:321–44.
16. Lanz RB, McKenna NJ, Onate SA, Albrecht U, Wong J, Tsai SY, Tsai MJ, O'Malley BW. A steroid receptor coactivator, SRA, functions as an RNA and is present in an SRC-1 complex. *Cell* 1999;**97**:17–27.
17. Hermanson O, Glass CK, Rosenfeld MG. Nuclear receptor coregulators: multiple modes of modification. *Trends Endocrinol Metabol* 2002;**13**:55–60.
18. Weigel NL, Moore NL. Steroid receptor phosphorylation: a key modulator of multiple receptor functions. *Mol Endocrinol* 2007;**21**:2311–19.

19. Shao D, Lazar MA. Modulating nuclear receptor function: may the phos be with you. *J Clin Invest* 1999;**103**:1617–18.
20. Tremblay A, Tremblay GB, Labrie F, Giguère V. Ligand-independent recruitment of SRC-1 by estrogen receptor b through phosphorylation of activation function AF-1. *Mol Cell* 1999;**3**:513–19.
21. Couse JF, Korach KS. Estrogen receptor null mice: what have we learned and where will they lead us?. *Endocr Rev* 1999;**20**:358–417.
22. Curtis SW, Washburn T, Sewall C, DiAugustine R, Lindzey J, Couse JF, Korach KS. Physiological coupling of growth factor and steroid receptor signaling pathways: estrogen receptor knockout mice lack estrogen-like response to epidermal growth factor. *Proc Natl Acad Sci USA* 1996;**93**:12,626–30.
23. Klotz DM, Hewitt SC, Ciana P, Raviscioni M, Lindzey JK, Foley J, Maggi A, DiAugustine RP, Korach KS. Requirement of estrogen receptor a in insulin-like growth factor-1 (IGF-1)-induced uterine responses and in vivo evidence for IGF-1/estrogen receptor cross-talk. *J Biol Chem* 2002;**277**:8531–7.
24. Kato S, Endoh H, Masuhiro Y, Kitamoto T, Uchiyama S, Sasaki H, Masushige S, Gotoh Y, Nishida E, Kawashima H, Metzger D, Chambon P. Activation of the estrogen receptor through phosphorylation by mitogen-activated protein kinase. *Science* 1995;**270**:1491–4.
25. Diamond MI, Miner JN, Yoshinaga SK, Yamamoto KR. Transcription factor interactions: selectors of positive or negative regulation from a single DNA element. *Science* 1990;**249**:1266–72.
26. Nissen RM, Yamamoto KR. The glucocorticoid receptor inhibits NFkB by interfering with serine-2 phosphorylation of the RNA polymerase II carboxy-terminal domain. *Genes Dev* 2000;**14**:2314–29.
27. Dufour CR, Wilson BJ, Huss JM, Kelly DP, Alaynick WA, Downes M, Evans RM, Blanchette M, Giguère V. Genone-wide orchestration of cardiac functions by orphan nuclear receptors ERRalpha and gamma. *Cell Metabolism* 2007;**5**:345–56.
28. Carroll JS, Meyer CA, Song J, Li W, Geistlinger TR, Eeckhoute J, Brodsky AS, Keeton EK, Fertuck KC, Hall GF, Wang Q, Bekiranov S, Sementchenko V, Fox EA, Silver PA, Gingeras TR, Liu XS, Brown M. Genome-wide analysis of estrogen receptor binding sites. *Nat Genet* 2006;**38**:1289–97.
29. Wang Q, Li W, Liu XS, Carroll JS, Janne OA, Keeton EK, Chinnaiyan AM, Pienta KJ, Brown M. A hierarchical network of transcription factors governs androgen receptor-dependent prostate cancer growth. *Mol Cell* 2007;**27**:380–92.
30. Massie CE, Adryan B, Barbosa-Morais NL, Lynch AG, Tran MG, Neal DE, Mills IG. New androgen receptor genomic targets show an interaction with the ETS1 transcription factor. *EMBO Rep* 2007;**8**:871–8.
31. Laganière J, Deblois G, Lefebvre C, Bataille AR, Robert F, Giguère V. Location analysis of estrogen receptor a target promoters reveals that FOXA1 defines a domain of the estrogen response. *Proc Natl Acad Sci USA* 2005;**102**:11,651–11,656,.
32. Carroll JS, Liu XS, Brodsky AS, Li W, Meyer CA, Szary AJ, Eeckhoute J, Shao W, Hestermann EV, Geistlinger TR, Fox EA, Silver PA, Brown M. Chromosome-wide mapping of estrogen receptor binding reveals long-range regulation requiring the forkhead protein FoxA1. *Cell* 2005;**122**:33–43.
33. Reichardt HM, Kaestner KH, Tuckermann J, Kretz O, Wessely O, Bock R, Gass P, Schmid W, Herrlich P, Angel P, Schütz G. DNA binding of the glucocorticoid receptor is not essential for survival. *Cell* 1998;**93**:531–41.
34. Kousteni S, Bellido T, Plotkin LI, O'Brien CA, Bodenner DL, Han L, Han K, DiGregorio GB, Katzenellenbogen JA, Katzenellenbogen BS, Roberson PK, Weinstein RS, Jilka RL, Manolagas SC. Nongenotropic, sex-nonspecific signaling through the estrogen or androgen receptors: dissociation from transcriptional activity. *Cell* 2001;**104**:719–30.
35. Revelli A, Massobrio M, Tesarik J. Nongenomic actions of steroid hormones in reproductive tissues. *Endocr Rev* 1998;**19**:3–17.
36. Watson CS, Gametchu B. Membrane-initiated steroid actions and the proteins that mediate them. *Proc Soc Exp Biol Med* 1999;**220**:9–19.
37. Ferrell Jr. JE Xenopus oocyte maturation: new lessons from a good egg. *Bioessays* 1999;**21**:833–42.
38. McEwen BS. Steroid hormone actions on the brain: when is the genome involved? *Horm Behav* 1994;**28**:396–405.
39. Moggs JG, Orphanides G. Estrogen receptors: orchestrators of pleiotropic cellular responses. *EMBO Reports* 2001;**2**:775–81.
40. Kelly MJ, Levin ER. Rapid actions of plasma membrane estrogen receptors. *Trends Endocrinol Metabol* 2001;**12**:152–6.
41. Levin ER. Integration of the extranuclear and nuclear actions of estrogen. *Mol Endocrinol* 2005;**19**:1951–9.
42. Vadlamudi RK, Kumar R. Functional and biological properties of the nuclear receptor coregulator PELP1/MNAR. *Nucl Recept Signal* 2007;**5**:e004.
43. Prossnitz ER, Arterburn JB, Sklar LA. GPR30: A G protein-coupled receptor for estrogen. *Mol Cell Endocrinol* 2007;**265–266**:138–42.
44. Boonyaratanakornkit V, Scott MP, Ribon V, Sherman L, Anderson SM, Maller JL, Miller WT, Edwards DP. Progesterone receptor contains a proline-rich motif that directly interacts with SH3 domains and activates c-Src family tyrosine kinases. *Mol Cell* 2001;**8**:269–80.
45. Boonyaratanakornkit V, Edwards DP. Receptor mechanisms mediating non-genomic actions of sex steroids. *Semin Reprod Med* 2007;**25**:139–53.
46. Giguère V, Yang N, Segui P, Evans RM. Identification of a new class of steroid hormone receptors. *Nature* 1988;**331**:91–4.
47. Wiley SR, Kraus RJ, Zuo F, Murray EE, Loritz K, Mertz JE. SV40 early-to-late switch involves titration of cellular transcriptional repressors. *Genes Dev* 1993;**7**:2206–19.
48. Sladek R, Bader J-A, Giguère V. The orphan nuclear receptor estrogen-related receptor a is a transcriptional regulator of the human medium-chain acyl coenzyme A dehydrogenase gene. *Mol Cell Biol* 1997;**17**:5400–9.
49. Vega RB, Kelly DP. A role for estrogen-related receptor a in the control of mitochondrial fatty acid β-oxidation during brown adipocyte differentiation. *J Biol Chem* 1997;**272**:31,693–9.
50. Luo J, Sladek R, Bader J-A, Rossant J, Giguère V. Placental abnormalities in mouse embryos lacking orphan nuclear receptor ERRβ. *Nature* 1997;**388**:778–82.
51. Giguère V. To ERR in the estrogen pathway. *Trends Endocrinol Metabol* 2002;**13**:220–5.
52. Tremblay GB, Bergeron D, Giguère V. 4-hydroxytamoxifen is an isoform-specific inhibitor of orphan estrogen-receptor-related (ERR) nuclear receptors β and γ. *Endocrinology* 2001;**142**:4572–5.
53. Tremblay GB, Kunath T, Bergeron D, Lapointe L, Champigny C, Bader J-A, Rossant J, Giguère V. Diethylstilbestrol regulates trophoblast stem cell differentiation as a ligand of orphan nuclear receptor ERRβ. *Genes Dev* 2001;**15**:833–8.
54. Coward P, Lee D, Hull MV, Lehmann JM. 4-Hydroxytamoxifen binds to and deactivates the estrogen-related receptor γ. *Proc Natl Acad Sci USA* 2001;**98**:8880–4.
55. Mootha VK, Handschin C, Arlow D, Xie X, St Pierre J, Sihag S, Yang W, Altshuler D, Puigserver P, Patterson N, Willy PJ, Schulman IG, Heyman RA, Lander ES, Spiegelman BM. ERRa and GABPAa/β specify PGC-1a-dependent oxidative phosphorylation

gene expression that is altered in diabetic muscle. *Proc Natl Acad Sci USA* 2004;**101**:6570–5.
56. Sonoda J, Laganière J, Mehl IR, Barish GD, Chong LW, Li X, Scheffler IE, Mock DC, Bataille AR, Robert F, Lee C-H, Giguère V, Evans RM. Nuclear receptor ERRa and coactivator PGC-1β are effectors of IFN-γ induced host defense. *Genes Dev* 2007;**21**:1909–20.
57. Alaynick WA, Kondo RP, Xie W, He W, Dufour CR, Downes M, Jonker JW, Giles W, Naviaux RK, Giguère V, Evans RM. ERRγ directs and maintains the transition to oxidative metabolism in the post-natal heart. *Cell Metabolism* 2007;**6**:16–24.
58. Huss JM, Imahashi K-I, Dufour C, Weinheimer CJ, Courtois M, Kovacs A, Giguère V, Murphy E, Kelly DP. The nuclear receptor ERRa is required for the bioenergetic and functional adaption to cardiac pressure overload. *Cell Metabolism* 2007;**6**:25–37.
59. Schreiber SN, Emter R, Hock MB, Knutti D, Cardenas J, Podvinec M, Oakeley EJ, Kralli A. The estrogen-related receptor alpha (ERRa) functions in PPARγ coactivator 1a (PGC-1a)-induced mitochondrial biogenesis. *Proc Natl Acad Sci USA* 2004;**101**:6472–7.
60. Schreiber SN, Knutti D, Brogli K, Uhlmann T, Kralli A. The transcriptional coactivator PGC-1 regulates the expression and activity of the orphan nuclear receptor ERRa. *J Biol Chem* 2003;**278**:9013–18.
61. Villena JA, Hock MB, Giguère V, Kralli A. Orphan nuclear receptor ERRa is essential for adaptive thermogenesis. *Proc Natl Acad Sci USA* 2007;**104**:1418–23.
62. Jordan VC, O'Malley BW. Selective estrogen-receptor modulators and antihormonal resistance in breast cancer. *J Clin Oncol* 2007;**25**:5815–24.
63. Tremblay GB, Tremblay A, Copeland NG, Gilbert DJ, Jenkins NA, Labrie F, Giguère V. Cloning, chromosomal localization and functional analysis of the murine estrogen receptor b. *Mol Endocrinol* 1997;**11**:353–65.
64. Norris JD, Paige LA, Christensen DJ, Chang CY, Huacani MR, Fan D, Hamilton PT, Fowlkes DM, McDonnell DP. Peptide antagonists of the human estrogen receptor. *Science* 1999;**285**:744–6.
65. Paech K, Webb P, Kuiper GGJM, Nilsson S, Gustafsson J-Å, Kushner PJ, Scanlan TS. Differential ligand activation of estrogen receptors ERa and ERb at AP1 sites. *Science* 1997;**277**:1508–10.
66. Shang Y, Brown M. Molecular determinants for the tissue specificity of SERMs. *Science* 2002;**295**:2465–8.

Chapter 6

FOXO Transcription Factors: Key Targets of the PI3K-Akt Pathway that Regulate Cell Proliferation, Survival, and Organismal Aging

Anne Brunet, Hien Tran and Michael E. Greenberg
Division of Neuroscience, Children's Hospital and Department of Neurobiology, Harvard Medical School, Boston, Massachusetts

INTRODUCTION

The development and integrity of multicellular organisms depend on the ability of each cell to integrate a wide range of external and internal cues and to trigger appropriate cellular responses. In recent years, Forkhead transcription factors of the FOXO subfamily have emerged as key players in controlling cell cycle progression, cell survival, detoxification, and DNA damage repair by integrating extracellular signals and triggering changes in gene expression. At the organismal level, FOXO transcription factors may play a critical role in the control of lifespan, a function that may be conserved throughout evolution.

IDENTIFICATION OF THE FOXO SUBFAMILY OF TRANSCRIPTION FACTORS

Forkhead transcription factors all share a conserved DNA binding domain (DBD) of 100 amino acids that folds into a winged helix structure termed the Forkhead box. The Forkhead family is evolutionarily conserved from yeast to mammals, and in humans comprises around 30 members that have been divided into 17 subgroups (FOX for "Forkhead box" A to Q) [1]. FOX transcriptional regulators play a wide range of roles during development, from organogenesis (FOXC) [2] to language and speech acquisition (FOXP) [3].

Among the large Forkhead family, the FOXO subgroup first attracted interest because all three members of this family, namely, FOXO1/FKHR, FOXO3a/FKHRL1, and FOXO4/AFX, were identified at the site of chromosomal translocations in human tumors. The chromosomal translocations involving FOXO family members result in the generation of a chimeric protein in which the transactivation domain of FOXOs is fused to the DBD of another transcription factor, creating a dysregulated and highly active transcriptional fusion protein (Table 6.1) [4–7]. These initial findings suggested that, when mutated, FOXO transcription factors might play a role in tumor development.

REGULATION OF FOXO TRANSCRIPTION FACTORS BY THE PI3K-AKT PATHWAY

In recent years, a series of genetic and biochemical studies revealed that FOXO family members are regulated by the PI3K-Akt pathway in response to growth factor stimulation. Binding of growth factors or insulin to their receptors triggers the activation of the phosphoinositide kinase (PI3K), which in turn is responsible for the activation of several serine/threonine kinases, including Akt or SGK (serum glucocorticoid inducible kinase) [8]. The PI3K-Akt pathway is conserved in the nematode *Caenorhabditis elegans*, where this pathway is activated by an insulin-like signal [9, 10]. Inactivation of the PI3K-Akt pathway during development of the nematode results in arrest in dauer, a long-lived stress resistant larval stage, whereas in the adult, inactivation of this pathway increases lifespan by two- to threefold. Strikingly, two independently conducted genetic screens found that all suppressor alleles of the PI3K-Akt pathway mutants map to the gene encoding a FOXO transcription factor termed DAF-16 [11, 12]. These results indicate that DAF-16 is a key target of the PI3K-Akt pathway.

Handbook of Cell Signaling, Three-Volume Set 2 ed.

TABLE 6.1 FOXO family members

New name	Old name	Chromosomal translocation	Fusion protein	Type of cancer	Expression				
FOXO1	FKHR	2–13	PAX3-FOXO1	Alveolar rhabdomyosarcoma	Ubiquitous but high in ovary		1–13	PAX7-FOXO1	
3FOXO2	3AF6q21	6–21	MLL-FOXO3a	Acute myeloblastic leukemia	Ubiquitous but high in FOXO3a	FKHRL1			brain and kidney
FOXO4	AFX	X-11	MLL-FOXO4	Acute myeloblastic leukemia	Placenta and muscle				

Several studies conducted in mammalian cells provide a molecular mechanism that appears to explain the genetic link between the PI3K-Akt pathway and FOXOs. These studies show that the protein kinase Akt phosphorylates for all FOXO family members at three key regulatory sites (Thr32, Ser253, and Ser315 for FOXO3a) that are conserved from *C. elegans* to mammals and are part of a perfect consensus sequence for Akt phosphorylation (RXRXX(S/T)) (Figure 6.1) [13–17]. Akt is not the only protein kinase that phosphorylates FOXOs at these regulatory sites; SGK also phosphorylates FOXO3a. However, the efficacy with which SGK and Akt phosphorylate the three phosphorylation sites of FOXO transcription factors differs. Whereas Thr32 is phosphorylated by both kinases, Akt preferentially phosphorylates Ser253, and SGK favors the phosphorylation of Ser315 [18].

The three FOXO regulatory sites are phosphorylated in response to a wide range of stimuli that activate Akt and SGK, including insulin-like growth factor 1 (IGF-1) [13], insulin [19], interleukin 3 [20], erythropoietin [21], epidermal growth factor [22], and transforming growth factor β [23]. The phosphorylation of FOXO transcription factors by Akt and SGK in response to growth factors triggers a change in the subcellular localization of FOXOs [13, 14] (Figure 6.2). In the absence of growth factors, when Akt and SGK are inactive, FOXOs are localized within the nucleus. When cells are exposed to growth factors, the PI3K-Akt/SGK cascade is activated and leads to the relocalization of FOXOs from the nucleus to the cytoplasm, away from FOXO target genes (Figure 6.2). Mutation analyses of the three regulatory sites have shown that the phosphorylation of each site contributes to the relocalization of FOXOs from the nucleus to the cytoplasm [18]. This apparent redundancy of phosphorylation may represent a way of modulating the extent of the relocalization of FOXOs to the cytoplasm in different cell types or in response to different combinations of signals.

OTHER REGULATORY PHOSPHORYLATION SITES IN FOXOS

In addition to the sites on FOXOs that are phosphorylated in response to growth factors, several other sites of phosphorylation have recently been identified (Figure 6.1). FOXO4 (AFX) is phosphorylated at two residues, Thr 447 and Thr 451, in response to the activation of the small G protein termed Ral, although the kinase directly responsible for the phosphorylation of these two threonines is not yet known [24]. Phosphorylation of Thr 447 and Thr 451 does not affect the subcellular localization of FOXO4, but instead appears to abolish FOXO4 transcriptional activity. Another study has reported that the MAP kinase family member DYRK catalyzes the phosphorylation of Ser 329 in FOXO1, a residue that is conserved in all FOXO family members [25]. However, the phosphorylation of Ser 329 is not modulated by growth factors and its role in the regulation of FOXO function is still unclear.

MECHANISM OF THE EXCLUSION OF FOXOS FROM THE NUCLEUS IN RESPONSE TO GROWTH FACTOR STIMULATION

The translocation of FOXOs from the nucleus to the cytoplasm appears to be the primary mechanism of regulation of these transcription factors in response to growth factors. Experiments using leptomycin B, a specific inhibitor of nuclear export, show that the translocation of FOXO transcription factors from the nucleus to the cytoplasm is mediated by a nuclear export signal (NES) dependent mechanism [14, 26, 27]. Mutation analyses have revealed that one (FOXO1) or two (FOXO3a) leucine-rich domains in the conserved C-terminal region of FOXOs function as NES [14, 26, 27] (Figure 6.1). Furthermore, the phosphorylated forms of FOXOs have been shown to specifically

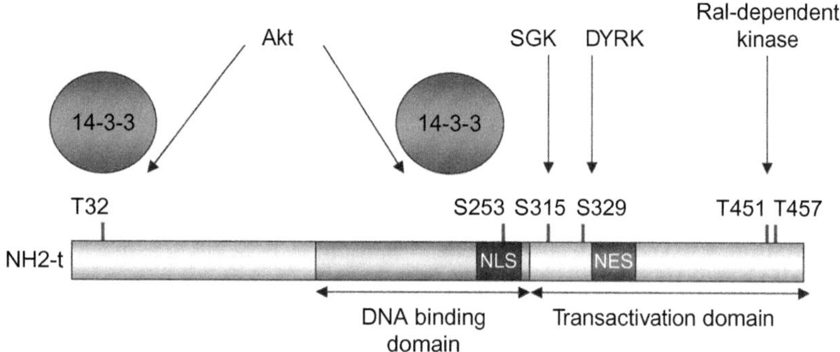

FIGURE 6.1 Domain structure and phosphorylation sites within FOXO family members.
Numbering for FOXO3a: T32, S253, S315; numbering for FOXO1, S329; numbering for FOXO4: T451, T457.

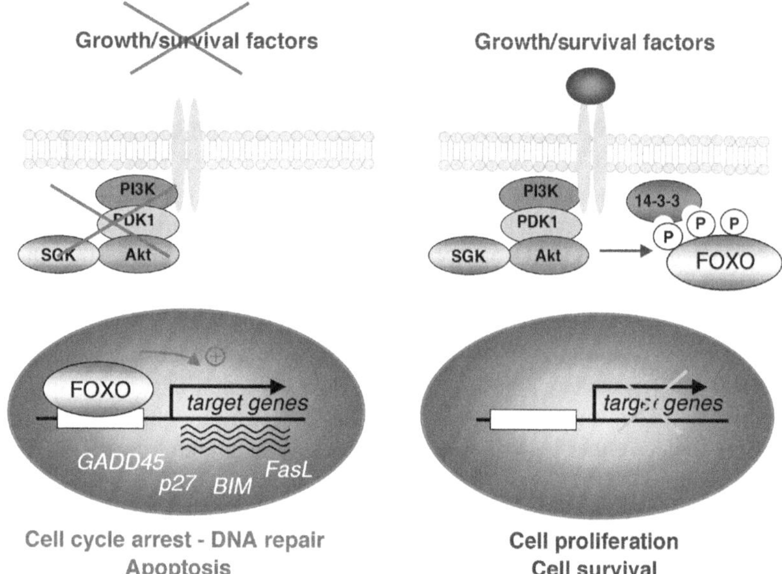

FIGURE 6.2 Regulation of FOXO family members by growth and survival factors.
In the absence of growth/survival factors, FOXO family members induce the expression of death genes, cell cycle arrest genes or stress response genes, including Fas ligand, the proapoptotic Bcl-2 family member BIM, the cell cycle inhibitor p27KIP1, and the growth arrest and DNA damage response gene GADD45. In the presence of growth/survival factors, the PI3K-Akt/SGK pathway is activated. Akt and SGK promote survival and cell cycle progression by phosphorylating and inhibiting FOXOs.

interact with 14-3-3 proteins [13], which appear to serve as chaperone molecules. The phosphorylation of the first FOXO regulatory site (Thr 32 in FOXO3a) creates a perfect 14-3-3 binding site, and the phosphorylation of the second FOXO regulatory site (Ser 253 in FOXO3a) – although it does not create a canonical 14-3-3 binding site – also participates in 14-3-3 binding [13]. Several mechanisms have been proposed to explain how 14-3-3 binding to FOXOs may promote the relocalization of FOXOs from the nucleus to the cytoplasm. One study suggests that 14-3-3 binding decreases the ability of FOXOs to bind DNA, releasing FOXOs from a nuclear DNA anchor and allowing the relocalization of FOXOs to the cytoplasm [28]. A second study suggests that 14-3-3 binding to FOXOs occurs within the nucleus and actively promotes the nuclear export of FOXOs, perhaps by inducing a conformational change in FOXOs that would expose the NES and allow interaction with Exportin/Crm1 [27].

Finally 14-3-3 binding to FOXOs has been shown to prevent the nuclear reimport of FOXOs by masking FOXOs' nuclear localization signal (NLS) [26, 29], consistent with a known function of 14-3-3 in regulating the subcellular localization of several of its other binding partners. These various mechanisms for regulating the translocation of FOXOs from the nucleus to the cytoplasm may serve as a fail-safe mechanism to ensure a complete sequestration of FOXOs in the cytoplasm, away from FOXO target genes in the nucleus. The sequestration of a transcription factor in the cytoplasm upon phosphorylation is a mechanism by which the activity of a variety of transcriptional regulators, including NFAT [30, 31] and the yeast transcription factor Pho4 [32], are controlled, and this mechanism may have been selected by evolution because it represents an efficient way to inhibit the function of a transcription factor.

TRANSCRIPTIONAL ACTIVATOR PROPERTIES OF FOXOS

Like other FOX family members, FOXOs bind to DNA as a monomer via the Forkhead box, a 100 amino acid region located in the central part of the molecule (Figure 6.1). The NMR structure of the DBD of FOXO4 has been solved and shows that this domain adopts a winged helix structure similar to those of the other FOX family members that have been solved, FOXA (HNF3γ) and FOXD (Genesis) [33]. The consensus recognition site for FOXOs on DNA has been determined by three independent groups [34–36] and the core motif, GTAAA(C/T)A, is very similar to that of other Forkhead transcription factors [37]. Both more extensive DNA sequence motifs in promoters of FOXO target genes and regions of FOXO proteins outside the Forkhead box are likely involved in conferring binding specificity to particular FOXO family members under different conditions of stimulation.

When present in the nucleus and bound to DNA, FOXOs act as potent activators of transcription. The phosphorylation of FOXOs by Akt or SGK appears to regulate the transcriptional activity of FOXOs primarily by excluding this transcription factor from the nucleus [13, 15, 17, 18, 38]. When the phosphorylated form of FOXOs is artificially maintained within the nucleus either by leptomycin B treatment or by mutation of the FOXOs' NES, FOXOs are transcriptionally active even in the presence of growth factors, indicating that the main function of phosphorylation at the growth factor regulated sites is to relocalize FOXOs to the cytoplasm rather than to directly affect the DNA binding or transcriptional activity of FOXOs [26]. Deletion analyses have revealed that the transactivation domain of FOXOs spans the C-terminal region of the molecule (Figure 6.1). In the chimeric protein PAX3-FOXO1 present in some human rhabdomyosarcomas, it is the transactivation domain of FOXO1 that is fused to PAX3 and that confers potent transactivation properties to PAX3 [39]. FOXOs' ability to bind to the transcriptional coactivator CBP (CREB binding protein) may provide a connection with the basal transcriptional machinery resulting in FOXO/CBP dependent changes in target gene expression [40].

FOXOS AND THE REGULATION OF APOPTOSIS

The expression of constitutively active mutants of FOXOs triggers cell death in several different cell types, ranging from primary neurons to lymphocytes [13, 17, 41–43]. FOXO induced apoptosis is dependent on the ability of FOXOs to bind to DNA and to induce transcription, indicating that a subset of target genes that FOXOs transactivate consists of death genes (Table 6.2). Thus, one way in which Akt and SGK promote cell survival is by sequestering FOXOs away from death genes. Several death genes contain FOXO binding sites in their promoters, including the genes encoding for death cytokines (Fas ligand, TNFα, CD30 ligand) or death cytokine receptors (Fas, TNFR). In particular, FOXO3a has been shown to bind to and activate the promoter for the Fas ligand gene [13, 44]. Consistent with the hypothesis that FOXOs regulate Fas ligand transcription, the inactivation of Akt has recently been shown to induce Fas ligand expression [45]. Because FOXO3a induced apoptosis is reduced in cells in which Fas ligand signaling is blocked [13], Fas ligand may relay in a paracrine manner the effects of FOXOs by triggering cellular pathways that culminate in apoptosis. FOXOs also appear to control the expression of several members of the Bcl-2 family, which are known to play a critical role in the balance between cell survival and cell death. FOXO3a induces the expression of Bim, a potent prodeath Bcl-2 family member, and apoptosis induced by the inactivation of the PI3K-Akt/SGK pathway is reduced in lymphocytes from Bim deficient mice, indicating that Bim may be an important target gene of FOXO3a that mediates cell death [43, 46]. In addition, FOXO4 indirectly downregulates the expression of the prosurvival Bcl-2 family member Bcl-xL via the induction of a transcriptional repressor Bcl-6 [47]. Thus, one way in which FOXOs appear to trigger apoptosis is by modulating the ratio of prodeath and prosurvival members of the Bcl-2 family.

FOXOS ARE KEY REGULATORS OF SEVERAL PHASES OF THE CELL CYCLE

In cycling cells such as fibroblasts, the main effect of the expression of FOXO family members is not apoptosis but cell cycle arrest at the G_1/S boundary [48]. One target gene that mediates FOXO induced cell cycle arrest is the Cdk inhibitor p27KIP1, which inhibits Cdk2 activity, thereby blocking cell cycle progression at the G_1/S transition [20, 48, 49] (Table 6.2). FOXOs induce p27 promoter activity, and FOXOs' ability to induce G_1 arrest is diminished in p27 deficient fibroblasts, suggesting that p27 is a critical FOXO target that mediates G_1 arrest [48]. The FOXO mediated arrest at the G_1/S transition may be a critical first step in the differentiation process because transgenic mice expressing a dominant negative form of FOXO1 have impaired T-cell differentiation [36].

Endogenous FOXO3a is localized to the nucleus in cells passing through the G_2 phase of the cell cycle, suggesting a role for FOXO3a at the G_2/M checkpoint [50]. Cells in which FOXO3a is activated in the S phase display a delay in their progression through the G_2 phase of the cell cycle. Microarray analyses led to the identification of several

TABLE 6.2 FOXO target genes

	FOXO genes activation	Gene function binding sites	Binding to induction	Promoter	Promoter
Cell death					
Fas ligand	Death cytokine	ICC	EMSA; ChIP	+	+
Bim	BH3 only Bcl-2 family member	NB; WB			
bNIP3	BH3 only Bcl-2 family member	Microarray			
Legumain	Cysteine protease	Microarray			
Bcl-6	Transcriptional repressor	Microarray; NB	EMSA	+	+
Cell cycle					
p27KIP1	Cdk inhibitor	WB; NB		+	+
WIP1	Phosphatase	Microarray			+
EXT1	Tumor suppressor	Microarray			+
Cyclin G_2	G_2 delay	Microarray			+
Polo-like kinase (PLK)	Mitosis control	WB	ChIP	+	+
Cyclin B	Cdc2 activator	WB	ChIP	+	+
Antioxidant					
Selenoprotein P	Detoxification of ROS	Microarray; NB			+
SOD3		NB			+
DNA repair					
GADD45	Growth arrest and	Microarray; NB; WB		+	+
PA26	DNA damage response	Microarray; NB			+
Metabolism					
Glucose-6-phosphatase	Glucose metabolism	NB	EMSA	+	+
Phosphoenolpyruvate carboxykinase	Glucose metabolism	NB		+	
IGF-BP1	Regulation of IGF-I levels	NB	EMSA	+	+

Note: ICC, immunocytochemistry; NB, Northern blot; WB, Western blot; EMSA, electrophoretic mobility shift assay; ChIP, chromatin immunoprecipitation.

FOXO3a target genes that may mediate FOXOs' effect at the G_2/M boundary (Table 6.2) [50]. As with the G_1/S checkpoint, the G_2/M checkpoint is critical in the cellular response to stress, in part to allow time for detoxification and the repair of damaged DNA. Indeed, when active, FOXO3a induces DNA repair [50]. One of the genes that may mediate both FOXO3a induced arrest at the G_2/M checkpoint and DNA repair is the growth arrest and DNA damage response gene 45 (GADD45), since the GADD45 promoter is activated by FOXO3a, and FOXO3a induced DNA repair is diminished in GADD45 deficient fibroblasts [50].

FOXO3a also promotes the exit from the M phase and allows the transition to the following G_1 phase [44]. FOXO3a binds to and induces the promoter of two genes that play a critical role in the exit from M phase, cyclin B and polo-like kinase (PLK) (Table 6.2) [44]. These findings are consistent with the observation that the activation of PI3K and Akt prevents the completion of the M phase [44].

As the two yeast Forkhead transcription factors Fkh1 and Fkh2 have been found to play a role in the completion of M phase [51], this function of FOXOs may have been conserved throughout evolution.

How can FOXOs induce both an arrest in G_1 and G_2 and promote the exit from the M phase? One possible explanation is that during normal cell cycles, FOXOs would be recruited to promoters of genes that promote the exit from the M phase. In contrast, under conditions of stress or absence of growth factors, FOXOs may be recruited to promoters of genes controlling cell cycle arrest at the G_1/S and G_2/M checkpoints, allowing repair of damaged DNA. Thus, FOXOs may integrate different extracellular cues, possibly by undergoing various posttranslational modifications, and elicit the appropriate cell cycle response by triggering a specific array of target genes.

FOXOS IN CANCER DEVELOPMENT: POTENTIAL TUMOR SUPPRESSORS

In mammals, FOXOs' ability to induce a G_1 arrest, a G_2 delay, DNA repair, and apoptosis makes it an attractive candidate as a tumor suppressor. Loss of FOXO function may lead to a decreased ability to induce cell cycle arrest, leading to tumor development. A decreased ability to repair damaged DNA due to the absence of FOXOs may result in genomic instability. Finally, in the absence of FOXOs, abnormal cells that would normally die may instead survive, resulting in tumor formation. FOXOs' ability to induce cell cycle arrest, DNA repair, and apoptosis are reminiscent of the functions of the tumor suppressor protein p53. In that respect, it is interesting to note that genes such as GADD45, WIP1, and PA26 (Table 6.2) that are induced in response to FOXOs have also been found to be regulated by p53 [52–54]. These observations raise the possibility that FOXOs and p53 may under some circumstances function in a cooperative manner.

One further link between the FOXO family and cancer is that all FOXO members had been initially characterized because of their presence at chromosomal breakpoints in cells from human tumors (Table 6.1). However the expression of these human tumor fusion proteins in transgenic mice is not sufficient to promote cancer [55]. This finding suggests that the human tumors may have arisen not only as a result of the chimeric molecules but perhaps also as a consequence of the loss of one FOXO allele. If FOXOs function as tumor suppressors, then the haplo insufficiency of one FOXO family member could be a contributing factor to the development of the tumor.

ROLE OF FOXOS IN THE RESPONSE TO STRESS AND ORGANISMAL AGING

All mutants of the nematode that lead to the activation of the FOXO transcription factor DAF-16 are not only long lived but also display resistance to oxidative stress, heat shock, and UV [9, 10, 56, 57]. This observation suggests that one way in which DAF-16 activity may lead to an increase in organismal lifespan is by augmenting the resistance of cells to various stresses. In the nematode, expression of the manganese dependent superoxide dismutase SOD3 is induced when DAF-16 is active [58]. The promoter of SOD3 contains a Forkhead binding site that is in a region conserved across species, suggesting that SOD may be a conserved target of FOXOs that detoxify cells of reactive oxygen species (ROS), thereby conferring resistance to oxidative stress [34]. Another gene that is induced when DAF-16 is active and contains FOXO binding sites in its promoter is the tyrosine kinase receptor OLD-1, which is distantly related to the mammalian PDGF receptor. Expression of OLD-1 increases nematode longevity in part by increasing the resistance to oxidative stress [59].

In mammals, FOXOs induce cell cycle arrest at two critical checkpoints that allow repair of damaged DNA [48, 50]. An organism's ability to respond to stress and in particular to induce detoxification and repair damaged DNA, has been shown to correlate with an increased longevity in many organisms [60]. The GADD45 gene, which in part mediates FOXO induced arrest and DNA repair [50], is upregulated in old mice [61]. These correlative evidence suggest that this pathway may be upregulated to protect against oxidative stress produced by metabolism as an organism ages. Thus, by upregulating a program of gene expression that protects cells against both internal and external cellular stresses, FOXO transcriptional regulators might promote longevity in mammals as well as in *C. elegans*.

Mosaic analyses in *C. elegans* indicate that the longevity phenotype of the nematode mutants that leads to DAF-16 activation is noncell autonomous [62]. In addition, neurons are the cell types that appear to be critical regulators of the aging phenotype in the nematode [63]. These findings, combined with the indications that DAF-16 may protect cells against stress, raise the possibility that a subset of neurons in the central nervous system may represent a "longevity control center" that might be more sensitive than other organs to oxidative stress [63]. It is tempting to speculate that factor(s) released from these neurons would convey the lifespan signal to the whole organism. It will be interesting to determine if a central neuronal regulator of life span also operates in mammals.

FOXOS AND THE REGULATION OF METABOLISM IN RELATION TO ORGANISMAL AGING

Reduction in nutrient intake consistently correlates with an increase in lifespan from *C. elegans* to mammals [60]. In the nematode, nutrient deprivation induces the relocalization of the DAF-16 to the nucleus [56], suggesting that one way starvation may result in increased lifespan is by relocalizing FOXO transcription factors to the nucleus where they may in turn induce target genes that lead to the cellular resistance to stress.

In addition, FOXOs appear to themselves regulate target genes that are involved in the control of metabolism (Table 6.2). FOXO1 induces the expression of glucose-6-phosphatase, a gene that catalyzes the hydrolysis of glucose-6-phosphate into glucose [64, 65]. FOXO1 also regulates the gene encoding phosphoenolpyruvate carboxykinase (PEPCK), which mediates the conversion of pyruvate to glucose, although the mechanism by which FOXO1 upregulates PEPCK expression is probably not via direct binding of FOXO1 to the promoter of the PEPCK gene [66, 67]. The regulation of glucose metabolism by FOXOs may be one indirect mechanism by which FOXOs increase organismal longevity. Indeed, switching the cell metabolism toward glucose production and away from the oxidative phosphorylation at the mitochondria may decrease the production of ROS by this organelle. Thus, FOXOs may extend organismal life span both by decreasing the production of ROS and by increasing detoxification and repair of DNA damage caused by ROS.

CONCLUSION

A series of recent studies have shown that FOXO transcription factors are regulated in a conserved manner throughout evolution by the PI3K-Akt pathway. In mammalian cells, FOXOs' functions include cell cycle regulation at various key checkpoints, apoptosis, repair of damaged DNA, and the regulation of glucose metabolism. How can the same transcription factor induce such a range of biological responses, responses that appear in some cases even to be antagonistic? Depending on their posttranslational modifications or interaction with protein partners, FOXOs may be recruited to particular subsets of promoters. It is possible that FOXOs may act as rheostats depending on the stress level: in low stress conditions, FOXOs may promote cell cycle arrest and DNA repair, whereas in high levels of stress, FOXOs may induce apoptosis. This graded response to stress stimuli would protect cells from damage, but also facilitate the elimination of terminally damaged cells. The result would be a suppression of tumor formation or an increase in the lifespan of the organism.

Understanding the molecular mechanisms that generate specificity among the FOXO family members will help to define the role of this family in various cellular responses. In addition, the generation of mice models in which FOXO family members are either inactivated or constitutively activated will help to uncover the respective contribution of the FOXO family members to tumor development and the control of lifespan in mammals.

ACKNOWLEDGEMENTS

We thank members of the Greenberg lab, in particular S. R. Datta, S. E. Ross, and A. J. Shaywitz, for their helpful comments on the manuscript. This work was supported by a Senior Scholars Award from the Ellison Foundation, NIH grant PO1-HD24926, and Mental Retardation Research Center grant NIHP30-HD18655 (MEG). AB was supported by a Goldenson Berenberg Fellowship. MEG acknowledges the generous contribution of the FM Kirby Foundation to the Division of Neuroscience.

REFERENCES

1. Kaestner KH, Knochel W, Martinez DE. Unified nomenclature for the winged helix/forkhead transcription factors. *Genes Dev* 2000;**15**:142–6.
2. Kume T, Deng K, Hogan BL. Murine forkhead/winged helix genes Foxc1 (Mf1) and Foxc2 (Mfh1) are required for the early organogenesis of the kidney and urinary tract. *Development* 2000;**127**:1387–95.
3. Lai CS, Fisher SE, Hurst JA, Vargha-Khadem F, Monaco AP. A forkhead-domain gene is mutated in a severe speech and language disorder. *Nature* 2001;**413**:519–23.
4. Galili N, et al. Fusion of a fork head domain gene to PAX3 in the solid tumour alveolar rhabdomyosarcoma. *Nat Genet* 1993;**5**:230–5.
5. Borkhardt A, et al. Cloning and characterization of AFX, the gene that fuses to MLL in acute leukemias with a t(X;11) (q13;q23). *Oncogene* 1997;**14**:195–202.
6. Hillion J, Le Coniat M, Jonveaux P, Berger R, Bernard OA. AF6q21, a novel partner of the MLL gene in t(6;11) (q21;q23), defines a forkhead transcriptional factor subfamily. *Blood* 1997;**90**:3714–19.
7. Davis RJ, D'Cruz CM, Lovell MA, Biegel JA, Barr FG. Fusion of PAX7 to FKHR by the variant t(1;13) (p36;q14) translocation in alveolar rhabdomyosarcoma. *Cancer Res* 1994;**54**:2869–72.
8. Datta SR, Brunet A, Greenberg ME. Cellular survival: a play in three Akts. *Genes Dev* 1999;**13**:2905–27.
9. Morris JZ, Tissenbaum HA, Ruvkun G. A phosphatidylinositol-3-OH kinase family member regulating longevity and diapause in Caenorhabditis elegans. *Nature* 1996;**382**:536–9.
10. Kimura KD, Tissenbaum HA, Liu Y, Ruvkun G. Daf-2, an insulin receptor-like gene that regulates longevity and diapause in Caenorhabditis elegans. *Science* 1997;**277**:942–6.
11. Lin K, Dorman JB, Rodan A, Kenyon C. Daf-16: An HNF-3/forkhead family member that can function to double the life-span of Caenorhabditis elegans. *Science* 1997;**278**:1319–22.
12. Ogg S, et al. The Fork head transcription factor DAF-16 transduces insulin-like metabolic and longevity signals in C. elegans. *Nature* 1997;**389**:994–9.
13. Brunet A, et al. Akt promotes cell survival by phosphorylating and inhibiting a Forkhead transcription factor. *Cell* 1999;**96**:857–68.
14. Biggs WHI, Meisenhelder J, Hunter T, Cavenee WK, Arden KC. Protein kinase B/Akt-mediated phosphorylation promotes nuclear exclusion of the winged helix transcription factor FKHR1. *Proc Natl Acad Sci USA* 1999;**96**:7421–6.
15. Kops GJ, et al. Direct control of the Forkhead transcription factor AFX by protein kinase B. *Nature* 1999;**398**:630–4.
16. Rena G, Guo S, Cichy S, Unterman TG, Cohen P. Phosphorylation of the transcription factor forkhead family member FKHR by protein kinase B. *J. Biol. Chem.* 1999;**274**:17,179–17,183.
17. Tang ED, Nunez G, Barr FG, Guan K-L. Negative regulation of the Forkhead transcription factor FKHR by Akt. *J Biol Chem* 1999;**274**:16,741–16,746.

18. Brunet A, et al. The protein kinase SGK mediates survival signals by phosphorylating the Forkhead transcription factor FKHRL1/FOXO3a. *Mol Cell Biol* 2001;**21**:952–65.
19. Nakae J, Park BC, Accili D. Insulin Stimulates Phosphorylation of the Forkhead Transcription Factor FKHR on Serine 253 through a Wortmannin-sensitive Pathway. *J Biol Chem* 1999;**274**:15,982–5.
20. Dijkers PF, et al. Forkhead transcription factor FKHR-L1 modulates cytokine-dependent transcriptional regulation of p27(KIP1). *Mol Cell Biol* 2000;**20**:9138–48.
21. Kashii Y, et al. A member of Forkhead family transcription factor, FKHRL1, is one of the downstream molecules of phosphatidylinositol 3-kinase-Akt activation pathway in erythropoietin signal transduction. *Blood* 2000;**96**:941–9.
22. Jackson JG, Kreisberg JI, Koterba AP, Yee D, Brattain MG. Phosphorylation and nuclear exclusion of the forkhead transcription factor FKHR after epidermal growth factor treatment in human breast cancer cells. *Oncogene* 2000;**19**:4574–81.
23. Shin I, Bakin AV, Rodeck U, Brunet A, Arteaga CL. Transforming growth factor beta enhances epithelial cell survival via Akt-dependent regulation of FKHRL1. *Mol Cell Biol* 2001;**12**:3328–39.
24. De Ruiter ND, Burgering BM, Bos JL. Regulation of the Forkhead transcription factor AFX by Ral-dependent phosphorylation of threonines 447 and 451. *Mol Cell Biol* 2001;**21**:8225–35.
25. Woods YL, et al. The kinase DYRK1A phosphorylates the transcription factor FKHR at Ser329 in vitro, a novel in vivo phosphorylation site. *Biochem J* 2001;**355**:597–607.
26. Brownawell AM, Kops GJ, Macara IG, Burgering BM. Inhibition of nuclear import by protein kinase B (Akt) regulates the subcellular distribution and activity of the forkhead transcription factor AFX. *Mol Cell Biol* 2001;**21**:3534–46.
27. Brunet A, et al. 14-3-3 Transits to the nucleus and actively participates in dynamic nucleo-cytoplasmic transport. *J Cell Biol* 2002;**156**:817–28.
28. Cahill CM, et al. PI-3 kinase signaling inhibits DAF-16 DNA binding and function via 14-3-3 dependent and 14-3-3 independent pathways. *J Biol Chem* 2000;**276**:13,402–13,410.
29. Rena G, Prescott AR, Guo S, Cohen P, Unterman TG. Roles of the forkhead in rhabdomyosarcoma (FKHR) phosphorylation sites in regulating 14-3-3 binding, transactivation and nuclear targeting. *Biochem J* 2001;**354**:605–12.
30. Beals CR, Sheridan CM, Turck CW, Gardner P, Crabtree GR. Nuclear export of NF-ATc enhanced by glycogen synthase kinase-3. *Science* 1997;**275**:1930–4.
31. Chow CW, Rincon M, Cavanagh J, Dickens M, Davis RJ. Nuclear accumulation of NFAT4 opposed by the JNK signal transduction pathway. *Science* 1997;**278**:1638–41.
32. Komeili A, O'Shea EK. Roles of phosphorylation sites in regulating activity of the transcription factor Pho4. *Science* 1999;**284**:977–80.
33. Weigelt J, Climent I, Dahlman-Wright K, Wikstrom M. Solution structure of the DNA binding domain of the human forkhead transcription factor AFX (FOXO4). *Biochemistry* 2001;**40**:5861–9.
34. Furuyama T, Nakazawa T, Nakano I, Mori N. Identification of the differential distribution patterns of mRNAs and consensus binding sequences for mouse DAF-16 homologues. *Biochem J* 2000;**349**:629–34.
35. Biggs III WH, Cavenee WK. Identification and characterization of members of the FKHR (FOXO) subclass of winged-helix transcription factors in the mouse. *Mamm Genome* 2001;**12**:416–25.
36. Leenders H, Whiffield S, Benoist C, Mathis D. Role of the forkhead transcription family member, FKHR, in thymocyte differentiation. *Eur J Immunol* 2000;**30**:2980–90.
37. Pierrou S, Hellqvist M, Samuelsson L, Enerback S, Carlsson P. Cloning and characterization of seven human forkhead proteins: binding site specificity and DNA bending. *EMBO J* 1994;**13**:5002–12.
38. Guo S, et al. Phosphorylation of serine 256 by protein kinase B disrupts transactivation by FKHR and mediates effects of insulin on IGF binding protein-1 promoter activity through a conserved insulin response sequence. *J Biol Chem* 1999;**274**:17,184–92.
39. Fredericks WJ, et al. The PAX3-FKHR fusion protein created by the t(2;13) translocation in alveolar rhabdomyosarcomas is a more potent transcriptional activator than PAX3. *Mol Cell Biol* 1995;**15**:1522–35.
40. Nasrin N, et al. DAF-16 recruits the CREB-binding protein coactivator complex to the insulin-like growth factor binding protein 1 promoter in HepG2 cells. *Proc Natl Acad Sci USA* 2000;**97**:10,412–17.
41. Takaishi H, et al. Regulation of nuclear translocation of forkhead transcription factor AFX by protein kinase B. *Proc Natl Acad Sci USA* 1999;**96**:11,836–41.
42. Dijkers PF, et al. Forkhead transcription factor FKHR-L1 modulates cytokine-dependent transcriptional regulation of p27(KIP1). *Mol Cell Biol* 2000;**20**:9138–48.
43. Dijkers PF, et al. FKHR-L1 can act as a critical effector of cell death induced by cytokine withdrawal: protein kinase B-enhanced cell survival through maintenance of mitochondrial integrity. *J Cell Biol* 2002;**156**:531–42.
44. Alvarez B, Martinez AC, Burgering BM, Carrera AC. Forkhead transcription factors contribute to execution of the mitotic programme in mammals. *Nature* 2001;**413**:744–7.
45. Suhara T, Kim HS, Kirshenbaum LA, Walsh K. Suppression of Akt signaling induces Fas ligand expression: involvement of caspase and Jun kinase activation in Akt-mediated Fas ligand regulation. *Mol Cell Biol* 2002;**22**:680–91.
46. Dijkers PF, Medemadagger RH, Lammers JJ, Koenderman L, Coffer PJ. Expression of the pro-apoptotic bcl-2 family member bim is regulated by the forkhead transcription factor FKHR-L1. *Curr Biol* 2000;**10**:1201–4.
47. Tang TT, et al. The forkhead transcription factor AFX activates apoptosis by induction of the BCL-6 transcriptional repressor. *J Biol Chem* 2002;**2**:2.
48. Medema RH, Kops GJ, Bos JL, Burgering BM. AFX-like Forkhead transcription factors mediate cell-cycle regulation by Ras and PKB through p27kip1. *Nature* 2000;**404**:782–7.
49. Nakamura N, et al. Forkhead transcription factors are critical effectors of cell death and cell cycle arrest downstream of PTEN. *Mol Cell Biol* 2000;**20**:8969–82.
50. Tran H, et al. DNA repair pathway stimulated by the Forkhead transcription factor FOXO3a (FKHRL1) through the GADD45 protein. *Science* 2002; **296**(5567):530–34.
51. Zhu G, et al. Two yeast forkhead genes regulate the cell cycle and pseudohyphal growth. *Nature* 2000;**406**:90–4.
52. Fiscella M, et al. Wip1, a novel human protein phosphatase that is induced in response to ionizing radiation in a p53-dependent manner. *Proc Natl Acad Sci USA* 1997;**94**:6048–53.
53. Velasco-Miguel S, et al. PA26, a novel target of the p53 tumor suppressor and member of the GADD family of DNA damage and growth arrest inducible genes. *Oncogene* 1999;**18**:127–37.
54. Kastan MB, et al. A mammalian cell cycle checkpoint pathway utilizing p53 and GADD45 is defective in ataxia-telangiectasia. *Cell* 1992;**71**:587–97.
55. Anderson MJ, Shelton GD, Cavenee WK, Arden KC. Embryonic expression of the tumor-associated PAX3-FKHR fusion protein interferes with the developmental functions of Pax3. *Proc Natl Acad Sci USA* 2001;**98**:1589–94.

56. Henderson ST, Johnson TE. Daf-16 integrates developmental and environmental inputs to mediate aging in the nematode Caenorhabditis elegans. *Curr Biol* 2001;**11**:1975–80.
57. Larsen PL. Aging and resistance to oxidative damage in Caenorhabditis elegans. *Proc Natl Acad Sci USA* 1993;**90**:8905–9.
58. Honda Y, Honda S. The daf-2 gene network for longevity regulates oxidative stress resistance and Mn-superoxide dismutase gene expression in Caenorhabditis elegans. *FASEB J* 1999;**13**:1385–93.
59. Murakami S, Johnson TE. The OLD-1 positive regulator of longevity and stress resistance is under DAF-16 regulation in Caenorhabditis elegans. *Curr Biol* 2001;**11**:1517–23.
60. Kirkwood TBL, Austad SN. Why do we age? *Nature* 2000;**408**:233–8.
61. Lee C-K, Klopp RG, Weindruch R, Prolla TA. Gene expression profile of aging and its retardation by caloric restriction. *Science* 1999;**285**:1390–3.
62. Apfeld J, Kenyon C. Cell nonautonomy of C. elegans daf-2 function in the regulation of diapause and life span. *Cell* 1998;**95**:199–210.
63. Wolkow CA, Kimura KD, Lee MS, Ruvkun G. Regulation of C. elegans life-span by insulinlike signaling in the nervous system. *Science* 2000;**290**:147–50.
64. Schmoll D, et al. Regulation of glucose-6-phosphatase gene expression by protein kinase Balpha and the forkhead transcription factor FKHR. Evidence for insulin response unit-dependent and -independent effects of insulin on promoter activity. *J Biol Chem* 2000;**275**:36,324–36,333.
65. Nakae J, Kitamura T, Silver DL, Accili D. The forkhead transcription factor Foxo1 (Fkhr) confers insulin sensitivity onto glucose-6-phosphatase expression. *J Clin Invest* 2001;**108**:1359–67.
66. Yeagley D, Guo S, Unterman T, Quinn PG. Gene- and activation-specific mechanisms for insulin inhibition of basal and glucocorticoid-induced insulin-like growth factor binding protein-1 and phosphoenolpyruvate carboxykinase transcription. Roles of forkhead and insulin response sequences. *J Biol Chem* 2001;**276**:33,705–33,710.
67. Hall RK, et al. Regulation of phosphoenolpyruvate carboxykinase and insulin-like growth factor-binding protein-1 gene expression by insulin. The role of winged helix/forkhead proteins. *J Biol Chem* 2000;**275**:30,169–30,175.

Chapter 7

The Multi-Gene Family of Transcription Factor AP-1

Peter Angel and Jochen Hess

Deutsches Krebsforschungszentrum, Division of Signal Transduction and Growth Control, Heidelberg

INTRODUCTION

Much of our present knowledge about transcription factors comes from the discovery and study of the activating protein-1 (AP-1) family. AP-1 has served to detect one of the decisive DNA binding motifs required for regulation of both basal and inducible transcription of several genes containing AP-1 sites (5′-TGA$^{G}/_{C}$TCA-3′), also known as TPA responsive elements (TRE), by a variety of extracellular signals. These include growth factors, cytokines, tumor promoters, such as the phorbol ester TPA (12-O-tetradecanoyl-phorbol-13-acetate) and carcinogens, for example UV irradiation and other DNA damaging agents. AP-1 collectively describes a group of structurally and functionally related members of the Jun (Jun (originally described as c-Jun), JunB, and JunD) and Fos (Fos (originally described as c-Fos), FosB, Fra-1, and Fra-2) protein family that form homo- and heterodimers (Jun-Jun and Jun-Fos). Additionally, members of the ATF protein subfamily form heterodimers predominantly with Jun proteins and can bind to TRE-like sequences.

The major intention of this chapter is to summarize our current knowledge of: (a) the characteristic structural features of AP-1 subunits, (b) mechanisms involved in modulation of AP-1 activity, and (c) function of AP-1 proteins in cellular processes using mouse genetics.

GENERAL STRUCTURE OF THE AP-1 SUBUNITS

According to their function in controlling gene expression AP-1 subunits are composed of a region responsible for binding to a specific DNA recognition sequence (DNA binding domain) and a second region that is required for transcriptional activation (transactivation domain, TAD) once the protein is bound to DNA.

The DNA binding domain of AP-1 proteins, also known as the "bZip" region, can be divided in two evolutionary conserved, independently acting domains: the "basic domain" ("b"), which is rich in basic amino acids and responsible for contacting the DNA and the "leucine-zipper" ("Zip") region characterized by heptad repeats of leucines forming a coiled-coil structure that is responsible for dimerization (Figure 7.1). In addition to the leucines other hydrophobic and charged amino acid residues within the leucine zipper region are responsible for specificity and stability of homo- or heterodimer formation between the various Jun, Fos, or ATF proteins. Jun-Jun and Jun-Fos dimers preferentially bind to the 7-bp motif 5'-TGA$^{G}/_{C}$TCA-3', whereas Jun-ATF dimers or ATF homodimers prefer to bind to a related, 8-bp consensus sequence 5'-T$^{G}/_{T}$ACNTCA-3' [1]. Therefore, the characteristics of the AP-1 DNA binding sites in promoters as well as the abundance of the individual AP-1 subunits are decisive for the selection of target genes.

In addition to the "classical" AP-1 members (Jun, Fos, and ATF), some other bZip proteins have been discovered some of which can heterodimerize with the core AP-1 subunits, e.g., Maf, Maf related proteins, Nrl, and Jun dimerizing partners (JDPs). Binding of AP-1 to DNA may also support binding of other transcription factors to adjacent or overlapping binding sites (composite elements) to allow the formation of "quaternary" complexes. The interaction of NF-AT and Ets proteins with the IL-2 and collagenase promoters, respectively, may serve as paradigms for this type of protein–protein interaction [2]. AP-1 has also been reported to associate with NFκB [3], the glucocorticoid receptor [4], Runx proteins [5], Hypoxia inducible factor 1α [6], the T-cell factor/lymphoid enhancing factor

Handbook of Cell Signaling, Three-Volume Set 2 ed.

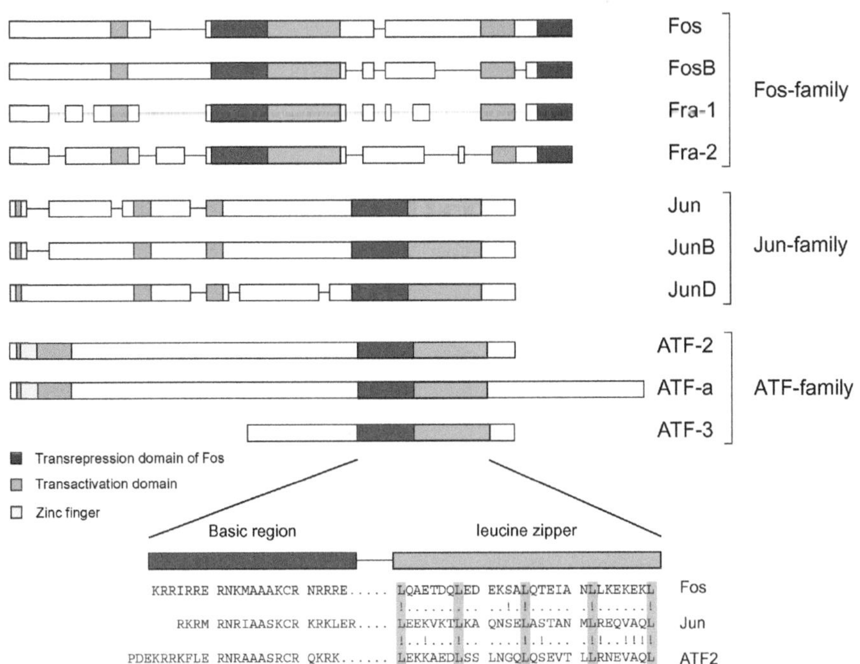

FIGURE 7.1 Structural organization of the Fos, Jun, and ATF proteins.
Conserved regions are highlighted. On the bottom, the amino acid sequences of the bZip region of Fos, Jun, and ATF-2, the most extensively analyzed members of the Fos, Jun, and ATF protein families are shown.

family member TCF4 [7], and many more. Recently, an atomic model for the interferon-β enhanceosome was published demonstrating a cooperative binding of Jun/ATF-2, IRF-3/IRF-7, and NFκB that critically depends on binding induced changes in DNA conformation and interactions with additional coactivators, such as CBP [8].

In contrast to the well-defined DNA binding domains of AP-1 proteins, the structural properties of the domains mediating transcriptional activation of target genes are yet ill defined. The TAD (and its function) can be transferred to heterologous DNA binding domains, which do not require heterodimerization, e.g. to that of the yeast transcription factor GAL4. Such chimeric proteins permitted to identify critical amino acids in TADs, and to recognize that the TADs of individual Jun, Fos, and ATF proteins differ in their transactivation potential. Under specific circumstances, some subunits, such as JunB, Fra-1, and Fra-2 proteins, may even act as repressors of AP-1 activity by competitive binding to AP-1 sites, or by forming inactive heterodimers with other AP-1 members (for details see [9]).

TRANSCRIPTIONAL AND POSTTRANSLATIONAL CONTROL OF AP-1 ACTIVITY

Given the aspects of subunit heterogeneity, their dimerization and DNA binding rules, the control of transcription of AP-1 subunit genes, the time course of subunit synthesis as well as degradation, and the regulation of their function become important issues for an evaluation of the effect on target gene expression and on cellular behavior. *Jun* and *Fos* genes represent the prototype of the class of "immediate early" genes, which are characterized by a rapid and transient activation of transcription in response to changes of environmental conditions, such as growth factors, cytokines, tumor promoters, carcinogens, and expression of certain oncogenes [10]. Since this type of regulation of promoter activity is also observed in the absence of ongoing protein synthesis, it is generally accepted that pre-existing factors, whose activity gets altered by changes in posttranslational modification (in particular phosphorylation), are responsible for the rapid regulation of promoter activity.

Most of our current knowledge on transcriptional activation of immediate early genes is derived from studies on deletion and point mutations of *Fos* and *Jun* promoters, combined with *in vitro* and *in vivo* footprinting analyses. The serum response element (SRE), which is bound by a ternary complex containing the transcription factor p67-SRF and p62-TCF, is required for the majority of extracellular stimuli including growth factors and phorbol esters. Changes in the phosphorylation pattern of SRF and, predominantly, TCF regulate *Fos* promoter activity by these stimuli. Other elements include the cAMP response element (CRE) and the Sis inducible enhancer (SIE) that is recognized by the STAT group of transcription factors. The transient induction of promoter activity by extracellular stimuli can be explained by increased phosphatase activity counteracting the function of upstream protein kinases, which address promoter associated transcription factors. On the other hand, negative auto-regulation by newly synthesized Fos protein may also play an important role [11].

Analysis of deletion mutants of the *Jun* promoter identified two TRE-like binding sites (Jun1, Jun2) that are recognized by Jun/ATF heterodimers or ATF homodimers, which are involved in transcriptional regulation in response to the majority of extracellular stimuli affecting *Jun* transcription. In response to G-protein coupled receptor activation, EGF, and other growth factors the TRE-like binding sites and an additional element in the *Jun* promoter recognized by MEF2 proteins cooperate in transcriptional control of the *Jun* gene. Similar to the factors binding to the *Fos* promoter, the activity of factors binding to the *Jun* promoter is regulated by their phosphorylation status [1, 10].

The most critical members of the class of protein kinases regulating the activity of AP-1 in response to extracellular stimuli are mitogen activated protein kinases (MAPKs; reviewed in [1, 12]). Depending on the type of stimuli these kinases can be dissected into three subgroups: (i) the extracellular signal regulated kinases (ERK-1 and -2), which are robustly activated by growth factors and phorbol esters, but only weakly activated by cytokines and cellular stress inducing stimuli (e.g., UV irradiation or chemical carcinogens), (ii) Jun-N-terminal kinases (JNK-1, -2, and -3), also known as stress activated kinases (SAPK), and (iii) structurally related p38 MAP kinases (p38α, -β, and -γ). Both, JNK and p38 kinases are strongly activated by cytokines and environmental stress, but are poorly activated by growth factors and phorbol ester. These kinases themselves are under strict control of upstream kinases and phosphatases that are part of individual signaling pathways initiated by specific classes of extra- and intracellular stimuli (e.g., growth factors, DNA damaging agents, and oncoproteins). This network, which exhibits a high degree of evolutionary conservation between yeast, *Drosophila*, and mammals is, however, far too complex to be discussed in greater detail in this review.

The JNK/SAPKs were originally identified by their ability to specifically phosphorylate the Jun protein at two positive regulatory sites (Ser-63 and Ser-73) residing within the TAD. Hyperphosphorylation of both sites is observed in response to stress stimuli and oncoproteins. The JNKs can also phosphorylate and regulate the transcriptional activity of other AP-1 proteins, including JunD and ATF-2. Notably, the nuclear protein Menin that is encoded by the tumor suppressor gene *MEN1* specifically interacts with JunD and inhibits JNK dependent phosphorylation of JunD, but also of Jun.

The same positive sites on ATF-2 also serve as phosphoacceptor sites for p38, while Ser-63 and Ser-73 of Jun are not affected by p38. Most likely, hyperphosphorylation of Jun and ATF proteins results in a conformational change of the TAD allowing more efficient interaction with cofactors, which facilitate and stabilize the connection with the RNA polymerase II/initiation complex to enhance transcription of target genes. In addition to enhanced transactivation, hyperphosphorylation of the Jun TAD also regulates protein stability by reducing ubiquitin dependent degradation [13, 14]. In contrast, in neurons the stability of Jun is regulated by the E3 ligase Fbw7, which ubiquitinates phosphorylated Jun protein and facilitates its degradation [15]. Additionally, GSK-3 can phosphorylate the C-terminus of Jun, at a site that is mutated in v-Jun, resulting in Fbw7 binding and proteasomal degradation [16]. Phosphorylation dependent changes in the half-life of Fos proteins have also been observed that predominantly depends on the activity of ribosomal S6 kinase (RSK) and ERK (reviewed in [1, 12]).

In addition to phosphorylation, other posttranslational modifications (e.g., oxidation of a cysteine in the basic region and sumoylation) and physical interaction with transcriptional cofactors (e.g., alphaNAC, Jab1, p300/CBP, TAF1, TAF4b, TAF7, Trip6, and WWOX), subunits of the chromatin remodeling complex (e.g., SWI/SNF and HDACs), and other types of cellular proteins (e.g., DexD/H-box RNA helicase RHII/Gu, BAF60a, and Lamin A/C) modulate AP-1 activity.

FUNCTION OF MAMMALIAN AP-1 SUBUNITS DURING EMBRYOGENESIS AND TISSUE HOMEOSTASIS: LESSONS FROM LOSS-OF-FUNCTION AND GAIN-OF-FUNCTION APPROACHES IN MICE

The generation of mice with genetic disruption and/or transgenic overexpression, as well as the availability of genetically defined mutant cells isolated from these animals represent a major breakthrough in our understanding of the regulatory functions of AP-1 subunits (Tables 7.1 and 7.2). The distinct phenotypes of the individual knockout mice, induced by defects in cells or tissues in which the subunit was particularly important or where its absence became first limiting, support the notion that AP-1 dimers exhibit specific and independent functions *in vivo*. As a general rule derived from all studies, the AP-1 subunits must be present in a complementary and coordinated manner in order to ensure proper development or physiology of the organism (reviewed in [1, 12]).

Conventional knockout approaches demonstrated that expression of JunD, Fos, and FosB is dispensable for normal embryogenesis, whereas loss of Jun, JunB, Fra-1, and Fra-2 results in embryonic lethality or postnatal death, excluding functional studies of tissue homeostasis in adult mice. Therefore, conditional tissue and cell-type specific ablation of AP-1 subunits using the Cre-loxP system or ectopic expression in transgenic mice has become a major tool to study their function in physiological and pathological processes *in vivo* (Tables 7.1 and 7.2).

TABLE 7.1 Knockout and knockin mouse models

Genotype	Phenotype	Affected tissues
$Jun^{-/-}$	Embryonic lethality at E12.5–13.5	Liver, heart
$Jun^{AA/AA}$ for Jun	Rescue of embryonic lethality; resistance to epileptic seizures and neuronal apoptosis induced by excitatory amino acid kainite; impaired skin and intestinal tumorigenesis	Liver, heart, CNS, skin, intestine
JunB for Jun	Rescue of embryonic lethality until birth	Liver, heart
JunD for Jun	Rescue of embryonic lethality until birth	Liver, heart
$Jun^{\Delta/\Delta}$ Alfp-Cre	Impaired postnatal hepatocyte proliferation and liver regeneration; increased hepatotoxicity in a mouse model of inflammatory liver disease; impaired liver tumorigenesis	Liver
$Jun^{\Delta/\Delta}$ Bal1-Cre	Malformation of axial skeleton	Skeleton
$Jun^{\Delta/\Delta}$ Col2a1-Cre	Increased apoptosis of notochordal cells, fusion of ventral bodies and scoliosis of axial skeleton	Skeleton
$Jun^{\Delta/\Delta}$ K5-Cre	Eye lid closure defect; impaired skin tumor development	Skin
$Jun^{\Delta/\Delta}$ Nestin-Cre	Impaired axonal regeneration	CNS
$Jun^{\Delta/\Delta}$ Villin-Cre	Impaired intestinal tumorigenesis	Intestine
$JunB^{-/-}$	Embryonic lethality at day E8.5 to E10	Extraembryonic tissue, placenta
$JunB^{\Delta/\Delta}$ Col1a2-Cre	Pronounced epidermal hyperplasia, disturbed differentiation and prolonged inflammation	Skin, immune system
$JunB^{\Delta/\Delta}$ LysM-Cre	Osteopetrosis	Skeleton
$JunB^{\Delta/\Delta}$ More-Cre	Osteopenia, myeloproliferative disease, and highly susceptible for B-cell leukemia	Immune system, skeleton
$JunB^{-/-}$ Ubi-JunB	Rescue of embryonic lethality; myeloproliferative disease; altered T-helper 2 cell differentiation and impaired allergen-induced airway inflammation; osteoporosis-like phenotype	Immune system, skeleton
$Jun^{\Delta/\Delta}$ $JunB^{\Delta/\Delta}$ K5-Cre-ER	Psoriasis-like phenotype and arthritic lesions	Skin, skeleton
$JunD^{-/-}$	Male sterility and growth retardation; cardiomyocyte hypertrophy and heart failure; hepatic stellate cell activation, liver fibrosis, and protection from ischemia/reperfusion injury; impaired T-helper cell differentiation; increased bone mass	Testis, heart, liver, immune system, skeleton
$Fos^{-/-}$	Osteopetrosis and accelerated light induced apoptosis of photoreceptor cells; impaired malignant progression of skin tumors	Skeleton, CNS, skin
$Fos^{-/-}$ H2Kb-Fra1	Rescue of osteopetrosis and photoreceptor cell apoptosis	Skeleton, CNS
$Fos^{\Delta/\Delta}$ Nestin-Cre	Impaired long-term memory and synaptic plasticity	CNS

(Continued)

Table 7.1 (Continued)

Genotype	Phenotype	Affected tissues
$FosB^{-/-}$	Nurturing defect	CNS, hypothalamus
$Fra1^{-/-}$	Embryonic lethality at E9.5–10.5	Extraembryonic tissue, placenta
$Fra1^{-/-}$ Ubi-JunB	Rescue of embryonic lethality	Extraembryonic tissue, placenta
$Fra1^{\Delta/\Delta}$ More-Cre	Osteopenia	Skeleton
Fra1 for Fos	Rescue of osteopetrosis and photoreceptor cell apoptosis	Skeleton, CNS
$Fra2^{-/-}$	Postnatal lethality and defective chondrocyte differentiation	Skeleton
$Fra2^{\Delta/\Delta}$ Coll2a1-Cre	Defective chondrocyte differentiation and kyphosis-like phenotype	Skeleton

Note: -/- conventional knockout, Δ/Δ Cre-induced conditional knockout.

TABLE 7.2 Transgenic mouse models

Genotype	Phenotype	Affected tissues
H2kb-Jun	None	None
K14-Jun(TAM67)	Impaired skin tumor development	Skin
UbiC-JunB	Inhibition of B-cell proliferation; increased bone mass	Immune system, skeleton
CD4-JunB	Altered T-helper cell differentiation	Immune system
UbiC-JunD	Reduced peripheral T- and B-cells and impaired T-cell activation	Immune system
H2Kb-Fos	Osteosarcoma	Skeleton
K5-tTA TetO-A-Fos	Impaired skin tumor development, but benign sebaceous adenomas	Skin
H2Kb-FosB	None	None
Tcrb-ΔFosB	Impaired T-cell differentiation	Immune system
NSE-ΔFosB	Osteosclerosis and impaired adipogenesis	Skeleton, fat tissue
H2Kb-Fra1	Osteosclerosis	Skeleton
CMV-Fra2	Ocular malformation	Anterior eye structure
H2Kb-Fra2	Increased bone mass	Skeleton

Jun-null embryos die at midgestation (E12.5 to 13.5) due to dysregulation in liver and heart development [17–19]. Mice with perinatal liver specific *Jun* deletion survive the specific postnatal gene inactivation in hepatocytes, but show an impaired liver regeneration in response to partial hepatectomy and increased hepatotoxicity in a mouse model of inflammatory liver disease [20, 21]. Using several double knockout mouse models revealed that Jun controls hepatocyte proliferation and liver regeneration by a p53/p21 dependent mechanism accompanied by repressing p38 activity [22]. Knockin mice expressing a mutated Jun protein in which the N-terminal phosphorylation sites at Ser-63 and Ser-73 were changed to alanines (JunAA mice) developed normally, were viable, and fertile as adults [23]. However, JunAA mice are resistant to epileptic seizures and neuronal apoptosis induced by the excitatory amino acid kainate and JunAA fibroblasts show proliferation defects as well as apoptotic defects upon stress induction [23]. By contrast, T-cell proliferation and differentiation appears to be independent of Jun N-terminal phosphorylation, whereas efficient T-cell receptor induced thymocyte apoptosis is affected [24]. Interestingly, knockin mice having *Jun* alleles replaced by *JunB* or *JunD* undergo normal embryogenesis and develop a normal liver and heart, indicating that spatial and temporal regulation of Jun protein expression may be more important than the coding sequence of the individual family member (reviewed in [1]). Detailed analysis of indicative marker genes showed that expression of genes regulated by Jun-Fos, but not those regulated by Jun-ATF dimers, are restored, thereby rescuing Jun dependent defects *in vivo* as well as in primary fibroblasts and fetal hepatoblasts *in vitro*.

Conditional tissue- and cell-type specific ablation highlighted further functions of Jun in adult tissues and showed that its deficiency results in axial skeleton malformation accompanied by accelerated apoptosis of notochord cells, fusion of ventral bodies, and scoliosis [25]. Furthermore, a specific inactivation in the central nervous system caused only minor defects in neurogenesis, but identified Jun as an important regulator of perineural sprouting, lymphocyte recruitment, and microglia activation after injury and axonal regeneration [26]. Jun also plays a major role in skin development and homeostasis as an important regulator of keratinocyte proliferation and differentiation through regulation of the epidermal growth factor receptor (EGFR) signaling [27, 28]. However, impaired proliferation of Jun deficient keratinocytes could be restored not only by the EGFR ligands heparin binding EGF (HB-EGF) and transforming growth factor-α (TGF-α), but also by other growth factors such as keratinocyte growth factor (KGF) and granulocyte-macrophage colony stimulating factor (GM-CSF). This indicates that AP-1 controls keratinocyte proliferation and differentiation by cell intrinsic genetic programs and in a paracrine manner. Indeed, in an organotypic *in vitro* coculture system composed of mouse fibroblasts and human keratinocytes, Jun and JunB regulate the synthesis of numerous fibroblast derived growth factors and cytokines, including KGF and GM-CSF, that control the balance between keratinocyte proliferation and differentiation [29, 30].

Targeted inactivation of JunB resulted also in embryonic lethality between days 8.5 and 10.0 due to multiple defects in extraembryonic tissues [31]. Affected cell types/organs comprise the trophoblast giant cells, yolk sac mesenthelium, and placental labyrinth. The observed phenotypes in JunB-null embryos is, at least in part, due to severe impairment of proper endothelial morphogenesis mediated by a cell autonomous endothelial JunB function and regulation of the vascular endothelial growth factor (VEGF), the core binding factor-β (CBF-β), and expression of matrix-metalloproteinases [32, 33]. The lethal phenotype can be rescued by a ubiquitously expressed *JunB* transgene [31]. In such rescued mice loss of *JunB* transgene expression in the myeloid lineage revealed a transplantable myeloproliferative disease resembling human chronic myeloid leukemia (see below).

Detailed analysis of mouse models with cell-type specific ablation or ectopic expression further elucidated specific functions of JunB in cells of the immune system, in bone development, and skin homeostasis. Hence, JunB regulates proper differentiation and function of T-helper cell [34, 35], inhibits proliferation and transformation of B-lymphoid cells [36], and is required for IgE mediated degranulation and cytokine release of mast cells [37]. Adult mice lacking JunB are osteopenic due to cell autonomous defects in osteoblasts and osteoclasts [38, 39], while JunB deficiency in the skin revealed pronounced epidermal hyperproliferation, disturbed differentiation, and prolonged inflammation upon wounding or treatment with phorbol ester [40]. Finally, an inducible epidermal deletion of JunB and Jun in adult mice leads to a phenotype resembling histological and molecular hallmarks of psoriasis, including arthritic lesions [41].

JunD deficient mice develop normally, but postnatal growth of homozygous *JunD*$^{-/-}$ animals is reduced and males develop age dependent defects in reproduction, hormone imbalance, and impaired spermatogenesis [42]. Analysis of mice lacking JunD or exhibit cardiomyocyte specific JunD overexpression revealed distinct functions of JunD in cardiac hypertrophy and heart failure, suggesting a major role in both adaptive–protective and maladaptive hypertrophy in heart, depending on its expression levels [43]. In liver tissue, JunD is implicated in hepatic stellate cells activation and liver fibrosis by regulation of tissue inhibitor of metalloproteinase-1 expression [44], but also protects liver from ischemia/reperfusion injury by dampening JNK-1 activation and c-Jun/AP-1 dependent gene transcription [45]. Moreover, JunD is part of a regulatory network, including TGF-α/EGFR signaling, that controls proliferation to prevent pathological progression in chronic renal disease [46]. Finally, JunD regulates lymphocyte proliferation and T-helper cell cytokine expression [47], and is implicated in estrogen depletion induced osteopenia via its action to suppress bone formation and to enhance bone resorption [48].

Fos-null mice are viable and fertile but suffer from severe osteopetrosis caused by lack of mature osteoclasts [49–51]. A broad range of experimental evidence support that Fos is a key player during osteoclastogenesis by regulation of distinct target genes, e.g. *Fosl1* (Fra-1), *Ifnb1* (interferon-β), and *Nfatc1* (nuclear factor of activated T-cells c1) [52–54]. Accordingly, Fos dependent functions in osteoclasts can be substituted by Fra-1, when the *Fos* locus is deleted and replaced by the *Fosl1* gene [55], or by gene transfer of an active form of NFAT in Fos deficient precursor cells [52].

Genetic deficiency in mice further highlighted a major role of Fos in molecular and physiological aspects of the central nervous system, including synaptic plasticity, long-term memory, and cell survival [56–58]. Again, replacement of the *Fos* locus by the *Fosl1* gene restored, at least in part, impairments caused by brain specific Fos depletion [59, 60].

Similar to Jun and JunB, Fra-1 expression is essential for embryonic development and null embryos died between E9.5 and 10.5 [61]. The vascularization of the placental labyrinth was impaired in null embryos suggesting that JunB and Fra-1, possibly as heterodimers, address common target genes responsible for the generation of a functional placental labyrinth. Accordingly, Fra-1 deficient embryos can be rescued from embryonic death by JunB overexpression [61].

Mice lacking Fra-1 due to conditional deletion develop osteopenia, whereas, mice with ectopic Fra-1 overexpression develop a progressive increase in bone mass leading to osteosclerosis [62, 63]. A similar phenotype is observed in transgenic mice expressing ΔFosB, a naturally occurring truncated form of FosB that arises from alternative splicing. These mice also show impaired adipogenesis, a phenotype that was not observed in Fra-1 transgenic mice [64].

In contrast to Fra-1, Fra-2 and FosB expression are dispensable for embryonic development. However, adult $FosB^{-/-}$ females nurture insufficiently [65], and pups lacking Fra-2 die shortly after birth showing reduced zones of hypertrophic chondrocytes and impaired matrix deposition in long bone growth plates [66]. Similar to Fra-2 deficient embryos, animals with conditional deletion of Fra-2 by a Coll2a1-Cre transgene display smaller growth plates and are growth retarded, supporting its role in chondrocyte differentiation and function [66].

FUNCTION OF MAMMALIAN AP-1 SUBUNITS DURING CANCER DEVELOPMENT AND PROGRESSION

Identification of Jun and Fos as cellular homologs of retroviral oncoproteins (v-Jun and v-Fos) together with the finding that growth factors, membrane bound, and cytoplasmic oncogenes, as well as tumor promoters permanently upregulate AP-1 abundance as part of their transforming capacity, immediately linked AP-1 to cellular growth control and neoplastic transformation (reviewed in [1]). Indeed, AP-1 is a key player in cell cycle control by the regulation of important cell cycle regulators such as cyclin D1, cyclin A, cyclin E, p53, $p21^{Cip1}$, $p16^{Ink4}$, $p19^{Arf}$, and many more [67–70]. Analysis of cell culture models that exhibit altered expression of individual AP-1 members revealed unique and crucial roles for each Jun protein, but functional redundancy among the Fos proteins. Again, genetically modified mouse models have contributed to our current understanding, how AP-1 family members affect the balance between cell proliferation, differentiation, and apoptosis in the context of neoplastic transformation and malignancy.

Genetic analysis of AP-1 function revealed that Fos overexpression induced chondro- and osteosarcoma formation in transgenic mice (reviewed in [71]). However, in the absence of RSK-2 or in JunAA mutants, Fos dependent tumor formation is impaired most likely due to decreased proliferation and increased apoptosis of transformed osteoblasts [72, 73]. RSK-2 can phosphorylate Fos at Ser-362 and Ser-374, allowing subsequent phosphorylation by ERK on Thr-325 and Thr-331. Both phosphorylation events stabilize the protein and modulate Fos induced cell transformation.

Fos protein expression and function also contribute to epithelial malignancy, since *Fos* deficient mice fail to undergo malignant progression of skin tumors in a transgenic model of oncogenic Ras and conditional expression of an *A-Fos* transgene, a dominant negative mutant that inhibits AP-1 DNA binding, in epidermal keratinocytes interferes with the development of characteristic benign or malignant squamous cell lesions during chemically induced skin carcinogenesis [74, 75]. Notably, these mice develop benign sebaceous adenomas.

The important role of AP-1 for malignant transformation of keratinocytes is further supported by the fact that expression of a dominant-negative Jun mutant (TAM67) in basal keratinocytes protects transgenic mice from UV- and chemically induced as well as HPV16 driven skin tumor formation (reviewed in [76]). Additionally, JunAA mutants and mice with an epidermal specific *Jun* knockout exhibit impaired skin tumor development in the *K5-SOS-F* transgenic tumor model [23, 27].

Mouse models for intestinal and hepatocellular carcinogenesis provided further evidence that AP-1 dependent genetic programs are implicated in epithelial malignancy [7, 77]. During chemically induced liver tumorigenesis, Jun prevents apoptosis by antagonizing p53 activity and thereby contributes to early stage hepatocellular cancer development [77]. Moreover, liver specific *Mapk14* deficient mice highlighted a new mechanism whereby p38α negatively regulates cell proliferation by antagonizing the JNK-Jun pathway during liver cancer development [78].

In contrast to Jun, JunB was identified as a potential tumor suppressor gene, at least in hematopoietic cells. In line with the observation that lymphoid cells from transgenic mice with JunB overexpression poorly respond to mitogenic stimuli [36], JunB acts as a gatekeeper for B-lymphoid leukemia by inhibition of cell proliferation and transformation [79]. JunB is also a critical downstream target of the tumor suppressor activity of PU.1 and reduced JunB expression is a common feature of acute myeloid leukemia [80]. Furthermore, JunB inactivation in postnatal mice resulted in a transplantable myeloproliferative disorder eventually progressing to blast crisis and resembling early human chronic myelogenous leukemia (CML) [81]. More detailed analysis provided experimental evidence that JunB regulates the number of hematopoietic stem cells in normal and leukemic hematopoiesis [82]. Loss of JunB expression is also evident in human CML patients, and the downregulation seems to be mediated by hypermethylation of the JunB promoter [83].

Despite a broad knowledge concerning genes that harbor putative AP-1 binding sites in their regulatory elements, only few direct AP-1 target genes have been identified, which are affected in mutant mice or cell lines derived thereof and critically contribute to cellular transformation and tumor formation *in vivo*. In addition to AP-1 target genes involved in the regulation of cell proliferation, differentiation, or apoptosis, the most well characterized AP-1 responsive genes in cancer are those implicated in signal transduction, (e.g., *Egfr*), chromatin remodeling (e.g., *Dmnt1*, *Hdac3*), invasion (e.g., *Mmp´s*, *uPa*), metastasis (e.g., *Cd44*, *Opn*), and angiogenesis (e.g., *Vegf*). Some of these target genes support the assumption that AP-1 is a key player in aggressive spread of malignant tumor cells and metastasis that is a major cause of death in cancer patients [1].

CONCLUSION

Despite the fact that AP-1 was identified more than a decade ago, it still maintains a lot of its mystery. Further research on tissue specific inactivation of AP-1 members and the identification of subunit specific target genes may yield an even more complex picture of function and regulation of AP-1 than exists at present.

ACKNOWLEDGEMENTS

We gratefully acknowledge Marina Schorpp-Kistner and Bettina Hartenstein for critical discussion and reading of the manuscript. The laboratory is supported by grants from the Deutsche Forschungsgemeinschaft, by the Cooperation Program in Cancer Research of the DKFZ and the Israeli Ministry of Science, Deutsche Krebshilfe, German Israeli Foundation for Scientific Research and Development, Helmholtz Society (Program SB-Cancer), and the German Ministry for Education and Research (National Genome Research Network, NGFN-2 and NGFN-Plus).

REFERENCES

1. Eferl R, Wagner EF. AP-1: a double-edged sword in tumorigenesis. *Nat Rev Cancer* 2003;**3**:859–68.
2. Macian F, Lopez-Rodriguez C, Rao A. Partners in transcription: NFAT and AP-1. *Oncogene* 2001;**20**:2476–89.
3. Stein B, Baldwin Jr AS, Ballard DW, Greene WC, Angel P, Herrlich P. Cross-coupling of the NF-kappa B p65 and Fos/Jun transcription factors produces potentiated biological function. *Embo J* 1993;**12**:3879–91.
4. Herrlich P. Cross-talk between glucocorticoid receptor and AP-1. *Oncogene* 2001;**20**:2465–75.
5. Hess J, Porte D, Munz C, Angel P. AP-1 and Cbfa/runt physically interact and regulate parathyroid hormone-dependent MMP13 expression in osteoblasts through a new osteoblast-specific element 2/AP-1 composite element. *J Biol Chem* 2001;**276**:20,029–38.
6. Laderoute KR, Calaoagan JM, Gustafson-Brown C, Knapp AM, Li GC, Mendonca HL, Ryan HE, Wang Z, Johnson RS. The response of c-jun/AP-1 to chronic hypoxia is hypoxia-inducible factor 1 alpha dependent. *Mol Cell Biol* 2002;**22**:2515–23.
7. Nateri AS, Spencer-Dene B, Behrens A. Interaction of phosphorylated c-Jun with TCF4 regulates intestinal cancer development. *Nature* 2005;**437**:281–5.
8. Panne D, Maniatis T, Harrison SC. An atomic model of the interferon-beta enhanceosome. *Cell* 2007;**129**:1111–23.
9. Wagner EF. AP-1 reviews. *Oncogene* 2001;**20**:2333–497.
10. Mechta-Grigoriou F, Gerald D, Yaniv M. The mammalian Jun proteins: redundancy and specificity. *Oncogene* 2001;**20**:2378–89.
11. Treisman R. Journey to the surface of the cell: Fos regulation and the SRE. *Embo J* 1995;**14**:4905–13.
12. Hess J, Angel P, Schorpp-Kistner M. AP-1 subunits: quarrel and harmony among siblings. *J Cell Sci* 2004;**117**:5965–73.
13. Treier M, Staszewski LM, Bohmann D. Ubiquitin-dependent c-Jun degradation in vivo is mediated by the delta domain. *Cell* 1994; **78**:787–98.
14. Musti AM, Treier M, Bohmann D. Reduced ubiquitin-dependent degradation of c-Jun after phosphorylation by MAP kinases. *Science* 1997;**275**:400–2.
15. Nateri AS, Riera-Sans L, Da Costa C, Behrens A. The ubiquitin ligase SCFFbw7 antagonizes apoptotic JNK signaling. *Science* 2004;**303**:1374–8.
16. Wei W, Jin J, Schlisio S, Harper JW, Kaelin Jr WG. The v-Jun point mutation allows c-Jun to escape GSK3-dependent recognition and destruction by the Fbw7 ubiquitin ligase. *Cancer Cell* 2005;**8**:25–33.
17. Hilberg F, Aguzzi A, Howells N, Wagner EF. c-jun is essential for normal mouse development and hepatogenesis. *Nature* 1993;**365**:179–81.
18. Johnson RS, van Lingen B, Papaioannou VE, Spiegelman BM. A null mutation at the c-jun locus causes embryonic lethality and retarded cell growth in culture. *Genes Dev* 1993;**7**:1309–17.
19. Eferl R, Sibilia M, Hilberg F, Fuchsbichler A, Kufferath I, Guertl B, Zenz R, Wagner EF, Zatloukal K. Functions of c-Jun in liver and heart development. *J Cell Biol* 1999;**145**:1049–61.
20. Behrens A, Sibilia M, David JP, Mohle-Steinlein U, Tronche F, Schutz G, Wagner EF. Impaired postnatal hepatocyte proliferation and liver regeneration in mice lacking c-jun in the liver. *Embo J* 2002; **21**:1782–90.

21. Hasselblatt P, Rath M, Komnenovic V, Zatloukal K, Wagner EF. Hepatocyte survival in acute hepatitis is due to c-Jun/AP-1-dependent expression of inducible nitric oxide synthase. *Proc Natl Acad Sci USA* 2007;**104**:17,105–17,110.
22. Stepniak E, Ricci R, Eferl R, Sumara G, Sumara I, Rath M, Hui L, Wagner EF. c-Jun/AP-1 controls liver regeneration by repressing p53/p21 and p38 MAPK activity. *Genes Dev* 2006;**20**:2306–14.
23. Behrens A, Sibilia M, Wagner EF. Amino-terminal phosphorylation of c-Jun regulates stress-induced apoptosis and cellular proliferation. *Nat Genet* 1999;**21**:326–9.
24. Behrens A, Sabapathy K, Graef I, Cleary M, Crabtree GR, Wagner EF. Jun N-terminal kinase 2 modulates thymocyte apoptosis and T cell activation through c-Jun and nuclear factor of activated T cell (NF-AT). *Proc Natl Acad Sci USA* 2001;**98**:1769–74.
25. Behrens A, Haigh J, Mechta-Grigoriou F, Nagy A, Yaniv M, Wagner EF. Impaired intervertebral disc formation in the absence of Jun. *Development* 2003;**130**:103–9.
26. Raivich G, Bohatschek M, Da Costa C, Iwata O, Galiano M, Hristova M, Nateri AS, Makwana M, Riera-Sans L, Wolfer DP, et al. The AP-1 transcription factor c-Jun is required for efficient axonal regeneration. *Neuron* 2004;**43**:57–67.
27. Zenz R, Scheuch H, Martin P, Frank C, Eferl R, Kenner L, Sibilia M, Wagner EF. c-Jun regulates eyelid closure and skin tumor development through EGFR signaling. *Dev Cell* 2003;**4**:879–89.
28. Li G, Gustafson-Brown C, Hanks SK, Nason K, Arbeit JM, Pogliano K, Wisdom RM, Johnson RS. c-Jun is essential for organization of the epidermal leading edge. *Dev Cell* 2003;**4**:865–77.
29. Szabowski A, Maas-Szabowski N, Andrecht S, Kolbus A, Schorpp-Kistner M, Fusenig NE, Angel P. c-Jun and JunB antagonistically control cytokine-regulated mesenchymal-epidermal interaction in skin. *Cell* 2000;**103**:745–55.
30. Florin L, Hummerich L, Dittrich BT, Kokocinski F, Wrobel G, Gack S, Schorpp-Kistner M, Werner S, Hahn M, Lichter P, et al. Identification of novel AP-1 target genes in fibroblasts regulated during cutaneous wound healing. *Oncogene* 2004;**23**:7005–17.
31. Schorpp-Kistner M, Wang ZQ, Angel P, Wagner EF. JunB is essential for mammalian placentation. *Embo J* 1999;**18**:934–48.
32. Licht AH, Pein OT, Florin L, Hartenstein B, Reuter H, Arnold B, Lichter P, Angel P, Schorpp-Kistner M. JunB is required for endothelial cell morphogenesis by regulating core-binding factor beta. *J Cell Biol* 2006;**175**:981–91.
33. Schmidt D, Textor B, Pein OT, Licht AH, Andrecht S, Sator-Schmitt M, Fusenig NE, Angel P, Schorpp-Kistner M. Critical role for NF-kappaB-induced JunB in VEGF regulation and tumor angiogenesis. *Embo J* 2007;**26**:710–19.
34. Hartenstein B, Teurich S, Hess J, Schenkel J, Schorpp-Kistner M, Angel P. Th2 cell-specific cytokine expression and allergen-induced airway inflammation depend on JunB. *Embo J* 2002;**21**:6321–9.
35. Li B, Tournier C, Davis RJ, Flavell RA. Regulation of IL-4 expression by the transcription factor JunB during T helper cell differentiation. *Embo J* 1999;**18**:420–32.
36. Szremska AP, Kenner L, Weisz E, Ott RG, Passegue E, Artwohl M, Freissmuth M, Stoxreiter R, Theussl HC, Parzer SB, et al. JunB inhibits proliferation and transformation in B-lymphoid cells. *Blood* 2003;**102**:4159–65.
37. Textor B, Licht AH, Tuckermann JP, Jessberger R, Razin E, Angel P, Schorpp-Kistner M, Hartenstein B. JunB is required for IgE-mediated degranulation and cytokine release of mast cells. *J Immunol* 2007; **179**:6873–80.
38. Hess J, Hartenstein B, Teurich S, Schmidt D, Schorpp-Kistner M, Angel P. Defective endochondral ossification in mice with strongly compromised expression of JunB. *J Cell Sci* 2003;**116**:4587–96.
39. Kenner L, Hoebertz A, Beil T, Keon N, Karreth F, Eferl R, Scheuch H, Szremska A, Amling M, Schorpp-Kistner M, et al. Mice lacking JunB are osteopenic due to cell-autonomous osteoblast and osteoclast defects. *J Cell Biol* 2004;**164**:613–23.
40. Florin L, Knebel J, Zigrino P, Vonderstrass B, Mauch C, Schorpp-Kistner M, Szabowski A, Angel P. Delayed wound healing and epidermal hyperproliferation in mice lacking JunB in the skin. *J Invest Dermatol* 2006;**126**:902–11.
41. Zenz R, Eferl R, Kenner L, Florin L, Hummerich L, Mehic D, Scheuch H, Angel P, Tschachler E, Wagner EF. Psoriasis-like skin disease and arthritis caused by inducible epidermal deletion of Jun proteins. *Nature* 2005;**437**:369–75.
42. Thepot D, Weitzman JB, Barra J, Segretain D, Stinnakre MG, Babinet C, Yaniv M. Targeted disruption of the murine junD gene results in multiple defects in male reproductive function. *Development* 2000;**127**:143–53.
43. Ricci R, Eriksson U, Oudit GY, Eferl R, Akhmedov A, Sumara I, Sumara G, Kassiri Z, David JP, Bakiri L, et al. Distinct functions of junD in cardiac hypertrophy and heart failure. *Genes Dev* 2005;**19**:208–13.
44. Smart DE, Green K, Oakley F, Weitzman JB, Yaniv M, Reynolds G, Mann J, Millward-Sadler H, Mann DA. JunD is a profibrogenic transcription factor regulated by Jun N-terminal kinase-independent phosphorylation. *Hepatology* 2006;**44**:1432–40.
45. Marden JJ, Zhang Y, Oakley FD, Zhou W, Luo M, Jia HP, McCray Jr PB, Yaniv M, Weitzman JB, Engelhardt JF. JunD protects the liver from ischemia/reperfusion injury by dampening AP-1 transcriptional activation. *J Biol Chem* 2008;**283**:6687–95.
46. Pillebout E, Weitzman JB, Burtin M, Martino C, Federici P, Yaniv M, Friedlander G, Terzi F. JunD protects against chronic kidney disease by regulating paracrine mitogens. *J Clin Invest* 2003;**112**:843–52.
47. Meixner A, Karreth F, Kenner L, Wagner EF. JunD regulates lymphocyte proliferation and T helper cell cytokine expression. *Embo J* 2004;**23**:1325–35.
48. Kawamata A, Izu Y, Yokoyama H, Amagasa T, Wagner EF, Nakashima K, Ezura Y, Hayata T, Noda M. JunD suppresses bone formation and contributes to low bone mass induced by estrogen depletion. *J Cell Biochem* 2008;**103**:1037–45.
49. Wang ZQ, Ovitt C, Grigoriadis AE, Mohle-Steinlein U, Ruther U, Wagner EF. Bone and haematopoietic defects in mice lacking c-fos. *Nature* 1992;**360**:741–5.
50. Johnson RS, Spiegelman BM, Papaioannou V. Pleiotropic effects of a null mutation in the c-fos proto-oncogene. *Cell* 1992;**71**:577–86.
51. Grigoriadis AE, Wang ZQ, Cecchini MG, Hofstetter W, Felix R, Fleisch HA, Wagner EF. c-Fos: a key regulator of osteoclast-macrophage lineage determination and bone remodeling. *Science* 1994;**266**:443–8.
52. Matsuo K, Galson DL, Zhao C, Peng L, Laplace C, Wang KZ, Bachler MA, Amano H, Aburatani H, Ishikawa H, et al. Nuclear factor of activated T-cells (NFAT) rescues osteoclastogenesis in precursors lacking c-Fos. *J Biol Chem* 2004;**279**:26,475–26,480.
53. Matsuo K, Owens JM, Tonko M, Elliott C, Chambers TJ, Wagner EF. Fosl1 is a transcriptional target of c-Fos during osteoclast differentiation. *Nat Genet* 2000;**24**:184–7.
54. Takayanagi H, Kim S, Matsuo K, Suzuki H, Suzuki T, Sato K, Yokochi T, Oda H, Nakamura K, Ida N, et al. RANKL maintains bone homeostasis through c-Fos-dependent induction of interferon-beta. *Nature* 2002;**416**:744–9.

55. Fleischmann A, Hafezi F, Elliott C, Reme CE, Ruther U, Wagner EF. Fra-1 replaces c-Fos-dependent functions in mice. *Genes Dev* 2000;**14**:2695–700.
56. Yasoshima Y, Sako N, Senba E, Yamamoto T. Acute suppression, but not chronic genetic deficiency, of c-fos gene expression impairs long-term memory in aversive taste learning. *Proc Natl Acad Sci USA* 2006;**103**:7106–11.
57. Fleischmann A, Hvalby O, Jensen V, Strekalova T, Zacher C, Layer LE, Kvello A, Reschke M, Spanagel R, Sprengel R, et al. Impaired long-term memory and NR2A-type NMDA receptor-dependent synaptic plasticity in mice lacking c-Fos in the CNS. *J Neurosci* 2003;**23**:9116–22.
58. Hafezi F, Steinbach JP, Marti A, Munz K, Wang ZQ, Wagner EF, Aguzzi A, Reme CE. The absence of c-fos prevents light-induced apoptotic cell death of photoreceptors in retinal degeneration in vivo. *Nat Med* 1997;**3**:346–9.
59. Gass P, Fleischmann A, Hvalby O, Jensen V, Zacher C, Strekalova T, Kvello A, Wagner EF, Sprengel R. Mice with a fra-1 knock-in into the c-fos locus show impaired spatial but regular contextual learning and normal LTP. *Brain Res Mol Brain Res* 2004;**130**:16–22.
60. Wenzel A, Iseli HP, Fleischmann A, Hafezi F, Grimm C, Wagner EF, Reme CE. Fra-1 substitutes for c-Fos in AP-1-mediated signal transduction in retinal apoptosis. *J Neurochem* 2002;**80**:1089–94.
61. Schreiber M, Wang ZQ, Jochum W, Fetka I, Elliott C, Wagner EF. Placental vascularisation requires the AP-1 component fra1. *Development* 2000;**127**:4937–48.
62. Eferl R, Hoebertz A, Schilling AF, Rath M, Karreth F, Kenner L, Amling M, Wagner EF. The Fos-related antigen Fra-1 is an activator of bone matrix formation. *Embo J* 2004;**23**:2789–99.
63. Jochum W, David JP, Elliott C, Wutz A, Plenk Jr H, Matsuo K, Wagner EF. Increased bone formation and osteosclerosis in mice overexpressing the transcription factor Fra-1. *Nat Med* 2000;**6**:980–4.
64. Sabatakos G, Sims NA, Chen J, Aoki K, Kelz MB, Amling M, Bouali Y, Mukhopadhyay K, Ford K, Nestler EJ, et al. Overexpression of DeltaFosB transcription factor(s) increases bone formation and inhibits adipogenesis. *Nat Med* 2000;**6**:985–90.
65. Brown JR, Ye H, Bronson RT, Dikkes P, Greenberg ME. A defect in nurturing in mice lacking the immediate early gene fosB. *Cell* 1996;**86**:297–309.
66. Karreth F, Hoebertz A, Scheuch H, Eferl R, Wagner EF. The AP1 transcription factor Fra2 is required for efficient cartilage development. *Development* 2004;**131**:5717–25.
67. Shaulian E, Karin M. AP-1 in cell proliferation and survival. *Oncogene* 2001;**20**:2390–400.
68. Shaulian E, Karin M. AP-1 as a regulator of cell life and death. *Nat Cell Biol* 2002;**4**:E131–6.
69. Andrecht S, Kolbus A, Hartenstein B, Angel P, Schorpp-Kistner M. Cell cycle promoting activity of JunB through cyclin A activation. *J Biol Chem* 2002;**277**:35,961–8.
70. Passegue E, Wagner EF. JunB suppresses cell proliferation by transcriptional activation of p16(INK4a) expression. *Embo J* 2000;**19**:2969–79.
71. Grigoriadis AE, Wang ZQ, Wagner EF. Fos and bone cell development: lessons from a nuclear oncogene. *Trends Genet* 1995;**11**:436–41.
72. Behrens A, Jochum W, Sibilia M, Wagner EF. Oncogenic transformation by ras and fos is mediated by c-Jun N-terminal phosphorylation. *Oncogene* 2000;**19**:2657–63.
73. David JP, Mehic D, Bakiri L, Schilling AF, Mandic V, Priemel M, Idarraga MH, Reschke MO, Hoffmann O, Amling M, et al. Essential role of RSK2 in c-Fos-dependent osteosarcoma development. *J Clin Invest* 2005;**115**:664–72.
74. Gerdes MJ, Myakishev M, Frost NA, Rishi V, Moitra J, Acharya A, Levy MR, Park SW, Glick A, Yuspa SH, et al. Activator protein-1 activity regulates epithelial tumor cell identity. *Cancer Res* 2006;**66**:7578–88.
75. Saez E, Rutberg SE, Mueller E, Oppenheim H, Smoluk J, Yuspa SH, Spiegelman BM. c-fos is required for malignant progression of skin tumors. *Cell* 1995;**82**:721–32.
76. Young MR, Yang HS, Colburn NH. Promising molecular targets for cancer prevention: AP-1, NF-kappa B and Pdcd4. *Trends Mol Med* 2003;**9**:36–41.
77. Eferl R, Ricci R, Kenner L, Zenz R, David JP, Rath M, Wagner EF. Liver tumor development. c-Jun antagonizes the proapoptotic activity of p53. *Cell* 2003;**112**:181–92.
78. Hui L, Bakiri L, Mairhorfer A, Schweifer N, Haslinger C, Kenner L, Komnenovic V, Scheuch H, Beug H, Wagner EF. p38alpha suppresses normal and cancer cell proliferation by antagonizing the JNK-c-Jun pathway. *Nat Genet* 2007;**39**:741–9.
79. Ott RG, Simma O, Kollmann K, Weisz E, Zebedin EM, Schorpp-Kistner M, Heller G, Zochbauer S, Wagner EF, Freissmuth M, et al. JunB is a gatekeeper for B-lymphoid leukemia. *Oncogene* 2007;**26**:4863–71.
80. Steidl U, Rosenbauer F, Verhaak RG, Gu X, Ebralidze A, Otu HH, Klippel S, Steidl C, Bruns I, Costa DB, et al. Essential role of Jun family transcription factors in PU.1 knockdown-induced leukemic stem cells. *Nat Genet* 2006;**38**:1269–77.
81. Passegue E, Jochum W, Schorpp-Kistner M, Mohle-Steinlein U, Wagner EF. Chronic myeloid leukemia with increased granulocyte progenitors in mice lacking junB expression in the myeloid lineage. *Cell* 2001;**104**:21–32.
82. Passegue E, Wagner EF, Weissman IL. JunB deficiency leads to a myeloproliferative disorder arising from hematopoietic stem cells. *Cell* 2004;**119**:431–43.
83. Yang MY, Liu TC, Chang JG, Lin PM, Lin SF. JunB gene expression is inactivated by methylation in chronic myeloid leukemia. *Blood* 2003;**101**:3205–11. Δ

Chapter 8

NFκB: A Key Integrator of Cell Signaling

John K. Westwick, Klaus Schwamborn and Frank Mercurio
Signal Research Division, Celgene, San Diego, California

Eukaryotic cells possess a number of distinct signal transduction pathways that couple environmental stimuli to specific changes in gene expression. One such pathway regulates the transcription factor NFκB, which is known to orchestrate the expression of a diverse array of genes essential in host defense, inflammatory, cell survival, and immune responses. NFκB is activated by a surprisingly broad array of cellular stimuli [1]. Likewise, NFκB has been shown to induce the expression of a large functionally diverse array of genes [2]. The seemingly promiscuous nature of NFκB biology raises the question as to how NFκB proteins elicit a specific transcriptional program in response to a given environmental challenge. The discovery of several key regulatory proteins within this pathway has provided insight into the mechanism by which the NFκB achieves specific coupling of these distinct cellular processes. As would be predicted of such a central signaling pathway, aberrant regulation of NFκB activation has been associated with the pathogenesis of several diseases, including autoimmunity, arthritis, asthma, and cancer [3]. This chapter summarizes the mechanisms of NFκB regulation and discusses emerging opportunities for target based therapeutic intervention in various disease settings.

NFκB was originally identified as a transcription factor required for B-cell specific gene expression [4]. However, it was quickly recognized that NFκB activity could be induced in most cell types in response to myriad stimuli, including proinflammatory cytokines, bacterial lipopolysaccharides, viral infection, DNA damage, oxidative stress, and chemotherapeutic agents (Figure 8.1) [1]. The classic experiment by Baeuerle and Baltimore, which demonstrated that NFκB exists in latent form in the cytosol of unstimulated cells and undergoes rapid translocation to the nucleus upon stimulation, set the stage for what is now the hallmark of NFκB regulation [5]. This paradigm provides a mechanism for NFκB to undergo rapid induction in response to cellular stress, resulting in the upregulation of NFκB target genes. A wide variety of genes are regulated by NFκB, including those encoding cytokines, chemokines, adhesion molecules, acute phase proteins, inducible effector enzymes, antimicrobial peptides, adaptive immune response, regulators of apoptosis, and cell proliferation (Figure 8.1). Remarkably, specific subsets of NFκB-responsive genes can be activated in a cell- and stimulus specific fashion. Biochemical and genetic characterization of molecular components that impinge on the NFκB signaling pathway has greatly facilitated our understanding of this process.

NFκB exists as a multi-gene family of proteins that can form stable homo- and heterodimeric complexes that vary in their DNA binding specificity and transcriptional activation potential. Five proteins belonging to the NFκB family have been identified in mammalian cells: RelA (p65), c-Rel, RelB, NFκB1 (p50 and its precursor p105), and NFκB2 (p52 and its precursor p100) (Figure 8.2) [1]. NFκB/Rel proteins share a highly conserved 300 amino acid N-terminal Rel homology domain (RHD) responsible for DNA binding, dimerization, and association with the IκB inhibitory proteins. The prototype NFκB complex comprises p50 and p65, but a variety of NFκB/Rel containing dimers are also known to exist. The p50/p65 complex displays strong transcriptional activation, whereas the p50/p50 and p52/p52 homodimers function to repress transcription of NFκB target genes [1]. Thus, the existence of a multi-gene family provides one tier of regulation by which the cell can fine-tune NFκB mediated gene expression. Moreover, knockout mice lacking distinct NFκB subunits display distinct phenotypes, further implicating specific NFκB dimers as activators of distinct sets of NFκB target genes [6–10].

NFκB resides in the cytoplasm in an inactive form by virtue of its association with a class of inhibitory proteins termed IκBs. Seven IκBs have been identified: IκBα, IκBβ, IκBε, IκBγ, Bcl3, NFκB1 precursor (p105), and the NFκB2 precursor (p100) (Figure 8.2) [11]. The IκB family members, having in common ankyrin repeat domains,

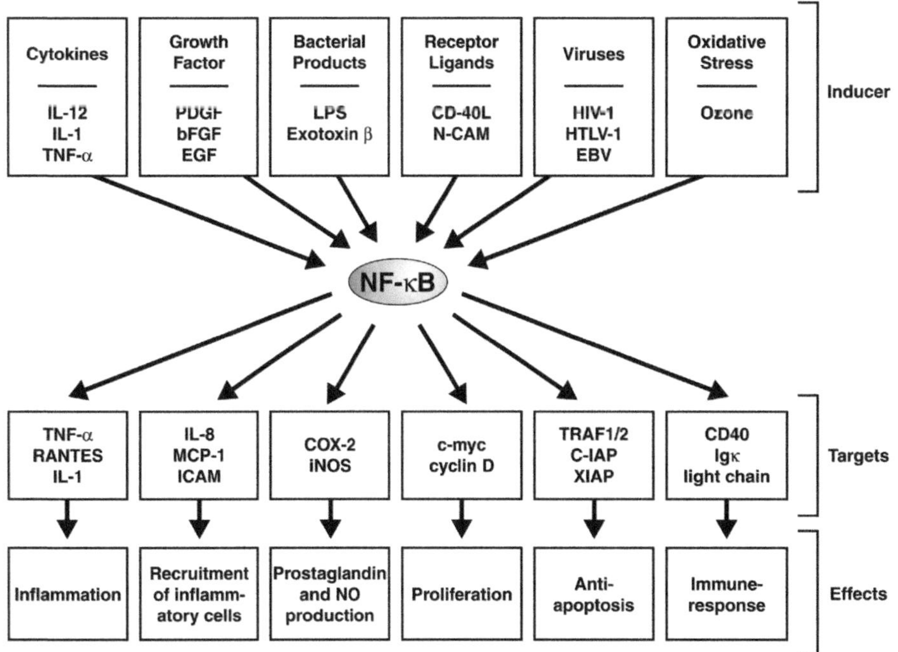

FIGURE 8.1 NFκB plays a central role in gene regulation.
Stimulus dependent activation of NFκB results in the modulation of specific subsets of NFκB target genes that are involved in distinct cellular processes. IL, interleukin; TNF, tumor necrosis factor; bFGF, basic fibroblast growth factor; PDGF, platelet derived growth factor; EGF, epidermal growth factor; LPS, lipopolysaccharide; CD-40L, CD 40 ligand; N-CAM, neural cell adhesion molecule; HIV, human immunodeficiency virus; HTLV, human T-cell leukemia/ lymphoma virus; EBV, Epstein-Barr virus; RANTES, regulated on activation, normal T cells expressed and secreted; MCP, monocyte chemoattractant protein; COX-2, cyclooxygenase-2; iNOS, inducible nitric oxide synthetase; TRAF1/2, TNF receptor associated factor; c-IAP, cellular inhibitors of apoptosis protein; XIAP, X-chromosome-linked inhibitor of apoptosis protein.

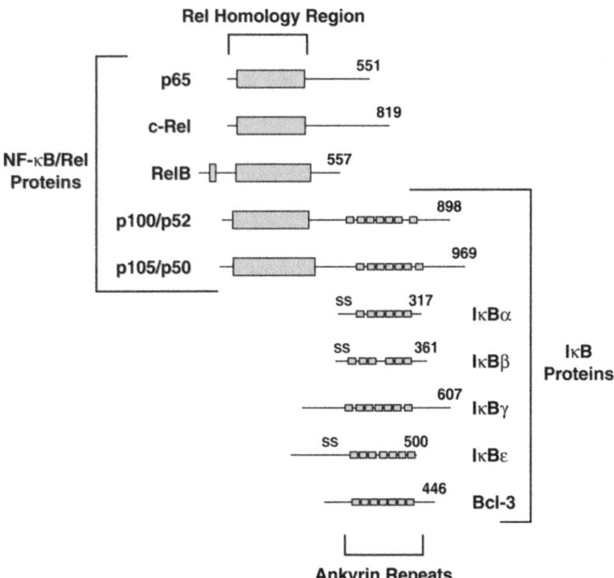

FIGURE 8.2 Schematic representation of members of the NFκB/Rel and IκB family of proteins.
The position of the Rel homology region and the ankyrin repeat domain are indicated in the figure. p100/p52 and p105/p50 precursor proteins comprise a unique subgroup of proteins that contain both a Rel homology and an ankyrin repeat domain. p100 and p105 proteins have been shown to function in an IκB-like capacity. The number of amino acids of each protein is indicated to the right. For RelB, the shaded box N terminal to the Rel homology region represents a putative leucine zipper region.

regulate the subcellular localization, and hence the DNA binding and transcriptional activity, of NFκB proteins. The basis for the cytoplasmic localization of the inactive NFκB : IκB complex is thought to be due to masking of the nuclear localization signals (NLS) on the NFκB subunits by the IκB proteins. Thus, IκB degradation would lead to unmasking of the NLS, allowing NFκB to undergo translocation to the nucleus (Figure 8.3). The IκBs display a preference for specific NFκB/Rel complexes, which may provide a means to regulate the activation of distinct Rel/NFκB complexes. Interestingly, NFκB induces the expression of IκBα; the newly synthesized IκB molecules enter the nucleus and remove NFκB from DNA [12–14]. The NFκB: IκB is then expelled from the nucleus as a result of potent nuclear export signals on IκB and p65 [15]. The IκBs provide yet another tier of regulatory complexity to modulate NFκB mediated gene expression.

Activation of NFκB is achieved primarily through the signal induced proteolytic degradation of IκB that is mediated by the ubiquitin/proteasome system [2]. The critical event that initiates the degradation of IκB is the stimulus dependent phosphorylation of IκB at specific N-terminal residues (S32/ S36 for IκBα, S19/S23 for IκBβ) [16, 17]. Mutation of IκBα at either of these residues was found to block stimulus dependent IκB phosphorylation, thereby preventing IκB degradation and subsequent activation of

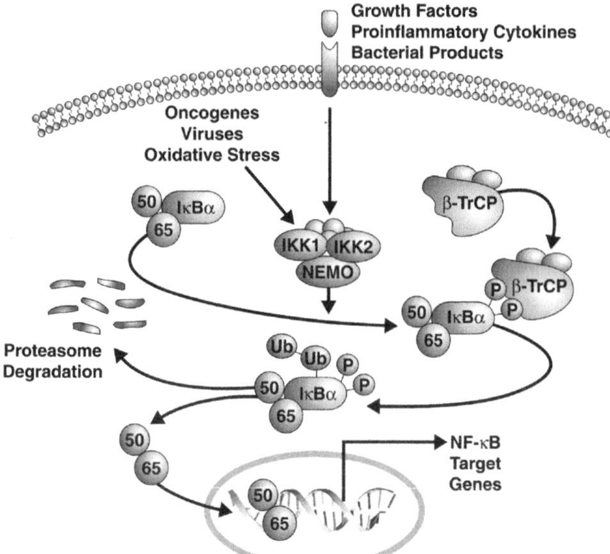

FIGURE 8.3 Schematic representation of components of the classical NFκB signal transduction pathway.
In response to extracellular signals, the IKK complex becomes activated by a poorly defined series of membrane proximal events. The activated IKK complex phosphorylates (P) IκBα at serines 32 and 36, leading to site specific ubiquitination (Ub) and degradation by the proteasome. The released NFκB complex (p50/p65) is then free to undergo nuclear translocation and activate transcription of NFκB target genes.

NFκB. Phosphorylation of IκB serves as a molecular tag leading to its rapid ubiquitination and subsequent degradation by the proteasome. β-TrCP, an F-box/WD containing component of the Skp1-Cullin-F-box (SCF) class of E3 ubiquitin ligases, is required for recognition of phosphorylated IκB and recruitment of the degradation machinery [18, 19].

Phosphorylation of IκBα on serines 32 and 36 is mediated by IκB kinases (IKKs), whose activity is induced by activators of the NFκB pathway [20–24]. IKK activity exists as large cytoplasmic multi-subunit complex (700–900 kDa) containing two kinase subunits, IKK1 (IKKα) and IKK2 (IKKβ), and a regulatory subunit, NEMO (IKKγ, IKKAP1, FIP3) [25–28]. Hence the "core" IKK complex comprises IKK1, IKK2, and NEMO. IKK1 and IKK2 are highly homologous kinases, both containing an N-terminal kinase domain and a C-terminal region with two protein interaction motifs, a leucine zipper (LZ), and a helix-loop-helix (HLH) motif. The LZ domain is responsible for dimerization of IKK1 and IKK2 and is essential for activity of the IKK complex. The IKK1/2 complex associates with NEMO through a short interaction motif located at the very C terminus of either catalytic subunit. Short peptides derived from the interaction motif can be used to disrupt the IKK complex and prevent its activation [29]. NEMO is thought to serve an important regulatory function by connecting the IKK complex to upstream activators through its C terminus, which contains a zinc finger motif (Figure 8.4). Further support for the regulatory function of NEMO is found by the observation that NEMO undergoes stimulus dependent interaction with components of the TNF receptor complex [30]. More recently, a report described two potentially novel components of the IKK complex, namely, Cdc37 and Hsp90 [31]. Apparently, formation of the core IKK complex with Cdc37/Hsp90 is required for TNF induced activation and recruitment of the core IKK complex from the cytoplasm to the membrane.

Sequence analysis revealed that both IKK1 and IKK2 contain a canonical MAP kinase kinase (MAPKK) activation loop motif. This region contains specific sites whose phosphorylation induces a conformational change that results in kinase activation. Phosphorylation within the activation loop typically occurs through the action of an upstream kinase or through transphosphorylation enabled by regulated proximity between two kinase subunits. IKK2 activation loop mutations, in which serines 177 and 181 were replaced with alanine, render the kinase refractory to stimulus dependent activation. In contrast, replacement of serine 177 and 181 with glutamic acid, to mimic phosphoserine, yielded a constitutively active kinase, and was capable of inducing NFκB mediated gene expression in the absence of cell stimulation [21]. The corresponding mutations in IKK1 did not interfere with NFκB activation in response to IL-1 or TNF, providing the first data suggesting that IKK2 plays a more prominent role in NFκB activation in response to proinflammatory cytokines [21, 23, 24]. Subsequent studies with IKK knockout mice further validate that IKK2, and not IKK1, is required for NFκB activation in response to most proinflammatory stimuli [32–35]. The mechanism by which diverse stimuli converge to activate the IKK complex remains unresolved. For example, a number of kinases have been reported to function in the capacity of an IKK kinase, including NIK, IKKι/ε, NAK/T2K/TBK, MEKK1, MEKK3, and Cot/TPL2 [36]. However, with the exception of MEKK3, no effect was observed on either IKK activation or induction of NFκB DNA binding activity in mice devoid of these kinases [37].

Biochemical and genetic analyses demonstrate that IKK1 and IKK2 have distinct cellular functions [2]. Disruption of the IKK2 gene results in embryonic lethality due to extensive liver apoptosis [32, 34, 38]. This phenotype is remarkably similar to that seen previously for RelA$^{-/-}$ mice [39]. Interestingly, IKK2$^{-/-}$ and RelA mice could be rescued by ablation of the TNF receptor I gene, which is consistent with the role of NFκB in preventing TNF induced hepatocyte apoptosis [34, 40]. In addition, IKK2$^{-/-}$ mouse embryonic fibroblasts were demonstrated to be refractory to activation of NFκB in response to inducers of NFκB, including TNF, IL-1, LPS, and dsRNA [34, 38]. In contrast, the IKK1$^{-/-}$ mice were born alive but died within 30 minutes, and IKK1$^{-/-}$ derived mouse embryonic fibroblasts display normal activation of IKK activity and induction of NFκB DNA binding in response to proinflammatory stimuli. These

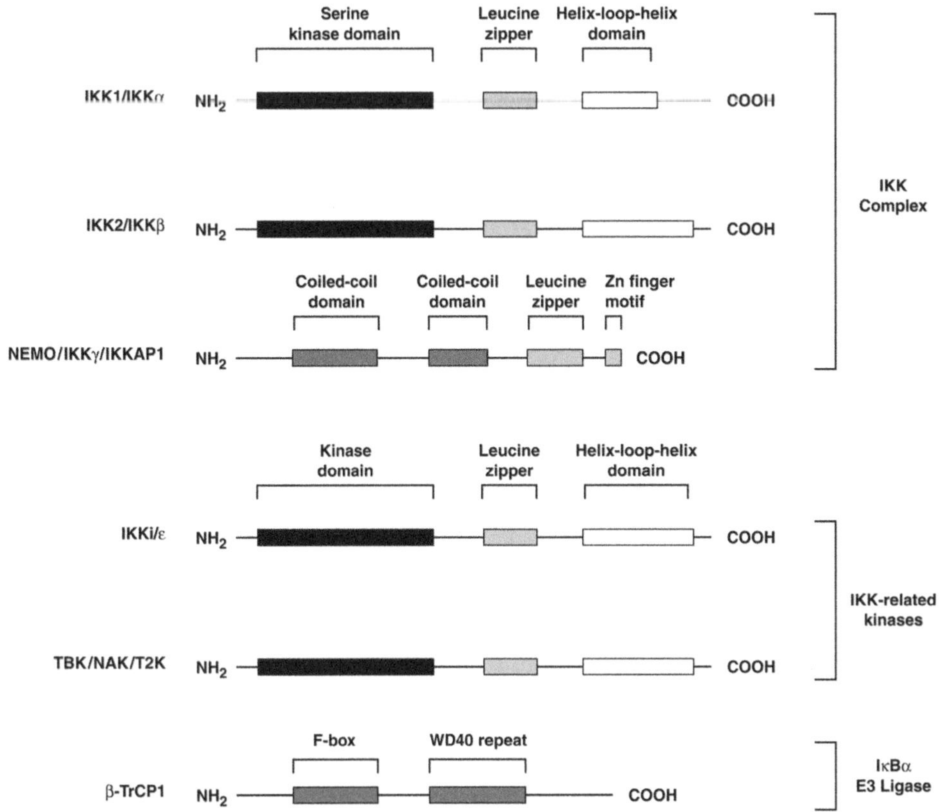

FIGURE 8.4 Schematic diagram showing the known subunits of the IKK complex, IKK-related kinases and β-TrCP, the IκBα E3 ubiquitin ligase. The name(s) of the protein is indicated to the left and the putative structural and functional motifs are indicated on top.

mice exhibit a plethora of developmental defects, the most striking of which is defective epidermal differentiation [32, 33]. Interestingly, IKK1 was found to play a prominent role in regulating keratinocyte differentiation, which is independent from its kinase activity or modulation of NFκB [32, 41].

More recently, a series of eloquent studies revealed a unique function of IKK1 in the lymphoid system [42]. Specifically, lethally irradiated mice that were reconstituted with $IKK^{-/-}$ hematopoietic stem cells displayed a diminution in B-cell maturation, germinal center formation, and antibody production, as well as defective splenic microarchitecture. Interestingly, it was discovered that $IKK1^{-/-}$B cells exhibit a specific deficiency in NFκB2/p100 processing [42]. Moreover, independent studies demonstrated that NIK induces ubiquitin dependent processing of NFκB2/p100 [43]. Subsequently, it was demonstrated that an inactive variant of IKK1 blocked NIK induced NFκB2/p100 processing. Together, these studies suggest that NIK mediated activation of IKK1 may lead to the phosphorylation and ubiquitin dependent processing of NFκB2/p100 (Figure 8.5). NFκB2/p100 is unique in that it contains both a Rel homology domain and an ankyrin domain and, consequently, possesses an intrinsic IκB-like capacity. Thus, enhanced proteolytic removal of the inhibitory C-terminal ankyrin domain will generate increased levels of transcriptionally active p52 containing NFκB complexes, perhaps leading to upregulation of a specific subset of NFκB target genes. It is quite intriguing that, although IKK1 and IKK2 display high similarity and form stable heterodimers, they modulate distinct signaling pathways with unique biological consequences.

NFκB proteins may play a role in any disease possessing one of the cytokines, including TNF, IFNβγ, IL-2, IL-6, IL-8, and IL-12, as a component of their pathophysiology (Figure 8.1). The likely diseases for an NFκB targeted therapeutic strategy may be those with a chronic, unresolved inflammatory component characterized by constitutively elevated systemic or local cytokine levels. Rheumatoid and osteoarthritis, inflammatory bowel disease, atherosclerosis, diabetes, multiple sclerosis, and cachexia, among other conditions, fall into this category. Note that the chronic inhibition of NFκB likely to be necessary to ameliorate these conditions may have unknown repercussions. However, agents such as glucocorticoids inhibit NFκB and are used chronically, so we may predict that long-term NFκB based therapy can be tolerated [44].

Despite the emphasis on NFκB's role as a regulator of cytokine gene expression and immune function, the initial application of therapeutic modulators of NFκB is likely to be in the field of oncology. If this is indeed the case, it will recapitulate the initial discovery of v-Rel proteins as

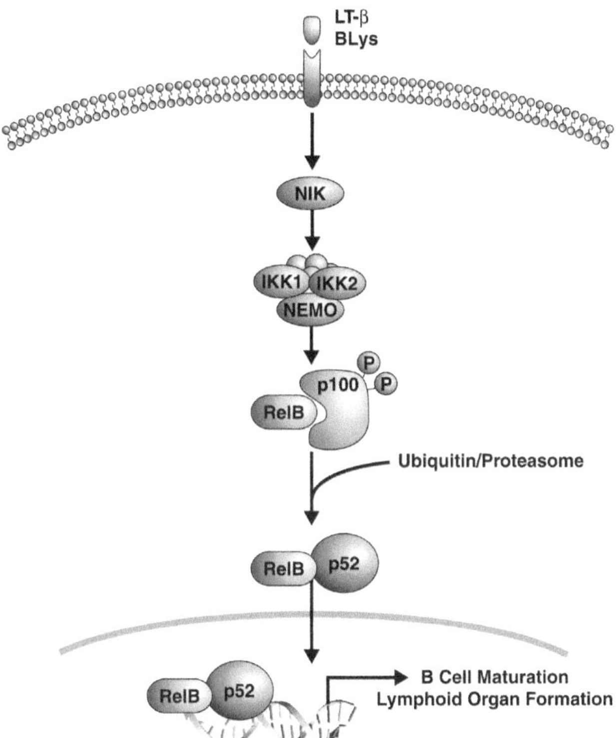

FIGURE 8.5 Activation by IKK1 is involved in an alternative NFκB signaling pathway.
Select members of the TNF family, namely, lymphotoxin-beta (LT-beta) and B-lymphocyte stimulator (BLyS), activate IKK1 through NIK (NFκB inducing kinase). Activation of this pathway leads to IKK1 dependent phosphorylation (P) and processing of NFκB2/p100. The active p52 : RelB heterodimer undergoes nuclear translocation and regulates specific NFκB target genes involved in B-cell maturation and lymphoid organ formation.

transforming oncogenes nearly 20 years ago [45]. While the genetics of viral Rel proteins were being worked out, simultaneous work was proceeding on the transcription factors (including NFκB) regulating immunoglobulin and MHC genes [4, 46]. It was several years before these divergent fields of research coalesced with the discovery that NFκB proteins are the cellular homologs of v-Rel [47, 48]. The functional roles of NFκB proteins, of course, extend far beyond regulation of immunoglobulins and are quite widely expressed.

More recent information has shown a wide variety of genetic aberrations of these genes in human neoplasms, from overexpression to rearrangement and amplification. The list of such observations is beyond the scope of this review, and details can be found elsewhere [49, 50]. Transforming viral proteins of SV40, EBV, adenovirus E1A, and HTLV-1 activate NFκB [51–54]. Regardless of the mechanism, it is clear that the NFκB transcription factor is constitutively active in a wide variety of human cancers, including but not limited to colon, gastric, pancreatic, ovarian, hepatocellular, breast, head and neck carcinomas, melanoma, lymphoblastic leukemias, and Hodgkin's disease [55–65].

As described earlier, NFκB activity is controlled to a large extent by the activity of the IKKs, and this represents the most tractable point for therapeutic intervention. However, other mechanisms for activation and avenues for intervention exist. For example, the DNA binding and transcriptional activity of NFκB proteins can be controlled by phosphorylation of multiple sites in both the Rel homology and transactivation domains [66–69]. Well characterized oncogenes such as Ras, Rac, and Bcr-Abl as well as integrin activated signals induce NFκB signaling both through the traditional IKK/IκB pathway and through direct phosphorylation of the p65 subunit of the transcription factor [66–70]. Recent results with agents selectively inhibiting the IKKs (e.g., PS-1145) or acting in part on other aspects of NFκB signaling (e.g., proteasome inhibitors, NSAIDs, and thalidomide) are beginning to show great promise in the treatment of various cancers [50, 71, 72].

From a functional standpoint, NFκB proteins have been suggested to play a role in cellular transformation via two general routes: direct stimulation of growth (cell cycle dysregulation) and/or inhibition of differentiation, and inhibition of programmed cell death (apoptosis). Both are likely to be involved to varying degrees based on the type of neoplasm, but the protection from apoptosis has garnered the most attention of late and may be the most important aspect from a therapeutic standpoint. With respect to direct cell cycle control, NFκB has been shown to upregulate cyclin D1 transcription, a crucial step in the G_1 to S phase cell cycle transition [73, 74]. NFκB activity peaks at multiple points in the mammalian cell cycle, including G_0/G_1, late G_1, and S phase [75]. In support of this concept, analysis of tumor cells treated with selective inhibitors of IKKs reveals a G_2/M phase blockade as well (unpublished observation), suggesting the existence of additional cell cycle targets.

NFκB activates central players in the programmed cell death pathways, including c-IAP-1 and -2, Bcl-Xl, and A1/Bfl-1 [76, 77]. NFκB activation of protective proteins holds the key to the promise of NFκB inhibitors as chemotherapeutic agents. Of particular importance is the observation that diverse classes of chemotherapeutic agents are potent activators of NFκB, a response that proceeds at least in part through the activation of IKKs. Agents known to activate NFκB include etoposide, CPT-11, and TRAIL [78–80]. *In vitro* and *in vivo* experiments using the IκB super-repressor, a mutant version of IκB resistant to stimulus induced degradation, have shown that inhibition of NFκB activity acts synergistically with chemotherapeutic agents to decrease tumor cell growth and tumor burden. Small molecule inhibitors of IKK2 show the same effect *in vitro* (unpublished observations). It is likely that agents previously used in chemotherapeutic regimens, such as glucocorticoids and proteasome inhibitors in multiple myeloma, are working at least in part through the inhibition of NFκB. It is interesting to note that specific inhibitors of IKK are only partially efficacious when used as monotherapy compared to a general proteasome inhibitor, suggesting other proteasome targets contribute to efficacy [71]. On the other hand, the side effect profile of more specific inhibitors is likely to be superior.

The development of potent and selective inhibitors of NFκB, targeting multiple aspects of NFκB signal transduction, is imminent. An important task for the near future is to determine the human diseases most likely to be resolved with these agents. Chronic inflammatory conditions are intriguing possibilities, but we must first define the consequences of long-term NFκB inhibition. The optimal tumor types to approach have not been completely defined, but are likely to include multiple myeloma [71] and one of the many neoplasms with chronically elevated NFκB (as described earlier). Tumors in which traditional chemotherapy elicits a robust (and presumably activating or protective) induction of NFκB will be likely targets for combination therapy. The molecular correlates of sensitivity to NFκB inhibitors, alone or in combination with other agents, need to be identified. For example, tumors lacking the PTEN tumor suppressor may be more sensitive to NFκB inhibitors [81]. It is clear that continued work on the basic aspects of NFκB biology is necessary. In concert with development of potent and specific modulators, these efforts have the potential to yield novel therapeutics for a wide variety of human diseases.

REFERENCES

1. Grossmann M, O'Reilly LA, Gugasyan R, et al. The anti-apoptotic activities of Rel and RelA required during B-cell maturation involve the regulation of Bcl-2 expression. *EMBO J* 2000;**19**:6351–60.
2. Karin M, Ben-Neriah Y. Phosphorylation meets ubiquitination: the control of NFκB activity. *Annu Rev Immunol* 2000;**18**:621–63.
3. Yamamoto Y, Gaynor RB. Therapeutic potential of inhibition of the NFkappaB pathway in the treatment of inflammation and cancer. *J Clin Invest* 2001;**107**:135–42.
4. Sen R, Baltimore D. Inducibility of kappa immunoglobulin enhancer-binding protein Nf-kappa B by a posttranslational mechanism. *Cell* 1986;**47**:921–8.
5. Baeuerle PA, Baltimore D. Activation of DNA-binding activity in an apparently cytoplasmic precursor of the NFkappa B transcription factor. *Cell* 1988;**53**:211–17.
6. Sha WC, Liou HC, Tuomanen EI, et al. Targeted disruption of the p50 subunit of NFkappa B leads to multifocal defects in immune responses. *Cell* 1995;**80**:321–30.
7. Kontgen F, Grumont RJ, Strasser A, et al. Mice lacking the c-rel proto-oncogene exhibit defects in lymphocyte proliferation, humoral immunity, and interleukin-2 expression. *Genes Dev* 1995;**9**:1965–77.
8. Beg AA, Baltimore D. An essential role for NFkappaB in preventing TNF-alpha-induced cell death. *Science* 1996;**274**:782–4.
9. Caamano JH, Rizzo CA, Durham SK, et al. Nuclear factor (NF)-kappa B2 (p100/p52) is required for normal splenic microarchitecture and B cell-mediated immune responses. *J Exp Med* 1998;**187**:185–96.
10. Franzoso G, Carlson L, Poljak L, et al. Mice deficient in nuclear factor (NF)-kappa B/p52 present with defects in humoral responses, germinal center reactions, and splenic microarchitecture. *J Exp Med* 1998;**187**:147–59.
11. Tam WF, Sen R. IkappaB family members function by different mechanisms. *J Biol Chem* 2001;**276**:7701–4.
12. Arenzana-Seisdedos F, Thompson J, Rodriguez MS, et al. Inducible nuclear expression of newly synthesized I kappa B alpha negatively regulates DNA-binding and transcriptional activities of NFkappa B. *Mol Cell Biol* 1995;**15**:2689–96.
13. Chiao PJ, Miyamoto S, Verma IM. Autoregulation of I kappa B alpha activity. *Proc Natl Acad Sci USA* 1994;**91**:28–32.
14. Sun SC, Ganchi PA, Ballard DW, et al. NFkappa B controls expression of inhibitor I kappa B alpha: Evidence for an inducible autoregulatory pathway. *Science* 1993;**259**:1912–15.
15. Harhaj EW, Sun SC. Regulation of RelA subcellular localization by a putative nuclear export signal and p50. *Mol Cell Biol* 1999;**19**:7088–95.
16. Brown K, Park S, Kanno T, et al. Mutual regulation of the transcriptional activator NFkappa B and its inhibitor, I kappa B-alpha. *Proc Natl Acad Sci USA* 1993;**90**:2532–6.
17. DiDonato J, Mercurio F, Rosette C, et al. Mapping of the inducible IkappaB phosphorylation sites that signal its ubiquitination and degradation. *Mol Cell Biol* 1996;**16**:1295–304.
18. Yaron A, Gonen H, Alkalay I, et al. Inhibition of NFkappa-B cellular function via specific targeting of the I-kappa-B-ubiquitin ligase. *EMBO J* 1997;**16**:6486–94.
19. Yaron A, Hatzubai A, Davis M, et al. Identification of the receptor component of the IkappaBalpha-ubiquitin ligase. *Nature* 1998; **396**:590–4.
20. DiDonato JA, Hayakawa M, Rothwarf DM, et al. A cytokine-responsive IkappaB kinase that activates the transcription factor NFkappaB. *Nature* 1997;**388**:548–54.
21. Mercurio F, Zhu H, Murray BW, et al. IKK-1 and IKK-2: Cytokine-activated IkappaB kinases essential for NFkappaB activation. *Science* 1997;**278**:860–6.
22. Regnier CH, Song HY, Gao X, et al. Identification and characterization of an IkappaB kinase. *Cell* 1997;**90**:373–83.
23. Woronicz JD, Gao X, Cao Z, et al. IkappaB kinase-beta: NFkappaB activation and complex formation with IkappaB kinase-alpha and NIK. *Science* 1997;**278**:866–9.
24. Zandi E, Rothwarf DM, Delhase M, et al. The IkappaB kinase complex (IKK) contains two kinase subunits, IKKalpha and IKKbeta, necessary for IkappaB phosphorylation and NFkappaB activation. *Cell* 1997;**91**:243–52.
25. Yamaoka S, Courtois G, Bessia C, et al. Complementation cloning of NEMO, a component of the IkappaB kinase complex essential for NFkappaB activation. *Cell* 1998;**93**:1231–40.
26. Rothwarf DM, Zandi E, Natoli G, et al. IKK-gamma is an essential regulatory subunit of the IkappaB kinase complex. *Nature* 1998;**395**:297–300.
27. Mercurio F, Murray BW, Shevchenko A, et al. IkappaB kinase (IKK)-associated protein 1, a common component of the heterogeneous IKK complex. *Mol Cell Biol* 1999;**19**:1526–38.
28. Li Y, Kang J, Friedman J, et al. Identification of a cell protein (FIP-3) as a modulator of NFkappaB activity and as a target of an adenovirus inhibitor of tumor necrosis factor alpha-induced apoptosis. *Proc Natl Acad Sci USA* 1999;**96**:1042–7.
29. May MJ, D'Acquisto F, Madge LA, et al. Selective inhibition of NFkappaB activation by a peptide that blocks the interaction of NEMO with the IkappaB kinase complex. *Science* 2000;**289**:1550–4.
30. Zhang SQ, Kovalenko A, Cantarella G, et al. Recruitment of the IKK signalosome to the p55 TNF receptor: RIP and A20 bind to NEMO (IKKgamma) upon receptor stimulation. *Immunity* 2000;**12**:301–11.

31. Chen G, Cao P, Goeddel DV. TNF-induced recruitment and activation of the IKK complex require Cdc37 and Hsp90. *Mol Cell* 2002;**9**:401–10.
32. Hu Y, Baud V, Delhase M, et al. Abnormal morphogenesis but intact IKK activation in mice lacking the IKKalpha subunit of IkappaB kinase. *Science* 1999;**284**:316–20.
33. Li Q, Lu Q, Hwang JY, et al. IKK1-deficient mice exhibit abnormal development of skin and skeleton. *Genes Dev* 1999;**13**:1322–8.
34. Li Q, Van Antwerp D, Mercurio F, et al. Severe liver degeneration in mice lacking the IkappaB kinase 2 gene. *Science* 1999;**284**:321–5.
35. Takeda K, Takeuchi O, Tsujimura T, et al. Limb and skin abnormalities in mice lacking IKKalpha. *Science* 1999;**284**:313–16.
36. Ghosh S, Karin M. Missing pieces in the NFkappaB puzzle. *Cell* 2002;**109**(Suppl.):S81–96.
37. Yang J, Lin Y, Guo Z, et al. The essential role of MEKK3 in TNF-induced NFkappaB activation. *Nat Immunol* 2001;**2**:620–4.
38. Li ZW, Chu W, Hu Y, et al. The IKKbeta subunit of IkappaB kinase (IKK) is essential for nuclear factor kappaB activation and prevention of apoptosis. *J Exp Med* 1999;**189**:1839–45.
39. Beg AA, Sha WC, Bronson RT, et al. Embryonic lethality and liver degeneration in mice lacking the RelA component of NFkappa B. *Nature* 1995;**376**:167–70.
40. Alcamo E, Mizgerd JP, Horwitz BH, et al. Targeted mutation of TNF receptor I rescues the RelA-deficient mouse and reveals a critical role for NFkappa B in leukocyte recruitment. *J Immunol* 2001;**167**:1592–600.
41. Hu Y, Baud V, Oga T, et al. IKKalpha controls formation of the epidermis independently of NFkappaB. *Nature* 2001;**410**:710–14.
42. Senftleben U, Cao Y, Xiao G, et al. Activation by IKKalpha of a second, evolutionary conserved, NFkappa B signaling pathway. *Science* 2001;**293**:1495–9.
43. Xiao G, Harhaj EW, Sun SC. NFkappaB-inducing kinase regulates the processing of NFkappaB2 p100. *Mol Cell* 2001;**7**:401–9.
44. Auphan N, DiDonato JA, Rosette C, et al. Immunosuppression by glucocorticoids: Inhibition of NFkappa B activity through induction of I kappa B synthesis. *Science* 1995;**270**:286–90.
45. Wilhelmsen KC, Eggleton K, Temin HM. Nucleic acid sequences of the oncogene v-rel in reticuloendotheliosis virus strain T and its cellular homolog, the proto-oncogene c-rel. *J Virol* 1984;**52**:172–82.
46. Baldwin Jr. AS, Sharp PA Two transcription factors, NFkappa B and H2TF1, interact with a single regulatory sequence in the class I major histocompatibility complex promoter. *Proc Natl Acad Sci USA* 1998;**85**:723–7.
47. Kieran M, Blank V, Logeat F, et al. The DNA binding subunit of NFkappa B is identical to factor KBF1 and homologous to the rel oncogene product. *Cell* 1990;**62**:1007–18.
48. Ghosh S, Gifford AM, Riviere LR, et al. Cloning of the p50 DNA binding subunit of NFkappa B: Homology to rel and dorsal. *Cell* 1990;**62**:1019–29.
49. Rayet B, Gelinas C. Aberrant rel/nfkb genes and activity in human cancer. *Oncogene* 1999;**18**:6938–47.
50. Yamamoto Y, Gaynor RB. Role of the NFkappaB pathway in the pathogenesis of human disease states. *Curr Mol Med* 2001;**1**: 287–96.
51. Hiscott J, Kwon H, Genin P. Hostile takeovers: viral appropriation of the NFkappaB pathway. *J Clin Invest* 2001;**107**:143–51.
52. Chu ZL, DiDonato JA, Hawiger J, et al. The tax oncoprotein of human T-cell leukemia virus type 1 associates with and persistently activates IkappaB kinases containing IKKalpha and IKKbeta. *J Biol Chem* 1998;**273**:15,891–15,894.
53. Yamaoka S, Inoue H, Sakurai M, et al. Constitutive activation of NFkappa B is essential for transformation of rat fibroblasts by the human T-cell leukemia virus type I Tax protein. *EMBO J* 1996;**15**:873–87.
54. Mosialos G. The role of Rel/NFkappa B proteins in viral oncogenesis and the regulation of viral transcription. *Semin Cancer Biol* 1997;**8**:121–9.
55. Izban KF, Ergin M, Huang Q, et al. Characterization of NFkappaB expression in Hodgkin's disease: Inhibition of constitutively expressed NFkappaB results in spontaneous caspase-independent apoptosis in Hodgkin and Reed-Sternberg cells. *Mod Pathol* 2001;**14**:297–310.
56. Romieu-Mourez R, Landesman-Bollag E, Seldin DC, et al. Roles of IKK kinases and protein kinase CK2 in activation of nuclear factor-kappaB in breast cancer. *Cancer Res* 2001;**61**:3810–18.
57. Arlt A, Vorndamm J, Breitenbroich M, et al. Inhibition of NFkappaB sensitizes human pancreatic carcinoma cells to apoptosis induced by etoposide (VP16) or doxorubicin. *Oncogene* 2001;**20**:859–68.
58. Arlt A, Vorndamm J, Muerkoster S, et al. Autocrine production of interleukin 1beta confers constitutive nuclear factor kappaB activity and chemoresistance in pancreatic carcinoma cell lines. *Cancer Res* 2002;**62**:910–16.
59. Sasaki N, Morisaki T, Hashizume K, et al. Nuclear factor-kappaB p65 (RelA) transcription factor is constitutively activated in human gastric carcinoma tissue. *Clin Cancer Res* 2001;**7**:4136–42.
60. Kordes U, Krappmann D, Heissmeyer V, et al. Transcription factor NFkappaB is constitutively activated in acute lymphoblastic leukemia cells. *Leukemia* 2000;**14**:399–402.
61. Tai DI, Tsai SL, Chang YH, et al. Constitutive activation of nuclear factor kappaB in hepatocellular carcinoma. *Cancer* 2000;**89**:2274–81.
62. Hinz M, Loser P, Mathas S, et al. Constitutive NFkappaB maintains high expression of a characteristic gene network, including CD40, CD86, and a set of antiapoptotic genes in Hodgkin/Reed-Sternberg cells. *Blood* 2001;**97**:2798–807.
63. Dhawan P, Richmond A. A novel NFkappa B-inducing kinase-MAPK signaling pathway up-regulates NFkappa B activity in melanoma cells. *J Biol Chem* 2002;**277**:7920–8.
64. Lind DS, Hochwald SN, Malaty J, et al. Nuclear factor-kappa B is upregulated in colorectal cancer. *Surgery* 2001;**130**:363–9.
65. Tamatani T, Azuma M, Aota K, et al. Enhanced IkappaB kinase activity is responsible for the augmented activity of NFkappaB in human head and neck carcinoma cells. *Cancer Lett* 2001;**171**:165–72.
66. Finco TS, Westwick JK, Norris JL, et al. Oncogenic Ha-Ras-induced signaling activates NFkappaB transcriptional activity, which is required for cellular transformation. *J Biol Chem* 1997;**272**:24,113–24,116.
67. Zhong H, SuYang H, Erdjument-Bromage H, et al. The transcriptional activity of NFkappaB is regulated by the IkappaB-associated PKAc subunit through a cyclic AMP-independent mechanism. *Cell* 1997;**89**:413–24.
68. Zhong H, Voll RE, Ghosh S. Phosphorylation of NFkappa B p65 by PKA stimulates transcriptional activity by promoting a novel bivalent interaction with the coactivator CBP/p300. *Mol Cell* 1998;**1**:661–71.
69. Zhong H, May MJ, Jimi E, et al. The Phosphorylation Status of Nuclear NFkappaB Determines Its Association with CBP/p300 or HDAC-1. *Mol Cell* 2002;**9**:625–36.
70. Reuther JY, Reuther GW, Cortez D, et al. A requirement for NFkappaB activation in Bcr-Abl-mediated transformation. *Genes Dev* 1998;**12**:968–81.
71. Hideshima T, Chauhan D, Richardson P, et al. NFkappa B as a therapeutic target in multiple myeloma. *J Biol Chem* 2002;**277**: 16,639–16,647.

72. Baldwin Jr. AS Series introduction: the transcription factor NFkappaB and human disease. *J Clin Invest* 2001;**107**:3–6.
73. Guttridge DC, Albanese C, Reuther JY, et al. NFkappaB controls cell growth and differentiation through transcriptional regulation of cyclin D1. *Mol Cell Biol* 1999;**19**:5785–99.
74. Joyce D, Albanese C, Steer J, et al. NFkappaB and cell-cycle regulation: the cyclin connection. *Cytokine Growth Factor Rev* 2001;**12**:73–90.
75. Ansari SA, Safak M, Del Valle L, et al. Cell cycle regulation of NFkappa b-binding activity in cells from human glioblastomas. *Exp Cell Res* 2001;**265**:221–33.
76. Wang CY, Mayo MW, Korneluk RG, et al. NFkappaB antiapoptosis: induction of TRAF1 and TRAF2 and c-IAP1 and c-IAP2 to suppress caspase-8 activation. *Science* 1998;**281**:1680–3.
77. Grumont RJ, Rourke IJ, O'Reilly LA, et al. B lymphocytes differentially use the Rel and nuclear factor kappaB1 (NFkappaB1) transcription factors to regulate cell cycle progression and apoptosis in quiescent and mitogen-activated cells. *J Exp Med* 1998;**187**:663–74.
78. Cusack Jr. JC, Liu R, Houston M, et al Enhanced chemosensitivity to CPT-11 with proteasome inhibitor PS-341: Implications for systemic nuclear factor-kappaB inhibition. *Cancer Res* 2001;**61**:3535–40.
79. Wang Jr. CY, Cusack JC, Liu R, et al Control of inducible chemoresistance: Enhanced anti-tumor therapy through increased apoptosis by inhibition of NFkappaB. *Nat Med* 1999;**5**:412–17.
80. Wang CY, Guttridge DC, Mayo MW, et al. NFkappaB induces expression of the Bcl-2 homologue A1/Bfl-1 to preferentially suppress chemotherapy-induced apoptosis. *Mol Cell Biol* 1999;**19**:5923–9.
81. Gustin JA, Maehama T, Dixon JE, et al. The PTEN tumor suppressor protein inhibits tumor necrosis factor-induced nuclear factor kappa B activity. *J Biol Chem* 2001;**276**:27,740–27,744.

Chapter 9

Ubiquitin-mediated Regulation of Protein Kinases in NFκB Signaling

Ming Xu[1,3] and Zhijian J. Chen[1,2,3]

[1] *Department of Molecular Biology*
[2] *Howard Hughes Medical Institute*
[3] *University of Texas Southwestern Medical Center, Dallas, Texas*

INTRODUCTION

It is well known that phosphorylation regulates protein ubiquitination and subsequent degradation by the proteasome. Accumulating evidence suggests that ubiquitination can also regulate protein phosphorylation through both proteasome-dependent and -independent mechanisms. Most of our understanding of the regulatory role of ubiquitin in protein kinase activation and inactivation is based on the study of the NFκB signaling pathways. In this chapter, we will discuss the current understanding of ubiquitin-mediated regulation of protein kinases in different pathways leading to NFκB activation, with particular focus on those emanating from tumor necrosis factor receptor (TNFR), interleukin-1 receptor (IL-1R), and Toll-like receptors (TLRs).

THE UBIQUITIN PATHWAY

Ubiquitin is a small and highly conserved protein ubiquitously expressed in all eukaryotic cells [1]. From yeast to human, each species has four ubiquitin genes, two of which encode ubiquitin fused to ribosomal subunits, while the other two encode linear polyubiquitin in which ubiquitin is linked to each other in tandem in a "head-to-tail" fashion [2]. The ubiquitin gene products are recognized and cleaved by ubiquitin C-terminal hydrolase (UCH) to generate monomeric 76-amino-acid ubiquitin.

Ubiquitination is a process that covalently attaches the C-terminus of ubiquitin to specific target proteins. This process consists of a cascade of three enzymatic reactions catalyzed by ubiquitin-activating enzymes (E1), ubiquitin-conjugating enzymes (E2), and ubiquitin–protein ligases (E3) [3, 4]. In the first step, the C-terminus of ubiquitin forms a thioester bond with E1 in an ATP-dependent manner. In the second step, the activated ubiquitin is transferred to a cysteine within the E2 catalytic center to form an E2~Ub thioester. In the last step, in the presence of E3, which binds to both E2 and substrates, the C-terminal carboxyl group of ubiquitin forms an isopeptide bond with the ε-NH_2 group of a lysine residue on the substrate.

Ubiquitin contains seven lysines, which can be conjugated by another ubiquitin to form polyubiquitin chains. The polyubiquitin chains with different lysine linkages have distinct functions. Whereas polyubiquitin chains linked through lysine 48 (K48) of ubiquitin primarily target a protein for proteasomal degradation, polyubiquitin chains linked through lysine 63 (K63) can function as a signal transducer in cellular processes such as protein kinase activation and DNA repair (Figure 9.1). Polyubiquitin chains linked through other lysines of ubiquitin have been reported to engage proteolytic as well as non-proteolytic functions [5]. Monoubiquitination normally does not target protein for degradation, but regulates processes such as vesicle trafficking, chromatin dynamics (through histone ubiquitination), and DNA repair [4]. The ubiquitination and polyubiquitination reactions can be reversed by de-ubiquitination enzymes (DUBs, also known as isopeptidase). Therefore, similar to phosphorylation, ubiquitination is a regulated, reversible, protein modification.

Ubiquitination and de-ubiquitination are complex biological reactions catalyzed by large families of enzymes. There are two ubiquitin E1s, ~50 E2s, ~700 E3s, and ~80 DUBs encoded by the human genome. All E2s contain a conserved UBC domain, harboring an invariant cysteine residue at the active site. The E3s for ubiquitination can be categorized into two subfamilies based on whether they contain a HECT (Homology to E6AP C-Terminus) or RING (Really Interesting New Gene) domain [6, 7]. The E3s that contain the HECT domain have a highly

Handbook of Cell Signaling, Three-Volume Set 2 ed.

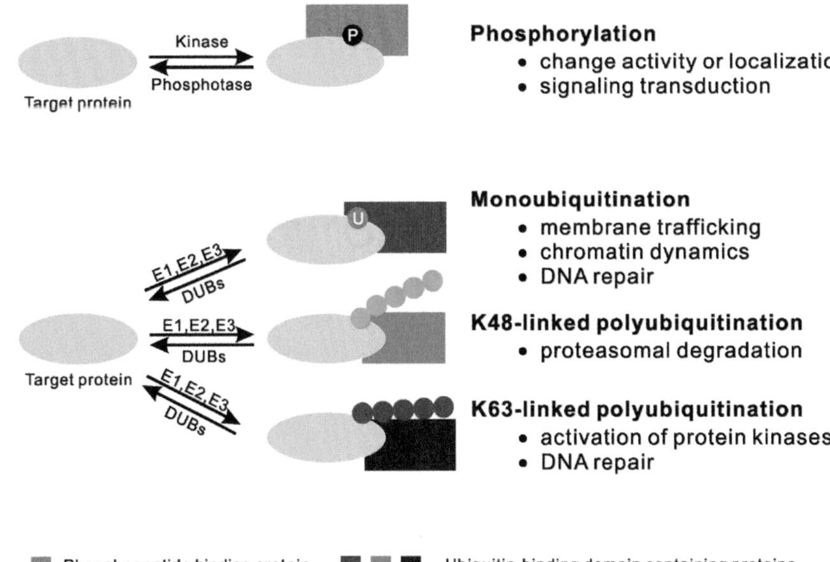

FIGURE 9.1 Protein phosphorylation and ubiquitination in cell signaling.
Both phosphorylation and ubiquitination are reversible forms of covalent modification that regulate diverse cellular functions. Both involve large families of enzymes: kinases, and phosphatases for phosphorylation and dephosphorylation; E1, E2, E3, and DUBs for ubiquitination and de-ubiquitination. Different types of ubiquitination, including monoubiquitination and polyubiquitination involving different lysines of ubiquitin, provide additional signaling specificity and they regulate cellular processes through proteasome-dependent and -independent mechanisms. The signals of ubiquitin and polyubiquitin chains are decoded by specialized ubiquitin-binding domains, much like phosphorylation signals, which are recognized by phosphopeptide-binding domains.

conserved cysteine residue that can relay ubiquitin from E2s to substrates. In contrast, the E3s containing the RING domain do not have a well-defined active site, although they can bind to E2s to facilitate the ubiquitin transfer from E2s to the target proteins. Some E3s can form multi-protein complexes. In these cases, a RING domain E3 lacking the intrinsic substrate-binding site is associated with another subunit that provides substrate recognition specificity. De-ubiquitination enzymes (DUBs) can be divided into five classes: ubiquitin-C-terminal hydrolases (UCH), ubiquitin-specific proteases (USPs), Machado-Joseph Disease Protein Domain Proteases (MJDs), ovarian tumor proteases (OTUs), and JAMM motif proteases (JAMMs). All of the DUBs are cysteine proteases except JAMMs, which are metalloproteases [8, 9].

In eukaryotic cells, many proteins contain motifs that recognize phosphorylated peptides. Similarly, the ubiquitin signal on target proteins can be recognized by ubiquitin-binding domains (UBD) embedded in many signaling proteins, generating a large variety of cellular outputs [10, 11]. A combination of biochemical and bioinformatics analyses has led to the identification of approximately 20 different types of UBDs, which vary in sizes, structure, and properties. Most of these UBDs bind to ubiquitin with weak affinity, and they contact a hydrophobic surface surrounding Ile-44 of ubiquitin. The weak interaction between ubiquitin and UBDs suggest that additional binding between ubiquitinated substrates and ubiquitin receptor must occur in order to achieve optimal signaling specificity and efficiency.

NFκB SIGNALING

NFκB is a heterodimeric transcription factor controlling the expression of a plethora of genes involved in inflammation, immunity, and apoptosis [12]. This heterodimer is composed of REL family members, including REL-A (also known as p65), c-REL, REL-B, p50, and p52. All subunits contain a REL-homology domain (RHD), which is responsible for DNA binding, dimerization, nuclear localization, and interaction with the inhibitory protein of the IκB family. In addition, REL-A, c-REL, and REL-B each has a transactivation domain (TAD), responsible for the transcriptional activity. The other two subunits, p50 and p52, are processed from their precursors, p105 and p100, respectively, and each must dimerize with a TAD-containing subunit to form a functional transactivator. The C-termini of p105 and p100 contain ankyrin repeats, which are also found in IκB proteins, including IκBα, IκBβ and IκBε. These ankyrin repeats bind to the RHD of NFκB and mask the nuclear localization signal, thereby sequestering NFκB in the cytoplasm.

The activation of NFκB can be divided into two schemes, canonical and non-canonical, depending on whether it involves IκB degradation or p100 processing [13]. In the canonical pathway, when cells are stimulated with proinflammatory cytokines such as Tumor Necrosis Factor α (TNFα) and Interleukin-1β (IL-1β), or microbial pathogens such as bacteria and viruses, a large protein kinase complex containing IKKα, IKKβ, and NEMO (also known as IKKγ or IKKAP) is rapidly activated. The IKK complex phosphorylates IκB proteins at two N-terminal serine resides, enabling binding to an E3 ligase complex consisting of SKP1, CUL1, ROC1, and the Fbox protein βTrCP (SCFβTrCP). This E3 complex, together with an E2 of the Ubc4/Ubc5 family, catalyzes polyubiquitination of IκB, which is selectively degraded by the 26S proteasome, allowing NFκB to enter the nucleus and turn on target genes [14, 15].

The non-canonical NFκB activation pathway results in the processing of p100 to the mature p52 subunit. Stimulation of B cells through CD40 or BAFF receptor, which belongs to the TNF receptor superfamily, results in the activation of NFκB activating kinase (NIK) [16]. NIK phosphorylates and activates IKKα, which in turn phosphorylates p100 at two serine residues at the C-terminus [17]. Phosphorylated p100 is polyubiquitinated by SCFβTrCP and processed by the proteasome. Interestingly, only the C-terminus of ubiquitinated p100 is degraded, leaving its RHD domain (p52) intact, which dimerizes with REL-B, translocates to the nucleus, and turns on genes required for B cell maturation and activation.

UBIQUITIN-MEDIATED ACTIVATION OF PROTEIN KINASES IN THE IL-1R AND TLR PATHWAYS

The role of ubiquitination in IKK activation was discovered even before IKK subunits were molecularly cloned [18]. In an effort to understand IκBα phosphorylation and ubiquitination, a large kinase complex (~700 kDa) capable of site-specific phosphorylation of IκBα was identified. Unexpectedly, this kinase complex could be activated *in vitro* by polyubiquitination in a manner independent of proteasomal degradation. Subsequent experiments showed that the ubiquitin E3 important for IKK activation is TNF receptor associated factor 6 (TRAF6), a RING domain-containing protein required for NFκB activation by interleukin-1 (IL-1) and Toll-like receptor (TLR) pathways [19–21]. TLRs are single transmembrane receptors that bind to structurally conserved molecules derived from microbial organisms including bacteria and viruses, and these receptors play a key role in innate immune responses [22]. Biochemical fractionation further identified two protein complexes that mediate IKK activation by TRAF6: an E2 complex consisting of Ubc13 and Uev1A, and a kinase complex composed of the TAK1 kinase and the adaptor proteins TAB1 and TAB2 [20, 23, 24]. Ubc13/Uev1A and TRAF6 catalyze the synthesis of K63-linked polyubiquitin chains on target proteins including TRAF6 itself. This polyubiquitination leads to the activation of the TAK1 kinase complex. The ubiquitin-mediated activation of TAK1 requires the regulatory subunits TAB2 and TAB3, which are homologous proteins containing a highly conserved zinc-finger domain at their C-termini [25]. The zinc-finger domain is an NZF-type ubiquitin-binding domain that binds preferentially to K63-linked polyubiquitin chains. Mutations of the NZF domain of TAB2 or TAB3 prevent the activation of TAK1 and IKK, whereas replacement of the NZF domain with another ubiquitin-binding domain restores activation of these kinases. These results indicate that polyubiquitin-binding of TAB2 and TAB3 is important for the activation of TAK1 and IKK.

Following the activation by K63 polyubiquitination, the TAK1 kinase complex phosphorylates IKKβ at two serine residues in the activation loop, resulting in the activation of the IKK complex [23]. This activation requires the essential regulatory subunit NEMO, which contains a novel ubiquitin-binding domain termed NUB (Nemo-ubiquitin binding) [26, 27]. The NUB domain binds preferentially to K63 polyubiquitin chains and its integrity is required for NFκB activation by TNFα and IL-1β. Importantly, several mutations within the NUB domain that disrupt its ubiquitin-binding have been found in patients inflicted with anhidrotic ectodermal dysplasia with immunodeficiency (EDA-ID) [28]. The NUB domain is conserved in several NFκB inhibitory proteins, including ABIN-1, -2, -3 and optineurin, and the ability of these proteins to inhibit NFκB requires their ubiquitin-binding function [29, 30]. More recently, the C-terminal zinc finger (ZF) domain of NEMO has also been found to bind ubiquitin, and mutations of the ubiquitin-binding residues in the ZF domain impair NFκB activation by TNFα [31].

A model that emerges from these studies is depicted in Figure 9.2. Stimulation of cells with ligands such as IL-1β and bacterial lipopolysaccharides (LPS) leads to the dimerization of the receptors, IL-1R and TLR4, respectively. These receptors contain a conserved cytoplasmic domain termed Toll/Interleukin-1 receptor (TIR) domain, which recruits the adaptor protein myeloid differentiation primary gene 88 (MyD88) and two protein kinases termed IL-1 receptor-associated kinase 1 (IRAK1) and IRAK4. IRAK4 phosphorylates IRAK1 to induce its autophosphorylation and dissociation from the receptor. In the cytoplasm, IRAK1 interacts with TIFA (TRAF6-interacting protein with a forkhead-associated domain) and TRAF6, leading to the oligomerization of TRAF6, which activates its ubiquitin ligase activity [32, 33]. In the presence of Ubc13/Uev1A, TRAF6 promotes the synthesis of K63-linked polyubiquitin chains, which are conjugated to itself as well as other target proteins such as IRAK1 and NEMO [23, 34, 35]. The polyubiquitin chains recruit and activate the TAK1 kinase complex through TAB2 and TAB3. The mechanism by which TAK1 is activated is not clearly understood, but may involve oligomerization and autophosphorylation of TAK1 as a result of polyubiquitin binding. The polyubiquitin chains also serve as a scaffold to recruit the IKK complex through NEMO, and this recruitment facilitates the phosphorylation of IKKβ by TAK1, leading to IKK

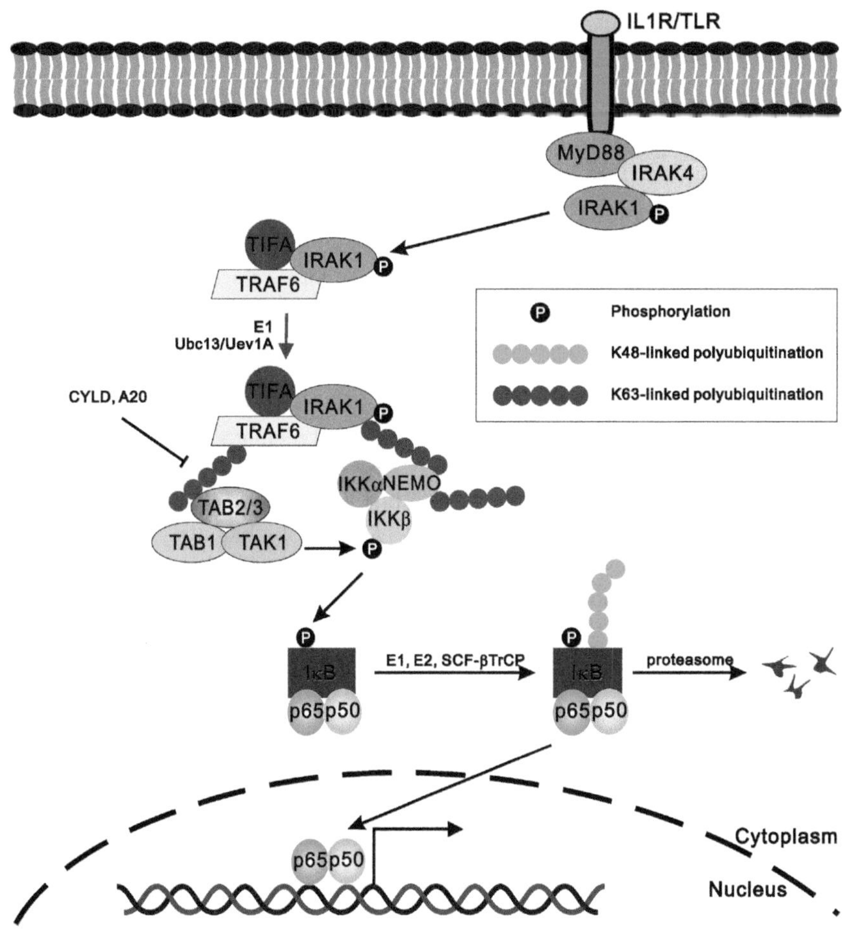

FIGURE 9.2 NFκB activation by IL1R/TLR.
Upon ligand binding, IL-1 receptor (IL1R) and Toll-like receptors (TLR) recruit the adaptor protein MyD88 and protein kinases IRAK4 and IKAK1. Phosphorylation of IRAK1 by IRAK4 leads to its translocation to the cytoplasm. IRAK1 associates with the adaptor protein TIFA and ubiquitin ligase TRAF6. In the presence of E1 and Ubc13/Uev1A, TRAF6 promotes K63-polyubiquitination of IRAK1 and TRAF6 itself. These polyubiquitination events lead to the recruitment of TAK1 and IKK complexes, activating TAK1, which in turn phosphorylates and activates IKK. IKK phosphorylates IκB proteins, which are then polyubiquitinated by the SCFβTrCP E3 complex. Ubiquitinated IκB is degraded by the proteasome, allowing NFκB to enter the nucleus to turn on downstream target genes. CYLD and A20, two de-ubiquitination enzymes, disassemble polyubiquitin chains, thereby inhibiting the activation of TAK1 and IKK.

activation. TAK1 also phosphorylates MAP kinase kinases (e.g., MKK6 and MKK7), leading to activation of Jun N-terminal kinase (JNK) and p38 kinase pathways.

UBIQUITIN-MEDIATED REGULATION OF NFκB AND APOPTOSIS IN THE TNFα PATHWAY

TNFα is a potent inflammatory cytokine produced by many cell types, especially macrophages [36]. It is synthesized as a single transmembrane protein (mTNFα), which is cleaved by the membrane-bound protease TNFα converting enzyme (TACE) to produce the soluble form (sTNFα). Two receptors, TNFR1 and TNFR2, bind to TNFα. TNFR1 is constitutively expressed in most tissues, and responds to both soluble and membrane-bound TNFα. In contrast, TNFR2 is only found in cells of the immune system, and is activated by membrane-bound TNFα. Most of our knowledge on TNFα signaling comes from TNFR1, and this will be the focus of this discussion (Figure 9.3).

TNFα forms a trimer, and its binding to TNFR1 induces trimerization of the receptor, which recruits a signaling complex (complex I) that includes TNFR-associated death domain protein (TRADD), TRAF2, TRAF5, receptor interacting protein kinase 1 (RIP1), and cellular inhibitor of apoptosis protein 1 and 2 (cIAP1 and cIAP2). RIP1 is rapidly modified by K63-linked polyubiquitin at a specific lysine (K377), and this ubiquitination is required for IKK activation [26, 37]. Although the TRAF and cIAP proteins recruited to the TNF receptor are RING domain proteins, only cIAP1 and cIAP2 have been shown to promote ubiquitination of RIP1 *in vitro* [38]. Polyubiquitinated RIP1 binds to the TAK1 and IKK complexes through the NZF domain

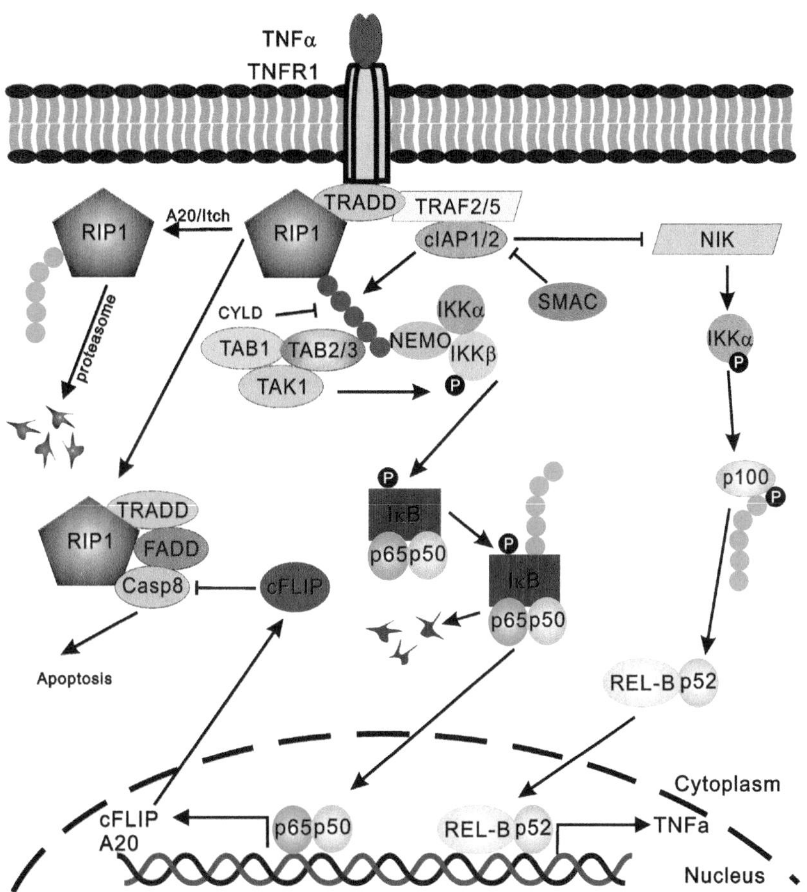

FIGURE 9.3 TNFα-induced NFκB and caspase activation.
Binding of TNFα to its receptor (TNFR) induces the assembly of complex I, which includes TRADD, RIP1, and the RING domain proteins TRAF2, TRAF5, cIAP1, and cIAP2. RIP1 is polyubiquitinated and then recruits the TAK1 kinase complex through the binding of polyubiquitin chains with TAB2 and TAB3. The polyubiquitin chains also recruit IKK by binding to NEMO, thus allowing TAK1 to phosphorylate and activate IKK. IKK activates NFκB, which induces target genes involved in immune and inflammatory responses, as well as negative inhibitors of NFκB and apoptosis, such as A20 and c-FLIP. TNFα-induced apoptosis is initiated by a death-inducing complex (complex II), which forms in the cytosol after TRADD and RIP1 dissociate from the receptor complex to interact with the death domain adaptor protein FADD and procaspase-8. In this complex, procaspase-8 undergoes autocatalytic cleavage to generate the mature caspase-8 to initiate apoptosis. The activation of caspase-8 is prevented by c-FLIP. A20 removes K63 polyubiquitin chains from RIP1, and it also functions with the HECT domain E3 ITCH to catalyze K48 polyubiquitination of RIP1, resulting in RIP1 degradation. The transition of RIP1 from complex I to complex II is facilitated by CYLD, which de-ubiquitinates RIP1, or by the loss of cIAPs, which is triggered by SMAC, an IAP-interacting protein released from the mitochondria. cIAPs also negatively regulate the non-canonical NFκB pathway through ubiquitination and degradation of NIK. In the absence of cIAPs, NIK is stabilized, leading to the activation of IKKα and processing of p100 to p52.

of TAB2/3 and the NUB domain of NEMO, respectively. The recruitment of TAK1 and IKK complexes leads to activation of these kinases through a mechanism similar to that involved in the IL-1R/TLR pathways.

After the initial association with the TNF receptor, TRADD and RIP1 leave the receptor and become associated with Fas-associated death domain protein (FADD), which binds to procaspase-8 to form complex II in the cytosol (Figure 9.3) [39]. Within this complex, procaspase-8 is activated by autocatalytic processing to generate mature caspase-8, which in turn cleaves procaspase-3, resulting in apoptotic cell death. Therefore, TNFα triggers two sequential signaling pathways that lead to NFκB activation and apoptosis, respectively. Normally, the apoptotic effect is suppressed because NFκB induces the expression of several anti-apoptotic proteins, including c-FLIP and cIAPs. c-FLIP binds to and inhibits caspase-8, thereby preventing TNFα-induced apoptosis. When new protein synthesis is inhibited by cyclohexamide, NFκB-induced expression of anti-apoptotic genes is blocked, whereas caspase activation is allowed to proceed to execute the apoptotic cell death program. c-FLIP is also targeted for degradation by the HECT domain ubiquitin ligase ITCH, whose activity is enhanced by JNK-catalyzed phosphorylation [40]. This JNK-mediated degradation of c-FLIP may contribute to the pro-apoptotic effect of prolonged JNK activation.

cIAPs prevent TNFα-induced apoptosis through their function as ubiquitin ligases, rather than inhibiting caspases directly. This ubiquitin-dependent function of cIAPs was uncovered through the use of small molecule mimetics of second mitochondrial activator of caspase (SMAC, also known as DIABLO), which is released from the mitochondrial matrix to the cytosol under apoptotic conditions [41–45]. SMAC binds to cIAPs, and triggers their autoubiquitination and degradation by the proteasome. In the presence of TNFα, loss of cIAPs facilitates the translocation of RIP1 from the receptor-associated complex I to the cytosolic complex II, where RIP1 promotes the activation of procaspase-8 in a manner that is resistant to inhibition by c-FLIP [46]. The kinase activity of RIP1 is required for TNFα-induced apoptosis and necrosis, but the substrate(s) is currently unknown. Ubiquitination status of RIP1 is an important determinant of whether a cell lives or dies in response to TNFα [47]. K63 polyubiquitination of RIP1 not only promotes NFκB activation, but also prevents the association of RIP1 with caspase-8. Inhibition of K63 polyubiquitination through inhibition or removal of Ubc13, TRAF2, or cIAPs, or mutation of the RIP1 ubiquitination sites, promotes the formation of the RIP1/FADD/caspase-8 complex. Conversely, RNAi of CYLD promotes RIP1 ubiquitination and inhibits its release from the receptor complex [46]. More recently, the ubiquitin-binding protein ABIN-1 has been found to prevent TNFα-induced apoptosis through inhibiting the binding of FADD to procaspase-8 [48]. This anti-apoptotic activity of ABIN-1 depends on its NUB domain, which binds to polyubiquitin chains.

In addition to their role in inhibiting the pro-apoptotic function of RIP1, cIAPs function as ubiquitin ligases that target the degradation of NIK, thereby inhibiting the induction of TNFα by the non-canonical pathway of NFκB activation [43, 49, 50]. This explains why degradation of cIAPs by the use of SMAC mimetics is sufficient to trigger apoptosis of some tumor cell lines capable of producing TNFα. Ubiquitination of NIK by cIAPs requires both TRAF2 and TRAF3, because B cells lacking TRAF2 or TRAF3 constitutively process p100 to p52 owing to the stabilization of NIK. Biochemical experiments suggest a model in which TRAF2 and TRAF3 function as adaptors that bridge cIAPs to NIK, permitting NIK degradation in unstimulated B cells [49]. When B cells are stimulated through members of the TNF receptor family, such as CD40 or BAFFR, TRAF2 is activated to catalyze K63 polyubiquitination of cIAP1 and cIAP2, which enhances their ubiquitin ligase activity and directs them to catalyze K48 polyubiquitination of TRAF3, resulting in TRAF3 degradation by the proteasome. The degradation of TRAF3 prevents NIK ubiquitination by cIAPs, leading to enhanced processing of p100.

cIAP1/2–induced degradation of TRAF3 also plays an important role in the activation of JNK by CD40 [51]. Stimulation of CD40 in B cells leads to the assembly of a receptor-associated protein complex consisting of several proteins involved in the ubiquitin pathway, including the E2 Ubc13; the E3s TRAF2, TRAF3, TRAF6, cIAP1, and cIAP2; and the ubiquitinated protein NEMO. In addition, the receptor complex contains protein kinases including TAK1 and MEKK1, which are required for IKK and JNK activation, respectively. MEKK1 also contains a RING-like domain and has E3 activity, but the function of this E3 activity is still unknown. While Ubc13, TRAF2, NEMO, and cIAP1/2 are all required for the receptor recruitment and activation of MEKK1, MEKK1 and JNK do not appear to be active at the receptor. Instead, TRAF3 must be ubiquitinated by cIAPs and degraded by the proteasome in order to release the MEKK1 complex to the cytosol, where the activation of MEKK1 and JNK takes place. It is not clear why these MAP kinases need to translocate to the cytosol to be activated. In contrast to the MAP kinases, activation of IKK occurs earlier at the receptor-associated complex and does not require proteasomal degradation of TRAF3. These results illustrate complex regulation of protein kinase activity by ubiquitin through both proteasome-dependent and -independent mechanisms. It remains to be seen whether MAPK activation in other signaling pathways requires a similar mechanism involving proteasome-dependent cytoplasmic translocation of signaling complexes.

DE-UBIQUITINATION ENZYMES PREVENT PROTEIN KINASES ACTIVATION IN THE NFκB PATHWAY

The identification of several de-ubiquitination enzymes that negatively regulate IKK activity further supports the role of polyubiquitination in protein kinase activation [52]. The first DUB known to inhibit IKK is CYLD, a tumor suppressor protein implicated in cylindromatosis, multiple familial trichoepithelioma, and multiple myeloma in humans [53–56]. CYLD contains three CAP-Gly domains at the N-terminus and a ubiquitin-specific protease (USP) domain at the C-terminus. CYLD specifically cleaves K63-, not K48-, linked polyubiquitin chains, and inhibits the activation of TAK1, IKK, and JNK in several pathways. A point mutation of the catalytic cysteine residue in CYLD abolishes its inhibitory effect on IKK activation, resulting in hyperactivation of NFκB. Remarkably, many disease-related mutations of CYLD have been found in the USP domain and impair its DUB activity, emphasizing the importance of de-ubiquitination in controlling NFκB activation and tumor formation [57]. Mice deficient in CYLD have been generated by different laboratories. Although the reported phenotypes of these mice are not identical, accumulation of ubiquitinated proteins, hyperactivation of IKK and excessive inflammation have been consistently observed [52].

A20 is another DUB that negatively regulates IKK activation [58, 59]. It contains an N-terminal OTU domain and several zinc fingers at the C-terminus [60, 61]. While the OTU domain removes the polyubiquitin chains from

TRAF6 and RIP1, one of the C-terminal zinc fingers has been reported to possess the ubiquitin E3 activity, which catalyzes K48-linked polyubiquitination on RIP1, leading to its degradation [62]. Consistent with its inhibitory effect on IKK activation, A20-deficient mice develop excessive inflammation in multiple organs, and cells lacking A20 display prolonged IKK and NFκB activation after TNFα stimulation [63]. However, unlike CYLD, which specifically cleaves K63-linked polyubiquitin chains, A20 can cleave polyubiquitin chains linked through K48 or K63 *in vitro* [64–66]. It is possible that other proteins may regulate the ubiquitin chain linkage specificity of A20. In fact, A20 has been shown to form a ubiquitin-editing complex with the ubiquitin ligase ITCH and ubiquitin-binding proteins that include TAX1BP1 and ABINs [67–69]. Genetic evidence suggests that TAX1BP1 and ITCH are required for A20-mediated RIP1 de-ubiquitination and degradation. TAX1BP1 contains a UBZ-type ubiquitin-binding domain, which is required for TRAF6 de-ubiquitination by A20 [67].

POLYUBIQUITINATION REGULATES PROTEIN KINASE ACTIVATION IN DIVERSE NFκB PATHWAYS

While the IL-1R/TLR and TNFR pathways have been most extensively studied, NFκB is known to be activated by a large variety of agents, virtually all of which activate NFκB through the IKK complex. Remarkably, although the upstream pathways leading to IKK activation are quite diverse, many of these pathways employ polyubiquitination as a mechanism to regulate IKK activation [14, 70]. These pathways include those emanating from the cell surface receptor (e.g., T cell receptor), as well as those that signal from within the cells (e.g. NOD-like receptors, RIG-I-like receptors, and DNA damage; Figure 9.4).

B and T cells, which carry out adaptive immune responses, express an extremely large repertoire of antigen receptors through somatic gene recombination and hypermutation. Each T cell receptor (TCR) is activated when a cognate antigenic peptide is presented by a major histocompatibility complex (MHC) on antigen presenting cells (APC). The engagement of TCR triggers a tyrosine phosphorylation cascade that leads to the activation of the serine/threonine kinase PKCθ. PKCθ recruits and activates a protein complex composed of the CARD domain proteins CARMA1 and BCL10 and a caspase-like protein MALT1 [71–73]. This complex, termed CBM (CARMA1–BCL10–MALT1), mediates the activation of IKK through a ubiquitin-dependent mechanism. MALT1 was reported to function as a ubiquitin ligase to catalyze K63 polyubiquitination of NEMO in conjunction with Ubc13/Uev1A [74]. Another report showed that MALT1 contains binding sites for TRAF2 and TRAF6, and that the binding of MALT1 to TRAF6 promotes the oligomerization of TRAF6, resulting in the activation of its ubiquitin ligase activity [75]. TRAF6 catalyzes auto-polyubiquitination as well as ubiquitination of NEMO. These events lead to activation of the TAK1 kinase complex, which in turn activates IKK. However, a recent report showed that a mutation of the ubiquitination site of NEMO in T cells does not impair IKK activation by TCR [76]. More recent studies showed that TRAF6 catalyzes the polyubiquitination of BCL10 and MALT1, both of which appear to be important for IKK activation by TCR [77, 78]. In support of a key role of K63 polyubiquitination in the TCR pathway, conditional deletion of Ubc13 in T cells abrogates the activation of TAK1, IKK, and MAP kinases by TCR stimulation [79]. However, deletion of Ubc13 in other cell types, including B cells and fibroblasts, impairs the activation of MAPK, but the effect on IKK activation is less profound [80, 81]. It is possible that other E2s, such as Ubc5, may substitute for the loss of Ubc13 in certain NFκB pathways [18]. TRAF6 and TRAF2 may also have redundant functions in the TCR pathway, as TRAF6-deficient T cells are still capable of activating NFκB, whereas RNAi of both TRAF6 and TRAF2 led to potent inhibition of NFκB by TCR signaling [75, 82].

NOD1 (nucleotide-binding oligomerization domain 1) and NOD2 are cytosolic proteins involved in host defense against intracellular bacteria [83]. Mutations of NOD2 have been associated with certain inflammatory bowel diseases, such as Crohn's. In additional to the NOD domain, NOD1 and NOD2 contain N-terminal CARD domains and C-terminal leucine-rich repeats (LRR), which may be involved (directly or indirectly) in recognition of bacterial peptidoglycans. The activation of NFκB by NOD1 and NOD2 requires RIP2 (also known as RICK or CARDIAK), a protein kinase that also contains a CARD domain. RIP2 interacts with NOD1 or NOD2 through the CARD–CARD interaction, and becomes oligomerized after NOD1/2 oligomerization induced by ligand binding. NOD1 and NOD2 promote K63 polyubiquitination of RIP2 at a specific lysine (K209) within the kinase domain [84, 85]. Polyubiquitination of RIP2 leads to the recruitment and activation of the TAK1 kinase complex, which then activates IKK [85–87]. TRAF2 and TRAF5 are required for NFκB activation by NOD1, whereas TRAF6 and Ubc13/Uev1A are required for maximal NFκB activation by NOD2 [84, 85]. However, it has not been demonstrated that the TRAF proteins can catalyze polyubiquitination of RIP2 *in vitro*. NOD2 and RIP2 can also promote the K63 polyubiquitination of NEMO at a specific lysine (K285), and mutations of NOD2 associated with Crohn's disease impair the ability of NOD2 to stimulate NEMO ubiquitination [88]. A20 de-ubiquitinates RIP2 and hence attenuates NFκB activation by both NOD1 and NOD2 [89].

Viral infection produces cytosolic viral RNA, which is detected by RIG-I-like RNA helicases (RLHs) [90–92]. RIG-I contains two CARD domains at the N-terminus, followed by an ATP-dependent RNA helicase domain and

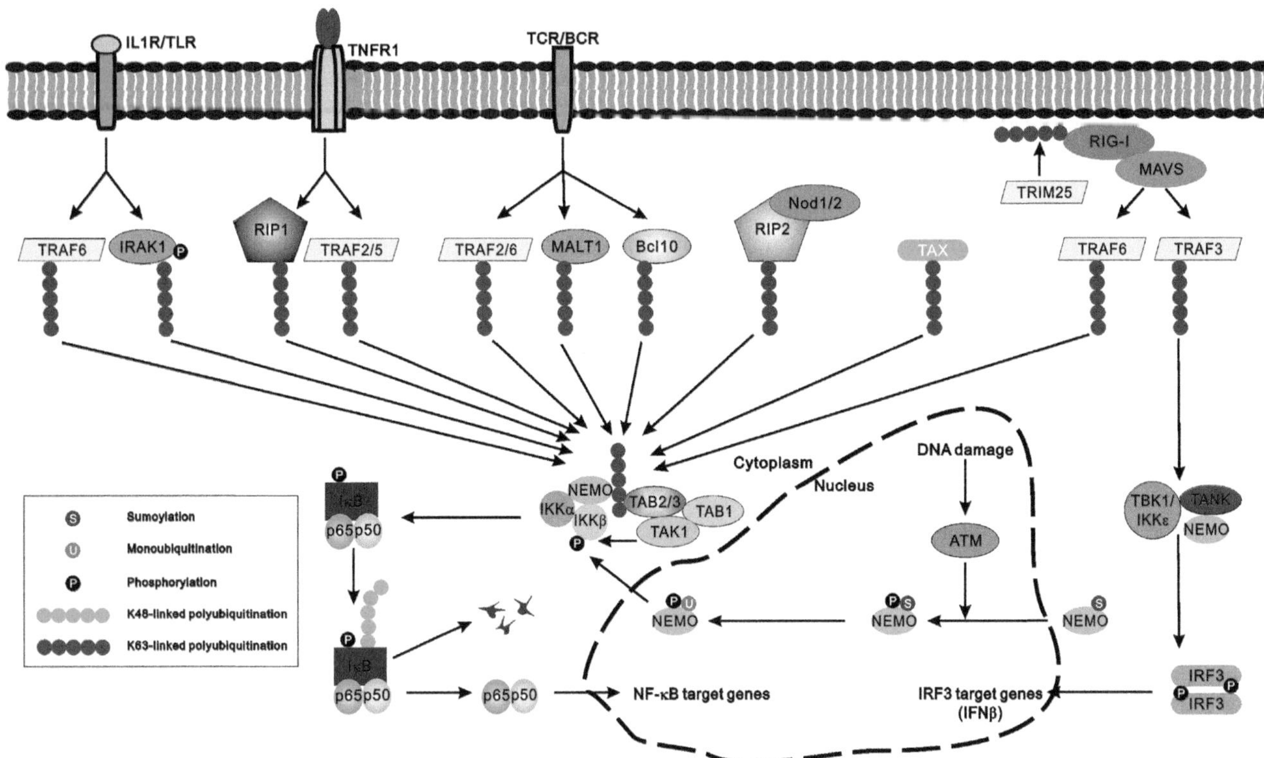

FIGURE 9.4 Expanding role of ubiquitin in diverse pathways leading to the activation of NFκB and IRF3.
NFκB is activated by a large variety of agents, including those that bind to cell surface receptors, including TNFR, IL-1R, TLR, and TCR, as well as those that signal through cytosolic receptors, including NOD1, NOD2 and RIG-I. In addition, viral proteins such as the Tax protein of human T cell leukemia virus (HTLV-1) and DNA damage in the nucleus can also activate NFκB. These different pathways converge on IKK to activate NFκB. The RIG-I pathway also activates the transcription factor IRF3 through its phosphorylation by the IKK-like kinases, TBK1 and IKKε. All of these pathways employ polyubiquitination as a common mechanism to activate TAK1 and IKK (see text for details).

a regulatory domain at the C-terminus [93]. It binds to double-stranded RNA as well as single-stranded RNA with 5′-triphosphate. The binding of viral RNA presumably induces a conformational change that exposes the N-terminal CARD domains of RIG-I. Following viral infection, a lysine within the second CARD domain (K172) of RIG-I is conjugated by K63-linked polyubiquitin chains [94]. This polyubiquitination is catalyzed by tripartite motif protein 25 (TRIM25), a RING domain-containing E3 ubiquitin ligase that binds to the CARD domains of RIG-I. CYLD de-ubiquitinates RIG-I, thereby inhibiting RIG-I signaling [95, 96]. Ubiquitination of RIG-I facilitates its binding to mitochondrial antiviral signaling protein (MAVS, also known as IPS-1, VISA or CARDIF) [96–99]. MAVS contains an N-terminal CARD domain, a proline-rich region, and a C-terminal transmembrane domain that is responsible for its mitochondrial localization [100]. MAVS activates NFκB through IKK, and another transcription factor, IRF3, through the IKK-related kinases TBK1 and IKKε. NFκB and IRF3 enter the nucleus and become parts of the enhanceosome complex that induces type-I interferons such as IFN-β [101].

The cytoplasmic portion of MAVS contains binding sites for TRAF2, TRAF3, and TRAF6. While TRAF2 and TRAF6 are likely to be important for IKK activation, TRAF3 is required for the activation of TBK1/IKKε [102]. TRAF3 forms a complex with TANK (TRAF family associated NFκB activator), which in turn binds to TBK1. The RING domain of TRAF3 is indispensable for TBK1 activation, suggesting that TRAF3 activates TBK1 through a ubiquitin-dependent mechanism [103]. In support of this model, the de-ubiquitination enzyme DUBA has recently been identified as a negative regulator of IRF3 [104]. DUBA contains the OTU-type DUB domain, which removes K63-linked polyubiquitin chains from TRAF3 and suppresses interferon induction.

DNA damage is known to activate NFκB, which might provide a survival mechanism while the DNA lesion is being repaired. A series of studies show that DNA damage in the nucleus activates the cytosolic IKK through sequential modification of NEMO by ubiquitin and the ubiquitin-like protein SUMO [105]. Although the majority of NEMO is associated with the catalytic subunits of IKK in the cytoplasm, a small population of NEMO is free of IKK, and this population is modified by SUMO-1 following DNA double-strand break (DSB). The sumoylation of NEMO is catalyzed by the SUMO E2, Ubc9, and a SUMO E3, PIASy [106]. This modification facilitates the nuclear import of NEMO. Inside the nucleus, NEMO is phosphorylated by the kinase ATM, which is activated following DSB. Subsequently, NEMO is monoubiquitinated at two specific lysines (K277 and K309) and then exported to the cytosol along with ATM [107]. The NEMO/ATM complex associates with the catalytic subunits of IKK in the cytosol, resulting in the activation of the IKK complex. The mechanism by which monoubiquitinated NEMO and ATM activate IKK is still not clear.

CONCLUSIONS AND PERSPECTIVES

Mounting evidence supports a key role of polyubiquitination in the activation of TAK1, IKK, and MAP kinases in immune and inflammatory pathways. It is remarkable that in some pathways, such as the TNF pathway, the majority of proteins recruited to the receptor are involved in the ubiquitin pathway. These proteins include ubiquitin E3s such as TRAFs and cIAPs; the ubiquitin E2 Ubc13; de-ubiquitination enzymes including A20 and CYLD; ubiquitin receptors including TAB2/3, NEMO, and TAX1BP1; and ubiquitination targets such as RIP1. Such an extensive utilization of the ubiquitination machinery may help to coordinate activation of multiple enzymes including IKK, MAP kinases, and caspases, providing a delicate balance between immune and inflammatory responses and cell death. A common mechanism underlying ubiquitin signaling in different pathways is that polyubiquitin chains serve as a scaffold to recruit protein kinase complexes that contain regulatory subunits harboring ubiquitin-binding domains (e.g., TAB2/3 in the TAK1 complex and NEMO in the IKK complex). It will be interesting to investigate whether other protein kinase complexes in the kinome contain ubiquitin-binding domains, and whether these kinases are regulated through ubiquitin-dependent mechanisms.

The regulatory role of ubiquitin is not limited to protein kinase activation. It is now clear that monoubiquitination and polyubiquitination regulate many cellular processes, including membrane trafficking, DNA repair, and gene transcription through mechanisms independent of protein degradation [108–110]. In almost all cases, ubiquitin signaling is mediated by a network of interactions between ubiquitinated proteins and ubiquitin-binding proteins. We envision that the pace of discovery of new ubiquitination targets, ubiquitin-binding proteins, and the ubiquitination enzymes involved in various signaling pathways will be greatly accelerated by technological advances such as mass spectrometry and high throughput screens, and the availability of resources such as genomics and proteomics. As evidenced by the success of proteasome inhibitors in the treatment of multiple myeloma, increased understanding of both proteolytic and non-proteolytic functions of ubiquitin should facilitate the development of potent and specific therapeutics for the treatment of various human diseases.

ACKNOWLEDGEMENTS

Research in our laboratory is supported by grants from the NIH and the Welch Foundation. Ming Xu is a postdoctoral fellow of the Leukemia and Lymphoma Society. Zhijian J. Chen is an Investigator at the Howard Hughes Medical Institute.

REFERENCES

1. Wilkinson KD. Ubiquitin: a Nobel protein. *Cell* 2004;**119**:741–5.
2. Finley D, Ciechanover A, Varshavsky A. Ubiquitin as a central cellular regulator. *Cell* 2004;**116**:S29–32. 22 pp. following S32.
3. Hershko A. Ubiquitin: roles in protein modification and breakdown. *Cell* 1983;**34**:11–12.
4. Pickart CM. Back to the future with ubiquitin. *Cell* 2004;**116**:181–90.
5. Haas AL, Wilkinson KD. DeTEKting ubiquitination of APC/C substrates. *Cell* 2008;**133**:570–2.
6. Bernassola F, Karin M, Ciechanover A, Melino G. The HECT family of E3 ubiquitin ligases: multiple players in cancer development. *Cancer Cell* 2008;**14**:10–21.
7. Petroski MD, Deshaies RJ. Function and regulation of cullin-RING ubiquitin ligases. *Nat Rev Mol Cell Biol* 2005;**6**:9–20.
8. Amerik AY, Hochstrasser M. Mechanism and function of de-ubiquitinating enzymes. *Biochim Biophys Acta* 2004;**1695**:189–207.
9. Nijman SM, Luna-Vargas MP, Velds A, et al. A genomic and functional inventory of de-ubiquitinating enzymes. *Cell* 2005;**123**:773–86.
10. Hicke L, Schubert HL, Hill CP. Ubiquitin-binding domains. *Nat Rev Mol Cell Biol* 2005;**6**:610–21.
11. Hurley JH, Lee S, Prag G. Ubiquitin-binding domains. *Biochem J* 2006;**399**:361–72.
12. Hayden MS, Ghosh S. Shared principles in NFκB signaling. *Cell* 2008;**132**:344–62.
13. Pomerantz JL, Baltimore D. Two pathways to NFκB. *Mol Cell* 2002;**10**:693–5.
14. Chen ZJ. Ubiquitin signalling in the NFκB pathway. *Nat Cell Biol* 2005;**7**:758–65.
15. Maniatis T. A ubiquitin ligase complex essential for the NFκB, Wnt/Wingless, and Hedgehog signaling pathways. *Genes Dev* 1999;**13**:505–10.
16. Xiao G, Harhaj EW, Sun SC. NFκB-inducing kinase regulates the processing of NFκB2 p100. *Mol Cell* 2001;**7**:401–9.
17. Senftleben U, Cao Y, Xiao G, et al. Activation by IKKα of a second, evolutionary conserved, NFκ B signaling pathway. *Science* 2001;**293**:1495–9.
18. Chen ZJ, Parent L, Maniatis T. Site-specific phosphorylation of IκBα by a novel ubiquitination-dependent protein kinase activity. *Cell* 1996;**84**:853–62.
19. Cao Z, Xiong J, Takeuchi M, Kurama T, Goeddel DV. TRAF6 is a signal transducer for interleukin-1. *Nature* 1996;**383**:443–6.
20. Deng L, Wang C, Spencer E, et al. Activation of the IκB kinase complex by TRAF6 requires a dimeric ubiquitin-conjugating enzyme complex and a unique polyubiquitin chain. *Cell* 2000;**103**:351–61.
21. Ishida T, Mizushima S, Azuma S, et al. Identification of TRAF6, a novel tumor necrosis factor receptor-associated factor protein that mediates signaling from an amino-terminal domain of the CD40 cytoplasmic region. *J Biol Chem* 1996;**271**:28,745–28,748.
22. Akira S, Uematsu S, Takeuchi O. Pathogen recognition and innate immunity. *Cell* 2006;**124**:783–801.

23. Wang C, Deng L, Hong M, Akkaraju GR, Inoue J, Chen ZJ. TAK1 is a ubiquitin-dependent kinase of MKK and IKK. *Nature* 2001;**412**:346–51.
24. Ninomiya-Tsuji J, Kishimoto K, Hiyama A, Inoue J, Cao Z, Matsumoto K. The kinase TAK1 can activate the NIK-I κB as well as the MAP kinase cascade in the IL-1 signalling pathway. *Nature* 1999;**398**:252–6.
25. Kanayama A, Seth RB, Sun L, et al. TAB2 and TAB3 activate the NFκB pathway through binding to polyubiquitin chains. *Mol Cell* 2004;**15**:535–48.
26. Ea CK, Deng L, Xia ZP, Pineda G, Chen ZJ. Activation of IKK by TNFα requires site-specific ubiquitination of RIP1 and polyubiquitin binding by NEMO. *Mol Cell* 2006;**22**:245–57.
27. Wu CJ, Conze DB, Li T, Srinivasula SM, Ashwell JD. Sensing of Lys 63-linked polyubiquitination by NEMO is a key event in NFκB activation [corrected]. *Nat Cell Biol* 2006;**8**:398–406.
28. Sebban H, Yamaoka S, Courtois G. Posttranslational modifications of NEMO and its partners in NFκB signaling. *Trends Cell Biol* 2006; **16**:569–77.
29. Wagner S, Carpentier I, Rogov V, et al. Ubiquitin binding mediates the NFκB inhibitory potential of ABIN proteins. *Oncogene* 2008;**27**:3739–45.
30. Zhu G, Wu CJ, Zhao Y, Ashwell JD. Optineurin negatively regulates TNFα- induced NFκB activation by competing with NEMO for ubiquitinated RIP. *Curr Biol* 2007;**17**:1438–43.
31. Cordier F, Grubisha O, Traincard F, Veron M, Delepierre M, Agou F. The zinc finger of NEMO is a functional ubiquitin-binding domain. *J Biol Chem* 2008, in press.
32. Ea CK, Sun L, Inoue J, Chen ZJ. TIFA activates IκB kinase (IKK) by promoting oligomerization and ubiquitination of TRAF6. *Proc Natl Acad Sci USA* 2004;**101**:15,318–323.
33. Takatsuna H, Kato H, Gohda J, et al. Identification of TIFA as an adapter protein that links tumor necrosis factor receptor-associated factor 6 (TRAF6) to interleukin-1 (IL-1) receptor-associated kinase-1 (IRAK-1) in IL-1 receptor signaling. *J Biol Chem* 2003;**278**:12,144–150.
34. Conze DB, Wu CJ, Thomas JA, Landstrom A, Ashwell JD. Lys63-linked polyubiquitination of IRAK-1 is required for interleukin-1 receptor- and toll-like receptor-mediated NFκB activation. *Mol Cell Biol* 2008;**28**:3538–47.
35. Windheim M, Stafford M, Peggie M, Cohen P. Interleukin-1 (IL-1) induces the Lys63-linked polyubiquitination of IL-1 receptor-associated kinase 1 to facilitate NEMO binding and the activation of IκBα kinase. *Mol Cell Biol* 2008;**28**:1783–91.
36. Tartaglia LA, Goeddel DV. Two TNF receptors. *Immunol Today* 1992;**13**:151–3.
37. Li H, Kobayashi M, Blonska M, You Y, Lin X. Ubiquitination of RIP is required for tumor necrosis factor α-induced NFκB activation. *J Biol Chem* 2006;**281**:13,636–643.
38. Bertrand MJ, Milutinovic S, Dickson KM, et al. cIAP1 and cIAP2 facilitate cancer cell survival by functioning as E3 ligases that promote RIP1 ubiquitination. *Mol Cell* 2008;**30**:689–700.
39. Micheau O, Tschopp J. Induction of TNF receptor I-mediated apoptosis via two sequential signaling complexes. *Cell* 2003;**114**:181–90.
40. Chang L, Kamata H, Solinas G, et al. The E3 ubiquitin ligase itch couples JNK activation to TNFα-induced cell death by inducing c-FLIP(L) turnover. *Cell* 2006;**124**:601–13.
41. Du C, Fang M, Li Y, Li L, Wang X. Smac, a mitochondrial protein that promotes cytochrome c-dependent caspase activation by eliminating IAP inhibition. *Cell* 2000;**102**:33–42.
42. Yang QH, Du C. Smac/DIABLO selectively reduces the levels of c-IAP1 and c-IAP2 but not that of XIAP and livin in HeLa cells. *J Biol Chem* 2004;**279**:16,963–970.
43. Varfolomeev E, Blankenship JW, Wayson SM, et al. IAP antagonists induce autoubiquitination of c-IAPs, NFκB activation, and TNFα-dependent apoptosis. *Cell* 2007;**131**:669–81.
44. Vince JE, Wong WW, Khan N, et al. IAP antagonists target cIAP1 to induce TNFα-dependent apoptosis. *Cell* 2007;**131**:682–93.
45. Petersen SL, Wang L, Yalcin-Chin A, et al. Autocrine TNFα signaling renders human cancer cells susceptible to Smac-mimetic-induced apoptosis.. *Cancer Cell* 2007;**2**:45–56.
46. Wang L, Du F, Wang X. TNFα induces two distinct caspase-8 activation pathways. *Cell* 2008;**133**:693–703.
47. O'Donnell MA, Legarda-Addison D, Skountzos P, Yeh WC, Ting AT. Ubiquitination of RIP1 regulates an NFκB-independent cell-death switch in TNF signaling. *Curr Biol* 2007;**17**:418–24.
48. Oshima S, Turer EE, Callahan JA, et al. ABIN-1 is a ubiquitin sensor that restricts cell death and sustains embryonic development. *Nature* 2008.
49. Vallabhapurapu S, Matsuzawa A, Zhang W, et al. Nonredundant and complementary functions of TRAF2 and TRAF3 in a ubiquitination cascade that activates NIK-dependent alternative NFκB signaling. *Nat Immunol* 2008;**9**:1364–70.
50. Zarnegar BJ, Wang Y, Mahoney DJ, et al. Noncanonical NFκB activation requires coordinated assembly of a regulatory complex of the adaptors cIAP1, cIAP2, TRAF2 and TRAF3 and the kinase NIK. *Nat Immunol* 2008;**9**:1371–8.
51. Matsuzawa A, Tseng PH, Vallabhapurapu S, et al. Essential cytoplasmic translocation of a cytokine receptor-assembled signaling complex. *Science* 2008;**321**:663–8.
52. Sun SC. De-ubiquitylation and regulation of the immune response. *Nat Rev Immunol* 2008;**8**:501–11.
53. Bignell GR, Warren W, Seal S, et al. Identification of the familial cylindromatosis tumour-suppressor gene. *Nat Genet* 2000;**25**:160–5.
54. Brummelkamp TR, Nijman SM, Dirac AM, Bernards R. Loss of the cylindromatosis tumour suppressor inhibits apoptosis by activating NFκB. *Nature* 2003;**424**:797–801.
55. Kovalenko A, Chable-Bessia C, Cantarella G, Israel A, Wallach D, Courtois G. The tumour suppressor CYLD negatively regulates NFκB signalling by de-ubiquitination. *Nature* 2003;**424**:801–5.
56. Trompouki E, Hatzivassiliou E, Tsichritzis T, Farmer H, Ashworth A, Mosialos G. CYLD is a de-ubiquitinating enzyme that negatively regulates NFκB activation by TNFR family members. *Nature* 2003;**424**:793–6.
57. Courtois G. Tumor suppressor CYLD: negative regulation of NFκB signaling and more. *Cell Mol Life Sci* 2008;**65**:1123–32.
58. Krikos A, Laherty CD, Dixit VM. Transcriptional activation of the tumor necrosis factor α-inducible zinc finger protein, A20, is mediated by κB elements. *J Biol Chem* 1992;**267**:17,971–976.
59. Coornaert B, Carpentier I, Beyaert R. A20: Central gatekeeper in inflammation and immunity. *J Biol Chem* 2008, in press.
60. Evans PC, Ovaa H, Hamon M, et al. Zinc-finger protein A20, a regulator of inflammation and cell survival, has de-ubiquitinating activity. *Biochem J* 2004;**378**:727–34.
61. Evans PC, Smith TS, Lai MJ, et al. A novel type of de-ubiquitinating enzyme. *J Biol Chem* 2003;**278**:23,180–186.
62. Wertz IE, O'Rourke KM, Zhou H, et al. De-ubiquitination and ubiquitin ligase domains of A20 downregulate NFκB signalling. *Nature* 2004;**430**:694–9.

63. Lee EG, Boone DL, Chai S, et al. Failure to regulate TNF-induced NFκB and cell death responses in A20-deficient mice. *Science* 2000;**289**:2350–4.
64. Komander D, Lord CJ, Scheel H, et al. The structure of the CYLD USP domain explains its specificity for Lys63-linked polyubiquitin and reveals a B box module. *Mol Cell* 2008;**29**:451–64.
65. Komander D, Barford D. Structure of the A20 OTU domain and mechanistic insights into de-ubiquitination.. *Biochem J* 2008;**409**:77–85.
66. Lin SC, Chung JY, Lamothe B, et al. Molecular basis for the unique de-ubiquitinating activity of the NFκB inhibitor A20. *J Mol Biol* 2008;**376**:526–40.
67. Iha H, Peloponese JM, Verstrepen L, et al. Inflammatory cardiac valvulitis in TAX1BP1-deficient mice through selective NFκB activation. *EMBO J* 2008;**27**:629–41.
68. Shembade N, Harhaj NS, Liebl DJ, Harhaj EW. Essential role for TAX1BP1 in the termination of TNFα-, IL-1- and LPS-mediated NFκB and JNK signaling. *EMBO J* 2007;**26**:3910–22.
69. Shembade N, Harhaj NS, Parvatiyar K, et al. The E3 ligase Itch negatively regulates inflammatory signaling pathways by controlling the function of the ubiquitin-editing enzyme A20. *Nat Immunol* 2008;**9**:254–62.
70. Krappmann D, Scheidereit C. A pervasive role of ubiquitin conjugation in activation and termination of IκB kinase pathways. *EMBO Rep* 2005;**6**:321–6.
71. Rawlings DJ, Sommer K, Moreno-Garcia ME. The CARMA1 signalosome links the signalling machinery of adaptive and innate immunity in lymphocytes. *Nat Rev Immunol* 2006;**6**:799–812.
72. Thome M. CARMA1, BCL-10 and MALT1 in lymphocyte development and activation. *Nat Rev Immunol* 2004;**4**:348–59.
73. van Oers NS, Chen ZJ. Cell biology. Kinasing and clipping down the NFκB trail. *Science* 2005;**308**:65–6.
74. Zhou H, Wertz I, O'Rourke K, et al. Bcl10 activates the NFκB pathway through ubiquitination of NEMO. *Nature* 2004;**427**:167–71.
75. Sun L, Deng L, Ea CK, Xia ZP, Chen ZJ. The TRAF6 ubiquitin ligase and TAK1 kinase mediate IKK activation by BCL10 and MALT1 in T lymphocytes. *Mol Cell* 2004;**14**:289–301.
76. Ni CY, Wu ZH, Florence WC, et al. Cutting edge: K63-linked polyubiquitination of NEMO modulates TLR signaling and inflammation in vivo. *J Immunol* 2008;**180**:7107–11.
77. Oeckinghaus A, Wegener E, Welteke V, et al. Malt1 ubiquitination triggers NFκB signaling upon T-cell activation. *EMBO J* 2007;**26**:4634–45.
78. Wu CJ, Ashwell JD. NEMO recognition of ubiquitinated Bcl10 is required for T cell receptor-mediated NFκB activation. *Proc Natl Acad Sci USA* 2008;**105**:3023–8.
79. Yamamoto M, Sato S, Saitoh T, et al. Cutting edge: pivotal function of Ubc13 in thymocyte TCR signaling. *J Immunol* 2006;**177**:7520–4.
80. Fukushima T, Matsuzawa S, Kress CL, et al. Ubiquitin-conjugating enzyme Ubc13 is a critical component of TNF receptor-associated factor (TRAF)-mediated inflammatory responses. *Proc Natl Acad Sci USA* 2007;**104**:6371–6.
81. Yamamoto M, Okamoto T, Takeda K, et al. Key function for the Ubc13 E2 ubiquitin-conjugating enzyme in immune receptor signaling. *Nat Immunol* 2006;**7**:962–70.
82. King CG, Kobayashi T, Cejas PJ, et al. TRAF6 is a T cell-intrinsic negative regulator required for the maintenance of immune homeostasis. *Nat Med* 2006;**12**:1088–92.
83. Kanneganti TD, Lamkanfi M, Nunez G. Intracellular NOD-like receptors in host defense and disease. *Immunity* 2007;**27**:549–59.
84. Hasegawa M, Fujimoto Y, Lucas PC, et al. A critical role of RICK/RIP2 polyubiquitination in Nod-induced NFκB activation. *EMBO J* 2008;**27**:373–83.
85. Yang Y, Yin C, Pandey A, Abbott D, Sassetti C, Kelliher MA. NOD2 pathway activation by MDP or Mycobacterium tuberculosis infection involves the stable polyubiquitination of Rip2. *J Biol Chem* 2007;**282**:36,223–229.
86. Windheim M, Lang C, Peggie M, Plater LA, Cohen P. Molecular mechanisms involved in the regulation of cytokine production by muramyl dipeptide. *Biochem J* 2007;**404**:179–90.
87. Abbott DW, Yang Y, Hutti JE, Madhavarapu S, Kelliher MA, Cantley LC. Coordinated regulation of Toll-like receptor and NOD2 signaling by K63-linked polyubiquitin chains. *Mol Cell Biol* 2007;**27**:6012–25.
88. Abbott DW, Wilkins A, Asara JM, Cantley LC. The Crohn's disease protein, NOD2, requires RIP2 in order to induce ubiquitinylation of a novel site on NEMO. *Curr Biol* 2004;**14**:2217–27.
89. Hitotsumatsu O, Ahmad RC, Tavares R, et al. The ubiquitin-editing enzyme A20 restricts nucleotide-binding oligomerization domain containing 2-triggered signals. *Immunity* 2008;**28**:381–90.
90. Pichlmair A, Reis e Sousa C. Innate recognition of viruses. *Immunity* 2007;**27**:370–83.
91. Seth RB, Sun L, Chen ZJ. Antiviral innate immunity pathways. *Cell Res* 2006;**16**:141–7.
92. Yoneyama M, Kikuchi M, Matsumoto K, et al. Shared and unique functions of the DExD/H-box helicases RIG-I, MDA5, and LGP2 in antiviral innate immunity. *J Immunol* 2005;**175**:2851–8.
93. Yoneyama M, Kikuchi M, Natsukawa T, et al. The RNA helicase RIG-I has an essential function in double-stranded RNA-induced innate antiviral responses. *Nat Immunol* 2004;**5**:730–7.
94. Gack MU, Shin YC, Joo CH, et al. TRIM25 RING-finger E3 ubiquitin ligase is essential for RIG-I-mediated antiviral activity. *Nature* 2007;**446**:916–20.
95. Friedman CS, O'Donnell MA, Legarda-Addison D, et al. The tumour suppressor CYLD is a negative regulator of RIG-I-mediated antiviral response. *EMBO Rep* 2008.
96. Zhang M, Wu X, Lee AJ, et al. Regulation of IκB kinase-related kinases and antiviral responses by tumor suppressor CYLD. *J Biol Chem* 2008;**283**:18,621–626.
97. Kawai T, Takahashi K, Sato S, et al. IPS-1, an adaptor triggering RIG-I- and Mda5-mediated type I interferon induction. *Nat Immunol* 2005;**6**:981–8.
98. Meylan E, Curran J, Hofmann K, et al. Cardif is an adaptor protein in the RIG-I antiviral pathway and is targeted by hepatitis C virus. *Nature* 2005;**437**:1167–72.
99. Xu LG, Wang YY, Han KJ, Li LY, Zhai Z, Shu HB. VISA is an adapter protein required for virus-triggered IFN-β signaling. *Mol Cell* 2005;**19**:727–40.
100. Seth RB, Sun L, Ea CK, Chen ZJ. Identification and characterization of MAVS, a mitochondrial antiviral signaling protein that activates NFκB and IRF 3. *Cell* 2005;**122**:669–82.
101. McWhirter SM, Tenoever BR, Maniatis T. Connecting mitochondria and innate immunity. *Cell* 2005;**122**:645–7.
102. Saha SK, Cheng G. TRAF3: a new regulator of type I interferons. *Cell Cycle* 2006;**5**:804–7.
103. Saha SK, Pietras EM, He JQ, et al. Regulation of antiviral responses by a direct and specific interaction between TRAF3 and Cardif. *EMBO J* 2006;**25**:3257–63.
104. Kayagaki N, Phung Q, Chan S, et al. DUBA: a de-ubiquitinase that regulates type I interferon production. *Science* 2007;**318**:1628–32.

105. Huang TT, Wuerzberger-Davis SM, Wu ZH, Miyamoto S. Sequential modification of NEMO/IKKγ by SUMO-1 and ubiquitin mediates NFκB activation by genotoxic stress. *Cell* 2003;**115**:565–76.

106. Mabb AM, Wuerzberger-Davis SM, Miyamoto S. PIASy mediates NEMO sumoylation and NFκB activation in response to genotoxic stress. *Nat Cell Biol* 2006;**8**:986–93.

107. Wu ZH, Shi Y, Tibbetts RS, Miyamoto S. Molecular linkage between the kinase ATM and NFκB signaling in response to genotoxic stimuli. *Science* 2006;**311**:1141–6.

108. Hurley JH. ESCRT complexes and the biogenesis of multivesicular bodies. *Curr Opin Cell Biol* 2008;**20**:4–11.

109. Huang TT, D'Andrea AD. Regulation of DNA repair by ubiquitylation. *Nat Rev Mol Cell Biol* 2006;**7**:323–34.

110. Weake VM, Workman JL. Histone ubiquitination: triggering gene activity. *Mol Cell* 2008;**29**:653–63.

Chapter 10

Transcriptional Regulation via the cAMP Responsive Activator CREB

Paul K. Brindle

Department of Biochemistry, St. Jude Children's Research Hospital, Memphis, Tennessee

THE CREB FAMILY OF TRANSCRIPTION FACTORS

There are three principal cAMP responsive transcription factors in mammals, the cAMP responsive element binding protein (CREB), activating transcription factor (ATF-1), and cAMP responsive element modulator tau (CREMτ) [1–5]. These closely related proteins are encoded by the *Creb1*, *Atf1*, and *Crem* genes in mice (*CREB1, ATF1,* and *CREM* in humans). There are excellent reviews that detail the structural isoforms produced from these three genes [3, 4], and additional online information about gene structure and phenotype is available from Ensembl, NCBI HomoloGene, and Mouse Genome Informatics [6–8].

DOMAIN STRUCTURE AND FUNCTION

Mammalian CREB has four main functional domains: the basic/leucine zipper domain (bZIP), the Kinase Inducible Domain (KID), and two glutamine-rich domains (Q1, and Q2 or CAD) (Figure 10.1). CREB, ATF-1, and CREM belong to the bZIP superfamily of transcription factors and *Creb1* or *Crem* homologs that contain bZIP and KID, the distinguishing domains for the CREB family, are found in insects and nematodes, but not fungi (source: NCBI HomoloGene). Generally, the bZIP provides dimerization and DNA binding activities, but more recently the CREB bZIP has also been shown to confer transactivation function by binding the TORC coactivators [9] (Figure 10.1).

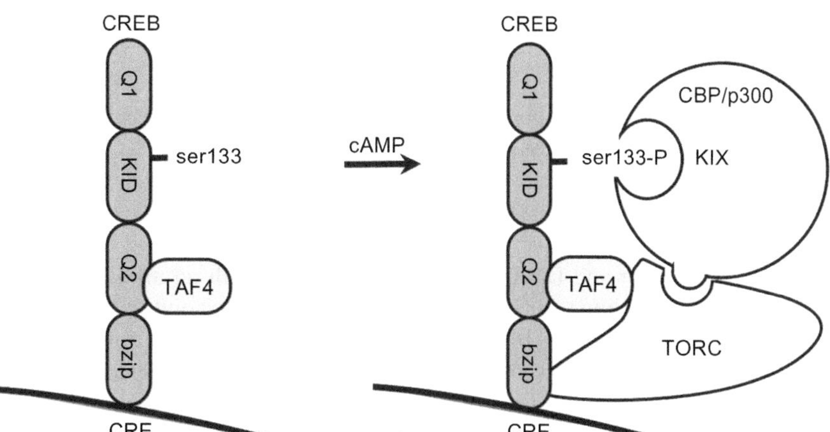

FIGURE 10.1 A schematic of CREB domain structure and the standard model of coactivator recruitment.
Under basal conditions CREB binds constitutively to DNA (heavy line) via a CRE sequence and interacts with the TFIID subunit TAF4 (other subunits omitted for clarity). Elevated intracellular cAMP actives protein kinase A (not shown) and leads to Ser133 phosphorylation (Ser133-P) and the shuttling of TORC into the nucleus (not shown). The KID domain of CREB binds the KIX domain of CBP/p300, and TORC binds the bZIP. Also indicated are additional protein–protein interactions between TAF4 and TORC [9], and CBP/p300 and TORC [17, 18]. The Q1, KID, Q2, and bZIP domains of CREB are shown.

Handbook of Cell Signaling, Three-Volume Set 2 ed.

The two glutamine-rich constitutive activation domains, Q1 and Q2, provide transactivation function in the absence of cAMP signaling and synergize with KID and bZIP to confer full activity to CREB in response to cAMP, however, ATF-1 lacks Q1, and Q2 is considered to be more critical for CREB function [4]. KID is the unique feature that distinguishes the activating isoforms of the CREB family from other bZIP transcription factors such as ATF-2 (CREB-2), ATF-3, ATF-4, Fos, and Jun. KID is an activation domain that is directly phosphorylated at CREB serine 133 (Ser133) by the cAMP dependent protein kinase A (PKA), an event that is necessary but not sufficient for the maximal transcriptional response to cAMP (Figure 10.1). About 300 different cell stimuli can lead to Ser133 phosphorylation [10], but CREB is typically not activated fully, if at all, except by cAMP and certain types of intracellular calcium signals [4].

Alternatively spliced CREM isoforms (α, β, γ, ε) that retain KID and bZIP but lack one or both of the Q domains, function as repressors or gene context dependent activators [11, 12]. The CREM ICER isoform retains only a bZIP domain and is alternatively transcribed in a CREB dependent manner from an intronic promoter. ICER is thought to function as a repressor that helps to attenuate cAMP responsive gene expression [13].

CREB is generally regarded as the predominant CREB family protein form in many cell types, and its expression in mice is more widespread than ATF-1 and especially CREM [14]. Indeed, *Creb1* null mice that lack all known functional isoforms die shortly after birth, but although *Atf1* null mice do not have an obvious phenotype, *Creb1* and *Atf1* double knockout embryos die before implantation, indicating that there is some functional redundancy between the two genes [14, 15]. *Crem* null mice survive but have defective spermiogenesis [16].

OVERVIEW OF CREB ACTIVATION

Unlike other classes of signal responsive transcription factors that are controlled by posttranslational modifications that affect nuclear localization (e.g., calcium regulated NFAT), dimerization and DNA binding (e.g., cytokine responsive STATs), and protein stability (e.g., hypoxia inducible factor, HIF), CREB is constitutively nuclear and DNA bound. In response to a signal (e.g., hormone binding to a G-protein coupled receptor), intracellular cAMP is produced by adenyl cyclase and cAMP binds to the two regulatory subunits of the tetrameric PKA holoenzyme in the cytoplasm, causing the disassociation and activation of the two catalytic subunits [4]. Following activation of PKA, CREB is phosphorylated on Ser133, and coactivators and the RNA polymerase II machinery are recruited to the target gene promoter (Figure 10.1). The coactivators CBP and p300 bind to Ser133 phosphorylated KID, whereas TORC coactivators shuttle to the nucleus from the cytoplasm in response to cAMP and bind to the bZIP domain. TORC and CBP/p300 also interact directly, which enhances their recruitment to CREB target genes [17, 18]. Thus, CREB belongs to a class of activators where signal responsive transactivation function is affected both directly (Ser133) and indirectly (TORC localization) by phosphorylation.

KEY PHOSPHORYLATION EVENTS

CREB target gene transcription in response to cAMP generally follows immediate-early (i.e., independent of new protein synthesis) and burst attenuation refractory kinetics (i.e., burst of transcription following phosphorylation of CREB and coactivator recruitment; attenuation of transcription due to protein phosphatase dependent dephosphorylation of CREB, and TORC degradation by the proteosome; refractory to further stimulation due to loss of the PKA catalytic subunit) [4, 19–21]. The rate of CREB Ser133 phosphorylation is limited by the diffusion of the catalytic subunit of PKA into the nucleus [22]. PKA also phosphorylates the salt inducible kinase (SIK) in the cytoplasm, inhibiting its ability to phosphorylate TORC2 at Ser171. TORC2 that lacks phospho-Ser171 is unable to bind its 14-3-3 protein anchor in the cytoplasm and enters the nucleus where it binds the CREB bZIP [23]. The calcium regulated phosphatase calcineurin also dephosphorylates TORC1 and TORC2, and promotes their nuclear translocation [23, 24]. Other sites on CREB that are phosphorylated can also affect its function. For example, phosphorylation of Ser142 in KID by casein kinase II or calcium/calmodulin dependent kinase II inhibits CREB recruitment of CBP and transactivation function [25, 26].

CREB TARGET GENES

There are an estimated 4,000 CREB target genes in humans based on bioinformatic analysis of full (TGACGTCA) and half-site (TGACG) CREs, but the majority are not induced by cAMP in a given cell type [27]. Functional CREs *in vivo* do not always match these particular full and half-site sequences, however, and CREB binds some CRE variants *in vitro* with an affinity comparable to the canonical CRE [28]. Most evolutionarily conserved CREs fall within 200 bps of the transcription start site [27]. In most instances, CREB is believed to bind constitutively to accessible CREs in the genome, and CREB modifying stimuli do not generally affect its nuclear localization, protein level, and DNA binding activity [4]. A regulated effect on binding DNA may occur under certain circumstances, such as for a subset of CRE sequences, in response to nitric oxide, and in cooperative binding to a viral CRE with the HTLV-1 protein Tax [3, 4, 29].

The pattern of CREB target gene expression varies between tissues, an effect partly related to the ability of CREB to bind the CREs of specific genes in different cell types [30]. Cytosine methylation of the central CpG dinucleotide of the CRE sequence is known to block CREB binding [4], and it is possible that tissue specific facultative heterochromatin blocks CREB access as well.

Recruitment of CBP to a CREB target gene correlates better with gene expression than does Ser133 phosphorylation, again indicating that CREB DNA binding and phosphorylation is generally insufficient to induce target gene expression [27]. Indeed, the interaction between CBP/p300 and TORC appears to be important in determining which genes respond to different stimuli that result in Ser133 phosphorylation (e.g., cAMP vs. stress pathways) [17]. In this light, individual CREB target genes dictate, by unknown means, the usage of distinct coactivation mechanisms required for a cAMP response [18]. This further suggests that the presence of coactivators at a promoter *per se* does not indicate their functional relevance for that gene [18]. Thus, CBP/p300 and TORC appear to have distinct biochemical roles that are not equally necessary for every CREB target gene, and which may contribute to signal and tissue dependent gene expression patterns.

CBP AND P300

The genes for CBP (*Crebbp*) and p300 (E1A binding protein p300, *Ep300*) encode highly related histone acetyltransferases that possess several conserved protein binding domains that bind a variety of transcriptional regulators and other proteins [31]. CBP/p300 are among the most highly connected proteins in the known mammalian protein interactome, with more than 340 described interaction partners [32]. Consistent with extensive roles in gene regulation, null mutations of *Crebbp* or *Ep300*, or mutations in their acetyltransferase domain, are lethal in mice [33–36]. The critical downstream functions (histone acetylation, protein adaptor function, etc.) of CBP/p300 that are required for CREB target gene transcription remain unclear, however.

Mice with point mutations in either CBP or p300 that block CREB binding to KIX are viable, although $Ep300^{KIX/KIX}$ mutant mice have hematopoietic abnormalities that appear to be largely due to disruption of c-Myb function, a hematopoietic transcription factor unrelated to CREB [37]. $Crebbp^{KIX/KIX}$ mice are mostly normal but have modest thymic hypoplasia and defects in learning and memory [37–40]. CBP KIX therefore appears to be limiting for important CREB functions in the brain, but another study showed it was mostly dispensable for CREB dependent gluconeogenic gene expression in the liver [41]. Mutation of Ser436 in the CBP CH1 domain, a serine that is not present in p300, increases CREB activity and liver gluconeogenesis in mice, suggesting that this residue negatively controls the interaction between CREB and CBP [42, 43]. Finally, mouse embryonic fibroblasts that have point mutations in the KIX domains of both CBP and p300 show that individual CREB target genes differ in their reliance on KIX [18]. Moreover, CBP/p300 recruitment in response to cAMP is only partially attenuated by this KIX mutation; a direct interaction between TORC and CBP/p300 appears to partially compensate for the KIX defect (Figure 10.1) [18]. These studies together indicate that the requirements for CBP/p300 are complex and vary between individual CREB target genes and possibly between tissues.

TORC

The three TORC family members, TORC1, TORC2, and TORC3 (or CREB regulated transcription coactivator, *Crtc1*, *Crtc2*, *Crtc3*) are unrelated to CBP and p300 [9, 44]. TORCs 1 and 2 are sequestered in the cytoplasm until activated by cAMP or calcium, whereupon they transit to the nucleus; TORC3 is constitutively nuclear [23]. They possess an N-terminal coiled-coil domain that binds the CREB bZIP, and they have a potent transactivation domain, but it is unknown if they possess enzymatic activities. TORCs appear to enhance the interaction of CREB with TFIID and CBP/p300 (Figure 10.1), which may account for their coactivation function [9, 17, 18].

TORC2 is an important regulator of liver gluconeogenesis during fasting [41]. The counter regulatory pancreatic hormones glucagon and insulin control liver glucose production at least in part by controlling TORC. During fasting conditions, glucagon induces cAMP levels in the liver, resulting in TORC translocation to the nucleus where it contributes to the expression of CREB target genes with roles in gluconeogenesis. In the fed state when liver gluconeogenesis is reduced, insulin induces SIK2 kinase activity, resulting in TORC2 phosphorylation at Ser171 and its cytoplasmic translocation whereupon it is degraded by the ubiquitin dependent 26S proteasome [21]. Metformin (Glucophage) is a widely prescribed drug for Type 2 diabetes that may reduce liver glucose output by inhibiting the TORC pathway [45].

OTHER COACTIVATORS AND INTERACTING PROTEINS

Besides its well studied coactivators, CBP/p300 and the TORCs, CREB has more than 25 described protein interaction partners [10]. One of the most important is TFIID, which binds constitutively via its TAF4 (TAFII130, TAFII135) subunit to the CREB glutamine-rich Q2 domain. Consistent with this, TAF4 (*Taf4a*) null mouse cells are deficient for CREB transactivation function [46]. Interestingly, a tissue specific coactivator, ACT, has been shown to coactivate CREB and CREM by binding KID independently of Ser133 phosphorylation and CBP [47]. In addition, the inhibition of some CREB target genes by the histone deacetylase inhibitor trichostatin A (TSA) indicates that deacetylases may also provide CREB coactivation function [48]. Other CBP/p300 dependent activators that do not bind KIX, such as hypoxia inducible factor (HIF), are also sensitive to TSA, suggesting that histone deacetylases may have broad roles both as coactivators and corepressors [49, 50].

QUESTIONS TO BE ADDRESSED

The CREB pathway is a paradigm for signal responsive transcription, yet our understanding of events downstream of CREB that lead to gene activation remains unclear. For instance, what are the detailed mechanisms by which CREB and its coactivators stimulate RNA Pol II activity? Why do individual CREB target genes in a single cell type recruit the same group of coactivators but require them differentially? And what role do coactivators, transcription factors, and epigenetics play in signal and tissue specific CREB target gene expression?

ACKNOWLEDGEMENTS

I apologize to those whose work was not discussed or cited directly because of space limitations. I would like to thank Lawryn Kasper for helpful comments on the manuscript. Support was provided by NIH Cancer Center (CORE) support grant P30 CA021765, and the American Lebanese Syrian Associated Charities of St. Jude Children's Research Hospital.

REFERENCES

1. De Cesare D, Sassone-Corsi P. Transcriptional regulation by cyclic AMP-responsive factors. *Prog Nucleic Acid Res Mol Biol* 2000;**64**:343–69.
2. Lonze BE, Ginty DD. Function and regulation of CREB family transcription factors in the nervous system. *Neuron* 2002;**35**:605–23.
3. Shaywitz AJ, Greenberg ME. CREB: a stimulus-induced transcription factor activated by a diverse array of extracellular signals. *Annu Rev Biochem* 1999;**68**:821–61.
4. Mayr B, Montminy M. Transcriptional regulation by the phosphorylation-dependent factor CREB. *Nat Rev Mol Cell Biol* 2001;**2**:599–609.
5. Quinn PG. Mechanisms of basal and kinase-inducible transcription activation by CREB. *Prog Nucleic Acid Res Mol Biol* 2002;**72**: 269–305.
6. Hubbard TJ, Aken BL, Beal K, et al. Ensembl 2007. *Nucleic Acids Res* 2007;**35**:D610–17.
7. Wheeler DL, Barrett T, Benson DA, et al. Database resources of the national center for biotechnology information. *Nucleic Acids Res* 2007;**35**:D5–D12.
8. Eppig JT, Blake JA, Bult CJ, et al. The mouse genome database (MGD): new features facilitating a model system. *Nucleic Acids Res* 2007;**35**:D630–7.
9. Conkright MD, Canettieri G, Screaton R, et al. TORCs: transducers of regulated CREB activity. *Mol Cell* 2003;**12**:413–23.
10. Johannessen M, Delghandi MP, Moens U. What turns CREB on?. *Cell Signal* 2004;**16**:1211–27.
11. Foulkes NS, Borrelli E, Sassone-Corsi P. CREM gene: use of alternative DNA-binding domains generates multiple antagonists of cAMP-induced transcription. *Cell* 1991;**64**:739–49.
12. Brindle P, Linke S, Montminy M. Protein-kinase-A-dependent activator in transcription factor CREB reveals new role for CREM repressors. *Nature* 1993;**364**:821–4.
13. Molina CA, Foulkes NS, Lalli E, Sassone-Corsi P. Inducibility and negative autoregulation of CREM: an alternative promoter directs the expression of ICER, an early response repressor. *Cell* 1993;**75**: 875–86.
14. Bleckmann SC, Blendy JA, Rudolph D, et al. Activating transcription factor 1 and CREB are important for cell survival during early mouse development. *Mol Cell Biol* 2002;**22**:1919–25.
15. Rudolph D, Tafuri A, Gass P, et al. Impaired fetal T cell development and perinatal lethality in mice lacking the cAMP response element binding protein. *Proc Natl Acad Sci U S A* 1998;**95**:4481–6.
16. Nantel F, Monaco L, Foulkes NS, et al. Spermiogenesis deficiency and germ-cell apoptosis in CREM-mutant mice. *Nature* 1996;**380**:159–62.
17. Ravnskjaer K, Kester H, Liu Y, et al. Cooperative interactions between CBP and TORC2 confer selectivity to CREB target gene expression. *Embo J* 2007;**26**:2880–9.
18. Xu W, Kasper LH, Lerach S, Jeevan T, Brindle PK. Individual CREB-target genes dictate usage of distinct cAMP-responsive coactivation mechanisms. *Embo J* 2007;**26**:2890–903.
19. Armstrong R, Wen W, Meinkoth J, et al. A refractory phase in cyclic AMP-responsive transcription requires down regulation of protein kinase A. *Mol Cell Biol* 1995;**15**:1826–32.
20. Hagiwara M, Alberts A, Brindle P, et al. Transcriptional attenuation following cAMP induction requires PP-1-mediated dephosphorylation of CREB. *Cell* 1992;**70**:105–13.
21. Dentin R, Liu Y, Koo SH, et al. Insulin modulates gluconeogenesis by inhibition of the coactivator TORC2. *Nature* 2007;**449**:366–9.
22. Hagiwara M, Brindle P, Harootunian A, et al. Coupling of hormonal stimulation and transcription via the cyclic AMP-responsive factor CREB is rated limited by nuclear entry of protein kinase A. *Mol Cell Biol* 1993;**13**:4852–9.
23. Screaton RA, Conkright MD, Katoh Y, et al. The CREB coactivator TORC2 functions as a calcium- and cAMP-sensitive coincidence detector. *Cell* 2004;**119**:61–74.
24. Bittinger MA, McWhinnie E, Meltzer J, et al. Activation of cAMP response element-mediated gene expression by regulated nuclear transport of TORC proteins. *Curr Biol* 2004;**14**:2156–61.
25. Parker D, Jhala US, Radhakrishnan I, et al. Analysis of an activator–coactivator complex reveals an essential role for secondary structure in transcriptional activation. *Mol Cell* 1998;**2**:353–9.
26. Sun P, Enslen H, Myung P, Maurer RA. Differential activation of CREB by Ca^{2+} /calmodulin-dependent protein kinases type II and type IV involves phosphorylation of a site that negatively regulates activity. *Gene Dev* 1994;**8**:2527–39.
27. Zhang X, Odom DT, Koo SH, et al. Genome-wide analysis of cAMP-response element binding protein occupancy, phosphorylation, and target gene activation in human tissues. *Proc Natl Acad Sci U S A* 2005;**102**:4459–64.
28. Benbrook DM, Jones NC. Different binding specificities and transactivation of variant CREs by CREB complexes. *Nucleic Acids Res* 1994;**22**:1463–9.

29. Riccio A, Alvania RS, Lonze BE, et al. A nitric oxide signaling pathway controls CREB-mediated gene expression in neurons. *Mol Cell* 2006;**21**:283–94.
30. Cha-Molstad H, Keller DM, Yochum GS, et al. Cell-type-specific binding of the transcription factor CREB to the cAMP-response element. *Proc Natl Acad Sci U S A* 2004;**101**:13,572–13,577,.
31. Goodman RH, Smolik S. CBP/p300 in cell growth, transformation, and development. *Genes Dev* 2000;**14**:1553–77.
32. Kasper LH, Fukuyama T, Biesen MA, et al. Conditional knockout mice reveal distinct functions for the global transcriptional coactivators CBP and p300 in T-cell development. *Mol Cell Biol* 2006;**26**:789–809.
33. Kung AL, Rebel VI, Bronson RT, et al. Gene dose-dependent control of hematopoiesis and hematologic tumor suppression by CBP. *Genes Dev* 2000;**14**:272–7.
34. Oike Y, Takakura N, Hata A, et al. Mice homozygous for a truncated form of CREB-binding protein exhibit defects in hematopoiesis and vasculo-angiogenesis. *Blood* 1999;**93**:2771–9.
35. Yao TP, Oh SP, Fuchs M, et al. Gene dosage-dependent embryonic development and proliferation defects in mice lacking the transcriptional integrator p300. *Cell* 1998;**93**:361–72.
36. Shikama N, Lutz W, Kretzschmar R, et al. Essential function of p300 acetyltransferase activity in heart, lung and small intestine formation. *Embo J* 2003;**22**:5175–85.
37. Kasper LH, Boussouar F, Ney PA, et al. A transcription-factor-binding surface of coactivator p300 is required for haematopoiesis. *Nature* 2002;**419**:738–43.
38. Oliveira AM, Abel T, Brindle PK, Wood MA. Differential role for CBP and p300 CREB-binding domain in motor skill learning. *Behav Neurosci* 2006;**120**:724–9.
39. Wood MA, Attner MA, Oliveira AM, et al. A transcription factor-binding domain of the coactivator CBP is essential for long-term memory and the expression of specific target genes. *Learn Mem* 2006;**13**:609–17.
40. Vecsey CG, Hawk JD, Lattal KM, et al. Histone deacetylase inhibitors enhance memory and synaptic plasticity via CREB: CBP-dependent transcriptional activation. *J Neurosci* 2007;**27**:6128–40.
41. Koo SH, Flechner L, Qi L, et al. The CREB coactivator TORC2 is a key regulator of fasting glucose metabolism. *Nature* 2005;**437**: 1109–14.
42. Zhou XY, Shibusawa N, Naik K, et al. Insulin regulation of hepatic gluconeogenesis through phosphorylation of CREB-binding protein. *Nat Med* 2004;**10**:633–7.
43. Hussain MA, Porras DL, Rowe MH, et al. Increased pancreatic beta-cell proliferation mediated by CREB binding protein gene activation. *Mol Cell Biol* 2006;**26**:7747–59.
44. Iourgenko V, Zhang W, Mickanin C, et al. Identification of a family of cAMP response element-binding protein coactivators by genome-scale functional analysis in mammalian cells. *Proc Natl Acad Sci U S A* 2003;**100**:12,147–12,152,.
45. Shaw RJ, Lamia KA, Vasquez D, et al. The kinase LKB1 mediates glucose homeostasis in liver and therapeutic effects of metformin. *Science* 2005;**310**:1642–6.
46. Mengus G, Fadloun A, Kobi D, et al. TAF4 inactivation in embryonic fibroblasts activates TGF beta signalling and autocrine growth. *Embo J* 2005;**24**:2753–67.
47. Fimia GM, De Cesare D, Sassone-Corsi P. CBP-independent activation of CREM and CREB by the LIM-only protein ACT. *Nature* 1999;**398**:165–9.
48. Fass DM, Butler JE, Goodman RH. Deacetylase activity is required for cAMP activation of a subset of CREB target genes. *J Biol Chem* 2003;**278**:43,014–43,019,.
49. Kasper LH, Brindle PK. Mammalian gene expression program resiliency: the roles of multiple coactivator mechanisms in hypoxia-responsive transcription. *Cell Cycle* 2006;**5**:142–6.
50. Kasper LH, Boussouar F, Boyd K, et al. Two transactivation mechanisms cooperate for the bulk of HIF-1-responsive gene expression. *Embo J* 2005;**24**:3846–58.

Chapter 11

The NFAT Family: Structure, Regulation, and Biological Functions

Fernando Macian[1], Fernando Cruz-Guilloty[2], Sonia Sharma[2] and Anjana Rao[2]

[1]*Department of Pathology, Albert Einstein College of Medicine, Bronx, New York*

[2]*Department of Pathology, Harvard Medical School and the Immune Disease Institute, Boston, Massachusetts*

INTRODUCTION

NFAT is a small but important family of transcription factors originally described in T cells [1–7]. It is now well established that NFAT proteins direct specific biological programs in a variety of cells and tissues. The NFAT family consists of five members. The primordial family member NFAT5 (TonEBP) is found in *Drosophila* while the genes encoding the four calcium regulated NFAT proteins, NFAT1–4 (also known as NFATc1–c4), appear to have emerged simultaneously early in the course of vertebrate evolution. The five NFAT proteins are classified into one family based on the sequence and structural similarity of their DNA binding domains: the degree of sequence identity is ~60–70 percent when NFAT1–4 are compared among themselves and ~40–50 percent when NFAT1–4 are compared with NFAT5.

The primordial NFAT family member NFAT5/TonEBP is expressed ubiquitously in mammalian cells and regulates the response to hypertonic stress [4, 8]. NFAT5 is also likely to be involved in regulating diverse other biological programs. Genetic analysis in *Drosophila* suggests that NFAT5 plays a role in the Ras signal transduction pathway, and this function may be conserved in evolution [9]. NFAT5 protein levels are increased by antigen receptor stimulation of T cells, suggesting a role for this protein in lymphocyte responses downstream of the TCR [10, 11]. NFAT5 is also activated by $\alpha 6 \beta 4$ integrin stimulation of carcinoma cells, suggesting a role in tumor metastasis [12].

Except for NFAT2, which is the only family member expressed at a specific stage of development of cardiac valves [13, 14], one or more calcium regulated NFAT proteins are expressed redundantly in many cell types of the embryo and the adult [3, 15]. Three structural features are common to all four calcium regulated NFAT proteins: an N-terminal transactivation domain, a highly phosphorylated regulatory domain that binds and is dephosphorylated by the calcium/calmodulin regulated phosphatase calcineurin, and a DNA binding domain that is monomeric in solution, forms dimers on certain κB-like DNA elements, and interacts with Fos and Jun proteins on "composite" NFAT:AP-1 DNA elements [1–5].

STRUCTURE AND DNA BINDING

The NFAT family is evolutionarily related to the Rel/NFκB family [6, 10, 16, 17]. The level of sequence identity between these two families is marginal (~17 percent) but the structural similarity is remarkable. The DNA binding domains of all NFAT and NFκB/Rel family members have two domains, an N-terminal specificity domain (~180 amino acids) involved in making base specific DNA contacts, and a C-terminal domain (~100 amino acids) involved in dimer formation [16, 18, 19]. Together these domains constitute the Rel homology region (RHR) common to all members of the extended NFAT/NFκB/Rel family. Consistent with this level of structural homology, NFAT proteins can also function as dimeric transcription factors at quasi-palindromic sites that resemble NFκB binding sites [20, 21]. Comparison of the structures of NFAT5/TonEBP and NFκB p50 dimers on DNA illustrates the striking structural similarity between the NFAT and NFκB families [17–19]. The C-terminal domains of NFAT5 and NFκB p50 utilize a similar interface to form the dimer contacts, but NFAT5 has an additional surface for dimer formation, which involves the N-terminal domain [17–19]. The structure of NFAT1 homodimers bound to κB-like sites, which contain two core NFAT sites separated by one or two nucleotides, in the *IL-8* promoter and the HIV-1 long terminal repeat have also been

Handbook of Cell Signaling, Three-Volume Set 2 ed.

described and closely resemble those of Rel dimer [20, 21]. The existence of a flexible linker region between the N- and C-terminal regions of the RHR domain, makes NFAT1 dimers more flexible than those formed by Rel proteins or NFAT5, which always act as homodimers [17, 22]. This flexible linker permits the RHR-C to acquire different conformations providing, thus, a dynamic surface for protein–protein interactions [22]. In part, this ability is possible because NFAT1 is monomeric in solution and forms complexes with partners only when it is bound to DNA [6]. The flexibility of the NFAT1 monomer bound to DNA should facilitate the formation of higher order NFAT transcriptional complexes [22]. Many of the dimer interface residues observed in the NFAT1 dimer are also conserved in NFAT2, NFAT3, and NFAT4, which suggest that these NFAT family members may also form dimers on κB-like sites and even opens the possibility of the formation of heterodimers containing two different NFAT proteins.

REGULATION

Activation by Receptors Coupled to Calcium Entry

The calcium regulated NFAT proteins (NFAT1–4) are activated by ligand binding to a variety of cell surface receptors [1–3, 7, 15]. The common feature of the receptors is their ability to activate phosphatidylinositol specific phospholipase C (PLC), thereby inducing calcium influx across the plasma membrane. In the immune system, the ability of immunoreceptors (T and B cell antigen receptors, Fcε receptors on mast cells, and Fcγ receptors on NK cells and monocytes) to activate NFAT is well documented [1, 4–7]. Stimulation of immune cells through these receptors activates a cascade of several tyrosine kinases, leading to tyrosine phosphorylation and activation of PLC-γ. In other cell types, NFAT activation has been shown to result from stimulation of G-protein coupled receptors leading to PLCβ activation, or stimulation of receptor tyrosine kinases leading to PLCγ activation. This leads to PIP2 hydrolysis and generation of IP3, which by binding to the IP3 receptor and depleting intracellular ER (endoplasmic reticulum) calcium stores, initiates the process of store operated calcium entry through the plasma membrane [23, 24].

NFAT dependent gene transcription is exquisitely sensitive to changes in intracellular calcium concentration ($[Ca^{2+}]_I$). Even in the continuous presence of stimulus, $[Ca^{2+}]_I$ levels may oscillate depending on specific parameters of receptor occupancy and desensitization [23]. $[Ca^{2+}]_I$ levels are thus modulated at two levels: amplitude and oscillation frequency, and NFAT activation is sensitive to both types of modulation [25–27]. This is well illustrated by T cells from two severe combined immunodeficiency patients with a primary defect in store operated calcium entry [28]. When activated, these cells show a small $[Ca^{2+}]_I$ spike resulting from store depletion but they lack the ability to sustain increased $[Ca^{2+}]_I$ levels for several hours as observed in wild-type T cells. The brief increase in $[Ca^{2+}]_I$ suffices for transient dephosphorylation and nuclear localization of NFAT, manifested by transcription of a small number of NFAT target genes [28, 29]. However, optimal activation of NFAT dependent genes requires sustained calcium/calcineurin signals [28, 29], which are most effectively elicited by slow $[Ca^{2+}]_I$ oscillations or low sustained $[Ca^{2+}]_I$ increases [25–27]. The differing sensitivities of NFAT target genes to intracellular free calcium levels are likely to reflect the multiple configurations of NFAT sites in gene regulatory regions: for instance the requirement for sustained activation could be either because NFAT itself needs to remain within a transcription complex for many hours, or because formation of a cooperative NFAT:AP-1 complex is important for gene transcription (*de novo* synthesis of Fos proteins is required for optimal AP-1 activation).

Identification of the Store Operated Calcium Entry Pathway

The major pathway for sustained intracellular calcium signaling in lymphocytes involves the calcium release activated calcium (CRAC) channels, and the interplay between CRAC channels and NFAT activation has been well documented [30, 31]. In fact, this interplay was used to identify the elusive CRAC channel, whose existence had been postulated for almost 20 years. The electrophysiological characteristics of the CRAC channel had been well studied but the molecular identity of the channel had remained unknown. The channel was finally identified almost simultaneously by three groups as a result of a major technological advance, the development of RNAi screens in *Drosophila* cells [32–34]. All three screens took advantage of the fact that *Drosophila* cells had previously been shown to possess a functional CRAC channel [34]. Two of the screens utilized high throughput assessment of store operated calcium entry [33, 34], whereas the third took advantage of the fact that calcium entry activates calcium regulated NFAT proteins and screened for candidates whose RNAi mediated depletion led to loss of nuclear translocation of an NFAT-GFP fusion protein [32]. All three screens identified the ER calcium sensor Stim, as well as a poorly annotated gene, *olf-186F*, encoding a protein that was later renamed Orai.

In addition to the RNAi screen, a separate genome-wide screening approach identified a point mutation in the human homolog ORAI1 as the underlying genetic defect in a family exhibiting severe combined immunodeficiency (SCID) secondary to loss of store operated calcium entry and downstream calcineurin/NFAT activation [32]. Genome-wide linkage analysis using single nucleotide polymorphism arrays for 23 individuals from the affected family identified a region on chromosome 12q24 that was strongly linked to the SCID gene defect. This region, encompassing approximately 6.5 Mb of genomic sequence, contained the gene for *tmem142a (orai1)*, the human paralog of *Drosophila orai* [32]. SCID patients are homozygous for a single C >T transition in the *ORAI1* gene, which causes a missense mutation of arginine-91 to tryptophan. ORAI1 is predicted to contain four transmembrane domains, and arginine-91 lies at the very beginning of the first transmembrane domain. Overexpression of wild-type ORAI1 in SCID T cells restored store operated calcium influx and the associated CRAC current (I)CRAC. Mutation of two conserved glutamates in the predicted first and third transmembrane portions of ORAI1 altered the electrophysiological characteristics of (I)CRAC, including the degree of calcium influx and ion selectivity, providing strong evidence that ORAI1 is a pore forming subunit of the CRAC channel [35].

Interaction of NFAT with Calcineurin and NFAT Kinases

Interaction with calcineurin is central to the calcium responsiveness of NFAT1–4. The major calcineurin docking site on NFAT is located at the N-terminus of the regulatory domain and has the consensus sequence PxIxIT [36]. Substitution of the PxIxIT sequence of NFAT1 with a higher affinity version obtained by peptide selection increases the basal $[Ca^{2+}]_I$ sensitivity of NFAT [37]. The principal calcineurin–NFAT contact between the PxIxIT motive in NFAT and the catalytic subunit of calcineurin has been mapped by photocrosslinking and the introduction of point mutations in calcineurin, and recently confirmed with the crystallographic structure of calcineurin complexed with a VIVIT peptide [38, 39]. This interaction requires only a PxIxIT sequence in the calcineurin substrate, and variations within this sequence easily modulate the affinity of this interaction, allowing calcineurin to adapt to and fine-tune its activity on a wide variety of substrates [38, 39]. In fact, recent studies analyzing the affinity of several calcineurin substrates in yeast have shown that calcineurin binds to them with varying affinities. The PxIxIT motif is critical in determining the binding affinities, which in turn define the calcium concentration dependence of the interaction and, therefore, the output signaling [40]. The surface of the NFAT–calcineurin interaction may not be limited to the PxIxIT motif, however: a second interacting region is present at the C-terminus of the regulatory domain [41, 42].

By mass spectrometric analysis, NFAT1 isolated from resting cells was shown to contain at least 21 phosphoserine residues, of which 14 were located in characteristic conserved sequence motifs in the regulatory domain [43]. Following cell stimulation, 13 of these residues were dephosphorylated by calcineurin. Dephosphorylation of five phosphoserines in one of these conserved motifs (SRR-1) sufficed for exposure of a nuclear localization signal in the regulatory domain, while dephosphorylation of all 13 residues was required for masking of a nuclear export sequence, complete nuclear localization, and full transcriptional activity [43]. In NFAT2, dephosphorylation of the conserved SPxx motifs has been correlated with increased DNA binding [44]. It is interesting that the gene for an endogenous calcineurin inhibitor, MCIP1, is itself an NFAT target, implying the existence of a negative feedback loop that downmodulates NFAT activity [45, 46].

When calcium entry is prevented or calcineurin activity is inhibited in stimulated cells, NFAT is rephosphorylated by NFAT kinases and rapidly leaves the nucleus. Multiple constitutively active kinases regulate the distinct serine motifs of NFAT. Classical biochemical fractionation and protein chromatography techniques demonstrated that casein kinase 1 (CK1) targets only the NFAT1 SRR-1 motif, the region that primarily controls nuclear import [47]. Glycogen synthase kinase 3 (GSK3), which does not phosphorylate SRR-1, specifically targets the SP-2 motif of NFAT and synergizes with CK1 to promote NFAT nuclear export. Phosphorylation of the SP-2 motif by GSK3 requires a priming phosphorylation, which can be achieved *in vitro* by protein kinase A (PKA) [48, 49]. To identify additional kinase regulators of NFAT import, a genome-wide RNA interference screen was performed in *Drosophila* cells, which contain the signaling pathways that regulate NFAT subcellular localization [50]. Genetic screening in *Drosophila* identified DYRK family kinases as negative regulators of NFAT through phosphorylation of the SP-3 motif, which primes for further phosphorylation by CK1 and GSK3 in the absence of PKA phosphorylation. Both DYRK and the endogenous calcineurin inhibitor MCIP1 lie within the Down's syndrome critical region of human chromosome 21, and synergize to prevent NFAT nuclear accumulation under conditions of overexpression [51]. Increased dosage of DYRK and MCIP1 in the context of chromosome 21 trisomy may reduce NFAT activity, and contribute to specific developmental defects associated with Down's syndrome such as immunodeficiency, placental vascular abnormalities, and heart disease [51].

The nuclear import and trafficking of NFAT is under further regulation through a previously unsuspected mechanism: a large scale functional analysis of 512 mouse noncoding RNAs (ncRNA) with homology to human genome sequences was performed using an RNA interference approach. Specific depletion of one, termed ncRNA repressor of NFAT (NRON) enhanced NFAT driven luciferase reporter activity in several human cell lines. Affinity purification followed by mass spectrometry analysis of NRON interacting protein identified multiple proteins, including nuclear transport factors, such as importin beta and importin alpha recyclers and scaffold/regulatory proteins such as IQGAP1 in the NRON complex that negatively regulates activation induced NFAT nuclear import [52].

With known protein components of the upstream signaling events leading to calcineurin and NFAT activation, it should be possible to dissect the complex networks required for NFAT function and how they affect each individual NFAT family member under different conditions and cell types.

TRANSCRIPTIONAL FUNCTIONS

All NFAT proteins contain intrinsic transactivation domains, and thus are *bona fide* transcription factors that can independently induce gene transcription. The large C-terminal region of NFAT5 contains a transactivation domain that functions in reporter assays when fused to the GAL4 DNA binding domain [10]. By the same criterion, NFAT1 and NFAT4 possess two transactivation domains, at their N- and C-termini respectively [53, 54]: both are strongly acidic and contain a DELDF[S/K] sequence that may serve as a docking site for the SAGA histone acetyltransferase complex [55]. Similar motifs are found in the corresponding regions of NFAT3 and the constitutive isoforms of NFAT2. The transactivation domain of NFAT2 has surprisingly been mapped to the beginning of its regulatory domain, overlapping a sequence utilized as the major docking site for calcineurin [56]. Constitutively active NFAT1 and NFAT2, which mimic the corresponding dephosphorylated proteins, have been shown to induce or potentiate target gene expression in resting cells [43, 57, 58].

Mass spectrometric analysis indicated that in stimulated cells, NFAT1 becomes phosphorylated on a serine residue in its transactivation domain [43, 59]. It remains to be determined if this modification is necessary for transcriptional function in the context of the full length protein. Pim1, Cot, and protein kinase C zeta have been suggested as candidate kinases, since when overexpressed, they enhance reporter activity dependent on the NFAT transactivation domain [60–62]. However, as previously suggested in a similar study using calmodulin dependent kinase II, such enhancement could arise from indirect effects on coactivator function rather than through direct NFAT phosphorylation [59].

As we have previously described, the RHR can adopt different structural conformations on DNA binding sites, which allows NFAT to interact with many different transcriptional partners. NFAT–Fos–Jun cooperation constitutes a major mechanism of NFAT dependent gene transcription. While the residues required for Fos–Jun contact are almost completely conserved in the DNA binding domains of all four calcium regulated NFAT proteins [63], they are absent from NFAT5 suggesting that the ability to cooperate with Fos and Jun was a relatively late evolutionary development [64]. NFAT has also been shown to cooperate functionally with many other families of transcription factors including GATA [65, 66], MEF2 [67], Maf [68], IRF4 [69], C/EBP [70], Oct [71], Egr [72], and Foxp3 [73]. Unlike the NFAT:AP-1 interaction, which cannot be detected in solution, some of these interactions appear to be mediated by direct protein–protein contact and are not cooperative on DNA. In at least one case (the NFAT–GATA interaction), the surface of interaction of these other transcription factors with NFAT is known not to overlap the surface of NFAT–AP-1 interaction: a mutant NFAT1 protein engineered to lack completely the ability to cooperate with AP-1 [74] was as or more effective than wild-type NFAT1 in its ability to synergize functionally with GATA3 in a transient reporter assay in T cells [75]. Transcriptional repressors, such as ICER [76], PPARγ [77], and p21^{snft} [78] have also been shown to interact with NFAT and inhibit its transcriptional activity.

The shortest isoform of NFAT2 (NFATc/A) is induced in a CsA sensitive manner by NFAT itself, in a process suggested to constitute a positive autoregulatory loop [46]. This protein is generated through utilization of a distinct inducible promoter that is preferentially coupled to the most proximal polyadenylation site [79]. As a result, NFATc/A lacks the entire C-terminal domain and contains an alternate N-terminal domain that is not highly acidic and lacks the SAGA interaction sequence DELDF[S/K]. It is not clear whether this protein would be transcriptionally active in the absence of partner proteins such as AP-1.

BIOLOGICAL PROGRAMS REGULATED BY NFAT

Originally described as a key regulator of T cell activation, the calcium regulated NFAT proteins have been implicated in a variety of gene expression programs, not only in cells of the immune system but also in many other cell types and tissues. These include cardiac hypertrophy, slow and fast twitch fiber differentiation, cardiac valve development, vascular patterning during embryogenesis, chondrocyte development, and adipose differentiation, among others.

Cooperation of NFAT proteins with members of the AP-1 family of transcription factors is known to play an important role in the establishment of a productive immune response [4, 7]. Activation of both transcription factors is achieved upon full stimulation of B and T cells through their antigen receptors and costimulatory molecules. Composite sites for NFAT and AP-1 have been described in promoters and enhancers of many cytokine genes [1]. Paradoxically, NFAT without AP-1 induces an unresponsive or tolerant state in lymphocytes [57]. This response is induced in T cells by TCR activation without engagement of costimulatory receptors. This results in increased $[Ca^{2+}]_I$ levels with minimal MAP kinase or IκB kinase activation, thereby leading to NFAT activation without appreciable AP-1 or NFκB activation. Under these conditions NFAT initiates a distinct gene expression profile associated with a profound block in TCR signaling [57]. Thus, NFAT proteins play a central role in the control of two opposite aspects of T cell function by activating two different genetic programs depending on the presence or absence of AP-1 cooperation.

Besides being a key regulator of mature T cell function, NFAT is also required in thymocyte development and T helper cell differentiation [7]. Studies performed on NFAT or calcineurin deficient mice have indicated that calcineurin/NFAT signaling regulates thymocyte proliferation and survival as well as the development of immature CD4-CD8-thymocytes into mature positively selected CD4+ or CD8+ T cells [80–82]. Cooperation between NFAT and STAT proteins is necessary for the expression of the T helper lineage specific transcription factors T-bet (Th1) and GATA-3 (Th2). NFAT is further required to cooperate with those transcription factors that, by recruiting histone modifying enzymes, induce the epigenetic changes that determine the specific pattern of cytokine expression in those two T helper cell populations [75, 83–85].

Another role for NFAT transcription factors has emerged in the response of memory T cells to antigen stimulation [86]. Naïve and memory CD4+ T cells differ in their cytokine expression profiles following TCR engagement, with memory T cells being capable of rapid cytokine (IL-2) production. This correlates with accumulated levels of NFAT1 and NFAT2 in memory cells, a previously unknown mechanism for immediate secondary protection against pathogenic challenge. The specific NFAT family members and/or isoforms that mediate memory T cell generation, maintenance, and function remain to be determined in detail, especially in the context of primary differentiation of CD4+ T cells into Th1, Th2, or Th17 cells, and CD8+ T cell differentiation into effector memory or central memory cells.

Recent evidence indicates that NFAT1/Foxp3 complexes are crucial for regulatory T cell (Treg) function. Cooperation of these two factors in Tregs is essential for their suppressor activity and is responsible for the downregulation of IL-2 transcription in these cells as well as for the upregulation of the expression of genes like CTLA-4, CD25, or GITR [73]. Store operated calcium entry and NFAT activity are also required for Treg development, as mice with selective ablation of the calcium sensor proteins STIM1 and STIM2 in T cells, which show severely impaired store operated calcium influx and NFAT activation, have reduced numbers of Tregs [87].

The role of NFAT in the immune system is not restricted to T cells. NFAT proteins are also expressed in other cells of the immune system, such as B cells, NK cells, and mast cells, where they have been shown to regulate the expression of cytokines, cell surface receptors, and immunoglobulins [88–90].

NFAT proteins are also expressed in skeletal, cardiac, and smooth muscle where they regulate development and differentiation [91–93]. Specific NFAT isoforms are expressed at different stages of skeletal muscle development and regulate progression from early precursors to mature myocytes [92, 94, 95]. NFAT proteins also control the expression of the myosin heavy chain and regulate muscle growth [95, 96].

Bone remodeling is also regulated by NFAT signaling. NFAT proteins control osteoclast and osteoblast differentiation, coupling bone formation and resorption during skeletal development and repair [97–99]. Furthermore, NFAT2 also regulates the expression of specific chemokines, such as CCL8, in osteoblasts, which attract osteoclast precursors to sites of osteoblast activation [99]. In osteoclasts, increased intracellular calcium promoted by RANKL signaling activates NFAT2, which promotes the expression of several osteoclast specific genes [100].

NFAT2 plays a key role in the development of the embryo's heart. Mice lacking this NFAT protein die in early embryonic stages with defects in the formation of valves and septum in the heart [13, 14]. NFAT proteins also cooperate with transcription factors of the GATA and MEF2 families in the adult heart, to regulate cardiac muscle hypertrophic responses [65, 101].

Angiogenesis is also regulated by NFAT proteins. NFAT3 and 4 are expressed in perivascular mesenchymal cells, which are implicated in the assembly of blood vessels during embryogenesis. The lack of NFAT signaling results in a disorganized and inappropriate growth of developing vessels [102]. In endothelial cells, Vascular Endothelial Growth Factor activates NFAT dependent transcription of endothelial genes like Cyclooxygenase-2 and promotes endothelial cell migration [103].

NFAT proteins modulate many other cellular functions. For instance, they regulate neuronal axon growth [104] and are also involved in insulin homeostasis, by controlling beta cell growth and regulating insulin signaling pathways and adipogenesis [105–107].

An important aspect of NFAT function involves its action on cell growth and the cell cycle, which should be addressed in a cell-type specific manner. For example, NFAT signaling has been implicated in the positive regulation of pancreatic beta-cell growth and function, promoting the expression of cell cycle regulators and beta-cell proliferation [105]. However, during T cell activation, NFAT acts as a repressor of cyclins A2, E, and B1 gene expression [108], suggesting a role in controlling cell cycle progression during T cell priming. This could explain the observed hyperproliferation in response to TCR stimulation in NFAT1 deficient T cells. NFAT2 is expressed in the stem cells located in the hair follicle bulge, where it represses CDK4 transcription and therefore maintains stem cell quiescence. Activation of those stem cells leads to downregulation of NFAT2, expression of CDK4, and cell division [109]. Therefore, the involvement of NFAT family members during cell growth and proliferation requires temporal and cell-type specific analysis.

THE PRIMORDIAL FAMILY MEMBER: NFAT5

As the most evolutionary conserved member of the NFAT family, NFAT5 provides important clues as to the ancestral functions for this family of transcription factors. While the mechanisms of NFAT5 regulation are incompletely understood, recent evidence has emerged to clarify its regulation by hypertonicity. In most cell types examined, NFAT5 is present in both the nucleus and the cytoplasm; however, in at least one subline of Jurkat T cells, NFAT is cytoplasmic and is translocated to the nucleus following hypertonic stimulation [8, 64]. Indeed, it was independently cloned in a yeast one-hybrid system as TonEBP (tonicity element binding protein) [8]. There are three main processes involved in the activation of NFAT5: nuclear translocation, enhanced transcriptional activity, and increased NFAT5 synthesis [110]. Isotonic conditions lead to both nuclear and cytoplasmic NFAT5 localization while exclusive accumulation in the nucleus or cytoplasm is observed under hypertonic or hypotonic challenges, respectively. This nucleocytoplasmic trafficking of NFAT5 is achieved by opposing effects of a nuclear localization sequence and a nuclear export sequence in response to changes in extracellular tonicity [111]. Additional regulatory domains are present in the C-terminus, including a large hypertonicity responsive transactivation domain composed of four subdomains, some of which can be specifically phosphorylated, which correlates with increased transcriptional activity [112]. While some kinases, such as p38, Fyn, and ATM, have been shown to have an effect on NFAT5 activity in studies using inhibitors, it remains to be determined whether they act by directly phosphorylating NFAT5 or indirectly through signaling cascades, and the actual phosphorylation sites within the protein itself have not been identified. Furthermore, the total amount of mature NFAT5 protein increases under hypertonic conditions and overexpression in HEK cells can activate NFAT5 dependent promoters in the absence of stimulation [112]. Detailed studies of stabilization of pre-existing mRNA, possible binding partners, and DNA regulatory elements in the NFAT5 gene should provide more insight into its regulation.

Functionally, NFAT5 activates a large number of target genes implicated in osmoprotective responses, including those encoding aldose reductase, the betaine transporter, and the inositol transporter [113]. With its wide range of cell type and tissue expression patterns, NFAT5 plays a prominent role in restoring intracellular osmotic balance under hypertonic conditions. However, it also has a number of functions independent of the hypertonic response, including embryonic development, integrin mediated cell migration, and T cell receptor (TCR) signaling. NFAT5 deficient mice have substantially reduced embryonic viability and increased perinatal lethality [114, 115], possibly due to a renal defect. In carcinoma cells, NFAT5 activity is induced in response to $\alpha6\beta4$ clustering and specific inhibition of NFAT5 results in reduced carcinoma invasion [12], suggesting a role for NFAT5 as an inducer of tumor metastasis. In human T cells, activation through the TCR increases NFAT5 levels and activates transcription from NFAT5 dependent reporters in a calcineurin dependent fashion [10, 11]. These results suggest an additional function for NFAT5 as a mediator of responses downstream of the TCR.

PERSPECTIVES

While a great deal has been discovered about the NFAT family in the 5 years since the founding family members were cloned, much still remains to be understood. It is clear that the individual family members can be independently regulated even in the same cell types, but the basis of this specificity is unclear. The genes regulated by NFAT1–4 in immune cells and by NFAT5 under hypertonic conditions in kidney cells are by large known, but the target genes for these proteins and their cell specific biological functions in other cell types remain to be identified.

REFERENCES

1. Rao A, Luo C, Hogan PG. Transcription factors of the NFAT family: regulation and function. *Annu Rev Immunol* 1997;**15**:707–47.
2. Crabtree GR. Generic signals and specific outcomes: signaling through Ca2+, calcineurin, and NF-AT. *Cell* 1999;**96**:611–4.
3. Crabtree GR, Olson EN. NFAT signaling: choreographing the social lives of cells. *Cell* 2002;**109**:S67–79.
4. Macian F, Lopez-Rodriguez C, Rao A. Partners in transcription: NFAT and AP-1. *Oncogene* 2001;**20**:2476–89.
5. Kiani A, Rao A, Aramburu J. Manipulating immune responses with immunosuppressive agents that target NFAT. *Immunity* 2000;**12**:359–72.
6. Hogan PG, Chen L, Nardone J, Rao A. Transcriptional regulation by calcium, calcineurin, and NFAT. *Genes Dev* 2003;**17**:2205–32.
7. Macian F. NFAT proteins: key regulators of T-cell development and function. *Nat Rev Immunol* 2005;**5**:472–84.
8. Miyakawa H, Woo SK, Dahl SC, Handler JS, Kwon HM. Tonicity-responsive enhancer binding protein, a rel-like protein that stimulates transcription in response to hypertonicity. *Proc Natl Acad Sci USA* 1999;**96**:2538–42.
9. Huang AM, Rubin GM. A misexpression screen identifies genes that can modulate RAS1 pathway signaling in Drosophila melanogaster. *Genetics* 2000;**156**:1219–30.
10. Lopez-Rodriguez C, Aramburu J, Jin L, Rakeman AS, Michino M, Rao A. Bridging the NFAT and NF-kappaB families: NFAT5 dimerization regulates cytokine gene transcription in response to osmotic stress. *Immunity* 2001;**15**:47–58.
11. Trama J, Lu Q, Hawley RG, Ho SN. The NFAT-related protein NFATL1 (TonEBP/NFAT5) is induced upon T cell activation in a calcineurin-dependent manner. *J Immunol* 2000;**165**:4884–94.
12. Jauliac S, Lopez-Rodriguez C, Shaw LM, Brown LF, Rao A, Toker A. The role of NFAT transcription factors in integrin-mediated carcinoma invasion. *Nat Cell Biol* 2002;**4**:540–4.
13. de la Pompa JL, Timmerman LA, Takimoto H, Yoshida H, Elia AJ, Samper E, et al. Role of the NF-ATc transcription factor in morphogenesis of cardiac valves and septum. *Nature* 1998;**392**:182–6.
14. Ranger AM, Grusby MJ, Hodge MR, Gravallese EM, de la Brousse FC, Hoey T, et al. The transcription factor NF-ATc is essential for cardiac valve formation. *Nature* 1998;**392**:186–90.

15. Horsley V, Pavlath GK. NFAT: ubiquitous regulator of cell differentiation and adaptation. *J Cell Biol* 2002;**156**:771–4.
16. Chen L, Rao A, Harrison SC. Signal integration by transcription-factor assemblies: interactions of NF-AT1 and AP-1 on the IL-2 promoter. *Cold Spring Harb Symp Quant Biol* 1999;**64**:527–31.
17. Stroud JC, Lopez-Rodriguez C, Rao A, Chen L. Structure of a TonEBP-DNA complex reveals DNA encircled by a transcription factor. *Nat Struct Biol* 2002;**9**:90–4.
18. Ghosh G, van Duyne G, Ghosh S, Sigler PB. Structure of NF-kappa B p50 homodimer bound to a kappa B site. *Nature* 1995;**373**:303–10.
19. Muller CW, Harrison SC. The structure of the NF-kappa B p50:DNA-complex: a starting point for analyzing the Rel family. *FEBS Lett* 1995;**369**:113–7.
20. Jin L, Sliz P, Chen L, Macian F, Rao A, Hogan PG, et al. An asymmetric NFAT1 dimer on a pseudo-palindromic kappa B-like DNA site. *Nat Struct Biol* 2003;**10**:807–11.
21. Giffin MJ, Stroud JC, Bates DL, von Koenig KD, Hardin J, Chen L. Structure of NFAT1 bound as a dimer to the HIV-1 LTR kappa B element. *Nat Struct Biol* 2003;**10**:800–6.
22. Stroud JC, Chen L. Structure of NFAT bound to DNA as a monomer. *J Mol Biol* 2003;**334**:1009–22.
23. Berridge MJ, Lipp P, Bootman MD. The versatility and universality of calcium signalling. *Nat Rev Mol Cell Biol* 2000;**1**:11–21.
24. Lewis RS. Calcium signaling mechanisms in T lymphocytes. *Annu Rev Immunol* 2001;**19**:497–521.
25. Dolmetsch RE, Lewis RS, Goodnow CC, Healy JI. Differential activation of transcription factors induced by Ca2+ response amplitude and duration. *Nature* 1997;**386**:855–8.
26. Dolmetsch RE, Xu K, Lewis RS. Calcium oscillations increase the efficiency and specificity of gene expression. *Nature* 1998;**392**: 933–6.
27. Li W, Llopis J, Whitney M, Zlokarnik G, Tsien RY. Cell-permeant caged InsP3 ester shows that Ca2+ spike frequency can optimize gene expression. *Nature* 1998;**392**:936–41.
28. Feske S, Giltnane J, Dolmetsch R, Staudt LM, Rao A. Gene regulation mediated by calcium signals in T lymphocytes. *Nat Immunol* 2001;**2**:316–24.
29. Feske S, Draeger R, Peter HH, Eichmann K, Rao A. The duration of nuclear residence of NFAT determines the pattern of cytokine expression in human SCID T cells. *J Immunol* 2000;**165**:297–305.
30. Feske S. Calcium signalling in lymphocyte activation and disease. *Nat Rev Immunol* 2007;**7**:690–702.
31. Hogan PG, Rao A. Dissecting ICRAC, a store-operated calcium current. *Trends Biochem Sci* 2007;**32**:235–45.
32. Feske S, Gwack Y, Prakriya M, Srikanth S, Puppel SH, Tanasa B, et al. A mutation in Orai1 causes immune deficiency by abrogating CRAC channel function. *Nature* 2006;**441**:179–85.
33. Vig M, Peinelt C, Beck A, Koomoa DL, Rabah D, Koblan-Huberson M, et al. CRACM1 is a plasma membrane protein essential for store-operated Ca2+ entry. *Science* 2006;**312**:1220–3.
34. Zhang SL, Yeromin AV, Zhang XH, Yu Y, Safrina O, Penna A, et al. Genome-wide RNAi screen of Ca(2+) influx identifies genes that regulate Ca(2+) release-activated Ca(2+) channel activity. *Proc Natl Acad Sci USA* 2006;**103**:9357–62.
35. Prakriya M, Feske S, Gwack Y, Srikanth S, Rao A, Hogan PG. Orai1 is an essential pore subunit of the CRAC channel. *Nature* 2006;**443**:230–3.
36. Aramburu J, Garcia-Cozar F, Raghavan A, Okamura H, Rao A, Hogan PG. Selective inhibition of NFAT activation by a peptide spanning the calcineurin targeting site of NFAT. *Mol Cell* 1998;**1**:627–37.
37. Aramburu J, Yaffe MB, Lopez-Rodriguez C, Cantley LC, Hogan PG, Rao A. Affinity-driven peptide selection of an NFAT inhibitor more selective than cyclosporin A. *Science* 1999;**285**:2129–33.
38. Li H, Zhang L, Rao A, Harrison SC, Hogan PG. Structure of calcineurin in complex with PVIVIT peptide: portrait of a low-affinity signalling interaction. *J Mol Biol* 2007;**369**:1296–306.
39. Li H, Rao A, Hogan PG. Structural delineation of the calcineurin-NFAT interaction and its parallels to PP1 targeting interactions. *J Mol Biol* 2004;**342**:1659–74.
40. Roy J, Li H, Hogan PG, Cyert MS. A conserved docking site modulates substrate affinity for calcineurin, signaling output, and in vivo function. *Mol Cell* 2007;**25**:889–901.
41. Liu J, Masuda ES, Tsuruta L, Arai N, Arai K. Two independent calcineurin-binding regions in the N-terminal domain of murine NF-ATx1 recruit calcineurin to murine NF-ATx1. *J Immunol* 1999;**162**: 4755–61.
42. Park S, Uesugi M, Verdine GL. A second calcineurin binding site on the NFAT regulatory domain. *Proc Natl Acad Sci USA* 2000;**97**: 7130–5.
43. Okamura H, Aramburu J, Garcia-Rodriguez C, Viola JP, Raghavan A, Tahiliani M, et al. Concerted dephosphorylation of the transcription factor NFAT1 induces a conformational switch that regulates transcriptional activity. *Mol Cell* 2000;**6**:539–50.
44. Neal JW, Clipstone NA. Glycogen synthase kinase-3 inhibits the DNA binding activity of NFATc. *J Biol Chem* 2001;**276**:3666–73.
45. Chuvpilo S, Jankevics E, Tyrsin D, Akimzhanov A, Moroz D, Jha MK, et al. Autoregulation of NFATc1/A expression facilitates effector T cells to escape from rapid apoptosis. *Immunity* 2002;**16**:881–95.
46. Zhou B, Cron RQ, Wu B, Genin A, Wang Z, Liu S, et al. Regulation of the murine Nfatc1 gene by NFATc2. *J Biol Chem* 2002;**277**: 10,704–10,711.
47. Okamura H, Garcia-Rodriguez C, Martinson H, Qin J, Virshup DM, Rao A. A conserved docking motif for CK1 binding controls the nuclear localization of NFAT1. *Mol Cell Biol* 2004;**24**:4184–95.
48. Beals CR, Sheridan CM, Turck CW, Gardner P, Crabtree GR. Nuclear export of NF-ATc enhanced by glycogen synthase kinase-3. *Science* 1997;**275**:1930–4.
49. Sheridan CM, Heist EK, Beals CR, Crabtree GR, Gardner P. Protein kinase A negatively modulates the nuclear accumulation of NF-ATc1 by priming for subsequent phosphorylation by glycogen synthase kinase-3. *J Biol Chem* 2002;**277**:48,664–48,676.
50. Gwack Y, Sharma S, Nardone J, Tanasa B, Iuga A, Srikanth S, et al. A genome-wide Drosophila RNAi screen identifies DYRK-family kinases as regulators of NFAT. *Nature* 2006;**441**:646–50.
51. Arron JR, Winslow MM, Polleri A, Chang CP, Wu H, Gao X, et al. NFAT dysregulation by increased dosage of DSCR1 and DYRK1A on chromosome 21. *Nature* 2006;**441**:595–600.
52. Willingham AT, Orth AP, Batalov S, Peters EC, Wen BG, Aza-Blanc P, et al. A strategy for probing the function of noncoding RNAs finds a repressor of NFAT. *Science* 2005;**309**:1570–3.
53. Luo C, Burgeon E, Rao A. Mechanisms of transactivation by nuclear factor of activated T cells-1. *J Exp Med* 1996;**184**:141–7.
54. Imamura R, Masuda ES, Naito Y, Imai S, Fujino T, Takano T, et al. Carboxyl-terminal 15-amino acid sequence of NFATx1 is possibly created by tissue-specific splicing and is essential for transactivation activity in T cells. *J Immunol* 1998;**161**:3455–63.
55. Massari ME, Grant PA, Pray-Grant MG, Berger SL, Workman JL, Murre C. A conserved motif present in a class of helix-loop-helix proteins activates transcription by direct recruitment of the SAGA complex. *Mol Cell* 1999;**4**:63–73.

56. Chuvpilo S, Avots A, Berberich-Siebelt F, Glockner J, Fischer C, Kerstan A, et al. Multiple NF-ATc isoforms with individual transcriptional properties are synthesized in T lymphocytes. *J Immunol* 1999;**162**:7294–301.
57. Macian F, Garcia-Cozar F, Im SH, Horton HF, Byrne MC, Rao A. Transcriptional mechanisms underlying lymphocyte tolerance. *Cell* 2002;**109**:719–31.
58. Porter CM, Clipstone NA. Sustained NFAT signaling promotes a Th1-like pattern of gene expression in primary murine CD4+ T cells. *J Immunol* 2002;**168**:4936–45.
59. Garcia-Rodriguez C, Rao A. Requirement for integration of phorbol 12-myristate 13-acetate and calcium pathways is preserved in the transactivation domain of NFAT1. *Eur J Immunol* 2000;**30**:2432–6.
60. de Gregorio R, Iniguez MA, Fresno M, Alemany S. Cot kinase induces cyclooxygenase-2 expression in T cells through activation of the nuclear factor of activated T cells. *J Biol Chem* 2001;**276**:27,003–27,009.
61. Rainio EM, Sandholm J, Koskinen PJ. Cutting edge: transcriptional activity of NFATc1 is enhanced by the Pim-1 kinase. *J Immunol* 2002;**168**:1524–7.
62. San-Antonio B, Iniguez MA, Fresno M. Protein kinase Czeta phosphorylates nuclear factor of activated T cells and regulates its transactivating activity. *J Biol Chem* 2002;**277**:27,073–27,080.
63. Chen L, Glover JN, Hogan PG, Rao A, Harrison SC. Structure of the DNA-binding domains from NFAT, Fos and Jun bound specifically to DNA. *Nature* 1998;**392**:42–8.
64. Lopez-Rodriguez C, Aramburu J, Rakeman AS, Rao A. NFAT5, a constitutively nuclear NFAT protein that does not cooperate with Fos and Jun. *Proc Natl Acad Sci USA* 1999;**96**:7214–19.
65. Molkentin JD, Lu JR, Antos CL, Markham B, Richardson J, Robbins J, et al. A calcineurin-dependent transcriptional pathway for cardiac hypertrophy. *Cell* 1998;**93**:215–28.
66. Musaro A, McCullagh KJ, Naya FJ, Olson EN, Rosenthal N. IGF-1 induces skeletal myocyte hypertrophy through calcineurin in association with GATA-2 and NF-ATc1. *Nature* 1999;**400**:581–5.
67. Youn HD, Chatila TA, Liu JO. Integration of calcineurin and MEF2 signals by the coactivator p300 during T-cell apoptosis. *EMBO J* 2000;**19**:4323–31.
68. Ho IC, Hodge MR, Rooney JW, Glimcher LH. The proto-oncogene c-maf is responsible for tissue-specific expression of interleukin-4. *Cell* 1996;**85**:973–83.
69. Rengarajan J, Mowen KA, McBride KD, Smith ED, Singh H, Glimcher LH. Interferon regulatory factor 4 (IRF4) interacts with NFATc2 to modulate interleukin 4 gene expression. *J Exp Med* 2002;**195**:1003–12.
70. Yang TT, Chow CW. Transcription cooperation by NFAT.C/EBP composite enhancer complex. *J Biol Chem* 2003;**278**:15,874–15,885.
71. Duncliffe KN, Bert AG, Vadas MA, Cockerill PN. A T cell-specific enhancer in the interleukin-3 locus is activated cooperatively by Oct and NFAT elements within a DNase I-hypersensitive site. *Immunity* 1997;**6**:175–85.
72. Decker EL, Nehmann N, Kampen E, Eibel H, Zipfel PF, Skerka C. Early growth response proteins (EGR) and nuclear factors of activated T cells (NFAT) form heterodimers and regulate proinflammatory cytokine gene expression. *Nucleic Acids Res* 2003;**31**:911–21.
73. Wu Y, Borde M, Heissmeyer V, Feuerer M, Lapan AD, Stroud JC, et al. FOXP3 controls regulatory T cell function through cooperation with NFAT. *Cell* 2006;**126**:375–87.
74. Macian F, Garcia-Rodriguez C, Rao A. Gene expression elicited by NFAT in the presence or absence of cooperative recruitment of Fos and Jun. *EMBO J* 2000;**19**:4783–95.
75. Avni O, Lee D, Macian F, Szabo SJ, Glimcher LH, Rao A. T(H) cell differentiation is accompanied by dynamic changes in histone acetylation of cytokine genes. *Nat Immunol* 2002;**3**:643–51.
76. Bodor J, Habener JF. Role of transcriptional repressor ICER in cyclic AMP-mediated attenuation of cytokine gene expression in human thymocytes. *J Biol Chem* 1998;**273**:9544–51.
77. Yang XY, Wang LH, Chen T, Hodge DR, Resau JH, DaSilva L, et al. Activation of human T lymphocytes is inhibited by peroxisome proliferator-activated receptor gamma (PPARgamma) agonists. PPARgamma co-association with transcription factor NFAT. *J Biol Chem* 2000;**275**:4541–4.
78. Iacobelli M, Wachsman W, McGuire KL. Repression of IL-2 promoter activity by the novel basic leucine zipper p21SNFT protein. *J Immunol* 2000;**165**:860–8.
79. Chuvpilo S, Zimmer M, Kerstan A, Glockner J, Avots A, Escher C, et al. Alternative polyadenylation events contribute to the induction of NF-ATc in effector T cells. *Immunity* 1999;**10**:261–9.
80. Gallo EM, Winslow MM, Cante-Barrett K, Radermacher AN, Ho L, McGinnis L, et al. Calcineurin sets the bandwidth for discrimination of signals during thymocyte development. *Nature* 2007;**450**:731–5.
81. Neilson JR, Winslow MM, Hur EM, Crabtree GR. Calcineurin B1 is essential for positive but not negative selection during thymocyte development. *Immunity* 2004;**20**:255–66.
82. Oukka M, Ho IC, de la Brousse FC, Hoey T, Grusby MJ, Glimcher LH. The transcription factor NFAT4 is involved in the generation and survival of T cells. *Immunity* 1998;**9**:295–304.
83. Ansel KM, Lee DU, Rao A. An epigenetic view of helper T cell differentiation. *Nat Immunol* 2003;**4**:616–23.
84. Agarwal S, Avni O, Rao A. Cell-type-restricted binding of the transcription factor NFAT to a distal IL-4 enhancer in vivo. *Immunity* 2000;**12**:643–52.
85. Agarwal S, Rao A. Modulation of chromatin structure regulates cytokine gene expression during T cell differentiation. *Immunity* 1998;**9**:765–75.
86. Dienz O, Eaton SM, Krahl TJ, Diehl S, Charland C, Dodge J, et al. Accumulation of NFAT mediates IL-2 expression in memory, but not naive, CD4+ T cells. *Proc Natl Acad Sci USA* 2007;**104**:7175–80.
87. Oh-Hora M, Yamashita M, Hogan PG, Sharma S, Lamperti E, Chung W, et al. Dual functions for the endoplasmic reticulum calcium sensors STIM1 and STIM2 in T cell activation and tolerance. *Nat Immunol* 2008;**9**:432–43.
88. Peng SL, Gerth AJ, Ranger AM, Glimcher LH. NFATc1 and NFATc2 together control both T and B cell activation and differentiation. *Immunity* 2001;**14**:13–20.
89. Aramburu J, Azzoni L, Rao A, Perussia B. Activation and expression of the nuclear factors of activated T cells, NFATp and NFATc, in human natural killer cells: regulation upon CD16 ligand binding. *J Exp Med* 1995;**182**:801–10.
90. Monticelli S, Solymar DC, Rao A. Role of NFAT proteins in IL13 gene transcription in mast cells. *J Biol Chem* 2004;**279**:36,210–36,218.
91. Horsley V, Friday BB, Matteson S, Kegley KM, Gephart J, Pavlath GK. Regulation of the growth of multinucleated muscle cells by an NFATC2-dependent pathway. *J Cell Biol* 2001;**153**:329–38.
92. Abbott KL, Friday BB, Thaloor D, Murphy TJ, Pavlath GK. Activation and cellular localization of the cyclosporine A-sensitive transcription factor NF-AT in skeletal muscle cells. *Mol Biol Cell* 1998;**9**:2905–16.
93. Schulz RA, Yutzey KE. Calcineurin signaling and NFAT activation in cardiovascular and skeletal muscle development. *Dev Biol* 2004;**266**:1–16.

94. Schulze M, Belema-Bedada F, Technau A, Braun T. Mesenchymal stem cells are recruited to striated muscle by NFAT/IL-4-mediated cell fusion. *Genes Dev* 2005;**19**:1787–98.
95. Delling U, Tureckova J, Lim HW, De Windt LJ, Rotwein P, Molkentin JD. A calcineurin-NFATc3-dependent pathway regulates skeletal muscle differentiation and slow myosin heavy-chain expression. *Mol Cell Biol* 2000;**20**:6600–11.
96. Wada H, Hasegawa K, Morimoto T, Kakita T, Yanazume T, Abe M, et al. Calcineurin-GATA-6 pathway is involved in smooth muscle-specific transcription. *J Cell Biol* 2002;**156**:983–91.
97. Zayzafoon M. Calcium/calmodulin signaling controls osteoblast growth and differentiation. *J Cell Biochem* 2006;**97**:56–70.
98. Takayanagi H. Mechanistic insight into osteoclast differentiation in osteoimmunology. *J Mol Med* 2005;**83**:170–9.
99. Winslow MM, Pan M, Starbuck M, Gallo EM, Deng L, Karsenty G, et al. Calcineurin/NFAT signaling in osteoblasts regulates bone mass. *Dev Cell* 2006;**10**:771–82.
100. Takayanagi H, Kim S, Koga T, Nishina H, Isshiki M, Yoshida H, et al. Induction and activation of the transcription factor NFATc1 (NFAT2) integrate RANKL signaling in terminal differentiation of osteoclasts. *Dev Cell* 2002;**3**:889–901.
101. Passier R, Zeng H, Frey N, Naya FJ, Nicol RL, McKinsey TA, et al. CaM kinase signaling induces cardiac hypertrophy and activates the MEF2 transcription factor in vivo. *J Clin Invest* 2000;**105**:1395–406.
102. Graef IA, Chen F, Chen L, Kuo A, Crabtree GR. Signals transduced by Ca(2+)/calcineurin and NFATc3/c4 pattern the developing vasculature. *Cell* 2001;**105**:863–75.
103. Hernandez GL, Volpert OV, Iniguez MA, Lorenzo E, Martinez-Martinez S, Grau R, et al. Selective inhibition of vascular endothelial growth factor-mediated angiogenesis by cyclosporin A: roles of the nuclear factor of activated T cells and cyclooxygenase 2. *J Exp Med* 2001;**193**:607–20.
104. Graef IA, Wang F, Charron F, Chen L, Neilson J, Tessier-Lavigne M, et al. Neurotrophins and netrins require calcineurin/NFAT signaling to stimulate outgrowth of embryonic axons. *Cell* 2003;**113**:657–70.
105. Heit JJ, Apelqvist AA, Gu X, Winslow MM, Neilson JR, Crabtree GR, et al. Calcineurin/NFAT signalling regulates pancreatic beta-cell growth and function. *Nature* 2006;**443**:345–9.
106. Kim HB, Kong M, Kim TM, Suh YH, Kim WH, Lim JH, et al. NFATc4 and ATF3 negatively regulate adiponectin gene expression in 3T3-L1 adipocytes. *Diabetes* 2006;**55**:1342–52.
107. Yang TT, Suk HY, Yang X, Olabisi O, Yu RY, Durand J, et al. Role of transcription factor NFAT in glucose and insulin homeostasis. *Mol Cell Biol* 2006;**26**:7372–87.
108. Caetano MS, Vieira-de-Abreu A, Teixeira LK, Werneck MB, Barcinski MA, Viola JP. NFATC2 transcription factor regulates cell cycle progression during lymphocyte activation: evidence of its involvement in the control of cyclin gene expression. *FASEB J* 2002;**16**:1940–2.
109. Horsley V, Aliprantis AO, Polak L, Glimcher LH, Fuchs E. NFATc1 balances quiescence and proliferation of skin stem cells. *Cell* 2008;**132**:299–310.
110. Aramburu J, Drews-Elger K, Estrada-Gelonch A, Minguillon J, Morancho B, Santiago V, et al. Regulation of the hypertonic stress response and other cellular functions by the Rel-like transcription factor NFAT5. *Biochem Pharmacol* 2006;**72**:1597–604.
111. Tong EH, Guo JJ, Huang AL, Liu H, Hu CD, Chung SS, et al. Regulation of nucleocytoplasmic trafficking of transcription factor OREBP/TonEBP/NFAT5. *J Biol Chem* 2006;**281**:23,870–23,879.
112. Lee SD, Colla E, Sheen MR, Na KY, Kwon HM. Multiple domains of TonEBP cooperate to stimulate transcription in response to hypertonicity. *J Biol Chem* 2003;**278**:47,571–47,577.
113. Burg MB, Kwon ED, Kultz D. Regulation of gene expression by hypertonicity. *Annu Rev Physiol* 1997;**59**:437–55.
114. Go WY, Liu X, Roti MA, Liu F, Ho SN. NFAT5/TonEBP mutant mice define osmotic stress as a critical feature of the lymphoid microenvironment. *Proc Natl Acad Sci USA* 2004;**101**:10,673–10,678.
115. Lopez-Rodriguez C, Antos CL, Shelton JM, Richardson JA, Lin F, Novobrantseva TI, et al. Loss of NFAT5 results in renal atrophy and lack of tonicity-responsive gene expression. *Proc Natl Acad Sci USA* 2004;**101**:2392–7.

Chapter 12

JAK-STAT Signaling

Li Song[1] **and Christian Schindler**[2]

[1]*Department of Microbiology, College of Physicians and Surgeons, Columbia University, New York*

[2]*Department of Microbiology and Medicine, College of Physicians and Surgeons, Columbia University, New York*

ABBREVIATIONS

STAT-Signal Transducer and Activator of Transcription; JAK – Janus Activated Kinase; GAS – Gamma-IFN Activation Site; ISRE – IFN Stimulated Response Elements; ISGF-3 – IFN Stimulating Gene Factor 3; GAF – Gamma-IFN activation Factor; IRF – IFN Regulatory Factor.

INTRODUCTION

Characterization of the ability of interferons (IFNs) to rapidly induce genes led to the discovery of the JAK-STAT signaling pathway. Subsequent studies determined that JAKs (Janus activated kinases) and STATs (signal transducers and activators of transcription) transduce signals for over 50 members of the four-helix bundle cytokine family (see Table 12.1). Although components of this pathway are found in more primitive eukaryotes, signaling through the JAK-STAT pathway expanded significantly as more sophisticated immune systems emerged. This review will provide a brief overview of this signaling paradigm and highlight more recent discoveries. Additional information can be found in a number of more comprehensive reviews [1–5].

THE JAK-STAT PARADIGM

The development of recombinant IFNs afforded investigators an early opportunity to determine how cytokines direct the rapid expression of target genes. This led to the identification of the founding members of the JAK and STAT families (see Figure 12.1). Subsequent studies identified a total of seven STATs and four JAKs, revealing how over 50 four-helix bundle cytokines direct their potent biological responses. These cytokines can be divided into five subgroups based on the receptors they bind and the STATs through which they signal (see Table 12.1). Of note, a number of receptor tyrosine kinases and G-protein coupled receptors have also been shown to activate STATs, but these signals do not appear to be pivotal in the corresponding biological responses [5].

IFNs can be divided into three distinct families, types I, II, and III (see Table 12.1, [3]). Type II IFN (IFN-γ; immune IFN) binds to a unique receptor consisting of IFN-γ receptor chain 1 (IFNGR1) and IFNGR2, which trigger a JAK-STAT signaling paradigm that is exploited by virtually all four-helix bundle cytokines (see Figure 12.2). The type I IFNs (IFN-Is; IFN-αs, IFN-β, IFN-ω, and IFN-τ) and type III IFNs (IFN-λs or IL-28a, b, and IL-29) are the only known exceptions, transducing signals through a distinct JAK-STAT signaling paradigm (see Figure 12.2).

Characteristic of four-helix bundle cytokines, IFN-γ's two symmetric faces each bind to one receptor chain. This drives a conformational change bringing Jak1 and Jak2, which stably associate with IFNGR1 and IFNGR2 respectively, into close apposition, allowing them to activate each other by transphosphorylation [6, 7]. Upon activation, these tyrosine kinases phosphorylate one or more specific receptor tyrosines, which are then recognized by signaling molecules. For IFN-γ and its receptor, this entails the phosphorylation of IFNGR1 tyrosine 440 and the subsequent, SH2 dependent recruitment of Stat1. Once at the receptor, Stat1 is phosphorylated on tyrosine 701, driving a conformational change from inactive to active homodimer. This conformational change is directed through a reciprocal interaction between the phosphotyrosyl residue of one STAT and the SH2 domain of the other [8–12]. The active Stat1 homodimers, which are released from the receptor, are also competent for nuclear translocation [13], where they bind to members of the GAS (gamma-IFN activation site) family of enhancers (the palindrome TTTCCNGGAAA; [14]), driving the expression of target genes (see Figure 12.2).

Handbook of Cell Signaling, Three-Volume Set 2 ed.

TABLE 12.1 JAK-STAT signaling by the four-helix bundle cytokines[1]

Ligands	JAKs	STATs
IFN family		
IFN-I (Type I)[2]	**Jak1**, Tyk2	**Stat1**, **Stat2**, Stats3, Stat4 (Stats5–6)
IFN-γ (Type II)	**Jak1**, **Jak2**	**Stat1**
IFN-γ (IL-28a,b,-29)	Jak1, Tyk2	**Stat1**, **Stat2**, Stat3
IL-10	**Jak1, Tyk2**	**Stat3**, Stat1
IL-19	Jak1, Jak2	**Stat3**, Stat1
IL-20	Jak1, Jak2	**Stat3**, Stat1
IL-22 (IL-TIF)	Jak1, Tyk2	**Stat3**, Stat1, (Stat5)
IL-24 (mda7)	Jak1, Jak2	**Stat3**, Stat1
IL-26 (AK155)	Jak1, Tyk2	**Stat3**, Stat1
gp130 family		
IL-6	**Jak1**, (Jak2)	**Stat3**, Stat1
IL-11	Jak1	**Stat3**, Stat1
LIF	**Jak1**, (Jak2)	**Stat3**, Stat1
CNTF	Jak1, (Jak2)	**Stat3**, Stat1
CLC/CLF[3]	Jak1, (Jak2)	**Stat3**, Stat1
NP	Jak1, (Jak2)	Stat3
CT-1	Jak1, (Jak2)	**Stat3**
OSM	Jak1, (Jak2)	**Stat3**, Stat1
IL-31	Jak1, (Jak2)	Stat3, Stat5, Stat1
G-CSF	Jak1, (Jak2)	**Stat3**
Leptin	Jak2	**Stat3**
IL-12 (p35+p40)	**Tyk2**, Jak2	**Stat4**
IL-23 (p19+p40)	**Tyk2**, Jak2	**Stat3, Stat4,** Stat1
IL-27[4] (p28+EBI3)	Jak2	**Stat1, Stat3,** Stat4, (Stat5)
IL-35 (p35+EBI3)		
γC family		
IL-2	**Jak1**, **Jak3**	**Stat5**, (Stat3)
IL-7	**Jak1**, **Jak3**	**Stat5**, (Stat3)

(*Continued*)

TABLE 12.1 *(Continued)*

TSLP[5]			Stat5
IL-9	Jak1, Jak3		**Stat5**, Stat3
IL-15	**Jak1**, **Jak3**	**Stat5**, (Stat3)	
IL-21	**Jak1, Jak3**		Stat3**, Stat5, (Stat1)**
IL-4	**Jak1**, **Jak3**	**Stat6**	
IL-13[5]	**Jak1**, **Jak2**		Stat6**, (Stat3)**
IL-3 family			
IL-3	**Jak2**		**Stat5**
IL-5	**Jak2**		**Stat5**
GM-CSF	**Jak2**		**Stat5**
Single chain family			
Epo	**Jak2**		**Stat5**
GH	**Jak2**		**Stat5**, (Stat3)
Prl	**Jak2**		**Stat5**
Tpo	**Jak2**		**Stat5**

[1]*Genetic and biochemical studies have determined that four-helix bundle cytokines transduce their signals through specific JAKs and STATs. Assignments with the highest level of confidence are shown in bold. Those with less confidence are shown in plain lettering, and those with the least confidence are shown in brackets.*
[2]*In humans this family consists of 12 IFN-αs, IFN-β,ω and Limitin.*
[3]*a.k.a. NNT-1/BSF-3.*
[4]*IL-30 is the p28 subunit of IL-27.*
[5]*Bind to related, but γC independent receptors.*

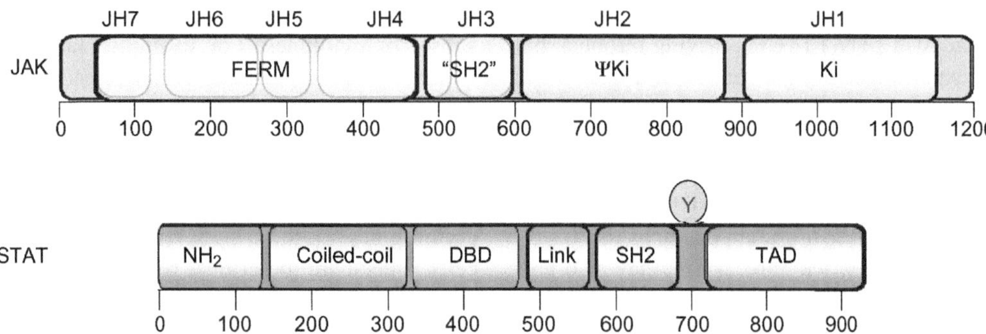

FIGURE 12.1 The JAK and STAT families.
Conserved motifs in the JAK family (top) include JAK Homology (JH) domains 1–7, which can be divided into the FERM (four point one, ezrin, radixin, moesin), SH2-related ("SH2"), pseudo-kinase (ΨKi) and kinase (Ki) domains. Conserved motifs in the STAT family include the amino terminal (NH_2), coiled-coil, DNA binding domain (DBD), Linker (Link), SH2, and transcriptional activation domains (TAD). STATs also feature a conserved tyrosine (Y) near residue 700 that is phosphorylated upon activation. See text for details.

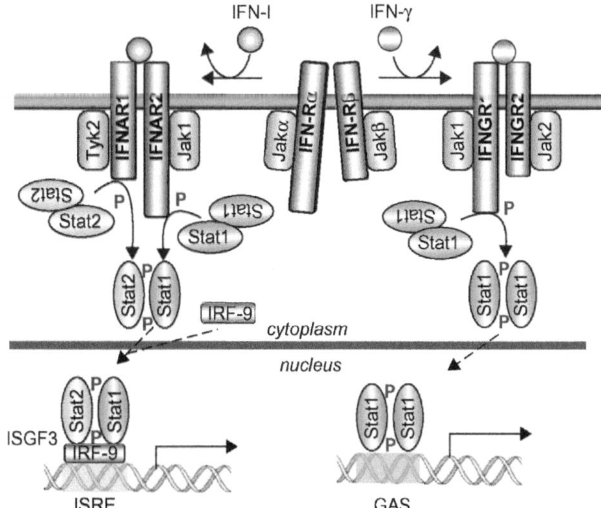

FIGURE 12.2 IFN-Is and IFN-g transduce signals through the two JAK-STAT paradigms.

JAKs (Jakα and Jakβ denote two generic JAKs) are preassociated with two chains of the type I and II IFN receptors, respectively (IFN-Rα and IFN-Rβ denote two resting, generic IFN receptor chains). Upon binding IFN-I, the IFN-α receptor (IFNAR) chains, IFNAR1 and IFNAR2, drive the Jak1/Tyk2 dependent activation of Stat1 and Stat2, culminating in the formation of ISGF-3 (Stat1 + Stat2 + IRF-9) and the rapid expression of ISRE driven target genes. In contrast, IFN-γ binds to two chains of the IFN-γ receptor (IFNGR1 and IFNGR2) directing the Jak1/Jak2 dependent activation of Stat1, which leads to the formation of active Stat1 homodimers and the expression of GAS driven genes.

Although IFN-I binding is thought to direct an analogous pattern of JAK activation, the tyrosine kinases associated with the two IFN-α receptor chains, IFNAR1 and IFNAR2, are Tyk2 and Jak1 respectively. This leads to phosphorylation of a receptor tyrosine and the IFNAR2 $Y^{510} > Y^{335}$ dependent recruitment/activation of Stat1 and Stat2 [15, 16]. Uniquely, Stat1/Stat2 heterodimers are not able to bind DNA directly, but rather associate with IRF-9 to form ISGF-3 (IFN Stimulated Gene Factor 3). This transcription factor binds to members of the ISRE (IFN Stimulated Response Element) family of enhancers (the direct repeat AGTTTN$_3$TTTCC), driving a robust, but more transient expression of target genes (see Figure 12.2; [17]). In addition, IFN-I dependent phosphorylation of Stat1 also leads to the formation of active Stat1 homodimers and the expression of GAS driven genes (not shown). Although IFN-λs bind to a receptor consisting of IL-28R1 and IL-10R2 (IL-10R2 is also a component of the IL-10, IL-22, and IL-26 receptors; see Table 12.1), they also promote the activation of ISGF-3 and Stat1 homodimers [3].

An important feature of the JAK-STAT signaling cascade is its transient nature, usually limited to less than a few hours. This decay process involves the activity of counter regulatory phosphatases and SOCS (suppressors of cytokine signaling) proteins, as well as a number of less well characterized regulators like PIAS, Nmi, and SLIM [18, 19]. Important phosphatases include SHP-1, SHP-2, PTP1B, TC-PTP, and PTP-BL [20–22]. Several appear to target receptors and JAKs, whereas others target the STATs [13, 23, 24]. This latter process is also associated with nuclear export. SOCS proteins, which are themselves STAT target genes, represent the most important specific negative regulators of the JAK-STAT signaling cascade (reviewed in [25]). For example, Socs-1 is an important negative regulator of STAT signals stimulated by the IFN, IL-12, and IL-4 cytokine families [17, 25–27], whereas Socs-3 exhibits more specificity for the IL-6 ligand family [25, 28]. Socs-2 is an important negative regulator of GH signaling [29].

THE JAK FAMILY

Members of the JAK family of tyrosine kinases, Jak1, Jak2, Jak3, and Tyk2, were initially identified as orphan tyrosine kinases and subsequently linked to cytokines through a genetic screen for IFN-I response [30–32]. They range in size from 120 to 140kDa and are widely expressed; except for Jak3, whose expression is restricted to lymphoid tissues (reviewed in [2, 5]). This family of kinases features seven conserved JAK homology (JH) domains, where the two-carboxy terminal JH regions represent the kinase (JH1/Ki) and pseudo kinase (JH2/Ψki) domains, respectively (see Figure 12.1). Analogous to other kinases, activation is driven by phosphorylation of critical tyrosines in the inactivation loop, which drives this loop out of the catalytic pocket. The four amino terminal JH domains (JH1–3 and half of JH4) constitute a FERM (four point one, ezrin, radixin, moesin) domain that mediates stable association with membrane proximal receptor motifs. An SH2 related domain ("SH2"; JH5 and half of JH4), of unknown function, lies between the pseudokinase and FERM domains. As illustrated in Figure 12.2, Jak1 stably associates with IFNAR2 and IFNGR1, Jak2 associates with IFNGR2, and Tyk2 associates with IFNAR1.

Gene targeting studies reveal that each JAK plays an important role in directing the biological response for a subset of cytokines. For example, Jak1 knockout mice, which feature a perinatal lethal phenotype, are defective in their response to cytokines from the IL-2, IL-6, IFN, and IL-10 families ([33]; see also Table 12.1). The Jak2 knockout mice exhibit an even more severe phenotype (i.e., lethality at E12.5), reflecting an important role in definitive erythropoiesis [34, 35]. *Ex vivo* studies on Jak2/fetal liver cells underscore the critical role this kinase plays in transducing signals for the IFN-γ, single chain, the IL-2, and IL-3 receptor families (see Table 12.1). Moreover, a single point mutation in the JH2 domain of Jak2, V617F, is associated with several myeloproliferative disorders, including the majority of cases of polycythemia vera, essential thrombocythemia, and primary myelofibrosis [36]. Thus, Jak2 inhibitors represent a promising therapeutic avenue for this group of patients [36].

The most unique member of this family is Jak3, which is only expressed in lymphoid tissues. Biochemical studies initially demonstrated a robust association between Jak3 and the common gamma chain (γC) from the IL-2 family of lymphoid predominant receptors (e.g., IL-2, IL-4, IL-7, IL-9, IL-15, and IL-21; see Table 12.1). Consistent with this, Jak3 and γC knockout mice both exhibit severe combined immunodeficiency (SCID)-like defects [37–39]. Thus, Jak3 has become another appealing drug target [36].

As mentioned above, Tyk2 was initially associated with IFN-I response, but subsequent biochemical studies implicated it in the response to IL-12, IL-23, as well as several members of the IL-6 and IL-10 receptor families. Subsequent analysis of Tyk2 deficient cells uncovered a remarkable divergence between humans and mice. Tyk2 knockout mice were characterized by modest defects in cytokine response and a proclivity toward type 2 (i.e., allergic) T-cell responses [40–42]. In contrast, Tyk2 deficient humans exhibited a severe allergic phenotype that was associated with impaired immunity to microbial infections [43]. Finally, studies exploring the response of Tyk2 knockout mice to LPS toxemia have highlighted a role for integrating the response to multiple cytokines during septic stress [44].

THE STAT FAMILY

Mammals express seven STAT proteins, which range in size from 750 to 900 amino acids. Both the chromosomal distribution of these STATs, as well as the identification of homologs in more primitive eukaryotes, suggested that this family arose from a single primordial gene [5, 45]. Duplications of this locus corresponded with an increasing need for cell-to-cell communication. Homologs, most closely related to Stat3 and Stat5, have been identified in model eukaryotes, including *Dictyostelium*, *C. elegans*, and *Drosophila* (reviewed in [5]). Whereas in *Drosophila* a single STAT transduces signals through a "classical" JAK-STAT pathway, the STAT homologs in *Dictyostelium* and *C. elegans* appear to signal through different pathway(s).

STATs can be divided into five structurally and functionally conserved domains (see Figure 12.1; [8, 11]). The amino terminal domain (NH_2; ~125 amino acids) is well conserved and promotes homotypic interactions between unphosphorylated STATs [9, 12]. The coiled-coil domain (amino acids ~135 to ~315) consists of a four-helix bundle that protrudes laterally (~80 Å) from the core. It has been shown to associate with a number of potentially important regulatory proteins and implicated in nuclear import/export [5, 13]. The DNA binding domain (DBD; amino acids ~320 to ~480) of activated STAT homodimers (except Stat2) recognize the palindromic GAS element and may also participate in the process of nuclear import/export [8, 11, 13, 46]. The Linker domain (amino acids ~480 to ~575) structurally translates the dimerization signal to the DNA binding motif. Studies suggest it regulates a process of basal (i.e., in resting cells) nuclear export [24]. The SH2 domain (amino acids ~575 to ~680) is the most highly conserved motif. It mediates specific recruitment to the appropriate receptor, as well as the formation of active STAT dimers [5]. The tyrosine activation motif is a conserved tyrosine near residue 700, which upon phosphorylation marks STAT activation. As detailed above, this phosphotyrosyl residue is recognized by a SH2 domain of the partner STAT to allow the formation of active STAT dimers [5]. The carboxy termini of all STATs vary considerably in both length and sequence. They encode the transcriptional activation domain (TAD), which is conserved between mouse and man for every STAT member (except Stat2), but diverges substantially between STATs [5]. These TADs include conserved serine phosphorylation sites that direct the recruitment of coactivators (e.g., CBP or MCM complex; [47, 48]) and in some cases regulate STAT stability [19, 49]. Finally, a number of native carboxy terminally truncated STAT isoforms direct unique programs of gene expression through their association with other transcription factors (e.g., Stat1β in ISGF-3 and Stat3β with c-jun; [50–52]).

Stat1

Consistent with its identification during the purification of ISGF-3 (Stat1 + Stat2 + IRF9; [53]) and gamma-IFN activating factor (GAF; [54]), Stat1 knockout mice exhibit profound defects in their biological response to type I, type II, and type III IFNs, but less so for other ligands [55, 56]. Intriguingly, Stat1 target genes appear to promote inflammation and antagonize proliferation. This contrasts the pro-proliferative and anti-inflammatory activities associated with Stat3 (see below). This raises the possibility that activation of both Stat1 and Stat3 by members of the IFN-I and IL-6 families (see Table 12.1) represents an effort to achieve a balanced response. Reflecting on its status as a founding STAT, both its structural and functional features are well conserved within this family (see Figures 12.1 and 12.2).

Stat2

Consistent with its identification during the purification of ISGF-3, Stat2 knockout mice are defective in their response to type I IFNs and likely type III IFNs [57]. Yet, in contrast to Stat1, Stat2 exhibits a number of unique features. This includes being the largest (850 amino acids in man; 925 amino acids in mouse) and most divergent STAT (e.g., the murine and human TADs are unrelated). Again, in contrast to all other STATs, there is no concrete evidence that active Stat2 homodimers form, or that Stat2 directly binds DNA. Rather, Stat2 exclusively heterodimerizes with Stat1 to form ISGF-3, where IRF-9 is responsible for DNA binding.

Stat3

Stat3 was initially identified as an IL-6 dependent transcription factor [58]. This prolific STAT is now known to transduce signals for the IL-6 family (IL-6, IL-11, IL-31, LIF, CNTF, CLC/CLF, NP, CT1, OSM), IL-10 family (IL-10, IL-19, IL-20, IL-22, IL-24, IL-26), as well as G-CSF, Leptin, IL-21, IL-27, and potentially IFN-Is (see Table 12.1; [2, 5, 59]). In addition, a number of growth factors and oncogenes have been shown to activate Stat3 *in vitro*. Consistent with these pervasive and pleiotropic effects, deletion of the Stat3 gene yields an early embryonic lethal phenotype (at E6.5–7.5 [60]). Tissue specific Stat3 deletions are associated with increased inflammation and decreased oncogenesis [61, 62]. Stat3 activity is associated with expression of anti-apoptotic/pro-survival genes (e.g., in head and neck cancers, mammary carcinomas, multiple myelomas, and other hematological malignancies; [62, 63]). Further supporting a role in cell growth, expression of a hyperactive Stat3 allele was associated with transformation and reduction in Stat3 activity (e.g., through dominant negative inhibitors, decoy oligonucleotides, RNA silencing, or genetic ablation) has been associated with tumor regression in a number of model systems [62, 64, 65]. However, Stat3's role in tumors is likely to be complex [64]. Moreover, the ability of Stat3 to regulate immune response is also likely to have an important effect on the development of cancer [63, 66, 67]. Reflecting this notion, exciting new evidence has emerged on the role Stat3 activating cytokines (e.g., IL-6, IL-10, IL-22, and IL-27) play in both the development of Th17 effector and regulatory T-cells [59].

Stat4

Initially identified through its homology to Stat1, its companion gene on murine chromosome 2, Stat4 was subsequently found to transduce signals for IL-12 (consisting of p40 + p35 subunits) and more recently IL-23 (consisting of p19 + p35 subunits; [68–70]). Specifically, Stat4 directs the IL-12 dependent polarization of naïve CD4 + T-cells towards IFN-γ secreting Th1 cells, and the IL-23 dependent polarization toward IL-17 secreting Th17 cells [59, 71]. Likewise, Stat4 plays an important role in the IL-12 dependent activation of IFN-γ secreting NK cells. More recent studies have underscored the ability of other cytokines to synergize with IL-12 stimulated Stat4 tyrosine phosphorylation through their capacity to promote Stat4 serine phosphorylation [72].

Stat5

Stat5a and Stat5b were initially identified as transcription factors that are activated by prolactin and IL-3 (reviewed in [2, 5]). These tandemly duplicated genes lie adjacent to Stat3 on murine chromosome 17 and exhibit the highest degree of homology with invertebrate STATs [45, 70]. Consistent with this ancient pedigree, Stat5a/b are also functionally pleiotropic, directing the biological response for the IL-3 (IL-3, IL-5, and GM-CSF), single chain (e.g., GH, Prl, Tpo, and Epo), and γC (i.e., the IL-2, IL-7, IL-9, IL-15, and IL-21) receptor families [2, 5]. Except in the biological response to Prl, which favors Stat5a, and GH, which favors Stat5b, these two STATs appear to be functionally redundant (they exhibit ~96 percent amino acid identity; [2, 5, 71]). Thus, a full deletion of both genes was required to demonstrate the important role Stat5 plays in directing normal erythropoiesis and lymphopoiesis [73].

Stat6

Initially identified as IL-4Stat, Stat6 was subsequently found to transduce signals for IL-13, whose receptor overlaps with that of IL-4 [2, 5, 71]. Analogous to Stat2, its neighbor on murine chromosome 12, Stat6 is more divergent in sequence. It also features a relatively large 150 amino acid TAD, which has been shown to interact with numerous transcriptional regulators [70, 74]. Intriguingly, Stat6 homodimers bind to a GAS element that features an additional central nucleotide, providing an opportunity to activate a distinct subset of genes. Stat6 knockout mice have underscored the critical role this STAT plays in directing the IL-4/IL-13 dependent polarization of naïve CD4 T-cells into Th2 effector cells, and in promoting B-cell functions (e.g., proliferation, maturation, MHC-II, and IgE expression), as well as mast cell activity [71].

A BRIGHT FUTURE

Characterization of IFN response, first described over 50 years ago, has both led to the identification of the JAK-STAT signaling cascade, as well as important insight into how more than 50 cytokines transduce their biological responses (see Table 12.1). Both gene targeting and developing pharmaceuticals will provide an opportunity to explore how this pathway regulates immune and non-immune homeostasis. This is likely not only to include evidence of receptor, JAK, and STAT modifications, but also other regulators that synergize with these defined components through both more traditional and non-traditional signaling pathways.

REFERENCES

1. Decker T, Muller M, Stockinger S. The yin and yang of type I interferon activity in bacterial infection. *Nat Rev Immunol* 2005;**5**(9):675–87.
2. Murray PJ. The JAK–STAT signaling pathway: input and output integration. *J Immunol* 2007;**178**(5):2623–9.

3. Pestka S, Krause CD, Sarkar D, Walter MR, Shi Y, et al. Interleukin-10 and related cytokines and receptors. *Annu Rev Immunol* 2004; **22**:929–79.
4. Levy Jr. DE, Darnell JE Stats: transcriptional control and biological impact. *Nat Rev Mol Cell Biol* 2002;**3**(9):651–62.
5. Kisseleva T, Bhattacharya S, Braunstein J, Schindler CW. Signaling through the JAK/STAT pathway, recent advances and future challenges. *Gene* 2002;**285**(1–2):1–24.
6. Remy I, Wilson IA, Michnick SW. Erythropoietin receptor activation by a ligand-induced conformation change. *Science* 1999; **283**(5404):990–3.
7. Stark GR, Kerr IM, Williams BR, Silverman RH, Schreiber RD. How cells respond to interferons. *Annu Rev Biochem* 1998;**67**:227–64.
8. Chen X, Vinkemeier U, Zhao Y, Jeruzalmi D, Darnell JE, Jr. et al Crystal structure of a tyrosine phosphorylated STAT-1 dimer bound to DNA. *Cell* 1998;**93**(5):827–39.
9. Mertens C, Zhong M, Krishnaraj R, Zou W, Chen X, et al. Dephosphorylation of phosphotyrosine on STAT1 dimers requires extensive spatial reorientation of the monomers facilitated by the N-terminal domain. *Genes Dev* 2006;**20**(24):3372–81.
10. Neculai D, Neculai AM, Verrier S, Straub K, Klumpp K, et al. Structure of the unphosphorylated STAT5a dimer. *J Biol Chem* 2005;**280**(49):40,782–7.
11. Becker S, Groner B, Muller CW. Three-dimensional structure of the Stat3beta homodimer bound to DNA. *Nature* 1998;**394**(6689): 145–51.
12. Mao X, Ren Z, Parker GN, Sondermann H, Pastorello MA, et al. Structural bases of unphosphorylated STAT1 association and receptor binding. *Mol Cell* 2005;**17**(6):761–71.
13. McBride KM, Reich NC. The ins and outs of STAT1 nuclear transport. *Sci STKE* 2003;**2003**(195):RE13.
14. Decker T, Kovarik P, Meinke A. GAS elements: A few nucleotides with a major impact on cytokine-induced gene expression. *J. Interferon Cytokine Res.* 1997;**17**(3):121–34.
15. Zhao W, Lee C, Piganis R, Plumlee C, de Weerd N, et al. A Conserved IFN-α Receptor Tyrosine Motif Directs the Biological Response to Type I IFNs. *J Immunol* 2008;**180**(8):5483–9.
16. Wagner TC, Velichko S, Vogel D, Rani MR, Leung S, et al. Interferon signaling is dependent on specific tyrosines located within the intracellular domain of IFNAR2c. Expression of IFNAR2c tyrosine mutants in U5A cells. *J Biol Chem* 2002;**277**(2):1493–9.
17. Zhao W, Cha EN, Lee C, Park CY, Schindler C. Stat2-Dependent Regulation of MHC Class II Expression. *J Immunol* 2007;**179**(1):463–71.
18. Shuai K. Modulation of STAT signaling by STAT-interacting proteins. *Oncogene* 2000;**19**(21):2638–44.
19. Tanaka T, Soriano MA, Grusby MJ. SLIM is a nuclear ubiquitin E3 ligase that negatively regulates STAT signaling. *Immunity* 2005;**22**(6):729–36.
20. Klingmueller U, Lorenz U, Cantley LC, Neel BG, Lodish HF. Specific recruitment of SH-PTP1 to the erythropoietin receptor causes inactivation of Jak2 and termination of proliferative signals. *Cell* 1995;**80**(5):729–38.
21. Mustelin T, Vang T, Bottini N. Protein tyrosine phosphatases and the immune response. *Nat Rev Immunol* 2005;**5**(1):43–57.
22. Nakahira M, Tanaka T, Robson BE, Mizgerd JP, Grusby MJ. Regulation of Signal Transducer and Activator of Transcription Signaling by the Tyrosine Phosphatase PTP-BL. *Immunity* 2007;**26**(2):163–76.
23. Vinkemeier U. Getting the message across, STAT! Design principles of a molecular signaling circuit. *J Cell Biol* 2004;**167**(2):197–201.
24. Bhattacharya S, Schindler C. Regulation of Stat3 nuclear export. *J Clin Invest* 2003;**111**(4):553–9.
25. Alexander WS, Hilton DJ. The role of suppressors of cytokine signaling (SOCS) proteins in regulation of the immune response. *Annu Rev Immunol* 2004;**22**:503–29.
26. Fenner JE, Starr R, Cornish AL, Zhang JG, Metcalf D, et al. Suppressor of cytokine signaling 1 regulates the immune response to infection by a unique inhibition of type I interferon activity. *Nat Immunol* 2006;**7**(1):33–9.
27. Rothlin CV, Ghosh S, Zuniga EI, Oldstone MB, Lemke G. TAM Receptors Are Pleiotropic Inhibitors of the Innate Immune Response. *Cell* 2007;**131**(6):1124–36.
28. Heinrich PC, Behrmann I, Haan S, Hermanns HM, Muller-Newen G, et al. Principles of interleukin (IL)-6-type cytokine signalling and its regulation. *Biochem J* 2003;**374**(Pt 1):1–20.
29. Greenhalgh CJ, Rico-Bautista E, Lorentzon M, Thaus AL, Morgan PO, et al. SOCS2 negatively regulates growth hormone action in vitro and in vivo. *J Clin Invest* 2005;**115**(2):397–406.
30. Firmbach-Kraft I, Byers M, Shows T, Dalla-Favera R, Krolewski JJ. tyk2, prototype of a novel class of non-receptor tyrosine kinase genes. *Oncogene* 1990;**5**(9):1329–36.
31. Wilks AF. Two putative protein-tyrosine kinases identified by application of the polymerase chain reaction. *Proc Nat Acad Sci, USA* 1989;**86**:1603–7.
32. Velazquez L, Fellous M, Stark GR, Pellegrini S. A protein tyrosine kinase in the interferon alpha/beta signaling pathway. *Cell* 1992;**70**(2):313–22.
33. Rodig SJ, Meraz MA, White JM, Lampe PA, Riley JK, et al. Disruption of the Jak1 gene demonstrates obligatory and nonredundant roles of the Jaks in cytokine-induced biologic responses. *Cell* 1998;**93**(3):373–83.
34. Neubauer H, Cumano A, Mueller M, Wu H, Huffstadt U, et al. Jak2 deficiency defines an essential developmental checkpoint in definitive hematopoiesis. *Cell* 1998;**93**(3):397–409.
35. Parganas E, Wang D, Stravopodis D, Topham DJ, Marine JC, et al. Jak2 is essential for signaling through a variety of cytokine receptors. *Cell* 1998;**93**(3):385–95.
36. Tefferi A. JAK and MPL mutations in myeloid malignancies. *Leuk Lymphoma* 2008;**49**(3):388–97.
37. Nosaka T, vanDeursen JM, Tripp RA, Thierfelder WE, Witthuhn BA, et al. Defective lymphoid development in mice lacking Jak3. *Science* 1995;**270**(5237):800–2.
38. Park SY, Saijo K, Takahashi T, Osawa M, Arase H, et al. Developmental defects of lymphoid cells in Jak3 kinase-deficient mice. *Immunity* 1995;**3**(6):771–82.
39. Thomis DC, Gurniak CB, Tivol E, Sharpe AH, Berg LJ. Defects in B lymphocyte maturation and T lymphocyte activation in mice lacking Jak3. *Science* 1995;**270**(5237):794–7.
40. Seto Y, Nakajima H, Suto A, Shimoda K, Saito Y, et al. Enhanced Th2 cell-mediated allergic inflammation in Tyk2-deficient mice. *J Immunol* 2003;**170**(2):1077–83.
41. Karaghiosoff M, Neubauer H, Lassnig C, Kovarik P, Schindler H, et al. Partial impairment of cytokine responses in Tyk2-deficient mice. *Immunity* 2000;**13**(4):549–60.
42. Shimoda K, Kato K, Aoki K, Matsuda T, Miyamoto A, et al. Tyk2 plays a restricted role in IFN alpha signaling, although it is required for IL-12-mediated T cell function. *Immunity* 2000;**13**(4):561–71.
43. Minegishi Y, Saito M, Morio T, Watanabe K, Agematsu K, et al. Human tyrosine kinase 2 deficiency reveals its requisite roles in multiple cytokine signals involved in innate and acquired immunity. *Immunity* 2006;**25**(5):745–55.

44. Karaghiosoff M, Steinborn R, Kovarik P, Kriegshauser G, Baccarini M, et al. Central role for type I interferons and Tyk2 in lipopolysaccharide-induced endotoxin shock. *Nat Immunol* 2003;**4**(5):471–7.
45. Miyoshi K, Cui Y, Riedlinger G, Lehoczky J, Zon L, et al. Structure of the mouse stat 3/5 locus: evolution from drosophila to zebrafish to mouse. *Genomics* 2001;**71**(2):150–5.
46. Meyer T, Marg A, Lemke P, Wiesner B, Vinkemeier U. DNA binding controls inactivation and nuclear accumulation of the transcription factor Stat1. *Genes Dev* 2003;**17**(16):1992–2005.
47. Decker T, Müller M, Kovarik P. Regulation of Stats by posttranslational modification. In: Seghal PB, Hirano T, Levy DE, editors. *Signal transducers and activators of transcription (Stats): Activation and biology*. Dordrecht: Kluwer Academic; 2003.
48. Ramsauer K, Farlik M, Zupkovitz G, Seiser C, Kroger A, et al. Distinct modes of action applied by transcription factors STAT1 and IRF1 to initiate transcription of the IFN-gamma-inducible gbp2 gene. *Proc Natl Acad Sci USA* 2007;**104**(8):2849–54.
49. Wang D, Moriggl R, Stravopodis D, Carpino N, Marine JC, et al. A small amphipathic alpha-helical region is required for transcriptional activities and proteasome-dependent turnover of the tyrosine-phosphorylated Stat5. *Embo J* 2000;**19**(3):392–9.
50. Maritano D, Sugrue ML, Tininini S, Dewilde S, Strobl B, et al. The STAT3 isoforms alpha and beta have unique and specific functions. *Nat Immunol* 2004;**5**(4):401–9.
51. Ivanov VN, Bhoumik A, Krasilnikov M, Raz R, Owen-Schaub LB, et al. Cooperation between STAT3 and c-jun suppresses Fas transcription. *Mol Cell* 2001;**7**(3):517–28.
52. Fu X-Y, Schindler C, Improta T, Aebersold R, Darnell JE. The proteins of ISGF-3, the IFN-a induced transcription activator, define a new family of signal transducers. *Proc Nat Acad Sci USA* 1992;**89**(16):7840–3.
53. Schindler Jr. C, Fu XY, Improta T, Aebersold R, Darnell JE Proteins of transcription factor ISGF-3: one gene encodes the 91-and 84-kDa ISGF-3 proteins that are activated by interferon alpha. *Proc Natl Acad Sci USA* 1992;**89**(16):7836–9.
54. Shuai K, Schindler C, Prezioso V, Darnell JE. Activation of transcription by IFN-g: Tyrosine phosphorylation of a 91-kDa DNA binding protein. *Science* 1992;**258**(5089):1808–12.
55. Durbin JE, Hackenmiller R, Simon MC, Levy DE. Targeted disruption of the mouse Stat1 gene results in compromised innate immunity to viral disease. *Cell* 1996;**84**(3):443–50.
56. Meraz MA, White JM, Sheehan KC, Bach EA, Rodig SJ, et al. Targeted disruption of the Stat1 gene in mice reveals unexpected physiologic specificity in the JAK-STAT signaling pathway. *Cell* 1996;**84**(3):431–42.
57. Park C, Li S, Cha E, Schindler C. Immune response in Stat2 knockout mice. *Immunity* 2000;**13**(6):795–804.
58. Akira S, Nishio Y, Inoue M, Wang X-J, Wei S, et al. Molecular cloning of APRF, a novel IFN-stimulated gene factor 3 p91-related transcription factor involved in the gp130-mediated signaling pathway. *Cell* 1994;**77**(1):63–71.
59. Jankovic D, Trinchieri G. IL-10 or not IL-10: that is the question. *Nat Immunol* 2007;**8**(12):1281–3.
60. Takeda K, Noguchi K, Shi W, Tanaka T, Matsumoto M, et al. Targeted disruption of the mouse Stat3 gene leads to early embryonic lethality. *Proc Natl Acad Sci USA* 1997;**94**(8):3801–4.
61. Takeda K, Akira S. STAT family of transcription factors in cytokine-mediated biological responses. *Cytokine Growth Factor Rev* 2000;**11**(3):199–207.
62. Inghirami G, Chiarle R, Simmons WJ, Piva R, Schlessinger K, et al. New and old functions of STAT3: a pivotal target for individualized treatment of cancer. *Cell Cycle* 2005;**4**(9):1131–3.
63. Naugler WE, Karin M. The wolf in sheep's clothing: the role of interleukin-6 in immunity, inflammation and cancer. *Trends Mol Med* 2008;**14**(3):109–19.
64. de la Iglesia N, Konopka G, Puram SV, Chan JA, Bachoo RM, et al. Identification of a PTEN-regulated STAT3 brain tumor suppressor pathway. *Genes Dev* 2008;**22**(4):449–62.
65. Bromberg JF, Wrzeszczynska MH, Devgan G, Zhao Y, Pestell RG, et al. Stat3 as an Oncogene. *Cell* 1999;**98**(3):295–303.
66. Schafer ZT, Brugge JS. IL-6 involvement in epithelial cancers. *J Clin Invest* 2007;**117**(12):3660–3.
67. Levy DE, Loomis CA. STAT3 signaling and the hyper-IgE syndrome. *N Engl J Med* 2007;**357**(16):1655–8.
68. Hunter CA. New IL-12-family members: IL-23 and IL-27, cytokines with divergent functions. *Nat Rev Immunol* 2005;**5**(7):521–31.
69. Zhong Z, Wen Z, Darnell JE. Stat3 and Stat4: Members of the family of signal transducers and activators of transcription. *Proc Nat Acad Sci USA* 1994;**91**:4806–10.
70. Copeland NG, Gilbert DJ, Schindler C, Zhong Z, Wen Z, et al. Distribution of the mammalian Stat gene family in mouse chromosomes. *Genomics* 1995;**29**(1):225–8.
71. Wurster AL, Tanaka T, Grusby MJ. The biology of Stat4 and Stat6. *Oncogene* 2000;**19**(21):2577–84.
72. Morinobu A, Gadina M, Strober W, Visconti R, Fornace A, et al. STAT4 serine phosphorylation is critical for IL-12-induced IFN-gamma production but not for cell proliferation. *Proc Natl Acad Sci USA* 2002;**99**(19):12,281–6.
73. Yao Z, Cui Y, Watford WT, Bream JH, Yamaoka K, et al. Stat5a/b are essential for normal lymphoid development and differentiation. *Proc Natl Acad Sci USA* 2006;**103**(4):1000–5.
74. Hebenstreit D, Wirnsberger G, Horejs-Hoeck J, Duschl A. Signaling mechanisms, interaction partners, and target genes of STAT6. *Cytokine Growth Factor Rev* 2006;**17**(3):173–88.

Part 2

Chromatin Remodeling

Chapter 13

Histone Acetylation Complexes

Tara L. Burke and Patrick A. Grant
Department of Biochemistry and Molecular Genetics, University of Virginia School of Medicine, Charlottesville, Virginia

INTRODUCTION

Dense chromatin structure poses a problem for the transcriptional machinery of eukaryotic cells. Histones are posttranslationally modified at certain residues and these covalent modifications can either allow or prevent access to the associated DNA. A number of covalent modifications occur on histones including acetylation, methylation, phosphorylation, ubiquitylation, and sumoylation. Histone acetylation is one of the most investigated modifications and, in 1964, the first evidence was published that acetylation correlated with transcriptional activation [1, 2]. Acetylation of lysine residues neutralizes the positive charge of the histone proteins, is a reversible process, and can occur at multiple conserved positions. Within the context of different combinations of modifications, acetylation can dictate the transcriptional activity of a gene.

Lysine acetyltransferases (KATs), formerly histone acetyltransferases or HATs, are enzymes that acetylate lysine residues on proteins, including histones [3]. The development of a successful assay to detect KAT activity had impeded their identification for years, but a novel in-gel assay has helped to link KAT activity with specific proteins [4]. The protein p55 in *Tetrahymena thermophilia* was shown to have close homology to the yeast KAT2 (formerly Gcn5) protein, which was also shown to have acetyltransferase activity [5]. KAT2 had previously been described as a transcriptional coactivator due to its ability to enable Gcn4 to promote maximal transcriptional levels in yeast [6]. These seminal studies were the first to directly link histone acetylation with gene activation.

One caveat realized soon after the discovery of the KAT activity of KAT2 was that while KAT2 could effectively acetylate free histones, it could not acetylate relevant nucleosomal histones in a chromatin context. It was soon discovered that KAT2 existed in high molecular weight multi-subunit complexes and that the proteins contained in these complexes helped KAT2 to recognize and acetylate histones within a native nucleosomal substrate. KAT2 was shown to be part of the ADA (alteration/deficiency in activation) and SAGA (Spt-Ada-Gcn5-acetyltransferase) complexes [7]. Since then, not only has KAT2 been shown to be part of other KAT complexes but a number of additional KAT complexes that contain other catalytic subunits have been discovered in various organisms. These complexes vary greatly in their protein subunit composition, substrate specificity, and function with each KAT often having overlapping and multiple roles [8] (Table 13.1). The additional complex subunits, the functional significance for which some is still unknown, regulate and assist in the acetylation of chromatin, contribute to the structural integrity of the complex, or play a role in additional processes. In addition to acetylation, these complexes regulate a number of other modifications via activity of other subunits, such as histone methylation, phosphorylation, and de-ubiquitylation. These complexes modify chromatin using multiple posttranslational modifications to create and maintain functional chromatin regions.

KAT complexes have been implicated in such processes as DNA transcription, repair, and replication and, more recently, the involvement of these complexes in the acetylation non-histone substrates has expanded the role of KATs in other areas of cellular signaling (Figure 13.1). Although the epigenetic role of KAT complexes is well established, we are just beginning to understand the immense diversity of roles these complexes play in nuclear function. This chapter will focus on the two main families of KAT complexes, GNAT and MYST, and an additional orphan group of KATs, highlighting specific complexes that exemplify the multiple roles these KAT activities play. Next, we will examine the role of acetylation in a larger context and

Handbook of Cell Signaling, Three-Volume Set 2 ed.

TABLE 13.1 Classification of GNAT lysine acetyltransferase complexes

	Catalytic subunit	Former name of catalytic subunit [3]	Histones modified	Complex subunits
GNAT				
SAGA (*Sc*)	KAT2	Gcn5	H2B/H3/H4	Tra1, Spt7, Spt8, Spt3, Spt20, Ada1, Ada2, Ada3, Sgf29, Sgf73, Ubp8, Sgf11, Taf5, Taf6, Taf9, Taf10, Taf12, Chd1, Sus1
SLIK (*Sc*)	KAT2	Gcn5	H2B/H3/H4	Tra1, Spt7, Spt3, Spt20, Ada1, Ada2, Ada3, Sgf29, Sgf73, Ubp8, Sgf11, Taf5, Taf6, Taf9, Taf10, Taf12, Rtg2, Chd1
ADA (*Sc*)	KAT2	Gcn5	H3	Ada2, Ada3, Sgf29, Ahc1, Ahc2
HAT-A2 (*Sc*)	KAT2	Gcn5	H3	Ada2, Ada3, Sgf29
SAGA (*Dm*)	KAT2	GCN5	H3	TRA1, SPT7, SPT3, ADA1, ADA2B, ADA3, SGF29, TAF5, TAF6, TAF9, TAF10B, TAF12, WDA
ATAC (*Dm*)	KAT2	GCN5	H3/H4	ADA2A, ADA3, ATAC1, HCF1
PCAF (*Hs*)	KAT2B	PCAF	H3/H4	PAF400, SPT3, ADA2, ADA3, TAF5L, TAF6L, TAF9, TAF10, TAF12, STAF36, STAF46
STAGA (*Hs*)	KAT2A	GCN5L	H3/H4	TRAPP, STAF65y, SPT3, STAF42, STAF54, ATXN7, TAF5L, TAF6L, TAF9, TAF10, TAF12, STAF36, STAF46
TFTC (*Hs*)	KAT2A	GCN5L	H3/H4	TRAPP, SPT3, ADA3, ATXN7, TAF5L, TAF6L, TAF9, TAF10, TAF12, TAF2, TAF4, TAF5, TAF6
Hpa2 (*Sc*)	KAT10	Hpa2	H3/H4	None to date
HATB (*Sc*)	KAT1	Hat1	H2A/H4	Hat2, Hif1
Elongator (*Sc*)	KAT9	Elp3	H3/H4	Elp1, Ep2, Elp4, Elp5, Elp6
Elongator (*Hs*)	KAT9	Elp3	H3/H4	Elp1, Elp3, Elp4, p38, p40

discuss how recognition of acetylation and other modification marks perpetuate epigenetic signaling within the nucleus and, finally, we will touch on the role of KAT complexes in human disease.

KAT CLASSIFICATION AND DIVERSITY

Lysine acetyltransferase complexes are classified based on their acetyltransferase catalytic domains. There are two main groups of KATs, the GNATs, named for KAT2 (*G*cn5) *N*-*a*cetyl*t*ransferases and the MYST family, named after the initial members of this group, KAT6B (*M*OZ), KAT6 (*Y*bfs/Sas3), KAT8 (*S*as2), and KAT5 (*T*ip60) (Table 13.1 and Table 13.2). Additionally there is an orphan group of acetyltransferases that do not contain true consensus KAT domains but possess intrinsic KAT activity (Table 13.2). Although the majority of the original KAT studies were conducted in yeast, there are novel and highly conserved KAT complexes in other organisms rapidly being discovered.

The GNAT Family of KAT Complexes

With a few exceptions, the GNAT family is made up primarily of complexes that contain either KAT2B, formerly PCAF or p300/CBP-associated factor, or KAT2 as their acetyltransferase. KAT2B was pulled out of a human cDNA screen as being 75 percent identical to yKAT2 [9]. Both KAT2 and KAT2B contain intrinsic acetyltransferase activity and possess a highly conserved bromodomain and acetyltransferase domain. Yeast KAT2 is part of four known multi-subunit complexes SAGA (Spt-Ada-Gcn5-acetyltransferase), SLIK/

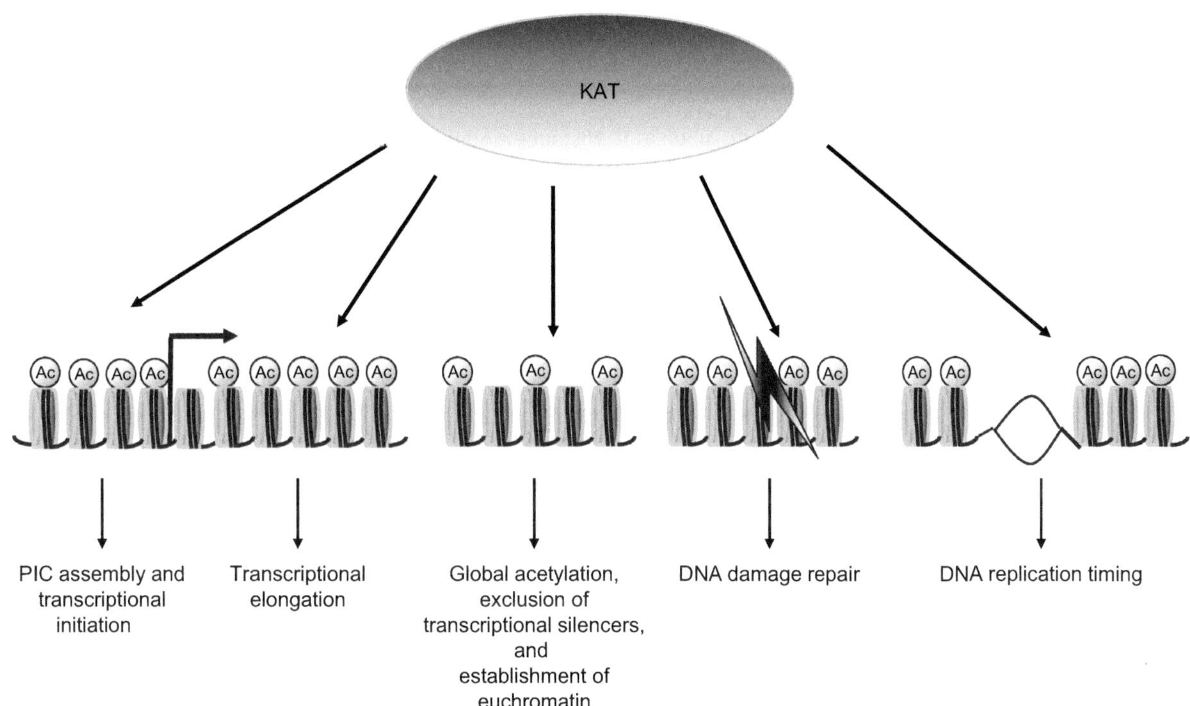

FIGURE 13.1 Roles of KAT complexes in chromatin dynamics.
Acetylation of chromatin by KAT complexes occurs at different stages during transcription. KATs, recruited by DNA transcriptional activators, acetylate the promoter proximal region initiating transcription by assisting in formation of the pre-initiation complex (PIC). Acetylation during transcriptional elongation occurs as RNAPII moves along the coding region of the gene helping to make DNA more accessible to the polymerase machinery. Global acetylation, a non-targeted function of KATs, occurs across chromatin, displaces silencing factors to establish a basic environment of euchromatin. Different KAT complexes are involved in repairing both nucleotide excisions and double-strand breaks. Acetylation of chromatin after DNA damage helps initiate repair of the site. Acetylation during the replication process participates in completing the assembly of the pre-replication complex.

SALSA (SAGA-like, or SAGA altered, Spt8 absent), ADA, and HAT-A2. These complexes contain a number of subunits, which are important in assisting proper and specific activity *in vivo*, and prior to being discovered as subunits in KAT complexes had been shown to be involved in transcription. Members of the GNAT family of KAT complexes have been shown to be involved in multiple processes including transcription initiation, elongation, histone disposition, DNA double-strand break repair, and global acetylation of chromatin [8, 10].

Two of the largest and highly studied GNAT KAT complexes are SAGA and SLIK/SALSA, two highly related complexes that share a number of subunits. SAGA is a 1.8 MDa complex and SLIK is 1.7 MDa in size [7, 11, 12]. These complexes function in the promoter region of genes to acetylate nucleosomes during transcription. In addition to KAT2, SAGA and SLIK contain a distinctive set of proteins that had previously been shown to be involved in transcription; the Spt, Ada, and TAF proteins. Before the isolation of these complexes, it had been suggested through a number of genetic studies that some of these proteins, for example Ada2 and KAT2, might indeed form a complex [13]. Suppressor of Ty (SPT) genes were isolated in a genetic screen for their ability to restore expression of genes disrupted by insertion of the Ty transposon [14]. These complexes minimally contain Spt3, Spt7, and Spt20/Ada5. The Adas (alteration/deficiency in activation) are another group of proteins contained in these complexes and include Ada1, Ada2, Ada3, Ada4/KAT2, and Ada5/Spt20. To date, all Ada genes are contained within SAGA and SLIK. Ada2, the first Ada gene product discovered, was identified for its ability to suppress VP16 toxicity in *S. cerevisiae* when mutated [15]. The TAF (TBP associated factor) proteins contained within these complexes, Taf5, Taf6, Taf9, Taf10, and Taf12, help recruit basal transcriptional machinery, serve as structural subunits and mediate nucleosomal KAT activity [16]. Thus, the KAT2 complexes, SAGA and SLIK, contain a wide range of transcriptional regulators, indicative of functions beyond simple histone acetylation.

The KAT2 related complexes are some of the most well studied KAT complexes. The existence of KAT2 in many different complexes, with different functions and lysine specificities, helps explain the important role of additional complex subunits. For example, KAT2, when part of SAGA or SLIK, acetylates lysines on H3 and H2B with an order of preference for H3 lysines (Lys 14>Lys 18>Lys 9/23) [17]. However, when KAT2 is part of the ADA complex KAT2 still acetylates H3 and H2B but only acetylates Lys 14 and Lys 18 with equal preference for both [17]. Since the dis-

TABLE 13.2 Classification of MYST and other lysine acetyltransferase complexes

	Catalytic subunit	Former name of catalytic subunit [3]	Histones modified	Complex subunits
MYST				
NuA4 (*Sc*)	KAT5	Esa1	H2A/H4	Tra1, Yng2, Yaf9, Eaf1, Eaf2, Eaf3, Eaf5, Eaf6, Eaf7, Epl1, Act1, Arp4
Piccolo NuA4 (*Sc*)	KAT5	Esa1	H2A/H4	Yng2, Epl1
NuA3 (*Sc*)	KAT6	Sas3	H3	Yng1, Taf14, Nto1
SAS (*Sc*)	KAT8	Sas2	H4	Sas4, Sas5
TIP60 (*Dm/Hs*)	KAT5	TIP60	H2A/H4	TRRAP, ING3, p400, BRD8, EPC1, EPC2, DMAP1, RUVBL1, MRG15, BAF53a, Actin, GAS41, MRGX, MRGBP, FLJ11730, YL1, TIP49a, TIP49b, TRCp120
HBO1 (*Hs*)	KAT7	HBO1	H3/H4	ING4/5, JADE1/2/3, EAF6
MOZ/MORF (*Hs*)	KAT6A/6B	MOZ/MORF	H3	ING5, BRPF1, EAF6
MSL (*Dm*)	KAT8	MOF	H4	MSL1, MSL2, MSL3, MLE, roX DNA
CLOCK	KAT13D	CLOCK	H3/H4	BMAL1
Other				
KAT3B/3A (*Hs*)	KAT3B/3A	p300/CBP	H2A/H2B/H3/ H4[1]	None to date
KAT11 (*Sc*)	KAT11	Rtt109	H3	Vps76p or Asf1
TFIIC (*Hs*)	KAT12	TFIIIC90	H2A/H3/H4	TFIII102, TFIII63
TFIID (*Sc/Dm/Hs*)	KAT4	TAF1	H3/H4	TAF2, TAF3, TAF4, TAF5, TAF6, TAF7, TAF8, TAF9, TAF10, TAF11, TAF12, TAF13, TAF14
ATCR/SRC-1 (*Hs*)	KAT13A/B	ATCR/SRC-1	H3/H4	SS-A, IKKβ, IKKγ

[1]*To date, only in vitro activity on all four histones has been observed.*

covery of SAGA in *S. cerevisiae* additional complexes, homologous to SAGA, have been described in humans and fruit flies. Human STAGA/TFTC and PCAF complexes have shown to have many of the same subunits and even STAGA/TFTC was shown to have a similar three-dimensional structure as yeast SAGA [18–20]. These complexes in human, in addition to facilitation of gene transcription via histone acetylation activity, have also been shown to have additional functions such as acetylation of non-histone substrates and DNA damage response. *Drosophila* KAT2 complexes, dSAGA and ATAC, are also very similar in structure and function to yeast SAGA [21, 22].

It is important to note that since many of these complexes are often large, consisting of many different subunits, it is not surprising that these complexes function to assist in transcription and other nuclear functions that go beyond acetylation of histones. The SAGA complex is a prime example of this due to its plethora of subunits, resulting in its extensive role in many nuclear functions [23]. For example, the subunit Tra1, which is a member of both KAT complexes SAGA and NuA4, is important for their interaction with transcriptional activators, selectively directing these complexes to the promoter regions of genes [24]. SAGA subunit, Sus1, which interacts with nuclear export machinery proteins Sac3-Thp1, is important for mRNA export [25]. Also, Ubp8 deubiquitinates Lys 123 of H2B which regulates the nucleosome H3 Lys 4 trimethylation status, a positive mark for active transcription, and thus assisting SAGA in transcriptional regulation independently of its acetylation activity [26–28].

Although the acetylation by these complexes is often referred to in a gene specific manner, these complexes also participate in global acetylation of chromatin. This process is not as well understood but has been shown to occur in yeast, fruit flies, and humans. Global acetylation by yKAT2 has been suggested to loosen chromatin condensation thus making formation of pre-initiation complexes at core promoters more accessible [29]. ySAGA has also been shown to acetylate nucleosomes present in transcribed open reading frames (ORFs) that may be a major part of the global acetylation function of KAT2 containing complexes [30]. Evidence of global acetylation is also present in *Drosophila* and humans. ATAC, dSAGA, and STAGA/TFTC all exhibit some form of global acetylation activity [30]. For example, *Drosophila* cells lacking KAT2 show drastic differences in deposited acetyl marks. Such extreme changes in acetylation, evident in total histone extracts, indicate an important role of KAT2 acetylation beyond regulation of genes at their respective promoters [31]. The converse is also true; when KAT2 is overexpressed, global acetylation also increases [32]. Thus, global acetylation by SAGA related complexes seems to be a well conserved, yet newly discovered function of these complexes. While much is to be learned about the significance of global acetylation, one likely function is in the marking of euchromatin.

In addition to KAT2 and KAT2B, there are also other GNAT family members. The highly conserved KAT, KAT9 (formerly Elongator protein 3, Elp3) is a component of the six subunit elongator complex, which is part of RNA polymerase II (RNAPII) holoenzyme and is involved in transcriptional elongation. KAT9 is the catalytic subunit of the elongator complex and is essential for the integrity of the holo-elongator complex [33]. Recombinant yeast KAT9 acetylates all four core histones *in vitro*, but when KAT9 is part of its complex it has preference for acetylating H3 Lys 14 and H4 Lys 8 [34, 35]. Human KAT9 was shown to have similar function and structure to yeast KAT9, and is able to complement the loss of KAT9 in yeast [36]. Since H3 and H4 are thought to greatly impede movement of RNAPII as it moves along the coding region of genes, acetylation of these two tails via acetyltransferases such as KAT9 may reduce this strain and participate in making transcription more efficient [37].

The KAT complex HATB is interesting in that its catalytic subunit, KAT1 (formerly Hat1) was the first KAT to be identified. Upon discovery, KAT1 was shown to be localized to the cytoplasm and acetylate newly synthesized histones in yeast. Subsequently, KAT1 has been shown to be present in both the cytoplasm and nucleus. KAT1 forms a complex with two other subunits, Hat2 and Hif1, called the HATB complex, which occurs exclusively in the nucleus [38]. Since Hif1 is a known histone chaperone, HATB, with its acetyltransferase and chaperone properties, may direct transfer and deposition of new H3-H4 tetramers during DNA synthesis. Additionally, yeast genetic studies have implicated the HATB complex in DNA double-strand break (DSB) repair and telomeric silencing; however, until recently little was known about how exactly HATB functions in DSB repair [39, 40]. The Parthun Laboratory showed that upon induction of a single DNA DSB, HATB is recruited to the sites of damage, associates with chromatin, and acetylates H4 on lysine 12 [39]. These findings point to a more complicated role of HATB acetylation in the nucleus beyond directed histone deposition.

The MYST Family of KAT Complexes

The MYST family is a diverse group of KATs in which all members contain a MYST homology domain (Table 13.2). This highly conserved domain, along with other sequence similarities, contains the acetyl-CoA binding region, also present in all KAT families. All members of the MYST family, with the exception of the yKAT5, formerly Esa1, also contain a C_2HC finger, which are important *in vivo* and *in vitro* for HAT activity [41]. While many MYST complexes have homologous complexes in other organisms, the MYST KATs themselves are much more diverse and vast than those of the GNAT family. In addition to transcriptional regulation, these KATs are known to be involved in a number of other processes such as development, DNA damage repair, DNA replication, and circadian rhythm [10, 42].

The male specific lethal (MSL) complex is found both in *Drosophila* and humans. The catalytic subunit in this complex is KAT8 (formerly MOF or Male absent on the first) and it acetylates H4 Lys 16 in a very specific manner with no preference for any other histone lysine. KAT8 was discovered in *Drosophila* and was shown to be involved in dosage compensation and required to equalize the expression of genes contained on the X chromosome between males and females [43]. In addition to dosage compensation, the acetylation of H4 Lys 16 by KAT8 has been implicated in a number of cellular processes including transcriptional regulation, establishment of chromatin boundaries, and DNA repair [44]. The acetylation of H4 Lys 16 is particularly interesting because this modification can alter chromatin formation alone such that 30 nm chromatin fiber formation is inhibited [45]. Additionally, knockdown of hKAT8 results in specific loss of H4 Lys 16, thus, bulk acetylation of this residue is largely due to hKAT8 [46]. Both *Drosophila* and human complexes contain similar subunits with the exception of the two non-coding RNAs; roX1 and roX2 that, to date, have only been found in the *Drosophila* complex. Recently, these RNAs were found to be important for function of the MSL complex; binding of roX RNAs alters the binding specificity thereby allowing the subunits MSL1/MSL2 to recognize specific features on the X chromosome [47].

It is also speculated that acetylation of H4 by MSL can recruit certain proteins. Multiple studies point to possible binding factors such as the methyltransferase KMT2A, formerly mixed lineage leukemia 1 (MLL1), where upon binding may help maintain transcriptionally active regions [46]. Also two studies, one in yeast with dKAT8 homolog, Sas2, and the other in *Drosophila* showed that acetylation of this residue inhibits binding of proteins involved in telomeric silencing and chromatin compaction [48, 49]. For example, Sas2 acetylates H4 Lys 16 that, in turn, inhibits binding of the telomeric gene silencing protein, Sir3, thus preventing the spread of heterochromatin beyond the telomeres [49].

KAT8 has also been implicated in DNA damage response. In yeast, upon induction of double-strand break via HO endonuclease in yeast, there is an increase of H4 acetylation, including lysine 16 [50]. Human cells with depleted KAT8 are kinetically slower at DNA repair and show more accumulation of specific protein markers specific for DNA DSBs [51]. Also, human cells depleted for KAT8 accumulate in the G_2/M cell cycle phase [51, 52]. These studies are suggestive of a role for KAT8 in cell cycle progression and DNA repair; however, more studies are needed to decipher the exact role H4 Lys 16 and KAT8 play in these processes.

The yeast KAT5 complex, NuA4, and its human and *Drosophila* counterpart, TIP60, are large multi-subunit complexes with many homologous subunits and functions. The yeast NuA4 complex is composed of 12 subunits with 6 that are essential for viability. yKAT5 is also part of a smaller, yet distinct, complex Piccolo composed of only three of NuA4 subunits. Piccolo, composed of KAT5, Epl1, and Yng2, acetylates global non-targeted chromatin [53]. NuA4 and TIP60 complexes acetylate H4 and H2A. The fact that KAT5 in yeast is an essential protein points to its role in multiple processes. KAT5 acts as a transcriptional activator through its recruitment to promoters of active protein encoding genes, especially ribosomal protein genes, and KAT5 acetylation is important at the *PHO5* promoter where it primes the site for transcriptional activation and chromatin remodeling [54–56]. yKAT5 also plays a vital role in cell cycle progression. Conditional mutants of yKAT5 fail to progress past the G_2/M checkpoint and is RAD9 dependent since RAD9 deletion rescues the G_2/M arrest. The recovery seen by RAD9 deletion is thought to be due to the role of KAT5 in DNA DSB repair, where acetylation of H4 by yKAT5 is required [57]. However, RAD9 deletion in yKAT5 mutants does not prevent loss of cell viability pointing toward the role of yKAT5 in other cell cycle regulatory functions. Additionally, yKAT5 functions in telomeric and rDNA silencing [58].

To date, the function of many of the subunits contained in this large complex have not been elucidated beyond their necessity for cell viability. Recently, with the use of a genome-wide SL-SGA analysis, the potential role of NuA4 has expanded to include its role in Golgi-vacuole vesicle mediated transport, while genetically linking the functions of other non-essential complex subunits to other cellular pathways [59]. Although this study was performed in yeast it helps to confirm and reiterate the multiple roles found for the human complex TIP60. In a parallel study, the role of the yeast subunit, Eaf1, was shown to be an important protein for maintaining complex integrity [60]. Additionally, since the human TIP60 complex contains homologs of NuA4 and SWR1 (an ATP dependent chromatin remodeling complex) subunits, this study suggests that the human complex contains a functional merge of these two separate complexes in yeast [60]. As such, TIP60 acetylates chromatin and mediates ATP dependent exchange of H2A variants. TIP60 also functions in many aspects of cellular function such as cell cycle checkpoint regulation, transcriptional regulation, DNA repair, and nuclear receptor regulation [61].

KAT7, formerly Hbo1/MYST2, is an acetyltransferase that is conserved from flies to humans and is important in cellular processes such as steroid dependent transcription and DNA replication. Through knockdown studies, it was revealed that KAT7 might also participate in acetylating the majority of H4. This KAT is known to be part of two highly related acetyltransferase complexes that contain members of the Jade and ING families [62]. KAT7 was pulled out of a yeast two hybrid study from its interaction with ORC1, a subunit of the origin replication complex and was later shown to interact with an additional DNA replication protein, MCM-2 [63, 64]. KAT7 is required for the chromatin loading of the MCM2-7 complex, allowing completion of the pre-replication complex assembly and initiation of DNA replication. Recent evidence shows that Hbo1 links p53 dependent stress signaling to DNA replication licensing, in that upon initiation of a p53 stress such as hyperosmotic stress, p53 interacts with Hbo1, Hbo1 acetylation activity is inhibited, loading of Mcm2-7 is prevented, and thus replication licensing does not occur [65].

Other KAT Complexes

There is an orphan class of KATs in which the acetyltransferases share little relation to the other classes of KATs (Table 13.2). While there seems to be some distantly related similarities of these KATs to either GNAT or MYST members, these KATs are separated into their own group. Also, the KATs in this group share no significant similarities or relationships among each other. However, like the other classes, the function of the acetyltransferases is diverse. For example, TAF1 is one of the subunits of the general transcriptional factor, TFIID, and is also shown have histone acetyltransferase activity [66]. Although the function of this HAT activity is not completely understood it is speculated that its KAT activity on histones, allows TBP to bind and initiate formation of the transcription pre-initiation complex.

One of the original members of this group is the KAT3B/KAT3A, formerly named p300/CBP. KAT3B and

KAT3A are structural and functional homologs and thus are often grouped together but have been shown to have some independent roles within the cell. KAT3B/KAT3A are known to acetylate both histones and non-histone substrates. *In vitro*, recombinant KAT3B/KAT3A can acetylate all four histones either as free histones or in nucleosomes; this ability is unique in that most acetyltransferases cannot efficiently acetylate nucleosomes alone and usually require additional binding partners. Although KAT3B/KAT3A have binding partners that regulate its activity either negatively or positively, KAT3B/KAT3A is thought to work alone and not need a multi-subunit complex [67]. Non-histone substrates of KAT3B/KAT3A are numerous and include HMG I(Y), p53, GATA-1, and HIV Tat and the acetylation of these proteins by KAT3B/KAT3A regulates the activity of each protein. For example, the acetylation of p53 near the DNA binding domain increases binding of p53 to DNA, thus increasing its activity [68].

KAT11, formerly named Rtt109, is a recently identified yeast acetyltransferase that has been shown to acetylate H3 Lys 56 when associated with either one of the histone chaperones, Vps75p or Asf1p [69]. KAT11 shares no obvious similarities to any of the known acetyltransferases, and, surprisingly, not even the highly conserved domain responsible for binding the cofactor acetyl-CoA contained in other acetyltransferase is present. Since binding of KAT11 to histones is weak, it is thought that by associating with either Vps75p or Asf1 the chaperones help direct KAT11 to its proper substrate for acetylation. What makes this acetylation event interesting is that the acetylated residue is not located in the amino-terminal tail but at the region where H3 makes contact with DNA. It is proposed that the acetylation of lysine 56 at this residue may break this interaction [70]. This elimination of DNA–histone contact is thought to help unwind the 30 nm fibers of DNA. Since mutation of all three proteins, along with H3 Lys 56, sensitize yeast to DNA damaging agents such as methyl methane sulfate and hydroxyurea, it is thought that this modification by KAT11 may assist in DNA DSB repair that may occur during replication, loosening up the DNA–histone contact at the stalled replication fork, and allowing the fork to regress and repair damage [70, 71].

BROMODOMAINS AND OTHER INTERPRETERS OF HISTONE MODIFICATIONS

Aside from the acetylation of lysine 16 on histone H4, most histone modifications are thought to function through the recruitment and stabilization of other proteins to chromatin. Addition of an acetyl group to histones is recognized by proteins containing a bromodomain [72]. Histone lysine acetylation creates a docking site for this protein module. Most bromodomains consist of a conserved four helical left-handed bundle, which is highly conserved [73]. Similar to lysine acetylation itself, this interaction between bromodomain containing proteins and acetylated lysines is highly dynamic and acts to perpetuate signaling within the nucleus. Transcription factors, chromatin associated proteins, including KATs and subunits within KAT complexes, contain bromodomains [74]. Some of these proteins include KAT2B, KAT2, Spt7, and KAT3A [23, 74]. For example, the SAGA complex contains two bromodomain containing proteins, KAT2 and Spt7, the latter being a SAGA subunit involved in complex stability. Although the significance of bromodomain present in Spt7 remains unknown, the bromodomain of KAT2 has been shown to be important for anchoring SAGA to acetylated nucleosomes, allowing SAGA to be stabilized and retained on chromatin by its own acetylation activity [23].

Bromodomains, along with other protein domains that recognize covalent modifications on histones such as chromodomains, WD40 repeats, tudor domains, and PHD fingers, help to interpret the histone code. These other domains are also present in protein subunits of chromatin modifying complexes, including some HAT complexes with many complexes containing a variety of different domains. For example, chromodomains, which recognize methylated histones are present in proteins of NuA4, Tip60, and SAGA [10]. Chromodomains have also been shown to recognize RNAs, as is the case the MSL complex. Both chromodomains of MSL-3 and KAT8 bind the non-coding RNAs present in the complex and are important for proper function [75]. Other domains such as WD40 and PHD domains bind methylated lysines and are also contained within HAT complexes. SAGA, along with chromodomain and bromodomain containing proteins also contain both a WD40 containing protein, Spt8, and tudor domain containing protein, Sgf29 [10]. Although the function of these domains await elucidation, they are thought to read histone marks and interpret their signaling, helping to effectively target and recruit proper HAT activity.

KATS AND DISEASE

It is not surprising that KATs, which function in many different facets of transcriptional regulation, are linked to human disease. Improper function of either the catalytic subunit or associated subunits can have serious effects on transcriptional regulation. Additionally, acetylation of non-histone substrates is coming to light as a vital regulatory process important for proper signaling cascades involved in transcription and other processes. Misregulation of these cascades has disastrous consequences that potentially lead to human disease, especially cancer. Some examples are highlighted below.

Spinocerebellar ataxia (SCA7) is an autosomal dominant neurodegenerative disorder caused by polyglutamine expansion in the Ataxin-7 protein. This disease manifests as retinal and brainstem degeneration and pigmentary macular dystrophy eventually leading to blindness. Ataxin-7 and its yeast homolog Sca7/Sgf73, are subunits of the SAGA family of transcriptional complexes. The pathogenic form of this protein still associates with the complex but there is a decreased association of other subunits such as Ada2, Ada3, and TAF12 [76]. Loss of association of these proteins with the complex results in a loss of Gcn5 mediated KAT activity *in vitro* [76]. In addition, polyglutamine expanded Sca7 also affects SAGA recruitment and pre-initiation complex formation (PIC) resulting in loss of these additional functions of SAGA suggesting that Sca7 also plays an important role in maintenance of complex integrity [77]. In humans, SAGA malfunction via polyglutamine expansion is seen at the promoter of the retinal specific gene, cone-rod homeobox (CRX), a transactivator of photoreceptor genes, where acetylation of H3 by STAGA is inhibited in a dominant negative manner leading to compromised expression of CRX target genes [78].

Since many of these HAT complexes are implicated in cell cycle regulation, DNA replication and transcriptional initiation it is understandable that many of these KATs are linked to oncogenesis. Also many KAT complex subunits and binding partners of KATs are tumor suppressors or oncogenes. For example, the Hbo1 complex contains the ING4 proteins, which is a known tumor suppressor and is involved in maintaining contact inhibition in cells [79]. Human TIP60 interacts with and acetylates the androgen receptor and has been implicated in prostate cancer [80]. TIP60 also contributes to transcription of genes regulated by tumor suppressors, p53, and Myc, and staining of human tumors with mutated p53 show decreased staining of TIP60 in the nuclei of various different cancer cells, suggesting that these two proteins function in parallel pathways. These data also suggest that loss of TIP60 may exacerbate cellular misregulation exhibited after loss of p53 [81]. The KATs KAT3B/KAT3A are also known tumor suppressors. A monoallelic mutation in KAT3A is known to cause Rubinstein-Taybi syndrome, which results in an increased risk of developing malignant tumors. Some epithelial cancers show biallelic mutations of the KAT3B and when this protein is overexpressed, growth of human carcinoma cells is suppressed [82]. Providing direct evidence that KATs are linked to cancer is the discovery that both of these proteins are translocated in leukemia. KAT6A, KAT6B, and the histone methyltransferase KMT2A, are three known fusion partners of KAT3B and KAT3A and the KAT domains of KAT3A and KAT3B are known to be necessary for the leukemogenic activity of these fused proteins [83]. For example, the KMT2A-CBP fusion protein causes aberrant acetylation at the loci of HOX genes leading to increased expression of these genes and development of leukemia [84].

CONCLUSION AND FUTURE DIRECTIONS

We have only touched on the diversity and complexity of KAT complexes and their role in nuclear function. Evident in this chapter is the ever expanding role of these enzymes in cellular signaling, their highly conserved functions among species, and their complex architecture. The interplay of acetylation with other posttranslational modifications on histones regulates chromatin structure and accessibility contributing to epigenetic control of DNA transcription, replication, and repair. The continued discovery of acetylation of non-histone proteins points toward a broader participation of acetylation as a regulator of protein signaling. As more and more roles of acetylation are discovered, distinct parallels between protein acetylation and phosphorylation can be drawn pointing toward an even more vast use of acetylation as a regulatory modification within the cell. Although to date, most of the KAT signaling is focused within the nucleus, it will be interesting to see if signaling via acetylation marks also becomes as common a method of protein signaling within the cytoplasm as well.

In addition, the role of acetylation in temporal changes such as circadian rhythm, development, and tissue specificity suggests that regulation of these catalytic activities and modifications are also critical processes. Understanding KAT function also proves to be an important task in understanding various human diseases. As described above, KAT misregulation can lead to human diseases and is involved in a number of cancers. To date, much attention has focused on the use of histone deacetylase (HDAC) inhibitors as treatment for cancer and other human diseases. However, exploring potential small molecule inhibitors for KAT activity should also be considered as it garners much therapeutic potential.

ACKNOWLEDGEMENTS

We apologize to colleagues whose work could not be cited due to space limitations. Research in the Grant Lab is funded from National Institutes of Health R01 Grant no. NS049065. TLB is supported by National Institutes of Health predoctoral cancer training grant no. 2 T32 CA009109-31A1.

REFERENCES

1. Allfrey VG, Faulkner R, Mirsky AE. Acetylation and Methylation of Histones and their Possible Role in the Regulation of Rna Synthesis. *Proc Natl Acad Sci USA* 1964;**51**:786–94.
2. Marmorstein R, Roth SY. Histone acetyltransferases: function, structure, and catalysis. *Curr Opin Genet Dev* 2001;**11**:155–61.
3. Allis CD, Berger SL, Cote J, et al. New nomenclature for chromatin-modifying enzymes. *Cell* 2007;**131**:633–6.

4. Brownell JE, Allis CD. An activity gel assay detects a single, catalytically active histone acetyltransferase subunit in Tetrahymena macronuclei. *Proc Natl Acad Sci USA* 1995;**92**:6364–8.
5. Brownell JE, Zhou J, Ranalli T, et al. Tetrahymena histone acetyltransferase A: a homolog to yeast Gcn5p linking histone acetylation to gene activation. *Cell* 1996;**84**:843–51.
6. Georgakopoulos T, Thireos G. Two distinct yeast transcriptional activators require the function of the GCN5 protein to promote normal levels of transcription. *EMBO J* 1992;**11**:4145–52.
7. Grant PA, Duggan L, Cote J, et al. Yeast Gcn5 functions in two multisubunit complexes to acetylate nucleosomal histones: characterization of an Ada complex and the SAGA (Spt/Ada) complex. *Genes Dev* 1997;**11**:1640–50.
8. Torok MS, Grant PA. Histone acetyltransferase proteins contribute to transcriptional processes at multiple levels. *Adv Protein Chem* 2004;**67**:181–99.
9. Yang XJ, Ogryzko VV, Nishikawa J, et al. A p300/CBP-associated factor that competes with the adenoviral oncoprotein. E1A. *Nature* 1996;**382**:319–24.
10. Lee KK, Workman JL. Histone acetyltransferase complexes: one size doesn't fit all. *Nat Rev Mol Cell Biol* 2007;**8**:284–95.
11. Pray-Grant MG, Schieltz D, McMahon SJ, et al. The novel SLIK histone acetyltransferase complex functions in the yeast retrograde response pathway. *Mol Cell Biol* 2002;**22**:8774–86.
12. Sterner DE, Belotserkovskaya R, Berger SL. SALSA, a variant of yeast SAGA, contains truncated Spt7, which correlates with activated transcription. *Proc Natl Acad Sci USA* 2002;**99**:11,622–11,627.
13. Marcus GA, Silverman N, Berger SL, et al. Functional similarity and physical association between GCN5 and ADA2: putative transcriptional adaptors. *EMBO J* 1994;**13**:4807–15.
14. Winston F, Chaleff DT, Valent B, Fink GR. Mutations affecting Ty-mediated expression of the HIS4 gene of Saccharomyces cerevisiae. *Genetics* 1984;**107**:179–97.
15. Berger SL, Pina B, Silverman N, et al. Genetic isolation of ADA2: a potential transcriptional adaptor required for function of certain acidic activation domains. *Cell* 1992;**70**:251–65.
16. Grant PA, Schieltz D, Pray-Grant MG, et al. A subset of TAF(II)s are integral components of the SAGA complex required for nucleosome acetylation and transcriptional stimulation. *Cell* 1998;**94**:45–53.
17. Grant PA, Berger SL. Histone acetyltransferase complexes. *Semin Cell Dev Biol* 1999;**10**:169–77.
18. Wu PY, Ruhlmann C, Winston F, Schultz P. Molecular architecture of the S. cerevisiae SAGA complex. *Mol Cell* 2004;**15**:199–208.
19. Brand M, Yamamoto K, Staub A, Tora L. Identification of TATA-binding protein-free TAFII-containing complex subunits suggests a role in nucleosome acetylation and signal transduction. *J Biol Chem* 1999;**274**:18,285–18,289.
20. Martinez E, Kundu TK, Fu J, Roeder RG. A human SPT3-TAFII31-GCN5-L acetylase complex distinct from transcription factor IID. *J Biol Chem* 1998;**273**:23,781–23,785.
21. Guelman S, Suganuma T, Florens L, et al. Host cell factor and an uncharacterized SANT domain protein are stable components of ATAC, a novel dAda2A/dGcn5-containing histone acetyltransferase complex in Drosophila. *Mol Cell Biol* 2006;**26**:871–82.
22. Kusch T, Guelman S, Abmayr SM, Workman JL. Two Drosophila Ada2 homologues function in different multiprotein complexes. *Mol Cell Biol* 2003;**23**:3305–19.
23. Baker SP, Grant PA. The SAGA continues: expanding the cellular role of a transcriptional co-activator complex. *Oncogene* 2007;**26**:5329–40.
24. Brown CE, Howe L, Sousa K, et al. Recruitment of HAT complexes by direct activator interactions with the ATM-related Tra1 subunit. *Science* 2001;**292**:2333–7.
25. Rodriguez-Navarro S, Fischer T, Luo MJ, et al. Sus1, a functional component of the SAGA histone acetylase complex and the nuclear pore-associated mRNA export machinery. *Cell* 2004;**116**:75–86.
26. Daniel JA, Torok MS, Sun ZW, et al. Deubiquitination of histone H2B by a yeast acetyltransferase complex regulates transcription. *J Biol Chem* 2004;**279**:1867–71.
27. Henry KW, Wyce A, Lo WS, et al. Transcriptional activation via sequential histone H2B ubiquitylation and deubiquitylation, mediated by SAGA-associated Ubp8. *Genes Dev* 2003;**17**:2648–63.
28. Santos-Rosa H, Schneider R, Bannister AJ, et al. Active genes are tri-methylated at K4 of histone H3. *Nature* 2002;**419**:407–11.
29. Imoberdorf RM, Topalidou I, Strubin M. A role for gcn5-mediated global histone acetylation in transcriptional regulation. *Mol Cell Biol* 2006;**26**:1610–16.
30. Nagy Z, Tora L. Distinct GCN5/PCAF-containing complexes function as co-activators and are involved in transcription factor and global histone acetylation. *Oncogene* 2007;**26**:5341–57.
31. Kikuchi H, Takami Y, Nakayama T. GCN5: a supervisor in all-inclusive control of vertebrate cell cycle progression through transcription regulation of various cell cycle-related genes. *Gene* 2005;**347**:83–97.
32. Knoepfler PS, Zhang XY, Cheng PF, et al. Myc influences global chromatin structure. *EMBO J* 2006;**25**:2723–34.
33. Petrakis TG, Wittschieben BO, Svejstrup JQ. Molecular architecture, structure-function relationship, and importance of the Elp3 subunit for the RNA binding of holo-elongator. *J Biol Chem* 2004;**279**:32,087–32,092.
34. Winkler GS, Kristjuhan A, Erdjument-Bromage H, et al. Elongator is a histone H3 and H4 acetyltransferase important for normal histone acetylation levels *in vivo*. *Proc Natl Acad Sci USA* 2002;**99**:3517–22.
35. Wittschieben BO, Otero G, de Bizemont T, et al. A novel histone acetyltransferase is an integral subunit of elongating RNA polymerase II holoenzyme. *Mol Cell* 1999;**4**:123–8.
36. Li F, Lu J, Han Q, et al. The Elp3 subunit of human Elongator complex is functionally similar to its counterpart in yeast. *Mol Genet Genomics* 2005;**273**:264–72.
37. Protacio RU, Li G, Lowary PT, Widom J. Effects of histone tail domains on the rate of transcriptional elongation through a nucleosome. *Mol Cell Biol* 2000;**20**:8866–78.
38. Ai X, Parthun MR. The nuclear Hat1p/Hat2p complex: a molecular link between type B histone acetyltransferases and chromatin assembly. *Mol Cell* 2004;**14**:195–205.
39. Qin S, Parthun MR. Histone H3 and the histone acetyltransferase Hat1p contribute to DNA double-strand break repair. *Mol Cell Biol* 2002;**22**:8353–65.
40. Kelly TJ, Qin S, Gottschling DE, Parthun MR. Type B histone acetyltransferase Hat1p participates in telomeric silencing. *Mol Cell Biol* 2000;**20**:7051–8.
41. Carrozza MJ, Utley RT, Workman JL, Cote J. The diverse functions of histone acetyltransferase complexes. *Trends Genet* 2003;**19**:321–9.
42. Avvakumov N, Cote J. The MYST family of histone acetyltransferases and their intimate links to cancer. *Oncogene* 2007;**26**:5395–407.
43. Hilfiker A, Hilfiker-Kleiner D, Pannuti A, Lucchesi JC. mof, a putative acetyl transferase gene related to the Tip60 and MOZ human genes and to the SAS genes of yeast, is required for dosage compensation in Drosophila. *EMBO J* 1997;**16**:2054–60.
44. Rea S, Xouri G, Akhtar A. Males absent on the first (MOF): from flies to humans. *Oncogene* 2007;**26**:5385–94.

45. Shogren-Knaak M, Ishii H, Sun JM, et al. Histone H4-K16 acetylation controls chromatin structure and protein interactions. *Science* 2006;**311**:844–7.
46. Dou Y, Milne TA, Tackett AJ, et al. Physical association and coordinate function of the H3 K4 methyltransferase MLL1 and the H4 K16 acetyltransferase MOF. *Cell* 2005;**121**:873–85.
47. Li F, Schiemann AH, Scott MJ. Incorporation of the noncoding roX RNAs alters the chromatin-binding specificity of the Drosophila MSL1/MSL2 complex. *Mol Cell Biol* 2008;**28**:1252–64.
48. Corona DF, Clapier CR, Becker PB, Tamkun JW. Modulation of ISWI function by site-specific histone acetylation. *EMBO Rep* 2002;**3**:242–7.
49. Suka N, Luo K, Grunstein M. Sir2p and Sas2p opposingly regulate acetylation of yeast histone H4 lysine16 and spreading of heterochromatin. *Nat Genet* 2002;**32**:378–83.
50. Tamburini BA, Tyler JK. Localized histone acetylation and deacetylation triggered by the homologous recombination pathway of double-strand DNA repair. *Mol Cell Biol* 2005;**25**:4903–13.
51. Taipale M, Rea S, Richter K, et al. hMOF histone acetyltransferase is required for histone H4 lysine 16 acetylation in mammalian cells. *Mol Cell Biol* 2005;**25**:6798–810.
52. Smith ER, Cayrou C, Huang R, et al. A human protein complex homologous to the Drosophila MSL complex is responsible for the majority of histone H4 acetylation at lysine. 16. *Mol Cell Biol* 2005;**25**:9175–88.
53. Boudreault AA, Cronier D, Selleck W, et al. Yeast enhancer of polycomb defines global Esa1-dependent acetylation of chromatin. *Genes Dev* 2003;**17**:1415–28.
54. Reid JL, Iyer VR, Brown PO, Struhl K. Coordinate regulation of yeast ribosomal protein genes is associated with targeted recruitment of Esa1 histone acetylase. *Mol Cell* 2000;**6**:1297–307.
55. Robert F, Pokholok DK, Hannett NM, et al. Global position and recruitment of HATs and HDACs in the yeast genome. *Mol Cell* 2004;**16**:199–209.
56. Nourani A, Utley RT, Allard S, Cote J. Recruitment of the NuA4 complex poises the PHO5 promoter for chromatin remodeling and activation. *EMBO J* 2004;**23**:2597–607.
57. Bird AW, Yu DY, Pray-Grant MG, et al. Acetylation of histone H4 by Esa1 is required for DNA double-strand break repair. *Nature* 2002;**419**:411–15.
58. Clarke AS, Samal E, Pillus L. Distinct roles for the essential MYST family HAT Esa1p in transcriptional silencing. *Mol Biol Cell* 2006;**17**:1744–57.
59. Mitchell L, Lambert JP, Gerdes M, et al. Functional dissection of the NuA4 histone acetyltransferase reveals its role as a genetic hub and that Eaf1 is essential for complex integrity. *Mol Cell Biol* 2008;**28**:2244–56.
60. Auger A, Galarneau L, Altaf M, et al. Eaf1 is the platform for NuA4 molecular assembly that evolutionarily links chromatin acetylation to ATP-dependent exchange of histone H2A variants. *Mol Cell Biol* 2008;**28**:2257–70.
61. Sapountzi V, Logan IR, Robson CN. Cellular functions of TIP60. *Int J Biochem Cell Biol* 2006;**38**:1496–509.
62. Doyon Y, Cayrou C, Ullah M, et al. ING tumor suppressor proteins are critical regulators of chromatin acetylation required for genome expression and perpetuation. *Mol Cell* 2006;**21**:51–64.
63. Iizuka M, Stillman B. Histone acetyltransferase HBO1 interacts with the ORC1 subunit of the human initiator protein. *J Biol Chem* 1999;**274**:23,027–23,034.
64. Burke TW, Cook JG, Asano M, Nevins JR. Replication factors MCM2 and ORC1 interact with the histone acetyltransferase HBO1. *J Biol Chem* 2001;**276**:15,397–15,408.
65. Iizuka M, Sarmento OF, Sekiya T, et al. Hbo1 Links p53 dependent stress signaling to DNA replication licensing. *Mol Cell Biol* 2008; **28**:140–53.
66. Mizzen CA, Yang XJ, Kokubo T, et al. The TAF(II)250 subunit of TFIID has histone acetyltransferase activity. *Cell* 1996;**87**:1261–70.
67. Iyer NG, Ozdag H, Caldas C. p300/CBP and cancer. *Oncogene* 2004;**23**:4225–31.
68. Gu W, Shi XL, Roeder RG. Synergistic activation of transcription by CBP and p53. *Nature* 1997;**387**:819–23.
69. Tsubota T, Berndsen CE, Erkmann JA, et al. Histone H3-K56 acetylation is catalyzed by histone chaperone-dependent complexes. *Mol Cell* 2007;**25**:703–12.
70. Ozdemir A, Masumoto H, Fitzjohn P, et al. Histone H3 lysine 56 acetylation: a new twist in the chromosome cycle. *Cell Cycle* 2006;**5**:2602–8.
71. Peterson CL. Genome integrity: a HAT needs a chaperone. *Curr Biol* 2007;**17**:R324–6.
72. Jacobson RH, Ladurner AG, King DS, Tjian R. Structure and function of a human TAFII250 double bromodomain module. *Science* 2000;**288**:1422–5.
73. Dhalluin C, Carlson JE, Zeng L, et al. Structure and ligand of a histone acetyltransferase bromodomain. *Nature* 1999;**399**:491–6.
74. Mujtaba S, Zeng L, Zhou MM. Structure and acetyl-lysine recognition of the bromodomain. *Oncogene* 2007;**26**:5521–7.
75. Akhtar A, Zink D, Becker PB. Chromodomains are protein-RNA interaction modules. *Nature* 2000;**407**:405–9.
76. McMahon SJ, Pray-Grant MG, Schieltz D, et al. Polyglutamine-expanded spinocerebellar ataxia-7 protein disrupts normal SAGA and SLIK histone acetyltransferase activity. *Proc Natl Acad Sci USA* 2005;**102**:8478–82.
77. Shukla A, Bajwa P, Bhaumik SR. SAGA-associated Sgf73p facilitates formation of the preinitiation complex assembly at the promoters either in a HAT-dependent or independent manner *in vivo*. *Nucleic Acids Res* 2006;**34**:6225–32.
78. Palhan VB, Chen S, Peng GH, et al. Polyglutamine-expanded ataxin-7 inhibits STAGA histone acetyltransferase activity to produce retinal degeneration. *Proc Natl Acad Sci USA* 2005;**102**:8472–7.
79. Kim S, Chin K, Gray JW, Bishop JM. A screen for genes that suppress loss of contact inhibition: identification of ING4 as a candidate tumor suppressor gene in human cancer. *Proc Natl Acad Sci USA* 2004;**101**: 16,251–16,256.
80. Brady ME, Ozanne DM, Gaughan L, et al. Tip60 is a nuclear hormone receptor coactivator. *J Biol Chem* 1999;**274**:17,599–17,604.
81. Gorrini C, Squatrito M, Luise C, et al. Tip60 is a haplo-insufficient tumour suppressor required for an oncogene-induced DNA damage response. *Nature* 2007;**448**:1063–7.
82. Suganuma T, Kawabata M, Ohshima T, Ikeda MA. Growth suppression of human carcinoma cells by reintroduction of the p300 coactivator. *Proc Natl Acad Sci USA* 2002;**99**:13,073–13,078.
83. Yang XJ. The diverse superfamily of lysine acetyltransferases and their roles in leukemia and other diseases. *Nucleic Acids Res* 2004;**32**:959–76.
84. Ayton PM, Cleary ML. Transformation of myeloid progenitors by MLL oncoproteins is dependent on Hoxa7 and Hoxa9. *Genes Dev* 2003; **17**:2298–307.

Chapter 14

Regulation of Histone Deacetylase Activities and Functions by Phosphorylation and Dephosphorylation

Edward Seto[1] and Xiang-Jiao Yang[2]

[1]*Molecular Oncology Program, H. Lee Moffitt Cancer Center and Research Institute, Tampa, Florida*

[2]*Molecular Oncology Group, Department of Medicine, McGill University Health Center and McGill Cancer Center, McGill University, Montréal, Québec, Canada*

INTRODUCTION

Histone deacetylases (HDACs) were originally identified as enzymes that catalyze the removal of acetyl moieties from the ε-amino groups of conserved lysine residues in the amino terminal tail of histones. The removal of this modification strengthens histone–DNA interactions and/or generateS specific docking surfaces for proteins that regulate chromatin folding and/or transcription. The discovery of HDAC enzymatic activity was first reported in 1969 [1] and, despite early awareness of the potential biological importance, histone deacetylation research was dormant for a long time. Many initial attempts to purify HDACs to homogeneity using conventional chromatography were unsuccessful. In 1996, the first *bona fide* histone deacetylase, HDAC1, was isolated and cloned [2] spurring an explosion of the HDAC field. Since that time, more than 6,000 papers have been published on this topic (compared to a total of 85 papers published until that point in time). Most discussions on eukaryotic transcriptional repression now refer to some aspects of histone deacetylation. We also now know that HDACs play crucial roles in many biological processes beyond histone and gene regulation.

In humans, HDACs are divided into four Classes [3, 4]: the Class I RPD3-like proteins (HDAC1, HDAC2, HDAC3, and HDAC8); the Class II HDA1-like proteins (HDAC4, HDAC5, HDAC6, HDAC7, HDAC9, and HDAC10); and the Class III SIR2-like proteins (SIRT1, SIRT2, SIRT3, SIRT4, SIRT5, SIRT6, and SIRT7). According to domain organization, the Class II enzymes can be further divided into two subclasses: IIa (HDAC4, HDAC5, HDAC7, and HDAC9) and IIb (HDAC6 and HDAC10). HDAC11, the unique member of the Class IV enzyme, shares sequence homology to the catalytic domains of both Class I and II HDAC enzymes [5]. The Class III proteins do not exhibit any sequence similarity to HDAC family members from Classes I, II, or IV, and differ from these members in that they require the cofactor NAD rather than zinc for deacetylase activity.

The current studies of HDACs can be divided into many areas. Several examples follow. First, many laboratories are engaged in determining whether a particular HDAC preferentially deacetylates some histones and whether that HDAC mediated deacetylation occurs preferentially on particular lysine residues of those histones. With the successful development of many antibodies that specifically recognize particular site specific histone acetylation, insights into this area are continuously evolving. Second, it is important to determine whether the deacetylase activity of HDACs *per se* is necessary for repression of gene transcription. This is a critical issue since many laboratories have found that some HDACs possess autonomous repressor functions that are independent of their deacetylase activity (e.g., see ref. [6]). A "one size fits all" model in which HDACs mediate transcriptional repression via deacetylation of core histones may not always be correct. Specifically, under what conditions is the targeting of HDACs to promoters the mechanism by which HDACs represses transcription? Are histones the physiological substrates for HDACs when they are targeted to promoters? Finally, an active area of HDAC research is to identify non-histone substrates of HDACs and to understand the functional consequences of non-histone deacetylation (e.g., see ref. [7]).

Although understanding the nature and mechanisms of histone and non-histone deacetylation is clearly an important

Handbook of Cell Signaling, Three-Volume Set 2 ed.

endeavor for current and future research, an equally important immediate need is to address the mechanisms by which HDACs themselves are regulated. Like many cellular enzymes, HDACs are mediators of cell signals. HDACs are downstream targets of signaling pathways, and signaling molecules (kinases in particular) regulate the activities of HDACs through multiple diverse pathways. HDACs in turn modulate different signaling pathways by deacetylation of histones and non-histones or through as yet unknown mechanisms. The focus of this review is on the regulation of mammalian HDAC activities and functions by phosphorylation and dephosphorylation.

REVERSIBLE PHOSPHORYLATION OF MAMMALIAN CLASS I HDACs

Complex, Multifaceted, Regulation of HDAC1 and HDAC2 Activities by Kinases and Phosphatases

The first detailed studies of Class I HDAC phosphorylation were reported independently from the laboratories of Stuart Schreiber [8] and Dalia Cohen [9]. With the goal of exploring the possibility of HDAC1 regulation by posttranslational modifications, Pflum *et al.* treated a cell lysate with alkaline phosphatase and visualized HDAC1 on a western blot with an anti-HDAC1 antibody. In the presence of phosphatase, HDAC1 migrated faster than untreated HDAC1 in a polyacrylamide gel, providing the first demonstration that HDAC1 might be phosphorylated. Subsequent *in vivo* [^{32}P]-orthophosphate labeling of cells, followed by immunoprecipitation with an anti-HDAC1 antibody, clearly confirmed that HDAC1 is posttranslationally modified by phosphorylation. Ion trap mass spectrometry analysis identified two phospho-acceptor sites on the carboxyl terminal of HDAC1, Ser421 and Ser423 (Figure 14.1 and Table 14.1). The functional relevance of HDAC1 phosphorylation came with the finding that mutation of Ser421 and Ser423 to alanines resulted in a significant reduction in enzymatic activity. Further, alanine substitution mutations disrupted protein complex formation of HDAC1 with RbAp48, MTA2, mSin3A, and CoREST. Glutamic acid and aspartic acid partially (but not completely) substituted for the phosphoserines, suggesting that in addition to charge, the unique property of phosphate contributes to HDAC1 function. Protein kinase CK2, but not CK1, phosphorylates HDAC1 *in vitro*. However, although endogenous HDAC1 and CK2 co-immunoprecipitates *in vivo*, the phosphorylation state of HDAC1 is not affected by DRB, a specific small molecule inhibitor of CK2 enzymatic activity.

Consistent with the data presented by Pflum *et al.*, *in vivo* labeling assays by Cai *et al.* demonstrated that both human and murine HDAC1 proteins are phosphorylated in cells. Also, assays using HDAC1 deletion mutants indicated that phosphorylation occurs in the C-terminal, consistent with the fact that HDAC1 is phosphorylated at Ser421 and Ser423. In the study by Cai *et al.*, it was shown that cAMP dependent kinase PKA and protein kinase CK2, but not PKC, cdc2, or MAP kinase, phosphorylated HDAC1 *in vitro*. In sharp contrast to the results presented by Pflum *et al.*, Cai *et al.* found that phosphorylation did not influence HDAC1 enzymatic activity. The use of different substrates, synthetic histone H4 N-terminal peptide by Cai *et al.* and histone isolated from HeLa cells by Pflum *et al.*, could not account for the different results because phosphorylated HDAC1 and HDAC2 were active toward a H4 peptide substrate in a separate study [10]. In a recent

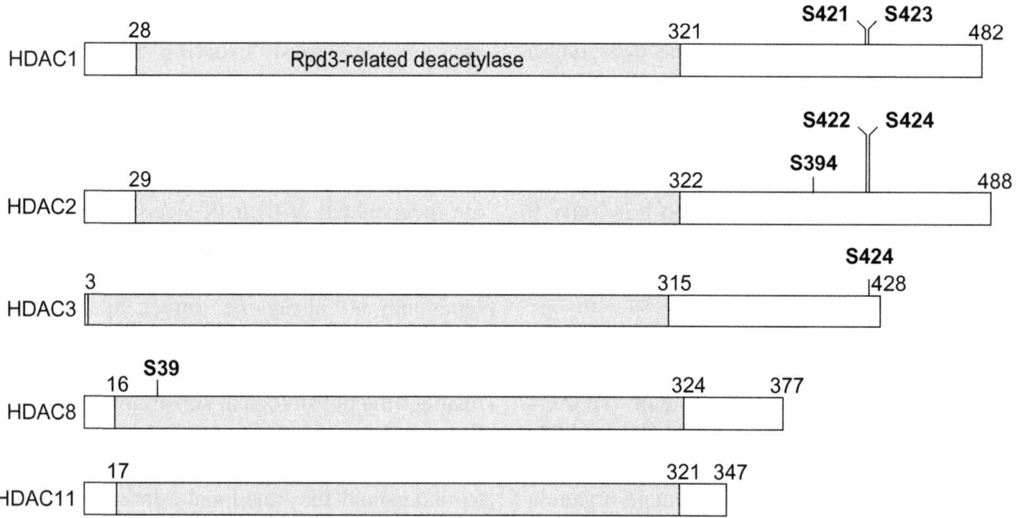

FIGURE 14.1 Phosphorylation sites on Class I HDACs.
Simplified representation of Class I HDAC proteins. Conserved "HDAC domains" are indicated with shaded areas. Position of known phosphorylation sites is indicated in bold.

TABLE 14.1 Kinases and phosphatases for Class I HDACs

	Kinase	Phospho-acceptor site	Consequence of phosphorylation	Phosphatase
HDAC1	CK2	S421	Increases deacetylase activity,	PP1
		S423	alters HDAC–protein complexes	
HDAC2	CK2	S394	Increases deacetylase activity,	PP1
		S422	alters HDAC–protein complexes	
		S424		
HDAC3	CK2	S424	Increases deacetylase activity	PP4
	PKcs	Not yet identified	Increases deacetylase activity	?
HDAC8	PKA	S39	Decreases deacetylase activity	?
HDAC11[a]	?	?	?	?

[a]HDAC11, a Class IV enzyme, is listed here for convenience. Although there is no report yet that HDAC11 is a phospho-protein, a casual inspection of the HDAC11 amino acid sequence reveals 12 (9 Ser, 2 Thr, and 1 Tyr) potential phosphorylation sites.

study, using limited trypsin proteolysis of HDAC1, it was discovered that phosphorylation probably does not induce a conformational change, but instead promoted protein associations to increase HDAC1 activity [11]. However, using phospho specific antibodies to Ser421 or Ser423, it was concluded that HDAC1 Ser421 or Ser423 phosphorylation is constitutive *in vivo* and dispensable for activity *in vitro* [12]. Also, HDAC1 treated with the calf intestinal phosphatase remained catalytically active *in vitro* [13].

HDAC2, like HDAC1, is a phospho-protein [14]. This came as no surprise, since human HDAC1 and HDAC2 are 85 percent identical in overall protein sequence, and Ser421 and Ser423 of HDAC1 are conserved in HDAC2 (Ser422 and Ser424). There are several similarities and differences between the phosphorylation of HDAC1 and HDAC2. First, interestingly, HDAC2 is phosphorylated at Ser394, in addition to Ser422 and Ser424. Although Pflum *et al.* did not identify phosphorylation in Ser393 of HDAC1 (corresponding to Ser394 of HDAC2), Cai *et al.* did report that at least one phosphorylation site exists between residues 387 and 409 of HDAC1 (corresponding to residues 388 to 410 of HDAC2). Additionally, a global analysis of protein phosphorylation sites did reveal that Ser393 of HDAC1 is phosphorylated [15]. Second, in contrast to HDAC1, which can be phosphorylated *in vitro* by CK2, PKA, and PKG, HDAC2 is phosphorylated exclusively by protein kinase CK2. Third, like HDAC1, HDAC2 phosphorylation promotes enzymatic activity and affected protein complex formation with mSin3 and Mi2. However, unlike HDAC1, phosphorylation of HDAC2 did not affect transcription repression when targeted to promoters via a Gal4 DNA binding domain in transient transfection, assayed with luciferase reporters. A recent study provided evidences that unmodified HDAC2 is associated with the coding region of transcribed genes, whereas phosphorylated HDAC2 is primarily recruited to promoters [16]. This may perhaps explain why phosphorylation had no effect on transcription when HDAC2 was targeted to a promoter using an HDAC2 protein fused to the Gal4 DNA binding domain.

CK2 is a serine/threonine protein kinase ubiquitously distributed in both the cytoplasm and nucleus of eukaryotic cells [17]. It is constitutively active and has been implicated in the regulation of many cellular processes, including DNA replication, basal and inducible transcription, and the regulation of cell growth and metabolism. CK2 has a broad range of substrates that include nuclear oncoproteins such as Myc, Myb, Jun, and Fos. It plays an important role in the transduction of extracellular signals to effector proteins in the nucleus and appears to be responsible for the phosphorylation of many growth related and cell cycle specific proteins. CK2 concentrations are increased in rapidly proliferating tumors, and studies have shown that dys regulated CK2 expression can be oncogenic. Although HDAC1 and HDAC2 are critical in cell cycle control, and there is circumstantial evidence that they might be involved in cancer development, whether HDAC1 and HDAC2 serve as mediators in the role of CK2 in oncogenesis remains to be determined. In this regard, it is interesting to note that transcription factors Sp1 and Sp3, which have key roles in estrogen induced proliferation and gene expression in estrogen dependent breast cancer cells with upregulated CK2, bind preferentially to phosphorylated HDAC2 [18].

Also, protein kinase CK2 is a key activator of HDAC1 and HDAC2 in hypoxia associated tumors [19].

Because results from mass spectrometry analysis show that HDAC1 was always doubly phosphorylated at Ser421 and Ser423, and the fact that HDAC1 migrates as a single band by PAGE, it was hypothesized that HDAC1 is constitutively phosphorylated and inaccessible to phosphatase enzymes [8]. Surprisingly, the work from Natalie Ahn's laboratory shows that treatment of cells with the protein phosphatase inhibitor okadaic acid led to the hyperphosphorylation of HDAC1 and HDAC2 [10]. The okadaic acid concentration range used is known to inhibit both PP1 and PP2A, and further analysis based on differential sensitivities to individual phosphatases indicates that hyperphosphorylation of HDAC1 and HDAC2 is most likely physiologically regulated by PP1. HDAC1 and HDAC2 were shown to exist in three distinct phosphorylation forms: unphosphorylated, basally phosphorylated, and hyperphosphorylated. Interestingly, the site(s) of basal phosphorylation in HDAC2 differ from the okadaic acid induced site(s) occupied in the hyperphosphorylated state. Consistent with the earlier studies [8, 14], phosphorylation of HDAC1 and HDAC2 correlated with an increase in enzymatic activities, and the increase is reversed upon phosphatase treatment. However, in contrast to other studies, HDAC1 and HDAC2 phosphorylation disrupted HDAC-mSin3A and HDAC–YY1 interactions, but had no effect on the HDAC–RbAp46/48 complex. These data clearly confirm the importance of HDAC1 and HDAC2 phosphorylation in protein complex formation. However, it also points to the possibility that basal phosphorylation of HDAC1 and HDAC2 dictates their ability to interact with certain cellular factors, but hyperphosphorylation of HDAC1 or HDAC2 can discriminate against complex formation.

Perhaps the most intriguing finding that came out of the study by Galasinski *et al.*, is that mitotic arrest resulted in hyperphosphorylation of HDAC2 without any change in HDAC1. Since HDAC2 phosphorylation was not changed in G1/S arrested cells, it was suggested that spindle checkpoint activation provides a physiological stimulus that leads to HDAC2 hyperphosphorylation. Indeed, an inventory of phosphorylation sites of spindle associated proteins purified from human mitotic spindles identified Ser394 phosphorylated HDAC2 [20].

Related to HDAC1 and HDAC2 regulation by phosphatases, HDAC1 co-immunoprecipitates with PP1 [21]. Although the conclusion of this study is that HDAC1 promotes dephosphorylation of the cAMP responsive element binding protein (CREB) via a stable interaction with PP1, it nevertheless strengthens the hypothesis that PP1 is a major regulator of HDAC1 phosphorylation status. In a separate study, affinity isolation of protein serine/threonine phosphatases on immobilized phosphatase inhibitor microcystin-LR identified HDAC1 (but not HDAC2) as one of the components of cellular phosphatase complexes [22]. PP1 activation correlates with changes in HDAC1 phosphorylation, dissociation of the HDAC1-PP1 complex, and an increase in HDAC1 activity [23]. Further, HDAC inhibitors can disrupt the HDAC1-PP1 complex [24].

HDAC3 is a Target Protein and a Key Mediator of Cell Signals

Like HDAC1 and HDAC2, HDAC3 is also phosphorylated in the C-terminal domain by the protein kinase CK2 [14, 25]. The phospho-acceptor site in HDAC3 is Ser424, which is a non-conserved residue among the Class I HDACs (Figure 14.1 and Table 14.1). Mutation of Ser424 of HDAC3 to alanine severely compromised enzymatic activity, reminiscent of HDAC1 and HDAC2 phosphorylation site mutations. However, unlike HDAC1 and HDAC2, HDAC3 associates with the catalytic and regulatory subunits of the protein serine/threonine phosphatase 4 complex ($PP4_c/PP4_{R1}$). Several lines of evidence suggest that HDAC3 is a PP4 substrate, and dephosphorylation of HDAC3 by PP4 downregulates HDAC3 enzymatic activity. First, HDAC3 serine phosphorylation decreased in cells that overexpressed PP4. Overexpression of PP4 had no effect on serine phosphorylation of HDAC1 and HDAC2, consistent with the notion that PP1 but not PP4 regulates HDAC1/2 phosphorylation. Second, treatment of cells with fostriecin under conditions that inhibited PP4, but not PP1 or PP2B, activity resulted in an elevation of HDAC3 serine phosphorylation. Finally, overexpression of $PP4_c$ resulted in a decrease in HDAC3 activity while knockdown of $PP4_c$ increased HDAC3 enzymatic activity. Taken together, these data strongly argue that HDAC3 phosphorylation status, and thus HDAC3 activity, is regulated by kinase and phosphatase. However, it is unclear how phosphorylation/dephosphorylation in itself alters HDAC3. An answer to this important question requires further investigations.

In addition to protein kinase CK2, HDAC3 has been reported to be phosphorylated *in vitro* by DNA-PKcs, a member of the phosphatidylinositol 3-kinase family, in the presence of the purified TR-RXR-corepressor complex [26]. Similar to the phosphorylation of HDAD3 by CK2, DNA-PKcs phosphorylation of HDAC3 caused a significant enhancement of HDAC3 activity. DNA-PK phosphorylates target proteins on serine or threonine residues, preferentially at the consensus sequence Ser/Thr-Gln [27]. Because HDAC3 is only serine (but not threonine or tyrosine) phosphorylated [25], if HDAC3 is truly a physiological substrate of DNA-PK, the site of phosphorylation may potentially be located at Ser54. However, results from an *in vivo* ^{32}P-labeling experiment indicate that HDAC3 phosphorylation is limited to the C-terminal of the protein [25]. Thus, whether DNA-PK is a physiologically relevant HDAC3 kinase and whether HDAC3 phosphorylation by

DNA-PK has any *in vivo* functional consequences remain to be determined.

Besides being a target protein of signal transduction, HDAC3 also modulates downstream signaling pathways. For example, HDAC3 interacts directly with MAPK11 (p38 β isoform), decreases MAPK11 phosphorylation, and inhibits the activity of the MAPK11 dependent transcription factor ATF2 [28]. Recently, it was established that HDAC3 forms a complex with A-kinase-anchoring proteins AKAP95 and HA95, which are targeted to mitotic chromosomes [29]. Although HDAC3 does not directly modify AKAP95/HA95 or kinases associated with AKAP95/HA95, deacetylation of histone H3 during mitosis requires AKAP95/HA95 along with HDAC3 to provide a hypoacetylated tail that allows phosphorylation by Aurora B kinase. HDAC3 and AKAP95/HA95, therefore, are required for normal mitotic progression in a signal dependent but transcriptional independent manner.

HDAC8 Activity is Negatively Regulated by the cAMP Dependent Kinase PKA

Although HDAC8 is a Class I deacetylase, there are many distinct differences between the phosphorylation of HDAC8 compared to other Class I enzymes. So far, a large body of evidence supports the role of phosphorylation as a positive regulatory mechanism for activation of HDAC1, HDAC2, and HDAC3 enzymatic activities. In contrast, a study by Lee *et al.* suggests that phosphorylation of HDAC8 inhibited the enzymatic activity of HDAC8 [30]. For example, mutation of the HDAC8 phosphorylation site to alanine enhanced HDAC8 activity, while an activator of HDAC8 phosphorylation caused a reduction in HDAC8 activity. HDAC8 is refractory to phosphorylation by protein kinase CK2, but instead is phosphorylated by PKA [14, 30]. Potential phosphatases that target HDAC8 have not yet been identified. Also, unlike other Class I HDACs, which are phosphorylated in their C-terminal ends, the phosphoacceptor site of HDAC8 is located in the N-terminal of HDAC8 at Ser39, a non-conserved residue among Class I HDACs. Crystal structures of HDAC8 reveal that Ser39 lies at a surface of HDAC8, roughly 20Å from the opening to the HDAC8 active site [31–33]. The prediction is that phosphorylation of Ser39 would lead to a major structural disruption of this region of the surface, which ultimately would negatively affect HDAC8's activity. A follow-up study to define the biological outcomes of HDAC8 phosphorylation found that phospho-HDAC8 binds to the EST1B protein [34]. Interestingly, by associating with EST1B, HDAC8 modulates the protein stability of EST1B. These results again underscored the importance of Class I HDAC phosphorylation, and furnished another example of a biological consequence of HDAC phosphorylation that is beyond histone modifications.

REVERSIBLE PHOSPHORYLATION OF MAMMALIAN CLASS II HDACs

Yeast Hda1 is the founding member of Class II HDACs [35]. In humans, there are six members of Class II enzymes [3, 36–38]. According to amino acid sequence similarity, they are further divided into two subgroups, which are often referred to as Classes IIa (HDAC4, 5, 7, and 9) and IIb (HDAC6 and 10). As illustrated in Figure 14.2, only their catalytic domains display sequence homology to yeast Hda1, indicating that additional domains and motifs have been acquired during evolution. As discussed below, these domains and motifs are crucial for the function and regulation of Class II HDACs in mammals.

Conserved Sequence Motifs on Class IIa HDACs

The Hda1 related domain of each mammalian Class IIa member is located in the C-terminal half. In addition to this domain, there is a long N-terminal extension and a short tail at the C-terminal end, neither of which shows any sequence similarity to yeast Hda1. Within the N-terminal extension, there are multiple conserved sequence motifs, including the well characterized MEF2 and 14-3-3 binding sites (Figure 14.2). While there is a single site for MEF2 binding, multiple sites confer the interaction of the HDACs with 14-3-3 proteins [39, 40]. Moreover, phosphorylation is required for the 14-3-3 interaction, thereby linking cellular signaling pathways to Class IIa HDAC regulation. Adjacent to the major 14-3-3 binding site is a conserved nuclear localization signal [41]. In addition to 14-3-3 binding sites, two other conserved motifs of HDAC4 have been found to be phosphorylated, i.e., Ser210 and 298 [42, 43]. While the sequence surrounding Ser298 is conserved among all four members of Class IIa, Ser210 of HDAC4 is conserved in HDAC5 and 9, but not in HDAC7. A conserved sequence element, tentatively called the SP box, is also conserved among all four HDACs [3]. Similar to Ser210, a sumoylation site is conserved among HDAC4, 5, and 9, but not in HDAC7. Therefore, multiple sequence elements are present in the N-terminal extension of Class IIa members. By contrast, only two conserved regions can be found in the C-terminal tail (Figure 14.2). While one of the regions is hydrophobic and serves as a nuclear export signal [44–46], the other is located at the C-terminal end and appears to be acidic. The function significance of this acidic region remains unclear.

Kinases and Phosphatases for Class IIa HDAC Phosphorylation

As alluded above, phosphorylation links cellular signaling to Class IIa HDAC regulation. Phosphorylation at 14-3-3 binding sites has been extensively characterized. HDAC4

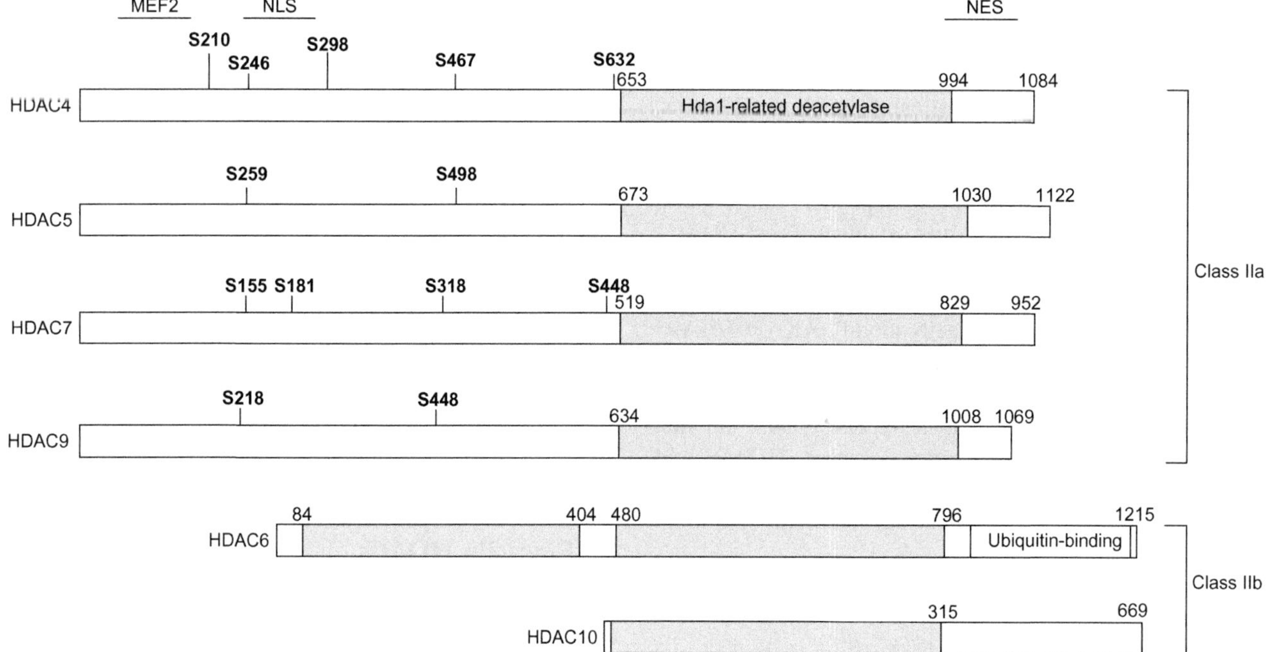

FIGURE 14.2 Phosphorylation sites on Class II HDACs.
Simplified representation of Class II HDAC proteins. Conserved "HDAC domains" are denoted with shaded areas. Position of known phosphorylation sites is indicated by bold. MEF2, MEF2 binding domain; NLS, nuclear localization signal; NES, nuclear export signal.

contains three 14-3-3 binding sites: Ser246, Ser467, and Ser632. All three sites are conserved in HDAC5, 7 and 9 (Figure 14.2), but only the sites equivalent to Ser246 and Ser467 of HDAC4 seem to be functional in HDAC5 and HDAC9 [47–49]. In HDAC7, a fourth site is also important (Figure 14.2) [40]. Mutational analysis revealed that the S246 site of HDAC4 is the most important [39, 41]. This site is conserved in both *C. elegans* and *D. melanogaster*. As listed in Table 14.2, six groups of kinases have been shown to phosphorylate the 14-3-3 binding sites: Ca^{2+}/calmodulin dependent kinases (CaMKs) [47, 50], protein kinase D [49, 51–55], microtubule affinity regulating kinases [40, 56], salt inducible kinases [57, 58], checkpoint kinase-1 (CHK1) [59], and AMPK [60] (Table 14.2). Several of these kinases have been shown only for some Class IIa members, so further experiments are needed to establish whether these kinases target all four members of Class IIa in humans.

Upon phosphorylation by these kinases, HDAC4, 5, 7, and 9 bind to 14-3-3 proteins and are subsequently subjected to nuclear export and cytoplasmic retention. This in turn inhibits the ability of the HDACs to function as transcriptional corepressors in the nucleus [39, 41]. In addition to nuclear export and cytoplasmic retention, phosphorylation by the kinases may lead to ubiquitination and proteasomal degradation [61]. Gene targeting in mice has linked the resultant phenotypes mainly to MEF2 transcription factors, indicating that these family members are major nuclear targets in different tissues [62]. This is consistent with the fact that the MEF2 binding site is conserved from *C. elegans* to humans. However, importance of the MEF2 link does not mean that links to other transcription factors are not important.

Among the Class IIa HDAC kinases, Ca^{2+}/calmodulin dependent kinase II δ (CaMKIIδ) appears to be exceptional [43]. It binds to a specific motif in HDAC4 and phosphorylates Ser210. This serine residue is conserved in HDAC5 and HDAC9, but CaMKIIδ is unable to bind and phosphorylate HDAC5. As the sequence of the CaMKIIδ binding motif is not conserved in HDAC9, it is unlikely that CaMKIIδ is able to bind and phosphorylate HDAC9. Therefore, kinase targeting is a prerequisite for CaMKIIδ to phosphorylate HDAC4. Whether other kinases also require physical targeting is an important issue awaiting further investigation.

For dephosphorylation, two phosphatases have been identified, protein phosphatase (PP)1β/myosin phosphatase targeting subunit-1 (MYPT1) [63] and PP2A [40, 42, 64, 65]. In addition to the 14-3-3 binding sites, PP2A targets Ser298 of HDAC4 [42]. Intriguingly, dephosphorylation of Ser298 is required for nuclear import of HDAC4. The underlying molecular mechanisms remain to be determined. Similarly, the mechanistic impact of Ser210 phosphorylation is also unclear. Therefore, in addition to those at 14-3-3 binding sites, other phosphorylation events are also important.

Phosphorylation of Class IIb HDACs

Compared to Class IIa members, much less is known about regulation of the Class IIb members, which in part is due to the lack of recognizable phosphorylation motifs in HDAC6

TABLE 14.2 Kinases and Phosphatases for Class II HDACs

	Kinase	Phospho-acceptor site	Consequence of phosphorylation	Phosphatase
HDAC4	CaMK	S210 (CaMKδB)	Increases export from nucleus to cytoplasm,	PP2A
		S246	promotes HDAC4-14-3-3 binding,	
		S467	derepression of HDAC4 target genes	
		S632		
	Not yet identified	S298		
HDAC5	CaMK	S259	Increases export from nucleus to cytoplasm,	PP2A
		S498	promotes HDAC5-14-3-3 binding	
	PKC/PKD	S259	Increases export from nucleus to cytoplasm,	
		S498	promotes HDAC5-14-3-3 binding,	
			derepression of HDAC5 target genes	
	AMPK	S259	Promotes HDAC5-14-3-3 binding,	
		S498	increases histone H3 acetylation,	
			enhances HDAC5 target gene expression	
	SIK1	S259	Increases export from nucleus to cytoplasm,	
		S498	enhances HDAC5 target gene expression	
	MARK2	Not yet identified	Not yet identified	
HDAC6	AurA	Not yet identified	Increases tubulin deacetylase activity	?
HDAC7	PKD	S155	Increases export from nucleus to cytoplasm,	PP1β
		S181	promotes HDAC7-14-3-3 binding,	MYPT1
		S318	derepression of HDAC7 target genes	
		S448		
HDAC9	CaMK	S218	Promotes HDAC9-14-3-3 binding,	?
(MITR)		S448	disrupts MEF2-HDAC9 binding	
HDAC10	?	?	?	?

and 10 [66]. Also different from Class IIa members, HDAC6 appears to be a cytoplasmic deacetylase (e.g., deacetylating α-tubulin) [67]. A recent study indicates that phosphorylation by Aurora A kinase activates the tubulin deacetylase activity of HDAC6 and promotes ciliary disassembly [68]. The site(s) of phosphorylation remains unknown. Regulation of HDAC6 and 10 by phosphorylation is a largely unchartered research area that remains to be explored.

CONCLUSION AND PERSPECTIVES

With the successful identification and cloning of 18 human HDACs in the last decade comes an increasing need not only to define their chemical and biological functions, but also to understand precisely how their activities are regulated and the exact molecular mechanisms by which they influence downstream processes. Proper coordination of cell signaling with histone/non-histone deacetylation is critical to precisely regulate both gene expression and a number of transcription independent events. Many HDAC phosphorylations are independent of protein synthesis and occur reversibly, consistent with signaling responses. Phosphorylations of many Class II HDACs are responsive to acute signals.

Compared to many target or mediator proteins of cell signals, our knowledge of how HDACs respond to signals or stimuli and process them into biochemical reactions are only beginning to accumulate slowly. Many areas are wide open for exploration. For example, the expression and activities of many kinases, and some phosphatases, have been reported to be regulated by HDAC inhibitors. As discussed, phosphorylations of some HDACs influence their ability to interact with cellular proteins. Conversely, there are ample examples where phosphorylation and dephosphorylation of some cellular proteins changed their ability to interact with HDACs. In addition to the HDAC kinases and phosphatases, there are HDAC interacting kinases and phosphatases that do not phosphorylate/dephosphorylate HDACs. Finally, there are a large number of reports that some HDAC inhibitors can synergize with certain kinases/ phosphatases to promote or inhibit gene expression. We predict that the number of proteins and molecules that participate in HDAC signal transductions may continue to grow. A complete understanding of HDACs in cell signals may eventually require the collaborations from multiple fields.

Although the discussions in this chapter is limited to HDACs from humans and mice, based on amino acid sequence analysis and perhaps intuition, it is clear that phosphorylation plays a critical role in the functions and activity of HDACs in other organisms. For example, early studies suggest that histone deacetylases are phospho-proteins in maize [69, 70]. *Xenopus* HDACm, which has 91 percent sequence homology to human HDAC1, is serine phosphorylated by CK2, and phosphorylation influences not only HDACm enzymatic activity but also its translocation to the nucleus [71, 72].

Because the role of phosphorylation in switching enzymes on or off has long been recognized, coupled with the fact that phosphorylation is the best known form of signal transduction, our discussions have centered on how phosphorylation/dephosphorylation regulates HDAC activities and functions. However, additional posttranslational modifications clearly exist in HDACs. For example, HDAC1 and HDAC4 are sumoylated [73, 74], and HDAC5, HDAC6, and HDAC7 are ubiquitinated [61, 75, 76]. Whether these modifications crosstalk with HDAC phosphorylation and function as part of the signaling network to control the activities of HDACs remain to be determined.

HDAC proteins are vital regulators of fundamental cellular events, such as cell cycle progression, differentiation, and tumorigenesis. Abnormal HDACs can contribute to many different human diseases including cancer [77, 78], neurodegenerative disorders [79], cardiac hypertrophy [80], and pulmonary diseases [81, 82]. HDAC inhibitors are currently in clinical trials for the treatment of leukemia and solid tumors [83, 84]. A thorough understanding of how cell signals, and phosphorylation/dephosphorylation in particular, modulate HDAC activities and functions may perhaps guide us into new alternative, and unique avenues for the development of better therapeutics and treatments for these diseases.

REFERENCES

1. Inoue A, Fujimoto D. Enzymatic deacetylation of histone. *Biochem Biophys Res Commun* 1969;**36**:146–50.
2. Taunton J, Hassig CA, Schreiber SL. A mammalian histone deacetylase related to the yeast transcriptional regulator Rpd3p. *Science* 1996;**272**:408–11.
3. Yang XJ, Seto E. The Rpd3/Hda1 family of lysine deacetylases: from bacteria and yeast to mice and men. *Nat Rev Mol Cell Biol* 2008;**9**:206–18.
4. Haigis MC, Guarente LP. Mammalian sirtuins: emerging roles in physiology, aging, and calorie restriction. *Genes Dev* 2006;**20**:2913–21.
5. Gao L, Cueto MA, Asselbergs F, Atadja P. Cloning and functional characterization of HDAC11, a novel member of the human histone deacetylase family. *J Biol Chem* 2002;**277**:25,748–55.
6. Kao HY, Downes M, Ordentlich P, Evans RM. Isolation of a novel histone deacetylase reveals that class I and class II deacetylases promote SMRT-mediated repression. *Genes Dev* 2000;**14**:55–66.
7. Kim SC, Sprung R, Chen Y, et al. Substrate and functional diversity of lysine acetylation revealed by a proteomics survey. *Mol Cell* 2006;**23**:607–18.
8. Pflum MK, Tong JK, Lane WS, Schreiber SL. Histone deacetylase 1 phosphorylation promotes enzymatic activity and complex formation. *J Biol Chem* 2001;**276**:47,733–41.
9. Cai R, Kwon P, Yan-Neale Y, et al. Mammalian histone deacetylase 1 protein is posttranslationally modified by phosphorylation. *Biochem Biophys Res Commun* 2001;**283**:445–53.
10. Galasinski SC, Resing KA, Goodrich JA, Ahn NG. Phosphatase inhibition leads to histone deacetylases 1 and 2 phosphorylation and disruption of corepressor interactions. *J Biol Chem* 2002;**277**:19,618–26.
11. Kamath N, Karwowska-Desaulniers P, Pflum MK. Limited proteolysis of human histone deacetylase 1. *BMC Biochem* 2006;**7**:22.
12. Karwowska-Desaulniers P, Ketko A, Kamath N, Pflum MK. Histone deacetylase 1 phosphorylation at S421 and S423 is constitutive in vivo, but dispensable in vitro. *Biochem Biophys Res Commun* 2007;**361**:349–55.
13. Schultz BE, Misialek S, Wu J, et al. Kinetics and comparative reactivity of human class I and class IIb histone deacetylases. *Biochemistry* 2004;**43**:11,083–91.
14. Tsai SC, Seto E. Regulation of histone deacetylase 2 by protein kinase CK2. *J Biol Chem* 2002;**277**:31,826–33.
15. Olsen JV, Blagoev B, Gnad F, et al. Global, in vivo, and site-specific phosphorylation dynamics in signaling networks. *Cell* 2006;**127**:635–48.
16. Sun JM, Chen HY, Davie JR. Differential distribution of unmodified and phosphorylated histone deacetylase 2 in chromatin. *J Biol Chem* 2007;**282**:33,227–36.

17. Faust M, Montenarh M. Subcellular localization of protein kinase CK2. A key to its function? *Cell Tissue Res* 2000;**301**:329–40.
18. Sun JM, Chen HY, Moniwa M, et al. The transcriptional repressor Sp3 is associated with CK2-phosphorylated histone deacetylase 2. *J Biol Chem* 2002;**277**:35,783–6.
19. Pluemsampant S, Safronova OS, Nakahama K, Morita I. Protein kinase CK2 is a key activator of histone deacetylase in hypoxia-associated tumors. *Int J Cancer* 2008;**122**:333–41.
20. Nousiainen M, Sillje HH, Sauer G, et al. Phosphoproteome analysis of the human mitotic spindle. *Proc Natl Acad Sci USA* 2006;**103**:5391–6.
21. Canettieri G, Morantte I, Guzman E, et al. Attenuation of a phosphorylation-dependent activator by an HDAC-PP1 complex. *Nat Struct Biol* 2003;**10**:175–81.
22. Brush MH, Guardiola A, Connor JH, et al. Deactylase inhibitors disrupt cellular complexes containing protein phosphatases and deacetylases. *J Biol Chem* 2004;**279**:7685–91.
23. Guo C, Mi J, Brautigan DL, Larner JM. ATM regulates ionizing radiation-induced disruption of HDAC1 : PP1 : Rb complexes. *Cell Sign* 2007;**19**:504–10.
24. Chen CS, Weng SC, Tseng PH, Lin HP. Histone acetylation-independent effect of histone deacetylase inhibitors on Akt through the reshuffling of protein phosphatase 1 complexes. *J Biol Chem* 2005;**280**:38,879–87.
25. Zhang X, Ozawa Y, Lee H, et al. Histone deacetylase 3 (HDAC3) activity is regulated by interaction with protein serine/threonine phosphatase 4. *Genes Dev* 2005;**19**:827–39.
26. Jeyakumar M, Liu XF, Erdjument-Bromage H, et al. Phosphorylation of thyroid hormone receptor-associated nuclear receptor corepressor holocomplex by the DNA-dependent protein kinase enhances its histone deacetylase activity. *J Biol Chem* 2007;**282**:9312–22.
27. Jackson SP. DNA damage detection by DNA dependent protein kinase and related enzymes. *Cancer Surv* 1996;**28**:261–79.
28. Mahlknecht U, Will J, Varin A, et al. Histone deacetylase 3, a class I histone deacetylase, suppresses MAPK11-mediated activating transcription factor-2 activation and represses TNF gene expression. *J Immunol* 2004;**173**:3979–90.
29. Li Y, Kao GD, Garcia BA, et al. A novel histone deacetylase pathway regulates mitosis by modulating Aurora B kinase activity. *Genes Dev* 2006;**20**:2566–79.
30. Lee H, Rezai-Zadeh N, Seto E. Negative regulation of histone deacetylase 8 activity by cyclic AMP-dependent protein kinase A. *Mol Cell Biol* 2004;**24**:765–73.
31. Somoza JR. Structural snapshots of human HDAC8 provide insights into the class I histone deacetylases. *Structure* 2004;**12**:1325–34.
32. Vannini A, Volpari C, Filocamo G, et al. Crystal structure of a eukaryotic zinc-dependent histone deacetylase, human HDAC8, complexed with a hydroxamic acid inhibitor. *Proc Natl Acad Sci U S A* 2004;**101**:15,064–9.
33. Vannini A, Volpari C, Gallinari P, et al. Substrate binding to histone deacetylases as shown by the crystal structure of the HDAC8-substrate complex. *EMBO Rep* 2007;**8**:879–84.
34. Lee H, Sengupta N, Villagra A, et al. Histone deacetylase 8 safeguards the human ever-shorter telomeres 1B (hEST1B) protein from ubiquitin-mediated degradation. *Mol Cell Biol* 2006;**26**:5259–69.
35. Rundlett SE, Carmen AA, Kobayashi R, et al. HDA1 and RPD3 are members of distinct yeast histone deacetylase complexes that regulate silencing and transcription. *Proc Natl Acad Sci USA* 1996; **93**:14,503–8.
36. Grozinger CM, Hassig CA, Schreiber SL. Three proteins define a class of human histone deacetylases related to yeast Hda1p.. *Proc Natl Acad Sci U S A* 1999;**96**:4868–73.
37. Verdel A, Khochbin S. Identification of a new family of higher eukaryotic histone deacetylases. Coordinate expression of differentiation-dependent chromatin modifiers. *J Biol Chem* 1999;**274**:2440–5.
38. Fischle W, Emiliani S, Hendzel MJ, et al. A new family of human histone deacetylases related to *Saccharomyces cerevisiae* HDA1p. *J Biol Chem* 1999;**274**:11,713–20.
39. Verdin E, Dequiedt F, Kasler HG. Class II histone deacetylases: versatile regulators. *Trends Genet* 2003;**19**:286–93.
40. Martin M, Kettmann R, Dequiedt F. Class IIa histone deacetylases: regulating the regulators. *Oncogene* 2007;**26**:5450–67.
41. Yang XJ, Grégoire S. Class II histone deacetylases: From sequence to function, regulation and clinical implication. *Mol Cell Biol* 2005;**25**:2873–84.
42. Paroni G, Cernotta N, Russo CD, et al. PP2A regulates HDAC4 nuclear import. *Mol Biol Cell* 2008;**19**:655–67.
43. Little GH, Bai Y, Williams T, Poizat C. Nuclear calcium/calmodulin-dependent protein kinase IIdelta preferentially transmits signals to histone deacetylase 4 in cardiac cells. *J Biol Chem* 2007;**282**:7219–31.
44. Wang AH, Yang XJ. Histone deacetylase 4 possesses intrinsic nuclear import and export signals. *Mol Cell Biol* 2001;**21**:5992–6005.
45. McKinsey TA, Zhang CL, Olson EN. Identification of a signal-responsive nuclear export sequence in class II histone deacetylases. *Mol Cell Biol* 2001;**21**:6312–21.
46. Gao C, Li X, Lam M, et al. CRM1 mediates nuclear export of HDAC7 independently of HDAC7 phosphorylation and association with 14-3-3s. *FEBS Lett* 2006;**580**:5096–104.
47. McKinsey TA, Zhang CL, Lu J, Olson EN. Signal-dependent nuclear export of a histone deacetylase regulates muscle differentiation. *Nature* 2000;**408**:106–11.
48. McKinsey TA, Zhang CL, Olson EN. Activation of the myocyte enhancer factor-2 transcription factor by calcium/calmodulin-dependent protein kinase-stimulated binding of 14-3- 3 to histone deacetylase 5. *Proc Natl Acad Sci USA* 2000;**97**:14,400–5.
49. Vega RB, Harrison BC, Meadows E, et al. Protein kinases C and D mediate agonist-dependent cardiac hypertrophy through nuclear export of histone deacetylase 5. *Mol Cell Biol* 2004;**24**:8374–85.
50. Linseman DA, Bartley CM, Le SS, et al. Inactivation of the myocyte enhancer factor-2 repressor histone deacetylase-5 by endogenous Ca(2+)/calmodulin-dependent kinase II promotes depolarization-mediated cerebellar granule neuron survival. *J Biol Chem* 2003;**278**:41,472–81.
51. Parra M, Kasler H, McKinsey TA, et al. Protein kinase D1 phosphorylates HDAC7 and Induces its nuclear export after TCR activation. *J Biol Chem* 2005;**280**:13,762–70.
52. Dequiedt F, Van Lint J, Lecomte E, et al. Phosphorylation of histone deacetylase 7 by protein kinase D mediates T cell receptor-induced Nur77 expression and apoptosis. *J Exp Med* 2005;**201**:793–804.
53. Harrison BC, Kim MS, van Rooij E, et al. Regulation of cardiac stress signaling by protein kinase D1. *Mol Cell Biol* 2006;**26**:3875–88.
54. Matthews SA, Liu P, Spitaler M, et al. Essential role for protein kinase D family kinases in the regulation of class II histone deacetylases in B lymphocytes. *Mol Cell Biol* 2006;**26**:1569–77.
55. Fielitz J, Kim MS, Shelton JM, et al. Requirement of protein kinase D1 for pathological cardiac remodeling. *Proc Natl Acad Sci USA* 2008;**105**:3059–63.
56. Chang S, Bezprozvannaya S, Li S, Olson EN. An expression screen reveals modulators of class II histone deacetylase phosphorylation. *Proc Natl Acad Sci USA* 2005;**102**:8120–5.
57. van der Linden AM, Nolan KM, Sengupta P. KIN-29 SIK regulates chemoreceptor gene expression via an MEF2 transcription factor and a class II HDAC. *EMBO J* 2007;**26**:358–70.

58. Berdeaux R, Goebel N, Banaszynski L, et al. SIK1 is a class II HDAC kinase that promotes survival of skeletal myocytes. *Nat Med* 2007;**13**:597–603.
59. Kim MA, Kim HJ, Brown AL, et al. Identification of novel substrates for human checkpoint kinase Chk1 and Chk2 through genome-wide screening using a consensus Chk phosphorylation motif. *Exp Mol Med* 2007;**39**:205–12.
60. McGee SL, van Denderen BJ, Howlett KF, et al. Amp-Activated Protein Kinase Regulates Glut4 Transcription by Phosphorylating Histone Deacetylase 5. *Diabetes* 2008;**57**:860–7.
61. Li X, Song S, Liu Y, et al. Phosphorylation of the histone deacetylase 7 modulates its stability and association with 14-3-3 proteins. *J Biol Chem* 2004;**279**:34,201–8.
62. Arnold MA, Kim Y, Czubryt MP, et al. MEF2C transcription factor controls chondrocyte hypertrophy and bone development. *Dev Cell* 2007;**12**:377–89.
63. Parra M, Mahmoudi T, Verdin E. Myosin phosphatase dephosphorylates HDAC7, controls its nucleocytoplasmic shuttling, and inhibits apoptosis in thymocytes. *Genes Dev* 2007;**21**:638–43.
64. Sucharov CC, Langer S, Bristow M, Leinwand L. Shuttling of HDAC5 in H9C2 cells regulates YY1 function through CaMKIV/PKD and PP2A. *Am J Physiol Cell Physiol* 2006;**291**:C1029–37.
65. Illi B, Russo CD, Colussi C, et al. Nitric Oxide Modulates Chromatin Folding in Human Endothelial Cells via PP2A Activation and Class II HDACs Nuclear Shuttling. *Circ Res* 2008;**102**:51–8.
66. Boyault C, Sadoul K, Pabion M, Khochbin S. HDAC6, at the crossroads between cytoskeleton and cell signaling by acetylation and ubiquitination. *Oncogene* 2007;**26**:5468–76.
67. Hubbert C, Guardiola A, Shao R, et al. HDAC6 is a microtubule-associated deacetylase. *Nature* 2002;**417**:455–8.
68. Pugacheva EN, Jablonski SA, Hartman TR, et al. HEF1-dependent Aurora A activation induces disassembly of the primary cilium. *Cell* 2007;**129**:1351–63.
69. Kolle D, Brosch G, Lechner T, et al. Different types of maize histone deacetylases are distinguished by a highly complex substrate and site specificity. *Biochemistry* 1999;**38**:6769–73.
70. Brosch G, Georgieva EI, Lopez-Rodas G, et al. Specificity of Zea mays histone deacetylase is regulated by phosphorylation. *J Biol Chem* 1992;**267**:20,561–4.
71. Smillie DA, Llinas AJ, Ryan JT, et al. Nuclear import and activity of histone deacetylase in Xenopus oocytes is regulated by phosphorylation. *J Cell Sci* 2004;**117**:1857–66.
72. Ryan J, Llinas AJ, White DA, et al. Maternal histone deacetylase is accumulated in the nuclei of Xenopus oocytes as protein complexes with potential enzyme activity. *J Cell Sci* 1999;**112**:2441–52.
73. David G, Neptune MA, DePinho RA. SUMO-1 modification of histone deacetylase 1 (HDAC1) modulates its biological activities. *J Biol Chem* 2002;**277**:23,658–63.
74. Kirsh O, Seeler JS, Pichler A, et al. The SUMO E3 ligase RanBP2 promotes modification of the HDAC4 deacetylase. *EMBO J* 2002;**21**:2682–91.
75. Hook SS, Orian A, Cowley SM, Eisenman RN. Histone deacetylase 6 binds polyubiquitin through its zinc finger (PAZ domain) and copurifies with deubiquitinating enzymes. *Proc Natl Acad Sci USA* 2002;**99**:13,425–30.
76. Seigneurin-Berny D, Verdel A, Curtet S, et al. Identification of components of the murine histone deacetylase 6 complex: link between acetylation and ubiquitination signaling pathways. *Mol Cell Biol* 2001;**21**:8035–44.
77. Cress WD, Seto E. Histone deacetylases, transcriptional control, and cancer. *J Cell Physiol* 2000;**184**:1–16.
78. Glozak MA, Seto E. Histone deacetylases and cancer. *Oncogene* 2007;**26**:5420–32.
79. Bodai L, Pallos J, Thompson LM, Marsh JL. Altered protein acetylation in polyglutamine diseases. *Curr Med Chem* 2003;**10**:2577–87.
80. Zhang CL, McKinsey TA, Chang S, et al. Class II histone deacetylases act as signal-responsive repressors of cardiac hypertrophy. *Cell* 2002;**110**:479–88.
81. Ito K, Ito M, Elliott WM, et al. Decreased histone deacetylase activity in chronic obstructive pulmonary disease. *N Engl J Med* 2005;**352**:1967–76.
82. Barnes PJ, Adcock IM, Ito K. Histone acetylation and deacetylation: importance in inflammatory lung diseases. *Eur Respir J* 2005;**25**:552–63.
83. Marks PA, Dokmanovic M. Histone deacetylase inhibitors: discovery and development as anticancer agents. *Expert Opin Investig Drugs* 2005;**14**:1497–511.
84. Palmieri C, Coombes RC, Vigushin DM. Targeted histone deacetylase inhibition for cancer prevention and therapy. *Prog Drug Res* 2005;**63**:147–81.

Chapter 15

Histone Methylation: Chemically Inert but Chromatin Dynamic

Johnathan R. Whetstine
Harvard Medical School and Massachusetts General Hospital Cancer Center, Charlestown, Massachusetts

INTRODUCTION

There is an ever increasing appreciation for the role that posttranslational modifications (PTMs) have on cellular responses to both extrinsic and intrinsic factors. One of the most well studied PTMs is phosphorylation. The phospho-group carries a negative charge, which can influence confirmation or interactions. However, dramatic consequences can also be associated with another chemically inert 15 dalton modification: the methyl group (CH3-). This chapter focuses on how this small inert molecule can dramatically change the chromatin environment and, in turn, the survival or demise of an organism. This chapter will unveil the complexity and dynamics associated with the methylation of arginines and lysines within histone tails. The following points will be addressed in this chapter: (1) histone methylation is very dynamic; (2) the degree of methylation has specific consequences; (3) the placement of methylated histones within the chromosomal environment directly impacts different DNA templated processes.

HISTORICAL PERSPECTIVE OF CHROMATIN AND HISTONE METHYLATION

Eukaryotic cells package their DNA into a nuclear structure called chromatin, which was first identified and named by Walther Flemming in 1882 because of its refractory nature and affinity for dyes [1]. Chromatin was later found to contain both nucleic acids and a series of acid soluble proteins that were termed "histone" by Albrecht Kossel [2]. The proper packaging of DNA into chromosomes is dependent on histones and other chromosomal proteins. In subsequent years a number of groups used electron microscopy and crystallography to determine the basic building block of chromatin. This building block is referred to as the nucleosome [3]. The nucleosome is composed of 146 base pairs of DNA wrapped around the histone octamer (two copies each of H2A, H2B, H3, and H4) [4]. In the early 1960s, increasing amounts of evidence were accumulating that PTMs such as acetylation and methylation occur on histone proteins. The data supported a role for these PTMs in RNA synthesis [5]. In 1964, Murray further demonstrated that methylation occurred on the e-amino group of lysine (K) [6]. A few years later, the guanidino group of arginine (R) [7, 8] was shown to be methylated. These methylation reactions were shown to be catalyzed by enzymes using S-adenosyl-L-methionine (SAM) as the methyl group donor [9, 10]. The arginine methyltransferases responsible for histone and other cellular protein methylation were discovered in the late 1980s [11]; however, it was not until 2000 that a landmark discovery uncovered the molecular identity of the first lysine specific histone methyltransferase [12]. The discovery of the lysine specific methyltransferases resulted in a flurry of studies that propelled our understanding of how histone methylation regulates nuclear processes and provided the foundation for the concept that the degree of histone methylation has specific consequences. The discovery of these methyltransferases also provided valuable insights into the fact that all the lysines and arginines within histones are not modified in the same manner and serve as specific substrates for these enzymes. This added an important level of regulation that is currently being investigated by many groups around the world.

Histone methylation is now recognized as an important modification linked to both transcriptional activation and repression [13]. Numerous lysine and arginine residues within the histones are methylated *in vivo*, especially within their tails (Figure 15.1a) [13–15]. For example, six of the lysine residues (K), including histone H3K4, 9, 27, 36, and 79,

Handbook of Cell Signaling, Three-Volume Set 2 ed.

(a)

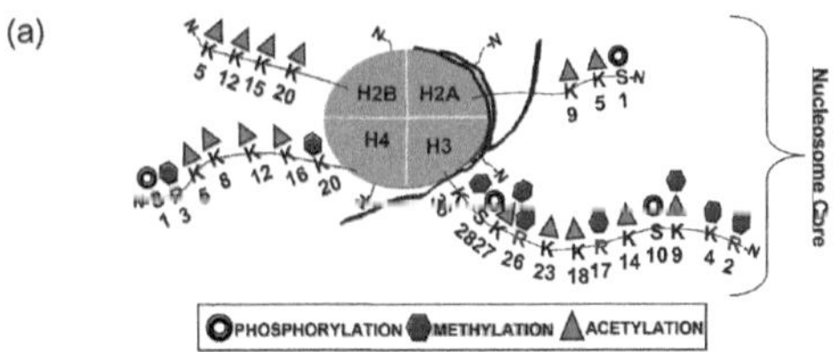

(b)

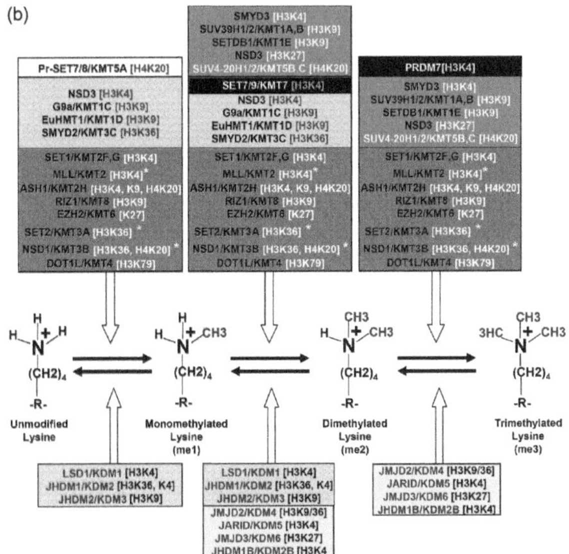

(c)

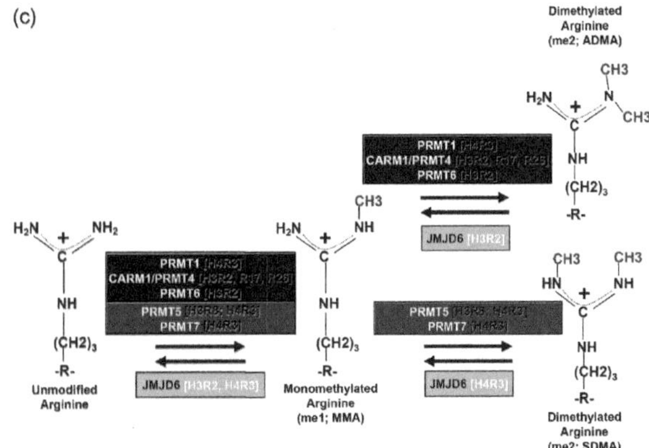

FIGURE 15.1 Multiple arginines and lysines are methylated in histones. (a) Schematic representation of the nucleosome. Some of the amino acids that ate posttranslationally modified are shown. The blue open circle represents phosphorylation. The red hexagon represents methylation. The green triangle represents acetylation. (b) Lysines are mono-, di-, and trimethylated by multiple histone methyltransferases (KMT). The different KMT are shown above the forward reactions. The histone lysine demethylases (KDM) remove methyl groups from the lysines. The different KDMs that have been discovered are shown below the reverse reactions. The sites the KMTs and KDMs modify are indicated next to their names. The degree of methylation these enzymes impact are based on their clustering (see color box code: pink, methylate from unmodified to trimethylated; light blue, unmodified to dimethylated; orange, monomethylation to trimethylation; white, unmodified to monomethylated; black, dimethylated to trimethylated). The methyltransferases with an * indicate the last reaction is not as strong. (c) Arginines are monomethylated (MMA) and symmetrically (SDMA) and asymmetrically (ADMA) methylated. The type I arginine methyltransferases are indicated in the blue box. The type II symmetrical methyltransferases are indicated in the red box. The only arginine demethylase is shown in the green box.

as well as histone H4K20, have been studied extensively and linked to chromatin and transcriptional regulation as well as DNA damage response (Figure 15.1a) [13, 16]. Lysine can be mono-, di- and trimethylated, while arginine can be both monomethylated and symmetrically or asymmetrically dimethylated (Figure 15.1b, c) [17, 18]. The numerous lysine and arginine residues within the histone tails, in conjunction with the various methylation levels that can be generated at each of these sites, provide tremendous regulatory potentials for chromatin modifications.

The ability to study the impact that specific lysines and arginines as well as their degree of methylation have on chromatin biology resulted from the development of highly specific antibodies that became available in early 2000 [17]. These critical tools, the combined use of genetics in model organisms, and the advances in genomic technologies have provided additional support to the notion that specific residues, their degree of methylation, and their position within the genome have very specific consequences. The importance of fine-tuning the degree of lysine methylation was further solidified in 2004 by the ground breaking discovery of the first *bona fide* histone lysine demethylase called lysine specific demethylase (LSD1, which recently was renamed KDM1) [19, 20]. Subsequent to this discovery, a flurry of studies reported additional demethylases that are capable of demethylating specific mono-, di-, and tri-methylated lysines as well as arginines [21, 22]. These studies emphasize the importance of keeping methylation finely tuned. The impact demethylation has on the concept of methylation state will be discussed later in this chapter.

ENZYMES REGULATING ARGININE AND LYSINE METHYLATION STATES

Histone Arginine Methyltransferases

In order to methylate the arginine or lysine amino acids, the methyltransferases use S-adenosylmethionine (SAM). The methyl transfer results in a methylated amino acid and the production of S-adenosyl-L-homocysteine (AdoHcy) (Figure 15.2) [9, 10]. The methyl group is transferred to either or both guanidino nitrogen on arginine, resulting in one of three states: N^G monomethylarginine (MMA), N^GN^G (asymmetric) dimethylarginine (ADMA), or N^GN^G (symmetric) dimethylarginine (SDMA) (Figure 15.1c). There are three classes of arginine methyltransferase that are conserved from yeast to human [18, 23, 24]. The type I enzymes (PRMT1, PRMT3, CRM1/PRMT4, PRMT6, and PRMT8) are responsible for both monomethylation (MMA) as well as asymmetric methylation (ADMA). The type II class of enzymes (PRMT5, PRMT7, and PRMT9/FBXO11) conducts monomethylation (MMA) and symmetric methylation (SDMA) (Figure 15.1c). PRMT7 is also classified as a type III arginine methyltransferase because of the ability to

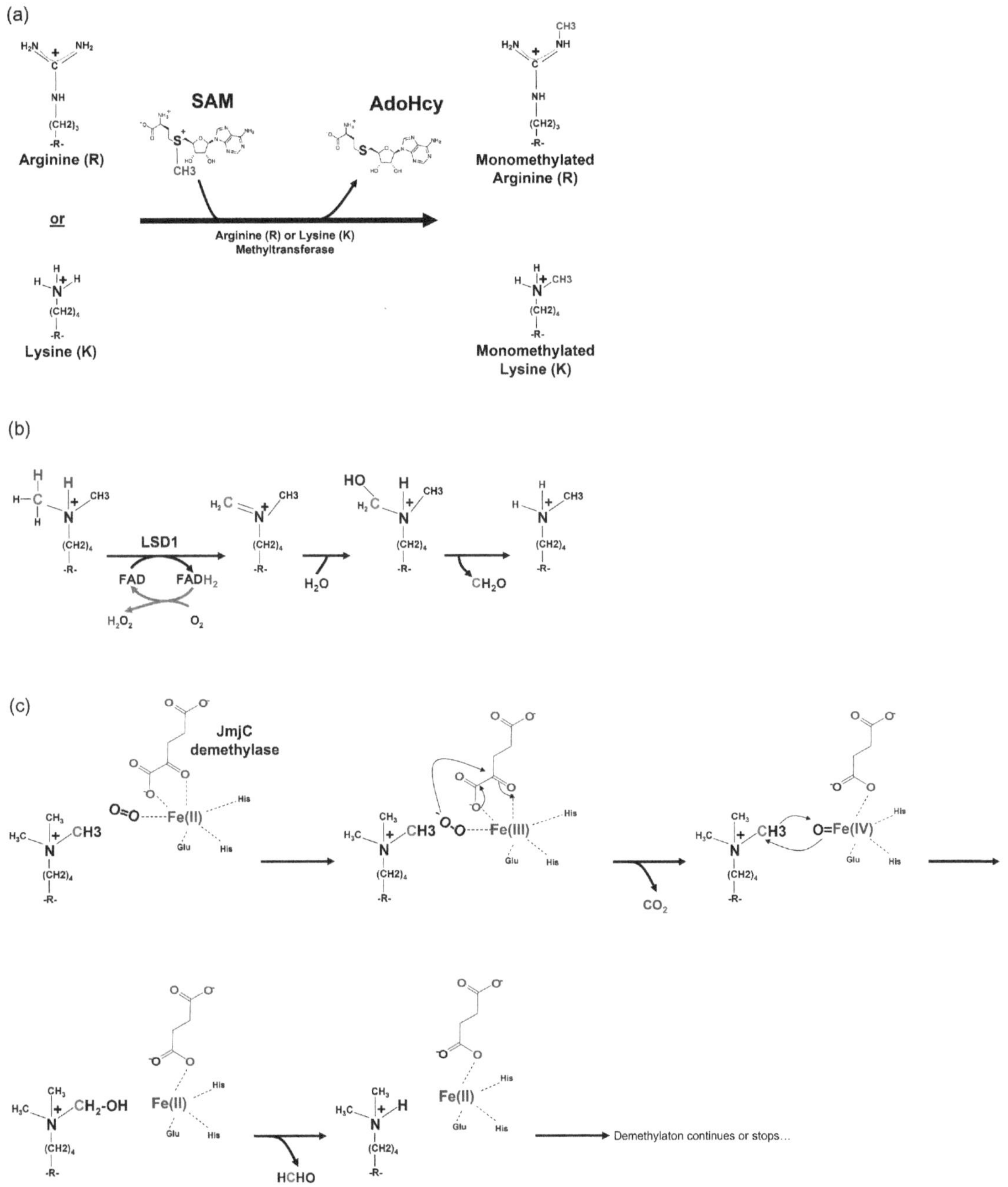

FIGURE 15.2 Histone methyltransferase and demethylase reaction mechanisms.
(a) The arginine and lysine methyltransferases use S-adenosylmethionine (SAM) as a cofactor to transfer the methyl group (red). These reactions can proceed to various degrees (see Figure 15.1). (b) The reaction scheme for LSD1 demethylating a dimethylated lysine to a monomethylated lysine. The reaction can proceed one more time to generate an unmodified lysine. This scheme was adapted from Shi *et al.* [19]. (c) The reaction scheme for JmjC histone demethylases. The reaction scheme depicts the demethylation of a trimethylated lysine. A complete description of this reaction is in [34]. The Fe(II) coordinates the molecular oxygen and α-ketoglutarate. Through electron transfers, CO_2 and succinate are generated. The highly reactive and unstable Fe(IV)-oxo intermediate is generated and hydroxylates the methyl group and spontaneous demethylation ensues. These mechanisms are reviewed in [23].

stop at monomethylation [18, 23, 24]. The arginine methyltransferases are known to methylate both histone and non-histone proteins. For example, CRM1/PRMT4 catalyzes the methylation of histone H3R2, R17, and R26. PRMT1 and 7 catalyze asymmetric and symmetric methylation of histone H4R3, respectively (Figure 15.1c). PRMT5 symmetrically methylates histone H3R8 and H4R3 [24, 25]. The methylation of H4R3 facilitates transcriptional activation, while the asymmetric methylation of H3R2 by CARM1 and PRMT6 results in decreased gene expression [26–31]. These examples demonstrate the unique roles that each of the modification events can have on the cell.

HISTONE LYSINE METHYLTRANSFERASES

Lysine specific methylation is extensively studied in chromatin biology. Therefore, our knowledge of the impact lysine methylation has on nuclear events and organismal development is rather advanced when compared to arginine methylation. Currently, there are two types of histone lysine methyltransferases. The first class of enzymes is defined by the presence of the SET domain, which is a 130 amino acid domain found in the suppressor of variegation (Su(var)3–9), enhancer of zeste (E(z)) and trithorax [32]. This class comprises the largest group of enzymes. The other class is the DOT1L/KMT4 related enzymes [20, 33]. They do not have a conserved SET domain. Regardless of SET or DOT1L methyltransferase domains, both enzyme groups use SAM to methylate lysines (Figure 15.2a). These enzymes do not always methylate from unmodified to trimethylated states. Some of the enzymes have preferences for catalyzing methylation from an unmethylated to a monomethylated state, while others have preferences for di- to trimethylation (Figure 15.1b) [34, 35]. Not only do these enzymes show specificity toward the degree of methylation but a high degree of lysine specificity within the histones. This level of specificity is functionally important, as it allows DNA templated processes such as transcriptional activation or repression to be tightly linked to the degree of methylation on certain lysines. A clear example of this can be seen in Figure 15.3. There are emerging hypotheses about how methyltransferases are able to have specificity at the level of methylation and choice of substrate. Most of the hypotheses are based on structure function studies that have emerged over the past few years. For a complete discussion on the structural constraints involved in the specificity of methylation on arginine and lysine residues see a review from Smith and Denu [23].

The discovery of the first histone lysine methyltransferase was the result of bioinformatics, genetics, and biochemical approaches. The similarity of the SET domain to a plant methyltransferase is what prompted Jenuwein and colleagues to ask whether Su(var)3–9 was in fact a *bona fide* histone lysine methyltransferase [12, 32]. This discovery resulted in a tremendous burst of research in the area of histone lysine methylation and the identification of many methyltransferases (Figure 15.1b) [34]. The discovery of methyltransferases started putting their methylation events in context. For example, Su(var)3–9 trimethylates H3K9 from a monomethylated state [36]. H3K9me3 is essential for heterochromatin assembly and maintenance [12, 36, 37]. Therefore, heterochromatin defects are observed in the Su(var)3–9 knockout mice [36]. This type of lysine and degree of methylation specificity has emerged for other enzymes. For example, the G9a histone H3K9 methyltransferase methylates from an unmethylated to dimethylated state, with a preference for mono- to dimethylation [38, 39]. Similar examples are observed for other lysine residues such as H3K4, K27, and H4K20 [20, 34]. H3K4 is methylated by enzymes that catalyze all degrees of methylation (Set1; [40, 41]) or that catalyze di- and/or trimethylation (SMYD3; [42]), or dimethylation from unmodified (SET7/9; [43]) (Figure 15.1b). The MLL1 enzyme (also known as ALL-1, see Gene ID 4297) methylates H3K4 from an unmodified state to a dimethylated state [44, 45]. However, when MLL1 is associated with the endogenous interacting proteins, there is an increase in the enzymatic activity that results in a trimethylated H3K4 [46]. This phenomenon has also been seen for the histone demethylases (discussed in [34]). This point is rather important because it demonstrates that these enzymes have specific activities, but their associated complexes or interacting partners could alter their target or degree of activity. These higher order protein interactions add additional levels of complexity and specificity to the system that will become clearer as researchers uncover the impact interacting partners have on enzyme function.

The main point of this section is to reiterate that there are a number of enzymes that methylate both arginines and lysines to different degrees and that this degree of methylation has specific and direct consequences within the nucleus. The consequence of specific modifications will be discussed in more detail later in this chapter.

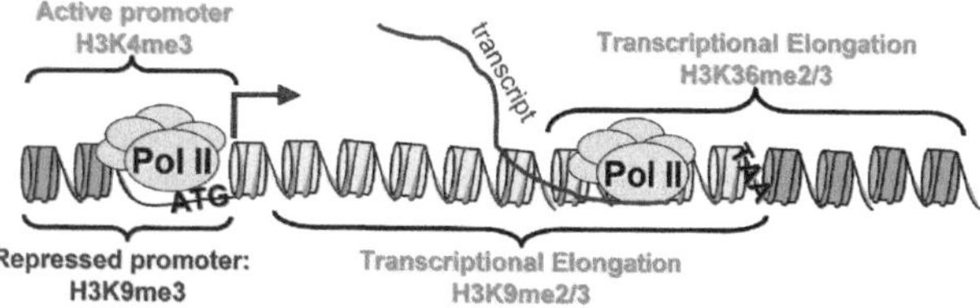

FIGURE 15.3 Schematic representation of how site, degree, and location of methylation can impact transcription.
The RNA polymerase II complex loads at active promoters where H3K4me3 peaks [40, 67–70]. Repressed promoters have increased H3K9me3 at their promoters [70, 75]; while, the coding regions of induced genes have H3K9me3 [70, 76]. This demonstrates how placement makes the difference. The elongating RNA polymerase II is associated with H3K36me3, which extends from the end of the open reading frame into the 3'-untranslated region [74].

HISTONE DEMETHYLASE ENZYMES

The field of chromatin biology was recently stimulated by a very important observation from Yang Shi and colleagues. Shi's group identified the first *bona fide* lysine specific demethylase, referred to as LSD1/KDM1 [19]. The discovery of this enzyme had two major contributions to the fields of protein methylation and chromatin biology. First, the discovery demonstrated that methylated lysines within proteins are, in fact, enzymatically reversible. Second, the discovery clearly documented that oxidation reactions are important for the demethylation of lysines or arginines. This knowledge fueled many laboratories to identify additional enzymes that use oxidation to reverse the methylation of lysines and arginines. These groups of enzymes are the focus of this section.

Lysine Specific Demethylases

FAD Dependent Amine Oxidase Histone Lysine Demethylases

Currently there are two classes of histone lysine demethylases. The first classes of enzymes are the FAD dependent amine oxidase enzymes that are restricted to di- and monodemethylation [34]. The second class of enzyme is the JmjC Fe(II) containing demethylases that are responsible for demethylating tri, di-, and monomethyl residues [21, 22, 47]. Like the histone methyltransferases, the demethylases have specificity at the level of the specific amino acids as well as the degree of methylation (Figure 15.1b) [34].

The enzymatic reversal of methylation remained elusive for ~30 years. In fact, many researchers started believing this enzymatic reaction did not occur. This belief was based on studies in the early 1970s that demonstrated a comparable turnover rate for bulk histones and the methyl groups on histone arginines and lysines in mammalian cells, suggesting methylation was only removed by bulk histone removal [48, 49]. In the same year, a separate study identified a low level of histone methyl group turnover (~2 percent per hour) [50]. The same research group also identified enzymatic activity from rat kidney extract; however, the enzyme responsible for this demethylation event remained elusive [51, 52]. Nevertheless, the lack of an enzyme seemed unlikely when considering all other PTMs are reversible.

The landmark finding of the histone lysine demethylase resulted from the purification of a corepressor complex called the C-terminal binding protein 1 complex (CtBP1) [53]. The CtBP1 protein complex, along with a number of other corepressors, contains the LSD1 protein, which has a flavin adenine dinucleotide (FAD) dependent amine oxidase domain (reviewed in [34]). The association of this protein with transcriptional corepressor complexes as well as the presence of the amine oxidase domain suggested a role in either polyamine metabolism or histone methylation. The pure enzyme did not have any affect on polyamines. However, when LSD1 was incubated with methylated peptides, site specific demethylation was observed for H3K4me2 and H3K4me1 (Figure 15.1b) [19].

The LSD1 related enzymes are not capable of removing methyl groups from trimethylated lysines. The demethylase chemistry requires oxidative cleavage of the α-carbon bond of the substrate to form an imine intermediate, which is hydrolyzed to form an aldehyde and amine. The cofactor FAD is reduced to $FADH_2$, which is then reoxidized by molecular oxygen, producing H_2O_2. The LSD1 mediated demethylation reactions generate the predicted reaction products (unmethylated histone peptides, formaldehyde, and H_2O_2) that were readily detected by multiple approaches [19]. The reaction scheme for LSD1 is described in Figure 15.2b.

JMJC Fe(II) Dependent Histone Lysine Demethylases

The discovery of LSD1 as a histone demethylase revealed the possibility that other enzymes capable of oxidative reactions could act as histone demethylases. The Fe(II) and α-ketoglutarate-dependent dioxygenases seemed to be likely candidates because they could use a radical attack to remove the methyl group. Furthermore, this reaction would allow for histone tridemethylation (Figure 15.2c). This prediction comes from the observations that ALKB dioxygenase proteins dealkylate DNA through radical hydroxylation [54–57]. This prediction was shown to be true when a JmjC family of Fe(II) and α-ketoglutarate dependent dioxygenases was shown to remove mono- and dimethylation from H3K36 [58]. Not long after this discovery, the first group of JmjC proteins capable of histone lysine tridemethylation was discovered [59–62]. The JmjC domain defines a group of proteins that contain a β-barrel structure that coordinates Fe(II) and α-ketoglutarate [63]. The JmjC domain is found in approximately 100 proteins from bacteria to eukaryotes and is the catalytic domain responsible for histone demethylation [58, 63]. Some of the JmjC containing proteins also contain the JmjN domain, which is only observed in the presence of the JmjC domain and seems to provide an important structural feature [64].

The initial biochemical studies and demethylase crystal structure of JMJD2A/KDM4A confirmed the requirements of Fe(II) and α-ketoglutarate as cofactors for demethylation [64]. The coordination of Fe(II) and α-ketoglutarate plays a critical role in generating the radical attack of the methyl groups. This radical attack is what distinguishes JmjC proteins from LSD1 related demethylases (Figure 15.2, compare b and c). The Fe(II) provides a resonance structure to the coordinated molecular oxygen, resulting in Fe(III) and a superoxide radical species that attacks the α-ketoglutarate and results in the production of CO_2 and an Fe(IV)-oxo intermediate (Fe[IV] =O; Figure 15.2c). The Fe(IV)-oxo intermediate removes a hydrogen from the methyl group, creates a free radical, and generates an Fe(III) hydroxide. The free radical on the substrate attacks the iron (III) hydroxide and becomes hydroxylated, creating a carbinolamine that spontaneously releases formaldehyde (Figure 15.2c). Unlike LSD1, the JmjC domain demethylases do not require a protonated lysine for demethylation; therefore, they are capable of demethylating not only mono- and dimethylated lysines but also trimethylated lysine residues (Figure 15.2c). However, several of the JmjC enzymes are not able to remove the mono- or trimethylation from lysines *in vitro* or *in vivo*. Future structure function studies will reveal valuable insights into this limitation.

The first JMJC demethylases to be discovered were the JHDM1A/B proteins (KDM2A/B; Figure 15.1b) [58]. These proteins as well as the H3K9me2 demethylases JMJD1A-C/KDM3A-C were identified in chromatography fractions screened for the ability to release formaldehyde from methylated histones [58, 65]. This discovery reiterated that formaldehyde is a common by-product from demethylase reactions, which was originally reported for LSD1 [19]. Using a candidate based approach that targeted about 50 enzymes with chromatin related motifs or nuclear localization signals, Whetstine *et al.* identified the first histone tridemethylase capable of removing both H3K9me3 and H3K36me3 [59]. This enzyme family is referred to as JMJD2A-D/KDM4A-D. This same family was subsequently identified with the fractionation approach and affinity purification [60–62]. Interestingly, this family of enzymes was unable to remove monomethylation on H3K9 or H3K36. This observation reiterates that methylation states are distinctly regulated.

Using the same candidate based approach, Whetstine and colleagues proceeded to identify the H3K4me3 and H3K27me3 demethylases. During this rapid expansion of our understanding of demethylation, many groups revealed the same proteins to be histone demethylases (reviewed in [21, 22]). The JARID/KDM5A-D family of enzymes removes the trimethylation from H3K4. This family of enzymes is composed of four members that have the ability to remove tri- and dimethylation but lack the ability to remove monomethylation [21, 22]. The H3K27me3 demethylases were subsequently discovered (JMJD3 class/KDM6) and shown to remove tri- and dimethylation. The JMJD3/KDM6A enzyme has the ability to demethylate K27 *in vitro* and *in vivo*. However, the JMJD3 homolog UTX/KDM6C lacks the ability to demethylate histones in cells overexpressing this protein. The other homolog UTY/KDM6B lacks any detectable demethylase activity *in vitro* or *in vivo*. The data from these studies demonstrate the specificity and unique activities associated with each protein, even when highly homologous [21, 22]. This same phenomenon was observed for the JMJD2/KDM4A-D family of K9/36me3 demethylases. JMJD2A-C generate dimethylated products; however, increased enzyme concentrations or overexpression can result in monomethylation. Interestingly, the JMJD2D/KDM4D homolog generates di- and monomethylated products for H3K9me3 at a range of enzyme concentrations [59]. The data presented in this section demonstrates that demethylases, like methyltransferases, are able to finely tune the methylation status. The fine-tuning of methylation will undoubtedly impact specific chromatin templated processes.

JMJC Fe(II) Dependent Histone Arginine Demethylases

The *bona fide* arginine demethylase remained elusive till October 2007. The JMJD6 enzyme (also known as PTDSR, Gene ID 23210) was shown to demethylate H3R2me2 (ADMA) and H4R3me2 (SDMA). JMJD6 has weak demethylase activity toward H4R3me1 in histones. Upon over expression in cells, JMJD6 was able to demethylate H3R2me2 and H4R3me2 (Figure 15.1c) [66]. Much like the histone lysine demethylases, these data suggest that the arginine demethylases will have specificity at the level of the site and degree of methylation. Future studies are needed to determine if this is a common property among other unknown arginine demethylases.

DEGREE AND LOCATION MATTER

Since the cell has developed a number of enzymes to balance methylation states in the forward and reverse reaction, one anticipates that there must be direct or specific consequences associated with these various methylation events. The exact interplay between lysines and arginines has not been completely resolved; however, there are several examples of how the specific sites can impact chromosomal structure and DNA templated processes (e.g., transcriptional events and DNA damage response) (Figure 15.3). The objective of this section is not to list all of the possible relationships but demonstrate that the site and degree of methylation has specific consequences.

The site and degree of R or K methylation have been demonstrated to be a hallmark of certain nuclear processes or chromosomal structures. For example, the trimethylation of H3K4 is found at the promoters of actively transcribed genes (Figure 15.3) [40, 67–70]. The enzymes that coordinate this activity aid in the activation of gene expression (e.g., Set1 or MLL) [40, 41, 44–46, 68]. However, enzymes that remove this methylation state or preclude this methylation from occurring will result in decreased gene expression and/or more silenced states within the genome such as heterochromatin. The demethylation of H3K4me3 does in fact repress transcription (discussed in [21, 22]). Consistent with the requirement of this site for active transcription, H3R2me2 (ADMA) results in the inhibition of the trimethylation of H3K4. In both yeast and human cells, H3R2me2 (ADMA) inhibits the binding of proteins that promote H3K4me3 and gene activation [27–31]. Kouzarides and colleagues demonstrated that H3R2me2 (ADMA) is found at heterochromatic regions and telomeric sites in the yeast genome. This study also demonstrated that H3R2me2 (ADMA) is localized within the coding regions and the 3'-untranslated regions of genes, which corresponds to regions void of H3K4me3 [29]. The human PRMT6 was also shown to be responsible for H3R2me2 (ADMA) and to interfere with H3K4me3 [28, 30]. These studies demonstrate how modifying the site and degree of one site can lead to distinct transcriptional consequences.

Methylation of lysines has also been shown to suppress gene expression. For example, the presence of H3K9me3 and HeK27me3 has been shown to suppress gene expression, to maintain the silent state of the X chromosome, and to establish and maintain heterochromatin formation [16]. H3K9me3 serves as a docking site for the heterochromatin protein 1 (HP1), which creates a more compacted chromatin state and results in gene silencing and heterochromatin establishment [12, 36, 37]. Therefore, altering the levels of the methyltransferase or demethylases that modify this methylation state can have dramatic consequences on gene expression and/or chromatin structure. SUV39H1/KMT1 knockout mice have decreased viability and develop cancer [36]. The mouse knockout cells have abnormal heterochromatin and increased genomic instability. Consistent with this observation, the H3K9/36me3 demethylase JMJD2C/KDM4C is amplified in several cancers and has been shown to alter HP1 localization and heterochromatin formation when overexpressed [61, 62, 71–73].

Methylation states are also important for protecting the genome from aberrant transcriptional events and for recognizing damaged DNA. The trimethylation of H3K36, which is catalyzed by Set2/KMT, is found across the 3'-end of the open reading frame of genes and into their 3'-untranslated region (Figure 15.3) [74]. The lack of this modification is rather deleterious because the site serves as a docking site for the RPD3S repressor complex that blocks spurious transcription within the coding region of genes [74]. Interestingly, if the H3K36 methyltransferase is localized to the promoter of genes, gene repression occurs, which is likely through the RPD3S recruitment. This example emphasizes the importance of methylated amino acid placement.

Previous studies have shown that H3K9me3 is found in repressed promoters and in heterochromatin; however, this mark has been found in the coding region of actively induced genes as well (Figure 15.3) [12, 36, 37, 75, 76]. This was originally observed at specific genes but recently was observed in several genes on a genome-wide level [70]. Interestingly, when the *C. elegans* member of the JMJD2/KDM4 H3K9/36me3 demethylases (JMJD-2) is depleted or deleted, increased levels of H3K9me3 and H3K36me3 are observed. All of the adult germline nuclei have increased levels of H3K9me3, while H3K36me3 levels increase at a very specific region on the X chromosome. The lack of this enzyme also results in increased double-strand breaks that trigger P53 dependent apoptosis [59]. The data from *C. elegans* emphasize that specific regions are regulated within the genome and that methylation balance is essential for genomic integrity. The need for balance and site specific control emphasizes why cells have developed many modifying enzymes that are capable of fine-tuning methylation at distinct loci in the genome. The more we uncover about the genomic localization of histone arginine and lysine methyltransferases and demethylases, the more we will appreciate how these enzymes regulate distinct genes, how these enzymes regulate chromatin templated processes, and how these enzymes help ward off disease.

Histone methylation plays an important role in marking areas in the genome that are damaged. The trimethylation of H4K20 serves as a hallmark of heterochromatin but it also plays an important role in docking DNA damage machinery when DNA double-strand breaks occur [16, 77]. This dual specificity reiterates the importance in balance and placement. Much like SUV39H1/2 knockout mice, mice lacking H4K20me3 have increased risks for cancer and the mouse embryonic fibroblast have increased genomic instability. The lack of the H4K20me3 does not alter H4K20me1, but these animals still have increased DNA instability. This is also true for the H3K9me3 knockout mice mentioned above. Consistent with these observations, primary tumors lost H4K20me3 [78]. Overall, these *in vivo* studies emphasize the requirement for certain sites as well as the degree of methylation so that organisms develop properly and ward off tumorigenesis. The need to balance methylation is not only required for inhibiting cancer risk but is also important in neurological development. For example, mutations in the H3K4me3 demethylase SmcX/KDM3C are associated with mental retardation [79, 80]. The same mutations found in mentally retarded patients result in decreased H3K4me3 demethylase activity *in vitro* and *in vivo* [80]. The lack of SmcX results in decreased neuronal survival and is required for dendritic morphogenesis [80]. From these few examples, one can appreciate the need to balance site, degree, and placement of arginine and lysine methylation within the histones.

In closing, this chapter highlighted the impact that an inert 15 dalton molecule can have on chromatin and organismal development. The cell has developed a number of enzymes and pathways to regulate the amount of methylation that occurs within the genome; therefore, understanding the intricate relationships between the methylation states, the enzymes that regulate the methylation levels, and how the enzymes determine their localization and degree of specificity will undoubtedly reveal important insights into developmental patterning, potential mechanisms that result in disease, but more importantly, how we might be able to intervene when histone methylation patterns are abnormal in disease.

REFERENCES

1. Flemming W. *Zellsubstanz, Kern und Zelltheilung*. Liepzig: Verlag Vogel; 1882.
2. Kossel A. Ueber die chemische Beschaffenheit des Zelkerns. Munchen. *Med Wochenschrift* 1911;**58**:65–9.
3. Olin DE, Olin AL. Chromatin history: our view from the bridge. *Nature Rev* 2003;**4**:809–14.
4. Kornberg RD, Lorch Y. Twenty-five years of the nucleosome, fundamental particle of the eukaryote chromosome. *Cell* 1999;**98**:285–94.
5. Allfrey VG, Faulkner R, Mirsky AE. Acetylation and Methylation of Histones and Their Possible Role in the Regulation of Rna Synthesis. *Proc Natl Acad Sci USA* 1964;**51**:786–94.

6. Murray K. The Occurrence of Epsilon-N-Methyl Lysine in Histones. *Biochemistry* 1964;**3**:10–15.
7. Paik WK, Kim S. Enzymatic methylation of protein fractions from calf thymus nuclei. *Biochem Biophys Res Commun* 1967;**29**:14–20.
8. Paik WK, Kim S. Enzymatic methylation of histones. *Arch Biochem Biophys* 1969;**134**:632–7.
9. Kim S, Paik WK. Studies on the origin of epsilon-N-methyl-L-lysine in protein. *J Biol Chem* 1965;**240**:4629–34.
10. Paik WK, Kim S. Protein methylation. *Science* 1971;**174**:114–19.
11. Ghosh SK, Paik WK, Kim S. Purification and molecular identification of two protein methylases I from calf brain. Myelin basic protein- and histone-specific enzyme. *J Biol Chem* 1988;**263**:19,024–19,033.
12. Rea S, Eisenhaber F, O'Carroll D, Strahl BD, Sun Z-W, Opravil S, Mechtier K, Ponting CP, Allis CD, Jenuwein T. Regulation of chromatin structure by site-specific histone H3 methyltransferases. *Nature* 2000;**406**:593–9.
13. Margueron R, Trojer P, Reinberg D. The key to development: interpreting the histone code? *Curr Opin Genet Dev* 2005;**15**:163–76.
14. Zhang L, Eugeni EE, Parthun MR, Freitas MA. Identification of novel histone post-translational modifications by peptide mass fingerprinting. *Chromosoma* 2003;**112**:77–86.
15. Zhang K, Siino JS, Jones PR, Yau PM, Bradbury EM. A mass spectrometric "Western blot" to evaluate the correlations between histone methylation and histone acetylation. *Proteomics* 2004;**4**:3765–75.
16. Martin C, Zhang Y. The diverse functions of histone lysine methylation. *Nat Rev Mol Cell Biol* 2005;**6**:838–49.
17. Bannister AJ, Kouzarides T. Histone methylation: recognizing the methyl mark. *Methods Enzymol* 2004;**376**:269–88.
18. Bedford M,T. Arginine methylation at a glance. *J Cell Sci* 2007; **120**:4243–6.
19. Shi Y, Lan F, Matson C, Mulligan P, Whetstine JR, Cole PA, Casero RA, Shi Y. Histone demethylation mediated by the nuclear amine oxidase homolog LSD1. *Cell* 2004;**119**:941–53.
20. Allis CD, Berger SL, Cote J, Dent S, Jenuwien T, Kouzarides T, Pillus L, Reinberg D, Shi Y, Shiekhattar R, Shilatifard A, Workman J, Zhang Y. New nomenclature for chromatin-modifying enzymes. *Cell* 2007;**131**:633–6.
21. Cloos PA, Christensen J, Agger K, Helin K. Erasing the methyl mark: histone demethylases at the center of cellular differentiation and disease. *Genes Dev* 2008;**22**:1115–40.
22. Lan F, Nottke AC, Shi Y. Mechanisms involved in the regulation of histone lysine demethylases. *Curr Opin Cell Biol* 2008;**20**:316–25.
23. Smith BC, Denu JM. Chemical mechanisms of histone lysine and arginine modifications. *Biochim Biophys Acta* 2008;**1789**:45–57.
24. Pal S, Sif S. Interplay between chromatin remodelers and protein arginine methyltransferases. *J Cell Physiol* 2007;**213**:306–15.
25. Wysocka J, Allis CD, Coonrod S. Histone arginine methylation and its dynamic regulation. *Front Biosci* 2006;**11**:344–55.
26. Wang H, Huang ZQ, Xia L, Feng Q, Erdjument-Bromage H, Strahl BD, Briggs SD, Allis CD, Wong J, Tempst P, Zhang Y. Methylation of histone H4 atarginine 3 facilitating transcriptional activation by nuclear hormone receptor. *Science* 2001;**293**:853–7.
27. van Ingen H, van Schaik FM, Wienk H, Ballering J, Rehmann H, Dechesne AC, Kruijzer JA, Liskamp RM, Timmers HT, Boelens R. Structural Insight into the Recognition of the H3K4me3 Mark by the TFIID Subunit TAF3. *Structure* 2008;**16**:1245–56.
28. Iberg AN, Espejo A, Cheng D, Kim D, Michaud-Levesque J, Richard S, Bedford MT. Arginine methylation of the histone H3 tail impedes effector binding. *J Biol Chem* 2008;**283**:3006–10.
29. Kirmizis A, Santos-Rosa H, Penkett CJ, Singer MA, Vermeulen M, Mann M, Bähler J, Green RD, Kouzarides T. Arginine methylation at histone H3R2 controls deposition of H3K4 trimethylation. *Nature* 2007;**449**:928–32.
30. Guccione E, Bassi C, Casadio F, Martinato F, Cesaroni M, Schuchlautz H, Lüscher B, Amati B. Methylation of histone H3R2 by PRMT6 and H3K4 by an MLL complex are mutually exclusive. *Nature* 2007;**449**:933–7.
31. Vermeulen M, Mulder KW, Denissov S, Pijnappel WW, van Schaik FM, Varier RA, Baltissen MP, Stunnenberg HG, Mann M, Timmers HT. Selective anchoring of TFIID to nucleosomes by trimethylation of histone H3 lysine 4. *Cell* 2007;**131**:58–69.
32. Jenuwein T. The epigenetic magic of histone lysine methylation. *FEBS J* 2006;**273**:121–35.
33. Okada Y, Feng Q, Lin Y, Jiang Q, Li Y, Coffield VM, Su L, Xu G, Zhang Y. hDOT1L links histone methylation to leukemogenesis. *Cell* 2005;**121**:167–78.
34. Shi Y, Whetstine JR. Dynamic Regulation of Histone Lysine Methylation by Demethylases. *Mol Cell* 2007;**25**:1–14.
35. Dillon SC, Zhang X, Trievel RC, Cheng X. The SET-domain protein superfamily: protein lysine methyltransferases. *Genome Biol* 2005;**6**:227.
36. Peters AH, O'Carroll D, Scherthan H, Mechtler K, Sauer S, Schofer C, Weipoltshammer K, Pagani M, Lachner M, Kohlmaier A, et al. Loss of the Suv39h histone methyltransferases impairs mammalian heterochromatin and genome stability. *Cell* 2001;**107**:323–37.
37. Nakayama J, Rice JC, Strahl BD, Allis CD, Grewal SI. Role of histone H3 lysine 9 methylation in epigenetic control of heterochromatin assembly. *Science* 2001;**292**:110–13.
38. Tachibana M, Sugimoto K, Fukushima T, Shinkai Y. Set domain-containing protein, G9a, is a novel lysine-preferring mammalian histone methyltransferase with hyperactivity and specific selectivity to lysines 9 and 27 of histone H3. *J Biol Chem.* 2001;**276**:25,309–25,317,.
39. Tachibana M, Sugimoto K, Nozaki M, Ueda J, Ohta T, Ohki M, Fukuda M, Takeda N, Niida H, Kato H, Shinkai Y. G9a histone methyltransferase plays a dominant role in euchromatic histone H3 lysine 9 methylation and is essential for early embryogenesis. *Genes Dev* 2002;**16**:1779–91.
40. Ng HH, Robert F, Young RA, Struhl K. Targeted recruitment of Set1 histone methylase by elongating Pol II provides a localized mark and memory of recent transcriptional activity. *Mol Cell* 2003;**11**:709–19.
41. Wysocka J, Myers MP, Laherty CD, Eisenman RN, Herr W. Human Sin3 deacetylase and trithorax-related Set1/Ash2 histone H3-K4 methyltransferase are tethered together selectively by the cell-proliferation factor HCF-1. *Genes Dev* 2003;**17**:896–911.
42. Hamamoto R, Furukawa Y, Morita M, Iimura Y, Silva FP, Li M, Yagyu R, Nakamura Y. SMYD3 encodes a histone methyltransferase involved in the proliferation of cancer cells. *Nat Cell Biol* 2004;**6**:731–40.
43. Kwon T, Chang JH, Kwak E, Lee CW, Joachimiak A, Kim YC, Lee J, Cho Y. Mechanism of histone lysine methyl transfer revealed by the structure of SET7/9-AdoMet. *EMBO J* 2003;**22**:292–303.
44. Milne TA, Briggs SD, Brock HW, Martin ME, Gibbs D, Allis CD, Hess JL. MLL Targets SET Domain Methyltransferase Activity to Hox Gene Promoters. *Mol Cell* 2002;**10**:1107–17.
45. Nakamura T, Mori T, Tada S, Krajewski W, Rozovskaia T, Wassell R, Dubois G, Mazo A, Croce CM, Canaani E. ALL-1 Is a Histone Methyltransferase that Assembles a Supercomplex of Proteins Involved in Transcriptional Regulation. *Mol Cell* 2002;**10**:1119–28.
46. Dou Y, Milne TA, Ruthenburg AJ, Lee S, Lee JW, Verdine GL, Allis CD, Roeder RG. Regulation of MLL1 H3K4 methyltransferase activity by its core components. *Nat Struct Mol Biol* 2006;**13**:713–19.
47. Klose RJ, Zhang Y. Regulation of histone methylation by demethylimination and demethylation. *Nat Rev Mol Cell Biol* 2007;**8**:307–18.

48. Byvoet P, Shepherd GR, Hardin JM, Noland BJ. The distribution and turnover of labeled methyl groups in histone fractions of cultured mammalian cells. *Arch Biochem Biophys* 1972;**148**:558–67.
49. Thomas G, Lange HW, Hempel K. [Relative stability of lysine-bound methyl groups in arginine-rich histones and their subfrations in Ehrlich ascites tumor cells in vitro]. *Hoppe Seylers Z Physiol Chem* 1972;**353**:1423–8.
50. Borun TW, Pearson D, Paik WK. Studies of histone methylation during the HeLa S-3 cell cycle. *J Biol Chem* 1972;**247**:4288–98.
51. Paik WK, Kim S. Enzymatic demethylation of calf thymus histones. *Biochem Biophys Res Commun* 1973;**51**:781–8.
52. Paik WK, Paik DC, Kim S. Historical review: the field of protein methylation. *Trends Biochem Sci* 2007;**32**:146–52.
53. Shi YJ, Sawada J-I, Sui GC, Affar EB, Whetstine J, Lan F, Ogawa H, Luke MP-S, Nakatani Y, Shi Y. Coordinated histone modifications mediated by a CtBP co-repressor complex. *Nature* 2003;**422**:735–8.
54. Falnes PO, Johansen RF, Seeberg E. AlkB-mediated oxidative demethylation reverses DNA damage in Escherichia coli. *Nature* 2002;**419**:178–82.
55. Trewick SC, Henshaw TF, Hausinger RP, Lindahl T, Sedgwick B. Oxidative demethylation by Escherichia coli AlkB directly reverts DNA base damage. *Nature* 2002;**419**:174–8.
56. Kubicek S, Jenuwein T. A crack in histone lysine methylation. *Cell* 2004;**119**:903–6.
57. Trewick SC, McLaughlin PJ, Allshire RC. Methylation: lost in hydroxylation? *EMBO Rep* 2005;**6**:315–20.
58. Tsukada Y, Fang J, Erdjument-Bromage H, Warren ME, Borchers CH, Tempst P, Zhang Y. Histone demethylation by a family of JmjC domain-containing proteins. *Nature* 2006;**439**:811–16.
59. Whetstine JR, Nottke A, Lan F, Huarte M, Smolikov S, Chen Z, Spooner E, Li E, Zhang G, Colaiacovo M, Shi Y. Reversal of histone lysine trimethylation by the JMJD2 family of histone demethylases. *Cell* 2006;**125**:467–81.
60. Fodor BD, Kubicek S, Yonezawa M, O'Sullivan RJ, Sengupta R, Perez-Burgos L, Opravil S, Mechtler K, Schotta G, Jenuwein T. Jmjd2b antagonizes H3K9 trimethylation at pericentric heterochromatin in mammalian cells. *Genes Dev* 2006;**20**:1557–62.
61. Klose RJ, Yamane K, Bae Y, Zhang D, Erdjument-Bromage H, Tempst P, Wong J, Zhang Y. The transcriptional repressor JHDM3A demethylates trimethyl histone H3 lysine 9 and lysine 36. *Nature* 2006;**442**:312–16.
62. Cloos PA, Christensen J, Agger K, Maiolica A, Rappsilber J, Antal T, Hansen KH, Helin K. The putative oncogene GASC1 demethylates tri- and dimethylated lysine 9 on histone H3. *Nature* 2006;**442**:307–11.
63. Clissold PM, Ponting CP. JmjC: cupin metalloenzyme-like domains in jumonji, hairless and phospholipase A2beta. *Trends Biochem Sci* 2001;**26**:7–9.
64. Chen Z, Zang J, Whetstine JR, Hong X, Davrazou F, Kutateladze TG, Simpson M, Mao Q, Pan C-H, Dai S, Hagman J, Hansen K, Shi Y, Zhang G. Structural insights into histone demethylation by JMJD2 family members. *Cell* 2006;**125**:691–702.
65. Yamane K, Toumazou C, Tsukada Y, Erdjument-Bromage H, Tempst P, Wong J, Zhang Y. JHDM2A, a JmjC-containing H3K9 demethylase, facilitates transcription activation by androgen receptor. *Cell* 2006; **125**:483–95.
66. Chang B, Chen Y, Zhao Y, Bruick RK. JMJD6 is a histone arginine demethylase. *Science* 2008;**318**:444–7.
67. Santos-Rosa H, Schneider R, Bannister AJ, Sherriff J, Bernstein BE, Emre NC, Schreiber SL, Mellor J, Kouzarides T. Active genes are tri-methylated at K4 of histone H3. *Nature* 2002;**419**:407–11.
68. Boa S, Coert C, Patterton HG. Saccharomyces cerevisiae Set1p is a methyltransferase specific for lysine 4 of histone H3 and is required for efficient gene expression. *Yeast* 2003;**20**:827–35.
69. Bernstein BE, Humphrey EL, Erlich RL, Schneider R, Bouman P, Liu JS, Kouzarides T, Schreiber SL. Methylation of histone H3 Lys 4 in coding regions of active genes. *Proc Natl Acad Sci USA* 2002; **99**:8695–700.
70. Mikkelsen TS, Ku M, Jaffe DB, Issac B, Lieberman E, Giannoukos G, Alvarez P, Brockman W, Kim TK, Koche RP, Lee W, Mendenhall E, O'Donovan A, Presser A, Russ C, Xie X, Meissner A, Wernig M, Jaenisch R, Nusbaum C, Lander ES, Bernstein BE. Genome-wide maps of chromatin state in pluripotent and lineage-committed cells. *Nature* 2007;**7153**:548–9.
71. Yang ZQ, Imoto I, Fukuda Y, Pimkhaokham A, Shimada Y, Imamura M, Sugano S, Nakamura Y, Inazawa J. Identification of a novel gene, GASC1, within an amplicon at 9p23-24 frequently detected in esophageal cancer cell lines. *Cancer Res* 2000;**60**:4735–9.
72. Ehrbrecht A, Müller U, Wolter M, Hoischen A, Koch A, Radlwimmer B, Actor B, Mincheva A, Pietsch T, Lichter P, Reifenberger G, Weber RG. Comprehensive genomic analysis of desmoplastic medulloblastomas: identification of novel amplified genes and separate evaluation of the different histological components. *J Pathol* 2006;**208**:554–63.
73. Hélias C, Struski S, Gervais C, Leymarie V, Mauvieux L, Herbrecht R, Lessard M. Polycythemia vera transforming to acute myeloid leukemia and complex abnormalities including 9p homogeneously staining region with amplification of MLLT3, JMJD2C, JAK2, and SMARCA2. *Cancer Genet Cytogenet* 2008;**180**:51–5.
74. Lee JS, Shilatifard AA. Site to remember: H3K36 methylation a mark for histone deacetylation. *Mutat Res* 2007;**618**:130–4.
75. Nielsen SJ, Schneider R, Bauer UM, Bannister AJ, Morrison A, O'Carroll D, Firestein R, Cleary M, Jenuwein T, Herrera RE, Kouzarides T. Rb targets histone H3 methylation and HP1 to promoters. *Nature* 2001;**412**:561–5.
76. Vakoc CR, Mandat SA, Olenchock BA, Blobel GA. Histone H3 lysine 9 methylation and HP1gamma are associated with transcription elongation through mammalian chromatin. *Mol Cell* 2005;**19**:381–91.
77. Sanders SL, Portoso M, Mata J, Bähler J, Allshire RC, Kouzarides T. Methylation of histone H4 lysine 20 controls recruitment of Crb2 to sites of DNA damage. *Cell* 2004;**119**:603–14.
78. Fraga MF, Ballestar E, Villar-Garea A, Boix-Chornet M, Espada J, Schotta G, Bonaldi T, Haydon C, Ropero S, Petrie K, Iyer NG, Pérez-Rosado A, et al. Loss of acetylation at Lys16 and trimethylation at Lys20 of histone H4 is a common hallmark of human cancer. *Nat Genet* 2005;**37**:391–400.
79. Jensen LR, Amende M, Gurok U, Moser B, Gimmel V, Tzschach A, Janecke AR, Tariverdian G, Chelly J, Fryns JP. Mutations in the JARID1C gene, which is involved in transcriptional regulation and chromatin remodeling, cause X-linked mental retardation. *Am J Hum Genet* 2005;**76**:227–36.
80. Iwase S, Lan F, Bayliss P, de la Torre-Ubieta L, Huarte M, Qi HH, Whetstine JR, Bonni A, Roberts TM, Shi Y. The X-linked mental retardation gene SMCX/JARID1C defines a family of histone H3 lysine 4 demethylases. *Cell* 2007;**128**:1077–88.

Chapter 16

Histone Phosphorylation: Chromatin Modifications that Link Cell Signaling Pathways to Nuclear Function Regulation

Priscilla Nga Ieng Lau and Peter Cheung
Department of Medical Biophysics, University of Toronto, and Division of Signaling Biology, Ontario Cancer Institute, Toronto, Ontario, Canada

INTRODUCTION

The human genome consists of three billion base pairs of DNA (roughly 2 m in length when stretched out), which has to be packaged into a nucleus of less than 10 μm in diameter. To achieve this greater than 10,000-fold compaction, genomic DNA is packaged into chromatin, a nucleoprotein structure that serves as the physiological template for all DNA related functions. The fundamental unit of chromatin is the nucleosome core particle, which is composed of 146 bp of DNA wrapped around a histone octamer containing two copies each of the four core histones – H2A, H2B, H3, and H4 [1]. The binding of histone H1 to linker DNA facilitates folding of the nucleosome arrays into 30 nm fibers, and the association of non-histone chromosomal proteins further compacts the chromatin fiber into higher order structures. The most condensed form of chromatin is seen in mitosis as metaphase chromosomes [2].

The organization of DNA within chromatin successfully solves the basic packaging problem; however, this poses another problem for the cell – the repressive nature of chromatin restricts the access of DNA by regulatory factors, and thus affects all DNA templated processes such as transcription, DNA replication, DNA repair, and recombination. To facilitate these essential cellular functions, eukaryotic cells have developed multiple strategies to modulate chromatin structure and to alleviate this repressive nature. For example, histone modifying enzymes add posttranslational modifications to specific amino acids on histones to alter the basic nucleosome structure and to elicit a variety of downstream effects [3]. In addition, ATP dependent chromatin remodeling enzymes mobilize and reposition nucleosomes to change accessibility of the chromatin fiber [4]. Finally, histone variants replace core histones at strategic positions within the genome to confer specialized functions [5–7]. All together, these mechanisms allow the chromatin template to undergo dynamic changes during cell growth and cell division to accommodate ongoing nuclear processes, and to alter gene expression profiles in response to extracellular stimuli.

All histones, particularly at their N-terminal tails, are heavily modified by a variety of posttranslational modifications that include acetylation, methylation, phosphorylation, ubiquitylation, sumoylation, and ADP-ribosylation. Although acetylation and methylation of histones have been discovered since the mid-1960s, progress in the last decade or so has greatly expanded our understanding of the diversity and complexity of covalent histone modifications (for review, see [2, 8, 9]). Not only has the list of modified histone residues grown considerably, collective efforts from many labs have identified diverse enzymes and pathways that target histones, and have elucidated elegant mechanistic and functional details associated with specific histone modifications. Phosphorylation is one of the most commonly occurring posttranslational modifications on proteins and specific phosphorylation sites have been identified on all four core histones, as well as the linker histone H1 (see Table 16.1 for a list of histone phosphorylation sites and histone kinases). Histone phosphorylation is often a direct outcome of activated intracellular signaling pathways, and functions to translate extracellular signals into appropriate nuclear biological outputs. In this chapter, we will highlight the diverse signaling pathways that converge onto chromatin, the kinases that phosphorylate specific residues on the different histones, and the various cellular processes that are regulated by this modification.

Handbook of Cell Signaling, Three-Volume Set 2 ed.

TABLE 16.1 Current list of histone phosphorylation sites and histone kinases

Histone	Phosphorylation site	Kinase	Organism
Transcriptional activation			
H3	S10	Snf1	*S. cerevisiae*
		JIL-1	*D. melanogaster*
		RSK1, MSK1/2, COT, IKKα, PKA, PIM1	Mammals
	T11	PRK1	Mammals
	S28	MSK1/2, MLTKα	Mammals
DNA damage response			
H2A	S129	Tec1, Mec1	*S. cerevisiae*
H2A.X	S139	ATM, ATR, DNA-PK	Mammals
H2B	S14	?	Mammals
H4	S1	CKII	*S. cerevisiae*
Mitosis			
H2A	S1	?	*C. elegans, D. melanogaster*, mammals
	T119	NHK-1	*D. melanogaster*
H3	T3	Haspin	Mammals
	S10	Ip11	*S. cerevisiae*
		NIMA	*A. nidulans*
		AIR-2	*C. elegans*
		Aurora B	Mammals
	T11	Dlk-ZIP	Mammals
	S28	Aurora B	Mammals
H3.3	S31	?	Mammals
CENP-A	S7	Aurora A/B	Mammals
H4	S1	?	*C. elegans, D. melanogaster*, mammals
Apoptosis			
H2B	S10	Ste20	*S. cerevisiae*
	S14	MST1	Mammals

HISTONE PHOSPHORYLATION AND TRANSCRIPTIONAL REGULATION

Transcription of eukaryotic genes requires the establishment of a chromatin environment permissive to binding of activators and transcription factors, as well as to the subsequent assembly of pre-initiation complexes [10]. These chromatin changes are brought about by multiple histone modifying enzymes and chromatin remodeling complexes, and they represent key regulatory steps in the transcription process. The roles of histone acetylation and methylation in this process have been well studied and summarized in a number of excellent reviews [11–13]. On the other hand, the mechanistic functions of histone phosphorylation in transcriptional regulation are not as clearly defined. Phosphorylation of H3 at multiple and distinct sites (S10, T11, S28) has been linked to the activation of immediate-early (IE), NFκB responsive, myc regulated and steroid hormone regulated genes. One common feature amongst these genes is their rapid induction in response to external stimuli. In that regard, much of the intracellular signaling pathways are mediated through highly dynamic phosphorylation cascades, and these cascades ultimately converge onto chromatin to phosphorylate H3 and to elicit appropriate changes in gene expression patterns.

H3 Phosphorylation and Transcriptional Activation

The first link between intracellular signaling pathways, gene activation, and H3 phosphorylation came from studies of IE gene induction in mouse fibroblasts [14]. Extracellular stimuli, such as growth factors, stress, or pharmacological agents, trigger the rapid and transient expression of the IE *c-fos* and *c-jun* genes in mammalian cells. Their induction strongly correlates with a transient phosphorylation of S10 and S28 on histone H3, as well as S6 on high mobility group protein HMGN1, and these phosphorylation events have been termed the "nucleosomal response" [15]. The kinetics of H3 phosphorylation closely mirrors the expression profiles of IE genes, suggesting that this histone modification is part of the activation process of these genes. A direct link between H3 phosphorylation and IE gene induction is confirmed by chromatin immunoprecipitation (ChIP) assays that show that phosphorylated H3S10 (H3S10ph) is physically associated with the promoter and coding regions of IE genes (c-*jun*, c-*fos*, c-*myc*) upon activation [16–19]. The association of phosphorylated H3S28 (H3S28ph) with specific genes has not been reported so far; however, it is not yet clear whether this reflects the biology or is due to technical challenges with the usage of the available antibodies in ChIP assays. Immunofluorescence studies of mitogen stimulated cells show that H3S10ph and H3S28ph both display punctuate staining throughout the nucleoplasm, and they are excluded from DAPI dense regions [20, 21]. Interestingly, the majority of the H3S10ph foci do not colocalize with the H3S28ph foci, suggesting that these are independent phosphorylation events. At present, the significance of this observation in terms of IE gene activation, and the functional differences between the H3 phosphorylated at these distinct sites are unknown.

The link between H3 phosphorylation and transcriptional activation is not only observed in mammals, but also in lower organisms as well. In budding yeast *Saccharomyces cerevisiae*, phosphorylation of H3S10 by the Snf1 kinase is required for the activation of *INO1* and *GAL1* genes in response to inositol starvation and galactose respectively [22, 23]. In *Drosophila*, heat shock treatment causes a rapid induction of heat shock genes and a concomitant global shutdown of transcription of all other genes. This process is best exemplified by the induction of heat shock puffs on the polytene chromosomes isolated from *Drosophila* larvae. In this system, heat shock induces a dramatic enrichment of H3S10ph at the transcriptionally active heat shock puffs, and a concomitant global loss of H3S10ph at all other gene loci [24]. Genetics studies further show that H3S10ph level in *Drosophila* is maintained by the opposing JIL-1 kinase and PP2A phosphatase [25, 26]. Together, these studies illustrate the link between H3 phosphorylation and rapid gene induction is evolutionarily conserved amongst different species.

MAP Kinase Pathway Mediated H3 Phosphorylation

It has long been known that mitogen and stress activate H3 phosphorylation respectively through the MAPK and p38 pathways. RSK2, a downstream kinase of ERK1/2, was the first kinase linked to H3S10 phosphorylation and IE gene expression [27]. RSK2 knockout mouse cells and human fibroblasts derived from the RSK2 deficient Coffin-Lowry syndrome (CLS) patients both exhibit drastic reduction in EGF induced H3-S10 phosphorylation and in c-*fos* expression [27–29]. Moreover, ectopic expression of RSK2 can restore mitogen stimulated H3 phosphorylation in CLS cells, suggesting that this kinase is required for the mitogen induced nucleosomal response. Subsequent studies have found that induction of H3 phosphorylation can be inhibited by H89, a chemical inhibitor that is specific for PKA and MSK1 but has no effect on RSK2, suggesting the involvement of additional kinases in this process [30]. Indeed, MSK1 and MSK2, two structurally and functionally similar kinases that lie downstream of ERK and p38 pathways, can phosphorylate H3 at S10 and S28 in response to mitogen and stress stimulation. Embryonic fibroblasts from MSK1/2 double knockout mice show severe reductions of H3 phosphorylation and IE gene induction, and reintroduction of the MSK2 gene alone is sufficient to rescue the H3 phosphorylation response in these cells [31]. These studies, therefore, implicate MSK2 as a key mediator of H3 phosphorylation in mammalian cells. However, it is more likely that, depending on the specific stimulus, multiple enzymes within the MAPK pathways can target H3. For example, arsenite induced H3 phosphorylation is mediated by AKT, ERK2, and RSK2, but not by MSK1 [32]. In addition, COT and MLTKα (both members of the MAPKKK family) are responsible for H3S10 and S28 phosphorylation, respectively, in UVB irradiated cells [33, 34]. Further studies will be needed to clarify the specific roles and contributions of each of these kinases in H3 phosphorylation and the nucleosomal response.

Other Kinase Pathways that Target H3

Besides the MAPK cascades, H3 phosphorylation is a common downstream event for other signaling pathways as well. For example, IKKα, one of the subunits of the IκB kinase complex in the NFκB pathway, directly phosphorylates H3 at the promoters of cytokine induced genes [35, 36]. Knockout mouse studies showed that IKKα is required for optimal TNF induced expression of NFκB regulated genes (IκB, IL-6, IL-8), as well as for H3 phosphorylation, and the recruitment of IKKα to these promoters in normal cells correlates with the induction of both H3 phosphorylation and gene expression. In addition, stimulation of ovarian granulosa cells with follicle stimulating hormone (FSH) induces H3 phosphorylation in a PKA dependent and MAPK independent manner [37]. This study implicates PKA as the direct H3 kinase associated with the activation of FSH responsive genes during granulosa cell differentiation. Finally, several other inducible genes have been associated with H3 phosphorylation, including collagenase I [38], interferon-β (IFN-β) [39], and retinoic acid receptor β (RARβ) [40]. In those cases, the responsible H3 kinases have not yet been identified.

H3 Phosphorylation and MYC Regulation

H3 phosphorylation has recently been linked to the regulation of MYC target genes through PIM1, a newly identified H3S10 kinase [41]. After VEGF-A stimulation, PIM1 co-localizes with c-MYC and H3S10ph at sites of active transcription. This is mediated through its interaction with the MYC-MAX dimer whereby PIM1 is recruited to the E-boxes of MYC target genes such as *FOSL1* (*FRA-1*) and *ID2*. MYC is recruited to two regions of the *FOSL1* gene: *FOSL1* upstream region, which contains a non-canonical E-box element, and *FOSL1* downstream enhancer, which contains a canonical E-box. Phosphorylation of H3S10 is detected at both sites, albeit with different kinetics. Phosphorylation at the upstream element occurs at an earlier time point and is mediated by MSK1/2, whereas PIM1 associates with the downstream enhancer and its recruitment correlates with H3S10ph at the canonical E-box and during the peak of transcriptional activation. Inhibition of PIM1 strongly reduces H3 phosphorylation and mRNA expression of *FOSL1* and *ID2*. The significance of having two independent H3 phosphorylation events, mediated by different H3S10 kinases, on the same gene is not clearly understood. It is of interest to note that the kinase activity of PIM1 and its recruitment to chromatin by MYC both contribute to MYC dependent transformation. Insofar as microarray expression analysis indicates that PIM1 is necessary in regulating expression of >200 (~20 percent) MYC targets, these findings together suggest that H3 phosphorylation may have a role in oncogenesis through MYC regulation.

H3T11 Phosphorylation and Hormone Regulated Gene Expression

Hormone regulated gene expression is a complex pathway that involves multiple modification steps, such as acetylation, arginine methylation, and lysine demethylation, on various histones. More recently, H3 phosphorylation has been implicated in this pathway as well. Instead of S10 or S28 phosphorylation, it is the phosphorylation of H3 at T11 that has been linked to the activation of hormone regulated genes [42]. PRK1 is a novel kinase that associates with the androgen receptor (AR) upon stimulation by the agonist R1881. ChIP assays show that PRK1 is recruited to the promoters of AR regulated genes in a hormone dependent manner. Moreover, promoter targeting of this kinase not only results in phosphorylation of H3T11, but leads to the acetylation of H3K9/K14 and demethylation of H3K9 as well. Co-expression of PRK1 and H3K9 demethylases LSD1 and/or JMJD2C induces strong activation of reporter gene, and *in vitro* enzymatic assays show that nucleosomes with H3T11ph are better substrates for JMJD2C. Inhibition of PRK1 completely abolishes all histone modification changes induced by R1881, and severely reduces expression of AR target genes. Moreover, phosphorylation of RNA Pol II CTD-S5, but not recruitment of RNA Pol II, is also impaired upon PRK1 inhibition. All together, these findings suggest that PRK1 and its phosphorylation of H3T11 is a pivotal step that triggers sequential histone modifications as well as the transition of the pre-initiation complex to the fully engaged polymerase complex.

Similar to the AR regulated genes, induction of progesterone target genes also involves phosphorylation of H3 [43]. Hormone stimulation activates the SRC/p21^{ras}/ERK pathway, and results in the phosphorylation and activation of progesterone receptor (PR) and MSK1. Phosphorylated PR, ERK, and MSK1 form a complex, which is then recruited to the MMTV promoter. These events result in H3S10 phosphorylation, H3K14 acetylation and displacement of HP1γ from the promoter. In addition, the chromatin remodeling complex Brg1 and histone acetyltransferase PCAF are also recruited upon hormone stimulation, and these changes combine to recruit RNA Pol II and activate hormone regulated genes.

DOWNSTREAM EFFECTS OF TRANSCRIPTION ASSOCIATED H3 PHOSPHORYLATION

Structural Effects of Histone Phosphorylation

The addition of acetyl or phosphate groups changes the net charge on histones, and these modifications have been proposed to alter histone–DNA or histone–histone interactions, and thereby affect the stability of chromatin compaction states. A native chemical ligation strategy has been developed to generate recombinant histones modified at specific residues, and biophysical analyses of nucleosomal arrays assembled with K16 acetylated H4 show that H4K16Ac inhibits the formation of 30 nm-like fibers *in vitro* [44]. Similar experiments using histones phosphorylated at specific sites could provide further insight into the direct effect of histone phosphorylation on chromatin compaction and structure. A study using hydroxyapatite dissociation chromatography to fractionate nucleosomes based on structural stability found that H3S28ph containing nucleosomes are more labile and destabilized compared to the H3S10ph containing nucleosomes [45]. In fact, the H3S10ph- and unphosphorylated H3 containing nucleosomes co-fractionate together, suggesting they have similar structural stability. It is interesting to note that phosphorylation at the different sites have different effects on nucleosome stability and it remains to be determined whether these physical differences are translated into functional differences as well.

Crosstalk Between H3 Phosphorylation and Other Modifications

The abundance of posttranslational modifications on histones has led to the idea that crosstalks can occur between different modifications on one or more histone tails [46]. A pre-existing modification can enhance or reduce the activity of another histone modifying enzyme toward its target residue, and histone phosphorylation has been shown to have synergistic or antagonistic effect on the acetylation and methylation on nearby residues [17, 47, 48]. Such a complex interplay between specific modifications can dictate the combinatorial modifications that occur together on histones, and thus encode information that can be translated into different cellular responses.

Insofar as H3 phosphorylation at S10 and acetylation at K9 or K14 have all been linked to transcriptional activation, and given the proximity of these modified residues on H3, several labs have questioned whether specific combinations of modifications occur together during gene activation. Antibodies generated to specifically recognize di-modified H3 (anti-H3K9ac/S10ph and anti-H3S10ph/K14ac) show that these specific H3 modification combinations do exist on the same H3 tail *in vivo* [17, 18]. The global levels of the phosphoacetylated H3 are increased upon mitogen stimulation of mammalian cells, and ChIP assays show that both combinations of di-modified H3 are associated with the IE gene promoters during transcriptional activation.

Additional studies further suggest that these modifications are mechanistically linked. For example, *in vitro* enzymatic assays show that several histone acetyltransferases

(HAT), including yeast Gcn5, human PCAF, and p300, preferentially acetylate H3S10ph peptide substrates over the unmodified form [17, 48]. Structural analyses suggest that this enzymatic preference is due to the phosphate dependent stabilization of the enzyme substrate complex [49]. In yeast, the functional coupling of H3 phosphorylation and acetylation is required for efficient activation of the *INO1* promoter upon inositol starvation [22]. Deletion of the H3 kinase *SNF1* gene, or replacing the endogenous H3 gene with an H3-S10A mutant both eliminated S10 phosphorylation in the mutant yeast strains, and also significantly lowered K14 acetylation levels at the *INO1* promoter. For specific genes in yeast, such as *INO1*, H3S10 phosphorylation is prerequisite for acetylation of K14 during gene activation.

In higher eukaryotes, due to the difficulties in genetic manipulation of histone genes, such detailed *in vivo* dissection of the crosstalk between H3 phosphorylation and acetylation has not been done. Nevertheless, the timing of H3S10ph was found to precede H3K14ac during the activation of genes such as IE genes [17], IFNβ [39], and NFκB regulated genes [35, 36]. In contrast, for other genes, such as the RARβ2 [40] and *Drosophila* heat shock genes [24], transcriptional activation is only accompanied by increased H3S10 phosphorylation without changes in the H3 acetylation levels. These suggest that phosphorylation and acetylation can be independently targeted to the same region of the genome and they coincide to produce phosphoacetylated H3 tails [19]. Further *in vivo* studies are still needed to clarify the functional relationship between H3S10 phosphorylation and K14 acetylation in mammalian cells.

Recruitment of Phospho-H3 Binding Proteins

Various modifications on histones elicit downstream effects by the modification dependent recruitment of effector binding proteins. Indeed, the paradigm of modification dependent protein–protein interactions is well established and best exemplified by the binding of SH2 domains to phosphotyrosines and FHA domains to phospho-serines/threonines [50]. Analogously, bromodomains and chromodomains have been found to respectively bind specific acetylated and methylated lysine residues [51]. These domains are present in many different histone modifying enzymes and ATP dependent chromatin remodeling enzymes, and acetylation or methylation of specific sites on histones play a crucial role in recruiting these enzymatic activities to the appropriate chromatin locations.

Phospho-histone binding proteins have also been identified: for example, several groups found that specific 14-3-3 isoforms bind to H3S10ph and H3S28ph peptides. An initial study reported that this interaction is phospho-dependent, but not affected by the acetylation status of nearby lysine residues (K9, K14) [52]. However, more detailed analyses in recent reports demonstrated that acetylated K9/K14 greatly enhances the binding affinity of 14-3-3 to H3S10ph peptides [53, 54]. Whether this holds true for 14-3-3/H3S28ph binding has not been tested. *In vivo* studies using ChIP assays further validate the findings from these *in vitro* pull-down assays. 14-3-3 is recruited to c-*jun*, c-*fos*, and *HDAC1* promoters upon gene activation, and this occurs concomitantly with phosphoacetylation of H3. Furthermore, RNAi knockdown of 14-3-3ε/ζ strongly reduces transcription of HDAC1, but has no effect on H3 phosphoacetylation levels, suggesting that 14-3-3 is a downstream effector of H3 phosphoacetylation. This interaction and function of 14-3-3 is also conserved in yeast. Bmh1 and Bmh2, the yeast homologs of 14-3-3, are essential for optimal transcriptional activation of the *GAL1* gene. Binding of Bmh1 to GAL1 promoter is dependent on H3S10 and K14 since H3 S10A, S10A/K14R, and K14R yeast mutants are all defective in Bmh1 recruitment. These findings provide further *in vivo* support that H3K14ac is important for 14-3-3 binding.

In addition to the recruitment of phospho binding proteins to chromatin, H3 phosphorylation can also function to repel other histone binding factors. HP1 isoforms bind specifically to H3K9me through their chromodomain, and this is a key step in heterochromatin assembly and gene silencing [55, 56]. Phosphorylation of H3S10 by Aurora B disrupts the HP1/H3K9me interaction, and causes the release of HP1 from chromosomes during mitosis [57, 58]. While this "phos/methyl switch" was first observed in mitosis, it also operates in the context of transcriptional activation in interphase cells. HP1γ displacement from promoter is concomitant with phosphorylation of H3S10 during progesterone receptor mediated gene activation and HDAC1 induction [43, 54]. H3 phosphorylation, therefore, can simultaneously recruit 14-3-3 and eject the transcriptional repressor HP1γ during transcription activation. These findings further illustrate the complexity of how histone modifications regulate chromatin binding of various proteins, as well as the importance of combinatorial histone modifications in the regulation of cellular processes.

HISTONE PHOSPHORYLATION IN RESPONSE TO DNA DAMAGE

The timely and efficient repair of DNA damage is essential for maintaining genome integrity and tumor suppression. Chromatin, in fact, has a major role in the DNA repair process by integrating signaling pathways that senses DNA damage with the coordinated assembly of DNA repair proteins. In mammals, this is specifically mediated through the phosphorylation of the H2A variant H2A.X at S139 (γH2A.X) [59]. Upon DNA damage, members of the PIKK family (ATM, ATR, DNA-PK) rapidly phosphorylate H2A.X located at sites of double-strand breaks (DSBs) [60–62]. DNA repair factors, such as BRCA1, 53BP1, and MRN complex, then co-localize with γH2A.X to form discrete IR induced nuclear foci (IRIF). Peptide pull-down assays found that BRCT domain of MDC1 directly binds to γH2A.X in a phospho-dependent manner, and *in vivo* studies confirmed that this interaction is critical for localization of MDC1 to IRIF and for the normal radiotolerance of cells [63]. Mice lacking H2A.X are radiation sensitive and exhibit genome instability [64, 65]. Moreover, they show defects in repair pathways and fail to recruit repair proteins to nuclear foci. Therefore, presence of H2A.X is required for coupling the DNA damage sensing pathways to the DNA repair response.

The γH2A.X response is conserved in yeast; however, since this organism does not have H2A variants, the equivalent phosphorylation event occurs on S129 of canonical H2A [66]. Exposure of yeast cells to DSB inducers also leads to H4 phosphorylation at S1 and this modification functions in the non-homologous end-joining (NHEJ) repair pathway [67]. In-gel kinase assays identified casein kinase II (CKII) as the kinase for H4S1, and temperature sensitive CKII mutants not only are defective for H4S1 phosphorylation, but are hyper-sensitive to DNA damaging agents. This DNA damage induced H4 phosphorylation is conserved in mammalian cells (personal observation); however, additional kinase besides CKII may be involved in the higher eukaryotic system.

HISTONE PHOSPHORYLATION AND MITOSIS

During mitosis, most histones become highly phosphorylated at multiple serine/threonine residues. These include: H2A/H4S1 [68], H2AT119 [69], H3T3 [70], H3S10 [71], H3T11 [72], H3S28 [73], H3.3S31 [74], and CENP-A S7 [75, 76]. H3 phosphorylation on S10 and S28 during mitosis is particularly interesting since these are the same phosphorylation sites associated with transcriptional activation. It is still unclear how these H3 phosphorylation sites are associated with contrasting functions – chromosome condensation during mitosis and chromatin relaxation during gene expression. Nevertheless, this paradox only further reinforces the idea that functional/biological outcomes are dictated by the overall combinations of modifications on histones rather than by individual modifications. The characteristics of H3S10 and S28 phosphorylation associated with mitosis and IE gene activation are quite distinct [77, 78]. During cell cycle progression, H3 phosphorylation begins in late G2 at the pericentromeric chromatin, and then spreads along the chromosome arms as mitosis proceeds. H3 is highly phosphorylated during mitosis, and it is commonly assumed that the majority of all H3 is phosphorylated during chromosome condensation (although a definitive quantification of the percentage of H3 phosphorylated during mitosis has not been done). In mitogen stimulated interphase cells, H3 phosphorylation is only observed in a very small fraction of total H3 and correlates with transcriptional activation of a small number of genes [79]. In addition, different kinases are responsible for H3 phosphorylation at the different cell cycle stages: mitotic H3 phosphorylation is mediated by Ipl1/Aurora B kinase [80, 81], whereas kinases from MAPK (mitogen activated protein kinase) pathways are largely responsible for H3 phosphorylation in interphase cells.

The functional significance of the high levels of histone phosphorylation during mitosis is currently not well understood. Earlier studies in *Tetrahymena* suggested that H3S10ph is required for proper chromosome condensation and segregation during mitosis [82]. However, this function does not appear to be conserved in yeast since the H3S10A mutant does not show any growth defects and is able to progress through the cell cycle normally [80]. Given that so many histones and histone sites are phosphorylated during mitosis, it is possible that some of these phosphorylation marks have redundant functions in the different organisms.

HISTONE PHOSPHORYLATION DURING APOPTOSIS

Chromatin condensation and DNA fragmentation are hallmarks of apoptosis, and therefore, apoptotic signaling cascades must target chromatin to mediate these effects. H2B phosphorylation (S14 in vertebrates and S10 in yeast) is the only histone modification that has been firmly linked to apoptosis so far. The caspase 3 activated MST1 kinase phosphorylates H2B at S14 both *in vitro* and *in vivo*, and H2B phosphorylation appears to be a general apoptotic marker in species ranging from frogs to human [83]. In *Saccharomyces cerevisiae*, hydrogen peroxide (H_2O_2) treatment induces an apoptotic-like cell death, and this treatment also induces H2B phosphorylation at S10 (the H2BS14 equivalent site in yeast) [84]. This phosphorylation event is mediated by the yeast MST1 homolog Ste20, suggesting a functional conservation between yeast and mammals. Importantly, H2B S10A mutant is resistant to H_2O_2 induced cell death and does not display DNA fragmentation or chromatin condensation. Moreover, the yeast phospho-mimic H2B S10E mutant shows severe growth defects and aberrant chromatin condensation even in the absence of H_2O_2. These results strongly suggest that H2B phosphorylation is functionally linked to cell death induced chromatin condensation in eukaryotic cells.

A unidirectional crosstalk pathway has been reported for H2BS10ph and H2BK11Ac in yeast [85]. H2BK11Ac is normally found in asynchronously growing yeast cells, and this modification has an inhibitory effect on the H_2O_2 induced H2BS10 phosphorylation. Death signal induces deacetylation of K11 through the Hos3 deacetylase, and this event in turn allows subsequent phosphorylation of S10 by Ste20. Although in yeast studies, H2B phosphorylation has been linked to chromatin condensation, the mechanism of this function is unknown. Phosphorylated H2B peptides form unusual peptide "aggregates" *in vitro* and such phosphohistone aggregation may lead to chromatin condensation during apoptosis [84]. Alternatively, analogous to phospho-H3 recruiting 14-3-3 or γH2A.X binding to MDC1, phosphorylated H2B may serve to recruit downstream effectors proteins that in turn condense chromatin during apoptosis.

HISTONE PHOSPHORYLATION AND HUMAN DISEASES

Given that chromatin plays such an important role in regulating multiple DNA templated processes, it is not surprising that perturbation of chromatin modifiers and histone modifying enzymes, including histone kinases, can lead to cancer and other diseases [86]. Mitogen induced H3 phosphorylation is mainly mediated via the Ras-MAPK (Raf-MEK-ERK) pathway, which is often dysregulated and constitutively activated in cancer cells. Ras transformed mouse fibroblasts display elevated levels of both H3S10ph and H3S28ph owing to increased MSK1 activity [16, 20, 87]. Many of the H3 kinases we introduced in this chapter, including MLTKα and COT, have been linked to neoplastic cell transformation or cancer development [33, 88]. *c-myc* is one of the most frequently mutated oncogenes in human cancers and PIM1 is known to cooperate with MYC in tumorigenesis. Indeed, PIM1 is overexpressed in MYC driven prostate tumors [89], and the oncogenic effects of this enzyme may be mediated through its H3 kinase activity [41]. PRK1 mediated H3T11 phosphorylation is involved in regulating AR dependent genes, and inhibition of PRK1 reduces androgen induced proliferation of a prostate cancer cell line [42]. High levels of PRK1 and H3T11ph are also detected in prostate cancer cells and thus, PRK1 has been proposed as a predictive tumor marker. Finally, the mitotic H3 kinase Aurora B is also overexpressed in many cancer cell lines. This leads to increased mitotic S10 phosphorylation and triggers chromosome instability and multi-nuclearity – features that are often seen in malignant cells [90]. Importantly, Aurora kinase inhibitors have shown potent anticancer activities in preclinical studies, suggesting that targeting histone kinases may have therapeutic effects.

As mentioned earlier, histone H2A variant H2A.X and its phosphorylation (γH2A.X) are critical to the DNA damage response. H2A.X is haploinsufficient and partial or complete loss of H2A.X in a $p53^{-/-}$ background ($H2AX^{+/-}$ $p53^{-/-}$ or $H2AX^{-/-}$ $p53^{-/-}$) predisposes mice to lymphomas and other cancers [91, 92]. Human *H2AX* (*H2AFX*) maps to chromosome 11 at 11q23, a region that commonly exhibits loss of heterozygosity (LOH) or deleted in human lymphoid and solid tumors. These observations even led some to classify H2A.X as a tumor suppressor gene. All together, these findings underscore the importance of histone phosphorylation in normal cell functions and growth.

CONCLUSIONS AND PERSPECTIVES

In this chapter, we discussed and highlighted some of the advances that functionally link histone phosphorylation to multiple cellular functions such as transcriptional activation, DNA damage response, mitosis, as well as apoptosis. Many of the upstream signaling cascades that lead to histone phosphorylation have been well defined and their correlative downstream effects have been documented (Figure 16.1). However, there are still many questions regarding

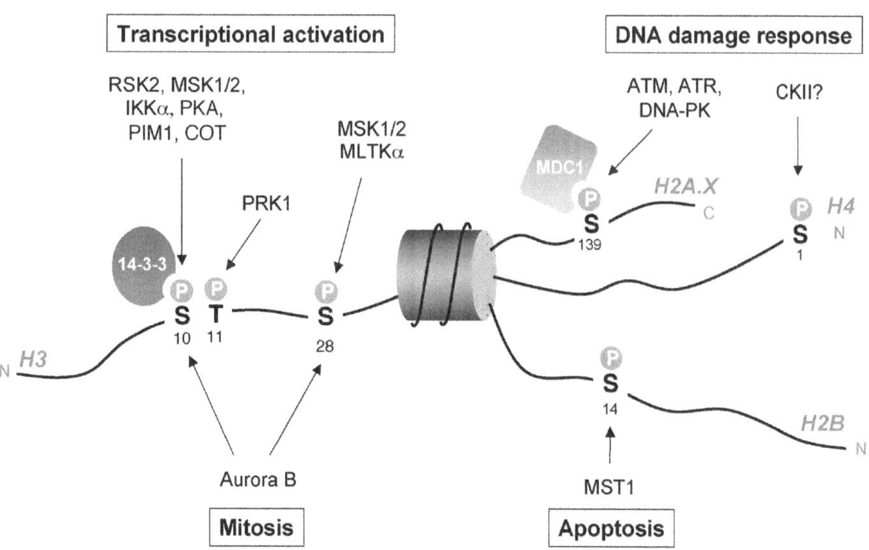

FIGURE 16.1 Phosphorylation on histone tails plays important roles in multiple cellular processes.
Known sites of phosphorylation associated with transcriptional activation, mitosis, DNA damage response, and apoptosis in mammalian cells, as well as the respective responsible kinases, are indicated.

mechanisms mediated by histone phosphorylation that need to be answered. Phosphorylation is a dynamic modification, and the H3 phosphorylation/dephosphorylation cycle linked to the activation/repression of IE genes occur within minutes to an hour. While the activation process has been studied extensively, the mechanisms that mediate H3 dephosphorylation and transcriptional repression have largely been neglected. Glc7/PP1 phosphatases in yeast and nematodes dephosphorylate H3S10 at the end of mitosis [80], whereas PP2A is responsible for genome-wide H3 dephosphorylation during heat shock response in *Drosophila* [26]. However, it is not yet known whether these phosphatases also mediate H3 dephosphorylation after IE gene induction. In addition, mechanisms by which kinases are targeted to specific promoters or sites of DNA damage are generally not known. In some cases, transcription factors (MYC) and nuclear receptors (PR, AR) help recruit PIM1, MSK1, and PRK1 to specific genes [41–43], but for the rest of the H3 kinases described in this chapter, their targeting mechanisms are unknown. Further dissection of the molecules involved in signaling to chromatin and how phosphorylation is targeted to specific loci will be areas for future investigations. Finally, in the case of H3 phosphorylation associated transcriptional activation, the mechanism of how H3 phosphorylation contributes to this process has not been defined. H3 acetylation can be functionally coupled to H3 S10 phosphorylation, but examples of transcriptional activation mediated only through H3 phosphorylation do exist. It is likely that H3 phosphorylation follows the paradigm exemplified by protein recruitment through histone acetylation or methylation, but so far only 14-3-3 has been shown to bind H3S10ph, and it is not yet clear how this protein functions in transcription. These and other questions relating to the mechanistic functions of histone phosphorylation are no doubt currently being investigated. These future advances will provide a more complete understanding of how phosphorylation of histones translates signal transduction pathways into specific biological responses.

REFERENCES

1. Luger K, Mader AW, Richmond RK, Sargent DF, Richmond TJ. Crystal structure of the nucleosome core particle at 2.8 A resolution. *Nature* 1997;**389**:251–60.
2. Felsenfeld G, Groudine M. Controlling the double helix. *Nature* 2003;**421**:448–53.
3. Strahl BD, Allis CD. The language of covalent histone modifications. *Nature* 2000;**403**:41–5.
4. Becker PB, Horz W. ATP-dependent nucleosome remodeling. *Annu Rev Biochem* 2002;**71**:247–73.
5. Kamakaka RT, Biggins S. Histone variants: deviants? *Genes Dev* 2005;**19**:295–310.
6. Sarma K, Reinberg D. Histone variants meet their match. *Nat Rev Mol Cell Biol* 2005;**6**:139–49.
7. Cheung P, Lau P. Epigenetic regulation by histone methylation and histone variants. *Mol Endocrinol* 2005;**19**:563–73.
8. Kouzarides T. Chromatin modifications and their function. *Cell* 2007;**128**:693–705.
9. Berger SL. The complex language of chromatin regulation during transcription. *Nature* 2007;**447**:407–12.
10. Li B, Carey M, Workman JL. The role of chromatin during transcription. *Cell* 2007;**128**:707–19.
11. Shahbazian MD, Grunstein M. Functions of site-specific histone acetylation and deacetylation. *Annu Rev Biochem* 2007;**76**:75–100.
12. Martin C, Zhang Y. The diverse functions of histone lysine methylation. *Nat Rev Mol Cell Biol* 2005;**6**:838–49.
13. Shilatifard A. Chromatin modifications by methylation and ubiquitination: implications in the regulation of gene expression. *Annu Rev Biochem* 2006;**75**:243–69.
14. Mahadevan LC, Willis AC, Barratt MJ. Rapid histone H3 phosphorylation in response to growth factors, phorbol esters, okadaic acid, and protein synthesis inhibitors. *Cell* 1991;**65**:775–83.
15. Thomson S, Mahadevan LC, Clayton AL. MAP kinase-mediated signalling to nucleosomes and immediate-early gene induction. *Semin Cell Dev Biol* 1999;**10**:205–14.
16. Chadee DN, Hendzel MJ, Tylipski CP, Allis CD, Bazett-Jones DP, Wright JA, Davie JR. Increased Ser-10 phosphorylation of histone H3 in mitogen-stimulated and oncogene-transformed mouse fibroblasts. *J Biol Chem* 1999;**274**:24,914–20,.
17. Cheung P, Tanner KG, Cheung WL, Sassone-Corsi P, Denu JM, Allis CD. Synergistic coupling of histone H3 phosphorylation and acetylation in response to epidermal growth factor stimulation. *Mol Cell* 2000;**5**:905–15.
18. Clayton AL, Rose S, Barratt MJ, Mahadevan LC. Phosphoacetylation of histone H3 on c-fos- and c-jun-associated nucleosomes upon gene activation. *EMBO J* 2000;**19**:3714–26.
19. Thomson S, Clayton AL, Mahadevan LC. Independent dynamic regulation of histone phosphorylation and acetylation during immediate-early gene induction. *Mol Cell* 2001;**8**:1231–41.
20. Dunn KL, Davie JR. Stimulation of the Ras-MAPK pathway leads to independent phosphorylation of histone H3 on serine 10 and 28. *Oncogene* 2005;**24**:3492–502.
21. Dyson MH, Thomson S, Inagaki M, Goto H, Arthur SJ, Nightingale K, Iborra FJ, Mahadevan LC. MAP kinase-mediated phosphorylation of distinct pools of histone H3 at S10 or S28 via mitogen- and stress-activated kinase 1/2. *J Cell Sci* 2005;**118**:2247–59.
22. Lo WS, Duggan L, Emre NC, Belotserkovskya R, Lane WS, Shiekhattar R, Berger SL. Snf1 a histone kinase that works in concert with the histone acetyltransferase Gcn5 to regulate transcription. *Science* 2001;**293**:1142–6.
23. Lo WS, Gamache ER, Henry KW, Yang D, Pillus L, Berger SL. Histone H3 phosphorylation can promote TBP recruitment through distinct promoter-specific mechanisms. *EMBO J* 2005;**24**:997–1008.
24. Nowak SJ, Corces VG. Phosphorylation of histone H3 correlates with transcriptionally active loci. *Genes Dev* 2000;**14**:3003–13.
25. Ivaldi MS, Karam CS, Corces VG. Phosphorylation of histone H3 at Ser10 facilitates RNA polymerase II release from promoter-proximal pausing in Drosophila. *Genes Dev* 2007;**21**:2818–31.
26. Nowak SJ, Pai CY, Corces VG. Protein phosphatase 2A activity affects histone H3 phosphorylation and transcription in Drosophila melanogaster. *Mol Cell Biol* 2003;**23**:6129–38.
27. Sassone-Corsi P, Mizzen CA, Cheung P, Crosio C, Monaco L, Jacquot S, Hanauer A, Allis CD. Requirement of Rsk-2 for epidermal growth factor-activated phosphorylation of histone H3. *Science* 1999;**285**:886–91.

28. De Cesare D, Jacquot S, Hanauer A, Sassone-Corsi P. Rsk-2 activity is necessary for epidermal growth factor-induced phosphorylation of CREB protein and transcription of c-fos gene. *Proc Natl Acad Sci U S A* 1998;**95**:12,202–7.
29. Bruning JC, Gillette JA, Zhao Y, Bjorbaeck C, Kotzka J, Knebel B, Avci H, Hanstein B, Lingohr P, Moller DE, Krone W, Kahn CR, Muller-Wieland D. Ribosomal subunit kinase-2 is required for growth factor-stimulated transcription of the c-Fos gene. *Proc Natl Acad Sci U S A* 2000;**97**:2462–7.
30. Thomson S, Clayton AL, Hazzalin CA, Rose S, Barratt MJ, Mahadevan LC. The nucleosomal response associated with immediate-early gene induction is mediated via alternative MAP kinase cascades: MSK1 as a potential histone H3/HMG-14 kinase. *EMBO J* 1999;**18**:4779–93.
31. Soloaga A, Thomson S, Wiggin GR, Rampersaud N, Dyson MH, Hazzalin CA, Mahadevan LC, Arthur JS. MSK2 and MSK1 mediate the mitogen- and stress-induced phosphorylation of histone H3 and HMG-14. *EMBO J* 2003;**22**:2788–97.
32. He Z, Ma WY, Liu G, Zhang Y, Bode AM, Dong Z. Arsenite-induced phosphorylation of histone H3 at serine 10 is mediated by Akt1, extracellular signal-regulated kinase 2, and p90 ribosomal S6 kinase 2 but not mitogen- and stress-activated protein kinase 1. *J Biol Chem* 2003;**278**:10,588–93.
33. Choi HS, Kang BS, Shim JH, Cho YY, Choi BY, Bode AM, Dong Z. Cot, a novel kinase of histone H3, induces cellular transformation through up-regulation of c-fos transcriptional activity. *Faseb J* 2008;**22**:113–26.
34. Choi HS, Choi BY, Cho YY, Zhu F, Bode AM, Dong Z. Phosphorylation of Ser28 in histone H3 mediated by mixed lineage kinase-like mitogen-activated protein triple kinase alpha. *J Biol Chem* 2005;**280**:13,545–53.
35. Anest V, Hanson JL, Cogswell PC, Steinbrecher KA, Strahl BD, Baldwin AS. A nucleosomal function for IkappaB kinase-alpha in NF-kappaB-dependent gene expression. *Nature* 2003;**423**:659–63.
36. Yamamoto Y, Verma UN, Prajapati S, Kwak YT, Gaynor RB. Histone H3 phosphorylation by IKK-alpha is critical for cytokine-induced gene expression. *Nature* 2003;**423**:655–9.
37. Salvador LM, Park Y, Cottom J, Maizels ET, Jones JC, Schillace RV, Carr DW, Cheung P, Allis CD, Jameson JL, Hunzicker-Dunn M. Follicle-stimulating hormone stimulates protein kinase A-mediated histone H3 phosphorylation and acetylation leading to select gene activation in ovarian granulosa cells. *J Biol Chem* 2001;**276**:40,146–55.
38. Martens JH, Verlaan M, Kalkhoven E, Zantema A. Cascade of distinct histone modifications during collagenase gene activation. *Mol Cell Biol* 2003;**23**:1808–16.
39. Agalioti T, Chen G, Thanos D. Deciphering the transcriptional histone acetylation code for a human gene. *Cell* 2002;**111**:381–92.
40. Lefebvre B, Ozato K, Lefebvre P. Phosphorylation of histone H3 is functionally linked to retinoic acid receptor beta promoter activation. *EMBO Rep* 2002;**3**:335–40.
41. Zippo A, De Robertis A, Serafini R, Oliviero S. PIM1-dependent phosphorylation of histone H3 at serine 10 is required for MYC-dependent transcriptional activation and oncogenic transformation. *Nat Cell Biol* 2007;**9**:932–44.
42. Metzger E, Yin N, Wissmann M, Kunowska N, Fischer K, Friedrichs N, Patnaik D, Higgins JM, Potier N, Scheidtmann KH, Buettner R, Schule R. Phosphorylation of histone H3 at threonine 11 establishes a novel chromatin mark for transcriptional regulation. *Nat Cell Biol* 2008;**10**:53–60.
43. Vicent GP, Ballare C, Nacht AS, Clausell J, Subtil-Rodriguez A, Quiles I, Jordan A, Beato M. Induction of progesterone target genes requires activation of Erk and Msk kinases and phosphorylation of histone H3. *Mol Cell* 2006;**24**:367–81.
44. Shogren-Knaak M, Ishii H, Sun JM, Pazin MJ, Davie JR, Peterson CL. Histone H4-K16 acetylation controls chromatin structure and protein interactions. *Science* 2006;**311**:844–7.
45. Sun JM, Chen HY, Espino PS, Davie JR. Phosphorylated serine 28 of histone H3 is associated with destabilized nucleosomes in transcribed chromatin. *Nucleic Acids Res* 2007;**35**:6640–7.
46. Latham JA, Dent SY. Cross-regulation of histone modifications. *Nat Struct Mol Biol* 2007;**14**:1017–24.
47. Rea S, Eisenhaber F, O'Carroll D, Strahl BD, Sun ZW, Schmid M, Opravil S, Mechtler K, Ponting CP, Allis CD, Jenuwein T. Regulation of chromatin structure by site-specific histone H3 methyltransferases. *Nature* 2000;**406**:593–9.
48. Lo WS, Trievel RC, Rojas JR, Duggan L, Hsu JY, Allis CD, Marmorstein R, Berger SL. Phosphorylation of serine 10 in histone H3 is functionally linked in vitro and in vivo to Gcn5-mediated acetylation at lysine 14. *Mol Cell* 2000;**5**:917–26.
49. Clements A, Poux AN, Lo WS, Pillus L, Berger SL, Marmorstein R. Structural basis for histone and phosphohistone binding by the GCN5 histone acetyltransferase. *Mol Cell* 2003;**12**:461–73.
50. Seet BT, Dikic I, Zhou MM, Pawson T. Reading protein modifications with interaction domains. *Nat Rev Mol Cell Biol* 2006;**7**:473–83.
51. Taverna SD, Li H, Ruthenburg AJ, Allis CD, Patel DJ. How chromatin-binding modules interpret histone modifications: lessons from professional pocket pickers. *Nat Struct Mol Biol* 2007;**14**:1025–40.
52. Macdonald N, Welburn JP, Noble ME, Nguyen A, Yaffe MB, Clynes D, Moggs JG, Orphanides G, Thomson S, Edmunds JW, Clayton AL, Endicott JA, Mahadevan LC. Molecular basis for the recognition of phosphorylated and phosphoacetylated histone h3 by 14-3-3. *Mol Cell* 2005;**20**:199–211.
53. Walter W, Clynes D, Tang Y, Marmostein R, Mellor J, Berger SL. 14-3-3 interaction with histone H3 involves dual modification pattern of phosphoacetylation. *Mol Cell Biol* 2008;**28**:2840–9.
54. Winter S, Simboeck E, Fischle W, Zupkovitz G, Dohnal I, Mechtler K, Ammerer G, Seiser C. 14-3-3 proteins recognize a histone code at histone H3 and are required for transcriptional activation. *EMBO J* 2008;**27**:88–99.
55. Bannister AJ, Zegerman P, Partridge JF, Miska EA, Thomas JO, Allshire RC, Kouzarides T. Selective recognition of methylated lysine 9 on histone H3 by the HP1 chromo domain. *Nature* 2001;**410**:120–4.
56. Lachner M, O'Carroll D, Rea S, Mechtler K, Jenuwein T. Methylation of histone H3 lysine 9 creates a binding site for HP1 proteins. *Nature* 2001;**410**:116–20.
57. Fischle W, Tseng BS, Dormann HL, Ueberheide BM, Garcia BA, Shabanowitz J, Hunt DF, Funabiki H, Allis CD. Regulation of HP1-chromatin binding by histone H3 methylation and phosphorylation. *Nature* 2005;**438**:1116–22.
58. Hirota T, Lipp JJ, Toh BH, Peters JM. Histone H3 serine 10 phosphorylation by Aurora B causes HP1 dissociation from heterochromatin. *Nature* 2005;**438**:1176–80.
59. Fillingham J, Keogh MC, Krogan NJ. GammaH2AX and its role in DNA double-strand break repair. *Biochem Cell Biol* 2006;**84**:568–77.
60. Burma S, Chen BP, Murphy M, Kurimasa A, Chen DJ. ATM phosphorylates histone H2AX in response to DNA double-strand breaks. *J Biol Chem* 2001;**276**:42,462–7.
61. Ward IM, Chen J. Histone H2AX is phosphorylated in an ATR-dependent manner in response to replicational stress. *J Biol Chem* 2001;**276**:47,759–62.

62. Burma S, Chen DJ. Role of DNA-PK in the cellular response to DNA double-strand breaks. *DNA Repair (Amst)* 2004;**3**:909–18.
63. Stucki M, Clapperton JA, Mohammad D, Yaffe MB, Smerdon SJ, Jackson SP. MDC1 directly binds phosphorylated histone H2AX to regulate cellular responses to DNA double-strand breaks. *Cell* 2005;**123**:1213–26.
64. Bassing CH, Chua KF, Sekiguchi J, Suh H, Whitlow SR, Fleming JC, Monroe BC, Ciccone DN, Yan C, Vlasakova K, Livingston DM, Ferguson DO, Scully R, Alt FW. Increased ionizing radiation sensitivity and genomic instability in the absence of histone H2AX. *Proc Natl Acad Sci U S A* 2002;**99**:8173–8.
65. Celeste A, Petersen S, Romanienko PJ, Fernandez-Capetillo O, Chen HT, Sedelnikova OA, Reina-San-Martin B, Coppola V, Meffre E, Difilippantonio MJ, Redon C, Pilch DR, Olaru A, Eckhaus M, Camerini-Otero RD, Tessarollo L, Livak F, Manova K, Bonner WM, Nussenzweig MC, Nussenzweig A. Genomic instability in mice lacking histone H2AX. *Science* 2002;**296**:922–7.
66. Downs JA, Lowndes NF, Jackson SP. A role for Saccharomyces cerevisiae histone H2A in DNA repair. *Nature* 2000;**408**:1001–4.
67. Cheung WL, Turner FB, Krishnamoorthy T, Wolner B, Ahn SH, Foley M, Dorsey JA, Peterson CL, Berger SL, Allis CD. Phosphorylation of histone H4 serine 1 during DNA damage requires casein kinase II in S. cerevisiae. *Curr Biol* 2005;**15**:656–60.
68. Barber CM, Turner FB, Wang Y, Hagstrom K, Taverna SD, Mollah S, Ueberheide B, Meyer BJ, Hunt DF, Cheung P, Allis CD. The enhancement of histone H4 and H2A serine 1 phosphorylation during mitosis and S-phase is evolutionarily conserved. *Chromosoma* 2004;**112**:360–71.
69. Aihara H, Nakagawa T, Yasui K, Ohta T, Hirose S, Dhomae N, Takio K, Kaneko M, Takeshima Y, Muramatsu M, Ito T. Nucleosomal histone kinase-1 phosphorylates H2A Thr 119 during mitosis in the early Drosophila embryo. *Genes Dev* 2004;**18**:877–88.
70. Dai J, Sultan S, Taylor SS, Higgins JM. The kinase haspin is required for mitotic histone H3 Thr 3 phosphorylation and normal metaphase chromosome alignment. *Genes Dev* 2005;**19**:472–88.
71. Wei Y, Mizzen CA, Cook RG, Gorovsky MA, Allis CD. Phosphorylation of histone H3 at serine 10 is correlated with chromosome condensation during mitosis and meiosis in Tetrahymena. *Proc Natl Acad Sci U S A* 1998;**95**:7480–4.
72. Preuss U, Landsberg G, Scheidtmann KH. Novel mitosis-specific phosphorylation of histone H3 at Thr11 mediated by Dlk/ZIP kinase. *Nucleic Acids Res* 2003;**31**:878–85.
73. Goto H, Tomono Y, Ajiro K, Kosako H, Fujita M, Sakurai M, Okawa K, Iwamatsu A, Okigaki T, Takahashi T, Inagaki M. Identification of a novel phosphorylation site on histone H3 coupled with mitotic chromosome condensation. *J Biol Chem* 1999;**274**:25,543–49.
74. Hake SB, Garcia BA, Kauer M, Baker SP, Shabanowitz J, Hunt DF, Allis CD. Serine 31 phosphorylation of histone variant H3.3 is specific to regions bordering centromeres in metaphase chromosomes. *Proc Natl Acad Sci U S A* 2005;**102**:6344–9.
75. Kunitoku N, Sasayama T, Marumoto T, Zhang D, Honda S, Kobayashi O, Hatakeyama K, Ushio Y, Saya H, Hirota T. CENP-A phosphorylation by Aurora-A in prophase is required for enrichment of Aurora-B at inner centromeres and for kinetochore function. *Dev Cell* 2003;**5**:853–64.
76. Zeitlin SG, Shelby RD, Sullivan KF. CENP-A is phosphorylated by Aurora B kinase and plays an unexpected role in completion of cytokinesis. *J Cell Biol* 2001;**155**:1147–57.
77. Nowak SJ, Corces VG. Phosphorylation of histone H3: a balancing act between chromosome condensation and transcriptional activation. *Trends Genet* 2004;**20**:214–20.
78. Prigent C, Dimitrov S. Phosphorylation of serine 10 in histone H3, what for?. *J Cell Sci* 2003;**116**:3677–85.
79. Barratt MJ, Hazzalin CA, Cano E, Mahadevan LC. Mitogen-stimulated phosphorylation of histone H3 is targeted to a small hyperacetylation-sensitive fraction. *Proc Natl Acad Sci U S A* 1994;**91**:4781–5.
80. Hsu JY, Sun ZW, Li X, Reuben M, Tatchell K, Bishop DK, Grushcow JM, Brame CJ, Caldwell JA, Hunt DF, Lin R, Smith MM, Allis CD. Mitotic phosphorylation of histone H3 is governed by Ipl1/aurora kinase and Glc7/PP1 phosphatase in budding yeast and nematodes. *Cell* 2000;**102**:279–91.
81. Goto H, Yasui Y, Nigg EA, Inagaki M. Aurora-B phosphorylates Histone H3 at serine28 with regard to the mitotic chromosome condensation. *Genes Cells* 2002;**7**:11–7.
82. Wei Y, Yu L, Bowen J, Gorovsky MA, Allis CD. Phosphorylation of histone H3 is required for proper chromosome condensation and segregation. *Cell* 1999;**97**:99–109.
83. Cheung WL, Ajiro K, Samejima K, Kloc M, Cheung P, Mizzen CA, Beeser A, Etkin LD, Chernoff J, Earnshaw WC, Allis CD. Apoptotic phosphorylation of histone H2B is mediated by mammalian sterile twenty kinase. *Cell* 2003;**113**:507–17.
84. Ahn SH, Cheung WL, Hsu JY, Diaz RL, Smith MM, Allis CD. Sterile 20 kinase phosphorylates histone H2B at serine 10 during hydrogen peroxide-induced apoptosis in S. cerevisiae. *Cell* 2005;**120**:25–36.
85. Ahn SH, Diaz RL, Grunstein M, Allis CD. Histone H2B deacetylation at lysine 11 is required for yeast apoptosis induced by phosphorylation of H2B at serine 10. *Mol Cell* 2006;**24**:211–20.
86. Wang GG, Allis CD, Chi P. Chromatin remodeling and cancer, Part I: Covalent histone modifications. *Trends Mol Med* 2007;**13**:363–72.
87. Drobic B, Espino PS, Davie JR. Mitogen- and stress-activated protein kinase 1 activity and histone h3 phosphorylation in oncogene-transformed mouse fibroblasts. *Cancer Res* 2004;**64**:9076–9.
88. Cho YY, Bode AM, Mizuno H, Choi BY, Choi HS, Dong Z. A novel role for mixed-lineage kinase-like mitogen-activated protein triple kinase alpha in neoplastic cell transformation and tumor development. *Cancer Res* 2004;**64**:3855–64.
89. Ellwood-Yen K, Graeber TG, Wongvipat J, Iruela-Arispe ML, Zhang J, Matusik R, Thomas GV, Sawyers CL. Myc-driven murine prostate cancer shares molecular features with human prostate tumors. *Cancer Cell* 2003;**4**:223–38.
90. Katayama H, Brinkley WR, Sen S. The Aurora kinases: role in cell transformation and tumorigenesis. *Cancer Metastasis Rev* 2003;**22**:451–64.
91. Bassing CH, Suh H, Ferguson DO, Chua KF, Manis J, Eckersdorff M, Gleason M, Bronson R, Lee C, Alt FW. Histone H2AX: a dosage-dependent suppressor of oncogenic translocations and tumors. *Cell* 2003;**114**:359–70.
92. Celeste A, Difilippantonio S, Difilippantonio MJ, Fernandez-Capetillo O, Pilch DR, Sedelnikova OA, Eckhaus M, Ried T, Bonner WM, Nussenzweig A. H2AX haploinsufficiency modifies genomic stability and tumor susceptibility. *Cell* 2003;**114**:371–83.

Chapter 17

Histone Variants: Signaling or Structural Modules?

Toyotaka Ishibashi, Andra Li and Juan Ausió

Department of Biochemistry and Microbiology, University of Victoria, Victoria, British Columbia, Canada

Abbreviations: 53BP1, p53 binding protein 1; ASF1, anti-silencing function 1 homolog A; ATM, ataxia-telangiectasia mutated; ATR, ATM and Rad3 related; AUT, acetic acid urea Triton X-100; BRCA1, breast cancer susceptibility protein 1; CHD1, (chromo)-helicase/ATPase-DNA binding protein 1; CENP-A, centromere protein A; Cid, centromere identifier; Cnp1p, centromere protein 1p; Cse4p, capping enzyme suppressor 4p; CTCF, CCCTC binding factor; DNA-PK, DNA protein kinase; DSB, double-stranded break; FACT, facilitates chromatin transcription; FRAP, fluorescence recovery after photobleaching; FRET, Forster resonance energy transfer; GADD45, growth arrest- and DNA damage inducible; GFP, green fluorescence protein; H2BFWT, H2B family, member W, testis specific; HDAC1, histone deacetylase 1; HIRA, histone regulator A; HPLC, high performance liquid chromatography; INO80, inosiol 80 containing complex; MDM2, mouse double minutes 2 protein; Mec1p, meiosis entry checkpoint 1 protein; mH2A, macroH2A; MRN, Mre11/Rad50/Nbs1; Msx1, msh homeobox 1; NAP-1, nucleosome assembly protein 1; NASP, nuclear autoantigenic sperm protein; NEF, nucleosome free region; NCP, nucleosome core particle; NHR, C-terminal non-histone region; NuA4, nucleosome acetyltransferase of H4; PAGE, polyacrylamide gel electrophoresis; PARP-1, poly(ADP-ribose) polymerase 1; PHO5, phosphatase5; PTM, posttranslational modification; SAGA, Spt-Ada-Gcn5-acetyltransferase; Sim3, silencing in the middle of the centromere-3; SWI/SNF, switch of the mating type/sucrose non-fermentable (chromatin remodeling complex); SWR1, Swi2/Snf2 related adenosine triphosphatase complex; Tel1p, telomere length 1; Tip60, Tat interactive protein; WHD, winged helix domain; Xi, inactive X chromosome.

INTRODUCTION

Histones are the main protein component of chromatin and they can be divided into two major groups: core histones and linker histones. Core histones: H2A, H2B, H3, and H4 have a rather simple structural organization in which a central folded domain, the histone fold [1], is flanked by two disordered N- and C-terminal "tails" which, in the case of H3 and H4, are very short. They associate to form a histone octamer, consisting of an H3-H4 tetramer and two H2A-H2B dimers. This constitutes the protein core around which approximately 145 bp of DNA wrap to form the most basic subunit of chromatin, the nucleosome core particle (NCP) [2]. Linker histones, or histones of the H1 family [3], consist of a similar structural organization but the central folded region consists of a winged helix domain [4]. Linker histones bind to the linker DNA, connecting adjacent NCPs in the chromatin fiber.

Both core histones and linker histones exhibit inter- and intra-specific amino acid sequence variability and each have their own set of variants. This variability can be very small, like between canonical H3.1/H3.2 and histone H3.3 where only a few amino acid differences exist among them, or it can be substantially large. This latter instance is the case with some of the histone H2A variants where functionally relevant variations in the C-terminal tail [5] can dramatically increase the size of the molecule, as in macro H2A. The different extent of sequence variation has led to the term homomorphous for the former and heteromorphous for the latter [6].

From an evolutionary point of view, histones have long been known to have evolved at different rates, with core histones being much more conserved than linker histones. Histones H3 and H4 stand among some of the most evolutionarily conserved proteins, with H2A and H2B occupying an intermediate position between them and linker histones [7]. Histone H4 is extremely conserved from yeast to humans and basically no variants of this histone exist. The evolutionary sequence variability is somehow reflected in the number of variants existing in the different major histone types. This is highest in H2A/H2B and in linker histones.

Histone variants have been reviewed extensively in recent years [8–14]; therefore, rather than provide an extensive review for each variant, this chapter is centered on some of the most recent achievements and we will direct

Handbook of Cell Signaling, Three-Volume Set 2 ed.

the reader to the appropriate reviews for more comprehensive information on each individual histone variant. We are going to focus our attention on structural and signaling aspects that contribute to the epigenetic role recently assigned to these variants.

It has been well documented that histone variants can modulate the highly dynamic nature of the nucleosome (Figure 17.1) [10, 15] to accommodate for different nuclear DNA metabolic needs such as transcription, replication, and repair. In this regard, H2A.Z has been shown to enhance the stability of NCP consisting exclusively of this H2A variant [16]. In contrast, NCPs consisting of a double histone variant combination (H2A.Z and H3.3) exhibited a highly destabilized nature (Figure 17.1) [17, 18]. Based on crystallographic data, initially it had been proposed that NCP hybrids consisting of variants of the same histone (i.e., H2A and H2A.Z) would probably not exist due to potential structural constraints [19, 20]. However, now it has been clearly demonstrated that it is possible to reconstitute such hybrid particles *in vitro* [15] and evidence has been provided for their existence in the nucleus [21]. The existence of such heterotypic single and multiple variant nucleosomes (see Figure 17.1) may add an additional layer of dynamic complexity, for which the consequences have not been completely explored. As discussed below, all of these conformational effects of histone variants may be enhanced further by the presence of specific posttranslational modifications (PTMs) to create a code of structures and signals with strong epigenetic implications.

H2A.Bbd IN SEARCH OF A FUNCTION

Histone H2A.Bbd (Barr body deficient) was first described in 2001 by Chadwick and Willard [22]. The name given to this variant was based on the observation that ectopically expressed tagged forms of the protein in mammalian cells were largely excluded from the inactive X chromosome [22]. Initially, it was identified from a search of human ESTs with distant homology to histone H2A. H2A.Bbd has only 48 percent identity compared with canonical H2A in humans. Because of this, and because of the interspecies homology (i.e., 54 percent identity between human and mouse H2A.Bbd), this variant has been identified in mammals only [23]. H2A.Bbd lacks both the C-terminal tail and the end of docking domain that are typical of the H2A

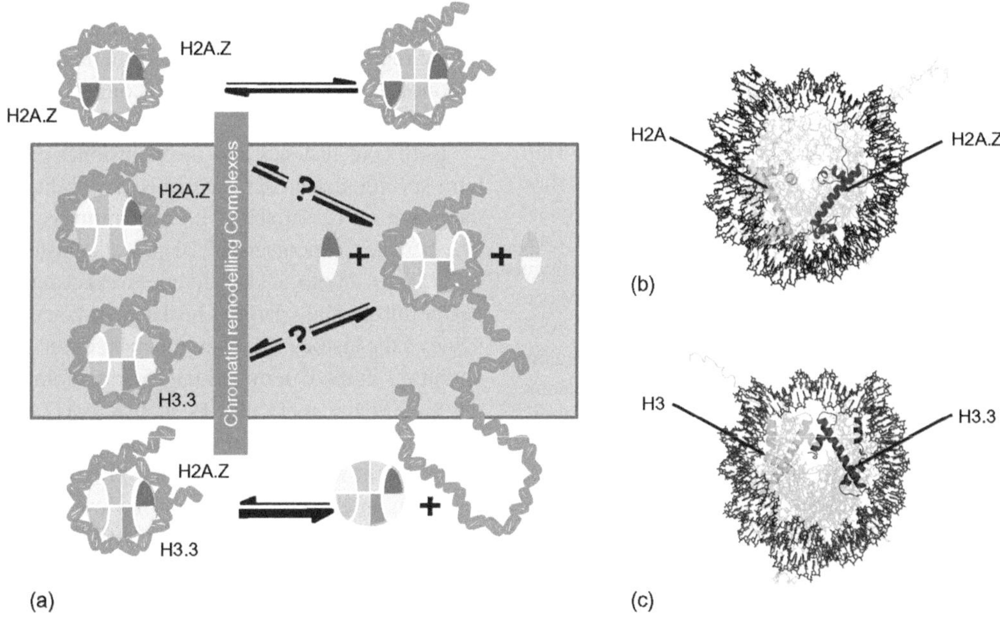

FIGURE 17.1 Histone variants affect the stability and dynamics of the nucleosome.
(a) Nucleosomes can consist of different combinations of histone variants. A double copy of a unique variant (H2A.Z) may be present leading to a homotypic single variant NCP (top). Most of the structurally characterized NCPs obtained by *in vitro* reconstitution occur in this way. H2A.Z reconstituted NCPs exhibit enhanced stability [16]. Hybrid nucleosomes consisting of only one copy of a histone variant (heterotypic single variant NCP) have been observed *in situ* for histone variants H3.3 [134] and H2A.Z [21] and prepared *in vitro* with H2A.Z [15] (middle). These types of histone variant hybrid particles may exhibit an altered stability. A reduced stability has been suggested in the case of H3.3 [17].The lower part shows an NCP with two different histone variants co-existing within the histone octamer (heterotypic multiple variant NCP). Evidence for the existence of highly unstable H2A.Z-H3.3 NCPs has been provided [17]. The thickness of the arrows schematically depict the association dynamics (resulting in different stabilities) of the corresponding NCPs. Chromatin remodeling complexes may enhance the structural NCP alterations introduced by the histone variants and their PTMs. The nomenclature of the histone variant containing nucleosomes followed here is that described in the paper by Bernstein and Hake [20].
(b–c) Three dimensional structure of heterotypic single variant H2A.Z and H3.3 NCPs, respectively. The variants and their canonical counterparts are represented in a ribbon form, for highlighting purposes, and the rest of the histones and DNA in the background are in a stick representation. The crystallographic coordinates of the canonical [29] and H2A.Z-containing [19] NCPs were used in this representation.

family. In humans, the N-terminus of this histone variant exhibits an unusual row of six arginines, which surprisingly is not present in mouse [23, 24].

Chromatin fractionation using cells with ectopic expression of H2A.Bbd showed that the variant is incorporated in nucleosomes that overlap with regions of the nucleus enriched in acetylated H4 [22]. GFP-H2A.Bbd was shown to exchange much more rapidly than GFP-H2A from chromatin as determined by FRAP assays [25]. H2A.Bbd-containing NCPs exhibit a relaxed conformation compared with the native NCPs under a broad range of salt concentrations as determined by analytical ultracentrifuge and by FRET [18]. In this nucleosome organization, the H2A.Bbd-containing histone core protects only 118+2 bp of DNA from digestion by micrococcal nuclease (compared to 147 bp in the H2A nucleosome) [18]. This is most likely the result of the flanking ends of the DNA at the entry and exit site of the NCP being released from their interaction with the core histones. This would result in a particle adopting a more open structure, which is highly reminiscent of what is observed in NCPs consisting of highly acetylated histones.

H2A.Bbd-containing nucleosomes are unstable [23], and NAP-1, that is a chaperone for H2A, can efficiently remove H2A.Bbd-H2B dimers from the nucleosome [26]. All of this suggests that the H2A.Bbd containing nucleosome is a rather unstable and quite open particle whose presence in dynamically active regions of the genome may circumvent the need for any remodeling. In agreement with this notion and strikingly in contrast with the H2A containing nucleosomes, H2A.Bbd containing nucleosomes cannot be remodeled by SWI/SNF [27, 28].

As for the molecular determinants of the unique structural conformation conferred by H2A.Bbd to the nucleosome, they are not completely elucidated. Initially, it was thought that the lack of the last 16–18 C-terminal residues of the protein, when compared to the canonical H2A counterpart, could be responsible. Indeed, the crystallographic structure of the nucleosome containing canonical core histones suggests that this region forms hydrogen bonds with the αN and α2 helices from the histone fold domain of histone H3 [29]. Despite this, it has been shown that it is the docking domain of H2A.Bbd and not its truncated C-terminal end that is responsible for the unusual conformation of the H2A.Bbd-NCP [18, 27]. Recently, it has also been pointed out that the histone fold of H2A.Bbd lacks several of the acidic amino acid residues that are responsible for the formation of an "acidic patch" in the nucleosome [29], which has been putatively involved in the folding of the chromatin fiber [30, 31]. Should arrays of contiguous H2A.Bbd nucleosomes exist *in vivo*, this would unfold the chromatin region encompassing these nucleosomes.

In closing this section, it is important to emphasize that the difficulty in generating highly specific antibodies against H2A.Bbd and its low abundance have made it extremely difficult to find a function for this protein. Determining the physiological function of this intriguing H2A variant is the most pressing issue to be addressed next. In this regard, the use of mammalian testes for this analysis may prove very rewarding as this is one of the tissues where the H2A.Bbd gene appears to be more abundantly transcribed [23].

H2A.X: DNA DAMAGE AND BEYOND

The role of histone H2A.X variant in DNA double-stranded break (DSB) repair has been an intensively studied topic in the last decade. This variant is characterized by the presence of a highly conserved SQE motif at the C-terminal tail of the protein. Histone H2A.X is immediately phosphorylated at the SQE motif upon DNA DSB damage and this phosphorylation plays an important role in the ensuing downstream repair events. This phosphorylated form is commonly denoted as γ-H2A.X. Such phosphorylation is carried out by three enzymes: ataxia-telangiectasia mutated (ATM), ATM and Rad3 related (ATR), and DNA protein kinase (DNA-PK) [32]. In *Sacharomyces cerevisiae,* homologs for ATM and ATR are Mec1p and Tel1p, respectively [33].

In mammals, histone H2A.X and the canonical histone H2A have 95 percent identity and the two histones differ in the sequence at their C-terminal tail. In contrast, histone H2Av, in *Drosophila*, shares properties with H2A.X-type and H2A.Z-type histone [34]. The similarity of H2Av to canonical histone H2A is lower in this instance, but it still retains a longer C-terminal end containing the characteristic H2A.X SQ motif. In yeast, H2A.X is the major H2A histone and is encoded by the *HTA1* and the *HTA2* genes that, in both instances, encode for H2A histones with the characteristic SQE sequences at their C-terminus [35]. Thus, it is not surprising that the phosphorylation of histone H2A.X at this SQ motif, as a mark for the presence of DNA DSB, has also been highly conserved.

Recently, new developments on histone H2A.X have shown that the DNA damage response involves more H2A.X PTMs. In addition, new evidence has brought back our attention to the role of histone H2A.X during mitosis, replication, meiosis, transcription, and apoptosis [36–38].

Histone H2A.X participates in DNA DSB repair by flanking the site of DNA DSB, signaling the recruitment of DNA DSB response machineries and eliminating the DNA DSB signal after DNA repair is completed. In addition to the recruitment of repair factors such as the MRN complex, 53BP1, and BRCA1 [39–41], H2A.X serves as a docking site for chromatin remodeling complexes. These, alter chromatin conformation to facilitate the access of DNA repair factors to the otherwise highly condensed chromatin with a broken DNA strand. Chromatin remodeling complexes that are known to associate with γ-H2A.X include: NuA4 [42], INO80 [43] and SWR1 [44] in yeast, Tip60 in *Drosophila* [45], and FACT in humans [46]. Association of γ-H2A.X with these complexes leads to the release of histone components to promote an "open" chromatin conformation, or leads to the exchange with other histone variants including the non-phosphorylated form of H2A.X upon completion of DNA repair.

In addition to the recruitment potential of γ-H2A.X, the variant itself may directly contribute to the chromatin alteration involved in DNA repair process [47]. The C-terminal tail of histone H2A.X is located at the entry and exit site of the DNA that wraps around the nucleosome. Phosphorylation of histone H2A.X at the SQE site introduces a significant increase in its negative charge of this region. This has the potential to reduce the H2A.X interaction with DNA and/or alter the DNA trajectory, hampering its ability to bind to linker histones [8]. Although the H2A.X has been shown to exhibit a very low diffusional mobility [48] this increases significantly upon induction of DNA repair by γ-irradiation [49]. The H2A.X phosphorylation associated with this process destabilizes the nucleosome, presumably facilitating its FACT mediated exchange in human cells [46].

In addition to phosphorylation, other PTMs on H2A.X are found to be important for DNA DSB repair, such as the acetylation and ubiquitination of H2A.X at K5 and K119, respectively, in human cells [50] that lead to the release of the modified histone H2A.X from the DSB sites, altering the chromatin structure surrounding this region. The H2A.X epigenetic code of DNA repair is quite thoroughly mapped in yeast [51]; however, the combinatorial modification of histone H2A.X in mammals awaits further investigation.

Histone H2A.X is not restricted to DNA repair; it is also involved in other cellular processes. Three populations of histone H2A.X have been described: DSB related and chromatin associated H2A.X, non-DSB related and chromatin associated H2A.X, and, most recently, soluble, non-nucleosomal associated H2A.X [37, 52, 53]. Although soluble, non-nucleosomal associated H2A.X has been found in both nucleus and cytoplasm [37, 53], it is important to note that the majority of histone H2A.X remains associated with chromatin. From the cytological point of view, the DSB associated H2A.X population, as described above, forms large foci at the DSB sites, and it is excluded from heterochromatin regions [52, 54]. In contrast, the other more abundant, non-DSB associated γ-H2A.X forms small foci that partition equally within euchromatin and heterochromatin [52]. The phosphorylation of H2A.X seems to be regulated through cell cycle where it reaches maximal at G2/M and then begins to decrease, reaching steady state immediately after cytokinesis, suggesting a role in cell cycle [52]. While the functional role of γ-H2A.X and its involvement in DNA DSB repair (large foci) has been the intense focus of research in the past decade, the biochemical nature of the recently described small foci are less well defined. The observation of these small foci could be partly related to the involvement of H2A.X in transcription described long ago [46, 55–57]. Additionally, they have been shown to participate in the replication of the inactive X chromosome [58].

Interestingly, the soluble, non-nucleosomal associated H2A.X, but not the canonical histone H2A, can sensitize cells to undergo apoptosis [37]. The eviction of H2A.X from chromatin by remodeling complexes, and the increase of H2A.X expression within the cells may cause the accumulation of soluble H2A.X. The apoptotic inducing property of the soluble H2A.X is not dependent on the phosphorylation of the histone [37]. In addition to all this, histone H2A.X is involved in the restoration of the double-strand breaks that occur during the reorganization of the meiotic nuclei in *Tetrahymena* [38] and, hence, it is not surprising that it accumulates in spermatogonia [59], likely as a result of its participation in the homologous recombination events that take place during this process.

Finally, H2A.X has been shown to accumulate in highly differentiated cells such as in the mature sperm of some mammals and in neurons. Whether its presence in human sperm [60] corresponds to residual histones from the spermiogenesis process or it has a role in post-fertilization events remains to be elucidated. Its accumulation in differentiating, mature rat brain cortical neurons is equally enigmatic [61, 62]. Whatever these functions, these examples underscore a multifaceted functional role for this variant that is far beyond DNA repair and that still remains to be explored.

H2A.Z FUNCTION AT A FLIP OF A COIN

From a functional perspective, H2A.Z (Htz1 in yeast) is one of the most studied histone variants in recent years [63]. H2A.Z is highly conserved amongst animals, plants, and yeast [64], and it has been shown to be necessary for survival in a broad range of organisms including *Trypanosoma* [65], *Tetrahymena* [66], and *Drosophila* [67]. Lack of this histone impairs vertebrate development in mouse [68] and *Xenopus* [69]. However, the function of this histone variant has been, and still remains, extremely controversial. H2A.Z has been involved in gene activation or inactivation (or both) (see [10, 63, 64] for reviews), genomic stability [70], and progression through the cell cycle [12].

The functional information gathered to date, points to an important physiological role of H2A.Z in the establishment and maintenance of chromatin boundaries. Indeed, it has been shown that H2A.Z protects euchromatin from the ectopic spread of heterochromatin [71]. Furthermore, the insulator binding protein, CTCF, is flanked by arrays of well positioned nucleosomes that are enriched in H2A.Z as well as several additional core histone PTMs [72]. Additionally, H2A.Z appears to participate in the demarcation of genes [73]. An enrichment of H2A.Z is found at the 5' ends of genes both in yeast [74] and in vertebrates [75] where the acetylated form is also present in certain enhancers [76]. It defines promoter boundary elements [77].

Accordingly, eukaryotic genes can be classified based on their promoter chromatin organization into low plasticity depleted proximal nucleosome promoters or high plasticity occupied proximal nucleosome promoters. The former are enriched with H2A.Z where it is possible that it serves to establish the boundaries of their highly characteristic nucleosome free region (NFR) [77]. While H2A.Z appears to be excluded from constitutive heterochromatin [78], it has been involved in the post-meiotic assembly of mammalian X and Y chromosomes into facultative heterochromatin [79] as well as pericentric heterochromatin during early development [80].

From a structural perspective, H2A.Z has been shown to increase the interaction of neighboring H2A.Z containing nucleosomes within the chromatin fiber [81] and to produce a moderate but reproducible increase in nucleosome stability [16, 19] while enhancing its mobility [82]. Histone H2A.Z destabilizes the H2A.Z-H2B dimer [16, 83]. However, the functional consequences of these observations remain unclear. Two PTMs of this variant, acetylation and ubiquitination, may be important in defining its multiple functions and providing further insight into the potential molecular mechanisms involved.

The presence of acetylated H2A.Z at active genes is a common feature in many organisms from yeast [84] to vertebrates [75] where H2A.Z containing nucleosomes often occupy a defined position in the promoters [74, 85] around the transcriptional start sites (TSS) [86–88]. In yeast, H2A.Z is acetylated at K3, K8, K10, and K14 within the N-terminal tail by NuA4 and SAGA histone acetyltransferases [84, 89, 90]. K14 of H2A.Z is the most acetylated residue, and strains where this residue is mutated exhibit specific defects in chromosome transmission without affecting transcription, telomeric silencing, or DNA repair [89]. Recently, NuA4 has been shown to affect histone H2A.Z incorporation/acetylation *in vivo*, which are both required to preset the *PHO5* promoter for activation in yeast [91]. This work hinted to the possibility that Tip60, a human ortholog of NuA4, may possibly be responsible for acetylation of human H2A.Z [91], which is known to be acetylated at K4, K7, and K11 [92–94].

H2A.Z has also been reported to be indiscriminately mono-ubiquitinated at sites K120, K121, or K125 of its C-terminal tail and the ubiquitinated forms are present in the inactive X chromosome of mammalian female cells. It has been postulated that this PTM provides a specific mark that distinguishes euchromatin from facultative heterochromatin. Nevertheless, it is unclear whether H2A.Z is ubiquitinated in *S. cerevisiae* [78] as mass spectrometry analysis showed the C-terminus Htz1 to be depleted of any PTMs [84].

Whether acetylation [90] or mono-ubiquitination [84] described above are the ultimate structural determinants of the bimodal activation/repressive action of H2A.Z remains to be established. In this regard, the enhanced translational mobility of the H2A.Z containing nucleosome [95] may be critical for its role in delineating boundaries both at promoter and intergenic regions. It has been shown that acetylation increases the rate of association–dissociation of H2A from chromatin [96] and weakens the interaction of H2A.Z with DNA, probably in conjunction with acetylation of the other core histone counterparts [16]. In this regard, it is interesting to note that histone H2A.Z acetylation appears to be critical for the prevention of heterochromatin spreading into euchromatin domains [90]. A combination of an unacetylatable H2A.Z and a mutant version of histone H4 that could not be acetylated by NuA4 was found to be lethal in yeast [90], suggesting a concerted action of H2A.Z acetylation and core histone acetylation are required. The molecular mechanism behind the mode of action of H2A.Z acetylation is not completely understood, but it is likely to involve a charge patch effect, which, in *Tetrahymena*, requires only the retention of one of the acetylatable lysines at the N-termini of the molecule [97]. The additional contribution of chromatin remodeling complexes, such as the SWR1 complex in yeast [98] or TIP60 in mammals [91], cannot be disregarded.

In the case of ubiquitination, the molecular mechanism by which uH2A.Z contributes to chromatin silencing is not known, as neither is that of ubiquitination of canonical H2A that has also been involved in silencing [99, 100]. In fact, ubiquitination of H2A has only minor structural effects, both at the level of the nucleosome and the chromatin fiber, where it slightly enhances the binding of nucleosomes to linker histones [101]. Whether this is sufficient to account for the ubiquitination repressive effects, or ubiquitination simply serves as a signal for recognition by other transacting inactivating factors, needs to be elucidated further. Another mechanism of repression by H2A.Z could be mediated by the interaction of the non-acetylated form of the variant with HP1a [102].

Finally, not only is H2A.Z indispensable for survival, a few papers have recently suggested a correlation of this variant with disease. It has been involved in breast cancer progression and associated with lymph node metastasis and decreased patient survival [57]. Also, at the molecular level, p21 transcription is repressed by H2A.Z deposition at the p53 binding sites of its promoter region. Other p53 responsive genes, such as GADD45, Bax, and MDM2 have their expression regulated by H2A.Z as well [103]. These interesting observations underscore the important epigenetic role of this variant.

MACRO H2A: PHOSPHORYLATION MATTERS

The significant differences in size and amino acid sequence between the primary structure of the canonical histone H2A and macroH2A (mH2A) allow this variant to be classified well within the realm of the heteromorphous type of histone variants [6, 8]. MacroH2A has an N-terminal H2A-like domain that is very similar (65 percent homology) to canonical H2A and a C-terminal non-histone region (NHR) that has been referred to as the macrodomain [104]. The macrodomain is related to a family of proteins that includes polynucleotide hydrolases and nictinamide dinucleotide metabolite binding proteins. The mH2A macrodomain binds to ADP-ribose [105], histone deacetylase 1 (HDAC1), and HDAC2, consequently, contributing to its association with inactive hypoacetylated chromatin [15]. The H2A-like domain and the macrodomain are linked by a lysine-rich, 30 amino acid hinge region resulting in the largest of all histone variants. The basic hinge region has very little secondary and tertiary structure and, hence, cannot be resolved in crystal structures [15]. The mH2A variants are conserved in vertebrates and absent in most invertebrates except for sea urchin [106].

The mH2A gene product only accounts for approximately 3 percent of the H2A population that is present in the cell [107]. Three mH2A isoforms, mH2A1.1, mH2A1.2, and mH2A.2 have been reported. The former two are alternatively spliced from the *H2AFY* gene; whereas, mH2A2 is encoded by a different gene [108]. The functional differences between the three isoforms are currently unknown. Overall, mH2A is associated with chromatin condensation. For example, mH2A is depleted on active genes, but concentrated on the female inactive X chromosome (Xi) and in the sex body of male spermatocytes in mammals [109–111]. In other vertebrates, mH2A is found ubiquitously distributed in the nucleus of their tissues [112], where it exhibits an unusually exclusive relationship with linker histones [5]. Its transcriptionally repressive role is also evident from its association with senescence associated heterochromatin foci and CpG-methylated domains, including imprinted loci [113, 114]. Furthermore, *in vitro* data showed that the presence of mH2A within a reconstituted nucleosome significantly impedes the binding of chromatin remodeling complexes and transcription factors and prevents transcription initiation by RNA polymerase II [115, 116]. While mH2A is present in heterochromatic regions within the nucleus and associated with transcriptional inactivation, posttranslationally modified mH2A has been shown to be involved in very different cellular processes.

Several PTMs on mH2A have been identified: poly-ADP ribosylation [5], ubiquitination, methylation [107] and, most recently, phosphorylation. MacroH2A.1 was found to be phosphorylated at T128 and S137 [20], both of which lie within the basic hinge region of mH2A. While the phosphorylation of T128 has been reported previously [107], the functional role of this modification was not clear. S137 is phosphorylated by Cdk complexes that are involved in the cell cycle process [20]. Interestingly, phosphorylation at S137 is excluded from the transcriptionally inactive Xi and enriched during mitosis [20], thus providing mH2A with a unique chromatin role beyond mammalian X chromosome inactivation [5]. Interestingly, ubiquitination has been proposed to have a contrasting effect by localizing mH2A.1 to the inactive X chromosome [117], whereupon de-ubiquitination and phosphorylation could act synergistically to alter the repressive nature of the molecule. The absence of S137 in mH2A.2 suggests a functional difference between this histone variant and the mH2A.1 isoforms. The specialized functions for each of the mH2A isoforms create a new level of regulation.

It has been suggested that the negative charge resulting from T128/S137 phosphorylation may disrupt the interaction of mH2A.1 with Xist RNA, consistent with the exclusion of the phosphorylated version of this variant from the Xi [20]; therefore, phosphorylation and ubiquitination of mH2A.1 could act as a switch that determines its chromosomal location.

At the gene level, the macrodomain of mH2A binds to, and suppresses, the activity of a transcriptional regulator, poly-ADP ribosylation polymerase (PARP-1), both *in vitro* and *in vivo* as observed at the *Hsp70.1* and *Hsp70.2* promoters [118, 119]. Interestingly, it has been hypothesized that a PTM at the NHR of mH2A1.1 (such as phosphorylation) may affect this interaction, thereby modulating the activity of these genes [119].

At the molecular level, the crystallographic structures of a nucleosome, consisting of the H2A-like region [15] and that of the macrodomain, have already been published [105]. Hydrodynamic characterization of *in vitro* mH2A reconstituted nucleosome indicates that the overall particle has a very asymmetric conformation as would be expected from the large protruding macrodomain [112]. The nucleosome core particle also exhibits enhanced stability to changes in the ionic strength of the media [5], a property that would be in agreement with the repressive properties of this variant. Furthermore, the region of DNA where the NHR exits the particle (close to the dyad axis of the NCP) exhibits an enhanced DNase I sensitivity, indicative of an extended conformation of this region. The contribution of the functionally antagonistic phosphorylation and ubiquitination of mH2A to the physical characteristic of the nucleosome core particle remains to be established.

H2B VARIANCE AND UNKNOWN PARTNERS

There are at least 17 predicted different isoforms of human H2B. Following the Swiss-Prot nomenclature, these include H2BA-F, H2BJ, H2BK, H2BN, H2BQ, H2BR, H2BS, H2B.X, Q5QNW6, Q6NWQ3 [94], and the testis specific TSH2B [120] and H2BFWT [121]. From a structural perspective they can be considered to be grouped into three major types: H2B.1, H2B.2, and tH2B. Histone H2B.1 putatively include the H2B isoforms whose N-terminal sequence starts with PEP(A/V/S/T) (H2BA-F, H2BK, H2BQ-S, and Q6NWQ3).

Histone H2B.2 would include the isoforms whose sequence starts with **PD**P(A/S) (H2BJ, H2BN, H2BX, and Q5QNW6). The human TSH2B isoform starts with **PEVS**. The amino acid sequence variability goes beyond that of the very N-terminal sequence [94] that is highlighted here for classification purposes only. Histone H2B.1 histones elute later than H2B.2 in reversed phase HPLC and, in acetic acid urea Triton-X100 (AUT) PAGE, H2B.1 exhibits a significantly higher mobility than H2B.2 and tH2B.

While many of these subtypes may have redundant functions, it is interesting to note that in rat, H2B.2 accumulates during rat brain development [61], whereas, it decreases in thymus during mouse development [122], suggesting that H2B.2 is the preferred variant for nuclear metabolic activity. Intriguingly though, H2B.2 appears to stabilize the interaction of the H2A-H2B dimer in the nucleosome [123]. These contrasting results point to the need of research focusing on H2B variability for which only very limited information is currently available. It is possible that the H2B.1 and H2B.2 isoforms may preferentially associate with some specific histone H2A variants in order to exert their functional–structural role. Indeed, it has been shown that, in *Trypanosome brucei*, a specialized H2BV pairs specifically with H2A.Z [65]. Again, research on this aspect is currently lacking.

In addition to the homomorphous variants, H2B also includes several heteromorphic forms that are usually found in the sperm of vertebrate and invertebrate organisms. The H2BFWT is a 175 amino acid long, TSH2B variant that participates in the telomere binding complex in the human sperm [121]. The protein has an extended N-terminal sequence and an unusual C-terminal end. The sperm H2B variants of invertebrates also exhibit long N-terminal regions that, in echinoderms, include sequences consisting of the repetitive motif SP(R/K)(R/K) [124] that confers specificity of binding to the minor groove of DNA [49]. The role of these N-terminal extensions, in the case of invertebrates, has been attributed to their involvement in the highly compacted chromatin of their sperm as a result of charge neutralization and interchromatin fiber interactions [125].

H3.3 PROVIDING TRANSCRIPTIONAL MEMORY

In humans, H3.3 and the canonical H3.2 and H3.1 differ only by 4 and 5 amino acids, respectively. The small difference in amino acid sequence between these variants constitutes a good example of homomorphous variation [6, 9]. Variation at structurally homologous sites within the H3.3 molecule appears to have occurred repeatedly throughout evolution, likely through a process of parallel evolution and hence this variant cannot be considered an independent lineage [14]. Interestingly, this variant appears to accumulate during development and it is the major histone H3 component of rat neuronal cells [61] and thymus in adult mice [122]. This strongly suggests that H3.3 is a marker of the accumulated past transcriptional activity of the tissue rather than an indicator of the overall activity of the tissue at a given stage in development.

Histone H3.3 is a replacement histone variant that participates in the histone turnover taking place in the nucleus throughout the entire cell cycle in contrast to H3.1/H3.2 whose expression is restricted to the S phase. H3.3 has been shown to play a major role in regulation of transcription. It is enriched in actively transcribing regions of the genome [126] where it replaces the canonical counterpart during transcriptional elongation [13, 127] and at promoter regions [20]. Of note, a paper by Ng and Gurdon, has shown that H3.3 provides epigenetic memory for the maintenance of transcription during subsequent cell cycles [128, 129]. Furthermore, the memory appears to depend on the presence of methylatable K4 of this variant [128].

At the molecular level, and despite the small sequence differences between H3.3 and the canonical H3 counterpart, a few recent papers suggest that the difference is large enough to at least partially exert a direct effect on the stability [17] and assembly–disassembly ability of the NCP [26]. Like with other histone variants, it is likely that these effects are enhanced by the presence of specific PTMs [130–133] and/or chaperone binding proteins. Recently, phosphorylation of S28 of H3.3 has been associated with destabilized nucleosomes in transcribed chromatin [96]. The H3.3 chaperones HIRA, ASF1A/ASF1B, and CHD1 are integral parts of complexes that participate in chromatin remodeling processes required for nuclear DNA metabolic activity [134–136] and for chromatin assembly in the early post-fertilization stages of development [137, 138].

CENP-A: SPLITTING NUCLEOSOMES IN *DROSOPHILA*

Centromere protein A (CENP-A) is a histone variant that replaces histone H3 in the chromatin regions of centromeres from where it co-purifies with nucleosomes [139, 140]. It has received different names in different organisms; for instance, Cse4p in yeast (*Sacharomyces cerevisiae*), Cnp1p in fission yeast (*Schizosaccharomyces pombe*), and Cid in *Drosophila.* Like H2A.Z, CENP-A is a histone variant that is indispensable for survival [141]. Several reviews on this variant have been published [14, 142–144].

The histone fold domain of this protein is quite conserved (63 percent identity with that of canonical H3 in humans) and required for targeting to centromeres [145] in contrast to the N-terminal region, which is very variable [14]. From the evolutionary point of view, this is one of the fastest evolving histone H3 variants and similarly to H3.3, it cannot be considered an independently evolving lineage [14, 146].

One of the most recent exciting developments with this variant has to do with its ability to lead to the formation of split nucleosomes in *Drosophila* [147] (hemisomes [148]) (see Figure 17.2a). The elusive occurrence of split nucleosomes has been haunting the chromatin field since the early prediction of Harold Weintraub and co-workers [149], which was followed by sporadic structural [150] and biochemical observations [151]. However, and despite the enormous structural flexibility of the histone core that can adopt different chiral conformations in the context of the NCP [148], direct convincing evidence for the existence of the hemisome was missing until now. It has been shown that it is possible to reconstitute full nucleosomes *in vitro* using recombinant CENP-A and the complementary core histone counterpart [152]. Nevertheless, such nucleosomes exhibited some DNA unwrapping, a noticeable destabilization of the CENP-A/H4 tetramer, and more rigidity [121, 153] when compared to canonical NCPs. As loop 1 and α2 helix of CENP-A (Figure 17.2b) have been shown to be the distinctive motif responsible for the targeting of this centromeric H3 variant to centromeres [153], it is highly likely that these structural motifs are responsible for the stiffness and instability of the reconstituted complexes as well as for the hemisome structures that have been observed *in vivo*. It has been proposed that the unusual composition of loop 1 of CENP-A results in a straightening of α2 helix of H4 which, in turn, would result in a strengthening of the H2B/H4 interface and the weakening of the CENP-A homodimerization domain relative to the canonical octameric NCP [154]. This would favor the formation of the CENP-A hemisomes that are observed *in vivo* (Figure 17.2). Such nucleosomal organization would result in the specialized decondensed chromatin organization [154].

The hemisome organization shown in Figure 17.2a would, in principle, preclude the interaction with linker histones, a notion that is further supported by experimental work carried out with the CENP-A nucleosomes reconstituted *in vitro* [121]. However, the occurrence of long, exposed linkers may allow for the interaction of other

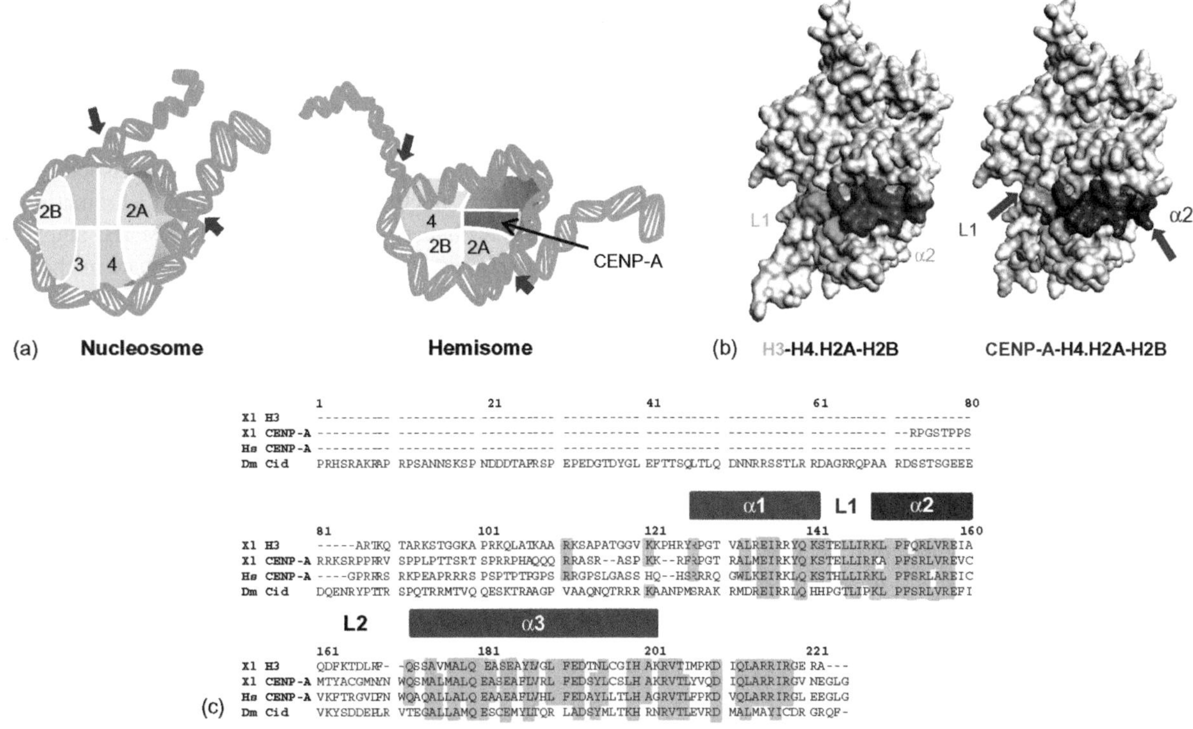

FIGURE 17.2 Centromeric protein A (CENP-A) is a highly specialized histone H3 variant that alters the conformation of the nucleosome. (a) Schematic representation of a canonical nucleosome and a hypothetical hemisome consisting of an H2A-H2B-H4-CENP-A tetramer [147]. The arrows point to the DNA protected by the protein cores that in the NCP would correspond to 145 bp and in the hemisome it would approximately protect 100–120 bp as experimentally observed in *Drosophila* [147]. (b) Three dimensional structure of a core histone tetramer consisting of H3-H2A-H2B-H4 in comparison to that of CENP-A-H2A-H2B-H4. The model is based on the crystallographic structure of NCPs reconstituted with *Xenopus laevis* recombinant histones [29]. The structure of *X .laevis* CENP-A was predicted from its sequence using the SWISS-MODEL [215] and it was next used in combination with PyMOL [216] to recreate its organization within the NCP. The regions of H3 and CENP-A corresponding to loop 1 and α-helix 2 of the histone fold domain of H3 and CENP-A are highlighted. In CENP-A, this region is sufficient to target this histone to the centromeres [153] and it has been held responsible for the unusual rigidity [153] and hemisome conformation of the *Drosophila* CENP-A/(Cid) containing nucleosomes [154]. The arrows point to structural differences in *Xenopus* CENP-A within this domain. (c) Sequence alignment of H3 from *X. laevis* (Xl) in comparison to the sequences of CENP-A from *X. laevis* (Xl), *Homo sapiens* (Hs), and the fruit fly *D. melanogaster* (Dm). Identical amino acids in the sequence are highlighted and the α-helices and loops of the histone fold are also indicated above the sequences. Note the highly divergent long N-terminal tail of *Drosophila* CENP-A (Cid). Although this region has not been involved in the formation of the hemisome, its contribution to the conformation of the Cid-NCP remains to be determined.

centromeric proteins such as CENP-B, which, in turn, could further associate with other centromeric proteins, such as CENP-C, CENP-H, CENP-I [155, 156], resulting in the formation of the kinetochore [157]. Also, H2A.Z has been shown to be a structural component of the centromere where it contributes to the highly compact organization that is characteristic of centromeric chromatin [21].

A long-standing question with this variant has been the mechanism(s) involved in its targeting to centromeres [158]. A recent report has shown that *de novo* centromere assembly of this variant is dependent on heterochromatin and RNA interference pathway in yeast [159] and in its deposition by Sim3, a histone chaperone related to human NASP [160], thus identifying NASP as a potential chaperone for this protein. NASP is a histone chaperone dimer [161] that, in vertebrates, has two distinct somatic and testes versions and it was originally described to be a chaperone for histones H1 and H3 [162, 163].

Despite the progress made in recent years with this variant, many unanswered questions remain. Can the hemisome structure that has been observed by the Henikoff's group in the fruit fly *Drosophila* [147, 154] be generalized to all eukaryotic organisms? What is the relation between these structures and the orderly sequence dependent nucleosome positioning pattern that is observed in fission yeast [164]? These are only two of the issues that await further analysis.

HISTONE H1: THE MICRO-HETEROGENEITY OF SPECIALIZED FUNCTION

As it has already been pointed out, the histones of the H1 family are also known by the generic name of linker histones. This is because, unlike the core histone variants, the members of this family bind to the linker DNA regions connecting adjacent nucleosomes in the chromatin fiber. In doing so, these proteins help to stabilize and dynamically modulate the folded state of the chromatin fiber [10, 165].

The histone H1 family consists of 11 non-allelic variants in humans [10]. These include the somatic forms H1.1, H1.2, H1.3, H1.4, H1.5 (nomenclature from [166]), H1°, H1.x [166–168]; and the germinal forms, male H1t [169], HILS1 [32, 39], H1T2 [170], and female H1foo [50] types. A similar situation is observed in other vertebrate organisms such as birds [171] and amphibians where a highly specific H1 variant (histone H5) is additionally present in the nucleated erythrocytes of these organisms [172]. As well, an egg specific histone B4 has been described in amphibians [173]. Histone H1 variants are also present in plants and include distinctive sets of somatic and "drought inducible" variants [174, 175]. In invertebrates and lower eukaryotes, the range of inter- and intra-specific amino acid sequence variation is very broad [176]. In the fruit fly *Drosophila melanogaster*, only one cysteine-containing variant appears to be present, whereas *D. virilis* contains three histone H1 variants [177] and the nematode *Caenorhabditis elegans* has at least five family members [178]. Yeast has minimal amounts of an unusual histone H1 (Hho1p) [179], which consists of two winged helix domains.

From an evolutionary perspective, the origin of the histones of the H1 family is older than that of core histones and can be traced down to eubacteria [176]. In metazoans, these proteins have evolved through a "birth-and-death" mechanism under a strong purifying selection [180, 181]. In the process, the interspecific conservation between the different variants has remained stronger than their intraspecific relation, providing evidence for their functional differentiation [182].

The histone H1 family also includes a unique group of highly specialized sperm specific PL-I proteins [183] that are the major chromosomal protein component of the sperm of several species of invertebrate [112] and vertebrate [184, 185] organisms. The PL-I proteins of vertebrates appear to be related to histone H1.x [186]. Histone H1.x [187] is one of the most recently characterized histone H1 variants, which is enriched in chromatin domains that are refractory to micrococcal nuclease digestion [167] and is required for mitotic progression [188].

While some of the members of the histone H1 family have a clear functional developmental component (H1foo, B4, H5, H1t), the function of the microheterogeneity of the somatic counterparts [3, 189–191], exemplified by the somatic mammalian forms, has remained quite elusive. In this regard, many knockout experiments (see [192] for a recent review) including those affecting specialized variants such as H1t [193–195] have suggested an extensive amount of functional redundancy for the different variants. The effect of these deletions for the long term survival of the organism has yet to be determined [196]. Indeed, triple knockout mutants have been shown to affect male fertility in mice [197]. Furthermore, there is an accumulating amount of data in support of the notion that the histone H1 subtypes may support specialized functions [198, 199]. The first source of evidence comes from their evolution as described above. This is further supported by the molecular evidence that indicates that the mammalian subtypes exhibit differential binding affinity for DNA and chromatin both *in vivo* [200] and *in vitro* [201].

A few examples that attest to the functional relevance of the histone H1 somatic micro-heterogeneity include human H1.2, which is involved in apoptosis induced by DNA double-strand breaks [202], and human endothelial specific H1.5, which participates in the transcriptional regulation of the Von Willebrandt factor, a glycoprotein involved in hemostasis [203]. In mice, recent work suggests that H1.2 and H1.4 may have a role in position effect variegation [204] and mouse H1.3 interacts with Msx1 to regulate muscle differentiation. Finally, a role for chicken H1 1R [205] in DNA damage response has been described [206].

The association of linker histones with chromatin is very dynamic [10, 15, 165, 207]. It is possible to imagine that the association of chromatin during replication with different histone H1 variants establishes chromatin domains that in themselves or upon PTM can provide domains with different states of folding (Figure 17.3a). It has been recently shown that chromatin folding dramatically decreases the accessibility of linker DNA [63, 208]. Therefore, the effect of histone H1 variability may exert an important role in the differential folding dynamics of chromatin along the genome.

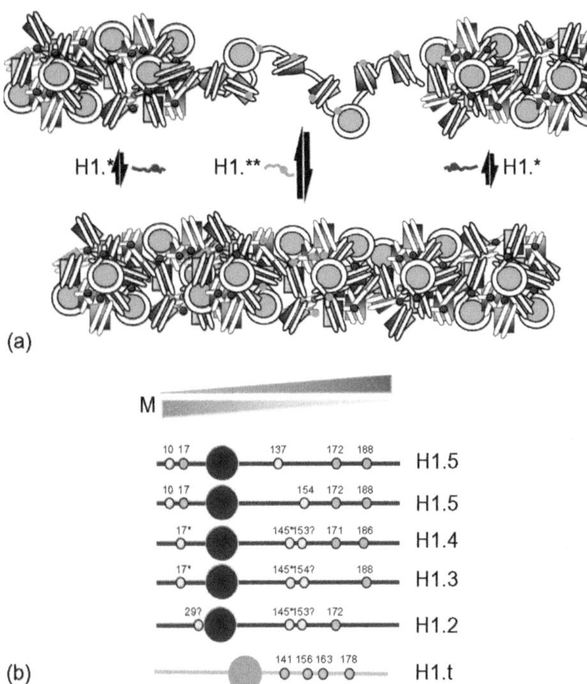

FIGURE 17.3 Phosphorylation enhancement of the structural and the functional potential of histone H1 variability.
(a) Different histone H1 variants denoted here by H1.* and H1.**, bind to the chromatin fiber with different affinities [201] leading to chromatin domains with different folding dynamics. (b) Phosphorylation of the somatic histone H1 variants (subtypes) takes place in a well orchestrated fashion [29] with a sequential and progressive increase in the extent of phosphates incorporated from the C- to the N-terminal end of the molecule during interphase (I). A similar wave of additional phosphorylation running in the opposite direction takes place during mitosis (M). The limited phosphorylation observed during interphase provides the different H1 somatic subtypes with chromatin binding characteristics that are quite different from those of the fully phosphorylated mitotic forms [212]. In contrast, the developmentally regulated testis specific H1t is globally phosphorylated at its C-terminal end [213]. The phosphorylation events that affect different residues in each variant add an additional layer of structural specificity to the chromatin dynamics shown in (a). In (b), the phosphates are represented by little circles with numbers above them denoting the residue (S/T) affected. In the somatic subtypes the sites were mainly based on those identified in [29] and in [217] (denoted in this case with an asterisk). The sites for H1t were obtained from [213]. The question marks point to potential sites that have not been yet identified. Note the distinct distribution of phosphorylated sites in different subtypes and also that in some instances as in H1.5 phosphorylation of T137 and T154 appear to be mutually exclusive [29].

We propose that the effect exerted on the folding of the chromatin fiber by the different histone H1 subtypes and variants operates in a synergistic and/or combinatorial way with their PTMs, in particular, phosphorylation [209]. In support of this, the human H1 subtypes have been shown to exhibit different phosphorylation patterns (Figure 17.3b) that are carefully synchronized throughout the cell cycle [29].

Phosphorylation has been shown to regulate the dynamic mobility and affinity with which linker histones bind to chromatin [210] and globally unfold the chromatin fiber [211]. Moreover, phosphorylation of the C-terminal domain of linker histones has been shown to alter its structure in a way that it is site specific and depends also on the overall level of phosphorylation [212]. Low levels of localized phosphorylation (observed during interphase) disrupt the α-helical content whereas hyperphosphorylation (of metaphase) results in a significant increase in β-structure providing the molecule with a higher aggregation capacity [212].

Phosphorylation of other developmentally regulated histone H1 variants, such as histone H1t [213], may also play an important role in the modulation of chromatin binding dynamics. However, in this instance, the structural effects of unfolding chromatin appear to be more dependent on the extent of phosphorylation than on the distribution pattern of the phosphorylated sites (Figure 17.3b).

CONCLUDING REMARKS

Throughout this chapter, we have discussed how histone variants can have both structural and signaling roles that can be directly mediated by the nature of the variability of the protein sequence itself and/or in combination with PTMs.

We have provided examples where core histone variants can have structural implications of their own at the levels of the NCP and the chromatin fiber. For instance, macroH2A and H2A.Z can both by themselves enhance the stability of the NCP [16, 18, 112] whereas H2A.Bbd can contribute to NCP unfolding and destabilization [18, 23, 27]. In other instances, some of these variants can operate synergistically as with H2A.Z-H3.3 to exert an enhanced destabilizing effect [17]. It is not clear yet how the existence of nucleosome hybrids in which histone variants co-exist with their canonical forms can also modulate the conformation and stability of the NCP and the chromatin fiber. Linker histone variants have also been shown to bind to chromatin with different affinities [201] and hence have subtle but noticeable effects on chromatin folding and dynamics [165]. It appears that the epigenetic contribution of histone variants is a combination of these subtle, yet important structural changes and a strong signaling component often exerted together through well defined individual PTMs. We have seen how ubiquitination of H2A.Z and mH2A participate in different forms of chromatin inactivation [78, 214], a trait they share with ubiquitinated canonical H2A [99]. In contrast, acetylated H2A.Z is targeted to transcriptionally active regions [78] and phosphorylation of mH2A allows this protein to be in chromatin regions other than those related to X chromosome inactivation [20]. In this way, acetylation/ubiquitination of H2A.Z and phosphorylation/ubiquitination of mH2A can operate as functional switches for the partitioning of nucleosomes between active and inactive regions of the genome. More strikingly, H3.3 K4 methylation has been recently shown to provide epigenetic transcriptional memory [128].

The dynamics with which the histone variants exchange from chromatin and interact with individual nucleosomes at localized regions of the genome are often modulated by specific chaperones (i.e., HIRA for H3.3, NASP for CENP-A, CHZ1 for H2A.Z) [15].

Despite all of what is already known about histone variants, the structural and functional roles of some of them remain controversial. This suggests that discerning the individual contribution of their structural and signaling components to the epigenetic changes they impart on chromatin may still be a question that requires more than what we currently know.

ACKNOWLEDGEMENTS

This work was supported in part by Canadian Institutes of Health Research (CIHR).

REFERENCES

1. Arents G, Moudrianakis EN. The histone fold: a ubiquitous architectural motif utilized in DNA compaction and protein dimerization. *Proc Natl Acad Sci* 1995;**92**:11,170–4.
2. van Holde KEChromatin. NY: Springer-Verlag; 1988.
3. Cole RD. Microheterogeneity in H1 histones and its consequences. *Int J Pept Protein Res* 1987;**30**:433–49.
4. Ramakrishnan V, Finch JT, Graziano V, Lee PL, Sweet RM. Crystal structure of globular domain of histone H5 and its implications for nucleosome binding. *Nature* 1993;**362**:219–23.
5. Ausió J, Abbott DW. The many tales of a tail: carboxyl-terminal tail heterogeneity specializes histone H2A variants for defined chromatin function. *Biochemistry* 2002;**41**:5945–9.
6. West MH, Bonner WM. Histone 2A, a heteromorphous family of eight protein species. *Biochemistry* 1980;**19**:3238–45.
7. Isenberg I. Histones. In: Busch H, editor. *The Cell Nucleus*. New York: Academic Press; 1978. p. 135–54.
8. Ausió J, Abbott DW, Wang X, Moore SC. Histone variants and histone modifications: a structural perspective. *Biochem Cell Biol* 2001;**79**:693–708.
9. Ausió J, Abbott DW. The role of histone variability in chromatin stability and folding. In: Zlatanova JaL S, editor. *New comprehensive biochemistry. Chromatin Structure and Dynamics: State-of-the-Art*. Amsterdam, The Netherlands: Elsevier; 2004. p. 241–90.
10. Ausió J. Histone variants: the structure behind the function. *Brief Funct Genomic Proteomic* 2006;**5**:228–43.
11. Pusarla RH, Bhargava P. Histones in functional diversification. Core histone variants. *FEBS J* 2005;**272**:5149–68.
12. Dhillon N, Oki M, Szyjka SJ, Aparicio OM, Kamakaka RT. H2A.Z functions to regulate progression through the cell cycle. *Mol Cell Biol* 2006;**26**:489–501.
13. Sarma K, Reinberg D. Histone variants meet their match. *Nat Rev Mol Cell Biol* 2005;**6**:139–49.
14. Malik HS, Henikoff S. Phylogenomics of the nucleosome. *Nat Struct Biol* 2003;**10**:882–91.
15. Chakravarthy S, Gundimella SK, Caron C, Perche PY, Pehrson JR, Khochbin S, Luger K. Structural characterization of the histone variant macroH2A. *Mol Cell Biol* 2005;**25**:7616–24.
16. Thambirajah AA, Dryhurst D, Ishibashi T, Li A, Maffey AH, Ausio J. H2A.Z stabilizes chromatin in a way that is dependent on core histone acetylation. *J Biol Chem* 2006;**281**:20,036–44.
17. Jin C, Felsenfeld G. Nucleosome stability mediated by histone variants H3.3 and H2A.Z. *Genes Dev* 2007;**21**:1519–29.
18. Bao Y, Konesky K, Park YJ, Rosu S, Dyer PN, Rangasamy D, Tremethick DJ, Laybourn PJ, Luger K. Nucleosomes containing the histone variant H2A.Bbd organize only 118 base pairs of DNA. *Embo J* 2004;**23**:3314–24.
19. Suto RK, Clarkson MJ, Tremethick DJ, Luger K. Crystal structure of a nucleosome core particle containing the variant histone H2A.Z. *Nat Struct Biol* 2000;**7**:1121–4.
20. Bernstein E, Muratore-Schroeder TL, Diaz RL, Chow JC, Changolkar LN, Shabanowitz J, Heard E, Pehrson JR, Hunt DF, Allis CD. A phosphorylated subpopulation of the histone variant macroH2A1 is excluded from the inactive X chromosome and enriched during mitosis. *Proc Natl Acad Sci U S A* 2008;**105**:1533–8.
21. Viens A, Mechold U, Brouillard F, Gilbert C, Leclerc P, Ogryzko V. Analysis of human histone H2AZ deposition in vivo argues against its direct role in epigenetic templating mechanisms. *Mol Cell Biol* 2006;**26**:5325–35.
22. Chadwick BP, Willard HF. A novel chromatin protein, distantly related to histone H2A, is largely excluded from the inactive X chromosome. *J Cell Biol* 2001;**152**:375–84.
23. Eirin-Lopez JM, Ishibashi T, Ausió J. H2A.Bbd: a quickly evolving hypervariable mammalian histone that destabilizes nucleosomes in an acetylation-independent way. *Faseb J* 2008;**22**:316–26.
24. Gonzalez-Romero R, Mendez J, Ausió J, Eirin-Lopez JM. Quickly evolving histones, nucleosome stability and chromatin folding: all about histone H2A.Bbd.. *Gene* 2008;**413**:1–7.
25. Gautier T, Abbott DW, Molla A, Verdel A, Ausió J, Dimitrov S. Histone variant H2ABbd confers lower stability to the nucleosome. *EMBO Rep* 2004;**5**:715–20.
26. Okuwaki M, Kato K, Shimahara H, Tate S, Nagata K. Assembly and disassembly of nucleosome core particles containing histone variants by human nucleosome assembly protein I. *Mol Cell Biol* 2005;**25**:10,639–51.
27. Doyen CM, Montel F, Gautier T, Menoni H, Claudet C, Delacour-Larose M, Angelov D, Hamiche A, Bednar J, Faivre-Moskalenko C, Bouvet P, Dimitrov S. Dissection of the unusual structural and functional properties of the variant H2A.Bbd nucleosome. *EMBO J* 2006;**25**:4234–44.
28. Montel F, Fontaine E, St-Jean P, Castelnovo M, Faivre-Moskalenko C. Atomic force microscopy imaging of SWI/SNF action: mapping the nucleosome remodeling and sliding. *Biophys J* 2007;**93**:566–78.
29. Luger K, Mader AW, Richmond RK, Sargent DF, Richmond TJ. Crystal structure of the nucleosome core particle at 2.8 A resolution. *Nature* 1997;**389**:251–60.
30. Zhou J, Fan JY, Rangasamy D, Tremethick DJ. The nucleosome surface regulates chromatin compaction and couples it with transcriptional repression. *Nat Struct Mol Biol* 2007;**14**:1070–6.

31. Dorigo B, Schalch T, Bystricky K, Richmond TJ. Chromatin fiber folding: requirement for the histone H4 N-terminal tail. *J Mol Biol* 2003;**327**:85–96.
32. Yang J, Yu Y, Hamrick HE, Duerksen-Hughes PJ. ATM, ATR and DNA-PK: initiators of the cellular genotoxic stress responses. *Carcinogenesis* 2003;**24**:1571–80.
33. Mallory JC, Petes TD. Protein kinase activity of Tel1p and Mec1p, two Saccharomyces cerevisiae proteins related to the human ATM protein kinase. *Proc Natl Acad Sci U S A* 2000;**97**:13,749–54.
34. Madigan JP, Chotkowski HL, Glaser RL. DNA double-strand break-induced phosphorylation of Drosophila histone variant H2Av helps prevent radiation-induced apoptosis. *Nucleic Acids Res* 2002;**30**:3698–705.
35. Redon C, Pilch DR, Rogakou EP, Sedelnikova O, Newrock K, Bonner WM. Histone H2A variants H2AX and H2AZ. *Curr Opin in Genet Dev* 2002;**12**:162–9.
36. Ismail IH, Hendzel MJ. The gamma-H2A.X: is it just a surrogate marker of double-strand breaks or much more?. *Environ Mol Mutagen* 2008;**49**:73–82.
37. Liu Y, Parry JA, Chin A, Duensing S, Duensing A. Soluble histone H2AX is induced by DNA replication stress and sensitizes cells to undergo apoptosis. *Mol Cancer* 2008;**7**:61.
38. Mochizuki K, Novatchkova M, Loidl J. DNA double-strand breaks, but not crossovers, are required for the reorganization of meiotic nuclei in Tetrahymena. *J Cell Sci* 2008;**121**:2148–58.
39. Bassing CH, Chua KF, Sekiguchi J, Suh H, Whitlow SR, Fleming JC, Monroe BC, Ciccone DN, Yan C, Vlasakova K, Livingston DM, Ferguson DO, Scully R, Alt FW. Increased ionizing radiation sensitivity and genomic instability in the absence of histone H2AX. *Proc Natl Acad Sci* 2002;**99**:8173–8.
40. Bassing CH, Alt FW. H2AX may function as an anchor to hold broken chromosomal DNA ends in close proximity. *Cell Cycle* 2004;**3**:149–53.
41. Stucki M, Jackson SP. gammaH2AX and MDC1: anchoring the DNA-damage-response machinery to broken chromosomes. *DNA Repair (Amst)* 2006;**5**:534–43.
42. Downs JA, Allard S, Jobin-Robitaille O, Javaheri A, Auger A, Bouchard N, Kron SJ, Jackson SP, Cote J. Binding of chromatin-modifying activities to phosphorylated histone H2A at DNA damage sites. *Mol Cell* 2004;**16**:979–90.
43. Morrison AJ, Highland J, Krogan NJ, Arbel-Eden A, Greenblatt JF, Haber JE, Shen X. INO80 and γ-H2AX interaction links ATP-dependent chromatin remodeling to DNA damage repair. *Cell* 2004;**119**:767–75.
44. van Attikum H, Fritsch O, Hohn B, Gasser SM. Recruitment of the INO80 complex by H2A phosphorylation links ATP-dependent chromatin remodeling with DNA double-strand break repair. *Cell* 2004;**119**:777–88.
45. Kusch T, Florens L, Macdonald WH, Swanson SK, Glaser RL, Yates 3rd JR, Abmayr SM, Washburn MP, Workman JL. Acetylation by Tip60 is required for selective histone variant exchange at DNA lesions. *Science* 2004;**306**:2084–7.
46. Heo K, Kim H, Choi SH, Choi J, Kim K, Gu J, Lieber MR, Yang AS, An W. FACT-Mediated Exchange of Histone Variant H2AX Regulated by Phosphorylation of H2AX and ADP-Ribosylation of Spt. *Mol Cell* 2008;**30**:86–97.
47. Downs JA, Lowndes NF, Jackson SP. A role for *Saccharomyces cerevisiae* histone H2A in DNA repair. *Nature* 2000;**408**:1001.
48. Siino JS, Nazarov IB, Zalenskaya IA, Yau PM, Bradbury EM, Tomilin NV. End-joining of reconstituted histone H2AX-containing chromatin in vitro by soluble nuclear proteins from human cells. *FEBS Lett* 2002;**527**:105–8.
49. Hamada N, Schettino G, Kashino G, Vaid M, Suzuki K, Kodama S, Vojnovic B, Folkard M, Watanabe M, Michael BD, Prise KM. Histone H2AX phosphorylation in normal human cells irradiated with focused ultrasoft X rays: evidence for chromatin movement during repair. *Radiat Res* 2006;**166**:31–8.
50. Ikura T, Tashiro S, Kakino A, Shima H, Jacob N, Amunugama R, Yoder K, Izumi S, Kuraoka I, Tanaka K, Kimura H, Ikura M, Nishikubo S, Ito T, Muto A, Miyagawa K, Takeda S, Fishel R, Igarashi K, Kamiya K. DNA damage-dependent acetylation and ubiquitination of H2AX enhances chromatin dynamics. *Mol Cell Biol* 2007;**27**:7028–40.
51. Moore JD, Yazgan O, Ataian Y, Krebs JE. Diverse roles for histone H2A modifications in DNA damage response pathways in yeast. *Genetics* 2007;**176**:15–25.
52. McManus KJ, Hendzel MJ. ATM-dependent DNA damage-independent mitotic phosphorylation of H2AX in normally growing mammalian cells. *Mol Biol Cell* 2005;**16**:5013–25.
53. Liu Y, Tseng M, Perdreau SA, Rossi F, Antonescu C, Besmer P, Fletcher JA, Duensing S, Duensing A. Histone H2AX is a mediator of gastrointestinal stromal tumor cell apoptosis following treatment with imatinib mesylate. *Cancer Res* 2007;**67**:2685–92.
54. Kim JA, Kruhlak M, Dotiwala F, Nussenzweig A, Haber JE. Heterochromatin is refractory to gamma-H2AX modification in yeast and mammals. *J Cell Biol* 2007;**178**:209–18.
55. Huang SY, Barnard MB, Xu M, Matsui S, Rose SM, Garrard WT. The active immunoglobulin kappa chain gene is packaged by non-ubiquitin-conjugated nucleosomes. *Proc Natl Acad Sci U S A* 1986;**83**:3738–42.
56. Allis CD, Richman R, Gorovsky MA, Ziegler YS, Touchstone B, Bradley WA, Cook RG. hv1 is an evolutionarily conserved H2A variant that is preferentially associated with active genes. *J Biol Chem* 1986;**261**:1941–8.
57. Hua S, Kallen CB, Dhar R, Baquero MT, Mason CE, Russell BA, Shah PK, Liu J, Khramtsov A, Tretiakova MS, Krausz TN, Olopade OI, Rimm DL, White KP. Genomic analysis of estrogen cascade reveals histone variant H2A.Z associated with breast cancer progression. *Mol Syst Biol* 2008;**4**:188.
58. Chadwick BP, Lane TF. BRCA1 associates with the inactive X chromosome in late S-phase, coupled with transient H2AX phosphorylation. *Chromosoma* 2005;**114**:432–9.
59. Meistrich ML, Bucci LR, Trostle-Weige PK, Brock WA. Histone variants in rat spermatogonia and primary spermatocytes. *Dev Biol* 1985;**112**:230–40.
60. Gatewood JM, Cook GR, Balhorn R, Schmid CW, Bradbury EM. Isolation of four core histones from human sperm chromatin representing a minor subset of somatic histones. *J Biol Chem* 1990;**265**:20,662–6.
61. Bosch A, Suau P. Changes in core histone variant composition in differentiating neurons: the roles of differential turnover and synthesis rates. *Eur J Cell Biol* 1995;**68**:220–5.
62. Pina B, Suau P. Changes in histones H2A and H3 variant composition in differentiating and mature rat brain cortical neurons. *Dev Biol* 1987;**123**:51–8.
63. Zlatanova J, Thakar A. H2A.Z: view from the top. *Structure* 2008;**16**:166–79.
64. Eirin-Lopez J, Ausió J. H2A.Z-Mediated Genome-Wide Chromatin Specialization. *Curr Genomics* 2007;**8**:59–66.
65. Lowell JE, Kaiser F, Janzen CJ, Cross GA. Histone H2AZ dimerizes with a novel variant H2B and is enriched at repetitive DNA in Trypanosoma brucei. *J Cell Sci* 2005;**118**:5721–30.
66. Liu X, Li B, Gorovsky MA. Essential and nonessential histone H2A variants in Tetrahymena thermophila. *Mol Cell Biol* 1996;**16**:4305–11.

67. Clarkson MJ, Wells JR, Gibson F, Saint R, Tremethick DJ. Regions of variant histone His2AvD required for Drosophila development. *Nature* 1999;**399**:694–7.
68. Iouzalen N, Moreau J, Mechali M. H2A.ZI, a new variant histone expressed during Xenopus early development exhibits several distinct features from the core histone H2A. *Nucleic Acids Res* 1996;**24**:3947–52.
69. Faast R, Thonglairoam V, Schulz TC, Beall J, Wells JR, Taylor H, Matthaei K, Rathjen PD, Tremethick DJ, Lyons I. Histone variant H2A.Z is required for early mammalian development. *Curr Biol* 2001;**11**:1183–7.
70. Krogan NJ, Baetz K, Keogh MC, Datta N, Sawa C, Kwok TC, Thompson NJ, Davey MG, Pootoolal J, Hughes TR, Emili A, Buratowski S, Hieter P, Greenblatt JF. Regulation of chromosome stability by the histone H2A variant Htz1, the Swr1 chromatin remodeling complex, and the histone acetyltransferase NuA4. *Proc Natl Acad Sci USA* 2004;**101**:13,513–18.
71. Meneghini MD, Wu M, Madhani HD. Conserved histone variant H2A.Z protects euchromatin from the ectopic spread of silent heterochromatin. *Cell* 2003;**112**:725–36.
72. Fu Y, Sinha M, Peterson CL, Weng Z. The insulator binding protein CTCF positions 20 nucleosomes around its binding sites across the human genome. *PLoS Genet* 2008;**4**:e1,000,138.
73. Arimbasseri AG, Bhargava P. Chromatin structure and expression of a gene transcribed by RNA polymerase III are independent of H2A.Z deposition. *Mol Cell Biol* 2008;**28**:2598–607.
74. Raisner RM, Madhani HD. Patterning chromatin: form and function for H2A.Z variant nucleosomes. *Curr Opin Genet Dev* 2006;**16**:119–24.
75. Bruce K, Myers FA, Mantouvalou E, Lefevre P, Greaves I, Bonifer C, Tremethick DJ, Thorne AW, Crane-Robinson C. The replacement histone H2A.Z in a hyperacetylated form is a feature of active genes in the chicken. *Nucleic Acids Res* 2005;**33**:5633–9.
76. Myers FA, Lefevre P, Mantouvalou E, Bruce K, Lacroix C, Bonifer C, Thorne AW, Crane-Robinson C. Developmental activation of the lysozyme gene in chicken macrophage cells is linked to core histone acetylation at its enhancer elements. *Nucleic Acids Res* 2006;**34**:4025–35.
77. Tirosh I, Barkai N. Two strategies for gene regulation by promoter nucleosomes. *Genome Res* 2008;**18**:1084–91.
78. Sarcinella E, Zuzarte PC, Lau PN, Draker R, Cheung P. Monoubiquitylation of H2A.Z distinguishes its association with euchromatin or facultative heterochromatin. *Mol Cell Biol* 2007;**27**:6457–68.
79. Greaves IK, Rangasamy D, Ridgway P, Tremethick DJ. H2A.Z contributes to the unique 3D structure of the centromere. *Proc Natl Acad Sci U S A* 2007;**104**:525–30.
80. Rangasamy D, Berven L, Ridgway P, Tremethick DJ. Pericentric heterochromatin becomes enriched with H2A.Z during early mammalian development. *Embo J* 2003;**22**:1599–607.
81. Fan JY, Rangasamy D, Luger K, Tremethick DJ. H2A.Z alters the nucleosome surface to promote HP1alpha-mediated chromatin fiber folding. *Mol Cell* 2004;**16**:655–61.
82. Flaus A, Martin DM, Barton GJ, Owen-Hughes T. Identification of multiple distinct Snf2 subfamilies with conserved structural motifs. *Nucleic Acids Res* 2006;**34**:2887–905.
83. Placek BJ, Harrison LN, Villers BM, Gloss LM. The H2A.Z/H2B dimer is unstable compared to the dimer containing the major H2A isoform. *Protein Sci* 2005;**14**:514–22.
84. Millar CB, Xu F, Zhang K, Grunstein M. Acetylation of H2AZ Lys 14 is associated with genome-wide gene activity in yeast. *Genes Dev* 2006;**20**:711–22.
85. Zhang H, Roberts DN, Cairns BR. Genome-wide dynamics of Htz1, a histone H2A variant that poises repressed/basal promoters for activation through histone loss. *Cell* 2005;**123**:219–31.
86. Raisner RM, Hartley PD, Meneghini MD, Bao MZ, Liu CL, Schreiber SL, Rando OJ, Madhani HD. Histone variant H2A.Z marks the 5' ends of both active and inactive genes in euchromatin. *Cell* 2005;**123**:233–48.
87. Albert I, Mavrich TN, Tomsho LP, Qi J, Zanton SJ, Schuster SC, Pugh BF. Translational and rotational settings of H2A.Z nucleosomes across the Saccharomyces cerevisiae genome. *Nature* 2007;**446**:572–6.
88. Liu CL, Kaplan T, Kim M, Buratowski S, Schreiber SL, Friedman N, Rando OJ. Single-nucleosome mapping of histone modifications in Scerevisiae. *PLoS Biol* 2005;**3**:e328.
89. Keogh MC, Mennella TA, Sawa C, Berthelet S, Krogan NJ, Wolek A, Podolny V, Carpenter LR, Greenblatt JF, Baetz K, Buratowski S. The Saccharomyces cerevisiae histone H2A variant Htz1 is acetylated by NuA4. *Genes Dev* 2006;**20**:660–5.
90. Babiarz JE, Halley JE, Rine J. Telomeric heterochromatin boundaries require NuA4-dependent acetylation of histone variant H2A.Z in Saccharomyces cerevisiae. *Genes Dev* 2006;**20**:700–10.
91. Auger A, Galarneau L, Altaf M, Nourani A, Doyon Y, Utley RT, Cronier D, Allard S, Cote J. Eaf1 is the platform for NuA4 molecular assembly that evolutionarily links chromatin acetylation to ATP-dependent exchange of histone H2A variants. *Mol Cell Biol* 2008;**28**:2257–70.
92. Boyne 2nd MT, Pesavento JJ, Mizzen CA, Kelleher NL. Precise characterization of human histones in the H2A gene family by top down mass spectrometry. *J Proteome Res* 2006;**5**:248–53.
93. Beck HC, Nielsen EC, Matthiesen R, Jensen LH, Sehested M, Finn P, Grauslund M, Hansen AM, Jensen ON. Quantitative proteomic analysis of post-translational modifications of human histones. *Mol Cell Proteomics* 2006;**5**:1314–25.
94. Bonenfant D, Coulot M, Towbin H, Schindler P, van Oostrum J. Characterization of histone H2A and H2B variants and their post-translational modifications by mass spectrometry. *Mol Cell Proteomics* 2006;**5**:541–52.
95. Flaus A, Rencurel C, Ferreira H, Wiechens N, Owen-Hughes T. Sin mutations alter inherent nucleosome mobility. *Embo J* 2004;**23**:343–53.
96. Higashi T, Matsunaga S, Isobe K, Morimoto A, Shimada T, Kataoka S, Watanabe W, Uchiyama S, Itoh K, Fukui K. Histone H2A mobility is regulated by its tails and acetylation of core histone tails. *Biochem Biophys Res Commun* 2007;**357**:627–32.
97. Ren Q, Gorovsky MA. Histone H2A.Z acetylation modulates an essential charge patch. *Mol Cell* 2001;**7**:1329–35.
98. Kobor MS, Venkatasubrahmanyam S, Meneghini MD, Gin JW, Jennings JL, Link AJ, Madhani HD, Rine J. A protein complex containing the conserved Swi2/Snf2-related ATPase Swr1p deposits histone variant H2A.Z into euchromatin. *PLoS Biol* 2004;**2**:E131.
99. Wang H, Wang L, Erdjument-Bromage H, Vidal M, Tempst P, Jones RS, Zhang Y. Role of histone H2A ubiquitination in Polycomb silencing. *Nature* 2004;**431**:873–8.
100. Baarends WM, Wassenaar E, van der Laan R, Hoogerbrugge J, Sleddens-Linkels E, Hoeijmakers JH, de Boer P, Grootegoed JA. Silencing of unpaired chromatin and histone H2A ubiquitination in mammalian meiosis. *Mol Cell Biol* 2005;**25**:1041–53.
101. Moore SC, Jason L, Ausió J. The elusive structural role of ubiquitinated histones. *Biochem Cell Biol* 2002;**80**:311–9.
102. Fan JY, Zhou J, Tremethick DJ. Quantitative analysis of HP1alpha binding to nucleosomal arrays. *Methods* 2007;**41**:286–90.

103. Gevry N, Chan HM, Laflamme L, Livingston DM, Gaudreau L. transcription is regulated by differential localization of histone H2A. Z. *Genes Dev* 2007;**21**:1869–81.
104. Pehrson JR, Fried VA. MacroH2A, a core histone containing a large nonhistone region. *Science* 1992;**257**:1398–400.
105. Kustatscher G, Hothorn M, Pugieux C, Scheffzek K, Ladurner AG. Splicing regulates NAD metabolite binding to histone macroH2A. *Nat Struct Mol Biol* 2005;**12**:624–5.
106. Pehrson JR, Fuji RN. Evolutionary conservation of histone macroH2A subtypes and domains. *Nucleic Acids Res* 1998;**26**:2837–42.
107. Chu F, Nusinow DA, Chalkley RJ, Plath K, Panning B, Burlingame AL. Mapping post-translational modifications of the histone variant MacroH2A1 using tandem mass spectrometry. *Mol Cell Proteomics* 2006;**5**:194–203.
108. Costanzi C, Pehrson JR. MACROH2A2, a new member of the MARCOH2A core histone family. *J Biol Chem* 2001;**276**: 21,776–4.
109. Changolkar LN, Pehrson JR. macroH2A1 histone variants are depleted on active genes but concentrated on the inactive X chromosome. *Mol Cell Biol* 2006;**26**:4410–20.
110. Hoyer-Fender S, Costanzi C, Pehrson JR. Histone macroH2A1.2 is concentrated in the XY-body by the early pachytene stage of spermatogenesis. *Exp Cell Res* 2000;**258**:254–60.
111. Richler C, Dhara SK, Wahrman J. Histone macroH2A1.2 is concentrated in the XY compartment of mammalian male meiotic nuclei. *Cytogenet Cell Genet* 2000;**89**:118–20.
112. Abbott DW, Laszczak M, Lewis JD, Su H, Moore SC, Hills M, Dimitrov S, Ausió J. Structural characterization of macroH2A containing chromatin. *Biochemistry* 2004;**43**:1352–9.
113. Zhang R, Poustovoitov MV, Ye X, Santos HA, Chen W, Daganzo SM, Erzberger JP, Serebriiskii IG, Canutescu AA, Dunbrack RL, Pehrson JR, Berger JM, Kaufman PD, Adams PD. Formation of MacroH2A-containing senescence-associated heterochromatin foci and senescence driven by ASF1a and HIRA. *Dev Cell* 2005;**8**:19–30.
114. Choo JH, Kim JD, Chung JH, Stubbs L, Kim J. Allele-specific deposition of macroH2A1 in imprinting control regions. *Hum Mol Genet* 2006;**15**:717–24.
115. Angelov D, Molla A, Perche PY, Hans F, Cote J, Khochbin S, Bouvet P, Dimitrov S. The histone variant macroH2A interferes with transcription factor binding and SWI/SNF nucleosome remodeling. *Mol Cell* 2003;**11**:1033–41.
116. Doyen CM, An W, Angelov D, Bondarenko V, Mietton F, Studitsky VM, Hamiche A, Roeder RG, Bouvet P, Dimitrov S. Mechanism of polymerase II transcription repression by the histone variant macroH2A. *Mol Cell Biol* 2006;**26**:1156–64.
117. Hernandez-Munoz I, Lund AH, van der Stoop P, Boutsma E, Muijrers I, Verhoeven E, Nusinow DA, Panning B, Marahrens Y, van Lohuizen M. Stable X chromosome inactivation involves the PRC1 Polycomb complex and requires histone MACROH2A1 and the CULLIN3/SPOP ubiquitin E3 ligase. *Proc Natl Acad Sci U S A* 2005;**102**:7635–40.
118. Nusinow DA, Hernandez-Munoz I, Fazzio TG, Shah GM, Kraus WL, Panning B. Poly(ADP-ribose) polymerase 1 is inhibited by a histone H2A variant, MacroH2A, and contributes to silencing of the inactive X chromosome. *J Biol Chem* 2007;**282**:12,851–9.
119. Ouararhni K, Hadj-Slimane R, Ait-Si-Ali S, Robin P, Mietton F, Harel-Bellan A, Dimitrov S, Hamiche A. The histone variant mH2A1.1 interferes with transcription by down-regulating PARP-1 enzymatic activity. *Genes Dev* 2006;**20**:3324–36.
120. Zalensky AO, Siino JS, Gineitis AA, Zalenskaya IA, Tomilin NV, Yau PM, Morton Bradbury E. Human Testis/Sperm-specific Histone H2B (hTSH2B) Molecular cloning and characterization. *J Biol Chem* 2002;**277**:43,474–80.
121. Li A, Maffey AH, Abbott WD, Conde e Silva N, Prunell A, Siino J, Churikov D, Zalensky AO, Ausió J. Characterization of nucleosomes consisting of the human testis/sperm-specific histone H2B variant (hTSH2B). *Biochemistry* 2005;**44**:2529–35.
122. Zweidler A. Histone Genes: Structure, Organization and Regulation. In: Stein GS, Stein JL, Marzluff WF, editors. *Histone genes: Structure, organization and regulation*. New York: Wiley and Sons; 1984. p. 339–71.
123. Nagaraja S, Delcuve GP, Davie JR. Differential compaction of transcriptionally competent and repressed chromatin reconstituted with histone H1 subtypes. *Biochim Biophys Acta* 1995;**1260**:207–14.
124. Poccia DL, Green GR. Packaging and unpackaging the sea urchin sperm genome. *Trends Biochem Sci* 1992;**17**:223–7.
125. Bavykin SG, Usachenko SI, Zalensky AO, Mirzabekov AD. Structure of nucleosomes and organization of internucleosomal DNA in chromatin. *J Mol Biol* 1990;**212**:495–511.
126. Mito Y, Henikoff JG, Henikoff S. Genome-scale profiling of histone H3.3 replacement patterns. *Nat Genet* 2005;**37**:1090–7.
127. Schwartz BE, Ahmad K. Transcriptional activation triggers deposition and removal of the histone variant H3.3. *Genes Dev* 2005; **19**:804–14.
128. Ng RK, Gurdon JB. Epigenetic memory of an active gene state depends on histone H3.3 incorporation into chromatin in the absence of transcription. *Nat Cell Biol* 2008;**10**:102–9.
129. Boudreault AA, Cronier D, Selleck W, Lacoste N, Utley RT, Allard S, Savard J, Lane WS, Tan S, Cote J. Yeast enhancer of polycomb defines global Esa1-dependent acetylation of chromatin. *Genes Dev* 2003;**17**:1415–28.
130. McKittrick E, Gafken PR, Ahmad K, Henikoff S. Histone H3.3 is enriched in covalent modifications associated with active chromatin. *Proc Natl Acad Sci U S A* 2004;**101**:1525–30.
131. Garcia-Ramirez M, Dong F, Ausió J. Role of the Histone "Tails" in the Folding of Oligonucleosomes Depleted of Histone H1. *J Biol Chem* 1992;**267**:19,587–95.
132. Loyola A, Bonaldi T, Roche D, Imhof A, Almouzni G. PTMs on H3 variants before chromatin assembly potentiate their final epigenetic state. *Mol Cell* 2006;**24**:309–16.
133. Hake SB, Allis CD. Histone H3 variants and their potential role in indexing mammalian genomes: The "H3 barcode hypothesis." *Proc Natl Acad Sci U S A* 2006;**103**:6428–35.
134. Tagami H, Ray-Gallet D, Almouzni G, Nakatani Y. Histone H3.1 and H3.3 complexes mediate nucleosome assembly pathways dependent or independent of DNA synthesis. *Cell* 2004;**116**:51–61.
135. Konev AY, Tribus M, Park SY, Podhraski V, Lim CY, Emelyanov AV, Vershilova E, Pirrotta V, Kadonaga JT, Lusser A, Fyodorov DV. CHD1 motor protein is required for deposition of histone variant H3.3 into chromatin in vivo. *Science (New York)* 2007;**317**:1087–90.
136. Galvani A, Courbeyrette R, Agez M, Ochsenbein F, Mann C, Thuret JY. In vivo study of the nucleosome assembly functions of ASF1 histone chaperones in human cells. *Mol Cell Biol* 2008;**28**:3672–85.
137. Loppin B, Bonnefoy E, Anselme C, Laurencon A, Karr TL, Couble P. The histone H3.3 chaperone HIRA is essential for chromatin assembly in the male pronucleus. *Nature* 2005;**437**:1386–90.
138. Torres-Padilla ME, Bannister AJ, Hurd PJ, Kouzarides T, Zernicka-Goetz M. Dynamic distribution of the replacement histone variant

H3.3 in the mouse oocyte and preimplantation embryos. *Int J Dev Biol* 2006;**50**:455–61.

139. Palmer DK, O'Day K, Trong HL, Charbonneau H, Margolis RL. Purification of the centromere-specific protein CENP-A and demonstration that it is a distinctive histone. *Proc Natl Acad Sci U S A* 1991;**88**:3734–8.
140. Palmer DK, O'Day K, Wener MH, Andrews BS, Margolis RL. A 17-kD centromere protein (CENP-A) copurifies with nucleosome core particles and with histones. *J Cell Biol* 1987;**104**:805–15.
141. Howman EV, Fowler KJ, Newson AJ, Redward S, MacDonald AC, Kalitsis P, Choo KH. Early disruption of centromeric chromatin organization in centromere protein A (Cenpa) null mice. *Proc Natl Acad Sci U S A* 2000;**97**:1148–53.
142. Sullivan KF. A solid foundation: functional specialization of centromeric chromatin. *Curr Opin Genet Dev* 2001;**11**:182–8.
143. Bird AW, Yu DY, Pray-Grant MG, Qiu Q, Harmon KE, Megee PC, Grant PA, Smith MM, Christman MF. Acetylation of histone H4 by Esa1 is required for DNA double-strand repair. *Nature* 2002;**419**:411–15.
144. Fukagawa T. Centromere DNA, proteins and kinetochore assembly in vertebrate cells. *Chromosome Res* 2004;**12**:557–67.
145. Sullivan KF, Hechenberger M, Masri K. Human CENP-A contains a histone H3 related histone fold domain that is required for targeting to the centromere. *J Cell Biol* 1994;**127**:581–92.
146. Cooper JL, Henikoff S. Adaptive evolution of the histone fold domain in centromeric histones. *Mol Biol Evol* 2004;**21**:1712–18.
147. Dalal Y, Wang H, Lindsay S, Henikoff S. Tetrameric structure of centromeric nucleosomes in interphase Drosophila cells. *PLoS Biol* 2007;**5**:e218.
148. Lavelle C, Prunell A. Chromatin polymorphism and the nucleosome superfamily: a genealogy. *Cell Cycle (Georgetown)* 2007;**6**:2113–9.
149. Weintraub H, Worcel A, Alberts B. A model for chromatin based upon two symmetrically paired half-nucleosomes. *Cell* 1976;**9**:409–17.
150. Oudet P, Germond JE, Bellard M, Spadafora C, Chambon P. Nucleosome structure. *Philos Trans R Soc Lond* 1978;**283**:241–58.
151. Lee MS, Garrard WT. Transcription-induced nucleosome "splitting": an underlying structure for DNase I sensitive chromatin. *EMBO J* 1991;**10**:607–15.
152. Yoda K, Ando S, Morishita S, Houmura K, Hashimoto K, Takeyasu K, Okazaki T. Human centromere protein A (CENP-A) can replace histone H3 in nucleosome reconstitution in vitro. *Proc Natl Acad Sci U S A* 2000;**97**:7266–71.
153. Zimmerman ES, Chen J, Andersen JL, Ardon O, DeHart JL, Blackett J, Choudhary SK, Camerini D, Nghiem P, Planelles V. Human immunodeficiency virus type 1 Vpr-mediated G2 arrest requires Rad17 and Hus1 and induces nuclear BRCA1 and γ-H2AX focus Formation. *Mol Cell Biol* 2004;**24**:9286–94.
154. Dalal Y, Furuyama T, Vermaak D, Henikoff S. Structure, dynamics, and evolution of centromeric nucleosomes. *Proc Natl Acad Sci U S A* 2007;**104**:15,974–81.
155. Ando S, Yang H, Nozaki N, Okazaki T, Yoda K. CENP-A, -B, and -C chromatin complex that contains the I-type alpha-satellite array constitutes the prekinetochore in HeLa cells. *Mol Cell Biol* 2002;**22**:2229–41.
156. Amor DJ, Kalitsis P, Sumer H, Choo KH. Building the centromere: from foundation proteins to 3D organization. *Trends Cell Biol* 2004;**14**:359–68.
157. Sugata N, Li S, Earnshaw WC, Yen TJ, Yoda K, Masumoto H, Munekata E, Warburton PE, Todokoro K. Human CENP-H multimers colocalize with CENP-A and CENP-C at active centromere: kinetochore complexes. *Hum Mol Genet* 2000;**9**:2919–26.
158. Durand-Dubief M, Ekwall K. Heterochromatin tells CENP-A where to go. *Bioessays* 2008;**30**:526–9.
159. Folco HD, Pidoux AL, Urano T, Allshire RC. Heterochromatin and RNAi are required to establish CENP-A chromatin at centromeres. *Science (New York)* 2008;**319**:94–7.
160. Dunleavy EM, Pidoux AL, Monet M, Bonilla C, Richardson W, Hamilton GL, Ekwall K, McLaughlin PJ, Allshire RC. A NASP (N1/N2)-related protein, Sim3, binds CENP-A and is required for its deposition at fission yeast centromeres. *Mol Cell* 2007;**28**:1029–44.
161. Finn RM, Browne K, Hodgson KC, Ausió J. sNASP, a histone H1-specific eukaryotic chaperone dimer that facilitates chromatin assembly. *Biophys J* 2008;**95**:1314–25.
162. O'Rand MG, Richardson RT, Zimmerman LJ, Widgren EE. Sequence and localization of human NASP: conservation of a Xenopus histone-binding protein. *Dev Biol* 1992;**154**:37–44.
163. Richardson RT, Batova IN, Widgren EE, Zheng LX, Whitfield M, Marzluff WF, O'Rand MG. Characterization of the histone H1-binding protein, NASP, as a cell cycle-regulated somatic protein. *J Biol Chem* 2000;**275**:30,378–86.
164. Song JS, Liu X, Liu XS, He X. A high-resolution map of nucleosome positioning on a fission yeast centromere. *Genome Res* 2008;**18**:1064–72.
165. Bustin M, Catez F, Lim JH. The dynamics of histone H1 function in chromatin. *Mol Cell* 2005;**17**:617–20.
166. Albig W, Meergans T, Doenecke D. Characterization of the H1.5 gene completes the set of human H1 subtype genes. *Gene* 1997;**184**:141–8.
167. Happel N, Schulze E, Doenecke D. Characterisation of human histone H1x. *Biol Chem* 2005;**386**:541–51.
168. Parseghian MH, Henschen AH, Krieglstein KG, Hamkalo BA. A proposal for a coherent mammalian histone H1 nomenclature correlated with amino acid sequences. *Protein Sci* 1994;**3**:575–87.
169. Seyedin SM, Kistler WS. Isolation and characterization of rat testis H1t. An H1 histone variant associated with spermatogenesis. *J Biol Chem* 1980;**255**:5949–54.
170. Martianov I, Brancorsini S, Catena R, Gansmuller A, Kotaja N, Parvinen M, Sassone-Corsi P, Davidson I. Polar nuclear localization of H1T2, a histone H1 variant, required for spermatid elongation and DNA condensation during spermiogenesis. *Proc Natl Acad Sci U S A* 2005;**102**:2808–13.
171. Coles LS, Robins AJ, Madley LK, Wells JR. Characterization of the chicken histone H1 gene complement. Generation of a complete set of vertebrate H1 protein sequences. *J Biol Chem* 1987;**262**:9656–63.
172. Neelin JM, Callahan PX, Lamb DC, Murray K. The Histones Of Chicken Erythrocyte Nuclei. *Can J Biochem Physiol* 1964;**42**:1743–52.
173. Smith RC, Dworkin-Rastl E, Dworkin MB. Expression of a histone H1-like protein is restricted to early *Xenopus* development. *Genes Dev* 1988;**2**:1284–95.
174. Przewloka MR, Wierzbicki AT, Slusarczyk J, Kuras M, Grasser KD, Stemmer C, Jerzmanowski A. The "drought-inducible" histone H1s of tobacco play no role in male sterility linked to alterations in H1 variants. *Planta* 2002;**215**:371–9.
175. Kosterin OE, Bogdanova VS, Gorel FL, Rozov SM, Trusov YA, Berdnikov VA. Histone H1 of the garden pea (*Pisum sativum* L.): composition, developmental changes, allelic polymorphism and inheritance. *Plant Science* 1994;**101**:189–202.

176. Kasinsky HE, Lewis JD, Dacks JB, Ausió J. Origin of H1 linker histones. *FASEB J* 2001;**15**:34–42.
177. Nagel S, Grossbach U. Histone H1 genes and histone gene clusters in the genus Drosophila. *J Mol Evol* 2000;**51**:286–98.
178. Sanicola M, Ward S, Childs G, Emmons SW. Identification of a Caenorhabditis elegans histone H1 gene family. Characterization of a family member containing an intron and encoding a poly(A)+ mRNA. *J Mol Biol* 1990;**212**:259–68.
179. Patterton HG, Landel CC, Landsman D, Peterson CL, Simpson RT. The biochemical and phenotypic characterization of Hho1p, the putative linker histone H1 of Saccharomyces cerevisiae. *J Biol Chem* 1998;**273**:7268–76.
180. Eirín-López JM, Gonzalez-Tizon AM, Martinez A, Mendez J. Birth-and-death evolution with strong purifying selection in the histone H1 multigene family and the origin of orphon H1 genes. *Mol Biol Evol* 2004;**21**:1992–2003.
181. Li A, Eirín-López JM, Ausió J. H2AX: tailoring histone H2A for chromatin-dependent genomic integrity. *Biochem Cell Biol* 2005; **83**:505–15.
182. Ponte I, Vidal-Taboada JM, Suau P. Evolution of the vertebrate H1 histone class: evidence for the functional differentiation of the subtypes. *Mol Biol Evol* 1998;**15**:702–8.
183. Eirín-López JM, Frehlick LJ, Ausió J. Protamines, in the footsteps of linker histone evolution. *J Biol Chem* 2006;**281**:1–4.
184. Saperas N, Chiva M, Casas MT, Campos JL, Eirín-López JM, Frehlick LJ, Prieto C, Subirana JA, Ausió J. A unique vertebrate histone H1-related protamine-like protein results in an unusual sperm chromatin organization. *FEBS J* 2006;**273**:4548–61.
185. Watson CE, Davies PL. The high molecular weight chromatin proteins of winter flounder sperm are related to an extreme histone H1 variant. *J Biol Chem* 1998;**273**:6157–62.
186. Frehlick LJ, Prado A, Calestagne-Morelli A, Ausió J. Characterization of the PL-I-Related SP2 Protein from Xenopus. *Biochemistry* 2007;**46**:12,700–8.
187. Yamamoto T, Horikoshi M. Cloning of the cDNA encoding a novel subtype of histone H1. *Gene* 1996;**173**:281–5.
188. Yamaguchi-Iwai Y, Sonoda E, Sasaki MS, Morrison C, Haraguchi T, Hiraoka Y, Yamashita YM, Yagi T, Takata M, Price C, Kakazu N, Takeda S. Mre11 is essential for the maintenance of chromosomal DNA in vertebrate cells. *EMBO J* 1999;**18**:6619–29.
189. Cole RD. A minireview of microheterogeneity in H1 histone and its possible significance. *Anal Biochem* 1984;**136**:24–30.
190. Khochbin S, Wolffe AP. Developmentally regulated expression of linker-histone variants in vertebrates. *Eur J Biochem/FEBS* 1994;**225**:501–10.
191. Saeki H, Ohsumi K, Aihara H, Ito T, Hirose S, Ura K, Kaneda Y. Linker histone variants control chromatin dynamics during early embryogenesis. *Proc Natl Acad Sci U S A* 2005;**102**:5697–702.
192. Izzo A, Kamieniarz K, Schneider R. The histone H1 family: specific members, specific functions? *Biol Chem* 2008;**389**:333–43.
193. Lin Q, Sirotkin A, Skoultchi AI. Normal spermatogenesis in mice lacking the testis-specific linker histone H1t. *Mol Cell Biol* 2000;**20**:2122–8.
194. Drabent B, Saftig P, Bode C, Doenecke D. Spermatogenesis proceeds normally in mice without linker histone H1t. *Histochem Cell Biol* 2000;**113**:433–42.
195. Fantz DA, Hatfield WR, Horvath G, Kistler MK, Kistler WS. Mice with a targeted disruption of the H1t gene are fertile and undergo normal changes in structural chromosomal proteins during spermiogenesis. *Biol Reprod* 2001;**64**:425–31.
196. Ausió J. Are linker histones (histone H1) dispensable for survival? *Bioessays* 2000;**22**:873–7.
197. Nayernia K, Drabent B, Meinhardt A, Adham IM, Schwandt I, Muller C, Sancken U, Kleene KC, Engel W. Triple knockouts reveal gene interactions affecting fertility of male mice. *Mol Reprod Dev* 2005;**70**:406–16.
198. Parseghian MH, Newcomb RL, Hamkalo BA. Distribution of somatic H1 subtypes is non-random on active vs. inactive chromatin II: distribution in human adult fibroblasts. *J Cell Biochem* 2001;**83**:643–59.
199. Parseghian MH, Newcomb RL, Winokur ST, Hamkalo BA. The distribution of somatic H1 subtypes is non-random on active vs. inactive chromatin: distribution in human fetal fibroblasts. *Chromosome Res* 2000;**8**:405–24.
200. Th'ng JP, Sung R, Ye M, Hendzel MJ. H1 family histones in the nucleus. Control of binding and localization by the C-terminal domain. *J Biol Chem* 2005;**280**:27,809–14.
201. Orrego M, Ponte I, Roque A, Buschati N, Mora X, Suau P. Differential affinity of mammalian histone H1 somatic subtypes for DNA and chromatin. *BMC Biology* 2007;**5**:22.
202. Konishi A, Shimizu S, Hirota J, Takao T, Fan Y, Matsuoka Y, Zhang L, Yoneda Y, Fujii Y, Skoultchi AI, Tsujimoto Y. Involvement of histone H1.2 in apoptosis induced by DNA double-strand breaks. *Cell* 2003;**114**:673–88.
203. Wang X, Peng Y, Ma Y, Jahroudi N. Histone H1-like protein participates in endothelial cell-specific activation of the von Willebrand factor promoter. *Blood* 2004;**104**:1725–32.
204. Alami R, Fan Y, Pack S, Sonbuchner TM, Besse A, Lin Q, Greally JM, Skoultchi AI, Bouhassira EE. Mammalian linker-histone subtypes differentially affect gene expression in vivo. *Proc Natl Acad Sci U S A* 2003;**100**:5920–5.
205. Shannon MF, Wells JR. Characterization of the six chicken histone H1 proteins and alignment with their respective genes. *J Biol Chem* 1987;**262**:9664–8.
206. Hashimoto H, Sonoda E, Takami Y, Kimura H, Nakayama T, Tachibana M, Takeda S, Shinkai Y. Histone H1 variant, H1R is involved in DNA damage response. *DNA Repair* 2007;**6**:1584–95.
207. Lever MA, Th'ng JP, Sun X, Hendzel MJ. Rapid exchange of histone H1.1 on chromatin in living human cells. *Nature* 2000;**408**:873–6.
208. Poirier MG, Bussiek M, Langowski J, Widom J. Spontaneous access to DNA target sites in folded chromatin fibers. *J Mol Biol* 2008;**379**:772–86.
209. Godde JS, Ura K. Cracking the enigmatic linker histone code. *J Biochem* 2008;**143**:287–93.
210. Contreras A, Hale TK, Stenoien DL, Rosen JM, Mancini MA, Herrera RE. The dynamic mobility of histone H1 is regulated by cyclin/CDK phosphorylation. *Mol Cell Biol* 2003;**23**:8626–36.
211. Horn PJ, Carruthers LM, Logie C, Hill DA, Solomon MJ, Wade PA, Imbalzano AN, Hansen JC, Peterson CL. Phosphorylation of linker histones regulates ATP-dependent chromatin remodeling enzymes. *Nat Struct Biol* 2002;**9**:263–7.
212. Roque A, Ponte I, Arrondo JL, Suau P. Phosphorylation of the carboxy-terminal domain of histone H1: effects on secondary structure and DNA condensation. *Nucleic Acids Res* 2008;**36**:4719–26.
213. Rose KL, Li A, Zalenskaya I, Zhang Y, Unni E, Hodgson KC, Yu Y, Shabanowitz J, Meistrich ML, Hunt DF, Ausió J. C-Terminal phosphorylation of murine testis-specific histone h1t in elongating spermatids. *J Proteome Res* 2008;**7**:4070–8.
214. Muyrers-Chen I, Hernandez-Munoz I, Lund AH, Valk-Lingbeek ME, van der Stoop P, Boutsma E, Tolhuis B, Bruggeman SW, Taghavi P,

Verhoeven E, Hulsman D, Noback S, Tanger E, Theunissen H, van Lohuizen M. Emerging roles of Polycomb silencing in X-inactivation and stem cell maintenance. *Cold Spring Harb Symp Quant Biol* 2004; **69**:319–26.

215. Schwede T, Kopp J, Guex N, Peitsch MC. SWISS-MODEL: An automated protein homology-modeling server. *Nucleic Acids Res* 2003;**31**:3381–5.

216. DeLano WL. The PyMOL Molecular Graphics System. San Carlos, CA: DeLano Scientific; 2002.

217. Wisniewski JR, Zougman A, Kruger S, Mann M. Mass spectrometric mapping of linker histone H1 variants reveals multiple acetylations, methylations, and phosphorylation as well as differences between cell culture and tissue. *Mol Cell Proteomics* 2007;**6**:72–87.

Chapter 18

Histone Ubiquitination

Vikki M. Weake and Jerry L. Workman
Stowers Institute for Medical Research, Kansas City, Missouri

THE MECHANISM OF UBIQUITINATION

Ubiquitination of target proteins involves a three-step enzymatic process (reviewed in [1]). The 76 amino acid protein ubiquitin is first activated by a ubiquitin activating enzyme (E1) in an ATP dependent process. An E2 ubiquitin conjugating enzyme then transfers activated ubiquitin via a thioester bond to a cysteine residue in its active site. Finally, a ubiquitin protein isopeptide ligase (E3) transfers the ubiquitin moiety from the E2 enzyme to a lysine residue in the target protein. Although E2 and E3 enzymes are often associated with each other within a protein complex, target specificity is largely regulated by the E3 ubiquitin ligase. These E3 ubiquitin ligases are characterized by the presence of either a Zn binding RING (*r*eally *i*nteresting *n*ew *g*ene) C3HC4 finger motif, or a HECT (*h*omologous to the *E*6AP *c*arboxyl *t*erminus) domain. Ubiquitination functions as a versatile signaling mark because substrates can be modified either by the addition of a single ubiquitin molecule (mono-ubiquitination), or by the conjugation of ubiquitin to preceding ubiquitin moieties (poly-ubiquitination). Whereas poly-ubiquitination through linkage of ubiquitin at Lys-48 regulates protein stability and targets substrate proteins for degradation via the 26S proteasome, mono-ubiquitination tends to act as a signaling mark and controls processes ranging from membrane transport to transcription [2]. Consistent with its role in signaling processes, mono-ubiquitination is reversible and the removal of ubiquitin from target proteins is catalyzed by ubiquitin specific proteases (UBPs in yeast; USPs in mammals) [3].

HISTONE UBIQUITINATION

The histone proteins within chromatin were found to be ubiquitinated over 25 years ago [4–6]. A unique histone-like chromosomal protein named A24 was isolated in these original studies that was soon identified as mono-ubiquitinated H2A [5–8]. H2A is mono-ubiquitinated at Lys-119 (ubH2A) in higher eukaryotes, although ubH2A is not detected in *Saccharomyces cerevisiae* [9–11]. Following the discovery of ubH2A, H2B was also found to be mono-ubiquitinated at Lys-120 (ubH2B) in mouse cells [4, 12]. This ubiquitinated residue corresponds to Lys-123 in *Saccharomyces cerevisiae*, Lys-119 in *S. pombe*, and Lys-143 in *Arabidopsis* [13–15].

MONO-UBIQUITINATION OF H2A

H2A ubiquitination is generally associated with transcriptional silencing and is catalyzed by two separate E3 ubiquitin ligases in humans: Ring1B (Ring2/Rnf2) and 2A-HUB [16–18]. These are likely to be the major H2A ubiquitin ligases in human cells, as knockdown of Ring1B largely reduces the level of ubH2A [18]. Both of these enzymes reside within large multi-subunit protein complexes. Ring1B (dRing/*Sex combs extra* in *Drosophila*; Table 18.1) is associated with three different repressive complexes: the Polycomb repressive complex 1 (PRC1), E2F-6.com-1, and the FBXL10-BcoR complex. These complexes include some shared components such as Ring1A, Bmi-1, and YAF2 (Figure 18.1) but act to repress different sets of target genes. The E2F-6.com-1 complex is involved in the silencing of E2F and Myc responsive genes in quiescent cells [19, 20], while the FBXL10-BcoR complex represses BCL-6 target genes and contains an H3 Lys-4 demethylase, KDM2B, that is involved in the repression of ribosomal RNA genes [20–23]. PRC1 represses *Hox* gene expression and has roles in X inactivation: ubH2A and PRC1 co-localize on the inactive X chromosome (Xi) in mouse, and knockdown of Ring1A and Ring1B depletes ubH2A from Xi [16, 18, 24–26]. Ring1B activity is likely to be functional within at least the PRC1 and FBXL10-BcoR complexes, as additional subunits within these complexes such as Bmi-1 and NSPC1 stimulate its H2A

Handbook of Cell Signaling, Three-Volume Set 2 ed.

TABLE 18.1 Enzymes involved in H2A and H2B ubiquitination/de-ubiquitination in different organisms[1]

	H2B ubiquitination		H2B de-ubiquitination		H2A ubiquitination		H2A de-ubiquitination
	E2	E3	transcription	Silencing	E2	E3	
S. cerevisiae	Rad6	Bre1	Ubp8	Ubp10 (Dot4)	–	–	–
S. pombe	Rhp6	Brl1 (Rfp2/ Spcc1919.15)			?		
		Brl2 (Rfp1/ Spcc970.10c)					
Drosophila	Dhr6[2]	Bre1 (CG10542)[3]	Nonstop	USP7		dRing (Sce)	
Mouse	mHR6A/ mHR6B						
Human	hHR6A/ hHR6B	RNF20	USP22			Ring1B (RING2/ RNF2)	Ubp-M (USP16)
	UbcH6 ?		USP3 ?			2A-HUB (hRUL138)	2A-DUB (MYSM1)
	Mdm2 ?						USP21
							USP3 ? USP22?
Arabidopsis		HUB1[4]		SUP32 (UBP26)			

[1]*Table modified from [121]*
[2]*See [64]*
[3]*See [89]*
[4]*See [122, 123]*

ubiquitination activity [16, 20, 25, 27]. Notably all three of the complexes have additional subunits that contain RING domains and could act as potential E3 ubiquitin ligases: Ring1A and Bmi-1. However, only Ring1B possesses *in vitro* E3 ubiquitin ligase activity specific for H2A [16, 18]. Nevertheless, Ring1A may be able to functionally substitute for Ring1B in some contexts such as in X inactivation in mice [26].

A second H2A specific ubiquitin ligase has also been identified in mammalian cells: 2A-HUB/hRUL138 [17]. 2A-HUB resides within the N-CoR/HDAC1/3 complex and represses transcription at a subset of chemokine gene promoters via inhibition of RNA polymerase II elongation. Several other E3 ubiquitin ligases show activity specific for H2A *in vitro*, although whether this is functional within a chromatin context still remains to be determined. These potential H2A ubiquitin ligases include the HECT domain protein LASU1, which was originally isolated as a testis-specific ubiquitin ligase activity that could ubiquitinate H2A *in vitro* in the presence of the UBC4-1 and UBC4-testis specific E2 conjugases [28, 29]. Despite the identification of several E3 ubiquitin ligases specific for H2A, the E2 enzyme responsible for H2A ubiquitination has yet to be identified. It is plausible that several different E2 conjugases may be able to fulfill this role *in vivo* as UbcH5a, b, c, and UbcH6 are all able to complement Ring1B H2A ubiquitination activity *in vitro* [27, 30].

UBIQUITINATION OF HISTONE H2A VARIANTS

In addition to its role in H2A ubiquitination, Ring1B also catalyzes mono-ubiquitination of the histone variant H2A.Z (ubH2A.Z) at Lys-120 or Lys-121 in mammalian cells [31]. As observed previously for ubH2A, ubH2A.Z is enriched on the inactive X chromosome (Xi) in female cells, while unmodified H2A.Z is excluded from Xi [31].

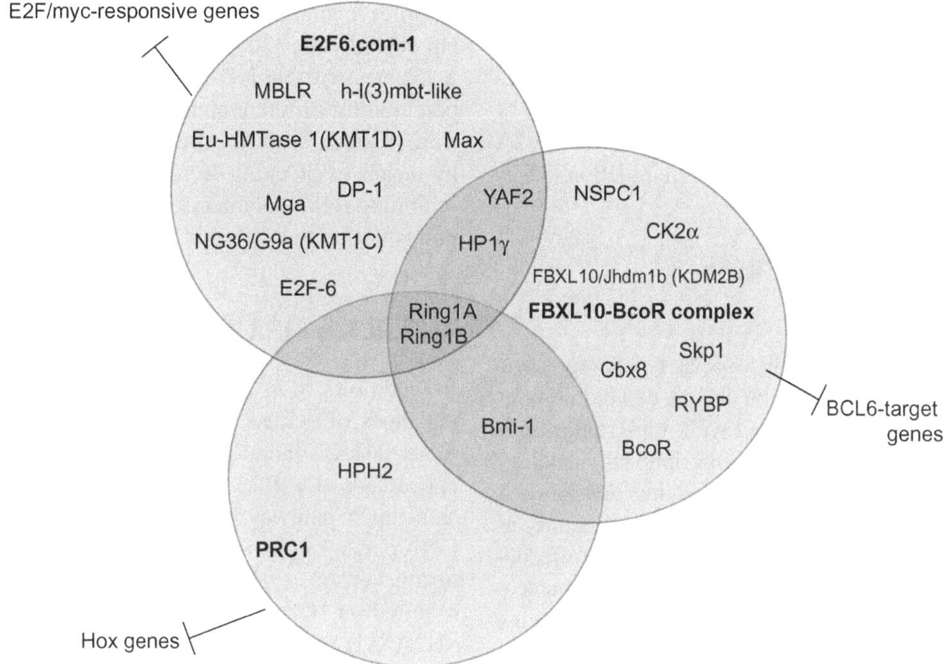

FIGURE 18.1 Shared components between the three repressive complexes containing the H2A ubiquitin ligase Ring1B.
The H2A ubiquitin ligase Ring1B is a component of three different repressive complexes: PRC1 localizes to trimethylated Lys-27 H3 and represses *Hox* gene transcription; E2F6.com-1 methylates Lys-9 H3 and binds E2F- and myc-response elements; the Fbxl10-BcoR complex contains the Lys-4 H3 demethylase KDM2B and represses BCL6-target genes.

Ubiquitination of H2A.Z is not required for its deposition on Xi, but may be a consequence of PRC1 activity on this chromosome [31]. Mass spectrometry analysis indicates that another histone variant, macroH2A1.2, is also mono-ubiquitinated at Lys-115 but the ubiquitin ligase responsible for this modification and its potential biological role have not been characterized [32].

DE-UBIQUITINATION OF UBH2A

De-ubiquitination of ubH2A is mediated by at least three different ubiquitin specific proteases: Ubp-M (USP16), 2A-DUB (KIAA1915/MYSM1), and USP21 [33–35] (reviewed in [36]). Some evidence also supports a role for USP3 and the ubH2B de-ubiquitinase USP22 in ubH2A de-ubiquitination *in vitro*, although the role of these in ubH2A de-ubiquitination in cells is not as well characterized [37–39].

Some early hints that Ubp-M might have a role in ubH2A de-ubiquitination came from studies showing that it possessed ubH2A de-ubiquitination activity *in vitro* on core histones, and that transient transfection of Ubp-M in human cells depleted ubH2A levels [40, 41]. Later work showed that Ubp-M could de-ubiquitinate nucleosomal ubH2A *in vitro* and that knockdown of Ubp-M increases ubH2A levels in cells, confirming its role in cells as an ubH2A de-ubiquitinase [33]. Ubp-M seems to act antagonistically to Ring1B within the PRC1 complex, as Ubp-M is required for expression of *Hox* genes [33]. Furthermore Ubp-M is also required for cell cycle progression and chromosome segregation during mitosis, and is essential for Aurora B mediated Ser-10 H3 phosphorylation [33]. These findings provide a potential explanation for the inverse correlation between Ser-10 H3 phosphorylation and ubH2A levels during the cell cycle; ubH2A is absent from isolated metaphase chromosomes [33, 42–44]. How is the cyclical activity of Ubp-M then regulated? Some hints come from the observation that Ubp-M is sequentially phosphorylated and dephosphorylated during the cell cycle, potentially by the cdc-2/cyclin B complex, which can phosphorylate Ubp-M *in vitro* [40]. A catalytically inactive form of Ubp-M fails to dissociate from mitotic chromosomes and remains bound throughout metaphase and anaphase, indicating that de-ubiquitination of ubH2A might be required for the release of Ubp-M [40].

The second of the H2A de-ubiquitinases, 2A-DUB, seems to counteract the effect of the 2A-HUB ubiquitin ligase [34]. 2A-DUB contains a JAMN/MPN+ domain, which can catalyze the hydrolysis of ubiquitin chain isopeptide bonds, and knockdown of 2A-DUB by siRNA in cell culture increases the global levels of ubH2A [34]. In contrast to the 2A-HUB ubiquitin ligase, which associates with a histone deacetylase (HDAC) complex, 2A-DUB interacts with the histone acetyltransferase (HAT) p300/CBP associated factor (PCAF/KAT2B) and is required for gene activation at a subset of promoters [34]. Moreover 2A-DUB preferentially de-ubiquitinates hyperacetylated nucleosomes *in vitro*, indicating some crosstalk between the HAT and de-ubiquitination activities of this complex [34]. The third of the H2A de-ubiquitinases, USP21, de-ubiquitinates ubH2A *in vitro* and is also involved in transcription initiation [35].

HOW DOES UBH2A REPRESS TRANSCRIPTION?

Ubiquitination of H2A is generally associated with repression of transcription, whereas both of the ubH2A de-ubiquitinases Ubp-M and 2A-DUB are associated with transcription activation. How does ubH2A mediate transcriptional silencing? There is some structural evidence indicating that ubH2A may directly influence higher order chromatin structure, as ubH2A enhances binding of the linker histone H1 to reconstituted nucleosomes *in vitro* [45]. Conversely, mononucleosomes purified from K119R H2A lack histone H1, indicating that ubH2A de-ubiquitination might result in dissociation of linker histones from core nucleosomes [34]. These findings are consistent with the structure of the nucleosome in which the C-terminus of H2A appears to interact with linker histones [46]. However, it is clear that ubH2A also directly inhibits some histone modifications associated with gene activation, and prevents recruitment of factors required for transcription elongation. Specifically, ubH2A inhibits MLL3 mediated di- and trimethylation of Lys-4 H3, repressing transcription initiation, but not elongation, *in vitro* [35]. The histone de-ubiquitinase USP21 is able to relieve this repression *in vitro* [35]. Furthermore, 2A-HUB mediated H2A ubiquitination inhibits recruitment of the Spt16 subunit of FACT to a subset of chemokine gene promoters, repressing transcription elongation [17]. Knockdown of 2A-HUB reduces the level of ubH2A at target promoters, concurrent with an enhancement of FACT recruitment and RNA polymerase II CTD phosphorylation [17]. This indicates that ubH2A might repress transcription elongation by blocking FACT recruitment. Whether this repression of FACT recruitment is mediated directly by ubH2A, or indirectly via crosstalk with Lys-4 H3 methylation or incorporation of histone H1, remains to be determined.

H2A ubiquitination does not appear to directly influence other modifications involved in transcription silencing such as Lys-27 H3 or Lys-9 H3 methylation [18, 35]. Rather, ubH2A functions downstream of these modifications, as loss of Lys-27 H3 methylation via knockdown of SUZ12 reduces the presence of Bmi-1, Ring1B, and ubH2A at silenced promoters [16]. However, several of the H2A ubiquitin ligase containing complexes contain subunits that catalyze additional repressive chromatin modifications. The E2F6.com-1 complex contains two Lys-9 H3 methyltransferases: Eu-HMTase1 (KMT1D) and NG36/G9a (KMT1C) [19]. Additionally, the FBXL10-BcoR complex contains the FBXL10/Jhdm1b (KDM2B) Lys-4 H3 demethylase [20, 21], and the 2A-HUB ubiquitin ligase associates with an HDAC complex [17]. Thus there is the potential for different combinations of these catalytic activities to have specific effects on transcription repression, by means of blocking recruitment of particular factors and inhibiting RNA polymerase II transcription at specific time points during transcription initiation and elongation.

THE ROLE OF UBH2A IN DNA REPAIR

In addition to its roles in transcription silencing, several lines of evidence indicate a role for ubH2A in DNA repair: H2A is mono-ubiquitinated at DNA lesions, and the appearance of ubH2A requires a functional nucleotide excision repair pathway [47]. The onset of ubH2A at sites of DNA damage coincides with phosphorylation of the variant histone H2AX (γH2AX), but this phosphorylation is not required for H2A ubiquitination [47]. However, similarly to γH2AX, H2A ubiquitination induced by DNA damage also requires the DNA damage signaling kinase ATR (ATM and Rad3 related) [47]. Several of the previously characterized E3 ubiquitin ligases specific for ubH2A have been implicated in this DNA damage induced ubiquitination including Ring1B, DDB1-CUL4^{DDB2}, and RNF8 [47–49]. However it is unclear which of these ubiquitin ligases is directly responsible for DNA damage induced H2A ubiquitination, and there is the possibility that these may act redundantly or target additional proteins other than H2A. For example, the potential H2A ubiquitin ligase RNF8 is recruited to DNA lesions through interactions with phosphorylated MDC1, which itself binds directly to γH2AX [49]. Although RNF8 shows moderate poly-ubiquitination activity on H2A *in vitro*, and the accumulation of both conjugated ubiquitin and ubH2A at DNA lesions requires RNF8, it is possible that RNF8 may ubiquitinate other substrates in addition to, or instead of, H2A [49]. RNF8 forms a complex with an E2 conjugase, UBC13, which catalyzes ubiquitination of γH2AX [50–53]. Thus whether RNF8 directly ubiquitinates H2A or γH2AX, or acts on both of these histones at sites of DNA damage, remains difficult to determine. It is clear however that RNF8, UBC13, and ubiquitin conjugates are all essential for the recruitment of proteins required at later stages of the DNA damage response [49, 50, 52]. It remains to be determined how ubH2A or potentially ubiquitinated γH2AX are involved in the recruitment of these late stage DNA damage response proteins.

MONO-UBIQUITINATION OF H2B

In contrast to ubH2A, ubH2B has roles in both transcription activation and in silencing, and is present in the yeast

S. cerevisiae that lacks detectable ubH2A [10]. H2B ubiquitination in *S. cerevisiae* is catalyzed by the Bre1 E3 ubiquitin ligase, which acts in combination with the Rad6 E2 ubiquitin conjugase [13, 54, 55]. Although Rad6 catalyzes both mono- and poly-ubiquitination of H2A and H2B *in vitro* [56–58], analysis of *S. cerevisiae* co-expressing a FLAG tagged H2B and HA tagged ubiquitin (FLAG-H2B:UB-HA) confirmed that Rad6 specifically mono-ubiquitinates only Lys-123 of H2B *in vivo* [13]. Bre1 and Rad6 are associated within a complex in *S. cerevisiae* that contains an additional protein, Lge1 that is also required for H2B ubiquitination [54]. An analogous complex (HULC) is present in *S. pombe* containing the Rad6 homolog Rhp6, the Bre1 homologs Brl1 (Rfp2/Spcc1919.15) and Brl2 (Rfp1/Spcc970.10c), and an addition protein Shf1 (Table 18.1; [14, 59]). In addition to its role in H2B ubiquitination, Rad6 interacts with other E3 ubiquitin ligases, such as Rad18, Ubr1, and Rad5, and has ubH2B independent functions in DNA damage repair and protein degradation [60, 61].

Similarly to *S. pombe*, there are two potential sequence homologs of Bre1 in humans: RNF20 and RNF40 [62]. These form a complex *in vivo* although only RNF20 affects ubH2B levels in cells [62, 63]. The two homologs of Rad6 in mammals, HR6A and HR6B [64, 65], can functionally rescue Lys-4 H3 methylation in *rad6Δ* yeast [61]. However, another ubiquitin conjugase, UbcH6, can form a complex *in vitro* with RNF20/RNF40 that specifically catalyzes mono-ubiquitination of nucleosomal H2B [63]. It is not yet clear whether any of these E2 enzymes interact with RNF20/RNF40 *in vivo*, and HR6A and HR6B might function redundantly as *Hrb6b* knockout mice are viable with wild-type levels of ubH2B, although male sterile [65, 66].

Additional E3 ubiquitin ligases have been implicated in H2B ubiquitination in human cells, for example Mdm2 [67]. Mdm2 has previously been demonstrated to ubiquitinate p53 and some of its cofactors, negatively regulating p53 activity (reviewed in [68]). A potential role for Mdm2 in H2B ubiquitination is supported by evidence showing Mdm2 interacts with core histones, and that overexpression of Mdm2 increases ubH2B levels [67]. However, this ubiquitination may not be specific for Lys-120 of H2B, as mutation of both Lys-120 and Lys-125 of H2B is required to prevent the increase in ubH2B caused by Mdm2 overexpression [67]. Other ubiquitin ligases including BRCA1 are able to catalyze ubiquitination of both H2A and H2B *in vitro*, but the relevance of these activities *in vivo* remains unclear [69–71].

H2B UBIQUITINATION REQUIRES FACTORS INVOLVED IN TRANSCRIPTION INITIATION AND ELONGATION

The initial recruitment of Rad6 and Bre1 to promoters during gene activation is dependent upon the interaction of Bre1 with activators such as p53 in humans and Gal4 in yeast [54, 55, 72] (Figure 18.2a). This recruitment however, is not sufficient for H2B ubiquitination. Instead mono-ubiquitination of H2B by Rad6 requires interactions with components of the transcription elongation machinery such as the PAF complex (reviewed in [73, 74]) (Figure 18.2b). Mutations in PAF, which associates with initiating and elongating RNA polymerase II, result in a global loss of ubH2B but do not affect Rad6 recruitment to target promoters [75, 76]. Furthermore, mutations disrupting the Bur1/Bur2 cyclin dependent protein kinase complex (BUR complex) also reduce ubH2B [77, 78]. It is possible that the requirement for the BUR complex may be indirect, as this complex is also required for PAF recruitment. However, phosphorylation of both Rad6 (Ser-120) in yeast and hHR6A in humans, by the BUR complex and CDK2 respectively, appear to stimulate the ubiquitin conjugase activity of at least CDK2 *in vitro* [78, 79].

In addition to its role in transcription initiation, some evidence suggests that H2B ubiquitination is involved in, and indeed dependent upon, transcription elongation. For example, deposition of ubH2B on a chromatin template in transcription assays is dependent on the addition of NTPs [80]. Moreover Rad6 associates with the elongating form of RNA polymerase II, and this interaction is dependent on both Bre1 and the PAF complex [81]. Sequential phosphorylation events on the C-terminal heptapeptide repeat sequences (CTD) of RNA polymerase II occur during the initial steps of transcription elongation. The CTD is initially phosphorylated at Ser-5 by Kin28 (CDK7 in humans), followed by Ctk1 (P-TEFb/CDK9 in humans) mediated phosphorylation at Ser-2 (reviewed in [82]). Mutations in Kin28 result in a loss of Ser-5 phosphorylated RNA polymerase II and eliminate ubH2B, indicating that the initial phosphorylation of the CTD is critical for Rad6/Bre1 activity [81]. Loss of Ctk1 has no effect on ubH2B, indicating that this modification functions downstream of Rad6 and Bre1 activity [81]. In addition to being regulated by transcription elongation, there is some evidence that ubH2B itself might directly stimulate transcription elongation by RNA polymerase II by assisting the histone chaperone FACT to displace an H2A/H2B dimer from the nucleosome core *in vitro* [80, 83]. Supporting this, the Spt16 subunit of FACT and ubH2B regulate histone deposition at the *GAL1* gene *in vivo* following the passage of elongating RNA polymerase II and functionally interact to repress cryptic transcription initiation [84]. Moreover, there is a reduction in the level of RNA polymerase II at the 3' end of coding sequences in the *htb1-K119R* mutant in *S. pombe* [14]. Furthermore, chromatin immunoprecipitation experiments in human cells using an antibody specific for ubH2B show that this modification is enriched across the entire open reading frame of highly transcribed genes, although it appears to be excluded from promoter regions [85].

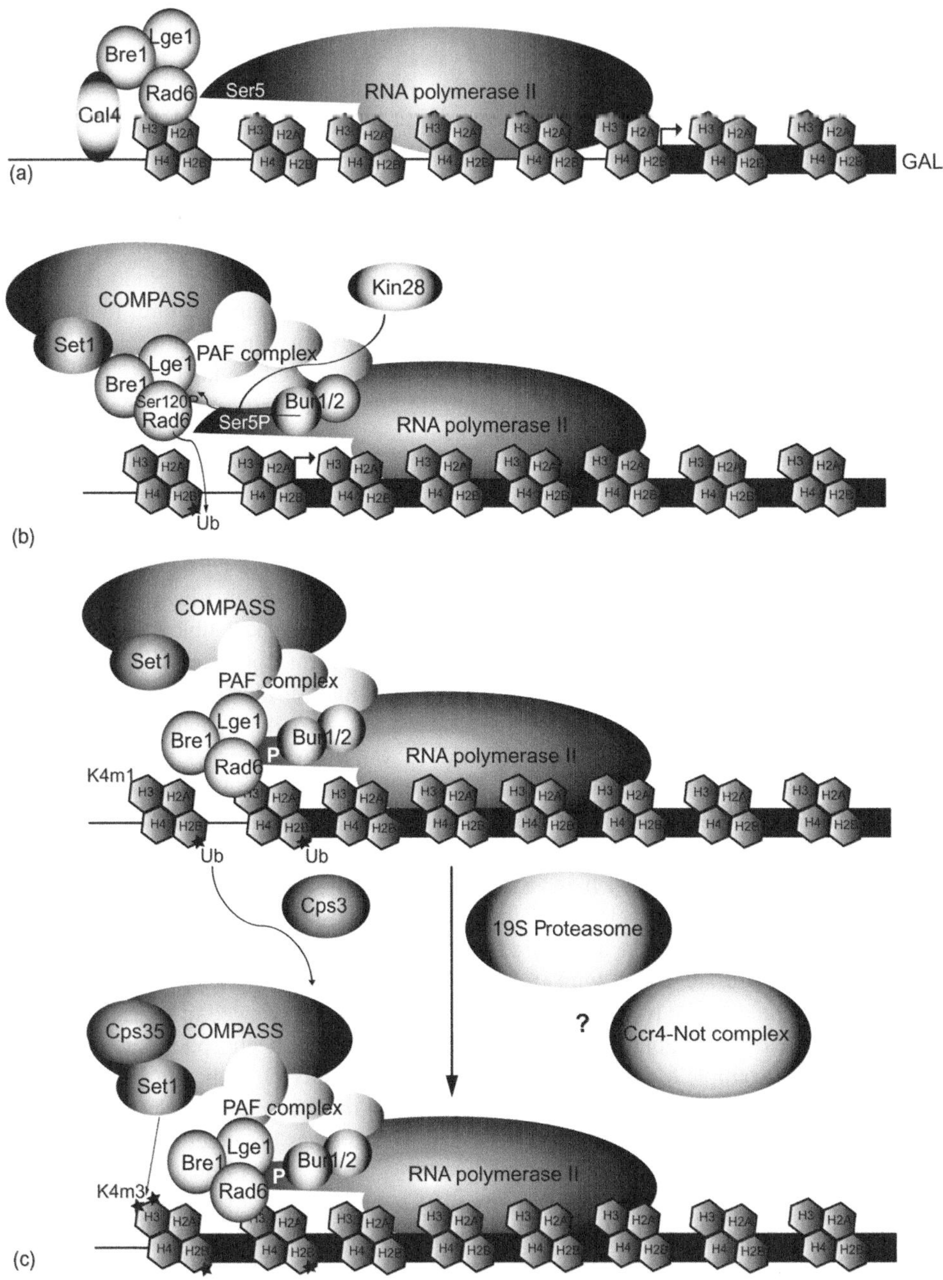

FIGURE 18.2 H2B ubiquitination requires early steps in transcription initiation and elongation. (a) The H2B ubiquitin ligase Bre1 interacts with acidic activators, such as Gal4, and recruits Rad6 and presumably its binding partner Lge1 to target promoters. This recruitment is not sufficient for H2B ubiquitination. (b) Mono-ubiquitination of H2B by Rad6/Bre1 requires interactions with the PAF complex, the BUR complex, and the elongating form of RNA polymerase II that has been phosphorylated on Ser-5 of the CTD by Kin28. The BUR complex phosphorylates Ser-120 of Rad6, which might stimulate its ubiquitin conjugase activity. (c) H2B ubiquitination is required for the recruitment of the Cps35 subunit of COMPASS, which activates di- and trimethylation of Lys-4 H3 by Set1. Cps35 recruitment is independent of COMPASS, indicating that this step is critical in linking H2B ubiquitination to trimethylation of Lys-4 H3. The 19S proteasome and Ccr4-Not complex also link H2B ubiquitination to Lys-4 H3 methylation by mechanisms that are less clear. This figure is reproduced from [121].

H2B UBIQUITINATION IS REQUIRED FOR PROCESSIVE LYS-4 H3 AND LYS-79 H3 METHYLATION

In contrast to H2A ubiquitination, which inhibits MLL mediated Lys-4 H3 methylation [35], Rad6/Bre1 mediated mono-ubiquitination of H2B is a prerequisite for methylation of both Lys-4 and Lys-79 of H3 [55, 61, 63, 76, 86–89]. Lys-79 H3 and Lys-4 H3 methylation are catalyzed by Dot1 and Set1 respectively in yeast [74, 90, 91]. Set1 is a component of the COMPASS complex, which is similar to the MLL complex in humans, and associates with the elongating form of RNA polymerase II (reviewed in [74]). Mutations affecting the enzymes responsible for H2B ubiquitination, or *htbK123R*, reduce global levels of methylated Lys-4 and Lys-79 H3. However, mutation of these methylated H3 residues has no reciprocal effect on ubH2B levels [61, 86].

More specifically, H2B ubiquitination is required for di- and tri-, but not mono-, methylation of Lys-4 and Lys-79 of H3 [92–94]. These effects can be differentiated further, as mutations disrupting the BUR complex reduce H2B ubiquitination but only affect Lys-4 H3 trimethylation [77]. How might mono-ubiquitination of H2B regulate processive methylation of H3? Recent studies using chemically ubiquitinated H2B indicate that nucleosomal ubH2B directly stimulates Lys-79 H3 methylation by human Dot1L (KMT4) [95]. The trimethylation activity of COMPASS is also regulated by ubH2B: H2B ubiquitination is required for incorporation of the Cps35 subunit of COMPASS into the complex, and this subunit is essential for di- and trimethylation activity *in vivo* [96] (Figure 18.2c). H2B ubiquitination is required for Cps35 recruitment to the *GAL1* promoter in yeast, which is reduced significantly in *rad6Δ* or *htbK123R* strains [96]. Furthermore, COMPASS purified from *rad6Δ* yeast lacks both the Cps35 subunit and the ability to di- and trimethylate Lys-4 H3 [96]. It remains to be determined how ubH2B regulates Cps35 recruitment and incorporation into COMPASS, but there are some hints that in yeast Cps35 might also have some involvement in Dot1 mediated Lys-79 H3 methylation [96].

THE 19S PROTEASOME AND THE CCR4-NOT COMPLEX LINK H2B UBIQUITINATION TO LYS-4 AND LYS-79 H3 METHYLATION

Several studies implicate the 19S regulatory complex of the proteasome and the Ccr4-Not complex in connecting H2B ubiquitination to Lys-4 and Lys-79 H3 methylation. The 19S regulatory complex of the proteasome has been shown to have roles in transcription independent of the 20S proteolytic complex [97]. This 19S regulatory complex is recruited to the *GAL1-10* promoter by the Gal4 activator in yeast [97] dependent upon Rad6 and H2B ubiquitination [98]. Temperature sensitive mutations in two of the ATPase subunits of the 19S proteasome, Rpt4 (Sug2) and Rpt6 (Sug1), reduce global levels of di- and trimethylated Lys-4 and Lys-79 H3 [98]. Intriguingly, the 19S proteasome has been shown to physically interact with components of SAGA and stimulate its activator dependent recruitment in an ATP dependent manner *in vitro* [99]. However, SAGA, which contains the H2B de-ubiquitinating enzyme Ubp8, is recruited independently of H2B ubiquitination [72, 100]. Thus it is unclear how the 19S proteasome links H2B ubiquitination and de-ubiquitination to Lys-4/Lys-79 H3 methylation.

The Ccr4-Not complex is involved in mRNA processing and degradation, and has roles in transcriptional repression (reviewed in [101]). Mutations in components of this complex including *not4*, *not5*, and the mRNA deadenylase *caf1*, specifically reduce trimethylated Lys-4 H3 without affecting global levels of any of the other H3 methylations including Lys-79 H3 [102, 103]. The Ccr4-Not complex interacts genetically and physically with components of the 19S regulatory complex of the proteasome and the BUR complex, indicating that these components may act together in linking H2B ubiquitination to Lys-4 H3 methylation [102, 103]. Intriguingly, the Not4 subunit of the Ccr4-Not complex contains a RING domain that may potentially act as an E3 ubiquitin ligase on as of yet uncharacterized substrate proteins [102].

DE-UBIQUITINATION OF UBH2B

Two separate ubiquitin specific proteases catalyze de-ubiquitination of H2B in *S. cerevisiae*: Ubp8 and Ubp10. The first of these, Ubp8, is a component of the Spt-Ada-Gcn5-acetyltransferase (SAGA) coactivator complex and is involved in transcription activation [104, 105]. In contrast the other ubH2B de-ubiquitinating enzyme, Ubp10 (Dot4), interacts with silencing factors such as Sir4 and is important for Sir mediated telomeric and rDNA silencing [106–109]. Ubp8 and Ubp10 appear to act on distinct pools of ubH2B within the cell, as deletion of both ubiquitin proteases results in a greater increase in the global level of ubH2B compared to either of the single deletions [106].

The first of these ubiquitin proteases, Ubp8, requires its association with several proteins within SAGA for de-ubiquitination activity: mutations disrupting the complex such as *spt20Δ*, or loss of components required for incorporation of Ubp8 into SAGA, such as Sgf11 and Sgf73, increase global levels of ubH2B [104, 105, 110–112]. Furthermore, recombinant Ubp8 alone is unable to de-ubiquitinate Ub-AMC (ubiquitin C-terminal 7-amido-4-methylcoumarin; a model substrate for de-ubiquitinating enzymes), whereas a subcomplex consisting of Ubp8, Sgf11, Sus1, and Sgf73 is able to de-ubiquitinate this substrate efficiently [112]. Orthologs of Ubp8 have recently been identified in higher eukaryotes: *Drosophila* Nonstop/Ubp8 and dSgf11/CG13379 are components of SAGA that are required for ubH2B de-ubiquitination [113]; human USP22 de-ubiquitinates ubH2B *in vitro* and associates with ATX7L3 within TFTC/STAGA [38, 114]. Notably Sus1 and Sgf73, which associate with the de-ubiquitination module within yeast SAGA, also interact with the Sac3-Thp1 mRNA export complex [112, 115]. The Sus1 homologs, E(y)2 and ENY2, have also been identified in *Drosophila* and human SAGA/STAGA respectively, indicating conservation of the link between ubH2B de-ubiquitination and mRNA export in higher eukaryotes [38, 116].

The second of the ubiquitin proteases, Ubp10, functions independently of Ubp8 and SAGA, even though *ubp10Δ* increases the global level of ubH2B to a level similar to that of *ubp8Δ* (109). Although Ubp10 has been shown to interact with silencing proteins, its function is not restricted to silenced regions as mutations disrupting its silencing function, but not its de-ubiquitination activity, still result in a global increase in ubH2B [106]. Furthermore Ubp8 and Ubp10 might function redundantly at some loci as many non-telomeric genes show increased levels of expression in the double deletion, in comparison to the deletion of either Ubp8 or Ubp10 alone [106]. Some of the silencing roles of ubH2B de-ubiquitination may be conserved in higher eukaryotes: the ubiquitin specific protease USP7 in *Drosophila* catalyzes de-ubiquitination of ubH2B *in vitro* and has been implicated in Polycomb mediated silencing [117]. However, it is not clear whether USP7 is orthologous to Ubp10, as phylogenetic analysis indicates that the Ubp10 homolog in *Drosophila* is an uncharacterized protein CG15817 [113]. An H2B de-ubiquitinase in *Arabidopsis* is also implicated in gene silencing: loss of SUP32/UBP26 increases global levels of ubH2B and is associated with a reduction in promoter Lys-9 H3 dimethylation and DNA methylation, concurrent with a loss of heterochromatic silencing of transgenes [15]. In human cells an additional ubiquitin protease, USP3, has also been implicated in both ubH2A and ubH2B de-ubiquitination [37].

DE-UBIQUITINATION OF UBH2B IS REQUIRED FOR LATER STAGES OF TRANSCRIPTION ELONGATION

The addition and removal of ubiquitin from H2B occurs in a sequential fashion that correlates with the phosphorylation events on the CTD of RNA polymerase II. Whereas H2B ubiquitination by Rad6/Bre1 requires the first of the CTD phosphorylations mediated by Kin28 [81], the H2B de-ubiquitinase Ubp8 is required for the second Ctk1 mediated CTD phosphorylation [118] (Figure 18.3). The Ser-2 phosphorylated CTD form of RNA polymerase II, Set2, and Ctk1 all fail to localize to the 5' end of the open reading frame in the absence of Ubp8 [118]. Removal of Bre1 rescues this defect, indicating that H2B ubiquitination prevents Ctk1 recruitment, although whether this is mediated directly remains unclear [118]. Thus H2B ubiquitination might function as a checkpoint at which RNA polymerase II pauses during transcription elongation. Notably, in ES cells Ring1B and H2A ubiquitination maintain RNA polymerase II in a Ser-5-phosphorylated poised state on genes that will be induced shortly after differentiation [119]. Therefore it is possible that H2A ubiquitination might substitute for some of the roles of H2B ubiquitination in these early steps in transcription elongation in higher eukaryotes.

CONCLUSION

Although histones were identified as substrates for ubiquitination over 25 years ago, the role of this modification in cellular signaling has only started to become clear in recent years. Ubiquitination of histone H2A and H2B have significantly different downstream effects, indicating that substrate targeting and recognition are regulated differentially within the context of the nucleosome. For example, these modifications have opposing effects on Lys-4 H3 methylation: whereas ubH2B is a prerequisite for Lys-4 H3 methylation, ubH2A inhibits MLL mediated methylation of this residue. Furthermore, while ubH2A inhibits transcription elongation and FACT recruitment, ubH2B interacts with FACT in restoring chromatin structure following passage of elongating RNA polymerase II. However, the lack of detectable ubH2A in *S. cerevisiae* raises the possibility that some of the functions mediated by ubH2B in yeast may have been reassigned to ubH2A in higher eukaryotes. In particular, it will be of interest to determine the extent of the involvement of ubH2B in silencing relative to ubH2A in *Drosophila* and mammalian cells.

The enzymes mediating both the addition and removal of ubiquitin from histones H2A and H2B tend to reside within large multi-subunit complexes that often contain additional catalytic activities. For example, both the histone H2B de-ubiquitinase, Ubp8, and the 2A-DUB H2A de-ubiquitinase associate with HAT complexes. It is interesting to speculate that the H2A/H2B ubiquitin ligases and de-ubiquitinases might have substrates in addition to histones as has been previously observed for other catalytic subunits within these complexes such as Gcn5/PCAF [120]. Moreover, it is likely that crosstalk between the modifications catalyzed by these different complexes might contribute to their downstream effects on transcription. The presence of different catalytic activities within the same complex raises several questions. Is there an order in which these modifications are catalyzed, and is the presence or removal of one modification, or even recruitment of an additional complex, required for the second activity? Furthermore, are these modifications catalyzed on the same or on adjacent nucleosomes? The recently published work describing the generation of recombinant ubH2B containing nucleosomes should greatly facilitate *in vitro* studies to examine some of these questions directly [95]. In addition, it is critical now to identify motifs responsible for the differential recognition of ubiquitinated H2A and H2B, in order to determine how these modifications exert their downstream effects. Histone ubiquitination provides an example of the versatility of this modification in signaling within the cell, and it is likely that many of the observations

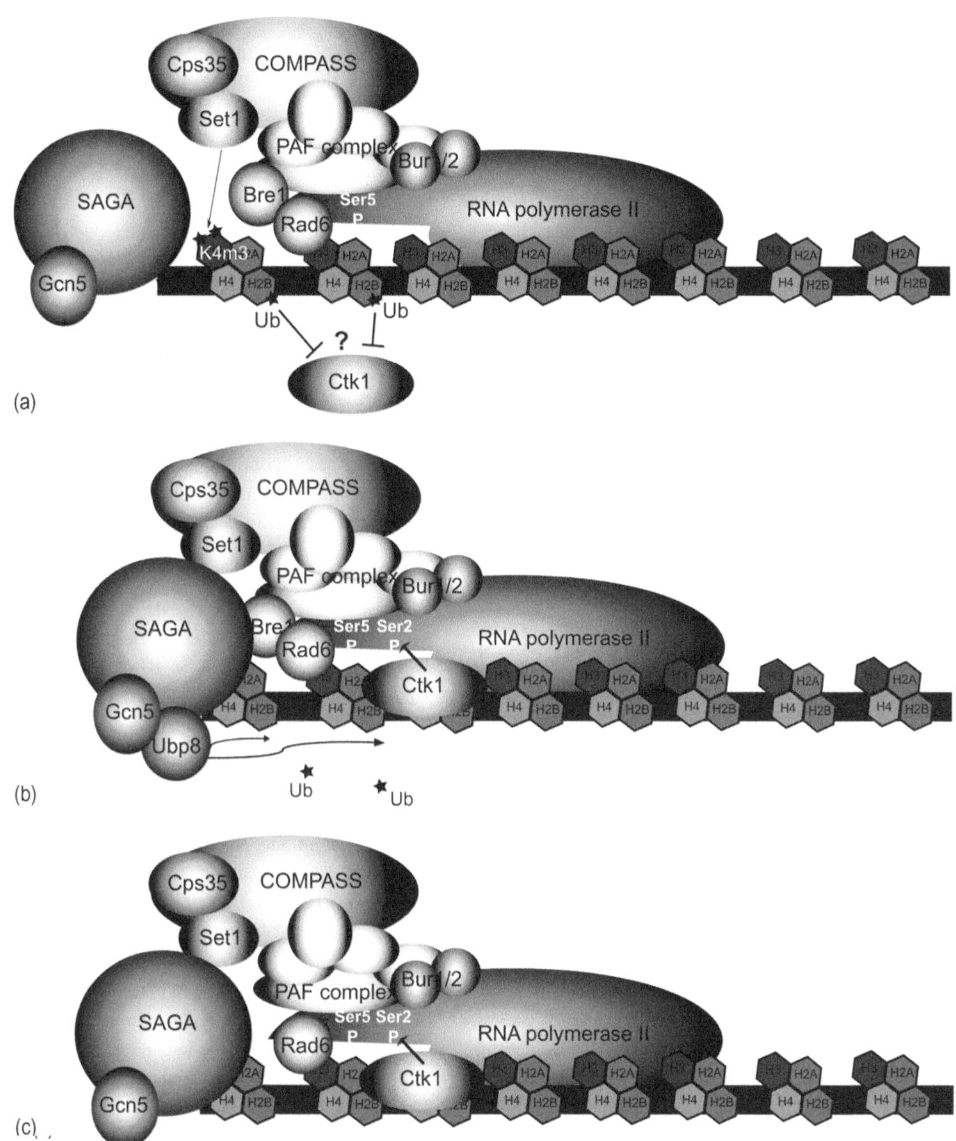

FIGURE 18.3 UbH2B de-ubiquitination is required for later stages of transcription elongation.
In the absence of the ubH2B de-ubiquitinase Ubp8, persistent ubH2B inhibits recruitment of the Ctk1 kinase. (b) Ubp8, within SAGA, de-ubiquitinates ubH2B enabling Ctk1 recruitment and phosphorylation of Ser-2 within the CTD of RNA polymerase II. Phosphorylation of Ser-2 provides a binding site for the Lys-36 H3 methyltransferase Set2, which functions in later steps of transcription elongation. (c) Deletion of the ubiquitin ligase Bre1 and the subsequent lack of ubH2B restores Ctk1 recruitment and Ser-2 phosphorylation even in yeast lacking Ubp8. This figure is reproduced from [121].

made with regard to mono-ubiquitination of histones will hold true for other pathways regulated by this modification.

ACKNOWLEDGEMENT

This chapter is based in part on an earlier review written by the authors [121].

REFERENCES

1. Hochstrasser M. Protein degradation or regulation: Ub the judge. *Cell* 1996;**84**:813–15.
2. Hicke L. Protein regulation by monoubiquitin. *Nat Rev* 2001; **2**:195–201.
3. Nijman SM, Luna-Vargas MP, Velds A, et al. A genomic and functional inventory of deubiquitinating enzymes. *Cell* 2005;**123**:773–86.
4. West MH, Bonner WM. Histone 2B can be modified by the attachment of ubiquitin. *Nucleic Acids Res* 1980;**8**:4671–80.
5. Goldknopf IL, Busch H. Isopeptide linkage between nonhistone and histone 2A polypeptides of chromosomal conjugate-protein A24. *Proc Natl Acad Sci USA* 1977;**74**:864–8.
6. Hunt LT, Dayhoff MO. Amino-terminal sequence identity of ubiquitin and the nonhistone component of nuclear protein A24. *Biochem Biophys Res Commun* 1977;**74**:650–5.
7. Ballal NR, Kang YJ, Olson MO, Busch H. Changes in nucleolar proteins and their phosphorylation patterns during liver regeneration. *J Biol Chem* 1975;**250**:5921–5.

8. Goldknopf IL, Taylor CW, Baum RM, et al. Isolation and characterization of protein A24, a "histone-like" non-histone chromosomal protein. *J Biol Chem* 1975;**250**:7182–7.
9. Bohm L, Crane-Robinson C, Sautiere P. Proteolytic digestion studies of chromatin core-histone structure. Identification of a limit peptide of histone H2A. *Eur J Biochem/FEBS* 1980;**106**:525–30.
10. Swerdlow PS, Schuster T, Finley D. A conserved sequence in histone H2A which is a ubiquitination site in higher eucaryotes is not required for growth in Saccharomyces cerevisiae. *Mol Cell Biol* 1990;**10**:4905–11.
11. Nickel BE, Davie JR. Structure of polyubiquitinated histone H2A. *Biochemistry* 1989;**28**:964–8.
12. Thorne AW, Sautiere P, Briand G, Crane-Robinson C. The structure of ubiquitinated histone H2B. *EMBO J* 1987;**6**:1005–10.
13. Robzyk K, Recht J, Osley MA. Rad6-dependent ubiquitination of histone H2B in yeast. *Science* 2000;**287**:501–4.
14. Tanny JC, Erdjument-Bromage H, Tempst P, Allis CD. Ubiquitylation of histone H2B controls RNA polymerase II transcription elongation independently of histone H3 methylation. *Genes Dev* 2007;**21**:835–47.
15. Sridhar VV, Kapoor A, Zhang K, et al. Control of DNA methylation and heterochromatic silencing by histone H2B deubiquitination. *Nature* 2007;**447**:735–8.
16. Cao R, Tsukada Y, Zhang Y. Role of Bmi-1 and Ring1A in H2A ubiquitylation and Hox gene silencing. *Mol Cell* 2005;**20**:845–54.
17. Zhou W, Zhu P, Wang J, et al. Histone H2A Monoubiquitination Represses Transcription by Inhibiting RNA Polymerase II Transcriptional Elongation. *Mol Cell* 2008;**29**:69–80.
18. Wang H, Wang L, Erdjument-Bromage H, et al. Role of histone H2A ubiquitination in Polycomb silencing. *Nature* 2004;**431**:873–8.
19. Ogawa H, Ishiguro K, Gaubatz S, et al. A complex with chromatin modifiers that occupies E2F- and Myc-responsive genes in G0 cells. *Science* 2002;**296**:1132–6.
20. Sanchez C, Sanchez I, Demmers JA, et al. Proteomics analysis of Ring1B/Rnf2 interactors identifies a novel complex with the Fbxl10/Jhdm1B histone demethylase and the Bcl6 interacting corepressor. *Mol Cell Proteomics* 2007;**6**:820–34.
21. Gearhart MD, Corcoran CM, Wamstad JA, Bardwell VJ. Polycomb group and SCF ubiquitin ligases are found in a novel BCOR complex that is recruited to BCL6 targets. *Mol Cell Biol* 2006;**26**:6880–9.
22. Frescas D, Guardavaccaro D, Bassermann F, et al. JHDM1B/FBXL10 is a nucleolar protein that represses transcription of ribosomal RNA genes. *Nature* 2007;**450**:309–13.
23. Allis CD, Berger SL, Cote J, et al. New nomenclature for chromatin-modifying enzymes. *Cell* 2007;**131**:633–6.
24. Fang J, Chen T, Chadwick B, et al. Ring1b-mediated H2A ubiquitination associates with inactive X chromosomes and is involved in initiation of X inactivation. *J Biol Chem* 2004;**279**:52,812–52,815.
25. Wei J, Zhai L, Xu J, Wang H. Role of Bmi1 in H2A ubiquitylation and Hox gene silencing. *J Biol Chem* 2006;**281**:22,537–22,544.
26. de Napoles M, Mermoud JE, Wakao R, et al. Polycomb group proteins Ring1A/B link ubiquitylation of histone H2A to heritable gene silencing and X inactivation. *Dev Cell* 2004;**7**:663–76.
27. Li Z, Cao R, Wang M, et al. Structure of a Bmi-1-Ring1B polycomb group ubiquitin ligase complex. *J Biol Chem* 2006;**281**:20,643–20,649.
28. Liu Z, Oughtred R, Wing SS. Characterization of E3Histone, a novel testis ubiquitin protein ligase which ubiquitinates histones. *Mol Cell Biol* 2005;**25**:2819–31.
29. Rajapurohitam V, Morales CR, El-Alfy M, et al. Activation of a UBC4-dependent pathway of ubiquitin conjugation during postnatal development of the rat testis. *Dev Biol* 1999;**212**:217–28.
30. Buchwald G, van der Stoop P, Weichenrieder O, et al. Structure and E3-ligase activity of the Ring-Ring complex of polycomb proteins Bmi1 and Ring1b. *EMBO J* 2006;**25**:2465–74.
31. Sarcinella E, Zuzarte PC, Lau PN, et al. Monoubiquitylation of H2A.Z distinguishes its association with euchromatin or facultative heterochromatin. *Mol Cell Biol* 2007;**27**:6457–68.
32. Chu F, Nusinow DA, Chalkley RJ, et al. Mapping post-translational modifications of the histone variant MacroH2A1 using tandem mass spectrometry. *Mol Cell Proteomics* 2006;**5**:194–203.
33. Joo HY, Zhai L, Yang C, et al. Regulation of cell cycle progression and gene expression by H2A deubiquitination. *Nature* 2007;**449**:1068–72.
34. Zhu P, Zhou W, Wang J, et al. A histone H2A deubiquitinase complex coordinating histone acetylation and H1 dissociation in transcriptional regulation. *Mol Cell* 2007;**27**:609–21.
35. Nakagawa T, Kajitani T, Togo S, et al. Deubiquitylation of histone H2A activates transcriptional initiation via trans-histone cross-talk with H3K4 di- and trimethylation. *Genes Dev* 2008;**22**:37–49.
36. Vissers JH, Nicassio F, van Lohuizen M, et al. The many faces of ubiquitinated histone H2A: insights from the DUBs. *Cell Div* 2008;**3**:8.
37. Nicassio F, Corrado N, Vissers JH, et al. Human USP3 is a chromatin modifier required for S phase progression and genome stability. *Curr Biol* 2007;**17**:1972–7.
38. Zhao Y, Lang G, Ito S, et al. A TFTC/STAGA Module Mediates Histone H2A and H2B Deubiquitination, Coactivates Nuclear Receptors, and Counteracts Heterochromatin Silencing. *Mol Cell* 2008;**29**:92–101.
39. Zhang XY, Pfeiffer HK, Thorne AW, McMahon SB. USP22, an hSAGA subunit and potential cancer stem cell marker, reverses the polycomb-catalyzed ubiquitylation of histone H2A. *Cell Cycle (Georgetown)* 2008;**7**:1522–4.
40. Cai SY. Babbitt RW and Marchesi VT: A mutant deubiquitinating enzyme (Ubp-M) associates with mitotic chromosomes and blocks cell division. *Proc Natl Acad Sci USA* 1999;**96**:2828–33.
41. Mimnaugh EG, Kayastha G, McGovern NB, et al. Caspase-dependent deubiquitination of monoubiquitinated nucleosomal histone H2A induced by diverse apoptogenic stimuli. *Cell Death Differ* 2001;**8**:1182–96.
42. Matsui SI, Seon BK, Sandberg AA. Disappearance of a structural chromatin protein A24 in mitosis: implications for molecular basis of chromatin condensation. *Proc Natl Acad Sci USA* 1979;**76**:6386–90.
43. Mueller RD, Yasuda H, Hatch CL, et al. Identification of ubiquitinated histones 2A and 2B in Physarum polycephalum. Disappearance of these proteins at metaphase and reappearance at anaphase. *J Biol Chem* 1985;**260**:5147–53.
44. Wu RS, Kohn KW, Bonner WM. Metabolism of ubiquitinated histones. *J Biol Chem* 1981;**256**:5916–20.
45. Jason LJ, Finn RM, Lindsey G, Ausio J. Histone H2A ubiquitination does not preclude histone H1 binding, but it facilitates its association with the nucleosome. *J Biol Chem* 2005;**280**:4975–82.
46. Luger K, Mader AW, Richmond RK, et al. Crystal structure of the nucleosome core particle at 2.8 A resolution. *Nature* 1997;**389**:251–60.
47. Bergink S, Salomons FA, Hoogstraten D, et al. DNA damage triggers nucleotide excision repair-dependent monoubiquitylation of histone H2A. *Genes Dev* 2006;**20**:1343–52.
48. Kapetanaki MG, Guerrero-Santoro J, Bisi DC, et al. The DDB1-CUL4ADDB2 ubiquitin ligase is deficient in xeroderma pigmentosum group E and targets histone H2A at UV-damaged DNA sites. *Proc Natl Acad Sci USA* 2006;**103**:2588–93.

49. Mailand N, Bekker-Jensen S, Faustrup H, et al. RNF8 ubiquitylates histones at DNA double-strand breaks and promotes assembly of repair proteins. *Cell* 2007;**131**:887–900.
50. Huen MS, Grant R, Manke I, et al. RNF8 transduces the DNA-damage signal via histone ubiquitylation and checkpoint protein assembly. *Cell* 2007;**131**:901–14.
51. Ikura T, Tashiro S, Kakino A, et al. DNA Damage-Dependent Acetylation and Ubiquitination of H2AX Enhances Chromatin Dynamics. *Mol Cell Biol* 2007;**27**:7028–40.
52. Kolas NK, Chapman JR, Nakada S, et al. Orchestration of the DNA-damage response by the RNF8 ubiquitin ligase. *Science* 2007;**318**:1637–40.
53. Wang B, Elledge SJ. Ubc13/Rnf8 ubiquitin ligases control foci formation of the Rap80/Abraxas/Brca1/Brcc36 complex in response to DNA damage. *Proc Natl Acad Sci USA* 2007;**104**:20,759–20,763.
54. Hwang WW, Venkatasubrahmanyam S, Ianculescu AG, et al. A conserved RING finger protein required for histone H2B monoubiquitination and cell size control. *Mol Cell* 2003;**11**:261–6.
55. Wood A, Krogan NJ, Dover J, et al. Bre1, an E3 ubiquitin ligase required for recruitment and substrate selection of Rad6 at a promoter. *Mol Cell* 2003;**11**:267–74.
56. Jentsch S, McGrath JP, Varshavsky A. The yeast DNA repair gene RAD6 encodes a ubiquitin-conjugating enzyme. *Nature* 1987;**329**:131–4.
57. Haas AL, Reback PB, Chau V. Ubiquitin conjugation by the yeast RAD6 and CDC34 gene products. Comparison to their putative rabbit homologs, E2(20K) AND E2(32K). *J Biol Chem* 1991;**266**:5104–12.
58. Sung P, Prakash S, Prakash L. The RAD6 protein of Saccharomyces cerevisiae polyubiquitinates histones, and its acidic domain mediates this activity. *Genes Dev* 1988;**2**:1476–85.
59. Zofall M, Grewal SI. HULC, a histone H2B ubiquitinating complex, modulates heterochromatin independent of histone methylation in fission yeast. *J Biol Chem* 2007;**282**:14,065–14,072.
60. Squazzo SL, O'Geen H, Komashko VM, et al. Suz12 binds to silenced regions of the genome in a cell-type-specific manner. *Genome Res* 2006;**16**:890–900.
61. Sun ZW, Allis CD. Ubiquitination of histone H2B regulates H3 methylation and gene silencing in yeast. *Nature* 2002;**418**:104–8.
62. Kim J, Hake SB, Roeder RG. The human homolog of yeast BRE1 functions as a transcriptional coactivator through direct activator interactions. *Mol Cell* 2005;**20**:759–70.
63. Zhu B, Zheng Y, Pham AD, et al. Monoubiquitination of human histone H2B: the factors involved and their roles in HOX gene regulation. *Mol Cell* 2005;**20**:601–11.
64. Koken M, Reynolds P, Bootsma D, et al. Dhr6, a Drosophila homolog of the yeast DNA-repair gene RAD6. *Proc Natl Acad Sci USA* 1991;**88**:3832–6.
65. Roest HP, van Klaveren J, de Wit J, et al. Inactivation of the HR6B ubiquitin-conjugating DNA repair enzyme in mice causes male sterility associated with chromatin modification. *Cell* 1996;**86**:799–810.
66. Baarends WM, Wassenaar E, Hoogerbrugge JW, et al. Increased phosphorylation and dimethylation of XY body histones in the Hr6b-knockout mouse is associated with derepression of the X chromosome. *J Cell Sci* 2007;**120**:1841–51.
67. Minsky N, Oren M. The RING domain of Mdm2 mediates histone ubiquitylation and transcriptional repression. *Mol Cell* 2004;**16**:631–9.
68. Coutts AS, La Thangue NB. Mdm2 widens its repertoire. *Cell Cycle (Georgetown)* 2007;**6**:827–9.
69. Chen A, Kleiman FE, Manley JL, et al. Autoubiquitination of the BRCA1*BARD1 RING ubiquitin ligase. *J Biol Chem* 2002;**277**:22,085–22,092.
70. Mallery DL, Vandenberg CJ, Hiom K. Activation of the E3 ligase function of the BRCA1/BARD1 complex by polyubiquitin chains. *EMBO J* 2002;**21**:6755–62.
71. Xia Y, Pao GM, Chen HW, et al. Enhancement of BRCA1 E3 ubiquitin ligase activity through direct interaction with the BARD1 protein. *J Biol Chem* 2003;**278**:5255–63.
72. Kao CF, Hillyer C, Tsukuda T, et al. Rad, 6 plays a role in transcriptional activation through ubiquitylation of histone H2B. *Genes Deve* 2004;**18**:184–95.
73. Osley MA. Regulation of histone H2A and H2B ubiquitylation. *Brief Funct Genomic Proteomic* 2006;**5**:179–89.
74. Shilatifard A. Chromatin modifications by methylation and ubiquitination: implications in the regulation of gene expression. *Annu Rev Biochem* 2006;**75**:243–69.
75. Wood A, Schneider J, Dover J, et al. The Paf1 complex is essential for histone monoubiquitination by the Rad6-Bre1 complex, which signals for histone methylation by COMPASS and Dot1p. *J Biol Chem* 2003;**278**:34,739–34,742.
76. Ng HH, Dole S, Struhl K. The Rtf1 component of the Paf1 transcriptional elongation complex is required for ubiquitination of histone H2B. *J Biol Chem* 2003;**278**:33,625–33,628.
77. Laribee RN, Krogan NJ, Xiao T, et al. BUR kinase selectively regulates H3 K4 trimethylation and H2B ubiquitylation through recruitment of the PAF elongation complex. *Curr Biol* 2005;**15**:1487–93.
78. Wood A, Schneider J, Dover J, et al. The Bur1/Bur2 complex is required for histone H2B monoubiquitination by Rad6/Bre1 and histone methylation by COMPASS. *Mol Cell* 2005;**20**:589–99.
79. Sarcevic B, Mawson A, Baker RT, Sutherland RL. Regulation of the ubiquitin-conjugating enzyme hHR6A by CDK-mediated phosphorylation. *EMBO J* 2002;**21**:2009–18.
80. Pavri R, Zhu B, Li G, et al. Histone H2B monoubiquitination functions cooperatively with FACT to regulate elongation by RNA polymerase II. *Cell* 2006;**125**:703–17.
81. Xiao T, Kao CF, Krogan NJ, et al. Histone H2B ubiquitylation is associated with elongating RNA polymerase II. *Mol Cell Biol* 2005;**25**:637–51.
82. Hartzog GA, Tamkun JW. A new role for histone tail modifications in transcription elongation. *Genes Dev* 2007;**21**:3209–13.
83. Laribee RN, Fuchs SM, Strahl BD. H2B ubiquitylation in transcriptional control: a FACT-finding mission. *Genes Dev* 2007;**21**:737–43.
84. Fleming AB, Kao CF, Hillyer C, et al. H2B ubiquitylation plays a role in nucleosome dynamics during transcription elongation. *Mol Cell* 2008;**31**:57–66.
85. Minsky N, Shema E, Field Y, et al. Monoubiquitinated H2B is associated with the transcribed region of highly expressed genes in human cells. *Nat Cell Biol* 2008;**10**:483–8.
86. Briggs SD, Xiao T, Sun ZW, et al. Gene silencing: trans-histone regulatory pathway in chromatin. *Nature* 2002;**418**:498.
87. Dover J, Schneider J, Tawiah-Boateng MA, et al. Methylation of histone H3 by COMPASS requires ubiquitination of histone H2B by Rad6. *J Biol Chem* 2002;**277**:28,368–28,371.
88. Ng HH, Xu RM, Zhang Y, Struhl K. Ubiquitination of histone H2B by Rad6 is required for efficient Dot1-mediated methylation of histone H3 lysine 79. *J Biol Chem* 2002;**277**:34,655–34,657.
89. Bray S, Musisi H, Bienz M. Bre1 is required for Notch signaling and histone modification. *Dev Cell* 2005;**8**:279–86.
90. Ng HH, Feng Q, Wang H, et al. Lysine methylation within the globular domain of histone H3 by Dot1 is important for telomeric silencing and Sir protein association. *Genes Dev* 2002;**16**:1518–27.
91. van Leeuwen F, Gafken PR, Gottschling DE. Dot1p modulates silencing in yeast by methylation of the nucleosome core. *Cell* 2002;**109**:745–56.

92. Dehe PM, Pamblanco M, Luciano P, et al. Histone H3 lysine 4 mono-methylation does not require ubiquitination of histone H2B. *J Mol Biol* 2005;**353**:477–84.
93. Schneider J, Wood A, Lee JS, et al. Molecular regulation of histone H3 trimethylation by COMPASS and the regulation of gene expression. *Mol Cell* 2005;**19**:849–56.
94. Shahbazian MD, Zhang K, Grunstein M. Histone H2B ubiquitylation controls processive methylation but not monomethylation by Dot1 and Set1. *Mol Cell* 2005;**19**:271–7.
95. McGinty RK, Kim J, Chatterjee C, et al. Chemically ubiquitylated histone H2B stimulates hDot1L-mediated intranucleosomal methylation. *Nature* 2008;**453**:812–16.
96. Lee JS, Shukla A, Schneider J, et al. Histone Crosstalk between H2B Monoubiquitination and H3 Methylation Mediated by COMPASS. *Cell* 2007;**131**:1084–96.
97. Gonzalez F, Delahodde A, Kodadek T, Johnston SA. Recruitment of a 19S proteasome subcomplex to an activated promoter. *Science* 2002;**296**:548–50.
98. Ezhkova E, Tansey WP. Proteasomal ATPases link ubiquitylation of histone H2B to methylation of histone H3. *Mol Cell* 2004;**13**:435–42.
99. Lee D, Ezhkova E, Li B, Pattenden SG, et al. The proteasome regulatory particle alters the SAGA coactivator to enhance its interactions with transcriptional activators. *Cell* 2005;**123**:423–36.
100. Shukla A, Stanojevic N, Duan Z, et al. Functional analysis of H2B-Lys-123 ubiquitination in regulation of H3-Lys-4 methylation and recruitment of RNA polymerase II at the coding sequences of several active genes in vivo. *J Biol Chem* 2006;**281**:19,045–19,054.
101. Collart MA, Timmers HT. The eukaryotic Ccr4-not complex: a regulatory platform integrating mRNA metabolism with cellular signaling pathways? *Prog Nucleic Acid Res Mol Biol* 2004;**77**:289–322.
102. Laribee RN, Shibata Y, Mersman DP, et al. CCR4/NOT complex associates with the proteasome and regulates histone methylation. *Proc Natl Acad Sci USA* 2007;**104**:5836–41.
103. Mulder KW, Brenkman AB, Inagaki A, et al. Regulation of histone H3K4 tri-methylation and PAF complex recruitment by the Ccr4-Not complex. *Nucleic Acids Res* 2007;**35**:2428–39.
104. Daniel JA, Torok MS, Sun ZW, et al. Deubiquitination of histone H2B by a yeast acetyltransferase complex regulates transcription. *J Biol Chem* 2004;**279**:1867–71.
105. Henry KW, Wyce A, Lo WS, et al. Transcriptional activation via sequential histone H2B ubiquitylation and deubiquitylation, mediated by SAGA-associated Ubp8. *Genes Dev* 2003;**17**:2648–63.
106. Gardner RG, Nelson ZW, Gottschling DE. Ubp10/Dot4p regulates the persistence of ubiquitinated histone H2B: distinct roles in telomeric silencing and general chromatin. *Mol Cell Biol* 2005;**25**:6123–39.
107. Kahana A, Gottschling DE. DOT4 links silencing and cell growth in Saccharomyces cerevisiae. *Mol Cell Biol* 1999;**19**:6608–20.
108. Calzari L, Orlandi I, Alberghina L, Vai M. The histone deubiquitinating enzyme Ubp10 is involved in rDNA locus control in Saccharomyces cerevisiae by affecting Sir2p association. *Genetics* 2006;**174**:2249–54.
109. Emre NC, Ingvarsdottir K, Wyce A, et al. Maintenance of low histone ubiquitylation by Ubp10 correlates with telomere-proximal Sir2 association and gene silencing. *Mol Cell* 2005;**17**:585–94.
110. Ingvarsdottir K, Krogan NJ, Emre NC, et al. H2B ubiquitin protease Ubp8 and Sgf11 constitute a discrete functional module within the Saccharomyces cerevisiae SAGA complex. *Mol Cell Biol* 2005;**25**:1162–72.
111. Lee KK, Florens L, Swanson SK, et al. The deubiquitylation activity of Ubp8 is dependent upon Sgf11 and its association with the SAGA complex. *Mol Cell Biol* 2005;**25**:1173–82.
112. Kohler A, Schneider M, Cabal GG, et al. Yeast Ataxin-7 links histone deubiquitination with gene gating and mRNA export. *Nat Cell Biol* 2008;**10**:707–15.
113. Weake VM, Lee KK, Guelman S, et al. SAGA-mediated H2B deubiquitination controls the development of neuronal connectivity in the Drosophila visual system. *EMBO J* 2008;**27**:394–405.
114. Zhang XY, Varthi M, Sykes SM, et al. The Putative Cancer Stem Cell Marker USP22 Is a Subunit of the Human SAGA Complex Required for Activated Transcription and Cell-Cycle Progression. *Mol Cell* 2008;**29**:102–11.
115. Rodriguez-Navarro S, Fischer T, Luo MJ, et al. Sus1, a functional component of the SAGA histone acetylase complex and the nuclear pore-associated mRNA export machinery. *Cell* 2004;**116**:75–86.
116. Kurshakova MM, Krasnov AN, Kopytova DV, et al. SAGA and a novel *Drosophila* export complex anchor efficient transcription and mRNA export to NPC. *EMBO J* 2007;**26**:4956–65.
117. van der Knaap JA, Kumar BR, Moshkin YM, et al. GMP synthetase stimulates histone H2B deubiquitylation by the epigenetic silencer USP7. *Mol Cell* 2005;**17**:695–707.
118. Wyce A, Xiao T, Whelan KA, et al. H2B ubiquitylation acts as a barrier to Ctk1 nucleosomal recruitment prior to removal by Ubp8 within a SAGA-related complex. *Mol Cell* 2007;**27**:275–88.
119. Stock JK, Giadrossi S, Casanova M, et al. Ring1-mediated ubiquitination of H2A restrains poised RNA polymerase II at bivalent genes in mouse ES cells. *Nat Cell Biol* 2007;**9**:1428–35.
120. Sakaguchi K, Herrera JE, Saito S, et al. DNA damage activates p53 through a phosphorylation-acetylation cascade. *Genes Dev* 1998;**12**:2831–41.
121. Weake VM, Workman JL. Histone ubiquitination: triggering gene activity. *Mol Cell* 2008;**29**:653–63.
122. Fleury D, Himanen K, Cnops G, et al. The Arabidopsis thaliana homolog of yeast BRE1 has a function in cell cycle regulation during early leaf and root growth. *Plant Cell* 2007;**19**:417–32.
123. Liu Y, Koornneef M, Soppe WJ. The absence of histone H2B monoubiquitination in the Arabidopsis hub1 (rdo4) mutant reveals a role for chromatin remodeling in seed dormancy. *Plant Cell* 2007;**19**:433–44.

Chapter 19

Chromatin Mediated Control of Gene Expression in Innate Immunity and Inflammation

Gioacchino Natoli
Department of Experimental Oncology, European Institute of Oncology (IEO), Milan, Italy

"How many genes will we find to be devoted to host defense? I am willing to bet that the genes that mediate defense against infection will make up a fair proportion of the human genome, somewhere between 1% and 10%, because host defense is such a fundamental property of life on Earth."

Charles A. Janeway (1943–2003)

INTRODUCTION

The genomes of multicellular eukaryotes have been built and organized around a few essential requirements. In addition to the faithful transmission of their genetic material to daughter cells (a necessity shared with unicellular organisms), multicellular eukaryotes had to deal with two additional major problems during their evolution: the maintenance of a complex metabolic equilibrium integrating the activities and specific requirements of different tissues and organs, and the need to defend themselves from microbes. Without efficient anti-microbial defense mechanisms, complex life on earth wouldn't have evolved at all and it's useful to remember that until the discovery of antibiotics, infections were the main cause of death also in the industrialized world. Indeed, the blueprints for the innate immune response were established in very early primitive life forms, and are now found throughout the animal and plant kingdoms [1].

Given these premises, it is not surprising at all that a large number of genes (although an estimate of the real number is impossible) have been positively selected and recruited to participate in various phases of the inflammatory response triggered by microbial stimuli. And it is not surprising that, because of the complexity of the response, the many phases in which it is deployed, and the many "flavors" in which it appears (depending on quality and intensity of the stimulus as well as the target organ), very elaborated mechanisms evolved to ensure that the expression of the induced genes is carefully and precisely organized so that each gene is expressed in response to specific stimuli and with kinetics and intensities that suit the peculiar function of its product(s). Data accumulated in the last years have strengthened the concept that chromatin is an essential substrate at which multiple signals are integrated to promote a perfectly choreographed expression of the genes involved in inflammatory transcriptional responses [2]. Although the current level of understanding of these mechanisms is far from complete, some concepts and ideas have resisted to experimental challenges and now represent accepted paradigms that will be the subject of this chapter.

INFLAMMATION AS A KINETICALLY COMPLEX TRANSCRIPTIONAL RESPONSE

Induction of an inflammatory response requires the activation of a complex gene expression program [3, 4]. Genes that are induced in response to microbes (or inflammatory stimuli in general) represent a highly heterogeneous set that includes chemokines devoted to recruit inflammatory cells to the site of inflammation (e.g., interleukin 8, IL8) or to recruit cells of the adaptive immune system for antigen presentation (e.g., Ccl5/Rantes), inflammatory cytokines that amplify and sustain the response (e.g., tumor necrosis factor α, and interleukin 1 α), antiapoptotic genes that promote cell survival in the inflammatory infiltrate, genes encoding proteins or enzymes involved in viral or bacterial recognition and killing (e.g., inducible nitric oxide synthase), and several others. Activation of these genes doesn't occur synchronously in response to stimulation; conversely, kinetics of induction and shut-off are extremely diverse, reflecting specific gene functions (Figure 19.1). For instance, among

Handbook of Cell Signaling, Three-Volume Set 2 ed.

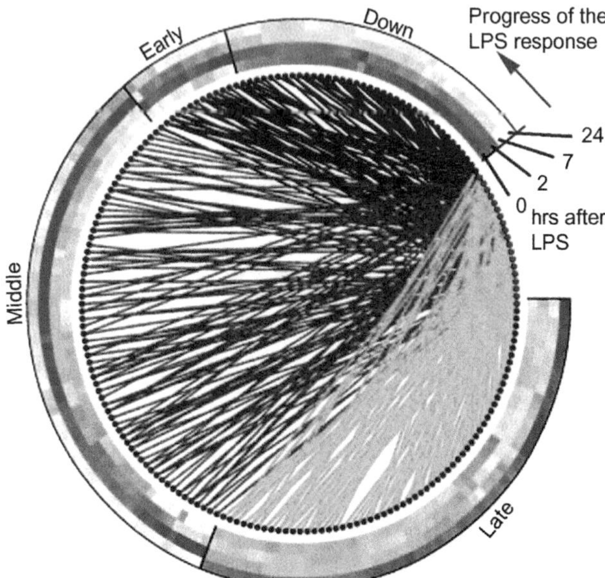

FIGURE 19.1 Dynamics of inflammatory gene expression in response to LPS stimulation.
Progress of the LPS response. A radial expression heat map for the microarray time-course data in LPS stilmulated macrophages (0, 2, 7, and 24 h treatment). Expression clusters are arranged to show the LPS response progress counterclockwise. Source [4]; Elsevier press.

the immediately inducible genes are those encoding chemoattractants for neutrophils (IL8), which represent the first line of cellular defense against microbes (and as such must be recruited immediately), while genes encoding for T cell chemoattractants (Ccl5/Rantes) or cytokines controlling T cell development (interleukin 12, IL12), are expressed at a later stage, as encounter with T cells occurs with slower kinetics and premature gene activation would be pointless.

CHROMATIN AND THE KINETIC CONTROL OF INFLAMMATORY RESPONSES

Mechanistic insights into how this kinetic complexity is achieved came from studies on Nuclear Factor kappa B (NFκB), a family of homo- and heterodimeric transcription factors directly controlling the expression of hundreds of inflammatory and immune response genes [5]. Activation of NFκB is a very rapid event invariably following exposure of cells to inflammatory stimuli (including microbial products, inflammatory cytokines, ultraviolet- and X-rays). In spite of this massive and rapid nuclear entry, only a fraction of NFκB target genes are immediately activated (heretofore indicated as "fast" genes), while several target genes bearing high affinity NFκB binding sites and absolutely requiring NFκB for induction remain in a "stand-by" state for hours before activation can occur ("slow" genes) [6, 7]. Importantly, slow genes include primary genes activated with slow kinetics (but not requiring new protein synthesis) and secondary genes (requiring new protein synthesis). Analysis of NFκB recruitment by chromatin immunoprecipitation revealed that the molecular basis for this delay was a postponed recruitment of NFκB to the slow genes [6, 7]. Delayed recruitment of NFκB to these genes does not depend on obvious differences in the underlying regulatory sequences, since both slow and fast genes are associated with high affinity NFκB binding sites. Moreover, genes that are activated with slow kinetics in one cell type (e.g., macrophages) may be activated with fast kinetics in a different cell type (e.g., fibroblasts). It is now clear that slow and cell-type specific recruitment kinetics depend on specific chromatin configurations of the target promoter. Biochemical evidence is represented by the differential acetylation and methylation state of fast and slow genes in unstimulated macrophages. Fast genes display a chromatin configuration typical of genes poised for immediate activation, including high levels of histone acetylation and trimethylation of histone H3 at lysine 4 (H3K4me3, a typical modification associated with transcription start sites that are active or poised for activation) [6, 8]. Recruitment of NFκB and gene induction is not associated with any obvious or dramatic change in this chromatin state and no further increase in acetylation and H3K4me3 can be detected. Conversely, slow genes are associated with undetectable or very low basal H3K4me3 and histone acetylation: in response to activation both modifications are slowly upregulated with kinetics that precede recruitment of NFκB, although their final levels will usually be lower than those found on fast genes. A simple scenario compatible with these data is that while the promoters of fast inflammatory genes are in a chromatin configuration permissive for immediate NFκB binding, promoters of slow genes must undergo biochemical changes before NFκB can be recruited. An inference of this model is that chromatin configuration at slow genes must negatively impact on the affinity of the NFκB sites in their promoters (otherwise these sites would presumably recruit NFκB as quickly as the fast genes). Another inference is that transcription factors different from NFκB must be able to bind the promoters of the slow genes at a stage when they are non-permissive for NFκB binding, and direct the biochemical changes leading to increased NFκB accessibility and eventually gene activation.

GENETIC DISSECTION OF CHROMATIN REMODELING AT INFLAMMATORY GENES

Genetic evidence supporting this model first came from the demonstration that activation of all the slow genes (with no apparent exception) absolutely requires the Swi/Snf chromatin remodeling complexes [9], multimolecular ATP dependent machines able to disrupt chromatin structure, displace nucleosomes, and promote transcription factor binding [10]. Specifically, simultaneous depletion of the two essential ATPase subunits of Swi/Snf (Brg1 and Brm) using RNA interference, resulted in the abrogation of slow gene induction, which was associated with defective or absent nucleosome remodeling. Conversely, fast genes were completely unaffected. A simple model compatible with these data would be that slow genes are associated with well positioned and stable nucleosomes that limit the accessibility of NFκB binding sites. However, the real scenario may be more complicated, as data support the idea that the non-permissive chromatin state of slow genes is actively enforced by an antagonistic chromatin remodeling complex, Mi2/Nurd, as indicated by the behavior of cells depleted of Mi2β/Chd4 (the helicase component Mi2/Nurd) [9]. Mi2/Nurd does not control the kinetics of induction of inflammatory genes but prevents their excessive induction, possibly by limiting Swi/Snf mediated remodeling. Therefore, the chromatin state of inflammatory genes is actively enforced by opposing chromatin remodeling activities, one (Swi/Snf) required for counteracting an inhibitory configuration that is non-permissive for gene activation, and the other (Mi2/Nurd) required to prevent excessive activation of slow inflammatory genes.

A refined genetic dissection of chromatin dependent control of inflammatory gene activation has been recently provided in lipopolisaccharide (LPS) stimulated macrophages. LPS acts via a cell surface receptor, toll-like receptor 4 (TLR4) that signals via two main adapters, MyD88 and TRIF [11]. MyD88$^{-/-}$ macrophages, in which NFκB activation is substantially normal, are unable to activate all secondary genes, and impaired activation is associated with absolute loss of nucleosome remodeling, lack of any increase in H3K4me3 after stimulation, as well as a complete lack of NFκB recruitment to slow genes [8]. Therefore, MyD88 is required to activate one or more factors that are in turn necessary for nucleosome remodeling, H3K4 trimethylation, and NFκB recruitment at secondary genes. A further dissection of the pathway came from the analysis of macrophages lacking IkBζ, a transcriptional coactivator encoded by a fast, primary response gene (Nfkbiz) and required for the activation of several inflammatory genes [12–14]. Absence of IkBζ compromises the induction of a subset of the slow genes induced in response to LPS. At the molecular level, Nfkbiz$^{-/-}$ macrophages were proficient in remodeling nucleosomes at secondary gene promoters, including those requiring IkBζ for induction. However, at these genes no increase in H3K4me3 and no NFκB recruitment were observed. Therefore IkBζ is specifically required at a step downstream of nucleosome remodeling and upstream of H3K4 trimethylation and NFκB recruitment [8]. A model integrating the data described above is presented in Figure 19.2.

Overall, these data also allow us to define a precise temporal sequence, supported by biochemical results and validated by genetic data, leading to the alleviation of chromatin dependent inhibition and to the induction of the slow inflammatory genes. The first event is the induction or activation of mediators of nucleosome remodeling. The identity of these factors has been elusive until now, but it is clear that their activation in response to LPS specifically requires MyD88. The second event is nucleosome remodeling that is essential but not sufficient for gene activation, which also requires further downstream signals leading to H3K4 trimethylation (presumably also H3 and H4 acetylation) and NFκB recruitment.

Since H3K4me3 deposition is dependent on histone methyl-transferase complexes associated with RNA polymerase II [15], it is unclear whether it is required for NFκB recruitment or, conversely, if its appearance follows NFκB recruitment and gene activation. The fact that some transcription factors have a marked preference for H3K4me3-positive sites may be in favor of the first possibility [16], but it's clear that definitive evidence is lacking.

In any case, NFκB recruitment to hundreds of inflammatory genes induced with slow kinetics is the last step of a complex multistep program whose final aim is to relieve NFκB binding sites from a negative control exerted by chromatin organization. Coordinated control of NFκB persistence in the nucleus and chromatin remodeling may also allow selective gene induction in response to alternative stimuli or to stimuli of different duration. For instance, a transient NFκB activation is incompatible with the induction of slow genes because remodeling will be completed only after the NFκB response has been terminated and the factor has been extruded from the nucleus [17]. Stimulus specificity likely arises as a consequence of the ability or inability of alternative stimuli signaling NFκB activation, to trigger pathways required for nucleosome remodeling [7].

The biochemistry of NFκB interaction with nucleosomes will be discussed next.

BINDING OF INFLAMMATORY TRANSCRIPTION FACTORS TO NUCLEOSOMAL DNA

While it is clear that *in vivo* nucleosomes and chromatin organization have a major impact on both the kinetics and the intensity of inflammatory gene induction, the molecular basis explaining how nucleosomes control inflammatory responses are only incompletely defined.

The biochemical and genetic data reported in the above sections clearly point to nucleosomes as the crucial controllers of the final and essential event in the induction of a large fraction of inflammatory genes, namely recruitment of NFκB.

In general, the assembly of DNA with core histones to generate nucleosomes, and the further folding of the nucleosomal chain into higher order fibers (with various degrees

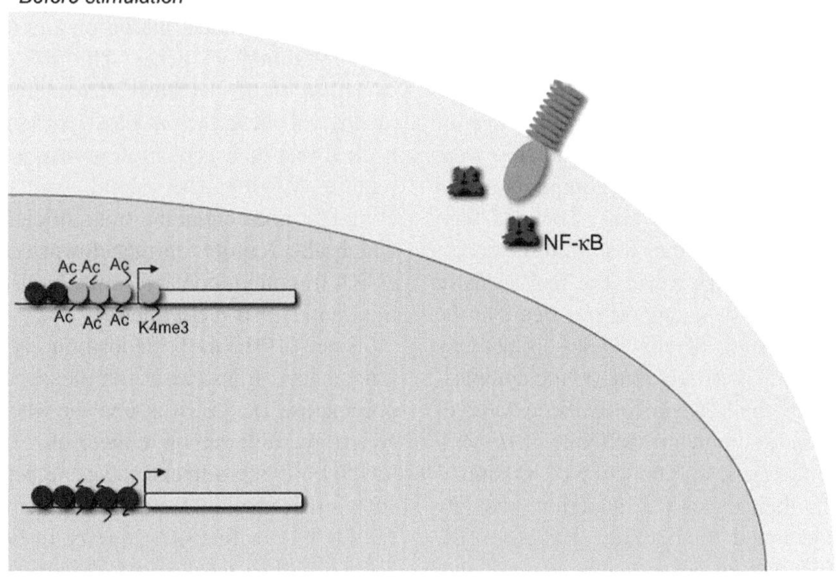

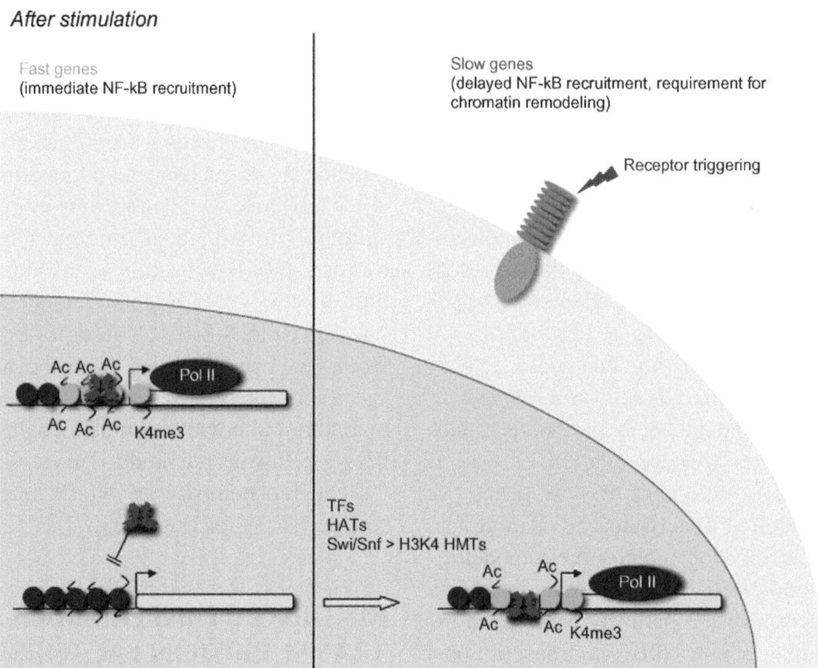

FIGURE 19.2 Chromatin mediated regulation of fast and slow inflammatory gene induction.
Above: chromatin state in unstimulated cells. Nucleosomal organization at a fast (up) and at a slow (down) gene. Only the fast gene shows features of transcriptional competence (like high level acetylation (Ac) and trimethylation of histone H3 at Lysine 4 (K4me3)). In unstimulated cells NFκB is cytoplasmic. Below: in response to receptor triggering, NFκB translocates to the nucleus and binds immediately to the genes poised for activation. Recruitment to the slow genes requires as yet unidentified transcription fractors (TFs) driving remodeling (by Swi/Snf) and histone modifications including acetylation (carried out by histone acetyl transferases, HAT) and H3 Lysine 4 methylation.

of compaction) impose a barrier to the recruitment of transcription factors to the underlying DNA [18]. However, because of their enormous structural differences, individual transcription factors greatly differ in their ability to contact high affinity binding sites embedded in a nucleosome. For instance the TATA-box binding protein has no measurable affinity for a site incorporated in a nucleosome [19]. The heat shock factor has a 100-fold higher affinity for a non-nucleosomal as compared to a nucleosomal site [20], while the affinity of the glucocorticoid receptor for a nucleosomal site is only two- to fivefold lower than that for a site in naked DNA [21].

Prior structural knowledge should allow predicting if a transcription factor is able or not to bind sites contained in a

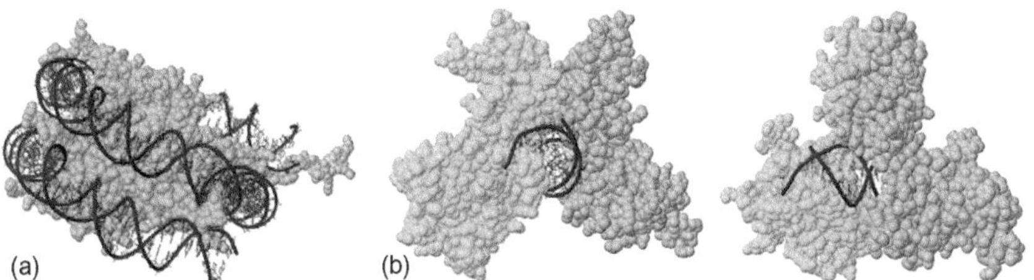

FIGURE 19.3 Structural views of the nucleosome and NFκB.
The nucleosome structure shows the DNA surrounding the histone octamer (a). The structure of an NFκB dimer (p65p50) crystallized on an NFκB site with front (left) and side (right) views. p50 is shown in green, p65 in gray, and DNA in red. Source [2]; Nature Publishing Group.

nucleosome. Although structural differences among various dimers exist, NFκB binds DNA with a butterfly-like shape (Figure 19.3), the two molecules of the dimer representing the two wings and the DNA the cylindrical body [22, 23]. The DNA is not completely encircled by the dimer; however, it is clear from the structure that the DNA binding cleft can't accommodate the surface of a nucleosome (Figure 19.3). Therefore, the simple structural prediction is that nucleosomal sites cannot be bound by NFκB, which would nicely fit the *in vivo* data discussed above. However, experimental data contradict this prediction, as incorporation of high affinity binding sites within a nucleosome causes only a marginal (if any) reduction in the affinity for NFκB, suggesting that nucleosomes are almost "transparent" to NFκB [24]. High affinity binding is detected independently of the position of the binding site relative to the dyad axis of the nucleosome: therefore binding doesn't occur as a consequence of the spontaneous uncoiling of nucleosomal DNA, because in this case a marked reduction of affinity for internally located sites (and a mild reduction for the peripheral ones) would be observed [25].

These data generate two main conceptual problems: first, the *in vivo* results described above cannot be explained if the incorporation of NFκB binding sites in nucleosomes has no effect on binding affinity whatsoever. Considering the strong genetic and biochemical evidence that nucleosomes *in vivo* shape the kinetics and intensity of the inflammatory response, we must assume that the nucleosomes controlling NFκB recruitment to (and activation of) slow inflammatory genes must have a special composition and/or biochemical properties (e.g., covalent modifications) that enable them to block NFκB binding.

For instance, incorporation of the histone variant macroH2A in nucleosomes abrogates NFκB binding [26] and in a specific case this has been shown to have regulatory consequences *in vivo* [27]. However, genomic distribution of macroH2A is still ill defined and whether it is commonly found in well positioned nucleosomes in promoters is still unknown.

Second, since nucleosomal sites can be efficiently bound by NFκB *in vitro* [24], and since this efficient binding is not compatible with the available structural information, we must conclude that the available structures of NFκB dimers should be considered snapshots of the final, stable conformation of the NFκB/DNA complex, but dynamic intermediates whose conformation is fully compatible with high affinity interactions with nucleosomal sites must exist.

A major piece of information that is still missing (and that is probably necessary to help solve the puzzle) is a detailed analysis of nucleosome distribution at the regulatory regions of inflammatory genes, which would clarify the relationship between individual nucleosomes and the underlying *cis*-regulatory sequences. The current understanding of the rules dictating nucleosome positions in higher eukaryotes is still primitive and available algorithms have limited predictive ability [28]. From an experimental point of view, the complexity of mammalian genomes hindered insofar the generation of extensive maps of nucleosome positions, although progresses in microarray based hybridization techniques and, more recently, high throughput sequencing led to detailed nucleosome maps in *Drosophila melanogaster* and yeast [29, 30]. Moreover, classical single gene analyses are complex and time consuming. Therefore, it is not surprising that in the specific case of inflammatory genes only anecdotic reports are available to date, which in some cases are compatible with the idea that nucleosomes may occlude NFκB binding sites. For instance, in macrophages a positioned nucleosome encompassing the promoter elements required for the induction of the *IL12b* gene is quickly remodeled in a manner that is independent of cRel, the NFκB subunit required for Il12b transcriptional activation [31, 32]. This observation is in line with the concept that different but coordinated signals lead to nucleosome remodeling and NFκB activation.

CONCLUSIONS

Nuclear translocation of inflammatory transcription factors can be dissociated from activation of specific genes because of chromatin regulated mechanisms controlling their recruitment to selected cognate binding sites in a highly promoter specific manner. Control of chromatin remodeling and binding site accessibility provides an additional regulatory layer ultimately shaping and tuning the ensuing transcriptional program in a manner that integrates information on the strength, duration, and nature of the stimulus. It can be assumed that a similar integration of signals and mechanisms operates in the vast majority of stimulus induced transcriptional responses, and from this point of view inflammatory gene regulation may provide paradigms and models of wider applicability.

We can anticipate that in the near future, the availability of detailed genome-wide chromatin maps describing nucleosome positions and their modifications in different basal and stimulated conditions, coupled with more accurate and complete information on transcription factor binding site affinities, will allow generating integrated and predictive models faithfully describing inducible transcriptional responses.

REFERENCES

1. Janeway Jr. CA, Medzhitov R Innate immune recognition. *Annu Rev Immunol* 2002;**20**:197–216.
2. Natoli G, Saccani S, Bosisio D, Marazzi I. Interactions of NF-kappaB with chromatin: the art of being at the right place at the right time. *Nat Immunol* 2005;**6**:439–45.
3. Nau GJ, Richmond JF, Schlesinger A, et al. Human macrophage activation programs induced by bacterial pathogens. *Proc Natl Acad Sci USA* 2002;**99**:1503–8.
4. Nilsson R, Bajic VB, Suzuki H, et al. Transcriptional network dynamics in macrophage activation. *Genomics* 2006;**88**:133–42.
5. Hayden MS, Ghosh S. Shared principles in NF-kappaB signaling. *Cell* 2008;**132**:344–62.
6. Saccani S, Pantano S, Natoli G. Two waves of nuclear factor kappaB recruitment to target promoters. *J Exp Med* 2001;**193**:1351–9.
7. Saccani S, Pantano S, Natoli G. p38-Dependent marking of inflammatory genes for increased NF-kappa B recruitment. *Nat Immunol* 2002;**3**:69–75.
8. Kayama H, Ramirez-Carrozzi VR, Yamamoto M, et al. Class-specific regulation of pro-inflammatory genes by MyD8 pathways and IkappaBzeta. *J Biol Chem* 2008;**283**:12,468–12,477.
9. Ramirez-Carrozzi VR, Nazarian AA, Li CC, et al. Selective and antagonistic functions of SWI/SNF and Mi-2beta nucleosome remodeling complexes during an inflammatory response. *Genes Dev* 2006;**20**:282–96.
10. Lusser A, Kadonaga JT. Chromatin remodeling by ATP-dependent molecular machines. *Bioessays* 2003;**25**:1192–200.
11. Akira S. TLR signaling. *Curr Top Microbiol Immunol* 2006;**311**:1–16.
12. Yamazaki S, Muta T, Takeshige K. A novel IkappaB protein, IkappaB-zeta, induced by proinflammatory stimuli, negatively regulates nuclear factor-kappaB in the nuclei. *J Biol Chem* 2001;**276**:27,657–27,662.
13. Haruta H, Kato A, Todokoro K. Isolation of a novel interleukin-1-inducible nuclear protein bearing ankyrin-repeat motifs. *J Biol Chem* 2001;**276**:12,485–12,488.
14. Yamamoto M, Yamazaki S, Uematsu S, et al. Regulation of Toll/IL-1-receptor-mediated gene expression by the inducible nuclear protein IkappaBzeta. *Nature* 2004;**430**:218–22.
15. Shilatifard A. Chromatin modifications by methylation and ubiquitination: implications in the regulation of gene expression. *Annu Rev Biochem* 2006;**75**:243–69.
16. Guccione E, Martinato F, Finocchiaro G, et al. Myc-binding-site recognition in the human genome is determined by chromatin context. *Nat Cell Biol* 2006;**8**:764–70.
17. Hoffmann A, Levchenko A, Scott ML, Baltimore D. The IkappaB-NF-kappaB signaling module: temporal control and selective gene activation. *Science* 2002;**298**:1241–5.
18. Kornberg RD, Lorch Y. Twenty-five years of the nucleosome, fundamental particle of the eukaryote chromosome. *Cell* 1999;**98**:285–94.
19. Imbalzano AN, Kwon H, Green MR, Kingston RE. Facilitated binding of TATA-binding protein to nucleosomal DNA. *Nature* 1994;**370**:481–5.
20. Taylor IC, Workman JL, Schuetz TJ, Kingston RE. Facilitated binding of GAL4 and heat shock factor to nucleosomal templates: differential function of DNA-binding domains. *Genes Dev* 1991;**5**:1285–98.
21. Perlmann T, Wrange O. Specific glucocorticoid receptor binding to DNA reconstituted in a nucleosome. *Embo J* 1988;**7**:3073–9.
22. Chen FE, Huang DB, Chen YQ, Ghosh G. Crystal structure of p50/p65 heterodimer of transcription factor NF-kappaB bound to DNA. *Nature* 1998;**391**:410–13.
23. Huxford T, Malek S, Ghosh G. Structure and mechanism in NF-kappa B/I kappa B signaling. *Cold Spring Harb Symp Quant Biol* 1999;**64**:533–40.
24. Angelov D, Lenouvel F, Hans F, et al. The histone octamer is invisible when NF-kappaB binds to the nucleosome. *J Biol Chem* 2004;**279**:42,374–42,382.
25. Anderson JD, Widom J. Sequence and position-dependence of the equilibrium accessibility of nucleosomal DNA target sites. *J Mol Biol* 2000;**296**:979–87.
26. Angelov D, Molla A, Perche PY, et al. The histone variant macroH2A interferes with transcription factor binding and SWI/SNF nucleosome remodeling. *Mol Cell* 2003;**11**:1033–41.
27. Agelopoulos M, Thanos D. Epigenetic determination of a cell-specific gene expression program by ATF-2 and the histone variant macroH2A. *EMBO J* 2006;**25**:4843–53.
28. Segal E, Fondufe-Mittendorf Y, Chen L, et al. A genomic code for nucleosome positioning. *Nature* 2006;**442**:772–8.
29. Yuan GC, Liu YJ, Dion MF, et al. Genome-scale identification of nucleosome positions in S. cerevisiae. *Science* 2005;**309**:626–30.
30. Mavrich TN, Jiang C, Ioshikhes IP, et al. Nucleosome organization in the Drosophila genome. *Nature* 2008;**453**:358–62.
31. Weinmann AS, Plevy SE, Smale ST. Rapid and selective remodeling of a positioned nucleosome during the induction of IL-12 p40 transcription. *Immunity* 1999;**11**:665–75.
32. Weinmann AS, Mitchell DM, Sanjabi S, et al. Nucleosome remodeling at the IL-12 p40 promoter is a TLR-dependent, Rel-independent event. *Nat Immunol* 2001;**2**:51–7.

Part 3

Stress Responses

Chapter 20

Complexity of Stress Signaling

Daniel R. Hyduke[1,2], Sally A. Amundson[3] and Albert J. Fornace Jr.[2,4]
[1]Department of Bioengineering, University of California-San Diego, La Jolla, California
[2]John B. Little Center for the Radiation Sciences and Environmental Health, Harvard School of Public Health, Boston, Massachusetts
[3]Center for Radiological Research, Columbia University Medical Center, New York
[4]Lombardi Comprehensive Cancer Center, and Department of Biochemistry and Molecular and Cellular Biology, Georgetown University, Washington, DC

ABBREVIATIONS

ATF6, activating transcription factor 6; BER, base excision repair; CGH, comparative genomic hybridization; ChIP, chromatin immunoprecipiation; DSB, double strand break; ER, endoplasmic reticulum; ERK, extracellular signal regulated kinase; IRE1, inositol requiring protein (enzyme) 1; JNK-c, Jun N-terminal kinase; LPS, lipopolysaccharide; MAPK, mitogen activated protein kinase; MRN, Mre11-Rad50-Nbs1; mtDNA, mitochondrial DNA; ncRNA, non-coding RNA; NER, nucleotide excision repair; NO, nitric oxide; PERK, PKR-like endoplasmic reticulum kinase; RNS, reactive nitrogen species; ROS, reactive oxygen species; SAPK, stress activated protein kinases; SG, stress granule; SNP, single nucleotide polymorphism; siRNA, small interfering RNA; sRNA, small RNA; TNF, tumor necrosis factor; UPR, unfolded protein response; XP, xeroderma pigmentosum.

INTRODUCTION

Life has evolved in an environment fraught with potential dangers. Successful organisms have developed strategies to meet the challenges posed by their environments. Many different kinds of stresses may be encountered by a cell or organism. These stresses arise from physical and chemical factors in the environment, toxic products of a competitor, or from normal metabolic processes. DNA damage is a universally encountered stress, and DNA damaging agents [1] are arguably the most important category of stressors. DNA contains the dynamic program for life and damage to DNA may alter the execution of this program and result in dire consequences, such as death or the runaway proliferation observed in cancer. Metabolic stress is another universally experienced stress – metabolism provides the building blocks for life, and events that disrupt the supply of these materials have a drastic impact upon the execution of the genomic program. Metabolic stresses can be broadly classified into two categories: (1) nutrient deprivation and (2) toxic by-products. Nutrient deprivation occurs when there are insufficient resources in the environment or when key metabolic pathways are inhibited. Toxic by-products, such as reactive oxygen species (ROS) and lactic acid, are generated during normal metabolism and will damage the cell if they are not cleared. In addition to preserving genetic integrity and staying alive, organisms must defend against a myriad of environmental stresses, such as osmotic stress, hypoxia (discussed by Chan et al.), heat shock (discussed by Morimoto and Westerheide), wounding, infection, and shear stress.

The cellular strategies used to combat these assaults on equilibrium are diverse; however, a number of strategies are relatively well-conserved through evolution (Table 20.1). Stress response networks are composed of sensors, transducers, and actuators: sensors detect the stress, and a signal is then transduced to the actuator. The whole response network can be contained in one cellular location or, in the case of eukaryotes, the signals can be exchanged between different cellular compartments (Figure 20.1). In multicellular organisms and symbiotic structures, such as biofilms [2–4], stress signals may be transmitted from cell to cell. Response networks may be composed of a small number of components with multifunctional abilities, such as the NsrR/HmpA system that is employed by *Escherichia coli* to combat nitric oxide (NO) [5]. NsrR senses NO and then transduces the signal by increasing the expression of the NO-oxidizer *hmpA*. Or, they can be complex interwoven networks with a variety of sensors, transducers, and actuators working concurrently to select the best response, such as the DNA damage sensing and response system

Handbook of Cell Signaling, Three-Volume Set 2 ed.

TABLE 20.1 Examples of conserved signaling proteins activated by different stress stimuli

Stress	Mammalian	*Drosophila*	*S. cerevisiae*
Ionizing radiation	ATM, CHK2	Mei-41	Mec1, Tel1, Rad53
	DNA-PKcs	Mei-41	Tor1
	p53	dp53	–
UV radiation	ATM, CHK2	Mei-41	Mec1, Tel1
	ATR, CHK1	Mei-41, grp	Esr1
	p38	p38b-P1	Hog1
	p53	dp53	–
UPR	PEK/PERK	Dmpek (E2K3)	Gcn2
	IRE1, ATF6, JNK	Bsk	Ire1
	HSF1	Hsf	Hsf1
Hypoxia	HIF1A	Sima	Swi1
	VHL	d-VHL	–
	p53	dp53	–
Heat shock	HSF1	Hsf	Hsf1
	Raf-1/ERK	Draf-1	Bck1

of eukaryotes (covered in the chapters by, Bennett and Resnick, Bradbury and Jackson, Lavin *et al.*, Anderson and Appella), where cells must decide amongst repair, senescence, and death. Molecular interaction maps (see online at: http://discover.nci.nih.gov/mim/index.jsp), pioneered by Kurt Kohn and others [6–8], provide an accessible framework that allows researchers to more easily navigate complex signaling networks.

Elegant studies in lower organisms (discussed in the chapters by Beuning and Walker, Demple, Bennett and Resnick, and Sogame and Abrams) have provided valuable insight for teasing apart complex stress response networks in higher eukaryotes. Specifically, studying the heat shock response in microbes and the unfolded protein response in yeast have facilitated the discovery and study of orthologous systems in mammals. Although much can be learned about complex networks by studying systems with similar functions in lower organisms, care must be taken when drawing conclusions by analogy – evolutionarily distinct solutions to problems are known to occur [9]. Furthermore, simple stress response modules are components of the greater genomic program and their initiation will have secondary effects that are not essential for countering the stress. These secondary effects can make it difficult to identify the core stress response elements, especially when the secondary effects result in global changes of transcription or remodeling of translation. This chapter attempts to provide an overview of these interwoven signaling pathways and processes, many of which will be presented in greater detail in the chapters that follow.

ORIGIN OF STRESS RESPONSE SIGNALS

Cells regularly encounter stressors, from the external environment or the by-products of cellular activity. These stresses may be detected in multiple cellular compartments and initiate signaling cascades that can permeate the cell, and can even be transduced to neighboring cells as is the case with the radiation induced bystander effect [10–13]. Stresses such as ionizing radiation and nitric oxide (NO) permeate the cell and can modify proteins or damage DNA, which will result in the activation of stress response components in multiple cellular compartments. Unlike an infection, where highly specific antibodies can target the invading pathogen, stress response networks do not necessarily respond to the stress

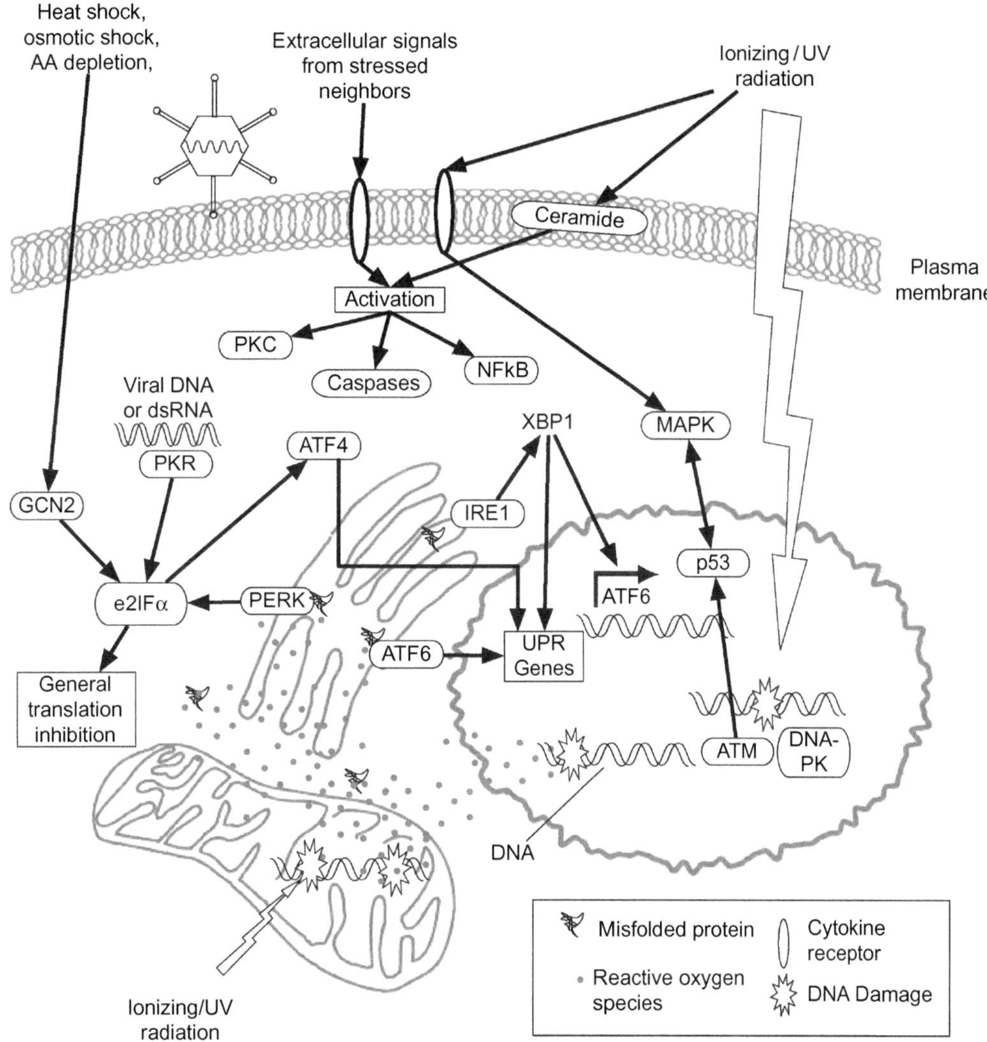

FIGURE 20.1 Major sites in the eukaryotic cell where stress signaling can originate, and some of the key signaling pathways activated in response.
Many of these pathways will be discussed in detail in the following chapters in this section. Figure not drawn to scale.

inducing agent but to the products of the agent's actions within the cell. In the case of ionizing radiation, cells detect by-products, such as reactive oxygen species (ROS) and subsequently activate ROS detoxification pathways; however, this will not mitigate all toxic effects of radiation.

Stress sensors and components may also have pleiotropic effects or be recycled for use in seemingly unrelated response programs. The ATM protein kinase, discussed in the chapter by Lavin *et al*, is present in both the nucleus and cytoplasm, and ATM is best known for its multiple roles in responding to various DNA damaging agents [14, 15]; however, there is evidence that it participates in other homeostatic processes [16]. The tyrosine kinase Abl, discussed in the chapter by Jean Wang, responds to distinct stresses in the nucleus and the cytoplasm: in the cytoplasm, Abl is involved in the regulation of actin assembly; while, in the nucleus, Abl regulates transcription in response to DNA damage and cell cycle signals. Additionally, c-Abl is believed to have cell type specific roles in signal transduction – [17] show that c-Abl plays a role in oxidative stress mediated neuronal apoptosis by cooperatively interacting with Cdk5 to stabilize p53.

In the case of some stresses, key components of the response network may be known but the initiating signals have yet to be elucidated. For example, Ink4a and Arf tumor suppressors are crucial entry points for oncogene induced senescence; however, the activating stimuli have yet to be identified (discussed in the chapter by Dimitry Bulavin). In eukaryotes, the division of labor amongst cellular compartments is a key contributor to the difficulty associated with identifying specific stress sensors – a stress response program may be initiated in a compartment other than the site of crucial damage. In this section, key stress signals in various cellular compartments are addressed and current knowledge of the sensing mechanisms are summarized when available.

The Nucleus

DNA damage has long been recognized as originating a stress signal. For instance, tracts of single stranded DNA at damaged sites in *Escherichia coli* are coated by the RecA protein, which initiates the bacterial SOS response to be discussed in detail in the chapter by Beuning and Walker. In mammals, ATM kinase, DNA dependent protein kinase (DNA-PK), and γ-H2AX foci are associated with double strand breaks (DSBs) [18–20].

ATM kinase activation by DSBs is an area of active investigation – with research indicating that ATM directly recognizes chromatin structural changes arising from DSBs [21] and that ATM can be recruited to DSBs by the Mre11-Rad50-Nbs1 (MRN) complex [22]. Regardless of the mode of activation, DSB induced activation of ATM results in phosphorylation of a number of downstream targets involved in signal transduction, including p53, Chk1, Chk2, Brca1, H2AX, NFκB, and Mdm2 [23–28]. Signaling through ATM is discussed further in the chapter by Lavin *et al.* whereas the cascades of events triggered specifically by double-strand breaks are covered in chapters by Bennett and Resnick and by Jackson and Bradbury.

DNA-PK is composed of the Ku70-Ku80 heterodimer, which binds to the DSB [29] then recruits and activates the catalytic subunit (DNA-PKcs). DNA-PK binding to broken DNA ends may initiate signaling to the cellular apoptosis machinery [30]. DNA-PKcs activates p53 *in vitro* [31], but the role of this activation *in vivo* has yet to be established [32]. A recent publication indicating that Ku70-Ku80 subunit may play a role in regulating ATM/ATR mediated phosphorylation of p53 [33] emphasizes the complexity of the DSB response network. The p53 response to DNA damage is reviewed in the chapter by Anderson and Appella.

DNA-PK phosporylates H2AX [28, 34, 35] and initiates the formation of γ-H2AX foci. γ-H2AX foci facilitate the assembly of DNA damage repair complexes [36], such as MRN [37], and prevent DSBs from developing into chromosomal translocations and breaks [38]. Suppression of genomic instability by H2AX is performed in synergy with ATM [39]. H2AX plays a key role in countering ROS induced genomic instability [39] and is discussed in the chapter by Dickey *et al.*

Single-strand breaks (SSBs) and damaged bases in DNA also result from many types of genotoxic stress. When a damaged site cannot be bypassed by the polymerase, stalled replication forks can result and DSBs and chromosomal aberrations may arise (reviewed in [40]). This in turn initiates a stress signal to prevent initiation of further DNA synthesis and to arrest cell cycle progression. In short, replication protein A (RPA) coats ssDNA at the stalled replication fork [41] then ATR interacting protein (ATRIP) recruits ATR [42], which phosphorylates CHK2 and subsequently leads to cell cycle arrest [43]. XPC, and other members of the xeroderma pigmentosum (XP) complementation group, detect damaged bases and facilitate nucleotide excision repair (NER) [44].

Although the DNA damage detection and repair system is efficient, somatic mutations accumulate over time and may lead to oncogenic stress by activating oncogenes or deactivating tumor suppressors [45]. As discussed by Dmitry Bulavin, these oncogenic mutations often result in defective cell cycle checkpoints and circumvent healthy apoptotic processes leading to unchecked proliferation.

Mitochondria

The mitochondria are a crucial center for stress detection and processing of stress signals arising in other cellular compartments [46]. Mitochondria activate apoptosis in response to external signals or stresses detected within the mitochondrion. Notably, p53 in the cytosol can localize to the mitochondria and instigate apoptosis [47–49]. Also, a role for mitochondria in Ca^{2+} signal processing is becoming apparent (reviewed in [50]); in particular, mitochondria can initiate an apoptotic program in response to Ca^{2+} released from the endoplasmic reticulum (ER) [51]. Mitochondria are important in ROS and reactive nitrogen species (RNS) signaling processes, and are believed to use these reactive molecules to amplify stress signals [11].

The best known mitochondrial function, respiration, is also a source of cellular stress. During respiration, ROS are chronically produced. During normal metabolism, superoxide is continuously produced at low levels and is rapidly converted into H_2O_2 by mitochondrial superoxide dismutases (SOD) – manganese SOD (Mn-SOD) in the matrix and copper/zinc SOD (Cu/Zn-SOD) in the intermembrane space. The H_2O_2 is subsequently detoxified by catalase and glutathione peroxidase. Mitochondrial DNA (mtDNA) is prone to damage by ROS and is slower to repair than nuclear DNA [52–54].

Metabolic and other stresses, such as ionizing and some types of ultraviolet (UV) radiation, may increase the mitochondrial ROS production rate beyond the detoxification capacity [55–57]. Elevated metabolic activity increases electron leakage from Complex I and subsequent ROS production [58]; nutrient deprivation leads to ROS production by Complex III [57]. Additionally, stress can increase the production of NO by mitochondrial NO synthase (mtNOS); NO rapidly reacts with superoxide to form peroxynitrite [59]. Peroxynitrite is a potent tyrosine nitrating agent that has been shown to inactivate mtSOD [60]. Inactivation of mtSOD will result in an amplification of ROS production. Excessive ROS production promotes shedding of cytochome C and initiation of apoptosis [61].

Mitochondria possess a DNA repair system; however, the stress sensors have yet to be elucidated and it is not as well characterized as the nuclear DNA damage repair system (reviewed in [62]). The mtDNA repair system is not as complex as the nuclear DNA repair system – there are systems for base excision repair (BER) and DSB repair and mismatch repair activity has been detected [63]; however, there is no solid evidence for a nucleotide excision repair (NER). Although the mitochondrial BER (mtBER) pathway has been relatively well characterized, the manner in which the mtDNA damage is sensed is unknown. p53 promotes mtBER activity but the significance of this interaction for mtDNA damage sensing has yet to be elucidated [64, 65].

The Cytoplasm

The cytoplasm is a dense gelatinous protein-filled matrix in which the organelles are embedded. It contains features that are involved in protein production, metabolism, and structural stability. Stresses such as ionizing radiation, ROS, RNS [66, 67], UV radiation, heat shock, osmotic shock, metabolic deprivation, and viral infection [68, 69] negatively impact a number of processes in the cytosol and the organelles. Nitrosative stress is sensed in the cytoplasm by S-nitrosylation of protein thiols; S-nitrosylation of GAPDH results in GADPH binding to Siah1 with subsequent translocation to the nucleus and initiation of apoptosis [67]. Oxidative stress that upsets the reducing ratio of the cytosol results in S-glutathionylation of various proteins also leading to apoptosis [70]. In response to UV radiation [47] and other stresses, p53 can activate BAX [71] to promote mitochondrial membrane permeabilization and subsequent apoptosis. However, the most notable stresses in the cytoplasm are those that impact translation and protein folding.

Essentially any stress that induces a transcriptional response will also invoke a translational response to facilitate production of proteins from stress induced mRNA. Translational reprogramming occurs mainly through inhibition of translation initiation [72]. In eukaryotes, this program in centered around a translation initiation factor eIF2a that stalls translation and initiates the formation of stress granules (SGs). The SGs serve as holding pens for mRNAs undergoing translation [73]. Shunting the transcriptome to SGs, allows the cell to focus resources on translating stress response proteins.

eIF2α is activated by the eIF2α kinases, GCN2, HRI, PERK, and PKR, which detect different stresses (reviewed in [74]). GCN2 responds to amino acid starvation [75] and UV radiation. HRI responds to heat shock, osmotic shock, or heme deficiency [76]. PKR is activated by double-stranded RNA and DNA that are present in the cytoplasm during viral infections [77]. PERK responds to protein misfolding in the ER [78] and mechanical stress [79]. Interestingly, activation of another translation initiation factor, eIF3h, has recently been shown to promote translation and oncogenesis [80].

Accumulation of misfolded and denatured proteins negatively impacts a variety of cellular processes (reviewed in [81]). Protein folding is a complex procedure with occasional failures, and proteins can be destabilized over time. Chaperones aid in the folding process and assist in clearing misfolded and denatured proteins. Stresses such as heat shock, ionizing radiation, oxidative stress, hypoxia, ROS, and RNS [82, 83] can increase the number of folding failures and unfolded proteins, subsequently initiating the protein misfolding responses. The heat shock response (discussed by Morimoto and Westerheide) [84, 85] is crucial to dealing with protein folding stresses in prokaryotes and in the cytoplasm of eukaryotes. In *E. coli*, heat shock protein 33 senses oxidative stress induced protein unfolding [86]. Eukaryotes also have a system for sensing protein folding stresses in the endoplasmic reticulum (ER) (reviewed in [87]).

Endoplasmic Reticulum Stress Signals and the Unfolded Protein Response

The ER is where secretory proteins are processed, a variety of posttranslational modifications are made, and the site of intracellular Ca^{2+} storage. Stresses affecting Ca^{2+} levels or the redox balance are the most notable ER stress signals that impact protein folding. Ca^{2+} is essential for a number of enzymes, including some involved in protein processing [88]. For example, insufficient Ca^{2+} interferes with glycoprotein processing [89]; UDP glycoprotein-glucosyltransferase identifies unprocessed glycoproteins [90] and schedules them for processing or degradation. The ER has a relatively oxidizing environment compared to the cytosol, to facilitate disulphide bond formation. GSH/GSSG in the ER is between 1:1 and 3:1 whereas the overall cellular ratio is at least 30:1 [91]. Because the ER has a lower ratio of GSH/GSSG compared to the cytoplasm, proteins in the ER will be more sensitive to ROS/RNS attack because there isn't as large a GSH pool to divert the reactive species from protein thiols. Regardless of the type of stress, most ER stresses induce some degree of protein misfolding and activate a program termed the unfolded protein response (UPR) to counter problems with protein folding and modification.

The key mammalian sensors that initiate the UPR are PKR-like endoplasmic reticulum kinase (PERK), activating transcription factor 6 (ATF6), and inositol requiring protein 1 (IRE1) [92]; they are transmembrane proteins with sensing regions in the ER interior. These sensors are inactive when bound by the chaperone BiP; an increase in the level of misfolded proteins will titrate BiP away from these sensors thus activating them [93]. The presence of three ER stress sensors allows mammalian systems to discriminate amongst various stresses that impact protein folding [94].

When responding to cellular stresses, the ER initiates both pro-survival and pro-apoptotic programs with the outcome (repair or death) being a function of stress mRNA and protein stabilities [95].

Both PERK and IRE1 are transmembrane receptor kinases that respond to ER stress. Although PERK and IRE1 have some overlapping functions, they activate different key components of the UPR. PERK phosphorylates eIF2α, which represses translation in general but increases translation of ATF4 [96]. ATF4 regulates a number of UPR, metabolic, and stress response genes such as FOS, JUN, NRF2, and GADD34 [97]. GADD34 promotes dephosphorylation of eIF2α thus serving as a negative feedback for translation repression [98]. In response to ER stress, IRE1 oligomerizes and undergoes autophosphorylation. Phophorylated IRE1 possesses an RNase activity that mediates non-conventional splicing of the yeast transcription factor HAC1 and metazoan XBP1 [99]; this splicing increases the activity of XBP1 and is essential for transcription of UPR genes [99, 100]. In yeast, IRE1's splicing activity appears to be highly specific for HAC1 [101]. XBP1 mRNA expression is positively regulated by the third ER stress sensor, ATF6 [99]. ATF6 is a membrane embedded transcription factor that is activated by ER stress [102]. Upon sensing ER stress, ATF6 is released from the ER membrane to the cytosol and is cleaved by serine proteases 1 and 2 into an active transcription factor, which localizes to the nucleus and increases transcription of genes essential to the UPR [103–105].

The Plasma Membrane

Exposure to toxins or infection can give rise to a stress signal originating in the plasma membrane. Engagement of cell surface receptors, such as the tumor necrosis factor (TNF) receptor superfamily or the interleukin 1 (IL-1) receptor may lead to activation of NFκB or initiation of the caspase cascade (reviewed in [106]). Activation of TNF receptors initiates hydrolysis of sphingomyelin to produce ceramide, which acts as a second messenger. Irradiation of enucleated cells can initiate ceramide signaling in the plasma membrane, clearly indicating signaling of radiation damage in the absence of DNA damage specific signals [107]. ROS may interact with membrane receptors, such as the TNFRs, and increase their sensitivity for their ligand or activate them in the absence of the ligand [108]. Additionally ROS, OH in particular, can initiate lipid peroxidation, which, if unchecked, will decrease membrane fluidity and interfere with membrane integrity [109]. 1-Cys peroxiredoxin scavenges peroxides and reduces peroxidized membrane phospholipids [109]. Membrane bound toll like receptors (TLRs) recognize various microbial components and can activate NFκB or the ERK, JNK, and p38 MAP kinases. For instance, lipopolysaccharide (LPS) from gram negative bacteria is sensed via TLR4 [110]. And the UV radiation response is mediated in part by growth factor and cytokine receptors [111–113].

Extracellular Signals

Cells also respond to stress or injury at a distance through communication with other cells. Cytokines, growth factors, and hormones are released in response to internal cues or external stressors, and their internalization or binding to receptors of other cells can set in motion signal cascades and alter transcriptional programs within the target cells. Such extracellular communication appears to be a basis of the bystander effect, where irradiation of one cell can result in altered protein expression and induction of damage end points such as micronuclei, mutation, genomic instability, and apoptosis in non-irradiated cells in the same culture [11, 114, 115]. Quite recently, bystander carcinogenesis has been demonstrated *in vivo* in the non-irradiated brains of partial-body irradiated mice [12]. The radiation bystander effect also appears to be cell type dependent [29].

Studies using gap junction inhibitors [116] or genetic manipulation of the gap junction protein connexin43 [117] have indicated a role for communication through gap junctions in bystander signaling. The stress sensed in the bystander effect may be partially due to ROS and RNS generation from damaged cells. Treating cells with the ROS scavenging enzymes superoxide dismutase or catalase [118–120] or with the NO scavenger c-PTIO [118] reduces micronuclei and γ-H2AX foci formation in bystander cells. As the levels of superoxide or NO required to damage DNA are relatively high [121, 122], it is likely that the damage is the result of a more reactive product. It has been speculated that this increase in ROS and RNS serves as an apoptotic signal that is transmitted to other mitochondria, where it is amplified by further ROS/RNS production [55]. ROS/RNS based amplification of DNA damage and other signals by mitochondria may play a role in the bystander effect, since cells depleted of functional mitochondria manifest a significantly weakened bystander response [13].

SIGNAL TRANSDUCTION

Once a stress situation has been recognized and signaling is initiated, numerous pathways exist for the amplification and direction of the message. Signal transduction can occur through precise protein to protein relays, or through less specific mechanisms where the concentration of a molecule is changed and sensors respond to the change in concentration gradient. The archetypical protein based method for conveying a highly specific signal is through phosphorelays where proteins pass a phosphate through a transduction pathway. Molecules whose gradients serve as a stress signal, or signal in a response network, include cAMP [123], cGMP [124], oxygen, ROS [125], and RNS [122, 126].

More recently, non-coding RNAs (ncRNAs) have been shown to be crucial signal transducers that facilitate degradation of mRNA [127].

Gradient signals are less specific with their effects being shaped by cellular architecture and fluid dynamics [126, 128, 129]. The relative gradients of these signals can also induce different responses [130, 131]. For example, O_2 concentration modulates NO bioavailability, and NO levels can influence the production of H_2O_2. Bruce Demple discusses how ROS gradients are employed in amplifying signals. Surprisingly, evidence is also accumulating for the employment of the free radical NO in protein relays [132, 133].

Protein Mediated Signal Transduction

Protein modification is a major mechanism for stress signal propagation within the cell, and perhaps the best studied modification is alteration of phosphorylation state. Other methods that proteins use for passing messages include acetylation [134, 135], ISGylation [136], nitrosylation [137], ROS generation [138], and sumolyation [139]. These protein signal transduction cascades serve as a network based decision-making process where multiple inputs may be fed to a number of components and a final decision is made.

p53 is the exemplar hub for processing and integrating multiple stress signals. In response to a variety of stresses, p53 can undergo a number of posttranslational modifications, which alter the transcriptional output and other cellular processes, as reviewed in detail in the chapter by Anderson and Appella. Briefly, at least 30 sites of regulatory posttranslational modification of p53 have been described, including sites of phosphorylation, acetylation, and sumoylation. Modifications of these sites vary in response to different stresses and the activation of different upstream regulatory pathways. For instance, distinct patterns of p53 modification have been reported in ionizing radiation treated, UV irradiated, or senescent human primary fibroblasts [140–142]. Agents inducing double-strand breaks, such as ionizing radiation, activate the ATM kinase pathway, resulting in phosphorylation of p53 at Ser15, Ser20, Ser9, and Ser46 [143], as well as activating a number of additional kinases phosphorylating further sites in p53. In contrast, agents such as UV radiation that induce bulky lesions in DNA activate the ATR [144], p38 MAPK [145], and JNK [146] kinase pathways, resulting in phosphorylations at Ser15 and Ser37, Ser33 and Ser46, and Thr81, respectively. Such modifications of p53 subsequently affect its interactions with other proteins and its sequence specific DNA binding and transactivation activities, resulting in differential gene transcrip-

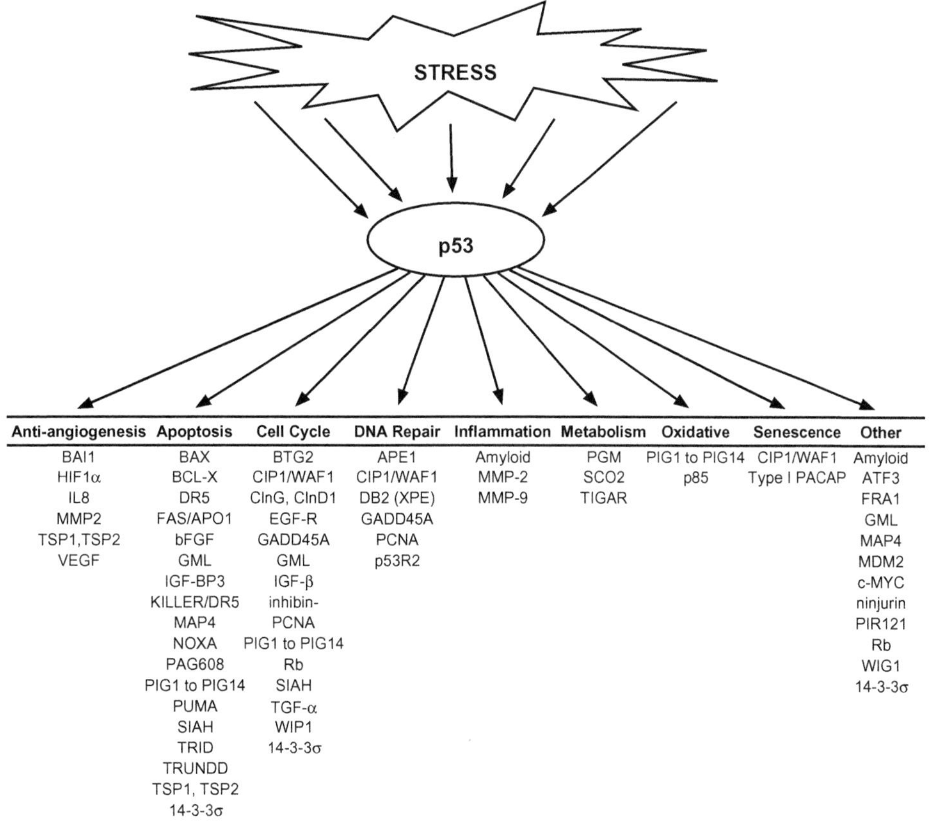

FIGURE 20.2 Diverse stress signals can be transduced by a variety of pathways through p53. This, in turn, results in the execution of different cellular programs, mediated in part by the activation or repression of transcription of a variety of genes, including the examples listed here.

tion profiles (Figure 20.2). Thus p53 activation can promote such diverse outcomes as apoptosis [147,148], reversible cell cycle arrest [149,150], or a permanent cell cycle arrest indistinguishable from senescence [151, 152]. The many different possible modifications and binding interactions of p53 serve as a means to tailor the specificity of p53 responses to suit the originating stress and the specific cellular environment.

MAP Kinase Pathways

Protein activation by phosphorylation is a common mechanism for intracellular signal transduction, and the mitogen activated protein kinase (MAPK) cascades provide a major example of this mechanism. The MAPK signaling cascades are three distinct but analogous pathways comprising the ERK-1/ERK-2 mitogen activated pathway and the stress activated protein kinases (SAPK), p38, and c-Jun N-terminal kinase (JNK). The MAPK/ERK pathway is activated primarily by mitogenic stimuli such as extracellular growth factors, as well as during physiological processes such as T cell activation in immune response [153]. Conversely, the SAPK/JNK and p38 pathways are activated predominantly by stresses, such as UV or ionizing radiation, heat shock, or inflammatory cytokines. There is also a regulatory phosphatase system for the downregulation and inactivation of each of these signaling cascades. The degree to which the different signaling cascades are activated and the duration of the activation appear to shape the cellular response and can result in altered gene transcription, cell cycle arrest, or apoptosis. The interplay of the various MAPK pathways is discussed in greater detail in the chapter by Böhmer, Weiss, and Herrlich.

Nitrosylation

NO mediated stress signaling is also an important component of many stress response systems [154]. The paradigmatic model for NO based signal transduction is binding to a metal center and altering enzymatic activity. The best known stress response that employs NO is modulation of blood flow in response to oxygen depletion. In response to decreases in oxygen, NO is produced in the endothelium and then diffuses to the vascular smooth muscle where it binds to the heme center of soluble guanylate cyclase and initiates a vasodilatory cascade.

In addition to reacting with heme centers, NO is known to modify protein thiols, and these modifications alter protein function [155, 156]. The most prominent modification is S-nitrosylation, which is the process of replacing cysteine's H^+ with NO^+. S-nitrosylation can induce protein misfolding [82], serve as a regulatory signal in apoptosis [132, 133] and chromatin remodeling [157], promote ISGylation [158], and impact a number of other biological processes [156]. Recent research has indicated that NO can also participate in relays reminiscent of phosphorelays. Thioredoxin (Trx) has been shown to have a transnitrosylase activity that allows it to transfer an NO from Trx-Cys73-SNO to Cys163 on caspase-3 (Casp-3) [132]. Trx and thioredoxin reductase can denitrosate Casp-3 in response to apoptotic signals, such as Fas [133].

Transcriptional and Post-Transcriptional Modifications

In addition to, and often consequent to, the modification of proteins, the cellular response to stress also alters the abundance of specific mRNA transcripts. By changing the transcripts and subsequently the proteins manufactured in response to specific stress situations, the cell makes components available to promote or protect against cell death, to repair damaged DNA, or to halt cell cycle progression and maintain damage checkpoints. Translation of stress specific mRNAs is facilitated by reprogramming translation and modifying mRNA levels [159].

Stress based translational reprogramming by eIF2α is described above and serves to reallocate translational resources away from the current transcriptome to the newly transcribed stress response components. More recently, small ncRNAs (sRNAs) have been shown to have a role in targeted remodeling of translation [160]. mRNA levels are modified by altering transcription factor activity and mRNA stability. mRNA stability is regulated by proteins and sRNAs. The chapter by Subramanya Srikantan and Myriam Gorospe covers proteins, such as HurR, that stabilize stress mRNAs, and proteins, such as TTP, KSRP, and BRF1, that facilitate mRNA degradation, as well as the upstream signaling pathways that signal shifts in mRNA degradation policies and general degradation policies. Proteins involved in regulating mRNA can be highly specific for a few transcripts, as is the case for BRF1, which facilitates the decay of TNFα and IL-3 [161].

sRNA regulation of mRNA translation and degradation is becoming known as an increasingly important component of response networks with at least 30 percent of genes subject to such regulation ([162, 163] and reviewed in [164]). Production of regulatory sRNAs in response to stress has a much shorter time lag than protein production, thus allowing for faster response dynamics. There are a number of classes of sRNAs that affect mRNA stability [127] with the two best known being: short interfering RNAs (siRNAs) [165, 166] and micro-RNAs (miRNAs) [160]. siRNAs and miRNAs are approximately 21–25 nucleotides and are produced from longer dsRNA or RNA hairpin loops, respectively (reviewed in [127, 160, 166]. siRNAs form dsRNA duplexes with their mRNA targets and facilitate degradation through dsRNA degradation pathways. Because they are perfect or near perfect matches for their targets they typically regulate a small number of genes. miRNAs bind to a much smaller sequence in their targets – the target binding sequence is located in nucleotides 2–7 at the miRNA's 5' end [167] and binding is enhanced by accessory features in the mRNA [168]. They can either activate transcript degradation or block translation.

The small target size for miRNAs gives them the potential to interact with a number of targets and convey signals that drastically alter the transcriptome and translation. In fact, the mir-34 miRNA family is transcriptionally regulated by p53, and its members facilitate the degradation of a number of genes employed in cell cycle progression [169] and promote apoptosis [150]. miR-34a regulates apoptosis by repressing SIRT1 (silent information regulator 1) mRNA stability [170]. miRNAs are also produced in response to other stresses [171], such as hypoxia [172]. As with other regulatory factors, miRNA expression appears to be cell-type dependent – the response to ionizing radiation is virtually non-existent in TK6 cell lines [171], whereas, there is a robust response in embryonic stem cells [173] and fibroblasts [174]. Also, miRNAs appear to be downregulated in cancers [175], and dysregulation of miRNAs' expression may be a contributing factor to some cancers [150].

ncRNA mediated signals can also affect other processes. sRNAs transduce signals to regulate transcription by promoting heterochromatin formation [176]. In addition, longer ncRNAs have been observed to play a role in regulating mRNA stability in response to stress. Starvation induces the expression of a long ncRNA that modulates Dicer [177]. Because of the diverse regulatory roles of ncRNAs it is essential to incorporate their activities into any systems level model of a stress response network.

SYSTEMS LEVEL DEDUCTIONS OF STRESS RESPONSE NETWORKS

From a systems perspective, organisms can be viewed as complex decision-making networks that receive information about the environmental state from a variety of input sensors and deliver their "best" decision to adapt to this state. Decisions made in response to stressors can induce a plethora of changes in the expression of mRNAs and proteins as well as dramatic alterations to the metabolite pool. The changes induced by the stressor may include a number of tangential secondary responses. These secondary effects may obscure the primary components of a stress response program. For example, Hog1 is a MAP kinase that was previously believed to counter osmotic stress by altering transcription; however, a recent study indicated that Hog1 counters osmotic stress by modulating glycerol metabolism [178] and the transcriptional activity may aid in preparing for extended stress. To identify the relevant components of a stress response and understand the global implications, it is prudent to collect finely detailed information about the cellular state and develop analytic tools to weed out the secondary effects.

Systems level tools allow us to acquire highly detailed information about cellular responses to stress. Advances in DNA sequencing technology [179,180] have made acquisition of the genomic program appear trivial. High throughput techniques have also been developed for taking snapshots of the whole transcriptome [181], metabolome [182, 183], and proteome [184, 185] thus providing researchers with mountains of information about cellular responses. The massive amount of descriptive information is just the first step in understanding systems level stress responses. Innovative algorithms [186–191] and experimental approaches [192–194] are also necessary to make sense of this information avalanche.

A great deal of progress has been made in untangling stress response networks in prokaryotes [195, 196], and decades of metabolic engineering research have provided us with tools for analyzing metabolic shifts [197–202]. For example, using regulatory information, transcriptome, and sequence data we [195] were able to deduce *E. coli*'s stress response network to NO and to identify a major mechanism of action for NO stress (Figure 20.3).

Yeast has provided a good eukaryotic model for early systems-level studies. In this organism, metabolic pathways, such as that for galactose utilization, can be systematically perturbed, and information from gene expression profiling and proteomic studies can be integrated with known molecular interactions and specific gene knockouts to suggest mechanisms of pathway regulation and interactions with other metabolic and cellular processes [203]. Similar approaches have also been applied to the study of DNA damage responses in yeast. By integrating global measurements of gene expression, transcription factor binding, genetic knockouts, and protein interaction data, new DNA damage specific transcription factor binding motifs have been identified [204].

Global genomic phenotyping is also feasible in *Saccharomyces cerevisiae*, and a complete set of genetically "bar coded" single gene deletion mutants is available for all non-essential genes in this organism. Genes conferring sensitivity or resistance to various stresses have been identified with this model system [205–207]. Genes contributing to surviving a stress were found to differ from the genes responding to the same stress in global expression assays [208], however. Genomic phenotyping studies of responses to methyl methanesulfonate, 4-nitroquinoline-N-oxide, *tert*-butyl hydroperoxide, and UV radiation have also been mapped onto the *S. cerevisiae* interactome, built from protein interaction information [209]. This has allowed the identification of the protein complexes and pathways that were most important for regulating the response to specific toxicants. Such a whole genome single gene knockout approach is not currently feasible for mammalian cells, but some investigators are working with high throughput approaches using siRNAs libraries [210] that may some day lead to analogous systematic genomic phenotyping in mammalian cells.

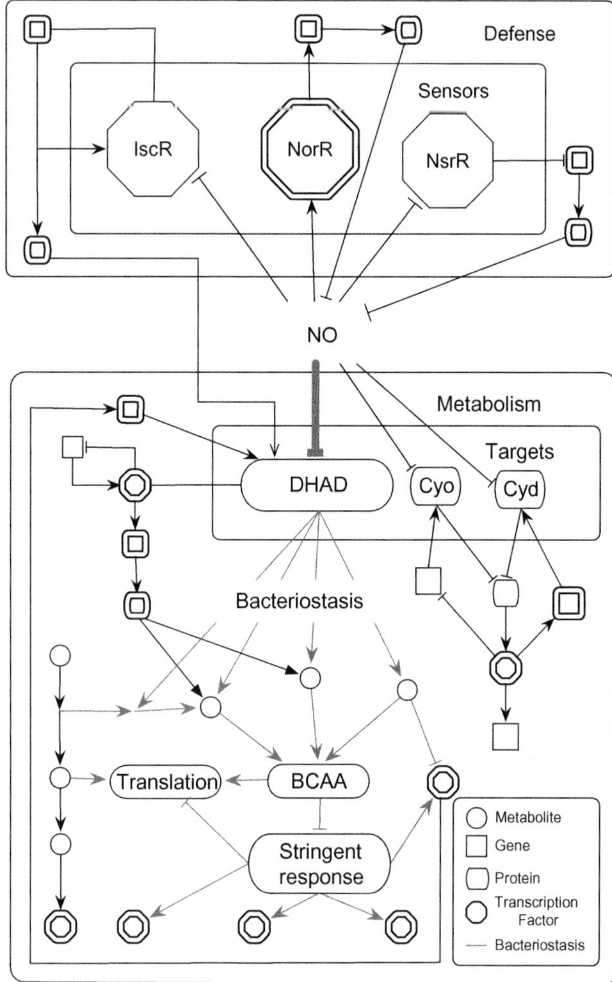

FIGURE 20.3 A network diagram of the essential NO response network in *E. coli*.
NO targets the metal prosthetics of the cytochrome oxidases (Cyo and Cyd) and the iron-sulfur cluster of dihydroxyacid dehydratase (DHAD). DHAD is an essential enzyme for branched chain amino acid biosynthesis, and its inhibition induces bacteriostasis. *E. coli* contains a defense system that senses NO and either facilitates NO detoxification or repair of NO mediated damage to DHAD [195].

Mammalian systems present more of a challenge to systems level analyses. The completion of the human genome sequencing projects [211, 212] was a humbling moment in science, due to the magnitude of the achievement and to the realization that the genome sequence is only the tip of the iceberg when it comes to unraveling the mysteries of human life. With the human genome sequence we in principle had the program for humans and should be able to unravel stress and other response networks. However, we were sorely lacking in knowledge regarding the running of the program. Only ~2 percent of the human genome codes for proteins with a significant fraction of the non-coding regions are believed to be used to precisely control execution of the genomic program [213]. Further layers of complexity are added by alternative splicing, additional transcriptional regulatory components, chromatin remodeling, and compartmentalization [164, 213, 214]. Endeavors such as the ENCODE project [214], have focused on exploring and richly annotating a subset of the genome, and they have revealed new insights into regulatory networks and evolutionary constraints on our genomic program. The Catalogue of Somatic Mutations in Cancer (COSMIC) project from the Cancer Genome Project at the Sanger Institute [215] has identified hot spots for the evolution of oncogenic stress.

Fortunately, recent advances in technologies [216] and algorithms [217–221] indicate that we will eventually be able to scale this mountain. Chromatin immunoprecipitation/DNA microarray (ChIP-chip) and sequencing technologies [216] have accelerated the discovery of regulatory interactions for transcription factors, such as p53 [222]. The development of techniques such as reverse engineering of regulatory networks attempts to build these vast amounts of data into coherent maps of mammalian molecular interactions [223–225]. Once such hybrid interactome maps are developed, they can be used to assess the perturbations resulting from exposure to various exogenous stresses or oncogenic stress in tumors, leading to the identification of key response mechanisms [223, 226].

"Omics" surveys of stress responses are providing a wealth of descriptive information. We and others have measured the transcriptional, or metabolomic, response to a variety of stresses including ionizing [183, 227–230] and UV [231, 232] irradiation, hypoxia [233], alkylating agents [234], and toxic metals [235–237]. Although some common factors emerge, such as increased expression of genes such as *CDKN1A* indicating activation of the p53 pathway, there is a profound heterogeneity in stress responsive genes. To understand this heterogeneity it is essential to develop integrated databases and tools for maximizing the benefits derived from these databases.

The NCI-60 and Integromic Approaches

The NCI Anti-Neoplastic Drug Screen (NCI-60) [238] is a panel of 60 human tumor cell lines derived from nine different tissue lineages, and is likely the most comprehensively characterized *in vitro* cancer cell model. The NCI-60 is the archetypal model for studying molecular mechanisms of disease and identifying therapeutic modes of action; lessons learned from studying the NCI-60 will be of immense benefit in the impending transition to systems level analyses of diseases *in vivo*. More than 100,000 chemicals and natural products have been screened in the NCI-60 as possible chemotherapy agents. This panel of 60 cell lines has also been the subject of extensive profiling of protein and gene expression, as well as studies of SNP profiles, CGH, DNA methylation, and mutation in specific genes [230, 239–246]. Correlations between gene expression or other cellular parameters and sensitivity to killing by many diverse agents have provided insight into mechanisms

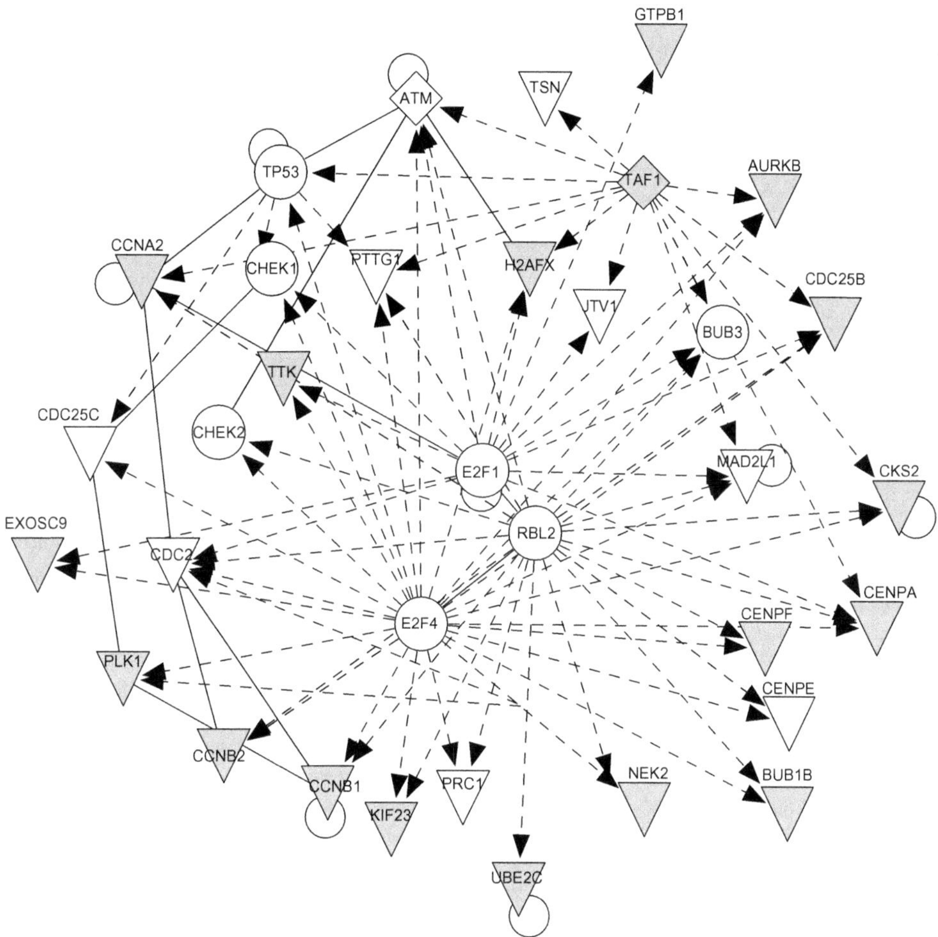

FIGURE 20.4 Using interaction maps to visualize high density array data facilitates the formation of hypotheses about the data. An interaction map of genes with most widespread downregulation across the NCI-60 in response to γ-ray exposure was created using interaction data from the BIND protein interaction database. A large number of the repressed genes interact with the transcription factors: E2F4, E2F1, RBL2, and TAF1. Experimental evidence indicated that E2F4 localized to the nucleus after irradiation [230]. Solid lines, protein–protein interactions; dashed lines with arrows, protein–DNA interactions. Diamonds, upregulated significantly in >10 percent of cell lines. Inverted triangles, downregulated significantly in >10 percent of cell lines. Gray significantly perturbed in >25 percent of cell lines.

underlying specific stress responses. This has translated into classification of mechanisms of drug action, and the approval of bortezomib and oxaliplatin as clinical chemotherapy agents [241, 245, 247].

We have employed transcriptome measurements and systems level information about protein–protein interactions to identify features in the NCI-60 that predict radiosensitivity [230]. Using interaction databases, such as BIND [248], and regulatory databases, such as TRED [249], it is possible to develop hypotheses about which transcription factors are responsible for the observed transcriptional perturbations. Analyzing the genes uniformly downregulated by ionizing radiation in the NCI-60 transcriptome data set we discovered that the transcription factor E2F4 is central to this response (Figure 20.4). The E2F transcription factor family had previously been shown to regulate replication and mitotic functions [250, 251].

Furthermore, to facilitate analysis of heterogenous stress responses, our group [252] and the Golub group [220] have developed databases of the transcriptional response of select cell lines to a wide variety of stresses. A key driving force behind developing these massive databases in a single lab is to provide the community with transcriptome data from relatively uniform data sets to analyze. These resources are already being used by other researchers in their investigations [253–256].

Some laboratories are even starting to survey transcriptome, proteome, metabolome, and genomic modifications in defined systems, to provide databases for integromics analyses [257–259]. Work from John Weinstein and collaborators [244, 246, 260–264] with the NCI-60 cell lines is starting to yield novel information about the network properties of cancer progression. Overall, systems level approaches hold great promise for unraveling mammalian stress response networks and should yield a substantial number of hypotheses about stress signaling that can be verified with traditional molecular biology techniques.

ACKNOWLEDGEMENT

DRH was supported by National Institutes of Health Grant 5T32CA009078.

REFERENCES

1. Li HH, Aubrecht J, Fornace Jr AJ. Toxicogenomics: overview and potential applications for the study of non-covalent DNA interacting chemicals. *Mutat Res* 2007;**623**(1–2):98–108.
2. Dietrich LE, Price-Whelan A, Petersen A, Whiteley M, Newman DK. The phenazine pyocyanin is a terminal signalling factor in the quorum sensing network of Pseudomonas aeruginosa. *Mol Microbiol* 2006;**61**(5):1308–21.
3. Dietrich LEP, Teal TK, Price-Whelan A, Newman DK. Redox-active antibiotics control gene expression and community behavior in divergent bacteria. *Science* 2008;**321**(5893):1203–6.
4. Lazdunski AM, Ventre I, Sturgis JN. Regulatory circuits and communication in Gram-negative bacteria. *Nat Rev Microbiol* 2004;**2**(7):581–92.
5. Bodenmiller DM, Spiro S. The yjeB (nsrR) gene of Escherichia coli encodes a nitric oxide-sensitive transcriptional regulator. *J Bacteriol* 2006;**188**(3):874–81.
6. Kohn KW. Molecular interaction map of the mammalian cell cycle control and DNA repair systems. *Mol Biol Cell* 1999;**10**(8):2703–34.
7. Kohn KW, Aladjem MI, Kim S, Weinstein JN, Pommier Y. Depicting combinatorial complexity with the molecular interaction map notation. *Mol Syst Biol* 2006;**2**:51.
8. Kohn KW, Aladjem MI, Weinstein JN, Pommier Y. Molecular interaction maps of bioregulatory networks: a general rubric for systems biology. *Mol Biol Cell* 2006;**17**(1):1–13.
9. Singh AH, Wolf DM, Wang P, Arkin AP. Modularity of stress response evolution. *Proc Natl Acad Sci USA* 2008;**105**(21):7500–5.
10. Nagasawa H, Little JB. Induction of sister chromatid exchanges by extremely low doses of alpha-particles. *Cancer Res* 1992;**52**(22):6394–6.
11. Hei TK, Zhou H, Ivanov VN, Hong M, Lieberman HB, Brenner DJ, Amundson SA, Geard CR. Mechanism of radiation-induced bystander effects: a unifying model. *J Pharm Pharmacol* 2008;**60**(8):943–50.
12. Mancuso M, Pasquali E, Leonardi S, Tanori M, Rebessi S, Di Majo V, Pazzaglia S, Toni MP, Pimpinella M, Covelli V, Saran A. Oncogenic bystander radiation effects in Patched heterozygous mouse cerebellum. *Proc Natl Acad Sci USA* 2008;**105**(34):12,445–12,450.
13. Zhou H, Ivanov VN, Lien YC, Davidson M, Hei TK. Mitochondrial function and nuclear factor-kappaB-mediated signaling in radiation-induced bystander effects. *Cancer Res* 2008;**68**(7):2233–40.
14. Barzilai A, Rotman G, Shiloh Y. ATM deficiency and oxidative stress: a new dimension of defective response to DNA damage. *DNA Repair (Amst)* 2002;**1**(1):3–25.
15. Rotman G, Shiloh Y. ATM: a mediator of multiple responses to genotoxic stress. *Oncogene* 1999;**18**(45):6135–44.
16. Weizman N, Shiloh Y, Barzilai A. Contribution of the Atm protein to maintaining cellular homeostasis evidenced by continuous activation of the AP-1 pathway in Atm-deficient brains. *J Biol Chem* 2003;**278**(9):6741–7.
17. Lee JH, Jeong MW, Kim W, Choi YH, Kim KT. Cooperative Roles of c-Abl and Cdk5 in Regulation of p53 in Response to Oxidative Stress. *J Biol Chem* 2008;**283**(28):19,826–19,835.
18. Morio T, Kim H. Ku, Artemis, and ataxia-telangiectasia-mutated: signalling networks in DNA damage. *Int J Biochem Cell Biol* 2008;**40**(4):598–603.
19. Shrivastav M, De Haro LP, Nickoloff JA. Regulation of DNA double-strand break repair pathway choice. *Cell Res* 2008;**18**(1):134–47.
20. Shroff R, Arbel-Eden A, Pilch D, Ira G, Bonner WM, Petrini JH, Haber JE, Lichten M. Distribution and dynamics of chromatin modification induced by a defined DNA double-strand break. *Curr Biol* 2004;**14**(19):1703–11.
21. Bakkenist CJ, Kastan MB. DNA damage activates ATM through intermolecular autophosphorylation and dimer dissociation. *Nature* 2003;**421**(6922):499–506.
22. Carson CT, Schwartz RA, Stracker TH, Lilley CE, Lee DV, Weitzman MD. The Mre11 complex is required for ATM activation and the G2/M checkpoint. *EMBO J* 2003;**22**(24):6610–20.
23. Banin S, Moyal L, Shieh S, Taya Y, Anderson CW, Chessa L, Smorodinsky NI, Prives C, Reiss Y, Shiloh Y, Ziv Y. Enhanced phosphorylation of p53 by ATM in response to DNA damage. *Science* 1998;**281**(5383):1674–7.
24. Canman CE, Lim DS, Cimprich KA, Taya Y, Tamai K, Sakaguchi K, Appella E, Kastan MB, Siliciano JD. Activation of the ATM kinase by ionizing radiation and phosphorylation of p53. *Science* 1998;**281**(5383):1677–9.
25. Cortez D, Wang Y, Qin J, Elledge SJ. Requirement of ATM-dependent phosphorylation of brca1 in the DNA damage response to double-strand breaks. *Science* 1999;**286**(5442):1162–6.
26. Khosravi R, Maya R, Gottlieb T, Oren M, Shiloh Y, Shkedy D. Rapid ATM-dependent phosphorylation of MDM2 precedes p53 accumulation in response to DNA damage. *Proc Natl Acad Sci USA* 1999;**96**(26):14,973–14,977.
27. Piret B, Schoonbroodt S, Piette J. The ATM protein is required for sustained activation of NF-kappaB following DNA damage. *Oncogene* 1999;**18**(13):2261–71.
28. Stiff T, O'Driscoll M, Rief N, Iwabuchi K, Lobrich M, Jeggo PA. ATM and DNA-PK function redundantly to phosphorylate H2AX after exposure to ionizing radiation. *Cancer Res* 2004;**64**(7):2390–6.
29. Vines AM, Lyng FM, McClean B, Seymour C, Mothersill CE. Bystander signal production and response are independent processes which are cell line dependent. *Int J Radiat Biol* 2008;**84**(2):83–90.
30. Wang S, Guo M, Ouyang H, Li X, Cordon-Cardo C, Kurimasa A, Chen DJ, Fuks Z, Ling CC, Li GC. The catalytic subunit of DNA-dependent protein kinase selectively regulates p53-dependent apoptosis but not cell-cycle arrest. *Proc Natl Acad Sci USA* 2000;**97**(4):1584–8.
31. Burma S, Kurimasa A, Xie G, Taya Y, Araki R, Abe M, Crissman HA, Ouyang H, Li GC, Chen DJ. DNA-dependent protein kinase-independent activation of p53 in response to DNA damage. *J Biol Chem* 1999;**274**(24):17,139–17,143.
32. Dip R, Naegeli H. More than just strand breaks: the recognition of structural DNA discontinuities by DNA-dependent protein kinase catalytic subunit. *FASEB J* 2005;**19**(7):704–15.
33. Tomimatsu N, Tahimic CG, Otsuki A, Burma S, Fukuhara A, Sato K, Shiota G, Oshimura M, Chen DJ, Kurimasa A. Ku70/80 modulates ATM and ATR signaling pathways in response to DNA double strand breaks. *J Biol Chem* 2007;**282**(14):10,138–10,145.
34. Burma S, Chen BP, Murphy M, Kurimasa A, Chen DJ. ATM phosphorylates histone H2AX in response to DNA double-strand breaks. *J Biol Chem* 2001;**276**(45):42,462–42,467.

35. Mukherjee B, Kessinger C, Kobayashi J, Chen BP, Chen DJ, Chatterjee A, Burma S. DNA-PK phosphorylates histone H2AX during apoptotic DNA fragmentation in mammalian cells. *DNA Repair (Amst)* 2006;**5**(5):575–90.
36. Fernandez-Capetillo O, Lee A, Nussenzweig M, Nussenzweig A. H2AX: the histone guardian of the genome. *DNA Repair (Amst)* 2004;**3**(8–9):959–67.
37. Celeste A, Petersen S, Romanienko PJ, Fernandez-Capetillo O, Chen HT, Sedelnikova OA, Reina-San-Martin B, Coppola V, Meffre E, Difilippantonio MJ, Redon C, Pilch DR, Olaru A, Eckhaus M, Camerini-Otero RD, Tessarollo L, Livak F, Manova K, Bonner WM, Nussenzweig MC, Nussenzweig A. Genomic instability in mice lacking histone H2AX. *Science* 2002;**296**(5569):922–7.
38. Franco S, Gostissa M, Zha S, Lombard DB, Murphy MM, Zarrin AA, Yan C, Tepsuporn S, Morales JC, Adams MM, Lou Z, Bassing CH, Manis JP, Chen J, Carpenter PB, Alt FW. H2AX prevents DNA breaks from progressing to chromosome breaks and translocations. *Mol Cell* 2006;**21**(2):201–14.
39. Zha S, Sekiguchi J, Brush JW, Bassing CH, Alt FW. Complementary functions of ATM and H2AX in development and suppression of genomic instability. *Proc Natl Acad Sci USA* 2008;**105**(27):9302–6.
40. Cimprich KA, Cortez D. ATR: an essential regulator of genome integrity. *Nat Rev Mol Cell Biol* 2008;**9**(8):616–27.
41. Fanning E, Klimovich V, Nager AR. A dynamic model for replication protein A (RPA) function in DNA processing pathways. *Nucleic Acids Res* 2006;**34**(15):4126–37.
42. Falck J, Coates J, Jackson SP. Conserved modes of recruitment of ATM, ATR and DNA-PKcs to sites of DNA damage. *Nature* 2005;**434**(7033):605–11.
43. Wang XQ, Redpath JL, Fan ST, Stanbridge EJ. ATR dependent activation of Chk2. *J Cell Physiol* 2006;**208**(3):613–19.
44. Batty D, Rapic'-Otrin V, Levine AS, Wood RD. Stable binding of human XPC complex to irradiated DNA confers strong discrimination for damaged sites. *J Mol Biol* 2000;**300**(2):275–90.
45. Jones S, Chen WD, Parmigiani G, Diehl F, Beerenwinkel N, Antal T, Traulsen A, Nowak MA, Siegel C, Velculescu VE, Kinzler KW, Vogelstein B, Willis J, Markowitz SD. Comparative lesion sequencing provides insights into tumor evolution. *Proc Natl Acad Sci USA* 2008;**105**(11):4283–8.
46. Kakkar P, Singh BK. Mitochondria: a hub of redox activities and cellular distress control. *Mol Cell Biochem* 2007;**305**(1–2):235–53.
47. Chipuk JE, Kuwana T, Bouchier-Hayes L, Droin NM, Newmeyer DD, Schuler M, Green DR. Direct activation of Bax by p53 mediates mitochondrial membrane permeabilization and apoptosis. *Science* 2004;**303**(5660):1010–14.
48. Leu JI, Dumont P, Hafey M, Murphy ME, George DL. Mitochondrial p53 activates Bak and causes disruption of a Bak-Mcl1 complex. *Nat Cell Biol* 2004;**6**(5):443–50.
49. Murphy ME, Leu JI, George DL. p53 moves to mitochondria: a turn on the path to apoptosis. *Cell Cycle* 2004;**3**(7):836–9.
50. Giorgi C, Romagnoli A, Pinton P, Rizzuto R. Ca2+ signaling, mitochondria and cell death. *Curr Mol Med* 2008;**8**(2):119–30.
51. Lao Y, Chang DC. Mobilization of Ca2+ from endoplasmic reticulum to mitochondria plays a positive role in the early stage of UV- or TNFalpha-induced apoptosis. *Biochem Biophys Res Commun* 2008;**373**(1):42–7.
52. Ballinger SW, Patterson C, Yan CN, Doan R, Burow DL, Young CG, Yakes FM, Van Houten B, Ballinger CA, Freeman BA, Runge MS. Hydrogen peroxide- and peroxynitrite-induced mitochondrial DNA damage and dysfunction in vascular endothelial and smooth muscle cells. *Circ Res* 2000;**86**(9):960–6.
53. Yakes FM, Van Houten B. Mitochondrial DNA damage is more extensive and persists longer than nuclear DNA damage in human cells following oxidative stress. *Proc Natl Acad Sci USA* 1997;**94**(2):514–19.
54. Santos JH, Hunakova L, Chen Y, Bortner C, Van Houten B. Cell sorting experiments link persistent mitochondrial DNA damage with loss of mitochondrial membrane potential and apoptotic cell death. *J Biol Chem* 2003;**278**(3):1728–34.
55. Leach JK, Van Tuyle G, Lin PS, Schmidt-Ullrich R, Mikkelsen RB. Ionizing radiation-induced, mitochondria-dependent generation of reactive oxygen/nitrogen. *Cancer Res* 2001;**61**(10):3894–901.
56. Paz ML, Gonzalez Maglio DH, Weill FS, Bustamante J, Leoni J. Mitochondrial dysfunction and cellular stress progression after ultraviolet B irradiation in human keratinocytes. *Photodermatol Photoimmunol Photomed* 2008;**24**(3):115–22.
57. Drose S, Brandt U. The mechanism of mitochondrial superoxide production by the cytochrome bc1 complex. *J Biol Chem* 2008;**283**(31):21,649–21,654.
58. Gredilla R, Sanz A, Lopez-Torres M, Barja G. Caloric restriction decreases mitochondrial free radical generation at complex I and lowers oxidative damage to mitochondrial DNA in the rat heart. *FASEB J* 2001;**15**(9):1589–91.
59. Huie RE, Padmaja S. The reaction of no with superoxide. *Free Radic Res Commun* 1993;**18**(4):195–9.
60. Yamakura F, Taka H, Fujimura T, Murayama K. Inactivation of human manganese-superoxide dismutase by peroxynitrite is caused by exclusive nitration of tyrosine 34 to 3-nitrotyrosine. *J Biol Chem* 1998;**273**(23):14,085–14,089.
61. Ott M, Gogvadze V, Orrenius S, Zhivotovsky B. Mitochondria, oxidative stress and cell death. *Apoptosis* 2007;**12**(5):913–22.
62. Stuart JA, Brown MF. Mitochondrial DNA maintenance and bioenergetics. *Biochim Biophys Acta* 2006;**1757**(2):79–89.
63. Mason PA, Matheson EC, Hall AG, Lightowlers RN. Mismatch repair activity in mammalian mitochondria. *Nucleic Acids Res* 2003;**31**(3):1052–8.
64. Chen D, Yu Z, Zhu Z, Lopez CD. The p53 pathway promotes efficient mitochondrial DNA base excision repair in colorectal cancer cells. *Cancer Res* 2006;**66**(7):3485–94.
65. Yoshida Y, Izumi H, Torigoe T, Ishiguchi H, Itoh H, Kang D, Kohno K. P53 physically interacts with mitochondrial transcription factor A and differentially regulates binding to damaged DNA. *Cancer Res* 2003;**63**(13):3729–34.
66. Pervin S, Tran AH, Zekavati S, Fukuto JM, Singh R, Chaudhuri G. Increased susceptibility of breast cancer cells to stress mediated inhibition of protein synthesis. *Cancer Res* 2008;**68**(12):4862–74.
67. Hara MR, Agrawal N, Kim SF, Cascio MB, Fujimuro M, Ozeki Y, Takahashi M, Cheah JH, Tankou SK, Hester LD, Ferris CD, Hayward SD, Snyder SH, Sawa A. S-nitrosylated GAPDH initiates apoptotic cell death by nuclear translocation following Siah1 binding. *Nat Cell Biol* 2005;**7**(7):665–74.
68. Takeuchi O, Akira S. Recognition of viruses by innate immunity. *Immunol Rev* 2007;**220**:214–24.
69. Ishikawa, H. & Barber, G. N. STING is an endoplasmic reticulum adaptor that facilitates innate immune signaling. *Nature*, *455* (7213), 674–678.
70. Townsend DM. S-glutathionylation: indicator of cell stress and regulator of the unfolded protein response. *Mol Interv* 2007;**7**(6):313–24.
71. Chipuk JE, Bouchier-Hayes L, Kuwana T, Newmeyer DD, Green DR. PUMA couples the nuclear and cytoplasmic proapoptotic function of p53. *Science* 2005;**309**(5741):1732–5.
72. Holcik M, Sonenberg N. Translational control in stress and apoptosis. *Nat Rev Mol Cell Biol* 2005;**6**(4):318–27.

73. Yamasaki S, Anderson P. Reprogramming mRNA translation during stress. *Curr Opin Cell Biol* 2008;**20**(2):222–6.
74. Wek RC, Jiang HY, Anthony TG. Coping with stress: eIF2 kinases and translational control. *Biochem Soc Trans* 2006;**34**(Pt 1):7–11.
75. Harding HP, Novoa I, Zhang Y, Zeng H, Wek R, Schapira M, Ron D. Regulated translation initiation controls stress-induced gene expression in mammalian cells. *Mol Cell* 2000;**6**(5):1099–108.
76. McEwen E, Kedersha N, Song B, Scheuner D, Gilks N, Han A, Chen JJ, Anderson P, Kaufman RJ. Heme-regulated inhibitor kinase-mediated phosphorylation of eukaryotic translation initiation factor 2 inhibits translation, induces stress granule formation, and mediates survival upon arsenite exposure. *J Biol Chem* 2005;**280**(17):16,925–16,933.
77. Dever TE. Gene-specific regulation by general translation factors. *Cell* 2002;**108**(4):545–56.
78. Wek RC, Cavener DR. Translational control and the unfolded protein response. *Antioxid Redox Signal* 2007;**9**(12):2357–71.
79. Mak BC, Wang Q, Laschinger C, Lee W, Ron D, Harding HP, Kaufman RJ, Scheuner D, Austin RC, McCulloch CA. Novel function of perk as a mediator of force-induced apoptosis. *J Biol Chem* 2008;**283**(34):23,462–23,472.
80. Zhang L, Smit-McBride Z, Pan X, Rheinhardt J, Hershey JW. An oncogenic role for the phosphorylated h-subunit of human translation initiation factor eIF3. *J Biol Chem* 2008;**283**(35):24,047–24,060.
81. Santucci R, Sinibaldi F, Fiorucci L. Protein folding, unfolding and misfolding: role played by intermediate States. *Mini Rev Med Chem* 2008;**8**(1):57–62.
82. Nakamura T, Lipton SA. Emerging roles of S-nitrosylation in protein misfolding and neurodegenerative diseases. *Antioxid Redox Signal* 2008;**10**(1):87–101.
83. Uehara T, Nakamura T, Yao D, Shi ZQ, Gu Z, Ma Y, Masliah E, Nomura Y, Lipton SA. S-nitrosylated protein-disulphide isomerase links protein misfolding to neurodegeneration. *Nature* 2006; **441**(7092):513–17.
84. Ananthan J, Goldberg AL, Voellmy R. Abnormal proteins serve as eukaryotic stress signals and trigger the activation of heat shock genes. *Science* 1986;**232**(4749):522–4.
85. Sherman MY, Goldberg AL. Cellular defenses against unfolded proteins: a cell biologist thinks about neurodegenerative diseases. *Neuron* 2001;**29**(1):15–32.
86. Ilbert M, Horst J, Ahrens S, Winter J, Graf PC, Lilie H, Jakob U. The redox-switch domain of Hsp33 functions as dual stress sensor. *Nat Struct Mol Biol* 2007;**14**(6):556–63.
87. Schroder M. Endoplasmic reticulum stress responses. *Cell Mol Life Sci* 2008;**65**(6):862–94.
88. Brostrom MA, Brostrom CO. Calcium dynamics and endoplasmic reticular function in the regulation of protein synthesis: implications for cell growth and adaptability. *Cell Calcium* 2003;**34**(4–5):345–63.
89. Schutzbach JS, Forsee WT. Calcium ion activation of rabbit liver alpha 1,2-mannosidase. *J Biol Chem* 1990;**265**(5):2546–9.
90. Taylor SC, Ferguson AD, Bergeron JJ, Thomas DY. The ER protein folding sensor UDP-glucose glycoprotein-glucosyltransferase modifies substrates distant to local changes in glycoprotein conformation. *Nat Struct Mol Biol* 2004;**11**(2):128–34.
91. Hwang C, Sinskey AJ, Lodish HF. Oxidized redox state of glutathione in the endoplasmic reticulum. *Science* 1992;**257**(5076):1496–502.
92. Kohno K. How transmembrane proteins sense endoplasmic reticulum stress. *Antioxid Redox Signal* 2007;**9**(12):2295–303.
93. Bertolotti A, Zhang Y, Hendershot LM, Harding HP, Ron D. Dynamic interaction of BiP and ER stress transducers in the unfolded-protein response. *Nat Cell Biol* 2000;**2**(6):326–32.
94. DuRose JB, Tam AB, Niwa M. Intrinsic capacities of molecular sensors of the unfolded protein response to sense alternate forms of endoplasmic reticulum stress. *Mol Biol Cell* 2006;**17**(7):3095–107.
95. Rutkowski DT, Arnold SM, Miller CN, Wu J, Li J, Gunnison KM, Mori K, Sadighi Akha AA, Raden D, Kaufman RJ. Adaptation to ER stress is mediated by differential stabilities of pro-survival and pro-apoptotic mRNAs and proteins. *PLoS Biol* 2006;**4**(11):e374.
96. Vattem KM, Wek RC. Reinitiation involving upstream ORFs regulates ATF4 mRNA translation in mammalian cells. *Proc Natl Acad Sci USA* 2004;**101**(31):11,269–11,274.
97. Harding HP, Zhang Y, Zeng H, Novoa I, Lu PD, Calfon M, Sadri N, Yun C, Popko B, Paules R, Stojdl DF, Bell JC, Hettmann T, Leiden JM, Ron D. An integrated stress response regulates amino acid metabolism and resistance to oxidative stress. *Mol Cell* 2003;**11**(3):619–33.
98. Kojima E, Takeuchi A, Haneda M, Yagi A, Hasegawa T, Yamaki K, Takeda K, Akira S, Shimokata K, Isobe K. The function of GADD34 is a recovery from a shutoff of protein synthesis induced by ER stress: elucidation by GADD34-deficient mice. *FASEB J* 2003;**17**(11):1573–5.
99. Yoshida H, Matsui T, Yamamoto A, Okada T, Mori K. XBP1 mRNA is induced by ATF6 and spliced by IRE1 in response to ER stress to produce a highly active transcription factor. *Cell* 2001;**107**(7):881–91.
100. Yoshida H, Nadanaka S, Sato R, Mori K. XBP1 is critical to protect cells from endoplasmic reticulum stress: evidence from Site-2 protease-deficient Chinese hamster ovary cells. *Cell Struct Funct* 2006;**31**(2):117–25.
101. Niwa M, Patil CK, DeRisi J, Walter P. Genome-scale approaches for discovering novel nonconventional splicing substrates of the Ire1 nuclease. *Genome Biol* 2005;**6**(1):R3.
102. Namba T, Ishihara T, Tanaka K, Hoshino T, Mizushima T. Transcriptional activation of ATF6 by endoplasmic reticulum stressors. *Biochem Biophys Res Commun* 2007;**355**(2):543–8.
103. Ye J, Rawson RB, Komuro R, Chen X, Dave UP, Prywes R, Brown MS, Goldstein JL. ER stress induces cleavage of membrane-bound ATF6 by the same proteases that process SREBPs. *Mol Cell* 2000;**6**(6):1355–64.
104. Haze K, Yoshida H, Yanagi H, Yura T, Mori K. Mammalian transcription factor ATF6 is synthesized as a transmembrane protein and activated by proteolysis in response to endoplasmic reticulum stress. *Mol Biol Cell* 1999;**10**(11):3787–99.
105. Wang Y, Shen J, Arenzana N, Tirasophon W, Kaufman RJ, Prywes R. Activation of ATF6 and an ATF6 DNA binding site by the endoplasmic reticulum stress response. *J Biol Chem* 2000;**275**(35):27,013–27,020.
106. Bowie A, O'Neill LA. Oxidative stress and nuclear factor-kappaB activation: a reassessment of the evidence in the light of recent discoveries. *Biochem Pharmacol* 2000;**59**(1):13–23.
107. Haimovitz-Friedman A, Kan CC, Ehleiter D, Persaud RS, McLoughlin M, Fuks Z, Kolesnick RN. Ionizing radiation acts on cellular membranes to generate ceramide and initiate apoptosis. *J Exp Med* 1994;**180**(2):525–35.
108. Ozsoy HZ, Sivasubramanian N, Wieder ED, Pedersen S, Mann DL. Oxidative stress promotes ligand-independent and enhanced ligand-dependent tumor necrosis factor receptor signaling. *J Biol Chem* 2008;**283**(34):23,419–23,428.
109. Courtois F, Seidman EG, Delvin E, Asselin C, Bernotti S, Ledoux M, Levy E. Membrane peroxidation by lipopolysaccharide and iron-ascorbate adversely affects Caco-2 cell function: beneficial role of butyric acid. *Am J Clin Nutr* 2003;**77**(3):744–50.

110. Lien E, Means TK, Heine H, Yoshimura A, Kusumoto S, Fukase K, Fenton MJ, Oikawa M, Qureshi N, Monks B, Finberg RW, Ingalls RR, Golenbock DT. Toll-like receptor 4 imparts ligand-specific recognition of bacterial lipopolysaccharide. *J Clin Invest* 2000;**105**(4):497–504.
111. Sachsenmaier C, Radler-Pohl A, Zinck R, Nordheim A, Herrlich P, Rahmsdorf HJ. Involvement of growth factor receptors in the mammalian UVC response. *Cell* 1994;**78**(6):963–72.
112. Rosette C, Karin M. Ultraviolet light and osmotic stress: activation of the JNK cascade through multiple growth factor and cytokine receptors. *Science* 1996;**274**(5290):1194–7.
113. Tobin D, van Hogerlinden M, Toftgard R. UVB-induced association of tumor necrosis factor (TNF) receptor 1/TNF receptor-associated factor-2 mediates activation of Rel proteins. *Proc Natl Acad Sci USA* 1998;**95**(2):565–9.
114. Iyer R, Lehnert BE, Svensson R. Factors underlying the cell growth-related bystander responses to alpha particles. *Cancer Res* 2000;**60**(5):1290–8.
115. Mothersill C, Seymour C. Radiation-induced bystander effects: past history and future directions. *Radiat Res* 2001;**155**(6):759–67.
116. Zhou H, Randers-Pehrson G, Waldren CA, Vannais D, Hall EJ, Hei TK. Induction of a bystander mutagenic effect of alpha particles in mammalian cells. *Proc Natl Acad Sci USA* 2000;**97**(5):2099–104.
117. Azzam EI, de Toledo SM, Little JB. Direct evidence for the participation of gap junction-mediated intercellular communication in the transmission of damage signals from alpha-particle irradiated to nonirradiated cells. *Proc Natl Acad Sci USA* 2001;**98**(2):473–8.
118. Yang H, Anzenberg V, Held KD. The time dependence of bystander responses induced by iron-ion radiation in normal human skin fibroblasts. *Radiat Res* 2007;**168**(3):292–8.
119. Yang H, Asaad N, Held KD. Medium-mediated intercellular communication is involved in bystander responses of X-ray-irradiated normal human fibroblasts. *Oncogene* 2005;**24**(12):2096–103.
120. Lyng FM, Maguire P, McClean B, Seymour C, Mothersill C. The involvement of calcium and MAP kinase signaling pathways in the production of radiation-induced bystander effects. *Radiat Res* 2006;**165**(4):400–9.
121. Olive PL, Johnston PJ. DNA damage from oxidants: influence of lesion complexity and chromatin organization. *Oncol Res* 1997;**9**(6–7):287–94.
122. Wink DA, Mitchell JB. Chemical biology of nitric oxide: Insights into regulatory, cytotoxic, and cytoprotective mechanisms of nitric oxide. *Free Radic Biol Med* 1998;**25**(4–5):434–56.
123. Tasken K, Aandahl EM. Localized effects of cAMP mediated by distinct routes of protein kinase A. *Physiol Rev* 2004;**84**(1):137–67.
124. Lucas KA, Pitari GM, Kazerounian S, Ruiz-Stewart I, Park J, Schulz S, Chepenik KP, Waldman SA. Guanylyl cyclases and signaling by cyclic GMP. *Pharmacol Rev* 2000;**52**(3):375–414.
125. Shah AM, Sauer H. Transmitting biological information using oxygen: reactive oxygen species as signalling molecules in cardiovascular pathophysiology. *Cardiovasc Res* 2006;**71**(2):191–4.
126. Lancaster JRJ. Diffusion of free nitric oxide. *Methods Enzymol* 1996;**268**:31–50.
127. Farazi TA, Juranek SA, Tuschl T. The growing catalog of small RNAs and their association with distinct Argonaute/Piwi family members. *Development* 2008;**135**(7):1201–14.
128. Liao JC, Hein TW, Vaughn MW, Huang KT, Kuo L. Intravascular flow decreases erythrocyte consumption of nitric oxide. *Proc Natl Acad Sci USA* 1999;**96**(15):8757–61.
129. Hyduke DR, Liao JC. Analysis of nitric oxide donor effectiveness in resistance vessels. *Am J Physiol Heart Circ Physiol* 2005;**288**(5):H2390–9.
130. Thomas DD, Ridnour LA, Espey MG, Donzelli S, Ambs S, Hussain SP, Harris CC, DeGraff W, Roberts DD, Mitchell JB, Wink DA. Superoxide fluxes limit nitric oxide-induced signaling. *J Biol Chem* 2006;**281**(36):25,984–25,993.
131. Zaccolo M, Movsesian MA. cAMP and cGMP signaling cross-talk: role of phosphodiesterases and implications for cardiac pathophysiology. *Circ Res* 2007;**100**(11):1569–78.
132. Mitchell DA, Marletta MA. Thioredoxin catalyzes the S-nitrosation of the caspase-3 active site cysteine. *Nat Chem Biol* 2005;**1**(3):154–8.
133. Benhar M, Forrester MT, Hess DT, Stamler JS. Regulated protein denitrosylation by cytosolic and mitochondrial thioredoxins. *Science* 2008;**320**(5879):1050–4.
134. Rahman I, Marwick J, Kirkham P. Redox modulation of chromatin remodeling: impact on histone acetylation and deacetylation, NF-kappaB and pro-inflammatory gene expression. *Biochem Pharmacol* 2004;**68**(6):1255–67.
135. Greer EL, Brunet A. FOXO transcription factors in ageing and cancer. *Acta Physiol (Oxf)* 2008;**192**(1):19–28.
136. Chang YG, Yan XZ, Xie YY, Gao XC, Song AX, Zhang DE, Hu HY. Different roles for two ubiquitin-like domains of ISG15 in protein modification. *J Biol Chem* 2008;**283**(19):13,370–13,377.
137. Sun J, Steenbergen C, Murphy E. S-nitrosylation: NO-related redox signaling to protect against oxidative stress. *Antioxid Redox Signal* 2006;**8**(9–10):1693–705.
138. Chen K, Kirber MT, Xiao H, Yang Y, Keaney JFJ. Regulation of ROS signal transduction by NADPH oxidase 4 localization. *J Cell Biol* 2008;**181**(7):1129–39.
139. Perry JJ, Tainer JA, Boddy MN. A SIM-ultaneous role for SUMO and ubiquitin. *Trends Biochem Sci* 2008;**33**(5):201–8.
140. Bulavin DV, Demidov ON, Saito S, Kauraniemi P, Phillips C, Amundson SA, Ambrosino C, Sauter G, Nebreda AR, Anderson CW, Kallioniemi A, Fornace Jr AJ, Appella E. Amplification of PPM1D in human tumors abrogates p53 tumor-suppressor activity. *Nat Genet* 2002;**31**(2):210–15.
141. Webley K, Bond JA, Jones CJ, Blaydes JP, Craig A, Hupp T, Wynford-Thomas D. Posttranslational modifications of p53 in replicative senescence overlapping but distinct from those induced by DNA damage. *Mol Cell Biol* 2000;**20**(8):2803–8.
142. Saito S, Yamaguchi H, Higashimoto Y, Chao C, Xu Y, Fornace Jr AJ, Appella E, Anderson CW. Phosphorylation site interdependence of human p53 post-translational modifications in response to stress. *J Biol Chem* 2003;**278**(39):37,536–37,544.
143. Saito S, Goodarzi AA, Higashimoto Y, Noda Y, Lees-Miller SP, Appella E, Anderson CW. ATM mediates phosphorylation at multiple p53 sites, including Ser(46), in response to ionizing radiation. *J Biol Chem* 2002;**277**(15):12,491–12,494.
144. Tibbetts RS, Brumbaugh KM, Williams JM, Sarkaria JN, Cliby WA, Shieh SY, Taya Y, Prives C, Abraham RT. A role for ATR in the DNA damage-induced phosphorylation of p53. *Genes Dev* 1999;**13**(2):152–7.
145. Bulavin DV, Saito S, Hollander MC, Sakaguchi K, Anderson CW, Appella E, Fornace Jr AJ. Phosphorylation of human p53 by p38 kinase coordinates N-terminal phosphorylation and apoptosis in response to UV radiation. *EMBO J* 1999;**18**(23):6845–54.
146. Buschmann T, Potapova O, Bar-Shira A, Ivanov VN, Fuchs SY, Henderson S, Fried VA, Minamoto T, Alarcon-Vargas D, Pincus MR, Gaarde WA, Holbrook NJ, Shiloh Y, Ronai Z. Jun NH2-terminal kinase phosphorylation of p53 on Thr-81 is important for p53 stabilization and transcriptional activities in response to stress. *Mol Cell Biol* 2001;**21**(8):2743–54.

147. Lowe SW, Ruley HE, Jacks T, Housman DE. p53-dependent apoptosis modulates the cytotoxicity of anticancer agents. *Cell* 1993;**74**(6):957–67.
148. Yonish-Rouach E, Resnitzky D, Lotem J, Sachs L, Kimchi A, Oren M. Wild-type p53 induces apoptosis of myeloid leukaemic cells that is inhibited by interleukin-6. *Nature* 1991;**352**(6333):345–7.
149. Agarwal ML, Agarwal A, Taylor WR, Stark GR. p53 controls both the G2/M and the G1 cell cycle checkpoints and mediates reversible growth arrest in human fibroblasts. *Proc Natl Acad Sci USA* 1995;**92**(18):8493–7.
150. Spurgers KB, Gold DL, Coombes KR, Bohnenstiehl NL, Mullins B, Meyn RE, Logothetis CJ, McDonnell TJ. Identification of cell cycle regulatory genes as principal targets of p53-mediated transcriptional repression. *J Biol Chem* 2006;**281**(35):25,134–25,142.
151. Bond JA, Wyllie FS, Wynford-Thomas D. Escape from senescence in human diploid fibroblasts induced directly by mutant p53. *Oncogene* 1994;**9**(7):1885–9.
152. Gire V, Wynford-Thomas D. Reinitiation of DNA synthesis and cell division in senescent human fibroblasts by microinjection of anti-p53 antibodies. *Mol Cell Biol* 1998;**18**(3):1611–21.
153. Weg-Remers S, Ponta H, Herrlich P, Konig H. Regulation of alternative pre-mRNA splicing by the ERK MAP-kinase pathway. *EMBO J* 2001;**20**(15):4194–203.
154. Ignarro LJ. *Nitric oxide : biology and pathobiology*. 1st ed San Diego: Academic Press; 2000.
155. Stamler JS, Simon I, Osborne JA, Mullins ME, Jaraki O, Michel T, Singel DJ, Loscalzo J. S-nitrosylation of proteins with nitric oxide: synthesis and characterization of biologically active compounds. *Proc Natl Acad Sci USA* 1992;**89**(1):444–8.
156. Hess DT, Matsumoto A, Kim SO, Marshall HE, Stamler JS. Protein S-nitrosylation: purview and parameters. *Nat Rev Mol Cell Biol* 2005;**6**(2):150–66.
157. Nott A, Watson PM, Robinson JD, Crepaldi L, Riccio A. S-nitrosylation of histone deacetylase 2 induces chromatin remodelling in neurons. *Nature* 2008;**455**(7211):411–15.
158. Okumura F, Lenschow DJ, Zhang DE. Nitrosylation of ISG15 prevents the disulfide bond mediated dimerization of ISG15 and contributes to effective ISGylation. *J Biol Chem* 2008;**283**(36):24,484–24,488.
159. Fan J, Yang X, Wang W, Wood WHR, Becker KG, Gorospe M. Global analysis of stress-regulated mRNA turnover by using cDNA arrays. *Proc Natl Acad Sci USA* 2002;**99**(16):10,611–10,616.
160. Bartel DP. MicroRNAs: genomics, biogenesis, mechanism, and function. *Cell* 2004;**116**(2):281–97.
161. Abdelmohsen K, Kuwano Y, Kim HH, Gorospe M. Posttranscriptional gene regulation by RNA-binding proteins during oxidative stress: implications for cellular senescence. *Biol Chem* 2008;**389**(3):243–55.
162. Selbach M, Schwanhausser B, Thierfelder N, Fang Z, Khanin R, Rajewsky N. Widespread changes in protein synthesis induced by microRNAs. *Nature* 2008;**455**(7209):58–63.
163. Baek D, Villen J, Shin C, Camargo FD, Gygi SP, Bartel DP. The impact of microRNAs on protein output. *Nature* 2008;**455**(7209):64–71.
164. Prasanth KV, Spector DL. Eukaryotic regulatory RNAs: an answer to the "genome complexity" conundrum. *Genes Dev* 2007;**21**(1):11–42.
165. Meister G, Tuschl T. Mechanisms of gene silencing by double-stranded RNA. *Nature* 2004;**431**(7006):343–9.
166. Golden DE, Gerbasi VR, Sontheimer EJ. An inside job for siRNAs. *Mol Cell* 2008;**31**(3):309–12.
167. Lewis BP, Shih IH, Jones-Rhoades MW, Bartel DP, Burge CB. Prediction of mammalian microRNA targets. *Cell* 2003;**115**(7):787–98.
168. Grimson A, Farh KK, Johnston WK, Garrett-Engele P, Lim LP, Bartel DP. MicroRNA targeting specificity in mammals: determinants beyond seed pairing. *Mol Cell* 2007;**27**(1):91 105.
169. He L, He X, Lim LP, de Stanchina E, Xuan Z, Liang Y, Xue W, Zender L, Magnus J, Ridzon D, Jackson AL, Linsley PS, Chen C, Lowe SW, Cleary MA, Hannon GJ. A microRNA component of the p53 tumour suppressor network. *Nature* 2007;**447**(7148):1130–4.
170. Yamakuchi M, Ferlito M, Lowenstein CJ. miR-34a repression of SIRT1 regulates apoptosis. *Proc Natl Acad Sci USA* 2008; **105**(36):13,421–13,426,.
171. Marsit CJ, Eddy K, Kelsey KT. MicroRNA responses to cellular stress. *Cancer Res* 2006;**66**(22):10,843–10,848.
172. Kulshreshtha R, Ferracin M, Wojcik SE, Garzon R, Alder H, Agosto-Perez FJ, Davuluri R, Liu CG, Croce CM, Negrini M, Calin GA, Ivan M. A microRNA signature of hypoxia. *Mol Cell Biol* 2007;**27**(5):1859–67.
173. Ishii H, Saito T. Radiation-induced response of micro RNA expression in murine embryonic stem cells. *Med Chem* 2006;**2**(6):555–63.
174. Maes OC, An J, Sarojini H, Wu H, Wang E. Changes in MicroRNA expression patterns in human fibroblasts after low-LET radiation. *J Cell Biochem* 2008;**105**(3):824–34.
175. Lu J, Getz G, Miska EA, Alvarez-Saavedra E, Lamb J, Peck D, Sweet-Cordero A, Ebert BL, Mak RH, Ferrando AA, Downing JR, Jacks T, Horvitz HR, Golub TR. MicroRNA expression profiles classify human cancers. *Nature* 2005;**435**(7043):834–8.
176. Yu W, Gius D, Onyango P, Muldoon-Jacobs K, Karp J, Feinberg AP, Cui H. Epigenetic silencing of tumour suppressor gene p15 by its antisense RNA. *Nature* 2008;**451**(7175):202–6.
177. Hellwig S, Bass BL. A starvation-induced noncoding RNA modulates expression of Dicer-regulated genes. *Proc Natl Acad Sci USA* 2008;**105**(35):12,897–12,902,.
178. Westfall PJ, Patterson JC, Chen RE, Thorner J. Stress resistance and signal fidelity independent of nuclear MAPK function. *Proc Natl Acad Sci USA* 2008;**105**(34):12,212–12,217.
179. Venter JC, Adams MD, Sutton GG, Kerlavage AR, Smith HO, Hunkapiller M. Shotgun sequencing of the human genome. *Science* 1998;**280**(5369):1540–2.
180. Shendure JA, Porreca GJ, Church GM. Overview of DNA sequencing strategies. *Curr Protoc Mol Biol* 2008. Chapter 7, Unit 7.1.
181. Schena M, Shalon D, Davis RW, Brown PO. Quantitative monitoring of gene expression patterns with a complementary DNA microarray. *Science* 1995;**270**(5235):467–70.
182. Walles M, Gauvin C, Morin PE, Panetta R, Ducharme J. Comparison of sub-2-microm particle columns for fast metabolite ID. *J Sep Sci* 2007;**30**(8):1191–9.
183. Patterson AD, Li H, Eichler GS, Krausz KW, Weinstein JN, Fornace Jr AJ, Gonzalez FJ, Idle JR. UPLC-ESI-TOFMS-based metabolomics and gene expression dynamics inspector self-organizing metabolomic maps as tools for understanding the cellular response to ionizing radiation. *Anal Chem* 2008;**80**(3):665–74.
184. Dominguez DC, Lopes R, Torres ML. Proteomics technology. *Clin Lab Sci* 2007;**20**(4):239–44.
185. Schneider-Mergener J. Synthetic Peptide and Protein Domain Arrays Prepared by the SPOT Technology. *Comp Funct Genomics* 2001;**2**(5):307–9.
186. Liao JC, Boscolo R, Yang YL, Tran LM, Sabatti C, Roychowdhury VP. Network component analysis: reconstruction of regulatory signals in biological systems. *Proc Natl Acad Sci USA* 2003;**100**(26):15,522–15,527.

187. Lee JK, Havaleshko DM, Cho H, Weinstein JN, Kaldjian EP, Karpovich J, Grimshaw A, Theodorescu D. A strategy for predicting the chemosensitivity of human cancers and its application to drug discovery. *Proc Natl Acad Sci USA* 2007;**104**(32):13,086–13,091.
188. Roth FP, Hughes JD, Estep PW, Church GM. Finding DNA regulatory motifs within unaligned noncoding sequences clustered by whole-genome mRNA quantitation. *Nat Biotechnol* 1998;**16**(10):939–45.
189. Getz G, Gal H, Kela I, Notterman DA, Domany E. Coupled two-way clustering analysis of breast cancer and colon cancer gene expression data. *Bioinformatics* 2003;**19**(9):1079–89.
190. Hyduke DR, Rohlin L, Kao KC, Liao JC. A software package for cDNA microarray data normalization and assessing confidence intervals. *OMICS* 2003;**7**(3):227–34.
191. Subramanian A, Tamayo P, Mootha VK, Mukherjee S, Ebert BL, Gillette MA, Paulovich A, Pomeroy SL, Golub TR, Lander ES, Mesirov JP. Gene set enrichment analysis: a knowledge-based approach for interpreting genome-wide expression profiles. *Proc Natl Acad Sci USA* 2005;**102**(43):15,545–15,550,.
192. Baetz K, McHardy L, Gable K, Tarling T, Reberioux D, Bryan J, Andersen RJ, Dunn T, Hieter P, Roberge M. Yeast genome-wide drug-induced haploinsufficiency screen to determine drug mode of action. *Proc Natl Acad Sci USA* 2004;**101**(13):4525–30.
193. Ronen M, Rosenberg R, Shraiman BI, Alon U. Assigning numbers to the arrows: parameterizing a gene regulation network by using accurate expression kinetics. *Proc Natl Acad Sci USA* 2002;**99**(16):10,555–10,560.
194. Winzeler EA, Shoemaker DD, Astromoff A, Liang H, Anderson K, Andre B, Bangham R, Benito R, Boeke JD, Bussey H, Chu AM, Connelly C, et al. Functional characterization of the S. cerevisiae genome by gene deletion and parallel analysis. *Science* 1999; **285**(5429):901–6.
195. Hyduke DR, Jarboe LR, Tran LM, Chou KJ, Liao JC. Integrated network analysis identifies nitric oxide response networks and dihydroxyacid dehydratase as a crucial target in Escherichia coli. *Proc Natl Acad Sci USA* 2007;**104**(20):8484–9.
196. Jarboe LR, Hyduke DR, Tran LM, Chou KJ, Liao JC. Determination of the Escherichia coli S-nitrosoglutathione response network using integrated biochemical and systems analysis. *J Biol Chem* 2008;**283**(8):5148–57.
197. Liao JC. Modelling and analysis of metabolic pathways. *Curr Opin Biotechnol* 1993;**4**(2):211–16.
198. Savinell JM, Palsson BO. Optimal selection of metabolic fluxes for in vivo measurement. I. Development of mathematical methods. *J Theor Biol* 1992;**155**(2):201–14.
199. Savageau MA. Biochemical systems analysis. II. The steady-state solutions for an n-pool system using a power-law approximation. *J Theor Biol* 1969;**25**(3):370–9.
200. Sorribas A, Savageau MA. Strategies for representing metabolic pathways within biochemical systems theory: reversible pathways. *Math Biosci* 1989;**94**(2):239–69.
201. Stephanopoulos G. Metabolic engineering. *Curr Opin Biotechnol* 1994;**5**(2):196–200.
202. Pramanik J, Keasling JD. Effect of Escherichia coli biomass composition on central metabolic fluxes predicted by a stoichiometric model. *Biotechnol Bioeng* 1998;**60**(2):230–8.
203. Ideker T, Thorsson V, Ranish JA, Christmas R, Buhler J, Eng JK, Bumgarner R, Goodlett DR, Aebersold R, Hood L. Integrated genomic and proteomic analyses of a systematically perturbed metabolic network. *Science* 2001;**292**(5518):929–34.
204. Workman CT, Mak HC, McCuine S, Tagne JB, Agarwal M, Ozier O, Begley TJ, Samson LD, Ideker T. A systems approach to mapping DNA damage response pathways. *Science* 2006;**312**(5776):1054–9.
205. Birrell GW, Giaever G, Chu AM, Davis RW, Brown JM. A genome-wide screen in Saccharomyces cerevisiae for genes affecting UV radiation sensitivity. *Proc Natl Acad Sci USA* 2001;**98**(22):12,608–12,613.
206. Game JC, Birrell GW, Brown JA, Shibata T, Baccari C, Chu AM, Williamson MS, Brown JM. Use of a genome-wide approach to identify new genes that control resistance of Saccharomyces cerevisiae to ionizing radiation. *Radiat Res* 2003;**160**(1):14–24.
207. Lee W, St Onge RP, Proctor M, Flaherty P, Jordan MI, Arkin AP, Davis RW, Nislow C, Giaever G. Genome-wide requirements for resistance to functionally distinct DNA-damaging agents. *PLoS Genet* 2005;**1**(2):e24.
208. Birrell GW, Brown JA, Wu HI, Giaever G, Chu AM, Davis RW, Brown JM. Transcriptional response of Saccharomyces cerevisiae to DNA-damaging agents does not identify the genes that protect against these agents. *Proc Natl Acad Sci USA* 2002;**99**(13): 8778–83.
209. Said MR, Begley TJ, Oppenheim AV, Lauffenburger DA, Samson LD. Global network analysis of phenotypic effects: protein networks and toxicity modulation in Saccharomyces cerevisiae. *Proc Natl Acad Sci USA* 2004;**101**(52):18,006–18,011.
210. Mousses S, Caplen NJ, Cornelison R, Weaver D, Basik M, Hautaniemi S, Elkahloun AG, Lotufo RA, Choudary A, Dougherty ER, Suh E, Kallioniemi O. RNAi microarray analysis in cultured mammalian cells. *Genome Res* 2003;**13**(10):2341–7.
211. Lander ES, Linton LM, Birren B, Nusbaum C, Zody MC, Baldwin J, Devon K, Dewar K, Doyle M, FitzHugh W, Funke R, Gage D, Harris K, et al. Initial sequencing and analysis of the human genome. *Nature* 2001;**409**(6822):860–921.
212. Venter JC, Adams MD, Myers EW, Li PW, Mural RJ, Sutton GG, Smith HO, Yandell M, Evans CA, Holt RA, Gocayne JD, Amanatides P, et al. The sequence of the human genome. *Science* 2001;**291**(5507):1304–51.
213. Frith MC, Pheasant M, Mattick JS. The amazing complexity of the human transcriptome. *Eur J Hum Genet* 2005;**13**(8):894–7.
214. Birney E, Stamatoyannopoulos JA, Dutta A, Guigo R, Gingeras TR, Margulies EH, Weng Z, Snyder M, Dermitzakis ET, Thurman RE, Kuehn MS, et al. Identification and analysis of functional elements in 1% of the human genome by the ENCODE pilot project. *Nature* 2007;**447**(7146):799–816.
215. Forbes SA, Bhamra G, Bamford S, Dawson E, Kok C, Clements J, Menzies A, Teague JW, Futreal PA, Stratton MR. The Catalogue of Somatic Mutations in Cancer (COSMIC). *Curr Protoc Hum Genet* 2008. Chapter 10, Unit 10.11.
216. Euskirchen GM, Rozowsky JS, Wei CL, Lee WH, Zhang ZD, Hartman S, Emanuelsson O, Stolc V, Weissman S, Gerstein MB, Ruan Y, Snyder M. Mapping of transcription factor binding regions in mammalian cells by ChIP: comparison of array- and sequencing-based technologies. *Genome Res* 2007;**17**(6):898–909.
217. Yang YL, Liao JC. Network component analysis of Saccharamyces cerevisiae stress response. *Conf Proc IEEE Eng Med Biol Soc* 2004;**4**:2937–40.
218. Sweet-Cordero A, Mukherjee S, Subramanian A, You H, Roix JJ, Ladd-Acosta C, Mesirov J, Golub TR, Jacks T. An oncogenic KRAS2 expression signature identified by cross-species gene-expression analysis. *Nat Genet* 2005;**37**(1):48–55.
219. Hieronymus H, Lamb J, Ross KN, Peng XP, Clement C, Rodina A, Nieto M, Du J, Stegmaier K, Raj SM, Maloney KN, Clardy J,

Hahn WC, Chiosis G, Golub TR. Gene expression signature-based chemical genomic prediction identifies a novel class of HSP90 pathway modulators. *Cancer Cell* 2006;**10**(4):321–30.

220. Lamb J, Crawford ED, Peck D, Modell JW, Blat IC, Wrobel MJ, Lerner J, Brunet JP, Subramanian A, Ross KN, Reich M, Hieronymus H, Wei G, et al. The Connectivity Map: using gene-expression signatures to connect small molecules, genes, and disease. *Science* 2006;**313**(5795):1929–35.

221. Wei G, Twomey D, Lamb J, Schlis K, Agarwal J, Stam RW, Opferman JT, Sallan SE, den Boer ML, Pieters R, Golub TR, Armstrong SA. Gene expression-based chemical genomics identifies rapamycin as a modulator of MCL1 and glucocorticoid resistance. *Cancer Cell* 2006;**10**(4):331–42.

222. Wei CL, Wu Q, Vega VB, Chiu KP, Ng P, Zhang T, Shahab A, Yong HC, Fu Y, Weng Z, Liu J, Zhao XD, Chew JL, Lee YL, Kuznetsov VA, et al. A global map of p53 transcription-factor binding sites in the human genome. *Cell* 2006;**124**(1):207–19.

223. Basso K, Margolin AA, Stolovitzky G, Klein U, Dalla-Favera R, Califano A. Reverse engineering of regulatory networks in human B cells. *Nat Genet* 2005;**37**(4):382–90.

224. Margolin AA, Wang K, Lim WK, Kustagi M, Nemenman I, Califano A. Reverse engineering cellular networks. *Nat Protoc* 2006;**1**(2):662–71.

225. Margolin AA, Nemenman I, Basso K, Wiggins C, Stolovitzky G, Dalla Favera R, Califano A. ARACNE: an algorithm for the reconstruction of gene regulatory networks in a mammalian cellular context. *BMC Bioinformatics* 2006;**7**(Suppl. 1):S7.

226. Mani KM, Lefebvre C, Wang K, Lim WK, Basso K, Dalla-Favera R, Califano A. A systems biology approach to prediction of oncogenes and molecular perturbation targets in B-cell lymphomas. *Mol Syst Biol* 2008;**4**:169.

227. Amundson SA, Bittner M, Chen YD, Trent J, Meltzer P, Fornace Jr AJ. Fluorescent cDNA microarray hybridization reveals complexity and heterogeneity of cellular genotoxic stress responses. *Oncogene* 1999;**18**:3666–72.

228. Amundson SA, Do KT, Shahab S, Bittner M, Meltzer P, Trent J, Fornace Jr AJ. Identification of potential mRNA biomarkers in peripheral blood lymphocytes for human exposure to ionizing radiation. *Radiat Res* 2000;**154**(3):342–6.

229. Amundson SA, Grace MB, McLeland CB, Epperly MW, Yeager A, Zhan Q, Greenberger JS, Fornace Jr AJ. Human in vivo radiation-induced biomarkers: gene expression changes in radiotherapy patients. *Cancer Res* 2004;**64**(18):6368–71.

230. Amundson SA, Do KT, Vinikoor LC, Lee RA, Koch-Paiz CA, Ahn J, Reimers M, Chen Y, Scudiero DA, Weinstein JN, Trent JM, Bittner ML, Meltzer PS, Fornace Jr AJ. Integrating global gene expression and radiation survival parameters across the 60 cell lines of the national cancer institute anticancer drug screen. *Cancer Res* 2008;**68**(2):415–24.

231. Murakami T, Fujimoto M, Ohtsuki M, Nakagawa H. Expression profiling of cancer-related genes in human keratinocytes following non-lethal ultraviolet B irradiation. *J Dermatol Sci* 2001;**27**(2):121–9.

232. Koch-Paiz Jr C, Amundson SA, Bittner ML, Meltzer P, Fornace AJ. Functional genomics of UV radiation responses in human cells. *Mutat Res* 2004;**549**(1–2):65–78.

233. Scandurro AB, Weldon CW, Figueroa YG, Alam J, Beckman BS. Gene microarray analysis reveals a novel hypoxia signal transduction pathway in human hepatocellular carcinoma cells. *Int J Oncol* 2001;**19**(1):129–35.

234. Jelinsky SA, Estep P, Church GM, Samson LD. Regulatory networks revealed by transcriptional profiling of damaged Saccharomyces cerevisiae cells: rpn4 links base excision repair with proteasomes. *Mol Cell Biol* 2000;**20**:8157–67.

235. Bouton CM, Hossain MA, Frelin LP, Laterra J, Pevsner J. Microarray analysis of differential gene expression in lead-exposed astrocytes. *Toxicol Appl Pharmacol* 2001;**176**(1):34–53.

236. Chen H, Liu J, Merrick BA, Waalkes MP. Genetic events associated with arsenic-induced malignant transformation: applications of cDNA microarray technology. *Mol Carcinog* 2001;**30**(2):79–87.

237. Momose Y, Iwahashi H. Bioassay of cadmium using a DNA microarray: genome-wide expression patterns of Saccharomyces cerevisiae response to cadmium. *Environ Toxicol Chem* 2001;**20**(10):2353–60.

238. Monks A, Scudiero D, Skehan P, Shoemaker R, Paull K, Vistica D, Hose C, Langley J, Cronise P, Vaigro-Wolff A, et al. Feasibility of a high-flux anticancer drug screen using a diverse panel of cultured human tumor cell lines. *J Natl Cancer Inst* 1991;**83**(11):757–66.

239. Amundson SA, Myers TG, Scudiero D, Kitada S, Reed JC, Fornace Jr AJ. An informatics approach identifying markers of chemosensitivity in human cancer cell lines. *Cancer Res* 2000; **60**:6101–10.

240. Ross DT, Scherf U, Eisen MB, Perou CM, Rees C, Spellman P, Iyer V, Jeffrey SS, Van dRM, Waltham M, Pergamenschikov A, et al. Systematic variation in gene expression patterns in human cancer cell lines. *Nat Genet* 2000;**24**(3):227–35.

241. Scherf U, Ross DT, Waltham M, Smith LH, Lee JK, Tanabe L, Kohn KW, Reinhold WC, Myers TG, Andrews DT, Scudiero DA, Eisen MB, et al. A gene expression database for the molecular pharmacology of cancer. *Nat Genet* 2000;**24**(3):236–44.

242. Staunton JE, Slonim DK, Coller HA, Tamayo P, Angelo MJ, Park J, Scherf U, Lee JK, Reinhold WO, Weinstein JN, Mesirov JP, Lander ES, Golub TR. Chemosensitivity prediction by transcriptional profiling. *Proc Natl Acad Sci USA* 2001;**98**(19):10,787–10,792.

243. Roschke AV, Tonon G, Gehlhaus KS, McTyre N, Bussey KJ, Lababidi S, Scudiero DA, Weinstein JN, Kirsch IR. Karyotypic complexity of the NCI-60 drug-screening panel. *Cancer Res* 2003;**63**(24):8634–47.

244. Bussey KJ, Chin K, Lababidi S, Reimers M, Reinhold WC, Kuo WL, Gwadry F, Ajay., Kouros-Mehr H, Fridlyand J, Jain A, Collins C, et al. Integrating data on DNA copy number with gene expression levels and drug sensitivities in the NCI-60 cell line panel. *Mol Cancer Ther* 2006;**5**(4):853–67.

245. Weinstein JN. Spotlight on molecular profiling: "Integromic" analysis of the NCI-60 cancer cell lines. *Mol Cancer Ther* 2006;**5**(11):2601–5.

246. Shankavaram UT, Reinhold WC, Nishizuka S, Major S, Morita D, Chary KK, Reimers MA, Scherf U, Kahn A, Dolginow D, Cossman J, et al. Transcript and protein expression profiles of the NCI-60 cancer cell panel: an integromic microarray study. *Mol Cancer Ther* 2007;**6**(3):820–32.

247. Shoemaker RH. The NCI60 human tumour cell line anticancer drug screen. *Nat Rev Cancer* 2006;**6**(10):813–23.

248. Alfarano C, Andrade CE, Anthony K, Bahroos N, Bajec M, Bantoft K, Betel D, Bobechko B, Boutilier K, Burgess E, Buzadzija K, Cavero R, et al. The biomolecular interaction network database and related tools 2005 update. *Nucleic Acids Res* 2005;**33**(Database issue):D418–24.

249. Jiang C, Xuan Z, Zhao F, Zhang MQ. TRED: a transcriptional regulatory element database, new entries and other development. *Nucleic Acids Res* 2007;**35**(Database issue):D137–40.

250. Ren B, Cam H, Takahashi Y, Volkert T, Terragni J, Young RA, Dynlacht BD. E2F integrates cell cycle progression with

DNA repair, replication, and G(2)/M checkpoints. *Genes Dev* 2002;**16**(2):245–56.

251. Ishida S, Huang E, Zuzan H, Spang R, Leone G, West M, Nevins JR. Role for E2F in control of both DNA replication and mitotic functions as revealed from DNA microarray analysis. *Mol Cell Biol* 2001;**21**(14):4684–99.
252. Amundson Jr SA, Do KT, Vinikoor L, Koch-Paiz CA, Bittner ML, Trent JM, Meltzer P, Fornace AJ. Stress-specific signatures: expression profiling of p53 wild-type and -null human cells. *Oncogene* 2005;**24**(28):4572–9.
253. Detours V, Delys L, Libert F, Weiss Solis D, Bogdanova T, Dumont JE, Franc B, Thomas G, Maenhaut C. Genome-wide gene expression profiling suggests distinct radiation susceptibilities in sporadic and post-Chernobyl papillary thyroid cancers. *Br J Cancer* 2007;**97**(6):818–25.
254. Wang SE, Xiang B, Guix M, Olivares MG, Parker J, Chung CH, Pandiella A, Arteaga CL. Transforming Growth Factor beta Engages TACE and ErbB3 To Activate Phosphatidylinositol-3 Kinase/Akt in ErbB2-Overexpressing Breast Cancer and Desensitizes Cells to Trastuzumab. *Mol Cell Biol* 2008;**28**(18):5605–20.
255. Creighton CJ, Casa A, Lazard Z, Huang S, Tsimelzon A, Hilsenbeck SG, Osborne CK, Lee AV. Insulin-like growth factor-I activates gene transcription programs strongly associated with poor breast cancer prognosis. *J Clin Oncol* 2008;**26**(25):4078–85.
256. Tirosh I, Barkai N. Two strategies for gene regulation by promoter nucleosomes. *Genome Res* 2008;**18**(7):1084–91.
257. Weinstein JN. Integromic analysis of the NCI-60 cancer cell lines. *Breast Dis* 2004;**19**:11–22.
258. Stylianou IM, Affourtit JP, Shockley KR, Wilpan RY, Abdi FA, Bhardwaj S, Rollins J, Churchill GA, Paigen B. Applying gene expression, proteomics and single-nucleotide polymorphism analysis for complex trait gene identification. *Genetics* 2008;**178**(3):1795–805.
259. Hirai MY, Yano M, Goodenowe DB, Kanaya S, Kimura T, Awazuhara M, Arita M, Fujiwara T, Saito K. Integration of transcriptomics and metabolomics for understanding of global responses to nutritional stresses in Arabidopsis thaliana. *Proc Natl Acad Sci USA* 2004;**101**(27):10,205–10,210,.
260. Blower PE, Chung JH, Verducci JS, Lin S, Park JK, Dai Z, Liu CG, Schmittgen TD, Reinhold WC, Croce CM, Weinstein JN, Sadee W. MicroRNAs modulate the chemosensitivity of tumor cells. *Mol Cancer Ther* 2008;**7**(1):1–9.
261. Blower PE, Verducci JS, Lin S, Zhou J, Chung JH, Dai Z, Liu CG, Reinhold W, Lorenzi PL, Kaldjian EP, Croce CM, Weinstein JN, Sadee W. MicroRNA expression profiles for the NCI-60 cancer cell panel. *Mol Cancer Ther* 2007;**6**(5):1483–91.
262. Reinhold WC, Reimers MA, Maunakea AK, Kim S, Lababidi S, Scherf U, Shankavaram UT, Ziegler MS, Stewart C, Kouros-Mehr H, Cui H, et al. Detailed DNA methylation profiles of the E-cadherin promoter in the NCI-60 cancer cells. *Mol Cancer Ther* 2007;**6**(2):391–403.
263. Stevens EV, Nishizuka S, Antony S, Reimers M, Varma S, Young L, Munson PJ, Weinstein JN, Kohn EC, Pommier Y. Predicting cisplatin and trabectedin drug sensitivity in ovarian and colon cancers. *Mol Cancer Ther* 2008;**7**(1):10–18.
264. Weinstein JN, Pommier Y. Transcriptomic analysis of the NCI-60 cancer cell lines. *C R Biol* 2003;**326**(10–11):909–20.

Chapter 21

Oxidative Stress and Free Radical Signal Transduction

Bruce Demple

Department of Pharmacological Sciences, Stony Brook University Medical Center Stony Brook, NY 11794

Department of Genetics and Complex Diseases, Harvard School of Public Health, Boston, Massachusetts

INTRODUCTION: REDOX BIOLOGY

During the past 25 years, considerable attention has been focused on issues related to the biology of free radicals. *Oxidative stress* [1] refers to an imbalance in the production of various reactive species (H_2O_2, superoxide, singlet oxygen, etc.) and the cellular antioxidant defenses that counteract them (glutathione (GSH), superoxide dismutase, catalase, etc.). Such imbalances extend from the levels of the reactive species themselves to the cellular capacity to handle damaged macromolecules. For example, the generation of oxidative DNA damage is balanced normally by cellular systems that correct these lesions; when DNA repair lags behind or damage production outstrips repair capacity, the accumulation of DNA lesions can eventually trigger cell cycle checkpoints, mutations, or cell death.

Sources of free radicals and reactive derivatives of oxygen and nitrogen are internal to the cell and also exogenous. A certain amount of autoxidation occurs in the electron transport chain and elsewhere in the cell to produce superoxide, the dismutation of which generates H_2O_2 [2]. Deficiencies in oxidative phosphorylation components can therefore increase the steady state level of reactive species. Environmental exposure produces oxidative stress through airway injury, irradiation, and compounds that disrupt cellular electron transfer pathways. In addition, actively generated free radicals can play physiological roles, including in cellular signaling pathways [3–7]. High level production of superoxide and another free radical, nitric oxide, contributes to immunological attack on pathogens and tumor cells [8–10]. Inflammation can provoke oxidative stress not only in the cells targeted by immunological attack, but also in "innocent bystanders" exposed to diffusible reactants such as H_2O_2 and NO. Inflammatory signals can also trigger production of these species in non-immune cells.

The main cellular macromolecules are all subject to oxidative damage, so it is not surprising to find that cells have evolved pathways to respond dynamically to oxidative stress by altering gene expression to offset the stress and its associated effects [5, 11, 12].

OXIDATIVE STRESS RESPONSES IN BACTERIA: SOME WELL-DEFINED MODELS OF REDOX SIGNAL TRANSDUCTION

Study of oxidative stress responses has focused especially on *Escherichia coli* and *Salmonella enterica* (*typhimurium*) as model systems amenable to combined genetic and biochemical analysis [7, 13, 14]. Studies in the Gram-positive bacterium *Bacillus subtilis* (e.g., [15]), in medically or agriculturally important organisms such as *Pseudomonas aeruginosa* [16], *Xanthomonas oryzae* [17], or *Agrobacterium tumefaciens* [18], and in organisms important for developing environmental bioremediation approaches such as *Pseudomonas putida* [19], are now yielding similarly fundamental insights.

RESPONSE TO SUPEROXIDE STRESS AND NITRIC OXIDE: SOXR PROTEIN

The SoxR regulatory system of *E. coli* was discovered by genetic analysis of strains with elevated defenses against superoxide generating agents such as paraquat and quinones [20, 21]. The system in *E. coli* and *S. enterica* works in two stages of transcriptional control; only the first stage involves direct redox signal transduction [22, 23]. In contrast, activated SoxR in *P. aeruginosa* and *P. putida* seems to act directly on the promoters of several genes without

Handbook of Cell Signaling, Three-Volume Set 2 ed.

the need for an intermediary transcription factor (see later in this section).

As described below, SoxR protein is activated in cells exposed to oxidative stress or nitric oxide; activated SoxR in *E. coli* and *S. enterica* then stimulates expression of just one gene, *soxS*. The resulting increase in the level of SoxS protein, a transcription activator of the AraC/XylS family [24], leads to the transcriptional upregulation of dozens of genes and the repression of dozens of others [25]. Many of the induced genes have obvious functions in oxidative defense (e.g., Mn-containing superoxide dismutase, or endonuclease IV for oxidative DNA damage), but many others (e.g., the *acrAB* encoded efflux pumps) serve broader functions [26] (Figure 21.1a).

The redox sensing SoxR protein is a homodimer of 17-kDa subunits, each of which contains a single [2Fe-2S] iron-sulfur cluster anchored to the protein's only four cysteine residues near the C-terminus [27]. In resting SoxR,

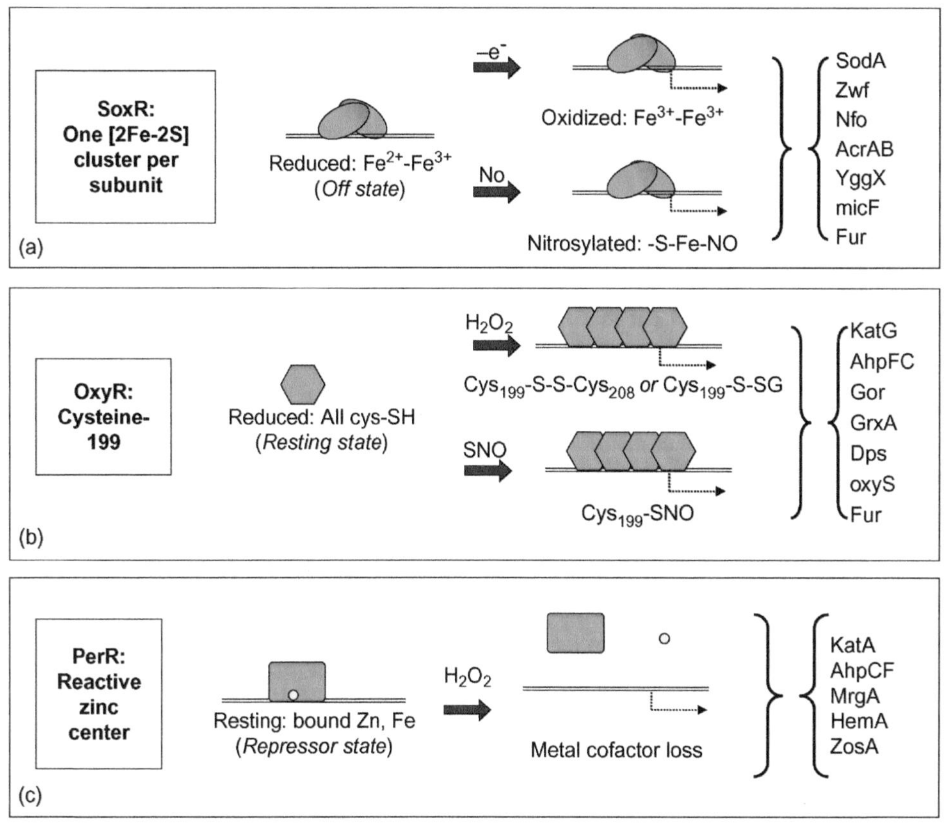

FIGURE 21.1 Representative systems of redox regulation.
Each box depicts the main aspects of one of the regulatory systems described in the text. In all three cases, transcriptional activation is indicated by a dotted, right-angle arrow at the target DNA (thin parallel lines). (a) The SoxR system of *E. coli*. Each 17 kDa subunit of the homodimer contains a [2Fe-2S] cluster, which is in the reduced state in the absence of free radical stress. Non-activated SoxR binds its DNA site in the *soxS* promoter with high affinity. Oxidative stress exerted by superoxide generating agents (such as paraquat) one-electron oxidizes the centers, which converts SoxR to a potent transcriptional activator without changing its DNA binding affinity. Alternatively, nitric oxide activates SoxR by reacting with the iron-sulfur centers to form dinitrosyl-iron complexes anchored via cysteine residues in the protein. Activation of SoxR triggers expression of the *soxS* gene, whose protein product (not shown) then activates the many genes of the regulon. Among the many functions stimulated by activated SoxR are manganese containing superoxide dismutase (SodA), glucose-6-phosphate dehydrogenase (Zwf), the DNA repair enzyme endonuclease IV (Nfo), the AcrAB efflux pump, and the YggX protein involved in maintenance of iron-sulfur clusters. In addition, secondary regulators are under SoxR control: the micF antisense RNA that downregulates OmpF expression; and Fur protein, which controls many genes involved in iron uptake. (b) The OxyR system of *E. coli*. Resting OxyR is in the reduced state and does not bind most of its DNA targets. Oxidation by H_2O_2 causes structural changes that convert the protein to a DNA binding tetramer. Cysteine-199 is the critical residue for this response and forms a disulfide, although there is controversy (see text) as to whether this involves cysteine-208 or a molecule of glutathione. A form of OxyR with a sulfenic acid based on cysteine-199 (not shown) has also been reported. In addition, nitrosothiols produce a different activated form of OxyR containing S-nitrosylated cysteine-199. Activation of OxyR stimulates the expression of many genes not activated by SoxR, although there is a limited overlap (e.g., Fur protein). The functions include catalase (KatG), alkyl hydroperoxide reductase (AhpFC), GSH reductase (Gor), glutaredoxin-1 (GrxA), and the protective DNA binding protein Dps. As seen with SoxR, secondary regulators control still more genes; in addition to the Fur repressor, the oxyS RNA acts post-transcriptionally on the expression of several other genes. (c) The PerR system of *B. subtilis*. The Fur homolog PerR is a repressor of genes that respond to H_2O_2. The active (repressing) form of PerR contains zinc and iron (small yellow circle in the cartoon); oxidation causes these metals to be lost from the protein and DNA binding affinity is lowered dramatically. Functions regulated by PerR include catalase (KatA), AhpFC, the protective DNA binding protein MrgA, HemA and other heme biosynthetic enzymes, and the ZosA protein involved in zinc uptake.

these clusters are in the reduced (+2) state, which can be detected *in vivo* using electron paramagnetic resonance spectroscopy [28, 29]. Upon cellular exposure to agents such as paraquat, which sets up a heavy flux of superoxide in the cell and consumes NADPH by cyclic reduction and autoxidation [30], the [2Fe-2S] centers are oxidized to the +3 state. This one-electron oxidation converts SoxR to a powerful activator of *soxS* transcription that elevates the soxS mRNA level up to 100-fold [6].

These events can be reproduced *in vitro* using purified SoxR protein [31, 32]. SoxR isolated under air has iron-sulfur clusters in the oxidized state, and the protein has full transcriptional activity specific for *soxS*. SoxR stimulates transcription by increasing the rate of open complex formation by RNA polymerase [33]. Chemical reduction using sodium dithionite yields the +2 state of the [2Fe-2S] clusters, which switches off the transcriptional activity. The activity is restored by re-oxidation [31, 32].

SoxR binds tightly to its *soxS* target site independently of the iron-sulfur centers [34] or their oxidation state [32] (Figure 21.1a). The SoxR binding site is centered between the −35 and −10 elements of the *soxS* promoter [33]. This promoter has an unusually long spacer of 19 bp (vs. the usual 17±1 bp) between the RNA polymerase recognition elements, each of which is a good match to consensus sequences. The wild-type *soxS* promoter has very weak basal activity, but mutant promoters with spacers of 18, 17, or 16 bp confer high *in vivo* expression independent of SoxR [35].

The model for SoxR transcription activation is analogous to that for the Hg^{2+} sensing MerR protein [36]: DNA structural remodeling that compensates for the excessively long spacer. The structure of the related BmrR protein in complex with DNA [37] is consistent with the helical distortion model for this family of transcription activators. In the absence of structural information, the SoxR polypeptide was "threaded" on to the structure of the homologous CueR (copper sensing) protein [38] to allow interpetation of a large collection of activating or inactivating mutations [39], which revealed the importance of a long α-helix in each subunit that forms much of the dimer interface [40]. Unfortunately, this approach provided no insight into the redox sensing iron-sulfur clusters. This shortcoming was finally overcome by the recent crystallographic analysis of the SoxR–DNA complex [41], which extends the DNA helical distortion model of transcriptional activation to this redox sensor, also providing the first glimpse of the detailed structure and orientation of the [2Fe-2S] clusters. The co-crystal structure provides some insight into the most exquisite question about SoxR: how is a change in the oxidation state of a distant [2Fe-2S] cluster transmitted through the protein dimer to exert a pronounced structural distortion of the DNA? An asymmetric distribution of charges around the metal centers may provide electrostatic force for protein movement, as suggested before the structural information was available [40]. A second recent study provided an unexpected finding for the complementary question: how are the energetics of DNA binding and the redox reactions linked? In that work [42], measurement of electron transfer to SoxR bound to DNA on a surface indicated that DNA binding shifted the midpoint potential (which is −285 mV for free SoxR) by nearly +500 mV, which would render the [2Fe-2S] clusters relatively oxidation resistant. That measurement would be consistent with the observation that SoxR is not significantly activated during normal aerobic growth in *E. coli*, *S. enterica*, *P. aeruginosa*, or *P. putida*. On the other hand, the *in vivo* oxidation kinetics of bulk SoxR, most of which would not be bound to the DNA site, exactly parallels transcriptional activation, and its re-reduction parallels the switching off of transcription [43]. One explanation might be that the free and DNA bound forms of SoxR are equilibrated in some way.

SoxR is also activated by nitric oxide, and the chemistry of this process is fundamentally different from the one-electron oxidation described above (Figure 21.1a). Nitric oxide directly modifies the [2Fe-2S] clusters to form dinitrosyl-iron-dicysteine complexes anchored to the protein [44]. These nitrosylated complexes are quite stable *in vitro*, which allowed NO-modified SoxR to be repurified for analysis. In *E. coli*, SoxR activation by NO may play a role in bacterial resistance to the cytotoxic attack of NO-generating macrophages [45, 46].

Both the oxidation and the nitrosylation of SoxR are very rapidly reversed in intact bacteria when the oxidative stress or NO exposure is stopped [44]. The former observation, together with maintenance of SoxR in the reduced form in the absence of stress, implies that the [2Fe-2S] centers are actively reduced, probably enzymatically. However, despite a report of a SoxR reductase activity [47] and a genetic study implicating an *rnf*-like gene cluster [48] in maintaining the reduced form, the biochemistry remains undefined.

The reversal of the dinitrosyl complexes in NO-treated SoxR to regenerate unmodified [2Fe-2S] clusters may be of even more general significance. Nitrosylation of various types of iron-sulfur clusters is a general feature of NO toxicity in all cell types [49]. Delineating the mechanism by which such protein damage is corrected (possibly by removal of the nitrosylated clusters and the resynthesis of new ones) should provide general insights into how cells handle this basic problem. Some studies indicate a cysteine dependent disassembly process for nitrosylated [2Fe-2S] centers in *E. coli* ferredoxin, which are replaced by fresh iron-sulfur centers using general cellular pathways [50]. The DNA repair protein endonuclease III of *E. coli* undergoes a similar reversible inactivation by nitric oxide, followed by reconstitution of its [4Fe-4S] clusters *in vivo* [51].

Recent studies in the human pathogen *P. aeruginosa* and the soil bacterium *P. putida* reveal a different regulatory arrangement for SoxR. No gene encoding a SoxS homolog was identified in either organism. Instead, SoxR itself directly regulates several genes by means of regulatory sites in each that are similar to those found for *soxS* in *E. coli* and *S. enterica*. In *P. aeruginosa*, the SoxR regulated genes do not seem to constitute antioxidant defenses, nor is their expression induced by oxidative stress [52]. Newman and coworkers suggested that the phenazine compound pyocyanin is a quorum sensing signal via SoxR in *P. aeruginosa* [53]. A more comprehensive study [54] of SoxR and likely SoxR binding sites in several hundred available bacterial genomes indicated that the SoxR–SoxS arrangement is restricted to enterobacteria, where it coordinates oxidative stress defenses. In most other species, including *P. aeruginosa* and *Streptomyces coelicolor*, SoxS was absent and likely SoxR binding sites were adjacent to just a few genes involved in small molecule metabolism and export [54]. These observations supported the idea that, in many bacterial species, the role of SoxR is at the population level, perhaps in regulating the formation of biofilms [55].

Despite its contrasting roles in the different species, *P. putida* SoxR can be expressed as a functional replacement for its *E. coli* counterpart [19], which indicates that the *Pseudomonas* protein can be maintained in the reduced state by the same cellular systems and that oxidation can activate its transcriptional activity. This observation reflects a strong conservation of DNA binding specificity and suggests that phenazine signaling might involve a redox mechanism.

RESPONSE TO H_2O_2 AND NITROSOTHIOLS: OXYR PROTEIN

The *oxyR* regulatory system was identified by genetic analysis of *S. enterica* and *E. coli* stains in which inducible resistance to H_2O_2 [56] was activated constitutively [57]. OxyR protein both senses oxidative stress and activates the >60 genes of the *oxyR* regulon [58]. OxyR protein is a member of the LysR family of transcriptional regulators [59], and *oxyR* homologs are widely represented in bacterial genomes [7].

Like SoxR, OxyR is a positive regulator that exists in a resting state in non-stressed cells [6, 7]. Upon cellular exposure to H_2O_2 or nitrosothiols [60, 61], the protein is activated to stimulate transcription of the regulon genes. Unlike SoxR, the activity of OxyR is regulated largely by changes in its DNA binding activity [62, 63], as is the case for many prokaryotic signal transduction systems (Figure 21.1b). The positioning of the protein upstream of the RNA polymerase binding site stimulates transcription through protein–protein contacts with the C-terminal domain of the polymerase α subunit [64]. The *oxyR* regulon includes antioxidant genes such as *katG* (HP-I catalase-hydroperoxidase), *gor* (GSH reductase), and the *ahpFC* operon (alkyl hydroperoxide reductase) [58].

The molecular basis for OxyR activation has been explored to yield a detailed model of both the redox activation and the stimulation of transcription [65]. However, the mechanism of redox sensing is still subject to some debate [61]. At the DNA binding level, redox activation converts OxyR from a mainly dimeric form to a tetramer, which binds consecutive repeats of a recognition element and bends the DNA [63] (Figure 21.1b).

The proposed mechanism of redox activation of *E. coli* OxyR by H_2O_2 is the formation of an intramolecular disulfide in the protein [65], and genetic and biochemical evidence supports this view. For example, of the six cysteine residues in the OxyR polypeptide, one (cysteine-199) is absolutely required for full activity in response to oxidative stress, and the other (cysteine-208) has a significant effect. Peptides have been isolated from oxidized OxyR that contain a disulfide bond linking cysteines 199 and 208 (although the other four cysteine residues of the protein had been replaced by alanines for this experiment). These observations led to a model in which oxidative activation of OxyR occurs through the formation of an intramolecular disulfide bond linking cysteine-199 and cysteine-208 (Figure 21.1b). The structural basis for this specificity is proposed to be both the proximity of the two cysteine residues, and a basic residue near cysteine-199 that increases its reactivity with oxidants [62].

This view has been challenged by studies [61] indicating that there are various other activated forms of OxyR lacking the cysteine-199/cysteine-208 disulfide. Instead, Kim *et al.* [61] identified a form with cysteine-199 as a sulfenic acid (-SOH) in air-oxidized OxyR, or with a mixed disulfide linking cysteine-199 and GSH [61]. Such mixed disulfides were detected biochemically as a prominent product in OxyR activated *in vivo* [61]. On the other hand, it should be noted that GSH is not necessary to activate OxyR *in vivo* [65]. The intramolecular disulfide would be formed by a mechanism analogous to the proposed mixed disulfide with GSH.

GSH and glutaredoxin-1 (a small, thioredoxin-like protein with a redox-active cysteine pair) evidently play physiological roles in restoring reduced OxyR following oxidative stress [4, 66]. The overall view is that these proteins participate in thiol/disulfide exchange reactions that yield reduced OxyR, ultimately at the expense of NADPH through the activity of GSH reductase. Note that the observed mixed GSH-OxyR disulfide may have been formed as an intermediate in this process. This control involving GSH would establish a negative feedback loop in the system, since the *gor* and *grx1* (glutaredoxin-1) genes are activated by OxyR [4, 66]. However, *gor* and *grx1* are not required to maintain OxyR in the reduced state in unstressed cells [65].

Nitrosothiols have also been shown to activate OxyR, which was proposed to occur by generating a cysteine nitrosothiol (-SNO) in the protein [60]. Some data support this view [61] and suggest that different activated forms (-SNO, -SOH, and -S-SG) have variable affinity and transcription activating capacity for different promoters (Figure 21.1b). The discrepancy with earlier studies [65] was rationalized by pointing out that they had employed mutant forms of OxyR with four of the six cysteines substituted by other amino acids [61]. These observations open the possibility that different stress signals might give rise to different patterns of gene activation even while operating on the same sensor/transducer protein. For the enteric SoxR/SoxS system, in contrast, different activated forms of SoxR would likely influence only the *level* of SoxS protein. Differences in the pattern of gene expression would then reflect the degree of SoxR activation, rather than changes in its specificity, although there is presumably a hierarchy among *soxRS* regulon genes with respect to their sensitivities to SoxS activation.

More recent data have been provided supporting the functional importance of an intramolecular disulfide in *E. coli* OxyR [67]. Very recently, it was shown that the H_2O_2 regulated OxyR homolog of the radioresistant organism *Deinococcus radiodurans* has just a single cysteine residue [68]. In this case, the reversible reaction appears to be oxidant dependent formation of a sulfenic acid at cysteine-210 [68]. Thus, various response pathways may be available to OxyR, as is the case for SoxR, which may provide functional flexibility for sensing different environmental stresses.

PARALLELS IN REDOX AND FREE RADICAL SENSING

Systems that seem to share the thiol/disulfide mechanism of redox sensing used by OxyR have now been identified in several contexts. A particularly interesting example is the *E. coli* Hsp33 chaperone, which is activated by H_2O_2 through the formation of intramolecular disulfide bonds [69]. This activation might provide for increased cellular capacity to handle oxidatively damaged proteins. The four cysteines involved are ligands for a zinc atom in the resting protein; oxidation eliminates the metal and activates the chaperone function [70]. A similar mechanism may govern the redox activation of the *Bacillus subtilis* Fur homolog PerR (Figure 21.1c), which governs an oxidative stress regulon [71]. Reversible disulfide bond formation (whether intramolecular or mixed) makes physiological sense as a regulatory mechanism because it links redox sensing to overall metabolism through NADPH and the oxidation state of intermediaries such as thioredoxin and glutaredoxin [7, 66].

New observations on Hsp33 indicate a more complicated picture, however. The chaperone is more efficiently activated when hydrogen peroxide is present at heat shock temperatures. A physical rationale for this observation is that two sets of disulfides must be formed in the protein: one pair of cysteines reacts readily when the Zn is oxidatively displaced, while the second disulfide requires the displacement of a linker region (heat shock), which stabilizes the active dimeric form [72]. Unexpectedly, Hsp33 is very efficiently activated by HOCl, a key product of neutrophils and the active ingredient of household bleach [73]. This effect is rationalized by noting that HOCl reacts much more rapidly with thiols than does H_2O_2, which would allow disulfide formation even during transient displacement of the Hsp33 linker region [73]. It remains to be seen whether there are eukaryotic counterparts of this mechanism.

Disulfide bond formation in eukaryotic regulatory proteins has also been shown. Comprehensive reviews of this and related questions for many mammalian systems have appeared recently [3, 5, 74], and the topic will not be further covered here. In plants, mRNA translation in *Clamydomonas reinhardtii* chloroplasts is governed by disulfide bond formation in a regulatory protein [75, 76]. Other examples will surely emerge, though the role of mixed disulfides needs further exploration.

Eukaryotic examples analogous to the SoxR mechanism remain to be described, although the facile nature of one-electron oxidation and reduction of an iron-sulfur cluster should lend itself to redox signaling in various contexts. Another prokaryotic example of redox sensing using an iron-sulfur cluster is the *E. coli* Fnr protein, which acts as a repressor of genes involved in aerobic metabolism. Fnr harbors a [4Fe-4S] cluster that is sensitive to oxygen; exposure to O_2 converts the single cluster to a pair of [2Fe-2S] clusters and releases the protein from the DNA [77]. For nitric oxide, NsrR regulatory proteins have recently been shown to contain iron-sulfur clusters that form mixed dinitrosyl-iron complexes upon reaction with nitric oxide, although the *S. coelicolor* protein was purified with [2Fe-2S] centers [78], while the *B. subtilis* protein was obtained with [4Fe-4S] centers [79]. It will have to be resolved whether this variation occurs naturally or is an artifact of producing and isolating recombinant proteins.

THEMES IN REDOX SENSING

The ability of cellular proteins to detect changes in redox status and the presence of potentially toxic free radicals is a fundamental aspect of biology. Known examples of redox regulated responses already include transcriptional, translational, and enzymatic control. Other key features might include feedback loops that return the system to a "ground state" (e.g., the role of reduced glutathione and glutaredoxin-1 in OxyR regulation) and the use of molecular "canaries in the coalmine": reactions that damage and inactivate most proteins actually switch on sensors such as OxyR, SoxR, and Hsp33.

ACKNOWLEDGEMENTS

Work in the author's laboratory has been supported by the National Cancer Institute (grants R01-CA37831, R01-CA82737, and Radiation Biology Training Grant T32-9078).

REFERENCES

1. Sies H. Oxidative stress: introduction. In: Sies H, editor. *Oxidative stress: oxidants and antioxidants*. London: Academic Press; 1991. p. xv–xxii.
2. Imlay JA. Cellular defenses against superoxide and hydrogen peroxide. *Annu Rev Biochem* 2008;**77**:755–76.
3. Brandes N, Schmitt S, Jakob U. Thiol-Based Redox Switches in Eukaryotic Proteins. *Antioxid Redox Signal* 2009. In press.
4. Carmel-Harel O, Storz G. Roles of the glutathione- and thioredoxin-dependent reduction systems in the Escherichia coli and saccharomyces cerevisiae responses to oxidative stress. *Annu Rev Microbiol* 2000;**54**:439–61.
5. Droge W. Free radicals in the physiological control of cell function. *Physiol Rev* 2002;**82**:47–95.
6. Hidalgo E, Demple B. Adaptive responses to oxidative stress: the soxRS and oxyR regulons. In: Lin ECC, Lynch AS, editors. *Regulation of gene expression in escherichia coli*. Austin, TX: R.G. Landes Co; 1996. p. 435–52.
7. Zheng M, Storz G. Redox sensing by prokaryotic transcription factors. *Biochem Pharmacol* 2000;**59**:1–6.
8. Babior BM. The respiratory burst oxidase. *Enzymol Relat Areas Mol Biol* 1992;**65**:49–65.
9. MacMicking J, Xie QW, Nathan C. Nitric oxide and macrophage function. *Annu Rev Immunol* 1997;**15**:323–50.
10. Spiro S. Regulators of bacterial responses to nitric oxide. *FEMS Microbiol Rev* 2007;**31**:193–211.
11. Finkel T. Redox-dependent signal transduction. *FEBS Lett* 2000;**476**:52–4.
12. Liu H, Colavitti R, Rovira II, Finkel T. Redox-dependent transcriptional regulation. *Circ Res* 2005;**97**:967–74.
13. Pomposiello PJ, Demple B. Redox-operated genetic switches: the SoxR and OxyR transcription factors. *Trends Biotechnol* 2001;**19**:109–14.
14. Pomposiello PJ, Demple B. Global adjustment of microbial physiology during free radical stress. *Adv Microb Physiol* 2002;**46**:319–41.
15. Hecker M, Volker U. General stress response of Bacillus subtilis and other bacteria. *Adv Microb Physiol* 2001;**44**:35–91.
16. Hassett DJ, Ma JF, Elkins JG, McDermott TR, Ochsner UA, West SE, Huang CT, Fredericks J, Burnett S, Stewart PS, McFeters G, Passador L, Iglewski BH. Quorum sensing in Pseudomonas aeruginosa controls expression of catalase and superoxide dismutase genes and mediates biofilm susceptibility to hydrogen peroxide. *Mol Microbiol* 1999;**34**:1082–93.
17. Sukchawalit R, Loprasert S, Atichartpongkul S, Mongkolsuk S. Complex regulation of the organic hydroperoxide resistance gene (ohr) from Xanthomonas involves OhrR, a novel organic peroxide-inducible negative regulator, and posttranscriptional modifications. *J Bacteriol* 2001;**183**:4405–12.
18. Saenkham P, Eiamphungporn W, Farrand SK, Vattanaviboon P, Mongkolsuk S. Multiple superoxide dismutases in Agrobacterium tumefaciens: functional analysis, gene regulation, and influence on tumorigenesis. *J Bacteriol* 2007;**189**:8807–17.
19. Park W, Pena-Llopis S, Lee Y, Demple B. Regulation of superoxide stress in Pseudomonas putida KT2440 is different from the SoxR paradigm in Escherichia coli. *Biochem Biophys Res Commun* 2006;**341**:51–6.
20. Greenberg JT, Monach P, Chou JH, Josephy PD, Demple B. Positive control of a global antioxidant defense regulon activated by superoxide-generating agents in Escherichia coli. *Proc Natl Acad Sci U S A* 1990;**87**:6181–5.
21. Tsaneva IR, Weiss B. soxR, a locus governing a superoxide response regulon in Escherichia coli K-12. *J Bacteriol* 1990;**172**:4197–205.
22. Nunoshiba T, Hidalgo E, Amabile C, Demple B. Two-stage control of an oxidative stress regulon: the Escherichia coli SoxR protein triggers redox-inducible expression of the soxS regulatory gene. *J Bacteriol* 1992;**174**:6054–60.
23. Wu J, Weiss B. Two-stage induction of the soxRS (superoxide response) regulon of Escherichia coli. *J Bacteriol* 1992;**174**:3915–20.
24. Li Z, Demple B. SoxS, an activator of superoxide stress genes in Escherichia coli. Purification and interaction with DNA. *J Biol Chem* 1994;**269**:18,371–18,377,.
25. Pomposiello PJ, Bennik MH, Demple B. Genome-wide transcriptional profiling of the Escherichia coli responses to superoxide stress and sodium salicylate. *J Bacteriol* 2001;**183**:3890–902.
26. White DG, Goldman JD, Demple B, Levy SB. Role of the acrAB Locus in Organic Solvent Tolerance Mediated by Expression of MarA, SoxS, or RobA in Escherichia coli. *J Bacteriol* 1997;**179**:6122–6.
27. Bradley TM, Hidalgo E, Leautaud V, Ding H, Demple B. Cysteine-to-alanine replacements in the Escherichia coli SoxR protein and the role of the [2Fe-2S] centers in transcriptional activation. *Nucleic Acids Res* 1997;**25**:1469–75.
28. Gaudu P, Moon N, Weiss B. Regulation of the soxRS Oxidative Stress Regulon. Reversible Oxidation of the Fe-S Centers of SoxR In Vivo. *J Biol Chem* 1997;**272**:5082–6.
29. Hidalgo E, Ding H, Demple B. Redox signal transduction: mutations shifting [2Fe-2S] centers of the SoxR sensor–regulator to the oxidized form. *Cell* 1997;**88**:121–9.
30. Kappus H, Sies H. Toxic drug effects associated with oxygen metabolism: redox cycling and lipid peroxidation. *Experientia* 1981;**37**:1233–41.
31. Ding H, Hidalgo E, Demple B. The redox state of the [2Fe-2S] clusters in SoxR protein regulates its activity as a transcription factor. *J Biol Chem* 1996;**271**:33,173–33,175.
32. Gaudu P, Weiss B. SoxR, a [2Fe-2S] Transcription Factor, is Active Only in Its Oxidized Form. *Proc Natl Acad Sci U S A* 1996;**93**:10,094–10,098,.
33. Hidalgo E, Bollinger Jr JM, Bradley TM, Walsh CT, Demple B. Binuclear [2Fe-2S] clusters in the Escherichia coli SoxR protein and role of the metal centers in transcription. *J Biol Chem* 1995;**270**:20,908–20,914,.
34. Hidalgo E, Demple B. An iron-sulfur center essential for transcriptional activation by the redox-sensing SoxR protein. *EMBO J* 1994;**13**:138–46.
35. Hidalgo E, Demple B. Spacing of promoter elements regulates the basal expression of the soxS gene and converts SoxR from a transcriptional activator into a repressor. *EMBO J* 1997;**16**:1056–65.
36. Summers AO. Untwist and shout: a heavy metal-responsive transcriptional regulator. *J Bacteriol* 1992;**174**:3097–101.
37. Heldwein EE, Brennan RG. Crystal structure of the transcription activator BmrR bound to DNA and a drug. *Nature* 2001;**409**:378–82.
38. Changela A, Chen K, Xue Y, Holschen J, Outten CE, O'Halloran TV, Mondragon A. Molecular basis of metal-ion selectivity and zeptomolar sensitivity by CueR. *Science* 2003;**301**:1383–7.
39. Chander M, Raducha-Grace L, Demple B. Transcription-defective soxR mutants of Escherichia coli: isolation and in vivo characterization. *J Bacteriol* 2003;**185**:2441–50.
40. Chander M, Demple B. Functional analysis of SoxR residues affecting transduction of oxidative stress signals into gene expression. *J Biol Chem* 2004;**279**:41,603–41,610.
41. Watanabe S, Kita A, Kobayashi K, Miki K. Crystal structure of the [2Fe-2S] oxidative-stress sensor SoxR bound to DNA. *Proc Natl Acad Sci U S A* 2008;**105**:4121–6.
42. Gorodetsky AA, Dietrich LE, Lee PE, Demple B, Newman DK, Barton JK. DNA binding shifts the redox potential of the transcription factor SoxR. *Proc Natl Acad Sci U S A* 2008;**105**:3684–9.

43. Ding H, Demple B. In vivo kinetics of a redox-regulated transcriptional switch. *Proc Natl Acad Sci U S A* 1997;**94**:8445–9.
44. Ding H, Demple B. Direct nitric oxide signal transduction via nitrosylation of iron- sulfur centers in the SoxR transcription activator. *Proc Natl Acad Sci U S A* 2000;**97**:5146–50.
45. Nunoshiba T, DeRojas-Walker T, Tannenbaum SR, Demple B. Roles of nitric oxide in inducible resistance of Escherichia coli to activated murine macrophages. *Infect Immun* 1995;**63**:794–8.
46. Nunoshiba T. T. deRojas-Walker, J.S. Wishnok, S.R. Tannenbaum, and B. Demple, Activation by nitric oxide of an oxidative-stress response that defends Escherichia coli against activated macrophages. *Proc Natl Acad Sci U S A* 1993;**90**:9993–7.
47. Kobayashi K, Tagawa S. Isolation of reductase for SoxR that governs an oxidative response regulon from Escherichia coli. *FEBS Lett* 1999;**451**:227–30.
48. Koo MS, Lee JH, Rah SY, Yeo WS, Lee JW, Lee KL, Koh YS, Kang SO, Roe JH. A reducing system of the superoxide sensor SoxR in Escherichia coli. *Embo J* 2003;**22**:2614–22.
49. Drapier J-C. Interplay between NO and [Fe-S] clusters: Relevance to biological systems. *Methods* 1997;**11**:319–29.
50. Yang W, Rogers PA, Ding H. Repair of nitric oxide-modified ferredoxin [2Fe-2S] cluster by cysteine desulfurase (IscS). *J Biol Chem* 2002;**277**:12,868–12,873,.
51. Rogers PA, Eide L, Klungland A, Ding H. Reversible inactivation of E. coli endonuclease III via modification of its [4Fe-4S] cluster by nitric oxide. *DNA Repair (Amst)* 2003;**2**:809–17.
52. Palma M, Zurita J, Ferreras JA, Worgall S, Larone DH, Shi L, Campagne F, Quadri LE. Pseudomonas aeruginosa SoxR does not conform to the archetypal paradigm for SoxR-dependent regulation of the bacterial oxidative stress adaptive response. *Infect Immun* 2005;**73**:2958–66.
53. Dietrich LE, Price-Whelan A, Petersen A, Whiteley M, Newman DK. The phenazine pyocyanin is a terminal signalling factor in the quorum sensing network of Pseudomonas aeruginosa. *Mol Microbiol* 2006;**61**:1308–21.
54. Dietrich LE, Teal TK, Price-Whelan A, Newman DK. Redox-active antibiotics control gene expression and community behavior in divergent bacteria. *Science* 2008;**321**:1203–6.
55. Demple B. Community organizers and (bio)filmmaking. *Nat Chem Biol* 2008;**4**:653–4.
56. Demple B, Halbrook J. Inducible repair of oxidative DNA damage in Escherichia coli. *Nature* 1983;**304**:466–8.
57. Christman MF, Morgan RW, Jacobson FS, Ames BN. Positive control of a regulon for defenses against oxidative stress and some heat-shock proteins in Salmonella typhimurium. *Cell* 1985;**41**:753–62.
58. Zheng M, Wang X, Templeton LJ, Smulski DR, LaRossa RA, Storz G. DNA microarray-mediated transcriptional profiling of the Escherichia coli response to hydrogen peroxide. *J Bacteriol* 2001;**183**:4562–70.
59. Christman MF, Storz G, Ames BN. OxyR, a positive regulator of hydrogen peroxide-inducible genes in Escherichia coli and Salmonella typhimurium, is homologous to a family of bacterial regulatory proteins. *Proc Natl Acad Sci U S A* 1989;**86**:3484–8.
60. Hausladen A, Privalle CT, Keng T, Deangelo J, Stamler JS. Nitrosative stress – Activation of the transcription factor OxyR. *Cell* 1996;**86**:719–28.
61. Kim SO, Merchant K, Nudelman R, Beyer Jr. WF, Keng T, DeAngelo J, Hausladen A, Stamler JS OxyR: a molecular code for redox-related signaling. *Cell* 2002;**109**:383–96.
62. Choi H, Kim S, Mukhopadhyay P, Cho S, Woo J, Storz G, Ryu S. Structural basis of the redox switch in the OxyR transcription factor. *Cell* 2001;**105**:103–13.
63. Toledano MB, Kullik I, Trinh F, Baird PT, Schneider TD, Storz G. Redox-dependent shift of OxyR-DNA contacts along an extended DNA-binding site: a mechanism for differential promoter selection. *Cell* 1994;**78**:897–909.
64. Tao K, Zou C, Fujita N, Ishihama A. Mapping of the OxyR protein contact site in the C-terminal region of RNA polymerase alpha subunit. *J Bacteriol* 1995;**177**:6740–4.
65. Zheng M, Åslund F, Storz G. Activation of the OxyR Transcription Factor by Reversible Disulfide Bond Formation. *Science* 1998;**279**:1718–21.
66. Åslund F, Beckwith J. The thioredoxin superfamily: redundancy, specificity, and gray-area genomics. *J Bacteriol* 1999;**181**:1375–9.
67. Lee C, Lee SM, Mukhopadhyay P, Kim SJ, Lee SC, Ahn WS, Yu MH, Storz G, Ryu SE. Redox regulation of OxyR requires specific disulfide bond formation involving a rapid kinetic reaction path. *Nat Struct Mol Biol* 2004;**11**:1179–85.
68. Chen H, Xu G, Zhao Y, Tian B, Lu H, Yu X, Xu Z, Ying N, Hu S, Hua Y. A novel OxyR sensor and regulator of hydrogen peroxide stress with one cysteine residue in Deinococcus radiodurans. *PLoS ONE* 2008;**3**:e1602.
69. Jakob U, Muse W, Eser M, Bardwell JC. Chaperone activity with a redox switch. *Cell* 1999;**96**:341–52.
70. Graumann J, Lilie H, Tang X, Tucker KA, Hoffmann JH, Vijayalakshmi J, Saper M, Bardwell JC, Jakob U. Activation of the redox-regulated molecular chaperone Hsp33: a two-step mechanism. *Structure* 2001;**9**:377–87.
71. Bsat N, Herbig A, Casillas-Martinez L, Setlow P, Helmann JD. Bacillus subtilis contains multiple Fur homologues: identification of the iron uptake (Fur) and peroxide regulon (PerR) repressors. *Mol Microbiol* 1998;**29**:189–98.
72. Ilbert M, Horst J, Ahrens S, Winter J, Graf PC, Lilie H, Jakob U. The redox-switch domain of Hsp33 functions as dual stress sensor. *Nat Struct Mol Biol* 2007;**14**:556–63.
73. Winter J, Ilbert M, Graf PC, Ozcelik D, Jakob U. Bleach activates a redox-regulated chaperone by oxidative protein unfolding. *Cell* 2008;**135**:691–701.
74. Janssen-Heininger YM, Mossman BT, Heintz NH, Forman HJ, Kalyanaraman B, Finkel T, Stamler JS, Rhee SG, van der Vliet A. Redox-based regulation of signal transduction: principles, pitfalls, and promises. *Free Radic Biol Med* 2008;**45**:1–17.
75. Fong CL, Lentz A, Mayfield SP. Disulfide bond formation between RNA binding domains is used to regulate mRNA binding activity of the chloroplast poly(A)-binding protein. *J Biol Chem* 2000;**275**:8275–8.
76. Kim J, Mayfield SP. Protein Disulfide Isomerase as a Regulator of Chloroplast Translational Activation. *Science* 1997;**278**:1954–7.
77. Khoroshilova N, Popescu C, Münck E, Beinert H, Kiley PJ. Iron-sulfur cluster disassembly in the FNR protein of Escherichia coli by O_2: [4Fe-4S] to [2Fe-2S] conversion with loss of biological activity. *Proc Natl Acad Sci USA* 1997;**94**:6087–92.
78. Tucker NP, Hicks MG, Clarke TA, Crack JC, Chandra G, Le Brun NE, Dixon R, Hutchings MI. The transcriptional repressor protein NsrR senses nitric oxide directly via a [2Fe-2S] cluster. *PLoS ONE* 2008;**3**:e3623.
79. Yukl ET, Elbaz MA, Nakano MM, Moenne-Loccoz P. Transcription Factor NsrR from Bacillus subtilis Senses Nitric Oxide with a 4Fe-4S Cluster. *Biochemistry* 2008;**47**:13,084–13,092.

Chapter 22

Double-Strand Break Recognition and its Repair by Non-Homologous End-Joining

Xiaoping Cui and Michael R. Lieber
University of Southern California Keck School of Medicine, USC Norris Comprehensive Cancer Center, Los Angeles, California

OVERVIEW OF NON-HOMOLOGOUS END-JOINING (NHEJ)

Double-strand breaks (DSBs) are the most challenging type of DNA damage to repair in cells, and efficient repair mechanisms are critical. In mammalian cells, the error-free homologous recombination (HR), and the error prone non-homologous end-joining (NHEJ) are the two major pathways for repair of DSBs. The HR pathway, in general, repairs the damaged strand by using an intact sister chromatid. In contrast, the NHEJ pathway includes steps involving editing of DNA ends by deleting or adding nucleotides at the ends. NHEJ is the primary DSB repair pathway used in mammalian cells. DSBs can be induced both physiologically and pathologically. In somatic cells, only two physiological forms of DSBs exist, and these are repaired by the NHEJ pathway. These are (1) V(D)J recombination occurring during early B and T cell development, and (2) class switch recombination in mature B cells (reviewed in [1, 2]). Pathological DSB can be induced in variety of ways, including ionizing radiation, free radicals, endogenous enzymatic errors (replication across a nick, topoisomerase failures, and inadvertent action by nucleases) (reviewed in [3, 4]) (Figure 22.1). In either condition, unrepaired or misrepaired DNA will cause irreversible loss of genetic information at the site of the break, possibly leading to cell death or cancer [5, 6].

In simplest terms, repair of DSBs is focused on joining the two ends back together. However, it is a much more complicated procedure than just simple ligation. The NHEJ repair pathway involves several proteins that are involved in its various steps (Figure 22.2). Initiation of the repair pathway starts by a protein receiving the "signal" of DNA breakage, and outcompeting the components of other pathways for binding to the broken DNA ends. Ku70/86 is likely the first protein complex in the NHEJ repair pathway that competes with protein components from other pathways, such as Rad51 from the HR pathway, to bind DNA ends, thereby initiating NHEJ [7–9]. Ku70/86 can then remain at the DNA end or slide inwards, thereby leaving space to recruit other proteins such as DNA dependent protein kinase (DNA-PKcs), a nuclease (Artemis), and the polymerases mu and lambda. These proteins can function in editing the DNA by deletion or addition of nucleotides to make ligatable ends. Finally, a ligase complex including XRCC4, ligaseIV and the recently defined XLF/Cernunnos [10, 11] complete the repair by ligating the ends together.

KINASE ACTIVATION AND AUTOPHOSPHORYLATION OF DNA-PK

The largest protein in the NHEJ pathway, DNA dependent protein kinase or DNA-PKcs, is 469 kDa. In fact, this is the largest protein kinase known in biology and one of the largest known polypeptides (thyroglobulin and titan are larger). At the carboxy terminus of DNA-PKcs is a kinase domain that is similar to the phosphotidylinositide-3, 4 kinase family (PI3, 4-kinase) and therefore is categorized as a member of the phosphatidylinositol-3 kinase related protein kinases (PIKKs) [12, 13]. Cells lacking DNA-PKcs are sensitive to ionizing irradiation as well as chemical compounds that induce double-strand breaks [14]. DNA-PKcs null animals are immunodeficient due to lack of active DNA-PKcs, and thus, lack of functional V(D)J recombination [15–19]. DNA-PKcs is a potential therapeutic target [5, 20].

Though the full list of roles of DNA-PKcs in NHEJ may not be complete, considerable evidence demonstrates at least two roles. First, DNA-PKcs serves as a "bridge" protein that recruits and connects with all the other NHEJ

Handbook of Cell Signaling, Three-Volume Set 2 ed.

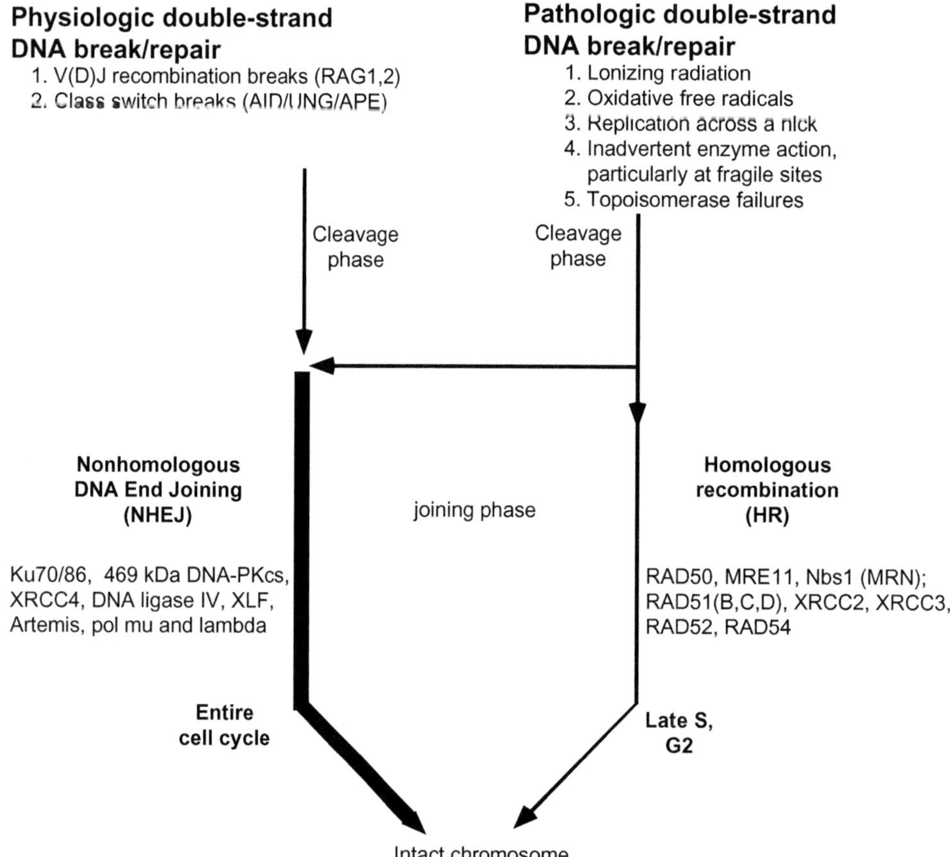

FIGURE 22.1 An introduction of DSBs and the two major repair pathways: NHEJ and HR.
Both physiological and pathological causes of DSBs are listed at the top. On the left, among physiological DSBs, V(D)J recombination is present in all true vertebrates and is initiated by an endonuclease complex composed of RAG1 and 2. Class switch recombination is only present among amphibians and land vertebrates and is initiated by a cytidine deaminase called activated induced deaminase (AID). Repair of these two physiological DSBs both rely on NHEJ. Pathological DNA damage can be repaired by either homologous recombination (HR) or NHEJ in multi-cellular eukaryotes. The protein components involved in each pathway of HR and NHEJ are listed. DSBs that arise in late S or G2 of the cell cycle are often repaired at long regions (>100 bp) of homology using homologous recombination (though single-strand annealing also can occur). NHEJ does not use long stretches of homology, but the processing of the DNA ends can, in a minority of cases, be influenced by alignment of a few nts of homology called terminal microhomology (typically 1–4 nt in length). NHEJ can function at any time during the cell cycle and therefore is the dominant pathway for the repair of double-strand breaks.

proteins at the broken DNA ends [21–24]. Second, the kinase activity of DNA-PKcs may alter some functions of other proteins and itself via phosphorylation at S/T sites. *In vitro* biochemistry assays have demonstrated that all of the known factors of NHEJ including DNA-PKcs itself, as well as some other proteins can be phosphorylated by DNA-PKcs. In these elegant mutagenesis studies, a number of phosphorylation sites have been identified and characterized on all of the proteins. However, mutations of the S/T sites on proteins including Ku70/86 (five S/T sites) [25], XRCC4 (nine S/T sites) [26], ligase IV (three S/T sites) [27], Artemis (14 S/T sites) [28], and XLF (two S sites) [29] showed no defect in the NHEJ repair efficiency. In fact, independent studies from various laboratories have demonstrated a role of kinase active DNA-PKcs in regulating nuclease activity, polymerase activity, and ligation activity in the process of NHEJ [30, 31]. Mutants of the S/T sites on DNA-PKcs showed defective NHEJ repair efficiency.

The two major clusters of autophosphorylation sites on DNA-PKcs itself are the ABCDE cluster (aa2609–2647) that contains six S/T sites [32–34] and PQR cluster (aa2023–2056) that contains five S sites [35] (Figure 22.3). Additional sites other than these two clusters include S3205 and T3950 [36]. To block phosphorylation of these sites, each S/T site was mutated to alanine. For either ABCDE or PQR cluster, individual S/T to A mutations did not affect DNA-PKcs's function, while alanine substitutions at all six sites of the ABCDE cluster resulted in a complete blocks of NHEJ repair and alanine substitutions at all five sites of the PQR cluster partially reduced NHEJ repair efficiency as well. Sequencing analysis on the repaired DNA ends revealed that mutations at both clusters affect the end processing. It was demonstrated in these studies that

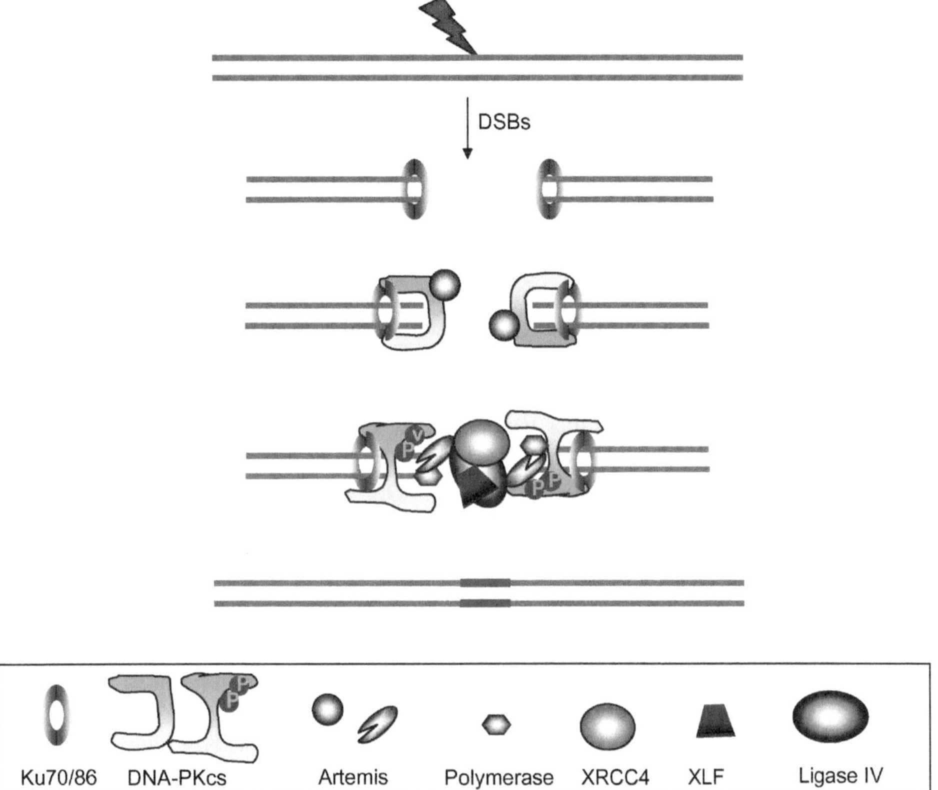

FIGURE 22.2 Autophosphorylation induced conformational changes of DNA-PKcs and the NHEJ repairing.
To initiate repair of double-strand breaks using NHEJ pathway, Ku70/86 heterdimers first recognize and bind the DNA ends of the DSBs. Binding of the Ku70/86 complex on DNA also improves the binding equilibrium of other NHEJ factors such as DNA-PKcs. After binding to the DNA ends, Ku70/86 slides inwards, leaving space for the nucleases, polymerases, and ligases to be recruited to the DNA ends. Although several steps of end processing are known to be involved before ligation and completion of the repair, it is still unclear how each enzyme is loaded on to the ends and no specific order for loading is thought to exist. However, *in vitro* and *in vivo* studies have clearly identified and confirmed the endonucleolytic activity of DNA-PKcs:Artemis complex. Two polymerases, mu and lamda, are also involved in the end processing by NHEJ. Finally, the ligation is done by the complex of XRCC4:XLF:ligaseIV. DNA-PK kinase activity is activated by binding to DNA, and therefore phophorylates its target proteins and itself. Autophosphorylation at the ABCDE cluster within DNA-PKcs results in an open conformation, which allows the nuclease, polymerase, and ligase to access the DNA ends. Autophosphorylation at the PQR sites changes DNA-PKcs to a more closed conformation, probably signaling completion of end processing. Mutagenesis studies on the autophosphorylation sites of DNA-PKcs have provided evidence that the status of DNA-PKcs autophosphorylation can lead to conformational changes of this scaffold protein, which therefore regulates the accessibility of the nuclease, polymerase, and ligase to the DNA ends. Moreover, the autophosphorylation status of DNA-PKcs may also regulate the DNA access of factors in the homologous recombination pathway. *In vitro* kinase assays have shown that including DNA-PKcs itself, six other proteins (Ku70, Ku86, Artemis, XRCC4, XLF, and LigaseIV) of NHEJ are also substrates of the DNA-PKcs kinase activity, but mutagenesis studies show no phenotype defects on any of these proteins except DNA-PKcs itself.

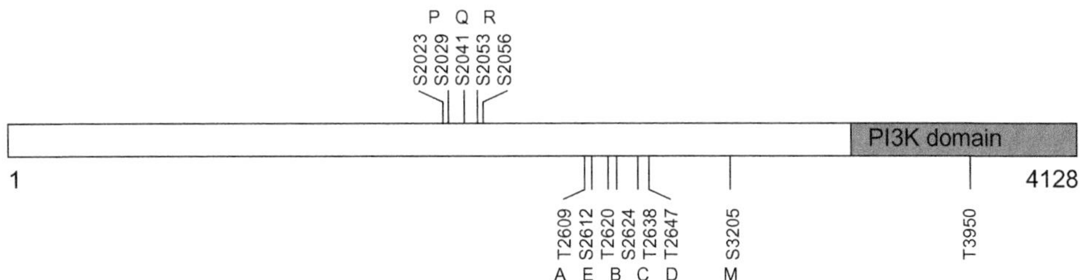

FIGURE 22.3 Schematic locations of the 13 identified and characterized autophosphorylation sites on DNA-PKcs.
The ABCDE cluster includes six serine (S) or threonine (T) autophosphorylation sites and is located between amino acid 2609 to 2647. The M site represents amino acid S3205. Each site of ABCDE is denoted as follows: A, T2609; B, T2620 and S2624; C, T2638; D, T2647; and E, S2612. The PQR cluster contains five serine sites, which are denoted as: P, S2023 and S2029; Q, S2041; R, S2053 and S2056.

blocking ABCDE sites resulted in less than two nucleotide losses at the ends of DSBs, while in wild-type cells, the average loss was about five to six nucleotides. Sequence analysis with the PQR to alanine mutants resulted in much longer deletions (average of 14 nucleotides). The endonucleolytic activity of the Artemis:DNA-PK complex confirmed that blocking the phosphorylation of the ABCDE cluster, resulted in no access to the DNA ends by Artemis and, thereby, no endonuclease activity [30]. Based on these experimental data, the authors suggested that autophosphorylation at the ABCDE cluster was more likely at the earlier stages of the process, and caused conformational changes that allowed other repair factors to gain access to the DNA ends. The autophosphorylation at the PQR cluster, on the other hand, is thought to occur at a later stage of the process and signals completion of end processing. Blocking this cluster then delays the completion of the NHEJ process, and modestly affects the repair efficiency. In summary, these clusters work in reciprocal ways in regulating end processing: while autophosphorylation at the ABCDE cluster promotes end processing, autophosphorylations at the PQR cluster may inhibit the ends processing.

Using purified proteins, biochemical assays are able to detect DNA-PKcs kinase activity initiated via binding to DNA and autophosphorylation of itself or substrate proteins. Autophosphorylation of DNA-PKcs correlates with loss of DNA-PK activity and dissociates from DNA, which might also be an indication of completion of the repair process. Although more than a dozen S/T sites have already been identified, and their potential roles during the process have been proposed, the mechanism of exactly when and how DNA-PKcs dissociates from the repaired DNA ends still remains unknown, and it is likely that additional autophosphorylation sites are involved and need to be identified. In fact, in the presence of double-stranded DNA and ATP, a DNA-PKcs mutant in which all 13 identified S/T sites are mutated to alanine could still undergo autophosphorylation. The level of this remaining autophosphorylation is about 40 percent of the wild-type level of autophosphorylation [35, 36].

DNA-PK MAY INFLUENCE THE BALANCE OF HR AND NHEJ DURING S PHASE

In the past decade, there has been considerable insight into the balance of HR and NHEJ from numerous labs. However, the interrelationship between the pathways and what regulates the pathway choice in the cells is still under active investigation. Among the hypotheses, two frequently suggested possible mechanisms of signaling have been proposed.

First, several studies point to the cell cycle as one factor that modulates the pathway choice. In this proposal, HR is favored during late S/G2, due to the proximity of a suitable homology donor, the sister chromatid. The NHEJ pathway, in contrast, does not rely on long (>25 bp) homologous sequences and is the primary repair pathway during G1 and early S, but can occur during late S/G2.

A second proposal is that the kinase activity and autophosphorylation status of DNA-PKcs regulate the pathway choice in the cells. This was an extension from another concept of mechanistic competition between HR and NHEJ [37]. In the earlier studies, it was shown that HR is increased in the absence of DNA-PKcs, but decreased when DNA-PKcs is complemented. More interestingly, with catalytically inactive (inactivated by DNA-PK specific inhibitors,IC86621,1-(2-hydroxy-4-morpholin-4-yl-phenyl)-ethanone) DNA-PKcs, the level of HR is also significantly reduced. Several studies also demonstrated that kinase inactive DNA-PKcs inhibits DNA repair pathways that are independent of DNA-PKcs [38, 39]. It is reasoned that inactivated DNA-PKcs binds to the DNA end and blocks the access of other factors including both the NHEJ factors and the HR repair factors. Detailed studies from a series of DNA-PKcs mutants or isoforms that either are kinase dead or lack certain phosphorylation sites on DNA-PKcs also suggested a possible role of DNA-PKcs in influencing pathway choice. These findings were part of the same studies of the two autophosphorylation clusters, as was discussed above and also other publications from independent investigators. An exchromosomal substrate was used to test homologous recombination efficiency in the cells with no DNA-PKcs, or complemented DNA-PKcs, or with either of the two clusters of S/T sites mutated to alanine. Using this substrate, the DSBs were introduced by expression of the enzyme, I-SceI, which can recognize and cut at its recognition sequence on the substrate. Just as the opposite roles of end processing demonstrated for the autophosphorylation of DNA-PKcs at the ABCDE cluster and PQR cluster, the same may also be true for their reciprocal roles in regulating the pathway choice. These results showed that while the PQR to alanine mutant improved HR efficiency, the ABCDE to alanine mutant reduced HR. The mechanism is also the same as discussed for end processing: the ABCDE to alanine mutant results in a more closed conformation for DNA-PKcs, which in turn blocks the DNA ends and blocks all the repair factors. But the PQR to alanine mutant has an open conformation, thereby allowing other factors including both NHEJ and HR proteins. Results from other mutants of DNA-PKcs, such as a kinase dead mutant, K>R, and kinase dead isoforms of DNA-PKcs, also showed that the kinase activity of DNA-PK is essential for wild-type repair efficiency of HR and NHEJ [40]. (Note that kinase dead DNA-PKcs decreases HR and NHEJ, but a complete lack of DNA-PKcs results in an increase in HR, presumably by permitting HR proteins better access. In contrast, a dead DNA-PKcs may obstruct HR proteins.) Taken together, DNA-PKcs plays a regulatory role that affects the pathway choice in the cell. This regulation role not only depends on the presence of the DNA-PKcs in the cells, but also the phosphorylation status and kinase activity of this protein.

LOCAL CHROMATIN STRUCTURE AT SITES OF NHEJ

Most histone octamers in the human genome consist of H2A, H2B, H3, and H4. However, one in every 10 octamers contains the histone variant H2AX rather than H2A. If an H2AX containing nucleosome is in the neighborhood of a DSB, then it becomes phosphorylated at the serine 139 within its C-terminus, and this phosphorylated form is called γ-H2AX. For homologous recombination, the distance of DNA processing is long, and it appears that zones containing a megabase of DNA can have H2AX phosphorylation. For NHEJ, the data is a bit less clear. Given that NHEJ is an extremely local event, then it might occur on a DSB in DNA wrapped around one nucleosome. If the NHEJ repair is fast enough, it is conceivable that no H2AX phosphorylation occurs.

This may seem at odds with some immunolocalization studies. One complexity of immunolocalization (focus) assays is that the confocal microscopy section thickness (0.2 microns) contains hundreds of DNA duplexes, some of which may be repaired by NHEJ and others that may be repaired by HR (reviewed in [41]).

To whatever extent that NHEJ does involve H2AX phosphorylation, there are data to support H2AX phosphorylation by either ATM or by DNA-PKcs [42–44]. When DNA-PKcs phosphorylates H2AX, this increases its vulnerability to the histone exchange factor called FACT (which consists of a heterodimer of Spt16 and SSRP1). Phosphorylated H2AX (γ-H2AX) is more easily exchanged out of the octamer, thereby leaving only a tetramer of $(H3)_2(H4)_2$ at the site, and this is more flexible, thereby perhaps permitting DNA repair factors to carry out their work [45].

PARP-1 is able to downregulate the activity of FACT by ADP-ribosylation of the Spt16 subunit of FACT. This may be able to shift the equilibrium of γ-H2AX and H2AX in the nucleosomes. That is, PARP-1 activation at a site of damage might shift the equilibrium toward retention of γ-H2AX in the region, perhaps thereby aiding in recruitment or retention of repair proteins.

Hence, FACT may initially act proximally at the most immediate nucleosome to leave an $(H3)_2(H4)_2$ tetramer at the site of damage for purposes of flexibility of the DNA. FACT may act more regionally (distally) to favor the retention of γ-H2AX for purposes of integrating the repair process with recruitment, retention, and cell cycle aspects.

REFERENCES

1. Lieber MR, Yu K, Raghavan SC. Roles of nonhomologous DNA end joining, V(D)J recombination, and class switch recombination in chromosomal translocations. *DNA Repair* 2006;**5**:1234–45.
2. Soulas-Sprauel P, Rivera-Munoz P, Malivert L, et al. V(D)J and immunoglobulin class switch recombinations: a paradigm to study the regulation of DNA end-joining. *Oncogene* 2007;**26**:7780–91.
3. Lieber MR. The mechanism of human nonhomologous DNA end joining. *J Biol Chem* 2008;**283**:1–5.
4. Jackson SP. Sensing and repairing DNA double-strand breaks. *Carcinogenesis* 2002;**23**:687–96.
5. Pastwa E, Malinowski M. Non-homologous DNA end joining in anticancer therapy. *Curr Cancer Drug Targets* 2007;**7**:243–50.
6. Burma S, Chen BP, Chen DJ. Role of non-homologous end joining (NHEJ) in maintaining genomic integrity. *DNA Repair (Amst)* 2006;**5**:1042–8.
7. Sonoda E, Hochegger H, Saberi A, et al. Differential usage of non-homologous end-joining and homologous recombination in double strand break repair. *DNA Repair (Amst)* 2006;**5**:1021–9.
8. Hochegger H, Dejsuphong D, Fukushima T, et al. Parp-1 protects homologous recombination from interference by Ku and Ligase IV in vertebrate cells. *EMBO J* 2006;**25**:1305–14.
9. Wang M, Wu W, Rosidi B, et al. PARP-1 and Ku compete for repair of DNA double strand breaks by distinct NHEJ pathways. *Nucleic Acids Res* 2006;**34**:6170–82.
10. Ahnesorg P, Smith P, Jackson SP. XLF interacts with the XRCC4-DNA ligase IV complex to promote nonhomologous end-joining. *Cell* 2006;**124**:301–13.
11. Buck D, Malivert L, deChasseval R, et al. Cernunnos, a novel nonhomologous end-joining factor, is mutated in human immunodeficiency with microcephaly. *Cell* 2006;**124**:287–99.
12. Beamish HJ, Jessberger R, Riballo E, et al. The C-terminal conserved domain of DNA-PKcs, missing in the SCID mouse, is required for kinase activity. *Nucleic Acids Res* 2000;**28**:1506–13.
13. Hartley KO, Gell D, Smith GC, et al. DNA-dependent protein kinase catalytic subunit: a relative of phosphatidylinositol 3-kinase and the ataxia telangiectasia gene product. *Cell* 1995;**82**:849–56.
14. Lees-Miller SP, Godbout R, Chan D, et al. Absence of p350 subunit of DNA-activated protein kinase from a radiosensitive human cell line. *Science* 1995;**267**:1183–5.
15. Kirschgessner C, Patel C, Evans J, et al. DNA-dependent kinase (p350) as a candidate gene for the murine SCID defect. *Science* 1995;**267**:1178–85.
16. Miller RD, Hogg J, Ozaki J, et al. Gene for the catalytic subunit of mouse DNA-PK maps to the scid locus. *Proc Natl Acad Sci USA* 1995;**92**:10,792–10,795.
17. Danska JS, Holland DP, Mariathasan S, et al. Biochemical and genetic defects in DNA-dependent protein kinase in murine scid lymphocytes. *Mol Cell Biol* 1996;**16**:5507–17.
18. Blunt T, Finnie NJ, Taccioli GE, et al. Defective DNA-Dependent Protein Kinase Activity is Linked to V(D)J Recombination and DNA Repair Defects Associated with the Murine scid Mutation. *Cell* 1995;**80**:813–23.
19. Blunt T, Gell D, Fox M, et al. Identification of a nonsense mutation in the carboxyl-terminal region of DNA-dependent protein kinase catalytic subunit in the scid mouse. *Proc Natl Acad Sci USA* 1996;**93**:10,285–10,290.
20. Salles B, Calsou P, Frit P, Muller C. The DNA repair complex DNA-PK, a pharmacological target in cancer chemotherapy and radiotherapy. *Pathol Biol (Paris)* 2006;**54**:185–93.
21. Chen L, Trujillo K, Sung P, Tomkinson AE. Interactions of the DNA ligase IV-XRCC4 complex with DNA ends and the DNA-dependent protein kinase. *J Biol Chem* 2000;**275**:26,196–26,205,.
22. Hsu HL, Yannone SM, Chen DJ. Defining interactions between DNA-PK and ligase IV/XRCC4. *DNA Repair (Amst)* 2002;**1**:225–35.

23. Han Z, Johnston C, Reeves WH, et al. Characterization of a Ku86 variant protein that results in altered DNA binding and diminished DNA-dependent protein kinase activity. *J Biol Chem* 1996;**271**:14,098–14,104.
24. Drouet J, Frit P, Delteil C, et al. Interplay between Ku, Artemis, and the DNA-dependent protein kinase catalytic subunit at DNA ends. *J Biol Chem* 2006;**281**:27,784–27,793.
25. Douglas P, Gupta S, Morrice N, et al. DNA-PK-dependent phosphorylation of Ku70/80 is not required for non-homologous end joining. *DNA Repair (Amst)* 2005;**4**:1006–18.
26. Yu Y, Wang W, Ding Q, et al. DNA-PK phosphorylation sites in XRCC4 are not required for survival after radiation or for V(D)J recombination. *DNA Repair (Amst)* 2003;**2**:1239–52.
27. Wang YG, Nnakwe C, Lane WS, et al. Phosphorylation and regulation of DNA ligase IV stability by DNA-dependent protein kinase. *J Biol Chem* 2004;**279**:37,282–37,290.
28. Ma Y, Pannicke U, Lu H, et al. The DNA-PKcs phosphorylation sites of human artemis. *J Biol Chem* 2005;**280**:33,839–33,846.
29. Yu Y, Mahaney BL, Yano KI, et al. DNA-PK and ATM phosphorylation sites in XLF/Cernunnos are not required for repair of DNA double strand breaks. *DNA Repair (Amst)* 2008;**7**:1680–92.
30. Goodarzi AA, Yu Y, Riballo E, et al. DNA-PK autophosphorylation facilitates Artemis endonuclease activity. *Embo J* 2006;**25**:3880–9.
31. Weterings E, Verkaik NS, Bruggenwirth HT, et al. The role of DNA dependent protein kinase in synapsis of DNA ends. *Nucleic Acids Res* 2003;**31**:7238–46.
32. Douglas P, Sapkota GP, Morrice N, et al. Identification of in vitro and in vivo phosphorylation sites in the catalytic subunit of the DNA-dependent protein kinase. *Biochem J* 2002;**368**:243–51.
33. Ding O, Reddy Y, Wang W, et al. Autophosphorylation of the catalytic subunit of the DNA-dependent protein kinase is required for efficient end processing during DNa double-strand break repair. *Mol Cell Biol* 2003;**23**:5836–48.
34. Reddy YV, Ding Q, Lees-Miller SP, et al. Non-homologous end joining requires that the DNA-PK complex undergo an autophosphorylation-dependent rearrangement at DNA ends. *J Biol Chem* 2004;**279**:39,408–39,413.
35. Cui X, Yu Y, Gupta S, et al. Autophosphorylation of DNA-dependent protein kinase regulates DNA end processing and may also alter double-strand break repair pathway choice. *Mol Cell Biol* 2005;**25**:10,842–10,852.
36. Meek K, Douglas P, Cui X, et al. trans Autophosphorylation at DNA-dependent protein kinase's two major autophosphorylation site clusters facilitates end processing but not end joining. *Mol Cell Biol* 2007;**27**:3881–90.
37. Allen C, Halbrook J, Nickoloff JA. Interactive competition between homologous recombination and non-homologous end joining. *Mol Cancer Res* 2003;**1**:913–20.
38. Udayakumar D, Bladen CL, Hudson FZ, Dynan WS. Distinct pathways of nonhomologous end joining that are differentially regulated by DNA-dependent protein kinase-mediated phosphorylation. *J Biol Chem* 2003;**278**:1631–5.
39. Perrault R, Wang H, Wang M, et al. Backup pathways of NHEJ are suppressed by DNA-PK. *J Cell Biochem* 2004;**92**:781–94.
40. Convery E, Shin EK, Ding Q, et al. Inhibition of homologous recombination by variants of the catalytic subunit of the DNA-dependent protein kinase (DNA-PKcs). *Proc Natl Acad Sci U S A* 2005;**102**:1345–50.
41. Friedberg EC, Walker GC, Siede W, et alDNA repair and mutagenesi. Washington D.C.: ASM Press; 2006 p. 1118.
42. Stiff T, O'Driscoll M, Rief N, et al. ATM and DNA-PK function redundantly to phosphorylate H2AX after exposure to ionizing radiation. *Cancer Res* 2004;**64**:2390–6.
43. Paull TT, Rogakou EP, Yamazaki V, et al. A critical role for histone H2AX in recruitment of repair factors to nuclear foci after DNA damage. *Curr Biol* 2000;**10**:886–95.
44. Wang H, Wang M, Bocker W, Iliakis G. Complex H2AX phosphorylation patterns by multiple kinases including ATM and DNA-PK in human cells exposed to ionizing radiation and treated with kinase inhibitors. *J Cell Physiol* 2005;**202**:492–502.
45. Heo K, Kim H, Choi SH, et al. FACT-mediated exchange of histone variant H2AX regulated by phosphorylation of H2AX and ADP-ribosylation of Spt16. *Mol Cell* 2008;**30**:86–97.

Chapter 23

ATM Mediated Signaling Defends the Integrity of the Genome

Martin F. Lavin[1,2], Magtouf Gatei[1], Philip Chen[1], Amanda Kijas[1] and Sergei Kozlov[1]
[1]*Queensland Cancer Fund Research Laboratory, Queensland Institute of Medical Research, Brisbane, Australia*
[2]*Department of Surgery, University of Queensland, Brisbane, Australia*

INTRODUCTION

The human genome is exposed to a variety of agents, both endogenous and exogenous, that threaten its integrity, compromise its ability to be expressed into RNA and protein and alter its capacity to be transmitted faithfully from one generation to the next. Several mechanisms have evolved to recognize these forms of DNA damage and signal them to the DNA repair machinery and the cell cycle checkpoints. These mechanisms maintain the integrity of the genomic material and minimize the risk of cancer and other pathologies. A number of human genetic disorders characterized by chromosomal instability and cancer predisposition has enhanced our understanding of the process of DNA damage recognition. One of these syndromes, ataxia-telangiectasia (A-T), has been a focal point because of the universal sensitivity to ionizing radiation and because of the central role the gene product involved plays in radiation signal transduction. A-T is characterized by immunodeficiency, neurodegeneration, radiosensitivity, meiotic defects, and cancer predisposition [1, 2]. Radiosensitivity has been described in A-T patients exposed to radiotherapy [3, 4] and in cells in culture [5, 6]. It seems likely that the basis of the radiosensitivity can be explained by a failure of A-T cells to respond appropriately to double-strand breaks (DSB) in DNA and the presence of residual breaks at longer times post-irradiation suggests that these cells are defective in repairing DNA DSB [7]. The presence of residual chromosomal breaks in A-T cells may lead to oxidative stress for which there is evidence in both humans and in *Atm* gene mutant mice [8–10]. Failure to activate ATM in response to DNA DSB gives rise to multiple cell cycle checkpoint defects in A-T cells [11, 12]. The first of these to be described was radioresistant DNA synthesis (RDS), which is manifested by reduced inhibition of DNA synthesis in response to IR exposure [13, 14]. This was the initial description of a cell cycle checkpoint defect in A-T cells. While the nature of the defect remains unclear it seems likely that a defect in Cdc25A phosphatase degradation, preventing dephosphorylation of the cyclin checkpoint kinase, Cdc2, allows DNA replication to proceed in the presence of DNA DSB [15]. The other pathway involved in the S-phase checkpoint, ATM-Mre11 (Mre11/Rad50/Nbs1) complex, SMC1, is also not activated in A-T cells [16, 17]. The defect in regulating the passage of cells from the G_1 to S phase can be explained primarily by a defective p53 response in irradiated A-T cells [18, 19] but it is also evident that ATM controls the G_1/S checkpoint through several intermediate phosphorylation reactions involving such intermediates as Chk2, MDM2, and MDMX [20]. ATM dependent phosphorylation events are also required for activation of the G_2/M checkpoint [21]. This is initiated by phosphorylation of Chk1 and Chk2, which inactivate Cdc25C phosphatise and prevent dephosphorylation of Cdk1-cyclinB kinase, which is required for progression into mitosis [22].

The gene defective in A-T was localized to chromosome 11q22-23 by Gatti *et al.* [23] and cloned by positional cloning, Savitsky *et al.* [24]. This gene was called ATM, ataxia- telangiectasia mutated. ATM is a member of a family of proteins that share a phosphatidylinositol 3-kinase (P13K) domain [25]. This group includes the catalytic subunit of DNA dependent protein kinase (DNA-PKcs), A-T, and rad3 related protein (ATR) suppressor with morphogenic effect on genitalia (SMG-1) and proteins in other organisms responsible for DNA damage recognition or cell cycle control [26]. ATM kinase is rapidly activated by ionizing radiation to phosphorylate a series of substrates involved in radiation signaling [27, 28]. The activation process has been largely elucidated in recent years and involves recruitment to a DNA DSB by the Mre11 complex

Handbook of Cell Signaling, Three-Volume Set 2 ed.

where it phosphorylates several proteins involved in stabilizing the recognition complex at or adjacent to the site of the break [29–31]. However, evidence also exists that ATM can be regulated at both the transcriptional and translational levels [32–34]. A more widespread role for ATM in events other than DNA damage recognition exists including receptor signaling, cellular proliferation, K^+ channel activity, and insulin signaling pathways [35–37]. Here we focus on the role of ATM in DNA damage recognition and its relationship to other DNA damage recognition systems, intermediates phosphorylated, and pathways activated to help coordinate the cellular response to radiation.

SENSING RADIATION DAMAGE IN DNA

DNA DSB arise normally during the process of V(D) J recombination in B cells during the process of V(D)J recombination and for T cell receptor rearrangements during meiosis and mitosis [38, 39]. On the other hand exposure of cells to IR and radiomimetic agents introduces potentially lethal DSB into DNA [40]. If left unrepaired they can cause the cell to die or lead to abnormal chromosomal rearrangements and translocations and a transformed phenotype [41–43]. In mammalian cells these breaks are repaired by three mechanisms: single-strand annealing, homologous recombination (HR), and non-homologous end-joining (NHEJ) [44]. The relative importance of these mechanisms varies with each stage of the cell cycle. NHEJ is primarily responsible for repair of DSBs in G_1 and early S phase, whereas HR predominates in G_2/M [45, 46]. The appearance of breaks in DNA is rapidly detected by several enzyme systems, as outlined in Figure 23.1. Poly(ADP-ribose) polymerase (PARP) responds rapidly to single-strand breaks in DNA by transferring ADP-ribose from NAD+ on to itself and to proteins involved in chromatin structure (histone proteins) and in DNA metabolism (topoisomerases, DNA replication factors) [47]. The presence of negatively charged poly(ADP-ribose) chains on these proteins alters their capacity to bind DNA and leads to their inactivation. As indicated above, NHEJ is responsible for initiating the process of DSB repair primarily in G_1 phase cells. Characterization of radiation sensitive mammalian mutants and the generation of gene disrupted mutant mice have identified DNA-PKcs, a heterodimer of Ku70 and Ku80, ligase IV, XRCC4, and XRCC4-like factor, Cerunnos, as central players in this process [48]. Mutants in any of these genes are hypersensitive to ionizing radiation, defective in DSB repair and V(D)J recombination, and have abnormalities in telomere maintenance [49–51]. The Ku heterodimer binds to the ends of a DSB and recruits and activates DNA-PKcs leading to the phosphorylation of a number of substrates implicated in DNA repair. There is also evidence that Ku recruits the XRCC4-ligase IV complex to DNA ends to complete the process of end-joining [52].

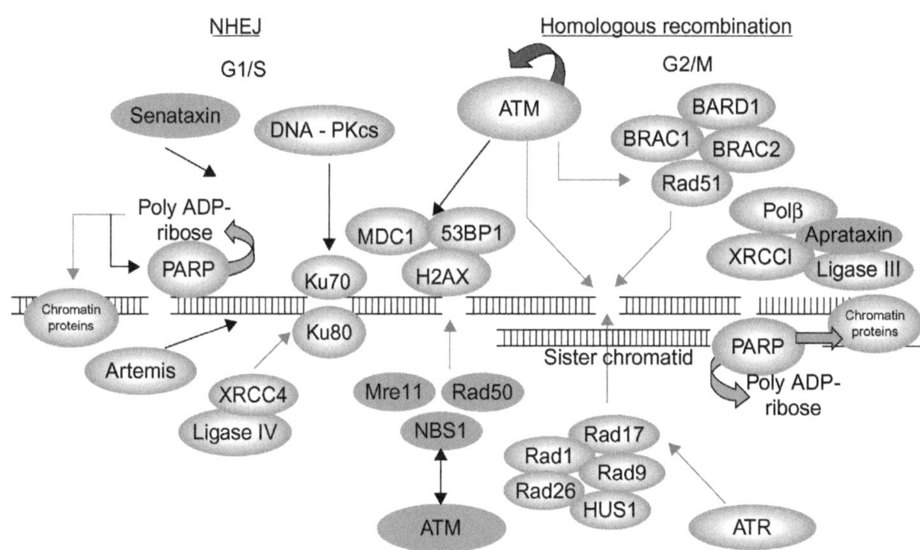

FIGURE 23.1 Detection of strand breaks in DNA.
PARP is rapidly activated by single- and double-strand breaks in DNA to poly ADP ribosylate a number of proteins including itself. Several other proteins are involved in this process including XRCC1, pol β, Ligase III, and aprataxin mutated in ataxia oculomotor apraxia type 1 (AOA1). DSBs are repaired by two major mechanisms: NHEJ and HR. Recognition of the DSBs involves the coordinated action of several proteins and complexes, including DNA-PKcs and the Ku heterodimer, that recognize free ends in DNA followed by recruitment of the DNA ligase IV/XRCC4 complex and XLF to seal the break. This occurs in concert with the Mre11/Rad50/Nbs1 complex, at least for some of these breaks. ATM also senses breaks in DNA and phosphorylates and activates a number of downstream effector proteins including Nbs1 and BRCA1. Evidence for indirect involvement of ATM through Rad51 and Mre11 also exists. Senataxin defective in AOA2 is implicated in the DNA DSB response during oxidative stress.

A second complex, Mre11/Rad50/Xrs2, was also shown to participate in NHEJ of breaks in *Saccharomyces cerevisiae* [53]. Homologs of Mre11 and Rad50 have been identified in mammalian cells and Nbs1 (nibrin); the gene defective in Nimegen breakage syndrome (NBS) appears to be the functional counterpart of Xrs2 [53, 54]. In response to radiation, the Mre11/Rad50/Nbs1 complex is localized rapidly to sites of DNA DSBs and associates in discrete foci [55, 56]. The Mre11 complex is also involved in homologous recombination, meiotic recombination, and telomere maintenance [57]. Cross-talk between the Mre11 complex and ATM, in recognizing DSB in DNA, is supported by overlap in phenotype between the three syndromes that arise when these genes are mutated: A-T (ATM), NBS (Nbs1), and A-T-like syndrome (Mre11) [24, 54, 58]. The MRN complex is rapidly localized to nuclear foci in response to radiation exposure of sites of DNA damage [55]. ATM, which phosphorylates Nbs1 for activation of the S-phase checkpoint is not required for association of the MRN complex with sites of damage [59]. The complex also binds tightly to chromatin in the absence of DNA damage during S phase [60]. The Mre11/Rad50 complex binds to DNA as a heterotetramer, tethering broken ends of a DSB [61]. The binding appears to be achieved through the two DNA binding motifs of Mre11 [62], which is arranged as a globular domain with Rad50 Walker A and B motifs (ATPase domains) and the bridging of DNA molecules is achieved through CXXC sequences in the middle of Rad50. Association with Rad50 stimulates both the exonuclease and endonuclease activities of Mre11 [63, 64] and Nbs1 stimulates its endonuclease activity [65]. These activities contribute to the processing of DNA DSB prior to repair.

ATM ACTIVATION AND RECRUITMENT OF DNA DAMAGE RESPONSE PROTEINS TO DNA DSB

ATM is primarily activated as a preexisting protein by DNA DSB [27, 28]. Agents that break DNA activate ATM so that the initiating signal may be a relaxation of chromatin superhelicity, which would be expected to be rapidly transmitted to ATM in complexes associated with chromatin (Figure 23.2). The initiating event may be a conformational change that would allow access to ATP and protein substrates [66]. This is supported by ATM activation by chloroquine, histone deacetylase inhibitors, and hypotonic buffers, all of which can alter chromatin without introducing DNA DSB [27]. However, this initial or "partial" activation

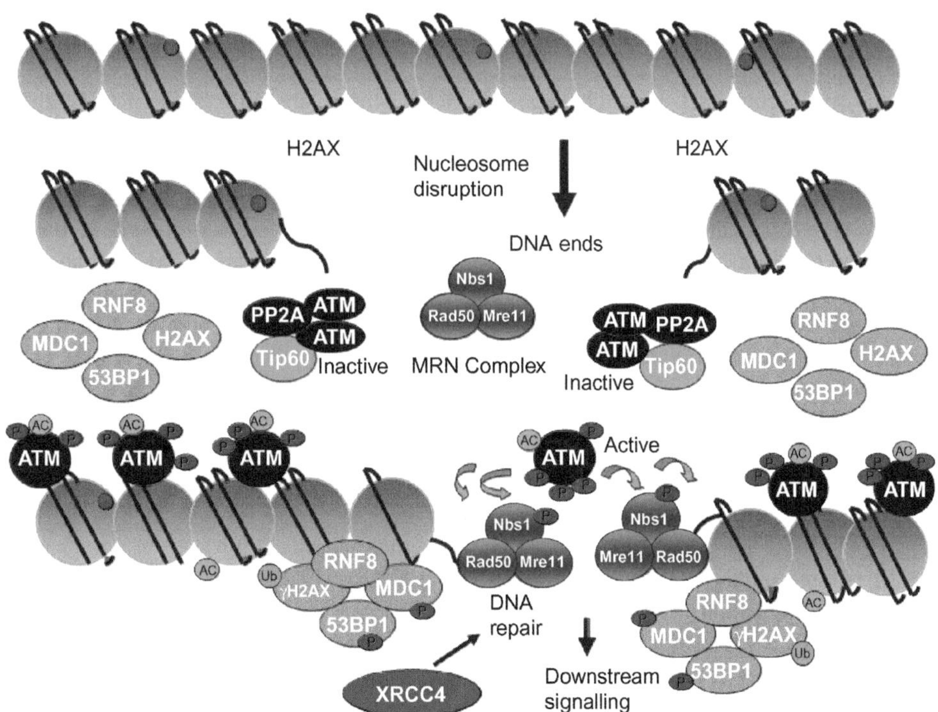

FIGURE 23.2 Sensing and signaling DNA DSB.
In response to DNA DSB chromatin relaxation occurs and a localized area of nucleosome disruption results allowing access to the DNA damage recognition complex. The MRN complex acts as the sensor for the break and recruits ATM, which is present in the nucleoplasm as an inactive dimer associated with PP2A phosphatase and Tip60 acetyltransferase. As ATM associates with the break so too do a number of other key proteins including H2AX, MDC1, 53BP1, and RNF8. These proteins are involved in stabilizing the DNA damage recognition apparatus. ATM is fully activated on recruitment and subsequently phosphorylates members of the MRN complex, some of these recognition proteins, as well as downstream substrates involved in cell cycle control and DNA repair.

of ATM does not cause it to appear in nuclear foci and its repertoire of substrate phosphorylation is very much restricted [67, 68]. For example in this state it can phosphorylate p53 but not H2AX and other substrates. For ATM to become active at least two posttranslational changes occur, phosphorylation status change and acetylation [27, 28, 69, 70]. There is evidence that both autophosphorylation and dephosphorylation contribute to the activation process. Bakkenist and Kastan [27] revealed that autophosphorylation on S1981 was crucial to the activation and recruitment of ATM to the site of damage. It soon became evident that this activation process was more complex with the description of two additional autophosphorylation sites on S367 and S1893 [28]. Phosphorylation site mutants for all three sites (S367A, S1893A, and S1981A) were defective in radiation induced signaling and failed to correct radiosensitivity or cell cycle checkpoint defects in A-T cells [28]. A second posttranslational modification, acetylation, was also shown to be induced by DNA DSB in parallel to autophosphorylation [70]. This was mapped to a single site, K3016, within the FATC domain and adjacent to the protein kinase domain of ATM [71]. Mutation at this site prevented radiation induced activation of ATM. While these events appear to play an important role in ATM activation in human cells this does not appear to be the case in Atm mouse mutants [72]. Use of a bacterial artificial chromosome reconstitution system to generate an Atm-S1987A mutant in an $Atm^{-/-}$ background showed a normal response to DNA damage. This included phosphorylation of Atm substrates, resistance to radiation, and normal cell cycle checkpoint activation. It will be of interest to compare this data in Atm-S1987A mice generated by the more conventional “knockin” approach.

As outlined above, the MRN complex is a sensor of DNA DSB and there is no requirement for ATM for localization of the complex to DNA damage [59]. Furthermore ATM is dependent on the MRN complex for its efficient activation in response to DNA DSB (Figure 23.2). ATLD and NBS cells are defective in ATM activation and downstream signaling [73]. ATM activation is also reduced in NBS cells expressing a mutant form of Nbs1 lacking the ATM binding site [74] and this is also the case for Nbs1 where the Mre11 interaction domain is deleted [75]. Degradation of Mre11 following adenovirus infection greatly reduced ATM autophosphorylation and abrogated the ATM dependent G2/M checkpoint [76]. *In vitro* studies show that purified MRN complex is required for stable association of ATM with DNA ends and stimulates its kinase activity [77]. Finally depletion of Mre11 in *Xenopus* extracts prevented ATM autophosphorylation and abrogated ATM dependent phosphorylation of H2AX [78]. Notwithstanding all these observations there is not an absolute requirement for the MRN complex for ATM activation. The data with ATLD and NBS cells reveal a retardation in ATM activation [73] and while ATM pS1981 does not appear in DNA damage induced foci in NBS cells it is diffusively present in the nucleus [67].

The recruitment and retention of DNA damage recognition and repair proteins to DNA DSB is a complex process involving multiple proteins, ATM dependent phosphorylation, as well as ubiquitination of some of those proteins [79, 80]. Once activated, ATM sits at the top of the cascade as a signal transducer of the response to DNA DSB phosphorylating several of the key players involved in the initiating events (Figure 23.2). It seems likely that phosphorylation of these proteins alters interactions with other effector proteins. These interactions are mediated through BRCA C-terminal repeat (BRCT) and forkhead associated (FHA) domains that recognize phosphopeptides [81, 82]. These domains are present in a number of proteins involved in the DNA damage response. Microlasers and charged peptide tracks determine the order of assembly of these proteins at the site of DNA damage [68, 83]. ATM dependent phosphorylation of the histone variant H2AX to produce γH2AX seems to be the initial signal for subsequent accumulation of DNA damage response proteins [84]. A second substrate, MDC1, binds to γH2AX via its BRCT domain and is a “master regulator” of the recognition and repair of DNA DSBs [85]. NBS1, as part of MRN, appears at the break with the same kinetics as MDC1, and it is also phosphorylated by ATM [86]. However, its retention on chromatin is not dependent on ATM phosphorylation but rather on casein kinase-2 phosphorylation of Ser-Asp-Thr (SDT) repeats in the N-terminus of MDC1 [87]. The RING-finger ubiquitin ligase, RNF8, also assembles at the DSB through interaction of its FHA domains with phosphorylated MDC1 (consensus sites for ATM phosphorylation) [88, 89]. RNF8 ubiquitylates H2A and facilitates the accumulation of 53BPI at the site of damage: both proteins are also substrates for ATM in response to DNA DSBs. Overall, ATM has a key role in phosphorylating various proteins that constitute the core components of the DNA damage recognition machinery and those responsible for amplification of the signal. These phosphorylations trigger the recruitment of RNF8, which ubiquitylates H2A (and γH2AX) to provide chromatin in the region of the break with the capacity to accumulate DNA damage response proteins to ensure repair of the break and maintenance of genome integrity [79].

ATM MEDIATED DOWNSTREAM SIGNALING

As outlined above, the DNA DSB sensor, the MRN complex, plays an important role together with other proteins in establishing the environment for ATM signaling (Figure 23.3). Once this role is accomplished the MRN complex itself becomes a target for ATM phosphorylation. One member of the complex, Nbs1, is stably maintained at the DNA DSB

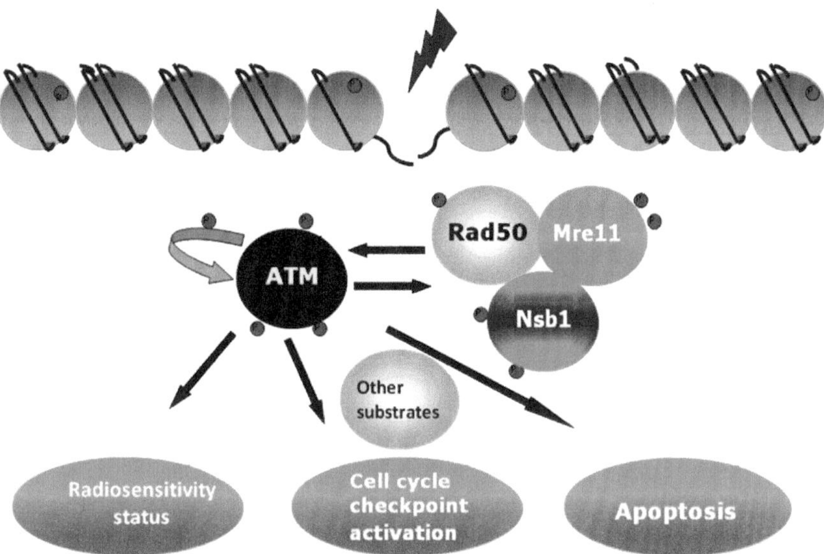

FIGURE 23.3 The MRN complex acts as an adaptor for ATM mediated signaling to protect the genome. ATM phosphorylates members of the complex that play a role in signaling through other substrates to cell survival, apoptosis, and cell cycle.

where it is phosphorylated by ATM on two sites, S278 and S343 [90–93]. These sites are functionally significant since mutant forms fail to correct the S-phase checkpoint defect in NBS cells [90–93]. However, these mutants did restore signaling to Chk2 and focus formation after DNA damage. Reports on the restoration of radioresistance in NBS cells by the phosphosite mutants are divided in their outcome [90–93]. These data are consistent with the results of Berkovich *et al.* [29] who showed that ATM is recruited to the MRN complex at the site of the break for maximal activation. This points to an adaptor or dependent role for Nbs1 in mediating the phosphorylation of downstream substrates in cell cycle control. Indeed ATM dependent phosphorylation of a number of substrates including FANCD2, Chk2, Chk1, and SMC1 is deficient in Nbs1 mutant cells [73, 94, 95]. It is likely that Mre11 and Rad50 also play adaptor roles in ATM dependent signaling since there is evidence for DNA damage induced phosphorylation of both proteins [96–99]. Furthermore an intact MRN complex is important not only for ATM recruitment and activation but also for downstream signaling [59, 60].

The first substrate described for ATM was p53 [100–102]. This was not surprising since it had been shown that the G_1/S checkpoint was defective in A-T cells and that this could be explained by a defect in the activation and stabilization of p53 [18, 19]. In addition, p53 was found to be phosphorylated *in vitro* by ATM. Exposure of cells to radiation led to phosphorylation of p53 on S15 and this was defective in A-T cells [100–102]. Subsequently, over 30 ATM or ATM dependent substrates have been described and in many cases the phosphorylation event has been shown to be functionally important (see Table 23.1). More recent data using large scale proteomic analysis of proteins phosphorylated in response to DNA damage, on consensus sites for ATM and ATR (S/TQ), have revealed that as many as 700 proteins are phosphorylated [96–99]. Not unexpectedly, this group of proteins is enriched for DNA damage response proteins and a number of new protein modules and networks were identified that were linked to this response. Linding *et al.* [97] described NetworkKIN that augments motif based predictions with the network context of protein kinases and phosphoproteins for the assignment of *in vivo* substrate specificity to improve predictability. Using this approach they showed that 53BPI and Rad50 were phosphorylated by CDK1 and ATM respectively after DNA damage. While the great majority of these substrates are mediators and effectors of the DNA damage response there is also evidence for cross-talk and cross-dependence amongst PIKK family members. ATM responds primarily to DNA DSB, whereas ATR is activated by stalled or collapsed DNA replication forks [103]. However, there is evidence that ATM functions upstream of ATR after exposure to radiation in S/G_2 phases of the cell cycle [104]. These data do not provide evidence that ATR is phosphorylated by ATM. On the other hand Stiff *et al.* [105] have shown that ATM phosphorylation on S1981 is ATR dependent after exposure to agents that cause DNA replication fork stalling. This was shown to be independent of ATM activity and did not require Nbs1 or Mre11, providing evidence that ATM is a substrate for ATR after UV damage. Similar to ATM, the catalytic subunit of DNA-PK, DNA-PKcs, is activated by autophosphorylation at several sites including S2056 and the T2609 cluster [106, 107]. Mutation of these sites reduces DNA DSB repair capacity and sensitizes the cells to radiation. While DNA-PKcs is responsible for pS2056 post-irradiation, phosphorylation at T2609 and other sites within the adjacent cluster is ATM dependent.

TABLE 23.1 Substrates for ATM kinase

Substrate	P sites	Activity	Role	Reference(s)
ATM	S1981 S1893 S367	Protein kinase	Activation	27, 28
DNA-PKcs	S2056 (ATM dependent) T2609 cluster	Protein kinase	Autophosphorylation Inactivation and dissociation from DNA Activation of DNA dsb repair	106, 107
Nbs1	S278 S343	Part of Mre11/Rad50/Nbs1 complex Sensor	Intra-S checkpoint Telomere maintenance	90–93
Rad9	S272	3′ to 5′ exonuclease	G1 checkpoint	107
H2AX	S139	Histone protein	Chromatin modification	Rogakou et al
53BP1	S25?	Binds DNA	Mitosis	Xia et al
Mdc1	S168?	BRCT domain	Master regulator	Goldberg et al
BRCA1	S1387 S1423	Tumor suppressor E3 ubiquitin ligase	S and G2/M checkpoints DNA repair	126, 127
BARD1	T714 T734	BRCA1 mediated tumor suppression, part of the RNA polII complex	inhibition of pre-mRNA polyadenylation and degradation of RNA polymerase II	Kim et al[33]
CtIP/ RBBP-8	S664 S745	BRCA1 functions modulator	?BRCA1 dissociation? GADD45 induction Repair	Li et al[34] Wu-Baer and Baer[35]
Chk1	S317 S345	Protein kinase	Cell cycle checkpoint control	93
Chk2	T68 S19, S33, and S35	Protein kinase	Cell cycle checkpoint control	95, 113
LKB1	T366	Protein kinase Tumor suppressor	Activation?	Sapkota et al[41]
Akt1/PKB	S473	Protein kinase	Activation	Viniegra et al[42]
PP2A	T21?	phosphatase	?Complex formation	Goodarzi et al[43]
p53	S15 S9, S46	tumor suppressor transcription factor	Checkpoints Activation	Khanna et al [44] Canman et al[45] Banin et al[46] Saito et al[47]
Mdm2/	S395	oncoprotein	p53 accumulation	Khosravi et al[48]
Hdm2/Sp		p53 ubiquitin ligase		Maya et al[49]
MdmX/	S403	ubiquitin ligase	Hdm2 mediated	Pereg et al[50]
Hdmx	S367 S342		ubiquitination	Chen et al[51]
BLM	T99	RecQ-like helicase	G2/M checkpoint dissociate from Top3alpha and PML	Beamish et al[52] Rao et al[53]
FANCD2	T691, S717 S222, S1401, S1404, S1418	chromosomal stability maintenance	Intra-S checkpoint	Tanaguichi et al[4] Ho et al[54] Nakanishi et al[55]

(Continued)

TABLE 23.1 (Continued)

SMC1	S957 S966	component of cohesin complex and centromere-kinetochore complex	Intra-S-checkpoint	16, 17
Mcm3	S535	helicase complex	Replication?	126
Xenopus Mcm2	S92	helicase complex	Replication checkpoints	Yoo et al
RPA32	T21	single-stranded DNA binding (SSB) protein	Replication	Oakley et al Block et al
E2F1	S31	transcription factor	stabilization of E2F1 apoptosis G1/S checkpoint?	Lin et al
ATF2	S490, S498	transcription factor	S-phase checkpoint	Bhoumik et al
TRF1/Pin2	S219	telomere binding	G2/M regulation Mitosis, apoptosis Radiation sensitivity	Kishi et al
KAP-1 (TIF1beta, KRIP-1 or TRIM28)	S824	corepressor of gene transcription	DSB induced chromatin decondensation/ relaxation	Ziv et al

CELL CYCLE CHECKPOINT ACTIVATION

G_1/S-Phase Checkpoint

The most widely described signaling pathway induced by ATM is that through p53 to arrest the passage of cells from G_1 to S phase [18–20]. ATM phosphorylates a series of proteins for the efficient and coordinated regulation of this pathway. ATM phosphorylates p53 on Ser15 and is required for Ser20 phosphorylation and Ser376 dephosphorylation [108]. P53 becomes transcriptionally activated to induce the cyclin kinase inhibitor p21/WAF1, which associates with cyclin E-Cdk2 to block its activity, to prevent phosphorylation of downstream substrates, and to delay the passage of cells from G_1 to S phase [109–111] (Figure 23.4). ATM phosphorylates Chk2 kinase on Thr68, which depends on the integrity of the FHA domain in Chk2 [112]. This phosphorylation is required for the subsequent autophosphorylation and activation of Chk2 [113], which enables it to exert its control on the G_1/S, S, and G_2/M checkpoints. This activation of Chk2 by ATM also requires Nbs1 since activation is defective in NBS cells [90, 93]. Whereas wild-type Nbs1 could complement this defect in NBS cells, a construct mutated in the ATM phosphorylation site Nbs1 (ser343) failed to do as well as a mutant form that abrogated the formation of the Mre11/Rad50/Nbs1 complex.

The involvment of ATM signaling at multiple levels in a single pathway is further illustrated by the capacity of activated ATM to phosphorylate both Mdm2, a negative regulator of p53, and its homolog Mdmx that represses p53 transactivation [20, 114]. Expression of Mdm2 is controlled by p53; Mdm2 in turn has a feedback effect, binding to p53, exporting it from the nucleus and promoting its degradation in the proteasome pathway [115, 116]. ATM dependent phosphorylation of Mdm2 is observed prior to p53 accumulation. Decreased reactivity of Mdm2 from irradiated cells to an anti-Mdm2 antibody, directed against an epitope containing Ser395, suggests that this is a site for phosphorylation *in vivo*, and phosphorylation at this site may reduce the capacity of Mdm2 to translocate p53 to the cytoplasm and ensure its stabilization [114]. ATM dependent phosphorylation of Mdmx occurs on at least three sites, one of which, S403, is directly phosphorylated by ATM [20]. These phosphorylations are important for Mdm2 mediated ubiquitination and degradation of Mdmx. Thus, it is evident that ATM acts directly on four different substrates and also has indirect effects on some of these, providing fine-tuning of p53 stabilization and activation for complex control of the G_1/S checkpoint (Figure 23.4).

S-Phase Checkpoint

A reduced inhibition of DNA synthesis in A-T cells in response to radiation, which is termed *radioresistant DNA synthesis* (RDS), represented the first report of a cell cycle anomaly [13, 14]. A functional link has also been established between ATM, the checkpoint signaling kinase, Chk2, and Cdc25A phosphatase and its downstream target for activation, Cdck2, in S phase [15]. Radiation induced degradation of Cdc25A requires both ATM and Chk2 mediated phosphorylation of Cdc25A or Ser123 and this prevents dephosphorylation of Cdk2, leading to a transient block in DNA replication (Figure 23.5). Exposure of A-T cells to radiation

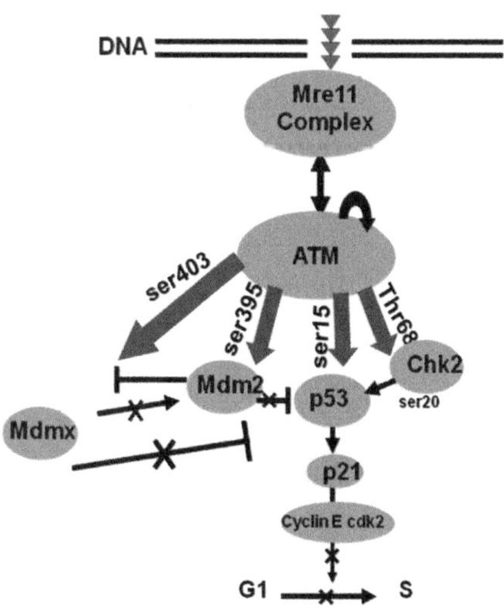

FIGURE 23.4 ATM activates the G_1/S checkpoint in response to ionizing radiation damage to DNA.
ATM is rapidly activated in response to DNA breaks, to phosphorylate itself, and four downstream effector proteins (p53, MDM2, MDMX, and Chk2) to achieve arrest of cells at the G_1/S checkpoint. It directly phosphorylates p53 on ser15, which may alter its transcriptional capacity. Stabilization of p53 is achieved by phosphorylating Chk2 on thr 68, which leads to its subsequent activation to, in turn, phosphorylate p53 on ser20 leading to stabilization. Negative regulation of MDM2 by ATM through phosphorylation on ser395 reduces its ability to bind to and mediate the ubiquitination and degradation of p53. ATM also phosphorylates MDMX leading to DNA damage induced degradation of this protein. Thus stabilization of p53 is ensured through these separate steps. When stabilized, p53 is capable of inducing p21 and other downstream genes to inhibit cyclin E-Cdk2 kinase activity and bring about G_1 arrest.

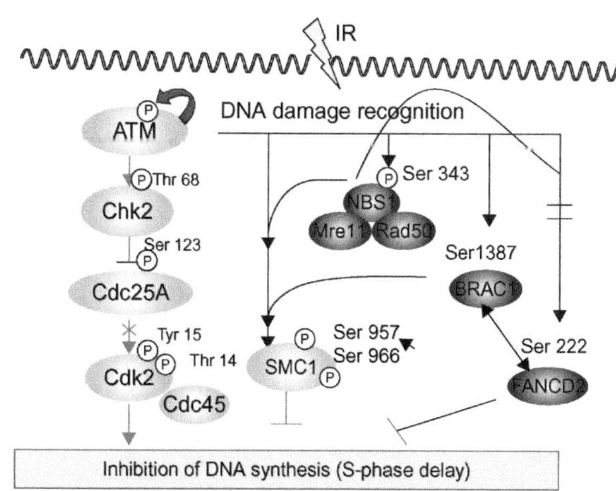

FIGURE 23.5 ATM activates the S-phase checkpoint.
Radioresistant DNA synthesis was first described in cells from A-T patients lacking ATM. This phenomenon is also observed in NBS where the ATM substrate Nbs1 is defective and in the A-T-like syndrome (ATLD) characterized by hypomorphic mutants in the Mre11 gene, a component of the Mre11/Rad50/Nbs1 complex. Radioresistant DNA synthesis is also observed when Chk2 is disrupted on its downstream pathway through Cdc25A/Cdk2. These observations point to an important role for ATM in inhibiting DNA synthesis through several separate but parallel pathways. In the first case ATM activates Chk2 as described for the G1/S checkpoint, which in this case destabilizes Cdc-25A by phosphorylation on Ser123, preventing its ability to remove the inhibitory Tyr 15/Thre 14 phosphorylations for activation of Cdk2, leading to inhibition of DNA synthesis. In a parallel but less well-described pathway, ATM activates the Mre11 complex and SMC1, a downstream target to inhibit DNA synthesis. Finally, the p53 activation pathway might also contribute through inhibition of Cdk2 kinase.

failed to cause an increase in Cdk2 Tyr 15 dephosphorylation or inhibition of cyclin E-Cdk2 kinase activity, consistent with the radioresistant DNA synthesis phenotype. Furthermore, Chk2 alleles defective in catalytic activity or ability to interact with Cdc25A had a dominant interfering effect and abrogated the S-phase checkpoint [15]. RDS is also evident in NBS and ATLD cells [54, 117]. RDS can be rescued by retroviral constructs expressing wild-type Nbs1 but not Ser343 phosphosile mutants. Thus ATM phosphorylation of Nbs1 is an important part of the mechanism to ensure inhibition of DNA replication in response to radiation damage.

The S-phase checkpoint is regulated by parallel pathways through Nbs1-Mre11 on the one hand and Chk2 on the other [118]. Concomitant interference with both of these pathways gave rise to RDS after exposure of cells to ionizing radiation. Thus it appears likely that ATM, by phosphorylating both Nbs1 and Chk2, triggers two pathways that inhibit distinct steps in DNA replication. Evidence for parallel pathways is further supported by the demonstration that the structural maintenance of chromosomes protein, SMC1, is a downstream effector in the ATM/Nbs1 branch of the S-phase checkpoint [16, 119]. The transitory delays in DNA synthesis post-irradiation also appears to be mediated through a calmodulin dependent regulatory cascade and this pathway is defective in A-T [120]. Beamish *et al.* [11] have shown that exposure of lymphoblastoid cells to radiation in the S phase causes a rapid inhibition of cyclin A-Cdk2 accompanied by markedly increased binding of p21/WAF1. In contrast, radiation did not inhibit cyclin kinase activity in A-T cells in S phase nor was there any significant change in cdk associated p21/WAF1 compared to unirradiated cells [11].

G_2/M Checkpoint

Arrest of cells in G_2 phase in response to radiation damage leads to suppression of the mitotic index to protect cell viability [121]. Consequently, abrogation of G_2 arrest sensitizes cells to radiation [122]. This checkpoint is defective in A-T cells as evidenced by a lesser delay of cells, irradiated in G_2 phase, in progressing into mitosis [123]. Defective G_2-phase delay in A-T cells was confirmed by labeling cells in S phase with BrdU, blocking their passage into mitosis with nocodazole prior to irradiation (in G_2 phase), and scoring for ability to enter the next G_1 phase [11]. This was also achieved by distinguishing G_2 phase cells from mitotic cells using histone H3 phosphospecific antibody [124]. As observed with cells irradiated in S phase, when A-T cells were exposed to radiation in

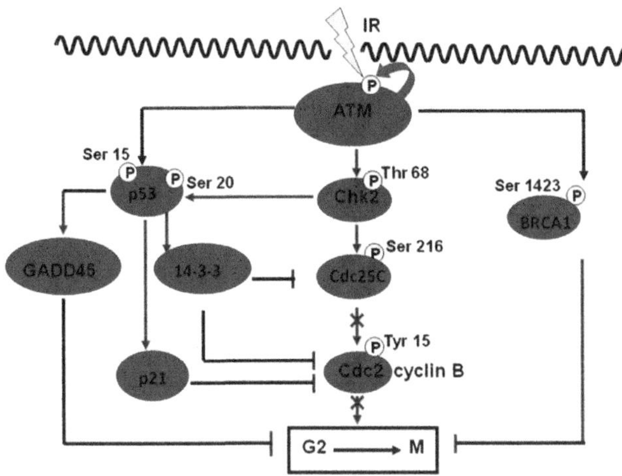

FIGURE 23.6 ATM involvement in G_2/M checkpoint control. The activation of the G2/M checkpoint is no less complex than the S-phase checkpoint. Again ATM is a central player, via chk2 phosphorylation of Cdc25C on ser216 in S phase, and in response to DNA damage the site is maintained in a phosphorylated state. Inactivation appears to be mediated by binding to 14-3-3 protein, a p53 downstream effector. Again in this case activation of the p53 pathway is prominent to induce 14-3-3 protein and p21/WAF1, which can bind to and inhibit Cdc2-cyclin B kinase blocking cells at the G_2/M checkpoint. Induction of GADD45 in this pathway may also interfere with Cdc2-cyclin B kinase. To add to the complexity mutations in BRCA1, including an ATM phosphorylation site (ser 1423), lead to a defective G_2/M checkpoint indicating that ATM also controls the G_2/M checkpoint through BRCA1.

G_2 phase, there was no inhibition of Cdc2-cyclin B kinase activity and, unlike that obtained with irradiated control cells, no increase in p21/WAF1 associated with Cdc2 was observed [11]. Again this checkpoint defect is compatible with a lack of ATM signaling in A-T cells to the cyclin–kinase complex. As with S and G_1 phase checkpoints, it appears that ATM influences G_2 arrest by more than one pathway (Figure 23.6). It is well established that BRCA1 is involved in the cellular response to radiation and this involves phosphorylation on several sites mediated by both ATM and ATR [125–127]. ATM interacts with BRCA1 and this molecule is a substrate for ATM kinase both *in vitro* and *in vivo* [125]. Deletion of exon 11 of BRCA1 leads to a defective G_2 checkpoint and extensive chromosomal abnormalities [128]. Because cells from NBS patients exhibited a normal radiation induced G_2/M checkpoint, it appears that ATM phosphorylation of Nbs1 does not play a significant role for this checkpoint. BRCA1 mediated induction of GADD45 also leads to G_2/M phase delay but only in response to microtubule poisons, not DNA damaging agents, revealing that it is ATM independent [129].

CONCLUDING REMARKS

It is evident that ATM plays a central role in maintaining the integrity of the genome and in preventing cancer and neurodegeneration. This role is largely in the recognition and signaling of DNA DSB. ATM is not the primary sensor of the break but it is at least partially activated by alteration of chromatin structure in the immediate region adjacent to the break. It is still not clear what is the actual stimulus for activation of ATM. The primary sensor of the DNA DSB is the MRN complex, which recruits ATM to the break where it is fully activated by autophosphorylation and acetylation at least in human cells. There exists some evidence that these events are less important in murine cells for ATM activation. Not surprisingly the requirement for the MRN complex for ATM activation is not absolute but when the complex is disrupted by mutations in Mre11 and Nbs1 defective downstream signaling through at least some substrates is evident. This impacts on cell cycle control and cell survival in response to DNA damage. While the MRN complex plays an important role upstream of ATM it also becomes a target for ATM dependent phosphorylation in downstream signaling. It is well established that ATM phosphorylates Nbs1, one member of that complex on two sites (S278, S343), and there is some evidence that these sites play a role in downstream signaling.

REFERENCES

1. Sedgwick RP, Boder E. Hereditary neuropathies and spinocerebellar atrophies. In: Vianney De Jong JMB, editor. *Hereditary neuropathies and spinocerebellar atrophies*. New York: Alan R Liss; 1991. p. 347–423.
2. Lavin MF, Shiloh Y. The genetic defect in ataxia-telangiectasia. *Ann Rev Immunol* 1997;**15**:177–202.
3. Gotoff SP, Amirmokri E, Liebner EJ. Radiation reaction in ataxia telangiectasia. *Am J Dis Child* 1967;**116**:557–8.
4. Morgan JL, Holcomb TM, Morrissey RW. Radiation reaction in ataxia telangiectasia. *Am J Dis Child* 1968;**116**:557–8.
5. Taylor AM, Harnden DG, Arlett CF, Harcourt SA, Lehmann AR, Stevens S, Bridges BA. Ataxia telangiectasia: A human mutation with abnormal radiation sensitivity. *Nature* 1975;**258**:427–9.
6. Chen PC, Lavin MF, Kidson C, Moss D. Identification of ataxia telangiectasia heterozygotes, a cancer prone population. *Nature* 1978;**274**: 484–6.
7. Foray N, Priestley A, Alsbeih G, Badie C, Capulas EP, Arlett CF, Malaise EP. Hypersensitivity of ataxia telangiectasia fibroblasts to ionizing radiation is associated with a repair deficiency of DNA double-strand breaks. *Int J Radiat Biol* 1997;**72**:271–83.
8. Barlow C, Hirotsune S, Paylor R, Liyanage M, Eckhaus M, Collins F, Shiloh Y, Crawley JN, Ried T, Tagle D, Wynshaw-Boris A. Atm-deficient mice: A paradigm of ataxia telangiectasia. *Cell* 1996;**86**:159–71.
9. Gatei M, Shkedy D, Khanna KK, Uziel T, Shiloh Y, Pandita TK, Lavin MF, Rotman G. Ataxia-telangiectasia: chronic activation of damage-responsive functions is reduced by alpha-lipoic acid. *Oncogene* 2001; **20**:289–94.
10. Rotman G, Shiloh Y. ATM: a mediator of multiple responses to genotoxic stress. *Oncogene* 2001;**18**:6135–44.
11. Beamish H, Lavin MF. Radiosensitivity in ataxia-telangiectasia: anomalies in radiation-induced cell cycle delay. *Int J Radiat Biol* 1994;**65**:175–84.
12. Morgan SE, Kastan MB. 53 and ATM: cell cycle, cell death, and cancer. *Adv Cancer Res* 1997;**71**:1–25.

13. Houldsworth J, Lavin MF. Effect of ionizing radiation on DNA synthesis in ataxia telangiectasia cells. *Nucleic Acids Res* 1980;**8**:3709–20.
14. Painter RB, Young BR. Radiosensitivity in ataxia-telangiectasia: a new explanation. *Proc Natl Acad Sci USA* 1980;**77**:7315–17.
15. Falck J, Mailand N, Syljuasen RG, Bartek J, Lukas J. The ATM-Chk2-Cdc25A checkpoint pathway guards against radioresistant DNA synthesis. *Nature* 2001;**410**:842–7.
16. Kim ST, Xu B, Kastan MB. Involvement of the cohesion protein, Smc1, in Atm-dependent and independent responses to DNA damage. *Genes Dev* 2002;**16**:560–70.
17. Yazdi PT, Wang Y, Zhao S, Patel N, Lee EY, Qin J. SMC1 is a downstream effector in the ATM/NBS1 branch of the human S-phase checkpoint. *Genes Dev* 2002;**16**:571–82.
18. Kastan MB, Zhan O, el-Deiry WS, Carrier F, Jacks T, Walsh WV, Plunkett BS, Vogelstein B, Fornace AJ Jr. A mammalian cell cycle checkpoint pathway utilising p53 and GADD45 is defective in ataxia-telangiectasia. *Cell* 1992;**71**:587–97.
19. Khanna KK, Lavin MF. Ionizing radiation and UV induction of p53 protein by different pathways in ataxia-telangiectasia cells. *Oncogene* 1993;**8**:3307–12.
20. Pereg Y, Shkedy D, de Graaf P, Meulmeester E, Edelson-Averbukh M, Salek M, Biton S, Teunisse AF, Lehmann WD, Jochemsen AG, Shiloh Y. Phosphorylation of Hdmx mediates its Hdm2- and ATM-dependent degradation in response to DNA damage. *Proc Natl Acad Sci USA* 2005;**102**:5056–61.
21. Lobrich M, Jeggo PA. The impact of a negligent B2/M checkpoint of genomic instability and cancer induction. *Nat Rev Cancer* 2007;**7**:861–9.
22. Lukas J, Lukas C, Bartek J. Mammalian cell cycle checkpoints: signalling pathways and their organization in space and time. *DNA Repair (Amst)* 2004;**3**:997–1007.
23. Gatti RA, Berkel I, Boder E, Braedt G, Charmley P, Concannon P, Ersoy F, Foroud T, Jaspers NG, Lange K. Localization of an ataxia-telangiectasia gene to chromosome 11q22-23. *Nature* 1998;**336**:577–80.
24. Savitsky K, Bar-Shira A, Gilad S, Rotman G, Ziv Y, Vanagaite L, Tagle DA, Smith S, Uziel T, Sfez S, Ashkenazi M, Pecker I, Harnik R, Patanjali SR, Simmons A, Frydman M, Sartiel A, Gatti RA, Chessa L, Sanal O, Lavin MF, Jaspers NGJ, Malcolm A, Taylor R, Arlett CF, Miki T, Weissman SM, Lovett M, Collins FS, Shiloh Y. A single ataxia-telengiectasia gene with a product simlinar to PI-3 kinase. *Science* 1995;**268**:1749–53.
25. Zakian VA. ATM-related genes: What do they tell us about functions of the human gene? *Cell* 1995;**82**:685–7.
26. Abrahams RT. Cell cycle checkpoint signalling through the ATM and ATR kinases. *Genes Dev* 2001;**15**:2177–96.
27. Bakkenist CJ, Kastan MB. DNA damage activates ATM through intermolecular autophosphorylation and dimer dissociation. *Nature* 2003;**421**:499–506.
28. Kozlov SV, Graham ME, Peng C, Chen P, Robinson PJ, Lavin MF. Involvement of novel autophosphorylation sites in ATM activation. *EMBO J* 2006;**25**:3504–14.
29. Berkovich Jr E, Monnat RJ, Kastan MB. Roles of ATM and NBS1 in chromatin structure modulation and DNA double-strand break repair. *Nat Cell Biol* 2007;**9**:683–90.
30. You Z, Bailis JM, Johnson SA, Dilworth SM, Hunter T. Rapid activation of ATM on DNA flanking double-strand breaks. *Nat Cell Biol* 2007;**9**:1311–18.
31. Soutoglou E, Dorn JF, Sengupta K, Jasin M, Nussenzweig A, Reid T, Danuser G, Misteli T. Positional stability of single double-strand breaks in mammalian cells. *Nat Cell Biol* 2007;**9**:675–82.
32. Fukao T, Kaneko H, Birrell G, Gatei M, Tashata H, Kasahara K, Cross S, Kedar P, Watters D, Khanna KK, Misko I, Kondo N, Lavin MF. ATM is upregulated during the mitogenic response in peripheral blood mononuclear cells. *Blood* 1999;**94**:1998–2006.
33. Gueven N, Keating KE, Chen P, Fukao T, Khanna KK, Watters D, Rodemann PH, Lavin MF. Epidermal growth factor sensitizes cells to ionizing radiation by down-regulating protein mutated in ataxia-telangiectasia. *J Biol Chem* 2001;**276**:8884–91.
34. Keating KE, Gueven N, Watters D, Rodemann HP, Lavin MF. Transcriptional downregulation of ATM by EGF is defective in ataxia-telangiectasia cells expressing mutant protein. *Oncogene* 2001;**20**:4281–90.
35. Khanna KK, Yan J, Watters D, Hobson K, Beamish H, Spring K, Shiloh Y, Gatti RA, Lavin MF. Defective signalling through the B cell antigen receptor in Epstein-Barr virus-transformed ataxia-telangiectasia cells. *J Biol Chem* 1997;**272**:9489–95.
36. Rhodes N, D'Souza T, Foster CD, Ziv Y, Kirsch DG, Shiloh Y, Kastan MB, Reinhart PH, Gilmer TM. Defective potassium currents in ataxia telangiectasia fibroblasts. *Genes Dev* 1998;**12**:3686–92.
37. Yang DO, Kastan MB. Participation of ATM in insulin signalling through phosphorylation of EIF-4E binding protein. *Nat Cell Biol* 2000;**2**:893–8.
38. Jeggo PA, Taccioli GE, Jackson SP. Menage a trios: Double-strand breaks repair, V(D)J recombination and DNA-PK. *Bioessays* 1995;**17**:949–57.
39. Haber JE. Partners and pathways: Repairing a double-strand break. *Trends Genet* 2000;**16**:259–64.
40. Ward JF. Molecular mechanisms of radiation-induced damage to nucleic acids. *Radiat Res* 1975;**86**:185–95.
41. Rooney S, Chaudhuri J, Alt FW. The role of the non-homologous end-joining pathway in lymphocyte development. *Immunol Rev* 2004;**200**:115–31.
42. Lees-Miller SP, Meek K. Repair of DNA double strand breaks by non-homologous end joining. *Biochimie* 2003;**85**:1161–73.
43. Thompson LH, Schild D. Recombinational DNA repair and human disease. *Mutat Res* 2002;**509**:49–78.
44. Jeggo P, Singleton B, Beamish H, Priestley A. Double-strand break rejoining by the Ku-dependent mechanism of non-homologous end-joining. *CR Acad Sci III* 1999;**322**:109–12.
45. Johnson RD, Jasin M. Sister chromatid gene conversion is a prominent double-strand break repair pathway in mammalian cells. *EMBO J* 2000;**19**:3398–407.
46. Sonoda E, Morrison C, Yamashita YM, Takata M, Takeda S. Reverse genetic studies of homologous DNA recombination using the chicken B-lymphocyte line, DT40. *Philos Trans R Soc Lond B Biol Sci* 2001;**356**:111–17.
47. Althaus FR, Richter C. ADP-ribosylation of proteins. enzymology and biological significance. *Mol Biol Biochem Biophys* 1987;**37**:1–237.
48. Weterings E, Chen DJ. The endless tale of non-homologous end-joining. *Cell Res* 2008;**18**:114–24.
49. Errani A, Smider V, Rathmell WK, He DM, Hendrickson EA, Zdzienicka MZ, Chu G. Ku86 defines the genetic defect and restores X-ray resistance and V(D)J recombination to complementation group 5 hamster cell mutants. *Mol Cell Biol* 1996;**16**:1519–26.
50. Jackson SP. DNA-dependent protein kinase. *Int J Biochem Cell Biol* 1997;**29**:935–8.
51. Grawunder U, Zimmer D, Fugmann S, Schwarz K, Lieber MR. DNA ligase IV is essential for V(D)J recombination and DNA double-strand break repair in human precursor lymphocytes. *Mol Cell* 1998;**2**:477–84.

52. McElhinny N, Snowden CM, McCarville J, Ramsden DA. Ku recruits the XRCC4-ligase IV complex to DNA ends. *Mol Cell Biol* 2000;**20**:2996–3003.
53. Petrini JH. The Mre11 complex and ATM: collaborating to navigate S phase. *Curr Opin Cell Biol* 2000;**12**:293–6.
54. Stewart GS, Maser RS, Stankovic T, Bressan DA, Kaplan MI, Jaspers NG, Raams A, Byrd PJ, Petrini JH, Taylor AM. The DNA double-strand break repair gene hMRE11 is mutated in individuals with an ataxia-telangiectasia-like disorder. *Cell* 1999;**99**:577–87.
55. Maser RE, Monsen KJ, Nelms BE, Petrini JH. hMre11 and hRad50 nuclear foci are induced during the normal cellular response to DNA double-strand breaks. *Mol Cell Biol* 1997;**17**:6087–96.
56. Nelms BE, Maser RS, MacKay JF, Lagally MG, Petrini JH. *In situ* visualization of DNA double-strand break repair in human fibroblasts. *Science* 1998;**280**:590–2.
57. Haber JE. The many interfaces of Mre11. *Cell* 1998;**95**:583–6.
58. Carney 3rd JP, Maser RS, Olivares H, Davis EM, Le Beau M, Yates JR, Hays L, Morgan WF, Petrini JH. The hMre11/hRad50 protein complex and Nijmegen breakage syndrome: linkage of double-strand break repair to the cellular DNA damage response. *Cell* 1998;**93**:477–86.
59. Mirzoeva OK, Petrini JH. DNA damage-dependent nuclear dynamics of the Mre11 complex. *Mol Cell Biol* 2001;**21**:281–8.
60. Mirzoeva OK, Petrini JH. DNA replication-dependent nuclear dynamics of the Mre11 complex. *Mol Cancer Res* 2003;**1**:207–18.
61. De Jager M, van Noort J, van Gent DC, Dekker C, Kanaar R, Wyman C. Human Rad50/Mre11 is a flexible complex that can tether DNA ends. *Mol Cell* 2001;**8**:1129–35.
62. Van den Bosch M, Bree RT, Lowndes NF. The MRN complex: coordinating and mediating the response to broken chromosomes. *EMBO Rep* 2003;**4**:844–9.
63. Paull TT, Gellert M. The 3' to 5' exonuclease activity of Mre11 facilitates repair of DNA double-strand breaks. *Mol Cell* 1999;**1**:969–79.
64. Trujillo KM, Sung P. DNA structure-specific nuclease activities in the Saccharomyces cerevisiae Rad50 Mre11 complex. *J Biol Chem* 2001;**276**:35,458–35,464.
65. Paull TT, Gellert M. Nbs1 potentiates ATP-driven DNA unwinding and endonuclease cleavage by the Mre11/Rad50 complex. *Genes Dev* 1999;**13**:1276–88.
66. Kozlov S, Gueven N, Keating K, Ramsay J, Lavin MF. ATP activates ataxia-telangiectasia mutated (ATM) in vitro. Importance of autophosphorylation. *J Biol Chem* 2003;**278**:9309–17.
67. Kitagawa R, Bakkenist CJ, McKinnon PJ, Kastan MB. Phosphorylation of SMC1 is a critical downstream event in the ATM-NBS1-BRCA1 pathway. *Genes Dev* 2004;**15**:1423–38.
68. Bekker-Jensen S, Lukas C, Kitagawa R, Melander F, Kastan MB, Bartek J, Lukas J. Spatial organization of the mammalian genome surveillance machinery in response to DNA strand breaks. *J Cell Biol* 2006;**173**:195–206.
69. Goodarzi AA, Yu Y, Riballo E, Douglas P, Walker SA, Ye R, Harer C, Marchetti C, Morrice N, Jeggo PA, Lees-Miller SP. DNA-PK autophosphorylation facilities Artemis endonuclease activity. *Embo J* 2006;**25**:3880–9.
70. Sun Y, Jiang X, Chen S, Fernandez N, Price BD. A role for the Tip60 histone acetyltransferase in the acetylation and activation of ATM. *Proc Natl Acad Sci* 2005;**102**:13,182–13,187.
71. Sun Y, Xu Y, Roy K, Price BD. DNA damage-induced acetylation of lysine 3016 of ATM activates ATM kinase activity. *Mol Cell Biol* 2007;**27**:8502–9.
72. Pellegrini M, Celeste A, Difilippantonio S, Guo R, Wang W, Feigenbaum L, Nussenzweig A. Autophosphorylation at serine 1987 is dispensable for murine Atm activation in vivo. *Nature* 2006;**443**: 222–5.
73. Uziel T, Lerenthal Y, Moyal L, Andegeko Y, Mittelman L, Shiloh Y. Requirement of the MRN complex for ATM activation by DNA damage. *EMBO J* 2003;**22**:5612–21.
74. Cerosaletti K, Wright J, Concannon P. Active role for nibrin in the kinetics of atm activation. *Mol Cell Biol* 2006;**26**:1691–9.
75. Horejsi Z, Falck J, Bakkenist CJ, Kastan MB, Lukas J, Bartek J. Distinct functional domains of Nbs1 modulate the timing and magnitude of ATM activation after low doses of ionizing radiation. *Oncogene* 2004;**23**:3122–7.
76. Carson CT, Schwartz RA, Stracker TH, Lilley CE, Lee DV, Weitzman MD. The Mre11 complex is required for ATM activation and the G2/M checkpoint. *EMBO J* 2003;**22**:6610–20.
77. Lee JH, Paull TT. Purification and biochemical characterization of ataxia-telangiectasia mutated and Mre11/Rad50/Nbs1. *Methods Enzymol* 2006;**408**:529–39.
78. Dupre A, Boyer-Chatenet L, Gautier J. Two-step activation of ATM by DNA and the Mre11-Rad50-Nbs1 complex. *Nat Struct Mol Biol* 2006;**13**:451–7.
79. Huen MS, Chen J. The DNA damage response pathways: at the crossroads of protein modifications. *Cell Res* 2008;**18**:8–16.
80. Lavin MF. Ataxia-telangiectasia: from a rare disorder to a paradigm for cell signalling and cancer. *Nat Rev Mol Cell Biol* 2008;**9**:759–69.
81. Glover JN, Williams RS, Lee MS. Interactions between BRCT repeats and phosphoproteins: tangled up in two. *Trends Biochem Sci* 2004;**29**:579–85.
82. Durocher D, Jackson SP. The FHA domain. *Febs Lett* 2002;**513**:58–66.
83. Jakob B, Rudolph JH, Gueven N, Lavin MF, Taucher-Scholz G. Live cell imaging of heavy-ion-induced responses by beamline microscopy. *Radiat Res* 2005;**163**:681–90.
84. Stucki M, Jackson SP. GammaH2AX and MDC1: anchoring the DNA-damage-response machinery to broken chromosomes. *DNA Repair* 2006;**5**:534–43.
85. Xie A, Hartlerode A, Stucki M, Odate S, Puget N, Kwok A, Nagaraju G, Yan C, Alt FW, Chen J, Jackson SP, Scully R. Distinct roles of chromatin-associated proteins MDC1 and 53BP1 in mammalian double-strand break repair. *Mol Cell* 2007;**28**:1045–57.
86. Chapman JR, Jackson SP. Phospho-dependent interactions between Mbs1 and MDC1 mediate chromatin retention of the MRN complex at sites of DNA damage. *EMBO Rep* 2008;**9**:795–801.
87. Melander F, Bekker-Jensen S, Falck J, Bartek J, Mailand M, Lukas J. Phosphorylation of SDT repeats in the MDC1 N Terminus triggers retention of NBS1 at the DNA damage-modified chromatin. *J Cell Biol* 2008;**131**:901–14.
88. Kolas NK, Chapman JR, Nakada S, Ylanko J, Chahwan R, Sweeney FD, Panier S, Mendez M, Wildenhain J, Thomson TM, Pelletier L, Jackson SP, Durocher D. Orchestration of the DNA-damage response by the RNF8 ubiquitin ligase. *Science* 2007;**318**:1637–40.
89. Mailand N, Bekker-Jensen S, Faustrup H, Melander F, Bartek J, Lukas C, Lukas J. RNF8 ubiquitylates histones at DNA double-strand breaks and promotes assembly of repair proteins. *Cell* 2007;**131**:887–900.
90. Lim DS, Kim ST, Xu B, Maser RS, Lin J, Petrini JH, Kastan MB. ATM phosphorylates p95/nbs1 in an S-phase checkpoint pathway. *Nature* 2000;**404**:613–17.
91. Wu X, Ranganathan V, Weisman DS, Heine WF, Ciccone DN, O'Neill TB, Crick KE, Pierce KA, Lane WS, Rathbun G, Livingston DM, Weaver DT. ATM phosphorylation of Nijmegen breakage syndrome protein is required in a DNA damage response. *Nature* 2000; **405**:477–82.

92. Zhao S, Weng YC, Yuan SS, Lin YT, Hsu HC, Lin SC, Gerbino E, Song MH, Zdzienicka MZ, Gatti RA, Shay JW, Ziv Y, Shiloh Y, Lee EY. Functional link between ataxia-telangiectasia and Nijmegen breakage syndrome gene products. *Nature* 2000;**405**:473–7.
93. Gatei M, Scott SP, Filippovich I, Soronika N, Lavin MF, Weber B, Khanna KK. Role for ATM in DNA damage-induced phosphorylation of BRCA1. *Cancer Res* 2000;**60**:3299–304.
94. Nakanishi K, Taniguchi T, Ranganathan V, New HV, Moreau LA, Stotsky M, Mathew CG, Kastan MB, Weaver DT, D'Andrea AD. Interaction of FANCD2 and NBS1 in the DNA damage response. *Nat Cell Biol* 2002;**4**:913–20.
95. Matsuoka S, Rotman G, Ogawa A, Shiloh Y, Tamai K, Elledge SJ. Ataxia telangiectasia-mutated phosphorylates *Chk2* in vivo and vitro. *Proc Natl Aced Sci USA* 2000;**97**:10,389–10,394.
96. Beausoleil SA, Jedrychowski M, Schwartz D, Elias JE, Villen J, Li J, Cohn MA, Cantley LC, Gygi SP. Large-scale characterisation of HeLa cell nuclear phosphoproteins. *Proc Natl Acad Sci USA* 2004;**101**:12,130–12,135.
97. Linding R, Jensen LJ, Ostheimer GJ, van Vugt MA, Jorgensen C, Miron IM, Diella F, Colwill K, Taylor L, Elder K, Metalnikov P, Nguyen V, Pasculescu A, Jin J, Park JG, Samson LD, Woodgett JR, Russell RB, Bork P, Yaffe MB, Pawson T. Systematic discovery of in vivo phosphorylation networks. *Cell* 2007;**129**:1415–26.
98. Matsuoka S, Baillif BA, Smogorzewska A, McDonald ER 3rd, Hurov KE, Luo J, Bakalarski CE, Zhao Z, Solimini N, Lerenthal Y, Shiloh Y, Gygi SP, Elledge SJ. ATM and ATR substrate analysis reveals extensive protein networks responsive to DNA damage. *Science* 2007;**316**:1160–6.
99. Dong Z, Zhong Q, Chen PL. The Nijmegen breakage syndrome protein is essential for Mre11 phosphorylation upon DNA damage. *J Biol Chem* 1999;**274**:19,513–19,516.
100. Banin S, Moyal L, Shieh S, Taya Y, Anderson CW, Chessa L, Smorodinsky NI, Prives C, Reiss Y, Shiloh Y, Ziv Y. Enhanced phosphorylation of p53 by ATM in response to DNA damage. *Science* 1998;**281**:1674–7.
101. Canman CE, Lim DS, Cimprich KA, Taya Y, Tamai K, Sakaguchi K, Appella E, Kastan MB, Siliciano JD. Activation of the ATM kinase by ionizing radiation and phosphorylation of p53. *Science* 1998;**281**:1677–9.
102. Khanna KK, Keating KE, Kozlov S, Scott S, Gatei M, Hobson K, Taya Y, Gabrielli B, Chan D, Lees-Miller SP, Lavin MF. ATM associates with and phosphorylates p53: mapping the region of interaction. *Nat Genet* 1998;**20**:398–400.
103. Zou L, Elledge SJ. Sensing DNA damage through ATRIP recognition of RPA-ssDNA complexes. *Science* 2003;**300**:1542–8.
104. Cuadrado M, Martinez-Pastor B, Murga M, Toledo LI, Gutierrez-Martinez P, Lopez E, Fernandez-Capetillo O. ATM regulates ATR chromatin loading in response to DNA double-strand breaks. *J Exp Med* 2006;**203**:297–303.
105. Stiff T, Walker SA, Cerosaletti K, Goodarzi AA, Petermann E, Concannon P, O'Driscoll M, Jeggo PA. ATR-dependent phosphorylation and activation of ATM in response to UV treatment or replication fork stalling. *EMBO J* 2006;**25**:5775–82.
106. Povirk LF, Zhou RZ, Ramsden DA, Lees-Miller SP, Valerie K. Phosphorylation in the serine/threonine 2609–2647 cluster promotes but is not essential for DNA-dependent protein kinase-mediated nonhomologous end joining in human whole-cell extracts. *Nucleic Acids Res* 2007;**35**:3869–78.
107. Chen X, Zhao R, Glick GG, Cortez D. Function of the ATR N-terminal domain revealed by an ATM/ATR chimera. *Exp Cell Res* 2007;**313**:1667–774.
108. Giaccia AJ, Kastan MB. The complexity of p53 modulation: Emerging patterns from divergent signals. *Genes Dev* 1998;**12**: 2973–83.
109. El-Deiry WS, Tokino T, Velculescu VE, Levy DB, Parsons R, Trent JM, Lin D, Mercer WE, Kinzler KW, Vogelstein B. WAF1, a potential mediator of p53 tumor suppression. *Cell* 1993;**75**:817–25.
110. Harper JW, Adami GR, Wei N, Keyomarsi K, Elledge SJ. The p21 Cdk-interacting protein Cip1 is a potent inhibitor of G1 cyclin-dependent kinases. *Cell* 1993;**75**:805–16.
111. Bartek J, Lukas J. Pathways governing G1/S transition and their response to DNA damage. *FEBS Lett* 2001;**490**:117–22.
112. Lee CH, Chung JH. The hCds1 (Chk2)-FHA domain is essential for a chain of phosphorylation events on hCds1 that is induced by ionizing radiation. *J Biol Chem* 2001;**276**:30,537–30,541.
113. Buscemi G, Savio C, Zannini L, Micciche F, Masnada D, Nakanishi M, Tauchi H, Komatsu K, Mitzutani S, Khanna K, Chen P, Concannon P, Chessa L, Delia D. Chk2 activation dependence on Nbs1 after DNA damage. *Mol Cell Biol* 2001;**21**:5214–22.
114. Khosravi R, Maya R, Gottlieb T, Oren M, Shiloh Y, Shkedy D. Rapid ATM-dependent phosphorylation of MDM2 precedes p53 accumulation in response to DNA damage. *Proc Natl Acad Sci USA* 1999;**96**:14,973–14,977.
115. Oren M. Regulation of the p53 tumor suppressor protein. *J Biol Chem* 1999;**274**:36,031–36,034.
116. Maya R, Balass M, Kim ST, Shkedy D, Leal JF, Shifman O, Moas M, Buschmann T, Ronai Z, Shiloh Y, Kastan MB, Katzir E, Oren M. ATM-dependent phosphorylation of 5: Role in p53 activation by DNA damage. *Genes Dev* 2001;**15**:1067–77.
117. Kleijer WJ, van der Kraan M, Los FJ, Jaspers NG. Prenatal diagnosis of ataxia-telangiectasia and Nijmegen Breakage syndrome by the assay of radioresistant DNA synthesis. *Int J Radiat Biol* 1994;**66**:S167–74.
118. Falck J, Petrini JH, Williams BR, Lukas J, Bartek J. The DNA damage damage-dependent intra-S phase checkpoint is regulated by parallel pathways. *Nat Genet* 2002;**30**:290–4.
119. Yazdi PT, Wang Y, Zhao S, Patel N, Lee E, Qin J. SMC1 is a downstream effector in ATM/NBS1 branch of the human S-phase checkpoint. *Genes Dev* 2002;**16**:571–82.
120. Mirzanyans R, Famulski KS, Enns L, Fraser M, Paterson MC. Characterisation of the signal transduction pathway mediating gamma ray-induced inhibition of DNA synthesis in human cells: Indirect evidence for involvement of calmodulin but not protein kinase C nor p53. *Oncogene* 1995;**11**:1597–605.
121. Zampetti-Besseler F, Scott D. Cell death, chromosome damage and mitotic delay in normal human, ataxia telangiectasia and retinoblastoma fibroblasts after x-irradiation. *Int J Radiat Biol Relat Stud Phys Chem Med* 1981;**39**:547–58.
122. Bache M, Pigorsch S, Dunst J, Wurl P, Meye A, Bartel F, Schmidt H, Rath FW, Taubert H. Loss of G2/M arrest correlates with radiosensitization in two human sarcoma cell lines with mutant p53. *Int J Cancer* 2001;**96**:110–17.
123. Scott D, Zampetti-Bosseler F. Cell cycle dependence of mitotic delay in X-irradiated normal and ataxia-telangiectasia fibroblasts. *Int J Radiat Biol Relat Stud Phys Chem Med* 1982;**42**:679–83.
124. Xu B, Kim ST, Lim DS, Kastan MB. Two molecularly distinct g(2)/m checkpoints are induced by ionizing irradiation. *Mol Cell Biol* 2002;**22**:1049–59.
125. Gatei M, Scott SP, Filippovich I, Sorokina N, Lavin MF, Weber B, Khanna KK. Role for ATM in DNA damage-induced phosphorylation of BRCA1. *Cancer Res* 2000;**60**:3299–304.

126. Cortez D, Wang Y, Qin J, Elledge SJ. Requirement of ATM-dependent phosphorylation of brca1 in the DNA damage response to double-strand breaks. *Science* 1999;**286**:1162–6.
127. Gatei M, Zhou BB, Hobson K, Scott S, Young D, Khanna KK. Ataxia telangiectasia mutated (ATM) kinase and ATM and Rad3 related kinase mediate phosphorylation of Brca1 at distinct and overlapping sites. In vivo assessment using phosphospecific antibodies. *J. Biol Chem* 2001;**276**:17,276–17,280.
128. Xu X, Wagner KU, Larson D, Weaver Z, Li C, Ried T, Hennighausen I, Wynshaw-Boris A, Deng CX. Conditional mutation of Brca1 in mammary epithelial cells results in blunted ductal morphogenesis and tumour formation. *Nat Genet* 1999;**22**:37–43.
129. Mullan PB, Quinn JE, Gilmore PM, McWilliams S, Andrews H, Gervin C, McCabe N, McKenna S, White P, Song YH, Maheswaran S, Liu E, Haber DA, Johnston PG, Harkin DP. BRCA1 and GADD45 mediated G2/M cell cycle arrest in response to antimicrotubule agents. *Oncogene* 2001;**20**:6123–31.

Chapter 24

Signaling to the p53 Tumor Suppressor through Pathways Activated by Genotoxic and Non-Genotoxic Stresses

Carl W. Anderson[1] and Ettore Appella[2]

[1]*Biology Department, Brookhaven National Laboratory, Upton, New York*

[2]*Laboratory of Cell Biology, National Cancer Institute, National Institutes of Health, Bethesda, Maryland*

INTRODUCTION

The product of the human p53 tumor suppressor gene is a 393 amino acid polypeptide that functions primarily as a homotetrameric transcription factor. p53 regulates the expression of genes that control cell cycle progression, the induction of apoptosis or senescence, DNA repair, and other functions that involve cellular responses to stress. Approximately 50 percent of all tumors harbor a p53 mutation, and loss of p53 function, either directly through mutation or indirectly through several mechanisms, plays a central role in the development of cancer [1–3]. The p53 protein has been highly conserved during evolution, and orthologs have been found in the sequenced genomes of most multi-cellular organisms from the animal kingdom including zebrafish, *Drosophila* and *Caenorhabditis elegans* [4], but not in unicellular yeasts. In vertebrates and mollusks, p53 is accompanied by family members p63 and p73, which are structurally similar and bind to similar DNA sequence response elements (RE), but, while they work with p53, p63 and p73 have evolved to accomplish different tasks from those of p53 [5]. Like p63 and p73, several isoforms of p53 were shown to be expressed in tumor cells [6].

The p53 protein normally is short lived and is present at low levels in unstressed mammalian cells; however, in response to both genotoxic and non-genotoxic stresses it accumulates in the nucleus where it binds to specific DNA sequences. Genomic approaches using serial analysis of gene expression or DNA microarrays have shown that p53 directly or indirectly induces or inhibits the expression of more than 1,000 genes including *CDKN1A* (*p21*, *WAF1*, *CIP1*), *GADD45*, *MDM2*, *IGFBP3*, and *BAX* that mediate cellular responses to stress (reviewed in [7]). These include the arrest of mammalian cells at either of two major cell cycle checkpoints, in G_1 near the border of S phase or in G_2 before mitosis, the initiation of p53 dependent apoptosis [8], or the induction of a permanent cell cycle arrest that is indistinguishable from senescence [9]. p53 also modulates DNA repair processes, and the arrest of cell cycle progression may provide time for the repair of DNA damage (reviewed in [10]). The biochemical links between p53 and cell cycle arrest, senescence, and apoptosis are cell- and stress-type dependent. These observations suggest that specific posttranslational modifications to the p53 protein, at least in part, determine cellular fate. In turn, specific modifications reflect the pathways that become activated in response to any particular stress condition.

In this chapter, we highlight recent studies on the pathways that modulate p53 stability and activity in response to genotoxic and non-genotoxic stresses through covalent posttranslational modifications to p53 including the phosphorylation of serine and threonines, the acetylation and methylation of lysines, the glycosylation of serine, or the attachment of regulatory polypeptides (Ubiquitin, SUMO1, Nedd8) to lysine residues. We also emphasize several of the many proteins that have been reported to interact with p53 to tailor its activity.

P53 PROTEIN STRUCTURE

The p53 polypeptide has the classical features of a sequence specific transcriptional activator that can be divided into several functionally distinct regions: an amino- (N) terminal region (Met1-Lys101, numbering for human p53) that interacts with regulatory proteins and the transcriptional

Handbook of Cell Signaling, Three-Volume Set 2 ed.

machinery, a central, sequence specific DNA binding domain (Thr102-Lys292), and a carboxyl-terminal tetramerization and regulatory domain (Gly293-Asp393) (Figure 24.1). The amino-terminal region, which is unstructured [11], comprises two independent transcription activation domains, TAD1 (Met1-Met40) and TAD2 (Asp41-Pro83) [12]. The highly conserved residues Met1-Met40 are required for most transactivation activity and interact with the transcription factors TFIID, TFIIH, several TAFs, the coactivator histone acetyltransferases KAT3A/KAT3B (CBP/p300; a new nomenclature for chromatin modifying enzymes recently was proposed [13]), and possibly KAT2B (PCAF), as well as the MDM2 ubiquitin ligase. Residues Glu17-Asn29 form an amphipathic helix that interacts directly with a hydrophobic cleft in the N-terminal domain of MDM2 [14], while residues Glu11-Leu26 are reported to function as a secondary nuclear export signal [15]. TAD2 largely overlaps with a proline-rich domain (PRD, Asp61-Leu94) first described by Walker and Levine [16] that has been shown to be important for p53 stability, transactivation ability, and the induction of transcription independent apoptosis. Residues Asp41-Ala79 of human p53 are not well conserved in length or sequence among other species, whereas residues Pro80-Val97 are highly conserved among mammals. The PRD contains several PXXP motifs that create potential binding sites for Src homology 3 (SH3) domain containing proteins, such as the corepressor Sin3a, which interacts with residues 61–75 to stabilize p53 [17], as well as two pS/T-P sites (phospho-serine or threonine followed by proline) that serve as potential binding sites

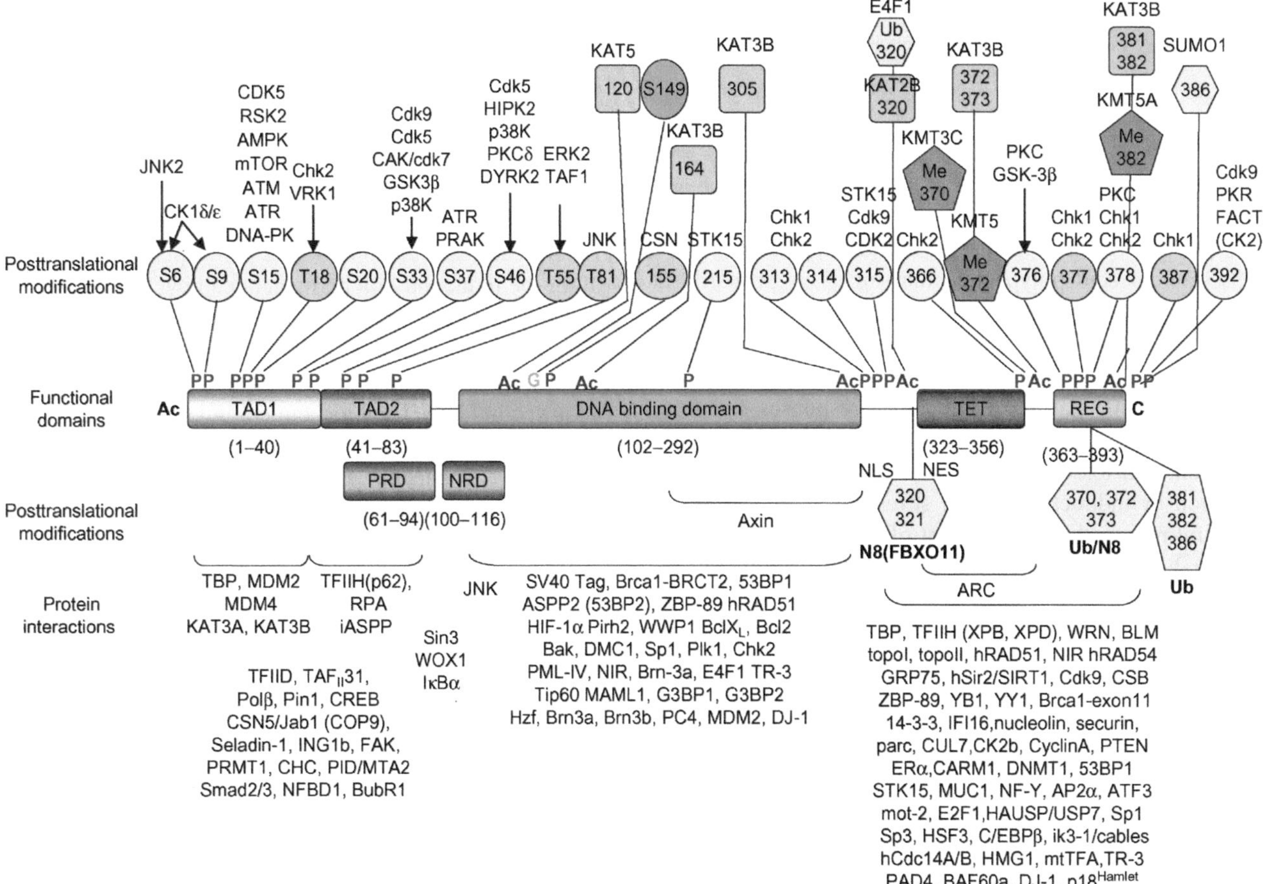

FIGURE 24.1 Protein domains, posttranslational modification sites, and proteins that interact with human p53.
The 393 amino acid human p53 polypeptide is represented schematically with postulated functional regions and domains indicated. Residues ~1–40 (TAD1) and 41–83 (TAD2) comprise independent tandem transactivation domains; residues ~61–94 represent a proline-rich domain (PRD); residues 33–80 are poorly conserved. Residues 100–116 constitute a recently described N-terminal repression domain that is required for repressing basal p53 activity in some cell types. Residues ~102–292 contain the central, sequence specific, DNA binding core region; residues 305–321 contain the primary bipartite nuclear localization signal (NLS); residues 323–356 comprise the tetramerization domain (TET), which contains a nuclear export signal within residues 339–350; residues 363–393 (REG) negatively regulate DNA binding by the central core to consensus recognition sites in oligonucleotides and interact in a sequence independent manner with single- and double-stranded nucleic acids but contribute positively to chromatin binding and transactivation *in vivo*. Posttranslational modification sites (P, phosphorylation; Ac, acetylation; G, glycosylation; Me, methylation, N8, neddylation; Ub, ubiquination) are indicated together with enzymes that can accomplish the modifications *in vitro*. The C-terminal six lysines (K370, K372, K373, K381, K382, and K386) can be ubiquitinated; K370, K372, and K373 are likely sites of attachment for the ubiquitin-like protein NEDD8; Lys386 may be modified by conjugation with SUMO1, a ubiquitin-like peptide. Interaction regions for selected proteins are indicated below the polypeptide. References are found in the text and recent reviews.

for the proline isomerase Pin1. Indeed, the Thr81-Pro82 site was recently shown to be required for Pin1 mediated isomerization of Pro82, which creates a binding site for the Chk2 protein kinase and allows phosphorylation of Ser20, leading to p53 stabilization [18]. Nevertheless, point mutants that eliminated the PXXP motifs or the Pin1 sites in the PRD of murine p53 had minimal effects on the stability or activity of murine p53 [19]. The PRD also provides binding sites for the p62/Tfb1 subunit of the basal transcription factor TFIIH [20], which is important for transcriptional initiation and elongation, and for the single-stranded DNA binding protein RPA [21], which is important for DNA repair. Recent structures showing these interactions provide insights into signaling mechanisms that activate p53 in response to DNA damage.

In contrast to the unstructured N- and C-terminal regions, the central core of p53 consists of an immunoglobulin-like β-sandwich that provides a scaffold for the DNA binding surface (reviewed in [22]). This surface is formed by two large loops, L2 (Lys164-Leu194, which includes a short helix Pro177-Glu180) and L3 (Met237-Pro250) that are stabilized by a zinc ion and a loop-sheet helix motif. Loss of the zinc ion, which is coordinated by three cysteines and histidine 179, destabilizes the protein and results in loss of sequence specific DNA binding. The majority (75 percent) of tumor derived p53 mutations are missense mutations that affect the central domain and block or alter sequence specific DNA binding either by destabilizing the domain, which is only marginally stable, or through changes to residues that directly contact DNA, e.g., Arg248 and Arg273 [23]. Six so-called hot-spot mutations (R175H, R248Q, R248W, R273H, R273C, and R282W) account for 20 percent of the somatic mutations found in human cancers [24]. Three-dimensional structures of the DNA binding domain free or bound to DNA have been determined both by X-ray crystallography and NMR, as well as structures of several mutant p53s in their DNA-free state. p53 binds through its core domain to two decameric half-site palindromes of the form 5'-RRRCWWGYYY-3' that may be separated by 0–13 bp [25], although for the majority of the ~100 known response elements that control transcription, the spacer length is zero [7]. Two core domain monomers bind to a half-site to form a symmetrical dimer, and two such dimers assemble to form the p53 tetramer. The human genome has over 10,000 predicted high affinity p53 binding sites and well over 200,000 weaker ones [26].

The carboxy-terminal region contains a nuclear localization signal (Thr312-Asp324), a tetramerization domain (Leu323-Gly356), and a basic 30 amino acid segment that binds RNA and certain DNA structures, including short single strands, four-way junctions, and insertions/deletions in a sequence independent manner (reviewed in [27]). Although reports have suggested that p53 regulates the translation of several mRNAs, including its own, through interactions with 5' mRNA regions, other studies have failed to show a preference for specific RNA primary sequences or secondary structures (reviewed in [28]), although p53 recently was reported to regulate the stability of plasminogen activator inhibitor-1 mRNA through interaction with a 70 nt sequence in the 3' UTR of the PAI-1 mRNA [29]. The unmodified C-terminal domain initially was thought to negatively regulate sequence specific binding by the core DNA binding domain [30]; however, more recently it has been shown to be required for efficient promoter activation possibly by promoting a search for response elements by facilitating diffusion along DNA [31–33]. The C-terminal region is highly modified and has been reported to interact with more than 50 proteins including components of the transcription apparatus and several other transcription factors (Figure 24.1). Only tetrameric p53 appears to be active as a transcription factor.

Although individual domains of p53 have been well characterized, only recently has the quaternary structure of full length p53 been addressed through electron microscopy, biophysical methods, and computational modeling [34]. The model that emerges for the free protein in solution is an elongated cross-shaped tetramer with extended N- and C-termini. Upon binding, p53 wraps around the DNA helix with all four N-termini pointing away from one face of the core domain DNA complex (reviewed in [22]).

POSTTRANSLATIONAL MODIFICATIONS TO P53

p53 activity is regulated through numerous posttranslational modifications that occur mainly in the N- and C-terminal regions, although several important modifications to the core DNA binding domain recently were identified (Figure 24.1). Western immunoblot experiments using monoclonal or affinity purified polyclonal antibodies that recognize specific modified sites in human or mouse p53, have identified a change in status at most of the 21 known phosphorylation sites and at eight identified acetylation sites in response to the treatment of cells with DNA damage inducing agents (reviewed in [35–37]). The N-terminal transactivation region of human p53 (Met1-Pro89) contains seven serines and two threonines; all, specifically Ser6, 9, 15, 20, 33, 37, 46, and Thr18 and 81, are phosphorylated or dephosphorylated in response to exposing cells to ionizing radiation or UV light. Thr55 is constitutively phosphorylated in unstressed cells and dephosphorylated after DNA damage through a mechanism involving recognition of phospho-Ser15 by the B56γ regulatory subunit of PP2A [38]. Serines 33 and 46 and Thr81 in the N-terminal region (as well as Ser315 in the C-terminal region) are each proximal to a proline residue and may regulate Pin1 mediated interconversion of the *cis-trans* proline conformations, which, in turn, regulate interactions with MDM2 and Chk2 and probably other binding partners [39].

In the central site specific DNA binding domain, three or four sites now are thought to be phosphorylated, two lysines can be acetylated, and Ser149 may be *O*-glycosylated. Thr155 and Thr150 or Ser149 are reported to be phosphorylated by the COP9 signalosome (CSN) associated kinase [40]. However, *O*-linked *N*-acetylglucosylation of Ser149 recently was reported to inhibit phosphorylation of Thr155, which resulted in increased p53 stability [41]. Ser215 can be phosphorylated by the mitotic kinase Aurora-A (STK15), and this phosphorylation inhibited DNA binding [42]. Lysine 120 can be acetylated by either KAT5 (Tip60) or KAT8 (hMOF) [43, 44], while Lys164 recently was shown to be acetylated by KAT3B [45].

In the C-terminal regulatory domain, Ser313, 314, 315, 366, 376, 378, 392, and Thr377 and 387 can be phosphorylated, Lys320, 372, 373, 381, and 382 can be acetylated, and lysines 370, 372, and 382 can be methylated (Figure 24.1). Lysines 370, 372, 373, 381, 382, and 386 are the major sites of p53 ubiquitination, and lysines 320, 321, 370, 372, and 373 also can be conjugated with the neural precursor cell-expressed developmentally downregulated 8 ubiquitin-like protein NEDD8 (reviewed in [46, 47]). Lys386 may be sumoylated in response to DNA damage. Ser376 and 378 were reported to be constitutively phosphorylated in unstressed cells, while Ser376, like Thr55, was dephosphorylated in response to ionizing radiation [48].

The *in vivo* biochemical and functional consequences of p53 posttranslational modifications have been difficult to assess; however, the ability to create "knockin" mutations that change single or several codons in the endogenous mouse p53 gene to ones that specify amino acids that cannot be modified or that mimic a modified residue recently has led to significant advances in our understanding [49, 50]. We now realize that several p53 modifications show dependencies, for example, phosphorylation interdependencies between several sites in the transactivation domain [51, 52], and methylation of mouse p53 Lys369 (human p53 K372) by the KMT5 (Set7/9) methytransferase recently was found to be required for p53 acetylation by Tip60 [53]. Furthermore, p53 modifications may exhibit synergies or exhibit redundancy. These possibilities are illustrated by analysis of the mouse Ser18 and Ser20 double knockin mutant compared to mice with the single amino acids changed [54] and by a recent analysis of cells with changes to eight p53 sites that are acetylated by KAT3B [45]. A table summarizing proposed functions for several of the p53 posttranslational modifications recently was published [37].

REGULATION OF P53 ACTIVITY

While the biochemical mechanisms that regulate p53 activity are complex and incompletely understood, it is widely believed that activation of p53 as a transcription factor involves two stages: p53 accumulation due to stabilization is sufficient to induce the transcription or repression of a subset of p53 regulated genes, but a full, cell specific transcriptional response also requires a cascade of posttranslational modifications that modulate activity.

P53 STABILIZATION

In response to a number of stress activated signaling pathways, p53 is stabilized and accumulates in the nucleus. In unstressed cells, p53 protein has a short half-life (~20 min) due to rapid, ubiquitin dependent degradation through the 26S proteosome. Ubiquitin is a 76 amino acid polypeptide that is transferred to lysine residues in proteins by ubiquitin ligases; multiple (poly-) ubiquitination targets proteins to the 26S proteosome complex where they are degraded. At least six cellular systems, MDM2, COP1, Pirh2, Topors, Synoviolin, and ARF-BP1, have been reported to target p53 for ubiquitination and degradation (reviewed in [47]). However, ubiquitination, like sumolyation and neddylation, also serves other functions and does not always lead to degradation of the modified protein [46, 55].

A number of viruses express proteins that induce p53 degradation so that cells enter a state in which the DNA viral genome can replicate and avoid virus induced apoptosis. Indeed, the first direct evidence for a role for ubiquitination in downregulating p53 came from studies on human papilloma viruses. The papillomavirus E6 protein, in conjunction with a ~100 kDa cellular HECT domain protein, E6AP, functions as a ubiquitin ligase in a manner similar to MDM2 [56]. As a consequence, papillomavirus-transformed human cells (e.g., HeLa) frequently have wild-type p53 genes but are functionally deficient for p53 activity. Similarly, the adenovirus E1B55K and E4orf6 proteins target p53 for degradation through a cullin containing complex that includes the RING finger protein Rbx1 [57].

In dividing cells, the primary cellular system that ubiquitinates p53 is the E3 ubiquitin ligase, MDM2 (murine double minute gene) [58]. This activity of MDM2 is vital as shown by the rescue of MDM2 knockout mice from embryonic lethality by deletion of p53; furthermore, p53 induces the transcription of MDM2 to form a negative feedback loop that regulates p53 accumulation [59]. MDM2 originally was identified on the basis of its ability to interact with p53. A cleft in the N-terminal domain of MDM2 (amino acids Thr26-Val108) binds to an amphipathic helix (Glu17-Asn29) in the transactivation domain at the N-terminus of p53 [14], and binding is required for subsequent ubiquitination of p53 at multiple C-terminal lysines (Figure 24.1). As noted above, p53 is phosphorylated at several N-terminal sites that reside in or near the MDM2 binding site by kinases activated through stress response pathways; this finding led to the hypothesis that phosphorylation of Ser15 and 37 in response to DNA damage might stabilize p53 by preventing its interaction with MDM2 [60]. Subsequently, phosphorylation of Thr18 and Ser20 were reported to negatively regulate the interaction of p53 with MDM2 [61–64]. Thr18 and Ser20 reside within the p53 N-terminal amphipathic helix that directly interacts with the N-terminus of MDM2, and Thr18 makes several stabilizing hydrogen bonds with neighboring residues that would be disrupted by phosphorylation. Consistent with the structural data and recent computational analysis, phosphorylation of Thr18, but not phosphorylation of Ser15, Ser20, or Ser37, was found to interfere directly with the interaction of an N-terminal p53 peptide with the N-terminal domain of MDM2 [63, 65]. Nevertheless, changing Ser15 to alanine was shown to significantly decrease the ability of p53 to activate transcription and induce apoptosis in both human [60] and mouse [66] systems, and changing Ser20 to alanine in human p53 abrogated stabilization in response to DNA damage [61]. Taken together, these results suggest that phosphorylation of p53 Ser15 and Ser20 may indirectly affect complex formation with MDM2, perhaps through increased competition for binding to the N-terminus of phosphorylated p53 by other factors. For example, KAT3A/KAT3B interacts with the N-terminus of p53, and binding is dramatically enhanced by phosphorylation of Ser15 [67, 68].

More recently, the interaction of p53 with MDM2 also has been found to involve both the conformation of the polyproline region (Figure 24.1), which depends on phosphorylation and Pin1 activity [18], and the central sequence specific DNA binding domain of p53 [69, 70]. Furthermore, like p53, MDM2 is extensively posttranslationally modified by phosphorylation, and these modifications both positively and negatively regulate MDM2 activity and its interaction with p53 [71]. MDM2 is modified by ATM and ATR in response to genotoxic stress signals, as well as by kinases regulated by ATM or ATR, and these signals also induce MDM2 degradation. Even in the absence of ATM, MDM2 can be inactivated in response to DNA double-strand breaks through a newly identified DNA-PK-Akt-GSK3β circuit [72]. Indeed, posttranslational modifications to MDM2 may be more important than modifications to p53 with respect to inhibiting p53 degradation. MDM2 activity also is regulated by several proteins with which it interacts, including $p14^{ARF}$, and the ribosomal proteins L5, L11, and L23 (see below).

MDM2 has been thought to regulate p53 through two mechanisms: first, by targeting it for degradation through ubiquitination, and second, by masking p53's access to the transcription machinery through binding to the N-terminal TAD1 (Figure 24.1). However, several recent studies, including one in which MDM2's ubiquitin ligase function was inactivated through a knockin mutation that changed a RING finger cysteine to alanine (C462A), indicate that the physical interaction between MDM2 and p53 is not sufficient to control p53 activity [73]. Instead, this job falls to the structurally related protein, MDM4 (MDMX) [74]. MDM4 was discovered in a screen for p53 binding proteins, and the MDM4 gene also is amplified or overexpressed in many cancers. Furthermore, like MDM2, loss of MDM4 expression also results in embryonic lethality that is rescued by deletion of p53 [75], indicating that MDM2 and MDM4 function in a coordinated manner. MDM2 and MDM4 also form heterodimers. However, unlike MDM2, MDM4 expression is not regulated by p53, and MDM4 does not have intrinsic ubiquitin ligase activity. MDM4 is posttranslationally modified following DNA damage, and DNA damage induces the rapid degradation of both MDM2 and MDM4. Although MDM2 was thought to be regulated by autoubiquitination, Itahana *et al.* [73] reported that the half-life of the ligase defective MDM2(C462A) was the same as that of the wild-type protein before and after DNA damage. Of note, the histone acetyltransferase KAT2B recently was reported to have intrinsic ubiquitination activity, and knockdown of KAT2B expression stabilized MDM2, suggesting that KAT2B may contribute to MDM2 ubiquitination and degradation [76].

p53 stability also can be modulated by de-ubiquitination. HAUSP (herpes virus associated ubiquitin specific proteinase), a de-ubiquitinating enzyme that acts on p53, and Daxx (death domain associated protein) recently were shown to have key roles in regulating MDM2 ubiquitination [77]. Under non-stress conditions, Daxx associated with HAUSP and MDM2, which stabilizes MDM2 and MDM4, directs MDM2 ligase activity toward p53. However, after DNA damage, HAUSP, Daxx, and p53 dissociate from MDM2, and this effect is mediated in part by ATM. The resulting MDM2–MDM4 complex is then degraded. Thus, the findings of Tang *et al.* [77] suggest that Daxx directs HAUSP toward MDM2, and possibly MDM4, which increase their activity toward p53.

As noted above, several other ubiquitin ligases, including Pirh2, COP1, and ARF-BP1 are reported to contribute to p53 degradation. The transcription of both COP1 and Pirh2 is activated in response to p53 thus forming negative feedback loops. In response to DNA damage, ATM phosphorylates COP1 on Ser387 and stimulates its autodegradation [78]. Pirh2 is regulated in a cell cycle specific manner, and its degradation is stimulated after phosphorylation by calmodulin dependent kinase II on Thr154 and Ser155 [79]. ARF-BP1 (also known as Mule, Huwe1, and UREB1) is a ~500 kDa HECT domain E3 ligase that interacts strongly with the tumor suppressor ARF (human $p14^{ARF}$ or murine $p19^{ARF}$), which inhibits ARF-BP1's activity [80]. ARF-BP1 is highly expressed in 80 percent of breast cancer cell lines while expression in normal breast cells is low. RNAi mediated knockdown of ARF-BP1 expression stabilized p53 and induced apoptosis, suggesting that ARF-BP1 plays a significant role in regulating p53 levels, whereas knockdown of COP1 or Pirh2 had more modest effects. However, overexpression or knockdown of ARF-BP1 in a mouse neuroblastoma cell line did not affect p53 levels

[81]. ARF-BP1 is phosphorylated by ATR or ATM following DNA damage, but its abundance does not change after DNA damage or during the cell cycle.

P53 ACTIVATION

Although early studies by Hupp *et al.* [30] indicated that p53 is synthesized in a latent form that is incompetent for sequence specific DNA binding, p53 accumulation in the absence of (known) posttranslational modifications, as shown by, e.g., studies using treatment with drugs that stabilize p53 [82], is sufficient to activate transcription of many p53 regulated genes. Indeed, p53 is reported to be constitutively bound to chromatin at some recognition sites *in vivo*, including sites in the *CDKN1A* ($p21^{WAF1}$), and *MDM2* promoters, and genotoxic stress caused only small increases in the amount of p53 bound to chromatin at these sites [83]. Nevertheless, it is widely believed that posttranslational modifications to p53 enhance its ability to induce or repress transcription from subsets of genes through several potential mechanisms including changes in response element recognition, attracting chromatin modifying proteins to p53 target sites, and modulation by proteins with which p53 interacts. These posttranslational modification induced effects are thought to modulate cellular fate – the decision to undergo temporary cell cycle arrest, senescence, or to undergo apoptosis in response to stress signaling. For example, KAT3B, which acetylates at least five p53 lysines (Figure 24.1), was shown to be necessary for p53 mediated G_1 arrest and the induction of $p21^{WAF1}$; furthermore, the absence of KAT3B promoted induction of PUMA transcription and apoptosis in response to the exposure of colorectal cancer cells to UV light [84]. Such changes also are shown by a recent study in which the endogenous gene of mouse p53 was changed so that codon 317 encoding lysine (human Lys320), which is acetylated by KAT2B [85], was changed to arginine, which cannot be acetylated [86]. Genome-wide microarray analyses showed that the expression of at least 150 of ~1400 genes that responded to treatment with ionizing radiation was significantly affected by this change, including a subset of genes involved in p53 mediated apoptosis. It should be noted, however, that Lys320 also is ubiquitinated in an E4F1 mediated process that does not promote degradation, and this modification and acetylation of Lys320 are mutually exclusive [55]. p53 oligo ubiquitinated at Lys320 was associated with chromatin in cell cycle arrested cells but not in cells induced to undergo apoptosis.

A role for p53 phosphorylation in integrating RTK/MAPK signaling with TGF-β signaling during development recently was described [87]. Fibroblast growth factor was shown to activate CK1δ/ε, which phosphorylate p53 on Ser 6 and 9; these modifications, in turn, induce binding of Smad2, which is induced by TGF-β and is required for deployment of the TGF-β cytostatic program. A second role for N-terminal phosphorylation is to promote the association between p53 and the coactivators KAT3A/KAT3B [68, 85]. In turn, KAT3A/KAT3B induces acetylation at about four sites in the C-terminal domain (Figure 24.1), and at a recently identified site in the central DNA binding domain [45]. Acetylation of C-terminal sites blocks ubiquitination of p53 by MDM2 at sites that are thought to be important for degradation and may directly prevent the interaction of p53 with MDM2 at promoters [45]. C-terminal acetylation also promotes p53 tetramerization and binding of the dual specificity phosphatase PTEN, which in turn promotes maintenance of high p53 acetylation through a phosphatase independent mechanism [88]. The promoter recruited KAT3A/KAT3B also affects the acetylation of histones near the p53 binding site [89, 90], and acetylation of p53 C-terminal residues also may serve to recruit coactivators [91].

The importance of p53 acetylation was further demonstrated by overexpression of the histone deacetylases HDAC-1, -2, or -3, or hSir2, which deacetylate p53 and inhibit the transcription of p53 target genes [92–94]. The targeting of HDACs by p53 to p53 repressed genes has been suggested as a mechanism of p53 mediated gene repression [95]. Indeed, acetylation of p53 C-terminal lysines recently was shown to be necessary for the recruitment of HDAC4 to NF-Y dependent repressed G_2/M promoters following DNA damage [96]. In this case, p53, which interacts with NF-Y, acts as an adaptor protein to promote gene repression.

Three C-terminal p53 residues, Lys370, Lys372, and Lys382 can be monomethylated by three different methyltransferases, KMT3C (SYMD2), KMT5 (SET9), and KMT5A (SET8), respectively, and these modifications also modulate p53 activity in response to DNA damage. The level of methylated Lys372 increases very rapidly after DNA damage, which promotes nuclear localization and increased stability of the p53 protein [97]. DNA damage does not affect the level of KMT5 protein, but its activity rapidly increases, suggesting that KMT5 activity may be regulated in response to stress signals. Methylation of K370 by KMT3C blocks p53 binding to DNA and represses transcriptional activation, but K370 methylation is inhibited in response to DNA damage by methylation of Lys372 by KMT5 [98]. Importantly, methylation of Lys372 by KMT5 recently was shown to be necessary for subsequent acetylation by KAT5 (Tip60) of p53 at Lys120 in the DNA binding domain [53]. Lys120 is a DNA contact residue, and in response to severe DNA damage, Lys120 acetylated p53 preferentially activates the transcription of proapoptotic genes [44]. Interestingly, increased dimethylation of Lys370 by an unknown methyltransferase is observed after DNA damage, and dimethyl-Lys370 is specifically bound by 53BP1 through its tandem tudor domains, which enhances transcriptional activation by p53 [99]. Through its

BRCT domains, 53BP1 also interacts with the p53 core DNA binding domain (Figure 24.1) [100]. Dimethylated-, but not monomethylated Lys370 can be demethylated by KDM1 (LSD1) [99]. Like monomethylation of Lys370, monomethylation of Lys382 by KMT5A also inhibits p53 DNA binding and prevents transcriptional activation of the p21^{WAF1} and PUMA promoters [101]; however, the level of Lys382 methylation decreases after DNA damage in parallel with an increase in Lys382 acetylation. KMT5A levels also decrease following DNA damage, suggesting that this methyltransferase also may be regulated by stress response signaling. The role of lysine methylation in p53 signaling is still developing, and new findings are expected.

To date there is no evidence that p53 is methylated at any of its 27 arginine residues, but it does bind two class I arginine methyltransferases, PRMT1 and CARM1, which function as coactivators that modulate the methylation of histones surrounding p53 target genes [102]. p53 also interacts with and recruits peptidylarginine deiminase 4 (PAD4), which converts histone arginine and methyl-arginine to citrulline, to the p21^{wAF1} promoter to repress transcription prior to exposure of cells to DNA damage inducing agents [103].

In response to severe DNA damage, p53 mediated apoptosis is enhanced by phosphorylation on Ser46 (Figure 24.1), apparently by one or more of five kinases, p38 MAPK [104], HIPK2 [105], DYRK2 [106], PKCδ [107], or CDK5 [108] (Figure 24.1). Phosphorylation of Ser46 in response to severe DNA damage induces the expression of p53AIP1 [109], a protein associated with mitochondrial membranes that induces the release of cytochrome c to initiate apoptosis. Less severe damage activates the ubiquitin E3 ligase Siah-1, which induces the degradation of HIPK2 during recovery from DNA damage [110].

Space precludes a discussion of many additional factors that are reported to affect p53 activity, and many questions remain. Where and when are modifications accomplished – in the cytoplasm, nucleoplasm, or on chromatin? What fraction of p53 is modified? What modifications coexist on the same p53 molecule – on the same p53 tetramer? Which are interdependent or temporally connected? Perhaps equally important, what modifications occur to p53 interacting proteins and to components of the transcriptional apparatus in response to stress signals? Taking a clue from MDM2, modifications to these components, which are much less well studied, may be as important as modifications to p53 in determining p53 responses [8, 111, 112].

ACTIVATION OF P53 BY GENOTOXIC STRESSES

Mammalian cells appear to have at least two interacting, but somewhat independent signaling pathways for activating p53 in response to genotoxic stress: one is activated by the presence of DNA double-strand breaks; the other is activated in response to lesions that inhibit DNA replication, such as pyrimidine dimers and bulky base adducts, that result in the formation of segments of single-stranded DNA. The effects of these pathways on p53 modifications are somewhat distinct (Figure 24.2) [51]. Although changes in gene expression, including genes known to be regulated by p53, can be detected in human lymphocytes or lymphoid derived cell lines after exposures to doses of ionizing radiation as low as 10 cGy [113, 114], most signaling studies employ "therapeutic" doses (2–10 Gy γ-rays, ~25 J/m^2 UV-C) at which cell survival is less than 0.1 percent but the induction of p53 posttranslational modifications is maximal and persistent. As genotoxic signaling pathways are described in detail in other chapters of this handbook, only brief summaries are presented below.

DNA DOUBLED-STRAND BREAKS

Treatment of cells with ionizing radiation or several radiomimetics (e.g., neocarzinostatin, bleomycin) induces DNA double-strand breaks (DSBs). Although the molecular mechanism(s) by which DSBs are recognized are incompletely known, key among the kinases activated in response to DNA breaks is ATM (A-T mutated), a protein kinase member of the phosphatidylinositol-3-kinase-like kinase (PIKK) family encoded by the gene mutated in the human genetic disorder ataxia-telangiectasia (A-T) [115]. ATM directly phosphorylates p53 at Ser15 and activates several other protein kinases that phosphorylate the N-terminal transactivation domain including Chk2 (Figure 24.3), which phosphorylates p53 at Ser20 and several other residues (Figure 24.1) [116]. The exact mechanism(s) by which ATM is activated in response to DSBs is controversial. In a landmark paper, Bakkenist and Kastan [117] reported that in undamaged cells ATM exists as an inactive dimer or oligomer; however, exposure of cells to ionizing radiation induced a rapid phosphorylation of Ser1981 that caused dimer dissociation and activation of the kinase. Autophosphorylation did not require direct binding to DNA breaks but was induced by changes in chromatin structure. Two protein complexes rapidly are recruited to DSBs: the MRN complex (Mre11, Nbs1, Rad50), which is involved in signaling and the activation of ATM [118, 119], and the Ku heterodimer (Ku70/Ku80). Ku recruits DNA-PKcs, the catalytic subunit of DNA-PK, a PIKK required for the repair of DSBs by non-homologous end-joining [120]. ATM interacts with and phosphorylates the MRN complex and the variant histone H2AX in the vicinity of DSBs, and the resulting foci can be detected with antibody specific for phosphorylated H2AX (γ-H2AX) or other proteins that colocalize at DSB sites. However, the majority of the rapidly activated ATM is free in the nucleoplasm, where it phosphorylates hundreds of different proteins [121]. In cells from A-T patients, the accumulation of p53, phosphorylation

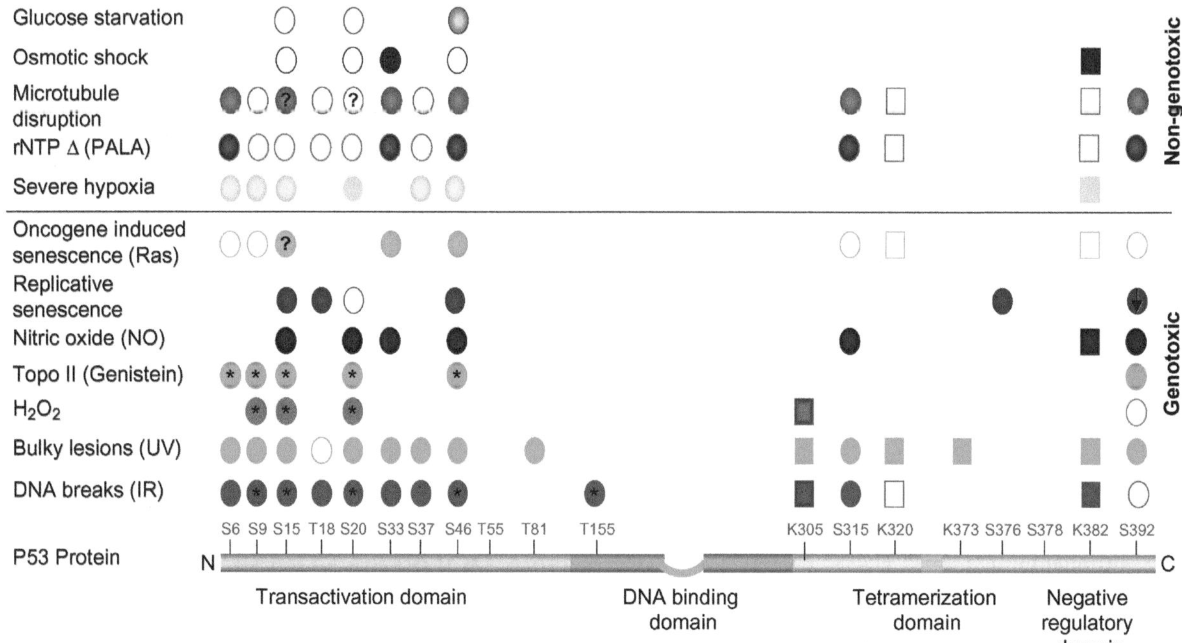

FIGURE 24.2 Posttranslational modifications to p53 in response to genotoxic and non-genotoxic stress.
The bar at the bottom represents the human 393 amino acid p53 polypeptide; functional regions are indicated. Selected posttranslational modification sites (Figure 24.1) are indicated above the bar; S, serine; T, threonine; K, lysine. Filled circles (phosphorylation) or squares (acetylation) indicate modification in response to the indicated stress (left); open symbols (or light gray) indicate no change in modification in response to stress. No symbol indicates the site has not been examined. An "*" denotes ATM dependent phosphorylations; a "?" indicates conflicting literature reports; a down arrow indicates treatment induced a decrease in site modification. The data were compiled from references in the text and our own published and unpublished work (e.g., [51]).

of several N-terminal sites (including Ser9, Ser15, Ser20, and Ser46), and the acetylation of Lys320 and Lys382 are delayed for several hours after the exposure of cells to ionizing radiation [116]. However, eventually these sites become phosphorylated, presumably as a consequence of the activation of ATR or DNA-PK.

Although the autophosphorylation model for ATM activation is attractive, and phospho-Ser1981 serves as a useful marker of activated ATM, Pellegrini *et al.* [122] reported that mice with a mutation that changed Ser1987 (equivalent to Ser1981 in human ATM) to alanine were not defective for ATM activation, indicating that autophosphorylation of this site was not necessary for ATM activation, at least in mice. However, Kozlov *et al.* [123] subsequently identified several additional ATM autophosphorylation sites, including Ser367 and Ser1893, and mutants that altered these sites were reported to be defective in ATM signaling. Recently Sun, 2007 [124] reported that Lys3016 in the highly conserved ATM FATC domain that is adjacent to the kinase domain is rapidly acetylated by KAT5 in response to DNA damage. Changing Lys3016 to arginine abolished upregulation of ATM's kinase activity by DNA damage, inhibited the conversion of inactive ATM dimers to active ATM monomers, and prevented ATM dependent phosphorylation of p53 and Chk2. Regardless of the exact mechanism of ATM activation, its activity is required for the rapid accumulation of p53 and p53 dependent activation of cell cycle checkpoints [125].

REPLICATION STRESS AND SINGLE-STRANDED DNA

A second DNA damage response pathway is dependent on the ATR (A-T and Rad3 related) PIKK and is activated in response to single-stranded DNA that is generated by the disruption of replication forks or the processing of bulky lesions, such as the pyrimidine dimers caused by UV-C (Figure 24.3) [126]. However, UV-A, which is more relevant to exposures on Earth, triggers activation of p53 through the ATM kinase, apparently as a consequence of the production of reactive oxygen species [127]. ATR also may be activated by transcriptional stress [128], presumably as a consequence of the generation of single-stranded DNA. ATR is found in a stable complex with ATRIP, and this complex binds to RPA coated single-stranded DNA. Like ATM, ATR can directly phosphorylate p53 on Ser15, and possibly Ser37, and it phosphorylates and activates Chk1, which then can phosphorylate additional p53 sites (Figure 24.1). In some circumstances, ATR can phosphorylate ATM on Ser1981 and activate it [129], and ATM can mediate the activation of Chk1 [130]. Thus, there can be considerable cross-talk between the ATM and ATR signaling pathways [131]. However, so far there is no evidence that ATR is activated through an autophosphorylation mechanism similar to ATM's activation, and, indeed, there do not appear to be serine-glutamine sites in locations equivalent to the three identified autophosphorylation sites

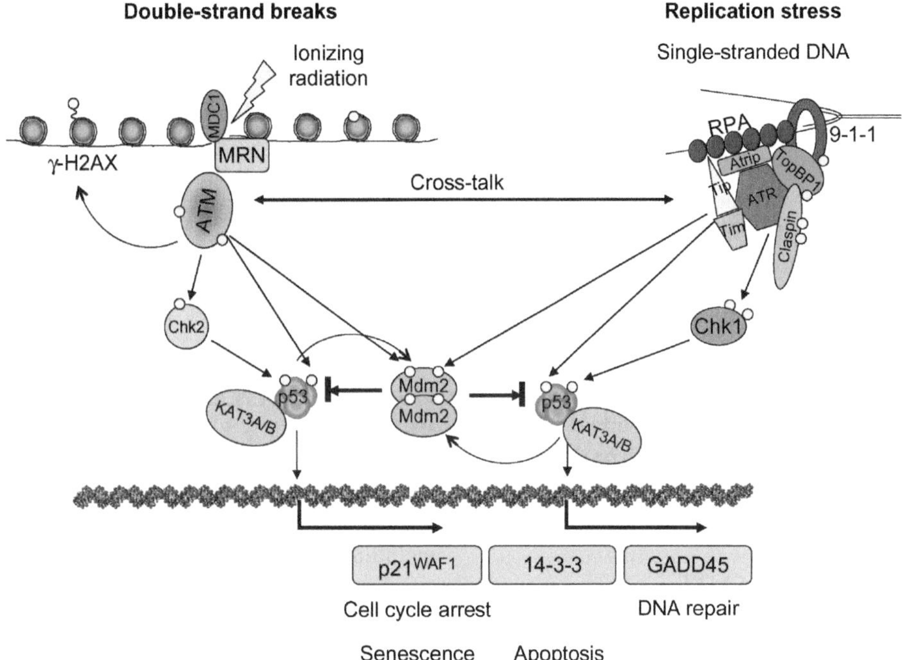

FIGURE 24.3 Pathways for p53 activation in response to genotoxic stress.
In response to DNA double-strand breaks, ATM interacts with chromatin and the MRN complex (Mre11, Nbs1, and Rad50), which is retained near the break site by MDC1. ATM is activated by conversion from an inactive dimer or oligomer through a process that involves autophosphorylation on several serines including Ser1981 (yellow dot) and acetylation on Lys3016 (not shown). ATM then phosphorylates numerous substrates including Ser15 near the N-terminus of p53, Ser395 of MDM2, and Ser68 on the effector checkpoint kinase Chk2. After Pin1 isomerizes p53 N-terminal prolines to create a docking site, Chk2 phosphorylates p53 Ser20 and at other residues. p53 phosphorylation enables binding of the histone acetyltransferase KAT3A/B (CBP/p300). In response to DNA damage, p53 also is methylated a C-terminal lysines as described in the text and Figure 24.1. The stabilized and activated p53 then induces or represses the transcription of numerous genes required for DNA repair, cell cycle arrest, senescence, or apoptosis. A parallel pathway involving ATR is activated in response to replication stress, which generates regions of single-stranded DNA. The ATR-ATRIP complex binds single-stranded DNA that has been coated by RPA. TopBP1 is recruited to RPA coated ss-dsDNA junctions by the phosphorylated 9-1-1 complex; it then interacts with the FATC domain of ATR and activates ATR's kinase activity. The Tim-Tipin (Tip) complex and Claspin are additional mediators that control ATR and Chk1 activation in response to DNA damage. ATR directly phosphorylates p53 Ser15 and, in the presence of Claspin, also phosphorylates the effector checkpoint kinase Chk1on Ser317 and Ser345, which activates it. Chk1 is then able to phosphorylate p53 on Ser20 and other sites.

in ATM. Instead, ATR is activated *in vitro* and *in vivo* by the binding to TopBP1 [132, 133]. The binding of claspin, a cell cycle regulated adapter protein, is then required to direct ATR activity to activate Chk1, which then is free to phosphorylate p53 on multiple residues (Figure 24.1). Tim and Tipin form a heterodimeric complex that coordinates the intra-S phase checkpoint response to UV induced DNA damage; Tipin binds RPA on DNA and the complex increases the efficiency of Chk1 activation by ATR [134].

REPLICATIVE SENESCENCE

Replicative senescence in human fibroblasts correlates with activation of p53 dependent transcription and was shown to be associated with increased phosphorylation at Ser15, Thr18, and probably Ser376, and with decreased phosphorylation at Ser392 [135]. It was inferred from the finding that no change occurred in staining with the DO-1 monoclonal antibody, the epitope for which includes Ser20, that phosphorylation on Ser20 was not induced. These results, in conjunction with findings showing that changes in p53 phosphorylation are abrogated in cells that have been immortalized by overexpression of telomerase, indicate that the above modifications may be the product of telomere erosion. Shortened or disrupted telomere structures signal to p53 via pathways partially shared with DNA damage responses [136].

ONCOGENE ACTIVATION

Oncogenes, such as Ras, c-Myc, or E1A, when activated or overexpressed, stabilize and activate p53, and depending on the cell type, induce senescence (Ras) or apoptosis (c-Myc or E1A) through hyperproliferative signaling pathways that activate ARF, the product of the alternative reading frame of the cell cycle regulatory gene INK4a (*CDKN2a*) [137, 138]. ARF, in turn, binds MDM2 and inhibits p53 degradation. Based on the observation that p53 Ser15 was not phosphorylated in response to adenovirus E1A expression, it was concluded that oncogenic activation of p53 occurs in the absence of DNA damage [139]. However, in normal fibroblasts, brief cMyc overexpression induced DNA damage prior to S phase that correlated with the induction of reactive oxygen species [140], raising the question of whether oncogenes activate p53 through DNA damage and whether the ability of oncogenes to promote either apoptosis or senescence correlates with different p53 posttranslational modifications. Ferbeyre *et al.* [141] reported that expression of oncogenic Ras induced phosphorylation of Ser15 in IMR90 cells. In contrast, Bulavin *et al.* [142] found p53 was phosphorylated at Ser33 and Ser46 but not at other N- or C-terminal sites, nor was it acetylated at Lys382 (Figure 24.2). However, "oncogenic stress" activates a DNA damage response that activates p53, perhaps through increased production of reactive oxygen species and/or disruption of normal DNA replication control [143, 144].

OTHER GENOTOXIC AGENTS

Because of their convenience, ionizing radiation and UV light commonly are used in the laboratory to produce two different forms of genotoxic damage, i.e., DNA double-strand breaks and pyrimidine dimers. Many other environmental, physiological, and therapeutic agents cause genotoxic damage that activate p53 through one or more signaling pathways. These include anti-cancer drugs such as adriamycin, topoisomerase inhibitors such as camptothecin, etoposide, and quercetin, DNA synthesis, and transcription inhibitors including aphidicolin, actinomycin D, and 5,6-dichloro-1-beta-D-ribofuranosylbenzimidazole (DRB), DNA cross-linking agents such as cisplatin and mitomycin C, and environmental chemicals including arsenite, cadmium, and chromate. Each of the above agents has been shown to induce p53 accumulation and its phosphorylation on Ser15 except DRB, which interferes with phosphorylation of the CTD domain of RNA polymerase II rather than with elongation [145].

ACTIVATION OF P53 BY NON-GENOTOXIC STRESSES

p53 can be activated or repressed in response to several physiological processes that are not associated with frank DNA damage, including hypoxia, low nucleoside triphosphate pool concentrations, low glucose or amino acid levels, microtubule disruption, heat and cold shock, the accumulation of unfolded proteins, nucleolar disruption, inflammation, nitrous oxide production, and mitotic spindle damage (see [146] for a review). Some of these processes may be mimicked by pharmacological agents that may or may not induce identical responses. The non-genotoxic stress pathways leading to alterations in p53 are complex, interacting, and incompletely characterized. Among the kinases that may directly phosphorylate p53 are AMPK and mTOR, both of which phosphorylate Ser15, and Aurora kinase A (also called STK15 or BTAK), which may phosphorylate Ser215 in the DBD and Ser315 in the C-terminal domain (Figure 24.1); other posttranslational modification pathways activated in response to non-genotoxic stress are less well studied. However, in many instances, p53 may simply be stabilized through effects that interfere with the targeting of p53 for destruction by MDM2. A few of these pathways are outlined below.

THE UNFOLDED PROTEIN RESPONSE – ER STRESS

A number of physiological or pathological conditions disrupt protein folding in the endoplasmic reticulum (ER), including glucose starvation, underglycosylation of glycosylated proteins, expression of mutant proteins, viral infection, extreme environmental conditions, and release of Ca2+ from the lumen, and these conditions lead to an imbalance between ER function and ER capacity. When chaperones of the heat shock family become saturated with unfolded proteins, cells respond by activating the unfolded protein response [147] pathway through the ER sensors AFT6, a transcription factor, and PERK and IRE1, two ER membrane associated protein kinases. The UPR enables cells to reduce the load of unfolded proteins in the ER by attenuating translation, increasing secretion and degradation, and increasing the synthesis of chaperones that promote folding (reviewed in [148, 149]). If ER homeostasis cannot be restored, then cells are induced to undergo apoptosis through activation of the JNK pathway.

ER stress is the only stress so far described that leads to p53 protein destabilization, which prevents p53 mediated apoptosis. As initially described by Qu *et al.* [150], ER stress leads to the activation of GSK-3β kinase, which binds p53 and mediates phosphorylation of human p53 on serines 315 and 376, both of which lie within nuclear localization signals, which leads to ubiquitination and increased cytoplasmic localization of p53. Although Ser376 does not lie in the normal GSK-3β kinase recognition sequence with a serine or threonine at the +4 position, GSK-3β kinase can directly phosphorylate Ser376 *in vitro*, but not Ser315, which may be phosphorylated by aurora kinase A [151]. The gene that encodes aurora kinase A (*AURKA*) is amplified in many cancers. Synoviolin (also called HRD1), an ER resident E3-ubiquitin ligase, has been shown to mediate p53 ubiquitination and target p53 for degradation [152].

HYPOXIA

Abnormal vasculature development causes hypoxia in most solid tumors; hypoxia also is an important pathophysiological feature of ischemic disorders. However, very low oxygen concentrations alone are not sufficient to induce DNA strand breaks [153, 154] although subsequent reoxygenation after periods of hypoxia does [155]. But recently it was reported that under very severe hypoxia conditions DNA damage may occur as a consequence of the repression or inhibition of DNA repair proteins including NBS1, MLH1, MSH2, and MSH6 (reviewed in [156]).

Extreme hypoxia (anoxia) leads to a rapid, reversible downregulation of protein synthesis by two distinct pathways. The first is the unfolded protein response [148, 149] discussed above; the second, which occurs only after prolonged anoxia is associated with disruption of the cap binding complex, eIF4F, which in turn inhibits recruitment of mRNA to the translation initiation complex at least partially through inhibition of the mTOR kinase (reviewed in [157]). However, hypoxia also selectively induces mRNA expression through one of the most responsive oxygen sensors in a cell, the hypoxia inducible transcription factor [158], which is composed of two subunits, HIF-1α and HIF-1β. HIF-1α is labile but is rapidly stabilized in response to falling oxygen concentrations through inhibition of proline hydroxylase activity. This activity is necessary for ubiquitination of HIF-1α by the von Hippel–Lindau E3 ubiquitin ligase (pVHL) and for subsequent proteosomal degradation [159].

While HIF-1α is stabilized under comparatively mild hypoxic conditions, p53 does not accumulate as a result of the downregulation of MDM2 until severe hypoxic/anoxic conditions are reached (0.2 percent oxygen) [160, 161] with concomitant phosphorylation of p53 Ser15, but not acetylation of Lys382 (Figure 24.2) [162]. Subsequent studies have shown that in response to hypoxia, p53 also becomes phosphorylated on Ser6, Ser9, Ser37, and Ser46 [154, 155]. HIF-1 also has been reported to stabilize p53 through a direct interaction with its DNA binding domain [163, 164]. However, in contrast to ionizing radiation, hypoxia treatment failed to induce the transcription of downstream effector mRNAs including GADD45, Bax, and $p21^{WAF1}$ [162], and, in contrast to DNA damage inducing agents, hypoxia primarily caused an association of p53 with mSin3a rather than p300. Consistent with this finding, p53 mediated transrepression was induced. A recent genome-wide analysis in oncogene transformed mouse embryo fibroblasts (MEFs) found that the association of p53 with its response elements in the promoters of several genes including *Cdkn1a* ($p21^{WAF1}$), *Mdm2*, *Perp*, and *Apaf-1* was not increased in response to hypoxia. Rather, hypoxia primarily induced p53 mediated repression, which required residues Leu25-Trp26 (equivalent to human p53 Leu22-Trp23), Gly53-Pro54 (human p53 Gly59-Pro60), the polyproline-rich region, and the DNA binding region [165].

Hypoxia induced replication arrest leads to activation of ATR, which acts to protect stalled replication forks, and to the ATR dependent phosphorylation of p53. Inhibition of ATR kinase activity reduced the hypoxia induced phosphorylation of p53 protein on Ser15 as well as p53 protein accumulation [153, 161]. ATM also becomes activated and phosphorylated on Ser1981 in response to hypoxia, but its activation does not require the MRN complex; rather, Ser1981 phosphorylation is mediated by ATR [129]. These data suggest that hypoxia could select for the loss of ATR dependent checkpoint controls, thus promoting cell transformation.

GLUCOSE DEPRIVATION – NUTRITIONAL STRESS

External signals provided by growth factors regulate homeostasis in part by activating signaling cascades that regulate nutrient uptake. In mammalian cells, mTOR, which like ATM and ATR belongs to the PIKK family, serves as a central integrator of signals arising from growth factors, nutrients, and cellular energy metabolism [166]. Energetically stressful conditions, such as glucose deprivation, activate AMP kinase (AMPK), which senses increases in the intracellular AMP:ATP ratio; it also may be activated by an upstream complex containing the LKB1 tumor suppressor protein. AMPK also is activated in response to hypoxia, heat and cold stress, and oxidative stress, each of which is tightly associated with an increase in the AMP:ATP ratio. AMPK activates catabolic pathways including autophagy and inhibits anabolic pathways to restore an energy balance through direct phosphorylation of metabolic enzymes; however, glucose limitation also may lead to cell cycle arrest, and glucose deprivation results in p53 mediated apoptosis. Imamura *et al.* [167] originally showed that pharmacological activation of AMPK inhibited cell growth and induced the accumulation of p53 and its phosphorylation at Ser15. Moreover, the sequence surrounding p53 Ser15 resembles the consensus recognition sequence for AMPK, suggesting that AMPK may directly phosphorylate p53 at this site. Subsequently, Jones *et al.* [168] showed that growth of MEFs in low (1 mM) glucose led to the activation of AMPK and to p53 phosphorylation at (mouse p53) Ser18. A constitutively active AMPK was shown to be sufficient to induce cell cycle arrest, and this arrest was dependent on p53 and its phosphorylation at Ser18, as MEFs with a p53 Ser18 to alanine substitution did not arrest. Furthermore, activated, immunoprecipitated AMPK was capable of phosphorylating p53 Ser15 (Figure 24.1).

In contrast to glucose limitation, glucose deprivation leads to apoptosis. Recently, Okoshi *et al.* [169] reported that starvation for glucose induced a p53 dependent apoptotic death in MEFs. Under these conditions, p53 accumulated and became phosphorylated at Ser46, a site phosphorylated after severe genotoxic damage that is associated with the transcriptional induction of pro-apoptotic genes, but not at

Ser15 or Ser20. Glucose deprivation resulted in the induction of pro-apoptotic Bax and AIP1, as well as the p53 gene itself, but hardly any change was observed in p21^{WAF1} expression levels. As indicated in Figure 24.1, several kinases are able to phosphorylate p53 Ser46, including Cdk5, HIPK2, p38MAPK, and PKCδ, but additional studies will be required to identify the kinase(s) responsible for Ser46 phosphorylation after glucose starvation.

While nutrient limitation activates p53 through non-genotoxic signaling pathways, activation of p53 by DNA damage also was shown to inhibit mTOR activity through activation of AMPK, which subsequently phosphorylates the TOR associated tuberous sclerosis TSC1/2 tumor suppressor complex [170]. Activation of AMPK by etoposide did not occur in p53−/− cells. How p53 activates AMPK is not clear, but interestingly, p53 is reported to associate with LKB1, an upstream kinase that activates AMPK, as shown by co-immune precipitation [171].

Loss of p53 results in the activation of NFκB, an upregulation of *Glut3* (a high affinity glucose transporter) and an increased rate of aerobic glycolysis that is typical of cancer cells (the Warburg effect) [172]. p53 activation also induces transcription of SCO2 (synthesis of cytochrome oxidase); thus, mitochondrial respiration also decreases following the loss of p53 from tumor cells [173].

RIBONUCLEOTIDE POOL IMBALANCE

Studies by Linke *et al.* [174] showed that p53 is activated in non-cycling normal human fibroblasts by the N-phosphoacetyl-L-aspartate induced (PALA) depletion of pyrimidine nucleotides in the absence of detectable DNA damage. In contrast to the G_1 arrest induced by DNA damage, that induced by PALA was readily reversible. PALA treatment induced a pattern of gene expression that was distinct from that induced by IR. Some genes, such as *MDGI*, a mammary derived growth inhibitor gene, were induced independent of p53 while for others, such as *TSG6*, a tumor necrosis factor stimulated gene, induction was p53 dependent [175]. In contrast to non-cycling cells, in cycling cells PALA induced DNA damage and activation of the ATR-Chk1 signaling pathway [176]. Phosphorylation at Ser6, 33, 46, 315, and 392 was observed in human HCT116 and A549 cells by 24 hours after treatment; little or no acetylation at Lys320 or Lys382 was observed [51].

Pharmacological inhibition of purine biosynthesis by glycinamide ribonucleotide formyltransferase (GART) induced p53 accumulation in human A549, MCF7, or HCT116 cells, but, surprisingly, p53 induced transcription of target genes, including p21^{WAF1}, was impaired [177]. The p53 accumulating in these cells was nuclear, but it was not phosphorylated on Ser6, 15, or 20, nor was it acetylated at Lys373 or 382. Recent studies showed that mycophenolic acid, a clinically used immunosuppressant, which inhibits *de novo* synthesis of guanine nucleotides through inhibition of inosine monophosphate dehydrogenase, stabilizes and activates p53 through inhibition of pre-rRNA synthesis and disruption of the nucleolus [177a], revealing a mechanistic connection between two non-genotoxic pathways that previously seemed unrelated (see below).

NUCLEOLAR AND RIBOSOMAL STRESS

The nucleolus is a specialized subnuclear compartment where ribosomal RNAs (rRNA) are synthesized, processed, modified, and associated with ribosomal proteins and 5S RNA before the subunits are exported to the cytoplasm [178]. Although the nucleolus is primarily associated with ribosome biogenesis, it is a highly dynamic structure, and several lines of evidence suggest that it has additional functions. Based on previous studies and their observation that p53 stabilization correlated tightly with nucleolar disruption, Rubbi and Milner proposed that the nucleolus is a stress sensor that is responsible for maintaining low levels of p53 protein [179]. Disruption of nucleolar function, in response to normal physiological processes as well as to various kinds of stress including DNA damage, invariably led to p53 stabilization and accumulation. While our current picture of p53 stabilization is more complex, it is clear that the nucleolus plays an important role in many stress responses [180, 181].

Initial studies showed that ribosomal protein L11 interacted with the central region of MDM2 and inhibited its activity, leading to p53 accumulation and activation in a manner similar to the previously described role of ARF [182, 183]. Like other ribosomal proteins, L11 is an abundant protein, which raised the questions why it did not inhibit MDM2 mediated p53 degradation normally and under what circumstances L11 might naturally regulate p53 activity? Bhat *et al.* [184] then showed that in addition to low levels of actinomycin D, which specifically inhibits RNA polymerase I (the polymerase that is responsible for rRNA synthesis), p53 dramatically increased in normal human fibroblasts after serum starvation or as they entered confluence. While MDM2 mainly resides in the nucleoplasm, L11 normally is found in the nucleolus or in the cytoplasm where it is associated with the 60S ribosomal subunit. However, under the above conditions, but not after exposure to UV or overexpression of E2F1, which induces p53 and growth arrest, L11 moved from the nucleolus to the nucleoplasm where it was free to interact with MDM2. Knockdown of L11 in serum starved cells attenuated growth arrest and p53 accumulation, showing that L11 was at least in part directly responsible for p53 stabilization.

Similarly, the ribosomal proteins L5 and L23 also interact with MDM2, activate p53 through inhibition of MDM2 mediated degradation without inducing N-terminal p53 phosphorylation, and induce a G_1 arrest [185–187].

It is interesting that L5, L11, and L23 each interacts with a different region of MDM2, and all three can interact with MDM2 simultaneously. As for L11, the interaction of L5 and L23 was enhanced by blocking RNA polymerase I transcription with low levels of actinomycin D, but the L23–MDM2 interaction was not induced by exposing cells to ionizing radiation [186]. L5, L11, and L23 do not bind MDM4, but L11, like ARF, stimulates ubiquitination of MDM4 by MDM2, thus inducing MDM4 degradation. MDM4 blocks p53 mediated transactivation; thus, MDM4 elimination is important for activation of the ribosomal stress response through p53. Several recent studies show that other nucleolar proteins also interact with MDM2, including nucleolin [188] and nucleostemin [189], and inhibit its activity, which in turn activates p53, while nucleophosmin affects p53 stability by interacting with ARF [190].

MICROTUBULE DISRUPTION

Activation of p53 also occurs in response to factors such as colcemid, nocodazole, and taxol that deregulate cell adhesion or microtubule architecture and dynamics. Taxol (Paclitaxel), one of the newer chemotherapy drugs commonly used to treat ovarian, breast, and head and neck cancers, inhibits microtubule depolymerization. After nocodazole treatment, which depolymerizes microtubules, quiescent human fibroblasts accumulated transcriptionally active p53 and arrested in G_1 with a 4 N DNA content [191]. Activation of p53 after colcemid treatment was accompanied by a moderate increase in phosphorylation at Ser15 and correlated with activation of Erk1/2 MAP kinases and the development of focal adhesions rather than disruption of the microtubule system [192]. Curiously, murine fibroblasts did not undergo the same response. Taxol and vincristine, but not nocodazole, were found to induce multi-site phosphorylation of p53 in several tumor derived human cell lines, including HCT-116 and RKO cells, and the pattern of p53 phosphorylation was distinct from that observed after DNA damage [51, 193]. Nevertheless, both nocodozole and taxol increased phosphorylation at Ser15 (Figure 24.2). Interestingly, microtubule inhibitor induced p53 stabilization and Ser15 phosphorylation did not occur in ATM deficient fibroblasts, nor in normal human dermal fibroblasts. Studies with ectopically expressed p53 phosphorylation site mutants indicated that several p53 amino-terminal residues, including Ser15 and Thr18, were required for the taxol mediated phosphorylation of p53 [193]. In contrast, Damia *et al.* [194] reported taxol induced p53 phosphorylation at Ser20 but not at Ser15 in HCT-116 cells. Ser20 phosphorylation was accompanied by increased Chk2 activity and was inhibited neither in A-T cells lines nor by wortmannin treatment. Thus, the signaling pathways that impinge on p53 after hypoxia, ribonucleotide depletion, or microtubule disruption, while still not well defined, appear distinct from the pathways induced by genotoxic stresses.

SETTING THRESHOLDS AND RESETTING ACTIVATION – P53 PHOSPHATASES

Protein modifications usually are reversible; thus, cellular enzymes potentially exist that can reverse each of the post-translational modifications discussed above. While these are less well studied than the kinases, transacetylases, methyltransferases, N-acetyl-glucosyl-transferases, and ubiquitin-like ligases that act on p53, at least four protein phosphatases are capable of dephosphorylating specific p53 sites *in vitro* (Table 24.1). Likewise, several deacetylases and the de-ubiquitinating enzyme HAUSP have been shown to use p53 as a substrate and, through knockout or knockdown experiments, to affect cellular responses to DNA damage and other stresses (reviewed in [36, 47]). As noted above, dimethylated-Lys370 can be demethylated by KDM1 [99]. Nevertheless, since modification stabilized p53 has a bulk half-life of only a few hours, the role of modification reversal enzymes in regulating p53 function is not necessarily obvious, although, that said, the lifetimes of modifications on chromatin bound p53, which presumably are the most important, are largely unknown. Furthermore, since each of these enzymes has many substrates in addition to p53, it is difficult, if not impossible, to sort out specific effects on p53 from overall effects on pathway responses.

Li and colleagues first cloned two human homologs of yeast CDC14, hCdc14A, and hCdc14B, and showed that both nuclear proteins interacted with the C-terminus of p53 and dephosphorylated Ser315 [195]. Subsequently Ljungman's lab showed that there is a strong bias for low Cdc14 expression in cancer cell lines, suggesting that high expression is not compatible with wild-type p53 [196]. Protein phosphatase 1 (PP1) has been reported to dephosphorylate three p53 residues, Ser15, Ser37, and Ser392,

TABLE 24.1 Dephosphorylaton of human p53 sites by phosphatases

Site	Phosphatase	References
Ser15	PP1; PM1D (Wip1)	198, 199, 211, 212
Ser37	PP1; PP2A	199, 203
Thr55	PP2A-B56γ	38, 202
Ser315	Cdc14A	195, 196
Ser392	PP1	197

and its inhibition was linked to changes in transcriptional and apoptotic activities [197–199]. PP1 forms stable complexes with small inhibitor proteins, such as I-2, that negatively regulate its activity. Somewhat counter intuitively, Tang *et al.* [200] recently showed that exposure of cells to IR induced the rapid dissociation of I-2 from PP1, which should activate PP1 and induce p53 dephosphorylation; this process required ATM. In response to IR, ATM phosphorylates I-2 on serine 43. PP1 also interacts with a subunit (PNUTS) that targets it to the nucleus; intriguingly, PNUTS is strongly induced in response to hypoxia [201], thus linking PP1 to non-genotoxic p53 responses. Increased PNUTS expression increased p53 nuclear localization, phosphorylation, and transcriptional activity as well as the ubiquitin dependent degradation of MDM2.

PP2A forms trimeric complexes with several regulatory subunits that alter its substrate specificity. The B56γ regulatory subunit has been shown to bind p53, and binding was found to be enhanced by the ATM dependent phosphorylation of Ser15 in response to DNA damage [38, 202]. B56γ expression also was increased after the exposure of cells to IR or UV. PP2A-B56γ dephosphorylates p53 Thr55, which when phosphorylated in normal cells by TAF1, promotes the MDM2 mediated degradation of p53. These results thus provide a mechanistic connection between the phosphorylation of Ser15 in response to stress and its stabilization. PP2A also has been implicated in the dephosphorylation of Ser37 and possibly Ser15 [203, 204]. Intriguingly, Thompson's group showed that another regulator of PP2A, α4, is an essential inhibitor of apoptosis, the deletion of which in mouse cells led to increased p53 phosphorylation at Ser18 and p53 dependent activation of the transcription of several pro-apoptotic genes [204].

PPM1D (WIP1), a member of the PP2C subfamily of metal dependent protein phosphatases [205], was first identified as a p53 inducible gene that was later shown to dephosphorylate and partially inactivate the p38 and JNK MAP kinases [206, 207]. PPM1D is overexpressed in 15–18 percent of primary human breast cancers and thus functions as an oncogene, while inhibition of PPM1D activity or deletion of the gene protected mice from tumors in several model systems [142, 208]. Recently, PPM1D was shown to dephosphorylate a number of threonine/serine-glutamine (S/TQ) sites that can be phosphorylated by the PIKK kinases ATM, ATR, and DNA-PK, including Ser1981 of ATM, which is associated with the autophosphorylation mediated activation of ATM, and p53 Ser15 [209–213]. *PPM1D* knockout mice had elevated levels of ATM and p53 phosphorylation on Ser15, and cells lacking PPM1D also were slower to return to normal p53 levels after DNA damage, suggesting that PPM1D has both a role in setting the threshold for p53 activation and in restoring normal p53 levels after DNA damage has been repaired. PPM1D also was shown to dephosphorylate Ser395 of MDM2, which stabilizes it, thus promoting p53 degradation [213]. Thus, PPM1D regulates p53 as well as components of the DNA damage signaling pathways that signal to p53.

PP2Cα and PP2Cβ, which like PPM1D are members of the PPM subfamily of phosphatases, also regulate p53 accumulation and activation; however, they do so through reducing the stability of MDM2, which results in an increase in the amount of p53 [205, 214].

CONCLUSIONS

Multiple, distinct signal transduction pathways clearly activate and modulate p53 dependent transcription in response to both genotoxic and non-genotoxic stresses. Although the key protein kinases that are likely to phosphorylate p53 in response to DNA damage have been identified, the identities of kinases that phosphorylate several important sites *in vivo* are still unknown. Furthermore, several sites are phosphorylated by more than one protein kinase. This complexity is augmented further by the facts that signaling pathway activation may be cell type and cell cycle dependent and that many signaling initiation events activate more than one pathway.

A fundamental question that remains only partially answered is what mechanism(s) contribute to the ability of different cells to interpret p53 activation in different ways. The activation of p53 by hypoxia or oncogenes clearly induces different effects than do responses to genotoxic stresses. Therefore, the pattern of posttranslational modifications may determine the selection of the subsets of target genes regulated in response to p53 activation, but a precise understanding of the mechanisms is not yet in hand. It is clear that the p53 protein forms complexes with many other cellular components and with particular nuclear structures (e.g., PML bodies). This characteristic may influence the degree of p53 activation and contribute to the heterogeneity of p53 dependent responses observed within a specific tissue. Analyses of modification patterns in different mouse tissues of knockin mutants should give insights as to the roles played by individual phosphorylation and acetylation sites in eliciting a molecular signaling outcome [49, 215]. While there is still much to learn, substantial progress in understanding the cause and effects of p53 posttranslational modifications is being made.

ACKNOWLEDGEMENTS

We thank W. Kaufmann, UNC, and S. Linn, UC-Berkeley, for useful suggestions. We apologize to those whose publications could not be cited due to space limitations. CWA was supported in part by the Low Dose Research Program of the Office of Biological and Environmental Research, Office of Science, U.S. Department of Energy. EA was supported by the Intramural Research Program of the NIH, National Cancer Institute.

REFERENCES

1. Hollstein M, Moeckel G, Hergenhahn M, Spiegelhalder B, Keil M, Werle-Schneider G, Bartsch H, Brickmann J. On the origins of tumor mutations in cancer genes: insights from the *p53* gene. *Mutat Res* 1998;**405**:145–54.
2. Soussi T, Wiman KG. Shaping genetic alterations in human cancer: the p53 mutation paradigm. *Cancer Cell* 2007;**12**:303–12.
3. Vogelstein B, Lane D, Levine AJ. Surfing the p53 network. *Nature* 2000;**408**:307–10.
4. Lu W-J, Abrams JM. Lessons from p53 in non-mammalian models. *Cell Death Differ* 2006;**13**:909–12.
5. Stiewe T. The p53 family in differentiation and tumorigenesis. *Nat Rev Cancer* 2007;**7**:165–8.
6. Bourdon J-C, Fernandes K, Murray-Zmijewski F, Liu G, Diot A, Xirodimas DP, Saville MK, Lane DP. p53 isoforms can regulate p53 transcriptional activity. *Genes Dev* 2005;**19**:2122–37.
7. Riley T, Sontag E, Chen P, Levine A. Transcriptional control of human p53-regulated genes. *Nat Rev Mol Cell Biol* 2008;**9**:402–12.
8. Meulmeester E, Jochemsen AG. p 53: a guide to apoptosis. *Curr Cancer Drug Targets* 2008;**8**:87–97.
9. Campisi J, d'Adda di Fagagna F. Cellular senescence: when bad things happen to good cells. *Nat Rev Mol Cell Biol* 2007;**8**:729–40.
10. Helton ES, Chen X. p53 modulation of the DNA damage response. *J Cell Biochem* 2007;**100**:883–96.
11. Dawson R, Müller L, Dehner A, Klein C, Kessler H, Buchner J. The N-terminal domain of p53 is natively unfolded. *J Mol Biol* 2003;**332**:1131–41.
12. Candau R, Scolnick DM, Darpino P, Ying CY, Halazonetis TD, Berger SL. Two tandem and independent sub-activation domains in the amino terminus of p53 require the adaptor complex for activity. *Oncogene* 1997;**15**:807–16.
13. Allis CD, Berger SL, Cote J, Dent S, Jenuwien T, Kouzarides T, Pillus L, Reinberg D, Shi Y, Shiekhattar R, Shilatifard A, Workman J, Zhang Y. New nomenclature for chromatin-modifying enzymes. *Cell* 2007;**131**:633–6.
14. Kussie PH, Gorina S, Marechal V, Elenbaas B, Moreau J, Levine AJ, Pavletich NP. Structure of the MDM2 oncoprotein bound to the p53 tumor suppressor transactivation domain. *Science* 1996;**274**:948–53.
15. Zhang Y, Xiong Y. A p53 amino-terminal nuclear export signal inhibited by DNA damage-induced phosphorylation. *Science* 2001;**292**:1910–15.
16. Walker KK, Levine AJ. Identification of a novel p53 functional domain that is necessary for efficient growth suppression. *Proc Natl Acad Sci U S A* 1996;**93**:15,335–15,340.
17. Zilfou JT, Hoffman WH, Sank M, George DL, Murphy M. The corepressor mSin3a interacts with the proline-rich domain of p53 and protects p53 from proteasome-mediated degradation. *Mol Cell Biol* 2001;**21**:3974–85.
18. Berger M, Stahl N, Del Sal G, Haupt Y. Mutations in proline 82 of p53 impair its activation by Pin1 and Chk2 in response to DNA damage. *Mol Cell Biol* 2005;**25**:5380–8.
19. Toledo F, Krummel KA, Lee CJ, Liu C-W, Rodewald L-W, Tang M, Wahl GM. A mouse p53 mutant lacking the proline-rich domain rescues Mdm4 deficiency and provides insight into the Mdm2-Mdm4-p53 regulatory network. *Cancer Cell* 2006;**9**:273–85.
20. Di Lello P, Jenkins LMM, Jones TN, Nguyen BD, Hara T, Yamaguchi H, Dikeakos JD, Appella E, Legault P, Omichinski JG. Structure of the Tfb1/p53 complex: insights into the interaction between the p62/Tfb1 subunit of TFIIH and the activation domain of p53. *Mol Cell* 2006;**22**:731–40.
21. Kaustov L, Yi G-S, Ayed A, Bochkareva E, Bochkarev A, Arrowsmith CH. p53 transcriptional activation domain: A molecular chameleon? *Cell Cycle* 2006;**5**:489–94.
22. Joerger AC, Fersht AR. Structural biology of the tumor suppressor p53. *Annu Rev Biochem* 2008;**77**:557–82.
23. Joerger AC, Fersht AR. Structure-function-rescue: the diverse nature of common p53 cancer mutants. *Oncogene* 2007;**26**:2226–42.
24. Petitjean A, Mathe E, Kato S, Ishioka C, Tavtigian SV, Hainaut P, Olivier M. Impact of mutant p53 functional properties on *TP53* mutation patterns and tumor phenotype: lessons from recent developments in the IARC TP53 database. *Hum Mutat* 2007;**28**:622–9.
25. El-Deiry WS, Kern SE, Pietenpol JA, Kinzler KW, Vogelstein B. Definition of a consensus binding site for p53. *Nat Genet* 1992;**1**:45–9.
26. Veprintsev DB, Fersht AR. Algorithm for prediction of tumour suppressor p53 affinity for binding sites in DNA. *Nucleic Acids Res* 2008;**36**:1589–98.
27. Kim E, Deppert W. The versatile interactions of p53 with DNA: when flexibility serves specificity. *Cell Death Differ* 2006;**13**:885–9.
28. Riley 3rd KJ-L, Maher LJ. p 53 RNA interactions: new clues in an old mystery. *RNA* 2007;**13**:1825–33.
29. Shetty S, Shetty P, Idell S, Velusamy T, Bhandary YP, Shetty RS. Regulation of plasminogen activator inhibitor-1 expression by tumor suppressor protein p53. *J Biol Chem* 2008;**283**:19,570–19,580,.
30. Hupp TR, Lane DP. Allosteric activation of latent p53 tetramers. *Curr Biol* 1994;**4**:865–75.
31. Liu Y, Lagowski JP, Vanderbeek GE, Kulesz-Martin MF. Facilitated search for specific genomic targets by p53 C-terminal basic DNA binding domain. *Cancer Biol Ther* 2004;**3**:1102–8.
32. McKinney K, Mattia M, Gottifredi V, Prives C. p 53 linear diffusion along DNA requires its C terminus. *Mol Cell* 2004;**16**:413–24.
33. Tafvizi A, Huang F, Leith JS, Fersht AR, Mirny LA, van Oijen AM. Tumor suppressor p53 slides on DNA with low friction and high stability. *Biophys J* 2008;**95**:L01–3.
34. Tidow H, Melero R, Mylonas E, Freund SMV, Grossmann JG, Carazo JM, Svergun DI, Valle M, Fersht AR. Quaternary structures of tumor suppressor p53 and a specific p53 DNA complex. *Proc Natl Acad Sci U S A* 2007;**104**:12,324–12,329.
35. Appella E, Anderson CW. Post-translational modifications and activation of p53 by genotoxic stresses. *Eur J Biochem* 2001;**268**:2764–72.
36. Bode AM, Dong Z. Post-translational modification of p53 in tumorigenesis. *Nat Rev Cancer* 2004;**4**:793–805.
37. Kruse JP, Gu W. SnapShot: p53 posttranslational modifications. *Cell* 2008;**133**:930–e931.
38. Shouse GP, Cai X, Liu X. Serine 15 phosphorylation of p53 directs its interaction with B56γ and the tumor suppressor activity of B56γ-specific protein phosphatase 2A. *Mol Cell Biol* 2008;**28**:448–56.
39. Braithwaite AW, Del Sal G, Lu X. Some p53-binding proteins that can function as arbiters of life and death. *Cell Death Differ* 2006;**13**:984–93.
40. Bech-Otschir D, Kraft R, Huang X, Henklein P, Kapelari B, Pollmann C, Dubiel W. COP9 signalosome-specific phosphorylation targets p53 to degradation by the ubiquitin system. *EMBO J* 2001;**20**:1630–9.
41. Yang WH, Kim JE, Nam HW, Ju JW, Kim HS, Kim YS, Cho JW. Modification of p53 with O-linked N-acetylglucosamine regulates p53 activity and stability. *Nat Cell Biol* 2006;**8**:1074–83.
42. Liu Q, Kaneko S, Yang L, Feldman RI, Nicosia SV, Chen J, Cheng JQ. Aurora-A abrogation of p53 DNA binding and transactivation activity by phosphorylation of serine 215. *J Biol Chem* 2004;**279**:52,175–82.
43. Tang Y, Luo J, Zhang W, Gu W. Tip60-dependent acetylation of p53 modulates the decision between cell-cycle arrest and apoptosis. *Mol Cell* 2006;**24**:827–39.

44. Sykes SM, Mellert HS, Holbert MA, Li K, Marmorstein R, Lane WS, McMahon SB. Acetylation of the p53 DNA-binding domain regulates apoptosis induction. *Mol Cell* 2006;**24**:841–51.
45. Tang Y, Zhao W, Chen Y, Zhao Y, Gu W. Acetylation is indispensable for p53 activation. *Cell* 2008;**133**:612–26.
46. Watson IR, Irwin MS. Ubiquitin and ubiquitin-like modifications of the p53 family. *Neoplasia* 2006;**8**:655–66.
47. Brooks CL, Gu W. p 53 Ubiquitination: Mdm2 and beyond. *Mol Cell* 2006;**21**:307–15.
48. Waterman MJF, Stavridi ES, Waterman JLF, Halazonetis TD. ATM-dependent activation of p53 involves dephosphorylation and association with 14-3-3 proteins. *Nat Genet* 1998;**19**:175–8.
49. Olsson A, Manzl C, Strasser A, Villunger A. How important are post-translational modifications in p53 for selectivity in target-gene transcription and tumour suppression? *Cell Death Differ* 2007;**14**:1561–75.
50. Toledo F, Lee CJ, Krummel KA, Rodewald L-W, Liu C-W, Wahl GM. Mouse mutants reveal that putative protein interaction sites in the p53 proline-rich domain are dispensable for tumor suppression. *Mol Cell Biol* 2007;**27**:1425–32.
51. Saito Jr S, Yamaguchi H, Higashimoto Y, Chao C, Xu Y, Fornace AJ, Appella E, Anderson CW. Phosphorylation site interdependence of human p53 post-translational modifications in response to stress. *J Biol Chem* 2003;**278**:37,536–44.
52. Dumaz N, Milne DM, Meek DW. Protein kinase CK1 is a p53-threonine 18 kinase which requires prior phosphorylation of serine 15. *FEBS Lett* 1999;**463**:312–16.
53. Kurash JK, Lei H, Shen Q, Marston WL, Granda BW, Fan H, Wall D, Li E, Gaudet F. Methylation of p53 by Set7/9 mediates p53 acetylation and activity in vivo. *Mol Cell* 2008;**29**:392–400.
54. Chao C, Herr D, Chun J, Xu Y. Ser18 and 23 phosphorylation is required for p53-dependent apoptosis and tumor suppression. *EMBO J* 2006;**25**:2615–22.
55. Le Cam L, Linares LK, Paul C, Julien E, Lacroix M, Hatchi E, Triboulet R, Bossis G, Shmueli A, Rodriguez MS, Coux O, Sardet C. E4F1 is an atypical ubiquitin ligase that modulates p53 effector functions independently of degradation. *Cell* 2006;**127**:775–88.
56. Talis AL, Huibregtse JM, Howley PM. The role of E6AP in the regulation of p53 protein levels in human papillomavirus (HPV)-positive and HPV-negative cells. *J Biol Chem* 1998;**273**:6439–45.
57. Querido E, Blanchette P, Yan Q, Kamura T, Morrison M, Boivin D, Kaelin WG, Conaway RC, Conaway JW, Branton PE. Degradation of p53 by adenovirus E4orf6 and E1B55K proteins occurs via a novel mechanism involving a Cullin-containing complex. *Genes Dev* 2001;**15**:3104–17.
58. Yang Y, Li C-CH, Weissman AM. Regulating the p53 system through ubiquitination. *Oncogene* 2004;**23**:2096–106.
59. Wu X, Bayle JH, Olson D, Levine AJ. The p53-mdm-2 autoregulatory feedback loop. *Genes & Development* 1993;**7**:1126–32.
60. Shieh S-Y, Ikeda M, Taya Y, Prives C. DNA damage-induced phosphorylation of p53 alleviates inhibition by MDM2. *Cell* 1997;**91**:325–34.
61. Chehab NH, Malikzay A, Stavridi ES, Halazonetis TD. Phosphorylation of Ser-20 mediates stabilization of human p53 in response to DNA damage. *Proc Natl Acad Sci U S A* 1999;**96**:13,777–82.
62. Craig AL, Burch L, Vojtesek B, Mikutowska J, Thompson A, Hupp TR. Novel phosphorylation sites of human tumour suppressor protein p53 at Ser^{20} and Thr^{18} that disrupt the binding of mdm2 (mouse double minute 2) protein are modified in human cancers. *Biochem J* 1999;**342**:133–41.
63. Sakaguchi K, Saito S, Higashimoto Y, Roy S, Anderson CW, Appella E. Damage-mediated phosphorylation of human p53 threonine 18 through a cascade mediated by a casein 1-like kinase. Effect on Mdm2 binding. *J Biol Chem* 2000;**275**:9278–83.
64. Unger T, Juven-Gershon T, Moallem E, Berger M, Vogt Sionov R, Lozano G, Oren M, Haupt Y. Critical role for Ser20 of human p53 in the negative regulation of p53 by Mdm2. *EMBO J* 1999;**18**:1805–14.
65. Lee HJ, Srinivasan D, Coomber D, Lane DP, Verma CS. Modulation of the p53-MDM2 interaction by phosphorylation of Thr18: a computational study. *Cell Cycle* 2007;**6**:2604–11.
66. Chao C, Saito S, Anderson CW, Appella E, Xu Y. Phosphorylation of murine p53 at Ser-18 regulates the p53 responses to DNA damage. *Proc Natl Acad Sci U S A* 2000;**97**:11,936–41.
67. Lambert PF, Kashanchi F, Radonovich MF, Shiekhattar R, Brady JN. Phosphorylation of p53 serine 15 increases interaction with CBP. *J Biol Chem* 1998;**273**:33,048–53.
68. Polley S, Guha S, Roy NS, Kar S, Sakaguchi K, Chuman Y, Swaminathan V, Kundu T, Roy S. Differential recognition of phosphorylated transactivation domains of p53 by different p300 domains. *J Mol Biol* 2008;**376**:8–12.
69. Kulikov R, Winter M, Blattner C. Binding of p53 to the central domain of Mdm2 is regulated by phosphorylation. *J Biol Chem* 2006;**281**:28,575–83.
70. Yu GW, Rudiger S, Veprintsev D, Freund S, Fernandez-Fernandez MR, Fersht AR. The central region of HDM2 provides a second binding site for p53. *Proc Natl Acad Sci U S A* 2006;**103**:1227–32.
71. Meek DW, Knippschild U. Posttranslational modification of MDM2. *Mol Cancer Res* 2003;**1**:1017–26.
72. Boehme KA, Kulikov R, Blattner C. p 53 stabilization in response to DNA damage requires Akt/PKB and DNA-PK. *Proc Natl Acad Sci U S A* 2008;**105**:7785–90.
73. Itahana K, Mao H, Jin A, Itahana Y, Clegg HV, Lindström MS, Bhat KP, Godfrey VL, Evan GI, Zhang Y. Targeted inactivation of Mdm2 RING Finger E3 ubiquitin ligase activity in the mouse reveals mechanistic insights into p53 regulation. *Cancer Cell* 2007;**12**:355–66.
74. Toledo F, Wahl GM. MDM2 and MDM4: p53 regulators as targets in anticancer therapy. *Int J Biochem Cell Biol* 2007;**39**:1476–82.
75. Parant J, Chavez-Reyes A, Little NA, Yan W, Reinke V, Jochemsen AG, Lozano G. Rescue of embryonic lethality in *Mdm4*-null mice by loss of *Trp53* suggests a nonoverlapping pathway with MDM2 to regulate p53. *Nat Genet* 2001;**29**:92–5.
76. Linares LK, Kiernan R, Triboulet R, Chable-Bessia C, Latreille D, Cuvier O, Lacroix M, Le Cam L, Coux O, Benkirane M. Intrinsic ubiquitination activity of PCAF controls the stability of the oncoprotein Hdm2. *Nat Cell Biol* 2007;**9**:331–8.
77. Tang J, Qu L-K, Zhang J, Wang W, Michaelson JS, Degenhardt YY, El-Deiry WS, Yang X. Critical role for Daxx in regulating Mdm2. *Nat Cell Biol* 2006;**8**:855–62.
78. Dornan D, Shimizu H, Mah A, Dudhela T, Eby M, O'Rourke K, Seshagiri S, Dixit VM. ATM engages autodegradation of the E3 ubiquitin ligase COP1 after DNA damage. *Science* 2006;**313**:1122–6.
79. Duan S, Yao Z, Hou D, Wu Z, Zhu W-G, Wu M. Phosphorylation of Pirh2 by calmodulin-dependent kinase II impairs its ability to ubiquitinate p53. *EMBO J* 2007;**26**:3062–74.
80. Chen D, Brooks CL, Gu W. ARF-BP1 as a potential therapeutic target. *Br J Cancer* 2006;**94**:1555–8.
81. Zhao X, Heng JI-T, Guardavaccaro D, Jiang R, Pagano M, Guillemot F, Iavarone A, Lasorella A. The HECT-domain ubiquitin ligase Huwe1 controls neural differentiation and proliferation by destabilizing the N-Myc oncoprotein. *Nat Cell Biol* 2008;**10**:643–53.
82. Thompson T, Tovar C, Yang H, Carvajal D, Vu BT, Xu Q, Wahl GM, Heimbrook DC, Vassilev LT. Phosphorylation of p53 on key serines is

dispensable for transcriptional activation and apoptosis. *J Biol Chem* 2004;**279**:53,015–22.
83. Kaeser MD, Iggo RD. Chromatin immunoprecipitation analysis fails to support the latency model for regulation of p53 DNA binding activity *in vivo*. *Proc Natl Acad Sci U S A* 2002;**99**:95–100.
84. Iyer NG, Chin S-F, Ozdag H, Daigo Y, Hu D-E, Cariati M, Brindle K, Aparicio S, Caldas C. p300 regulates p53-dependent apoptosis after DNA damage in colorectal cancer cells by modulation of PUMA/p21 levels. *Proc Natl Acad Sci U S A* 2004;**101**:7386–91.
85. Sakaguchi K, Herrera JE, Saito S, Miki T, Bustin M, Vassilev A, Anderson CW, Appella E. DNA damage activates p53 through a phosphorylation–acetylation cascade. *Genes Dev* 1998;**12**:2831–41.
86. Chao C, Wu Z, Mazur SJ, Borges H, Rossi M, Lin T, Wang JYJ, Anderson CW, Appella E, Xu Y. Acetylation of mouse p53 at lysine 317 negatively regulates p53 apoptotic activities after DNA damage. *Mol Cell Biol* 2006;**26**:6859–69.
87. Cordenonsi M, Montagner M, Adorno M, Zacchigna L, Martello G, Mamidi A, Soligo S, Dupont S, Piccolo S. Integration of TGF-β and Ras/MAPK signaling through p53 phosphorylation. *Science* 2007;**315**:840–3.
88. Li AG, Piluso LG, Cai X, Wei G, Sellers WR, Liu X. Mechanistic insights into maintenance of high p53 acetylation by PTEN. *Mol Cell* 2006;**23**:575–87.
89. Kaeser MD, Iggo RD. Promoter-specific p53-dependent histone acetylation following DNA damage. *Oncogene* 2004;**23**:4007–13.
90. Wang P, Yu J, Zhang L. The nuclear function of p53 is required for PUMA-mediated apoptosis induced by DNA damage. *Proc Natl Acad Sci U S A* 2007;**104**:4054–9.
91. Barlev NA, Liu L, Chehab NH, Mansfield K, Harris KG, Halazonetis TD, Berger SL. Acetylation of p53 activates transcription through recruitment of coactivators/histone acetyltransferases. *Mol Cell* 2001;**8**:1243–54.
92. Luo J, Nikolaev AY, Imai S, Chen D, Su F, Shiloh A, Guarente L, Gu W. Negative control of p53 by Sir2alpha promotes cell survival under stress. *Cell* 2001;**107**:137–48.
93. Vaziri H, Dessain SK, Eaton EN, Imai SI, Frye RA, Pandita TK, Guarente L, Weinberg RA. *hSIR2*^{SIRT1} functions as an NAD-dependent p53 deacetylase. *Cell* 2001;**107**:149–59.
94. Langley E, Pearson M, Faretta M, Bauer U-M, Frye RA, Minucci S, Pelicci PG, Kouzarides T. Human SIR2 deacetylates p53 and antagonizes PML/p53-induced cellular senescence. *EMBO J* 2002; **21**:2383–96.
95. Murphy M, Ahn J, Walker KK, Hoffman WH, Evans RM, Levine AJ, George DL. Transcriptional repression by wild-type p53 utilizes histone deacetylases, mediated by interaction with mSin3a. *Genes Dev* 1999;**13**:2490–501.
96. Basile V, Mantovani R, Imbriano C. DNA damage promotes histone deacetylase 4 nuclear localization and repression of G_2/M promoters, via p53 C-terminal lysines. *J Biol Chem* 2006;**281**:2347–57.
97. Ivanov 3rd GS, Ivanova T, Kurash J, Ivanov A, Chuikov S, Gizatullin F, Herrera-Medina EM, Rauscher F, Reinberg D, Barlev NA. Methylation–acetylation interplay activates p53 in response to DNA damage. *Mol Cell Biol* 2007;**27**:6756–69.
98. Huang J, Perez-Burgos L, Placek BJ, Sengupta R, Richter M, Dorsey JA, Kubicek S, Opravil S, Jenuwein T, Berger SL. Repression of p53 activity by Smyd2-mediated methylation. *Nature* 2006;**444**:629–32.
99. Huang J, Sengupta R, Espejo AB, Lee MG, Dorsey JA, Richter M, Opravil S, Shiekhattar R, Bedford MT, Jenuwein T, Berger SL. p53 is regulated by the lysine demethylase LSD1. *Nature* 2007; **449**:105–8.
100. Derbyshire DJ, Basu BP, Serpell LC, Joo WS, Date T, Iwabuchi K, Doherty AJ. Crystal structure of human 53BP1 BRCT domains bound to p53 tumour suppressor. *EMBO J* 2002;**21**:3863–72.
101. Shi X, Kachirskaia I, Yamaguchi H, West LE, Wen H, Wang EW, Dutta S, Appella E, Gozani O. Modulation of p53 function by SET8-mediated methylation at lysine 382. *Mol Cell* 2007;**27**:636–46.
102. An W, Kim J, Roeder RG. Ordered cooperative functions of PRMT1, p300, and CARM1 in transcriptional activation by p53. *Cell* 2004;**117**:735–48.
103. Li P, Yao H, Zhang Z, Li M, Luo Y, Thompson PR, Gilmour DS, Wang Y. Regulation of p53 target gene expression by peptidylarginine deiminase 4. *Mol Cell Biol* 2008;**28**:4745–58.
104. Bulavin Jr DV, Saito S, Hollander MC, Sakaguchi K, Anderson CW, Appella E, Fornace AJ. Phosphorylation of human p53 by p38 kinase coordinates N-terminal phosphorylation and apoptosis in response to UV radiation. *EMBO J* 1999;**18**:6845–54.
105. Calzado MA, Renner F, Roscic A, Schmitz ML. HIPK2: a versatile switchboard regulating the transcription machinery and cell death. *Cell Cycle* 2007;**6**:139–43.
106. Taira N, Nihira K, Yamaguchi T, Miki Y, Yoshida K. DYRK2 is targeted to the nucleus and controls p53 via Ser46 phosphorylation in the apoptotic response to DNA damage. *Mol Cell* 2007;**25**:725–38.
107. Yoshida K, Liu H, Miki Y. Protein Kinase C δ regulates Ser^{46} phosphorylation of p53 tumor suppressor in the apoptotic response to DNA damage. *J Biol Chem* 2006;**281**:5734–40.
108. Lee J-H, Kim H-S, Lee S-J, Kim K-T. Stabilization and activation of p53 induced by Cdk5 contributes to neuronal cell death. *J Cell Sci* 2007;**120**:2259–71.
109. Oda K, Arakawa H, Tanaka T, Matsuda K, Tanikawa C, Mori T, Nishimori H, Tamai K, Tokino T, Nakamura Y, Taya Y. *p53AIP1*, a potential mediator of p53-dependent apoptosis, and its regulation by Ser-46-phosphorylated p53. *Cell* 2000;**102**:849–62.
110. Winter M, Sombroek D, Dauth I, Moehlenbrink J, Scheuermann K, Crone J, Hofmann TG. Control of HIPK2 stability by ubiquitin ligase Siah-1 and checkpoint kinases ATM and ATR. *Nat Cell Biol* 2008;**10**:812–24.
111. Espinosa JM. Mechanisms of regulatory diversity within the p53 transcriptional network. *Oncogene* 2008;**27**:4013–23.
112. Das S, Boswell SA, Aaronson SA, Lee SW. p53 promoter selection: Choosing between life and death. *Cell Cycle* 2007;**7**.
113. Fachin AL, Mello SS, Sandrin-Garcia P, Junta CM, Donadi EA, Passos GAS, Sakamoto-Hojo ET. Gene expression profiles in human lymphocytes irradiated *in vitro* with low doses of gamma rays. *Radiat Res* 2007;**168**:650–65.
114. Amundson Jr SA, Lee RA, Koch-Paiz CA, Bittner ML, Meltzer P, Trent JM, Fornace AJ. Differential responses of stress genes to low dose-rate gamma irradiation. *Mol Cancer Res* 2003;**1**:445–52.
115. Shiloh Y. The ATM-mediated DNA-damage response: taking shape. *Trends Biochem Sci* 2006;**31**:402–10.
116. Saito S, Goodarzi AA, Higashimoto Y, Noda Y, Lees-Miller SP, Appella E, Anderson CW. ATM mediates phosphorylation at multiple p53 sites, including Ser^{46}, in response to ionizing radiation. *J Biol Chem* 2002;**277**:12,491–4.
117. Bakkenist CJ, Kastan MB. DNA damage activates ATM through intermolecular autophosphorylation and dimer dissociation. *Nature* 2003;**421**:499–506.
118. Rupnik A, Grenon M, Lowndes N. The MRN complex. *Curr Biol* 2008;**18**:R455–7.
119. Lee J-H, Paull TT. Activation and regulation of ATM kinase activity in response to DNA double-strand breaks. *Oncogene* 2007;**26**:7741–8.

120. Ward I, Chen J. Early events in the DNA damage response. *Curr Top Dev Biol* 2004;**63**:1–35.
121. Matsuoka 3rd S, Ballif BA, Smogorzewska A, McDonald ER, Hurov KE, Luo J, Bakalarski CE, Zhao Z, Solimini N, Lerenthal Y, Shiloh Y, Gygi SP, Elledge SJ. ATM and ATR substrate analysis reveals extensive protein networks responsive to DNA damage. *Science* 2007;**316**:1160–6.
122. Pellegrini M, Celeste A, Difilippantonio S, Guo R, Wang W, Feigenbaum L, Nussenzweig A. Autophosphorylation at serine 1987 is dispensable for murine Atm activation in vivo. *Nature* 2006;**443**:222–5.
123. Kozlov SV, Graham ME, Peng C, Chen P, Robinson PJ, Lavin MF. Involvement of novel autophosphorylation sites in ATM activation. *EMBO J* 2006;**25**:3504–14.
124. Sun Y, Xu Y, Roy K, Price BD. DNA damage-induced acetylation of lysine 3016 of ATM activates ATM kinase activity. *Mol Cell Biol* 2007;**27**:8502–9.
125. Kastan Jr MB, Zhan Q, El-Deiry WS, Carrier F, Jacks T, Walsh WV, Plunkett BS, Vogelstein B, Fornace AJ. A mammalian cell cycle checkpoint pathway utilizing p53 and *GADD45* is defective in ataxia-telangiectasia. *Cell* 1992;**71**:587–97.
126. Cimprich KA, Cortez D. ATR: an essential regulator of genome integrity. *Nat Rev Mol Cell Biol* 2008;**9**:616–27.
127. Zhang Y, Ma W-Y, Kaji A, Bode AM, Dong Z. Requirement of ATM in UVA-induced signaling and apoptosis. *J Biol Chem* 2002;**277**:3124–31.
128. Ljungman M. The transcription stress response. *Cell Cycle* 2007;**6**:2252–7.
129. Stiff T, Walker SA, Cerosaletti K, Goodarzi AA, Petermann E, Concannon P, O'Driscoll M, Jeggo PA. ATR-dependent phosphorylation and activation of ATM in response to UV treatment or replication fork stalling. *EMBO J* 2006;**25**:5775–82.
130. Gatei M, Sloper K, Sörensen C, Syljuäsen R, Falck J, Hobson K, Savage K, Lukas J, Zhou B-B, Bartek J, Khanna KK. Ataxia-telangiectasia-mutated (ATM) and NBS1-dependent phosphorylation of Chk1 on Ser-317 in response to ionizing radiation. *J Biol Chem* 2003;**278**:14,806–11.
131. Hurley PJ, Bunz F. ATM and ATR: components of an integrated circuit. *Cell Cycle* 2007;**6**:414–17.
132. Kumagai A, Lee J, Yoo HY, Dunphy WG. TopBP1 activates the ATR-ATRIP complex. *Cell* 2006;**124**:943–55.
133. Mordes DA, Glick GG, Zhao R, Cortez D. TopBP1 activates ATR through ATRIP and a PIKK regulatory domain. *Genes Dev* 2008;**22**:1478–89.
134. Ünsal-Kaçmaz K, Chastain PD, Qu P-P, Minoo P, Cordeiro-Stone M, Sancar A, Kaufmann WK. The human Tim/Tipin complex coordinates an Intra-S checkpoint response to UV that slows replication fork displacement. *Mol Cell Biol* 2007;**27**:3131–42.
135. Webley K, Bond JA, Jones CJ, Blaydes JP, Craig A, Hupp T, Wynford-Thomas D. Posttranslational modifications of p53 in replicative senescence overlapping but distinct from those induced by DNA damage. *Mol Cell Biol* 2000;**20**:2803–8.
136. d'Adda di Fagagna F. Living on a break: cellular senescence as a DNA-damage response. *Nat Rev Cancer* 2008;**8**:512–22.
137. Sherr CJ. The *INK4a/ARF* network in tumour suppression. *Nat Rev Mol Cell Biol* 2001;**2**:731–7.
138. Sherr CJ. Divorcing ARF and p53: an unsettled case. *Nat Rev Cancer* 2006;**6**:663–73.
139. de Stanchina E, McCurrach ME, Zindy F, Shieh SY, Ferbeyre G, Samuelson AV, Prives C, Roussel MF, Sherr CJ, Lowe SW. E1A signaling to p53 involves the p19(ARF) tumor suppressor. *Genes Dev* 1998;**12**:2434–42.
140. Vafa O, Wade M, Kern S, Beeche M, Pandita TK, Hampton GM, Wahl GM. c-myc can induce DNA damage, increase reactive oxygen species, and mitigate p53 function. A mechanism for oncogene-induced genetic instability. *Mol Cell* 2002;**9**:1031–44.
141. Ferbeyre G, de Stanchina E, Lin AW, Querido E, McCurrach ME, Hannon GJ, Lowe SW. Oncogenic *ras* and p53 cooperate to induce cellular senescence. *Mol Cell Biol* 2002;**22**:3497–508.
142. Bulavin Jr DV, Demidov ON, Saito S, Kauraniemi P, Phillips C, Amundson SA, Ambrosino C, Sauter G, Nebreda AR, Anderson CW, Kallioniemi A, Fornace AJ, Appella E. Amplification of *PPM1D* in human tumors abrogates p53 tumor-suppressor activity. *Nat Genet* 2002;**31**:210–15.
143. Mallette FA, Ferbeyre G. The DNA damage signaling pathway connects oncogenic stress to cellular senescence. *Cell Cycle* 2007;**6**:1831–6.
144. Di Micco R, Fumagalli M, Cicalese A, Piccinin S, Gasparini P, Luise C, Schurra C, Garre M, Nuciforo PG, Bensimon A, Maestro R, Pelicci PG, d'Adda di Fagagna F. Oncogene-induced senescence is a DNA damage response triggered by DNA hyper-replication. *Nature* 2006;**444**:638–42.
145. Ljungman M, O'Hagan HM, Paulsen MT. Induction of ser15 and lys382 modifications of p53 by blockage of transcription elongation. *Oncogene* 2001;**20**:5964–71.
146. Levine AJ, Feng Z, Mak TW, You H, Jin S. Coordination and communication between the p53 and IGF-1-AKT-TOR signal transduction pathways. *Genes Dev* 2006;**20**:267–75.
147. Zhang F, Hamanaka RB, Bobrovnikova-Marjon E, Gordan JD, Dai M-S, Lu H, Simon MC, Diehl JA. Ribosomal stress couples the unfolded protein response to p53-dependent cell cycle arrest. *J Biol Chem* 2006;**281**:30,036–45.
148. Zhao L, Ackerman SL. Endoplasmic reticulum stress in health and disease. *Curr Opin Cell Biol* 2006;**18**:444–52.
149. Wu J, Kaufman RJ. From acute ER stress to physiological roles of the Unfolded Protein Response. *Cell Death Differ* 2006;**13**:374–84.
150. Qu L, Huang S, Baltzis D, Rivas-Estilla A-M, Pluquet O, Hatzoglou M, Koumenis C, Taya Y, Yoshimura A, Koromilas AE. Endoplasmic reticulum stress induces p53 cytoplasmic localization and prevents p53-dependent apoptosis by a pathway involving glycogen synthase kinase-3β. *Genes Dev* 2004;**18**:261–77.
151. Katayama H, Sasai K, Kawai H, Yuan Z-M, Bondaruk J, Suzuki F, Fujii S, Arlinghaus RB, Czerniak BA, Sen S. Phosphorylation by aurora kinase A induces Mdm2-mediated destabilization and inhibition of p53. *Nat Genet* 2004;**36**:55–62.
152. Yamasaki S, Yagishita N, Sasaki T, Nakazawa M, Kato Y, Yamadera T, Bae E, Toriyama S, Ikeda R, Zhang L, Fujitani K, Yoo E, Tsuchimochi K, Ohta T, Araya N, Fujita H, Aratani S, Eguchi K, Komiya S, Maruyama I, Higashi N, Sato M, Senoo H, Ochi T, Yokoyama S, Amano T, Kim J, Gay S, Fukamizu A, Nishioka K, Tanaka K, Nakajima T. Cytoplasmic destruction of p53 by the endoplasmic reticulum-resident ubiquitin ligase "Synoviolin.". *EMBO J* 2006;**26**:113–22.
153. Hammond EM, Kaufmann MR, Giaccia AJ. Oxygen sensing and the DNA-damage response. *Curr Opin Cell Biol* 2007;**19**:680–4.
154. Hammond EM, Giaccia AJ. The role of p53 in hypoxia-induced apoptosis. *Biochem Biophys Res Commun* 2005;**331**:718–25.
155. Hammond EM, Dorie MJ, Giaccia AJ. ATR/ATM targets are phosphorylated by ATR in response to hypoxia and ATM in response to reoxygenation. *J Biol Chem* 2003;**278**:12,207–13.

156. Huang LE, Bindra RS, Glazer PM, Harris AL. Hypoxia-induced genetic instability: a calculated mechanism underlying tumor progression. *J Mol Med* 2007;**85**:139–48.
157. van den Beucken T, Koritzinsky M, Wouters BG. Translational control of gene expression during hypoxia. *Cancer Biol Ther* 2006;**5**:749–55.
158. Ke Q, Costa M. Hypoxia-inducible factor-1 (HIF-1). *Mol Pharmacol* 2006;**70**:1469–80.
159. Maxwell PH, Wiesener MS, Chang G-W, Clifford SC, Vaux EC, Cockman ME, Wykoff CC, Pugh CW, Maher ER, Ratcliffe PJ. The tumour suppressor protein VHL targets hypoxia-inducible factors for oxygen-dependent proteolysis. *Nature* 1999;**399**:271–5.
160. Alarcón R, Koumenis C, Geyer RK, Maki CG, Giaccia AJ. Hypoxia induces p53 accumulation through MDM2 down-regulation and inhibition of E6-mediated degradation. *Cancer Res* 1999;**59**:6046–51.
161. Hammond EM, Denko NC, Dorie MJ, Abraham RT, Giaccia AJ. Hypoxia links ATR and p53 through replication arrest. *Mol Cell Biol* 2002;**22**:1834–43.
162. Koumenis C, Alarcon R, Hammond E, Sutphin P, Hoffman W, Murphy M, Derr J, Taya Y, Lowe SW, Kastan M, Giaccia A. Regulation of p53 by hypoxia: dissociation of transcriptional repression and apoptosis from p53-dependent transactivation. *Mol Cell Biol* 2001;**21**:1297–310.
163. Hansson LO, Friedler A, Freund S, Rüdiger S, Fersht AR. Two sequence motifs from HIF-1alpha bind to the DNA-binding site of p53. *Proc Natl Acad Sci U S A* 2002;**99**:10,305–10,309,.
164. An WG, Kanekal M, Simon MC, Maltepe E, Blagosklonny MV, Neckers LM. Stabilization of wild-type p53 by hypoxia-inducible factor 1alpha. *Nature* 1998;**392**:405–8.
165. Hammond EM, Mandell DJ, Salim A, Krieg AJ, Johnson TM, Shirazi HA, Attardi LD, Giaccia AJ. Genome-wide analysis of p53 under hypoxic conditions. *Mol Cell Biol* 2006;**26**:3492–504.
166. Hay N, Sonenberg N. Upstream and downstream of mTOR. *Genes Dev* 2004;**18**:1926–45.
167. Imamura K, Ogura T, Kishimoto A, Kaminishi M, Esumi H. Cell cycle regulation via p53 phosphorylation by a 5'-AMP activated protein kinase activator, 5-aminoimidazole- 4-carboxamide-1-beta-D-ribofuranoside, in a human hepatocellular carcinoma cell line. *Biochem Biophys Res Commun* 2001;**287**:562–7.
168. Jones RG, Plas DR, Kubek S, Buzzai M, Mu J, Xu Y, Birnbaum MJ, Thompson CB. AMP-activated protein kinase induces a p53-dependent metabolic checkpoint. *Mol Cell* 2005;**18**:283–93.
169. Okoshi R, Ozaki T, Yamamoto H, Ando K, Koida N, Ono S, Koda T, Kamijo T, Nakagawara A, Kizaki H. Activation of AMP-activated protein kinase induces p53-dependent apoptotic cell death in response to energetic stress. *J Biol Chem* 2008;**283**:3979–87.
170. Feng Z, Zhang H, Levine AJ, Jin S. The coordinate regulation of the p53 and mTOR pathways in cells. *Proc Natl Acad Sci U S A* 2005;**102**:8204–9.
171. Karuman P, Gozani O, Odze RD, Zhou XC, Zhu H, Shaw R, Brien TP, Bozzuto CD, Ooi D, Cantley LC, Yuan J. The Peutz-Jegher gene product LKB1 is a mediator of p53-dependent cell death. *Mol Cell* 2001;**7**:1307–19.
172. Kawauchi K, Araki K, Tobiume K, Tanaka N. p 53 regulates glucose metabolism through an IKK-NFκB pathway and inhibits cell transformation. *Nat Cell Biol* 2008;**10**:611–18.
173. Matoba S, Kang J-G, Patino WD, Wragg A, Boehm M, Gavrilova O, Hurley PJ, Bunz F, Hwang PM. p53 regulates mitochondrial respiration. *Science* 2006;**312**:1650–3.
174. Linke SP, Clarkin KC, Di Leonardo A, Tsou A, Wahl GM. A reversible, p53-dependent G_0/G_1 cell cycle arrest induced by ribonucleotide depletion in the absence of detectable DNA damage. *Genes Dev* 1996;**10**:934–47.
175. Seidita G, Polizzi D, Costanzo G, Costa S, Di Leonardo A. Differential gene expression in p53-mediated G_1 arrest of human fibroblasts after gamma-irradiation or *N*-phosphoacetyl-L-aspartate treatment. *Carcinogenesis* 2000;**21**:2203–10.
176. Hastak K, Paul RK, Agarwal MK, Thakur VS, Amin ARMR, Agrawal S, Sramkoski RM, Jacobberger JW, Jackson MW, Stark GR, Agarwal ML. DNA synthesis from unbalanced nucleotide pools causes limited DNA damage that triggers ATR-CHK1-dependent p53 activation. *Proc Natl Acad Sci U S A* 2008;**105**:6314–19.
177. Bronder JL, Moran RG. A defect in the p53 response pathway induced by *de novo* purine synthesis inhibition. *J Biol Chem* 2003;**278**:48,861–48,871.
177a. Sun X-X, Dai M-S, Lu H. Mycophenolic acid activation of p53 requires ribosomal proteins L5 and L11. *J Biol Chem* 2008;**283**:12,387–92.
178. Boisvert F-M, van Koningsbruggen S, Navascués J, Lamond AI. The multifunctional nucleolus. *Nat Rev Mol Cell Biol* 2007;**8**: 574–85.
179. Rubbi CP, Milner J. Disruption of the nucleolus mediates stabilization of p53 in response to DNA damage and other stresses. *EMBO J* 2003;**22**:6068–77.
180. Olson MO. Sensing cellular stress: another new function for the nucleolus? *Sci STKE* 2004:pe10.
181. Kurki S, Peltonen K, Laiho M. Nucleophosmin, HDM2 and p53: players in UV damage incited nucleolar stress response. *Cell Cycle* 2004;**3**:976–9.
182. Lohrum MAE, Ludwig RL, Kubbutat MHG, Hanlon M, Vousden KH. Regulation of HDM2 activity by the ribosomal protein L11. *Cancer Cell* 2003;**3**:577–87.
183. Zhang Y, Wolf GW, Bhat K, Jin A, Allio T, Burkhart WA, Xiong Y. Ribosomal protein L11 negatively regulates oncoprotein MDM2 and mediates a p53-dependent ribosomal-stress checkpoint pathway. *Mol Cell Biol* 2003;**23**:8902–12.
184. Bhat KP, Itahana K, Jin A, Zhang Y. Essential role of ribosomal protein L11 in mediating growth inhibition-induced p53 activation. *EMBO J* 2004;**23**:2402–12.
185. Dai M-S, Lu H. Inhibition of MDM2-mediated p53 ubiquitination and degradation by ribosomal protein L5. *J Biol Chem* 2004;**279**:44,475–44,482.
186. Dai M-S, Zeng SX, Jin Y, Sun X-X, David L, Lu H. Ribosomal protein L23 activates p53 by inhibiting MDM2 function in response to ribosomal perturbation but not to translation inhibition. *Mol Cell Biol* 2004;**24**:7654–68.
187. Jin A, Itahana K, O'Keefe K, Zhang Y. Inhibition of HDM2 and activation of p53 by ribosomal protein L23. *Mol Cell Biol* 2004; **24**:7669–80.
188. Saxena A, Rorie CJ, Dimitrova D, Daniely Y, Borowiec JA. Nucleolin inhibits Hdm2 by multiple pathways leading to p53 stabilization. *Oncogene* 2006;**25**:7274–88.
189. Dai M-S, Sun X-X, Lu H. Aberrant expression of nucleostemin activates p53 and induces cell cycle arrest via inhibition of MDM2. *Mol Cell Biol* 2008;**28**:4365–76.
190. Korgaonkar C, Hagen J, Tompkins V, Frazier AA, Allamargot C, Quelle FW, Quelle DE. Nucleophosmin (B23) targets ARF to nucleoli and inhibits its function. *Mol Cell Biol* 2005;**25**:1258–71.

191. Khan SH, Wahl GM. p 53 and pRb prevent rereplication in response to microtubule inhibitors by mediating a reversible G_1 arrest. *Cancer Res* 1998;**58**:396–401.
192. Sablina AA, Chumakov PM, Levine AJ, Kopnin BP. p53 activation in response to microtubule disruption is mediated by integrin-Erk signaling. *Oncogene* 2001;**20**:899–909.
193. Stewart ZA, Tang LJ, Pietenpol JA. Increased p53 phosphorylation after microtubule disruption is mediated in a microtubule inhibitor- and cell-specific manner. *Oncogene* 2001;**20**:113–24.
194. Damia G, Filiberti L, Vikhanskaya F, Carrassa L, Taya Y, D'incalci M, Broggini M. Cisplatinum and taxol induce different patterns of p53 phosphorylation. *Neoplasia* 2001;**3**:10–16.
195. Li L, Ljungman M, Dixon JE. The human Cdc14 phosphatases interact with and dephosphorylate the tumor suppressor protein p53. *J Biol Chem* 2000;**275**:2410–14.
196. Paulsen MT, Starks AM, Derheimer FA, Hanasoge S, Li L, Dixon JE, Ljungman M. The p53-targeting human phosphatase hCdc14A interacts with the Cdk1/cyclin B complex and is differentially expressed in human cancers. *Mol Cancer* 2006;**5**:25.
197. Long X, Wu G, Gaa ST, Rogers TB. Inhibition of protein phosphatase-1 is linked to phosphorylation of p53 and apoptosis. *Apoptosis* 2002;**7**:31–9.
198. Haneda M, Kojima E, Nishikimi A, Hasegawa T, Nakashima I, Isobe K. Protein phosphatase 1, but not protein phosphatase 2A, dephosphorylates DNA-damaging stress-induced phospho-serine 15 of p53. *FEBS Lett* 2004;**567**:171–4.
199. Li DW-C, Liu J-P, Schmid PC, Schlosser R, Feng H, Liu W-B, Yan Q, Gong L, Sun S-M, Deng M, Liu Y. Protein serine/threonine phosphatase-1 dephosphorylates p53 at Ser-15 and Ser-37 to modulate its transcriptional and apoptotic activities. *Oncogene* 2006;**25**:3006–22.
200. Tang X, Hui Z-G, Cui X-L, Garg R, Kastan MB, Xu B. A novel ATM-dependent pathway regulates protein phosphatase 1 in response to DNA damage. *Mol Cell Biol* 2008;**28**:2559–66.
201. Lee S-J, Lim C-J, Min J-K, Lee J-K, Kim Y-M, Lee J-Y, Won M-H, Kwon Y-G. Protein phosphatase 1 nuclear targeting subunit is a hypoxia inducible gene: its role in post-translational modification of p53 and MDM2. *Cell Death Differ* 2007;**14**:1106–16.
202. Li H-H, Cai X, Shouse GP, Piluso LG, Liu X. A specific PP2A regulatory subunit, B56γ, mediates DNA damage-induced dephosphorylation of p53 at Thr55. *EMBO J* 2007;**26**:402–11.
203. Dohoney KM, Guillerm C, Whiteford C, Elbi C, Lambert PF, Hager GL, Brady JN. Phosphorylation of p53 at serine 37 is important for transcriptional activity and regulation in response to DNA damage. *Oncogene* 2004;**23**:49–57.
204. Kong M, Fox CJ, Mu J, Solt L, Xu A, Cinalli RM, Birnbaum MJ, Lindsten T, Thompson CB. The PP2A-associated protein alpha4 is an essential inhibitor of apoptosis. *Science* 2004;**306**:695–8.
205. Lammers T, Lavi S. Role of type 2C protein phosphatases in growth regulation and in cellular stress signaling. *Crit Rev Biochem Mol Biol* 2007;**42**:437–61.
206. Fiscella M, Zhang H, Fan S, Sakaguchi K, Shen S, Mercer WE, Vande Woude GF, O'Connor PM, Appella E. Wip1, a novel human protein phosphatase that is induced in response to ionizing radiation in a p53-dependent manner. *Proc Natl Acad Sci U S A* 1997;**94**:6048–53.
207. Takekawa M, Adachi M, Nakahata A, Nakayama I, Itoh F, Tsukuda H, Taya Y, Imai K. p53-inducible Wip1 phosphatase mediates a negative feedback regulation of p38 MAPK-p53 signaling in response to UV radiation. *EMBO J* 2000;**19**:6517–26.
208. Bulavin Jr DV, Phillips C, Nannenga B, Timofeev O, Donehower LA, Anderson CW, Appella E, Fornace AJ. Inactivation of the Wip1 phosphatase inhibits mammary tumorigenesis through p38 MAPK-mediated activation of the $p16^{Ink4a}$-$p19^{Arf}$ pathway. *Nat Genet* 2004;**36**:343–50.
209. Shreeram S, Demidov ON, Hee WK, Yamaguchi H, Onishi N, Kek C, Timofeev ON, Dudgeon C, Fornace AJ, Anderson CW, Minami Y, Appella E, Bulavin DV. Wip1 phosphatase modulates ATM-dependent signaling pathways. *Mol Cell* 2006;**23**:757–64.
210. Shreeram Jr S, Hee WK, Demidov ON, Kek C, Yamaguchi H, Fornace AJ, Anderson CW, Appella E, Bulavin DV. Regulation of ATM/p53-dependent suppression of myc-induced lymphomas by Wip1 phosphatase. *J Exp Med* 2006;**203**:2793–9.
211. Yamaguchi H, Durell SR, Chatterjee DK, Anderson CW, Appella E. The Wip1 phosphatase PPM1D dephosphorylates SQ/TQ motifs in checkpoint substrates phosphorylated by PI3K-like kinases. *Biochemistry* 2007;**46**:12,594–12,603.
212. Lu X, Nannenga B, Donehower LA. PPM1D dephosphorylates Chk1 and p53 and abrogates cell cycle checkpoints. *Genes Dev* 2005;**19**:1162–74.
213. Lu X, Ma O, Nguyen T-A, Jones SN, Oren M, Donehower LA. The Wip1 phosphatase acts as a gatekeeper in the p53-Mdm2 autoregulatory loop. *Cancer Cell* 2007;**12**:342–54.
214. Ofek P, Ben-Meir D, Kariv-Inbal Z, Oren M, Lavi S. Cell cycle regulation and p53 activation by protein phosphatase 2Cα. *J Biol Chem* 2003;**278**:14,299–14,305,.
215. Johnson TM, Attardi LD. Dissecting p53 tumor suppressor function *in vivo* through the analysis of genetically modified mice. *Cell Death Differ* 2006;**13**:902–8.

Chapter 25

The p53 Master Regulator and Rules of Engagement with Target Sequences

Alberto Inga[1], Jennifer J. Jordan[2,3], Daniel Menendez[2], Veronica De Sanctis[4] and Michael A. Resnick[2]

[1]*Unit of Molecular Mutagenesis and DNA repair, Department of Epidemiology and Prevention, National Institute for Cancer Research, IST, Genoa, Italy*

[2]*Laboratory of Molecular Genetics, National Institute of Environmental Health Sciences, NIH, Research Triangle Park, North Carolina*

[3]*Curriculum in Genetics and Molecular Biology, University of North Carolina, Chapel Hill, North Carolina*

[4]*Centre for Integrative Biology, CIBIO, University of Trento, Italy*

INTRODUCTION

The p53 protein is a master regulator of cellular responses to stress. It functions as a tetrameric, sequence specific transcription factor at the hub of various distinct transcriptional programs. There are several tight but flexible controls of p53 activity that tailor induced responses to the type and duration of cellular and microenvironmental perturbations. These controls appear to be dependent on cell type and state [1–8]. p53 is a prominent tumor suppressor and is impaired in nearly all human tumors, primarily by somatically acquired p53 mutations. The vast majority of tumor associated p53 mutations are missense, resulting in single amino acid changes in the large immunoglobulin-like folded DNA binding domain of the protein that provides for sequence specific recognition/interaction with DNA [9, 10].

In the previous chapter, Anderson and Appella [11] described the highly complex epigenetic/posttranslational modification code affecting protein stability, subcellular compartmentalization, and protein–protein interactions that could modulate selective DNA binding and transactivation. In this chapter, the response element sequences (REs) that provide for p53 recruitment to target promoters is addressed, focusing on the intrinsic potential of wild-type and mutant p53 to transactivate from various REs as a source of transactivation selectivity. Emphasis is placed on establishing "rules of engagement" using transcription assays in the budding yeast *Saccharomyces cerevisiae* and subsequently in human cells to determine functional p53 REs. These rules increase opportunities for understanding the p53 pathway using various biochemical, cellular, functional-genomic, and computational approaches. A brief summary is included about other sequence specific transcription factors and p53 recruited cofactors that contribute to establish selective transactivation from subsets of target genes in response to specific cell signals.

THE p53 INDUCED TRANSCRIPTIONAL NETWORK: GENES, BIOLOGICAL FUNCTIONS, AND THE COMPLEXITY OF TARGET SELECTION

Consistent with its tumor suppressor function, p53 activity was shown to be inducible by signaling pathways responding to DNA damage. The first established p53 cellular functions dependent on sequence specific transactivation were linked to cell cycle control, primarily at the G1/S checkpoint [12] and induction of apoptosis [13, 14]. The growing list of established p53 responsive genes has linked p53 to other biological processes, including senescence [15], ATP production [16, 17], autophagy [18], mTOR signaling [19, 20], inflammation and angiogenesis [21, 22], suntan response [23], DNA repair [24, 25], and, more recently, the modulation of a number of microRNAs [26]. While microRNAs contribute to p53 protein regulation via feedback loops, the primary negative feedback loop involves p53 dependent induction of MDM2, with modulation by MDMX [7, 27, 28], that affects p53 protein stability, nuclear localization, and function at target promoters.

Bioinformatic predictions of p53 target genes based on the presence and the "quality score" of p53 REs in gene regulatory regions suggest that the approximate 100 validated p53 target genes represent only a small fraction of

Handbook of Cell Signaling, Three-Volume Set 2 ed.

direct targets [29–32]. Genome-wide p53 binding results [33–35] and projections from smaller scale but high resolution studies [36] indicate that the number of sites bound by p53 in the genome is in the thousands. Although binding need not lead necessarily to transactivation these results suggest that the complexity of the p53 network is far from being fully described. While not addressed in this review, transcriptional repression can be included in p53 mediated responses [37], as well as transcription independent activities [38]. In addition to sequence specificity, there are additional factors including, for instance, chromatin context that can influence p53 specific activity [6, 8, 39–41]. There is still a significant gap in our understanding of the transactivation selectivity process that provides for induction of specific subsets of p53 target genes, resulting in distinct cellular responses to individual stress conditions.

YEAST AS AN *IN VIVO* TEST TUBE TO STUDY WILD-TYPE AND MUTANT p53 TRANSACTIVATION POTENTIAL TOWARD DEFINED RESPONSE ELEMENT SEQUENCES

p53 can bind as a tetramer to a variety of REs related to the degenerate consensus RRRCWWGYYY(N)RRRCWWGYYY (N = 0 to 13; W = A or T), where the REs frequently contain non-consensus bases (see below and [8]). The sequence specific interaction of wild-type p53 with cognate target sequences is a necessary and important first step in the complex process leading to differential regulation of downstream targets. While not addressed in this review, cancer hot-spot mutant p53s that have single amino acid changes in the DNA binding domain, which prevent sequence specific binding, may result in gain-of-function properties. Such mutants can affect cellular gene expression by being recruited to transcriptional complexes via protein–protein interactions favored by aberrant high expression and nuclear localization [42].

At the DNA level, p53 regulation of target genes is dependent on the RE sequence, number of REs, distance of the RE from the transcriptional start site, and intrinsic binding affinity of p53 toward an RE [43]. However, the mechanisms of regulation from the various promoter REs targeted by p53 remain unclear, particularly the relationship between p53 binding to target RE sequence and transactivation. Understanding the interaction of p53 with various RE sequences under conditions that render all other factors (such as chromatin, transcriptional cofactors, and the landscape of other transcription factor binding sites in the promoter) constant have helped to elucidate how different p53–DNA interactions alter the potential for transactivation and possibly influence other factors that affect transactivation.

With its ease of genetic manipulation and cost effective measures as a research tool, budding yeast has become a prominent model system to study various human diseases [44]. Fundamental aspects of cancer, including, but not limited to DNA replication, cell cycle checkpoints, and mechanisms of drug resistance, have been addressed in this smaller eukaryote, sometimes referred to as an "honorary mammal" due to its conservation of genes, signaling, and cellular pathways [45–48]. Yeast has emerged as an *in vivo* test tube that provides constant conditions that are often not attainable in mammalian systems to assess the transactivation capacity of p53, as well as other sequence specific master regulators including NKX2-5, HAND1, ER, and NFκB [49–52]. In contrast to *in vitro* biochemical assays, the yeast assays have the advantage of comparing p53 alleles (WT or mutant) in an isogenomic chromatin environment for transcriptional activity toward individual REs. Functional status of the transcription factor from specific binding elements can be evaluated using various reporters (including the beta-galactosidase gene, *HIS3*, *URA3*, *ADE2*, the firefly luciferase, or GFP) that are integrated into the genome or are plasmid based [29, 43, 53–58].

Several studies have utilized an approach referred to as FASAY (functional analysis of separated alleles in yeast) that exploits the exquisite proficiency for homologous recombination of yeast to achieve high efficiency capture of PCR amplified p53 cDNAs from tumor samples by *in vivo* cloning into an expression plasmid. A p53 dependent reporter gene present in the yeast strain enables the identification of transcriptionally defective p53 mutants [59, 60]. Using reporter genes containing different p53 REs derived from human p53 target genes involved in various biological pathways (e.g., cell cycle arrest and apoptosis), some of the p53 mutations associated with cancer were found to be at least partially functional toward some, but not necessarily all the REs tested; the responses might depend on the experimental conditions, such as temperature [58, 61, 62]. Similar findings were obtained using mammalian cell based assays or, in some cases knock-in mouse models [63–65]. Several mutations (i.e., T150I, G199R, R202S, and S215C) even appeared to be silent, or indistinguishable from WT p53 in terms of transactivation capabilities [66], which might have been due to the high levels of p53 expression as discussed below.

Yeast based assays have also been developed to screen for p53 mutations that could act as an intragenic suppressor of transcriptionally inactive cancer hot-spot mutations [67] and for proteins that could modulate p53 transactivation potential [68, 69]. Other studies and systems have been designed for evaluating the relative transactivation potential of wild-type p53 toward many target REs [70] or for isolating p53 alleles exhibiting altered functions [71].

A RHEOSTATABLE YEAST PROMOTER FOR CONTROLLED, INDUCIBLE EXPRESSION

All of the above yeast-based approaches were based on high expression of p53 proteins in yeast using a constitutive promoter such as *ADH1* or *PGK1* or the inducible *GAL1* promoter (uninduced or fully induced). The high expression allowed alleles that were active or inactive for function to be distinguished, but it did not allow for the discrimination between silent or subtle, altered-function mutations [72].

Since binding and transactivation of wild-type and mutant p53 to various REs will depend on the amount of p53 in the cell, we developed a system for variable induction of p53. Initially, variable p53 expression was provided within a plasmid based system [73]. Subsequently, the components were integrated into the chromosomes of diploid cells to better allow for a matrix of assessments of mutant p53's transactivation capacity toward many REs at various levels of p53 [74, 75]. Within these systems, the p53 coding sequence is placed under the control of the *GAL1* promoter that provides for inducible (i.e., "rheostatable") control of p53 expression through variation in the amount of galactose in the medium. Importantly, expression from the *GAL1* promoter in yeast displays a graded transcriptional response (rather than an on/off or binary response) when galactose is supplemented in the media, allowing for a broad range of activity from the promoter [76, 77] from most cells in the population.

This rheostatable promoter allows for the functional assessment of a sequence specific transcription factor at various protein concentrations from a single copy of a target RE upstream of a reporter [43, 49, 74], thereby providing the opportunity to address the responsiveness of individual REs to wild-type and mutant p53 proteins. This can be viewed as an opportunity for *in vivo* assessment of the biochemical properties of p53 and REs. Importantly, the rheostatable promoter can unmask subtle transactivational differences between WT p53 and several p53 missense mutations that had previously been determined to be equivalent to WT in transactivation capacity under conditions of high expression [66] (discussed below).

RULES OF p53 TRANSACTIVATION REVEALED BY YEAST-BASED ASSAYS

Using the rheostatable p53 system, it has been possible to establish the contribution of each RE sequence and level of p53 protein to transactivation. Transactivation capacity was assessed as the minimal level of p53 induction (i.e., amount of galactose) required for a detectable increase in reporter output from a given RE:reporter cassette. Initially, the reporter was based on change in colony color (i.e., RE:*ADE2*) where increases in p53 would result in colonies changing from red to white due to RE-mediated transcription of the *ADE2* gene. With this assay, large unpredicted differences in activation by WT p53 were observed between REs that differ by only small changes in sequence [78]. This has led to rules of p53 transactivation that describe elements within the canonical RE sequence that determine not only whether a RE is functional, but also the strength of p53 transactivation [72, 75] from an RE (Figure 25.1). Interestingly, the number of matches in the canonical consensus sequence (RRRCWWGYYY $N_{(0-13)}$ RRRCWWGYYY) does not correlate with transcriptional activity; however, functionality requires that there be no more than four mismatches from the consensus sequence [43]. Importantly, these *in vivo* measurements of transactivation differ from statistically based predictions of binding energy of p53 toward REs [79].

Within the decamer half-site (RRRCWWGYYY), changes in the conserved C and G at the fourth and seventh position greatly decreases transactivation. In addition, the nucleotides in the central core CWWG motif strongly influence the ability of WT p53 to transactivate from a RE, where CATG is most active, followed by CAAG and CTTG. Changing the CATG to CTAG decreases transactivation ~20-fold [74]. These observations suggested that an intrinsic property of the core CATG is required for strong levels of transactivation. Binding studies agree with this finding and indicate that the flexibility of the CATG sequence increases the affinity of p53 toward an RE [80]. Upon binding to the RE, p53 induces a conformational change in the DNA by bending the RE. The angle is dependent upon the particular RE, where REs containing CATG cores are associated with larger bending angles in comparison to REs containing other core motifs such as CTTG [81, 82]. Molecular dynamic simulations investigating p53 core domain tetramer–DNA interactions confirm that alterations of the core CWWG motifs modify the degree of bending, with p53 bound to CATG showing the greatest curvature, CAAG and CTTG moderate curvature, and CTAG having little or no bend [83]. Importantly, the simulations showed that the CATG core motif had the greatest ability to maintain interactions with the p53 protein. The capability of RE sequences to form secondary stem/loop structures has also been proposed to modulate p53 binding/quaternary conformation and transactivation potential [84].

Cooperation with cofactors *in vivo* may provide opportunities to enhance p53 interactions and transactivation with REs (discussed below). Proteins such as HMG can augment p53 binding to REs by inducing bends in the DNA [85]. In addition to the central core sequence contributing to the strength of transactivation, the two inner purines and pyrimidines that flank the core affect transactivation more than the outer nucleotides with GG/CC being the most active and AG/CT the least active [33, 34, 86]. This latter result indicates the existence of pairwise or more complex interactions between the nucleotides and binding of p53. Such interactions are usually overlooked in computational predictions of DNA binding probabilities such as the commonly used Position Weight Matrices that assume that each nucleotide within a RE contributes independently to the binding/activity score of the full sequence [87, 88].

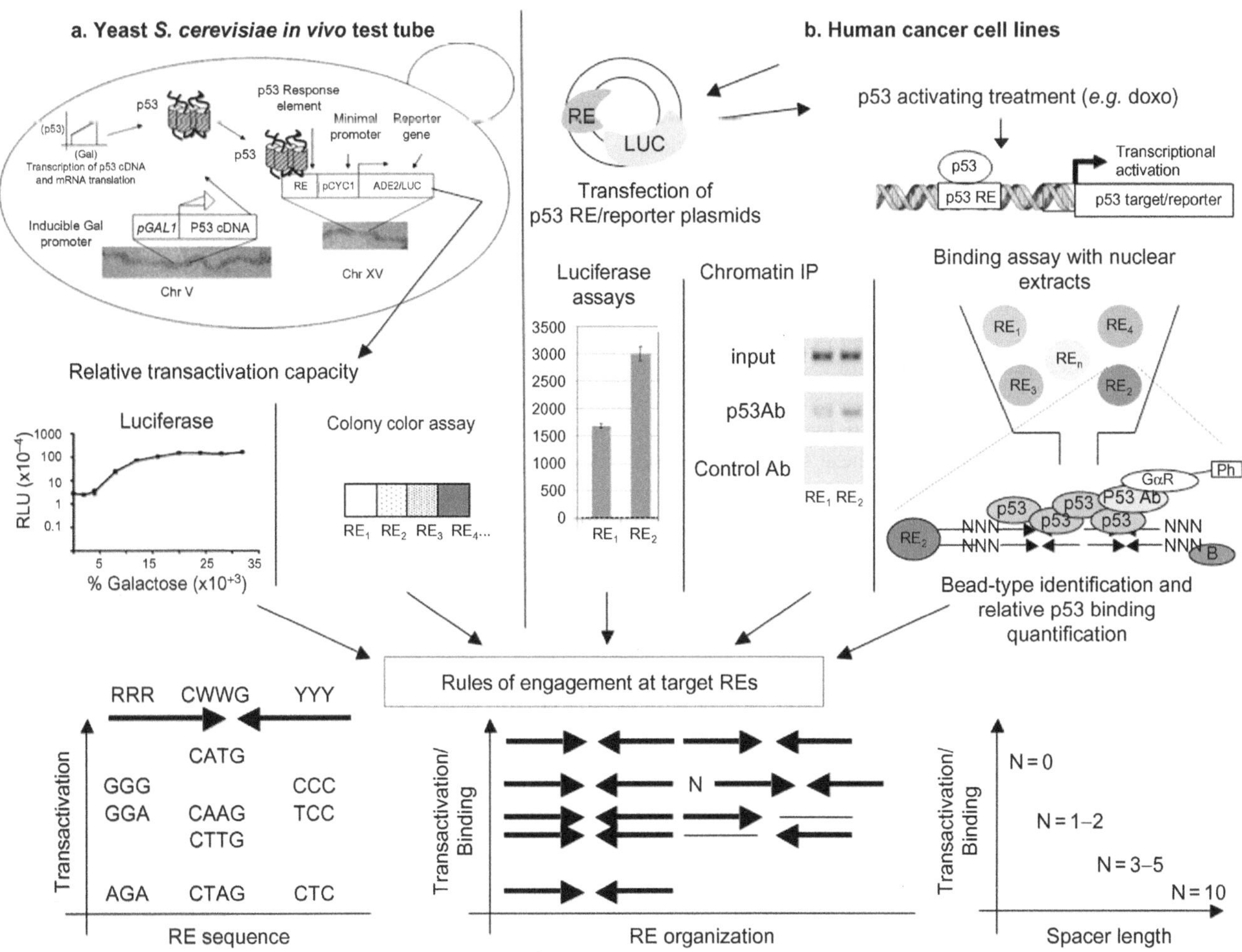

FIGURE 25.1 Yeast and mammalian model systems to study "rules of engagement" for the p53 master regulator. a. *Assays in the budding yeast S. cerevisiae* provide an *in vivo* test tube to determine relative transactivation capacity of p53. The integrated diploid system allows for a sensitive evaluation of the contribution of RE sequence, binding motif, and p53 expression on functionality and regulation within the p53 transcriptional network under constant, isogenomic conditions. Only the p53 mutation of interest and the RE sequence are varied between strains. The chromosomal position for all of the human derived REs is identical and the number of p53 molecules/cell can be varied over 100-fold using an integrated "rheostatable" *GAL1 : p53* promoter system for p53 expression that is sensitive to levels of galactose (Gal) in the medium. Evaluation of transactivation capacity can be based on colony color (*ADE2*) or enzyme based (luciferase, LUC) quantitative luminescence. The "*delitto perfetto*" *in vivo* mutagenesis approach is utilized for rapid inclusion of target REs upstream of a reporter and the development of mutant p53s [146]. b. *Human cancer cell lines.* The ability of p53 to transactivate from transfected RE-luciferase reporter plasmids or endogenous genes under stimulated or non-stimulated conditions can be assessed using luciferase assays while p53 binding to target REs can be measured using chromatin immunoprecipitation (ChIP) assays. The semi *in vitro* microsphere binding assay can be employed to examine p53 in nuclear extracts. This assay utilizes a Luminex microsphere technology to assess simultaneously the differential binding of p53 to multiple target sequences [96]. Key issues that have been addressed with these systems include (i) what constitutes a functional RE for transactivation, (ii) how do changes in the RE sequence affect transactivation, (iii) how does the organization of the binding motif alter transactivation, (iv) the importance of p53 level on transactivation, and (v) the relationship between *in vitro* binding and *in vivo* transactivation. Transactivation and binding assays have shown that p53 can transactivate from non-canonical response elements such as 1/2- and 3/4-sites and that the spacer (N) separating two half-sites in a full RE can have a large effect both on transactivation and DNA binding potential. Arrows in the lower part of the figure indicate the relative orientation of the pentameric sequence motifs where each pentamer corresponds to a p53 monomer binding site.

SPACERS AFFECT p53 TRANSACTIVATION POTENTIAL IN THE YEAST-BASED ASSAY

The length of nucleotide spacer (N) between decamer half-sites in the p53 consensus site has little impact on p53 interactions with REs in various computation algorithms [31, 32]. However, analysis of established RE sequences has shown that the majority of the p53 REs lack a spacer [6, 79]. In a comparison of REs of target genes subject to transcriptional activation or repression by p53, the former were primarily found to lack or contain a small spacer and genes associated with repression had spacers of various lengths approximately equally distributed between 4 and 15 nucleotides [6].

While consensus REs determined from *in vitro* binding studies contain spacers of up to 15 nucleotides, spacers of varying lengths can differentially impact the *in vivo* functionality of p53. Based on transactivation from REs containing spacers of 4, 13, and 14 nucleotides [29], spacers that would result in half-sites appearing on opposing faces of the DNA helix were inferred to reduce transactivation. Furthermore, altering the sequence and the length of the spacer within the DDB2 gene (p48) from T to GG was found to significantly reduce binding [89]. Other studies indicated that small spacers could confer repression by p53 from specific response elements as demonstrated with the 3 nt spacer in the RE of the survivin promoter [90].

The rheostatable diploid yeast system was employed in a comprehensive study to address the influence of spacer length upon p53 transcriptional functionality when present between adjacent full-site REs or between decamer half-sites within single REs (Figure 25.1). In agreement with previous results [91, 92], large spacers between two full-site REs lead to synergistic transactivation by p53. This could be due to stacking of tetramers to enhance transactivation from REs that have been brought into close proximity due to looping of the intervening DNA. However, a spacer located between the decamers of a full-site RE, can strongly decrease the strength of transactivation. The insertion of a single nucleotide placed between the half-sites of the strongly transactivated p21-5' RE was shown to decrease transactivation by greater than 50 percent at high levels of p53 expression. Subsequent increases in the spacer length up to 10 nucleotides revealed a continual decrease in transactivation at higher levels of p53 expression. At lower levels of p53, transactivation was abolished even for a few nucleotides.

Thus, spacers could be an important factor in determining responsiveness of RE targets at varying levels of p53 and could be a factor in the evolution of REs. This concept was illustrated by the ability to significantly enhance the relatively weak p53 dependent transactivation from the RE of the TIGAR gene (TP53 induced glycolysis and apoptosis regulator). The two half-sites of the natural RE are perfectly matched to the consensus and contain a 2 nucleotide spacer. Removal of the spacer greatly enhances p53 induced transactivation [75].

SPACER EFFECTS ON p53 BINDING AND TRANSACTIVATION IN MAMMALIAN CELL ASSAYS

While results from the yeast system describe the *potential* role that RE sequence may play in the p53 master regulatory network, further assays in mammalian cells are needed to fully characterize the impact of target binding sequence and organization. Additional promoter elements, chromatin structure, epigenetic landscape, and transcriptional cofactors in mammalian cells are expected to influence p53 mediated transactivation [6, 8, 93] (see Figure 25.2). Transient transfection and ChIP assays with human SaOS2 (osteosarcoma) cells to assess p53 transactivation and promoter occupancy from vectors containing the p21-5' RE with spacers of increasing length demonstrated effects comparable to those found with the yeast based system. These findings in yeast and human cells support the view that assessment of the impact of spacer needs to be evaluated *in vivo*.

These findings appear to differ from predictions based on crystal studies of p53 bound to REs where spacer has a limited effect and the specific sequence of an RE dictates the exact protein–DNA interaction [94]. Similar interpretations were drawn using a quaternary structure of human p53 in solution [86]. An alternative 3-D model based on cryoelectron microscopy proposes the best fit p53 RE sequence has a spacer length of 6 nucleotides [95]. The contradicting views of 3-D structures suggest different potential binding modes for the p53 tetramer and indicate that the optimal conformation of p53 protein bound to a DNA response element in terms of transactivation activity remains to be determined.

From comparisons between *in vitro* binding and *in vivo* transactivation, levels of transactivation are not predictable without direct assessment of function *in vivo* under conditions where p53 levels of expression can be varied. This implies that factors other than just sequence and sequence organization influence transactivation by p53 in a manner that is distinguishable from simple DNA binding. Thus, while *in vitro* biochemical assays may predict the potential for binding under ideal conditions with purified components, p53 functionality must be assessed within the context of a cellular environment. Along this line, a semi *in vitro* fluorescent microsphere binding assay based on Luminex FlexMap technology has recently been developed that provides opportunities to measure *in vitro* binding simultaneously to many different REs or more complex sequence motifs using nuclear extracts [96]. This microsphere-based assay has been employed to address the impact of spacers in REs upon p53 binding [75]. The p21-5' REs containing spacers of increasing length were tagged to unique identifier sequences on colored microspheres and then incubated with nuclear cellular extracts from lymphoblastoid cells in which p53 was induced by doxorubicin treatment. In this system, increases in spacer length >1 nucleotide reduced binding [75]; p53 binding to an RE with a 3 nt or greater spacer was comparable to binding to individual p21-5' half-site REs.

Genome wide approaches have been utilized to address sites of p53 binding and potential p53 target genes. Chromatin immunoprecipitation along with a paired-end ditag sequencing strategy (ChIP-PET) was employed to determine an unbiased genome-wide map of p53 transcription factor binding sites in human HCT116 colorectal cancer cells treated

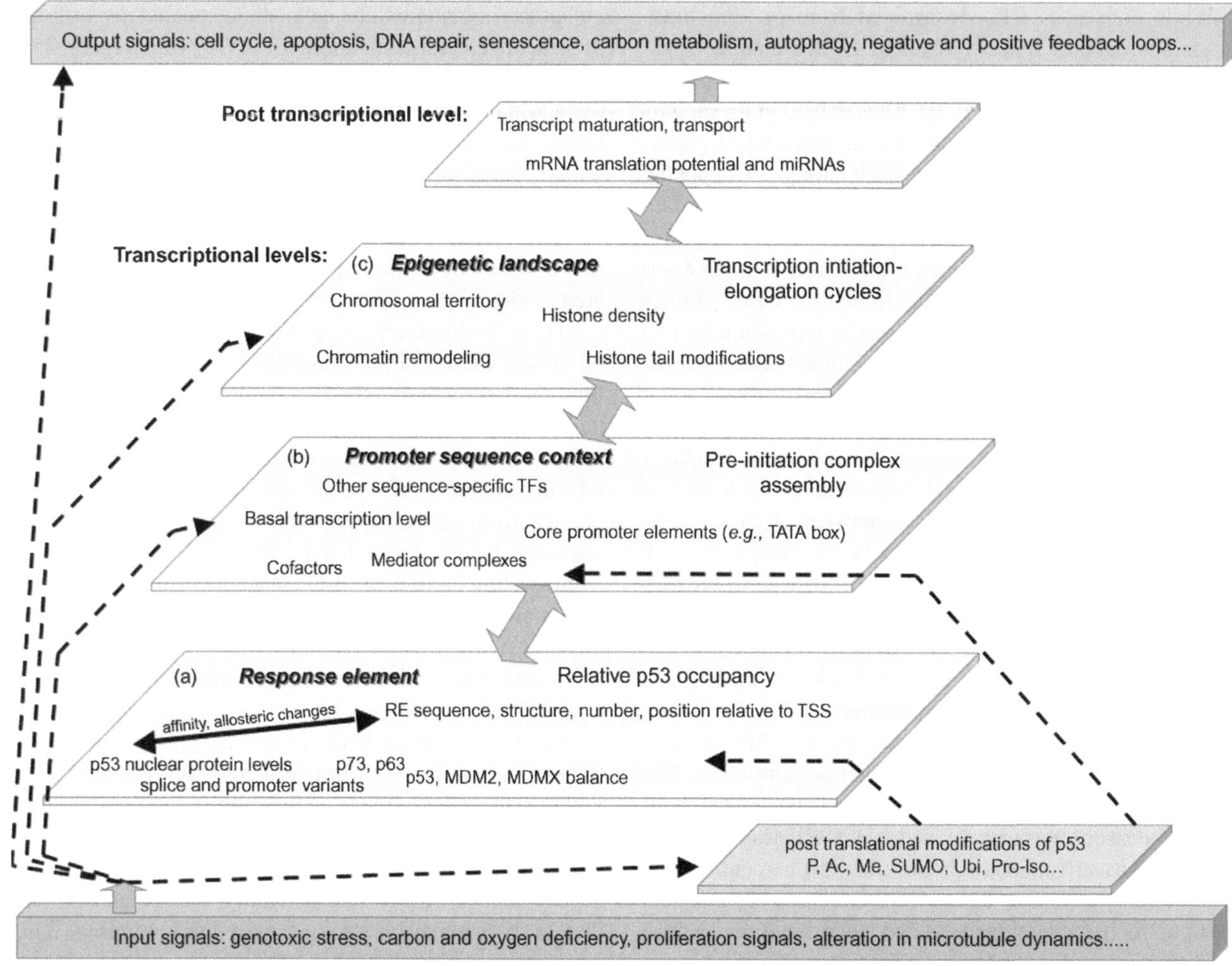

FIGURE 25.2 The p53 transcriptional network of stress responses.
Many cellular perturbations can activate the p53 transcriptional network, mainly via signaling pathways resulting in a complex web of posttranslational modification of the p53 protein. As shown and described in the text, signaling pathways can also activate other sequence specific transcription factors that can modulate common target genes cooperatively or independently from p53. The activation of p53 as a transcription factor is here divided conceptually into three highly connected levels of transcription (a–c). The interaction of p53 protein with response element sequences provides the first step (a) in which the intrinsic affinity of p53 for different REs and the regulation of p53 nuclear levels contribute to p53 binding at target promoter sites. In the next step (b), p53 occupancy can directly affect the recruitment and assembly of the transcriptional machinery. This process can be greatly influenced by the interaction/availability of cofactors and mediator complexes, the contribution from other sequence specific transcription factors, and specific features of the individual target core promoter sites. In step (c), various changes at the chromatin level, including posttranslational modifications of histone tails and remodeling of nucleosomes, in part mediated by p53 recruited histone modifying enzymes, can result in the initiation and elongation phases of transcription and influence the reloading of the transcription machinery for sustained gene expression changes. Additional "post-transcriptional level" controls also contribute to shaping the outcome of p53 activation, including the effect of regulatory non-coding RNAs on the stability and translation efficiency of mRNA. The output of p53 activation changes the transcription of hundreds (and possibly thousands) of genes with different functions. Contributing to this complexity are the negative and positive feedback loops connecting input and outputs, such as p53 dependent transcription of MDM2 and of microRNAs that in turn can target p53 cofactor proteins.

with 5-fluorouracil (5-FU) [34]. In the model developed to identify potential p53 consensus sites, 236 out of 288 motif sequences were found to lack any spacer between half-sites; 27 REs contained a spacer of 1 nucleotide and the remaining 25 had larger spacers. In a genome-wide ChIP-on-chip analysis with U2OS osteosarcoma cells treated with actinomycin D, Smeenk *et al.* (33) showed that ~80 percent of 1546 sites bound by p53 with high confidence levels were canonical full-sites that lacked a spacer (0 nucleotides) between decamer half-sites. Overall, it appears that based on *in vivo* transactivation and binding data the maximal allowable spacer for a functional RE is confined to <3 nucleotides.

NON-CANONICAL 3/4- AND 1/2-SITE RES EXPAND THE p53 TRANSCRIPTIONAL NETWORK

Extensive *in vivo* functional assays within the yeast and mammalian systems have found that p53 can transactivate from non-canonical sequences consisting of 1/2 and 3/4 RE sites (a 3/4-site is defined as a 1/4-site directly adjacent to a 1/2-site or a 1/2-site separated from a 1/4 -site by a 5 nt spacer) [75]. The structural requirements for p53 tetramer formation are similar for transactivation from both the full-site REs and the non-canonical elements. Interestingly, re-evaluation of transactivational activity from REs originally classified as canonical full-site REs including PIDD, APAF1, PCNA, and 14-3-3σ indicates that many biologically relevant REs are actually non-canonical 3/4-site REs and that non-canonical sequences commonly occur in the genome.

Half-site p53 REs exhibit limited transactivation potential that is less than for 3/4-sites when examined in functional assays [75]. The transactivation "rules" established for full-site REs [74], such as the higher activity with the CATG core sequence, also appear to be conserved for half-sites. Importantly, while the impact of spacer on transactivation constrains the number of functional p53 target sequences, the ability of p53 to function from non-canonical REs expands the universe of possible p53 downstream target genes.

IMPACT OF 1/2-SITE RES IN THE p53 TRANSCRIPTIONAL NETWORK

p53 half-sites could serve many roles in the genome. For example, half-sites could affect chromatin accessibility to transcriptional machinery (i.e., loosening the chromatin or recruiting and sequestering chromatin modifiers) [97]. They could also affect levels of available p53. Furthermore, the p53 half-sites may also bring together different transcriptional networks. For example, a half-site RE can be a necessary, but not sufficient, *cis*-element to confer p53 responsiveness to a promoter [72]. In the case of the VEGF receptor FLT1, a SNP in the FLT1 promoter creates a p53 half-site that is located near an estrogen receptor (ER) response element (ERE). The interaction of p53 and ER at their corresponding target sequences can greatly enhance p53 mediated transactivation [98]. Thus, p53 mediated stimulation of FLT1 can be accomplished through two sets of environmental factors: cellular stress such as DNA damage and ER ligands [98, 99]. Such transcriptional cooperation may be utilized in fine-tuning of p53 regulated responses and may render activity from non-canonical 1/2-site REs more dependent on availability of cooperating transcription factors or additional cofactors as well as levels of nuclear p53.

EVOLUTIONARY DEVELOPMENT OF p53 RES

Recent studies addressing evolutionary aspects in the p53 transcriptional network provide evidence of widespread evolutionary turnover of p53 binding sites and of species-specific genes or branches entering or being dropped from the network [100, 101]. This contrasts with findings for two other stress response transcription factors, the binding sites for NRF2 and NFκB, where both exhibit high inter-specific conservation suggesting strong purifying natural selection [100]. Findings such as these lead to the question of whether the p53 transcriptional network expanded during evolution to incorporate additional functions with the growing complexity of the organism or adapting to specific environments, as found for several DNA metabolism genes [100, 101].

There are several examples of weaker p53 response elements being evolutionarily conserved in spite of an overall high evolutionary turnover of p53 REs [100, 101]. This finding suggests that these non-optimal elements can be selected for as they may provide the opportunity for adaptive regulation in response to stress signals via quantitative or qualitative (posttranslational) modifications of p53 proteins or through other factors involved in transcriptional regulation [101].

Interestingly, a mechanism for spreading p53 regulatory modules in the genome of primates has been identified that relies on functional p53 REs present in specific families of LTR class I retroelements. These retroelements are primate specific and were active in transposition around the split in the appearance of New and Old World monkeys [102]. An analysis based on the results of global p53 binding to DNA (34) revealed putative p53 binding sites in different classes of other transposable elements, including several primate-specific Alu subfamilies. Notably, sequence analysis of the evolutionary precursors of those Alu elements revealed that the p53 REs likely arose from deamination of methylated cytosines in CpG sequences resulting in the creation of strongly p53 responsive CATG core sequence elements [103].

YEAST BASED FUNCTIONAL CLASSIFICATION OF p53 MUTANT ALLELES ASSOCIATED WITH CANCER

The rheostatable system has proven useful for characterizing p53 functional mutations associated with cancer [43]. The ability to discriminate between wild-type and mutant transactivation capacities at various p53 expression levels has allowed for the identification of altered function mutations that not only affect the spectrum of target genes regulated but the strength to which these targets are transactivated. Functional fingerprints defining the ability of p53 mutants to transactivate from a panel of human-derived REs relative to WT p53 have been determined and used to classify the mutants into groups according to their retained transactivation capacities at various levels of p53 expression. Interestingly, a subset of mutants was found to be indistinguishable from WT in terms of transactivation capacity at high levels of expression, but displayed altered function at lower levels of expression [43, 75].

Functional classification of p53 missense mutations has been employed to explore whether p53 transactivation capacity can correlate with specific clinical features of the associated tumors or add prognostic value to the simple assessment of p53 status (i.e., the presence or absence of a p53 mutation). A significant correlation between loss-of-function p53 mutations and poor clinical responses has been observed for familial cancer cases associated with germline p53 mutants [9, 104]. However, in a large study of sporadic breast cancers, functional classification did not add prognostic value [105]. In that study, the functional classification of mutations was derived from analysis of the transactivation potential of ~2300 different missense p53 mutations towards a set of eight different REs using a functional assay in yeast that relies on constitutive expression of p53 alleles [56] and a fluorescent reporter. When classification of p53 mutant functionality obtained from that screen was compared with results obtained with the rheostatable system described above, there was up to 30 percent discordance for mutants classified as partial or altered function, suggesting that further studies are needed to assess the relationship between functional status of p53 mutants and treatment response evaluation [147].

CONTRIBUTIONS TO THE RULES OF ENGAGEMENT BY p53 HOMOLOGS AND OTHER SEQUENCE SPECIFIC TRANSCRIPTION FACTORS

p53 family. Although the discovery of p53 homolog genes in mammalian genomes occurred several years after the identification of p53 [106, 107], it is now established that most vertebrate genomes harbor a p53 family consisting of three genes: p53, p63, and p73, which have a complex array of isoforms [108]. Even though both p63 and p73 are involved in different cellular processes from p53, based on the phenotypes from knockout mice [109, 110], both p63 and p73 are targets of stress signaling pathways, particularly those responding to DNA damage, and could participate at various levels in the p53 mediated responses. For example, in murine cells a genetic approach indicated that the lack of p63 or p73, and particularly the deletion of both genes, resulted in impaired or diminished activity of p53 as a transcription factor, especially toward apoptotic genes [111, 112].

Functional interactions with other transcription factors can modulate p53 mediated responses, and can be dependent on adjacent or closely located cognate REs at common target promoters or on physical interactions. While not covered in this review, cross-talk also occurs between p53 and SMAD2/3 (TGF-beta signaling) [28, 113], NF-Y [114–116], ER [98, 99, 117–119], and NFκB [120–125].

p53 COFACTORS CONTRIBUTE TO PROMOTER SELECTIVITY

Like many other transcriptional activators, p53 can modulate gene transcription through interaction with coactivators and corepressors and these interactions have been shown to influence the promoter selectivity by p53. For example, ASPP1 and ASPP2 (apoptosis stimulating protein of p53) were shown to enhance p53 binding toward some apoptotic REs, such as Bax [112, 126], an activity that is inhibited by iASPP [127]. The human cellular apoptosis susceptibility protein hCAS/CSE1L was found to be associated with a subset of p53 target genes involved in the apoptotic process such as PIG3 [128]. Whereas, HZF (hematopoietic zinc finger) was shown to bind the p53 DNA binding domain resulting in enhanced p53 binding specifically toward REs associated with cell cycle arrest genes [129].

Interestingly, several of the known coactivators and corepressors of p53 transactivation are histone modifying enzymes and local changes in histone acetylation and methylation on p53 target genes have been shown to be p53 dependent [130–134]. p53 recruited histone acetyltransferases include the ubiquitous CBP/p300 [135], the STAGA complex (containing GCN5) [136], and MOZ [137]. p53 was also shown to interact with two arginine methylases *in vivo*, PRMT1 and CARM1 [138]. An ordered recruitment and activation of histone acetyltransferases and methyltransferases appears to be necessary to mediate the fundamental changes in the chromatin structure associated with p53 dependent transcription [138]. Several subunits of the ATP dependent nucleosome remodeling complex SWI/SNF have also been shown to be recruited at promoter sites by p53 [139, 140]. Although not covered in this review, p53 could function as a chromatin accessibility factor not only to induce gene transcription but also to modulate other aspects of DNA metabolism including DNA repair and replication [97, 141, 142].

CONCLUSIONS

Through a combination of transcription and DNA binding assays in yeast and human cell systems it has been possible to address intrinsic functional properties of p53 interactions with target REs. There is a wide opportunity for transactivation specificity (Figure 25.1) [72], suggesting that changes in RE sequences can be a source of inter-individual or evolutionary diversification in p53 mediated responses [74, 80, 101]. Along this line, results from functional as well as DNA binding assays have led to the development of *in silico* approaches for the identification of single nucleotide polymorphisms (SNPs) at p53 REs that are predicted to be functionally significant [32, 55, 143]. A challenge for the near future will be to integrate information on individual SNPs within the p53 transcriptional network to predict haplotypes that would confer higher or lower cancer risk or impact tumor responsiveness to therapies.

In a broader sense, the growing knowledge of the multiple layers and players (mediator complexes, cofactors, transcription factors, epigenetic modifications) contributing to transactivation selectivity by p53, along with a more detailed understanding of which subsets of p53 induced responses are most effective in tumor suppression [144, 145], can generate new targets and strategies for therapeutic intervention. Along this line, functional assessment of mutant p53 alleles may provide valuable information in tailoring therapeutic protocols.

ACKNOWLEDGEMENTS

We regret that due to limitations on length many important contributions to the field could not be referenced in this review. We thank Dr. Omari Bandele for critical reading and comments on the manuscript. We are thankful for the financial support from the NIH intramural research program (to DM and MAR Z01 ES065079), the Department of Defense Breast Cancer Research Program Predoctoral Traineeship Award (BC051212) (to JJ), the Associazione Italiana per la Ricerca sul Cancro, AIRC (to AI) and the Fondazione Pezcoller (to VDS).

REFERENCES

1. Harris SL, Levine AJ. The p53 pathway: positive and negative feedback loops. *Oncogene* 2005;**24**:2899–908.
2. Ko LJ, Prives C. p53: puzzle and paradigm. *Genes Dev* 1996;**10**: 1054–72.
3. Levine AJ, Hu W, Feng Z. The P53 pathway: what questions remain to be explored? *Cell Death Differ* 2006;**13**:1027–36.
4. Olivier M, Petitjean A, Marcel V, et al. Recent advances in p53 research: an interdisciplinary perspective. *Cancer Gene Ther* 2009;**16**:1–12.
5. Vogelstein B, Lane D, Levine AJ. Surfing the p53 network. *Nature* 2000;**408**:307–10.
6. Riley T, Sontag E, Chen P, Levine A. Transcriptional control of human p53-regulated genes. *Nat Rev Mol Cell Biol* 2008;**9**:402–12.
7. Toledo F, Wahl GM. Regulating the p53 pathway: *in vitro* hypotheses, in vivo veritas. *Nat Rev Cancer* 2006;**6**:909–23.
8. Espinosa JM. Mechanisms of regulatory diversity within the p53 transcriptional network. *Oncogene* 2008;**27**:4013–23.
9. Petitjean A, Mathe E, Kato S, et al. Impact of mutant p53 functional properties on TP53 mutation patterns and tumor phenotype: lessons from recent developments in the IARC TP53 database. *Hum Mutat* 2007;**28**:622–9.
10. Cho Y, Gorina S, Jeffrey PD, Pavletich NP. Crystal structure of a p53 tumor suppressor-DNA complex: understanding tumorigenic mutations [see comments]. *Science* 1994;**265**:346–55.
11. Anderson CW, Appella E. Signaling of the p53 Tumor Suppressor through pathways activated by genotoxic and nongenotoxic stresses. In: Bradshaw RA, Dennis EA, editors. *Handbook of cell signalling*. Oxford: Elsevier; 2009.
12. Kastan MB, Onyekwere O, Sidransky D, et al. Participation of p53 protein in the cellular response to DNA damage. *Cancer Res* 1991;**51**:6304–11.
13. Yonish-Rouach E, Grunwald D, Wilder S, et al. p53-mediated cell death: relationship to cell cycle control. *Mol Cell Biol* 1993;**13**:1415–23.
14. Shaw P, Bovey R, Tardy S, et al. Induction of apoptosis by wild-type p53 in a human colon tumor-derived cell line. *PNAS* 1992;**89**:4495–9.
15. Cosme-Blanco W, Shen MF, Lazar AJ, et al. Telomere dysfunction suppresses spontaneous tumorigenesis *in vivo* by initiating p53-dependent cellular senescence. *EMBO Rep* 2007;**8**:497–503.
16. Matoba S, Kang JG, Patino WD, et al. p53 regulates mitochondrial respiration. *Science* 2006;**312**:1650–3.
17. Bensaad K, Tsuruta A, Selak MA, et al. TIGAR, a p53-inducible regulator of glycolysis and apoptosis. *Cell* 2006;**126**:107–20.
18. Crighton D, Wilkinson S, O'Prey J, et al. DRAM, a p53-induced modulator of autophagy, is critical for apoptosis. *Cell* 2006;**126**:121–34.
19. Levine AJ, Feng Z, Mak TW, et al. Coordination and communication between the p53 and IGF-1-AKT-TOR signal transduction pathways. *Genes Dev* 2006;**20**:267–75.
20. Budanov AV, Karin M. p53 target genes sestrin1 and sestrin2 connect genotoxic stress and mTOR signaling. *Cell* 2008;**134**:451–60.
21. Bian J, Sun Y. Transcriptional activation by p53 of the human type IV collagenase (gelatinase A or matrix metalloproteinase 2) promoter. *Molecular and cellular biology* 1997;**17**:6330–8.
22. Kunz C, Pebler S, Otte J, von der Ahe D. Differential regulation of plasminogen activator and inhibitor gene transcription by the tumor suppressor p53. *Nucleic Acids Res* 1995;**23**:3710–17.
23. Cui R, Widlund HR, Feige E, et al. Central role of p53 in the suntan response and pathologic hyperpigmentation. *Cell* 2007;**128**: 853–64.
24. Chen J, Sadowski I. Identification of the mismatch repair genes PMS2 and MLH1 as p53 target genes by using serial analysis of binding elements. *PNAS* 2005;**102**:4813–18.
25. Adimoolam S, Ford JM. p53 and DNA damage-inducible expression of the xeroderma pigmentosum group C gene. *PNAS* 2002;**99**:12,985–12,990.
26. He L, He X, Lowe SW, Hannon GJ. microRNAs join the p53 network: another piece in the tumour-suppression puzzle. *Nat Rev Cancer* 2007;**7**:819–22.
27. Toledo F, Wahl GM. MDM2 and MDM4: p53 regulators as targets in anticancer therapy. *Int J Biochem Cell Biol* 2007;**39**:1476–82.
28. Wang SE, Narasanna A, Whitell CW, et al. Convergence of p53 and transforming growth factor beta (TGFbeta) signaling on activating expression of the tumor suppressor gene maspin in mammary epithelial cells. *J Biol Chem* 2007;**282**:5661–9.
29. Tokino T, Thiagalingam S, el-Deiry WS, et al. p53 tagged sites from human genomic DNA. *Hum Mol Genet* 1994;**3**:1537–42.
30. Barenco M, Tomescu D, Brewer D, Callard R, Stark J, Hubank M. Ranked prediction of p53 targets using hidden variable dynamic modeling. *Genome Biol* 2006;**7**:R25.
31. Hoh J, Jin S, Parrado T, et al. The p53MH algorithm and its application in detecting p53-responsive genes. *PNAS* 2002;**99**:8467–72.
32. Veprintsev DB, Fersht AR. Algorithm for prediction of tumour suppressor p53 affinity for binding sites in DNA. *Nucleic Acids Res* 2008;**36**:1589–98.
33. Smeenk L, van Heeringen SJ, Koeppel M, et al. Characterization of genome-wide p53-binding sites upon stress response. *Nucleic Acids Res* 2008;**36**:3639–54.
34. Wei CL, Wu Q, Vega VB, et al. A global map of p53 transcription-factor binding sites in the human genome. *Cell* 2006;**124**:207–19.
35. Hearnes JM, Mays DJ, Schavolt KL, et al. Chromatin immunoprecipitation-based screen to identify functional genomic binding sites for sequence-specific transactivators. *Mol Cell Biol* 2005;**25**:10,148–10,158,.
36. Kaneshiro K, Tsutsumi S, Tsuji S, et al. An integrated map of p53-binding sites and histone modification in the human ENCODE regions. *Genomics* 2007;**89**:178–88.
37. Ho J, Benchimol S. Transcriptional repression mediated by the p53 tumour suppressor. *Cell Death Differ* 2003;**10**:404–8.

38. Moll UM, Wolff S, Speidel D, Deppert W. Transcription-independent pro-apoptotic functions of p53. *Curr Opin Cell Biol* 2005;**17**:631–6.
39. Murray-Zmijewski F, Slee EA, Lu X. A complex barcode underlies the heterogeneous response of p53 to stress. *Nat Rev Mol Cell Biol* 2008;**9**:702–12.
40. Laptenko O, Prives C. Transcriptional regulation by p53: one protein, many possibilities. *Cell Death Differ* 2006;**13**:951–61.
41. Vousden KH. Outcomes of p53 activation: spoilt for choice. *J Cell Sci* 2006;**119**:5015–20.
42. Weisz L, Oren M, Rotter V. Transcription regulation by mutant p53. *Oncogene* 2007;**26**:2202–11.
43. Resnick MA, Inga A. Functional mutants of the sequence-specific transcription factor p53 and implications for master genes of diversity. *PNAS* 2003;**100**:9934–9.
44. Bassett Jr DE, Boguski MS, Hieter P. Yeast genes and human disease. *Nature* 1996;**379**:589–90.
45. Palermo C, Walworth NC. Yeast as a model system for studying cell cycle checkpoints. In: Nitiss JL, Heitman J, editors. *Yeast as a tool in cancer research.* Dordrecht, Netherlands: Springer; 2007. p. 179–89.
46. Rogojina AT, Zhengsheng L, Nitiss KC, Nitiss JL. Using yeast tools to dissect the action of anticancer drugs: mechanisms of enzyme inhibition and cell killing by agents targeting DNA topoisomerases. In: Nitiss JL, Heitman J, editors. *Yeast as a tool in cancer research.* Dordrecht, Netherlands: Springer; 2007. p. 409–27.
47. Resnick MA, Cox BS. Yeast as an honorary mammal. *Mutat Res* 2000;**451**:1–11.
48. Guarente L. UASs and enhancers: common mechanism of transcriptional activation in yeast and mammals. *Cell* 1988;**52**:303–5.
49. Inga A, Reamon-Buettner SM, Borlak J, Resnick MA. Functional dissection of sequence-specific NKX2-5 DNA binding domain mutations associated with human heart septation defects using a yeast-based system. *Hum Mol Genet* 2005;**14**:1965–75.
50. Reamon-Buettner SM, Ciribilli Y, Inga A, Borlak J. A loss-of-function mutation in the binding domain of HAND1 predicts hypoplasia of the human hearts. *Hum Mol Genet* 2008;**17**:1397–405.
51. Epinat JC, Whiteside ST, Rice NR, Israel A. Reconstitution of the NF-kappa B system in Saccharomyces cerevisiae for isolation of effectors by phenotype modulation. *Yeast* 1997;**13**:599–612.
52. Balmelli-Gallacchi P, Schoumacher F, Liu JW, et al. A yeast-based bioassay for the determination of functional and non-functional estrogen receptors. *Nucleic Acids Res* 1999;**27**:1875–81.
53. Thukral SK, Lu Y, Blain GC, et al. Discrimination of DNA binding sites by mutant p53 proteins. *Mol Cell Biol* 1995;**15**:5196–202.
54. Flaman JM, Frebourg T, Moreau V, et al. A simple p53 functional assay for screening cell lines, blood, and tumors. *PNAS* 1995;**92**:3963–7.
55. Tomso DJ, Inga A, Menendez D, et al. Functionally distinct polymorphic sequences in the human genome that are targets for p53 transactivation. *PNAS* 2005;**102**:6431–6.
56. Kato S, Han SY, Liu W, et al. Understanding the function-structure and function-mutation relationships of p53 tumor suppressor protein by high-resolution missense mutation analysis. *PNAS* 2003;**100**:8424–9.
57. Ishioka C, Englert C, Winge P, et al. Mutational analysis of the carboxy-terminal portion of p53 using both yeast and mammalian cell assays *in vivo*. *Oncogene* 1995;**10**:1485–92.
58. Brachmann RK, Vidal M, Boeke JD. Dominant-negative p53 mutations selected in yeast hit cancer hot spots. *PNAS* 1996;**93**:4091–5.
59. Camplejohn RS, Rutherford J. p53 functional assays: detecting p53 mutations in both the germline and in sporadic tumours. *Cell Prolif* 2001;**34**:1–14.
60. Smardova J. FASAY: a simple functional assay in yeast for identification of p53 mutation in tumors. *Neoplasma* 1999;**46**:80–8.
61. Di Como CJ, Prives C. Human tumor-derived p53 proteins exhibit binding site selectivity and temperature sensitivity for transactivation in a yeast-based assay. *Oncogene* 1998;**16**:2527–39.
62. Flaman JM, Robert V, Lenglet S, et al. Identification of human p53 mutations with differential effects on the bax and p21 promoters using functional assays in yeast. *Oncogene* 1998;**16**:1369–72.
63. Ludwig RL, Bates S, Vousden KH. Differential activation of target cellular promoters by p53 mutants with impaired apoptotic function. *Mol Cell Biol* 1996;**16**:4952–60.
64. Friedlander P, Haupt Y, Prives C, Oren M. A mutant p53 that discriminates between p53-responsive genes cannot induce apoptosis. *Mol Cell Biol* 1996;**16**:4961–71.
65. Iwakuma T, Lozano G. Crippling p53 activities via knock-in mutations in mouse models. *Oncogene* 2007;**26**:2177–84.
66. Campomenosi P, Monti P, Aprile A, et al. p53 mutants can often transactivate promoters containing a p21 but not Bax or PIG3 responsive elements. *Oncogene* 2001;**20**:3573–9.
67. Baroni TE, Wang T, Qian H, et al. A global suppressor motif for p53 cancer mutants. *PNAS* 2004;**101**:4930–5.
68. Wang T, Kobayashi T, Takimoto R, et al. hADA3 is required for p53 activity. *EMBO J* 2001;**20**:6404–13.
69. Yousef AF, Xu GW, Mendez M, et al. Coactivator requirements for p53-dependent transcription in the yeast Saccharomyces cerevisiae. *Int J Cancer* 2008;**122**:942–6.
70. Qian H, Wang T, Brachmann RK. Not all p53 DNA binding sites are created equal. *Proc Am Assoc Cancer Res* 2002;**43**:1141.
71. Freeman J, Schmidt S, Scharer E, Iggo R. Mutation of conserved domain II alters the sequence specificity of DNA binding by the p53 protein. *EMBO J* 1994;**13**:5393–400.
72. Menendez D, Inga A, Jordan JJ, Resnick MA. Changing the p53 master regulatory network: ELEMENTary, my dear Mr Watson. *Oncogene* 2007;**26**:2191–201.
73. Inga A, Resnick MA. Novel human p53 mutations that are toxic to yeast can enhance transactivation of specific promoters and reactivate tumor p53 mutants. *Oncogene* 2001;**20**:3409–19.
74. Inga A, Storici F, Darden TA, Resnick MA. Differential transactivation by the p53 transcription factor is highly dependent on p53 level and promoter target sequence. *Mol Cell Biol* 2002;**22**:8612–25.
75. Jordan JJ, Menendez D, Inga A, et al. Noncanonical DNA motifs as transactivation targets by wild type and mutant p53. *PLoS Genet* 2008;**4**:e1,000,104.
76. Biggar SR, Crabtree GR. Cell signaling can direct either binary or graded transcriptional responses. *EMBO J* 2001;**20**:3167–76.
77. Ramsey SA, Smith JJ, Orrell D, et al. Dual feedback loops in the GAL regulon suppress cellular heterogeneity in yeast. *Nat Genet* 2006;**38**:1082–7.
78. Inga A, Nahari D, Velasco-Miguel S, et al. A novel p53 mutational hotspot in skin tumors from UV-irradiated Xpc mutant mice alters transactivation functions. *Oncogene* 2002;**21**:5704–15.
79. Zeng J, Yan J, Wang T, et al. Genome wide screens in yeast to identify potential binding sites and target genes of DNA-binding proteins. *Nucleic Acids Res* 2008;**36**:e8.
80. Weinberg RL, Veprintsev DB, Bycroft M, Fersht AR. Comparative binding of p53 to its promoter and DNA recognition elements. *J Mol Biol* 2005;**348**:589–96.
81. Balagurumoorthy P, Sakamoto H, Lewis MS, et al. Four p53 DNA-binding domain peptides bind natural p53-response elements and bend the DNA. *PNAS* 1995;**92**:8591–5.

82. Nagaich AK, Appella E, Harrington RE. DNA bending is essential for the site-specific recognition of DNA response elements by the DNA binding domain of the tumor suppressor protein p53. *J Biol Chem* 1997;**272**:14,842–14,849.
83. Pan Y, Nussinov R. p53-Induced DNA bending: the interplay between p53-DNA and p53-p53 interactions. *J Phys Chem* 2008;**112**: 6716–24.
84. Kim E, Deppert W. The versatile interactions of p53 with DNA: when flexibility serves specificity. *Cell Death Differ* 2006;**13**: 885–9.
85. McKinney K, Prives C. Efficient specific DNA binding by p53 requires both its central and C-terminal domains as revealed by studies with high-mobility group 1 protein. *Mol Cell Biol* 2002; **22**:6797–808.
86. Tidow H, Melero R, Mylonas E, et al. Quaternary structures of tumor suppressor p53 and a specific p53 DNA complex. *PNAS* 2007;**104**:12,324–12,329.
87. Ma B, Pan Y, Zheng J, et al. Sequence analysis of p53 response-elements suggests multiple binding modes of the p53 tetramer to DNA targets. *Nucleic Acids Res* 2007;**35**:2986–3001.
88. Li L, Liang Y, Bass RL. GAPWM: a genetic algorithm method for optimizing a position weight matrix. *Bioinformatics* 2007; **23**:1188–94.
89. Tan T, Chu G. p53 Binds and activates the xeroderma pigmentosum DDB2 gene in humans but not mice. *Mol Cell Biol* 2002;**22**:3247–54.
90. Hoffman WH, Biade S, Zilfou JT, et al. Transcriptional repression of the anti-apoptotic survivin gene by wild type p53. *J Biol Chem* 2002;**277**:3247–57.
91. Jackson P, Mastrangelo I, Reed M, et al. Synergistic transcriptional activation of the MCK promoter by p53: tetramers link separated DNA response elements by DNA looping. *Oncogene* 1998;**16**:283–92.
92. Stenger JE, Tegtmeyer P, Mayr GA, et al. p53 oligomerization and DNA looping are linked with transcriptional activation. *EMBO J* 1994;**13**:6011–20.
93. Murray-Zmijewski F, Lane DP, Bourdon JC. p53/p63/p73 isoforms: an orchestra of isoforms to harmonise cell differentiation and response to stress. *Cell Death Differ* 2006;**13**:962–72.
94. Kitayner M, Rozenberg H, Kessler N, et al. Structural basis of DNA recognition by p53 tetramers. *Mol Cell* 2006;**22**:741–53.
95. Okorokov AL, Sherman MB, Plisson C, et al. The structure of p53 tumour suppressor protein reveals the basis for its functional plasticity. *EMBO J* 2006;**25**:5191–200.
96. Chorley BN, Wang X, Campbell MR, et al. Discovery and verification of functional single nucleotide polymorphisms in regulatory genomic regions: current and developing technologies. *Mutat Res* 2008;**659**:147–57.
97. Allison SJ, Milner J. Remodelling chromatin on a global scale: a novel protective function of p53. *Carcinogenesis* 2004;**25**:1551–7.
98. Menendez D, Inga A, Snipe J, et al. A single-nucleotide polymorphism in a half-binding site creates p53 and estrogen receptor control of vascular endothelial growth factor receptor 1. *Mol Cell Biol* 2007;**27**:2590–600.
99. Menendez D, Krysiak O, Inga A, et al. A SNP in the flt-1 promoter integrates the VEGF system into the p53 transcriptional network. *PNAS* 2006;**103**:1406–11.
100. Horvath MM, Wang X, Resnick MA, Bell DA. Divergent evolution of human p53 binding sites: cell cycle versus apoptosis. *PLoS Genet* 2007;**3**:e127.
101. Jegga AG, Inga A, Menendez D, et al. Functional evolution of the p53 regulatory network through its target response elements. *PNAS* 2008;**105**:944–9.
102. Wang T, Zeng J, Lowe CB, et al. Species-specific endogenous retroviruses shape the transcriptional network of the human tumor suppressor protein p53. *PNAS* 2007;**104**:18,613–18,618.
103. Zemojtel T, Kielbasa SM, Arndt PF, et al. Methylation and deamination of CpGs generate p53-binding sites on a genomic scale. *Trends Genet* 2008;**25**:63–6.
104. Monti P, Ciribilli Y, Jordan J, et al. Transcriptional functionality of germ line p53 mutants influences cancer phenotype. *Clin Cancer Res* 2007;**13**:3789–95.
105. Olivier M, Langerod A, Carrieri P, et al. The clinical value of somatic TP53 gene mutations in 1,794 patients with breast cancer. *Clin Cancer Res* 2006;**12**:1157–67.
106. Lane DP, Crawford LV. T antigen is bound to a host protein in SV40-transformed cells. *Nature* 1979;**278**:261–3.
107. Linzer DI, Levine AJ. Characterization of a 54K dalton cellular SV40 tumor antigen present in SV40-transformed cells and uninfected embryonal carcinoma cells. *Cell* 1979;**17**:43–52.
108. Yang A, Kaghad M, Caput D, McKeon F. On the shoulders of giants: p63, p73 and the rise of p53. *Trends Genet* 2002;**18**:90–5.
109. Yang A, Schweitzer R, Sun D, et al. p63 is essential for regenerative proliferation in limb, craniofacial and epithelial development. *Nature* 1999;**398**:714–18.
110. Yang A, Walker N, Bronson R, et al. p73-deficient mice have neurological, pheromonal and inflammatory defects but lack spontaneous tumours. *Nature* 2000;**404**:99–103.
111. Flores ER, Sengupta S, Miller JB, et al. Tumor predisposition in mice mutant for p63 and p73: evidence for broader tumor suppressor functions for the p53 family. *Cancer Cell* 2005;**7**:363–73.
112. Flores ER, Tsai KY, Crowley D, et al. p63 and p73 are required for p53-dependent apoptosis in response to DNA damage. *Nature* 2002;**416**:560–4.
113. Cordenonsi M, Dupont S, Maretto S, et al. Links between tumor suppressors: p53 is required for TGF-beta gene responses by cooperating with Smads. *Cell* 2003;**113**:301–14.
114. Lecona E, Barrasa JI, Olmo N, et al. Upregulation of annexin A1 expression by butyrate in human colon adenocarcinoma cells: role of p53, NF-Y, and p38 mitogen-activated protein kinase. *Mol Cell Biol* 2008;**28**:4665–74.
115. Benatti P, Basile V, Merico D, et al. A balance between NF-Y and p53 governs the pro- and anti-apoptotic transcriptional response. *Nucleic Acids Res* 2008;**36**:1415–28.
116. Di Agostino S, Strano S, Emiliozzi V, et al. Gain of function of mutant p53: the mutant p53/NF-Y protein complex reveals an aberrant transcriptional mechanism of cell cycle regulation. *Cancer Cell* 2006;**10**:191–202.
117. Molinari AM, Bontempo P, Schiavone EM, et al. Estradiol induces functional inactivation of p53 by intracellular redistribution. *Cancer Res* 2000;**60**:2594–7.
118. Liu W, Konduri SD, Bansal S, et al. Estrogen receptor-alpha binds p53 tumor suppressor protein directly and represses its function. *J Biol Chem* 2006;**281**:9837–40.
119. Liu G, Schwartz JA, Brooks SC. Estrogen receptor protects p53 from deactivation by human double minute-2. *Cancer Res* 2000;**60**:1810–14.
120. Ikeda A, Sun X, Li Y, et al. p300/CBP-dependent and -independent transcriptional interference between NF-kappaB RelA and p53. *Biochem Biophys Res Commun* 2000;**272**:375–9.

121. Kawauchi K, Araki K, Tobiume K, Tanaka N. Activated p53 induces NF-kappaB DNA binding but suppresses its transcriptional activation. *Biochem Biophys Res Commun* 2008;**372**:137–41.
122. Webster GA, Perkins ND. Transcriptional cross talk between NF-kappaB and p53. *Mol Cell Biol* 1999;**19**:3485–95.
123. Chen F, Castranova V. Nuclear factor-kappaB, an unappreciated tumor suppressor. *Cancer Res* 2007;**67**:11,093–11,098.
124. Ryan KM, Ernst MK, Rice NR, Vousden KH. Role of NF-kappaB in p53-mediated programmed cell death. *Nature* 2000;**404**:892–7.
125. Karin M. Nuclear factor-kappaB in cancer development and progression. *Nature* 2006;**441**:431–6.
126. Samuels-Lev Y, O'Connor DJ, Bergamaschi D, et al. ASPP proteins specifically stimulate the apoptotic function of p53. *Mol Cell* 2001;**8**:781–94.
127. Bergamaschi D, Samuels Y, O'Neil NJ, et al. iASPP oncoprotein is a key inhibitor of p53 conserved from worm to human. *Nat Genet* 2003;**33**:162–7.
128. Tanaka T, Ohkubo S, Tatsuno I, Prives C. hCAS/CSE1L associates with chromatin and regulates expression of select p53 target genes. *Cell* 2007;**130**:638–50.
129. Das S, Raj L, Zhao B, et al. Hzf Determines cell survival upon genotoxic stress by modulating p53 transactivation. *Cell* 2007;**130**:624–37.
130. Ard PG, Chatterjee C, Kunjibettu S, et al. Transcriptional regulation of the mdm2 oncogene by p53 requires TRRAP acetyltransferase complexes. *Mol Cell Biol* 2002;**22**:5650–61.
131. Espinosa JM, Emerson BM. Transcriptional Regulation by p53 through Intrinsic DNA/Chromatin Binding and Site-Directed Cofactor Recruitment. *Mol Cell* 2001;**8**:57–69.
132. Espinosa JM, Verdun RE, Emerson BM. p53 functions through stress- and promoter-specific recruitment of transcription initiation components before and after DNA damage. *Mol Cell* 2003;**12**:1015–27.
133. Kaeser MD, Iggo RD. Chromatin immunoprecipitation analysis fails to support the latency model for regulation of p53 DNA binding activity *invivo*. *Proc Nat Acad Sci USA* 2002;**99**:95–100.
134. Vrba L, Junk DJ, Novak P, Futscher BW. p53 induces distinct epigenetic states at its direct target promoters. *BMC Genomics* 2008;**9**:486.
135. Chan HM, La Thangue NB. p300/CBP proteins: HATs for transcriptional bridges and scaffolds. *J Cell Sci* 2001;**114**:2363–73.
136. Gamper AM, Roeder RG. Multivalent binding of p53 to the STAGA complex mediates coactivator recruitment after UV damage. *Mol Cell Biol* 2008;**28**:2517–27.
137. Rokudai S, Aikawa Y, Tagata Y, et al. Monocytic Leukemia Zinc Finger (MOZ) Interacts with p53 to Induce p21 Expression and Cell-cycle Arrest. *J Biol Chem* 2009;**284**:237–44.
138. An W, Kim J, Roeder RG. Ordered cooperative functions of PRMT1, p300, and CARM1 in transcriptional activation by p53. *Cell* 2004;**117**:735–48.
139. Lee D, Kim JW, Seo T, et al. SWI/SNF complex interacts with tumor suppressor p53 and is necessary for the activation of p53-mediated transcription. *J Biol Chem* 2002;**277**:22,330–22,337.
140. Oh J, Sohn DH, Ko M, et al. BAF60a interacts with p53 to recruit the SWI/SNF complex. *J Biol Chem* 2008;**283**:11,924–11,934.
141. Rubbi CP, Milner J. p53 is a chromatin accessibility factor for nucleotide excision repair of DNA damage. *EMBO J* 2003;**22**:975–86.
142. Iizuka M, Sarmento OF, Sekiya T, et al. Hbo1 Links p53-dependent stress signaling to DNA replication licensing. *Mol Cell Biol* 2008;**28**:140–53.
143. Resnick MA, Tomso D, Inga A, et al. Functional diversity in the gene network controlled by the master regulator p53 in humans. *Cell cycle* 2005;**4**:1026–9.
144. Christophorou MA, Martin-Zanca D, Soucek L, et al. Temporal dissection of p53 function *in vitro* and *in vivo*. *Nat genet* 2005;**37**:718–26.
145. Christophorou MA, Ringshausen I, Finch AJ, et al. The pathological response to DNA damage does not contribute to p53-mediated tumour suppression. *Nature* 2006;**443**:214–17.
146. Storici F, Resnick MA. The delitto perfetto approach to *in vivo* site-directed mutagenesis and chromosome rearrangements with synthetic oligonucleotides in yeast. *Methods Enzymol* 2006;**409**:329–45.
147. Jordan, Jennifer J, Diversity within the master regulatory p53 transcriptional network: Impact of sequence, binding motifs and mutations. Dissertations & Theses at University of North Carolina at Chapel Hill, 2008, Chapter IV, 127–72.

Chapter 26

The Heat Shock Response and the Stress of Misfolded Proteins

Richard I. Morimoto and Sandy D. Westerheide

Department of Biochemistry, Molecular Biology and Cell Biology, Rice Institute for Biomedical Research, Northwestern University, Evanston, Illinois

INTRODUCTION

The heat shock response (HSR) is an inducible molecular response to a disruption of protein homeostasis (proteostasis) that results in the elevated expression of cytoprotective genes that protect the proteome (Figure 26.1). The four general categories of environmental and physiological regulators of the HSR include: (1) environmental stress, such as heat shock, amino acid analogs, drugs, oxidative stress, toxic chemicals, heavy metals, and pharmacologically active small molecules; (2) cell growth and developmental conditions, including cell cycle, growth factors, development, differentiation, and activation by certain oncogenes; (3) pathology and disease, such as neuroendocrine stress, tissue injury and repair, fever, inflammation, infection, ischemia and reperfusion, and cancer; and (4) diseases of protein conformation including Huntington's disease, Alzheimer's disease, Parkinson's disease, and ALS. For each of these categories, the various conditions indicated are typically associated with the elevated expression of one or more heat shock and stress induced proteins through activation of heat shock factor (HSF) and the HSR.

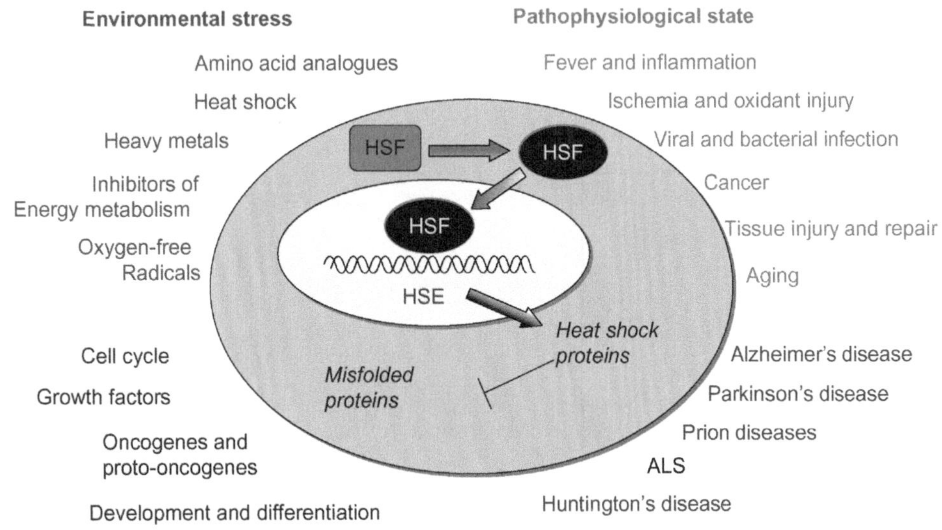

FIGURE 26.1 Conditions that activate the heat shock response.
The expression of heat shock (stress responsive) genes is regulated by diverse exposures to environmental and physiological stress, conditions of cell growth, and development, in response to conditions of altered pathophysiological states, and protein aggregation diseases, that is, conditions that are associated with the appearance of misfolded proteins. The expression of heat shock proteins including molecular chaperones is regulated by the heat shock transcription factor HSF1. Molecular chaperones function to prevent and repair protein damage through interactions with the misfolded proteins.

Handbook of Cell Signaling, Three-Volume Set 2 ed.

Common to these diverse stress conditions are acute and chronic challenges to proteostasis that influence protein synthesis, folding, translocation, assembly, disassembly, and degradation. Consequently, an increased flux of non-native protein intermediates, if left unprotected, can misfold to form protein aggregates and other toxic protein species. The HSR, through the elevated synthesis of molecular chaperones and proteases, responds rapidly and precisely to the intensity and duration of specific environmental and physiological stress signals to restore proteostasis and prevent further protein damage [1–4]. Transient exposure to intermediate elevated temperatures or lower levels of chemical and environmental stress has cytoprotective effects against sustained, normally lethal, exposures to stress [5]. This reveals a valuable cell and organismal survival strategy that "a little stress is good."

TRANSCRIPTIONAL REGULATION OF THE HEAT SHOCK RESPONSE

Simultaneous with the rapid transcriptional induction of genes encoding molecular chaperones and other cytoprotective genes, exposure to heat shock results in transcriptional repression of many genes [6–9] by inhibition of RNA polymerase II [10]. The stress induced transcription of heat shock genes is regulated by a family of heat shock transcription factors (HSFs 1–4) [11]. HSFs are highly conserved and are required for normal cell growth and development in addition to their central importance in stress adaptation, survival, and disease. In humans, four HSF genes (HSF1, 2, 3, and 4) are expressed of which HSF1 is essential for the heat shock response [12], HSF2 is developmentally regulated and important for neuronal specification [13–19], and HSF4 is important for lens crystallins [20, 21]. HSF1 as the principal stress activated factor binds to heat shock elements (HSEs) consisting of multiple contiguous inverted repeats of the pentamer sequence nGAAn located in the promoter regions of all heat shock responsive genes in eukaryotes [22–24].

The regulation of HSF1 activity is complex and involves a multi-step pathway (Figure 26.2) that responds to diverse stress conditions associated with the appearance of misfolded proteins [25]. In addition to heat shock, heavy metals, and oxidants, stress induced effects on intracellular redox status activates HSF1 [26]. There are also different stress specific activation pathways that have been characterized in addition to a proposed role for the non-coding RNA HSR1, which has a temperature sensing domain that could affect HSF1 activity [27–29]. Regardless of the specific

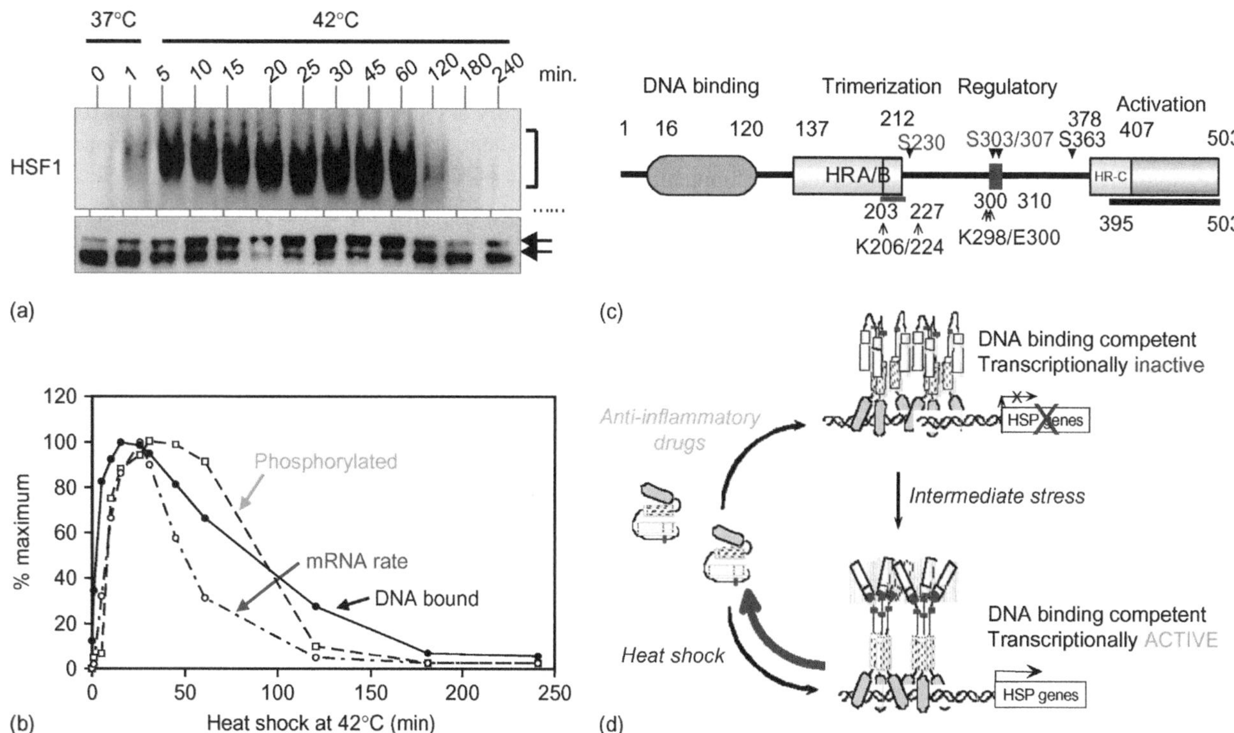

FIGURE 26.2 Regulation of the heat shock transcriptional response.
(a) Top panel – HSF1 DNA binding activity, as detected by electromobility shift assay, is induced to maximal levels within minutes of exposure of HeLa cells to heat shock. Thereafter, HSF1 DNA binding activity attenuates. Bottom panel – HSF1, detected by western blot analysis, is present prior to heat shock and undergoes a change in electrophoretic mobility due to phosphorylation. (b) Comparison of the kinetics of HSF1 DNA binding activity, transcription of the endogenous Hsp70 gene, and levels of HSF1 phosphorylation during heat shock. Physical map of HSF1 indicating the location of the DNA binding domain, HR AB and HR C heptad repeat domains, the negative regulatory domain with key amino acid residues indicated, and the transcriptional transactivation domain. (d) HSF1 prior to heat shock exists as a monomer that is the rapidly activated to a DNA-binding competent trimer. Exposure to anti-inflammatory drugs also activates HSF1 to a transcriptionally inert trimer that can be further activated by heat shock.

characteristics of the stress signal, the molecular response is that HSF1, which normally exists as an inert negatively regulated monomer in the cytoplasmic or nuclear compartments, becomes activated to a homotrimer that accumulates in the nucleus. DNA binding activity is acquired upon trimerization and results in occupancy of HSEs by HSF1. Many HSF1 target genes, including hsp70, hsp90, hsp40, and the small heat shock genes are also transcribed constitutively due to the presence of binding sites for other transcription factors or the binding of low basal levels of HSF1. HSF1 is negatively regulated by feedback control through transient interactions with the chaperones Hsp70 and Hsp90 [30, 31]. Thus, the HSR involves a multi-step pathway with both positive and negative regulators.

HSF1 is composed of multiple domains for DNA binding, trimer formation, negative regulation, and transcriptional transactivation [32]. The DNA binding domain (DBD), located at the N-terminus, is highly conserved and contains a helix-turn-helix motif first identified from the crystal structure of the *Kluyveromyces lactis* HSF DBD [33]. Upon activation, HSF1 forms a trimer by interaction of the N-terminal hydrophobic heptad repeats (HR-A/B) [34]. During stress conditions, these heptads form a leucine zipper coiled-coil allowing trimer formation, while during non-stress conditions, a C-terminal heptad sequence (HR-C) is thought to form an intramolecular coiled-coil with HR-A/B to suppress spontaneous HSF1 trimerization [35, 36]. Located at the extreme C-terminus of HSF1 is a bipartite activation domain (AD1 and AD2) composed of acidic and hydrophobic residues [37, 38]. Under non-stress conditions, the activity of AD1 and AD2 is suppressed by the negative regulatory domain (RD) [37–39].

In vivo studies on transcription of the *Drosophila* hsp70 gene have shown that HSF1 binds to HSEs in the promoter within seconds upon activation by heat shock, and accumulates to maximal levels by two minutes [40]. HSF1 DNA binding is regulated at two steps, upon activation and in attenuation when HSF1 trimers dissociate from the HSEs and hsp gene transcription arrests. HSF1 is extensively posttranslationally modified by phosphorylation, sumoylation and acetylation [41–47]. HSF1 is constitutively phosphorylated and becomes more extensively phosphorylated upon exposure to stress [44]. Among the many sites for constitutive HSF1 phosphorylation, residues S303, S307, and S308 have important roles in the negative regulation of HSF1 activity, whereas sites of inducible phosphorylation (S230, S326, and S419) promote HSF1 transcriptional activity [43, 46, 48]. Sumoylation of HSF1 on lysine 298 is dependent upon phosphorylation on serines 303 and 307 and has an inhibitory effect on transcriptional activity [42]. Additionally, acetylation occurs on multiple lysines within HSF1, with acetylation at K80 within the DBD causing inhibition of HSF1 DNA binding [47]. The regulation of HSF1 by posttranslational modification allows fine-tuning to ensure precise and tight regulatory control.

Studies on the HSR in *Drosophila* have provided an exceptional model for transcriptional regulation and led to an understanding of elongation control through a paused RNA polymerase II [49]. Upon heat shock, HSF1 cooperates with the paused polymerase to recruit elongation factors including P-TEFb, a heterodimer of the kinase Cdk9 and Cyclin T, for maturation of the polymerase into an elongating complex [50]. HSF1 transcriptional activity is also modulated through its interactions with accessory transcription factors, coactivators, and mRNA processing factors. Chromatin within the hsp70 promoter is remodeled by recruitment of the remodeling complexes Mediator and FACT (Facilitates Chromatin Transcription) [51, 52]. The association of HSF1 with mRNA processing components was shown in *Drosophila* by the recruitment of exosome factors [53] and in human cells by the recruitment of symplekin, a protein known to form a complex with mRNA polyadenylation factors [54]. HSF1 activity in human cells is further regulated by an intranuclear equilibrium between HSF1 localization to HSEs of target promoters and to nuclear stress bodies (nSBs) that assemble on satellite III repeats located at the 9q11–q12 heterochromatic region of primate genomes [55]. Within these satellite III repetitive sequences are thousands of copies of redundant, degenerate HSEs that are actively transcribed [56]. The function of these transcripts and their association with the HSR is unknown, but may turn out to have important positive or negative regulatory function as these sites also accumulate RNA polymerase II, acetylated histones, and a variety of RNA processing factors [57]. Recent genome-wide mRNA array studies using yeast and mammalian cells have shown that many genes besides chaperones are induced by heat stress, including genes involved in protein degradation, transport, signal transduction, cytoskeletal maintenance, and metabolism [58, 59]. It has been suggested that approximately 3 percent of the total genomic loci in yeast are occupied by HSF upon heat shock [58]. HSF1 binds to many, but not all, HSE containing promoters and can function both to activate and repress gene transcription [60]. Thus, HSF1 activation regulates a network of genes required to respond to stress. The complex regulatory system of HSF1 allows it to respond appropriately to the precise needs of the cell.

CHAPERONE FUNCTION IN NORMAL AND DISEASE STATES

The family of molecular chaperones are highly conserved, abundant in growing cells, and can attain concentrations of 5–20 percent of total cellular protein [1, 3, 61]. Chaperones are classified by molecular size: Hsp100, Hsp90, Hsp70, Hsp60, Hsp40, and small Hsps, and are distributed in all subcellular compartments (Table 26.1). Biochemical studies on chaperones have established common properties

TABLE 26.1 Nomenclature, location, and function of molecular chaperones

Family	Organism	Chaperones	Location	Functions
HSP 100	*E. coli*	ClpA, B, C	Cytosol	Role in stress tolerance; resolubilization of heat inactivated proteins from insoluble aggregates
	S. cerevisiae	HSP104	Cytosol	
HSP90	*E. coli*	HtpG	Cytosol	Role in signal transduction (e.g., interaction with steriod hormone receptors, kinases, phosphatases); maintains proteins in soluble intermediate states (assembly competent); autoregulation of the HSR; role in cell cycle and proliferation
	S. cerevisiae	HSP83	Cytosol/nucleus	
	Mammals	HSP90	Cytosol/nucleus	
		GRP94	ER	
HSP70	*E. coli*	DnaK	Cytosol	Roles in lambda phage replication; autoregulation of the HSR; interaction with nascent polypeptides; functions in interorganellar transport; roles in signal transduction; refolds and maintains denatured proteins *in vitro*; role in cell cycle and proliferation; anti-apoptotic activity; potential antigen-presenting molecule in tumor cells
	S. cerevisiae	Ssa 1–4	Cytosol	
		Ssb 1,2	Cytosol	
		Ssc1	Mitochondria	
		Kar2	ER	
	Mammals	Hsp70	Cytosol/nucleus	
		BIP	ER	
		mHSP70	Mitochondria	
HSP60	*E. coli*	groEL	Cytosol	Refolds and prevents aggregation of denatured proteins *in vitro*; may facilitate protein degradation by acting as a cofactor in proteolytic systems; role in the assembly of bacteriophages, mitochondria, and Rubisco in chloroplasts
	S. cerevisiae	HSP60	Mitochondria	
	Plants	Cpn60	*Chloroplast*	
	Mammals	HSP60	Mitochondria	
HSP40	*E. coli*	dnaJ	Cytosol	Essential co-chaperone activity with HSP70 proteins to enhance of ATPase activity and substrate release
	S. cerevisiae	Ydj1/Sis1	Cytosol/nucleus	
	Mammals	Hdj1	Cytosol/nucleus	
Small HSPs	*E. coli*	Ibp A and B	Cytosol	Suppresses aggregation and heat inactivation of protein *in vitro*; confers thermotolerance
	S. cerevisiae	HSP27	Cytosol	
	Mammals	αA and αB- crystallins	Cytosol	

of Hsp104, Hsp90, Hsp70, the small Hsps, immunophilins (FKPB52 and CyP40), the steroid aporeceptor protein p23, and Hip (Hsp70 and Hsp90 interacting protein) to prevent the *in vitro* aggregation of model protein substrates and to maintain substrates in intermediate folded states that are competent for subsequent refolding to the native state [62–65]. A distinction among proteins that exhibit the properties of chaperones is that folding to the native state requires the activity of a specific subset of ATP binding chaperones such as Hsp90, Hsp70, or TriC/Hsp60/GroEL (Figure 26.3). Cycles of nucleotide binding and hydrolysis, associated with the binding and release of the protein substrate, are regulated by co-chaperones such as p23 or the immunophilins, which enhance Hsp90, dnaJ/Hsp40, and Hip, or Bag proteins that regulate the nucleotide state of Hsp70, or Hsp10/groES that stimulates Hsp60/groEL [62, 66, 67]. The association of chaperones with different co-chaperone partner proteins thus influences the folded state of the substrate and consequently its activities (Figure 26.3).

Although some proteins can refold spontaneously, *in vitro* and perhaps *in vivo*, large multi-domain proteins readily misfold and form aggregates. The challenge, *in vivo*, within the densely packed environment of the cell is to ensure that nascent polypeptides fold, translocate, and assemble as multimeric complexes, and that non-native intermediates that accumulate during normal biosynthesis or are enhanced due to mutations or stress are efficiently captured and directed to a triage process for refolding or clearance. Molecular chaperones of the Hsp104, Hsp90, Hsp70, Hsp60, and small Hsp class are highly efficient and capture non-native intermediates and, together with co-chaperones and ATP, facilitate the folded native state (Figure 26.3) [3]. The Hsp70 chaperones are highly efficient to recognize hydrophobic residue-rich stretches in polypeptides that are transiently exposed in intermediates and become confined to the hydrophobic core as the protein folds to the native state [68]. This contrasts with the Hsp60/GroEL chaperonin, that creates a protected environment with the properties of a "protein folding test tube" in which non-native proteins undergo rounds of ATP binding and hydrolysis that drives substrate binding and release to the native state [69, 70]. Common to these chaperone interactions is their ability to shift the equilibrium of protein folding toward on-pathway events and to minimize the appearance of off-pathway, aggregation prone species.

The ability of a cell to know whether to grow, divide, differentiate, or die is influenced by extracellular signals and the ability to properly recognize and respond to these signals. The cell may receive these extracellular signals in different forms such as soluble hormones, small peptides, or proteins attached to neighboring cells. Cellular receptors receiving the signals transmit the extracellular information to the nucleus through cascades of protein–protein interactions and biochemical reactions. Chaperones of the Hsp90 and Hsp70 family and their co-chaperones have been found to interact with key cell signaling molecules, including intracellular receptors, tyrosine- and serine/threonine kinases, cell cycle regulators, and cell death regulators [71–74]. Decreasing the levels of functional Hsp90 in *Drosophila*, by genetic mutation or by treatment with the Hsp90 inhibitor geldanamycin, causes developmental abnormalities [75]. Likewise, increasing the levels of Hsp70, by overexpression or upon heat shock, has growth inhibitory effects on mammalian tissue culture cells and in *Drosophila* salivary gland cells, whereas expression of a dominant negative form of Hsp70 causes developmental defects in *Drosophila* [76–78]. However, what remains less well understood is whether this represents a general strategy of the cell to link specific signaling pathways with cell stress sensing events.

Interestingly, cells that have lost their ability to properly regulate cell growth, such as tumor cells, often express high levels of multiple heat shock proteins compared to their normal parental cells [79]. Depletion of Hsp90 by geldanamycin, or of Hsp70 by antisense methodology in transformed cells, but not in their non-transformed counterparts, causes either an arrest of cell growth or progression into cell death [80, 81]. Tumor cells seem to have become dependent on elevated levels of Hsps, although the beneficial reasons for this have yet to be clearly established. One possibility is the ability of chaperones to suppress and buffer mutations that accumulate during the transformation

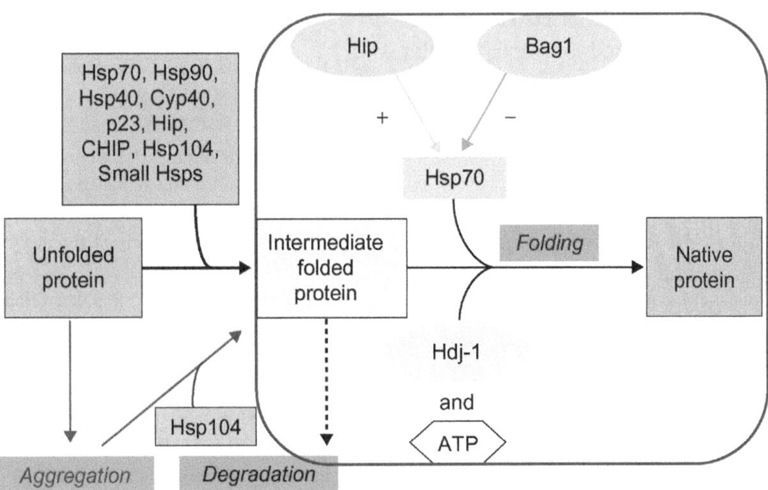

FIGURE 26.3 Chaperone networks and the regulation of protein conformation.
The biochemical fate of an unfolded protein is schematically presented. In the absence of chaperones, unfolded proteins are prone to aggregation. The presence of the molecular chaperones Hsp70, Hsp90, Hsp40, Cyp40, p23, Hip, CHIP Hsp104, or small Hsps suppress aggregation "holding" and result in an intermediate folded state. The chaperone Hsp 104 can also dissociate aggregates or target the damaged proteins for subsequent degradation. Proteins in an intermediate folded state can be refolded to the native state by Hsp70 and ATP, a process that is regulated by the co-chaperones Hdj-1, Hip, and Bag1.

process, thus allowing otherwise misfolded proteins to acquire various activities that influence cell viability and enhanced cell growth. This is exemplified by the relationship between p53 and Hsp90, where mutant forms of p53, but not wild-type p53, depend on Hsp90 for their normal levels and function [82]. Consistent with the dependence of cancer cells on high expression levels of chaperones for growth, it was found that mice lacking HSF1 are protected from tumors induced by mutations in Ras or p53 [83].

Human neurodegenerative disorders represent a clear example of a large class of diseases associated with the accumulation of misfolded and damaged proteins that lead to the appearance of protein aggregates, fibrils, or plaques. These include inherited disorders caused by CAG/polyglutamine expansion as occurs in Huntington's disease (HD), Kennedy's disease, the spinocerebellar ataxias SCA1, SCA2, MJD (SCA3), SCA6, and SCA7, dentorubral-pallidoluysian atrophy (DPRLA), Alzheimer's disease, and prion disease [84]. A hallmark of these diseases is the expression of abnormal proteins that form a predominant β-sheet conformation that self-associates to form insoluble aggregates in neuronal cells or in the extracellular space [85–87].

Indeed the HSR and heat shock proteins have been implicated in these polyglutamine expansion misfolding diseases. Studies with mammalian tissue culture cells and the nematode *C. elegans* have established that the HSR is activated in cells expressing polyglutamine expansion containing proteins [88, 89]. Moreover, the co-localization of several heat shock proteins, including the Hsp40 family members Hdj-1 and Hdj-2, Hsp70, and ubiquitin, with polyglutamine aggregates in mouse tissues and tissue culture cells has suggested a direct relationship between these heat shock proteins and polyglutamine diseases [90]. Recently, evidence has accumulated that heat shock proteins could play important roles in cytoprotection based on observations that overexpression of Hdj-1, Hdj-2, or Hsp70 reduces polyglutamine aggregate formation and prevents cellular degeneration [91–93]. These observations offer interesting possibilities to develop therapeutic strategies based on activation of the stress response or selectively increasing the levels of individual chaperones.

In addition to a central role for HSF1 in the stress response, HSF1 is essential for the lifespan prolonging effects of the insulin signaling pathway [94–96]. HSF1 and the insulin signaling pathway are linked in *C. elegans*, as HSF1 mutants have a similar reduction in lifespan as DAF-16 mutants [95, 96]. Moreover, HSF1 and DAF-16 regulate common target genes, including sHsps, which suggests a mechanism by which these stress transcription factors coordinately regulate lifespan and metabolism. These observations provide interesting insights into the aging process and suggest that age related changes in proteostasis could be a basis for neurodegenerative disease and cancer. For example, *C. elegans* with an extended lifespan show a dramatic reduction in polyQ aggregation and toxicity characteristic of Huntington's disease [97]. Therefore, slowing the aging process could have the added benefit of combating many diseases of aging.

Because targeting the HSR and molecular chaperones could have therapeutic value, much interest has focused on finding small molecule regulators of HSF1. These compounds could be of benefit to many disease states, including diseases associated with trauma, neurodegenerative disease and cancer. Of the several known small molecule activators and inhibitors of HSF1 [98], the first compound to be identified as the result of a high throughput screen is celastrol [99]. Celastrol is currently being tested for its therapeutic potential and mechanism of action [100–105]. Ongoing screens for additional modulators of the HSR may identify other therapeutically useful compounds and may also provide new insights into the mechansim of activation of HSF1.

ACKNOWLEDGEMENTS

RM is supported by research grant from the National Institutes of General Medical Sciences, the National Institutes of Aging, National Institutes for Neurological Diseases and Stroke (NIH), the Huntington's Disease Society of America Coalition for the Cure, the ALS Association, and the Daniel F. and Ada L. Rice Foundation.

REFERENCES

1. Morimoto RI, et al. The biology of heat-shock proteins and molecular chaperones. In: Morimoto RI, Tissieres A, Georgopoulos C, editors. *cold spring harbor laboratory press*. NY: Cold Spring Harbor; 1994. p. 417–55.
2. Bukau B. *Molecular chaperones and folding catalysis, regulation, cellular function and mechanism*. Amsterdam: Harwood Academic Publishers; 1999.
3. Hartl FU. Molecular chaperones in cellular protein folding. *Nature* 1996;**381**(6583):571–9.
4. Downes, C.P., C.R. Wolf, and D.P. Lane. 1999, *Cellular Responses to Stress*. Portland Press: London.
5. Parsell DA, Lindquist S. The biology of heat shock proteins and molecular chaperones. In: Morimoto RI, Tissieres A, Georgopoulos C, editors. Cold Spring Harbor Laboratory Press. NY: Cold Spring Harbor; 1994. p. 457–94.
6. Findly RC, Pederson T. Regulated transcription of the genes for actin and heat-shock proteins in cultured Drosophila cells. *J Cell Biol* 1981;**88**(2):323–8.
7. Gilmour DS, Lis JT. In vivo interactions of RNA polymerase II with genes of Drosophila melanogaster. *Mol Cell Biol* 1985;**5**(8):2009–18.
8. O'Brien T, Lis JT. Rapid changes in Drosophila transcription after an instantaneous heat shock. *Mol Cell Biol* 1993;**13**(6):3456–63.
9. Sonna LA, et al. Effect of acute heat shock on gene expression by human peripheral blood mononuclear cells. *J Appl Physiol* 2002;**92**(5):2208–20.

10. Mariner PD, et al. Human alu rna is a modular transacting repressor of mrna transcription during heat shock. *Mol Cell* 2008;**29**(4): 499–509.
11. Akerfelt M, et al. Heat shock factors at a crossroad between stress and development. *Ann N Y Acad Sci* 2007;**1113**:15–27.
12. McMillan DR, et al. Targeted disruption of heat shock transcription factor 1 abolishes thermotolerance and protection against heat-inducible apoptosis. *J Biol Chem* 1998;**273**(13):7523–8.
13. Chang Y, et al. Role of heat-shock factor 2 in cerebral cortex formation and as a regulator of p35 expression. *Genes Dev* 2006; **20**(7):836–47.
14. Kallio M, et al. Brain abnormalities, defective meiotic chromosome synapsis and female subfertility in HSF2 null mice. *Embo J* 2002;**21**(11):2591–601.
15. Ostling P, et al. Heat shock factor 2 (HSF2) contributes to inducible expression of hsp genes through interplay with HSF1. *J Biol Chem* 2007;**282**(10):7077–86.
16. Sistonen L, Sarge KD, Morimoto RI. Human heat shock factors 1 and 2 are differentially activated and can synergistically induce hsp70 gene transcription. *Mol Cell Biol* 1994;**14**(3):2087–99.
17. Sistonen L, et al. Activation of heat shock factor 2 during hemin-induced differentiation of human erythroleukemia cells. *Mol Cell Biol* 1992;**12**(9):4104–11.
18. Nakai., A et al., unpublished data.
19. Wang G, et al. Targeted disruption of the heat shock transcription factor (hsf)-2 gene results in increased embryonic lethality, neuronal defects, and reduced spermatogenesis. *Genesis* 2003;**36**(1):48–61.
20. Fujimoto M, et al. HSF4 is required for normal cell growth and differentiation during mouse lens development. *Embo J* 2004;**23**(21): 4297–306.
21. Nakai A, et al. HSF4, a new member of the human heat shock factor family which lacks properties of a transcriptional activator. *Mol Cell Biol* 1997;**17**(1):469–81.
22. Amin J, Ananthan J, Voellmy R. Key features of heat shock regulatory elements. *Mol Cell Biol* 1988;**8**(9):3761–9.
23. Xiao H, Lis JT. Germline transformation used to define key features of heat-shock response elements. *Science* 1988;**239**(4844):1139–42.
24. Kroeger PE, Morimoto RI. Selection of new HSF1 and HSF2 DNA-binding sites reveals difference in trimer cooperativity. *Mol Cell Biol* 1994;**14**(11):7592–603.
25. Ananthan J, Goldberg AL, Voellmy R. Abnormal proteins serve as eukaryotic stress signals and trigger the activation of heat shock genes. *Science* 1986;**232**(4749):522–4.
26. Ahn SG, Thiele DJ. Redox regulation of mammalian heat shock factor 1 is essential for Hsp gene activation and protection from stress. *Genes Dev* 2003;**17**(4):516–28.
27. Hahn JS, Thiele DJ. Activation of the Saccharomyces cerevisiae heat shock transcription factor under glucose starvation conditions by Snf1 protein kinase. *J Biol Chem* 2004;**279**(7):5169–76.
28. Shamovsky I, et al. RNA-mediated response to heat shock in mammalian cells. *Nature* 2006;**440**(7083):556–60.
29. Thomson S, et al. Distinct stimulus-specific histone modifications at hsp70 chromatin targeted by the transcription factor heat shock factor-1. *Mol Cell* 2004;**15**(4):585–94.
30. Shi Y, Mosser DD, Morimoto RI. Molecular chaperones as HSF1-specific transcriptional repressors. *Genes Dev* 1998;**12**(5):654–66.
31. Zou J, et al. Repression of heat shock transcription factor HSF1 activation by HSP90 (HSP90 complex) that forms a stress-sensitive complex with HSF1. *Cell* 1998;**94**(4):471–80.
32. Pirkkala L, Nykanen P, Sistonen L. Roles of the heat shock transcription factors in regulation of the heat shock response and beyond. *Faseb J* 2001;**15**(7):1118–31.
33. Harrison CJ, Bohm AA, Nelson HC. Crystal structure of the DNA binding domain of the heat shock transcription factor. *Science* 1994;**263**(5144):224–7.
34. Sorger PK, Nelson HC. Trimerization of a yeast transcriptional activator via a coiled-coil motif. *Cell* 1989;**59**(5):807–13.
35. Rabindran SK, et al. Regulation of heat shock factor trimer formation: role of a conserved leucine zipper. *Science* 1993;**259**(5092):230–4.
36. Zuo J, et al. Activation of the DNA-binding ability of human heat shock transcription factor 1 may involve the transition from an intramolecular to an intermolecular triple-stranded coiled-coil structure. *Mol Cell Biol* 1994;**14**(11):7557–68.
37. Green M, et al. A heat shock-responsive domain of human HSF1 that regulates transcription activation domain function. *Mol Cell Biol* 1995;**15**(6):3354–62.
38. Shi Y, Kroeger PE, Morimoto RI. The carboxyl-terminal transactivation domain of heat shock factor 1 is negatively regulated and stress responsive. *Mol Cell Biol* 1995;**15**(8):4309–18.
39. Newton EM, et al. The regulatory domain of human heat shock factor 1 is sufficient to sense heat stress. *Mol Cell Biol* 1996;**16**(3):839–46.
40. Boehm AK, et al. Transcription factor and polymerase recruitment, modification, and movement on dhsp70 in vivo in the minutes following heat shock. *Mol Cell Biol* 2003;**23**(21):7628–37.
41. Guettouche T, et al. Analysis of phosphorylation of human heat shock factor 1 in cells experiencing a stress. *BMC Biochem* 2005;**6**:4.
42. Hietakangas V, et al. Phosphorylation of serine 303 is a prerequisite for the stress-inducible SUMO modification of heat shock factor 1. *Mol Cell Biol* 2003;**23**(8):2953–68.
43. Holmberg CI, et al. Phosphorylation of serine 230 promotes inducible transcriptional activity of heat shock factor 1. *Embo J* 2001; **20**(14):3800–10.
44. Holmberg CI, et al. Multisite phosphorylation provides sophisticated regulation of transcription factors. *Trends Biochem Sci* 2002;**27**(12):619–27.
45. Hong Y, et al. Regulation of heat shock transcription factor 1 by stress-induced SUMO-1 modification. *J Biol Chem* 2001;**276**(43): 40,263–40,267.
46. Kim SA, et al. Polo-like kinase 1 phosphorylates heat shock transcription factor 1 and mediates its nuclear translocation during heat stress. *J Biol Chem* 2005;**280**(13):12,653–7.
47. Westerheide, S.D., Anckar, J., Stevens, S., et al., *Stress-inducible regulation of heat shock factor 1 by the deacetylase SIRT1*. Science, 2009. **323**(5917): p. 1063–6.
48. Guettouche T, et al. Analysis of phosphorylation of human heat shock factor 1 in cells experiencing a stress. *BMC Biochem* 2005;**6**(1):4.
49. Lis J. Promoter-associated pausing in promoter architecture and postinitiation transcriptional regulation. *Cold Spring Harb Symp Quant Biol* 1998;**63**:347–56.
50. Lis JT, et al. P-TEFb kinase recruitment and function at heat shock loci. *Genes Dev* 2000;**14**(7):792–803.
51. Park JM, et al. Mediator, not holoenzyme, is directly recruited to the heat shock promoter by HSF upon heat shock. *Mol Cell* 2001; **8**(1):9–19.
52. Saunders A, et al. Tracking FACT and the RNA polymerase II elongation complex through chromatin in vivo. *Science* 2003;**301**(5636):1094–6.
53. Andrulis ED, et al. The RNA processing exosome is linked to elongating RNA polymerase II in Drosophila. *Nature* 2002;**420**(6917): 837–41.

54. Xing H, et al. HSF1 modulation of Hsp70 mRNA polyadenylation via interaction with symplekin. *J Biol Chem* 2004;**279**(11):10,551–10,555.
55. Jolly C, et al. HSF1 transcription factor concentrates in nuclear foci during heat shock: relationship with transcription sites. *J Cell Sci* 1997;**110**(Pt 23):2935–41.
56. Jolly C, et al. Stress-induced transcription of satellite III repeats. *J Cell Biol* 2004;**164**(1):25–33.
57. Sandqvist A, Sistonen L. Nuclear stress granules: the awakening of a sleeping beauty? *J Cell Biol* 2004;**164**(1):15–17.
58. Hahn JS, et al. Genome-wide analysis of the biology of stress responses through heat shock transcription factor. *Mol Cell Biol* 2004;**24**(12):5249–56.
59. Murray JI, et al. Diverse and specific gene expression responses to stresses in cultured human cells. *Mol Biol Cell* 2004;**15**(5):2361–74.
60. Trinklein ND, et al. The role of heat shock transcription factor 1 in the genome-wide regulation of the mammalian heat shock response. *Mol Biol Cell* 2004;**15**(3):1254–61.
61. Gething MJ. Molecular chaperones and protein folding catalysts. New York: Oxford University Press; 1997.
62. Freeman BC, Toft DO, Morimoto RI. Molecular chaperone machines: chaperone activities of the cyclophilin Cyp-40 and the steroid aporeceptor-associated protein p23. *Science* 1996;**274**(5293):1718–20.
63. Jakob U, et al. Transient interaction of Hsp90 with early unfolding intermediates of citrate synthase. Implications for heat shock in vivo. *J Biol Chem* 1995;**270**(13):7288–94.
64. Pratt WB, Welsh MJ. Chaperone functions of the heat shock proteins associated with steroid receptors. *Semin Cell Biol* 1994;**5**(2):83–93.
65. Smith DF. *Molecular chaperones and protein-folding catalysts*. New York: Oxford University Press; 1997. p. 518–521.
66. Freeman ML, et al. Characterization of a signal generated by oxidation of protein thiols that activates the heat shock transcription factor. *J Cell Physiol* 1995;**164**(2):356–66.
67. Goloubinoff P, Gatenby AA, Lorimer GH. GroE heat-shock proteins promote assembly of foreign prokaryotic ribulose bisphosphate carboxylase oligomers in Escherichia coli. *Nature* 1989;**337**(6202):44–7.
68. Rudiger S, et al. Substrate specificity of the DnaK chaperone determined by screening cellulose-bound peptide libraries. *Embo J* 1997;**16**(7):1501–7.
69. Horwich AL, et al. Folding in vivo of bacterial cytoplasmic proteins: role of GroEL. *Cell* 1993;**74**(5):909–17.
70. Mayhew M, et al. Protein folding in the central cavity of the GroEL-GroES chaperonin complex. *Nature* 1996;**379**(6564):420–6.
71. Morimoto RI. Dynamic remodeling of transcription complexes by molecular chaperones. *Cell* 2002;**110**(3):281–4.
72. Xu Y, Lindquist S. Heat-shock protein hsp90 governs the activity of pp60v-src kinase. *Proc Natl Acad Sci USA* 1993;**90**(15):7074–8.
73. Dittmar KD, et al. The role of DnaJ-like proteins in glucocorticoid receptor.hsp90 heterocomplex assembly by the reconstituted hsp90.p60.hsp70 foldosome complex. *J Biol Chem* 1998;**273**(13):7358–66.
74. Sato S, Fujita N, Tsuruo T. Modulation of Akt kinase activity by binding to Hsp90. *Proc Natl Acad Sci USA* 2000;**97**(20):10,832–10,837.
75. Rutherford SL, Lindquist S. Hsp90 as a capacitor for morphological evolution. *Nature* 1998;**396**(6709):336–42.
76. Elefant F, Palter KB. Tissue-specific expression of dominant negative mutant Drosophila HSC70 causes developmental defects and lethality. *Mol Biol Cell* 1999;**10**(7):2101–17.
77. Feder JH, et al. The consequences of expressing hsp70 in Drosophila cells at normal temperatures. *Genes Dev* 1992;**6**(8):1402–13.
78. Song J, Takeda M, Morimoto RI. Bag1-Hsp70 mediates a physiological stress signalling pathway that regulates Raf-1/ERK and cell growth. *Nat Cell Biol* 2001;**3**(3):276–82.
79. Jaattela M. Heat shock proteins as cellular lifeguards. *Ann Med* 1999;**31**(4):261–71.
80. Nylandsted J, et al. Selective depletion of heat shock protein 70 (Hsp70) activates a tumor-specific death program that is independent of caspases and bypasses Bcl-2. *Proc Natl Acad Sci USA* 2000;**97**(14):7871–6.
81. Whitesell L, et al. Inhibition of heat shock protein HSP90-pp60v-src heteroprotein complex formation by benzoquinone ansamycins: essential role for stress proteins in oncogenic transformation. *Proc Natl Acad Sci USA* 1994;**91**(18):8324–8.
82. Blagosklonny MV, et al. Mutant conformation of p53 translated in vitro or in vivo requires functional HSP90. *Proc Natl Acad Sci USA* 1996;**93**(16):8379–83.
83. Dai C, et al. Heat shock factor 1 is a powerful multifaceted modifier of carcinogenesis. *Cell* 2007;**130**(6):1005–18.
84. Kakizuka A. Protein precipitation: a common etiology in neurodegenerative disorders? *Trends Genet* 1998;**14**(10):396–402.
85. Jackson GS, et al. Reversible conversion of monomeric human prion protein between native and fibrilogenic conformations. *Science* 1999;**283**(5409):1935–7.
86. Jimenez JL, et al. Cryo-electron microscopy structure of an SH3 amyloid fibril and model of the molecular packing. *Embo J* 1999;**18**(4):815–21.
87. Scherzinger E, et al. Huntingtin-encoded polyglutamine expansions form amyloid-like protein aggregates *in vitro* and *in vivo*. *Cell* 1997;**90**(3):549–58.
88. Satyal SH, et al. Polyglutamine aggregates alter protein folding homeostasis in Caenorhabditis elegans. *Proc Natl Acad Sci USA* 2000;**97**(11):5750–5.
89. Wyttenbach A, et al. Effects of heat shock, heat shock protein 40 (HDJ-2), and proteasome inhibition on protein aggregation in cellular models of Huntington's disease. *Proc Natl Acad Sci USA* 2000;**97**(6):2898–903.
90. Cummings CJ, et al. Chaperone suppression of aggregation and altered subcellular proteasome localization imply protein misfolding in SCA1. *Nat Genet* 1998;**19**(2):148–54.
91. Krobitsch S, Lindquist S. Aggregation of huntingtin in yeast varies with the length of the polyglutamine expansion and the expression of chaperone proteins. *Proc Natl Acad Sci USA* 2000;**97**(4):1589–94.
92. Warrick JM, et al. Suppression of polyglutamine-mediated neurodegeneration in Drosophila by the molecular chaperone HSP70. *Nat Genet* 1999;**23**(4):425–8.
93. Kazemi-Esfarjani P, Benzer S. Genetic suppression of polyglutamine toxicity in Drosophila. *Science* 2000;**287**(5459):1837–40.
94. Garigan D, et al. Genetic analysis of tissue aging in Caenorhabditis elegans: a role for heat-shock factor and bacterial proliferation. *Genetics* 2002;**161**(3):1101–12.
95. Hsu AL, Murphy CT, Kenyon C. Regulation of aging and age-related disease by DAF-16 and heat-shock factor. *Science* 2003;**300**(5622):1142–5.
96. Morley JF, Morimoto RI. Regulation of longevity in Caenorhabditis elegans by heat shock factor and molecular chaperones. *Mol Biol Cell* 2004;**15**(2):657–64.
97. Morley JF, et al. The threshold for polyglutamine-expansion protein aggregation and cellular toxicity is dynamic and influenced by aging in Caenorhabditis elegans. *Proc Natl Acad Sci USA* 2002;**99**(16):10,417–10,422.

98. Westerheide SD, Morimoto RI. Heat shock response modulators as therapeutic tools for diseases of protein conformation. *J Biol Chem* 2005;**280**(39):33,097–33,100.
99. Westerheide SD, et al. Celastrols as inducers of the heat shock response and cytoprotection. *J Biol Chem* 2004;**279**(53):56,053–56,060.
100. Cleren C, et al. Celastrol protects against MPTP- and 3-nitropropionic acid-induced neurotoxicity. *J Neurochem* 2005;**94**(4):995–1004.
101. Kiaei M, et al. Celastrol blocks neuronal cell death and extends life in transgenic mouse model of amyotrophic lateral sclerosis. *Neurodegener Dis* 2005;**2**(5):246–54.
102. Chow AM, Brown IR. Induction of heat shock proteins in differentiated human and rodent neurons by celastrol. *Cell Stress Chaperones* 2007;**12**(3):237–44.
103. Zhang YQ, Sarge KD. Celastrol inhibits polyglutamine aggregation and toxicity though induction of the heat shock response. *J Mol Med* 2007;**85**(12):1421–8.
104. Trott A, et al. Activation of heat shock and antioxidant responses by the natural product celastrol: transcriptional signatures of a thiol-targeted molecule. *Mol Biol Cell* 2008;**19**(3):1104–12.
105. Mu TW, et al. Chemical and biological approaches synergize to ameliorate protein-folding diseases. *Cell* 2008;**134**(5):769–81.

Chapter 27

Hypoxia Mediated Signaling Pathways

Denise A. Chan, Albert C. Koong and Amato J. Giaccia

Division of Radiation Biology, Department of Radiation Oncology, Stanford University School of Medicine, Stanford, California

INTRODUCTION

Molecular oxygen is a key element in the evolution of higher life forms. Oxygen levels must be maintained within a fairly narrow window to respond to the metabolic demands of different tissues and to be compatible with viability. Oxygen tensions that are too high or too low can lead to diseases of the cardiovascular, central nervous, and pulmonary systems. Thus, oxygen homeostasis must be exquisitely controlled.

Oxygen homeostasis is deregulated in several pathological settings. Hypoxia, or low oxygen conditions, exists in virtually all solid tumors [1]. As tumor cells rapidly expand, their growth outpaces the ability of the existing vasculature to supply both nutrients and oxygen, resulting in tumor hypoxia [2]. Cells experiencing hypoxia can cope with the low oxygen tension by either increasing delivery of oxygen or adapting to the low oxygen level. For example, hypoxic tumor cells can induce angiogenesis and decrease their proliferation rate [3,4]. Both of these pathological processes contribute to the aggressive nature of hypoxic tumor cells and to their resistance against conventional cancer therapies, such as cytotoxic chemotherapy and radiation therapy [5]. As anticancer drugs are generally administered intravenously, these agents must diffuse out of the blood vessel to reach the tumor cells. Drug delivery to the tumor is determined in part by the distance away from the blood vessel as well as the leakiness of the newly formed vasculature [6]. Moreover, chemotherapeutic agents are less active on hypoxic cells since hypoxic cells proliferate slower than well-perfused tissues. Similarly, the efficacy of radiation treatment is directly correlated with oxygenation status of a tumor. Molecular oxygen is necessary to form the cytotoxic double-stranded breaks in DNA that cause cell death. In addition to its important role in therapeutic resistance, hypoxia also functions as a physiological selection pressure to select for cells with increased metastatic potential. Tumor hypoxia poses a significant obstacle in the treatment of cancer and is an independent indicator of poor prognosis [2, 7–9]. Thus, hypoxia mediated signaling pathways play important roles in the development of cancer and may potentially be exploited in the treatment of cancer.

HIF REGULATION

How is oxygen homeostasis achieved? At both the molecular and cellular levels, in physiological and pathological settings, oxygen homeostasis is controlled primarily through the function of a heterodimeric transcription factor, the hypoxia inducible factor-1 (HIF-1) [10, 11]. Since HIF plays a critical role in an essential process, tight regulation of HIF is necessary. HIF is composed of an oxygen-labile α subunit and a constitutive β subunit. The HIF-α subunit is divided into several functional domains. Both HIF-α and HIF-β are members of the basic helix-loop-helix Per, AhR, and Sim (bHLH-PAS) family of transcription factors. Dimerization with HIF-β and binding to hypoxia responsive elements (HREs) of target DNA sequences is achieved through these bHLH-PAS domains. HIF-α contains amino- and carboxy-terminal transactivation domains (N-TAD and C-TAD), which interact with the transcriptional coactivators p300/CBP. The recruitment of transcriptional coactivators is responsible for the transcriptional activity of HIF. Thus, the transactivation domains of HIF-α serves as a potential means of regulation.

The transcriptional activity of HIF is regulated by at least two steps, protein stability and recruitment of transactivation coactivators (Figure 27.1). Protein stability is the initial and the predominant means of HIF activity regulation. HIF-α contains an oxygen dependent degradation

Handbook of Cell Signaling, Three-Volume Set 2 ed.

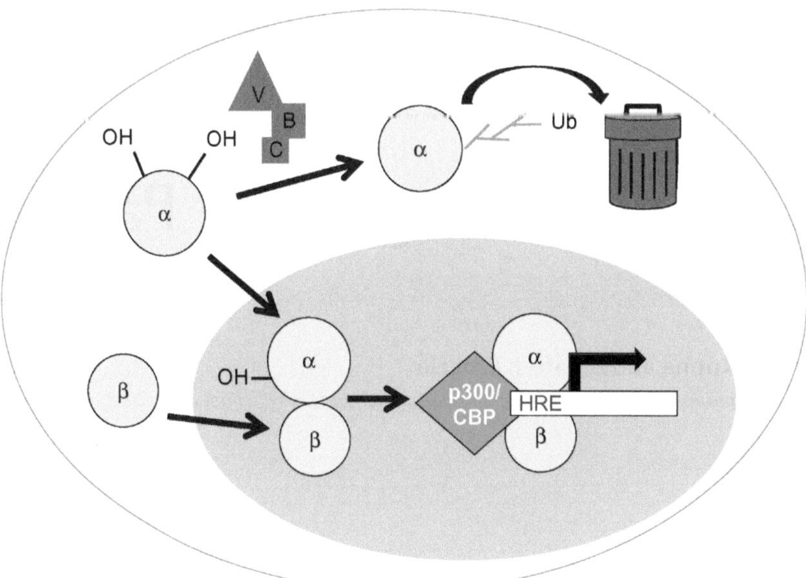

FIGURE 27.1 Regulation of the HIF signaling pathway.
Tight regulation of HIF is achieved through posttranslational modification of the oxygen-labile α-subunit. The hydroxylation of the α-subunit under normal oxygenated conditions allows the VHL-ElonginB-ElonginC (VBC) complex to interact. The VBC is an E3 ubiquitin ligase that adds ubiquitin ladders, which marks client proteins to the proteasome for degradation. The α-subunit is also hydroxylated on a third residue, which prevents the interaction with p300/CBP. Under hypoxic conditions, this posttranslational modification is impaired and the α-subunit binds the β-subunit. Together this heterodimeric transcription factor recruits the p300/CBP transcriptional coactivators to the hypoxia responsive elements of target genes, driving transcription of a myriad processes.

domain (ODD), responsible for the oxygen-lability of HIF-α. In the presence of oxygen, HIF-1α is hydroxylated on two highly conserved proline residues, prolines 402 and 564, both found within the ODD [12–15]. A family of 4-prolyl hydroxylases, distinct from the structurally important collagen-4-prolyl hydroxylases, modifies these prolines [16]. These enzymes catalyze a hydroxylation reaction, which utilizes molecular oxygen and 2-oxoglutarate as substrates and iron and ascorbate as cofactors. These posttranslational modifications allow the von Hippel–Lindau (VHL) tumor suppressor protein to recognize HIF-α. VHL functions in conjunction with Elongins B and C as an E3 ubiquitin ligase, covalently attaching ubiquitin modifiers to substrate proteins, such as HIF-α [17]. Ubiquitin ladders mark the respective client proteins to the proteasome for degradation. Under normal oxygenated conditions, HIF-α protein is constantly turned over due to hydroxylation, recognition by VHL, and subsequent destruction by the proteasome. In contrast, the 4-prolyl hydroxylases have reduced activity under hypoxia, as oxygen is a requisite substrate of the hydroxylation reaction. When HIF-α is not hydroxylated, it escapes recognition by VHL, resulting in protein stabilization. Once stabilized HIF-α dimerizes with its binding partner, HIF-β (also known as ARNT). Together the HIF-α/β dimer binds to a core sequence of 5'-RCGTG-3' in the enhancer elements of target genes to initiate gene transcription.

Another layer of regulation on HIF transactivation activity occurs at the level of recruitment of coactivators. A third hydroxylation occurs on HIF-α in the C-terminal transactivation domain on asparagine residue 803 [18]. This hydroxylation reaction is catalyzed by a separate asparaginyl hydroxylase, factor inhibiting HIF (FIH) [19]. Hydroxylation of this residue under normoxic conditions prevents interaction with p300, which in turn prevents gene transcription. This modification further demonstrates the fastidious regulation needed to ensure appropriate expression of HIF under hypoxic conditions.

HIF SIGNALING AND METASTASIS

Other transcription factors are induced by hypoxia, including AP-1 and NFκB, but HIF signaling controls the vast majority of hypoxia regulated genes [20, 21]. A wide variety of HIF target genes have been identified, including those involved in an array of many cellular processes. HIF has been shown to regulate cellular metabolism, genomic instability, cell proliferation and viability, tissue remodeling, differentiation, angiogenesis, and recently, metastasis [22] (Figure 27.2).

HIF signaling occurs at different stages of metastasis to promote the dissemination of primary tumor cells to distant sites. In order to metastasize and colonize a distant

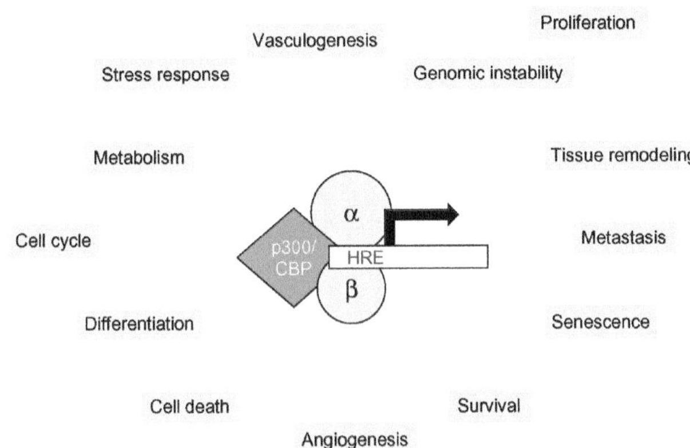

FIGURE 27.2 Diagram depicting processes that are regulated by HIF signaling.

site, a tumor cell must first detach from the primary tumor. These renegade cells degrade both the basement membrane and the extracellular matrix to move into either the lymphatic of vascular systems, while hiding from the immune system. These cells then exit the circulation, homing in on its target site.

Remodeling of extracellular matrix proteins and disruption of cell–cell adhesion are important steps in the metastatic process and numerous ECM proteins are in fact hypoxia inducible and HIF targets. One such ECM protein is lysyl oxidase (LOX). LOX, through controlling focal adhesion kinase activity, regulates cell migration [23]. In a retrospective expression analysis of a breast cancer study and head and neck tumors, high LOX levels were associated with hypoxia, lower metastasis free disease, and overall survival rates. Importantly, in an orthotopic breast cancer metastasis model, disruption of LOX through pharmacological, genetic, or with a neutralizing antibody decreased metastatic potential, indicating that HIF signaling through LOX could be a new therapeutic target.

Another key step in the development of metastasis is the epithelial – mesenchymal transition (EMT), whereby epithelial cells lose cell contacts and polarity and phenotypically become more like undifferentiated connective tissue. A hallmark of the EMT is the loss or repression of E-cadherin expression. E-cadherin is a transmembrane glycoprotein that maintains epithelial cell–cell adhesion contacts and is negatively regulated by HIF [24–26]. More recently, HIF has also been shown to regulate EMT through the expression of TWIST, a transcription factor whose targets are involved in gastrulation and mesodermal specification [27].

Additional HIF targets have been identified that regulate cell migration and homing of cells to distant sites. CXCR-4, a chemokine receptor, and its ligand, stromal derived factor-1 (SDF-1) act in concert to determine the distinct anatomical location of where a metastasis will colonize depending on the tissue of origin of the primary tumor [28–30]. For example, breast cancer frequently metastasizes to the lung and liver. CXCR4 levels were elevated in both cell lines and breast tumors when compared to primary mammary epithelial cells. SDF-1 expression increased migration and invasion of breast cancer cell lines and similarly, chemotaxis was enhanced in response to liver- or lung-derived protein extracts. These results suggest HIF signaling from the metastatic site helps determine whether a particular organ will be colonized. Altogether, HIF via its downstream effectors play essential roles in both the development of a primary tumor as well as the metastatic potential of a tumor.

UNFOLDED PROTEIN RESPONSE

Prolonged periods of hypoxia can activate additional non-HIF signaling pathways, including the unfolded protein response (UPR). The UPR is a cellular stress response that is induced by the accumulation of unfolded proteins in the endoplasmic reticulum (ER) in order to process misfolded proteins. This is achieved twofold: by stopping protein translation to prevent further bottleneck and through the induction of chaperone proteins that function in the ER to assist in folding, trafficking, and secretion of proteins [31] (Figure 27.3). The UPR is highly conserved throughout evolution. It has been well characterized in yeast and more recently, a greater understanding has been elucidated in humans. The only ER stress sensor ortholog in yeast and humans is the inositol requiring kinase 1 (IRE1). Through homodimerization and transautophosphorylation, IRE1 becomes activated. The endoribonuclease activity of IRE1 splices the pre-mRNA of the X-box binding protein (XBP1) transcription factor. Excision of a 252 base pair intron results in translation of XBP1, which has many target genes involved in ER protein maturation [32].

A second ER stress sensor structurally related to IRE1 is the protein kinase-like endoplasmic reticulum kinase (PERK). One of the initial response steps to activated UPR is to prevent further translation of proteins by activating PERK. PERK activation occurs rapidly under severe hypoxia on the order of minutes, while under milder hypoxia PERK activation occurs on the order of hours. Through oligomerization and autophosphorylation, PERK self-activates, resulting in translation inhibition by directly phosphorylating serine 51 of eIF2α, a key initiator of the mRNA translation machinery [33]. Activation of PERK is important to survive under hypoxic conditions as PERK-deficient cells undergo apoptosis following hypoxia, resulting in a lower survival rate compared to genetically matched cells that have the PERK pathway intact [34]. These PERK-deficient cells are less tumorigenic, likely in part due to higher apoptosis in the hypoxic regions of the tumor. Thus, one cellular response to chronic conditions of hypoxic stress is to limit the load of misfolded proteins.

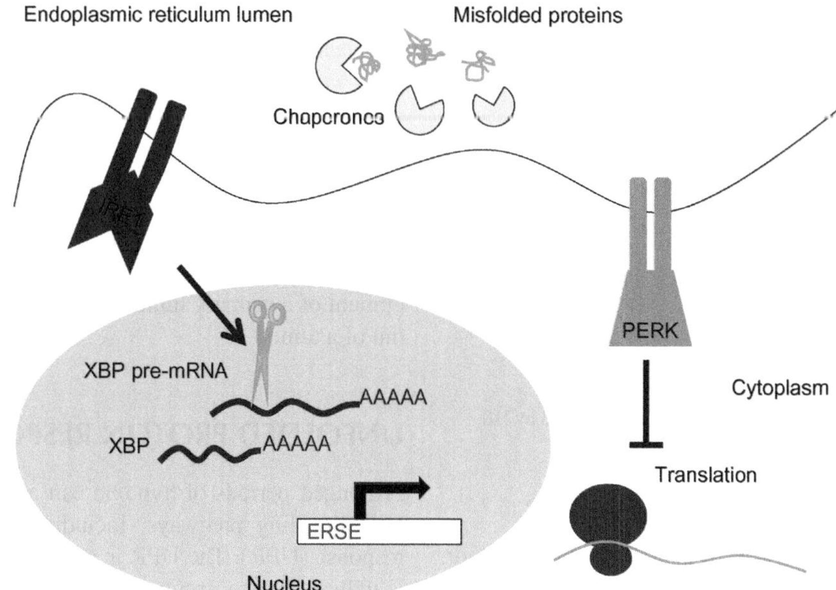

FIGURE 27.3 The unfolded protein response in hypoxia.
Under hypoxic conditions, misfolded proteins accumulate in the endoplasmic reticulum, resulting in activation of two ER stress sensors. IRE1, a heterodimer, potentiates a signal to splice the pre-mRNA of XBP, a transcription factor, which drives the transcription of genes with ER stress elements, including chaperone proteins to help properly fold the malfolded proteins. Similarly, PERK activation inhibits translation, preventing the further accumulation of misfolded proteins.

CONCLUSIONS

Hypoxia mediated signaling pathways play important roles in the development of cancer and the response of tumors to conventional therapies. These pathways, including HIF signaling cascade and the unfolded protein response, represent potential, new avenues for targeted treatments. By gaining a better understanding of these signaling pathways, we may gain a better understanding of how to exploit this common feature of solid tumors for therapeutic benefit.

REFERENCES

1. Brown JM, Giaccia AJ. The unique physiology of solid tumors: opportunities (and problems) for cancer therapy. *Cancer Res* 1998;**58**(7):1408–16.
2. Brizel DM, Scully SP, Harrelson JM, Layfield LJ, Bean JM, Prosnitz LR, Dewhirst MW. Tumor oxygenation predicts for the likelihood of distant metastases in human soft tissue sarcoma. *Cancer Res* 1996;**56**(5):941–3.
3. Folkman J. The role of angiogenesis in tumor growth. *Semin Cancer Biol* 1992;**3**(2):65–71.
4. Shweiki D, Itin A, Soffer D, Keshet E. Vascular endothelial growth factor induced by hypoxia may mediate hypoxia-initiated angiogenesis. *Nature* 1992;**359**(6398):843–5.
5. Kallman RF, Dorie MJ. Tumor oxygenation and reoxygenation during radiation therapy: their importance in predicting tumor response. *Int J Radiat Oncol Biol Phys* 1986;**12**(4):681–5.
6. Tannock IF. The relation between cell proliferation and the vascular system in a transplanted mouse mammary tumour. *Br J Cancer* 1968;**22**(2):258–73.
7. Hockel M, Knoop C, Schlenger K, Vorndran B, Baussmann E, Mitze M, Knapstein PG, Vaupel P. Intratumoral pO2 predicts survival in advanced cancer of the uterine cervix. *Radiother Oncol* 1993;**26**(1):45–50.
8. Hockel M, Schlenger K, Aral B, Mitze M, Schaffer U, Vaupel P. Association between tumor hypoxia and malignant progression in advanced cancer of the uterine cervix. *Cancer Res* 1996;**56**(19):4509–15.
9. Fyles AW, Milosevic M, Wong R, Kavanagh MC, Pintilie M, Sun A, Chapman W, Levin W, Manchul L, Keane TJ, Hill RP. Oxygenation predicts radiation response and survival in patients with cervix cancer. *Radiother Oncol* 1998;**48**(2):149–56.
10. Jiang BH, Rue E, Wang GL, Roe R, Semenza GL. Dimerization, DNA binding, and transactivation properties of hypoxia-inducible factor 1. *J Biol Chem* 1996;**271**(30):17,771–17,778.
11. Jiang BH, Zheng JZ, Leung SW, Roe R, Semenza GL. Transactivation and inhibitory domains of hypoxia-inducible factor 1alpha. Modulation of transcriptional activity by oxygen tension. *J Biol Chem* 1997;**272**(31):19,253–19,260.
12. Ivan M, Kondo K, Yang H, Kim W, Valiando J, Ohh M, Salic A, Asara JM, Lane WS, Kaelin Jr WG. HIFalpha targeted for VHL-mediated destruction by proline hydroxylation: implications for O2 sensing. *Science* 2001;**292**(5516):464–8.
13. Jaakkola P, Mole DR, Tian YM, Wilson MI, Gielbert J, Gaskell SJ, Kriegsheim A, Hebestreit HF, Mukherji M, Schofield CJ, Maxwell PH, Pugh CW, Ratcliffe PJ. Targeting of HIF-alpha to the von Hippel–Lindau ubiquitylation complex by O2-regulated prolyl hydroxylation. *Science* 2001;**292**(5516):468–72.
14. Masson N, Willam C, Maxwell PH, Pugh CW, Ratcliffe PJ. Independent function of two destruction domains in hypoxia-inducible factor-alpha chains activated by prolyl hydroxylation. *Embo J* 2001;**20**(18):5197–206.

15. Chan DA, Sutphin PD, Yen SE, Giaccia AJ. Coordinate regulation of the oxygen-dependent degradation domains of hypoxia-inducible factor 1 alpha. *Mol Cell Biol* 2005;**25**(15):6415–26.
16. Epstein AC, Gleadle JM, McNeill LA, Hewitson KS, O'Rourke J, Mole DR, Mukherji M, Metzen E, Wilson MI, Dhanda A, Tian YM, Masson N, Hamilton DL, Jaakkola P, Barstead R, Hodgkin J, Maxwell PH, Pugh CW, Schofield CJ, Ratcliffe PJC. elegans EGL-9 and mammalian homologs define a family of dioxygenases that regulate HIF by prolyl hydroxylation. *Cell* 2001;**107**(1):43–54.
17. Ohh M, Park CW, Ivan M, Hoffman MA, Kim TY, Huang LE, Pavletich N, Chau V, Kaelin WG. Ubiquitination of hypoxia-inducible factor requires direct binding to the beta-domain of the von Hippel–Lindau protein. *Nat Cell Biol* 2000;**2**(7):423–7.
18. Lando D, Peet DJ, Whelan DA, Gorman JJ, Whitelaw ML. Asparagine hydroxylation of the HIF transactivation domain a hypoxic switch. *Science* 2002;**295**(5556):858–61.
19. Lando D, Peet DJ, Gorman JJ, Whelan DA, Whitelaw ML, Bruick RK. FIH-1 is an asparaginyl hydroxylase enzyme that regulates the transcriptional activity of hypoxia-inducible factor. *Genes Dev* 2002;**16**(12):1466–71.
20. Laderoute KR, Calaoagan JM, Gustafson-Brown C, Knapp AM, Li GC, Mendonca HL, Ryan HE, Wang Z, Johnson RS. The response of c-jun/AP-1 to chronic hypoxia is hypoxia-inducible factor 1 alpha dependent. *Mol Cell Biol* 2002;**22**(8):2515–23.
21. Koong AC, Chen EY, Giaccia AJ. Hypoxia causes the activation of nuclear factor kappa B through the phosphorylation of I kappa B alpha on tyrosine residues. *Cancer Res* 1994;**54**(6):1425–30.
22. Chan DA, Giaccia AJ. Hypoxia, gene expression, and metastasis. *Cancer Metastasis Rev* 2007;**26**(2):333–9.
23. Erler JT, Bennewith KL, Nicolau M, Dornhofer N, Kong C, Le QT, Chi JT, Jeffrey SS, Giaccia AJ. Lysyl oxidase is essential for hypoxia-induced metastasis. *Nature* 2006;**440**(7088):1222–6.
24. Krishnamachary B, Zagzag D, Nagasawa H, Rainey K, Okuyama H, Baek JH, Semenza GL. Hypoxia-inducible factor-1-dependent repression of E-cadherin in von Hippel–Lindau tumor suppressor-null renal cell carcinoma mediated by TCF3, ZFHX1A, and ZFHX1B. *Cancer Res* 2006;**66**(5):2725–31.
25. Esteban MA, Tran MG, Harten SK, Hill P, Castellanos MC, Chandra A, Raval R, O'Brien TS, Maxwell PH. Regulation of E-cadherin expression by VHL and hypoxia-inducible factor. *Cancer Res* 2006;**66**(7):3567–75.
26. Evans AJ, Russell RC, Roche O, Burry TN, Fish JE, Chow VW, Kim WY, Saravanan A, Maynard MA, Gervais ML, Sufan RI, Roberts AM, Wilson LA, Betten M, Vandewalle C, Berx G, Marsden PA, Irwin MS, Teh BT, Jewett MA, Ohh M. VHL promotes E2 box-dependent E-cadherin transcription by HIF-mediated regulation of SIP1 and snail. *Mol Cell Biol* 2007;**27**(1):157–69.
27. Yang MH, Wu MZ, Chiou SH, Chen PM, Chang SY, Liu CJ, Teng SC, Wu KJ. Direct regulation of TWIST by HIF-1alpha promotes metastasis. *Nat Cell Biol* 2008;**10**(3):295–305.
28. Muller A, Homey B, Soto H, Ge N, Catron D, Buchanan ME, McClanahan T, Murphy E, Yuan W, Wagner SN, Barrera JL, Mohar A, Verastegui E, Zlotnik A. Involvement of chemokine receptors in breast cancer metastasis. *Nature* 2001;**410**(6824):50–6.
29. Staller P, Sulitkova J, Lisztwan J, Moch H, Oakeley EJ, Krek W. Chemokine receptor CXCR4 downregulated by von Hippel–Lindau tumour suppressor pVHL. *Nature* 2003;**425**(6955):307–11.
30. Ceradini DJ, Kulkarni AR, Callaghan MJ, Tepper OM, Bastidas N, Kleinman ME, Capla JM, Galiano RD, Levine JP, Gurtner GC. Progenitor cell trafficking is regulated by hypoxic gradients through HIF-1 induction of SDF-1. *Nat Med* 2004;**10**(8):858–64.
31. Wouters BG, Koritzinsky M. Hypoxia signalling through mTOR and the unfolded protein response in cancer. *Nat Rev Cancer* 2008;**8**(11):851–64.
32. Romero-Ramirez L, Cao H, Nelson D, Hammond E, Lee AH, Yoshida H, Mori K, Glimcher LH, Denko NC, Giaccia AJ, Le QT, Koong AC. XBP1 is essential for survival under hypoxic conditions and is required for tumor growth. *Cancer Res* 2004;**64**(17):5943–7.
33. Koumenis C, Naczki C, Koritzinsky M, Rastani S, Diehl A, Sonenberg N, Koromilas A, Wouters BG. Regulation of protein synthesis by hypoxia via activation of the endoplasmic reticulum kinase PERK and phosphorylation of the translation initiation factor eIF2alpha. *Mol Cell Biol* 2002;**22**(21):7405–16.
34. Bi M, Naczki C, Koritzinsky M, Fels D, Blais J, Hu N, Harding H, Novoa I, Varia M, Raleigh J, Scheuner D, Kaufman RJ, Bell J, Ron D, Wouters BG, Koumenis C. ER stress-regulated translation increases tolerance to extreme hypoxia and promotes tumor growth. *EMBO J* 2005;**24**(19):3470–81.

Chapter 28

Regulation of mRNA Turnover by Cellular Stress

Subramanya Srikantan and Myriam Gorospe
Laboratory of Cellular and Molecular Biology, National Institute on Aging-IRP, Baltimore, Maryland

INTRODUCTION

Changes in gene expression patterns following cellular damage are controlled at multiple levels through a variety of regulatory mechanisms. In addition to extensively studied transcriptional events, there is increasing recognition that gene expression is also potently modulated by less well understood post-transcriptional processes, which include pre-mRNA splicing and maturation (3' polyadenylation, 5' capping), followed by mRNA export to the cytoplasm, subcytoplasmic transport, turnover, and translation [1–3]. In this chapter, we will review the process of regulated mRNA turnover, one of the most influential mechanisms effecting changes in the patterns of expressed mRNAs in response to stress [4]. By changing the half-lives of mRNAs, which typically range from 20 min to 24 h [5], mammalian cells can quickly allow certain subsets of mRNAs to accumulate (*mRNA stabilization*) and rapidly eliminate other subsets of mRNAs (*mRNA decay*). Among the genes whose expression is regulated through altered mRNA turnover are many encoding T cell activation molecules, cell cycle regulators, cell survival proteins, and stress response factors. Given their pivotal role in many physiologic and pathologic conditions, including conditions of stress, there is mounting interest in understanding the mechanisms that govern mRNA stability and mRNA degradation.

Broadly speaking, the process of mRNA turnover involves the association of *trans* acting factors that recognize specific *cis* elements (RNA sequences) on the target mRNA. Two main sets of *trans* binding factors have been identified: RNA binding proteins (RBPs), and microRNAs [6, 7]. In the case of stress-regulated gene expression, the involvement of RBPs has been investigated in significant detail, as described in this chapter, while studies of the participation of microRNAs have only recently begun [8–10] and will not be reviewed here.

RNP complexes responsible for controlling mRNA stability following stress are regulated by many of the signaling pathways described in detail in other chapters of this *Handbook*. They include several signal transduction events that also regulate transcription factor activity [11], such as those leading to the activation of mitogen-activated protein kinases (MAPKs) (comprising the extracellular signal-regulated kinases (ERK), the c-Jun N-terminal kinases (JNK), and p38), phosphoinositide 3-kinase (PI3K) and protein kinase B (PKB, also called Akt), the glycogen synthase kinase (GSK)3β, protein kinases A (PKA) and C (PKC), as well as ataxia-telangiectasia mutated (ATM) kinase and downstream checkpoint kinases (Chk1, Chk2); stress-causing and proliferative stimuli can also repress other pathways such as that controlling the AMP activated protein kinase (AMPK). Examples of the influence of these stress-regulated pathways upon RNP function will be discussed later in this chapter.

RNA-BINDING PROTEINS CONTROLLING mRNA TURNOVER

Many RNA-binding proteins have been described that selectively recognize and bind to specific instability-conferring regions of labile mRNAs. Many stress-regulated mRNAs are the targets of RBPs, which can positively or negatively alter their half-life. These RBPs constitute a heterogeneous family of proteins whose functions are tightly interconnected, as they often bind to similar RNA sequences. In addition to modulating mRNA half-life, they often influence the *translation* status of bound mRNA; accordingly, the name TTR-RBP was recently introduced to designate the group of mRNA *t*urnover and *t*ranslation *r*egulatory RBPs [12]. In this regard, different TTR-RBPs with affinity for the same target mRNA can display competitive

Handbook of Cell Signaling, Three-Volume Set 2 ed.

or cooperative interactions among them, as reported for AUF1 (AU binding factor 1) and TIAR (related to the T-cell intracellular antigen-1 (TIA-1)), as well as for HuR and AUF1 [29]. Some TTR-RBPs can influence several distinct processes, such as mRNA translation and turnover, as described for NF90 and HuR [13–17]. In addition, some TTR-RBPs such as AUF1 can promote and reduce mRNA stability [18, 19]. Finally, TTR-RBPs can regulate each other's expression levels through binding to cognate mRNAs, although this regulation has not been examined in response to stress [12]. Despite the well-recognized overlap between mRNA stability and translation, this chapter will focus primarily on the regulation of mRNA turnover.

HuR. Mammalian cells encode four members of the Hu/elav family of TTR-RBPs: the ubiquitous HuR (HuA), and the primarily neuronal HuB (Hel-N1), HuC, and HuD. HuR, the most extensively studied member of this family, is predominantly nuclear, but its influence upon the expression of target mRNAs is linked to its localization in the cytoplasm. A variety of stresses, such as alkylating agents (methylmethane sulfonate), oxidants (H_2O_2, arsenite), and ultraviolet irradiation (UVC), are potent inducers of the translocation of HuR to the cytoplasm, where it functions as a promoter of mRNA stability and can also modulate mRNA translation. As with other Hu/elav proteins, HuR contains three RNA recognition motifs (RRM) through which it binds with high affinity and specificity to U-rich and AU-rich sequences in a variety of mRNAs. Many stress-regulated mRNAs are targets of HuR, including those that encode c-fos, p21, cyclins A2, B1, and D1, iNOS, granulocyte macrophage colony stimulating factor (GM-CSF), vascular endothelial growth factor (VEGF), sirtuin 1 (SIRT1), tumor necrosis factor (TNF)α, bcl-2, mcl-1, cyclooxygenase (COX)-2, γ-glutamylcysteine synthetase heavy subunit (γ-GCSh), urokinase plasminogen activator (uPA) and its receptor (uPAR), and interleukin (IL)-3 [20–32]. Treatment with H_2O_2 was found to enhance the binding of HuR to target mRNAs including p21 and MAPK phosphatase-1 (MKP-1), thereby elevating transcript half-life and abundance [24, 31, 106]. On the other hand, H_2O_2 treatment *reduced* the binding of HuR to SIRT1, cyclin A2, and cyclin B1 mRNAs [31], associated with a reduction in mRNA abundance following treatment with the oxidant.

AUF1. The AUF1 TTR-RBP comprises a family of four proteins that arise from alternative splicing (p37, p40, p42, p45), shuttle between the nucleus and the cytoplasm, and contain two RRMs [33–37]. Many studies using cultured cells that expressed different levels of AUF1, as well as studies of AUF1-deficient mice have demonstrated that AUF1 functions as a decay promoting TTR-RBP [18, 29, 38–41]. Nonetheless, in some instances AUF1 was also shown to promote mRNA stabilization [19, 38, 40, 42] or translation [43]. The levels of many AUF1 target mRNAs are regulated following stress, including those that encode p21, cyclin D1, c-myc, c-fos, GM-CSF, TNFα, IL-3, parathyroid hormone (PTH), and growth arrest- and DNA damage-inducible (GADD)45α mRNAs [18, 19, 29, 38, 40, 41, 44]. While AUF1's role in the stress regulated stability of these mRNAs has not been studied in detail, several examples have emerged in recent years. For example, in response to treatments with the stress agents prostaglandin A_2 or UVC, AUF1 was shown to bind to the cyclin D1 mRNA and to reduce its stability [44, 45]; similarly, the cellular response to bacterial lipopolisaccharide (LPS) implicated AUF1 in the degradation of target mRNAs encoding TNFα, IL-1β, and cyclooxygenase (COX)-2 [46, 47].

TTP. This predominantly cytoplasmic TTR-RBP is strongly inducible [48, 49] and promotes the decay of target mRNAs in a variety of systems [49–58]. TTP binds numerous stress regulated mRNAs such as those that codify GM-CSF, c-fos, TNFα, COX-2, IL-3, and IL-10 [49, 52–55, 59–62], many of them shared targets of other TTR-RBPs. In one specific example, TTP was implicated in the response to oxidative stress and inflammation seen in animals fed high glucose diets; these animals had elevated TTP levels in liver and skeletal muscle, linked to a reduction in TNFα levels [63, 64].

KSRP. In a phosphorylation-dependent manner, KSRP recruits labile mRNAs to the exosome (discussed below), wherein mRNAs are degraded [56, 61]. KSRP was proposed to carry out this function in competition with the binding of TTR-RBPs that promote mRNA stability such as HuR. Indeed, several KSRP target mRNAs are also shared with other TTR-RBPs, including the stress-regulated c-fos, c-jun, inducible nitric oxide synthase (iNOS), TNFα, and IL-2 mRNAs [49, 61, 65], although the specific involvement of KSRP in the stress response has not been studied.

BRF1. The decay of TNFα and IL-3 mRNAs is accelerated by BRF1 [63, 66]. The stability of both cytokine mRNAs is regulated by oxidative damage, even though the involvement of BRF1 in their degradation by oxidative stress also awaits experimental testing [67].

CUGBP1. This TTR-RBP associates with widespread GU-rich elements in the mRNAs encoding stress regulated c-jun, junB, and TNF receptor 1B and promoted their decay [68]. CUGBP1 was also found to promote the deadenylation of c-jun mRNA, thereby reducing its stability [69].

NF90, aCP1, RNPC1, CUGBP2, PAIP2, and nucleolin. Additional TTR-RBPs have also been shown to stabilize target mRNAs. While their functions remain to be studied in detail, they were found to stabilize many target mRNAs that encode stress-regulated gene products, such as p21, IL-2, p53, tyrosine hydroxylase, bcl-2, and COX-2 mRNAs [15, 70–74]. The involvement of these TTR-RBPs in stress responses also remains to be investigated in depth.

mRNA DECAY DETERMINANTS AND DEGRADATION MACHINERIES

cis Elements Regulating mRNA Turnover

Certain mRNA stability-promoting structures are ubiquitous, like the *5' end cap*, a 7 methylguanosine triphosphate (m^7GpppG) structure that protects messages from general 5'→3' exonucleases (Figure 28.1). The 5'-cap can be removed in a potentially regulatable fashion by a decapping nuclease (DCP1/2) [75]. The *3' end polyadenylate* or *poly(A) tail*, protects the 3' terminus from degradation by 3'→5' exonucleases. The 3' poly(A) tail is bound by a poly(A) RNA-binding protein (PABP) that protects it. Accordingly, steps leading to poly(A) removal are also subject to regulation; indeed, poly(A) shortening appears to be a rate-limiting step in the decay of many mammalian mRNAs. In addition, at least one sequence (CU-rich) was shown to confer unusually high *stability* to transcripts bearing it (e.g., α-globin, β-globin, and α-collagen mRNAs) [76–78].

By contrast, the sequences that enhance mRNA decay appear to be more variable. In some cases, these are found in the 5' untranslated region (UTR), as described for interleukin (IL)-2 mRNA [79], and in the coding region, as shown for the c-fos and c-myc mRNAs [80, 81]. However, the majority of specific regulatory sequences reside within the 3'UTR. The best understood examples of 3'UTR mediated stability are the IRE (iron responsive element), which confers stability to transferrin mRNA, and AREs (AU-rich and U-rich elements) present in many mRNAs, such as those encoding cytokines (interferons, interleukins, TNFα), cell cycle-regulatory proteins (p21, cyclin A, cyclin B1, cdc25), growth factors (GM-CSF, VEGF), and oncoproteins (c-fos, c-myc) [82–84]. AREs often (but not always) comprise one or several copies of the pentamer AUUUA, generally within a U-rich sequence [82]. More recently a GU-rich element (GRE) widely found in mRNAs was also identified as conferring mRNA instability [68]. However, many additional sequences that modulate mRNA stability have been reported that do not resemble any of these RNA elements.

Among the plausible mechanisms whereby associations of RNA-binding proteins with turnover-determining sequences can affect mRNA turnover are the following: (1) RNA-binding proteins may mask/unmask sites of endonucleolytic cleavage; (2) they may enhance or reduce the activity of potential decapping nucleases; (3) they may elevate or diminish the protective influence of PABPs; and (4) they may direct mRNAs to or away from sites in the cell where degradation machineries are at work (below). While the influence of stress upon these mechanisms remains to be investigated in depth, two general mRNA turnover pathways have been described. *Deadenylation-independent mRNA decay* is initiated through the endonucleolytic cleavage of a mRNA, followed by subsequent exonuclease digestion of each fragment through the combined action of 5'→3' and 3'→5' exonucleases. *Deadenylation-dependent decay*, believed to be the major pathway, is initiated by the removal of the poly(A) tract, followed by decapping and degradation by both 5'→3' and 3'→5' exonucleases. This

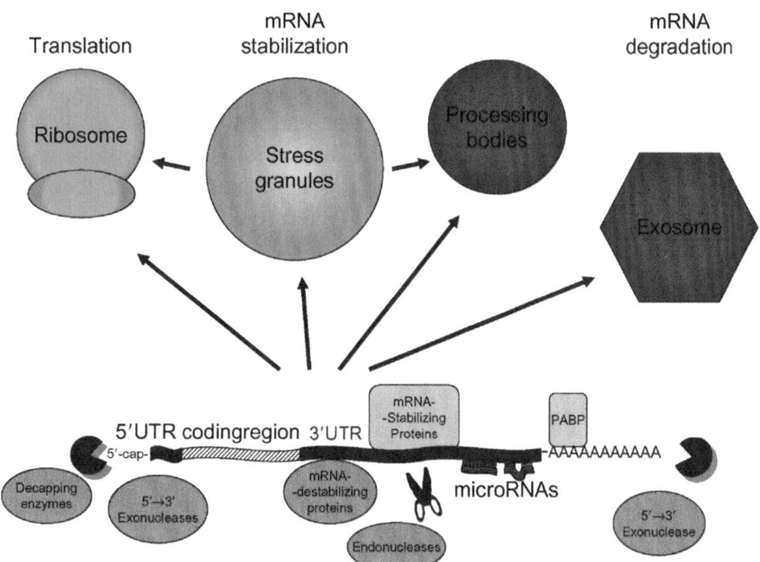

FIGURE 28.1 Regulation of mRNA turnover.
Mammalian mRNA turnover can be influenced by a number of events. Events favoring mRNA stability include its association with RNA binding proteins and poly(A) binding protein (PABP), as well as the recruitment of the mRNA to cytoplasmic sites such as stress granules. Events favoring mRNA degradation include its association with decay promoting proteins or microRNAs, its recruitment to P-bodies or the exosome, as well as its enzymatic digestion by endonucleases and exonucleases. Arrows indicate the recruitment of mRNAs to specialized cytoplasmic sites that modify its turnover and translation status.

is the pathway of degradation of most ARE containing mRNAs. For a review on this topic, see [1].

Degradation Machineries

Besides the molecular factors that dictate mRNA turnover (mRNA sequences and TTR-RBPs), three cellular entities have been identified in the regulation of mRNA stability: the *exosome* and *processing (P)-bodies*, two structures where mRNAs containing instability sequences are digested, and *stress granules*, wherein mRNAs appear to be protected from degradation (Figure 28.1). The exosome is a large multi-protein complex responsible for 3'→5' degradation of labile mRNAs following their recruitment by decay promoting TTR-RBPs [61, 65, 85, 86]. For mRNAs recruited to the exosome by TTP and KSRP, the activity of the deadenylase PARN (poly(A) ribonuclease) is also required for transcript decay, suggesting that the turnover of TTP and KSRP target mRNAs involves the combined action of deadenylation factors and exosome-associated ribonucleases [55, 65]. P-bodies (PBs) are cytoplasmic structures that contain proteins involved in mRNA decapping and 5'→3' degradation [87–90]. PBs are functionally linked to stress granules (SGs), cytoplasmic foci that form transiently in response to a variety of damaging agents, and represent sites of translational repression [91]. As mRNAs are dynamically transported between these two cytosolic domains, SGs have been proposed to serve as transient storage sites for untranslated mRNAs until molecular decisions are made to direct the mRNA to polysomes for translation or to PBs for degradation [90, 92]. Such molecular decisions involve TTR–RBPs that regulate mRNA turnover (such as TTP, KSRP, HuR), translational regulatory factors (TIA-1, TIAR), as well as the metabolic state of the cell and the extent of cellular injury and repair [91].

STRESS-ACTIVATED SIGNALING MOLECULES THAT REGULATE mRNA TURNOVER

As mentioned above, many recent studies have identified signal transduction pathways that regulate RNP complex activity. We discuss several prominent pathways and provide examples. Table 28.1 lists several major stress-regulated pathways known to influence TTR-RBP activity and mRNA stability.

p38, PKCa, Cdk1, AMPK, and Chk2 Regulate HuR Function

In response to various stress agents (H_2O_2, arsenite, UVC) HuR is exported to the cytoplasm [24, 31]. At least four kinases have been implicated in the nuclear export of HuR by ROS-producing agents: p38, AMPK, PKCα, and Cdk1 (cyclin-dependent kinase 1 or Cdc2). Activation of the MAPK p38 in response to oxidants such as sulindac and taxanes was linked to increases in the cytoplasmic localization of HuR [93, 94], which in turn increased HuR-COX-2 mRNA and HuR-γ-GCSh mRNA complexes and enhanced the production of the encoded proteins. In this paradigm, the effectors of p38 action remain to be identified; whether or not HuR is a direct phosphorylation substrate for p38 as well as the specific transport proteins that participate in this transport process are unknown. Treatment with UVC inhibited AMPK, a kinase that phosphorylated importin α1 and promoted its acetylation, in turn enhancing the nuclear import of HuR [95]. At least partly by inhibiting AMPK, UVC led to an increase in cytoplasmic HuR and to the stabilization of target mRNAs encoding p21, cyclin A2, and cyclin B1 [95], although HuR itself did not appear to be a phosphorylation substrate for AMPK.

Other kinases influence HuR subcellular localization by phosphorylating HuR itself. Treatment with an ATP analog activated PKCα, which phosphorylated HuR at residues S158 and S221 and thereby promoted the export of HuR to the cytoplasm, where its target COX-2 mRNA became more stable [96]. Whether the elevated ATP levels contributed to increasing cytoplasmic HuR levels by concomitantly inhibiting AMPK remains to be tested. Our unpublished studies reveal that HuR is an *in vivo* substrate of Cdk1. Cdk1 phosphorylated HuR at residue S202 and this modification increased HuR's association with a novel nuclear ligand, the molecular chaperone 14-3-3. According to our findings, Cdk1 phosphorylated HuR at S202, thereby retaining it in the nucleus in association with 14-3-3 and hindering its post-transcriptional function and anti-apoptotic influence. UVC treatment inhibits Cdk1, in turn decreasing the levels of phosphorylated HuR and augmenting its abundance in the cytoplasm [106]. It is interesting to note that both PKCα and Cdk1 phosphorylate HuR within its hinge region, either at the HuR nucleocytoplasmic shuttling (HNS) domain (S221) or proximal to it (S202), suggesting that local perturbations in charge or conformation within this region perturbs HuR shuttling and consequently its function.

While H_2O_2 treatment increased the cytoplasmic abundance of HuR, it unexpectedly triggered the dissociation of HuR from several targets including the SIRT1 and cyclin D1 mRNAs; in turn, their stability was reduced [31]. The loss of these RNPs was attributed to H_2O_2-mediated activation of Chk2, a kinase that phosphorylates HuR within RRM1 (S88), RRM2 (T118), and between both RRMs (S100). Point mutation of these residues revealed a complex pattern of HuR binding, with S100 appearing important for [HuR-SIRT1 mRNA] dissociation after H_2O_2

TABLE 28.1 Stress-triggered signaling pathways that modulate mRNA turnover[1]

Stress signaling pathway	Downstream TTR-RBP	Influence on TTR-RBP function	Influence on bound mRNA	Target mRNAs
ATM/Chk2	HuR (S88, S100, T118)	Binding to mRNA changes	stability, translation	c-fos, p21, cyclins A2/B1/D1, iNOS, GM-CSF, VEGF, SIRT1, TNFα, IL-3, bcl-2, mcl-1, COX-2, γ-GCSh, uPA, uPAR, MKP-1, p53, ProTα, CAT-1, cytochrome c, MKP-1, p27, IGF-IR, Wnt5a, TNFα
PKC	HuR (S158, S221)	Cytoplasmic localization		
AMPK	[HuR]	Nuclear localization		
Cdk1	HuR(S202)	Cytoplasmic localization		
p38	[HuR]	Cytoplasmic localization		
p38	KSRP (T692)	No binding to mRNA	↑ stability	c-fos, c-jun, iNOS, TNFα, IL-2
PI3K/Akt	KSRP (S193)	Binding to 14-3-3, exclusion from exosome		
PI3K/Akt	BRF1 (S92, S203)	Binding to 14-3-3, exclusion from exosome	↑ stability	TNFα, IL-3
MK2	TTP (S52, S218)	Binding to 14-3-3, exclusion from SG	↑ stability	GM-CSF, c-fos, TNFα, COX-2, IL-3, IL-10
ALK	AUF1 (?)	Binding to mRNA prevented	↑ stability	cyclin D1, c-myc, c-fos, TS, GM-CSF, TNFα, IL-1β, IL-3, PTH, p16, p21, GADD45α, COX-2
PKA	AUF1 (S87)	Binding to mRNA prevented		
GSK3β	AUF1 (S83)	Binding to mRNA prevented		

[1]*Stress-regulated signaling cascades ("Stress signaling pathways") and their influence on TTR-RBP function ("Downstream TTR-RBPs"). Specific amino acid residues phosphorylated in TTR-RBPs are indicated: [HuR] has not been reported to be a direct phosphorylation substrate of the corresponding stress-regulated kinases. "Influence on TTR-RBP function," effect of individual TTR-RBP modifications by stress-regulated signaling pathways. "Influence on bound mRNA," changes in target mRNA stability and/or translation as a result of TTR-RBP modification. "Target mRNAs," mRNAs identified as being regulated through mRNA stability and/or translation by the corresponding TTR-RBPs.*

treatment, but S88 and T118 appeared necessary for HuR to bind efficiently to other target mRNAs [31].

MAPKAPK2 (MK2) Affects TTP Function

Arsenite treatment prevented the degradation of short-lived mRNAs forming RNP complexes with TTP [63]. This inhibitory influence was attributed to the phosphorylation of TTP by MAPKAPK2 (MK2) at S52 and S178, which triggered the association of TTP with 14-3-3. The resulting [TTP-14-3-3] complex was excluded from SGs and the bound mRNAs became stable. A similar mechanism was proposed to mediate the stabilization of TNFα in response to LPS treatment [63].

PKA, GSK3β, and ALK Affect AUF1 Activity

$p40^{AUF1}$ was shown to be phosphorylated at residues S83 and S87 [97, 98]. As determined by *in vitro* kinase assays, S87 was phosphorylated by protein kinase A (PKA) and S83 by glycogen synthase kinase (GSK)3β [99]. These modifications altered the ability of AUF1 to bind to target mRNAs *in vitro*, but have not yet been studied in cells responding to stress agents. More recently, AUF1 was shown to be an *in vitro* substrate of the anaplastic lymphoma kinase (ALK) and was found to be hyperphosphorylated in cells expressing a chimeric protein (nucleophosmin-ALK or NPM-ALK) linked to anaplastic large cell lymphoma [100]. AUF1 phosphorylation in these cells was correlated with increased stability of several AUF1 target mRNAs encoding proteins with roles in cell proliferation and survival [100].

Akt Affects BRF1 Function

After phosphorylation by Akt at residues S92 and S203, BRF1 was still capable of binding to target mRNAs, as described for TTP [63]. However, phosphorylated BRF1 formed [BRF1-14-3-3] complexes that could not be recruited to the exosome and thus the labile, BRF target mRNAs were stabilized [101, 102]. Instead, phosphorylated BRF1 was shown to associate with the cytoskeleton [102]. Despite the broad involvement of Akt in stress responses, the influence of cellular damage upon the Akt mediated BRF1 phosphorylation and subsequent mRNA stabilization also awaits experimental testing.

CONCLUDING REMARKS

Given that steady state mRNA levels are essential for determining cellular protein levels, the regulation of mRNA turnover critically influences the cell's ability to respond to situations of stress. As reviewed in this chapter, the involvement of TTR-RBPs in regulating mRNA half-life is firmly established and we are making strides in understanding the stress-regulated signaling pathways that modulate the abundance, location, and function of the resulting RNP complexes. Among the immediate challenges facing the field are the systematic identification of TTP-RBP target mRNAs and a more comprehensive knowledge of the stress pathways that control TTR-RBP levels, distribution, and function. Many exciting developments in this area will be presented by microRNAs, another major class of mRNA-interacting molecules. With the expectation that microRNAs will critically regulate gene expression during stress [103], there is mounting interest in the specific mechanisms whereby microRNAs affect gene expression [104] and the various ways in which microRNAs and RBPs may interact to modulate in gene expression patterns [105]. The development of suitable animal models in which to study the effect of post-transcriptional regulators will also critically advance our understanding of physiologic changes in gene expression patterns that arise in response to stress. As our knowledge in these areas of gene expression advances, so will our ability to intervene in a broad range of pathological processes arising from aberrant gene regulation.

ACKNOWLEDGEMENTS

The authors are supported by the Intramural Research Program of the National Institute on Aging, National Institutes of Health.

REFERENCES

1. Mitchell P, Tollervey D. mRNA stability in eukaryotes. *Curr Opin Genet Dev* 2000;**10**:193–8.
2. Orphanides G, Reinberg D. A unified theory of gene expression. *Cell* 2002;**108**:439–51.
3. Moore MJ. From birth to death: the complex lives of eukaryotic mRNAs. *Science* 2005;**309**:1514–18.
4. Fan J, Yang X, Wang W, Wood 3rd WH, Becker KG, Gorospe M. Global analysis of stress-regulated mRNA turnover by using cDNA arrays. *Proc Natl Acad Sci USA* 2002;**99**:10,611–10,616.
5. Peltz SW, Brewer G, Bernstein P, Hart PA, Ross J. Regulation of mRNA turnover in eukaryotic cells. *Crit Rev Eukaryot Gene Expr* 1991;**1**:99–126.
6. Valencia-Sanchez MA, Liu J, Hannon GJ, Parker R. Control of translation and mRNA degradation by miRNAs and siRNAs. *Genes Dev* 2006;**20**:515–24.
7. Keene JD. RNA regulons: coordination of post-transcriptional events. *Nat Rev Genet* 2007;**8**:533–43.
8. Bhattacharyya SN, Habermacher R, Martine U, Closs EI, Filipowicz W. Relief of microRNA-mediated translational repression in human cells subjected to stress. *Cell* 2006;**125**:1111–24.
9. Sunkar R, Kapoor A, Zhu JK. Posttranscriptional induction of two Cu/Zn superoxide dismutase genes in Arabidopsis is mediated by downregulation of miR398 and important for oxidative stress tolerance. *Plant Cell* 2006;**18**:2051–65.
10. Wu L, Fan J, Belasco JG. MicroRNAs direct rapid deadenylation of mRNA. *Proc Natl Acad Sci USA* 2006;**103**:4034–9.
11. Martindale JL, Holbrook NJ. Cellular response to oxidative stress: signaling for suicide and survival. *J Cell Physiol* 2002;**192**:1–15.
12. Pullmann R, Kim HH, Abdelmohsen K, Lal A, Martindale JL, Yang X, Gorospe M. Analysis of Stability and Translation Regulatory RBP Expression Through Binding to Cognate mRNAs. *Mol Cell Biol* 2007;**27**:6265–78.
13. Xu YH, Grabowski GA. Molecular cloning and characterization of a translational inhibitory protein that binds to coding sequences of human acid beta-glucosidase and other mRNAs. *Mol Genet Metab* 1999;**68**:441–54.
14. Xu YH, Busald C, Grabowski GA. Reconstitution of TCP80/NF90 translation inhibition activity in insect cells. *Mol Genet Metab* 2000;**70**:106–15.
15. Shim J, Lim HR, Yates J, Karin M. Nuclear export of NF90 is required for interleukin-2 mRNA stabilization. *Mol Cell* 2002;**10**:1331–44.
16. Gorospe M. HuR in the mammalian genotoxic response: Post-transcriptional multitasking. *Cell Cycle* 2003;**2**:412–14.
17. Xu YH, Leonova T, Grabowski GA. Cell cycle dependent intracellular distribution of two spliced isoforms of TCP/ILF3 proteins. *Mol Genet Metab* 2003;**80**:426–36.
18. Brewer G. An A+U-rich element RNA-binding factor regulates c-myc mRNA stability in vitro. *Mol Cell Biol* 1991;**11**:2460–6.
19. Xu N, Chen C, Shyu A-B. Versatile role for hnRNPD isoforms in the differential regulation of cytoplasmic mRNA turnover. *Mol Cell Biol* 2001;**21**:6960–71.

20. Fan XC, Steitz JA. Overexpression of HuR, a nuclear-cytoplasmic shuttling protein, increases the in vivo stability of ARE containing mRNAs. *EMBO J* 1998;**17**:3448–60.
21. Levy N, Chung S, Furneaux H, Levy A. Hypoxic stabilization of vascular endothelial growth factor mRNA by the RNAbinding protein HuR. *J Biol Chem* 1998;**273**:6417–23.
22. Peng SS, Chen CY, Xu N, Shyu A-B. RNA stabilization by the AU-rich element binding protein, HuR, an ELAV protein. *EMBO J* 1998;**17**:3461–70.
23. Wang W, Caldwell MC, Lin S, Furneaux H, Gorospe M. HuR regulates cyclin A and cyclin B1 mRNA stability during cell proliferation. *EMBO J* 2000;**19**:2340–50.
24. Wang W, Furneaux H, Cheng H, Caldwell MC, Hutter D, Liu Y, Holbrook NJ, Gorospe M. HuR Regulates p21 mRNA stabilization by Ultraviolet Light. *Mol Cell Biol* 2000;**20**:760–9.
25. Ming XF, Stoecklin G, Lu M, Looser R, Moroni C. Parallel and independent regulation of interleukin-3 mRNA turnover by phosphatidylinositol 3-kinase and p38 mitogen-activated protein kinase. *Mol Cell Biol* 2001;**21**:5778–89.
26. Chen CY, Xu N, Shyu AB. Highly selective actions of HuR in antagonizing AU-rich element-mediated mRNA destabilization. *Mol Cell Biol* 2002;**22**:7268–78.
27. Sengupta S, Jang BC, Wu MT, Paik JH, Furneaux H, Hla T. The RNA-binding protein HuR regulates the expression of cyclooxygenase-2. *J Biol Chem* 2003;**278**:25,227–25,233.
28. Tran H, Maurer F, Nagamine Y. Stabilization of urokinase and urokinase receptor mRNAs by HuR is linked to its cytoplasmic accumulation induced by activated mitogen-activated protein kinase-activated protein kinase 2. *Mol Cell Biol* 2003;**23**:7177–88.
29. Lal A, Mazan-Mamczarz K, Kawai T, Yang X, Martindale JL, Gorospe M. Concurrent versus individual binding of HuR and AUF1 to common labile target mRNAs. *EMBO J* 2004;**23**:3092–102.
30. Song IS, Tatebe S, Dai W, Kuo MT. Delayed mechanism for induction of gamma-glutamylcysteine synthetase heavy subunit mRNA stability by oxidative stress involving p38 mitogen-activated protein kinase signaling. *J Biol Chem* 2005;**280**:28,230–28,240.
31. Abdelmohsen K, Pullmann R, Lal A, Kim HH, Galban S, Yang X, Blethrow JD, Walker M, Shubert J, Gillespie DA, Furneaux H, Gorospe M. Phosphorylation of HuR by Chk2 regulates SIRT1 expression. *Mol Cell* 2007;**25**:543–57.
32. Abdelmohsen K, Lal A, Kim HH, Gorospe M. Posttranscriptional orchestration of an anti-apoptotic program by HuR. *Cell Cycle* 2007;**6**:1288–92.
33. Zhang W, Wagner BJ, Ehrenman K, Schaefer AW, DeMaria CT, Crater D, DeHaven K, Long L, Brewer G. Purification, characterization, and cDNA cloning of an AU-rich element RNA-binding protein, AUF1. *Mol Cell Biol* 1993;**13**:7652–65.
34. Laroia G, Cuesta R, Brewer G, Schneider RJ. Control of mRNA decay by heat shock-ubiquitin-proteasome pathway. *Science* 1999;**284**:499–502.
35. Loflin P, Chen CY, Shyu A-B. Unraveling a cytoplasmic role for hnRNP D in the in vivo mRNA destabilization directed by the AU-rich element. *Genes Dev* 1999;**13**:1884–97.
36. Shyu AB, Wilkinson MF. The double lives of shuttling mRNA binding proteins. *Cell* 2000;**102**:135–8.
37. Sarkar B, Lu JY, Schneider RJ. Nuclear import and export functions in the different isoforms of the AUF1/heterogeneous nuclear ribonucleoprotein protein family. *J Biol Chem* 2003;**278**:20,700–20,707.
38. Sela-Brown A, Silver J, Brewer G, Naveh-Many T. Identification of AUF1 as a parathyroid hormone mRNA 3'-untranslated region-binding protein that determines parathyroid hormone mRNA stability. *J Biol Chem* 2000;**275**:7424–9.
39. Sarkar B, Xi Q, He C, Schneider RJ. Selective degradation of AU-rich mRNAs promoted by the p37 AUF1 protein isoform. *Mol Cell Biol* 2003;**23**:6685–93.
40. Raineri I, Wegmueller D, Gross B, Certa U, Moroni C. Roles of AUF1 isoforms, HuR and BRF1 in ARE-dependent mRNA turnover studied by RNA interference. *Nucleic Acids Res* 2004;**32**:1279–88.
41. Fialcowitz EJ, Brewer BY, Keenan BP, Wilson GM. A hairpin-like structure within an AU-rich mRNA-destabilizing element regulates trans-factor binding selectivity and mRNA decay kinetics. *J Biol Chem* 2005;**280**:22,406–22,417.
42. Puig S, Askeland E, Thiele DJ. Coordinated remodeling of cellular metabolism during iron deficiency through targeted mRNA degradation. *Cell* 2005;**120**:99–110.
43. Liao B, Hu Y, Brewer G. Competitive binding of AUF1 and TIAR to MYC mRNA controls its translation. *Nat Struct Mol Biol* 2007;**14**:511–18.
44. Lal A, Abdelmohsen K, Pullmann R, Kawai T, Yang X, Galban S, Brewer G, Gorospe M. Posttranscriptional derepression of GADD45a by genotoxic stress. *Mol Cell* 2006;**22**:117–28.
45. Lin S, Wang W, Wilson GM, Yang X, Brewer G, Holbrook NJ, Gorospe M. Downregulation of cyclin D1 expression by prostaglandin A2 is mediated by enhanced cyclin D1 mRNA turnover. *Mol Cell Biol* 2000;**20**:7903–13.
46. Cok SJ, Acton SJ, Sexton AE, Morrison AR. Identification of RNA-binding proteins in RAW 264.7 cells that recognize a lipopolysaccharide-responsive element in the 3-untranslated region of the murine cyclooxygenase-2 mRNA. *J Biol Chem* 2004;**279**:8196–205.
47. Lu JY, Sadri N, Schneider RJ. Endotoxic shock in AUF1 knockout mice mediated by failure to degrade proinflammatory cytokine mRNAs. *Genes Dev* 2006;**20**:3174–84.
48. Carballo E, Lai WS, Blackshear PJ. Feedback inhibition of macrophage tumor necrosis factor-alpha production by tristetraprolin. *Science* 1998;**281**:1001–5.
49. Linker K, Pautz A, Fechir M, Hubrich T, Greeve J, Kleinert H. Involvement of KSRP in the post-transcriptional regulation of human iNOS expression-complex interplay of KSRP with TTP and HuR. *Nucleic Acids Res* 2005;**33**:4813–27.
50. Carballo E, Gilkeson GS, Blackshear PJ. Bone marrow transplantation reproduces the tristetraprolin-deficiency syndrome in recombination activating gene-2 mice. Evidence that monocyte/macrophage progenitors may be responsible for TNFalpha overproduction. *J Clin Invest* 1997;**100**:986–95.
51. Carballo E, Lai WS, Blackshear PJ. Evidence that tristetraprolin is a physiological regulator of granulocyte-macrophage colony-stimulating factor messenger RNA deadenylation and stability. *Blood* 2000;**95**:1891–9.
52. Lai WS, Carballo E, Strum JR, Kennington EA, Phillips RS, Blackshear PJ. Evidence that tristetraprolin binds to AU-rich elements and promotes the deadenylation and destabilization of tumor necrosis factor alpha mRNA. *Mol Cell Biol* 1999;**19**:4311–23.
53. Lai WS, Carballo E, Thorn JM, Kennington EA, Blackshear PJ. Interactions of CCCH zinc finger proteins with mRNA. Binding of tristetraprolin-related zinc finger proteins to Au-rich elements and destabilization of mRNA. *J Biol Chem* 2000;**275**:17,827–17,837.
54. Lai WS, Kennington EA, Blackshear PJ. Interactions of CCCH zinc finger proteins with mRNA: non-binding tristetraprolin mutants exert an inhibitory effect on degradation of AU-rich element containing mRNAs. *J Biol Chem* 2002;**277**:9606–13.

55. Lai WS, Kennington EA, Blackshear PJ. Tristetraprolin and its family members can promote the cell-free deadenylation of AU-rich element-containing mRNAs by poly(A) ribonuclease. *Mol Cell Biol* 2003;**23**:3798–812.
56. Briata P, Forcales SV, Ponassi M, Corte G, Chen CY, Karin M, Puri PL, Gherzi R. p38-dependent phosphorylation of the mRNA decay-promoting factor KSRP controls the stability of select myogenic transcripts. *Mol Cell* 2005;**20**:891–903.
57. Lykke-Andersen J, Wagner E. Recruitment and activation of mRNA-decay enzymes by two ARE-mediated decay activation domains in the proteins TTP and BRF-1. *Genes Dev* 2005;**19**:351–61.
58. Ogilvie RL, Abelson M, Hau HH, Vlasova I, Blackshear PJ, Bohjanen PR. Tristetraprolin down-regulates IL-2 gene expression through AU-rich element-mediated mRNA decay. *J Immunol* 2005;**174**:953–61.
59. Stoecklin G, Ming XF, Looser R, Moroni C. Somatic mRNA turnover mutants implicate tristetraprolin in the interleukin-3 mRNA degradation pathway. *Mol Cell Biol* 2000;**20**:3753–63.
60. Stoecklin G, Tenenbaum SA, Mayo T, Chittur SV, George AD, Baroni TE, Blackshear PJ, Anderson P. Genome-wide analysis identifies interleukin-10 mRNA as target of tristetraprolin. *J Biol Chem* 2008;**283**:11,689–11,699.
61. Chen CY, Gherzi R, Ong SE, Chan EL, Raijmakers R, Pruijn GJ, Stoecklin G, Moroni C, Mann M, Karin M. AU binding proteins recruit the exosome to degrade ARE-containing mRNAs. *Cell* 2001;**107**:451–64.
62. Sawaoka H, Dixon DA, Oates JA, Boutaud O. Tristetraprolin binds to the 30-untranslated region of cyclooxygenase-2 mRNA. A polyadenylation variant in a cancer cell line lacks the binding site. *J Biol Chem* 2003;**278**:13,928–13,935.
63. Stoecklin G, Stubbs T, Kedersha N, Wax S, Rigby WF, Blackwell TK, Anderson P. MK2-induced tristetraprolin: 14-3-3 complexes prevent stress granule association and ARE-mRNA decay. *EMBO J* 2004;**23**:1313–24.
64. Cao H, Kelly MA, Kari F, Dawson HD, Urban JF, Coves S, Roussel AM, Anderson RA. Green tea increases anti-inflammatory tristetraprolin and decreases pro-inflammatory tumor necrosis factor mRNA levels in rats. *J Inflamm* 2007;**5**:1.
65. Gherzi R, Lee KY, Briata P, Wegmuller D, Moroni C, Karin M, Chen CY. A KH-domain RNA-binding protein, KSRP, promotes ARE-directed mRNA turnover by recruiting the degradation machinery. *Mol Cell* 2004;**14**:571–83.
66. Lai WS, Blackshear PJ. Interactions of CCCH zinc finger proteins with mRNA: tristetraprolin-mediated AU-rich element dependent mRNA degradation can occur in the absence of a poly(A) tail. *J Biol Chem* 2001;**276**:23,144–23,154.
67. Ming XF, Kaiser M, Moroni C. c-jun N-terminal kinase is involved in AUUUA-mediated interleukin-3 mRNA turnover in mast cells. *EMBO J* 1998;**17**:6039–48.
68. Vlasova IA, Tahoe NM, Fan D, Larsson O, Rattenbacher B, Sternjohn JR, Vasdewani J, Karypis G, Reilly CS, Bitterman PB, Bohjanen PR. Conserved GU-Rich Elements Mediate mRNA Decay by Binding to CUG-Binding Protein 1. *Mol Cell* 2008;**29**:263–70.
69. Paillard L, Legagneux V, Maniey D, Osborne HB. c-Jun ARE targets mRNA deadenylation by an EDEN-BP (embryo deadenylation element-binding protein)-dependent pathway. *J Biol Chem* 2002;**277**:3232–5.
70. Paulding WR, Czyzyk-Krzeska MF. Hypoxia-induced regulation of mRNA stability. *Adv Exp Med Biol* 2000;**475**:111–21.
71. Mukhopadhyay D, Houchen CW, Kennedy S, Dieckgraefe BK, Anant S. Coupled mRNA stabilization and translational silencing of cyclooxygenase-2 by a novel RNA binding protein, CUGBP2. *Mol Cell* 2003;**11**:113–26.
72. Sengupta TK, Bandyopadhyay S, Fernandes DJ, Spicer EK. Identification of nucleolin as an AU-rich element binding protein involved in bcl-2 mRNA stabilization. *J Biol Chem* 2004;**279**:10,855–10,863.
73. Shi L, Zhao G, Qiu D, Godfrey WR, Vogel H, Rando TA, Hu H, Kao PN. NF90 regulates cell cycle exit and terminal myogenic differentiation by direct binding to the 30-untranslated region of MyoD and p21WAF1/CIP1 mRNAs. *J Biol Chem* 2005;**280**:18,981–18,989.
74. Shu L, Yan W, Chen X. RNPC1, an RNA-binding protein and a target of the p53 family, is required for maintaining the stability of the basal and stress-induced p21 transcript. *Genes Dev* 2006;**20**:2961–72.
75. Simon, E., Camier, S., and Séraphin, B. New insights into the control of mRNA decapping. *Trends Biochem Sci* **31:** 241–243.
76. Kiledjian M, Wang X, Liebhaber SA. Identification of two KH domain proteins in the -globin mRNP stability complex. *EMBO J* 1995;**14**:4357–64.
77. Yu J, Russell JE. Structural and functional analysis of an mRNP complex that mediates the high stability of human β-globin mRNA. *Mol Cell Biol* 2001;**21**:5879–88.
78. Lindquist JN, Parsons CJ, Stefanovic B, Brenner DA. Regulation of 1(I) collagen messenger RNA decay by interactions with CP at the 3'-untranslated region. *J Biol Chem* 2004;**279**:23,822–23,829.
79. Chen C-Y, Del Gatto-Konczak F, Wu Z, Karin M. Stabilization of Interleukin-2 mRNA by the c-Jun NH2-Terminal Kinase Pathway. *Science* 1998;**280**:1945–9.
80. Shyu AB, Greenberg ME, Belasco JG. The c-fos transcript is targeted for decay by two distinct mRNA degradation pathways. *Genes Dev* 1989;**3**:60–72.
81. Wisdom R, Lee W. The protein-coding region of c-myc mRNA contains a sequence that specifies rapid mRNA turnover and induction by protein synthesis inhibitors. *Genes Dev* 1991;**5**:232–43.
82. Chen CY, Shyu AB. AU-rich elements: characterization and importance in mRNA degradation. *Trends Biochem Sci* 1995;**20**:465–70.
83. Lagnado CA, Brown CY, Goodall GJ. AUUUA is not sufficient to promote poly(A) shortening and degradation of an mRNA: the functional sequence within AU-rich elements may be UUAUUUA(U/A)(U/A). *Mol Cell Biol* 1994;**14**:7984–95.
84. Xu N, Chen C-YA, Shyu A-B. Modulation of the fate of cytoplasmic mRNA by AU-rich elements: key sequence features controlling mRNA deadenylation and decay. *Mol Cell Biol* 1997;**17**:4611–21.
85. Butler JS. The yin and yang of the exosome. *Trends Cell Biol* 2002;**12**:90–6.
86. Mukherjee D, Gao M, O'Connor JP, Raijmakers R, Pruijn G, Lutz CS, Wilusz J. The mammalian exosome mediates the efficient degradation of mRNAs that contain AU-rich elements. *EMBO J* 2002;**21**:165–74.
87. Ingelfinger D, Arndt-Jovin DJ, Luhrmann R, Achsel T. The human LSm1-7 proteins colocalize with the mRNA-degrading enzymes Dcp1/2 and Xrnl in distinct cytoplasmic foci. *RNA* 2002;**8**:1489–501.
88. Eystathioy T, Jakymiw A, Chan EK, Seraphin B, Cougot N, Fritzler MJ. The GW182 protein colocalizes with mRNA degradation associated proteins hDcp1 and hLSm4 in cytoplasmic GW bodies. *RNA* 2003;**9**:1171–3.
89. Cougot N, Babajko S, Seraphin B. Cytoplasmic foci are sites of mRNA decay in human cells. *J Cell Biol* 2004;**165**:31–40.
90. Kedersha N, Anderson P. Stress granules: sites of mRNA triage that regulate mRNA stability and translatability. *Biochem Soc Trans* 2002;**30**:963–9.
91. Kedersha N, Stoecklin G, Ayodele M, Yacono P, Lykke-Andersen J, Fitzler MJ, Scheuner D, Kaufman RJ, Golan DE, Anderson P. Stress

granules and processing bodies are dynamically linked sites of mRNP remodeling. *J Cell Biol* 2005;**169**:871–84.

92. Wilczynska A, Aigueperse C, Kress M, Dautry F, Weil D. The translational regulator CPEB1 provides a link between dcp1 bodies and stress granules. *J Cell Sci* 2005;**118**:981–92.
93. Subbaramaiah K, Marmo TP, Dixon DA, Dannenberg AJ. Regulation of cyclooxgenase-2 mRNA stability by taxanes: evidence for involvement of p38, MAPKAPK-2, and HuR. *J Biol Chem* 2003;**278**:37,637–37,647.
94. Song IS, Tatebe S, Dai W, Kuo MT. Delayed mechanism for induction of gamma-glutamylcysteine synthetase heavy subunit mRNA stability by oxidative stress involving p38 mitogen-activated protein kinase signaling. *J Biol Chem* 2005;**280**:28,230–28,240.
95. Wang W, Fan J, Yang X, Fürer S, Lopez de Silanes I, von Kobbe C, Guo J, Georas SN, Foufelle F, Hardie DG, Carling D, Gorospe M. AMP-activated kinase regulates cytoplasmic HuR. *Mol Cell Biol* 2002;**22**:3425–36.
96. Doller A, Huwiler A, Müller R, Radeke HH, Pfeilschifter J, Eberhardt W. Protein Kinase C alpha-dependent phosphorylation of the mRNA-stabilizing factor HuR: implications for posttranscriptional regulation of cyclooxygenase-2. *Mol Biol Cell* 2007;**18**:2137–48.
97. Wilson GM, Lu J, Sutphen K, Suarez Y, Sinha S, Brewer B, Villanueva-Feliciano EC, Ysla RM, Charles S, Brewer G. Phosphorylation of p40AUF1 regulates binding to A+U-rich mRNA-destabilizing elements and protein-induced changes in ribonucleoprotein structure. *J Biol Chem* 2003;**278**:33,039–33,048.
98. Wilson GM, Lu J, Sutphen K, Sun Y, Huynh Y, Brewer G. Regulation of A+U-rich element-directed mRNA turnover involving reversible phosphorylation of AUF1. *J Biol Chem* 2003;**278**:33,029–33,038.
99. Tolnay M, Juang Y-T, Tsokos GC. Protein kinase A enhances, whereas glycogen synthase kinase-3 beta inhibits, the activity of the exon 2-encoded transactivator domain of heterogeneous nuclear ribonucleoprotein D in a hierarchical fashion. *Biochem J* 2002;**363**:127–36.
100. Fawal M, Armstrong F, Ollier S, Dupont H, Touriol C, Monsarrat B, Delsol G, Payrastre B, Morello D. A "liaison dangereuse" between AUF1/hnRNPD and the oncogenic tyrosine kinase NPM-ALK. *Blood* 2006;**108**:2780–8.
101. Schmidlin M, Lu M, Leuenberger SA, Stoecklin G, Mallaun M, Gross B, Gherzi R, Hess D, Hemmings BA, Moroni C. The ARE-dependent mRNA-destabilizing activity of BRF1 is regulated by protein kinase B. *EMBO J* 2004;**23**:4760–9.
102. Benjamin D, Schmidlin M, Min L, Gross B, Moroni C. BRF1 protein turnover and mRNA decay activity are regulated by protein kinase B at the same phosphorylation sites. *Mol Cell Biol* 2006;**26**:9497–507.
103. Leung AK, Sharp PA. microRNAs: a safeguard against turmoil?. *Cell* 2007;**130**:581–5.
104. Filipowicz W, Bhattacharyya SN, Sonenberg N. Mechanisms of post-transcriptional regulation by microRNAs: are the answers in sight? *Nat Rev Genet* 2008;**9**:102–14.
105. George AD, Tenenbaum SA. MicroRNA modulation of RNA-binding protein regulatory elements. *RNA Biol* 2006;**3**:57–9.
106. Kim HH, Abdelmohsen K, Lal A, Pullmann R Jr, Yang X, Galban S, Srikantan S, Martindale JL, Blethrow J, Shokat KM, Gorospe M. Nuclear HuR accumulation through phosphorylation by Cdkl. *Genes Dev* 2008;**22**:1804–15.

Chapter 29

Oncogenic Stress Responses

Dmitry V. Bulavin
Institute of Molecular and Cell Biology, Proteos, Singapore

INTRODUCTION

Cancer is a major cause of death in older mammals. This raises the question whether cancer had a selective pressure during evolution to create cancer protective mechanisms. One could argue that it would be unlikely for such mechanisms to evolve since cancer is primarily a post-reproductive disease. Evidence described in this review suggests that there are in fact multiple mechanisms that efficiently prevent cancer that would otherwise occur from the daily onslaught of oncogenic mutations that long lived mammals are subjected to.

The origin of cancer is debatable, with accumulating evidence pointing toward the involvement of tissue specific stem cells and early progenitors [1, 2]. This model requires that stem cells convert into cancer stem cells at the earliest stages of tumorigenesis, a task that stem cells are prohibited from fulfilling due to their very nature [3–5]. Stem cells, while dividing continuously during their lifespan, rarely develop carcinogenic mutations suggesting that their genome is extremely well protected. This protection is achieved through several mechanisms, including their ability to activate robust apoptosis [6]. As such, upon acquiring genetic changes that lead to the activation of oncogenes, stem cells would turn on a program of self-elimination [5]. If an oncogene-expressing stem cell survives, however, the oncogene would be passed on to progenitors, and to more differentiated cell types. At this stage of tumorigenesis, additional mechanisms are engaged to prevent the formation of full-blown cancer.

It has been known for many years that there are numerous small neoplastic lesions in some tissues that rarely become cancerous. These lesions, such as melanocytes nevi [7], stop growing appreciably once they reach a certain size. The absence of mitotic figures in these benign tumors is puzzling, and suggests that at a certain stage the oncogene-driven mitogenic signals cease to evoke a proliferative response. A potential explanation for this phenomenon comes from recent work that identified an intriguing mechanism of growth cessation of pre-malignant or benign neoplasms. The first indication of this mechanism came from *in vitro* studies, which showed that under certain conditions oncogenes can elicit, paradoxically, a new type of cellular response called senescence [8–10]. Senescence is a specific form of stable growth arrest provoked by diverse stresses, including the enforced expression of cancer-promoting genes in cultured cells. This oncogene-induced senescence (OIS) is linked to well-characterized pathways in cultured cells; but whether OIS is an authentic anticancer process *in vivo* or simply an artifact of enforced oncogene expression in cells experiencing culture shock remains controversial [11]. Recent studies, however, show that OIS occurs *in vivo* in diverse pre-cancerous tissues from both human and mouse [12–15]. A compelling feature of these studies is the consistency of OIS in response to a variety of cancer-promoting mutations. Another important finding is that *in vivo* senescence is molecularly heterogeneous, requiring different pathways in different cell types and occurring in response to different oncogenic insults. Our present knowledge on cellular senescence is discussed below.

DOWNSTREAM EFFECTORS OF OIS

Primary human and murine cells (predominantly fibroblasts) have proven essential in the analysis of oncogene-induced cell cycle arrest and in the identification of underlying molecular mechanisms. Expression of activated oncogenes, such as Ha-ras, triggers an irreversible growth arrest state with many similarities to the cellular senescence

Handbook of Cell Signaling, Three-Volume Set 2 ed.

seen in late-passage primary fibroblasts [16]. This premature senescence response is considered to be protective, since it removes cells with aberrant oncogene expression from the pool of growing cells.

In terms of molecular mechanisms, the Ink4a and Arf tumor suppressor proteins are crucial gatekeepers for OIS in both human and rodent cells [17–19]. The binding of the Ink4a proteins to Cdk4 and Cdk6 induces an allosteric change that abrogates the binding of these kinases to D-type cyclins, and inhibits Cdk4/6-mediated phosphorylation of retinoblastoma (Rb) family members [20]. Expression of Ink4a maintains Rb family proteins in a hypophosphorylated state, which in turn promotes binding of Rb to E2F to induce cell cycle arrest. The tumor suppressor activity of Arf is largely ascribed to its ability to regulate p53 in response to aberrant growth or oncogenic stress [21]. Arf binds to and inactivates Mdm2, which in turn negatively regulates p53 [22]. One mechanism proposed to explain how Mdm2 regulates p53 is that it acts as an E3 ubiquitin ligase to target p53 for proteosomal degradation [23].

The signaling pathways that are engaged to activate Ink4a and Arf in the course of OIS are still not fully understood, and could be different depending on a cell types and the oncogene (Figure 29.1). Notably, the best studied of these molecular signals are those that induce the Erk MAPK pathway [24, 25]. A few studies have suggested how Ras activation might lead to increased Ink4a/Arf expression, including Erk mediated activation of Ets1/2 to induce *Ink4a* and Jun mediated activation of the transcription factor Dmp1 to induce *Arf* expression [26–28]. Additionally, a few repressors of *Ink4a/Arf* expression have been identified. For example, the T box proteins and the polycomb group (PcG) genes (Bmi1, Cdx7, Mel18) reportedly repress the *Ink4a/Arf* genes [29–32]. Others identified a DNA replication origin ($RD^{Ink/Arf}$) in close proximity to the *Ink4a/Arf* locus that appeared to transcriptionally repress it in a manner dependent on Cdc6 [33]. Whether one or several of these mechanisms take place during OIS to activate *Ink4a/Arf* expression, however, remains unknown.

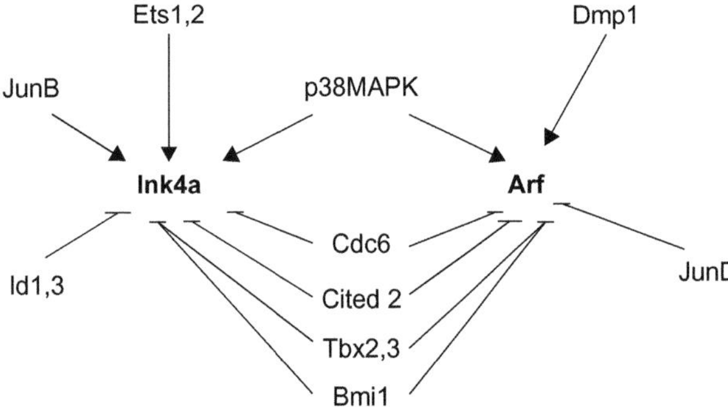

FIGURE 29.1 Multiple regulators of Ink4a and Arf expression.
Signaling pathways that are potentially engaged to activate Ink4a/Arf expression in the course of OIS are shown. These signaling pathways could be dependent on cell type and/or the type of oncogenic insult.

Both Rb and p53 are critical regulators of OIS downstream of Ink4a/Arf [8, 20]. The activities of Rb and p53 are significantly increased during OIS. Inactivation of either protein results in the reversal of the senescence phenotype in mouse embryo fibroblasts with subsequent re-entry into the cell cycle. This result suggests that Rb and p53 together play an important role in both initiating and maintaining senescence [34, 35]. Interestingly, targeted inactivation of all Rb family genes (Rb, p107, and p130) immortalizes mouse cells despite the presence of high levels of p53, suggesting that Rb family proteins play a role downstream of the p53 pathway in cellular senescence [36]. In human primary fibroblasts, however, the situation is different. In these cells, senescence is primarily dependent on Rb and to lesser extent on p53 [37, 38]. In human senescent cells, once Rb is fully engaged, particularly by its activator Ink4a, cell cycle arrest becomes irreversible and is no longer overcome by the subsequent inactivation of Rb and p53 [39, 40]. Interestingly, subsequent inactivation of either Rb or p53 reinitiates DNA synthesis, but not full cell cycle re-entry in human senescent cells. This supports the notion that additional mechanisms besides Rb and p53 are engaged to activate the senescent state in human fibroblasts. This Rb-and p53-independent cell cycle block, which seems to be more specific in human cells, likely acts as a second barrier to cellular immortalization, and may help explain the remarkable stability of the senescent cell cycle arrest in human cells.

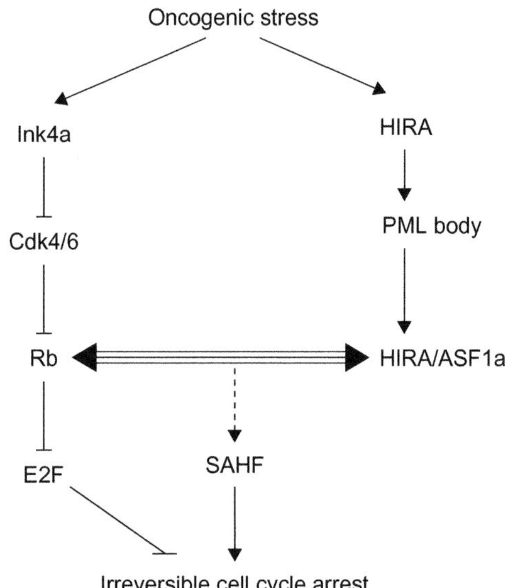

FIGURE 29.2 The cooperative formation of SAHF through the Rb and HIRA/ASF1a pathways.
Rb and HIRA/ASF1a complexes are mutually interdependent, and play a major role in the formation of SAHF. Dashed lines indicate that the molecular details of SAHF formation remain to be determined.

An important characteristic of cellular senescence is the stability of the phenotype (Figure 29.2). It has been proposed that the irreversibility of senescence is mediated by dramatic changes in chromatin through the formation of senescence-induced heterochromatic foci (SAHF) [41]. SAHF are assembled via a sequence involving PML nuclear bodies and the pro-senescence chromatin regulators histone repressor A (HIRA), heterochromatin protein 1(HP1), and anti-silencing function 1a (ASF1a) [42–44]. Formation of SAHF is highly dependent upon the integrity of the Rb and p53 pathways. In turn, assembly of SAHF by constitutive activation of Rb via ectopic expression of Ink4a requires the HIRA/ASF1a pathway. This suggests that the Rb and HIRA/ASF1a pathways are mutually dependent and cooperate to form SAHF. Consistent with this idea, the Rb tumor suppressor protein and the HIRA and ASF1a histone chaperones are all well-established regulators of chromatin structure and function. For example, Rb recruits the histone methylase Suv39h1 to chromatin, which is required for oncogene-induced formation of heterochromatin in murine T cells [45]. One simple model to describe the cooperation between Rb and HIRA/ASF1a pathways is that Rb initiates heterochromatin formation at its target gene promoters, and then HIRA and ASF1a generate the more extensive domains of heterochromatin characteristic of SAHF.

As for molecular mechanisms of SAHF formation, a role for Wnt dependent signaling was recently proposed to delay OIS by regulating the recruitment of HIRA to PML bodies [46]. It was shown that this recruitment is dependent on serine 697 of HIRA, which is most likely phosphorylated by GSK3β, a key effector of the repressed Wnt signaling pathway. How Wnt2 signaling is inactivated during the early steps of senescence, however, remains uncertain, and it would be of interest to know whether p38MAPK and/or ATM kinases play any role in this suppression.

The OIS response is complex and there are examples of cell types that use neither p53 nor Ink4a to activate OIS [47]. Ras activation causes growth arrest in some ovarian epithelial cells in the presence of E6, suggesting that p53 does not play a major role in activation of OIS in these cells. In addition, there was no accumulation of Ink4a at the protein level. Therefore, it appears that inducible expression of Ras leads to growth arrest in some ovarian epithelial cells through an entirely p53- and Ink4a-independent mechanism. This growth arrest was dependent on one of the members of the MAPK family of protein kinases, p38MAPK, which is discussed next.

p38MAPK SIGNALING AND OIS

Activation of MAPK, including p38 as measured by its constitutive phosphorylation, was previously observed in primary human tumors [48–50]. This unexpected activation, apparently occurring without an external stressor, raises the question of the origin of the stimulus involved in this response as well as the significance of the activation. Although it cannot be completely excluded that phosphorylation of p38MAPK favors cancer cell proliferation under certain conditions, this may reflect the activation of signaling pathways that are engaged to restrain aberrant cell cycle progression in the presence of activated oncogenes. These initial observations are now supported by the emerging role of p38MAPK in the regulation of OIS [51–54].

In addition to activation of the MEK1-ERK1/2 pathway, which is believed to be a major trigger of OIS in the presence of oncogenic Ras, there is increasing evidence that the other two MAPK pathways, p38 MAP kinase and JNK, play important roles [52, 55]. Sequential activation of the ERK and then p38 pathways reportedly contributes to Ha-ras V12-induced premature senescence (Figure 29.3) [56]. This study argued that Ha-ras activation of ERK and JNK can have growth stimulatory properties, while premature senescence relies on the subsequent activation of p38MAPK. Growth arrest triggered by constitutively active forms of MEK1, MKK3, or MKK6 (upstream activators of p38MAPK) was blocked by inhibition of p38MAPK, which highlights the pivotal role of p38 as the cellular "brake" after oncogenic stress. In another study [57], pharmacologic inhibition of p38MAPK, together with Raf-dependent activation of ERK, was sufficient to mimic the morphologic and growth transformation characteristics caused by oncogenic Ras. In addition, in human fibroblasts, growth arrest after expression of Ha-ras was prevented

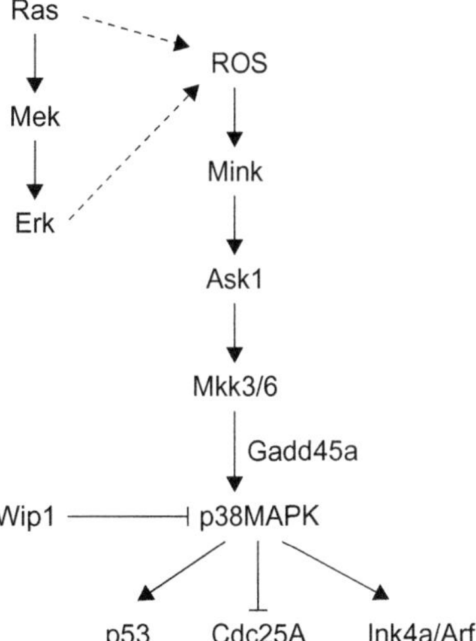

FIGURE 29.3 Activation of p38MAPK during OIS.
Activation of p38MAPK is shown using the Ras induced OIS as an example. Activation of Ras leads to accumulation of reactive oxygen species (ROS) through several pathways by an as yet-to-be determined mechanism (dashed lines). This subsequently activates the MINK-Ask1-p38MAPK signaling pathway. Gadd45a is required for proper activation of p38MAPK in the presence of activated Ras. In turn, p38MAPK could halt cell cycle progression through multiple mechanisms.

by pharmacological inhibition of p38MAPK [53, 54]. Interestingly, disruption of the *Gadd45a* gene, a positive regulator of p38MAPK, was sufficient to abrogate Ha-ras-induced activation of p38 [52]. This observation correlated with the inability of *Gadd45a* deficient MEFs to undergo permanent cell cycle arrest, and further highlights the role of p38MAPK in OIS. Studies with the p53 inducible Wip1 phosphatase, a repressor of p38MAPK, also supports the contention that p38 is an important component mediating oncogene-induced responses. The Wip1 phosphatase mediates a negative feedback regulation of p53 through the inactivation p38MAPK (a potent regulator of p53 [51]) and is induced after UV radiation, but not after introduction of Ha-ras [58]. Without induction of Wip1, Ha-ras likely triggers a positive feedback loop involving p38 MAPK, p53, and Gadd45a that leads to irreversible growth arrest.

Introduction of Ha-ras stimulates all three major branches of the MAPK pathway: ERK, JNK, and p38. The activities of Jnk and p38MAPK, however, are induced relatively slowly over a few days [52]. The delayed response likely indicates that the induction is indirect. One of the potential mediators in this delayed activation was shown to be MINK kinase, whose activity is increased at both the transcriptional and protein levels in the presence of activated Ras (Figure 29.3) [47]. MINK was previously reported to activate both the JNK and p38MAPK stress kinase pathways through downstream MAP kinase kinases, such as MAP3K5 (Ask1) [59]. Thus, MINK may represent a missing link between prolonged Raf/ERK activation and p38MAPK induction.

The nature of the intermediate steps linking ERK activation and MINK induction are not presently known in detail; however, they are sensitive to conditions that reduce the level of reactive oxygen species (ROS) produced after Ras activation [47]. Oxidative stress is necessary and sufficient for the activation of MINK downstream of Ras and in turn activated Ras has long been associated with elevated levels of ROS, both in the induction of cell proliferation and, at higher ROS levels, cellular senescence.

p38MAPK-mediated growth inhibition engages multiple pathways to restrain cell cycle progression [51]. The ability of p38MAPK to target different components of the Ink4a/cyclin D/Rb and Arf/p53 pathways further supports its potential role in regulating OIS [60]. Induction of Ink4a in the presence of Ras is believed to be regulated through the MEK1-ERK1/2 pathway (discussed above). Others have proposed that this process is also dependent on p38MAPK (Figure 29.1). This is consistent with the fact that overexpression of active MKK3 or MKK6, which lie upstream of p38MAPK, is sufficient to activate *Ink4a* transcription [56]. Activation of Arf-dependent signals, a primary mechanism for Ras-induced senescence in MEFs, can also be controlled by p38MAPK. Activation of p38MAPK in Wip1 deficient cells contributes to Arf (as well as Ink4a) upregulation and causes Wip1 deficient MEFs to exit the cell cycle prematurely [60]. This pathway also turned out to be critical in rendering Wip1 deficient MEFs resistant to oncogene-induced transformation.

Activation of a senescence-like arrest in human cells could be abrogated after disruption of Ink4a alone, however p38MAPK has additional growth inhibiting properties. Overexpression of MKK6 could halt cell proliferation in the U2OS human tumor cell line, even though this cell line expresses neither Arf nor Ink4a [61]. These examples show that p38MAPK can negatively regulate cell growth by affecting multiple regulatory pathways. Overall, multiple findings support the model that p38MAPK dependent signaling, in cooperation with the MEK1-ERK1/2 pathway, contributes to activation of oncogene induced premature senescence in certain primary cells.

DNA DAMAGE RESPONSE AND OIS

Similar to the findings for constitutive activation of p38MAPK in certain human primary cancers, the key initial observation that the DNA damage response could be activated by oncogenes came from the analysis of primary cell lines and clinical specimens of large subsets of human breast and lung carcinomas [62]. The presence of the activated Thr68-phosphorylated form of the checkpoint kinase Chk2 found in these clinical samples suggested that some events that occur in tumors, but not in adjacent normal tissues, lead to constitutive activation of the DNA damage response. At first, these observations were difficult to interpret because advanced tumors are genetically unstable, which in turn could lead to activation of the DNA damage response. However, studies from several laboratories recently showed that activation of the DNA damage response is a direct and immediate consequence of activation of different oncogenes, including RasV12, Stat5, E2F1, Mos, and Cdc6 [63–65]. Collectively these papers demonstrated that different members of the DNA damage signaling pathway (ATM, ATR, Chk1, and Chk2) were implicated in OIS; downregulating them with RNA interference allowed cells to bypass senescence in the context of inactivation of the Rb tumor suppressor pathway. Importantly, in human clinical specimens from early stages of colorectal and urinary bladder cancers, markers of the DNA damage response correlate with other markers of senescence, such as the appearance of oncogene-induced DNA damage foci [63, 64]. These observations indicate that activation of the DNA damage response contributes to OIS together with other mechanisms discussed earlier.

Mechanistically, several models have been promulgated to explain the activation of the DNA damage response machinery in the presence of activated oncogenes (Figure 29.4). Data supporting a replication stress model have been provided by several groups. DiMicco *et al.* (2006) reported Ras-induced, replication associated stress marked by DNA re-replication

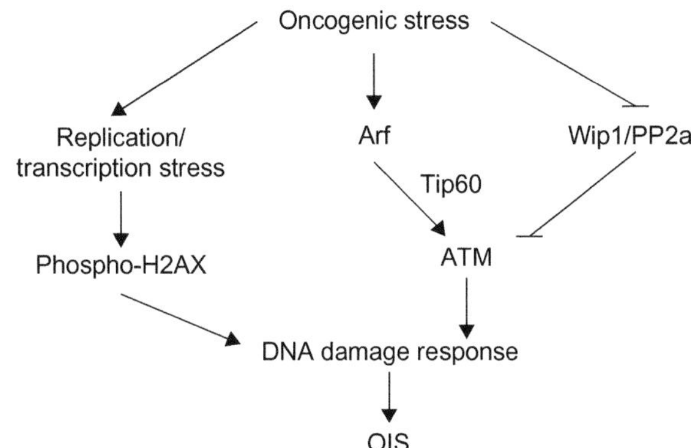

FIGURE 29.4 OIS as a response to DNA damage signals. Oncogenes may induce DNA damage (marked by H2AX phosphorylation) by forcing an aberrant DNA replication process where some replication forks arrest as a result of an unknown mechanism. Another possibility is that oncogenes induce the activation of ATM-dependent signaling without causing DNA damage through the Tip60-and/or Wip1/PP2A-dependent pathways.

suggesting that some replication origins fired more than once [63]. In addition, Bartkova *et al.* (2006) found that aberrant, premature termination of replication fork progression occurred in the presence of several oncogenes [64]. Such premature fork arrest can result in fork collapse and DNA breakage, consistent with the observed double-strand breaks and the activation of the DNA damage response.

The data supporting a replication stress model do not rule out a contribution from other potential sources of activation for the DNA damage response machinery during the early stages of tumorigenesis. Arguably, the activation of ATM-dependent signaling could occur in the absence of apparent DNA damage. Recent analysis of Arf overexpression revealed an increase in ATM-dependent signaling to p53, which likely occurs through a Tip60-dependent acetylation of ATM [66, 67]. Similarly, overexpression of E2F1, another efficient inducer of Arf expression, resulted in the phosphorylation of p53 at serine 15 via ATM without noticeable H2AX phosphorylation [68]. In another study, introduction of an activated Ras into IMR90 human fibroblasts resulted in phosphorylation of p53 at ATM site Ser46, which was efficiently mitigated by ATM's phosphatase, Wip1 [58]. In turn, deletion or downregulation of Wip1 could activate ATM-dependent signaling in both mouse and human cells without inducing DNA damage [69,70]. Because Wip1 controls several key autophosphorylation sites on ATM in concert with another phosphatase, PP2A [69, 71], it would be of great importance to know whether both phosphatases are engaged in activation of OIS.

CONCLUDING REMARKS

The identification of cellular senescence as an important *in vivo* process has fostered a field of extensive research. Nevertheless, many questions remain unanswered. For example, little is known about the ultimate fate of senescent cells *in vivo* and how they might control neighboring cells, such as stem cells, as a part of local tissue microenvironment. Senescent fibroblasts are known to secrete high levels of several matrix metalloproteinases, epithelial growth factor, and inflammatory cytokines (reviewed in [9]). These may in turn stimulate chronic tissue remodeling, local inflammation, and even enhance the proliferation of cells that harbor pre-neoplastic mutations. Another important and obvious question is the role of senescent cells in aging. As tissues have a fairly constant number of cells, the accumulation of non-dividing senescent cells may impact tissue renewal and repair, contributing to organismal aging. Addressing these important questions opens up interesting prospects for a better understanding of the molecular pathways and the role of cellular senescence in cancer and aging.

REFERENCES

1. Cho RW, Clarke MF. Recent advances in cancer stem cells. *Curr Opin Genet Dev* 2008;**18**:48–53.
2. Dalerba P, Cho RW, Clarke MF. Cancer stem cells: models and concepts. *Annu Rev Med* 2007;**58**:267–84.
3. Bach SP, Renehan AG, Potten CS. Stem cells: the intestinal stem cell as a paradigm. *Carcinogenesis* 2000;**21**:469–76.
4. Potten CS, Owen G, Booth D. Intestinal stem cells protect their genome by selective segregation of template DNA strands. *J Cell Sci* 2002;**115**:2381–8.
5. Demidov ON, Timofeev O, Lwin HN, Kek C, Appella E, Bulavin DV. Wip1 phosphatase regulates p53-dependent apoptosis of stem cells and tumorigenesis in the mouse intestine. *Cell Stem Cell* 2007;**1**:180–90.
6. Potten CS, Ellis JR. Adult small intestinal stem cells: identification, location, characteristics, and clinical applications. *Ernst Schering Res Found Workshop* 2006;**60**:81–98.
7. Pollock PM, Harper UL, Hansen KS, Yudt LM, Stark M, Robbins CM, Moses TY, Hostetter G, Wagner U, Kakareka J, Salem G, Pohida T, Heenan P, Duray P, Kallioniemi O, Hayward NK, Trent JM, Meltzer PS. High frequency of BRAF mutations in nevi. *Nat Genet* 2003;**33**:19–20.
8. Serrano M, Lin AW, McCurrach ME, Beach D, Lowe SW. Oncogenic ras provokes premature cell senescence associated with accumulation of p53 and p16INK4a. *Cell* 1997;**88**:593–602.
9. Campisi J, d'Adda dF. Cellular senescence: when bad things happen to good cells. *Nat Rev Mol Cell Biol* 2007;**8**:729–40.
10. Campisi J. Senescent cells, tumor suppression, and organismal aging: good citizens, bad neighbors. *Cell* 2005;**120**:513–22.
11. Sherr CJ, DePinho RA. Cellular senescence: mitotic clock or culture shock?. *Cell* 2000;**102**:407–10.
12. Michaloglou C, Vredeveld LC, Soengas MS, Denoyelle C, Kuilman T, van der Horst CM, Majoor DM, Shay JW, Mooi WJ, Peeper DS. BRAFE600-associated senescence-like cell cycle arrest of human naevi. *Nature* 2005;**436**:720–4.
13. Chen Z, Trotman LC, Shaffer D, Lin HK, Dotan ZA, Niki M, Koutcher JA, Scher HI, Ludwig T, Gerald W, Cordon-Cardo C,

Pandolfi PP. Crucial role of p53-dependent cellular senescence in suppression of Pten-deficient tumorigenesis. *Nature* 2005;**436**:725–30.

14. Collado M, Gil J, Efeyan A, Guerra C, Schuhmacher AJ, Barradas M, Benguria A, Zaballos A, Flores JM, Barbacid M, Beach D, Serrano M. Tumour biology: senescence in premalignant tumours. *Nature* 2005;**436**:642.
15. Braig M, Lee S, Loddenkemper C, Rudolph C, Peters AH, Schlegelberger B, Stein H, Dorken B, Jenuwein T, Schmitt CA. Oncogene-induced senescence as an initial barrier in lymphoma development. *Nature* 2005;**436**:660–5.
16. Serrano M, Lin AW, McCurrach ME, Beach D, Lowe SW. Oncogenic ras provokes premature cell senescence associated with accumulation of p53 and p16INK4a. *Cell* 1997;**88**:593–602.
17. Sherr CJ. The INK4a/ARF network in tumour suppression. *Nat Rev Mol Cell Biol* 2001;**2**:731–7.
18. Sherr CJ, Weber JD. The ARF/p53 pathway. *Curr Opin Genet Dev* 2000;**10**:94–9.
19. Narita M, Lowe SW. Executing cell senescence. *Cell Cycle* 2004;**3**:244–6.
20. Sherr CJ, McCormick F. The RB and p53 pathways in cancer. *Cancer Cell* 2002;**2**:103–12.
21. Sherr CJ, Weber JD. The ARF/p53 pathway. *Curr Opin Genet Dev* 2000;**10**:94–9.
22. Weber JD, Taylor LJ, Roussel MF, Sherr CJ, Bar-Sagi D. Nucleolar Arf sequesters Mdm2 and activates p53. *Nat Cell Biol* 1999;**1**:20–6.
23. Haupt Y, Maya R, Kazaz A, Oren M. Mdm2 promotes the rapid degradation of p53. *Nature* 1997;**387**:296–9.
24. Lin AW, Barradas M, Stone JC, van Aelst L, Serrano M, Lowe SW. Premature senescence involving p53 and p16 is activated in response to constitutive MEK/MAPK mitogenic signaling. *Genes Dev* 1998;**12**:3008–19.
25. Zhu J, Woods D, McMahon M, Bishop JM. Senescence of human fibroblasts induced by oncogenic Raf. *Genes Dev* 1998;**12**:2997–3007.
26. Huot TJ, Rowe J, Harland M, Drayton S, Brookes S, Gooptu C, Purkis P, Fried M, Bataille V, Hara E, Newton-Bishop J, Peters G. Biallelic mutations in p16(INK4a) confer resistance to Ras- and Ets-induced senescence in human diploid fibroblasts. *Mol Cell Biol* 2002;**22**:8135–43.
27. Ohtani N, Zebedee Z, Huot TJ, Stinson JA, Sugimoto M, Ohashi Y, Sharrocks AD, Peters G, Hara E. Opposing effects of Ets and Id proteins on p16INK4a expression during cellular senescence. *Nature* 2001;**409**:1067–70.
28. Sreeramaneni R, Chaudhry A, McMahon M, Sherr CJ, Inoue K. Ras-Raf-Arf signaling critically depends on the Dmp1 transcription factor. *Mol Cell Biol* 2005;**25**:220–32.
29. Jacobs JJ, Keblusek P, Robanus-Maandag E, Kristel P, Lingbeek M, Nederlof PM, van Welsem T, van de Vijver MJ, Koh EY, Daley GQ, Van Lohuizen M. Senescence bypass screen identifies TBX2, which represses Cdkn2a (p19(ARF)) and is amplified in a subset of human breast cancers. *Nat Genet* 2000;**26**:291–9.
30. Gil J, Bernard D, Martinez D, Beach D. Polycomb CBX7 has a unifying role in cellular lifespan. *Nat Cell Biol* 2004;**6**:67–72.
31. Kranc KR, Bamforth SD, Braganca J, Norbury C, van Lohuizen M, Bhattacharya S. Transcriptional coactivator Cited2 induces Bmi1 and Mel18 and controls fibroblast proliferation via Ink4a/ARF. *Mol Cell Biol* 2003;**23**:7658–66.
32. Jacobs JJ, Kieboom K, Marino S, DePinho RA, van Lohuizen M. The oncogene and Polycomb-group gene bmi-1 regulates cell proliferation and senescence through the ink4a locus. *Nature* 1999;**397**:164–8.
33. Gonzalez S, Klatt P, Delgado S, Conde E, Lopez-Rios F, Sanchez-Cespedes M, Mendez J, Antequera F, Serrano M. Oncogenic activity of Cdc6 through repression of the INK4/ARF locus. *Nature* 2006; **440**:702–6.
34. Sage J, Miller AL, Perez-Mancera PA, Wysocki JM, Jacks T. Acute mutation of retinoblastoma gene function is sufficient for cell cycle re-entry. *Nature* 2003;**424**:223–8.
35. Dirac AM, Bernards R. Reversal of senescence in mouse fibroblasts through lentiviral suppression of p53. *J Biol Chem* 2003;**278**:11,731–11,734.
36. Sage J, Mulligan GJ, Attardi LD, Miller A, Chen S, Williams B, Theodorou E, Jacks T. Targeted disruption of the three Rb-related genes leads to loss of G(1) control and immortalization. *Genes Dev* 2000;**14**:3037–50.
37. Beausejour CM, Krtolica A, Galimi F, Narita M, Lowe SW, Yaswen P, Campisi J. Reversal of human cellular senescence: roles of the p53 and p16 pathways. *EMBO J* 2003;**22**:4212–22.
38. Brookes S, Rowe J, Ruas M, Llanos S, Clark PA, Lomax M, James MC, Vatcheva R, Bates S, Vousden KH, Parry D, Gruis N, Smit N, Bergman W, Peters G. INK4a-deficient human diploid fibroblasts are resistant to RAS-induced senescence. *EMBO J* 2002;**21**:2936–45.
39. Beausejour CM, Krtolica A, Galimi F, Narita M, Lowe SW, Yaswen P, Campisi J. Reversal of human cellular senescence: roles of the p53 and p16 pathways. *EMBO J* 2003;**22**:4212–22.
40. Dai CY, Enders GH. p 16 INK4a can initiate an autonomous senescence program. *Oncogene* 2000;**19**:1613–22.
41. Narita M, Nunez S, Heard E, Narita M, Lin AW, Hearn SA, Spector DL, Hannon GJ, Lowe SW. Rb-mediated heterochromatin formation and silencing of E2F target genes during cellular senescence. *Cell* 2003;**113**:703–16.
42. Zhang R, Chen W, Adams PD. Molecular dissection of formation of senescence-associated heterochromatin foci. *Mol Cell Biol* 2007;**27**:2343–58.
43. Ye X, Zerlanko B, Zhang R, Somaiah N, Lipinski M, Salomoni P, Adams PD. Definition of pRB- and p53-dependent and -independent steps in HIRA/ASF1a-mediated formation of senescence-associated heterochromatin foci. *Mol Cell Biol* 2007;**27**:2452–65.
44. Zhang R, Poustovoitov MV, Ye X, Santos HA, Chen W, Daganzo SM, Erzberger JP, Serebriiskii IG, Canutescu AA, Dunbrack RL, Pehrson JR, Berger JM, Kaufman PD, Adams PD. Formation of MacroH2A-containing senescence-associated heterochromatin foci and senescence driven by ASF1a and HIRA. *Dev Cell* 2005;**8**:19–30.
45. Braig M, Lee S, Loddenkemper C, Rudolph C, Peters AH, Schlegelberger B, Stein H, Dorken B, Jenuwein T, Schmitt CA. Oncogene-induced senescence as an initial barrier in lymphoma development. *Nature* 2005;**436**:660–5.
46. Ye X, Zerlanko B, Kennedy A, Banumathy G, Zhang R, Adams PD. Downregulation of Wnt signaling is a trigger for formation of facultative heterochromatin and onset of cell senescence in primary human cells. *Mol Cell* 2007;**27**:183–96.
47. Nicke B, Bastien J, Khanna SJ, Warne PH, Cowling V, Cook SJ, Peters G, Delpuech O, Schulze A, Berns K, Mullenders J, Beijersbergen RL, Bernards R, Ganesan TS, Downward J, Hancock DC. Involvement of MINK, a Ste20 family kinase, in Ras oncogene-induced growth arrest in human ovarian surface epithelial cells. *Mol Cell* 2005;**20**:673–85.
48. Elenitoba-Johnson KS, Jenson SD, Abbott RT, Palais RA, Bohling SD, Lin Z, Tripp S, Shami PJ, Wang LY, Coupland RW, Buckstein R, Perez-Ordonez B, Perkins SL, Dube ID, Lim MS. Involvement of multiple signaling pathways in follicular lymphoma transformation: p38-mitogen-activated protein kinase as a target for therapy. *Proc Natl Acad Sci USA* 2003;**100**:7259–64.

49. Greenberg AK, Basu S, Hu J, Yie TA, Tchou-Wong KM, Rom WN, Lee TC. Selective p38 activation in human non-small cell lung cancer. *Am J Respir Cell Mol Biol* 2002;**26**:558–64.
50. Miki H, Yamada H, Mitamura K. Involvement of p38 MAP kinase in apoptotic and proliferative alteration in human colorectal cancers. *Anticancer Res* 1999;**19**:5283–91.
51. Bulavin DV, Fornace Jr AJ. p38 MAP kinase's emerging role as a tumor suppressor. *Adv Cancer Res* 2004;**92**:95–118.
52. Bulavin DV, Kovalsky O, Hollander MC, Fornace Jr AJ. Loss of oncogenic H-ras-induced cell cycle arrest and p38 mitogen-activated protein kinase activation by disruption of Gadd45a. *Mol Cell Biol* 2003;**23**:3859–71.
53. Han J, Sun P. The pathways to tumor suppression via route p38. *Trends Biochem Sci* 2007;**32**:364–71.
54. Sun P, Yoshizuka N, New L, Moser BA, Li Y, Liao R, Xie C, Chen J, Deng Q, Yamout M, Dong MQ, Frangou CG, Yates JR, III, Wright PE, Han J. PRAK is essential for ras-induced senescence and tumor suppression. *Cell* 2007;**128**:295–308.
55. Iwasa H, Han J, Ishikawa F. Mitogen-activated protein kinase p38 defines the common senescence-signalling pathway. *Genes Cells* 2003;**8**:131–44.
56. Wang W, Chen JX, Liao R, Deng Q, Zhou JJ, Huang S, Sun P. Sequential activation of the MEK-extracellular signal-regulated kinase and MKK3/6-p38 mitogen-activated protein kinase pathways mediates oncogenic ras-induced premature senescence. *Mol Cell Biol* 2002;**22**:3389–403.
57. Pruitt K, Pruitt WM, Bilter GK, Westwick JK, Der CJ. Raf-independent deregulation of p38 and JNK mitogen-activated protein kinases are critical for Ras transformation. *J Biol Chem* 2002;**277**:31,808–31,817.
58. Bulavin DV, Demidov ON, Saito S, Kauraniemi P, Phillips C, Amundson SA, Ambrosino C, Sauter G, Nebreda AR, Anderson CW, Kallioniemi A, Fornace Jr AJ, Appella E. Amplification of PPM1D in human tumors abrogates p53 tumor-suppressor activity. *Nat Genet* 2002;**31**:210–15.
59. Dan I, Watanabe NM, Kobayashi T, Yamashita-Suzuki K, Fukagaya Y, Kajikawa E, Kimura WK, Nakashima TM, Matsumoto K, Ninomiya-Tsuji J, Kusumi A. Molecular cloning of MINK, a novel member of mammalian GCK family kinases, which is up-regulated during post-natal mouse cerebral development. *FEBS Lett* 2000;**469**:19–23.
60. Bulavin DV, Phillips C, Nannenga B, Timofeev O, Donehower LA, Anderson CW, Appella E, Fornace Jr AJ. Inactivation of the Wip1 phosphatase inhibits mammary tumorigenesis through p38 MAPK-mediated activation of the p16(Ink4a)-p19(Arf) pathway. *Nat Genet* 2004;**36**:343–50.
61. Haq R, Brenton JD, Takahashi M, Finan D, Finkielsztein A, Damaraju S, Rottapel R, Zanke B. Constitutive p38HOG mitogen-activated protein kinase activation induces permanent cell cycle arrest and senescence. *Cancer Res* 2002;**62**:5076–82.
62. DiTullio Jr RA, Mochan TA, Venere M, Bartkova J, Sehested M, Bartek J, Halazonetis TD. 53BP1 functions in an ATM-dependent checkpoint pathway that is constitutively activated in human cancer. *Nat Cell Biol* 2002;**4**:998–1002.
63. Di Micco R, Fumagalli M, Cicalese A, Piccinin S, Gasparini P, Luise C, Schurra C, Garre' M, Nuciforo PG, Bensimon A, Maestro R, Pelicci PG, d'Adda dF. Oncogene-induced senescence is a DNA damage response triggered by DNA hyper-replication. *Nature* 2006;**444**:638–42.
64. Bartkova J, Rezaei N, Liontos M, Karakaidos P, Kletsas D, Issaeva N, Vassiliou LV, Kolettas E, Niforou K, Zoumpourlis VC, Takaoka M, Nakagawa H, Tort F, Fugger K, Johansson F, Sehested M, Andersen CL, Dyrskjot L, Orntoft T, Lukas J, Kittas C, Helleday T, Halazonetis TD, Bartek J, Gorgoulis VG. Oncogene-induced senescence is part of the tumorigenesis barrier imposed by DNA damage checkpoints. *Nature* 2006;**444**:633–7.
65. Mallette FA, Gaumont-Leclerc MF, Ferbeyre G. The DNA damage signaling pathway is a critical mediator of oncogene-induced senescence. *Genes Dev* 2007;**21**:43–8.
66. Li Y, Wu D, Chen B, Ingram A, He L, Liu L, Zhu D, Kapoor A, Tang D. ATM activity contributes to the tumor-suppressing functions of p14ARF. *Oncogene* 2004;**23**:7355–65.
67. Eymin B, Claverie P, Salon C, Leduc C, Col E, Brambilla E, Khochbin S, Gazzeri S. p14ARF activates a Tip60-dependent and p53-independent ATM/ATR/CHK pathway in response to genotoxic stress. *Mol Cell Biol* 2006;**26**:4339–50.
68. Powers JT, Hong S, Mayhew CN, Rogers PM, Knudsen ES, Johnson DG. E2F1 uses the ATM signaling pathway to induce p53 and Chk2 phosphorylation and apoptosis. *Mol Cancer Res* 2004;**2**:203–14.
69. Shreeram S, Hee WK, Demidov ON, Kek C, Yamaguchi H, Fornace Jr AJ, Anderson CW, Appella E, Bulavin DV. Regulation of ATM/p53-dependent suppression of myc-induced lymphomas by Wip1 phosphatase. *J Exp Med* 2006;**203**:2793–9.
70. Shreeram S, Demidov ON, Hee WK, Yamaguchi H, Onishi N, Kek C, Timofeev ON, Dudgeon C, Fornace AJ, Anderson CW, Minami Y, Appella E, Bulavin DV. Wip1 phosphatase modulates ATM-dependent signaling pathways. *Mol Cell* 2006;**23**:757–64.
71. Goodarzi AA, Jonnalagadda JC, Douglas P, Young D, Ye R, Moorhead GB, Lees-Miller SP, Khanna KK. Autophosphorylation of ataxia-telangiectasia mutated is regulated by protein phosphatase 2A. *EMBO J* 2004;**23**:4451–61.

Chapter 30

Ubiquitin and FANC Stress Responses

Stacy A. Williams and Gary M. Kupfer
Departments of Pediatrics and Pathology, Yale University School of Medicine, New Haven, Connecticut

INTRODUCTION

First described in 1927, Fanconi anemia (FA) is a rare genetic disorder affecting less than 1 in 100,000 people. Although FA patients display a wide variety of clinical phenotypes, the disease can generally be characterized by congenital abnormalities, bone marrow failure, and an increased susceptibility to cancer and leukemia. FA can be inherited in an autosomal or X-linked recessive manner, as one of the 13 genes associated with the disorder is located on the X chromosome, while the remaining 12 are scattered amongst the autosomes [1–3]. Diagnosis of FA can be performed prenatally or postnatally using chromosome breakage analysis. Taking advantage of their hypersensitivity to DNA crosslinking agents such as diepoxybutane (DEB) or mitomycin C (MMC), cells derived from FA patients can be distinguished from non-FA cells by their characteristic increase in chromosome breakage upon treatment with drug [4]. This cellular phenotype, observed when any of the 13 FA genes are mutated, has led to the belief that the proteins encoded by these genes represent components of a novel DNA repair pathway specific for inter-strand crosslinks (ICLs) [2]. Coupled with the connection of this disease to genomic instability and cancer, the molecular biology of Fanconi anemia therefore embodies a unique area of research that can be exploited to provide fresh insights into tumorigenesis and potential therapies.

COMPONENTS OF THE FANCONI ANEMIA PATHWAY

FA patients have been divided into 13 different complementation groups based on somatic cell hybridization studies using DNA crosslinker sensitivity as an assay to determine genetic complementation [5]. Over a decade has passed since the cloning of the first Fanconi anemia gene, FANCC [6], and to date all 13 genes have been identified. However, as there are very few recognizable structural domains in the proteins encoded by these genes, and many appear to only be conserved among vertebrates, elucidation of the function of the FA proteins has proven difficult. A summary of the FA proteins and their known functions is provided in Table 30.1. Since cellular sensitivity to DNA crosslinking agents results when any of the FA genes are mutated, it is generally accepted that the proteins function in a common pathway specific for the repair of ICLs known as the FA pathway [2]. Additionally, since the discovery that three FA genes are in fact previously identified familial breast cancer genes, the pathway is often referred to as the FA-BRCA pathway [1]. FANCD1 is now known to be BRCA2, a DNA repair protein that functions in homologous recombination by regulating the recruitment of RAD51 [7]. FANCJ, or BRIP1/BACH1, interacts with BRCA1 and has been shown to have DNA helicase activity [8, 9]. Lastly, FANCN, also known as PALB2, is required for BRCA2 nuclear localization [10, 11]. FA proteins can be divided into three different subgroups based mainly on their requirement for the mono-ubiquitylation of FANCD2: the FA core complex, FANCD2 and its partner FANCI, and downstream effectors.

THE FA CORE COMPLEX AND ACTIVATION OF THE FA PATHWAY

Eight of the 13 FA proteins (FANCA, B, C, E, F, G, L, and M) make up what is known as the FA core complex, all of which must be present for core complex assembly and stability. Using positional and complementation based

Handbook of Cell Signaling, Three-Volume Set 2 ed.

TABLE 30.1 Components of the Fanconi anemia pathway

FA gene	Protein size (kDa)	Function	Required for D2 ubiquitylation?
FANCA	163	Core complex; phosphorylated after damage	Y
FANCB	95	Core complex	Y
FANCC	63	Core complex	Y
FANCD1/ BRCA2	380	Regulates RAD51 recruitment	N
FANCD2	155, 162	Phosphorylated and mono-ubiquitylated after damage	Y
FANCE	60	Core complex; binds FANCD2; phosphorylated after damage	Y
FANCF	42	Core complex	Y
FANCG/XRCC9	68	Core complex; phosphorylated after damage	Y
FANCI	140, 147	Binds FANCD2; phosphorylated and mono-ubiquitylated after damage	Y
FANCJ/BRIP1/BACH1	140	Phosphorylated after DNA damage; DNA helicase/ATPase; binds BRCA1	N
FANCL/PHF9	43	Core complex; E3 ubiquitin ligase activity	Y
FANCM/Hef	250	Core complex; phosphorylated after damage; branch migration activity; DNA binding	Y
FANCN/PALB2	140	Regulates BRCA2 nuclear localization	N
FAAP24	24	Core complex; binds FANCM	Y
FAAP100	100	Core complex; binds FANCB and FANCL	Y

expression cloning techniques, five components of the core complex were identified first (FANCA, C, E, F, and G) [12, 13] and were later shown to be required for the mono-ubiquitylation of FANCD2 [14]. Purification of five additional components of the core complex known as FAAPs, or Fanconi anemia associated polypeptides, eventually led to the identification of FANCL, FANCB [15], and FANCM. The core complex represents a key component of the FA pathway, supported by the fact that approximately 90 percent of all FA patients have a mutation in one of these eight genes [2]. Current models prescribe a few significant functions for the core complex in ICL repair including binding to DNA, and its role in the activation of the FA pathway as the E3 ubiquitin ligase responsible for the mono-ubiquitylation of FANCD2 and FANCI (see Figure 30.1).

The FA core complex readily localizes to chromatin in response to DNA damage induced by a crosslinking agent or upon entry into S phase [16–17]. However, the exact mechanism by which the FA pathway is activated following DNA damage still remains unknown. Treatment with a DNA crosslinking agent results in a covalent linkage between the DNA strands, preventing the separation of these strands that is necessary during replication or transcription. When a replication fork stalls, the resulting persistent single-stranded DNA is recognized by the protein ATRIP, which in turn activates the checkpoint kinase Ataxia-telangiectasia and Rad3 related

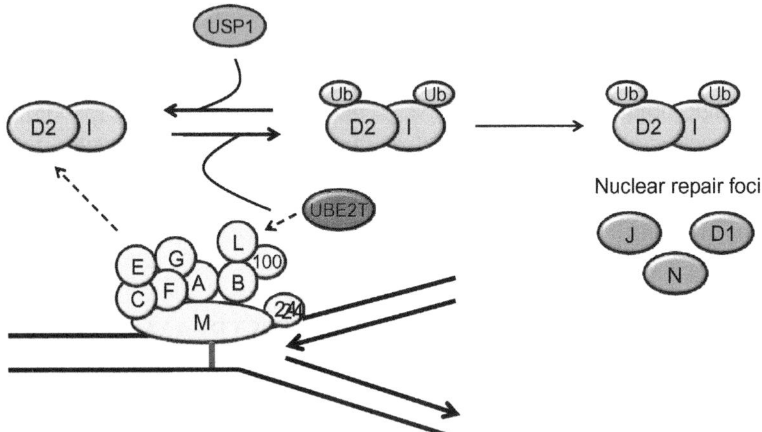

FIGURE 30.1 Current model of the regulation of mono-ubiquitylation of FANCD2 and FANCI. FANCM recruits the FA core complex to chromatin upon damage with a DNA inter-strand crosslinker such as mitomycin C. The entire core complex cooperates to mono-ubiquitylate FANCD2 and FANCI. FANCE binds FANCD2 directly and may serve to recruit the substrate, while FANCL, acting as the E3 ligase, catalyzes the ubiquitin reaction along with the E2 enzyme, UBE2T. Mono-ubiquitylation of FANCD2 and FANCI causes these proteins to localize to nuclear repair foci with FANCJ, FANCN, and FANCD1 as well as other DNA repair factors. USP1 de-ubiquitylates FANCD2 and FANCI to deactivate the FA pathway once the lesion has been repaired.

protein, or ATR. It is known that ATR is required for activation of the FA pathway [18], and, along with its downstream substrate CHK1, is responsible for phosphorylating four of the eight core complex proteins including FANCA, E, G, and M [19–22]. Recent studies using *Xenopus laevis* extracts show activation of the FA pathway, as determined by mono-ubiquitylation of FANCD2, in the presence of linear or branched double-stranded DNA structures as well as circular DNA containing no double-stranded ends [23]. Specific recognition of the lesion and subsequent recruitment of the core complex may therefore be afforded by FANCM and FAAP24.

FANCM was first identified as the protein that corresponded to a 250kDa FAAP with sequence similarity to the known DNA repair proteins Hef and XPF. This protein was shown to have ATP dependent DNA translocase activity, indicating a possible role as the engine responsible for motoring the core complex along DNA strands [22]. FAAP24, a core complex associated protein that interacts directly with FANCM, has been shown to specifically interact with DNA structures resembling recombination intermediates. Furthermore, a purified FANCM/FAAP24 complex recognized these intermediates *in vitro*, while a C-terminal portion of FANCM alone did not, leading to the theory that FAAP24 confers specific recognition of lesions to the FA core complex [24]. However, recent studies have shown that purified full-length FANCM is capable of binding preferentially to branched DNA structures such as Holliday junctions and replication forks with a high affinity [25, 26]. Additionally, FANCM could dissociate these structures via branch migration (as opposed to the unwinding activity of a helicase) and this branch migration activity was functional even in the absence of FAAP24 [26]. Other recent reports have indicated that although FANCM is not essential for core complex assembly and stability, it is required for the chromatin loading of the core complex in a DNA damage and cell cycle dependent manner [17]. Taken together, these data suggest a significant role for FANCM in recruitment of core complex to chromatin and subsequent repair of DNA lesions.

The observation that all of the components of the FA core complex are required for the mono-ubiquitylation of FANCD2 has led to a widely accepted model in which the core complex functions as the E3 ubiquitin ligase responsible for FANCD2 and FANCI mono-ubiquitylation. The discovery of FANCL provided some evidence to support this model, as FANCL contains both a PHD-type zinc finger, a motif associated with E3 ubiquitin ligase activity, and several WD40 repeats, which are known to mediate protein–protein interactions. Moreover, purified FANCL was shown to have autopolyubiquitin ligase activity *in vitro* [27]; however, *in vitro* mono-ubiquitylation of FANCD2 and FANCI by core complex or FANCL alone has yet to be shown. A direct interaction between FANCL and FANCD2 has yet to be shown as well, which has led some to suggest that the entire core complex acts as the E3 ligase. Since FANCE is known to stably associate with FANCD2 *in vitro* [28], this protein could be responsible for recruiting the substrate while FANCL functions as the catalytic subunit. One last component of the FA core complex is FAAP100, which is required for stability of the core complex and has been shown to directly interact with FANCB and FANCL in a subcomplex, the significance of which is largely unknown [29].

MONO-UBIQUITYLATION OF FANCD2 AND FANCI

A complete and functional core complex is required for the mono-ubiquitylation of FANCD2 and its partner FANCI, considered the next key step in the FA pathway. Mono-ubiquitylation reactions typically involve three types of enzymes that function to modify proteins through the covalent attachment of a single ubiquitin monomer. First, through an ATP dependent reaction, the ubiquitin molecule is activated by an E1 enzyme. This activated ubiquitin is then transferred to an E2 enzyme, or ubiquitin conjugating protein. Lastly, the E3 ubiquitin ligase interacts with both the E2 protein and the substrate, and catalyzes the exchange of the molecule. Therefore, the E3 enzyme is particularly important since it functions both as the ligase and provides substrate specificity. Once mono-ubiquitylated, the modified protein often changes function or undergoes relocalization within the cell [30]. Since FANCL contains a PHD finger motif and has demonstrated autopolyubiquitin ligase activity *in vitro*, it is thought that this represents the component of the core complex that catalyzes the ubiquitin reaction [27]. FANCD2, however, is only known to interact directly with FANCE [28], suggesting that the entire core complex cooperates to perform the reaction. A recent study also identified the putative E2 enzyme in the FA pathway, UBE2T, which interacts with FANCL and is required for FANCD2 mono-ubiquitylation [31]. However, mono-ubiquitylation of FANCD2 by FANCL and UBE2T directly has yet to be shown *in vitro*, leaving the possibility that other E2 and E3 enzymes could be involved in this process.

FANCD2 is mono-ubiquitylated on lysine 561 after exposure to DNA damaging agents such as mitomycin C and IR [14], or upon entry into S phase [32]. Once mono-ubiquitylated, FANCD2 readily localizes to chromatin. Studies using ubiquitin fusion proteins in chicken DT40 cells have shown that the attachment of a single ubiquitin molecule is crucial for localization to chromatin and correction of sensitivity to crosslinking agents. A ubiquitin molecule was fused to chicken FANCD2 carrying a mutation at its normal site of mono-ubiquitylation (K563R). Expression of this fusion protein in a mutant FANCD2 background resulted in constitutive presence in the chromatin fraction, while expression of the same FANCD2 mutant without the fused ubiquitin remained in the soluble fraction. Additionally, fusion of a ubiquitin containing a point mutation that disrupts interactions between ubiquitin and ubiquitin binding motifs to the chicken FANCD2-K563R resulted in reduced chromatin localization, indicating that the ubiquitin serves as the cue for chromatin localization. Lastly, fusion of FANCD2-K563R to histone H2B and expression of this protein in mutant cells corrected the sensitivity to MMC, suggesting that mono-ubiquitylation of FANCD2 serves only to alter the localization of this protein [33].

The recent identification of FANCI as a binding partner and paralog of FANCD2 has led some to refer to this group as the ID complex. FANCI is known to be mono-ubiquitylated on lysine 523 in response to several different DNA damaging agents or upon entry into S phase, and this activation depends both on the core complex and mono-ubiquitylation of FANCD2. Conversely, mono-ubiquitylation of FANCD2 depends on the ubiquitylation of FANCI. Upon activation, FANCI localizes to chromatin at sites of DNA damage with its binding partner FANCD2 [34–35]. Using chicken DT40 cells, multiple phosphorylation sites on FANCI have also recently been identified that are essential for mono-ubiquitylation and focus formation of FANCI and FANCD2 after DNA damage [36].

DOWNSTREAM EFFECTORS AND INTERACTIONS WITH OTHER DNA REPAIR PROTEINS

The three remaining FA proteins, FANCD1, FANCJ, and FANCN, are not required for the mono-ubiquitylation of FANCD2 and are therefore considered either downstream or components of a different pathway. FANCD2 is known to co-localize in nuclear repair foci containing FANCD1 and FANCN, and has been shown to co-immunoprecipitate with FANCD1 [37–38], providing strong evidence for participation of these proteins in a common pathway. As discussed above, all three downstream effector proteins have been identified as breast cancer susceptibility genes that function in homologous recombination.

Although much is still unknown about the exact mechanism by which an ICL is processed, it is known that many general DNA repair factors localize to repair foci and/or interact with FA proteins, including, among others, ERCC1/XPF, BRCA1, RAD51, and MLH1. Inter-strand crosslinks are complex lesions and restoration of a functional replication fork is thought to involve several other DNA repair systems including homologous recombination (HR), translesion synthesis (TLS), and nucleotide excision repair (NER). The first step in ICL repair involves incision of the crosslink, which is thought to be performed by the endonucleases Mus81-Eme1 and ERCC1/XPF, or mismatch repair proteins (see Figure 30.2). Incision of the crosslink creates a double-strand break, and subsequent repair of the lesion and restoration of the replication fork could be completed through TLS, HR, or a combination of both pathways. Translesion polymerases, or specialized polymerases containing a more accommodating active site that can replicate through bulky lesions, may be recruited to bypass the unhooked crosslink. Alternatively, HR may be used to copy the missing genetic information from the homologous chromosome. Restart of the replication fork could also be accomplished through HR using the broken strand as the invading strand [39]. The FA pathway appears to function subsequent to the creation of the double-strand break. Using a modified comet assay, the repair kinetics of photoactivated psoralen crosslinks were studied in FA mutant cells. Mutant cells were capable of restoring DNA migration in this assay with kinetics similar to that of wild-type cells, signifying that initial incision functions were intact in

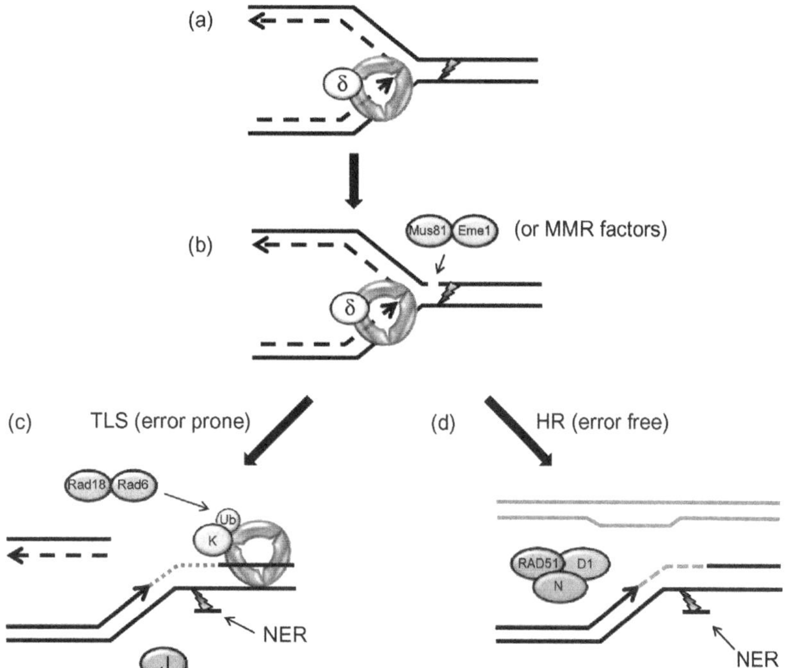

FIGURE 30.2 ICL repair may involve several different DNA repair pathways.
(a) When a replication fork encounters an ICL, the replication machinery, including PCNA and a replicative polymerase such as polymerase delta, will stall at the lesion. (b) It is thought that the endonuclease Mus81-Eme1 plays a role in at least one of the incisions surrounding the crosslink, while the other incision may be formed by either ERCC1/XPF or mismatch repair proteins. A double-strand break is formed when the crosslink is incised. Subsequent repair of the lesion and replication fork restart is thought to be accomplished through the translesion synthesis (TLS) pathway, homologous recombination (HR), or a combination of both. (c) TLS polymerases could be recruited through the mono-ubiquitylation of PCNA by Rad18/Rad6 in order to allow for replication (generally considered error prone) past the unhooked crosslink. FANCJ may be involved in clearing the DNA to allow for proper loading of TLS machinery. Nucleotide excision repair (NER) factors are then likely recruited to remove the unhooked crosslink and repair the resulting gap. (d) Alternatively, the homologous chromosome could be used to copy the missing genetic information in what is generally considered an error-free pathway. Use of the HR pathway for repair is thought to involve FANCD1, FANCN, and RAD51. The replication fork may be restarted through invasion of the broken strand using the HR pathway as well. FANCJ could be involved in resolving recombination intermediates to establish a functional replication fork.

FA mutant cells prior to S phase. Additionally, formation of γH2AX foci, which appear early on during double-strand break formation, also occurred with wild-type kinetics in FA mutant cells, suggesting that the FA pathway is involved in repair of ICLs once the double-strand break has been established [40]. Furthermore, phosphorylation of H2AX (γH2AX) is required for recruitment of FANCD2 to chromatin. FANCD2 was shown to co-precipitate with γH2AX in a BRCA1- and DNA damage dependent manner [41].

Evidence that the FA pathway may cooperate with translesion synthesis proteins in the repair of ICLs was provided by one key report using chicken DT40 cells. Strains that were deficient in either of the TLS polymerases Rev1 or Rev3 were epistatic to FANCC mutant strains with regards to sensitivity to crosslinking agents as measured by cell viability and chromosome breakage assays. Moreover, Rev1 co-localized with FANCD2 in nuclear foci upon replication arrest induced by X-rays, hydroxyurea, and thymidine [42]. Although a direct interaction between FA and TLS proteins has yet to be reported, it has been suggested that TLS polymerases may be recruited to the mono-ubiquitylated ID complex in a manner similar to the "polymerase switch" mechanism known to occur when PCNA becomes mono-ubiquitylated by the E3 ubiquitin ligase RAD18 [2].

Based on a number of different studies, a role for the FA pathway in homologous recombination repair is also possible. Perhaps most convincing is the fact that FANCD1, FANCJ, and FANCN have been identified as the breast cancer susceptibility proteins BRCA2, BACH1/BRIP1, and PALB2, which were initially found to perform essential functions in HR. Repair of a double-strand break (DSB) by homologous recombination occurs in four general steps [43]. First, the ends of the DSB are resected to generate a single-stranded DNA filament. This filament invades the homologous sequence and is followed by synthesis of new DNA starting from the invading strand. Once the missing information is copied, the recombination intermediate can be resolved to yield error-free repair of the damaged strand, or a restarted replication fork as is likely in ICL repair. In this system, it is known that FANCN/PALB2 is required for FANCD1/BRCA2 nuclear localization [10, 11].

FANCD1/BRCA2 is vital to homologous recombination repair, as it is responsible for the regulation of the recruitment of RAD51, a protein that nucleates the single-stranded filament that invades the homologous sequence [7]. FANCJ/BACH1/BRIP1 interacts with BRCA1, and HR of an I-SceI induced double-strand break was found to be disrupted in cells where FANCJ is depleted [44]. In a similar manner, the induction of I-SceI induced double-strand breaks was used to demonstrate that HR is disrupted in human cells deficient in FANCA, FANCG, and FANCD2 [45]. Despite all of this support, contradictory reports do exist that argue that the FA pathway functions separately from HR. One such report suggests that BRCA2 and RAD51 function in separate pathways due to the additive effect observed with MMC crosslinker sensitivity when both proteins are depleted [46]. More work will ultimately need to be done to resolve the precise role of the FA pathway in homologous recombination repair.

Additional studies on FANCJ have begun to provide more insights into the enzymatic function and repair roles of this protein. It has been shown that purified recombinant FANCJ preferentially binds to and unwinds model replication fork substrates [47], suggesting a role for this helicase in the restart of replication forks during ICL repair. RPA, the single-stranded DNA binding protein that is known for its ability to activate ATR through its interaction with ATRIP, also directly interacts with FANCJ in a DNA damage dependent manner. Moreover, RPA was capable of stimulating the unwinding of model replication fork substrates by FANCJ, further implicating the helicase as a major player in DNA repair [48]. In addition, FANCJ interacts with MLH1 and PMS2 [49], also known as the mismatch repair complex MutLa, indicating a possible role for mismatch repair proteins in ICL repair.

DE-UBIQUITYLATION OF FANCD2 AND FANCI

Once an inter-strand crosslink has been repaired and the replication fork has been restarted, the FA pathway must be deactivated to allow for normal replication to occur. The de-ubiquitylating enzyme, USP1, known to be the de-ubiquitylase for PCNA [50], is also responsible for removing the ubiquitin moiety from FANCD2 and FANCI [34, 51]. The de-ubiquitylation of the ID complex is essential for efficient repair of ICLs, as ablation of USP1 results in hypersensitivity to crosslinking agents, and this hypersensitivity is the result of persistence of the mono-ubiquitylated form of FANCD2 [52]. Though not yet entirely understood, it is clear that negative regulation of the FA pathway and recycling of the ID complex is accomplished through this enzyme and is a necessary step in the repair of ICLs.

A recent report adds yet another layer to the regulation of the FA pathway. Purification of a protein complex containing USP1 yielded a novel interacting protein, UAF1. Furthermore, presence of UAF1 stabilizes USP1, which is known to undergo autocleavage through an unknown mechanism. Binding of UAF1 to USP1 also stimulates its enzymatic activity over 30-fold, and is required for de-ubiquitylation of FANCD2. Lastly, transcription of the *USP1* gene rapidly shuts off after DNA damage, followed by a decrease in protein levels [53]. These results suggest a multi-tiered system of regulation of the FA pathway, which is important for efficient repair of inter-strand crosslinks.

NON-REPAIR FUNCTIONS OF THE FA PATHWAY

Many DNA repair proteins are multi-functional, and FA proteins are likely no exception. Several studies support a role for FA proteins in the cellular response to oxidative stress. FA mutant cells display increased levels of a DNA lesion that results from oxidative damage, 8-hydroxydeoxyguanosine [54–55]. Furthermore, FANCC and FANCG were shown to interact with proteins involved in oxygen metabolism. NADPH cytochrome P450 reductase, or RED, is an enzyme involved in the oxidative metabolism of drugs such as mitomycin C. FANCC interacts with RED and attenuates its activity, suggesting a regulatory role for FANCC in oxidative metabolism [56]. FANCC also interacts with glutathione S-transferase P1-1 (GSTP1), an enzyme that catalyzes the detoxification of oxygen metabolites. GSTP1 activity is increased in the presence of FANCC upon initiation of apoptosis, again supporting a role for FANCC in the regulation of enzymes involved in oxidative metabolism [57]. Lastly, FANCG interacts with cytochrome P450 2E1 (CYP2E1), an enzyme responsible for production of reactive oxygen intermediates. As FANCG mutant cells display higher levels of CYP2E1, it was suggested that FANCG plays a role in the regulation of this enzyme, and consequently oxygen metabolism [58].

The high incidence of bone marrow failure in FA patients implies a role for FA proteins in hematopoeisis. Interferon γ (IFN-γ) and tumor necrosis factor α (TNFα) are known negative regulators of hematopoeisis. In one study, bone marrow mononuclear cells from FA patients showed significant overexpression of both IFN-γ and TNFα as compared to healthy controls [59], indicating a possible role for FA proteins in the regulation of negative modulators of hematopoeisis. Binding of IFN-γ to the IFN receptor results in the downstream phosphorylation of STAT1, and this leads to the regulation of the IFN-γ responsive genes. FANCC has been shown to interact with STAT1, and although phosphorylation of STAT1 is defective in FANCC mutant cells [60], IFN-γ inducible genes are constitutively expressed in these cells [61]. Taken together, these data support a role for FANCC in modulation of the expression of IFN-γ inducible genes through a

STAT1 independent mechanism, as well as a general role for FA proteins in the regulation of hematopoeisis.

CONCLUSION

Although much remains to be elucidated about the true functions of the FA pathway and the mechanism of interstrand crosslink repair, a great deal of effort over the last few decades has led to a more complete view of the players involved and their possible roles. All 13 FA proteins have now been cloned, and are known to comprise a unique type of signaling pathway that utilizes ubiquitin as a cue essential for certain types of DNA repair. Exploiting the molecular biology of Fanconi anemia can provide novel insights into tumorigenesis and build paths leading to new areas of drug discovery.

REFERENCES

1. D'Andrea AD, Grompe M. The Fanconi anemia/BRCA Pathway. *Nat Rev Cancer* 2003;**3**:23–34.
2. Wang W. Emergence of a DNA-damage response network consisting of Fanconi anaemia and BRCA proteins. *Nat Rev Genet* 2007;**8**:735–48.
3. Collins N, Kupfer G. Molecular pathogenesis of fanconi anemia. *Int J Hematol* 2005;**82**:176–83.
4. Auerbach AD. Diagnosis of Fanconi anemia by diepoxybutane analysis. *Curr Protoc Hum Genet* 2003:1. Chapter 8: Unit 8.7.
5. Strathdee CA, Duncan AMV, Buchwald M. Evidence for at least four Fanconi anaemia genes including FACC on chromosome 9. *Nat Genet* 1992;**1**:196–8.
6. Strathdee CA, Gavish H, Shannon WR, Buchwald M. Cloning of cDNAs for Fanconi's anemia by functional complementation. *Nature* 1992;**356**:763–7.
7. Howlett NG, Taniguchi T, Olson S, Cox B, Waisfisz Q, De Die-Smulders C, Persky N, Grompe M, Joenje H, Pals G, Ikeda H, Fox EA, D'Andrea AD. Biallelic inactivation of BRCA2 in Fanconi anemia. *Science* 2002;**297**:606–9.
8. Levitus M, Waisfisz Q, Godthelp BC, de Vries Y, Hussain S, Wiegant WW, Elghalbzouri-Maghrani E, Steltenpool J, Rooimans MA, Pals G, Arwert F, Mathew CG, Zdzienicka MZ, Hiom K, De Winter JP, Joenje H. The DNA helicase BRIP1 is defective in Fanconi anemia complementation group J. *Nat Genet* 2005;**37**:934–5.
9. Levran O, Attwooll C, Henry RT, Milton KL, Neveling K, Rio P, Batish SD, Kalb R, Velleuer E, Barral S, Ott J, Petrini J, Schindler D, Hanenberg H, Auerbach AD. The BRCA1-interacting helicase BRIP1 is deficient in Fanconi anemia. *Nat Genet* 2005;**37**:931–3.
10. Xia B, Dorsman JC, Ameziane N, de Vries Y, Rooimans MA, Sheng Q, Pals G, Errami A, Gluckman E, Llera J, Wang W, Livingston DM, Joenje H, de Winter JP. Fanconi anemia is associated with a defect in the BRCA2 partner PALB2. *Nat Genet* 2007;**39**:159–61.
11. Reid S, Schindler D, Hanenberg H, Barker K, Hanks S, Kalb R, Neveling K, Kelly P, Seal S, Freund M, Wurm M, Batish SD, Lach FP, Yetgin S, Neitzel H, Ariffin H, Tischkowitz M, Mathew CG, Auerbach AD, Rahman N. Biallelic mutations in PALB2 cause Fanconi anemia subtype FA-N and predispose to childhood cancer. *Nat Genet* 2007; **39**:162–4.
12. Kupfer GM, Näf D, Suliman A, Pulsipher M, D'Andrea AD. The Fanconi anaemia proteins, FAA and FAC, interact to form a nuclear complex. *Nat Genet* 1997;**17**:487–90.
13. Medhurst AL, Huber PA, Waisfisz Q, de Winter JP, Mathew CG. Direct interactions of the five known Fanconi anaemia proteins suggest a common functional pathway. *Hum Mol Genet* 2001;**10**:423–9.
14. Garcia-Higuera I, Taniguchi T, Ganesan S, Meyn MS, Timmers C, Hejna J, Grompe M, D'Andrea AD. Interaction of the Fanconi anemia proteins and BRCA1 in a common pathway. *Mol Cell* 2001;**7**:249–62.
15. Meetei AR, Levitus M, Xue Y, Medhurst AL, Zwaan M, Ling C, Rooimans MA, Bier P, Hoatlin M, Pals G, de Winter JP, Wang W, Joenje H. X-linked inheritance of Fanconi anemia complementation group B. *Nat Genet* 2004;**36**:1219–24.
16. Qiao F, Moss A, Kupfer GM. Fanconi anemia proteins localize to chromatin and the nuclear matrix in a DNA damage- and cell cycle-regulated manner. *J Biol Chem* 2001;**276**:23,391–23,396.
17. Kim JM, Kee Y, Gurtan A, D'Andrea AD. Cell cycle-dependent chromatin loading of the Fanconi anemia core complex by FANCM/FAAP24. *Blood* 2008;**10**:5215–22.
18. Andreassen PR, D'Andrea AD, Taniguchi T. ATR couples FANCD2 monoubiquitination to the DNA-damage response. *Genes Dev* 2004;**18**:1958–63.
19. Yamashita T, Kupfer GM, Naf D, Suliman A, Joenje H, Asano S, D'Andrea AD. The fanconi anemia pathway requires FAA phosphorylation and FAA/FAC nuclear accumulation. *Proc Natl Acad Sci USA* 1998;**95**:13,085–13,090.
20. Wang X, Kennedy RD, Ray K, Stuckert P, Ellenberger T, D'Andrea AD. Chk1-mediated phosphorylation of FANCE is required for the Fanconi anemia/BRCA pathway. *Mol Cell Biol* 2007;**27**:3098–108.
21. Qiao F, Mi J, Wilson JB, Zhi G, Bucheimer NR, Jones NJ, Kupfer GM. Phosphorylation of fanconi anemia (FA) complementation group G protein, FANCG, at serine 7 is important for function of the FA pathway. *J Biol Chem* 2004;**279**:46,035–46,045.
22. Meetei AR, Medhurst AL, Ling C, Xue Y, Singh TR, Bier P, Steltenpool J, Stone S, Dokal I, Mathew CG, Hoatlin M, Joenje H, de Winter JP, Wang W. A human ortholog of archaeal DNA repair protein Hef is defective in Fanconi anemia complementation group M. *Nat Genet* 2005;**38**:958–63.
23. Sobeck A, Stone S, Hoatlin ME. DNA structure-induced recruitment and activation of the Fanconi anemia pathway protein FANCD2. *Mol Cell Biol* 2007;**27**:4283–92.
24. Ciccia A, Ling C, Coulthard R, Yan Z, Xue Y, Meetei AR, Laghmani el H, Joenje H, McDonald N, de Winter JP, Wang W, West SC. Identification of FAAP24, a Fanconi anemia core complex protein that interacts with FANCM. *Mol Cell* 2007;**9**:331–43.
25. Xue Y, Li Y, Guo R, Ling C, Wang W. FANCM of the Fanconi anemia core complex is required for both monoubiquitination and DNA repair. *Hum Mol Genet* 2008;**11**:1641–52.
26. Gari K, Décaillet C, Stasiak AZ, Stasiak A, Constantinou A. The Fanconi anemia protein FANCM can promote branch migration of Holliday junctions and replication forks. *Mol Cell* 2008;**1**:141–8.
27. Meetei AR, de Winter JP, Medhurst AL, Wallisch M, Waisfisz Q, van de Vrugt HJ, Oostra AB, Yan Z, Ling C, Bishop CE, Hoatlin ME, Joenje H, Wang W. A novel ubiquitin ligase is deficient in Fanconi anemia. *Nat Genet* 2003;**35**:165–70.
28. Pace P, Johnson M, Tan WM, Mosedale G, Sng C, Hoatlin M, de Winter J, Joenje H, Gergely F, Patel KJ. FANCE: the link between Fanconi anaemia complex assembly and activity. *EMBO J* 2002;**21**:3414–23.
29. Ling C, Ishiai M, Ali AM, Medhurst AL, Neveling K, Kalb R, Yan Z, Xue Y, Oostra AB, Auerbach AD, Hoatlin ME, Schindler D, Joenje H,

de Winter JP, Takata M, Meetei AR, Wang W. FAAP100 is essential for activation of the Fanconi anemia-associated DNA damage response pathway. *EMBO J* 2007;**26**:2104–14

30. Di Fiore PP, Polo S, Hofmann K. When ubiquitin meets ubiquitin receptors: a signalling connection. *Nat Rev Mol Cell Biol* 2003;**6**:491–7.
31. Machida YJ, Machida Y, Chen Y, Gurtan AM, Kupfer GM, D'Andrea AD, Dutta A. UBE2T is the E2 in the Fanconi anemia pathway and undergoes negative autoregulation. *Mol Cell* 2006;**23**:589–96.
32. Taniguchi T, Garcia-Higuera I, Andreassen PR, Gregory RC, Grompe M, D'Andrea AD. S-phase-specific interaction of the Fanconi anemia protein, FANCD2, with BRCA1 and RAD51. *Blood* 2002;**100**:2414–20.
33. Matsushita N, Kitao H, Ishiai M, Nagashima N, Hirano S, Okawa K, Ohta T, Yu DS, McHugh PJ, Hickson ID, Venkitaraman AR, Kurumizaka H, Takata M. A FancD2-monoubiquitin fusion reveals hidden functions of Fanconi anemia core complex in DNA repair. *Mol Cell* 2005;**9**:841–7.
34. Smogorzewska A, Matsuoka S, Vinciguerra P, McDonald ER 3rd, Hurov KE, Luo J, Ballif BA, Gygi SP, Hofmann K, D'Andrea AD, Elledge SJ. Identification of the FANCI protein, a monoubiquitinated FANCD2 paralog required for DNA repair. *Cell* 2007;**129**:289–301.
35. Sims AE, Spiteri E, Sims RJ 3rd, Arita AG, Lach FP, Landers T, Wurm M, Freund M, Neveling K, Hanenberg H, Auerbach AD, Huang TT. FANCI is a second monoubiquitinated member of the Fanconi anemia pathway. *Nat Struct Mol Biol* 2007;**14**:564–7.
36. Ishiai M, Kitao H, Smogorzewska A, Tomida J, Kinomura A, Uchida E, Saberi A, Kinoshita E, Kinoshita-Kikuta E, Koike T, Tashiro S, Elledge SJ, Takata M. FANCI phosphorylation functions as a molecular switch to turn on the Fanconi anemia pathway. *Nat Struct Mol Biol* 2008;**11**:1138–46.
37. Wang X, Andreassen PR, D'Andrea AD. Functional interaction of monoubiquitinated FANCD2 and BRCA2/FANCD1 in chromatin. *Mol Cell Biol* 2004;**24**:5850–62.
38. Xia B, Sheng Q, Nakanishi K, Ohashi A, Wu J, Christ N, Liu X, Jasin M, Couch FJ, Livingston DM. Control of BRCA2 cellular and clinical functions by a nuclear partner, PALB2. *Mol Cell* 2006;**22**:719–29.
39. Dronkert ML, Kanaar R. Repair of DNA interstrand cross-links. *Mutat Res* 2001;**486**:217–47.
40. Rothfuss A, Grompe M. Repair kinetics of genomic interstrand DNA cross-links: evidence for DNA double-strand break-dependent activation of the Fanconi anemia/BRCA pathway. *Mol Cell Biol* 2004;**24**:123–34.
41. Bogliolo M, Lyakhovich A, Callén E, Castellà M, Cappelli E, Ramírez MJ, Creus A, Marcos R, Kalb R, Neveling K, Schindler D, Surrallés J. Histone H2AX and Fanconi anemia FANCD2 function in the same pathway to maintain chromosome stability. *EMBO J* 2007;**5**:1340–51.
42. Niedzwiedz W, Mosedale G, Johnson M, Ong CY, Pace P, Patel KJ. The Fanconi anaemia gene FANCC promotes homologous recombination and error-prone DNA repair. *Mol Cell* 2004;**15**:607–20.
43. Sung P, Klein H. Mechanism of homologous recombination: mediators and helicases take on regulatory functions. *Nat Struct Mol Cell Biol* 2006;**7**:739–50.
44. Litman R, Peng M, Jin Z, Zhang F, Zhang J, Powell S, Andreassen PR, Cantor SB. BACH1 is critical for homologous recombination and appears to be the Fanconi anemia gene product FANCJ. *Cancer Cell* 2005;**8**:255–65.
45. Nakanishi K, Yang YG, Pierce AJ, Taniguchi T, Digweed M, D'Andrea AD, Wang ZQ, Jasin M. Human Fanconi anemia monoubiquitination pathway promotes homologous DNA repair. *Proc Natl Acad Sci USA* 2005;**102**:1110–15.
46. Ohashi A, Zdzienicka MZ, Chen J, Couch FJ. Fanconi anemia complementation group D2 (FANCD2) functions independently of BRCA2- and RAD51-associated homologous recombination in response to DNA damage. *J Biol Chem* 2005;**280**:14,877–14,883.
47. Gupta R, Sharma S, Sommers JA, Jin Z, Cantor SB, Brosh Jr RM. Analysis of the DNA substrate specificity of the human BACH1 helicase associated with breast cancer. *J Biol Chem* 2005;**27**:25,450–25,460.
48. Gupta R, Sharma S, Sommers JA, Kenny MK, Cantor SB, Brosh Jr RM. FANCJ (BACH1) helicase forms DNA damage inducible foci with replication protein A and interacts physically and functionally with the single-stranded DNA-binding protein. *Blood* 2007;**7**:2390–8.
49. Peng M, Litman R, Xie J, Sharma S, Brosh Jr RM, Cantor SB. The FANCJ/MutLalpha interaction is required for correction of the cross-link response in FA-J cells. *EMBO J* 2007;**13**:3238–49.
50. Huang TT, Nijman SM, Mirchandani KD, Galardy PJ, Cohn MA, Haas W, Gygi SP, Ploegh HL, Bernards R, D'Andrea AD. Regulation of monoubiquitinated PCNA by DUB autocleavage. *Nat Cell Biol* 2006;**4**:339–47.
51. Nijman SM, Huang TT, Dirac AM, Brummelkamp TR, Kerkhoven RM, D'Andrea AD, Bernards R. The deubiquitinating enzyme USP1 regulates the Fanconi anemia pathway. *Mol Cell* 2005;**17**:331–9.
52. Oestergaard VH, Langevin F, Kuiken HJ, Pace P, Niedzwiedz W, Simpson LJ, Ohzeki M, Takata M, Sale JE, Patel KJ. Deubiquitination of FANCD2 is required for DNA crosslink repair. *Mol Cell* 2007;**28**:798–809.
53. Cohn MA, Kowal P, Yang K, Haas W, Huang TT, Gygi SP, D'Andrea AD. A UAF1-containing multisubunit protein complex regulates the Fanconi anemia pathway. *Mol Cell* 2007;**28**:786–97.
54. Takeuchi T, Morimoto K. Increased formation of 8-hydroxydeoxyguanosine, an oxidative DNA damage, in lymphoblasts from Fanconi's anemia patients due to possible catalase deficiency. *Carcinogenesis* 1993;**6**:1115–20.
55. Degan P, Bonassi S, De Caterina M, Korkina LG, Pinto L, Scopacasa F, Zatterale A, Calzone R, Pagano G. In vivo accumulation of 8-hydroxy-2'-deoxyguanosine in DNA correlates with release of reactive oxygen species in Fanconi's anaemia families. *Carcinogenesis* 1995;**4**:735–41.
56. Kruyt FA, Hoshino T, Liu JM, Joseph P, Jaiswal AK, Youssoufian H. Abnormal microsomal detoxification implicated in Fanconi anemia group C by interaction of the FAC protein with NADPH cytochrome P450 reductase. *Blood* 1998;**9**:3050–6.
57. Cumming RC, Lightfoot J, Beard K, Youssoufian H, O'Brien PJ, Buchwald M. Fanconi anemia group C protein prevents apoptosis in hematopoietic cells through redox regulation of GSTP1. *Nat Med* 2001;**7**:814–20.
58. Futaki M, Igarashi T, Watanabe S, Kajigaya S, Tatsuguchi A, Wang J, Liu JM. The FANCG Fanconi anemia protein interacts with CYP2E1: possible role in protection against oxidative DNA damage. *Carcinogenesis* 2002;**1**:67–72.
59. Dufour C, Corcione A, Svahn J, Haupt R, Poggi V, Béka'ssy AN, Scimè R, Pistorio A, Pistoia V. TNF-alpha and IFN-gamma are overexpressed in the bone marrow of Fanconi anemia patients and TNF-alpha suppresses erythropoiesis *in vitro*. *Blood* 2003;**6**:2053–9.
60. Pang Q, Fagerlie S, Christianson TA, Keeble W, Faulkner G, Diaz J, Rathbun RK, Bagby GC. The Fanconi anemia protein FANCC binds to and facilitates the activation of STAT1 by gamma interferon and hematopoietic growth factors. *Mol Cell Biol* 2000;**13**:4724–35.
61. Fagerlie SR, Diaz J, Christianson TA, McCartan K, Keeble W, Faulkner GR, Bagby GC. Functional correction of FA-C cells with FANCC suppresses the expression of interferon gamma-inducible genes. *Blood* 2001;**10**:3017–24.

Chapter 31

Stress and γ-H2AX

Jennifer S. Dickey, Christophe E. Redon, Asako J. Nakamura, Brandon J. Baird, Olga A. Sedelnikova and William M. Bonner

Laboratory of Molecular Pharmacology, Center for Cancer Research, National Cancer Institute, National Institutes of Health, Bethesda, Maryland

INTRODUCTION

Stress occurs when a cell or organism is subjected to a non-optimal environment. Some of these disturbances will lead to cell death while others have been shown to contribute to increased levels of cellular damage. Though many types of cell damage can be reversed, those that target the genome may create permanent changes, contributing to genomic instability and cancer. One type of genomic damage that is a serious threat to cell health is the DNA double-stranded break (DSB) because of its propensity to induce irreversible genomic rearrangements, which, if not fatal, may promote cellular oncogenesis [1]. To combat threats to genome integrity, living systems have evolved a highly regulated, intricate system of repair pathways. These systems attempt to rectify DNA lesions and arrest the cell cycle, which decreases interference with ongoing DNA repair [2,3]. Other related pathways induce senescence or activate apoptosis to remove damaged cells from the population. This DNA damage response (DDR) machinery helps maintain the health of the cell population and prevent cancer [4–6].

A key protein in the DDR machinery is histone H2AX [7]. H2AX becomes phosphorylated at a serine four residues from the carboxy terminus upon the induction of any DNA DSB regardless of whether the break is induced by endogenous or exogenous agents [8]. Many hundreds of H2AX molecules in the chromatin flanking the break site become phosphorylated during the first 30 minutes following break induction, creating a focus at each break site that can be visualized with an antibody against phosphorylated H2AX (γ-H2AX) [3, 4]. These properties make γ-H2AX an excellent tool to observe DSB formation and repair [9]. In this review we discuss many stresses that may induce DSB formation, highlight the mechanisms of various cellular responses, and describe methods of DSB detection. Readers are also referred to other recent reviews [10,11].

STRESS INDUCES DNA DSB DAMAGE AND γ-H2AX FORMATION

Any deviation from normal physiological conditions can generate cell stress leading to DNA damage including DSBs. However, even in an optimum environment, DSBs can originate as part of normal cellular processes. One potential damage source is the presence of reactive oxygen species (ROS) in cells. ROS are present in all eukaryotes, and either a lack or an excess of ROS can be stressful [12]. ROS are produced by mitochondria as part of their normal function [13]. Peroxisomes, inflammatory responses, chlorinated compounds, metal ions, and phorbol esters are also sources of ROS [14–16]. Ionizing rays and particles interact with water to generate clusters of ROS [17], which may interact with DNA often leading to a DSB [18, 19]. On the whole, ROS are estimated to be responsible for about 5000 DNA single-stranded lesions per cell per day, mostly in replicating cells, about 1 percent of which may lead to DSBs [20, 21].

In addition to being a source of DSBs, ROS can oxidize protein components of the DDR pathways, indirectly leading to elevated genotoxicity [22]. Other stresses that may interfere with the DDR machinery include hypoxia, electromagnetic waves, decreased pH, heat, and hyperosmolarity. These factors can impact DSB levels regardless of their source [23–29]. Chronic hypoxia, frequently experienced by tumor cells, may also lead to increased genomic instability through downregulation of DDR systems [25, 30].

Table 31.1 categorizes sources of DSBs on the basis of their origin (endogenous or exogenous) and their properties (chemical/physical or biological). ROS are found in all four of these situations, thus they are not listed in Table 31.1. Figure 31.1, likewise illustrates the various causes of DNA DSBs.

Handbook of Cell Signaling, Three-Volume Set 2 ed.

TABLE 31.1 Sources of DSBs[1]

	Chemical/physical	Biological
Endogenous	Cellular radioactive constituents	Meiosis V(D)J, CSR Telomere erosion Replication/ transcription errors Mutations Apoptosis
Exogenous	IR-radiomimetic drugs Chemotherapeutic agents UV Electromagnetic waves	Bystander effect signals Viral infection

[1]DSBs are classified according to their origin.

Endogenous Sources of DSBs

Physical/chemical endogenous sources of DSBs include naturally abundant radioactive species such as K^{39}, C^{14}, and Rb^{87} among others. It is estimated that these elements account for about 8000 disintegrations per second per 60kg person, which can potentially result in DSBs. Walter LD, David MJ, Glenn ST. Modern Nuclear Chemistry 2005.

Biological sources of DSBs can be classified as programmed or accidental. DSB formation and resealing are essential steps in certain biological processes, including homologous recombination during meiosis and gene rearrangement during immune system development [31–33]. RAG1/RAG2 and AID induce DSBs in V(D)J recombination and class switch recombination (CSR) respectively, while SPO11 induces DSBs in meiosis [33–35]. Since these breaks are programmed, their presence would not be expected to be stressful, but they do make the cell more susceptible to chromosome translocations [31].

Accidental DSBs can arise during replication or transcription. If a polymerase stalls, the resulting DNA structure will be recognized by an endonuclease, which creates a SSB, and then a DSB near the block [36], leading to the transient appearance of γ-H2AX foci [37]. Situations that cause polymerase stalling include dNTP pool alterations, changes in DNA replication frequency or the presence of a blocking DNA lesion [38]. Incomplete or aberrant DNA repair is another common source of endogenous DNA DSBs. Base excision repair (BER), mismatch repair (MMR), and nucleotide excision repair (NER) have all been linked to increased formation of DSBs and γ-H2AX when these processes have been delayed or altered [1, 2, 39, 40]. Repair intermediates can be converted to DSBs when encountered by replication or transcription forks. Mutations in DNA damage repair proteins result in DSBs and elevated γ-H2AX formation, leading to genomic instability and cancer [41–43]. Examples of these deleterious events include mutations in MYH, a BER DNA glycosylase that has been linked with increased colon cancer risk [44]. In addition, mutations in 7 of the 30 proteins that participate in NER have been linked to the cancer predisposing disorder xeroderma pigmentosum [45, 46]. Finally, hereditary non-polyposis colorectal cancer is caused by mutations in the MMR machinery [47].

Telomere shortening is an endogenous programmed process that can be accelerated by stresses such as increased levels of ROS [48]. Depending on the initial length of the telomeres in an organism, this shortening may lead to telomere dysfunction [49]. This in turn generates an exposed DNA double-stranded end that forms a γ-H2AX focus [49, 50]. Eroded telomeres may play a role in aging and in preventing tumorigenesis [41, 50–53], but may also be a site of chromosome end-to-end fusions, leading to unstable genomes [54].

Exogenous Sources of DSBs

DSBs may arise from a variety of exogenous chemical/physical sources. Ionizing radiation (IR) has been studied for decades. In addition to naturally occurring radioactive isotopes and cosmic radiation, radiation sources include nuclear bombs and accidents, as well as medical X-rays and other diagnostic scanning procedures. Also the majority of cancer therapies involve IR of some sort [55]. Radiation causes DSBs through both direct and indirect mechanisms. Direct interaction of a radioactive ray or particle with a DNA double helix may result in a DNA lesion. A DSB can also be created if the ray or particle interacts with other cellular constituents, commonly water, to create a burst of ROS close to the double helix. This cluster of radicals may also create local multiply damaged sites (LMDS) [56] containing a variety of DNA lesions including DSBs and oxidative clusters [57].

Many chemotherapeutic drugs function by creating breaks in the genome through a variety of means. Topoisomerase I and II poisons function by stabilizing the enzyme–DNA complex formed when topoisomerase creates a break in DNA in order to untangle or decatenate the genetic material [58, 59]. Other chemicals including cisplatin function by adducting or inducing DNA base modifications that also create breaks upon replication fork progression [60]. These drugs as well as DNA modifications following alkylation or oxidation can induce replication and/or transcription stresses, leading to DSBs and γ-H2AX formation [36, 59].

DSBs can also be induced by non-ionizing radiation. Exposure to UVB or UVC rays can lead to the formation of thymine dimers and other 6–4 photoproducts that likewise can disrupt replication and transcription leading to DSB

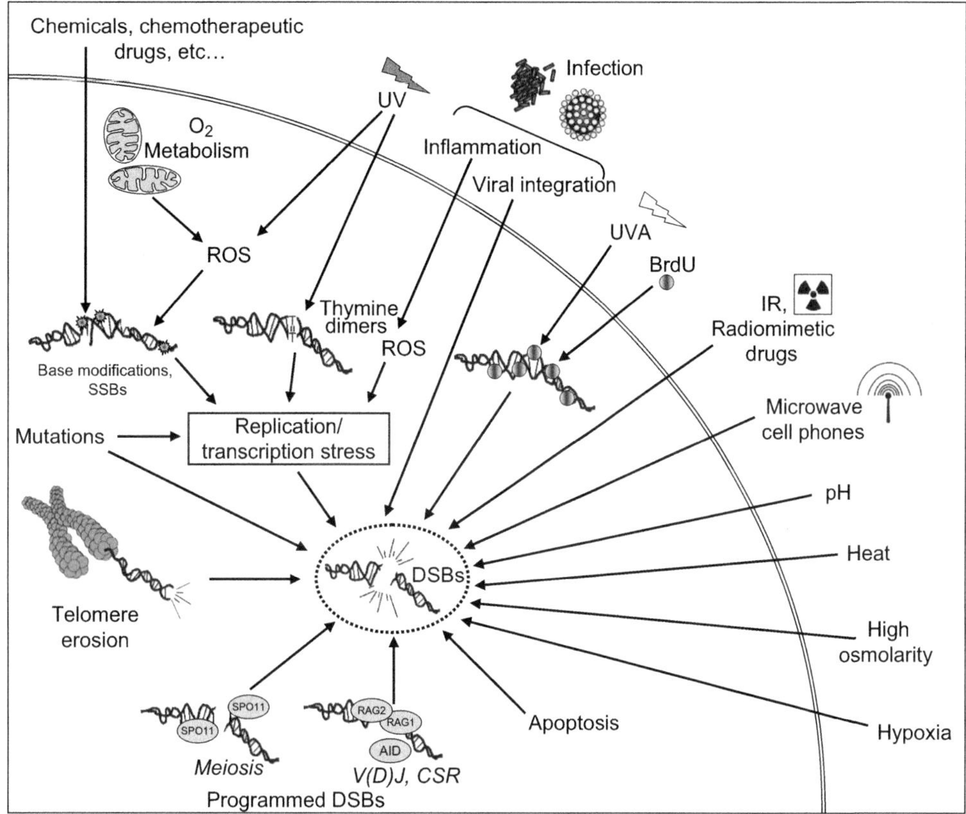

FIGURE 31.1 Factors that cause DNA DSBs.
Natural endogenous processes can cause DNA DSBs. Examples of these processes include ROS generated through normal oxygen metabolism, mutations in DNA repair proteins, telomere erosion, meiosis, V(D)J recombination, CSR, and apoptosis. Many DNA DSBs also arise from replication or transcription stresses caused in part by base modifications generated by ROS, UV, and drugs. Likewise, exogenous cellular stresses including chemicals, UV, viral or bacterial infection, photolysis, IR, radiomimetic compounds, electromagnetic waves from cell phones, changes in pH, heat, and osmolarity as well as hypoxia have been shown to cause DNA DSBs.

formation [61, 62]. UVA rays generally do not directly create DNA lesions; however, in the presence of the DNA intercalating agent Hoescht dye and BrdU, DSBs may be formed [63, 64].

Physical trauma can lead to increased levels of DSBs [65]. When normal human fibroblasts are grown to complete confluence and then scratched in a small line with a sterile rubber policeman, a slightly less than twofold induction of γ-H2AX foci is seen in cells near the wound 8 hours later. The number of foci declined after overnight incubation coincident with sealing of the scratch by cellular growth (Figure 31.2).

Biological sources of DSBs can also be external to the cell. For example, viruses such as the murine γ-herpes virus 68 actively induces γ-H2AX through expression of the viral kinase, orf36, in macrophages [66]. Orf36 is important for viral replication in infected animals, and orf36, H2AX, and ATM are all critical for efficient viral replication in primary macrophages [66]. Retroviral integration also introduces DSBs that promote the formation of γ-H2AX [67].

Another source of cellular DNA DSBs may be from chemicals released by stressed neighboring cells, either through gap junctions or the media. This effect, known as the bystander effect, involves cells that have not themselves been subjected to any kind of stress or damage, but which contact or share the environment with injured cells [18, 68–71]. The factors leading to this phenomenon can be emitted either by cells with damaged DNA [68], microbeam targeted cytoplasm [69, 70], or cells with genomic instability, including cancer cells [71]. These observations indicate a common mechanism of bystander effect induction, which has been shown to be partially caused by deregulated redox homeostasis involving various reactive species including nitric oxide (NO) [70, 72–75]. DNA damage has been suggested to be the key starting point in the cascade of reactions leading to bystander biological responses [76–78]. As DNA SSBs and oxidative clusters can be converted to DNA DSBs upon collision with the replication fork, [16] replicative cells are the most vulnerable to bystander signaling [79]. Thus, rapidly proliferating tissues in the body should be considered to be the primary oxidative stress targets [16]. DNA DSB repair proteins including γ-H2AX have recently proved themselves to be sensitive markers for detection of bystander cells [69, 73, 80–83]. For

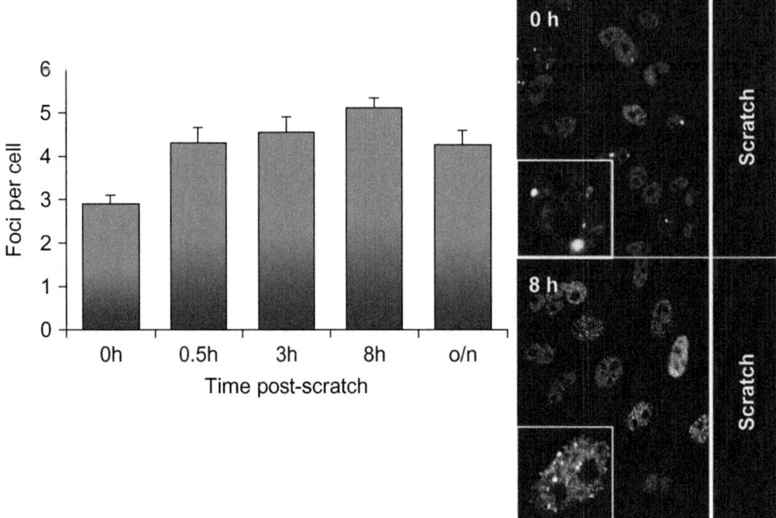

FIGURE 31.2 Wound formation by scratching induces increased γ-H2AX foci formation.
Quantification of the average number of γ-H2AX foci per cell in normal human fibroblasts after a confluent layer of cells was scratched at time zero is shown. Error bars represent the standard error of the mean for at least 100 cell nuclei counted. Representative images on the right show cells at the border of the scratch (white line) at time zero (top) and 8 hrs post-scratch (bottom). Insets show single magnified cells.

example, in marked contrast to the γ-H2AX focal dynamics in targeted cells [84], their bystander neighbors exhibit delayed increases in γ-H2AX focal incidences, which reach maxima by 12–48 hours post-IR, gradually decreasing over several days [73, 85, 86]. In bystander cell populations of microbeam irradiated 3-D human artificial tissue systems, increased levels of DNA DSBs were followed by loss of nuclear DNA methylation, increased formation of micronuclei, apoptosis, and premature senescence [86]. The temporal order of these events suggests that bystander DNA DSBs are an upstream event and initiate the full realization of the bystander effect.

The Stress Responses in H2AX-KO Mice

The H2AX null mouse provides some insights into the role of H2AX in stress. Because these mice are immune deficient, a germ free setting is necessary for survival, but they are otherwise viable [87]. H2AX null mouse embryonic fibroblasts (MEFs) demonstrated high sensitivity to various stresses including IR [87], hydroxyurea (HU), mitomycin C (MMC) [88], methyl methane sulfonate (MMS) [89], UVA [90], camptothecin [91], and H_2O_2 [92]. In addition, H2AX deficient mice showed increased sensitivity to 7 Gy whole body IR exposure [87], indicating that H2AX plays an important role for *in vitro* cell survival and for *in vivo* survival after trauma. Also, ATM and H2AX double deficient embryonic stem (ES) cells show impaired ROS regulation, a high sensitivity to oxidative stress, and a more severe genomic instability phenotype than either ATM or H2AX single deficient ES cells [92], indicating that H2AX has increased importance in other genetic backgrounds. H2AX null ES cells showed normal ROS levels, thus H2AX may be essential for the repair of damage induced by excessive ROS which are generated in the ATM null background.

ROLE OF γ-H2AX IN DNA DAMAGE REPAIR PATHWAYS

Immediately after DSB induction, members of the phosphatidylinositol-3 kinase (PIKKs) family, which includes ATM, ATR, and DNA-PK, are activated and phosphorylate H2AX along with other proteins involved in DNA repair (Figure 31.3) [11]. ATM, ATR, and DNA-PK, which have specialized and overlapping functions, are recruited to the sites of DNA damage through analogous mechanisms [93]. The kinase responsible for H2AX phosphorylation depends on the type of damage induced. Replication induced DSBs activate primarily ATR via replication protein A (RPA), while the MRN complex (MRE11, RAD50, NBS1) recruits ATM to IR induced DSB sites [94–96]. DNA-PK, which has a minor role in H2AX phosphorylation, acts in concert with both ATM and ATR [8]. During the 15–30 minutes following the formation of a DNA DSB, hundreds of γ-H2AX molecules are formed along the chromatin leading away from the break [3]. The phosphorylation of H2AX is required for the accumulation and the retention of DNA damage repair and signaling proteins at DSB sites. Some of these factors include the MRN complex, ATM, MDC1, BRCA1, 53BP1, MCPH1/BRIT1, chromatin remodeling complexes, and cohesion (Figure 31.3a) [11].

The assembly of DDR factors at DSB sites follows distinct kinetics, suggesting that the formation of the DSB induced repair complex is highly regulated and hierarchical

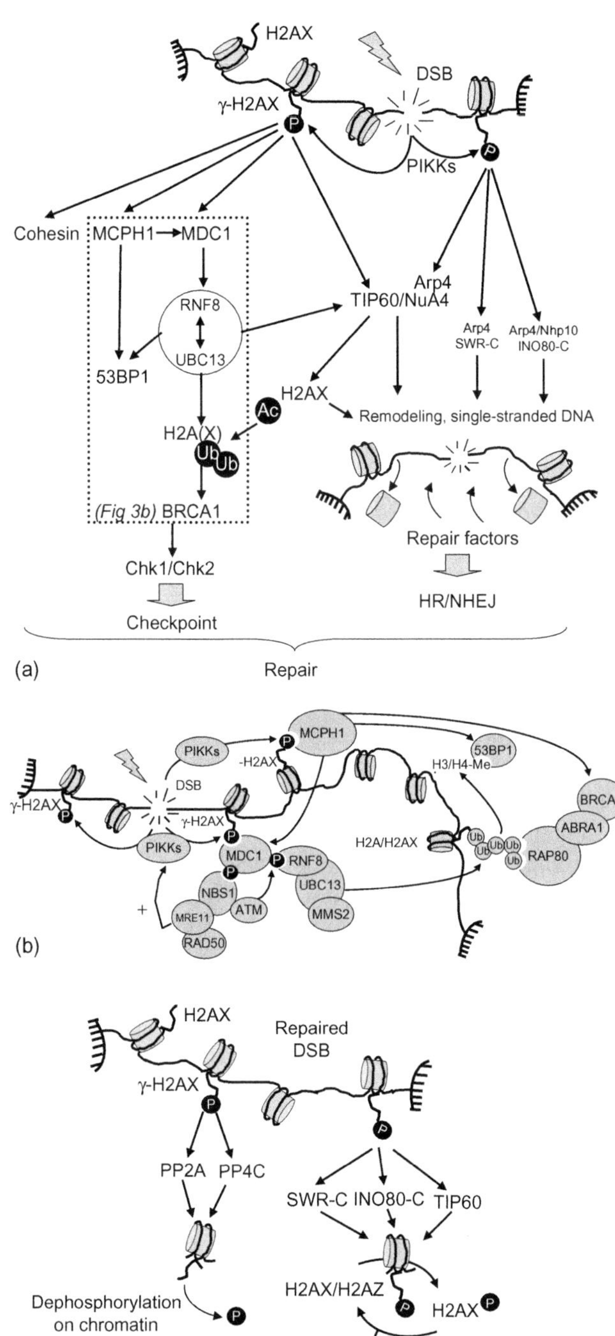

FIGURE 31.3 Model for the role of γ-H2AX in DNA damage repair. (a) The induction of a DNA DSB in chromatin results in the rapid phosphorylation of H2AX (γ-H2AX) by the PIKKs ATM, ATR, and/or DNA-PK. The phosphorylation of H2AX is a signal for recruitment of cohesin, chromatin remodeling complexes, and other DDR factors. The chromatin remodeling factors include the histone acetyl transferase TIP60 and its yeast homolog NuA4, and the yeast ATP dependent chromatin remodeling complexes INO80-C and SWR-C. The yeast complexes bind γ-H2AX through Arp4 or Nhp10, one of their constitutive subunits. The recruitment of these complexes is time dependent; the NuA4 complex is recruited before the association of INO80-C and SWR-C. Remodeling the chromatin around a DSB allows the recruitment of DDR factors and the appearance of single-stranded DNA necessary for homologous recombination. (b) Schematic model of the role of γ-H2AX in the assembly of DDR factors surrounding a DSB site. Figure 31.3b is a detailed view of the inset from Figure 31.3a. To simplify, only one side of a DSB is shown. The phosphorylation of H2AX allows the formation of the γ-H2AX–MDC1–MRN complex, which in turn both inhibits γ-H2AX dephosphorylation and increases phosphorylation by ATM of H2AX and other factors including MDC1. Phosphorylation of MDC1 recruits RNF8 and results in ubiquitination of H2A and H2AX. Ubiquitinated H2A/H2AX provides a direct signal for RAP80 and the pathway is then completed by BRCA1, which binds RAP80 via ABRA1. The phosphorylation of H2AX also allows the recruitment of MCPH1. Both H2A/H2AX ubiquitination and MCPH1 recruitment by γ-H2AX are involved in 53BP1 and MDC1 retention at the DSB site, probably by altering chromatin structure and unmasking methylated H3 and/or H4. (c) After repair of the DSB, γ-H2AX is removed. The removal of γ-H2AX from the chromatin is performed by the PP2A and PP4C phosphatases that hydrolyze the γ-phosphate or the INO80-C and SWR-C complexes that mediate the exchange of γ-H2AX with non-phosphorylated H2AX/H2AZ. Yeast phosphatase Pph3 is hypothesized to dephosphorylate γ-H2AX only after it has been removed from chromatin.

(Figure 31.3b). In the initial steps of the DNA damage response, MDC1 binds the phospho-serine of γ-H2AX through its BRCT domains. The role of MDC1 is both to prevent the premature dephosphorylation of H2AX [97] and to act as a mediator to recruit other DDR factors to DSBs [98, 99]. MDC1 is phosphorylated on a cluster of conserved repeat motifs by casein kinase 2. This phosphorylation then allows the interaction of the MDC1–MRN complex via NBS1 [98]. The γ-H2AX–MDC1–MRN complex increases the local concentration of ATM, allowing phosphorylation of adjacent H2AX and accumulation of MDC1 at the DSB site resulting in a signal amplification loop (reviewed in [99]). The γ-H2AX–MDC1–MRN complex also promotes MDC1 phosphorylation by ATM and its recognition by RNF8, a member of the ring finger containing nuclear factors [100]. The recruitment of RNF8 allows the ubiquitylation of H2A(X) by the ubiquitin conjugating enzyme UBC13, which, in turn, provides docking sites for the ubiquitin ligase domain of RAP80 [100–102].

At later stages of repair, the recruitment of BRCA1 is initiated by ABRA1, a mediator protein that can simultaneously bind to RAP80 and the BRCT domain of BRCA1 [100, 101]. BRCA1 plays a key role in genome integrity by modulating checkpoints and DNA repair. The phosphorylation of H2AX is also essential for 53BP1 accumulation at DSB sites [103, 104]. Similar to BRCA1, 53BP1 is a key transducer of DNA damage and cell cycle checkpoint signaling. 53BP1 contains two BRCT motifs that interact with p53 and two tandem tudor domains which bind methylated histones H3 and H4 with high affinity for H4K20me2 [105]. Like BRCA1, its accumulation at DSB sites is dependant on RNF8 [101]. However, while the link between RNF8, UBC13, and BRCA1 is well established, the relationship between RNF8 and 53BP1 is not well understood. One possibility is that the chromatin remodeling engineered by the DDR factors, including RNF8, could expose H4K20me2, which is thought to be masked by stacked nucleosomes, allowing 53BP1 accumulation

[105–107]. Indeed, the recruitment of RNF8 by H2AX and MDC1 could stimulate the binding of TRRAP-TIP60 histone acetyltransferase (HAT) to the DSB flanking regions or enhance its HAT activity [101]. Finally, the presence of a DSB seems to facilitate the association between TIP60 and UBC13, which then regulates H2AX acetylation on lysine 5. H2AX acetylation is necessary prior to its ubiquitylation by UBC13 and its release from chromatin [108].

Chromatin remodeling complexes also participate in the repair of DNA damage (reviewed in [109]). The human histone acetyltransferase TIP60 complex and its yeast counterpart the NuA4 complex, interact with H2AX and its phosphorylated form [108, 110]. The yeast NuA4 subunit Arp4 is recruited to a DSB [111] and binds the yeast γ-H2AX homolog *in vitro* (also named γ-H2AX in this review)[110]. Arp4 is also a component of the yeast chromatin remodeling complex INO80-C and SWR-C [109]. Act1, another component of NuA4, INO80-C, and SWR-C complexes also co-purifies with γ-H2AX [109]. INO80-C is recruited to DSBs in a γ-H2AX dependent manner [112] via Arp4 or through Nhp10 [113]. SWR-C, exchanges the histone variant H2AZ with H2A in chromatin [114]. SWR-C binds γ-H2AX in peptide pull-down experiments [113] probably through the Arp4 and/or Act1 [109], although it is not clear how this complex is recruited to the DSB site [110]. The chromatin remodeling complexes may function to relax chromatin, allowing the DDR factors access to the break site. They also may be involved in regulating homologous recombination mediated DSB repair. Finally they may intervene in the removal of γ-H2AX from chromatin [109].

H2AX phosphorylation is also critical for the recruitment of BRIT1/MCPH1 to DSB sites that are mediated by the BRIT1/MCPH1 C-terminal BRCT domains [115]. BRIT1/MCPH1 [116, 117] is another putative protein mediator, involved in chromatin condensation [118], cancer prevention [119, 120] and cell cycle checkpoint regulation by participating in the accumulation of MDC1, RPA, phospho-Rad17, and 53BP1 at DSB sites [115, 121–123] and/or its interaction with BRCA1, NBS1, CHK1, 53BP1, MDC1, ATR, RPA, ATM [121, 124].

γ-H2AX plays a key role in sister chromatid homologous recombination [125, 126], probably by facilitating the interaction between sister chromatids [125]. This hypothesis is strongly corroborated in yeast by the fact that γ-H2AX is essential for cohesin loading at the DSB site [127] and that γ-H2AX is in a same genetic pathway for post-replication repair as proteins involved in chromatid cohesion [128]. The loading of cohesin serves to hold sister chromatids together from the time of their replication until the metaphase–anaphase transition [129]. In addition, a role in DNA repair is suggested by observations from yeast and mammals [130–132].

The γ-H2AX foci dissolve after DNA repair is completed (Figure 31.3c). Several studies have shown a role for the PP2A and PP4 family of phosphatases in γ-H2AX dephosphorylation [133–135]. Moreover, the yeast INO80-C, SWR-C, and TIP60 remodeling complexes are implicated in removing phosphorylated H2AX from the chromatin by histone exchange [110, 112, 113, 136, 137]. In fact, it is suggested that the two mechanisms may work together, with the chromatin remodeling complexes removing γ-H2AX near the DSB sites allowing access to specific repair factors while the phosphatase complexes could mediate direct dephosphorylation at a distal site [133]. Yeast phosphatase Pph3 is hypothesized to dephosphorylate γ-H2AX only after it has been removed from chromatin [134].

γ-H2AX, A MARKER TO MONITOR CELL STRESS AND A PROTEIN INVOLVED IN STRESS SIGNALING

As mentioned above, H2AX phosphorylation is induced by various cellular stresses and this modification is one of the most rapid and sensitive cellular responses to DNA damage. Therefore the detection of γ-H2AX is a very powerful tool to monitor the effects of stress *in vitro* and *in vivo*. In addition, detection of overall γ-H2AX levels yields useful information concerning the amount of stress in the cellular environment. Several mouse monoclonal and rabbit polyclonal γ-H2AX antibodies are commercially available that can be used for immunoblotting, flow cytometry (FACS), and immunocyto- and immunohistochemistry [11, 138, 139]. Assay methods using the antibody for γ-H2AX can be divided into two types, those counting γ-H2AX foci in images of cells and tissues, and those measuring overall γ-H2AX protein levels.

The numbers of γ-H2AX foci have been found to correlate closely with the numbers of DSBs and increase linearly with stress dose [3, 9, 140]. In general, the γ-H2AX focus formation assay by immunocyto- or immunohistochemistry is very sensitive [141], detecting responses to as little as 1.2 mGy, equivalent to an average of 0.2 foci per cell [142]. However there are limitations to counting foci, especially at high doses of stress. Once γ-H2AX foci begin to overlap within the nuclear volume, one γ-H2AX focus may contain more than one DSB [143].

Other assays measure total amounts of γ-H2AX in a cell population by various immunoblotting procedures or on a per cell basis by FACS. FACS analysis of γ-H2AX has a detection level of 0.1–10 Gy and yields cell cycle information that is useful when studying drugs that preferentially act on S-phase cells [55, 144, 145]. Immunoblotting procedures are performed on extracted proteins from cells or tissues. As the amount of H2AX can vary from 2–20 percent of total H2A in different cells and tissues and since the amount of γ-H2AX formed per DSB is a percentage of total H2AX [146], total H2AX levels should also be determined when comparing different cell lines. Detection of γ-H2AX has been applied successfully to many human materials including peripheral blood mononuclear cells, tissues, and skin biopsies to detect DNA damage produced by various stresses [41, 53, 86, 141, 142, 147, 148].

CONCLUSIONS

Not only do living organisms need to be able to thrive in an optimal environment, they must be able to survive in the presence of multiple stresses. Proteins that may be expendable in a laboratory setting are essential to life in a stressful world. Organisms that have the most robust machinery to handle stresses are likely to survive in a variety of environments. It is accepted that the maintenance of genome integrity is an essential task, but the complexity of this task may still be underestimated. Elucidating the components required to overcome various harsh conditions, is an ongoing goal of stress research.

ACKNOWLEDGEMENTS

This work was supported by the intramural funding of the National Cancer Institute.

REFERENCES

1. McKinnon PJ, Caldecott KW. DNA strand break repair and human genetic disease. *Annu Rev Genomics Hum Genet* 2007;**8**:37–55.
2. Hakem R. DNA-damage repair; the good, the bad, and the ugly. *EMBO J* 2008;**27**:589–605.
3. Rogakou EP, Boon C, Redon C, Bonner WM. Megabase chromatin domains involved in DNA double-strand breaks *in vivo*. *J Cell Biol* 1999;**146**:905–16.
4. Paull TT, Rogakou EP, Yamazaki V, et al. A critical role for histone H2AX in recruitment of repair factors to nuclear foci after DNA damage. *Curr Biol* 2000;**10**:886–95.
5. Shroff R, Arbel-Eden A, Pilch D, et al. Distribution and dynamics of chromatin modification induced by a defined DNA double-strand break. *Curr Biol* 2004;**14**:1703–11.
6. Berkovich E, Monnat Jr RJ, Kastan MB. Roles of ATM and NBS1 in chromatin structure modulation and DNA double-strand break repair. *Nat Cell Biol* 2007;**9**:683–90.
7. Bartova E, Krejci J, Harnicarova A, et al. Histone modifications and nuclear architecture: a review. *J Histochem Cytochem* 2008; **56**:711–21.
8. Kurz EU, Lees-Miller SP. DNA damage-induced activation of ATM and ATM-dependent signaling pathways. *DNA Repair (Amst)* 2004;**3**:889–900.
9. Sedelnikova OA, Rogakou EP, Panyutin IG, Bonner WM. Quantitative detection of (125)IdU-induced DNA double-strand breaks with gamma-H2AX antibody. *Radiat Res* 2002;**158**:486–92.
10. Kinner A, Wu W, Staudt C, Iliakis G. Gamma-H2AX in recognition and signaling of DNA double-strand breaks in the context of chromatin. *Nucleic Acids Res* 2008;**36**:5678–94.
11. Bonner WM, Redon CE, Dickey JS, et al. g-H2AX and Cancer. *Nat Rev Cancer* 2008;**8**:957–67.
12. Salganik RI. The benefits and hazards of antioxidants: controlling apoptosis and other protective mechanisms in cancer patients and the human population. *J Am Coll Nutr* 2001;**20**:464S–472S. discussion 473S–475S.
13. Gruber J, Schaffer S, Halliwell B. The mitochondrial free radical theory of ageing: where do we stand? *Front Biosci* 2008;**13**:6554–7659.
14. Klaunig JE, Kamendulis LM. The role of oxidative stress in carcinogenesis. *Annu Rev Pharmacol Toxicol* 2004;**44**:239–267.
15. Ott M, Gogvadze V, Orrenius S, Zhivotovsky B. Mitochondria, oxidative stress and cell death. *Apoptosis* 2007;**12**:913–922.
16. Hussain SP, Hofseth LJ, Harris CC. Radical causes of cancer. *Nat Rev Cancer* 2003;**3**:276–285.
17. Hagen U. Mechanisms of induction and repair of DNA double-strand breaks by ionizing radiation: some contradictions. *Radiat Environ Biophys* 1994;**33**:45–61.
18. Bonner WM. Phenomena leading to cell survival values which deviate from linear-quadratic models. *Mutat Res* 2004;**568**:33–39.
19. Kielbassa C, Roza L, Epe B. Wavelength dependence of oxidative DNA damage induced by UV and visible light. *Carcinogenesis* 1997;**18**:811–816.
20. Tanaka T, Halicka HD, Huang X, et al. Constitutive histone H2AX phosphorylation and ATM activation, the reporters of DNA damage by endogenous oxidants. *Cell Cycle* 2006;**5**:1940–1945.
21. Pollycove M, Feinendegen LE. Radiation-induced versus endogenous DNA damage: possible effect of inducible protective responses in mitigating endogenous damage. *Hum Exp Toxicol* 2003;**22**:290–306. discussion 307,315–317,319–323.
22. Berlett BS, Stadtman ER. Protein oxidation in aging, disease, and oxidative stress. *J Biol Chem* 1997;**272**:20,313–20,316.
23. Takahashi A, Matsumoto H, Nagayama K, et al. Evidence for the involvement of double-strand breaks in heat-induced cell killing. *Cancer Res* 2004;**64**:8839–8845.
24. Kultz D, Chakravarty D. Hyperosmolality in the form of elevated NaCl but not urea causes DNA damage in murine kidney cells. *Proc Natl Acad Sci U S A* 2001;**98**:1999–2004.
25. To KK, Sedelnikova OA, Samons M, et al. The phosphorylation status of PAS-B distinguishes HIF-1alpha from HIF-2alpha in NBS1 repression. *EMBO J* 2006;**25**:4784–4794.
26. Xiao H, Li TK, Yang JM, Liu LF. Acidic pH induces topoisomerase II-mediated DNA damage. *Proc Natl Acad Sci USA* 2003;**100**:5205–5210.
27. Belyaev IY, Hillert L, Protopopova M, et al. 915 MHz microwaves and 50 Hz magnetic field affect chromatin conformation and 53BP1 foci in human lymphocytes from hypersensitive and healthy persons. *Bioelectromagnetics* 2005;**26**:173–184.
28. Belyaev IY, Markova E, Hillert L, et al. Microwaves from UMTS/GSM mobile phones induce long-lasting inhibition of 53BP1/gamma-H2AX DNA repair foci in human lymphocytes. *Bioelectromagnetics* 2009;**30**:129–141.
29. Markova E, Hillert L, Malmgren L, et al. Microwaves from GSM mobile telephones affect 53BP1 and gamma-H2AX foci in human lymphocytes from hypersensitive and healthy persons. *Environ Health Perspect* 2005;**113**:1172–1177.
30. Bristow RG, Hill RP. Hypoxia and metabolism. Hypoxia, DNA repair and genetic instability. *Nat Rev Cancer* 2008;**8**:180–192.
31. Edry E, Melamed D. Class switch recombination: a friend and a foe. *Clin Immunol* 2007;**123**:244–251.
32. Soulas-Sprauel P, Rivera-Munoz P, Malivert L, et al. V(D)J and immunoglobulin class switch recombinations: a paradigm to study the regulation of DNA end-joining. *Oncogene* 2007;**26**:7780–7791.

33. Chicheportiche A, Bernardino-Sgherri J, de Massy B, Dutrillaux B. Characterization of Spo11-dependent and independent phospho-H2AX foci during meiotic prophase I in the male mouse. *J Cell Sci* 2007;**120**:1733–1742.
34. Oettinger MA, Schatz DG, Gorka C, Baltimore D. RAG-1 and RAG-2, adjacent genes that synergistically activate V(D)J recombination. *Science* 1990;**248**:1517–1523.
35. Revy P, Muto T, Levy Y, et al. Activation-induced cytidine deaminase (AID) deficiency causes the autosomal recessive form of the Hyper-IgM syndrome (HIGM2). *Cell* 2000;**102**:565–575.
36. Takahashi A, Ohnishi T. Does gammaH2AX foci formation depend on the presence of DNA double strand breaks? *Cancer Lett* 2005;**229**:171–179.
37. Burhans WC, Weinberger M. DNA replication stress, genome instability and aging. *Nucleic Acids Res* 2007;**35**:7545–7556.
38. Furuta T, Takemura H, Liao ZY, et al. Phosphorylation of histone H2AX and activation of Mre11, Rad50, and Nbs1 in response to replication-dependent DNA double-strand breaks induced by mammalian DNA topoisomerase I cleavage complexes. *J Biol Chem* 2003;**278**:20,303–20,312.
39. Hofseth LJ, Khan MA, Ambrose M, et al. The adaptive imbalance in base excision-repair enzymes generates microsatellite instability in chronic inflammation. *J Clin Invest* 2003;**112**:1887–1894.
40. Nowosielska A, Marinus MG. DNA mismatch repair-induced double-strand breaks. *DNA Repair (Amst)* 2008;**7**:48–56.
41. Sedelnikova OA, Horikawa I, Redon C, et al. Delayed kinetics of DNA double-strand break processing in normal and pathological aging. *Aging Cell* 2008;**7**:89–100.
42. Yu T, MacPhail SH, Banath JP, et al. Endogenous expression of phosphorylated histone H2AX in tumors in relation to DNA double-strand breaks and genomic instability. *DNA Repair (Amst)* 2006;**5**:935–946.
43. Sharma S, Brosh RM. Human RECQ1 Is a DNA Damage Responsive Protein Required for Genotoxic Stress Resistance and Suppression of Sister Chromatid Exchanges. *PLoS ONE* 2007;**2**:e1297.
44. Cheadle JP, Sampson JR. Exposing the MYtH about base excision repair and human inherited disease. *Hum Mol Genet* 2003;**12**(Spec No 2):R159–R165.
45. Kraemer KH, Patronas NJ, Schiffmann R, et al. Xeroderma pigmentosum, trichothiodystrophy and Cockayne syndrome: a complex genotype-phenotype relationship. *Neuroscience* 2007;**145**:1388–1396.
46. Andressoo JO, Hoeijmakers JH, Mitchell JR. Nucleotide excision repair disorders and the balance between cancer and aging. *Cell Cycle* 2006;**5**:2886–2888.
47. Rustgi AK. The genetics of hereditary colon cancer. *Genes Dev* 2007;**21**:2525–2538.
48. von Zglinicki T. Oxidative stress shortens telomeres. *Trends Biochem Sci* 2002;**27**:339–344.
49. Takai H, Smogorzewska A, de Lange T. DNA damage foci at dysfunctional telomeres. *Curr Biol* 2003;**13**:1549–1556.
50. Nakamura AJ, Chiang YJ, Hathcock KS, et al. Both telomeric and non-telomeric DNA damage are determinants of mammalian cellular senescence. *Epigenetics & Chromatin* 2008;**1**:6.
51. d'Adda di Fagagna F, Reaper PM, Clay-Farrace L, et al. A DNA damage checkpoint response in telomere-initiated senescence. *Nature* 2003;**426**:194–198.
52. Herbig U, Jobling WA, Chen BP, et al. Telomere shortening triggers senescence of human cells through a pathway involving ATM, p53, and p21(CIP1), but not p16(INK4a). *Mol Cell* 2004;**14**:501–513.
53. Sedelnikova OA, Horikawa I, Zimonjic DB, et al. Senescing human cells and ageing mice accumulate DNA lesions with unrepairable double-strand breaks. *Nat Cell Biol* 2004;**6**:168–170.
54. van Steensel B, Smogorzewska A, de Lange T. TRF2 protects human telomeres from end-to-end fusions. *Cell* 1998;**92**:401–413.
55. Ismail IH, Wadhra TI, Hammarsten O. An optimized method for detecting gamma-H2AX in blood cells reveals a significant inter-individual variation in the gamma-H2AX response among humans. *Nucleic Acids Res* 2007;**35**:e36.
56. Brenner DJ, Ward JF. Constraints on energy deposition and target size of multiply damaged sites associated with DNA double-strand breaks. *Int J Radiat Biol* 1992;**61**:737–748.
57. Georgakilas AG. Processing of DNA damage clusters in human cells: current status of knowledge. *Mol Biosyst* 2008;**4**:30–35.
58. Fortune JM, Osheroff N. Topoisomerase II as a target for anticancer drugs: when enzymes stop being nice. *Prog Nucleic Acid Res Mol Biol* 2000;**64**:221–253.
59. Pommier Y, Topoisomerase I. inhibitors: camptothecins and beyond. *Nat Rev Cancer* 2006;**6**:789–802.
60. Sorenson CM, Eastman A. Mechanism of cis-diamminedichloroplatin um(II)-induced cytotoxicity: role of G2 arrest and DNA double-strand breaks. *Cancer Res* 1988;**48**:4484–4488.
61. Pfeifer GP, You YH, Besaratinia A. Mutations induced by ultraviolet light. *Mutat Res* 2005;**571**:19–31.
62. Garinis GA, Mitchell JR, Moorhouse MJ, et al. Transcriptome analysis reveals cyclobutane pyrimidine dimers as a major source of UV-induced DNA breaks. *EMBO J* 2005;**24**:3952–3962.
63. Wondrak GT, Jacobson MK, Jacobson EL. Endogenous UVA-photosensitizers: mediators of skin photodamage and novel targets for skin photoprotection. *Photochem Photobiol Sci* 2006;**5**:215–237.
64. Limoli CL, Giedzinski E, Bonner WM, Cleaver JE. UV-induced replication arrest in the xeroderma pigmentosum variant leads to DNA double-strand breaks, gamma-H2AX formation, and Mre11 relocalization. *Proc Natl Acad Sci USA* 2002;**99**:233–238.
65. Takahashi A, Aoshiba K, Nagai A. Apoptosis of wound fibroblasts induced by oxidative stress. *Exp Lung Res* 2002;**28**:275–284.
66. Tarakanova VL, Leung-Pineda V, Hwang S, et al. Gamma-herpesvirus kinase actively initiates a DNA damage response by inducing phosphorylation of H2AX to foster viral replication. *Cell Host Microbe* 2007;**1**:275–286.
67. Daniel R, Ramcharan J, Rogakou E, et al. Histone H2AX is phosphorylated at sites of retroviral DNA integration but is dispensable for postintegration repair. *J Biol Chem* 2004;**279**:45,810–45,814.
68. Mothersill C, Seymour C. Radiation-induced bystander effects: are they good, bad or both? *Med Confl Surviv* 2005;**21**:101–110.
69. Tartier L, Gilchrist S, Burdak-Rothkamm S, et al. Cytoplasmic Irradiation Induces Mitochondrial-Dependent 53BP1 Protein Relocalization in Irradiated and Bystander Cells. *Cancer Res* 2007;**67**:5872–5879.
70. Shao C, Folkard M, Michael BD, Prise KM. Targeted cytoplasmic irradiation induces bystander responses. *Proc Natl Acad Sci USA* 2004; **101**:13,495–13,500,.
71. Nagar S, Morgan WF. The death-inducing effect and genomic instability. *Radiat Res* 2005;**163**:316–323.
72. Shao C, Stewart V, Folkard M, et al. Nitric oxide-mediated signaling in the bystander response of individually targeted glioma cells. *Cancer Res* 2003;**63**:8437–8442.
73. Sokolov MV, Smilenov LB, Hall EJ, et al. Ionizing radiation induces DNA double-strand breaks in bystander primary human fibroblasts. *Oncogene* 2005;**24**:7257–7265.
74. Shao C, Folkard M, Prise KM. Role of TGF-beta1 and nitric oxide in the bystander response of irradiated glioma cells. *Oncogene* 2008;**27**:434–440.
75. Han W, Wu L, Chen S, et al. Constitutive nitric oxide acting as a possible intercellular signaling molecule in the initiation of radiation-

induced DNA double strand breaks in non-irradiated bystander cells. *Oncogene* 2007;**26**:2330–2339.
76. Kashino G, Suzuki K, Matsuda N, et al. Radiation induced bystander signals are independent of DNA damage and DNA repair capacity of the irradiated cells. *Mutat Res* 2007;**619**:134–138.
77. Prise KM, Folkard M, Kuosaite V, et al. What role for DNA damage and repair in the bystander response? *Mutat Res* 2006;**597**:1–4.
78. Burdak-Rothkamm S, Rothkamm K, Prise KM. ATM acts downstream of ATR in the DNA damage response signaling of bystander cells. *Cancer Res* 2008;**68**:7059–7065.
79. Burdak-Rothkamm S, Short SC, Folkard M, et al. ATR-dependent radiation-induced gamma H2AX foci in bystander primary human astrocytes and glioma cells. *Oncogene* 2007;**26**:993–1002.
80. Sokolov MV, Dickey JS, Bonner WM, Sedelnikova OA. gamma-H2AX in bystander cells: not just a radiation-triggered event, a cellular response to stress mediated by intercellular communication. *Cell Cycle* 2007;**6**:2210–2212.
81. Yang H, Asaad N, Held KD. Medium-mediated intercellular communication is involved in bystander responses of X-ray-irradiated normal human fibroblasts. *Oncogene* 2005;**24**:2096–2103.
82. Zhang Y, Zhou J, Held KD, et al. Deficiencies of double-strand break repair factors and effects on mutagenesis in directly gamma-irradiated and medium-mediated bystander human lymphoblastoid cells. *Radiat Res* 2008;**169**:197–206.
83. Hu B, Wu L, Han W, et al. The time and spatial effects of bystander response in mammalian cells induced by low dose radiation. *Carcinogenesis* 2006;**27**:245–251.
84. Cucinotta FA, Pluth JM, Anderson JA, et al. Biochemical kinetics model of DSB repair and induction of gamma-H2AX foci by non-homologous end joining. *Radiat Res* 2008;**169**:214–222.
85. Smilenov LB, Hall EJ, Bonner WM, Sedelnikova OAA. microbeam study of DNA double-strand breaks in bystander primary human fibroblasts. *Radiat Prot Dosimetry* 2006;**122**:256–259.
86. Sedelnikova OA, Nakamura A, Kovalchuk O, et al. DNA double-strand breaks form in bystander cells after microbeam irradiation of three-dimensional human tissue models. *Cancer Res* 2007;**67**:4295–4302.
87. Celeste A, Petersen S, Romanienko PJ, et al. Genomic instability in mice lacking histone H2AX. *Science* 2002;**296**:922–927.
88. Bogliolo M, Lyakhovich A, Callen E, et al. Histone H2AX and Fanconi anemia FANCD2 function in the same pathway to maintain chromosome stability. *Embo J* 2007;**26**:1340–1351.
89. Meador JA, Zhao M, Su Y, et al. Histone H2AX is a critical factor for cellular protection against DNA alkylating agents. *Oncogene* 2008;**27**:5662–5671.
90. Lu C, Zhu F, Cho YY, et al. Cell apoptosis: requirement of H2AX in DNA ladder formation, but not for the activation of caspase-3. *Mol Cell* 2006;**23**:121–132.
91. Pilch DR, Sedelnikova OA, Redon C, et al. Characteristics of gamma-H2AX foci at DNA double-strand breaks sites. *Biochem Cell Biol* 2003;**81**:123–129.
92. Zha S, Sekiguchi J, Brush JW, et al. Complementary functions of ATM and H2AX in development and suppression of genomic instability. *Proc Natl Acad Sci USA* 2008;**105**:9302–9306.
93. Morio T, Kim H. Ku, Artemis, and ataxia-telangiectasia-mutated: signalling networks in DNA damage. *Int J Biochem Cell Biol* 2008;**40**:598–603.
94. Durocher D, Jackson SP. DNA-PK, ATM and ATR as sensors of DNA damage: variations on a theme? *Curr Opin Cell Biol* 2001;**13**:225–231.
95. Yang J, Yu Y, Hamrick HE, Duerksen-Hughes PJ. ATM, ATR and DNA-PK: initiators of the cellular genotoxic stress responses. *Carcinogenesis* 2003;**24**:1571–1580.
96. Shiloh Y. ATM and ATR: networking cellular responses to DNA damage. *Curr Opin Genet Dev* 2001;**11**:71–77.
97. Stucki M, Clapperton JA, Mohammad D, et al. MDC1 directly binds phosphorylated histone H2AX to regulate cellular responses to DNA double-strand breaks. *Cell* 2005;**123**:1213–1226.
98. Chapman JR, Jackson SP. Phospho-dependent interactions between NBS1 and MDC1 mediate chromatin retention of the MRN complex at sites of DNA damage. *EMBO Rep* 2008;**9**:795–801.
99. Yan J, Jetten AM. RAP80 and RNF8, key players in the recruitment of repair proteins to DNA damage sites. *Cancer Lett* 2008;**271**:179–190.
100. Huen MS, Grant R, Manke I, et al. RNF8 transduces the DNA-damage signal via histone ubiquitylation and checkpoint protein assembly. *Cell* 2007;**131**:901–914.
101. Mailand N, Bekker-Jensen S, Faustrup H, et al. RNF8 ubiquitylates histones at DNA double-strand breaks and promotes assembly of repair proteins. *Cell* 2007;**131**:887–900.
102. Kolas NK, Chapman JR, Nakada S, et al. Orchestration of the DNA-damage response by the RNF8 ubiquitin ligase. *Science* 2007;**318**:1637–1640.
103. Celeste A, Fernandez-Capetillo O, Kruhlak MJ, et al. Histone H2AX phosphorylation is dispensable for the initial recognition of DNA breaks. *Nat Cell Biol* 2003;**5**:675–679.
104. Ward IM, Minn K, Jorda KG, Chen J. Accumulation of checkpoint protein 53BP1 at DNA breaks involves its binding to phosphorylated histone H2AX. *J Biol Chem* 2003;**278**:19,579–19,582.
105. Botuyan MV, Lee J, Ward IM, et al. Structural basis for the methylation state-specific recognition of histone H4-K20 by 53BP1 and Crb2 in DNA repair. *Cell* 2006;**127**:1361–1373.
106. Nakamura TM, Du LL, Redon C, Russell P. Histone H2A phosphorylation controls Crb2 recruitment at DNA breaks, maintains checkpoint arrest, and influences DNA repair in fission yeast. *Mol Cell Biol* 2004;**24**:6215–6230.
107. Sanders SL, Portoso M, Mata J, et al. Methylation of histone H4 lysine 20 controls recruitment of Crb2 to sites of DNA damage. *Cell* 2004;**119**:603–614.
108. Ikura T, Tashiro S, Kakino A, et al. DNA damage-dependent acetylation and ubiquitination of H2AX enhances chromatin dynamics. *Mol Cell Biol* 2007;**27**:7028–7040.
109. Fillingham J, Keogh MC, Krogan NJ. GammaH2AX and its role in DNA double-strand break repair. *Biochem Cell Biol* 2006;**84**:568–577.
110. Downs JA, Allard S, Jobin-Robitaille O, et al. Binding of chromatin-modifying activities to phosphorylated histone H2A at DNA damage sites. *Mol Cell* 2004;**16**:979–990.
111. Bird AW, Yu DY, Pray-Grant MG, et al. Acetylation of histone H4 by Esa1 is required for DNA double-strand break repair. *Nature* 2002;**419**:411–415.
112. van Attikum H, Fritsch O, Hohn B, Gasser SM. Recruitment of the INO80 complex by H2A phosphorylation links ATP-dependent chromatin remodeling with DNA double-strand break repair. *Cell* 2004;**119**:777–788.
113. Morrison AJ, Highland J, Krogan NJ, et al. INO80 and gamma-H2AX interaction links ATP-dependent chromatin remodeling to DNA damage repair. *Cell* 2004;**119**:767–775.
114. Kobor MS, Venkatasubrahmanyam S, Meneghini MD, et al. A protein complex containing the conserved Swi2/Snf2-related ATPase Swr1p deposits histone variant H2A.Z into euchromatin. *PLoS Biol* 2004;**2**:E131.
115. Wood JL, Singh N, Mer G, Chen J. MCPH1 functions in an H2AX-dependent but MDC1-independent pathway in response to DNA damage. *J Biol Chem* 2007;**282**:35,416–35,423.

116. Wang YQ, Su B. Molecular evolution of microcephalin, a gene determining human brain size. *Hum Mol Genet* 2004;**13**:1131–1137.
117. Evans PD, Anderson JR, Vallender EJ, et al. Reconstructing the evolutionary history of microcephalin, a gene controlling human brain size. *Hum Mol Genet* 2004;**13**:1139–1145.
118. Trimborn M, Bell SM, Felix C, et al. Mutations in microcephalin cause aberrant regulation of chromosome condensation. *Am J Hum Genet* 2004;**75**:261–266.
119. Bartek J. Microcephalin guards against small brains, genetic instability, and cancer. *Cancer Cell* 2006;**10**:91–93.
120. Chaplet M, Rai R, Jackson-Bernitsas D, et al. BRIT1/MCPH1: a guardian of genome and an enemy of tumors. *Cell Cycle* 2006;**5**:2579–2583.
121. Rai R, Dai H, Multani AS, et al. BRIT1 regulates early DNA damage response, chromosomal integrity, and cancer. *Cancer Cell* 2006;**10**:145–157.
122. Xu X, Lee J, Stern DF. Microcephalin is a DNA damage response protein involved in regulation of CHK1 and BRCA1. *J Biol Chem* 2004;**279**:34,091–34,094.
123. Lin SY, Rai R, Li K, et al. BRIT1/MCPH1 is a DNA damage responsive protein that regulates the Brca1-Chk1 pathway, implicating checkpoint dysfunction in microcephaly. *Proc Natl Acad Sci USA* 2005;**102**:15,105–15,109.
124. Alderton GK, Galbiati L, Griffith E, et al. Regulation of mitotic entry by microcephalin and its overlap with ATR signalling. *Nat Cell Biol* 2006;**8**:725–733.
125. Xie A, Puget N, Shim I, et al. Control of sister chromatid recombination by histone H2AX. *Mol Cell* 2004;**16**:1017–1025.
126. Xie A, Hartlerode A, Stucki M, et al. Distinct roles of chromatin-associated proteins MDC1 and 53BP1 in mammalian double-strand break repair. *Mol Cell* 2007;**28**:1045–1057.
127. Unal E, Arbel-Eden A, Sattler U, et al. DNA damage response pathway uses histone modification to assemble a double-strand break-specific cohesin domain. *Mol Cell* 2004;**16**:991–1002.
128. Redon C, Pilch DR, Bonner WM. Genetic analysis of Saccharomyces cerevisiae H2A serine 129 mutant suggests a functional relationship between H2A and the sister-chromatid cohesion partners Csm3-Tof1 for the repair of topoisomerase I-induced DNA damage. *Genetics* 2006;**172**:67–76.
129. Koshland DE, Guacci V. Sister chromatid cohesion: the beginning of a long and beautiful relationship. *Curr Opin Cell Biol* 2000;**12**:297–301.
130. Birkenbihl RP, Subramani S. Cloning and characterization of rad21 an essential gene of Schizosaccharomyces pombe involved in DNA double-strand-break repair. *Nucleic Acids Res* 1992;**20**:6605–6611.
131. Heo SJ, Tatebayashi K, Kato J, Ikeda H. The RHC21 gene of budding yeast, a homologue of the fission yeast rad21+ gene, is essential for chromosome segregation. *Mol Gen Genet* 1998;**257**:149–156.
132. Sonoda E, Matsusaka T, Morrison C, et al. Scc1/Rad21/Mcd1 is required for sister chromatid cohesion and kinetochore function in vertebrate cells. *Dev Cell* 2001;**1**:759–770.
133. Chowdhury D, Keogh MC, Ishii H, et al. gamma-H2AX dephosphorylation by protein phosphatase 2A facilitates DNA double-strand break repair. *Mol Cell* 2005;**20**:801–809.
134. Keogh MC, Kim JA, Downey M, et al. A phosphatase complex that dephosphorylates gammaH2AX regulates DNA damage checkpoint recovery. *Nature* 2006;**439**:497–501.
135. Chowdhury D, Xu X, Zhong X, et al. A PP4-phosphatase complex dephosphorylates gamma-H2AX generated during DNA replication. *Mol Cell* 2008;**31**:33–46.
136. Kusch T, Florens L, Macdonald WH, et al. Acetylation by Tip60 is required for selective histone variant exchange at DNA lesions. *Science* 2004;**306**:2084–2087.
137. van Attikum H, Gasser SM. ATP-dependent chromatin remodeling and DNA double-strand break repair. *Cell Cycle* 2005;**4**:1011–1014.
138. Pilch DR, Redon C, Sedelnikova OA, Bonner WM. Two-dimensional gel analysis of histones and other H2AX-related methods. *Methods Enzymol* 2004;**375**:76–88.
139. Nakamura A, Sedelnikova OA, Redon C, et al. Techniques for gamma-H2AX detection. *Methods Enzymol* 2006;**409**:236–250.
140. Rothkamm K, Lobrich M. Evidence for a lack of DNA double-strand break repair in human cells exposed to very low x-ray doses. *Proc Natl Acad Sci USA* 2003;**100**:5057–5062.
141. Redon C, Dickey JS, Bonner WM, Sedelnikova O. g-H2AX as a biomarker of DNA damage induced by ionizing radiation in human peripheral blood lymphocytes and artificial skin. *Advances in Space Research* 2009;**43**:1171–1178.
142. Lobrich M, Rief N, Kuhne M, et al. In vivo formation and repair of DNA double-strand breaks after computed tomography examinations. *Proc Natl Acad Sci USA* 2005;**102**:8984–8989.
143. Scherthan H, Hieber L, Braselmann H, Meineke V, Zitzelsberger H. Accumulation of DSBs in gamma-H2AX domains fuel chromosomal aberrations. *Biochem Biophys Res Commun* 2008;**371**:694–697.
144. Tanaka T, Huang X, Halicka HD, et al. Cytometry of ATM activation and histone H2AX phosphorylation to estimate extent of DNA damage induced by exogenous agents. *Cytometry A* 2007;**71**:648–661.
145. Olive PL, Banath JP, Keyes M. Residual gammaH2AX after irradiation of human lymphocytes and monocytes in vitro and its relation to late effects after prostate brachytherapy. *Radiother Oncol* 2008;**86**:336–346.
146. Rogakou EP, Pilch DR, Orr AH, et al. DNA double-stranded breaks induce histone H2AX phosphorylation on serine 139. *J Biol Chem* 1998;**273**:5858–5868.
147. Qvarnstrom OF, Simonsson M, Johansson KA, et al. DNA double strand break quantification in skin biopsies. *Radiother Oncol* 2004;**72**:311–317.
148. Rothkamm K, Balroop S, Shekhdar J, et al. Leukocyte DNA damage after multi-detector row CT: a quantitative biomarker of low-level radiation exposure. *Radiology* 2007;**242**:244–251.

Section C

Signaling to/from Intracellular Compartments

Chapter 32

Regulation of mRNA Turnover

Ann-Bin Shyu and Chyi-Ying A. Chen
Department of Biochemistry and Molecular Biology, University of Texas, Medical School at Houston, Houston, Texas

INTRODUCTION

Regulation of mRNA turnover is important for controlling the abundance of cellular transcripts and thus the levels of protein expression (for reviews, e.g., [1, 2]). RNA polymerase II transcripts, with the exception of histone mRNAs, contain a m^{7}G-cap structure at the 5′ termini and a poly(A) tail at the 3′ termini. The m^{7}G-cap, in conjunction with the cap binding protein complex, renders the mRNA 5′ termini resistant to 5′ to 3′ exonucleases [3]. The 3′ poly(A) tail along with the poly(A) binding protein (PABP) provides protection from ribonucleases' attack at the 3′ end of an mRNA [4]. Moreover, the 5′ cap/cap binding complex and the 3′ poly(A)/PABP complex can interact with each other to form a closed loop that enhances translation initiation and provides an effective means to protect mRNA ends from exonuclease attack [5].

In theory, mRNA degradation in the cytoplasm can be initiated by deadenylation (i.e., removal of the 3′ poly(A) tail), decapping (i.e., removal of the 5′ cap structure), or endonucleolytic cleavage within the message to generate 5′ and 3′ fragments. A newly created end would then be unprotected from subsequent exonuclease attack. However, mRNA decay pathways in eukaryotic cells involve multiple steps and are more complicated than simply creating free ends by the three theoretical ways followed by exonucleolytic digestion [1]. In this chapter, we describe what is known about how mRNA decay can be triggered and highlight some recent studies on regulation of mRNA turnover in mammalian cells.

CURRENT MODEL FOR MRNA DECAY IN MAMMALIAN CELLS

In eukaryotes, two major pathways for cytoplasmic mRNA decay begin with deadenylation with the predominant poly(A) nuclease being Ccr4-Caf1 complex [1]. Deadenylation primarily leads to decapping by Dcp1-Dcp2 complex at the 5′ end followed by the exonucleolytic digestion of the RNA body by Xrn1. Alternatively, mRNA body can also be degraded from the 3′ end after deadenylation by a large complex called exosome, especially when the 5′ to 3′ decay pathway is compromised. Deadenylation independent pathways of mRNA degradation also exist. For example, when an aberrant mRNA loses its stop codon and becomes a non-stop mRNA, exosomes are able to completely degrade the mRNA from the very 3′ end of the poly(A) tail [6, 7]. Also, in yeast, nonsense codons usually trigger rapid Dcp1-Dcp2 mediated decapping followed by 5′ to 3′ Xrn1 digestion of the RNA body before the poly(A) tail is significantly removed [8]. Nevertheless, mammalian nonsense mediated decay (NMD) is triggered by deadenylation but not by decapping [9]. Thus, deadenylation is the first major step triggering general mRNA decay in mammalian cells.

Recently, our group discovered that mammalian deadenylation is biphasic and is mediated by the consecutive activity of two different poly(A) nuclease complexes [10]. Poly(A) tails are first shortened to ~110 nucleotides (nt) by Pan2-Pan3 poly(A) nuclease complex. In the second phase of deadenylation, a complex containing Ccr4 and Caf1 catalyzes further shortening of the poly(A) tail to oligo(A). Ccr4 and Caf1 have complementary roles in this step. Decapping by the Dcp1-Dcp2 complex may occur during and/or after the second phase of deadenylation [10]. Our observations suggest that in mammalian cells Dcp2 directed decapping functions as a fail-safe mechanism to initiate mRNA decay if the first step, deadenylation, is compromised. Since deadenylation but not decapping is a reversible process, it is plausible that in mammals, deadenylation serves as an important checkpoint before an mRNA is committed to elimination during embryogenesis and cell growth and differentiation.

Handbook of Cell Signaling, Three-Volume Set 2 ed.

DEADENYLATION: THE FIRST MAJOR STEP TRIGGERING MRNA DECAY

Both translation and stability of mRNAs are affected by deadenylation [11], which occurs as soon as poly(A)$^+$ mRNAs arrive in the cytoplasm. Computational modeling of eukaryotic mRNA turnover indicates that changes in levels of mRNA are highly leveraged to the rate of deadenylation [12]. The importance of deadenylation in mRNA turnover can be observed in all major mRNA decay pathways yet recognized in mammals, including decay directed by AU-rich elements (AREs) in the 3′ UTR [13], the rapid decay mediated by destabilizing elements in protein coding regions [14], the surveillance mechanism that detects and degrades nonsense containing mRNA [9], and the default decay pathway for stable messages such as β-globin mRNA [15].

The importance of deadenylation in regulation of gene expression is further emphasized by two recent developments. First, a group of abundant, evolutionarily conserved small silencing RNAs termed microRNAs (miRNAs), are capable of promoting decay of their mRNA targets by accelerating deadenylation, thereby achieving gene silencing (e.g., [16]). Second, non-translatable mRNA-protein complexes (mRNPs) are found in RNA processing bodies (P-bodies), a newly discovered cytoplasmic domain related to translation repression and mRNA decay (see below) [17, 18]. Because one major consequence of deadenylation is the formation of non-translatable mRNPs [5, 11], this observation suggests that deadenylation leads to a remodeling of mRNPs necessary for them to subsequently enter existing or form new P-bodies.

REGULATION OF DEADENYLATION BY A PROTEIN THAT INTERACTS WITH BOTH POLY(A) NUCLEASE(S) AND PABP

In spite of the importance of deadenylation, relatively little is known about the mechanisms that control this process and its participating factors. Several observations indicate that PABP has potential to regulate the activities of poly(A) nucleases and decapping enzymes (e.g., [19, 20]). Thus, cytoplasmic PABP may play a key role in deadenylation by modulating the activities of Pan2-Pan3 and Ccr4-Caf1 poly(A) nucleases.

PABPs are known to play crucial roles in mRNA metabolism through their binding to the 3′ poly(A) tails of newly synthesized and mature mRNAs [4, 21]. The overall structure of PABP is highly conserved in eukaryotes and consists of four RNA recognition motifs (RRMs) connected to a large carboxy-terminal helical domain [4]. In addition to its highly conserved overall domain structure, the importance of PABP in mRNA decay and translation is also revealed by its ability to interact with at least 11 different proteins known to be involved in mRNA decay and translation [21, 22]. One key feature in its C-terminal region is the presence of a peptide interacting PABC domain that recruits proteins containing a highly specific PABP interacting motif 2 (PAM-2 motif) to the 3′ poly(A)/PABP complex of mRNPs [23]. In the cytoplasm, PABPs facilitate formation of a "closed loop" structure of the mRNP particle that is crucial for PABPs to promote translation initiation, termination, possibly recycling of ribosomes, and mRNA stability [5]. In many cases, mRNA decay occurs concurrently with the termination of PABP's role in enhancing translation initiation [4, 11]. Thus, PABP's role in mRNA stability is more complicated than merely protecting the 3′ poly(A) tail from being attacked by the 3′ to 5′ exonuclease.

Given that PABPs can interact with an array of binding factors involved in mRNA translation and decay, the interaction between poly(A) nucleases and PABP as well as factors that mediate or prohibit such an interaction could form the molecular basis for regulation of deadenylation. In order to identify potential regulatory factors involved in mammalian deadenylation, literature and database searches were carried out with a focus on proteins that have the potential to interact with a poly(A) nuclease and PABP. A family of anti-proliferation genes termed the *tob/btg* family [24] emerged from the searches because they contain a highly conserved N-terminal domain that can interact with the Caf1 poly(A) nuclease. In humans, *tob (tob1)* and *tob2* also encode a C-terminal domain containing two putative PABP interacting PAM2 motifs. Increasing evidence suggests that Tob proteins are involved in negative control of cell growth and can function as tumor suppressors (reviewed in [24]). Moreover, Tob is highly expressed in anergic T cell clones and in unstimulated peripheral blood T lymphocytes [25]. The ability of Tob to maintain T cell quiescence is thought to be due to its modulation of transcription [25]. Despite the fact that Tob proteins have been known for a decade to function in anti-proliferation and potentially in transcriptional control, the biochemical and molecular mechanisms by which they exert their functions remains unclear.

Using Tet-off transcriptional pulsing approach to investigate mRNA decay in mouse NIH3T3 fibroblasts [26], we recently found that ectopic expression of Tob1 or Tob2 enhanced mRNA deadenylation *in vivo* [27]. Results from GST pull-down, co-immunoprcipitation (IP), and gel-shift experiments demonstrate that Tob proteins can simultaneously interact with the Ccr4-Caf1 complex and the poly(A)-PABP complex [27]. Combining with mutagenesis studies, we further identified the motifs on Tob proteins and on PABP that are necessary for their interaction. When the PABP interaction motif on Tob was mutated, the mutant Tob not only could not interact with PABP effectively but also could not enhance deadenylation [27]. Thus, interaction with PABP is necessary for the deadenylation enhancing effect of Tob.

Immunoblotting has detected endogenous Tob proteins in NIH3T3 cells in G0/G1 phase but not in G2 and M phases [28]. Intriguingly, we found in NIH3T3 cells that Tob2 localizes to P-bodies, cytoplasmic domains where translationally repressed mRNPs may accumulate or be degraded [27], in a manner correlating with its cell cycle regulated expression. Our observations suggest that Tob proteins may achieve their anti-proliferation function in the G0/G1 phase by promoting deadenylation *via* their ability to deliver the Ccr4-Caf1 poly(A) nuclease complex to the 3′ poly(A)-PABP complex, thus facilitating remodeling of mRNPs and their subsequent movement into P-bodies. These findings reveal a novel role for Tob proteins in regulating cytoplasmic deadenylation. We identify a new mechanism by which the fate of mammalian mRNA is controlled at the deadenylation step by a protein that interacts with both poly(A) nuclease(s) and PABP and has the potential to direct poly(A) shortened mRNA intermediates to P-bodies.

A MECHANISM FOR TRANSLATIONALLY COUPLED MRNA TURNOVER

The linkage between translation and mRNA turnover plays an important role in regulating gene expression. The poly(A) tail has been shown to stimulate both cap dependent and cap independent translation initiation in the cytoplasm [5]. On the other hand, a large set of observations point to an important role for translation in the mRNA decay process [11, 29]. Here, we use the mRNA decay mediated by the c-*fos* major coding region determinant of instability (mCRD) as an example to describe a novel mechanism by which interplay between translation and mRNA turnover determines the lifespan of an mRNA in the cytoplasm.

The c-*fos* mCRD is located in the central portion of the protein coding region of the c-*fos* proto oncogene transcript [30]. While it has been shown that loss of poly(A) tail profoundly reduces the efficiency of translation initiation, the rapid decay directed by the mCRD requires ribosome transit, and deadenylation is a necessary first step coupled to translation [15, 30]. These seemingly contradictory observations were reconciled by our findings that a minimal spacer sequence must be maintained between the *c-fos* mCRD and the poly(A) tail to allow the formation of a unique mCRD complex, whose mRNA destabilizing function is modulated by ribosome transit [14].

To elucidate the mechanism underlying the mCRD mediated RNA decay, our group identified UNR, a cold shock domain containing RNA binding protein, as a factor that binds directly to both mCRD and PABP [31]. We detected a physical interaction between the poly(A)/PABP complex and the mCRD/UNR complex both *in vitro* and *in vivo*. Interference with this interaction diminishes the mCRD destabilizing function. Moreover, knocking down UNR *via* RNA interference (RNAi) resulted in retardation of deadenylation of the mCRD containing mRNA. To test the role of translation initiation in mCRD mediated RNA decay, we introduced a strong hairpin structure (hp) into the 5′ UTR to block translation initiation of a mCRD containing reporter mRNA. Polysome profile analysis by sucrose gradient fractionation confirmed the blockage of translation initiation, and Northern blot analysis showed that deadenylation of this mRNA was dramatically impeded, leading to stabilization of the mRNA. We further identified human Ccr4 as a poly(A) nuclease that associates with UNR and is involved in mCRD mediated mRNA decay. Collectively, our findings support a model in which mCRD/UNR serves as a "landing/assembly" platform for formation of a deadenylation/decay mRNP complex involving PABP and the Ccr4 poly(A) nuclease. The deadenylation/decay complex is kept in a dormant state prior to translation. Ribosome transit is required in order to alter this interaction and activate the nuclease, leading to deadenylation and subsequent decay of the RNA body. This novel mechanism represents another illustration of the importance of mRNP remodeling in regulating mRNA turnover.

THE INVOLVEMENT OF RNA PROCESSING BODIES (P-BODIES) IN REGULATION OF MRNA TURNOVER

P-bodies are newly discovered cytoplasmic foci that contain proteins known to function in mRNA metabolism (reviewed in [17, 18]). These foci are also referred to as GW bodies because they carry GW182 proteins that are involved in miRNA mediated translation repression [32]. Translationally repressed mRNPs, factors involved in translational repression including miRNA mediated translational silencing proteins (such as Argonaute proteins and Rck/p54), Dcp1/2 decapping enzymes, and a 5′ to 3′ exonuclease Xrn1 are also found in these foci. In contrast, ribosomes, PABPs, and all translation initiation factors, with the exception of eIF4E (the mRNA 5′ cap binding protein), are not found in P-bodies. Thus, P-bodies are considered to be sites where translationally repressed mRNAs in conjunction with the 5′ to 3′ decay machinery and other translation repressors may be sequestered and/or degraded. In addition, mRNAs subject to NMD [33] and miRNA targeted mRNA silencing are also linked to P-bodies [18, 34]. Global inhibition of miRNA biogenesis and general mRNA decay strongly decreases the number and size of P-bodies in HeLa cells [35]. Nevertheless, mRNAs accumulating in P-bodies are not necessarily destined for degradation. mRNAs can exit P-bodies and re-enter translation following general reactivation of cellular protein synthesis or in response to environmental stimuli [36, 37]. One possibility is that concentrating repressed mRNPs in P-bodies facilitates additional mRNP remodeling steps that reinforce this repression for long term storage in a repressed form. In other cases, these remodeling events may trigger more efficient mRNA degradation. Sequestration in P-bodies may also provide a rapid means to prevent accidental translation of aberrant mRNAs, such as nonsense containing transcripts, prior to degradation.

CONCLUDING REMARKS

It has become clear from numerous studies that mRNA turnover is not simply a "default" pathway secondary to the regulated process of transcription. Instead, mRNA turnover is a complex set of cellular processes regulated by mechanisms independent from transcription. Once transcribed, mRNAs associate with a host of protein factors throughout their lifetimes, some of which are stably bound while others are subject to dynamic exchange. Individual mRNP components may serve as adaptors that allow mRNAs to interface with the machineries mediating their subcellular localization, translation, or decay [38, 39]. Thus, mRNP remodeling plays a critical role in determining the fate of an mRNA in the cytoplasm although mechanisms of mRNP remodeling remain unclear. For example, as discussed above, studies in the c-*fos* mCRD suggest that protein–mRNA interaction and mRNP remodeling are involved in recognizing and degrading the message [14]. Also, because P-bodies can function as temporary storage sites for translationally repressed mRNPs, their numbers and morphology appear to change depending on cellular mRNA decay and translation activity (reviewed in [17, 18]). It is unclear what changes in mRNP composition are necessary for mRNPs to enter existing P-bodies, nucleate formation of P-bodies, or be released from P-bodies. An important area of future work will be to decipher the dynamic changes in mRNP composition throughout the lifetimes of different mRNAs under various conditions (e.g., during poly(A) shortening, exiting the translating pool, etc.) and also to learn how the process of mRNP remodeling is regulated during cell development, growth, and differentiation. Since many human diseases (e.g., cancers, autoimmune diseases, allergic inflammation, etc.) are associated with an alteration in the level of gene expression controlled by mRNA stability, studies on these crucial events for regulation of mRNA turnover have the potential to unravel new mechanisms underlying these conditions and thus facilitate the development of novel therapeutic agents.

REFERENCES

1. Parker R, Song H. The enzymes and control of eukaryotic mRNA turnover. *Nat Strut Mol Biol* 2004;**11**:121–7.
2. Wilusz CW, Wormington M, Peltz SW. The cap-to-tail guide to mRNA turnover. *Nat Rev Mol Cell Biol* 2001;**2**:237–46.
3. Muhlrad D, Decker CJ, Parker R. Deadenylation of the unstable mRNA encoded by the yeast MFA2 gene leads to decapping followed by 5′→3′ digestion of the transcript. *Genes Dev* 1994;**8**:855–66.
4. Mangus DA, Evans MC, Jacobson A. Poly(A)-binding proteins: multifunctional scaffolds for the post-transcriptional control of gene expression. *Genome Biol* 2003;**4**:233.
5. Jacobson A. Poly(A) metabolism and translation: the colsed-loop model. In: Hershey JWB, Mathews MB, Sonenberg N, editors. *Tanslational control*. Plainview, NY: Cold Spring Harbor Laboratory Press; 1996. p. 451–80.
6. Frischmeyer PA, van Hoof A, O'Donnell K, et al. An mRNA surveillance mechanism that eliminates transcripts lacking termination codons. *Science* 2002;**295**:2258–61.
7. van Hoof A, Frischmeyer PA, Dietz HC, Parker R. Exosome-mediated recognition and degradation of mRNAs lacking a termination codon. *Science* 2002;**295**:2262–4.
8. Muhlrad D, Parker R. Premature translational termination triggers mRNA decapping. *Nature* 1994;**370**:578–81.
9. Chen C-YA, Shyu A-B. Rapid deadenylation triggered by a nonsense codon precedes decay of the RNA body in a mammalian cytoplasmic nonsense-mediated decay pathway. *Mol Cell Biol* 2003;**23**:4805–13.
10. Yamashita A, Chang TC, Yamashita Y, et al. Concerted action of poly(A) nucleases and decapping enzyme in mammalian mRNA turnover. *Nat Struct Mol Biol* 2005;**12**:1054–63.
11. Jacobson A, Peltz SW. Interrelationships of the pathways of mRNA decay and translation in eukaryotic cells. *Annu Rev Biochem* 1996;**65**:693–739.
12. Gao D, Parker R. Computational modeling and experimental analysis of nonsense-mediated decay in yeast. *Cell* 2003;**113**:545–53.
13. Chen AC-Y, Shyu A-B. AU-rich elements: characterization and importance in mRNA degradation. *Trends Biochem Sci* 1995;**20**:465–70.
14. Grosset C, Chen C-YA, Xu N, et al. A mechanism for translationally coupled mRNA turnover: interaction between the poly(A) tail and a c-fos RNA coding determinant via a protein complex. *Cell* 2000;**103**:29–40.
15. Shyu A-B, Belasco JG, Greenberg MG. Two distinct destabilizing elements in the c-fos message trigger deadenylation as a first step in rapid mRNA decay. *Genes Dev* 1991;**5**:221–32.
16. Wu L, Fan J, Belasco JG. MicroRNAs direct rapid deadenylation of mRNA. *PNAS* 2006;**103**:4034–9.
17. Eulalio A, Behm-Ansmant I, Izaurralde E. P bodies: at the crossroads of post-transcriptional pathways. *Nat Rev Mol Cell Biol* 2007;**8**:9–22.
18. Parker R, Sheth U. P bodies and the control of mRNA translation and degradation. *Mol Cell* 2007;**25**:635–46.
19. Tucker M, Staples RR, Valencia-Sanchez MA, et al. Ccr4p is the catalytic subunit of a Ccr4p/Pop2p/Notp mRNA deadenylase complex in Saccharomyces cerevisiae. *EMBO J* 2002;**21**:1427–36.
20. Uchida N, Hoshino S-I, Katada T. Identification of a human cytoplasmic poly(A) nuclease complex stimulated by poly(A)-binding protein. *J Biol Chem* 2004;**279**:1383–91.
21. Kahvejian A, Roy G, Sonenberg N. The mRNA closed-loop model: the function of PABP and PABP-interacting proteins in mRNA translation. *Cold Spring Harb Symp Quant Biol* 2001;**66**:293–300.
22. Albrecht M, Lengauer T. Survey on the PABC recognition motif PAM2. *Biochem Biophys Res Commun* 2004;**316**:129–38.
23. Kozlov G, Trempe J-F, Khaleghpour K, et al. Structure and function of the C-terminal PABC domain of human poly(A)-binding protein. *PNAS* 2001;**98**:4409–13.
24. Matsuda S, Rouault J-P, Magaud J-P, Berthet C. In search of a function for the TIS21/PC3/BTG1/TOB family. *FEBS Letters* 2001;**497**:67–72.

25. Tzachanis D, Freeman GJ, Hirano N, et al. Tob is a negative regulator of activation that is expressed in anergic and quiescent T cells. *Nat Immunol* 2001;**2**:1174–82.
26. Xu N, Loflin P, Chen C-YA, Shyu A-B. A broader role for AU-rich element-mediated mRNA turnover revealed by a new transcriptional pulse strategy. *Nucl Acids Res* 1998;**26**:558–65.
27. Ezzeddine N, Chang T-C, Zhu W, et al. Human TOB, an antiproliferative transcription factor, Is a poly(A)-binding protein-dependent positive regulator of cytoplasmic mRNA deadenylation. *Mol Cell Biol* 2007;**27**:7791–801.
28. Suzuki T, K-Tsuzuku J, Ajima R, et al. Phosphorylation of three regulatory serines of Tob by Erk1 and Erk2 is required for Ras-mediated cell proliferation and transformation. *Genes Dev* 2002;**16**:1356–70.
29. Ross J. mRNA stability in mammalian cells. *Microbiol Rev* 1995;**59**:423–50.
30. Schiavi SC, Wellington CL, Shyu AB, et al. Multiple elements in the c-fos protein-coding region facilitate mRNA deadenylation and decay by a mechanism coupled to translation. *J Biol Chem* 1994;**269**:3441–8.
31. Chang T-C, Yamashita A, Chen C-YA, et al. UNR, a new partner of poly(A)-binding protein, plays a key role in translationally coupled mRNA turnover mediated by the c-fos major coding-region determinant. *Genes Dev* 2004;**18**:2010–23.
32. Eystathioy T, Jakymiw A, Chan EKL, et al. The GW182 protein colocalizes with mRNA degradation associated proteins hDcp1 and hLSm4 in cytoplasmic GW bodies. *RNA* 2003;**9**:1171–3.
33. Sheth U, Parker R. Targeting of aberrant mRNAs to cytoplasmic processing bodies. *Cell* 2006;**125**:1095–109.
34. Nilsen TW. Mechanisms of miroRNA-mediated gene regulation in animal cells. *Trends Genet* 2007;**23**:243–9.
35. Cougot N, Babajko S, Seraphin B. Cytoplasmic foci are sites of mRNA decay in human cells. *J Cell Biol* 2004;**165**:31–40.
36. Bhattacharyya SN, Habermacher R, Martine U, et al. Relief of microRNA-mediated translation repression in human cells subjected to stress. *Cell* 2006;**125**:1111–24.
37. Brengues M, Teixeira D, Parker R. Movement of eukaryotic mRNAs between polysomes and cytoplasmic processing bodies. *Science* 2005;**310**:486–9.
38. Dreyfuss G, Kim VN, Kataoka N. Messenger-RNA-binding proteins and the messages they carry. *Nat Rev Mol Cell Biol* 2003;**3**:195–205.
39. Moore MJ. From birth to death: the complex lives of eukaryotic mRNAs. *Science* 2005;**309**:1514–18.

Chapter 33

Signaling to Cytoplasmic Polyadenylation and Translation

Jong Heon Kim[1,2] and Joel D. Richter[1]
[1]Program in Molecular Medicine, University of Massachusetts Medical School, Worcester, Massachusetts
[2]Research Institute, National Cancer Center, Goyang, Gyeonggi, Korea

INTRODUCTION

Early animal development is programmed in part by mRNAs inherited by the fertilized egg at the time of fertilization. In vertebrates, oocytes are arrested at the end of meiotic prophase I, particularly at the diplotene or diakinesis stage. In this state, they reside in the ovary; they do not divide but grow by taking up yolk from the bloodstream. Upon receipt of hormonal signaling from surrounding follicle cells, the oocytes re-enter the meiotic divisions (a process known as oocyte maturation) and are ovulated and subsequently fertilized. This oocyte maturation process, especially from the frog *Xenopus laevis*, is particularly amenable for examining mRNA translational control. It is characterized by several salient features: translational activation of many dormant mRNAs, poly(A) elongation of several of the newly activated mRNAs, general stability of all mRNAs, and inhibited transcription. In addition, a number of kinase signaling cascades have been elucidated in maturing oocytes including those involving MAP kinase, cdk1, Aurora A, polo-like kinase, and a host of others. For these reasons, as well as the fact that maturing oocytes can be gathered in large numbers and that several of the molecular processes noted above such as polyadenylation and translation are robust even in cell free extracts, makes them ideally suited for studying the signaling events that lead to polyadenylation induced translation.

THE BIOCHEMISTRY OF CYTOPLASMIC POLYADENYLATION

When *Xenopus* oocytes are incubated in a simple salt solution containing progesterone, they mature *in vitro* to second meiotic metaphase. During this time, several dormant mRNAs that have short poly(A) tails, usually about 30–50 bases, undergo poly(A) extension, usually to ~150–200 bases. Following polyadenylation, translation ensues [1]. A large number of factors control polyadenylation and translation, the salient features of which are illustrated in Figure 33.1.

In *Xenopus* oocytes, gonadotropic hormones emanating from the hypothalamus stimulate the synthesis of progesterone in follicles cells that surround the oocyte. Progesterone is then secreted from these cells and interacts with an oocyte surface associated receptor: this event stimulates a cascade of events that culminates in the cytoplasmic polyadenylation and translation of several mRNAs and, ultimately, meiotic progression to metaphase II. To understand how the signaling events induce polyadenylation and translation it is first necessary to describe the most proximal factors that regulate this process. The RNAs that are regulated by this process contain one or more cytoplasmic polyadenylation elements (CPEs) in their 3′ untranslated regions (3′UTRs); the sequence of the CPE is UUUUUAU, or a slight variation thereof. Following synthesis, probably all nuclear pre-mRNAs, irrespective of whether they harbor a CPE, acquire a long poly(A) tail, probably many hundreds of bases in length [2]. After nuclear export, the CPE containing RNAs become associated with a large number of factors that include: (1) CPEB (CPE binding protein), an RNA recognition motif and zinc finger containing protein that binds the CPE [3, 4], (2) CPSF (cleavage and polyadenylation specificity factor), a group of four proteins that bind the AAUAAA polyadenylation hexanucleotide [5], which is also necessary for cytoplasmic polyadenylation, (3) symplekin, a scaffold protein upon which other factors assemble [6], (4) Gld2 (germline development 2), an unusual poly(A) polymerase that is a member of the DNA nucleotidyltransferase superfamily [6], (5) PARN (poly(A) specific ribonucleases), a deadenylating enzyme [2], (6) ePAB (embryonic poly(A) binding protein) [7], and (7)

Handbook of Cell Signaling, Three-Volume Set 2 ed.

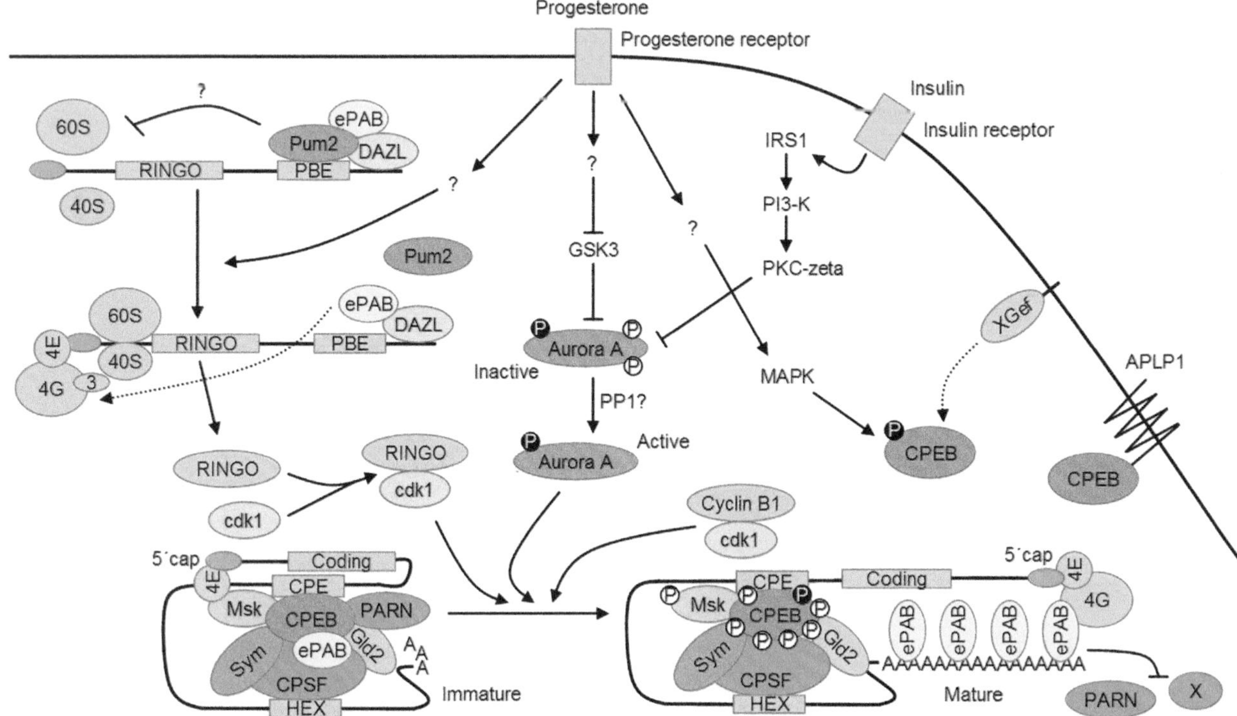

FIGURE 33.1 Signaling events leading to cytoplasmic polyadenylation.
Progesterone, secreted from follicles cells, interacts with the progesterone receptor associated with the cell surface. In an unknown mechanism signals cause pumilio 2 (Pum2), which is associated with the proteins DAZL and ePAB, to dissociate from its binding site, the PBE (pumilio binding element) in the RINGO mRNA 3′UTR, thereby releasing this mRNA from translational inhibition. ePAB may facilitate the association of eIF4G (4G) with the cap binding factor eIF4E (4E). This would allow eIF3 (3) to bring the 40S ribosomal subunit to the end of the mRNA. Newly synthesized RINGO associates with cdk1 and phosphorylates CPEB on six sites. These events induce the dissociation of ePAB from CPEB and its interaction with the newly elongated poly(A) tail (see below). ePAB protects the poly(A) tail from degradation by PARN or other deadenylating enzymes (denoted as factor X). Progesterone signaling also represses the kinase GSK-3, which in oocytes binds and phosphorylates Aurora A, causing an Aurora A inhibitory auto-phosphorylation. The phosphatase PP1 may remove the inhibitory phosphate, thereby leading to active Aurora A, which in turn phosphorylates CPEB serine 174. This phosphorylation causes the dissociation of PARN, a deadenylating enzyme, from a ribonucleoprotein complex that includes not only CPEB, a CPE containing RNA, but also Gld2, a poly(A) polymerase, CPSF, a group of four proteins that bind the AAUAAA polyadenylation hexanucleotide (HEX), symplekin (Sym), a scaffold-like protein, and maskin (Msk), which represses translation by binding not only CPEB, but the cap binding factor eIF4E. In addition to progesterone, insulin can also signal to cytoplasmic polyadenylation [13]. Insulin signals through the insulin receptor substrate 1 (IRS1), phosphoinositol 3 kinase (PI3-K), and the zeta isoform of protein kinase C (PKC-zeta). PKC-zeta inhibits the Aurora A inhibitory phosphorylation, thereby activating this kinase so that it can phosphorylate CPEB serine 174. MAP kinase can also phosphorylate CPEB serine 174, which may be facilitated by a guanine nucleotide exchange factor (XGef) [14]. Finally, some CPEB is also associated with membranes through an interaction with the intracellular domain of APLP1 (amyloid precursor like protein 1) [41].

Maskin, the most proximal protein that controls translation [8–10]. Curiously, both Gld2 and PARN, which are anchored to the ribonucleoprotein complex via CPEB, are both active, but because PARN activity is more robust, the poly(A) tails are shortened to about 50 nucleotides or so. However, because Gld2 is active, it continues to add adenosine residues to the end of the RNA only to be removed by PARN [2]. This apparently energetically wasteful reaction could possibly take place in oocytes over a period of many months. Translational dormancy is conferred to these "lightly" polyadenylated RNAs by maskin. Maskin binds both CPEB and the cap binding factor eIF4E. By associating with eIF4E, maskin precludes the eIF4E-eIF4G interaction; it is this step that maskin regulates translation since it is through eIF4E that eF4G brings the 40S ribosomal subunit to the 5′ end of the mRNA [11].

SIGNALING TO POLYADENYLATION

Progesterone stimulation of its cell surface receptor [12] is the switch that changes short poly(A) tails and translational repression to long poly(A) tails and translational activation. An early event following progesterone stimulation is the dissociation of the kinase GSK-3 from another kinase, Aurora A. This is a complicated affair, for GSK-3 phosphorylates Aurora A, which causes an auto-inhibitory phosphorylation. Upstream signaling events cause GSK-3 to dissociate from Aurora A, thereby alleviating the auto-inhibitory phosphorylation and resulting in Aurora A activation [13]. Aurora A (and possibly MAP kinase, [14]) phosphorylates CPEB on serine 174 [15], which causes an enhanced interaction between CPEB and CPSF [16], and more importantly, to expel PARN from the RNP complex [2].

The result of this PRN expulsion is Gld2 catalyzed polyadenylation, which occurs by default [2].

Prior to polyadenylation, maskin association with CPEB and the cap binding factor eIF4E precludes translation specifically of CPE containing mRNAs. Following polyadenylation, the newly elongated poly(A) tail associates with ePAB, which in turn binds eIF4G. This interaction apparently increases the affinity of eIF4G with eIF4E [17]; maskin is thus displaced from eIF4E and translation initiation proceeds [9].

Maskin also undergoes a series of cdk1 catalyzed phosphorylation events that help it dissociate from eIF4E [18]. While during oocyte maturation, polyadenylation and ePAB are most important for inducing maskin dissociation from eIF4E [9], in early embryogenesis, maskin phosphorylation is particularly essential [10]. Maskin is also phosphorylation by PKC, which is necessary for anchoring it to centrosomes ([18]; see also [19]).

A HIERARCHY OF TRANSLATION CONTROL

One of the first translational control events involves the mRNA encoding RINGO (*r*apid *in*ducer of G_2/M transition in *o*ocytes). RINGO was identified initially in a screen for mRNAs that can induce oocyte maturation [20]. While oocytes contain very little RINGO protein, they do contain dormant RINGO mRNA that becomes activated upon progesterone initiated signaling. RINGO appears to resemble, in a functional sense, cyclin B1, in that it can bind and activate the protein kinase cdk1 [20]. The RINGO 3′UTR is bound by three proteins: pumilio 2 (Pum2), deleted in azospermia-like (DAZL), and ePAB [21]. Among these, Pum2 may be the most important as its interaction with a pumilio binding element (PBE) within the 3′UTR of RINGO mRNA is necessary for translational repression in oocytes [21]. Although the mechanism of Pum2 mediated repression is not known, Pum2 dissociates from RINGO mRNA following progesterone stimulation, which allows it to be translated. Presumably, RINGO protein interacts with and activates at least a small portion of the available cdk1. What does cdk1 phosphorylate? Recall that Aurora A phosphorylation of CPEB causes the expulsion of PARN from the cytoplasmic polyadenylation complex. The RINGO/cdk1 complex also phosphorylates CPEB, but on six other residues [22]. These events cause ePAB to dissociate from CPEB and to bind the newly elongated poly(A) tail; here, it protects the poly(A) tail from deadenylation and binds eIF4G to stimulate translation [7].

Perhaps the first RNA to undergo CPEB controlled polyadenylation encodes mos, a serine/threonine kinase [23, 24]. Mos activates a kinase cascade that culminates in the formation of cyclin B1/cdk1, which then is responsible for most of the manifestations of oocyte maturation. Mos, or signals downstream from it, are also required for the polyadenylation of cycin B1 [25, 26]. It is not known how mos functions in cyclin B1 polyadenylation.

POLYADENYLATION IN MAMMALIAN OOCYTES

At least in growing mouse oocytes and in *Xenopus* embryos, Aurora A phosphorylation is reversible. In the mouse during the protracted period of oogenesis, Aurora A phosphorylates CPEB at pachytene in first meiotic prophase [27]; as the cells enter diplotene, PP1 dephosphorylates CPEB and it remains in this unphosphorylated state until maturation, when it is again phosphorylated by Aurora A [28]. This regulated CPEB phosphorylation controls the polyadenylation and translation of synaptonemal complex protein mRNAs during pachytene [27], and of growth differentiation factor-9 (GDF-9) mRNA [29].

During the cell cycle in early *Xenopus* embryos, polyadenylation is regulated; that is, at least one CPE containing mRNA is polyadenylated during M phase (i.e., mitosis), and deadenylated during interphase. Perhaps not surprisingly, CPEB is phosphorylated as cells begin to enter M phase, presumably by Aurora A, and is dephosphorylated as cells exit M phase [30]. It is not know whether Aurora A activity, or perhaps the PP1 phosphatase, is regulated during the cell cycle.

SIGNALING TO POLYADENYLATION IN THE BRAIN

CPEB is present not only in germs cells and embryos, but in the central nervous system as well. In neurons, CPEB is present at postsynaptic sites and in dendrites where, in response to synaptic activity, it stimulates polyadenylation and translation [31–36]. CPEB induced translation appears to influence synaptic plasticity and hippocampal dependent memories [37, 38] (the hippocampus is a part of the brain that is important for learning and memory). In neurons, there are a number of neurotransmitters, but only one that seems to induce polyadenylation in *N*-methyl-D-aspartate (NMDA), which causes calcium influx into synapses [32]. Two kinases, Aurora A [19] and calcium/calmodulin dependent protein kinase 2 (CaMKII) [39, 40] can phosphorylate CPEB in neurons and induce polyadenylation (Figure 33.2).

CONCLUSIONS

By virtue of its activity to control specific mRNA translation, CPEB plays a central role in germ cell and embryo development and in brain functions such as memory consolidation and retrieval. Hence the signaling events that lead to

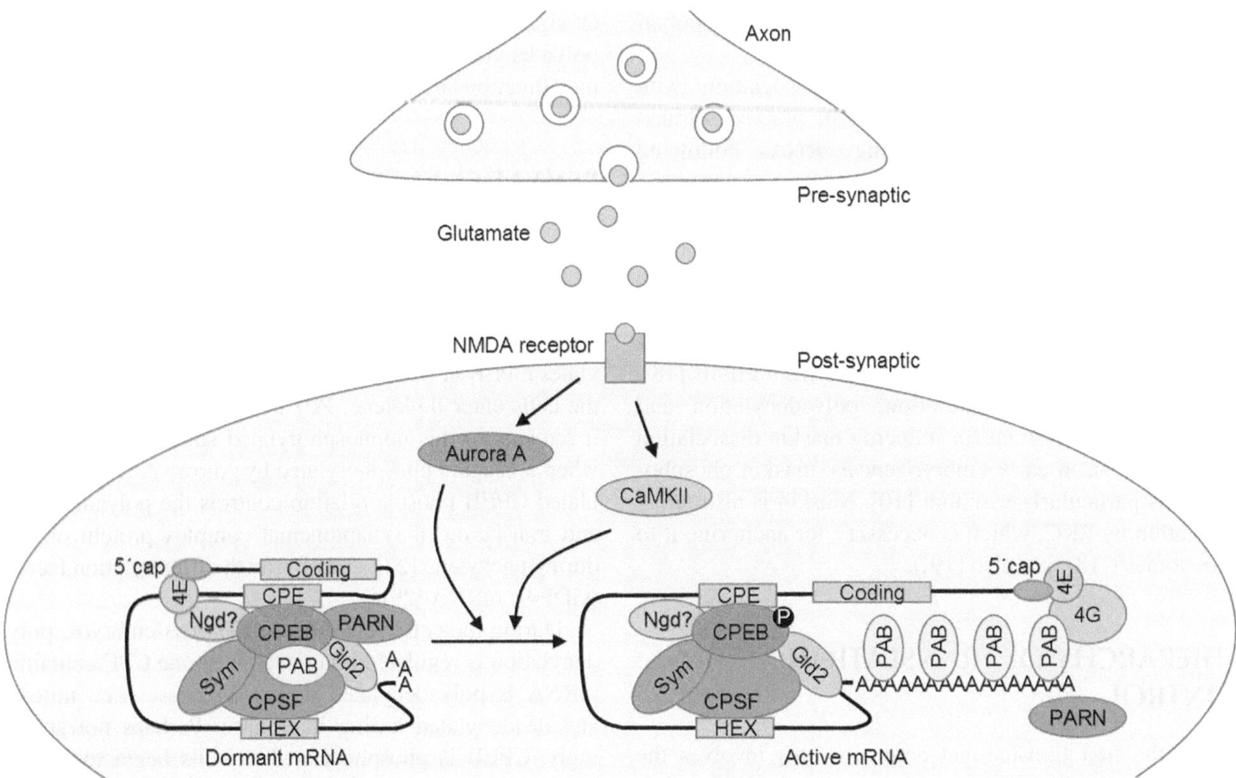

FIGURE 33.2 At synapses, glutamate, a neurotransmitter, is released from axon terminals, and interacts with the *N*-methyl-D-aspartate (NMDA) receptor.
Calcium, which enters the postsynaptic region through the NMDA receptors, then activates Aurora A and CaMKII protein kinases. These enzymes both have the capability of phosphorylating CPEB, which we surmise leads to polyadenylation and translation similar to that that occurs in *Xenopus* oocytes. However, neuroguidin (Ngd), rather than maskin, is most likely the factor that bridges CPEB and eIF4E, and thus is the most proximal regulator of translation [42].

CPEB phosphorylation are necessary for this translational control. While Aurora A, MAP kinase, and CaMKll have specific CPEB phosphorylating activity (i.e., serine 174 in *Xenopus* or threonine 171 in mammals), they may respond similarly in some cells, but perhaps uniquely in other cells under different circumstances. That is, in oocytes Aurora A and MAP kinase phosphorylate CPEB serine 174, but in the brain, Aurora A and CaMKII phosphorylate CPEB threonine 171. In other cells, perhaps only Aurora A, or MAP kinase, or CAMKII can carry out this CPEB modification. Thus, different pathways can converge on CPEB to modulate specific biological responses.

REFERENCES

1. Richter JD. CPEB: a life in translation. *Trends Biochem Sci* 2007;**32**:279–85.
2. Kim JH, Richter JD. Opposing polymerase-deadenylase activities regulate cytoplasmic polyadenylation. *Mol Cell* 2006;**24**:173–83.
3. Hake LE, Richter JD. CPEB is a specificity factor that mediates cytoplasmic polyadenylation during *Xenopus* oocyte maturation. *Cell* 1994;**79**:617–27.
4. Hake LE, Mendez R, Richter JD. Specificity of RNA binding by CPEB: requirement for RNA recognition motifs and a novel zinc finger. *Mol Cell Biol* 1998;**18**:685–93.
5. Bilger A, Fox CA, Wahle E, Wickens M. Nuclear polyadenylation factors recognize cytoplasmic polyadenylation elements. *Genes Dev* 1994;**8**:1106–16.
6. Barnard DC, Ryan K, Manley JL, Richter JD. Symplekin and xGLD-2 are required for CPEB-mediated cytoplasmic polyadenylation. *Cell* 2004;**119**:641–51.
7. Kim JH, Richter JD. RINGO/cdk1 and CPEB mediate poly(A) tail stabilization and translational regulation by ePAB. *Genes Dev* 2007;**21**:2571–9.
8. Stebbins-Boaz B, Cao Q, de Moor CH, Mendez R, Richter JD. Maskin is a CPEB-associated factor that transiently interacts with elF-4E. *Mol Cell* 1999;**4**:1017–27.
9. Cao Q, Richter JD. Dissolution of the maskin-eIF4E complex by cytoplasmic polyadenylation and poly(A)-binding protein controls cyclin B1 mRNA translation and oocyte maturation. *EMBO J* 2002;**21**:3852–62.
10. Cao Q, Kim JH, Richter JD. CDK1 and calcineurin regulate Maskin association with eIF4E and translational control of cell cycle progression. *Nat Struct Mol Biol* 2006;**13**:1128–34.
11. Richter JD, Sonenberg N. Regulation of cap-dependent translation by eIF4E inhibitory proteins. *Nature* 2005;**433**:477–80.
12. Martinez S, Grandy R, Pasten P, Montecinos H, Montecino M, Olate J, Hinrichs MV. Plasma membrane destination of the classical *Xenopus laevis* progesterone receptor accelerates progesterone-induced oocyte maturation. *J Cell Biochem* 2006;**99**:853–9.
13. Sarkissian M, Mendez R, Richter JD. Progesterone and insulin stimulation of CPEB-dependent polyadenylation is regulated by aurora a and glycogen synthase kinase-3. *Genes Dev* 2004;**18**:48–61.

14. Keady BT, Kuo P, Martinez SE, Yuan L, Hake LE. MAPK interacts with XGef and is required for CPEB activation during meiosis in *Xenopus* oocytes. *J Cell Sci* 2007;**120**:1093–103.
15. Mendez R, Hake LE, Andresson T, Littlepage LE, Ruderman JV, Richter JD. Phosphorylation of CPE binding factor by Eg2 regulates translation of c-*mos* mRNA. *Nature* 2000;**404**:302–7.
16. Mendez R, Murthy KG, Ryan K, Manley JL, Richter JD. Phosphorylation of CPEB by Eg2 mediates the recruitment of CPSF into an active cytoplasmic polyadenylation complex. *Mol Cell* 2000;**6**:1253–9.
17. von der Haar T, Gross JD, Wagner G, McCarthy JE. The mRNA cap-binding protein eIF4E in post-transcriptional gene expression. *Nat Struct Mol Biol* 2004;**11**:503–11.
18. Barnard DC, Cao Q, Richter JD. Differential phosphorylation controls Maskin association with eukaryotic translation initiation factor 4E and localization on the mitotic apparatus. *Mol Cell Biol* 2005;**25**:7605–15.
19. Groisman I, Huang YS, Mendez R, Cao Q, Theurkauf W, Richter JD. CPEB, maskin, and cyclin B1 mRNA at the mitotic apparatus: implications for local translational control of cell division. *Cell* 2000;**103**:435–47.
20. Ferby I, Blazquez M, Palmer A, Eritja R, Nebreda AR. A novel $p34^{cdc2}$-binding and activating protein that is necessary and sufficient to trigger G_2/M progression in *Xenopus* oocytes. *Genes Dev* 1999;**13**:2177–89.
21. Padmanabhan K, Richter JD. Regulated Pumilio-2 binding controls RINGO/Spy mRNA translation and CPEB activation. *Genes Dev* 2006;**20**:199–209.
22. Mendez R, Barnard D, Richter JD. Differential mRNA translation and meiotic progression require Cdc2-mediated CPEB destruction. *EMBO J* 2002;**21**:1833–44.
23. Sheets MD, Wu M, Wickens M. Polyadenylation of c-*mos* mRNA as a control point in *Xenopus* meiotic maturation. *Nature* 1995;**374**:511–16.
24. Stebbins-Boaz B, Hake LE, Richter JD. CPEB controls the cytoplasmic polyadenylation of cyclin, Cdk2 and c-*mos* mRNAs and is necessary for oocyte maturation in *Xenopus*. *EMBO J* 1996;**15**:2582–92.
25. de Moor CH, Richter JD. The Mos pathway regulates cytoplasmic polyadenylation in *Xenopus* oocytes. *Mol Cell Biol* 1997;**17**:6419–26.
26. Ballantyne Jr. S, Daniel DL, Wickens M A dependent pathway of cytoplasmic polyadenylation reactions linked to cell cycle control by c-*mos* and CDK1 activation. *Mol Biol Cell* 1997;**8**:1633–48.
27. Tay J, Richter JD. Germ cell differentiation and synaptonemal complex formation are disrupted in CPEB knockout mice. *Dev Cell* 2001;**1**:201–13.
28. Tay J, Hodgman R, Sarkissian M, Richter JD. Regulated CPEB phosphorylation during meiotic progression suggests a mechanism for temporal control of maternal mRNA translation. *Genes Dev* 2003;**17**:1457–62.
29. Racki WJ, Richter JD. CPEB controls oocyte growth and follicle development in the mouse. *Development* 2006;**133**:4527–37.
30. Groisman I, Jung MY, Sarkissian M, Cao Q, Richter JD. Translational control of the embryonic cell cycle. *Cell* 2002;**109**:473–83.
31. Wu L, Wells D, Tay J, Mendis D, Abbott MA, Barnitt A, Quinlan E, Heynen A, Fallon JR, Richter JD. CPEB-mediated cytoplasmic polyadenylation and the regulation of experience-dependent translation of alpha-CaMKII mRNA at synapses. *Neuron* 1998;**21**:1129–39.
32. Huang YS, Jung MY, Sarkissian M, Richter JD. *N*-methyl-D-aspartate receptor signaling results in aurora kinase-catalyzed CPEB phosphorylation and alpha CaMKII mRNA polyadenylation at synapses. *EMBO J* 2002;**21**:2139–48.
33. Wells DG, Dong X, Quinlan EM, Huang YS, Bear MF, Richter JD, Fallon JR. A role for the cytoplasmic polyadenylation element in NMDA receptor-regulated mRNA translation in neurons. *J Neurosci* 2001;**21**:9541–8.
34. Shin CY, Kundel M, Wells DG. Rapid, activity-induced increase in tissue plasminogen activator is mediated by metabotropic glutamate receptor-dependent mRNA translation. *J Neurosci* 2004;**24**:9425–33.
35. Du L, Richter JD. Activity-dependent polyadenylation in neurons. *RNA* 2005;**11**:1340–7.
36. Richter JD, Lorenz LJ. Selective translation of mRNAs at synapses. *Curr Opin Neurobiol* 2002;**12**:300–4.
37. Alarcon JM, Hodgman R, Theis M, Huang YS, Kandel ER, Richter JD. Selective modulation of some forms of schaffer collateral-CA1 synaptic plasticity in mice with a disruption of the CPEB-1 gene. *Learn Mem* 2004;**11**:318–27.
38. Berger-Sweeney J, Zearfoss NR, Richter JD. Reduced extinction of hippocampal-dependent memories in CPEB knockout mice. *Learn Mem* 2006;**13**:4–7.
39. Atkins CM, Nozaki N, Shigeri Y, Soderling TR. Cytoplasmic polyadenylation element-binding protein-dependent protein synthesis is regulated by calcium/calmodulin-dependent protein kinase II. *J Neurosci* 2004;**24**:5193–201.
40. Atkins CM, Davare MA, Oh MC, Derkach V, Soderling TR. Bidirectional regulation of cytoplasmic polyadenylation element-binding protein phosphorylation by Ca^{2+}/calmodulin-dependent protein kinase II and protein phosphatase 1 during hippocampal long-term potentiation. *J Neurosci* 2005;**25**:5604–10.
41. Cao Q, Huang YS, Kan MC, Richter JD. Amyloid precursor proteins anchor CPEB to membranes and promote polyadenylation-induced translation. *Mol Cell Biol* 2005;**25**:10,930–9.
42. Jung MY, Lorenz L, Richter JD. Translational control by neuroguidin, a eukaryotic initiation factor 4E and CPEB binding protein. *Mol Cell Biol* 2006;**26**:4277–87.

Chapter 34

Translation Control and Insulin Signaling

Anand Selvaraj and George Thomas

Department of Cancer and Cell Biology, Genome Research Institute, University of Cincinnati, College of Medicine, Cincinnati, Ohio

INTRODUCTION

Insulin controls a number of key anabolic responses in specific target tissues including liver, adipose, and muscle. Among these responses is the activation and maintenance of high rates of protein synthesis, the most energy consuming cellular anabolic process [1]. Importantly, it has been known for some time that the stimulation of protein synthesis by insulin is independent of, and not secondary to increased in glucose or amino acid transport into the cell [2]. It is thought that the loss of insulin induced protein synthesis significantly contributes to the cessation of growth and weight loss associated with untreated Type 1 *diabetes mellitus* [3]. Given the detrimental consequences of dysregulation of insulin function, it is not surprising that mechanisms have evolved to closely modulate translation rates as a function of time and demand [4]. In addition, regulation of gene expression at the level of translation allows the cell to rapidly respond to sudden changes in the external milieu. The use of mammalian cell culture systems and animal models has led to the identification of a number of signaling components involved in insulin mediated translational control. The intracellular signal transduction cascades utilize downstream effectors to modulate the specific activities of key translational components. Insulin stimulates global protein synthesis by increasing both the rates of translational initiation and elongation [5], but also by selectively triggering the upregulation of ribosome biogenesis, to increase translational capacity [6, 7].

THE INSULIN SIGNALING PATHWAY

Binding of insulin to the tetrameric insulin receptor triggers a conformational change in the receptor that leads to inter-autophosphorylation at specific tyrosine residues. These phosphorylated residues act as docking sites for proteins containing either phosphotyrosine binding (PTB) or Src-homology 2 (SH2) adapter motifs. A major insulin receptor adapter molecule is insulin receptor substrates (IRS). Upon insulin stimulation, IRS proteins, through their PTB domain, bind to the phosphorylated receptor. After their recruitment to the receptor, IRS proteins are tyrosine phosphorylated by the insulin receptor, which forms a second set of recognition adapter motifs for the binding of key signaling molecules [8].

One of the insulin responsive signaling cascades is initiated by the binding of son of sevenless (SOS) to IRS proteins through growth factor receptor bound protein 2 (GRB2). This binding leads to activation of SOS, which drives Ras from the GDP bound inactive state to the active GTP bound state, by functioning as a guanine nucleotide exchange factor (GEF). GTP bound Ras regulates multiple pathways including both the phosphatidylinositol 3-kinase (Class 1 PI3K, see below) and the Raf-MAPK cascade. A number of kinases, including MAP kinase signal integrating kinase 1 (MnK1) and p90 ribosomal S6 kinase (p90 RSK) act downstream of the Raf-MAPK cascade to mediate specific anabolic processes [9]. As mentioned above, a second signaling cascade is initiated by the binding of the p85 adapter of the phosphatidylinositol 3-kinase (Class 1 PI3K) to IRS proteins. PI3K signaling is involved in the regulation of cell growth, survival, and metabolism. The recruitment of PI3K to the membrane by IRS stimulates the production of the lipid second messenger phosphatidylinositol-3,4,5-trisphosphate (PtdIns (3,4,5) P3) from PtdIns (4,5)-P2, recruits protein kinase B (PKB/Akt) through its Plextrin Homology (PH) domain, and this is followed by the sequential phosphorylation and activation of PKB/Akt by phosphatidylinositol dependent kinase (PDK)1 and mammalian Target of Rapamycin Complex2 (mTOR Complex2) [10, 11]. The major negative regulator of this step appears to be the lipid phosphatase and tensing homolog (PTEN). PTEN converts PIP_3 to PIP_2 resulting in a reduction in

Handbook of Cell Signaling, Three-Volume Set 2 ed.

PKB/Akt's recruitment to the cell membrane and its subsequent activation [12]. Activated PKB/Akt has a number of downstream substrates including glycogen synthase kinase 3 (GSK3) and tuberin, also known as the tuberous sclerosis complex 2 (TSC2) protein.

The phosphorylation of the TSC2, a tumor suppressor that exists in complex with hamartin, also known as tuberous sclerosis complex 1 (TSC1) protein, results in its dissociation from TSC1 and its degradation [13]. This releases the small GTPase Ras homolog enriched in brain (Rheb) from the inhibitory GTPase activating protein (GAP) activity of TSC2, thus driving Rheb into the GTP bound active state. Initial studies suggested that GTP bound Rheb mediates the ability of mTOR Complex1 (raptor-mTOR) to signal to downstream substrates, such as ribosomal protein S6 kinase 1 (S6K1), by directly altering mTOR Complex1 activity [14]. However recent reports suggest that Rheb directly binds to the isoleucine-proline isomerase FKBP38, an endogenous inhibitor of mTOR Complex1, and prevents its association with mTOR Complex1 in a GTP dependent manner [15]. The anti-fungal macrolide, rapamycin, blocks mTOR Complex1 by forming an inhibitory complex with the immunophilin FKBP12. mTOR Complex1 belongs to an evolutionarily conserved nutrient sensitive pathway (for review see [16]). It is thought that the insulin signaling pathway merged with the nutrient sensitive mTOR Complex1 pathway to regulate protein synthetic rates as a function of translational precursor availability [17,18]. The PI3K-TSC-mTOR Complex1 branch of the insulin signaling pathway is the major effector of the translational apparatus (Figure 34.1).

INSULIN SIGNALING AND REGULATION OF TRANSLATION INITIATION

Translation initiation is a highly regulated, dynamic stepwise process. It is classically separated into four steps (Figure 34.2): (1) formation of the 43S (S stands for Svedberg unit) pre-initiation complex, which contains eukaryotic initiation factor 2-methionine transfer RNA (eIF2-Met-tRNAi-GTP), the 40S ribosomal subunit, eIF3, and eIF1A, (2) recruitment of the 43S pre-initiation complex to the 5' cap of the messenger RNA (mRNA) by eIF4F leading to the formation of the 48S initiation complex, (3) scanning of the 5' untranslated region (5' UTR) of the mRNA by the eIF 4F complex until reaching

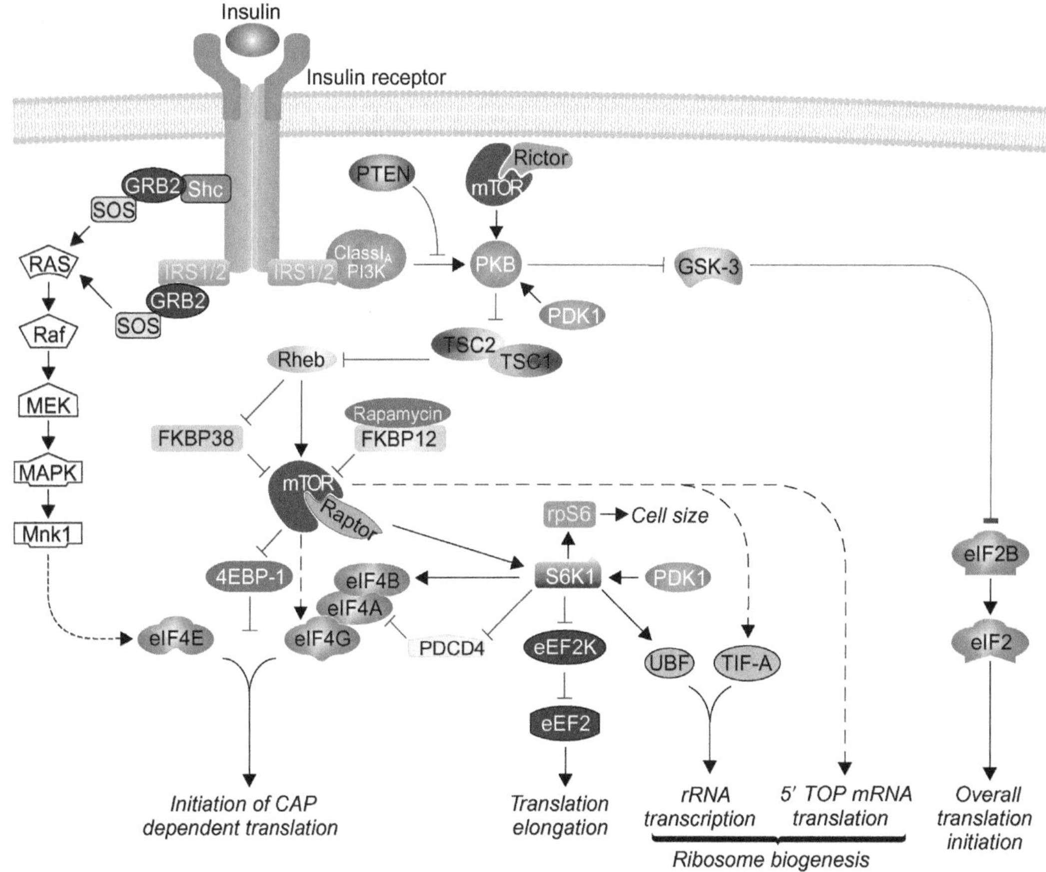

FIGURE 34.1 Schematic representation of insulin responsive signaling cascades and their targets in translational machinery. Arrows represent a positive input while bars represent a negative input. Broken arrows indicate unclear and/or indirect relationship. mTOR Complex1 is represented as mTOR-Raptor and mTOR Complex2 as mTOR-Rictor.

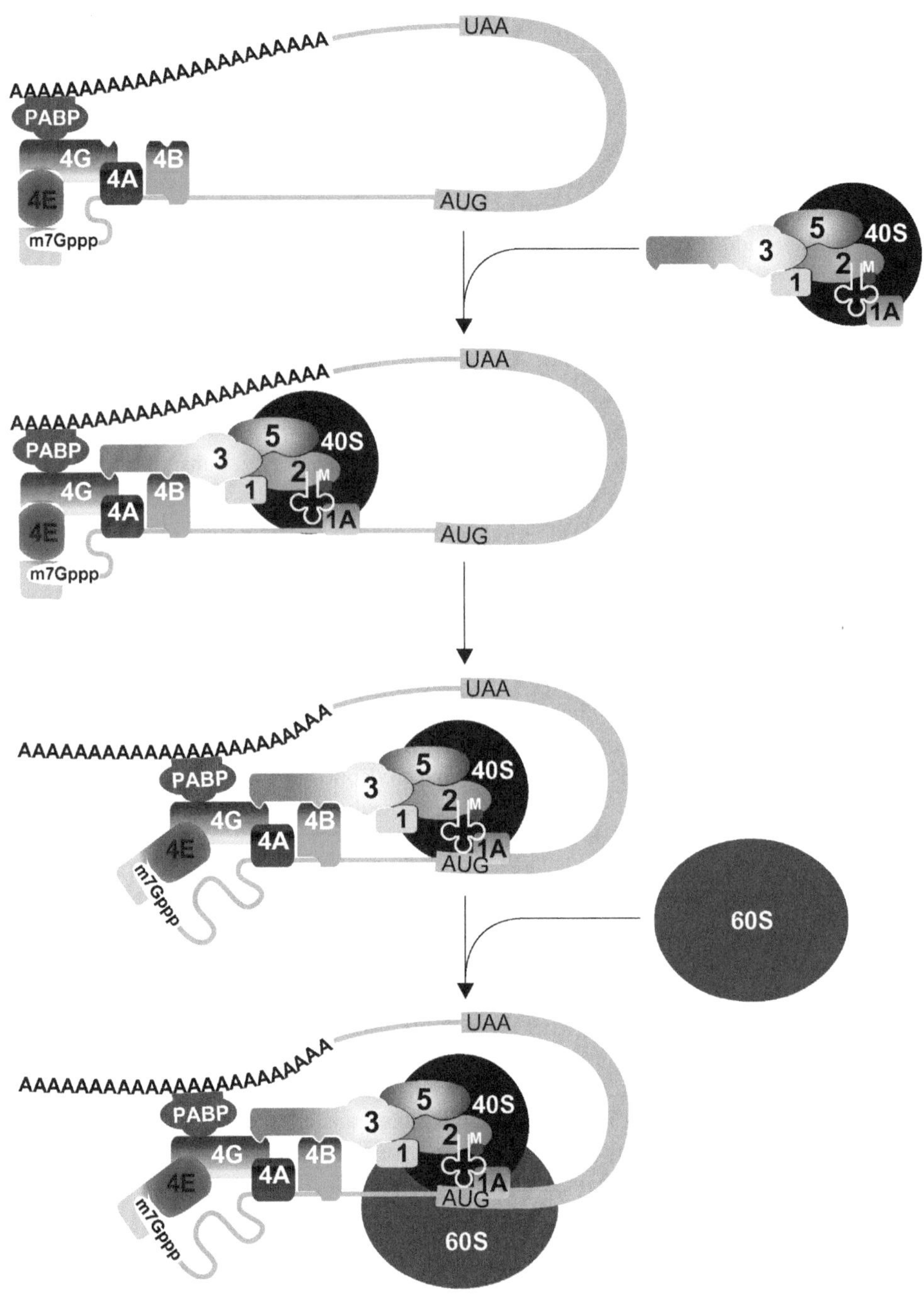

FIGURE 34.2 The translation initiation pathway.
This is separated into four steps: (1) formation of the 43S pre-initiation complex, which contains eukaryotic initiation factor 2-methionine transfer RNA (eIF2-Met-tRNAi-GTP), the 40S ribosomal subunit, eIF3, and eIF1A, (2) recruitment of the 43S pre-initiation complex to the 5' cap of the messenger RNA (mRNA) by eIF 4F leads to the formation of the 48S initiation complex, (3) scanning of the 5' untranslated region (5' UTR) of the mRNA by the eIF 4F complex until the AUG start codon, and (4) recruitment of the 60S subunit to assemble a complete 80s ribosome.

the AUG start codon, and (4) recruitment of the 60S subunit to assemble a complete 80S ribosome [19].

eIF2-Met-tRNAi-GTP binds the 40S subunit as a ternary complex. The GDP bound form of eIF2 generated by each initiation cycle cannot bind Met-tRNAi and requires exchange of GDP for GTP by the guanine nucleotide exchange factor, eIF2B. eIF2B is inactivated via phosphorylation by GSK3 at a conserved serine residue, S540. Insulin signaling regulates this first step of translation initiation by controlling the phosphorylation state of eIF2B. Insulin induced activation of PKB/Akt causes phosphorylation and inactivation of GSK3 leading to eIF2B dephosphorylation and consequent activation [20]. In parallel, different forms of cellular stress lead to phosphorylation of the eIF2 α subunit, which converts GDP bound eIF2 into an inhibitor of eIF2B. Thus, phosphorylation of eIF2 constitutes a major means of regulating translation [21]. Although the mechanism is not known, insulin treatment leads to dephosphorylation of the α subunit of eIF2 to promote translation [22,23].

All nuclear mRNA transcripts contain a 5'-terminal 7-methyl guanosine, or cap. The 5' cap of mRNA is bound by eIF4E, which is part of the eIF4F complex. The eIF4F complex contains a number of additional proteins including, eIF4G, eIF4A, and the polyA binding protein (PABP). This complex is critical for directing the binding of the 43S pre-initiation complex to the mRNA to form 48S initiation complex. eIF4E, which is the cap binding protein, is a limiting factor in eIF4F complex formation and is regulated by insulin through two distinct signaling pathways. The first involves a family of small repressor proteins termed eIF4E binding proteins (4E-BPs). 4E-BPs act as molecular mimics of the eIF4E binding region of eIF4G and, hence, are competitive inhibitors of eIF4G binding to eIF4E. Insulin induced phosphorylation of 4E-BP1 disrupts its binding to eIF4E, which allows eIF4E to interact with eIF4G to promote increased translation by the eIF4F complex. Insulin induced hierarchical phosphorylation of the 4E-BPs is mediated by mTOR Complex1 [24]. In parallel, the activation of a second signal transduction pathway involving the phosphorylation of eIF4E has been proposed as a mechanism to increase eIF4E's affinity toward the mRNA cap structure [25]. Insulin treatment leads, through the Ras-MAPK pathway, to the activation of MnK1, which docks to eIF4G to phosphorylate eIF4E at S209 [26–28]. Although the phosphorylation of eIF4E correlates with increased translation rates, the role of phosphorylation in the binding of the 5' cap has been controversial [29,30].

The scaffolding protein eIF4G acts to bring together the components of the 4F initiation complex and also directly interacts with eIF3 of the 43S pre-initiation complex. Insulin induces the phosphorylation of eIF4G in an mTOR Complex1 dependent manner, but the kinase that mediates this phosphorylation event and the functional significance of this event is not known [31]. eIF4A is an mRNA helicase that acts in unison with eIF4B and mRNA binding protein to unwind the 5' secondary structure of mRNA. Upon insulin treatment, S6K1, a critical downstream effector of mTOR Complex1, phosphorylates eIF4B at S422, stimulating its recruitment to the pre-initiation complex [32]. In addition, S6K1 phosphorylates the programmed cell death protein 4 (PDCD4), a tumor suppressor gene at S67, leading to its degradation [33]. PDCD4 binds to eIF4A and is thought to prevent translation by competing with eIF4G for binding to eIF4A or by inhibiting eIF4A helicase activity [33].

Recent biochemical studies suggest the existence of a yet larger initiation complex, as eIF3 has been reported to act as a scaffolding platform that binds to mTOR Complex1 in an insulin dependent manner. Insulin stimulation recruits mTOR Complex1 to the pre-initiation complex leading to phosphorylation and displacement of S6K1. Following the dissociation, full activation of S6K1 is achieved, leading to the phosphorylation of eIF4B and ribosomal protein S6 (S6) [34]. As mentioned above, phosphorylated eIF4B is recruited to the pre-initiation complex and with eIF4A forms an efficient RNA helicase complex. Although *in vitro* biochemical experiments support a role for S6 phosphorylation in translation initiation, genetic analysis based on S6 mutants failed to reveal a critical role in regulating overall rates of translation [35].

INSULIN SIGNALING AND REGULATION OF TRANSLATION ELONGATION

In mammalian cells, transit of the 80S ribosome along the mRNA, termed translational elongation, requires two elongation factors: eukaryotic elongation factors 1 and 2 (eEF1 and eEF2). eEF2 mediates the GTP dependent translocation of the ribosome along the mRNA. Phosphorylation of eEF2 within the GTP binding domain by eEF2 kinase (eEF2K) impairs its ability to bind to the ribosome, thus leading to its inactivation. Insulin treatment decreases the activity of eEF2K resulting in dephosphorylation of eEF2, which leads to accelerated rates of elongation. Insulin induced phosphorylation and inactivation of eEF2K is, in part, mediated by S6K1 [36,37].

INSULIN SIGNALING AND RIBOSOME BIOGENESIS

Apart from activating several translational factors to promote translation, insulin signaling also regulates ribosome biogenesis leading to increased translational capacity in the cell. It does so by two apparent mechanisms, (1) increasing the translation of ribosomal proteins and (2) increasing ribosomal RNA (rRNA) production.

The effect of insulin on global translation of most mRNAs is modest, but a set of transcripts that harbors a 5' terminal oligo pyrimidine (5' TOP) at their transcriptional start site is significantly affected [38]. The most representative members of the 5' TOP family are ribosomal proteins and the elongation factor family of transcripts. The 5' TOP represents a cis-repressive element and the repression is relieved upon insulin addition in mammalian cells. The PI3K- mTOR Complex1 pathway appears to be critical for the regulation of 5' TOPs [39,40]. The mechanism of regulation of 5' TOP translation by mTOR Complex1 is not known. Initial studies suggested a role of S6K1 and S6 phosphorylation downstream of mTOR Complex1 in the insulin mediated regulation of 5' TOPs [41], although later loss-of-function analysis suggested that mTOR Complex1 regulates 5' TOPs independent of S6K1 and S6 phosphorylation [35, 42].

Insulin has been shown to activate rDNA transcription in myoblasts suggesting another mechanism for insulin mediated regulation of ribosomal biogenesis [43]. Moreover, recent studies suggest that mTOR Complex1 via S6K1 plays a critical role in rDNA transcription by regulating upstream binding factor (UBF) and transcription initiation factor I A (TIF-IA). Regulation of UBF is mediated through the phosphorylation of the carboxy-terminal activation domain, which enhances the interaction of UBF with the basal rDNA transcription factor selectivity factor 1 (SL1), also known as transcription initiation factor I B (TIF-IB) [44]. Regulation of TIF-1 A is mediated by mTOR Complex1 dependent phosphorylation and localization. Moreover, rapamycin treatment impairs the formation of the RNA polymerase 1 (RNA pol 1) transcription initiation complex by disrupting the association of TIF-1 A with both RNA pol 1 and SL1 [45]. Interestingly, insulin has also been shown to induce translation of the myc oncogene in an mTOR Complex1 dependent manner [46,47]. The myc gene encodes a transcription factor and has been known to drive transcription of ribosomal protein genes as well as the rRNA and tRNA genes leading to increased translational capacity in the cell [48].

CONCLUDING REMARKS

Signal transduction cascades initiated by insulin alter the function of initiation and elongation factors through reversible phosphorylation. To date there is no evidence of any role for insulin in translation termination. However, as described above, insulin regulates multiple steps in translation initiation and also in translation elongation. Although many aspects of insulin signaling and translation are unfolding rapidly, the mechanism of 5' TOP regulation, and the functions of eIF4E and eIF4G phosphorylation remain to be resolved.

ACKNOWLEDGEMENTS

We thank G. Doerman for the preparations of figures, M. Daston for editing, and members of Kozma/Thomas laboratory for discussions and insights concerning this chapter. A.S is supported by an ORISE fellowship from Air Force Research Laboratory. G.T is supported by NIH grants R01DK078019 and R01DK073802

REFERENCES

1. Buttgereit F, Brand MD. A hierarchy of ATP-consuming processes in mammalian cells. *Biochem J* 1995;**312**:163–7.
2. Wool IG, Krahl ME. An effect of insulin on peptide synthesis independent of glucose or amino-acid transport. *Nature* 1959;**183**:1399–400.
3. Christian JF, Lawrence JC. Control of Protein Synthesis by Insulin. In: Alan R, Saltiel Jeffrey EP, editors. *Mechanism of Insulin Action*. New York: Springer; 2007. p. 71–89.
4. Hershey JWB. Translational control in mammalian cells. *Annual Review of Biochemistry* 1991;**60**:717–55.
5. Proud CG. Regulation of protein synthesis by insulin. *Biochem Soc Trans* 2006;**34**:213–16.
6. Antonetti DA, Kimball SR, Horetsky RL, Jefferson LS. Regulation of rDNA transcription by insulin in primary cultures of rat hepatocytes. *J Biol Chem* 1993;**268**:25,277–25,284.
7. Hammond ML, Merrick W, Bowman LH. Sequences mediating the translation of mouse S16 ribosomal protein mRNA during myoblast differentiation and in vitro and possible control points for the in vitro translation. *Genes & Development* 1991;**5**:1723–36.
8. Whitehead JP, Clark SF, Urso B, James DE. Signalling through the insulin receptor. *Curr Opin Cell Biol* 2000;**12**:222–8.
9. Goalstone ML, Draznin B. What does insulin do to Ras? *Cell Signal* 1998;**10**:297–301.
10. Fayard E, Tintignac LA, Baudry A, Hemmings BA. Protein kinase B/Akt at a glance. *J Cell Sci* 2005;**118**:5675–8.
11. Fruman DA, Meyers RE, Cantley LC. Phosphoinositide kinases. *Annu Rev Biochem* 1998;**67**:481–507.
12. Maehama T, Dixon JE. PTEN: a tumour suppressor that functions as a phospholipid phosphatase. *Trends Cell Biol* 1999;**9**:125–8.
13. Manning BD, Cantley LC. AKT/PKB signaling: navigating downstream. *Cell* 2007;**129**:1261–74.
14. Avruch J, Hara K, Lin Y, et al. Insulin and amino-acid regulation of mTOR signaling and kinase activity through the Rheb GTPase. *Oncogene* 2006;**25**:6361–72.
15. Bai X, Ma D, Liu A, et al. Rheb activates mTOR by antagonizing its endogenous inhibitor, FKBP38. *Science* 2007;**318**:977–80.
16. Dann SG, Selvaraj A, Thomas G. mTOR Complex1-S6K1 signaling: at the crossroads of obesity, diabetes and cancer. *Trends Mol Med* 2007;**13**:252–9.
17. Dennis PB, Thomas G. Quick guide: target of rapamycin. *Curr Biol* 2002;**12**:R269.
18. Hara K, Yonezawa K, Weng QP, Kozlowski MT, Belham C, Avruch J. Amino acid sufficiency and mTOR regulate p70 S6 kinase and eIF-4E BP1 through a common effector mechanism. *J Biol Chem* 1998;**273**:14,484–14,494.
19. Hershey JWB, Mathews MB, Sonenberg N. *Translational Control*Cold Spring Harbor Laboratory Press. NY: Cold Spring Harbor; 1996.
20. Welch H, Eguinoa A, Stephens LR, Hawkins PT. Protein kinase B and rac are activated in parallel within a phosphatidylinositide 3OH-kinase-controlled signaling pathway. *J Biol Chem* 1998;**273**: 11,248–11,256.
21. Dever TE. Gene-specific regulation by general translation factors. *Cell* 2002;**108**:545–56.
22. Sullivan JM, Alousi SS, Hikade KR, et al. Insulin induces dephosphorylation of eukaryotic initiation factor 2alpha and restores protein synthesis in vulnerable hippocampal neurons after transient brain ischemia. *J Cereb Blood Flow Metab* 1999;**19**:1010–19.
23. Towle CA, Mankin HJ, Avruch J, Treadwell BV. Insulin promoted decrease in the phosphorylation of protein synthesis initiation factor eIF-2. *Biochem Biophys Res Commun* 1984;**121**:134–40.
24. Richter JD, Sonenberg N. Regulation of cap-dependent translation by eIF4E inhibitory proteins. *Nature* 2005;**433**:477–80.
25. Minich WB, Balasta ML, Gross DJ, Rhoads RE. Chromatographic resolution of in vivo phosphorylated and nonphosphorylated eukaryotic translation initiation factor eIF-4E: Increased cap affinity of the phosphorylated form. *P Natl A Sci USA* 1994;**91**:7668–72.
26. Diggle TA, Moule SK, Avison MB, et al. Both rapamycin-sensitive and -insensitive pathways are involved in the phosphorylation of the factor-4E-binding protein (4E-BP1) in response to insulin in rat epididymal fat-cells. *Biochemical Journal* 1996;**316**:447–53.

27. Waskiewicz AJ, Johnson JC, Penn B, Mahalingam M, Kimball SR, Cooper JA. Phosphorylation of the cap-binding protein eukaryotic translation initiation factor 4E by protein kinase Mnk1 in vivo. *Mol Cell Biol* 1999;**19**:1871–80.
28. Pyronnet S, Imataka H, Gingras AC, Fukunaga R, Hunter T, Sonenberg N. Human eukaryotic translation initiation factor 4G (eIF4G) recruits mnk1 to phosphorylate eIF4E. *Embo J* 1999;**18**:270–9.
29. Scheper GC, Proud CG. Does phosphorylation of the cap-binding protein eIF4E play a role in translation initiation? *Eur J Biochem* 2002;**269**:5350–9.
30. Knauf U, Tschopp C, Gram H. Negative regulation of protein translation by mitogen-activated protein kinase-interacting kinases 1 and 2. *Mol Cell Biol* 2001;**21**:5500–11.
31. Raught B, Gingras AC, Gygi SP, et al. Serum-stimulated, rapamycin-sensitive phosphorylation sites in the eukaryotic translation initiation factor 4GI. *Embo J* 2000;**19**:434–44.
32. Raught B, Peiretti F, Gingras AC, et al. Phosphorylation of eucaryotic translation initiation factor 4B Ser422 is modulated by S6 kinases. *Embo J* 2004;**23**:1761–9.
33. Dorrello NV, Peschiaroli A, Guardavaccaro D, et al. S6K1- and betaTRCP-mediated degradation of PDCD4 promotes protein translation and cell growth. *Science* 2006;**314**:467–71.
34. Holz MK, Ballif BA, Gygi SP, Blenis J. mTOR and S6K1 mediate assembly of the translation preinitiation complex through dynamic protein interchange and ordered phosphorylation events. *Cell* 2005;**123**:569–80.
35. Ruvinsky I, Meyuhas O. Ribosomal protein S6 phosphorylation: from protein synthesis to cell size. *Trends Biochem Sci* 2006;**31**:342–8.
36. Redpath NT, Foulstone EJ, Proud CG. Regulation of translation elongation factor-2 by insulin via a rapamycin-sensitive signalling pathway. *EMBO Journal* 1996;**15**:2291–7.
37. Wang X, Li W, Williams M, Terada N, Alessi DR, Proud CG. Regulation of elongation factor 2 kinase by p90(RSK1) and p70 S6 kinase. *Embo J* 2001;**20**:4370–9.
38. Gingras AC, Raught B, Sonenberg N. Regulation of translation initiation by FRAP/mTOR. *Genes Dev* 2001;**15**:807–26.
39. Meyuhas O. Synthesis of the translational apparatus is regulated at the translational level. *Eur J Biochem* 2000;**267**:6321–30.
40. Fumagalli S, Thomas G. S6 phosphorylation and signal transduction. In: Sonenberg N, Hershey JWB, Mathews MB, editors. *Translational control of gene expression*. NY: Cold Spring Harbor Laboratory Press, Cold Spring Harbor; 2000. p. 695–717.
41. Jefferies HBJ, Fumagalli S, Dennis PB, Reinhard C, Pearson RB, Thomas G. Rapamycin suppresses 5'TOP mRNA translation through inhibition of $p70^{s6k}$. *EMBO Journal* 1997;**12**:3693–704.
42. Pende M, Um SH, Mieulet V, et al. S6K1(−/−)/S6K2(−/−) mice exhibit perinatal lethality and rapamycin-sensitive 5'-terminal oligopyrimidine mRNA translation and reveal a mitogen-activated protein kinase-dependent S6 kinase pathway. *Mol Cell Biol* 2004;**24**:3112–24.
43. Hammond ML, Bowman LH. Insulin stimulates the translation of ribosomal proteins and the transcription of rDNA in mouse myoblasts. *J Biol Chem* 1988;**263**:17,785–17,791.
44. Hannan KM, Brandenburger Y, Jenkins A, et al. mTOR-dependent regulation of ribosomal gene transcription requires S6K1 and is mediated by phosphorylation of the carboxy-terminal activation domain of the nucleolar transcription factor UBF. *Mol Cell Biol* 2003;**23**:8862–77.
45. Mayer C, Grummt I. Ribosome biogenesis and cell growth: mTOR coordinates transcription by all three classes of nuclear RNA polymerases. *Oncogene* 2006;**25**:6384–91.
46. Mendez R, Myers Jr MG, White MF, Rhoads RE. Stimulation of protein synthesis, eukaryotic translation initiation factor 4E phosphorylation , and PHAS-I phosphorylation by insulin requires insulin receptor substrate 1 and phosphatidylinositol 3-kinase. *Mol Cell Biol* 1996;**16**:2857–64.
47. West MJ, Stoneley M, Willis AE. Translational induction of the c-myc oncogene via activation of the FRAP/TOR signalling pathway. *Oncogene* 1998;**17**:769–80.
48. Oskarsson T, Trumpp A. The Myc trilogy: lord of RNA polymerases. *Nat Cell Biol* 2005;**7**:215–17.

Chapter 35

Signaling Pathways that Mediate Translational Control of Ribosome Recruitment to mRNA

Ryan J.O. Dowling and Nahum Sonenberg
Department of Biochemistry, Rosalind and Morris Goodman Cancer Centre, McGill University, Montreal, Quebec, Canada

INTRODUCTION

Translational control plays an important role in the regulation of gene expression in eukaryotes. It allows for immediate cellular responses to physiological stimuli without the need for *de novo* mRNA synthesis, and is important for several critical cellular processes including growth, proliferation, and differentiation. mRNA translation, in particular initiation, is tightly regulated by a number of mechanisms. In particular, the phosphorylation of translation initiation factors that belong to the eIF4 family plays a key role in the recruitment of ribosomes to mRNA. The signaling pathways that mediate these phosphorylation events, their effect on initiation factor activity, and their significance in the regulation of translation initiation is described in this chapter. In addition, the role of mRNA translation in cancer is discussed.

TRANSLATION INITIATION

The process of translation is divided into three stages: initiation, elongation, and termination. Initiation is the rate limiting step of translation under most circumstances and, as a result, is a primary target of translational control. Translation initiation is a complex, highly ordered process that culminates in the assembly of the 80S ribosome at the initiation codon of an mRNA. The rate limiting step in translation initiation is thought to be the formation of the eIF4F (eukaryotic initiation factor 4F) complex, which mediates the recruitment of ribosomal subunits to the initiation codon of an mRNA [1]. eIF4F consists of three subunits: eIF4E, which interacts with the 7-methylguanosine "cap" (m^7GpppX, where X is any nucleotide and m is a methyl group) found on the 5′ end of all nuclear transcribed cellular mRNAs, the helicase, eIF4A, and eIF4G [2, 3]. eIF4G is a large scaffolding protein that bridges the ribosome and the mRNA by interacting with eIF3, which binds the 40S ribosome. The 40S ribosome with its associated initiation factors scans the 5′untranslated region (UTR) until it encounters an initiation codon (AUG or a cognate thereof). Once the initiation codon is encountered, the 60S ribosome joins to form the active 80S ribosome.

A subset of mRNAs can be translated by a mechanism that does not involve recognition of the 5′ cap. This process, known as cap independent translation, does not require the 5′ cap structure or eIF4E, but instead involves binding of the 40S ribosomal subunit to an internal ribosomal entry site (IRES). The IRES is found in some viral RNAs including those of poliovirus (*Picornaviridae*), hepatitis C virus (hepacivirus genus, *Flaviviridae*), and human immunodeficiency virus (*Retroviridae*) [4]. A small number of cellular mRNAs contain IRESes, the majority of which encode proteins that are involved in cellular growth, apoptosis, and cell cycle regulation [2]. For example, the mRNAs encoding VEGF (vascular endothelial growth factor), FGF-2 (fibroblast growth factor-2), PDGF-2 (platelet derived growth factor-2), c-myc, and ODC (ornithine decarboxylase) contain IRES elements [5–9]. Cap independent translation allows for the selective expression of proteins during times when global protein synthesis is suppressed such as mitosis, virus infection, and apoptosis. During poliovirus infection, eIF4G is cleaved by a viral protease leading to an overall decline in

Handbook of Cell Signaling, Three-Volume Set 2 ed.

host cell translation [10]. However, translation of the viral mRNA continues owing to IRES mediated translation [4]. Likewise, the cellular mRNA encoding ODC is selectively translated at the G2/M phase of the cell cycle despite the general inhibition of protein synthesis that occurs as cells enter mitosis [5].

THE eIF4F COMPLEX

eIF4F comprises the cap binding protein, eIF4E, the helicase eIF4A, and the large scaffolding protein eIF4G (Figure 35.1). eIF4E was identified by its ability to crosslink to the mRNA cap structure, and was purified by affinity chromatography using sepharose coupled m^7GDP [11, 12]. eIF4A is an RNA helicase that is responsible for unwinding 5′UTR mRNA secondary structure, which is encountered by the ribosome during mRNA scanning [13]. The ATPase and helicase activities of eIF4A are stimulated by the RNA binding proteins, eIF4B and eIF4H [14–16]. The large scaffolding protein eIF4G is a critical component of the eIF4F complex. It contains binding sites for the other eIF4F members, eIF4E and eIF4A, as well as the ribosome associated eIF3, and the poly-A binding protein (PABP) [13]. The interaction between eIF4G and eIF3 is crucial for recruitment of the ribosome to the mRNA in mammalian cells (but not in yeast), whereas eIF4G's interaction with PABP is important for circularization of the mRNA [17, 18]. eIF4G also contains a binding site for the Mnk kinases (mitogen activated protein kinase interacting kinase; Mnk1, 2), which phosphorylate eIF4E on Ser209 [19, 20]. The phosphorylation of eIF4E is induced by a variety of growth promoting stimuli including hormones, growth factors, and mitogens; however, phosphorylation of eIF4E also occurs in response to cellular stress [13, 21, 22]. It was reported that eIF4E phosphorylation stimulates mRNA translation [22–24]. In accordance with this, phosphorylation of eIF4E on Ser209 is important for development and growth in *Drosophila* [25], and plays an important role in tumorigenesis in mice [26]. However, there are also reports that eIF4E phosphorylation inhibits mRNA translation [27, 28].

REGULATION OF TRANSLATION INITIATION BY mTOR

The PI3K/AKT/mTOR signaling pathway (Figure 35.2) is critical for the regulation of cellular growth, proliferation, and survival. The serine/threonine protein kinase mTOR (also known as FKBP12-rapamycin associated protein (FRAP), or rapamycin and FKBP12 target (RAFT)) is a major effector of PI3K signaling. Two mTOR containing complexes exist in mammalian cells; mTORC1 and mTORC2 (Figure 2). mTORC1 is rapamycin-sensitive, regulated by nutrients and growth factors, and composed of mTOR, raptor (regulatory associated protein of TOR), PRAS40 (Proline-rich AKT substrate 40 kDa), and mLst8 (also known as GβL). The major downstream targets of mTORC1, the 4E-BPs and S6Ks, are well characterized and will be discussed below. mTORC2 is rapamycin-insensitive and composed of mTOR, mLst8, rictor (rapamycin insensitive component of TOR), PRR5 (Proline-rich protein 5, also known as Protor) and mSIN1 (mammalian stress-activated protein kinase (SAPK)-interacting protein). mTORC2 is responsible for phosphorylating Ser 473 in AKT. Thus, mTORC2 controls the activity of mTORC1. In addition, it also controls cytoskeletal organization. Due to the role of mTORC1 in the phosphorylation of translational regulators, this chapter will mainly focus on mTORC1 signalling. mTOR integrates signals from nutrients, growth factors, hormones, cellular energy stores, oxygen levels, and other cues to control a variety of important cellular processes including growth, proliferation,

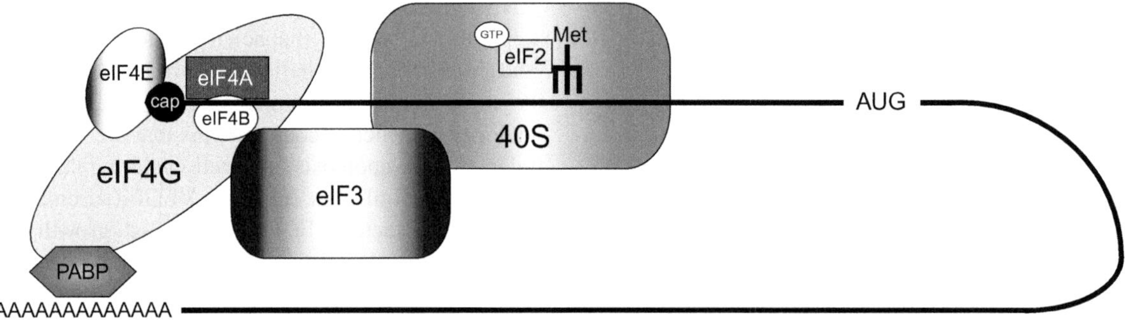

FIGURE 35.1 eIF4F complex formation.
The eIF4F complex comprises the cap binding protein, eIF4E, the helicase eIF4A, and the large scaffolding protein eIF4G and is responsible for recruiting ribosomal subunits to the initiation codon of an mRNA. eIF4G contains binding sites for eIF4E, eIF4A, as well as eIF3 and PABP. The interaction between eIF4G and eIF3 bridges the 40S ribosome and the mRNA, while eIF4G's interaction with PABP is important for circularization of the mRNA. Note: the size of the translation components is not to scale.

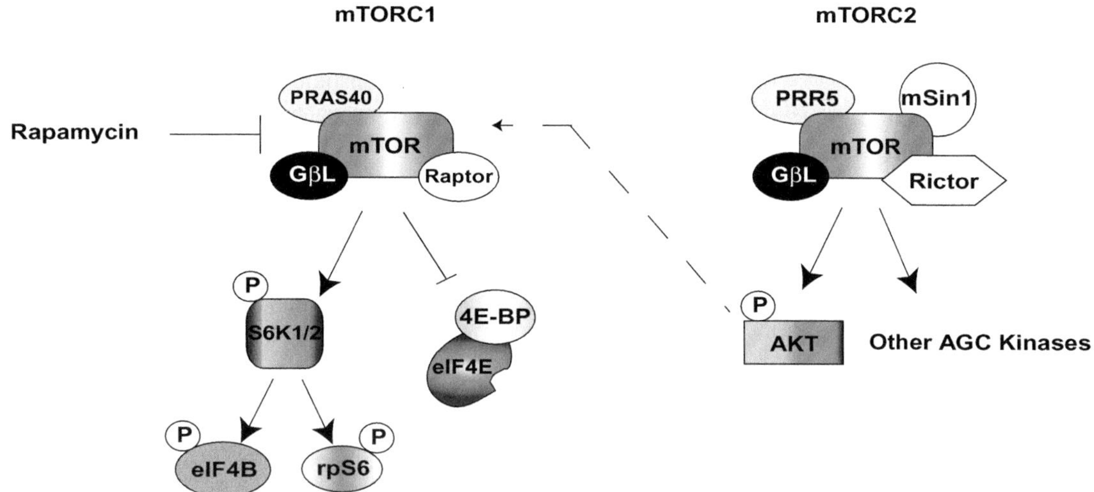

FIGURE 35.2 mTORC1 and mTORC2 signalling.
mTORC1 is composed of mTOR, raptor, PRAS40 (proline-rich AKT substrate 40 kDa and GβL (mLst8) and signals downstream to proteins involved in the regulation of mRNA translation. mTORC2 is composed of mTOR, GβL (mLst8), rictor, PRR5 (protor) and mSin1 and regulates AKT as well as other AGC family kinases. mTORC2 is implicated in the control of cytoskeletal organization.

differentiation, transcription, cytoskeletal organization, autophagy, and mRNA translation [29–32]. A key mechanism by which mTOR coordinates these processes is the regulation of mRNA translation initiation via phosphorylation of its two major downstream targets: the eIF4E binding proteins (4E-BPs) and the ribosomal protein S6 kinases (S6K1 and S6K2) (see below). In mammalian cells, mTOR signaling to its major downstream effectors is activated in response to growth factors and mitogens binding to cellular surface receptors (Figure 35.2). Upon activation by growth factors, PI3K phosphorylates phosphatidylinositol-4,5-biphosphate (PIP2) to convert it to phosphatidylinositol-3,4,5-triphosphate (PIP3). The pleckstrin homology domain bearing proteins AKT and PDK1, which are serine/threonine kinases, are recruited to the plasma membrane by PIP3, where PDK1 phosphorylates and activates AKT. AKT phosphorylates and inactivates TSC2 (tuberous sclerosis complex), a subunit of the TSC1/2 (hamartin/tuberin) complex that negatively regulates mTOR signaling. TSC2 is the GAP (GTPase activating protein) for the small GTPase, Rheb (Ras homolog enriched in brain), which activates mTOR in its GTP bound form. When inactivated by phosphorylation, TSC2 no longer hydrolyzes Rheb-GTP, leading to the accumulation of Rheb-GTP and activation of mTOR [33].

The 4E-BPs are a family of inhibitory proteins (4E-BP1, 2, and 3 in mammals) that negatively control the assembly of the eIF4F complex by competing with eIF4G for binding to eIF4E [34]. The binding of the 4E-BPs to eIF4E is regulated by mTOR- mediated phosphorylation. Hypophosphorylated 4E-BP1 (the best characterized member of the 4E-BPs) binds to eIF4E with high affinity and inhibits the formation of the eIF4F complex, which leads to a suppression of cap dependent translation. Upon stimulation by growth factors, nutrients, or hormones, mTOR phosphorylates 4E-BP1, leading to its release from eIF4E, and a subsequent increase in cap dependent translation [3, 35]. The phosphorylation of 4E-BP1 occurs in an ordered, hierarchical manner. mTOR directly phosphorylates 4E-BP1 on Thr37 and Thr46. The phosphorylation of these residues acts as a priming event, which is necessary for the subsequent phosphorylation of residues Thr70 and Ser65, and the ultimate release of 4E-BP1 from eIF4E [3, 36].

Phosphorylation of S6K1 by mTOR enhances its kinase activity toward its downstream substrates including the 40S ribosomal protein S6, eIF4B [37], and SKAR (S6K1 Aly/REF-like target) [38]. eIF4B is an RNA binding protein that stimulates the activity of eIF4A, and plays an important role in the recruitment of ribosomes to the mRNA. eIF4B is phosphorylated on Ser422 by S6K and the kinase RSK (p90 ribosomal protein S6 kinase) in response to insulin and growth factors. Phosphorylation of eIF4B leads to enhanced translation, likely due to increased interaction between phosphorylated eIF4B and eIF3 [37, 39]. The modular scaffolding protein eIF4G is also phosphorylated in an mTOR dependent manner. For example, eIF4G phosphorylation increases in response to serum, insulin, and growth factors, and this phosphorylation is sensitive to treatment with the mTOR inhibitor, rapamycin [32, 40]. The biological significance of eIF4G phosphorylation is unclear; however, phosphorylation may induce a conformational change in eIF4G, leading to an alteration of its activity in mRNA translation.

TRANSLATION AND CANCER

Due to its role in cellular growth and proliferation, it is not surprising that the control of mRNA translation is implicated in the development of cancer. It is thought that elevated translation contributes to tumorigenesis via increased synthesis of growth factors, antiapoptotic factors, and cell cycle regulators. Many of these proteins including survivin, cyclin D1, TGF-β (transforming growth factor), CDK4 (cyclin dependent kinase), and IGF-II (insulin-like growth factor), are encoded by a subset of mRNAs that contain long, highly structured 5′UTRs [41–44]. Efficient translation of these mRNAs is thought to require high levels of the eIF4F complex, which through the actions of its eIF4A helicase, facilitates unwinding of the extensive mRNA secondary structure. Therefore, increased expression of translation initiation factors, particularly eIF4E, is believed to promote tumor growth and progression. eIF4E is the least abundant translation initiation factor in mammalian cells [45], and overexpression of eIF4E results in transformation and oncogenesis. Overexpression of eIF4E transforms immortalized rodent fibroblasts [46] and in cooperation with the immortalizing genes E1A or *myc*, causes the malignant transformation of primary embryo fibroblasts [47]. Ectopic expression of eIF4E also caused the transformation of human mammary epithelial cells, enabling clonal expansion and anchorage-independent growth [48]. In mice, overexpression of eIF4E promotes the development of lymphomas, angiosarcomas, lung adenocarcinomas, and hepatocellular adenomas [49, 50]. Consistent with its role in tumorigenesis, eIF4E levels are elevated in a number of human cancers including those of the colon, breast, bladder, lung, and prostate [42, 51, 52]. Consequently, eIF4E has emerged as a target for anticancer therapy. For example, antisense oligonucleotides against eIF4E suppressed tumor growth in mice without causing toxicity and are currently under investigation in phase I clinical trials (Eli Lilly Co.) in human patients [53, 54]. Also, reduction of eIF4E levels by RNA interference inhibited the growth of head and neck squamous carcinoma cells as well as cell lines derived from breast tumors [55, 56]. The other eIF4F members are also implicated in human cancer, as eIF4A and eIF4G levels are elevated in melanoma and squamous cell lung carcinoma, respectively [57–59].

While eIF4E acts as an oncoprotein and contributes to tumorigenesis, the 4E-BPs act as tumor suppressors by inhibiting eIF4F complex formation and translation initiation. In fact, overexpression of 4E-BP1 or 4E-BP2 in NIH 3T3 cells that have been transformed by eIF4E, *ras*, or *src*, caused a reversion of the transformed phenotype [60]. In addition, overexpression of a non-phosphorylatable 4E-BP1, which constitutively binds to and inhibits eIF4E, reduced the tumorigenicity of breast cancer cells [48]. In humans, high levels of phosphorylated 4E-BP1 in ovarian, breast, and prostate tumors have been correlated with malignant progression and poor prognosis [61–63].

Translational control plays a key role in the regulation of cellular proliferation and the development of cancer via mTOR. A number of tumor suppressors (LKB1, PTEN, and TSC2) negatively regulate PI3K/AKT/mTOR signaling, and loss or inactivation of these proteins results in increased phosphorylation of S6K and 4E-BP1 and a cellular environment that is conducive to transformation [64–66]. Consistent with these data, mutational loss or inactivation of LKB1, PTEN, and TSC2 is implicated in a variety of human cancers including those of the breast, prostate, brain, colon, and kidney [67, 68]. Consequently, compounds that inhibit mTOR are being explored as anticancer therapies. For instance, rapamycin is a specific mTOR inhibitor, which suppresses the growth of a variety of cancer cell lines both *in vitro* and *in vivo*. As a result, a number of rapamycin derivatives (rapalogs) were developed for the treatment of cancer in humans. Three rapamycin analogs are currently under investigation in phase I–III clinical trials for the treatment of cancer: CCI-779 (temsirolimus, torisel; Wyeth-Ayerst, PA, USA), RAD001 (everolimus, certican; Novartis, Switzerland), and AP23573 (deforolimus; Ariad Pharmaceuticals, MA, USA) [67, 69, 70]. Importantly, CCI-779 (temsirolimus, torisel) was approved by the U.S. Food and Drug Administration in May 2007 for the treatment of patients with advanced renal cell carcinoma [69]. In addition, RAD001 (everolimus) was also FDA approved for the treatment of advanced renal cell carcinoma in March 2009.

CONCLUSIONS

Translation initiation is a highly ordered process that is regulated primarily by phosphorylation of initiation factors, in particular those that are involved in 5′ mRNA cap recognition and eIF4F complex formation. Increased expression or differential phosphorylation of these initiation factors leads to changes in cellular translation rates, which can result in drastic changes in growth, proliferation, differentiation, and survival. For example, increased translation as a result of overexpression of eIF4E [51], or inappropriate activation of mTOR signaling to the 4E-BPs [67] is implicated in a variety of human cancers. Therefore, eukaryotic cells have evolved intricate mechanisms to regulate mRNA translation initiation. A number of significant advancements have been made in the understanding of the mechanisms underlying translational control of gene expression. However, a great deal of work still needs to be performed. For example, the functional significance of eIF4E phosphorylation is still unclear. Future studies focusing on the role of eIF4E phosphorylation in translation initiation and its potential involvement in malignant transformation are required. In addition, mRNA translation has emerged as a target for anticancer therapy. Despite the recent success of anticancer therapies designed to inhibit translation, either by targeting eIF4E directly [54] or suppressing mTOR [67, 69], it is imperative that we continue to study the mechanisms of translational control in order to fully understand the process of translation initiation and its role in human disease.

REFERENCES

1. Pestova TV, Lorsch JR, Hellen CUT. The mechanism of translation initiation in eukaryotes. In: Mathews MB, Sonenberg N, Hershey JWB, editors. *Translational control in biology and medicine*. Cold Spring Harbor, NY: Cold Spring Harbor Laboratory Press; 2007. p. 87–128.
2. Holcik M, Sonenberg N. Translational control in stress and apoptosis. *Nat Rev Mol Cell Biol* 2005;**6**:318–27.
3. Gingras AC, Gygi SP, Raught B, Polakiewicz RD, Abraham RT, Hoekstra MF, Aebersold R, Sonenberg N. Regulation of 4E-BP1 phosphorylation: a novel two-step mechanism. *Genes Dev* 1999;**13**:1422–37.
4. Doudna JA, Sarnow P. Translation initiation by viral internal ribosome entry sites. In: Mathews MB, Sonenberg N, Hershey JWB, editors. *Translational control in biology and medicine*. Cold Spring Harbor, NY: Cold Spring Harbor Laboratory Press; 2007. p. 129–53.
5. Pyronnet S, Pradayrol L, Sonenberg N. A cell cycle-dependent internal ribosome entry site. *Mol Cell* 2000;**5**:607–16.
6. Stein I, Itin A, Einat P, Skaliter R, Grossman Z, Keshet E. Translation of vascular endothelial growth factor mRNA by internal ribosome entry: implications for translation under hypoxia. *Mol Cell Biol* 1998;**18**:3112–19.
7. Vagner S, Gensac MC, Maret A, Bayard F, Amalric F, Prats H, Prats AC. Alternative translation of human fibroblast growth factor 2 mRNA occurs by internal entry of ribosomes. *Mol Cell Biol* 1995;**15**:35–44.
8. Bernstein J, Sella O, Le SY, Elroy-Stein O. PDGF2/c-sis mRNA leader contains a differentiation-linked internal ribosomal entry site (D-IRES). *J Biol Chem* 1997;**272**:9356–62.
9. Nanbru C, Lafon I, Audigier S, Gensac MC, Vagner S, Huez G, Prats AC. Alternative translation of the proto-oncogene c-myc by an internal ribosome entry site. *J Biol Chem* 1997;**272**:32,061–32,066.
10. Etchison D, Milburn SC, Edery I, Sonenberg N, Hershey JW. Inhibition of HeLa cell protein synthesis following poliovirus infection correlates with the proteolysis of a 220,000-dalton polypeptide associated with eucaryotic initiation factor 3 and a cap binding protein complex. *J Biol Chem* 1982;**257**:14,806–14,810.
11. Sonenberg N, Morgan MA, Merrick WC, Shatkin AJ. A polypeptide in eukaryotic initiation factors that crosslinks specifically to the 5′-terminal cap in mRNA. *Proc Natl Acad Sci USA* 1978;**75**:4843–7.
12. Sonenberg N, Rupprecht KM, Hecht SM, Shatkin AJ. Eukaryotic mRNA cap binding protein: purification by affinity chromatography on sepharose-coupled m7GDP. *Proc Natl Acad Sci USA* 1979;**76**:4345–9.
13. Gingras AC, Raught B, Sonenberg N. eIF4 initiation factors: effectors of mRNA recruitment to ribosomes and regulators of translation. *Annu Rev Biochem* 1999;**68**:913–63.
14. Rozen F, Edery I, Meerovitch K, Dever TE, Merrick WC, Sonenberg N. Bidirectional RNA helicase activity of eucaryotic translation initiation factors 4A and 4F. *Mol Cell Biol* 1990;**10**:1134–44.
15. Altmann M, Muller PP, Wittmer B, Ruchti F, Lanker S, Trachsel H. A Saccharomyces cerevisiae homologue of mammalian translation initiation factor 4B contributes to RNA helicase activity. *Embo J* 1993;**12**:3997–4003.
16. Rogers Jr. GW, Komar AA, Merrick WC eIF4A: the godfather of the DEAD box helicases. *Prog Nucleic Acid Res Mol Biol* 2002;**72**:307–31.
17. Wells SE, Hillner PE, Vale RD, Sachs AB. Circularization of mRNA by eukaryotic translation initiation factors. *Mol Cell* 1998;**2**:135–40.
18. Wakiyama M, Imataka H, Sonenberg N. Interaction of eIF4G with poly(A)-binding protein stimulates translation and is critical for Xenopus oocyte maturation. *Curr Biol* 2000;**10**:1147–50.
19. Pyronnet S, Imataka H, Gingras AC, Fukunaga R, Hunter T, Sonenberg N. Human eukaryotic translation initiation factor 4G (eIF4G) recruits mnk1 to phosphorylate eIF4E. *Embo J* 1999;**18**:270–9.
20. Waskiewicz AJ, Flynn A, Proud CG, Cooper JA. Mitogen-activated protein kinases activate the serine/threonine kinases Mnk1 and Mnk2. *Embo J* 1997;**16**:1909–20.
21. Morley SJ, McKendrick L. Involvement of stress-activated protein kinase and p38/RK mitogen-activated protein kinase signaling pathways in the enhanced phosphorylation of initiation factor 4E in NIH 3T3 cells. *J Biol Chem* 1997;**272**:17,887–17,893.
22. Scheper GC, Proud CG. Does phosphorylation of the cap-binding protein eIF4E play a role in translation initiation? *Eur J Biochem* 2002;**269**:5350–9.
23. Ishida M, Ishida T, Nakashima H, Miho N, Miyagawa K, Chayama K, Oshima T, Kambe M, Yoshizumi M. Mnk1 is required for angiotensin II-induced protein synthesis in vascular smooth muscle cells. *Circ Res* 2003;**93**:1218–24.
24. Fraser CS, Pain VM, Morley SJ. Cellular stress in xenopus kidney cells enhances the phosphorylation of eukaryotic translation initiation factor (eIF)4E and the association of eIF4F with poly(A)-binding protein. *Biochem J* 1999;**342**:519–26.
25. Lachance PE, Miron M, Raught B, Sonenberg N, Lasko P. Phosphorylation of eukaryotic translation initiation factor 4E is critical for growth. *Mol Cell Biol* 2002;**22**:1656–63.
26. Wendel HG, Silva RL, Malina A, Mills JR, Zhu H, Ueda T, Watanabe-Fukunaga R, Fukunaga R, Teruya-Feldstein J, Pelletier J, Lowe SW. Dissecting eIF4E action in tumorigenesis. *Genes Dev* 2007;**21**:3232–7.
27. Knauf U, Tschopp C, Gram H. Negative regulation of protein translation by mitogen-activated protein kinase-interacting kinases 1 and 2. *Mol Cell Biol* 2001;**21**:5500–11.
28. Ross G, Dyer JR, Castellucci VF, Sossin WS. Mnk is a negative regulator of cap-dependent translation in Aplysia neurons. *J Neurochem* 2006;**97**:79–91.
29. Sarbassov DD, Ali SM, Sabatini DM. Growing roles for the mTOR pathway. *Curr Opin Cell Biol* 2005;**17**:596–603.
30. Yang Q, Guan KL. Expanding mTOR signaling. *Cell Res* 2007;**17**:666–81.
31. Wullschleger S, Loewith R, Hall MN. TOR signaling in growth and metabolism. *Cell* 2006;**124**:471–84.

32. Hay N, Sonenberg N. Upstream and downstream of mTOR. *Genes Dev* 2004;**18**:1926–45.
33. Inoki K, Li Y, Zhu T, Wu J, Guan KL. TSC2 is phosphorylated and inhibited by Akt and suppresses mTOR signalling. *Nat Cell Biol* 2002;**4**:648–57.
34. Gingras AC, Raught B, Sonenberg N. Regulation of translation initiation by FRAP/mTOR. *Genes Dev* 2001;**15**:807–26.
35. Pause Jr. A, Belsham GJ, Gingras AC, Donze O, Lin TA, Lawrence JC, Sonenberg N Insulin-dependent stimulation of protein synthesis by phosphorylation of a regulator of 5′-cap function. *Nature* 1994;**371**:762–7.
36. Gingras AC, Raught B, Gygi SP, Niedzwiecka A, Miron M, Burley SK, Polakiewicz RD, Wyslouch-Cieszynska A, Aebersold R, Sonenberg N. Hierarchical phosphorylation of the translation inhibitor 4E-BP1. *Genes Dev* 2001;**15**:2852–64.
37. Shahbazian D, Roux PP, Mieulet V, Cohen MS, Raught B, Taunton J, Hershey JW, Blenis J, Pende M, Sonenberg N. The mTOR/PI3K and MAPK pathways converge on eIF4B to control its phosphorylation and activity. *Embo J* 2006;**25**:2781–91.
38. Richardson CJ, Broenstrup M, Fingar DC, Julich K, Ballif BA, Gygi S, Blenis J. SKAR is a specific target of S6 kinase 1 in cell growth control. *Curr Biol* 2004;**14**:1540–9.
39. Holz MK, Ballif BA, Gygi SP, Blenis J. mTOR and S6K1 mediate assembly of the translation preinitiation complex through dynamic protein interchange and ordered phosphorylation events. *Cell* 2005;**123**:569–80.
40. Raught B, Gingras AC, Gygi SP, Imataka H, Morino S, Gradi A, Aebersold R, Sonenberg N. Serum-stimulated, rapamycin-sensitive phosphorylation sites in the eukaryotic translation initiation factor 4GI. *Embo J* 2000;**19**:434–44.
41. Mamane Y, Petroulakis E, Martineau Y, Sato TA, Larsson O, Rajasekhar VK, Sonenberg N. Epigenetic activation of a subset of mRNAs by eIF4E explains its effects on cell proliferation. *PLoS ONE* 2007;**2**: e242.
42. Clemens MJ, Bommer UA. Translational control: the cancer connection. *Int J Biochem Cell Biol* 1999;**31**:1–23.
43. Nielsen FC, Ostergaard L, Nielsen J, Christiansen J. Growth-dependent translation of IGF-II mRNA by a rapamycin-sensitive pathway. *Nature* 1995;**377**:358–62.
44. Willis AE. Translational control of growth factor and proto-oncogene expression. *Int J Biochem Cell Biol* 1999;**31**:73–86.
45. Duncan R, Milburn SC, Hershey JW. Regulated phosphorylation and low abundance of HeLa cell initiation factor eIF-4F suggest a role in translational control. Heat shock effects on eIF-4F. *J Biol Chem* 1987;**262**:380–8.
46. Lazaris-Karatzas A, Montine KS, Sonenberg N. Malignant transformation by a eukaryotic initiation factor subunit that binds to mRNA 5′ cap. *Nature* 1990;**345**:544–7.
47. Lazaris-Karatzas A, Sonenberg N. The mRNA 5′ cap-binding protein, eIF-4E, cooperates with v-myc or E1A in the transformation of primary rodent fibroblasts. *Mol Cell Biol* 1992;**12**:1234–8.
48. Avdulov S, Li S, Michalek V, Burrichter D, Peterson M, Perlman DM, Manivel JC, Sonenberg N, Yee D, Bitterman PB, Polunovsky VA. Activation of translation complex eIF4F is essential for the genesis and maintenance of the malignant phenotype in human mammary epithelial cells. *Cancer Cell* 2004;**5**:553–63.
49. Ruggero D, Montanaro L, Ma L, Xu W, Londei P, Cordon-Cardo C, Pandolfi PP. The translation factor eIF-4E promotes tumor formation and cooperates with c-Myc in lymphomagenesis. *Nat Med* 2004;**10**:484–6.
50. Wendel HG, De Stanchina E, Fridman JS, Malina A, Ray S, Kogan S, Cordon-Cardo C, Pelletier J, Lowe SW. Survival signalling by Akt and eIF4E in oncogenesis and cancer therapy. *Nature* 2004;**428**:332–7.
51. Mamane Y, Petroulakis E, Rong L, Yoshida K, Ler LW, Sonenberg N. eIF4E: from translation to transformation. *Oncogene* 2004;**23**:3172–9.
52. De Benedetti A, Harris AL. eIF4E expression in tumors: its possible role in progression of malignancies. *Int J Biochem Cell Biol* 1999;**31**:59–72.
53. Graff JR, Konicek BW, Carter JH, Marcusson EG. Targeting the eukaryotic translation initiation factor 4E for cancer therapy. *Cancer Res* 2008;**68**:631–4.
54. Graff JR, Konicek BW, Vincent TM, Lynch RL, Monteith D, Weir SN, Schwier P, Capen A, Goode RL, Dowless MS, Chen Y, Zhang H, Sissons S, Cox K, McNulty AM, Parsons SH, Wang T, Sams L, Geeganage S, Douglass LE, Neubauer BL, Dean NM, Blanchard K, Shou J, Stancato LF, Carter JH, Marcusson EG. Therapeutic suppression of translation initiation factor eIF4E expression reduces tumor growth without toxicity. *J Clin Invest* 2007;**117**:2638–48.
55. Oridate N, Kim HJ, Xu X, Lotan R. Growth inhibition of head and neck squamous carcinoma cells by small interfering RNAs targeting eIF4E or cyclin D1 alone or combined with cisplatin. *Cancer Biol Ther* 2005;**4**:318–23.
56. Dong K, Wang R, Wang X, Lin F, Shen JJ, Gao P, Zhang HZ. Tumor-specific RNAi targeting eIF4E suppresses tumor growth, induces apoptosis and enhances cisplatin cytotoxicity in human breast carcinoma cells. *Breast Cancer Res Treat* 2008;**113**:443–56.
57. Bauer C, Brass N, Diesinger I, Kayser K, Grasser FA, Meese E. Overexpression of the eukaryotic translation initiation factor 4G (eIF4G-1) in squamous cell lung carcinoma. *Int J Cancer* 2002;**98**:181–5.
58. Bauer C, Diesinger I, Brass N, Steinhart H, Iro H, Meese EU. Translation initiation factor eIF-4G is immunogenic, overexpressed, and amplified in patients with squamous cell lung carcinoma. *Cancer* 2001;**92**:822–9.
59. Eberle J, Krasagakis K, Orfanos CE. Translation initiation factor eIF-4A1 mRNA is consistently overexpressed in human melanoma cells in vitro. *Int J Cancer* 1997;**71**:396–401.
60. Rousseau D, Gingras AC, Pause A, Sonenberg N. The eIF4E-binding proteins 1 and 2 are negative regulators of cell growth. *Oncogene* 1996;**13**:2415–20.
61. Armengol G, Rojo F, Castellvi J, Iglesias C, Cuatrecasas M, Pons B, Baselga J, Ramon y Cajal S. 4E-binding protein 1: a key molecular "funnel factor" in human cancer with clinical implications. *Cancer Res* 2007;**67**:7551–5.
62. Castellvi J, Garcia A, Rojo F, Ruiz-Marcellan C, Gil A, Baselga J, Ramon y Cajal S. Phosphorylated 4E binding protein 1: a hallmark of cell signaling that correlates with survival in ovarian cancer. *Cancer* 2006;**107**:1801–11.
63. Rojo F, Najera L, Lirola J, Jimenez J, Guzman M, Sabadell MD, Baselga J, Ramon y Cajal S. 4E-binding protein 1, a cell signaling hallmark in breast cancer that correlates with pathologic grade and prognosis. *Clin Cancer Res* 2007;**13**:81–9.
64. Shaw RJ, Bardeesy N, Manning BD, Lopez L, Kosmatka M, DePinho RA, Cantley LC. The LKB1 tumor suppressor negatively regulates mTOR signaling. *Cancer Cell* 2004;**6**:91–9.
65. Neshat MS, Mellinghoff IK, Tran C, Stiles B, Thomas G, Petersen R, Frost P, Gibbons JJ, Wu H, Sawyers CL. Enhanced sensitivity of PTEN-deficient tumors to inhibition of FRAP/mTOR. *Proc Natl Acad Sci USA* 2001;**98**:10,314–10,319.
66. Jaeschke A, Hartkamp J, Saitoh M, Roworth W, Nobukuni T, Hodges A, Sampson J, Thomas G, Lamb R. Tuberous sclerosis complex tumor

suppressor-mediated S6 kinase inhibition by phosphatidylinositide-3-OH kinase is mTOR independent. *J Cell Biol* 2002;**159**:217–24.
67. Petroulakis E, Mamane Y, Le Bacquer O, Shahbazian D, Sonenberg N. mTOR signaling: implications for cancer and anticancer therapy. *Br J Cancer* 2006;**94**:195–9.
68. Guertin DA, Sabatini DM. An expanding role for mTOR in cancer. *Trends Mol Med* 2005;**11**:353–61.
69. Abraham RT, Eng CH. Mammalian target of rapamycin as a therapeutic target in oncology. *Expert Opin Ther Targets* 2008;**12**:209–22.
70. Faivre S, Kroemer G, Raymond E. Current development of mTOR inhibitors as anticancer agents. *Nat Rev Drug Discov* 2006;**5**:671–88.

Chapter 36

Nuclear and Cytoplasmic Functions of Abl Tyrosine Kinase

Jean Y. J. Wang
Moores Cancer Center, Division of Hematology-Oncology, Department of Medicine, University of California at San Diego, La Jolla, California

INTRODUCTION

The *Abl* gene encodes a non-receptor tyrosine kinase that is conserved through evolution. *Abl* was discovered by virtue of its identity to the oncogene of Abelson murine leukemia virus (A-MuLV), which encodes a Gag-Abl fusion protein that drives the development of lymphoma. The oncogenic potential of *Abl* is also activated in human chronic myelogenous leukemia (CML) through the formation of BCR-ABL fusion protein via chromosomal translocation. The oncogenic function of Gag-Abl and BCR-ABL requires the Abl tyrosine kinase activity. This is best demonstrated by the clinical success in treating CML with Abl kinase inhibitors, i.e., imatinib and dasatinib [1]. The constitutively activated BCR-ABL tyrosine kinase resides exclusively in the cytoplasm of transformed cells where it stimulates a large number of tyrosine kinase regulated signaling pathways. By contrast, wild-type Abl kinase activity is tightly regulated, it shuttles between the cytoplasmic and the nuclear compartments, and its overproduction does not lead to cell transformation. The mammalian *Abl* gene is expressed in all cell types examined, including embryonic stem cells and the mature sperms. Inactivation of the *Abl* gene in mice causes neonatal lethality and a series of low penetrant phenotypes including lymphopenia, osteoporosis, defects of splenocytes in responding to bacterial lipopolysaccaride, and resistance to hyperoxia induced retinopathy. The mammalian genomes contain an *Abl* related gene (*Arg*). The *Arg* knockout mice are healthy and fertile. However, the double knockout of *Abl* and *Arg* causes early embryonic lethality. Thus, *Abl* and *Arg* have redundant functions during early embryonic development, whereas *Abl* also has unique functions that cannot be compensated by *Arg*. This chapter focuses on the signaling functions of the mammalian Abl tyrosine kinase, with an emphasis on the role of nuclear Abl in cellular response to genotoxic stress.

FUNCTIONAL DOMAINS OF ABL

N-Terminal Region: Kinase Function

The mammalian Abl protein consists of a series of functional modules (Figure 36.1). The N-terminus of Abl is variable, encoded by two alternative 5′-exons. This variable domain is followed by the Src homology (SH) domains 3, 2, and the tyrosine kinase domain. The Abl SH3 domain exerts a negative effect on the catalytic activity. This inhibitory effect is mediated by an intramolecular interaction between the SH3 domain and the kinase N-lobe, similar to that found in the inhibited conformation of the Src tyrosine kinase [2]. The inhibited conformation involves a second intramolecular interaction between the SH2 domain and the kinase C-lobe. This interaction requires the kinase C-lobe binding to a myristoyl group, which can be donated by the N-terminal myristoylation in the mouse type-IV (human type Ib) Abl variant [2]. The Abl SH3 and SH2 domains can also engage in intermolecular interactions with other cellular proteins, including activators, inhibitors, or substrates of the Abl kinase (Table 36.1).

C-Terminal Region: Localization Cues

The C-terminal region of Abl specifies its subcellular localization. The mammalian Abl contains three nuclear localization signals (NLSs) (Figure 36.1). Disruption of all three NLSs is required to prevent Abl from entering the nucleus. The mammalian Abl also contains a nuclear export signal (NES) that binds to exportin-1. Mutation of the NES or inhibition of exportin-1 by leptomycin-B causes the nuclear accumulation of Abl. In proliferating cells, Abl undergoes nucleo-cytoplasmic shuttling such that its subcellular distribution at steady state is determined by a dynamic equilibrium between nuclear import and export.

Handbook of Cell Signaling, Three-Volume Set 2 ed.

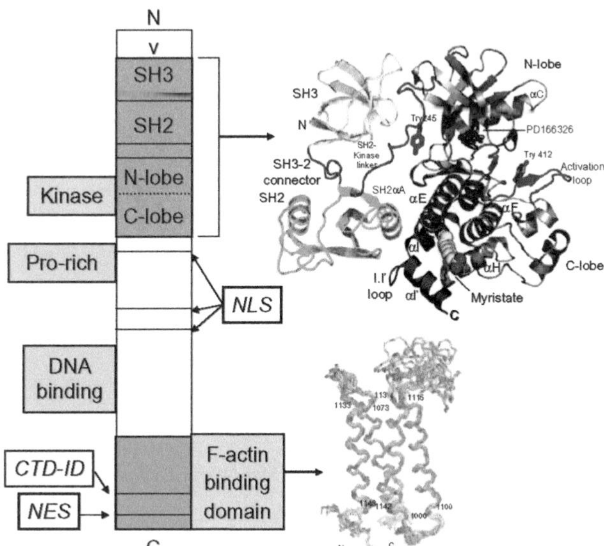

FIGURE 36.1 Functional domains of mammalian Abl.
The variable N terminus (V) is encoded by two different 5′-exons driven by two different promoters. The murine type I (human Ia) exon encodes 26 amino acids. The murine type IV (human Ib) exon encodes 45 amino acids. SH3, Src homology 3; SH2, Src homology 2: the crystal structure of the SH3-SH2 kinase domain is shown in the conformation of an autoinhibited assembly [2]; NLS, nuclear localization signal, of which there are three in Abl; NES, nuclear export signal, which is embedded in the F-actin binding domain: the DNA binding domain interacts with A/T-rich DNA with a distorted structure. Proline-rich (Pro-Rich) motifs that bind to SH3 domains of several adaptor proteins are found immediately C-terminal to the kinase domain. CTD-ID, binding site for the CTD of RNA polymerase II: the CTD-ID is conserved in mammalian Abl and Arg, it is also embedded in the F-actin binding domain. The NMR structure of the F-actin binding domain at the very C-terminus is shown [40].

In the cytoplasm, Abl associates with the actin cytoskeleton through a C-termimal F-actin binding domain (Figure 36.1). Purified mammalian Abl protein is an active kinase, which can be inhibited by F-actin *in vitro* and this inhibition is abolished by mutation of the F-actin binding domain [3]. The cytoplasmic Abl plays a role in actin dependent cellular processes, including cell adhesion, cell spreading, cell migration, and axon guidance [3].

In the nucleus, Abl associates with the chromatin. The mammalian Abl protein directly binds DNA through a DNA binding domain that prefers double-stranded DNA with A/T-rich sequences (Figure 36.1). The nuclear Abl interacts with a number of transcription factors, including p53, p73, CREB, c-Jun, RXFI, and E2F-1, either directly or indirectly through scaffolding proteins. Furthermore, Abl directly interacts with the C-terminal repeated domain (CTD) of the catalytic subunit of RNA polymerase II. The nuclear Abl also interacts with DNA damage signaling proteins, including ATM, RAD51, RAD52, WRN, UV-DDB1, and topoisomerase-1 (Table 36.1). Thus, nuclear Abi regulates gene expression and DNA damage response.

Functional Interactions between the N- and C-Terminal Regions of Abl

Functional interactions between the N- and the C-terminal regions of Abl are suggested by the following observations: (a) F-actin binding to the Abl C-terminus exerts a negative effect on its kinase activity only when the Abl SH2 domain is intact [4]; (b) the three NLSs in the C-terminal region are subjected to regulation by the conformation of the kinase domain such that the constitutively active BCR-ABL does not undergo nuclear import but its NLSs can be reactivated by imatinib, which binds to the kinase N-lobe [5]. The full extent of the interactions between the N- and C-regions of Abl are presently unclear and will require the solution of the three-dimensional structure of the full length protein.

PROTEINS THAT INTERACT WITH ABL

The multiple functional domains of Abl direct its interaction with a large number of proteins, selected examples of which are listed in Table 36.1, and briefly discussed below.

SH3 Binding Proteins

The Abl SH3 binding proteins are mostly activators and/or substrates of the Abl kinase (Table 36.1). As mentioned earlier, the SH3 domain engages in an intramolecular interaction to impose an inactive conformation on Abl kinase [2]. Therefore, SH3 binding proteins that disrupt the SH3-N-lobe assembly will disrupt the autoinhibited conformation to activate the Abl kinase. Several SH3 binding Abl substrates were initially identified as "inhibitors" of Abl kinase. Those conclusions were biased by the experiments, where substrate competition for binding to the Abl SH3 domain was misinterpreted as inhibition of the kinase catalytic activity. Another cause for the biased conclusion might be the use of truncated proteins that could have exerted a dominant negative effect. However, it is conceivable that a protein may bind to the Abl SH3 domain but does not interfere with the autoinhibitory assembly. Furthermore, an SH3 binding protein may even stabilize the autoinhibited assembly to enforce kinase inhibition. The fact that an Abl SH3 binding protein, PAG (Table 36.1), has been shown to inhibit kinase activity supports the concept of co-inhibition. In this concept, co-inhibitors are proteins that enforce the autoinhibitory assembly to prevent Abl kinase activation [6].

The SH3 domain is not found in the Gag-Abl oncoprotein of A-MuLV, accounting for the constitutive activation of this oncogenic tyrosine kinase. Mutations that disrupt the SH3-N-lobe assembly in Abl can lead to the constitutive activation of kinase activity and the acquisition of oncogenic activity, albeit of weaker potency than Gag-Abl. While the SH3 domain is dispensable for the signaling functions of oncogenic Abl tyrosine kinases, it is critical for the regulation

TABLE 36.1 Proteins that bind to the mammalian Abl

Binding domain in Abl	Protein	Function	Effect of or on Abl	
SH3	Abi-1, Abi-2	F-actin assembly	Abl substrates	19
	ATM	DNA damage signaling	Abl activator	20, 21
	Cables	adaptor	Abl substrate	22
	CSB	transcription and DNA repair	Abl substrate	23
	Mena	F-actin assembly	Abl substrate	24
	NMDA receptor -NR2D subunit	neuronal signaling	Abl inhibitor	25
	p73	transcription factor	Abl substrate	26, 27
	PAG	peroxiredoxin	Abl inhibitor	28
	RIN1	adaptor	Abl activator	29
	Scramblase-1	lipid bilayer remodeling	Abl substrate	30
	Topoisomerase-1	topoisomerase	Abl substrate	14
	WAVE	F-actin assembly	Abl substrate	19
	WRN	helicase	Abl substrate	31
	YAP	transcription factor	Abl substrate	32
SH2	Ptyr-Cbl	adaptor	Abl substrate	33
	Ptyr-Eph receptor	axon guidance	Abl substrate	34
	Ptyr-Dok1, Dok2	adaptors	Abl substrates	8
	Ptyr-CTD of RNA polymerase II	transcription	Abl substrate	35
	Ptyr-RIN1	adaptor	Abl substrate	29
Kinase	PAG	peroxiredoxin	Abl inhibitor	28
	RB	transcription corepressor	Abl inhibitor	36
Proline-rich motifs	Abi-1, Abi-2	F-actin assembly	Abl substrates	37
	Crk, CrkL	adaptor	Abl substrates	38
	Nck	adaptor	Abl activator	33
	PSTPIP1	adaptor	Abl substrate	39

of Abl kinase and it is likely to also be important in constraining the kinase function toward specific substrates.

SH2 Binding Proteins

The SH2 domain of Abl interacts with phosphorylated tyrosine in the consensus sequence (P) YXXP. The kinase domain of Abl also prefers the YXXP motif in its substrates. The coordinated recognition of substrates by the kinase and the SH2 domain is important for catalysis for two reasons. First, the SH2 domain can mediate a continued association of Abl with a substrate. This is particularly important in the processive phosphorylation of the multiple tyrosines in the p130cas adaptor protein or the CTD of RNA polymerase II. Second, the SH2 domain may increase the catalytic efficiency by extracting the product from the kinase domain. The Abl SH2 domain plays an essential role in the oncogenic functions of Gag-Abl and BCR-ABL, in keeping with its positive role in catalysis. In addition, the Abl SH2 domain could serve as an adaptor to bring Abl kinase to specific signaling complexes. For example, the Abl SH2 domain binds to tyrosine phosphorylated EphB2 receptors (Table 36.1). Thus, Abl SH2 binding proteins are substrates and/or recruiters of Abl kinase in signal transduction.

Kinase Domain Binding Proteins

The substrates of Abl must directly interact with the kinase domain. In addition, two cellular inhibitors of the Abl kinase have been shown to bind Abl through its kinase domain. These are the retinoblastoma tumor suppressor (RB) and PAG (Table 36.1). RB binds to the ATP binding lobe of the Abl kinase domain, and this interaction inhibits Abl kinase activity. *In vitro*, RB can inhibit Abl, Gag-Abl, and BCR-ABL kinase. *In vivo*, only the nuclear Abl is inhibited by RB. PAG is inducibly expressed under oxidative stress, and it catalyzes the destruction of H_2O_2. PAG binds Abl through the SH3 domain and the kinase domain to inhibit Abl kinase (Table 36.1). Thus, the Abl kinase domain binds substrates as well as co-inhibitors that function to enforce the autoinhibited conformation and to sequester inactive Abl in specific protein complexes [6].

Proline-Rich Motif Binding Proteins

The C-terminal region of Abl contains a series of proline-rich motifs that bind to proteins with SH3 domains (Table 36.1). The majority of these proteins are substrates of Abl, suggesting this proline-rich linker (PRL) may function as another substrate binding site. The PRL appears to also participate in the regulation of Abl kinase activity. This is suggested by the observations that either phosphorylation or protein binding to the PRL can override the inhibitory effect of F-actin on Abl kinase activity [4]. Furthermore, the PRL may have an adaptor function that is independent of kinase activity. The idea that Abl has kinase independent biological functions is supported by a recent report showing that kinase defective Abl can regulate nuclear excision repair in response to UV irradiation [7], although the specific role of the Abl PRL was not investigated in this experimental system.

ABL IN SIGNAL TRANSDUCTION

Cytoplasmic Signaling Function of Abl

In the cytoplasm, Abl transduces a variety of extracellular signals to regulate F-actin dynamics (Figure 36.2). Platelet derived growth factor (PDGF) and epidermal growth factor (EGF) activate Abl kinase to stimulate membrane ruffling in fibroblasts, and this ruffling response is likely to involve Abl dependent activation of the WAVE complex, including the tyrosine phosphorylation of WAVE itself and Abi-1/2 in the complex, to stimulate actin polymerization. Growth factors also activate an Abl dependent pathway leading to the stimulation of Rac-GTP, which is another activator of the WAVE complex. Thus, the current literature strongly

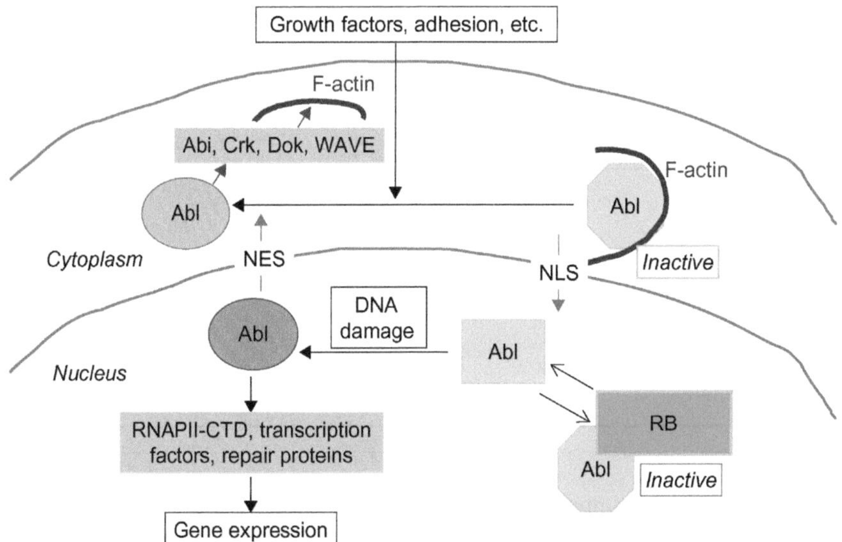

FIGURE 36.2 Signaling functions of nuclear and cytoplasmic Abl kinase.
Cytoplasmic Abl transduces a variety of extracellular signals to regulate F-actin assembly. Abl stimulates membrane ruffling in response to PDGF and EGF. Abl stimulates F-actin microspikes in response to ECM. On the other hand, Abl inhibits cell spreading, cell migration, and chemotaxis, most likely because Abl promotes the apical compartmentalization of the F-actin polymerization machinery. F-actin directly binds to the Abl protein and functions as a co-inhibitor to enforce the autoinhibitory assembly of the Abl SH3-SH2 kinase domains shown in Figure 36.1. The reciprocal regulation between Abl and F-actin may provide a self-limiting mechanism in regulating the dynamic structure of F-actin. Nuclear translocation of Abl is mediated by three functional NLSs. The nuclear Abl directly interacts with the retinoblastoma tumor suppressor (RB), which also functions as a co-inhibitor to block Abl activation. Phosphorylation of RB during cell cycle progression or degradation of RB following caspase activation releases Abl and allows it to be further activated. Genotoxins such as ionizing radiation and cisplatin activate nuclear Abl kinase through ATM. Activation of nuclear Abl leads to the tyrosine phosphorylation of RNA polymerase II CTD, transcription factors, and DNA repair proteins to regulate gene expression and to promote apoptosis. The Abl NLS function is inactivated upon the activation of Abl kinase, and thus preventing cytoplasmically activated Abl from entering the nucleus. DNA damage stimulates a transient accumulation of Abl in the nucleus. Nuclearly activated Abl can be exported to the cytoplasm, and may transduce DNA damage signal to the cytoplasm.

supports a role for Abl in transducing growth factor signals to stimulate the WAVE complex and actin polymerization (Figure 36.2). Cell adhesion to fibronectin also activates Abl kinase to stimulate F-actin assembly through a pathway that requires the tyrosine phosphorylation of p62Dok-1 [8]. Interestingly, however, Abl kinase *inhibits* cell spreading on fibronectin despite the stimulation of actin polymerization. This paradox may be explained by the observation that Abl promotes the sequestration of Rac-GTP at the apical surface, away from the cell matrix contact [9]. As a result, Abl kinase prolongs the apical ruffling response to fibronectin at the expense of lamellipodial spreading on fibronectin. The negative effect of Abl on cell spreading can account for the inhibition of cell migration and chemotaxis by Abl kinase (Figure 36.2). Taken together, current evidence suggests that the cytoplasmic Abl kinase activates actin polymerization and compartmentalizes this process to form apical ruffles and microspikes. It is interesting to note that F-actin itself is a co-inhibitor of the Abl kinase [3, 6]. The reciprocal regulation between Abl and F-actin can therefore provide a self-limiting mechanism to control the dynamics of actin filament formation in response to growth factor and cell adhesion signals (Figure 36.2).

Nuclear Signaling Function of Abl

In the nucleus, the Abl kinase transduces DNA damage signal to regulate transcription and DNA repair. Furthermore, nuclear Abl kinase promotes the apoptotic response to DNA damage [10].

Regulation of Nuclear Abl Kinase

The retinoblastoma tumor suppressor protein, RB, is a nuclear co-inhibitor of Abl kinase; it binds the kinase domain to prevent Abl activation in quiescent and early G_1 cells (Table 36.1). Upon the phosphorylation or the degradation of RB, nuclear Abl kinase is released and can then be further activated by DNA damage or tumor necrosis factor (TNF) [11]. DNA damaging agents also stimulate the nuclear accumulation of Abl kinase through mechanisms that require ATM [12] and/or JNK [13]. The mismatch repair (MMR) proteins, which recognize cisplatin crosslinked DNA, are required for cisplatin to activate Abl. The ATM kinase, which is activated by double-stranded breaks is required for ionizing radiation to activate Abl. The DNA lesions induced by IR and cisplatin are processed by different repair pathways, hence, DNA lesions are likely to activate the nuclear Abl kinase through multiple different mechanisms.

Regulation of Transcription

The CTD of RNA polymerase II is a substrate of the nuclear Abl tyrosine kinase. The CTD is composed of heptad repeats with the consensus sequence YSPTSPS. The CTD is phosphorylated at the Y1, S2, S5, and S7 positions during transcription elongation to regulate co-transcriptional processes such as RNA capping, splicing, and poly-adenylation. CTD phosphorylation regulates the recruitment of various nuclear proteins to the elongating RNA polymerase II complex and thus constitutes a CTD code for co-transcriptional regulation of gene expression. The effects of CTD-Y1 phosphorylation on gene expression are currently being elucidated and this line of investigation will further delineate the nuclear function of the Abl tyrosine kinase. The Abl protein has an intrinsic DNA binding activity that is mapped to the C-terminal region, which binds A/T-rich DNA with distorted structure. While its intrinsic DNA binding function does not support sequence specific interactions, Abl can be recruited to specific promoter complexes through interactions with sequence specific transcription factors, e.g., p53, p73, and the RB–E2F1 complex. Therefore, the nuclear Abl kinase may only phosphorylate the CTD of elongating RNA polymerase II at specific gene loci, dependent upon its association with gene specific transcription factors. As a result, the Abl targeted genes are likely to be variable and subject to regulation by mechanisms that control the formation of specific promoter complexes.

Regulation of DNA Repair

A role for Abl in DNA repair is indicated by its tyrosine phosphorylation of several DNA metabolic enzymes and repair proteins, including RAD51, RAD52, WRN, topoisomerase-1, and CSB (Table 36.1). However, the biological effects of Abl kinase on DNA repair are presently unclear. As mentioned above, the negative effect of Abl on nuclear excision repair in response to UV irradiation does not require its kinase activity [7]; this is consistent with the fact that UV does not induce the nuclear accumulation or the activation of Abl [12]. In the case of topoisomerase-1, tyrosine phosphorylation by Abl is associated with a stimulation of enzyme activity, the enhanced formation of camptothecin-enzyme adduct, and the sensitization of cells to the toxic effect of camptothecin [14]. Taken together, current evidence suggests that Abl can interact with DNA repair proteins, leading to kinase dependent or kinase independent regulation of their functions. Given the fact that repair proteins also transduce DNA damage signals [15], Abl may regulate both the repair and the signaling functions of proteins such as WRN and CSB.

Regulation of Apoptosis

One of the biological consequences resulting from the activation of nuclear Abl kinase is apoptosis [10]. The pro-apoptotic function of Abl has been associated with the downstream activation of p53 and p73, which are related transcription factors that stimulate the expression of pro-apoptotic genes such as Puma. In particular, the Abl kinase is an essential upstream activator of p73, because p73 does activate apoptosis in Abl deficient cells. Through the generation of an *AblμNLS* allele, which expresses an exclusively cytoplasmic Abl, in the mouse embryonic stem (ES) cells, we have shown that nuclear Abl promotes cisplatin and etoposide induced ES cell apoptosis through Bax and the release of cytochrome C from the mitochondria [12]. Furthermore, Abl knockout partially rescues the ectopic apoptosis in the developing central nervous system of RB deficient mouse embryos [16]. These results provide *in vivo* evidence for the pro-apoptotic function of nuclear Abl.

The BCR-ABL tyrosine kinase also possesses pro-apoptotic function, despite the fact that it is an oncoprotein. The BCR-ABL tyrosine kinase is an exclusively cytoplasmic protein, although it contains the three NLSs of ABL. Interestingly, the NLS function is reactivated when the BCR-ABL tyrosine kinase binds to the inhibitor imatinib [5]. The nuclear BCR-ABL can be trapped by a second drug, leptomycin B, which covalently inactivates exportin-1 to block the nuclear export of BCR-ABL [5]. Upon removal of imatinib, the nuclear BCR-ABL regains kinase activity to induce apoptosis [5]. These results are consistent with the conclusion that the pro-apoptotic function of Abl kinase is dependent on it being in the nucleus of a cell. Hence, conversion of Abl into an oncoprotein requires not only the activation of its tyrosine kinase activity but also the inhibition of its nuclear import to abolish its pro-apoptotic function.

The nuclear Abl also transduces death inducing signals other than those generated by DNA damage. In particular, the Abl/p73 pro-apoptotic pathway contributes to cell killing induced by TNF [11], it also mediates the death of cerebella neurons in a mouse model for the human Niman-Picks Syndrome caused by a defect in the NPC1 protein that regulates cholesterol homeostasis [17]. While the nuclear Abl kinase can promote apoptosis, this may not be the only biological consequence of nuclear Abl activation. For example, the maximal activation of nuclear Abl tyrosine kinase is achieved by IR at a dose (2 Gy) that does not induce apoptosis in most cell types. Nuclear Abl kinase can also phosphorylate MyoD to inhibit myogenic differentiation in response to DNA damage [18]. Thus, activation of the nuclear Abl kinase is necessary but not sufficient to cause cell death.

FUTURE PROSPECTS

A framework for the signaling functions of Abl has emerged from (a) the delineation of its modular domains, (b) the identification of cellular phenotypes associated with Abl kinase activation, and (c) the identification of Abl substrates. The cytoplasmic Abl transduces extracellular signals to regulate actin dynamics. The nuclear Abl transduces cell cycle, DNA damage, and TNF signals to regulate transcription, DNA repair, and apoptosis. With this framework, it is now feasible to elucidate further the mechanistic steps in the various Abl dependent signaling pathways. Of particular importance is the regulated nucleo-cytoplasmic shuttling of Abl. It appears that activated Abl kinase is prohibited from entering the nucleus, as demonstrated by the exclusive cytoplasmic localization of Gag-Abl and BCR-ABL. This regulatory mechanism may provide a firewall to prevent cytoplasmically activated Abl from regulating transcription, DNA repair, and apoptosis. By contrast, nuclearly activated Abl is readily exported back to the cytoplasm [12], indicating that nuclear Abl may transmit DNA damage signals into the cytosol. Given the fact that DNA damage activates mitochondrial dependent apoptotic pathways, the nuclearly activated Abl kinase may directly carry the damage signal to the cytoplasm to activate apoptosis.

The current evidence suggests that Abl tyrosine kinase does not function as a master switch, but instead, it functions as an integrator in signal transduction. The biological outputs from the activated Abl kinase is modulated by its subcellular locations, the nature of the upstream signals, the cell types, and the developmental stages of the cells. For example, cytoplasmic Abl kinase stimulates membrane ruffling in response to PDGF, and F-actin microspikes in response to fibronectin. Although both phenotypic outcomes involve the ability of Abl to regulate F-actin assembly, the pathways linking Abl to ruffles involve the activation of Rac-GTP whereas those leading microspikes do not require Rac activity. Likewise, nuclear Abl kinase can be recruited to different promoter complexes through an array of protein–protein interactions that will influence which gene expression events are regulated by Abl in response to DNA damage. Because the signaling functions of Abl are highly context dependent, it is important not to extrapolate the function of Abl deduced in one biological system to another. In other words, a biological effect of Abl found under one biological context might not be readily generalizable to all other experimental systems. Despite these possible complexities, it is clear that Abl regulates F-actin, transcription, DNA repair, and apoptosis. The precise mechanisms by which these functions of Abl are utilized under specific biological conditions will await further investigation.

ACKNOWLEDGEMENT

Work in the author's lab has been supported by grants from the National Institutes of Health (CA43054 and CA58320).

REFERENCES

1. O'Hare T, Corbin AS, Druker BJ. Targeted CML therapy: controlling drug resistance, seeking cure. *Curr Opin Genet Dev* 2006;**16**(1):92–9.
2. Nagar B, Hantschel O, Young MA, Scheffzek K, Veach D, Bornmann W, et al. Structural basis for the autoinhibition of c-Abl tyrosine kinase. *Cell* 2003;**112**(6):859–71.

3. Woodring PJ, Hunter T, Wang JY. Regulation of F-actin-dependent processes by the Abl family of tyrosine kinases. *J Cell Sci* 2003;**116**(Pt 13):2613–26.
4. Woodring PJ, Hunter T, Wang JY. Mitotic phosphorylation rescues Abl from F-actin-mediated inhibition. *J Biol Chem* 2005;**280**(11): 10,318–10,325,.
5. Vigneri P, Wang JY. Induction of apoptosis in chronic myelogenous leukemia cells through nuclear entrapment of BCR-ABL tyrosine kinase. *Nat Med* 2001;**7**(2):228–34.
6. Wang JY. Controlling Abl: auto-inhibition and co-inhibition? *Nat Cell Biol* 2004;**6**(1):3–7.
7. Chen X, Zhang J, Lee J, Lin PS, Ford JM, Zheng N, et al. A kinase-independent function of c-Abl in promoting proteolytic destruction of damaged DNA binding proteins. *Mol Cell* 2006;**22**(4):489–99.
8. Woodring PJ, Meisenhelder J, Johnson SA, Zhou GL, Field J, Shah K, et al. c-Abl phosphorylates Dok1 to promote filopodia during cell spreading. *J Cell Biol* 2004;**165**(4):493–503.
9. Jin H, Wang JY. Abl tyrosine kinase promotes dorsal ruffles but restrains lamellipodia extension during cell spreading on fibronectin. *Mol Biol Cell* 2007;**18**(10):4143–54.
10. Wang JYJ, Minami Y, Zhu J. Abl and Cell Death. In: Koleske AJ, editor. *Abl Family Kinases in Development and Disease*. Georgetown, TX: Landes Bioscience; 2006. p. 26–47.
11. Chau BN, Chen TT, Wan YY, DeGregori J, Wang JY. Tumor necrosis factor alpha-induced apoptosis requires p73 and c-ABL activation downstream of RB degradation. *Mol Cell Biol* 2004;**24**(10):4438–47.
12. Preyer M, Shu CW, Wang JY. Delayed activation of Bax by DNA damage in embryonic stem cells with knock-in mutations of the Abl nuclear localization signals. *Cell Death Differ* 2007;**14**(6):1139–48.
13. Yoshida K, Yamaguchi T, Natsume T, Kufe D, Miki Y. JNK phosphorylation of 14-3-3 proteins regulates nuclear targeting of c-Abl in the apoptotic response to DNA damage. *Nat Cell Biol* 2005;**7**(3):278–85.
14. Yu D, Khan E, Khaleque MA, Lee J, Laco G, Kohlhagen G, et al. Phosphorylation of DNA topoisomerase I by the c-Abl tyrosine kinase confers camptothecin sensitivity. *J Biol Chem* 2004;**279**(50):51,851–51,861,
15. Wang JY, Cho SK. Coordination of repair, checkpoint, and cell death responses to DNA damage. *Adv Protein Chem* 2004;**69**:101–35.
16. Borges HL, Hunton IC, Wang JY. Reduction of apoptosis in Rb-deficient embryos via Abl knockout. *Oncogene* 2007;**26**(26):3868–77.
17. Alvarez AR, Klein A, Castro J, Cancino GI, Amigo J, Mosqueira M, et al. Imatinib therapy blocks cerebellar apoptosis and improves neurological symptoms in a mouse model of Niemann-Pick type C disease. *Faseb J* 2008;**22**(10):3617–27.
18. Puri PL, Bhakta K, Wood LD, Costanzo A, Zhu J, Wang JY. A myogenic differentiation checkpoint activated by genotoxic stress. *Nat Genet* 2002;**32**(4):585–93.
19. Leng Y, Zhang J, Badour K, Arpaia E, Freeman S, Cheung P, et al. Abelson-interactor-1 promotes WAVE2 membrane translocation and Abelson-mediated tyrosine phosphorylation required for WAVE2 activation. *Proc Natl Acad Sci U S A* 2005;**102**(4):1098–103.
20. Baskaran R, Chiang GG, Mysliwiec T, Kruh GD, Wang JY. Tyrosine phosphorylation of RNA polymerase II carboxyl-terminal domain by the Abl-related gene product. *J Biol Chem* 1997;**272**(30):18,905–18,909,.
21. Shafman T, Khanna KK, Kedar P, Spring K, Kozlov S, Yen T, et al. Interaction between ATM protein and c-Abl in response to DNA damage. *Nature* 1997;**387**(6632):520–3.
22. Rhee J, Buchan T, Zukerberg L, Lilien J, Balsamo J. Cables links Robo-bound Abl kinase to N-cadherin-bound beta-catenin to mediate Slit-induced modulation of adhesion and transcription. *Nat Cell Biol* 2007;**9**(8):883–92.
23. Imam SZ, Indig FE, Cheng WH, Saxena SP, Stevnsner T, Kufe D, et al. Cockayne syndrome protein B interacts with and is phosphorylated by c-Abl tyrosine kinase. *Nucleic Acids Res* 2007;**35**(15):4941–51.
24. Tani K, Sato S, Sukezane T, Kojima H, Hirose H, Hanafusa H, et al. Abl interactor 1 promotes tyrosine 296 phosphorylation of mammalian enabled (Mena) by c-Abl kinase. *J Biol Chem* 2003;**278**(24):21,685–21,692,
25. Glover RT, Angiolieri M, Kelly S, Monaghan DT, Wang JY, Smithgall TE, et al. Interaction of the N-methyl-D-aspartic acid receptor NR2D subunit with the c-Abl tyrosine kinase. *J Biol Chem* 2000;**275**(17):12,725–12,729,.
26. Agami R, Blandino G, Oren M, Shaul Y. Interaction of c-Abl and p73alpha and their collaboration to induce apoptosis. *Nature* 1999;**399**(6738):809–13.
27. Gong Jr JG, Costanzo A, Yang HQ, Melino G, Kaelin WG, Levrero M, et al. The tyrosine kinase c-Abl regulates p73 in apoptotic response to cisplatin-induced DNA damage. *Nature* 1999;**399**(6738):806–9.
28. Wen ST, Van Etten RA. The PAG gene product, a stress-induced protein with antioxidant properties, is an Abl SH3-binding protein and a physiological inhibitor of c-Abl tyrosine kinase activity. *Genes Dev* 1997;**11**(19):2456–67.
29. Hu H, Bliss JM, Wang Y, Colicelli J. RIN1 is an ABL tyrosine kinase activator and a regulator of epithelial-cell adhesion and migration. *Curr Biol* 2005;**15**(9):815–23.
30. Sun J, Zhao J, Schwartz MA, Wang JY, Wiedmer T, Sims PJ. c-Abl tyrosine kinase binds and phosphorylates phospholipid scramblase 1. *J Biol Chem* 2001;**276**(31):28,984–28,990,
31. Cheng WH, von Kobbe C, Opresko PL, Fields KM, Ren J, Kufe D, et al. Werner syndrome protein phosphorylation by abl tyrosine kinase regulates its activity and distribution. *Mol Cell Biol* 2003;**23**(18):6385–95.
32. Levy D, Adamovich Y, Reuven N, Shaul Y. Yap1 phosphorylation by c-Abl is a critical step in selective activation of proapoptotic genes in response to DNA damage. *Mol Cell* 2008;**29**(3):350–61.
33. Miyoshi-Akiyama T, Aleman LM, Smith JM, Adler CE, Mayer BJ. Regulation of Cbl phosphorylation by the Abl tyrosine kinase and the Nck SH2/SH3 adaptor. *Oncogene* 2001;**20**(30):4058–69.
34. Yu HH, Zisch AH, Dodelet VC, Pasquale EB. Multiple signaling interactions of Abl and Arg kinases with the EphB2 receptor. *Oncogene* 2001;**20**(30):3995–4006.
35. Baskaran R, Escobar SR, Wang JY. Nuclear c-Abl is a COOH-terminal repeated domain (CTD)-tyrosine (CTD)-tyrosine kinase-specific for the mammalian RNA polymerase II: possible role in transcription elongation. *Cell Growth Differ* 1999;**10**(6):387–96.
36. Welch PJ, Wang JY. A C-terminal protein-binding domain in the retinoblastoma protein regulates nuclear c-Abl tyrosine kinase in the cell cycle. *Cell* 1993;**75**(4):779–90.
37. Shi Y, Alin K, Goff SP. Abl-interactor-1, a novel SH3 protein binding to the carboxy-terminal portion of the Abl protein, suppresses v-abl transforming activity. *Genes Dev* 1995;**9**(21):2583–97.
38. Feller SM. Crk family adaptors-signalling complex formation and biological roles. *Oncogene* 2001;**20**(44):6348–71.
39. Cong F, Spencer S, Cote JF, Wu Y, Tremblay ML, Lasky LA, et al. Cytoskeletal protein PSTPIP1 directs the PEST-type protein tyrosine phosphatase to the c-Abl kinase to mediate Abl dephosphorylation. *Mol Cell* 2000;**6**(6):1413–23.
40. Hantschel O, Wiesner S, Guttler T, Mackereth CD, Rix LL, Mikes Z, et al. Structural basis for the cytoskeletal association of Bcr-Abl/c-Abl. *Mol Cell* 2005;**19**(4):461–73.

Chapter 37

The SREBP Pathway: Gene Regulation through Sterol Sensing and Gated Protein Trafficking

Arun Radhakrishnan[1], Li-Ping Sun[1], Peter J. Espenshade[2], Joseph L. Goldstein[1] and Michael S. Brown[1]

[1]*Department of Molecular Genetics, University of Texas Southwestern Medical Center at Dallas, Dallas, Texas*

[2]*Department of Cell Biology, Johns Hopkins University School of Medicine, Baltimore, Maryland*

Cholesterol is an essential component of mammalian cell membranes. Sufficient levels of cellular cholesterol are required for the integrity and impermeability of the plasma membrane, for the proper assembly of cell surface lipid rafts and caveolae, and for the posttranslational modification of at least one protein, the morphogen Hedgehog [1, 2]. However, too much unesterified cholesterol is toxic to cells. Thus, levels of intracellular cholesterol must be tightly regulated [3]. Here, the cell faces a fundamental problem: how does a cell measure the concentration of insoluble, membrane embedded cholesterol and then appropriately adjust its levels? This chapter outlines the molecular mechanism that mammalian cells utilize to maintain proper cholesterol homeostasis.

Mammalian cells obtain cholesterol from two sources. Cholesterol can be synthesized *de novo* from acetyl-CoA or taken up in the form of lipoprotein particles by the low density lipoprotein (LDL) receptor [3]. Cells maintain cholesterol homeostasis through a feedback regulatory mechanism that acts at the level of transcription to coordinately control the input of sterols from these two sources [4]. When cells are depleted of sterols, transcription of genes required for the uptake and synthesis of cholesterol, such as the LDL receptor and 3-hydroxy-3-methylglutaryl coenzyme A (HMG CoA) reductase, increases. Conversely, when intracellular cholesterol levels are high, transcription of these target genes decreases, and sterol levels fall. Analysis of the LDL receptor gene identified a *cis*-acting, positive regulatory sequence in the promoter that mediates transcription in the absence of sterols and is silenced by sterols [5]. This sterol regulatory element (SRE) was used as a bait in the purification and subsequent cDNA cloning of a family of membrane bound transcription factors, designated sterol regulatory element binding proteins (SREBPs) [5, 6].

SREBPS: MEMBRANE BOUND TRANSCRIPTION FACTORS

SREBPs transmit information to the nucleus about the sterol content of membranes [4]. This information is generated through the process of regulated intramembrane proteolysis (Rip) [7]. Newly synthesized SREBPs are inserted into the membranes of the ER and nuclear envelope in a hairpin orientation such that the NH_2- and COOH-termini project into the cytosol (Figure 37.1). These termini are separated by two transmembrane segments that surround a short ~30 amino acid luminal loop. The NH_2-terminal domain of ~480 amino acids is a transcription factor of the basic helix-loop-helix-leucine zipper family, while the COOH-terminal domain of ~580 residues performs a regulatory function. Three SREBP proteins, encoded by two genes, are present in humans, hamsters, and mice [8]. *SREBP1* encodes two isoforms (SREBP-1a and SREBP-1c) through the use of alternate promoters that generate different first exons that are spliced to a common second exon. *SREBP-2* encodes a single protein. SREBP-2 preferentially activates genes involved in cholesterol biosynthesis, whereas SREBP-1 activates genes required for fatty acid synthesis. To date, in mammalian cells SREBPs have been shown to directly activate more than 30 genes involved in the synthesis and uptake of cholesterol, fatty acids, triglycerides, and phospholipids (for a detailed review, see [9]). The SREBP pathway has also been identified and studied in cells of non-vertebrates including yeast, worms, and insects (for a recent comprehensive review, see [10]). This chapter focuses on the mammalian system.

When mammalian cells are depleted of sterols, SREBPs are activated by two sequential proteolytic cleavage events that release the NH_2-terminal transcription factor from the

Handbook of Cell Signaling, Three-Volume Set 2 ed.

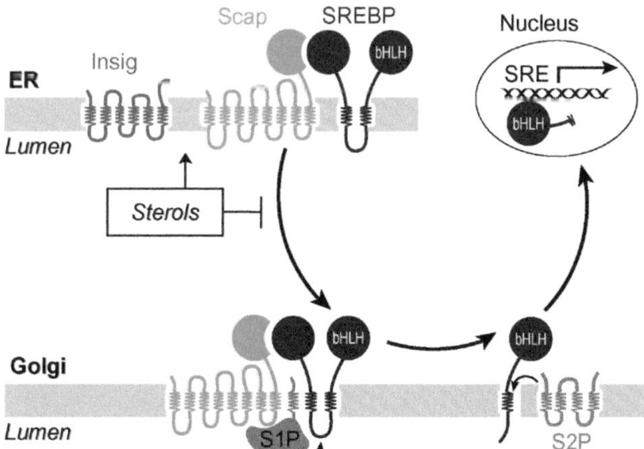

FIGURE 37.1 The SREBP pathway.
When mammalian cells are deprived of sterols, Scap escorts SREBPs from ER to Golgi. Two Golgi proteases (S1P and S2P) then sequentially cleave SREBP, releasing the active NH_2-terminal transcription factor domain, which travels to the nucleus and activates genes involved in cholesterol synthesis and uptake. High sterol levels trigger the binding of Scap to an ER retention protein, Insig. Transport of SREBP to Golgi and subsequent transcriptional activation is then blocked.

membrane, allowing it to enter the nucleus and activate transcription of target genes (Figure 37.1). The first cleavage occurs in the luminal loop of SREBP at Site-1 and is catalyzed by a transmembrane protease, called Site-1 protease (S1P) [11]. Cleavage by this 1052-amino acid, subtilisin related protease occurs after the consensus sequence Arg-X-X-Leu (RXXL) and separates the molecule into two halves [12]. S1P action is not restricted to proteins involved in cholesterol homeostasis. It also cleaves and activates two transcription factors involved in the ER stress response, ATF6 [13] and CREBH [14], and it participates in the processing of Lassa virus glycoprotein precursor GP-C, a step required for production of infectious virus [15].

Following cleavage by S1P, the NH_2-terminal domain of SREBP remains bound to the membrane until a 519-amino acid, membrane bound zinc metalloprotease, the Site-2 protease (S2P), cleaves SREBP within the first transmembrane segment [16, 17]. The liberated transcription factor now enters the nucleus and activates target gene transcription. Sterols control the activation of SREBPs by regulating cleavage at Site-1. When cholesterol accumulates, cleavage of SREBP by S1P is blocked. Cleavage by S2P is not regulated by sterols, but requires prior cleavage by S1P [18].

SCAP: STEROL SENSOR AND ESCORTER OF SREBP FROM ER TO GOLGI

Cleavage of SREBP at Site-1 requires a 1276-amino acid, polytopic membrane protein called Scap [19]. Scap is divided into two domains. The NH_2-terminal 730 amino acids contain eight transmembrane segments that attach Scap to membranes of the ER and nuclear envelope [20]. The COOH-terminal 546 amino acids contain five copies of the WD-40 motif, which form β-propeller structures that mediate protein–protein interactions [21]. This domain of Scap forms a tight complex with the COOH-terminal regulatory domain of SREBP, and this interaction is essential for cleavage of SREBP by S1P (Figure 37.1) [22]. Chinese hamster ovary (CHO) cells lacking Scap fail to process SREBP at Site-1 and show reduced levels of SREBP precursor protein, indicating that Scap is also required for stability of SREBP [23].

Scap functions as a sterol sensor in the SREBP pathway. The sterol sensing activity localizes to transmembrane segments 2–6 (~170 amino acids) in the NH_2-terminal domain of the protein (Figure 37.1). Mutant CHO cells that contain single amino acid substitutions in this sterol sensing domain (Y298C, L315F, and D443N) fail to sense sterols and continue to process SREBPs even in the presence of high levels of sterols [19, 24, 25].

Genetic, biochemical, and live-cell microscopy studies have established the following mechanism for sterol mediated regulation of the processing of SREBP (Figure 37.1). In the absence of sterols, Scap escorts SREBP from ER to Golgi where S1P cleaves SREBP at Site-1 and initiates release of the NH_2-terminal transcription factor. Following cleavage of SREBP, Scap recycles to the ER [26]. In the presence of sterols, the Scap•SREBP complex remains in the ER and is compartmentally separated from active S1P. The initial evidence for this model emerged from studies of the glycosylation state of Scap. In wild-type CHO cells cultured in the presence of sterols, the *N*-linked carbohydrate chains of Scap are sensitive to digestion with endoglycosidase H, indicating that the protein resides in the ER. However, when cells are depleted of sterols, *N*-linked carbohydrates on Scap become resistant to endoglycosidase H treatment, suggesting that Scap travels to the Golgi, but only in the absence of sterols [24].

The cloning and characterization of S1P provided further support for this model (Figure 37.1). S1P is synthesized as an inactive zymogen (S1P-A). Autocatalytic processing of S1P at two sites in the NH_2-terminus of the protein leads to the removal of an inhibitory prosegment and production of the active form of the enzyme (S1P-C) [27]. Glycosylation and immunolocalization studies demonstrated that S1P-C resides in the Golgi. Therefore, in order to be processed at Site-1, SREBP must move from ER to Golgi. Scap is the key to this movement.

Additional evidence for the role of Scap in the trafficking of SREBP comes from experiments using Scap deficient mutant CHO cells (SRD-13A cells) that possess two copies of a non-functional *Scap* gene [23]. These cells fail to process SREBPs, and they are therefore unable to synthesize cholesterol. Relocalization of active S1P from the Golgi to the ER bypasses the requirement of Scap for processing of SREBP in Scap deficient SRD-13A cells [28]. This relocalization can be accomplished experimentally in two ways: (1) by treatment of SRD-13A cells with the fungal metabolite, brefeldin A, which results in fusion of the Golgi with the ER, and (2) by transfection of SRD-13A cells with a cDNA encoding a fusion protein of S1P (lacking its membrane spanning segment) attached to the ER retrieval sequence, KDEL. In both of these situations, cleavage of SREBP occurs constitutively and is not inhibited by sterols [28]. Considered together, these results demonstrate that Scap functions to escort SREBP from ER to Golgi and that sterols act on Scap to block this transport.

Transport of SREBP from ER to Golgi requires Scap mediated clustering of SREBP into coat protein complex II (COPII) coated vesicles [29, 30]. This clustering occurs by the general mechanism that was elucidated primarily by studies in yeast [31, 32]. The COPII coat consists of five proteins: Sar1, Sec23/Sec24 complex, and Sec13/Sec31 complex. Newly synthesized membrane proteins and secreted proteins destined for export (called cargo proteins) are clustered into COPII coated vesicles in a reaction initiated when an ER resident protein, Sec12, stimulates the exchange of GTP for GDP on Sar1. This exchange triggers the binding of Sar1-GTP to ER membranes. There, the Sar1-GTP complex provides a platform for recruitment of Sec23/Sec24, followed by Sec13/Sec31. The Sec23/Sec24 complex selects certain membrane cargo proteins that have sequences recognized by Sec24, which thereby clusters them into budding vesicles [32–35]. In the case of Scap, the COPII binding site has been identified through mutagenesis and biochemical studies as a hexapeptide sequence, Met-Glu-Leu-Ala-Asp-Leu (MELADL), located in the cytoplasmic loop between transmembrane helices 6 and 7 [36, 37]. Sterols block the binding of Sec23/Sec24 complex to MELADL, thus retaining Scap•SREBP in the ER.

Indirect evidence suggested that the sterol dependent retention of Scap in the ER required a Scap binding protein that is present in limiting amounts [38]. The initial clue came from the observation that overexpression of Scap leads to non-inhibitable SREBP processing as though a protein required for ER retention is overwhelmed by saturating levels of Scap [19]. Supporting evidence came from competition experiments involving overexpression of a segment of Scap, designated Scap(TM1–6), that contains the sterol sensing domain [38]. Expression of this segment prevented the retention of the endogenous Scap•SREBP complex in the ER, apparently by competing for binding to the putative retention protein. Moreover, a shorter segment of Scap, designated Scap(TM1–5), which has an incomplete sterol sensing domain failed to release the endogenous Scap•SREBP complex from its retention site in the ER. A mutant version of Scap(TM1–6) containing the Y298C substitution that confers sterol resistance also failed to release the Scap•SREBP complex [38]. The truncated Scap(TM1–6) was then used as bait in the purification of a family of two membrane bound ER retention proteins, called Insigs [39].

INSIG: STEROL SENSOR AND ER RETENTION PROTEIN

The exit of Scap•SREBP complex from the ER in the presence of sterols is blocked by the interaction of Scap with Insigs. Mammalian cells contain two isoforms of Insig, called Insig-1 and Insig-2 [39–41]. Human Insig-1 and Insig-2 contain 277 and 225 amino acids, respectively. Both Insigs are extremely hydrophobic. Topology studies suggest that most of the protein consists of six transmembrane helices separated by short hydrophilic loops [42]. Only short sequences at the NH_2- and COOH-termini project into the cytoplasm. These sequences contain the major differences between the two Insigs. The NH_2-terminal sequence of Insig-2 is 50 residues shorter than the sequence in Insig-1. The membranous regions of the two proteins are 85 percent identical. Both Insigs bind to Scap in the presence of sterols, and both retain the Scap•SREBP complex in the ER [25, 39]. Furthermore, the three sterol resistant Scaps (Y298C, L315F, and D443N) do not bind Insigs and thus render cells resistant to the effect of sterols blocking SREBP processing [25, 39, 43].

The requirement of the Scap•Insig interaction for sterol regulation was confirmed in studies of SRD-15 cells, a line of mutant CHO cells that lacks both Insig proteins. In these cells, sterols do not block Scap•SREBP transport from the ER. Regulation is restored by transfecting cells with cDNAs encoding either of the two Insig isoforms, Insig-1 or Insig-2 [44].

The mechanism by which sterol regulated formation of Scap•Insig complex ultimately results in ER retention of SREBP is illustrated in Figure 37.2. When Insig is bound to Scap in the presence of sterols, the MELADL sequence on Scap is no longer accessible to the COPII proteins, and thus transport of SREBP to the Golgi is blocked (Figure 37.2). Insig does not simply sequester the MELADL signal by binding to it, inasmuch as the Fab fragment of anti-MELADL can bind to Scap even when Insig is bound to it. Instead, Insig acts by inducing a conformational change in Scap, which moves the MELADL sequence to a new location where it is no longer accessible to the COPII machinery [37]. Sterols thus block the transport of SREBP from ER to Golgi by stabilizing a complex between Scap and Insig (Figure 37.2).

The trigger for sterol sensing could be direct or indirect, either mechanism producing conformation changes leading to Scap•Insig complex formation. A direct mechanism would involve sterol binding to an ER membrane receptor; an indirect mechanism would involve sterols perturbing the physical properties of the ER membrane. Biochemical studies using purified detergent solubilized recombinant

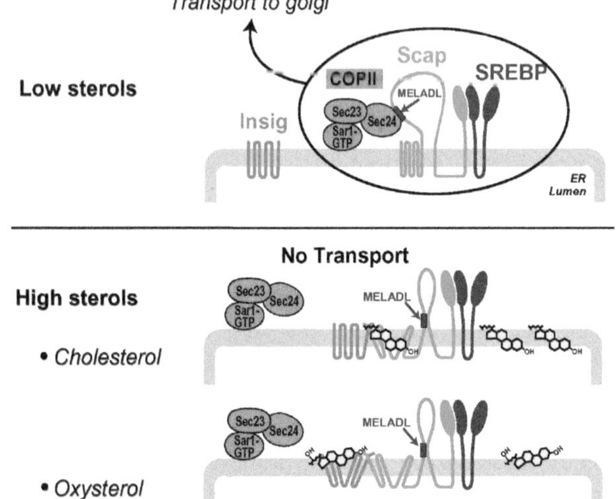

FIGURE 37.2 Differential mechanisms by which cholesterol and oxysterols trigger Insig binding to Scap.
Top panel: when sterol levels are low, Scap escorts SREBPs from ER to Golgi by binding to Sec24, a component of the Sarl•Sec23/Sec24 complex of the COPII protein coat. Once in the Golgi, the SREBPs are proteolytically processed to generate their nuclear forms that activate genes for cholesterol synthesis and uptake. Bottom panel: two classes of sterols, cholesterol and oxysterols, negatively regulate ER-to-Golgi transport of SREBP. Regulation is initiated by one of two mechanisms: cholesterol binding to Scap or oxysterols binding to Insig. Thereafter, the actions of both sterols converge. Conformation changes in Scap and Insig caused by binding to their respective ligands trigger the formation of a Scap•Insig complex, which prevents the binding of Scap to COPII proteins, thereby halting transport of SREBPs to Golgi.

proteins have revealed that Scap and Insigs sense sterols through direct binding interactions [45, 46].

SCAP AND INSIG: TWO SENSORS FOR TWO CLASSES OF STEROLS

It has been known for the last 30 years that cellular cholesterol homeostasis is regulated in a negative feedback fashion not only by cholesterol, but also by oxygenated derivatives of cholesterol, termed oxysterols [47–50]. We now know that this feedback system is mediated by the SREBP pathway.

As a sterol regulator, cholesterol can be derived either from endogenous synthesis or from the receptor mediated endocytosis of LDL. The regulatory pool of cholesterol is presumably a minor component of the bulk cholesterol that is crucial for structural integrity of membranes. Oxysterols are synthesized by specific hydroxylases that act on cholesterol. In addition to their regulatory role, oxysterols function in the export of excess cholesterol from brain and lung [51], and they also are intermediates in the synthesis of bile acids [52]. Although oxysterols are potent feedback regulators of cholesterol homeostasis, they make up only a minute fraction of total sterols in various tissues and in blood, present at concentrations 10^4 to 10^6-fold less than that of cholesterol [51, 53].

In vitro binding studies using the purified recombinant membrane domain of Scap (transmembrane helices 1–8) in detergent micelles show that [^{3}H]cholesterol binds directly to Scap [45] in a saturable and specific manner. Scap behaves like a standard receptor for cholesterol, although the kinetics of the binding reaction are unusual in that both the ligand and the receptor must be added in separate detergent micelles. In this *in vitro* reaction, the rate limiting step is the transfer of the cholesterol from the donor micelles to the acceptor micelles that contain Scap. The binding of cholesterol to Scap causes a conformational change that induces Scap to bind to Insig. The conformational change can be monitored by a change in the tryptic cleavage pattern of Scap. In membranes from sterol depleted cells, Arg-505 of Scap is inaccessible to trypsin. Addition of cholesterol, either to living cells or to isolated membranes *in vitro*, causes Arg-505 to become exposed so that it is cleavable by trypsin [54]. Surprisingly, oxysterols such as 25-hydroxycholesterol also induce Scap to bind to Insig [43], but they do not bind to Scap *in vitro* [52], nor do they induce a trypsin detectable conformational change in Scap in the absence of Insigs [43].

Oxysterols act by binding to Insig, triggering Insigs to bind to Scap. *In vitro* binding studies using purified recombinant Insig-2 in detergent micelles show that [^{3}H]25-hydroxycholesterol binds directly to Insig-2 in a saturable and specific manner [46]. For technical reasons, the sterol binding studies have been performed with Insig-2, which is easier to produce and to purify than Insig-1. However, all functional evidence indicates that Insig-1 and Insig-2 function in the same manner. The notion that 25-HC acts by binding to Insigs is supported by the finding that 25-HC has very little ability to cause a conformational change in Scap or to block SREBP processing in Insig deficient cells [37]. When SREBP-2 cleavage is measured in intact cells, the presence of Insig-1 increases the sensitivity to cholesterol by 13-fold, but it increases the sensitivity to 25-HC by 500-fold, essentially an all-or-none effect [37].

In remarkable reciprocity with Scap, Insig-2 does not bind cholesterol or any of the other Scap ligands [46]. Thus, Scap and Insig together cover the entire range of sterol molecules that regulate SREBP processing in mammalian cells. Binding of either cholesterol to Scap or oxysterols to Insig triggers the formation of Scap•Insig complex and prevents COPII proteins from gaining access to the MELADL sequence (Figure 37.2).

FUTURE CHALLENGES

It is noteworthy that the molecular switch that controls sterol metabolism in animal cells is a simple hexapeptide targeting signal (MELADL) in a single membrane protein (Scap).

Although much has been learned about the SREBP regulatory system, many exciting questions remain. How do Scap and Insig discriminate between cholesterol and oxysterols? What is the nature of the conformational change upon sterol binding that allows complex formation? How does formation of the Scap•Insig complex abrogate binding to COPII proteins? The answers to these questions await detailed molecular structures of Scap, Insig, and the Scap•Insig complex.

Other outstanding questions deal with membrane biochemistry. How does the ER membrane receive information about the cholesterol content in the plasma membrane? Are there cytosolic factors that transmit this information to the SREBP system? In detergent solutions, recombinant Scap shows nanomolar affinity for cholesterol; however, the concentration (area fraction) of cholesterol in membranes is many orders of magnitude higher. One possibility is that the chemical activity of cholesterol in the ER membrane is normally low, owing to its strong interaction with phospholipids (i.e., the concentration of free cholesterol is very low), and only when the cholesterol : phospholipid ratio exceeds a threshold value does free cholesterol become available to Scap. These fine details will be fully understood only when the entire system is reconstituted into chemically defined liposomes, a challenging task that is currently being undertaken.

ACKNOWLEDGEMENTS

This work was supported by research grants from the National Institutes of Health (HL20948 and HL077588) and the Perot Family Foundation. AR was a recipient of a fellowship from the Jane Coffin Childs Foundation for Medical Research. PJE is the recipient of a Burroughs Wellcome Fund Career Award in the Biomedical Sciences.

REFERENCES

1. Anderson RGW. The caveolae membrane system. *Annu Rev Biochem* 1998;**67**:199–225.
2. Mann RK, Beachy PA. Cholesterol modification of proteins. *Biochim Biophys Acta* 2000;**1529**:188–202.
3. Brown MS, Goldstein JL. A receptor-mediated pathway for cholesterol homeostasis. *Science* 1986;**232**:34–47.
4. Brown MS, Goldstein JL. A proteolytic pathway that controls the cholesterol content of membranes, cells, and blood. *Proc Natl Acad Sci USA* 1999;**96**:11,041–8.
5. Wang X, Briggs MR, Hua X, Yokoyama C, Goldstein JL, Brown MS. Nuclear protein that binds sterol regulatory element of LDL receptor promoter: II. Purification and characterization. *J Biol Chem* 1993;**268**:14,497–504.
6. Yokoyama C, Wang X, Briggs MR, Admon A, Wu J, Hua X, Goldstein JL, Brown MS. SREBP-1, a basic helix-loop-helix leucine zipper protein that controls transcription of the LDL receptor gene. *Cell* 1993;**75**:187–97.
7. Brown MS, Ye J, Rawson RB, Goldstein JL. Regulated intramembrane proteolysis: a control mechanism conserved from bacteria to humans. *Cell* 2000;**100**:391–8.
8. Brown MS, Goldstein JL. The SREBP pathway: Regulation of cholesterol metabolism by proteolysis of a membrane-bound transcription factor. *Cell* 1997;**89**:331–40.
9. Horton JD, Goldstein JL, Brown MS. SREBPs: activators of the complete program of cholesterol and fatty acid synthesis in the liver. *J Clin Invest* 2002;**109**:1125–31.
10. Espenshade P, Hughes AL. Regulation of sterol synthesis in eukaryotes. *Ann Rev Genet* 2007;**41**:401–27.
11. Sakai J, Rawson RB, Espenshade PJ, Cheng D, Seegmiller AC, Goldstein JL, Brown MS. Molecular identification of the sterol-regulated luminal protease that cleaves SREBPs and controls lipid composition of animal cells. *Mol Cell* 1998;**2**:505–14.
12. Duncan EA, Brown MS, Goldstein JL, Sakai J. Cleavage site for sterol-regulated protease localized to a Leu-Ser bond in lumenal loop of sterol regulatory element binding protein-2. *J Biol Chem* 1997;**272**:12,778–85.
13. Ye J, Rawson RB, Komuro R, Chen X, Dave UP, Prywes R, Brown MS, Goldstein JL. ER stress induces cleavage of membrane-bound ATF6 by the same proteases that process SREBPs. *Mol Cell* 2000;**6**:1355–64.
14. Zhang K, Shen X, Wu J, Sakaki K, Saunders T, Rutkowski DT, Back SH, Kaufman RJ. Endoplasmic reticulum stress activates cleavage of CREBH to induce a systemic inflammatory response. *Cell* 2006;**124**:587–99.
15. Lenz O, ter Meulen J, Klenk HD, Seidah NG, Garten W. The Lassa virus glycoprotein precursor GP-C is proteolytically processed by subtilase SK1-1/S1P. *Proc Natl Acad Sci USA* 2001;**98**:12,701–5.
16. Rawson RB, Zelenski NG, Nijhawan D, Ye J, Sakai J, Hasan MT, Chang T-Y, Brown MS, Goldstein JL. Complementation cloning of *S2P*, a gene encoding a putative metalloprotease required for intramembrane cleavage of SREBPs. *Mol Cell* 1997;**1**:47–57.
17. Duncan EA, Davé UP, Sakai J, Goldstein JL, Brown MS. Second-site cleavage in sterol regulatory element-binding protein occurs at transmembrane junction as determined by cysteine panning. *J Biol Chem* 1998;**273**:17,801–9.
18. Sakai J, Duncan EA, Rawson RB, Hua X, Brown MS, Goldstein JL. Sterol-regulated release of SREBP-2 from cell membranes requires two sequential cleavages, one within a transmembrane segment. *Cell* 1996;**85**:1037–46.
19. Hua X, Nohturfft A, Goldstein JL, Brown MS. Sterol resistance in CHO cells traced to point mutation in SREBP cleavage activating protein (SCAP). *Cell* 1996;**87**:415–26.
20. Nohturfft A, Brown MS, Goldstein JL. Topology of SREBP cleavage-activating protein, a polytopic membrane protein with a sterol-sensing domain. *J Biol Chem* 1998;**273**:17,243–50.
21. Smith TF, Gaitatzes C, Saxena K, Neer EJ. The WD repeat: a common architecture for diverse functions. *Trends Biochem Sci* 1999;**24**:181–5.
22. Sakai J, Nohturfft A, Goldstein JL, Brown MS. Cleavage of sterol regulatory element binding proteins (SREBPs) at site-1 requires interaction with SREBP cleavage-activating protein. Evidence from *in vivo* competition studies. *J Biol Chem* 1998;**273**:5785–93.
23. Rawson RB, DeBose-Boyd RA, Goldstein JL, Brown MS. Failure to cleave sterol regulatory element-binding proteins (SREBPs) causes cholesterol auxotrophy in Chinese hamster ovary cells with genetic absence of SREBP cleavage-activating protein. *J Biol Chem* 1999;**274**:28,549–56.
24. Nohturfft A, Brown MS, Goldstein JL. Sterols regulate processing of carbohydrate chains of wild-type SREBP cleavage-activating protein (SCAP), but not sterol-resistant mutants Y298C or D443N. *Proc Natl Acad Sci USA* 1998;**95**:12,848–53.

25. Yabe D, Xia Z-P, Adams CM, Rawson RB. Three mutations in sterol-sensing domain of SCAP block interaction with insig and render SREBP cleavage insensitive to sterols. *Proc Natl Acad Sci USA* 2002;**99**:16,672–7.
26. Nohturfft A, DeBose-Boyd RA, Scheek S, Goldstein JL, Brown MS. Sterols regulate cycling of SREBP cleavage-activating protein (SCAP) between endoplasmic reticulum and Golgi. *Proc Natl Acad Sci USA* 1999;**96**:11,235–40.
27. Espenshade PJ, Cheng D, Goldstein JL, Brown MS. Autocatalytic processing of Site-1 protease removes propeptide and permits cleavage of sterol regulatory element-binding proteins. *J Biol Chem* 1999;**274**:22,795–804.
28. DeBose-Boyd RA, Brown MS, Li W-P, Nohturfft A, Goldstein JL, Espenshade PJ. Transport-dependent proteolysis of SREBP: Relocation of Site-1 protease from Golgi to ER obviates the need for SREBP transport to Golgi. *Cell* 1999;**99**:703–12.
29. Nohturfft A, Yabe D, Goldstein JL, Brown MS, Espenshade PJ. Regulated step in cholesterol feedback localized to budding of SCAP from ER membranes. *Cell* 2000;**102**:315–23.
30. Espenshade PJ, Li W-P, Yabe D. Sterols block binding of COPII proteins to SCAP, thereby controlling SCAP sorting in ER. *Proc Natl Acad Sci USA* 2002;**99**:11,694–9.
31. Antonny B, Schekman R. ER export: public transportation by the COPII coach. *Curr Opin Cell Biol* 2001;**13**:438–43.
32. Barlowe C. Molecular recognition of cargo by the COPII complex: a most accommodating coat. *Cell* 2003;**114**:395–7.
33. Aridor M, Weissman J, Bannykh SI, Nuoffer C, Balch WE. Cargo selection by the COPII budding machinery during export from the ER. *J Cell Biol* 1998;**141**:61–70.
34. Mossessova E, Bickford LC, Goldberg J. SNARE selectivity of the COPII coat. *Cell* 2003;**114**:483–95.
35. Lee MCS, Miller EA, Goldberg J, Orci L, Schekman R. Bi-directional protein transport between the ER and Golgi. *Annu Rev Cell Biol* 2004;**20**:87–123.
36. Sun L-P, Li L, Goldstein JL, Brown MS. Insig required for sterol-mediated inhibition of Scap/SREBP binding to COPII proteins *in vitro*. *J Biol Chem* 2005;**280**:26,483–90.
37. Sun L-P, Seemann J, Brown MS, Goldstein JL. Sterol-regulated transport of SREBPs from endoplasmic reticulum to Golgi: Insig renders sorting signal in Scap inaccessible to COPII proteins. *Proc Natl Acad Sci USA* 2007;**104**:6519–26.
38. Yang T, Goldstein JL, Brown MS. Overexpression of membrane domain of SCAP prevents sterols from inhibiting SCAP/SREBP exit from endoplasmic reticulum. *J Biol Chem* 2000;**275**:29,881–6.
39. Yang T, Espenshade PJ, Wright ME, Yabe D, Gong Y, Aebersold R, Goldstein JL, Brown MS. Crucial step in cholesterol homeostasis: sterols promote binding of SCAP to INSIG-1, a membrane protein that facilitates retention of SREBPs in ER. *Cell* 2002;**110**:489–500.
40. Yabe D, Brown MS, Goldstein JL. Insig-2, a second endoplasmic reticulum protein that binds SCAP and blocks export of sterol regulatory element-binding proteins. *Proc Natl Acad Sci USA* 2002;**99**:12,753–8.
41. Goldstein JL, DeBose-Boyd RA, Brown MS. Protein sensors for membrane sterols. *Cell* 2006;**124**:35–46.
42. Feramisco JD, Goldstein JL, Brown MS. Membrane topology of human Insig-1, a protein regulator of lipid synthesis. *J Biol Chem* 2004;**279**:8487–96.
43. Adams CM, Reitz J, DeBrabander JK, Feramisco JD, Brown MS, Goldstein JL. Cholesterol and 25-hydroxycholesterol inhibit activation of SREBPs by different mechanisms, both involving SCAP and Insigs. *J Biol Chem* 2004;**279**:52,772–80.
44. Lee PCW, Sever N, DeBose-Boyd RA. Isolation of sterol-resistant Chinese hamster ovary cells with genetic deficiencies in both Insig-1 and Insig-2. *J Biol Chem* 2005;**280**:25,242–49.
45. Radhakrishnan A, Sun L-P, Kwon HJ, Brown MS, Goldstein JL. Direct binding of cholesterol to the purified membrane region of SCAP: mechanism for a sterol-sensing domain. *Mol Cell* 2004;**15**:259–68.
46. Radhakrishnan A, Ikeda Y, Kwon HJ, Brown MS, Goldstein JL. Sterol-regulated transport of SREBPs from endoplasmic reticulum to Golgi: Oxysterols block transport by binding to Insig. *Proc Natl Acad Sci USA* 2007;**104**:6511–8.
47. Kandutsch AA, Chen HW. Inhibition of sterol synthesis in cultured mouse cells by 7-hydroxycholesterol, 7-hydroxycholesterol, and 7-ketocholesterol. *J Biol Chem* 1973;**248**:8408–17.
48. Brown MS, Goldstein JL. Suppression of 3-hydroxy-3-methylglutaryl coenzyme A reductase activity and inhibition of growth of human fibroblasts by 7-ketocholesterol. *J Biol Chem* 1974;**249**:7306–14.
49. Goldstein JL, Faust JR, Brunschede GY, Brown MS. Steroid requirements for suppression of HMG CoA reductase activity in cultured human fibroblasts. In: Kritchevsky D, Paoletti R, Holmes WL, editors. *Lipids, Lipoproteins, and Drugs*. New York: Plenum Publishing Corp.; 1975. p. 77–84.
50. Chang T-Y, Limanek JS. Regulation of cytosolic acetoacetyl coenzyme A thiolase, 3-hydroxy-3-methylglutaryl coenzyme A synthase, 3-hydroxy-3-methylglutaryl coenzyme A reductase, and mevalonate kinase by low density lipoprotein and by 25-hydroxycholesterol in Chinese hamster ovary cells. *J Biol Chem* 1980;**255**:7787–95.
51. Bjorkhem I. Do oxysterols control cholesterol homeostasis? *J Clin Invest* 2002;**110**:725–30.
52. Russell DW. Oxysterol biosynthetic enzymes. *Biochim Biophys Acta* 2000;**1529**:126–35.
53. Lund EG, Diczfalusy U. Quantitation of receptor ligands by mass spectrometry. Analysis of nuclear receptor ligands. *Meth Enzymol.* 2003;**364**:24–37.
54. Brown AJ, Sun L, Feramisco JD, Brown MS, Goldstein JL. Cholesterol addition to ER membranes alters conformation of SCAP, the SREBP escort protein that regulates cholesterol metabolism. *Mol Cell* 2002;**10**:237–45.

Chapter 38

Ubiquitination/Proteasome

Daniel Kornitzer[1] and Aaron Ciechanover[2]

[1]Department of Molecular Microbiology, Bruce Rappaport Faculty of Medicine, Technion-Israel Institute of Technology, Haifa, Israel

[2]Vascular and Tumor Biology Research Center, Bruce Rappaport Faculty of Medicine, Technion-Israel Institute of Technology, Haifa, Israel

PROTEIN DEGRADATION AND THE UBIQUITIN/PROTEASOME SYSTEM

Although most cellular proteins are long lived, a large class exists of normally short lived proteins. Within this class one mainly finds regulatory factors, of which transcription factors (TFs) constitute the largest group. One explanation for the short half-life of many regulatory factors is that it enables the rapid modulation of the steady-state concentration of the protein: the levels of constitutively short lived proteins will respond much faster to changes in their rate of synthesis than that of long lived ones. Additionally, many factors, rather than being constitutively short lived, can be conditionally stabilized or degraded in response to various stimuli. The question how extracellular stimuli ultimately can determine the stability of specific TFs is central to the understanding of many signaling pathways.

The ubiquitin/proteasome pathway is the principal cellular system for selective protein degradation of normally short lived proteins and of damaged or unfolded proteins (see reference [1] for a more extensive review). The ubiquitin system covalently ligates the conserved, 76 amino acids protein ubiquitin to target proteins by creating an isopeptide bond linking the terminal carboxyl group of the ubiquitin polypeptide to an ε amino group of an internal lysine of the target polypeptide (or occasionally, to the α amino group of the target polypeptide chain). An internal lysine residue of the first ubiquitin adduct can then serve as acceptor for an additional ubiquitin moiety, eventually yielding a crosslinked chain of up to several tens of ubiquitin molecules. The polyubiquitin chain serves as a tag for recognition and destruction of the target protein by the 26S proteasome, a large, multicatalytic cytoplasmic protease. The ubiquitination reaction requires a number of enzymes and recognition factors that act sequentially: the ubiquitin activating enzyme, or E1; a ubiquitin conjugating enzyme, or E2; and a ubiquitin-protein ligase, or E3. Ubiquitin is initially conjugated in an ATP-requiring reaction to E1 via a thiolester bond, then transferred via a trans-esterification reaction to a cysteine residue in the active site of the E2, which finally transfers the activated ubiquitin moiety to an amino group of the target protein (see Figure 38.1). The E3 – or ubiquitin-protein ligase – is responsible for substrate recognition; it serves to bring together the E2 and the substrate in a single complex, thereby allowing ubiquitination to occur. Ubiquitin ligases form a heterogeneous group of proteins, that can be divided into two main subcategories: the more numerous and versatile RING finger domain containing ligases, and the HECT domain containing ligases. With the HECT domain ligases, the catalytic cascade of ubiquitination includes an additional transthiolation step in which the activated ubiquitin is transferred from the E2 to a cysteine residue on the E3 prior to its conjugation to the target.

The main site of regulation of the various ubiquitination reactions is at the level of the ubiquitin ligase–substrate interaction, which can be modulated either by modification of the substrate, or by modulation of ubiquitin ligase activity. In the next sections, we will review pathways where the ubiquitin system plays a role in the regulation of TF activity in response to extracellular signals.

REGULATION OF UBIQUITINATION BY SUBSTRATE MODIFICATION

Stimulation of Ubiquitination by Substrate Phosphorylation

Modification of proteins by the addition of phosphate groups can modulate their function in a multitude of ways, including

Handbook of Cell Signaling, Three-Volume Set 2 ed.

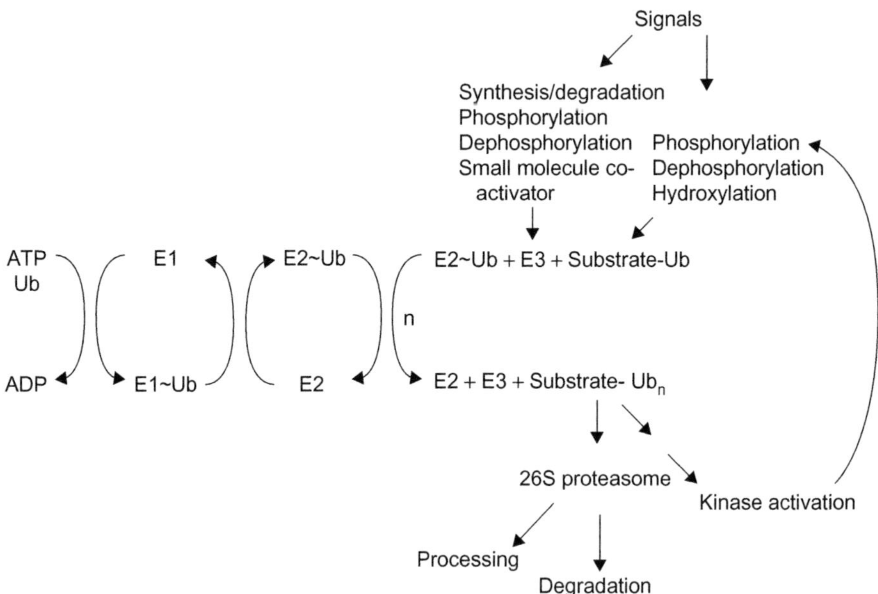

FIGURE 38.1 Schematic representation of the ubiquitination reaction cycle.
Known points of regulation of the ubiquitin system by signaling pathways are indicated. See text for details.

by enabling interactions with other proteins. In the event that such an interaction occurs with a proteolytic system, phosphorylation will result in degradation of the protein. A number of ubiquitin ligases were described that require phosphorylation of the substrate prior to ubiquitination. The known ubiquitin ligases requiring prior phosphorylation of the substrate are primarily of the SCF class. SCF ubiquitin ligases form a complex of at least four proteins: the three components that were initially isolated, which gave the complex its name (SCF stands for *S*kp1, *C*dc53 or *C*ullin, and *F*-box protein [2, 3]) and a subsequently identified RING finger domain containing component, Rbx1 or Roc1 (reviewed in [4]). The F-box protein is the variable component of the complex. Given that SCF complexes with different F-box proteins exhibit different substrate affinities, the F-box proteins are thought to carry the substrate recognition function of the complex (see [5] for a review). The SCF complex containing a specific F-box protein is marked with a superscript, e.g., SCF^{CDC4}.

Several TFs are known to date to be regulated by SCF mediated ubiquitination, e.g., Gcn4 [6], Tec1 [7, 8], and Met4 [9] in yeast, and NFκB [10], β-catenin [11], ATF4 [12], and E2-F1 [13] in mammalian cells. Some of these factors are regulated at the level of protein stability by signaling pathways that modulate the phosphorylation state of the substrate, and therefore recognition by the SCF complex.

IκBα. One of the most thoroughly investigated signaling pathways is that of NFκB activation (reviewed in [10]). IκBα inhibits NFκB by sequestering it in the cytoplasm. Upon stimulation of the pathway, IκBα becomes phosphorylated at two specific serine residues, whereupon it becomes a substrate for ubiquitination by the $SCF^{\beta-TrCP}$ complex. Interestingly, ubiquitination plays additional roles in the NFκB pathway: at the level of processing of the NFκB precursor, p105 (see below), and at the level of activation of the IκBα kinase (also see below).

Gcn4. Another example of phosphorylation being required for degradation is that of the yeast TF Gcn4. Gcn4 degradation depends on SCF^{CDC4} and on phosphorylation at a single specific site by a cyclin dependent kinase, Pho85 [6] in conjunction with the Pho85 cyclin Pcl5 [14], or alternatively, on phosphorylation on an undefined site by another cyclin dependent kinase, Srb10 [15]. Regulation of Gcn4 turnover by the availability of nutrients or, more generally, by the protein biosynthetic capacity of the cell, is mediated by regulation of Pho85 activity [6, 14].

β-catenin. Turnover of the mammalian transcriptional coactivator β-catenin depends on multiple phosphorylation by glycogen synthase kinase 3β (GSK3β), which leads to ubiquitination of β-catenin by $SCF^{\beta-TrCP}$ (reviewed in [11]). Interestingly, phosphorylation of β-catenin by GSK3β requires preliminary phosphorylation of β-catenin at a single residue, Ser45, by Casein Kinase I. Thus, both kinases are necessary for GSK3β phosphorylation and degradation. Activation of the Wnt pathway leads to inhibition of β-catenin phosphorylation, and consequently to stabilization of β-catenin. Whether the Wnt pathway directly inhibits Casein Kinase I or GSK3β is still somewhat unclear [16, 17].

Tec1. In the yeast *S. cerevisiae*, the Ste12 TF is involved in both the mating response, induced by activation of the MAP kinase cascade in the presence of pheromone, and filamentous growth, induced by activation of the MAP kinase cascade under nitrogen starvation conditions. Filamentation requires the combined activity of Ste12 with an additional TF, Tec1. The MAP kinase Fus3 can activate Ste12 under both conditions, whereas an alternative MAP kinase, Kss1, activates Ste12 under filamentation conditions only.

Inappropriate filamentation in the presence of mating pheromone is prevented in the following way: Fus3 – but not Kss1 – phosphorylates Tec1 at a specific site; phosphorylated Tec1 is then recognized by the SCF^{CDC4} ubiquitin ligase, and rapidly degraded. This ensures the strict segregation of the two alternative developmental programs, mating and filamentation, in spite of a common signal transduction pathway [8].

Inhibition of Ubiquitination by Substrate Phosphorylation

p53. The tumor suppressor protein p53 is regulated at several levels, including degradation by the ubiquitin system. p53 ubiquitination requires mdm2, a RING finger containing ubiquitin ligase [18]. Genotoxic stress leads to phosphorylation of p53 at several serine residues, including serine 20. Phosphorylation of serine 20 was found to stabilize the protein, probably by preventing its interaction with mdm2 [19]. Thus, in the case of p53, phosphorylation exerts a stabilizing effect on the protein.

Stimulation of Ubiquitination by Substrate Hydroxylation

HIF-1α. Although phosphorylation is the most common protein modification, other types of modification could, in principle, regulate the interaction between a protein and the ubiquitin system. The global transcriptional regulator of the hypoxic response, HIF-1α, provides an example of regulation by an alternative type of modification. HIF-1α is rapidly degraded under normoxic conditions, but stabilized under hypoxic conditions. Degradation of HIF-1α depends on ubiquitination by a ubiquitin ligase complex, the VBC complex (*V*on Hippel/Lindau protein, Elongin *B*, and Elongin *C*), which resembles the SCF complex in general architecture, and which contains the Von Hippel–Lindau tumor suppressor protein as substrate recognition component [20–22]. However, rather than phosphorylation, hydroxylation of a specific proline residue in HIF-1α was found to be responsible for the interaction of the protein with the VBC complex [23, 24]. This hydroxylation occurs only under normoxic conditions, thereby explaining the dependence of HIF-1α degradation on the oxygen concentration. A specific proline hydroxylase, the activity of which is directly regulated by the oxygen concentration, is responsible for the modification of HIF-1α [25, 26]. This pathway thus represents a case where a very short signaling cascade links the stimulus (oxygen concentration) to the response (HIF-1α degradation).

REGULATION OF UBIQUITIN LIGASE ACTIVITY

Signaling pathways can modulate degradation of proteins by modification of the substrate, as shown above. In principle, the same effect could be achieved by modification of the ubiquitination system in response to extracellular signals. The ubiquitin-protein ligase, being the substrate recognition component of the ubiquitin system, would *a priori* be the preferred site of regulation. Regulation of the ubiquitin ligase is best established in the case of the E3 *A*naphase *P*romoting *C*omplex/*C*yclosome (APC/C), responsible for the degradation of the regulators that govern, among other processes, the various mitotic transitions. Regulation of APC/C activity by signals emanating from the cell cycle machinery is essential for orderly cell cycle progression. However, it is not known whether the APC/C also responds to extracellular signals (reviewed in [27, 28]).

F-box protein stability. Activity of SCF type ligases can be modulated at the level of regulation of the F-box component, the substrate recognition component of the complex. Many F-box proteins were found to be intrinsically unstable [29], and therefore could be tightly regulated at the level of their synthesis. The TF E2F-1 accumulates in late G1 and is rapidly degraded in S phase, following ubiquitination by the SCF^{SKP2} complex. Regulation of E2F-1 degradation was shown to be due to the cell cycle dependent synthesis of the F-box protein SKP2 [13]. Another example is that of the yeast TF Met4, which is regulated by the SCF^{MET30} ubiquitin ligase [9]. Met30 is transcriptionally regulated by Met4 [9], but it is also regulated at the level of protein stability by the availability of methionine, so that in the absence of methionine, Met30 is destabilized and disappears from the cell, leading to increased Met4 activity [30]. Here again, regulation appears to be exerted by the concentration of an intracellular metabolite, rather than by extracellular signals.

Modulation of F-box protein–substrate interaction: SCF^{TIR1} and auxin. The plant hormone auxin exerts its effect by regulating the degradation of a family of transcriptional repressors, the Aux/IAA proteins, by the ubiquitin ligase SCF^{TIR1}. Degradation of the Aux/IAAs activates the ARF TFs, ultimately causing the effects of auxin on cell division, cell elongation, and differentiation [31]. Only recently has the mechanism of SCF^{TIR1} regulation by auxin been elucidated: crystal structure of the F-box protein TIR1 with the Aux/IAA substrate revealed that auxin fills a hydrophobic cavity in the substrate interacting face of TIR1, and acts as "molecular glue," enhancing the binding of Aux/IAA to the ubiquitin ligase [32]. This constitutes probably the most direct possible mechanism for specific regulation of the ubiquitin system by an extracellular signal.

PROCESSING OF TFS BY THE UBIQUITIN SYSTEM

NFκB. The p50 subunit of the TF NFκB is generated by proteolytic processing of a 105kD precursor, p105. Processing is mediated by the proteasome, in what constituted the first instance of partial degradation of a ubiquitinated protein [33].

The proteasome degrades the C-terminal half of the precursor, but arrests degradation and releases the mature N-terminal half because of the presence of a "stop transfer signal" of ill-defined nature in the middle of the precursor [34]. The only salient feature of the signal is that it is rich in glycine residues; how such a sequence would inhibit complete degradation of a protein is unclear, but it is notable that a similar glycine-rich region is found in the viral protein EBNA-1, where in contrast to p105, it confers complete protection of the protein against degradation by the proteasome. p105 processing can occur in two modes at least, constitutive and stimulated. The constitutive mode depends probably on signals near the glycine-rich region, and on an undefined ubiquitin ligase [34]. The second, stimulated mode of processing, depends on phosphorylation of residues in the C-terminus of p105 by the IκB kinase. The phosphorylated precursor is then recognized and ubiquitinated by $SCF^{\beta-TrCP}$, and processed or degraded by the proteasome [35, 36]. This second mode ensures that upon activation of the NFκB pathway, not only are NFκB molecules formerly sequestered in the cytosol recruited to the nucleus upon degradation of IκB, but new NFκB molecules are rapidly generated.

Spt23/Mga2. The yeast TFs Spt23 and Mga2 present another example of processing by the ubiquitin/proteasome system [37]. These two proteins are required for expression of *OLE1*, encoding a fatty acid desaturase essential for the synthesis of the monounsaturated fatty acids, palmitoleic and oleic acid. Regulated synthesis of these compounds is essential for the maintenance of proper membrane fluidity. Spt23 and Mga2 are normally membrane-bound due to a C-terminal transmembrane domain. The N-terminal domain of the TF is released from its membrane anchor via ubiquitination and selective degradation of the C-terminal domain by the proteasome. This processing is inhibited by addition of unsaturated fatty acids to the medium. The simplest hypothesis is that in this case, the signal for processing of the precursor is not extracellular, but rather consists of variations in membrane fluidity or thickness, which depend on the ratio of unsaturated to saturated fatty acids. These yeast proteins have in common with mammalian p105, an ankyrin domain in the C-terminal part of the protein that is removed during processing [37]. Interestingly, since the C-terminal end of Spt23 and Mga2's precursors are anchored in the membrane, proteasomal processing cannot initiate at the chain's end, but rather must initiate at an internal site in the polypeptide chain, presumably via introduction of a polypeptide loop into the 20S proteasome cavity, followed by endoproteolytic cleavage, and bidirectional degradation of the polypeptide [38].

MODULATION OF KINASE ACTIVITY BY UBIQUITINATION

The ubiquitin system is best known for its role in protein degradation; however, examples are emerging of nondestructive roles for ubiquitin. In signal transduction, an example for such a nondestructive role is the function of ubiquitination in the activation of the IκB kinase (IKK) (reviewed in [39]), the kinase responsible for the phosphorylation that induces IκB ubiquitination and degradation (see above). The NFκB pathway is activated by proinflammatory factors such as interleukin-1 that bind to their cognate receptors and ultimately lead to phosphorylation and degradation of IκB, thereby releasing NFκB to the nucleus [10].

One of the first proteins in the NFκB pathway is TRAF6, a signal transducer that links receptor activation to activation of IKK. TRAF6 activation depends on a complex of two ubiquitin conjugating enzymes, Ubc13/Uev1A, that catalyze the polyubiquitination of TRAF6 following receptor activation and (probably) TRAF6 oligomerization [40]. TRAF6, a RING domain protein, serves as its own ubiquitin ligase. The fact that polyubiquitinated TRAF6 escapes destruction is probably due to the nature of the ubiquitin chain: whereas "classical" ubiquitin chains are crosslinked via lysine 48 of ubiquitin, the Ubc13/Uev1A ubiquitin conjugating enzymes catalyze the formation of a lysine 63 crosslinked chain, which is probably not recognized by the proteasome. Lys 63 polyubiquitinated TRAF6 recruits and activates TAK1, a kinase that phosphorylates and activates IKK. TAK1 recruitment by TRAF6 requires TAB2/TAB3, adaptor proteins that binds both TAK1 and the polyubiquitin chain on TRAF6. The activation mechanism of IKK upon binding to TAB2/TAB3–$TRAF6^{Ubn}$ remains, however, unclear.

ROLE OF UBIQUITINATION/ PROTEASOME IN TF ACTIVITY

Role of proteasome in transcription. The proteasome subunits Rpt4 and Rpt6, two of the ATPases of the 19S regulatory subcomplex of the 26S proteasome, were originally identified genetically as *SUG1* and *SUG2*, genes involved in the activity of the yeast TF Gal4 [41]. The 19S subcomplex is responsible for recognizing the ubiquitinated proteins, unfolding the substrate protein, and threading the resulting polypeptide chain into the catalytic chamber of the proteasome 20S subcomplex [42]. Initially, the involvement of the proteasome in transcription was thought to reflect the role of the proteasome in degradation of TFs. Subsequently, however, biochemical evidence has accumulated that places the 19S subcomplex [43], and even the whole proteasome [44], at the site of actively transcribed genes, suggesting a more direct role for the proteasome in transcription initiation and elongation. The biochemical role of the proteasome in transcription is not entirely clear yet, but the superfluity of the 20S subcomplex for the effect of the 19S on transcription *in vitro* [45] argues that it is unrelated to its proteolytic activity, and may be related to the chaperone/unfoldase activity of the 19S subcomplex.

Ubiquitin as a potentiator of TF activity. The observation that activation domains of TFs often overlap with ubiquitination signals, and that conversely, cyclin degradation signals could serve as transcription activation domains, has spawned the theory that ubiquitination may itself potentiate the activity of TFs [46]. Further evidence for this possibility was provided by the observation that in yeast, the activity of the artificial TF, LexA-VP16, was increased in the presence of SCF^{MET30}, the ubiquitin ligase responsible for LexA-VP16 degradation [47]. Fusion of a single ubiquitin unit to LexA-VP16 was able to restore transcriptional activity even in the absence of MET30 [47], suggesting that monoubiquitination of a TF is sufficient to stimulate its activity. Since monoubiquitination is not sufficient for substrate recognition by the proteasome, this observation implies that the two opposing effects of ubiquitination on TFs – activation vs. degradation – may depend on the length of the ubiquitin chain added, activation by monoubiquitination being a prelude to degradation by polyubiquitination (or inactivation by polyubiquitination even in the absence of degradation; see [48]). An alternative model for the role of the ubiquitin/proteasome on TF activity posited that TFs are somehow inactivated during the process of transcriptional activators, and degradation is required for the turnover of "spent" TF on the promoter [49]. Evidence was put forward to show that both ubiquitination and proteasome proteolytic activity are required for the activity of the yeast transcription factors Gcn4, Gal4, and Ino2/4 *in vivo* [50]. Some of these conclusions have, however, been disputed [51]; and more generally, interpretation of experiments that rely on interference with proteolytic pathways *in vivo* must take potential pleiotropic effects of such interference into account. Thus, the mechanism(s) by which ubiquitination activates TFs is still elusive.

CONCLUSION

In the interplay between cellular signal transduction and the ubiquitin system, the latter was often thought as lying near the bottom of the signaling cascade, in the role of executioner of phosphorylated proteins, merely carrying out the verdict of the kinases. However, not only have we seen instances where the ubiquitination complex is itself the recipient of the signal, but in the case of the NFκB pathway, ubiquitination contributes to signaling by modulating the activity of kinases. In addition, the proteasome, and possibly ubiquitination, can contribute to TF activity in a nonproteolytic fashion. In summary, the versatility of the ubiquitin/proteasome system, both in its proteolytic and nonproteolytic functions, is nowhere better displayed than in regulation of cellular transcription.

ACKNOWLEDGEMENTS

Work in the authors' laboratories is supported by grants from the Israel Science Foundation, the Ministry of Health's Chief Scientist Office, the U.S.–Israel Binational Science Foundation, the German–Israeli Foundation for Scientific R&D, the German–Israeli Project Cooperation, a European Community TMR grant, and CapCure Israel.

REFERENCES

1. Hershko A, Ciechanover A. The ubiquitin system. *Annu Rev Biochem* 1998;**67**:425–79.
2. Skowyra D, Craig KL, Tyers M, Elledge SJ, Harper JW. F-box proteins are receptors that recruit phosphorylated substrates to the SCF ubiquitin-ligase complex. *Cell* 1997;**91**:209–19.
3. Feldman RM, Correll CC, Kaplan KB, Deshaies RJ. A complex of Cdc4p, Skp1p, and Cdc53p/cullin catalyzes ubiquitination of the phosphorylated CDK inhibitor Sic1p. *Cell* 1997;**91**:221–30.
4. Deshaies RJ. SCF and Cullin/Ring H2-based ubiquitin ligases. *Annu Rev Cell Dev Biol* 1999;**15**:435–67.
5. Patton EE, Willems AR, Tyers M. Combinatorial control in ubiquitin-dependent proteolysis: don't Skp the F-box hypothesis. *Trends Genet* 1998;**14**:236–43.
6. Meimoun A, Holtzman T, Weissman Z, McBride HJ, Stillman DJ, Fink GR, Kornitzer D. Degradation of the transcription factor Gcn4 requires the kinase Pho85 and the SCF(CDC4) ubiquitin-ligase complex. *Mol Biol Cell* 2000;**11**:915–27.
7. Bao MZ, Schwartz MA, Cantin GT, Yates 3rd JR, Madhani HD. Pheromone-dependent destruction of the Tec1 transcription factor is required for MAP kinase signaling specificity in yeast. *Cell* 2004;**119**:991–1000.
8. Chou S, Huang L, Liu H. Fus3-regulated Tec1 degradation through SCFCdc4 determines MAPK signaling specificity during mating in yeast. *Cell* 2004;**119**:981–90.
9. Rouillon A, Barbey R, Patton EE, Tyers M, Thomas D. Feedback-regulated degradation of the transcriptional activator Met4 is triggered by the SCF(Met30)complex. *Embo J* 2000;**19**:282–94.
10. Karin M, Ben-Neriah Y. Phosphorylation meets ubiquitination: the control of NF-κB activity. *Annu Rev Immunol* 2000;**18**:621–63.
11. Kimelman D, Xu W. beta-catenin destruction complex: insights and questions from a structural perspective. *Oncogene* 2006;**25**:7482–91.
12. Lassot I, Segeral E, Berlioz-Torrent C, Durand H, Groussin L, Hai T, Benarous R, Margottin-Goguet F. ATF4 degradation relies on a phosphorylation-dependent interaction with the SCF(betaTrCP) ubiquitin ligase. *Mol Cell Biol* 2001;**21**:2192–202.
13. Marti A, Wirbelauer C, Scheffner M, Krek W. Interaction between ubiquitin-protein ligase SCFSKP2 and E2F-1 underlies the regulation of E2F-1 degradation. *Nat Cell Biol* 1999;**1**:14–19.
14. Shemer R, Meimoun A, Holtzman T, Kornitzer D. Regulation of the transcription factor Gcn4 by Pho85 cyclin PCL5. *Mol Cell Biol* 2002;**22**:5395–404.
15. Chi Y, Huddleston MJ, Zhang X, Young RA, Annan RS, Carr SA, Deshaies RJ. Negative regulation of Gcn4 and Msn2 transcription factors by Srb10 cyclin-dependent kinase. *Genes Dev* 2001;**15**:1078–92.
16. Amit S, Hatzubai A, Birman Y, Andersen JS, Ben-Shushan E, Mann M, Ben-Neriah Y, Alkalay I. Axin-mediated CKI phosphorylation of beta-catenin at Ser 45: a molecular switch for the Wnt pathway. *Genes Dev* 2002;**16**:1066–76.
17. Liu C, Li Y, Semenov M, Han C, Baeg GH, Tan Y, Zhang Z, Lin X, He X. Control of beta-catenin phosphorylation/degradation by a dual-kinase mechanism. *Cell* 2002;**108**:837–47.
18. Haupt Y, Maya R, Kazaz A, Oren M. Mdm2 promotes the rapid degradation of p53. *Nature* 1997;**387**:296–9.
19. Unger T, Juven-Gershon T, Moallem E, Berger M, Vogt Sionov R, Lozano G, Oren M, Haupt Y. Critical role for Ser20 of human p53 in the negative regulation of p53 by Mdm2. *Embo J* 1999; **18**:1805–14.
20. Kamura T, Sato S, Iwai K, Czyzyk-Krzeska M, Conaway RC, Conaway JW. Activation of HIF1alpha ubiquitination by a reconstituted

von Hippel–Lindau (VHL) tumor suppressor complex. *Proc Natl Acad Sci USA* 2000;**97**:10,430–10,435.

21. Maxwell PH, Wiesener MS, Chang GW, Clifford SC, Vaux EC, Cockman ME, Wykoff CC, Pugh CW, Maher ER, Ratcliffe PJ. The tumour suppressor protein VHL targets hypoxia-inducible factors for oxygen-dependent proteolysis. *Nature* 1999;**399**:271–5.
22. Ohh M, Park CW, Ivan M, Hoffman MA, Kim TY, Huang LE, Pavletich N, Chau V, Kaelin WG. Ubiquitination of hypoxia-inducible factor requires direct binding to the beta-domain of the von Hippel–Lindau protein. *Nat Cell Biol* 2000;**2**:423–7.
23. Ivan Jr M, Kondo K, Yang H, Kim W, Valiando J, Ohh M, Salic A, Asara JM, Lane WS, Kaelin WG. HIFalpha targeted for VHL-mediated destruction by proline hydroxylation: implications for O2 sensing. *Science* 2001;**292**:464–8.
24. Jaakkola P, Mole DR, Tian YM, Wilson MI, Gielbert J, Gaskell SJ, Kriegsheim A, Hebestreit HF, Mukherji M, Schofield CJ, Maxwell PH, Pugh CW, Ratcliffe PJ. Targeting of HIF-alpha to the von Hippel–Lindau ubiquitylation complex by O2-regulated prolyl hydroxylation. *Science* 2001;**292**:468–72.
25. Epstein AC, Gleadle JM, McNeill LA, Hewitson KS, O'Rourke J, Mole DR, Mukherji M, Metzen E, Wilson MI, Dhanda A, Tian YM, Masson N, Hamilton DL, Jaakkola P, Barstead R, Hodgkin J, Maxwell PH, Pugh CW, Schofield CJ, Ratcliffe PJ. C. elegans EGL-9 and mammalian homologs define a family of dioxygenases that regulate HIF by prolyl hydroxylation. *Cell* 2001;**107**:43–54.
26. Bruick RK, McKnight SL. A conserved family of prolyl-4-hydroxylases that modify HIF. *Science* 2001;**294**:1337–40.
27. Fang G, Yu H, Kirschner MW. Control of mitotic transitions by the anaphase-promoting complex. *Philos Trans R Soc Lond B Biol Sci* 1999;**354**:1583–90.
28. Hershko A. Mechanisms and regulation of the degradation of cyclin B. *Philos Trans R Soc Lond B Biol Sci* 1999;**354**:1571–5. discussion 1575–1576.
29. Zhou P, Howley PM. Ubiquitination and degradation of the substrate recognition subunits of SCF ubiquitin-protein ligases. *Mol Cell* 1998;**2**:571–80.
30. Smothers DB, Kozubowski L, Dixon C, Goebl MG, Mathias N. The abundance of met30p limits SCF(Met30p) complex activity and is regulated by methionine availability. *Mol Cell Biol* 2000;**20**:7845–52.
31. Woodward AW, Bartel B. Auxin: regulation, action, and interaction. *Ann Bot (Lond)* 2005;**95**:707–35.
32. Tan X, Calderon-Villalobos LI, Sharon M, Zheng C, Robinson CV, Estelle M, Zheng N. Mechanism of auxin perception by the TIR1 ubiquitin ligase. *Nature* 2007;**446**:640–5.
33. Palombella VJ, Rando OJ, Goldberg AL, Maniatis T. The ubiquitin-proteasome pathway is required for processing the Nf-kappaB1 precursor protein and the activation of Nf-kappaB. *Cell* 1994;**78**:773–85.
34. Orian A, Schwartz AL, Israel A, Whiteside S, Kahana C, Ciechanover A. Structural motifs involved in ubiquitin-mediated processing of the NF- kappaB precursor p105: roles of the glycine-rich region and a downstream ubiquitination domain. *Mol Cell Biol* 1999;**19**:3664–73.
35. Orian A, Gonen H, Bercovich B, Fajerman I, Eytan E, Israel A, Mercurio F, Iwai K, Schwartz AL, Ciechanover A. SCF(beta)(-TrCP) ubiquitin ligase-mediated processing of NF-kappaB p105 requires phosphorylation of its C-terminus by IkappaB kinase. *Embo J* 2000;**19**:2580–91.
36. Heissmeyer V, Krappmann D, Hatada EN, Scheidereit C. Shared pathways of IkappaB kinase-induced SCF(betaTrCP)-mediated ubiquitination and degradation for the NF-kappaB precursor p105 and IkappaBalpha. *Mol Cell Biol* 2001;**21**:1024–35.
37. Hoppe T, Matuschewski K, Rape M, Schlenker S, Ulrich HD, Jentsch S. Activation of a membrane-bound transcription factor by regulated ubiquitin/proteasome-dependent processing. *Cell* 2000;**102**:577–86.
38. Piwko W, Jentsch S. Proteasome-mediated protein processing by bidirectional degradation initiated from an internal site. *Nat Struct Mol Biol* 2006;**13**:691–7.
39. Chen ZJ. Ubiquitin signalling in the NF-kappaB pathway. *Nat Cell Biol* 2005;**7**:758–65.
40. Wang C, Deng L, Hong M, Akkaraju GR, Inoue J, Chen ZJ. TAK1 is a ubiquitin-dependent kinase of MKK and IKK. *Nature* 2001;**412**:346–51.
41. Swaffield JC, Bromberg JF, Johnston SA. Alterations in a yeast protein resembling HIV Tat-binding protein relieve requirement for an acidic activation domain in GAL4. *Nature* 1992;**357**:698–700.
42. Glickman MH, Ciechanover A. The ubiquitin-proteasome proteolytic pathway: destruction for the sake of construction. *Physiol Rev* 2002;**82**:373–428.
43. Gonzalez F, Delahodde A, Kodadek T, Johnston SA. Recruitment of a 19S proteasome subcomplex to an activated promoter. *Science* 2002;**296**:548–50.
44. Gillette TG, Gonzalez F, Delahodde A, Johnston SA, Kodadek T. Physical and functional association of RNA polymerase II and the proteasome. *Proc Natl Acad Sci USA* 2004;**101**:5904–9.
45. Ferdous A, Gonzalez F, Sun L, Kodadek T, Johnston SA. The 19S regulatory particle of the proteasome is required for efficient transcription elongation by RNA polymerase II. *Mol Cell* 2001;**7**:981–91.
46. Salghetti SE, Muratani M, Wijnen H, Futcher B, Tansey WP. Functional overlap of sequences that activate transcription and signal ubiquitin-mediated proteolysis. *Proc Natl Acad Sci USA* 2000;**97**:3118–23.
47. Salghetti SE, Caudy AA, Chenoweth JG, Tansey WP. Regulation of transcriptional activation domain function by ubiquitin. *Science* 2001;**293**:1651–3.
48. Kaiser P, Flick K, Wittenberg C, Reed SI. Regulation of transcription by ubiquitination without proteolysis: Cdc34/SCF(Met30)-mediated inactivation of the transcription factor Met4. *Cell* 2000;**102**:303–14.
49. Lipford JR, Deshaies RJ. Diverse roles for ubiquitin-dependent proteolysis in transcriptional activation. *Nat Cell Biol* 2003;**5**:845–50.
50. Lipford JR, Smith GT, Chi Y, Deshaies RJ. A putative stimulatory role for activator turnover in gene expression. *Nature* 2005;**438**:113–16.
51. Nalley K, Johnston SA, Kodadek T. Proteolytic turnover of the Gal4 transcription factor is not required for function in vivo. *Nature* 2006;**442**:1054–7.

Chapter 39

Regulating Endoplasmic Reticulum Function through the Unfolded Protein Response

Alicia A. Bicknell and Maho Niwa

University of California San Diego, Division of Biological Sciences, La Jolla, California

INTRODUCTION

Secreted proteins and proteins that reside on the cell surface or within the secretory pathway begin their maturation process in the endoplasmic reticulum (ER). These proteins are synthesized in the cytosol, then targeted and translocated into the ER as nascent peptides [1]. Upon entry into the ER lumen, nascent proteins associate with ER chaperones and protein modification enzymes to fold into their native functional structures. Because misfolded proteins may be toxic to the cell, only properly folded proteins are allowed to exit the ER to arrive at their final cellular or extracellular destinations. Incompletely folded proteins are retained in the ER for further processing and permanently misfolded proteins are marked within the ER for proteasome mediated degradation [2–5]. Thus, the ER serves as a master regulator for the complex and error prone process of protein maturation, quality control, and trafficking. Furthermore, the ER must match its capacity for protein processing with the cell's dynamic need for protein synthesis, dictated by developmental and environmental cues.

The ER regulates its own protein processing capacity through an inter-organelle signaling pathway termed the unfolded protein response (UPR). The UPR is activated in the ER lumen when molecular sensors detect an accumulation of unfolded proteins that exceeds the ER's folding capacity, a condition known as ER stress. These sensors initiate a series of signaling events that reduce the influx of unfolded proteins into the ER and increase the ER's ability to properly process proteins and degrade those proteins that are permanently misfolded [6]. UPR regulation also extends beyond the ER, increasing the efficiency of post-ER protein processing steps to broadly enhance the cell's secretory capacity [2, 7]. Under certain conditions, such as prolonged ER stress, or ER stress that cannot be reversed, the UPR pathway also has the ability to induce apoptotic cell death [8].

MOLECULAR SENSORS

In the mammalian ER, there are at least three molecular sensors that initiate UPR activation: IRE1, PERK, and ATF6. Each of these sensors is a transmembrane protein with a luminal domain that detects ER stress and a cytosolic domain that activates a downstream UPR signal. In *S. cerevisiae*, there is no homolog of PERK or ATF6; Ire1p is the only known UPR component at the ER membrane.

IRE1 is a bifunctional kinase/endoribonuclease ER-transmembrane protein that has two isoforms in mammalian cells, IRE1α and IRE1β. When its N-terminal luminal domain detects ER stress, IRE1 dimerizes or oligomerizes within the ER membrane [9, 10]. Subsequently, its cytosolic kinase domain undergoes trans-autophosphorylation, thus causing the activation of its cytosolic endoribonuclease domain. IRE1's nuclease then cleaves a single UPR specific intron from the mRNA encoding Hac1p in yeast, or XBP-1 in mammalian cells [11–14]. In yeast, once this UPR specific intron is removed from *HAC1*, the two exons are joined by tRNA ligase to produce the spliced form of the transcript [15]. The *HAC1* mRNA splicing mechanism resembles that of tRNA splicing, such that a 2′ phosphate remains at the splice junction after ligation [16]. This phosphate is subsequently removed by a 2′ phosphotransferase (Tpt1p) [17]. In mammalian cells, Xbp-*1* exons are also joined following cleavage by IRE1, but the ligase that joins these exons has not yet been identified. Mouse embryonic fibroblasts (MEFs) lacking *Tpt1* show no *Xbp-1* splicing

Handbook of Cell Signaling, Three-Volume Set 2 ed.

phenotype [18], suggesting that either tRNA ligase uses a different mechanism to ligate *Xbp-1* in mammalian cells, or that tRNA ligase is not involved in mammalian UPR splicing. In addition to utilizing a unique mechanism, UPR specific mRNA splicing is also unique in that it occurs in the cytoplasm [19] independently of the spliceosome.

This *HAC1*/*Xbp-1* splicing event is a key regulatory step in the UPR signal transduction pathway. Once spliced, *Xbp-1* and *HAC1* are translated to become transcription factors that bind to promoters of target genes and activate a broad UPR specific transcriptional program that helps cells cope with ER stress [2, 7]. XBP-1 binds at least three promoter elements with differing affinities; it binds strongly to the unfolded protein response element (UPRE) and more weakly to ER stress elements I and II (ERSE and ERSE-II) [20]. Hac1p binds and activates the yeast UPRE-1, UPRE-2, and UPRE-3 [21, 22]. However, only the spliced form of each protein can serve this function. In yeast, the UPR intron is inhibitory to translation of *HAC1* [19, 23], so the unspliced protein has never been detected in the cell. In mammalian cells, splicing of *Xbp-1* mRNA results in a translational frame shift causing the production of a second form of the XBP-1 protein, with a unique C-terminus. This spliced form of XBP-1 is a far more potent transcriptional activator than the unspliced form [24]. In addition, the *Xbp-1* and *HAC1* genes contain an ERSE and UPRE, respectively within their own promoters, allowing the encoded proteins to potentiate the UPR signal by inducing their own transcription [25]. Furthermore, in yeast, the splicing of *HAC1* leads to an increase in the level of Gcn4p protein by an unknown mechanism. Gcn4p is a transcription factor that activates, along with Hac1p, all three yeast UPREs [22].

In mammalian cells, IRE1 has the additional function of recruiting the protein TRAF2 to the ER membrane [26]. TRAF2, in turn, recruits and activates ASK1 [27]. ASK1 is a MAP kinase kinase kinase that ultimately activates the MAP kinase, JNK [28]. JNK is a well known stress response protein that has the power to regulate multiple transcription factors [29]. However, the precise function of JNK activation during ER stress is not yet clear.

PERK, the second ER sensor in mammalian cells, is a type-I ER-transmembrane kinase that also senses protein folding demands in the ER through its N-terminal domain [30, 31]. Once activated by ER stress, PERK oligomerizes and autophosphorylates, thus activating its cytosolic kinase domain. PERK then phosphorylates the αsubunit of eukaryotic translation initiation factor 2 (eIF2α), which rapidly shuts down translation in the cell [32–34]. Translation attenuation during the UPR reduces the influx of newly synthesized proteins into the ER, thus alleviating ER stress. In addition, when translation efficiency drops, cyclin D1 levels rapidly diminish due to an intrinsically high turnover rate [35]. This leads to a G1 phase cell cycle arrest, which is thought to expand the window of time for the cell to decide between adapting to the stress or undergoing apoptosis [36].

Although eIF2α phosphorylation signals a global decline in translation, it actually increases the translation of the transcription factor, ATF4 [37], which goes on to transcribe a second set of UPR responsive genes [38]. In addition to eIF2α, activated PERK phosphorylates the transcription factor Nrf2, thus allowing it to enter the nucleus [39] where it presumably regulates UPR dependent gene expression. Currently, PERK is only known to phosphorylate eIF2α and Nrf2. However, multiple forms of phosphorylated PERK accumulate under certain UPR inducing conditions [40], suggesting that this kinase may have additional undiscovered substrates. Future work will characterize these different forms of PERK and identify their unique cellular targets, if such targets exist.

ATF6, the third UPR molecular sensor, exists as two isoforms, ATF6α and ATF6β. It is a type-I ER-transmembrane transcription factor that is required for the activation of many UPR target genes [41, 42]. When its luminal domain senses ER stress, ATF6 moves into the Golgi, where it is accessible to Site 1 Protease (S1P) and Site 2 Protease (S2P). These proteases sequentially cleave ATF6, liberating the soluble N-terminal domain into the cytosol [43]. Upon release, this domain moves into the nucleus where it binds to ERSE and ERSE-II promoter elements to regulate a third branch of the UPR specific transcriptional program [44, 45].

HOW MOLECULAR SENSORS DETECT ER STRESS

Although their downstream signals vary, IRE1, PERK, and ATF6 share the ability to self-activate upon sensing unfolded proteins in the ER. Precisely how these sensors detect unfolded proteins is currently an area of active research, which focuses primarily on understanding the sensors' ER luminal domains. Despite little sequence homology among their luminal domains, IRE1, PERK, and ATF6 share some features of their unfolded protein sensing mechanisms.

Shortly after the discovery of Ire1, it was proposed that a negative regulator might release Ire1's luminal domain during ER stress, thus causing its activation [46]. Subsequently, it was discovered that the luminal domains of all three molecular sensors bind to the ER resident chaperone BiP in unstressed cells. Upon exposure to ER stress, they release BiP with kinetics that correlate well with their own activation [47]. Therefore, an initial model regarded BiP as the proximal sensor of ER stress, with its binding and release comprising the UPR activation switch for all three sensors. While this model was attractive, several recent findings suggest that the mechanism of unfolded protein sensing may not rely solely upon a simple model of BiP binding and release. *IRE1* deletion studies found that *ire1* mutants that cannot bind BiP are not constitutively active [48]. Furthermore, a recent study monitored the kinetics of each sensor's activation during different types of ER stress. This study revealed that either BiP binds each sensor with a different affinity, or BiP release is not the rate limiting step for UPR sensor activation [40]. Therefore, although it seems evident that BiP binding has a role to play in regulating the UPR pathway, we still do not fully understand the nature of this role, and whether this role is the same or different for each of the molecular sensors.

Recently, Credle *et al.* solved the crystal structure of the yeast Ire1p core luminal domain, providing another clue about Ire1's activation mechanism [49]. The crystal structure shows that two monomers of Ire1p, when joined together, create a deep groove that is reminiscent of the peptide binding pocket of the major histocompatibility complex (MHC). Since the MHC binds a wide variety of peptides, this structural feature strongly suggests that Ire1's luminal domain directly binds misfolded proteins. Although it has been shown that residues lining the MHC-like groove are required for unfolded protein detection, a direct peptide binding function for the groove has not yet been verified. Furthermore, if Ire1 does bind misfolded peptides, future work will need to address how this binding interplays with Ire1's binding of BiP to regulate activity.

Similarly detailed studies of PERK and ATF6 activation mechanisms also remain to be done. Like IRE1, PERK and ATF6 release BiP from their luminal domains during UPR activation [47]. However, BiP release may play a more prominent role in the activation of these sensors, as deletion of the BiP binding site in their luminal domains is sufficient for constitutive activation [50, 51]. In addition, it is possible that PERK directly binds unfolded peptides, as secondary structure predictions indicate that PERK's luminal domain folds similarly to IRE1 [49].

Furthermore, in addition to BiP release and possible peptide binding, posttranslational modifications within the luminal domains of UPR sensors may also play a role in their activation. In response to ER stress, ATF6 becomes hypoglycosylated and its disulfide bonds reduced. Reducing ATF6's intermolecular disulfide bonds converts it from an oligomer to a monomer, a conversion that is necessary but not sufficient for its transport to the Golgi and efficient cleavage by S1P [52]. Hypo-glycosylation of ATF6 also contributes to its Golgi transport and activation, although the precise reasons for this are not yet clear [53]. A potential role for luminal modifications of PERK and IRE1 has not been thoroughly examined. However, mutating four conserved cysteines in PERK's luminal domain [54] or one conserved glycosylation site in IRE1's luminal domain [55] had no impact on the ability of either sensor to detect unfolded proteins.

Although we do not yet have a complete understanding of how the three UPR sensors detect unfolded proteins, emerging evidence indicates that slight differences in this sensing mechanism allow each sensor to respond with different sensitivities to specific types of ER stress. The use of pharmacological agents such as DTT or the ER calcium importin inhibitor thapsigargin has been instrumental in revealing these sensitivity differences. For example, breaking disulfide bonds in the ER rapidly activates ATF6 and slowly activates PERK mediated eIF2α phosphorylation, whereas depleting calcium in the ER quickly phosphorylates eIF2α and slowly activates ATF6 [40]. Various physiological conditions of ER stress impose unique types of protein folding load on the ER, and each activated sensor elicits a different, but overlapping, transcriptional program. Therefore, the distinct ER sensing mechanisms may be adapted to differentially activate each sensor according to the specific type of stress present in the ER lumen, thus allowing a fine-tuning of the mammalian UPR pathway.

DOWNREGULATING THE UPR

The three sensors of the UPR pathway induce a wide array of overlapping, but non-identical, physiological changes. Some of these changes promote adaptation to ER stress and survival, while others lead to apoptosis. Downregulation of each branch of the pathway is therefore critical in achieving the appropriate cell fate following ER stress. Although we still have a limited understanding of how the cells shut off the UPR, it is clear that each branch is downregulated by several mechanisms, each of which is a potential point for modulation of the pathway.

As autophosphorylation of Ire1 is critical for its activation, attenuation of Ire1 is likely to involve phosphatase activity. Two phosphatases, Ptc2p [56] and Dcr2p [57], have been implicated in the dephosphorylation and downregulation of yeast Ire1p, but an IRE1 phosphatase in mammalian cells has not yet been found. Mammalian cells do, however, have an interesting mechanism of downregulation the IRE1-XBP-1 UPR branch. During the induction phase of the UPR, XBP-1 upregulates its own transcription. The combined kinetics of IRE1 activation and *Xbp-1* transcription, coupled with the distinct degradation rates of the spliced and unspliced proteins, leads to an immediate accumulation of spliced XBP-1 protein, followed by a gradual increase in the unspliced protein. During the late stages of UPR induction, unspliced XBP-1 dimerizes with the spliced form, causing it to be removed from the nucleus and degraded by the proteasome [58], thus shutting down XBP-1 mediated transcription and turning off the pathway. However, such a mechanism is unlikely to exist in yeast, as the intron in unspliced *HAC1* mRNA prevents its translation.

Reminiscent of XBP-1 downregulation by its own unspliced isoform, ATF6α mediated transcription may be downregulated by the activation of ATF6β. Both isoforms of ATF6 are activated by ER stress, but ATF6α has a much shorter half-life than ATF6β [59]. *In vitro*, the two proteins can compete for binding to the ERSE, but ATF6α is a much stronger transcription factor than ATF6β. These data suggest a model in which the ERSE is primarily bound by

ATF6α during UPR induction, and target gene transcription is therefore strongly activated. During later UPR stages, because ATF6β is more stable than ATF6α, it begins to outcompete ATF6α for ERSE binding. Since ATF6β is a weak transcription factor, this binding reduces the transcription of ERSE containing genes [60]. The ATF6 branch of the UPR pathway is also downregulated upstream of transcription, at the level of ATF6 proteolytic cleavage. NUCB1 is a Golgi localized protein that gets induced by ATF6 during UPR activation. Upon increased expression in the Golgi, NUCB1 interferes with the S1P-ATF6 interaction, thus causing reduced ATF6 cleavage and nuclear translocation, and shutting off the ATF6 pathway by a negative feedback mechanism [61].

Like ATF6, PERK can be inhibited both at the level of its own activation, and at the downstream event that it regulates, in this case eIF2α phosphorylation. To directly inhibit PERK activity, ATF6 activates the transcription of $P58^{IPK}$. $\mathrm{P58}^{\mathrm{IPK}}$ is a DnaJ family protein that binds and inhibits PERK's kinase domain, causing the downregulation of eIF2α phosphorylation, and the reversal of translational inhibition [62, 63]. In addition, ATF4 activates the transcription of *GADD34*, a subunit of PP1c phosphatase, which directly dephosphorylates eIF2α during the later stages of ER stress, thus allowing cells to resume normal translation rates [64–66]. PERK inactivation and recovery of protein synthesis are crucial to the adaptive function of the other two branches of the UPR pathway, as these branches rely upon an increased transcription and translation of protein folding enzymes, as well as the translation of spliced XBP-1. In the absence of functional GADD34, when protein synthesis rates do not recover, spliced *Xbp-1* does not accumulate and XBP-1 target genes are not expressed.

CELLULAR EFFECTS OF UPR INDUCTION

Broadly, the UPR is defined as a pathway that senses ER stress, and elicits cellular changes in response to this stress. Most of these cellular changes are adaptive, allowing the cell to cope with the toxic conditions imposed by ER stress. As part of this adaptive response, the UPR increases the cell's protein folding and secretory capacity, adjusts the protein folding load in the ER, and degrades potentially toxic misfolded proteins. When adaptation is not possible or not desired, the UPR can also induce apoptosis. These adaptive and apoptotic responses are discussed in detail below.

Increase in Secretory Capacity

To help cells cope with an insufficiency in secretory protein processing, the UPR utilizes all three of its transcription factors (XBP1/Hac1p, ATF4, ATF6) to expand the cellular machinery that produces mature secretory proteins. Specifically, UPR target genes increase the ER's physical volume, as well as its protein folding activity, and also act to enhance post-ER protein processing functions.

In both yeast and mammals, ectopic expression of the spliced, active form of *HAC1/Xbp-1* transcription factor has been shown to expand the volume of the rough ER [7, 67, 68]. In yeast, it is clear that the UPR pathway transcriptionally activates many phospholipid and inositol metabolism genes [2]. Although it is presumed that these genes are responsible for the lipid biogenesis that the ER's physical expansion requires, the exact mechanism for generating extra ER membrane is not yet known. In mammals, expression of spliced *Xbp-1* increases the production of phosphatidyl choline, the primary phospholipid component of ER membranes [68, 69]. This lipid production is accompanied by the XBP-1 mediated production of choline cytidylyltransferase, a rate limiting enzyme of the phosphatidyl choline biosynthetic pathway [69, 70] Presumably, activation of this enzyme is at least partially responsible for the dramatic expansion of the ER that is induced during the UPR.

In addition to expanding the ER's physical volume, the UPR also increases the production of enzymes that reside in the ER and facilitate the process of protein folding. Chaperones, co-chaperones, disulfide bond catalyzing enzymes, and glycosylation enzymes are all transcriptional targets of UPR activation in yeast and mammalian systems [2, 7, 71]. Chaperone proteins fall into two main classes: heat shock family chaperones and chaperone lectins (carbohydrate binding proteins). Heat shock family proteins, which include BiP (Grp78) and Grp94, recognize hydrophobic regions of unfolded proteins and assist these proteins in achieving their appropriate conformation through a process that utilizes repeated rounds of ATP hydrolysis and ADP exchange. Co-chaperones, such as ERdj4, assist in ATP hydrolysis, and are required for proper chaperone function [72, 73]. The second class of chaperones, the chaperone lectins, includes the ER membrane protein, calnexin, and the luminal protein, calreticulin. These proteins bind to carbohydrate moieties of unfolded glycoproteins to promote their proper folding and exit from the ER [74].

Secretory protein maturation also requires the formation of specific intramolecular disulfide bonds. Disulfide bonds are formed within the ER by a series of reactions that ultimately uses molecular oxygen to oxidize free thiol groups on cysteine residues. These reactions are catalyzed by the ER resident oxidoreductases, ERO1 and PDI (protein disulfide isomerase), both of which are strongly induced by the UPR pathway [75–78].

Finally, the UPR is responsible for inducing a large number of ER resident glycosylation enzymes. Glycosylation within the ER promotes proper protein folding in two key ways. Firstly, glycosylation moieties are often an intrinsic component of a protein's native conformation, as they can help stabilize the structure of a protein, or increase its solubility. Secondly, glycans are used to tag unfolded proteins, so that they can be recognized by lectin chaperones to promote proper folding, or, if they remain unfolded for too long, glycan-tagged proteins are recognized and processed by the cell's protein degradation machinery [74].

Although the signal to stimulate UPR activity is initiated specifically within the lumen of the ER, the cell takes a very broad approach in responding to this signal. In addition to increasing the ER's protein folding capacity, the UPR appears to enhance the processing and trafficking of secretory proteins after they exit the ER. Genes involved in ER to Golgi transport, Golgi function, Golgi to ER transport, and exocytosis are all activated by UPR transcription factors in response to ER stress [2, 70, 79].

Adjustment of Protein Folding Load

After exiting the nucleus, mRNAs associate with ribosomal subunits in the cytosol to form an initiation complex and begin translation. During translation, those nascent proteins that are destined for the secretory pathway reveal a signal sequence that emerges from the ribosome and is recognized by the signal recognition particle (SRP). The SRP associates with the SRP receptor on the ER membrane to deliver the peptide and its associated ribosome to the translocon. Here, translocation of the secretory protein proceeds co-translationally into the ER lumen [1].

To cope with high levels of ER stress, the UPR adjusts the influx of nascent proteins into the ER. Thus, any step in the process of protein production and translocation is a potential point of regulation by the UPR. To date, studies have shown that mRNA stability, translation, and translocation are altered during ER stress. Modulating these points of protein production can stop the arrival of nascent proteins to the ER, where the protein folding machinery is already overwhelmed. In addition, this modulation might free ribosomes and translocons to engage in the production of chaperones and other proteins critical for re-establishment of ER homeostasis.

mRNA stability

A genome-wide study, examining steady state mRNA levels and transcriptional changes in mammalian cells [80], has revealed that more than 800 transcripts become destabilized in response to ER stress. The purpose and mechanism for this destabilization are not yet clear, but one clue may come from a study in *Drosophila* [81]. Here, it has been shown that IRE1 mediates the degradation of a specific subset of mRNAs, in a manner that is independent of XBP-1. Specifically, IRE1 promotes the internal cleavage of certain transcripts, generating RNA fragments that are subject to degradation by housekeeping machinery. It is not yet known how IRE1 signals this event, but one intriguing possibility is that IRE1's ribonuclease domain directly catalyzes the endonucleolytic cleavage of the transcripts in question. In this case, IRE1 might simply act as a nonspecific nuclease, and regulation of mRNA stability might occur via IRE1 activation and mRNA recruitment to the ER membrane, where IRE1 resides.

Genome-wide analysis shows that IRE1 mediated mRNA degradation specifically targets transcripts that encode plasma membrane and other secreted proteins, but spares transcripts that directly promote protein folding within the ER [80], suggesting that the purpose of degradation is to reduce the load of unfolded protein substrates in the ER. How the cell achieves this specificity is not yet clear. Interestingly, the ER targeting signal sequence is necessary for transcript degradation. This suggests that the degradation machinery targets only those transcripts that are being translated at the ER membrane, providing a mechanism for a selection of secretory proteins. However, the means of escape for protein folding enzymes, which also contain a signal sequence and are translated at the ER membrane, has not yet been explored.

Translation

To reduce the influx of unfolded proteins into the ER, the cell also takes the approach of downregulating translation during ER stress. In mammalian cells, the UPR is able to inhibit translation in at least two key ways. First, IRE1β promotes the cleavage of 28S ribosomal RNA during ER stress, causing a moderate decline in protein synthesis [82]. Although cleavage does require a functional IRE1 ribonuclease domain, it is not yet known whether IRE1 directly cleaves rRNA, or indirectly promotes the cleavage event.

Second, as described in the introduction, PERK activation mediates a translation block during ER stress. When phosphorylated by PERK kinase, eIF2α is unable to exchange its associated GDP for GTP [83]. Since this exchange is an obligate step in translation initiation, PERK dependent translation inhibition can be quite dramatic, reducing translation to 10 percent under certain conditions of ER stress. This inhibition is very rapid, preceding the slower transcriptional branches of the UPR pathway, and is therefore considered the "first line of defense" against ER stress [84]. *Perk* knockout cells are severely impaired in ER stress survival, and this defect can be rescued by external means of translation inhibition [85].

Budding yeast do not have a *Perk* homolog. However, they do have eIF2α and several eIF2α kinases that are able to repress translation. One of these kinases, Gcn2p is implicated to function during the UPR [22]. Therefore, it is surprising that the yeast UPR does not appear to repress translation [86]. We do not yet understand why PERK mediated translational repression is crucial to the mammalian UPR, but dispensable for budding yeast.

PERK dependent translational inhibition was initially thought to equally affect mRNAs translated in the cytosol and those translated at the ER membrane. However, recent studies suggest that translation in each compartment is actually affected differently by ER stress [87]. In the cytosol, ER stress causes mRNAs to move from a strongly translated polyribosomal fraction, to a non-ribosome associated fraction. By contrast, mRNAs at the ER shift from large polyribosomes to smaller polyribosomes and 80S monosomes. Furthermore, ribosomes at the ER membrane remain ER associated after induction of the UPR. Thus, ER stress decreases translation in both the cytosol and the ER, but a low level of translation is maintained at the ER membrane. During ER stress, *Bip*, *Xbp-1*, and *Atf4* all continue to be translated at the ER. This suggests that while reduced translation serves to decrease the protein folding load in the ER, the low level of translation at the ER is maintained for the preferential production of those proteins that help the ER resume homeostasis. This preferential translation at the ER membrane extends to proteins, such as XBP-1 and ATF4, which do not contain a signal sequence and are not destined to translocate into the ER [87]. How such mRNAs are targeted to the ER membrane, and why this targeting promotes escape from translational repression remain open and provocative questions.

Translocation

In addition to targeting nascent secretory proteins to the translocon, the signal sequence of a nascent peptide actually interacts with the translocon to promote protein translocation into the ER lumen [88]. Some signal sequences promote efficient translocation, while others are less efficient, and this variation in translocational ability is conserved [89]. One purpose of this signal sequence variation is to provide another means of coupling the influx of proteins into the ER to conditions within the cell. Under conditions of ER stress, proteins containing an inefficient signal sequence undergo co-translocational degradation, whereas efficiently translocated proteins remain unaffected. This process, termed "pre-emptive quality control," may protect the cell from the toxic effects of protein aggregation, as those proteins that tend to aggregate in the ER have a weaker signal sequence, and are therefore substrates for degradation [90, 91]. It stands to reason that proteins that reside in the ER and enhance protein folding might contain strong signal sequences, which would render them resistant to co-translocational degradation during ER stress, but this has not yet been demonstrated.

Degradation of Terminally Misfolded Proteins

ER Associated Degradation

Despite the fact that protein folding enzymes outnumber folding substrates in the unstressed ER [92], the protein folding process is error prone and a large percentage of secretory proteins permanently misfold. If allowed to escape the ER, these misfolded proteins could be highly toxic to the cell. They are therefore retained in the ER and subsequently recognized and degraded by the ER associated degradation (ERAD) pathway [93]. During ER stress, the ER's protein folding machinery is overwhelmed, and permanently misfolded proteins accumulate to especially high levels. The UPR enables the ER to safely clear these accumulated substrates by enhancing ERAD activity [2]. This function of the UPR is vital, as overexpression of a misfolded ERAD substrate has little impact on wild-type cells, but is lethal to yeast that do not have an intact UPR [86]. Furthermore, even in the absence of stress, simultaneously disabling both the ERAD and UPR pathways is synthetically lethal [2].

The details of the ERAD pathway are discussed in several recent reviews [94], so we will only briefly describe its general steps. ERAD begins when components within the ER lumen or ER membrane recognize an unfolded ER substrate to be permanently misfolded, thus distinguishing it from a legitimate folding intermediate that still could achieve its appropriate conformation. This recognition process is complex and incompletely understood. In mammalian cells, EDEM appears to recognize misfolded glycosylated proteins, and HERP helps recognize misfolded proteins that are not glycosylated [95]. In yeast, Yos9p [96, 97] and EDEM's homolog Htm1p [98] have both been implicated in misfolded glycoprotein recognition. Following recognition, misfolded substrates are delivered to an unknown channel, where they are retrotransolocated into the cytoplasm. Once retrotranslocation has begun, the HRD1/HRD3 E3 ubiquitin ligase complex catalyzes the poly-ubiquitination of the misfolded protein [99, 100]. The protein is then completely extracted from the ER by the p97 ATPase complex (Cdc48p in yeast), and delivered to the cytosolic proteasome for degradation [101].

During ER stress in yeast, Hac1p is known to activate *HRD1*, *HRD3*, and *UBC7*, an E2 ubiquitin conjugating enzyme required for Hrd1p/Hrd3p function [2]. In the mammalian UPR, XBP1 has been shown to activate *Edem*, *Herp*, *Derlin1* (which is involved in retrotranslocation), and *Hrd1* [7, 70, 71]. ATF6 activates *Herp* and *Hrd1* [71]. Thus, the UPR takes a broad approach, increasing the efficiency of multiple steps in the ERAD pathway, in order recognize and destroy proteins that cannot achieve their appropriate conformation.

Autophagy

Autophagy is a mechanism of bulk degradation, whereby large portions of the cytosol and its resident organelles are engulfed by a double lipid bi-layer. The resulting subcompartment, termed the autophagosome, ultimately fuses with the lysosome (or vacuole in yeast) so that its components can be degraded and recycled. Autophagy is best understood as a means of enduring starvation conditions, when recycled cellular content can replace nutrient supplementation to provide for the cell's most essential functions. During nutrient deprivation in yeast, the cell initiates autophagy by transcribing the ubiquitin-like gene, *ATG8*. Once translated, Atg8p undergoes a series of modifications, including lipidation, which are required for autophagosome formation. After lipidation, Atg8p co-localizes with, and probably helps to form pre-autophagosomal structures (PASs) in the cytosol. These PASs nucleate autophagosomes, which ultimately fuse with the vacuole to facilitate the degradation of autophagosomal content [102].

Recently, it has become evident that autophagy is also activated during ER stress in both yeast and mammals [67, 103–105]. This process, termed "ER-phagy," is thought to help the cell withstand the stress. Although we are just beginning to understand how ER-phagy helps the cell cope with ER stress, current data suggest that it supplements the ERAD pathway in clearing unfolded proteins from the ER [106, 107]. In addition, it might be used to counterbalance the membrane expansion that is induced by the UPR [67]. According to this theory, upon activation of ER-phagy, ER membranes and their resident misfolded proteins are engulfed and delivered to the lysosome to be degraded. This engulfment might occur indiscriminately, or subcompartments of the ER might be designated for misfolded protein targeting, and ultimate disposal. In support of this idea, ER-phagy induced autophagosomes in yeast cells contain membranes that are derived from the ER. Furthermore, inhibiting autophagy interferes with the disposal of certain misfolded mutant secretory proteins [106, 107] and causes the accumulation of potentially toxic protein aggregates [105, 108].

Although little is known about the autophagic mechanism during ER stress, it does appear to overlap somewhat with starvation induced autophagy. During ER stress, Atg8p is induced and lipidated [104], although other autophagic modifications to the protein have not been examined. Atg8p then co-localizes with cytosolic structures, which appear similar to PASs, although they are more numerous than the PASs formed during starvation. It is thought that these Atg8p containing structures help nucleate autophagosomes during ER stress because they are juxtaposed to ER induced autophagosomes and because *ATG8* is required for autophagosome formation during ER stress [67]. Although there are some similarities between ER-phagy and starvation induced autophagy, the autophagosomes produced by each process differ in their content, and therefore must be derived by processes that are somehow distinct. Future studies will seek to understand the similarities between ER-phagy and starvation induced autophagy, as well as the defining mechanistic features of each process.

Precisely how ER stress triggers the activation of autophagy is also poorly understood. In yeast, *IRE1* and *HAC1* are not required for the induction of Atg8p. However, ectopic expression of the spliced form of *HAC1* is sufficient to induce accumulation of Atg8p protein, but not sufficient to induce PAS formation. This suggests that both *HAC1* dependent and *HAC1* independent pathways emanate from the ER to regulate various aspects of the autophagic process in yeast [67]. However, the precise nature of this regulation remains to be uncovered. In mouse and human cell lines, *Ire1−/−* cells are defective in inducing autophagy during ER stress. However, IRE1 appears to trigger autophagy through its phosphorylation of JNK, rather than by splicing *Xbp-1* [103, 105]. Furthermore, expression of dominant negative *Perk*, or an unphosphorylatable mutant of *eIF2α* can inhibit ER induced autophagy in certain cell types, suggesting that PERK also has the ability to promote autophagy during ER stress [108]. However, the mechanism linking PERK to the autophagic pathway is completely unknown.

Apoptosis

In addition to providing a wide array of adaptive responses to help the cell survive ER stress, the UPR can also trigger apoptotic cell death. This outcome is presumably reserved for cases of ER stress when adaptation is not possible or not desired. ER induced apoptosis depends upon the activation of proapoptotic Bcl-2 family members, BAX and BAK [109, 110], which reside in the cytosol, mitochondria, and ER. In response to various apoptotic signals, these proteins oligomerize and insert themselves into the outer mitochondrial membrane, where they trigger cytochrome c release and caspase activation. During ER stress, inhibiting mitochondrial membrane permeabilization and cytochrome c release reduces apoptosis, indicating that the UPR utilizes this mitochondrial apoptotic pathway [111]. Some evidence also suggests that BAX and BAK can promote apoptosis directly from their position within the ER. During ER stress, BAX and BAK assume their activated forms within the ER membrane and caspase 12, which is ER localized, becomes activated in a *Bax*/*Bak* dependent manner. Furthermore, targeting BAK to the ER is sufficient to activate caspase 12 and trigger apoptosis [112]. However, the mitochondrial pathway, rather than the ER specific pathway, is probably the dominant means of inducing apoptosis during ER stress, as caspase 12 knockout MEFs are only mildly resistant to UPR induced apoptosis [113].

During ER stress, the UPR signals the activation of BAX and BAK in at least two key ways. Firstly, all three UPR transcription factors induce the expression of *Chop*, a transcription factor that is partly required for ER induced apoptosis [114]. Upon expression, CHOP downregulates the transcription of *Bcl-2*, a BAX/BAK inhibitor [115] and increases the expression of the BAX activating protein, DR5 [116]. Secondly, during ER stress, IRE1 activates a MAP kinase pathway that results in JNK activation. Cells that cannot activate JNK during ER stress are partially resistant to apoptosis [27]. It is not yet known how JNK activates apoptosis during ER stress. However, in response to other apoptotic stimuli, JNK has been shown to inhibit the BAX/BAK inhibitors BCL-2 and BCl-xL and activate the proapoptotic proteins, BIM and BMF. Furthermore, JNK activates several transcription factors, which activate a wide variety of genes including some that induce apoptosis. When apoptosis is the appropriate response to ER stress, the UPR pathway clearly relies upon CHOP and JNK activation to communicate this to the apoptotic machinery. However, it remains to be seen whether these two pathways account for all UPR induced apoptosis, or whether additional pathways exist to promote apoptosis during ER stress.

In response to an unfolded protein stimulus in the ER, the UPR can activate numerous adaptive responses and apoptotic signals, whose combined effects either allow the cell to survive ER stress, or apoptotically kill the cell. We do not yet understand the exact conditions that induce apoptosis over survival, nor do we know the mechanism of detecting these conditions and integrating them into a cell fate decision. This mechanism must depend upon conditions within the ER, including the type, severity, and duration of ER stress, which are probably sensed through the differential activation and repression of the three molecular UPR sensors. In addition, a cell's fate during ER stress depends upon its broader physiological environment, including cytosolic conditions, cell type, and developmental cues. Current research, focusing on the interplay between cellular context and UPR induction, will shed light on how the UPR arrives at cell fate decisions, and how these decisions are customized to promote survival of the organism.

PHYSIOLOGICAL UPR

The fundamental circuitry and logic of the UPR pathway were initially discerned with the help of pharmacological agents that cause extensive misfolding of bulk proteins in the ER and maximally activate all three branches of the pathway. However, it is now clear that the UPR has been refined to help the cell cope with a wide array of physiological situations that are more subtle than these pharmacological conditions of widespread protein misfolding. As expected, many secretory cell types invoke the UPR pathway to cope with their unusually high protein folding load. Surprisingly, the UPR pathway has also been shown to be active in non-secretory cells, and to perform non-secretory functions. Furthermore, emerging evidence suggests that the UPR can sometimes serve a housekeeping function, maintaining an appropriately equipped ER, even when levels of ER stress are low. As discussed below, the tissue specific nature of the UPR's involvement in cell function highlights the complexity and flexibility of the UPR pathway. UPR activation is also triggered by a variety of different diseases, in some cases contributing to the pathology of the disease, and in some cases helping cells cope with the disease. However, the role of the UPR in various disease conditions has been the subject of several excellent reviews [117–119] and will not be discussed here.

A pathway specialized for secretion

Tissues that primarily function in protein secretion, such as the liver, pancreas, salivary glands, and skeletal secretory cells, exhibit the most dramatic requirement for UPR components. In these cell types, the UPR is thought to respond to an increased secretory load by expanding the ER's function. Lacking this ability to expand, UPR deficient cells continue to experience extreme ER stress and ultimately undergo apoptosis.

Ire1α−/− and *Xbp-1−/−* mice die during development at E12.5 with small, apoptotic livers [120, 121], and *Xbp-1−/−* lethality can be rescued by expressing *Xbp-1* selectively in hepatocytes [122]. Since enzyme secretion is a primary function of the liver, it has been proposed that the UPR's role in hepatocytes is to support their secretory activity, but the precise nature of this role remains to be discovered.

Mice that express *Xbp-1* only in the liver survive until birth, but soon develop defects of the exocrine pancreas and salivary glands [122], suggesting that the IRE1/XBP-1 branch of the UPR is active in these tissues. This idea is further supported by the fact that wild-type mice highly express *Ire1α* and *Xbp-1* in the pancreas and salivary glands [123, 124]. In addition to higher expression levels, *Xbp-1* splicing has been detected in the mouse pancreas [125] and *Drosophila* salivary glands [126], indicating activation of IRE1 in these cells. Furthermore, pancreatic acinar cells and salivary gland cells of *Xbp-1* knockout mice contain much less ER than their wild-type counterparts, and fail to secrete important digestive enzymes. As a result, these mice die of malnutrition soon after birth [122]. Studies also suggest that PERK plays a role in the function of specialized secretory cells; PERK is activated in the exocrine pancreas and, like *Xbp-1−/−* cells, *Perk−/−* pancreatic cells have an abnormal ER and reduced secretion of digestive enzymes [127, 128].

Xbp-1 and *Perk* mRNA are also highly expressed in chondrocytes and osteoblasts of the developing mouse skeleton [124]. These two cell types are the major secretory cells of the skeletal system; chondrocytes are responsible for the secretion of cartilage proteins and osteoblasts secrete bone matrix proteins during development. Probably due to defects in the secretory function of these two cell types, *Perk−/−* mice have abnormal skeletal development, with reduced collagen production and defective bone matrix secretion [127]. PERK appears to perform a similar function in the human skeletal system, as individuals with a genetic disorder caused by loss of PERK function also suffer from poor skeletal development [129].

Studies in the above secretory tissues have refined our understanding of *in vivo* UPR signaling. It is now clear that activation of a specific branch of the UPR does not necessarily induce that pathway's full repertoire of downstream effects. For example, in pancreatic acinar cells, XBP-1 is activated, but only a subset of XBP-1's target genes are transcriptionally induced [122]. This may be because target genes differ in their sensitivities to XBP-1 activation. In this case, the level of ER stress and the resulting level of XBP-1 activity would dictate the specific transcriptional program that is appropriate for a given cell type. Alternatively, the UPR pathway might intersect with other pathways in order to refine its physiological effects on the cell. For example, certain cell types might induce transcriptional activators or repressors that act on the promoters of UPR target genes to enhance or dampen their transcription.

Inducing the UPR to Halt Protein Production

The tissues described thus far illustrate the ability of the UPR pathway to enhance protein secretion. In these cases, to allow for protein production, the cell must invoke some means of preventing or overcoming the translation attenuation that is normally caused by PERK activation. By contrast, in pancreatic β cells, translation attenuation is precisely the purpose of UPR induction. β cells detect high glucose levels in the bloodstream, and respond by secreting insulin. Following secretion, these cells must synthesize large quantities of insulin to replenish their stores and prepare for the next glucose stimulation. This process of insulin replacement is controlled by PERK mediated eIF2α phosphorylation. In their basal state, β cells activate PERK to maintain high levels of eIF2α phosphorylation and low levels of translation. After glucose stimulation, eIF2α becomes dephosphorylated, allowing large amounts of insulin to be produced and directed to the secretory pathway. When the appropriate amount of insulin has been produced, PERK becomes reactivated by the abundance of insulin in the ER, and shuts off insulin production [127]. As a result of their inability to control insulin production, the β cells of *Perk−/−* mice experience significant ER stress, causing them to undergo apoptosis around 4 weeks after birth. Consequently, β cells are depleted from the pancreas, causing deficient insulin production, hyperglycemia, and diabetes in *Perk* knockout mice [127, 128, 130]. Similarly, in humans, mutations in Perk have been linked to Wolcott-Rallinson syndrome, which is marked by early onset diabetes [129].

Plasma Cells: Two Phases of UPR

In the peripheral immune system, mature B cells that encounter antigen differentiate to become antibody secreting plasma cells. A role for the UPR in this process has been postulated for a long time, but only recently has experimental evidence supported such a role [131]. As plasma cells differentiate, *Xbp-1* becomes transcribed [132] and spliced [133] and ATF6 becomes cleaved and activated [134]. Ultimately, chaperones and other trafficking genes are transcriptionally activated [71], and the volume of the ER expands [7], thus allowing the plasma cell to produce and secrete large amounts of Ig. Proving the importance of XBP-1 mediated transcription in plasma cell differentiation, *Xbp-1−/−* B cells proliferate normally, but do not differentiate into plasma cells and do not secrete Ig. The expression of spliced *Xbp-1* restores Ig secretion in *Xbp1−/−* B cells, whereas the expression of an unspliceable *Xbp-1* does not. Furthermore, expression of *Xbp-1* in an activated B cell line is sufficient to drive cells toward a plasma cell differentiation program [132]. We do not yet know the significance of ATF6 activation in plasma cells, as *Atf6* knockout mice have only recently been generated [135, 136]. More work will need to be done to characterize the plasma cell differentiation program in hematopoietic stem cells taken from these mice. Interestingly, PERK phosphorylation is not detected during plasma cell differentiation, suggesting that PERK is not involved in this process. This is probably because translational inhibition would counteract the plasma cell's objective of generating large amounts of Ig [137].

These data all point to a classical model of UPR activation, whereby increased production of Ig crowds the ER and stimulates certain branches of the UPR pathway thus expanding the cell's capacity to secrete Ig. In support of this model, B cells deleted for Ig heavy chain produce less XBP-1 protein than wild-type cells during differentiation [133]. However, further studies uncovered several surprising aspects of the plasma cell UPR that argue against a simple classical model of activation. First, Ig production is not absolutely required for UPR induction during plasma cell differentiation. In cells deleted for heavy chain, some XBP-1 protein is produced during differentiation. Second, primary B cells that are activated through their B cell receptor are primed to initiate plasma cell differentiation, but do not differentiate and do not produce Ig. However, these cells do induce a modest amount of *Xbp-1* splicing, activate the transcription of chaperones, and expand their ER [138]. Finally, the initiation of *Xbp-1* splicing, ATF6 cleavage, and chaperone expression during the course of plasma cell differentiation actually precede the production of Ig light and heavy chains [134, 139]. Thus, in plasma cells the UPR appears to be active before an increased secretory load would mandate its activation.

Therefore, it is currently thought that the UPR is activated in two phases during plasma cell differentiation. First, in an "anticipatory phase," XBP-1 and ATF6 are activated prior to Ig production, to induce an initial expansion of the secretory pathway. Second, once Ig production begins, this imposes an increased secretory load that signals further activation of the UPR and further expansion of the cell's secretory capacity. The discovery of an anticipatory UPR was unexpected, and the mechanism of activation during this phase is still entirely unknown. Perhaps, the UPR sensors have adjustable sensitivities, and within certain physiological settings such as the developing plasma cell, the threshold for activation is reduced. This would allow UPR induction in the plasma cell when unfolded protein levels are still relatively low. Another possibility is that certain types of unfolded proteins, which are specifically present in the developing plasma cell, can signal UPR activation even if they are not abundant and the ER is not overloaded.

Beyond the UPR's Secretory Function

In yeast, the UPR pathway activates the transcription of 381 genes, 208 of which have a known function. Half of these functionally characterized UPR target genes function in protein processing, secretion, or lipid metabolism, and are presumably activated to expand the cell's secretory capacity during ER stress. The other half of these UPR target genes act in non-secretory processes such as signaling, gene regulation, metabolism, and DNA repair [2]. Similarly, of the hundreds of genes bound by XBP-1 in skeletal myotubes, plasma cells, and pancreatic β cells, approximately 40 percent perform non-secretory physiological functions, including cell growth and differentiation, RNA processing, signal transduction, and gene regulation [79].

Interestingly, in the mammalian system, this extra-secretory ability of the UPR manifests itself in a tissue specific manner. A growing number of cell types require UPR signaling for their specific differentiation programs and to perform their normal, non-secretory physiological functions. For example, in the developing liver, XBP-1 induces the expression of α1-antitrypsin, α-fetoprotein, transthyretin, and apoplipoprotein A1, all of which promote hepatic proliferation, but not secretory function [120]. During neuronal development, *Xbp-1* is spliced and this splicing promotes axonal growth and branching [140]. When expressed in myoblasts, spliced XBP-1 actually inhibits myotube differentiation through its activation of the myogenesis inhibitor, *Mist1* [79].

Reconstitution of *Rag2−/−* mice with *Ire1−/−* hematopoietic stem cells has revealed that *Ire1* is required in the pro B cell stage of early B lymphopoiesis for VDJ recombination, a process that does not require secretion [121]. Furthermore, in the developing plasma cell, spliced XBP-1 induces a dramatic expansion of the ER and the broader secretory pathway, but it also regulates the increase in cell size, mitochondrial mass, and lysosomal content that are characteristic of terminally differentiated plasma cells [7]. Moreover, spliced XBP-1 activates the transcription [141] and translation [69] of Ig heavy chain in the plasma cell, thus promoting Ig secretion even before it enters the secretory pathway. XBP-1 also enhances IL6 production [133], decreases CD44 expression, and increases Syndecan-1 expression [132], all changes that mark plasma cell differentiation but do not enhance secretory function.

Proper tissue development often depends upon reduced proliferation or apoptosis of selected cells. Surprisingly, in some cases, the UPR is responsible for triggering this inhibition in tissue growth. For example, during mammary acinar morphogenesis, loss of adhesion activates the PERK pathway, which downregulates translation. As a result, cell proliferation is reduced, CHOP is activated, and apoptosis is modestly enhanced. Mammary acinar cells expressing a dominant negative *Perk* form abnormally large and amorphous acini, indicating that UPR induced growth inhibition is a key component of mammary morphogenesis [142]. Similarly, during the development of a mature muscle fiber, myoblasts fuse to form multi-nucleated myotubes, a process that requires a modest degree of myoblast apoptosis. ATF6 and caspase 12 are activated specifically in those myoblasts that are undergoing apoptosis, and preventing ATF6 cleavage prevents apoptosis and proper differentiation of the myotube. This indicates that the apoptosis signal that is necessary for proper myotube development is generated by the UPR [143]. Therefore, in the mammary acinus and myotube, the UPR is selectively activated, not to enhance secretory function or to relieve ER stress, but to control cell growth and promote proper tissue differentiation. The surprising ability of the UPR to perform in this capacity emphasizes current gaps in our understanding of the upstream signals that activate the UPR and how cells can tailor downstream UPR signals to their specific physiological needs.

The UPR as a Housekeeping Pathway

The UPR is best known for its role in responding to conditions of extreme ER stress, such as those imposed by the production of massive amounts of secretory protein or by pharmacological agents that cause widespread misfolding. Recently, it has become evident that the UPR also has the ability to regulate ER function under normal conditions, when ER stress levels are low. For example, during unstressed growth in diploid yeast, basal *HAC1* mRNA splicing appears to inhibit meiosis and pseudohyphal growth, two differentiation programs that are only induced in response to nitrogen starvation [144]. Basal *HAC1* splicing also supports efficient cytokinesis in yeast. In the absence of ER stress, *hac1Δ* cells display a modest cytokinesis defect, and this defect is exacerbated when ER stress is applied [145]. This implies that during normal cell growth, basal UPR signaling provides a higher level of ER function, which is required for efficient cytokinesis. Furthermore, in unstressed skeletal muscle cells, XBP-1 binds the promoters of 118 genes, most of which are involved in protein folding and trafficking [79]. Since skeletal muscle cells have no apparent secretory function, and presumably do not experience ER stress, this XBP-1 binding represents a basal UPR activity, which most likely helps these cells maintain normal ER function.

The discovery that the UPR functions in unstressed cells has two possible non-mutually exclusive explanations. Firstly, constitutive low level UPR activity may enhance the ER's basal function in a constant manner. In this case, the UPR would not have to be modulated to achieve its function, it would simply provide a constant level of support to the ER. If this is true, cells without a functional UPR would always have decreased ER function, but this might only be evident when examining processes that require maximal ER capacity. Secondly, the UPR might monitor basal fluctuations in ER functional demand, becoming slightly active when the cell calls for incremental increases in ER capacity, and then shutting off when this need is met. In this way, the cell could fine-tune the ER to its changing environment, and perhaps even avoid the dangers of sudden and extreme ER stress.

PERSPECTIVES

In the past, researchers have taken drastic measures to induce the UPR pathway and detect its activity. These measures included adding DTT to cells to reduce all protein disulfide bonds, adding tunicamycin to inhibit N-linked glycosylation, adding thapsigargin to inhibit ER calcium import, or overexpressing a misfolded mutant protein. Because these methods strongly activate all three branches of the UPR pathway and induce the UPR's full repertoire of cellular consequences, they allowed scientists to identify the pathway's molecular components and describe their breadth of cellular effects. Furthermore, the ability to detect UPR activity only after causing widespread protein misfolding allowed an appreciation of the pathway's potential as a "stress response pathway," inactive until a stress is encountered, then activated to cope with conditions of ER stress. In this review, we have highlighted some of the more recently discovered roles for the UPR in normal cell physiology. Physiological studies have begun to reveal that the specific nature of the UPR's activation can be quite different for different cell types, and that the pathway's function can range from stress response to general housekeeping.

During physiological instances of its activation, the UPR can be modulated on several different levels to achieve results that are fine-tuned to specific cellular contexts. One way that the UPR can be modulated is at the level of sensor activation. Although pharmacological agents lead to rapid and strong induction of all three UPR signaling branches, many physiological studies, including one study that measured several different levels of PERK activity in different mouse tissues [127], indicate that activation of a given branch within a cell is not an "all or none" event. Only by moving away from strong external stressors have we begun to appreciate the pathway's potential for achieving intermediate levels of activity, operating more like a dimmer switch than an on/off switch. Furthermore, in many physiological cases, each UPR signaling branch appears to be tuned separately. For example, in the plasma cell, XBP-1 and ATF6 are activated strongly [132–134], but PERK activity has not been detected [137]. Another layer of physiological fine-tuning of the UPR occurs subsequent to activation, where individual UPR branches can selectively activate portions of their downstream response. For example, pancreatic acinar cells induce *Xbp-1* splicing, but only activate a subset of XBP-1's target genes. Therefore, under physiological conditions the cell can utilize several strategies to modulate the UPR pathway according to its specific requirements, thus lending the pathway enormous versatility and flexibility. One of the challenges now is to dissect the precise mechanisms that allow this modulation.

Certain physiological instances of UPR activation confirm the pathway's role as a stress response pathway that becomes active when it senses high levels of unfolded proteins in the ER. For example, high levels of collagen synthesis in chondrocytes or Ig production in plasma cells can activate the UPR pathway, which in turn increases the secretory capacity of these cell types and allows them to achieve their secretory function. However, more recent studies indicate that the UPR pathway also has a role to play in cells that do not appear to be experiencing ER stress. For example, even before they begin to produce Ig, plasma cells seem to induce the UPR pathway [134, 138, 139], indicating that some mechanism must be in place to induce the pathway when stress levels are low. Even in some non-secretory cell types, the UPR pathway is active at low levels. This basal UPR activity probably performs the housekeeping function of helping the ER sustain its basal protein folding capacity. As of yet, this basal UPR activity has only been detected in a few circumstances. For example, in the unstressed skeletal muscle, XBP-1 binds to a subset of its promoters [79], and during unstressed growth in budding yeast, cytokinesis requires UPR signaling [145]. However, studies of the UPR's potential housekeeping function have been hindered by a lack of sensitive methods to detect low levels of UPR activation. It is likely that the development of more sensitive assays will uncover a broad role for basal UPR signaling in helping many cell types maintain normal cellular function. This housekeeping role of the UPR, though subtle, may be a key aspect of the UPR's physiological function. In fact, other signal transduction pathways might also serve housekeeping functions that have been overlooked due to subtle activation levels. The model of a single pathway serving both a housekeeping function and a stress response function could be utilized for multiple aspects of cellular regulation, and may turn out to be a common mode of signal transduction.

REFERENCES

1. Brown JD, Ng DTW, Ogg SC, Walter P. Targeting pathways to the endoplasmic reticulum membrane. *Cold Spring Harb Symp Quant Biol* 1995;**60**:23–30.
2. Travers KJ, Patil CK, Wodicka L, et al. Functional and genomic analyses reveal an essential coordination between the unfolded protein response and ER-associated degradation. *Cell* 2000;**101**:249–58.
3. Ng DT, Spear ED, Walter P. The unfolded protein response regulates multiple aspects of secretory and membrane protein biogenesis and endoplasmic reticulum quality control. *J Cell Biol* 2000;**150**:77–88.

4. Plemper RK, Wolf DH. Retrograde protein translocation: ERADication of secretory proteins in health and disease. *Trends Biochem Sci* 1999; **24**:266–70.
5. Gardner RG, Shearer AG, Hampton RY. In vivo action of the HRD ubiquitin ligase complex: Mechanisms of endoplasmic reticulum quality control and sterol regulation. *Mol Cell Biol* 2001;**21**:4276–91.
6. Patil C, Walter P. Intracellular signaling from the endoplasmic reticulum to the nucleus: the unfolded protein response in yeast and mammals. *Curr Opin Cell Biol* 2001;**13**:349–55.
7. Shaffer AL, Shapiro-Shelef M, Iwakoshi NN, et al. XBP1, downstream of Blimp-1, expands the secretory apparatus and other organelles, and increases protein synthesis in plasma cell differentiation. *Immunity* 2004;**21**:81–93.
8. Larsson O, Carlberg M, Zetterberg A. Selective killing induced by an inhibitor of N-linked glycosylation. *J Cell Sci* 1993;**106**:299–3070.
9. Cox JS, Shamu CE, Walter P. Transcriptional induction of genes encoding endoplasmic reticulum resident proteins requires a transmembrane protein kinase. *Cell* 1993;**73**:1197–206.
10. Shamu CE, Walter P. Oligomerization and phosphorylation of the Ire1p kinase during intracellular signaling from the endoplasmic reticulum to the nucleus. *Embo J* 1996;**15**:3028–39.
11. Cox JS, Walter P. A novel mechanism for regulating activity of a transcription factor that controls the unfolded protein response. *Cell* 1996;**87**:391–404.
12. Sidrauski C, Walter P. The transmembrane kinase Ire1p is a site-specific endonuclease that initiates mRNA splicing in the unfolded protein response. *Cell* 1997;**90**:1031–9.
13. Mori K, Kawahara T, Yanagi H, Yura T. ER stress-induced mRNA splicing permits synthesis of transcription factor Hac1p/Ern4p that activates the unfolded protein response. *Mol Biol Cell* 1997;**8**:2056.
14. Mori K, Kawahara T, Yoshida H, et al. Signalling from endoplasmic reticulum to nucleus: transcription factor with a basic-leucine zipper motif is required for the unfolded protein-response pathway. *Genes Cells* 1996;**1**:803–17.
15. Sidrauski C, Cox JS, Walter P. tRNA ligase is required for regulated mRNA splicing in the unfolded protein response. *Cell* 1996;**87**:405–13.
16. Abelson J, Trotta CR, Li H. tRNA splicing. *J Biol Chem* 1998;**273**:12,685–8.
17. Culver GM, McCraith SM, Consaul SA, et al. A 2′-phosphotransferase implicated in tRNA splicing is essential in Saccharomyces cerevisiae. *J Biol Chem* 1997;**272**:13,203–10.
18. Harding HP, Lackey JG, Hsu HC, et al. An intact unfolded protein response in Trpt1 knockout mice reveals phylogenic divergence in pathways for RNA ligation. *Rna* 2008;**14**:225–32.
19. Ruegsegger U, Leber JH, Walter P. Block of HAC1 mRNA translation by long-range base pairing is released by cytoplasmic splicing upon induction of the unfolded protein response. *Cell* 2001;**107**:103–14.
20. Yamamoto K, Yoshida H, Kokame K, et al. Differential contributions of ATF6 and XBP1 to the activation of endoplasmic reticulum stress-responsive cis-acting elements ERSE, UPRE and ERSE-II. *J Biochem* 2004;**136**:343–50.
21. Mori K, Sant A, Kohno K, et al. A 22 bp cis-acting element is necessary and sufficient for the induction of the yeast KAR2 (BiP) gene by unfolded proteins. *Embo J* 1992;**11**:2583–93.
22. Patil CK, Li H, Walter P. Gcn4p and novel upstream activating sequences regulate targets of the unfolded protein response. *PLoS Biol* 2004;**2**:E246.
23. Chapman RE, Walter P. Translational attenuation mediated by an mRNA intron. *Curr Biol* 1997;**7**:850–9.
24. Yoshida H, Matsui T, Yamamoto A, et al. XBP1 mRNA is induced by ATF6 and spliced by IRE1 in response to ER stress to produce a highly active transcription factor. *Cell* 2001;**107**:881–91.
25. Ogawa N, Mori K. Autoregulation of the HAC1 gene is required for sustained activation of the yeast unfolded protein response. *Genes Cells* 2004;**9**:95–104.
26. Urano F, Wang X, Bertolotti A, et al. Coupling of stress in the ER to activation of JNK protein kinases by transmembrane protein kinase IRE1. *Science* 2000;**287**:664–6.
27. Nishitoh H, Matsuzawa A, Tobiume K, et al. ASK1 is essential for endoplasmic reticulum stress-induced neuronal cell death triggered by expanded polyglutamine repeats. *Genes Dev* 2002;**16**:1345–55.
28. Nishitoh H, Saitoh M, Mochida Y, et al. ASK1 is essential for JNK/SAPK activation by TRAF2. *Mol Cell* 1998;**2**:389–95.
29. Bogoyevitch MA, Kobe B. Uses for JNK: the many and varied substrates of the c-Jun N-terminal kinases. *Microbiol Mol Biol Rev* 2006;**70**:1061–95.
30. Harding HP, Zhang Y, Ron D. Protein translation and folding are coupled by an endoplasmic-reticulum-resident kinase. *Nature* 1999;**397**:271–4.
31. Shi YG, Vattem KM, Sood R, et al. Identification and characterization of pancreatic eukaryotic initiation factor 2 alpha-subunit kinase, PEK, involved in translational control. *Mol Cell Biol* 1998;**18**:7499–509.
32. Hinnebusch AG. Translational regulation of yeast GCN4. A window on factors that control initiator-trna binding to the ribosome. *J Biol Chem* 1997;**272**:21,661–4.
33. Srivastava SP, Kumar KU, Kaufman RJ. Phosphorylation of eukaryotic translation initiation factor 2 mediates apoptosis in response to activation of the double-stranded RNA-dependent protein kinase. *J Biol Chem* 1998;**273**:2416–23.
34. Scheuner D, Song B, McEwen E, et al. Translational control is required for the unfolded protein response and in vivo glucose homeostasis. *Mol Cell* 2001;**7**:1165–76.
35. Brewer JW, Diehl JA. PERK mediates cell-cycle exit during the mammalian unfolded protein response. *Proc Natl Acad Sci U S A* 2000;**97**:12,625–30.
36. Niwa M, Walter P. Pausing to decide. *Proc Natl Acad Sci U S A* 2000;**97**:12,396–7.
37. Harding HP, Novoa I, Zhang YH, et al. Regulated translation initiation controls stress-induced gene expression in mammalian cells. *Mol Cell* 2000;**6**:1099–108.
38. Harding HP, Zhang YH, Zeng HQ, et al. An integrated stress response regulates amino acid metabolism and resistance to oxidative stress. *Mol Cell* 2003;**11**:619–33.
39. Cullinan SB, Zhang D, Hannink M, et al. Nrf2 is a direct PERK substrate and effector of PERK-dependent cell survival. *Mol Cell Biol* 2003;**23**:7198–209.
40. DuRose JB, Tam AB, Niwa M. Intrinsic capacities of molecular sensors of the unfolded protein response to sense alternate forms of endoplasmic reticulum stress. *Mol Biol Cell* 2006;**17**:3095–107.
41. Wang Y, Shen JS, Arenzana N, et al. Activation of ATF6 and an ATF6 DNA binding site by the endoplasmic reticulum stress response. *J Biol Chem* 2000;**275**:27,013–20.
42. Haze K, Yoshida H, Yanagi H, et al. Mammalian transcription factor ATF6 is synthesized as a transmembrane protein and activated by proteolysis in response to endoplasmic reticulum stress. *Mol Biol Cell* 1999;**10**:3787–99.

43. Ye J, Rawson RB, Dave UP, et al. Site-1 protease (S1P) and site-2 protease (S2P), the two proteases that cleave membrane-bound SREBPs, also cleave ATF6 upon ER stress. *Mol Biol Cell* 2000;**11**:1511.
44. Kokame K, Kato H, Miyata T. Identification of ERSE-II, a new cis-actin element responsible for the ATF6-dependent mammalian unfolded protein response. *J Biol Chem* 2001;**276**:9199–205.
45. Yoshida H, Haze K, Yanagi H, et al. Identification of the cis-acting endoplasmic reticulum stress response element responsible for transcriptional induction of mammalian glucose-regulated proteins: Involvement of basic leucine zipper transcription factors. *J Biol Chem* 1998;**273**:33,741–9.
46. Shamu CE, Cox JS, Walter P. The unfolded-protein-response pathway in yeast. *Trends Cell Biol* 1994;**4**:56–60.
47. Bertolotti A, Zhang Y, Hendershot LM, et al. Dynamic interaction of BiP and ER stress transducers in the unfolded-protein response. *Nat Cell Biol* 2000;**2**:326–32.
48. Kimata Y, Oikawa D, Shimizu Y, et al. A role for BiP as an adjustor for the endoplasmic reticulum stress-sensing protein Ire1. *J Cell Biol* 2004;**167**:445–56.
49. Credle JJ, Finer-Moore JS, Papa FR, et al. Inaugural Article: On the mechanism of sensing unfolded protein in the endoplasmic reticulum. *Proc Natl Acad Sci U S A* 2005;**102**:18,773–84.
50. Shen JS, Chen X, Hendershot L, Prywes R. ER stress regulation of ATF6 localization by dissociation of BiP/GRP78 binding and unmasking of golgi localization signals. *Dev Cell* 2002;**3**:99–111.
51. Ma K, Vattem KM, Wek RC. Dimerization and release of molecular chaperone inhibition facilitate activation of eukaryotic initiation factor-2 kinase in response to endoplasmic reticulum stress. *J Biol Chem* 2002;**277**:18,728–35.
52. Nadanaka S, Okada T, Yoshida H, Mori K. Role of disulfide bridges formed in the luminal domain of ATF6 in sensing endoplasmic reticulum stress. *Mol Cell Biol* 2007;**27**:1027–43.
53. Hong M, Luo S, Baumeister P, et al. Underglycosylation of ATF6 as a Novel Sensing Mechanism for Activation of the Unfolded Protein Response. *J Biol Chem* 2004;**279**:11,354–63.
54. Ma K, Vattem KM, Wek RC. Dimerization and release of molecular chaperone inhibition facilitate activation of eukaryotic initiation factor-2 kinase in response to endoplasmic reticulum stress. *J Biol Chem* 2002;**277**:18,728–35.
55. Liu CY, Schroder M, Kaufman RJ. Ligand-independent dimerization activates the stress response kinases IRE1 and PERK in the lumen of the endoplasmic reticulum. *J Biol Chem* 2000;**275**:24,881–5.
56. Welihinda AA, Tirasophon W, Green SR, Kaufman RJ. Protein serine/threonine phosphatase Ptc2p negatively regulates the unfolded-protein response by dephosphorylating Ire1p kinase. *Mol Cell Biol* 1998;**18**:1967–77.
57. Guo J, Polymenis M. Dcr2 targets Ire1 and downregulates the unfolded protein response in Saccharomyces cerevisiae. *EMBO Rep* 2006;**7**:1124–7.
58. Yoshida H, Oku M, Suzuki M, Mori K. pXBP1(U) encoded in XBP1 pre-mRNA negatively regulates unfolded protein response activator pXBP1(S) in mammalian ER stress response. *J Cell Biol* 2006;**172**:565–75.
59. Thuerauf DJ, Morrison LE, Hoover H, Glembotski CC. Coordination of ATF6-mediated transcription and ATF6 degradation by a domain that is shared with the viral transcription factor, VP16. *J Biol Chem* 2002;**277**:20,734–9.
60. Thuerauf DJ, Marcinko M, Belmont PJ, Glembotski CC. Effects of the Isoform-specific Characteristics of ATF6alpha and ATF6beta on Endoplasmic Reticulum Stress Response Gene Expression and Cell Viability. *J Biol Chem* 2007;**282**:22,865–78.
61. Tsukumo Y, Tomida A, Kitahara O, et al. Nucleobindin 1 Controls the Unfolded Protein Response by Inhibiting ATF6 Activation. *J Biol Chem* 2007;**282**:29,264–72.
62. Yan W, Frank CL, Korth MJ, et al. Control of PERK eIF2alpha kinase activity by the endoplasmic reticulum stress-induced molecular chaperone P58IPK. *Proc Natl Acad Sci U S A* 2002;**99**:15,920–5.
63. van Huizen R, Martindale JL, Gorospe M, Holbrook NJ. P58IPK, a novel endoplasmic reticulum stress-inducible protein and potential negative regulator of eIF2alpha signaling. *J Biol Chem* 2003;**278**:15,558–64.
64. Novoa I, Zeng HQ, Harding HP, Ron D. Feedback inhibition of the unfolded protein response by GADD34-mediated dephosphorylation of eIF2 alpha. *J Cell Biol* 2001;**153**:1011–21.
65. Novoa I, Zhang YH, Zeng HQ, et al. Stress-induced gene expression requires programmed recovery from translational repression. *Embo J* 2003;**22**:1180–7.
66. Ma Y, Hendershot LM. Delineation of a negative feedback regulatory loop that controls protein translation during endoplasmic reticulum stress. *J Biol Chem* 2003;**278**:34,864–73.
67. Bernales S, McDonald KL, Walter P. Autophagy Counterbalances Endoplasmic Reticulum Expansion during the Unfolded Protein Response. *PLoS Biol* 2006;**4**:e423.
68. Sriburi R, Jackowski S, Mori K, Brewer JW. XBP1: a link between the unfolded protein response, lipid biosynthesis, and biogenesis of the endoplasmic reticulum. *J Cell Biol* 2004;**167**:35–41.
69. Tirosh B, Iwakoshi NN, Glimcher LH, Ploegh HL. XBP-1 specifically promotes IgM synthesis and secretion, but is dispensable for degradation of glycoproteins in primary B cells. *J Exp Med* 2005;**202**:505–16.
70. Sriburi R, Bommiasamy H, Buldak GL, et al. Coordinate regulation of phospholipid biosynthesis and secretory pathway gene expression in XBP-1(S)-induced endoplasmic reticulum biogenesis. *J Biol Chem* 2007;**282**:7024–34.
71. Lee AH, Iwakoshi NN, Glimcher LH. XBP-1 regulates a subset of endoplasmic reticulum resident chaperone genes in the unfolded protein response. *Mol Cell Biol* 2003;**23**:7448–59.
72. Kleizen B, Braakman I. Protein folding and quality control in the endoplasmic reticulum. *Curr Opin Cell Biol* 2004;**16**:343–9.
73. Ni M, Lee AS. ER chaperones in mammalian development and human diseases. *FEBS Lett* 2007;**581**:3641–51.
74. Helenius A, Aebi M. Roles of N-linked glycans in the endoplasmic reticulum. *Annu Rev Biochem* 2004;**73**:1019–49.
75. Tu BP, Weissman JS. Oxidative protein folding in eukaryotes: mechanisms and consequences. *J Cell Biol* 2004;**164**:341–6.
76. Frand AR, Kaiser CA. The ERO1 gene of yeast is required for oxidation of protein dithiols in the endoplasmic reticulum. *Mol Cell* 1998;**1**:161–70.
77. Pollard MG, Travers KJ, Weissman JS. Ero1p: a novel and ubiquitous protein with an essential role in oxidative protein folding in the endoplasmic reticulum. *Mol Cell* 1998;**1**:171–82.
78. Cabibbo A, Pagani M, Fabbri M, et al. ERO1-L, a human protein that favors disulfide bond formation in the endoplasmic reticulum. *J Biol Chem* 2000;**275**:4827–33.
79. Acosta-Alvear D, Zhou Y, Blais A, et al. XBP1 controls diverse cell type- and condition-specific transcriptional regulatory networks. *Mol Cell* 2007;**27**:53–66.
80. Kawai T, Fan J, Mazan-Mamczarz K, Gorospe M. Global mRNA stabilization preferentially linked to translational repression during the endoplasmic reticulum stress response. *Mol Cell Biol* 2004;**24**:6773–87.

81. Hollien J, Weissman JS. Decay of endoplasmic reticulum-localized mRNAs during the unfolded protein response. *Science* 2006;**313**:104–7.
82. Iwawaki T, Hosoda A, Okuda T, et al. Translational control by the ER transmembrane kinase/ribonuclease IRE1 under ER stress. *Nat Cell Biol* 2001;**3**:158–64.
83. Clemens MJ. Regulation of eukaryotic protein synthesis by protein kinases that phosphorylate initiation factor eIF-2. *Mol Biol Rep* 1994;**19**:201–10.
84. Ron D. Translational control in the endoplasmic reticulum stress response. *J Clin Invest* 2002;**110**:1383–8.
85. Harding HP, Zhang Y, Bertolotti A, et al. Perk is essential for translational regulation and cell survival during the unfolded protein response. *Mol Cell* 2000;**5**:897–904.
86. Spear ED, Ng DT. Stress tolerance of misfolded carboxypeptidase Y requires maintenance of protein trafficking and degradative pathways. *Mol Biol Cell* 2003;**14**:2756–67.
87. Stephens SB, Dodd RD, Brewer JW, et al. Stable Ribosome Binding to the Endoplasmic Reticulum Enables Compartment-specific Regulation of mRNA Translation. *Mol Biol Cell* 2005;**16**:5819–31.
88. Osborne AR, Rapoport TA, van den Berg B. Protein translocation by the Sec61/SecY channel. *Annu Rev Cell Dev Biol* 2005;**21**:529–50.
89. Kim SJ, Mitra D, Salerno JR, Hegde RS. Signal sequences control gating of the protein translocation channel in a substrate-specific manner. *Dev Cell* 2002;**2**:207–17.
90. Kang SW, Rane NS, Kim SJ, et al. Substrate-Specific Translocational Attenuation during ER Stress Defines a Pre-Emptive Quality Control Pathway. *Cell* 2006;**127**:999–1013.
91. Oyadomari S, Yun C, Fisher EA, et al. Cotranslocational degradation protects the stressed endoplasmic reticulum from protein overload. *Cell* 2006;**126**:727–39.
92. Marquardt T, Hebert DN, Helenius A. Post-translational folding of influenza hemagglutinin in isolated endoplasmic reticulum-derived microsomes. *J Biol Chem* 1993;**268**:19,618–25.
93. Hampton RY. ER-associated degradation in protein quality control and cellular regulation. *Curr Opin Cell Biol* 2002;**14**:476–82.
94. Nakatsukasa K, Brodsky JL. The Recognition and Retrotranslocation of Misfolded Proteins from the Endoplasmic Reticulum. *Traffic* 2008;**9**:861–70.
95. Okuda-Shimizu Y, Hendershot LM. Characterization of an ERAD Pathway for Nonglycosylated BiP Substrates, which Require Herp. *Mol Cell* 2007;**28**:544–54.
96. Carvalho P, Goder V, Rapoport TA. Distinct ubiquitin-ligase complexes define convergent pathways for the degradation of ER proteins. *Cell* 2006;**126**:361–73.
97. Denic V, Quan EM, Weissman JS. A luminal surveillance complex that selects misfolded glycoproteins for ER-associated degradation. *Cell* 2006;**126**:349–59.
98. Jakob CA, Bodmer D, Spirig U, et al. Htm1p, a mannosidase-like protein, is involved in glycoprotein degradation in yeast. *EMBO Rep* 2001;**2**:423–30.
99. Bays NW, Gardner RG, Seelig LP, et al. Hrd1p/Der3p is a membrane-anchored ubiquitin ligase required for ER-associated degradation. *Nat Cell Biol* 2001;**3**:24–9.
100. Gardner RG, Swarbrick GM, Bays NW, et al. Endoplasmic reticulum degradation requires lumen to cytosol signaling. Transmembrane control of Hrd1p by Hrd3p. *J Cell Biol* 2000;**151**:69–82.
101. Ye Y, Meyer HH, Rapoport TA. The AAA ATPase Cdc48/p97 and its partners transport proteins from the ER into the cytosol. *Nature* 2001;**414**:652–6.
102. Suzuki K, Ohsumi Y. Molecular machinery of autophagosome formation in yeast, Saccharomyces cerevisiae. *FEBS Lett* 2007;**581**:2156–61.
103. Ogata M, Hino S, Saito A, et al. Autophagy is activated for cell survival after endoplasmic reticulum stress. *Mol Cell Biol* 2006;**26**:9220–31
104. Yorimitsu T, Nair U, Yang Z, Klionsky DJ. Endoplasmic reticulum stress triggers autophagy. *J Biol Chem* 2006;**281**:30,299–304.
105. Ding WX, Ni HM, Gao W, et al. Linking of autophagy to ubiquitin-proteasome system is important for the regulation of endoplasmic reticulum stress and cell viability. *Am J Pathol* 2007;**171**:513–24.
106. Kruse KB, Brodsky JL, McCracken AA. Characterization of an ERAD gene as VPS30/ATG6 reveals two alternative and functionally distinct protein quality control pathways: one for soluble Z variant of human alpha-1 proteinase inhibitor (A1PiZ) and another for aggregates of A1PiZ. *Mol Biol Cell* 2006;**17**:203–12.
107. Kamimoto T, Shoji S, Hidvegi T, et al. Intracellular inclusions containing mutant alpha1-antitrypsin Z are propagated in the absence of autophagic activity. *J Biol Chem* 2006;**281**:4467–76.
108. Kouroku Y, Fujita E, Tanida I, et al. ER stress (PERK/eIF2alpha phosphorylation) mediates the polyglutamine-induced LC3 conversion, an essential step for autophagy formation. *Cell Death Differ* 2007;**14**:230–9.
109. Wei MC, Zong WX, Cheng EH, et al. Proapoptotic BAX and BAK: a requisite gateway to mitochondrial dysfunction and death. *Science* 2001;**292**:727–30.
110. Hetz C, Bernasconi P, Fisher J, et al. Proapoptotic BAX and BAK modulate the unfolded protein response by a direct interaction with IRE1alpha. *Science* 2006;**312**:572–6.
111. Boya P, Cohen I, Zamzami N, et al. Endoplasmic reticulum stress-induced cell death requires mitochondrial membrane permeabilization. *Cell Death Differ* 2002;**9**:465–7.
112. Zong WX, Li C, Hatzivassiliou G, et al. Bax and Bak can localize to the endoplasmic reticulum to initiate apoptosis. *J Cell Biol* 2003;**162**:59–69.
113. Nakagawa T, Zhu H, Morishima N, et al. Caspase-12 mediates endoplasmic-reticulum-specific apoptosis and cytotoxicity by amyloid-beta. *Nature* 2000;**403**:98–103.
114. Zinszner H, Kuroda M, Wang X, et al. CHOP is implicated in programmed cell death in response to impaired function of the endoplasmic reticulum. *Genes Dev* 1998;**12**:982–95.
115. McCullough KD, Martindale JL, Klotz LO, et al. Gadd153 sensitizes cells to endoplasmic reticulum stress by down-regulating Bcl2 and perturbing the cellular redox state. *Mol Cell Biol* 2001;**21**:1249–59.
116. Yamaguchi H, Wang HG. CHOP is involved in endoplasmic reticulum stress-induced apoptosis by enhancing DR5 expression in human carcinoma cells. *J Biol Chem* 2004;**279**:45,495–502.
117. Marciniak SJ, Ron D. Endoplasmic reticulum stress signaling in disease. *Physiol Rev* 2006;**86**:1133–49.
118. Zhang K, Kaufman RJ. The unfolded protein response: a stress signaling pathway critical for health and disease. *Neurology* 2006;**66**:S102–9.
119. Zhao L, Ackerman SL. Endoplasmic reticulum stress in health and disease. *Curr Opin Cell Biol* 2006;**18**:444–52.
120. Reimold AM, Etkin A, Clauss I, et al. An essential role in liver development for transcription factor XBP-1. *Genes Dev* 2000;**14**:152–7.
121. Zhang K, Wong HN, Song B, et al. The unfolded protein response sensor IRE1alpha is required at 2 distinct steps in B cell lymphopoiesis. *J Clin Invest* 2005;**115**:268–81.
122. Lee AH, Chu GC, Iwakoshi NN, Glimcher LH. XBP-1 is required for biogenesis of cellular secretory machinery of exocrine glands. *Embo J* 2005;**24**:4368–80.

123. Tirasophon W, Welihinda AA, Kaufman RJ. A stress response pathway from the endoplasmic reticulum to the nucleus requires a novel bifunctional protein kinase/endoribonuclease (Ire1p) in mammalian cells. *Genes Dev* 1998;**12**:1812–24.
124. Clauss IM, Gravallese EM, Darling JM, et al. In situ hybridization studies suggest a role for the basic region-leucine zipper protein hXBP-1 in exocrine gland and skeletal development during mouse embryogenesis. *Dev Dyn* 1993;**197**:146–56.
125. Iwawaki T, Akai R, Kohno K, Miura M. A transgenic mouse model for monitoring endoplasmic reticulum stress. *Nat Med* 2004;**10**:98–102.
126. Souid S, Lepesant JA, Yanicostas C. The xbp-1 gene is essential for development in Drosophila. *Dev Genes Evol* 2007;**217**:159–67.
127. Zhang P, McGrath B, Li S, et al. The PERK eukaryotic initiation factor 2 alpha kinase is required for the development of the skeletal system, postnatal growth, and the function and viability of the pancreas. *Mol Cell Biol* 2002;**22**:3864–74.
128. Harding HP, Zeng H, Zhang Y, et al. Diabetes mellitus and exocrine pancreatic dysfunction in perk−/− mice reveals a role for translational control in secretory cell survival. *Mol Cell* 2001;**7**:1153–63.
129. Delepine M, Nicolino M, Barrett T, et al. EIF2AK3, encoding translation initiation factor 2-alpha kinase 3, is mutated in patients with Wolcott-Rallison syndrome. *Nat Genet* 2000;**25**:406–9.
130. Ozcan U, Cao Q, Yilmaz E, et al. Endoplasmic reticulum stress links obesity, insulin action, and type 2 diabetes. *Science* 2004;**306**:457–61.
131. Calfon M, Zeng H, Urano F, et al. IRE1 couples endoplasmic reticulum load to secretory capacity by processing the XBP-1 mRNA. *Nature* 2002;**415**:92–6.
132. Reimold AM, Iwakoshi NN, Manis J, et al. Plasma cell differentiation requires the transcription factor XBP-1. *Nature* 2001;**412**:300–7.
133. Iwakoshi NN, Lee AH, Vallabhajosyula P, et al. Plasma cell differentiation and the unfolded protein response intersect at the transcription factor XBP-1. *Nat Immunol* 2003;**4**:321–9.
134. Gass JN, Gifford NM, Brewer JW. Activation of an unfolded protein response during differentiation of antibody-secreting B cells. *J Biol Chem* 2002;**277**:49,047–54.
135. Yamamoto K, Sato T, Matsui T, et al. Transcriptional Induction of Mammalian ER Quality Control Proteins Is Mediated by Single or Combined Action of ATF6alpha and XBP1. *Dev Cell* 2007;**13**:365–76.
136. Wu J, Rutkowski DT, Dubois M, et al. ATF6alpha Optimizes Long-Term Endoplasmic Reticulum Function to Protect Cells from Chronic Stress. *Dev Cell* 2007;**13**:351–64.
137. Gass JN, Jiang HY, Wek RC, Brewer JW. The unfolded protein response of B-lymphocytes: PERK-independent development of antibody-secreting cells. *Mol Immunol* 2008;**45**:1035–43.
138. Skalet AH, Isler JA, King LB, et al. Rapid B cell receptor-induced unfolded protein response in nonsecretory B cells correlates with pro- versus antiapoptotic cell fate. *J Biol Chem* 2005;**280**:39,762–71.
139. van Anken E, Romijn EP, Maggioni C, et al. Sequential waves of functionally related proteins are expressed when B cells prepare for antibody secretion. *Immunity* 2003;**18**:243–53.
140. Hayashi A, Kasahara T, Iwamoto K, et al. The role of brain-derived neurotrophic factor (BDNF)-induced XBP1 splicing during brain development. *J Biol Chem* 2007;**282**:34,525–34.
141. Shen Y, Hendershot LM. Identification of ERdj3 and OBF-1/BOB-1/OCA-B as Direct Targets of XBP-1 during Plasma Cell Differentiation. *J Immunol* 2007;**179**:2969–78.
142. Sequeira SJ, Ranganathan AC, Adam AP, et al. Inhibition of proliferation by PERK regulates mammary acinar morphogenesis and tumor formation. *PLoS ONE* 2007;**2**:e615.
143. Nakanishi K, Sudo T, Morishima N. Endoplasmic reticulum stress signaling transmitted by ATF6 mediates apoptosis during muscle development. *J Cell Biol* 2005;**169**:555–60.
144. Schroder M, Chang JS, Kaufman RJ. The unfolded protein response represses nitrogen-starvation induced developmental differentiation in yeast. *Genes Dev* 2000;**14**:2962–75.
145. Bicknell AA, Babour A, Federovitch CM, Niwa M. A novel role in cytokinesis reveals a housekeeping function for the unfolded protein response. *J Cell Biol* 2007;**177**:1017–27.

Chapter 40

Protein Quality Control in the Endoplasmic Reticulum

Yuki Okuda-Shimizu, Ying Shen and Linda Hendershot

Department of Genetics and Tumor Cell Biology, St. Jude Children's Research Hospital, Memphis, Tennessee

INTRODUCTION

Proteins destined for secretion or cell surface expression are synthesized in the endoplasmic reticulum (ER). In addition to the complexity of protein folding in any organelle, the oxidizing, calcium rich ER environment and high concentration of unfolded proteins pose further dangers and constraints to the process. Quality control measures exist to ensure that nascent polypeptide chains are specifically prevented from traveling further along the secretory pathway until they have completed their folding or assembly. Proteins that cannot achieve a proper conformation are recognized and removed from the ER for degradation by the 26S proteasome. The same chaperones that aid protein folding are also important for targeting misfolded proteins for degradation. It remains unclear how unfolded nascent proteins are distinguished from proteins that cannot fold. Recent studies reveal a functional and perhaps even physical organization of resident ER proteins that could provide a means of separating these contradictory functions.

ER QUALITY CONTROL

The ER is the site of synthesis of proteins destined for the cell surface or secretion. Although the concentration of nascent unfolded polypeptides in this organelle can be extremely high, particularly in secretory tissues like the pancreas, liver, and plasma cells, it has been estimated that >95 percent of nascent chains fold and assemble rapidly and correctly through the participation of chaperones and folding enzymes. In mammalian cells, proteins are co-translationally translocated into the ER through a protein channel called the translocon as unfolded, extended polypeptide chains (Figure 40.1). These proteins immediately encounter molecular chaperones that are situated on the luminal face of the translocon and can begin to fold and assemble into multimeric complexes even before they are completely synthesized. In addition to preventing their aggregation and helping them achieve their correct conformation, the chaperones also retain incompletely matured proteins in the ER by remaining associated until they are completely folded. Proteins that ultimately fail to mature properly are targeted for intracellular degradation, which is also achieved in part through the action of molecular chaperones [1,2]. The process of recognizing the folding status of newly synthesized proteins, allowing proteins that have reached their proper tertiary structure to move forward through the secretory pathway, and disposing of those that do not, is collectively referred to as ER quality control [3].

UNIQUE ENVIRONMENT OF THE ER

Although proteins folding in the ER encounter all the same constraints and risks faced by proteins synthesized in any organelle, the unusual ER environment poses additional perils and provides unique assistance for allowing the protein to reach a proper conformation. The ER contains millimolar concentrations of calcium used in many signal transduction pathways, but calcium can bind and neutralize negatively charged amino acids that often provide side chain interactions that initiate and stabilize protein folding and assembly intermediates. In addition, many secretory pathway proteins are modified by N-linked glycans as soon as they enter the ER. These highly charged moieties restrict and direct the available folding pathways for a given sequence of amino acids. Inhibiting glycosylation can dramatically affect protein folding and secretion [4]. Finally, unlike the cytosol, the ER possesses an environment that fosters disulfide bond formation between juxtaposed cysteine residues.

Handbook of Cell Signaling, Three-Volume Set 2 ed.

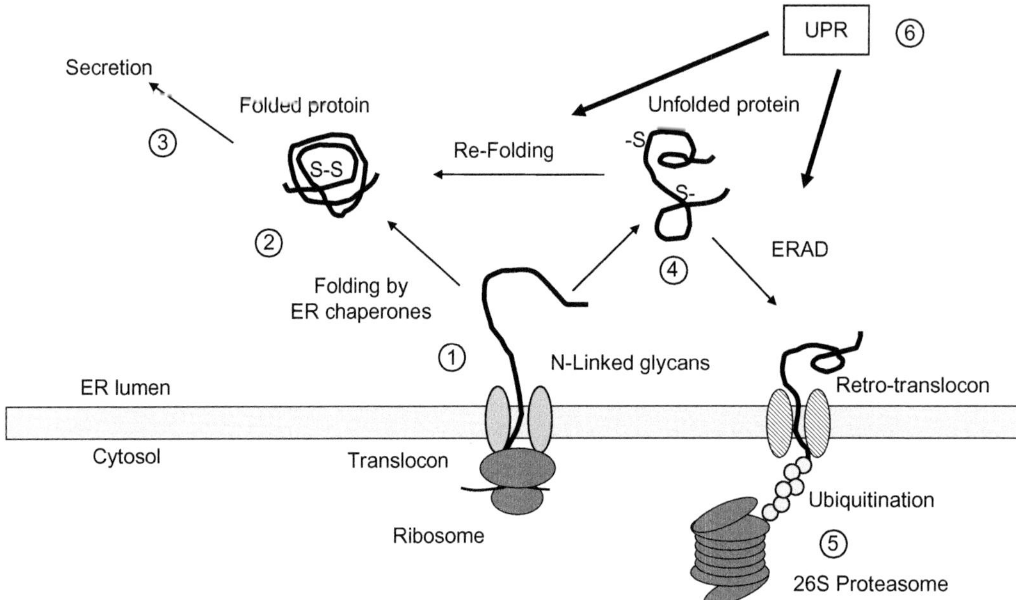

FIGURE 40.1 The ER quality control system.
(1) Proteins enter the ER co-translationally through a protein channel called the translocon as an extended polypeptide chain where they immediately encounter resident ER proteins known as molecular chaperones. (2) They are often modified by N-linked glycans and begin fold even before their synthesis is complete. This folding is assisted by ER chaperones and folding enzymes and is stabilized by the formation of disulfide bonds. (3) If folding is successful, the newly synthesized protein is transported further along the secretory pathway for eventual secretion or delivery to more distal compartments along the pathway. (4) Proteins that fail to reach a proper mature conformation are identified and transported back across the ER membrane through a protein channel known as the retrotranslocon. (5) Once in the cytosol, these proteins are ubiquitylated and degraded by the 26S proteosome. (6) If the concentration of unfolded proteins becomes too high in this organelle, the UPR is activated, which increases molecular chaperones to prevent the aggregation and enhance the refolding of these proteins while at the same time upregulating ERAD components in order to dispose of these proteins.

Previously it was believed that the oxidizing environment of the ER was sufficient to directly promote disulfide bond formation. It is now clear that the oxidation of thiols or rearrangements of non-native disulfide bonds is catalyzed by enzymes, which transfer oxidizing equivalents via a protein relay. Ero1p (Ero1Lα and Ero1β in mammals) accepts an electron directly from molecular oxygen. Oxidized, Ero1p transfers its disulfide bond to an oxidoreductase (discussed below), which in turn catalyzes the formation of disulfide bonds in substrates [5]. Disulfide bond formation is reversible, and the same catalytic proteins can often act to reduce disulfide bonds depending on their oxidation status [6]. The demands on this system can be particularly high in secretory cells. For example, plasma cells synthesize thousands of IgM pentamers per second. This requires the formation of ~20 disulfide bonds per monomer or ~100 disulfide bonds per pentamer, which means that ~100,000 disulfide bonds are produced per second and this does not include off-pathway products that must be reduced and reoxidized!

MOLECULAR CHAPERONES AND FOLDING ENZYME

Two major chaperone families exist in the ER. BiP is the resident Hsp70 homolog and is an essential protein that is involved in all major ER functions [7]. Mice that are null for BiP die at embryonic day 3.5 [8], and the subtilase cytotoxin, which is produced by certain Shiga toxigenic strains of *Escherichia coli*, kills cells by cleaving and inactivating BiP [9]. Like all Hsp70 proteins, BiP possesses an N-terminal ATPase domain that is crucial to its nucleotide dependent chaperoning activity and a C-terminal substrate binding domain that interacts with unfolded polypeptides. BiP preferentially binds to peptides containing aromatic and hydrophobic amino acids in alternating positions, suggesting that peptides bind in an extended conformation, with the side chains of alternating residues pointing into a cleft on the BiP molecule [10]. In the ATP-bound conformation, BiP associates with and releases polypeptides rapidly. Hydrolysis of ATP by the nucleotide binding domain triggers a conformational change in the substrate binding domain, which stabilizes BiP's binding to the substrate. Release of ADP and rebinding of ATP reopens the substrate binding domain, thus initiating substrate release and allowing it to fold. The BiP's ATP hydrolysis cycle is regulated by co-chaperones, including DnaJ homologs that accelerate ATP hydrolysis and nucleotide exchanging factors that release ADP to allow BiP to start another cycle.

Six ER localized DnaJ homologs have been identified, which bind to BiP and stimulate its ATPase activity [11]. ERdj1 is a transmembrane protein that interacts with ribosomes via its C-terminal domain in the cytosol and recruits BiP through its J domain in the ER, suggesting that ERdj1 plays a role in coupling translation and translocation [12]. ERdj2 is a Sec63 homolog [13] that in yeast plays a role in translocation of nascent polypeptides into the ER. Mutations in ERdj2 have been linked to polycystic liver disease in humans [14]. ERdj3 binds directly to a number of substrates [15,16] and participates in their folding. ERdj4 is normally expressed at very low levels but is strongly upregulated by ER stress [17] and may protect cells against apoptosis [18]. ERdj5 is an abundant, stress inducible protein with four thioredoxin-like domains in addition to a J domain [19]. Although its function is unknown, it is likely to play some role in disulfide bond chemistry. Finally ERdj6, also known as p58^{IPK}, protects cells from ER stress via its protein processing function [20]. The concentration of BiP in dog pancreatic microsomes was estimated to be ~5 μM, which is sufficient to simultaneously bind to all known ER localized ERdjs, which range from nearly undetectable (ERdj4) to ~2 μM (ERdj2 and ERdj5) [11]. The ATPase activity of BiP is also regulated by nucleotide exchange factors, including BAP/Sil1p [21] and GRP170 [22]. Mutations in the *BAP/Sil1* gene have been linked to an autosomal recessive form of Marinesco-Sjogren syndrome; a multisystem disorder that is characterized by cerebellar ataxia [23,24]. The *Sil1* null or "woozy" mouse is similarly affected by neurodegeneration [25].

The other major chaperone family is calnexin (CNX), a transmembrane, monovalent lectin binding chaperone, and its soluble homolog, calreticulin (CRT), which aid glycoprotein folding in the ER. Core glycans are added co-translationally to Asn-X-Ser/Thr sequences as the polypeptide enters the ER [4]. Processing of the N-linked sugar moiety ($Glc_3Man_9GlcNAc_2$) begins almost immediately. Once the two outermost of the three glucose residues are removed by glucosidase I and II, CNX/CRT binds the monoglucosylated form of the glycan [26]. In the case of the influenza hemaglutinin protein, this can occur when as few as 30 amino acids have entered the ER lumen, which suggests that both CNX and the glucosidases function in very close proximity to the translocon [27]. Dissociation of the glycoprotein from CNX/CRT requires removal of the remaining glucose on the core glycan by glucosidase II. Mutations in the β subunit of glucosidase II have been associated with an autosomal dominant form of polycystic liver disease [28], thus tying both the BiP and the CNX system to this disease. After the nascent glycoprotein is released from CNX, three possible fates await it. First, if properly folded, it is ready to exit the ER. Second, if it is incompletely folded, UDP-glucosyl transferase binds to unfolded regions on the protein and to the N-linked glycan and transfers a single glucose back to the core sugars, thereby recreating the signal for CXN/CRT association and allowing it additional opportunities to fold [29]. Finally, if proper folding cannot be achieved, the glycoprotein will be targeted for degradation as described below.

While disulfide bonds can certainly serve to stabilize proper tertiary structures, the correct cysteines that are involved in the disulfide bond can occur at distal sites on the linear sequence or on completely separate subunits [1,30]. Thus, the potential for the formation of incorrect disulfide bonds and the need to resolve these can be quite high. To minimize the formation of incorrect disulfide bonds, it appears that chaperones work together with disulfide bond catalysis that are characterized by containing the thioredoxin active site, C-X-X-C. At least 10 potential disulfide bond isomerases have been discovered in the mammalian ER, where they assist in the formation, reduction, and isomerization of disulfide bonds during protein synthesis [31]. At least three of these (i.e., PDI, ERp72, and CaBP1/P5) are components of a large multi-chaperone complex that contains BiP [32], and ERp57 is associated with CXN and CRT [33]. Both PDI and ERp57 have been shown to form mixed disulfide bonds with nascent proteins in mammalian cells [34], which is an intermediate in disulfide bond formation. A recent study demonstrated that Ig light chains must be at least partially reduced in order to be retrotranslocated and degraded, which strongly implicates this group of proteins in both folding and degradation [35].

DISPOSAL OF UNFOLDED AND MISFOLDED PROTEIN

The identification of proteins that do not fold or assemble properly and ensuring that they are disposed of instead of being allowed to travel further along the secretory pathway is defined as ER quality control (Figure 40.1). Early studies by McCracken and Brodsky supported a rather heretical idea that at least some of these proteins were actually retrotranslocated across the ER membrane to the cytosol and degraded by the 26S proteasome via a process known as ER associated degradation (ERAD) [1]. As currently understood [3], the ERAD process begins with the recognition of a substrate as being misfolded or unfolded. The substrate is transported across the ER membrane through a proteinaceous channel called the retrotranslocon. In the cytosol, it is poly-ubiquitylate and degraded by the 26S proteasome [36]. In *S. cerevisiae*, there are three major ERAD subpathways defined by which region of the protein is misfolded (i.e., luminal, transmembrane, or cytosolic). These pathways utilize distinct ubiquitin–ligase complexes and vary in their reliance on Der1p and Usa1p [37]. Proteins with misfolded ER-luminal domains rely on the Hrd1p/Hrd3p ubiquitin ligase, which binds Der1p via the linker protein Usa1p. Der1p possesses four transmembrane domains and is thought to constitute part of the channel for retrotranslocation of this group of substrates [38]. Disposal of substrates with misfolded intramembrane domains is independent of Usa1p and Der1p, and membrane proteins with misfolded cytosolic domains are directly targeted to the Doa10p ubiquitin ligase. All three ERAD subpathways converge at the Cdc48p AAA ATPase complex, which uses the energy of hydrolysis to extract the substrate [39]. The accumulation of unfolded proteins in the ER triggers a signal transduction cascade known as the UPR, which serves both to upregulate ER chaperones to prevent the aggregation of these proteins and to aid in their refolding and to increase the degradative capacity of the cell to dispose of misfolded proteins.

The existence of a similar ERAD pathway for mammalian proteins was first shown for the MHC Class I protein, which is an integral membrane protein that is extracted from the ER membrane, deglycosylated, and ubiquitylate [40]. Mammalian equivalents of a number of the yeast proteins involved in ERAD have been identified, including three different Der1p homologs, Derlin-1-3 [41–43]. The ERAD pathway in mammalian cells has been best described for glycoproteins, which interact with the CXN/CRT chaperone family during folding. These chaperones monitor the cyclic processing of N-linked glycans that occurs on unfolded glycoproteins [29]. If mannosidases trim the glycan to an eight mannose form, the protein is no longer a substrate for reglucosylation by UDP-glucosyltransferase and can now be recognized by an ER-degradation enhancing alpha-mannosidase-like protein (EDEM), which is associated with retrotranslocons containing Derlin-2 and Derlin-3 [44,45]. The degradation of the misfolded CXN/CRT substrate, α1-antitrypsin $\text{Null}_{\text{Hong Kong}}$ (AAT NHK) [46] is accelerated by EDEM overexpression and requires either Derlin-2 or 3, but its turnover is not affected when Derlin-1 levels are reduced [43]. The extraction of a number of glycoprotein substrates requires the activity of the Cdc48/p97-Ufd1-Npl4 AAA ATPase complex [47] and is followed by ubiquitylate, which is catalyzed by a growing number of E3 ligases [48].

It has been much less clear how unfolded, non-glycosylated proteins that utilize BiP are recognized and targeted for degradation. Very recently it was shown that some BiP substrates interact with a subset of ERAD components (i.e., Herp and Derlin-1) that are not used to dispose of glycosylated AAT mutants, but their turnover is also dependent on several common ERAD components including the E3 ligase Hrd1 and the p97 complex [35]. Herp, the mammalian counterpart of yeast Usa1p, is a UPR inducible ER membrane protein that is almost entirely cytosolically disposed [49]. It contains a ubiquitin-like domain and binds to both ubiquitinated proteins and to the 26S proteasome [35], suggesting that it may play a role in targeting this group of ERAD substrates to the proteasome. There is currently no functional ortholog of EDEM, which would serve to identify misfolded proteins in the ER and target them to the Derlin-1 containing retrotranslocon. This study is the first to suggest that unlike entry into the ER, pathways for retrotranslocation may be substrate specific in mammalian cells.

ER SUBCOMPARTMENTS

It is unclear how the ER is able to separate its various functions, particularly those of protein folding versus protein degradation. Recent fluorescence staining data [50], crosslinking studies [32], and proteomic analyses of ER fractions [51] have suggested that there may be a spatial separation of some resident ER proteins, which could then lead to a physical separation of functions. For instance, both the BiP and calnexin chaperone systems are associated with an oxidoreductase that can either form or reduce disulfide bonds in the substrate protein depending on its oxidation status. This would provide the chaperone systems with the ability to aid in both the folding of nascent proteins as well as the disposal of misfolded ones. Changes in their ability to oxidize (for folding) or reduce (for degradation) a substrate could conceivably be achieved by regulating the concentration of molecular oxygen, Ero1, or as yet to be identified oxidoreductases within subregions of the ER.

In summary, the ER represents a major and particularly complex site of protein synthesis. Quality control systems exist in order to ensure that only properly matured proteins are permitted to leave the ER for further destinations along the secretory pathway. Proteins that fail the checks provided are identified by components of ER quality control, targeted for retrotranslocation, and ultimately destroyed by the 26S proteasome. While much progress has been made in recent years in identifying the components of ER quality control and understanding how they work, some of the most basic elements of this system remain largely elusive, including the mechanism for distinguishing between unfolded proteins that are in the process of folding and those that cannot fold.

REFERENCES

1. Werner ED, Brodsky JL, McCracken AA. Proteasome-dependent endoplasmic reticulum-associated protein degradation: an unconventional route to a familiar fate. *Proc Natl Acad Sci USA* 1996;**93**:13,797–13,801,.
2. Jarosch E, Lenk U, Sommer T. Endoplasmic reticulum-associated protein degradation. *Int Rev Cytol* 2003;**223**:39–81.
3. Ellgaard L, Helenius A. Quality control in the endoplasmic reticulum. *Nat Rev Mol Cell Biol* 2003;**4**:181–91.
4. Kornfeld R, Kornfeld S. Assembly of asparagine-linked oligosaccharides. *Annu Rev Biochem* 1985;**54**:631–64.
5. Frand AR, Cuozzo JW, Kaiser CA. Pathways for protein disulphide bond formation. *Trends Cell Biol* 2000;**10**:203–10.
6. Tsai B, Ye Y, Rapoport TA. Retro-translocation of proteins from the endoplasmic reticulum into the cytosol. *Nat Rev Mol Cell Biol* 2002;**3**:246–55.
7. Hendershot LM. The ER chaperone BiP is a master regulator of ER function. *Mt Sinai J Med* 2004;**71**:289–97.
8. Luo S, Mao C, Lee B, Lee AS. GRP78/BiP is required for cell proliferation and protecting the inner cell mass from apoptosis during early mouse embryonic development. *Mol Cell Biol* 2006;**26**:5688–97.

9. Paton AW, Beddoe T, Thorpe CM, Whisstock JC, Wilce MC, Rossjohn J, Talbot UM, Paton JC. AB5 subtilase cytotoxin inactivates the endoplasmic reticulum chaperone BiP. *Nature* 2006;**443**:548–52.
10. Blond-Elguindi S, Cwirla SE, Dower WJ, Lipshutz RJ, Sprang SR, Sambrook JF, Gething MJ. Affinity panning of a library of peptides displayed on bacteriophages reveals the binding specificity of BiP. *Cell* 1993;**75**:717–28.
11. Weitzmann A, Baldes C, Dudek J, Zimmermann R. The heat shock protein 70 molecular chaperone network in the pancreatic endoplasmic reticulum: a quantitative approach. *FEBS J* 2007;**274**:5175–87.
12. Dudek J, Volkmer J, Bies C, Guth S, Muller A, Lerner M, Feick P, Schafer KH, Morgenstern E, Hennessy F, Blatch GL, Janoscheck K, Heim N, Scholtes P, Frien M, Nastainczyk W, Zimmermann R. A novel type of co-chaperone mediates transmembrane recruitment of DnaK-like chaperones to ribosomes. *EMBO J* 2002;**21**:2958–67.
13. Skowronek MH, Rotter M, Haas IG. Molecular characterization of a novel mammalian DnaJ-like Sec63p homolog. *Biol Chem* 1999;**380**:1133–8.
14. Davila S, Furu L, Gharavi AG, Tian X, Onoe T, Qian Q, Li A, Cai Y, Kamath PS, King BF, Azurmendi PJ, Tahvanainen P, Kaariainen H, Hockerstedt K, Devuyst O, Pirson Y, Martin RS, Lifton RP, Tahvanainen E, Torres VE, Somlo S. Mutations in SEC63 cause autosomal dominant polycystic liver disease. *Nat Genet* 2004;**36**:575–7.
15. Yu M, Haslam RH, Haslam DB. HEDJ, an Hsp40 co-chaperone localized to the endoplasmic reticulum of human cells. *J Biol Chem* 2000;**275**:24,984–24,992.
16. Shen Y, Hendershot LM. ERdj3, a stress-inducible endoplasmic reticulum DnaJ homologue, serves as a cofactor for BiP's interactions with unfolded substrates. *Mol Biol Cell* 2005;**16**:40–50.
17. Shen Y, Meunier L, Hendershot LM. Identification and characterization of a novel endoplasmic reticulum (ER) DnaJ homologue, which stimulates ATPase activity of BiP in vitro and is induced by ER stress. *J Biol Chem* 2002;**277**:15,947–15,956.
18. Kurisu J, Honma A, Miyajima H, Kondo S, Okumura M, Imaizumi K. MDG1/ERdj4, an ER-resident DnaJ family member, suppresses cell death induced by ER stress. *Genes Cells* 2003;**8**:189–202.
19. Cunnea PM, Miranda-Vizuete A, Bertoli G, Simmen T, Damdimopoulos AE, Hermann S, Leinonen S, Huikko MP, Gustafsson JA, Sitia R, Spyrou G. ERdj5, an endoplasmic reticulum (ER)-resident protein containing DnaJ and thioredoxin domains, is expressed in secretory cells or following ER stress. *J Biol Chem* 2003;**278**:1059–66.
20. Rutkowski DT, Kang SW, Goodman AG, Garrison JL, Taunton J, Katze MG, Kaufman RJ, Hegde RS. The role of p58IPK in protecting the stressed endoplasmic reticulum. *Mol Biol Cell* 2007;**18**:3681–91.
21. Chung KT, Shen Y, Hendershot LM. BAP, a mammalian BiP associated protein, is a nucleotide exchange factor that regulates the ATPase activity of BiP. *J Biol Chem* 2002;**277**:47,557–47,563.
22. Weitzmann A, Volkmer J, Zimmermann R. The nucleotide exchange factor activity of Grp170 may explain the non-lethal phenotype of loss of Sil1 function in man and mouse. *FEBS Lett* 2006;**580**:5237–40.
23. Anttonen AK, Mahjneh I, Hamalainen RH, Lagier-Tourenne C, Kopra O, Waris L, Anttonen M, Joensuu T, Kalimo H, Paetau A, Tranebjaerg L, Chaigne D, Koenig M, Eeg-Olofsson O, Udd B, Somer M, Somer H, Lehesjoki AE. The gene disrupted in Marinesco-Sjogren syndrome encodes SIL1, an HSPA5 cochaperone. *Nat Genet* 2005;**37**:1309–11.
24. Senderek J, Krieger M, Stendel C, Bergmann C, Moser M, Breitbach-Faller N, Rudnik-Schoneborn S, Blaschek A, Wolf NI, Harting I, North K, Smith J, Muntoni F, Brockington M, Quijano-Roy S, Renault F, Herrmann R, Hendershot LM, Schroder JM, Lochmuller H, Topaloglu H, Voit T, Weis J, Ebinger F, Zerres K. Mutations in SIL1 cause Marinesco-Sjogren syndrome, a cerebellar ataxia with cataract and myopathy. *Nat Genet* 2005;**37**:1312–14.
25. Zhao L, Longo-Guess C, Harris BS, Lee JW, Ackerman SL. Protein accumulation and neurodegeneration in the woozy mutant mouse is caused by disruption of SIL1, a cochaperone of BiP. *Nat Genet* 2005;**37**:974–9.
26. Helenius A, Aebi M. Roles of N-linked glycans in the endoplasmic reticulum. *Annu Rev Biochem* 2004;**73**:1019–49.
27. Daniels R, Kurowski B, Johnson AE, Hebert DN. N-linked glycans direct the cotranslational folding pathway of influenza hemagglutinin. *Mol Cell* 2003;**11**:79–90.
28. Li A, Davila S, Furu L, Qian Q, Tian X, Kamath PS, King BF, Torres VE, Somlo S. Mutations in PRKCSH cause isolated autosomal dominant polycystic liver disease. *Am J Hum Genet* 2003;**72**:691–703.
29. Hebert DN, Foellmer B, Helenius A. Glucose trimming and reglucosylation determine glycoprotein association with calnexin in the endoplasmic reticulum. *Cell* 1995;**81**:425–33.
30. Braakman I, Hoover Litty H, Wagner KR, Helenius A. Folding of influenza hemagglutinin in the endoplasmic reticulum. *J Cell Biol* 1991;**114**:401–11.
31. Kleizen B, Braakman I. Protein folding and quality control in the endoplasmic reticulum. *Curr Opin Cell Biol* 2004;**16**:343–9.
32. Meunier L, Usherwood YK, Chung KT, Hendershot LM. A subset of chaperones and folding enzymes form multiprotein complexes in endoplasmic reticulum to bind nascent proteins. *Mol Biol Cell* 2002;**13**:4456–69.
33. High S, Lecomte FJ, Russell SJ, Abell BM, Oliver JD. Glycoprotein folding in the endoplasmic reticulum: a tale of three chaperones?. *FEBS Lett* 2000;**476**:38–41.
34. Molinari M, Helenius A. Glycoproteins form mixed disulphides with oxidoreductases during folding in living cells. *Nature* 1999;**402**:90–3.
35. Okuda-Shimizu Y, Hendershot LM. Characterization of an ERAD pathway for nonglycosylated BiP substrates, which requires Herp. *Mol Cell* 2007;**28**:1–11.
36. Meusser B, Hirsch C, Jarosch E, Sommer T. ERAD: the long road to destruction. *Nat Cell Biol* 2005;**7**:766–72.
37. Ismail N, Ng DT. Have you HRD? Understanding ERAD is DOAble!. *Cell* 2006;**126**:237–9.
38. Knop M, Finger A, Braun T, Hellmuth K, Wolf DH. Der1, a novel protein specifically required for endoplasmic reticulum degradation in yeast. *EMBO J* 1996;**15**:753–63.
39. Rabinovich E, Kerem A, Frohlich KU, Diamant N, Bar-Nun S. AAA-ATPase p97/Cdc48p, a cytosolic chaperone required for endoplasmic reticulum-associated protein degradation. *Mol Cell Biol* 2002;**22**:626–34.
40. Wiertz EJ, Jones TR, Sun L, Bogyo M, Geuze HJ, Ploegh HL. The human cytomegalovirus US11 gene product dislocates MHC class I heavy chains from the endoplasmic reticulum to the cytosol. *Cell* 1996;**84**:769–79.
41. Ye Y, Shibata Y, Yun C, Ron D, Rapoport TA. A membrane protein complex mediates retro-translocation from the ER lumen into the cytosol. *Nature* 2004;**429**:841–7.
42. Lilley BN, Ploegh HL. A membrane protein required for dislocation of misfolded proteins from the ER. *Nature* 2004;**429**:834–40.
43. Oda Y, Okada T, Yoshida H, Kaufman RJ, Nagata K, Mori K. Derlin-2 and Derlin-3 are regulated by the mammalian unfolded protein response and are required for ER-associated degradation. *J Cell Biol* 2006;**172**:383–93.

44. Oda Y, Hosokawa N, Wada I, Nagata K. EDEM as an acceptor of terminally misfolded glycoproteins released from calnexin. *Science* 2003;**299**:1394–7.
45. Molinari M, Calanca V, Galli C, Lucca P, Paganetti P. Role of EDEM in the release of misfolded glycoproteins from the calnexin cycle. *Science* 2003;**299**:1397–400.
46. Cabral CM, Liu Y, Sifers RN. Dissecting glycoprotein quality control in the secretory pathway. *Trends Biochem Sci* 2001;**26**:619–24.
47. Ye Y, Meyer HH, Rapoport TA. The AAA ATPase Cdc48/p97 and its partners transport proteins from the ER into the cytosol. *Nature* 2001;**414**:652–6.
48. Kostova Z, Tsai YC, Weissman AM. Ubiquitin ligases, critical mediators of endoplasmic reticulum-associated degradation. *Semin Cell Dev Biol* 2007;**18**:770–9.
49. Kokame K, Agarwala KL, Kato H, Miyata T. Herp, a new ubiquitin-like membrane protein induced by endoplasmic reticulum stress. *J Biol Chem* 2000;**275**:32,846–32,853.
50. Kamhi-Nesher S, Shenkman M, Tolchinsky S, Fromm SV, Ehrlich R, Lederkremer GZ. A novel quality control compartment derived from the endoplasmic reticulum. *Mol Biol Cell* 2001;**12**:1711–23.
51. Gilchrist A, Au CE, Hiding J, Bell AW, Fernandez-Rodriguez J, Lesimple S, Nagaya H, Roy L, Gosline SJ, Hallett M, Paiement J, Kearney RE, Nilsson T, Bergeron JJ. Quantitative proteomics analysis of the secretory pathway. *Cell* 2006;**127**:1265–81.

Chapter 41

Protein Quality Control in Peroxisomes: Ubiquitination of the Peroxisomal Targeting Signal Receptors

Chris Williams and Ben Distel

Department of Medical Biochemistry, Academic Medical Center, University of Amsterdam, Amsterdam, The Netherlands

INTRODUCTION

The ability to import folded, cofactor bound, and even oligomeric proteins makes peroxisomes unique when compared to subcellular organelles such as mitochondria and chloroplasts [1]. Proteins destined for the peroxisomal matrix begin their journey in the cytosol, where they are synthesized on free polyribosomes. Peroxisomal sorting, like the sorting into other subcellular compartments, relies on targeting signals, in this case a peroxisomal targeting signal (PTS). To date, two signals have been identified: the PTS1 and the PTS2. Of these two signals, the PTS1 is by far the most common and is a C-terminal tripeptide related to the canonical S-K-L sequence of firefly luciferase [2]. The PTS2 however, is less common and is an N-terminal nona-peptide with the consensus (R/K)(L/V/I)X_5(H/Q)(L/A) [3]. Proteins equipped with a PTS are recognized and bound in the cytosol by a cycling receptor. The cycling receptor for PTS1 proteins, peroxin 5 (Pex5p) is a bi-domain protein. The N-terminal region is involved in the docking and recycling of the receptor [4, 5], while the C-terminal region contains seven tetratricopeptide repeats (TPRs), which specifically interact with the PTS1 sequence [6, 7]. Peroxin 7 (Pex7p) on the other hand, is a WD-40 repeat containing protein and the cycling receptor for PTS2 proteins [8,9]. A number of "helper proteins" have been identified for the PTS2 pathway. These co-receptors, known as the Pex20p family and consisting of the yeast proteins Pex18p, Pex20p, and Pex21p, assist Pex7p in the import of PTS2 proteins [10, 11]. The members of the Pex20p family have a similar domain structure to that of the N-terminal region of Pex5p. Indeed, expression of a chimeric protein consisting of Pex18p fused to the TPR domains of Pex5p can rescue the PTS1 protein import defect of *pex5Δ* cells, indicating that Pex18p fulfills the same function as the N-terminal region of Pex5p [4].

During a typical import cycle (Figure 41.1), the receptor recognizes and binds the PTS protein in the cytosol (I), transports it to the peroxisomal membrane (II), aids in the translocation of the cargo protein into the peroxisomal matrix (III), and recycles to the cytosol for another round of import (IV) (for review see [12]). Around 12 peroxins (the precise number depending on the organism) play important roles in the receptor cycle. A complex, consisting of the peroxisomal membrane proteins (PMPs) Pex13p and Pex14p (with Pex17p in yeast) is responsible for receptor docking [13–17]. A separate complex, consisting of the really interesting new gene (RING) domain containing proteins Pex2p, Pex10p, and Pex12p as well as the intra-peroxisomal protein Pex8p are involved in the translocation process, but the individual role of each component is not fully understood [18]. In addition, the ubiquitin conjugating enzyme Pex4p (alternate name Ubc10p), together with its membrane anchor PMP Pex22p may also play a role in this step or, alternatively, may be involved, together with the AAA (ATPase associated with various cellular activities) proteins Pex1p and Pex6p and the PMP Pex15p, in receptor recycling [19–21].

Recent efforts in the peroxisome field are aimed at understanding how such a complex import cycle, involving many different steps, may be regulated. Remarkably, the import of PTS proteins does not require a membrane potential or an energy source, such as ATP. The recycling of the receptors, however, does require ATP hydrolysis and there is compelling evidence that the AAA proteins Pex1p and Pex6p are involved in this ATP dependent step [20–22]. Are their other potential regulators of the import process? Over the last few years, it has become clear that

Handbook of Cell Signaling, Three-Volume Set 2 ed.

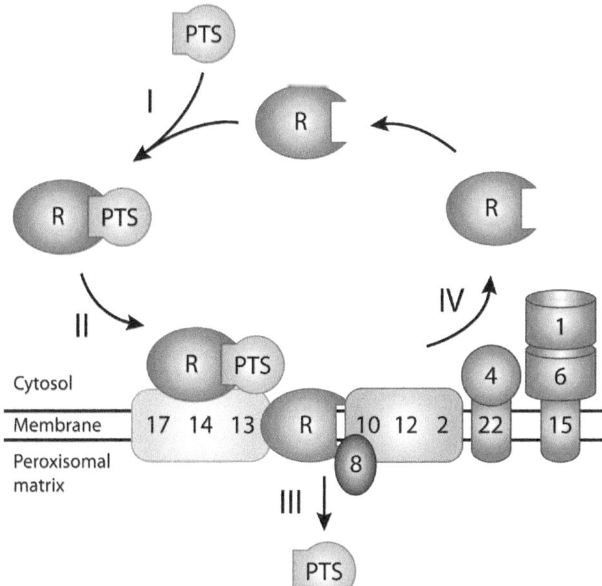

FIGURE 41.1 Model for the import of peroxisomal matrix proteins and receptor recycling.
Peroxisomal matrix proteins containing a peroxisomal targeting signal (PTS) are synthesized in the cytosol and recognized by an associated cycling receptor (I). The receptor–cargo complex then docks on the peroxisomal membrane (II). Next, the PTS cargo is dissociated from the receptor and translocated into the peroxisome (III) and the receptor is recycled to the cytosol for another round of import (IV). R represents the cycling receptors and numbers indicate specific peroxins. See text for details.

ubiquitination of peroxins plays a key role in the regulation of peroxisomal protein import.

Ubiquitination is the attachment of ubiquitin to a substrate protein. Ubiquitination plays a key role in a wide range of cellular events, including protein degradation, DNA repair, cell cycle control, multi-vesicular body sorting, and endocytosis, amongst others [23]. The attachment of ubiquitin is an ATP dependent process and requires the action of three distinct enzymes. First, the ubiquitin activating enzyme (E_1) activates ubiquitin in an ATP dependent manner. Next, the activated ubiquitin is transferred to the active site cysteine residue of an ubiquitin conjugating enzyme (UBC or E_2). Finally, with the aid of an ubiquitin ligase (E_3), ubiquitin is conjugated to the substrate protein, usually to an NH_2 group of a lysine residue. In turn, the ubiquitin itself can become a substrate for ubiquitination, resulting in the formation of an ubiquitin chain. The attachment of a chain consisting of at least four ubiquitin moieties, linked through lysine 48 (K_{48}) of ubiquitin, is often referred to as poly-ubiquitination and targets the modified substrate for 26S proteasome mediated degradation [24]. The attachment of less than four ubiquitin moieties, usually linked via other lysine residues in ubiquitin can be referred to as mono-ubiquitination and is for the non-proteolytic regulation of the modified substrates [25].

This chapter discusses the ubiquitination of the PTS (co-) receptors and addresses the role of each of the proteins involved in this process together with the implications of receptor ubiquitination on peroxisomal matrix protein import.

PTS (CO-) RECEPTOR UBIQUITINATION: CONUNDRUM AND CONFUSION

More than a decade ago, the E_2 enzyme Pex4p (Ubc10p) was identified as an essential factor in PTS mediated peroxisomal protein import [26]. However, it took many years before the first ubiquitinated peroxin, the PTS2 co-receptor Pex18p from the yeast *Saccharomyces cerevisiae*, was identified [27]. Since then, other ubiquitinated peroxins have been identified, for example the PTS1 receptor Pex5p, as well as other members of the Pex20p family of PTS2 co-receptors [28–31]. Nevertheless, the situation is far more complex than would, at first, appear. For example, in the absence of the presumed E_2 enzyme Pex4p, the PTS (co-) receptors are still ubiquitinated. Also, the fate of the ubiquitinated peroxins can vary between rapid degradation and accumulation, depending on the type of ubiquitination and the organism under study.

S. CEREVISIAE PEX18P IS DEGRADED IN AN UBIQUITIN DEPENDENT MANNER

The yeast *S. cerevisiae* is unique in that, unlike the other yeast species that only possess one PTS2 co-receptor, Pex20p, it contains two partially redundant proteins, Pex18p and Pex21p that are required for PTS2 import [10, 11]. The expression of Pex18p is dramatically upregulated when cells are grown on oleic acid, a carbon source requiring active peroxisomes for its metabolism. This regulation at the transcriptional level is combined with a rapid turnover of the mature protein that is dependent on ubiquitin and the E_2 enzymes Ubc4p and Ubc5p. It is therefore not surprising that Pex18p is ubiquitinated, via the conjugation of one or two ubiquitin moieties. A direct involvement for Ubc4/5p in this process, however, was not shown [27]. Interestingly, Pex18p is stabilized in certain peroxin deletion strains, including *pex14Δ*, *pex1Δ*, and, significantly, *pex4Δ*. The observation that both Pex4p and Ubc4/5p are required for Pex18p turnover led the authors to propose a model where Pex18p is sequentially ubiquitinated by Pex4p and Ubc4/5p, respectively, ultimately leading to proteasomal degradation of the protein [32]. Additional data concerning the ubiquitination of Pex18p have, however, not been forthcoming. We are therefore left with several unanswered questions, such as is the turnover of Pex18p essential for its function and, crucially, if so why? Further developments in this field are eagerly awaited.

S. CEREVISIAE PEX5P, TWO DISTINCT UBIQUITINATION EVENTS, TWO DISTINCT FUNCTIONS

Pex5p is Poly-Ubiquitinated by Ubc4p on Two Lysine Residues

In contrast to Pex18p, rapid turnover of Pex5p is only observed in cells that lack certain late acting peroxins. In the yeast *Pichia pastoris*, cells deleted for either Pex4p or Pex22p show a strong reduction in the levels of Pex5p. This effect is also observed in a *pex1Δ* and *pex6Δ* deletion strain, although to a lesser extent [19]. Similarly, in mammalian cells where either Pex1p or Pex6p are absent, Pex5p levels are reduced [14]. The reasons for the reduction remained unclear until data on *S. cerevisiae* Pex5p showed that the levels of this protein were not severely reduced in cells lacking one of these same peroxins but instead, ubiquitinated forms of the protein accumulated in the cell [28–30]. Later, it was also shown that certain Pex5p mutants are poly-ubiquitinated, in much the same way as in a *pex4Δ* strain [33]. The ubiquitination pattern in these strains varies and can be divided into two subgroups. In the first group, consisting of *pex4Δ*, *pex22Δ*, and certain Pex5p mutants, two ubiquitinated species were observed, corresponding to the attachment of one or two ubiquitin moieties. In the second group, containing *pex1Δ*, *pex6Δ*, and *pex15Δ*, three and sometimes four ubiquitinated species were present, consistent with the attachment of three to four ubiquitin moieties [28–30, 33]. The presence of such an ubiquitin "ladder" is reminiscent of poly-ubiquitination, which, as stated above, targets modified substrates for degradation via the 26S proteasome [24]. The appearance of a ladder can be explained by the fact that each ubiquitin molecule in the chain is attached separately and not transferred as a chain of ubiquitins in bulk [34]. Ubiquitin chain elongation can be efficiently blocked by mutation of the lysine residue present at position 48 in ubiquitin to an arginine (Ub-K_{48}R). Indeed, the expression of Ub-K_{48}R in *pex4Δ* or *pex1Δ* cells results in a significant decrease in the larger ubiquitinated Pex5p species, indicating that Pex5p is poly-ubiquitinated at a single site, rather than at a number of different sites [28]. Nevertheless, poly-ubiquitination for 26S proteasome mediated degradation usually requires the attachment of a chain of more than four ubiquitin moieties, which is not often observed in the *S. cerevisiae pex* deletion strains. Does this then mean that the purpose of poly-ubiquitination of Pex5p in *S. cerevisiae* is not its degradation, or, alternatively, that the ubiquitination machinery in this organism is not efficient, leading to relatively short ubiquitin chains and failure to degrade the protein? Conclusions in this area are difficult as most of the data concerning the poly-ubiquitination of Pex5p come from *S. cerevisiae*. One exception, however, is the methylotrophic yeast *Hansenula polymorpha*, where Pex5p abundance is severely reduced in a *pex4Δ* strain, much the same as in *P. pastoris*. Expression of the ubiquitin K_{48}R mutant stabilizes Pex5p levels and results in the formation of a higher molecular weight species of Pex5p, corresponding to the addition of a single ubiquitin moiety, suggesting that Pex5p is modified but, due to the inability of the Ub-K_{48}R to form a ubiquitin chain, is no longer degraded [35]. It remains to be seen whether the poly-ubiquitination of Pex5p is also responsible for the apparent degradation observed in mammalian and *P. pastoris* cells lacking one of the late acting peroxins. Data concerning the ubiquitination of the PTS2 co-receptor Pex20p in *P. pastoris* suggests that, at least for this organism, this is indeed the case (see below). Although reduction in Pex5p levels is not observed in *S. cerevisiae*, mutants deficient in proteasome function cause an apparent build up of poly-ubiquitinated Pex5p [28, 29]. These results, however, must be treated with caution as they were obtained with temperature sensitive mutants grown at the restrictive temperature of 37°C, conditions where others have observed that in wild-type cells, poly-ubiquitination of Pex5p already occurs [36].

It seems rather contradictory that in the absence of the E_2 enzyme Pex4p (Ubc10p), Pex5p is ubiquitinated [28–30]. *S. cerevisiae* contains 13 E_2 enzymes, of which Ubc9p and Ubc12p are specific for the ubiquitin like proteins SUMO (small ubiquitin related modifier) and Nedd8p (neural precursor cell-expressed developmentally down-regulated), respectively. The question then became which one of these E_2 enzymes was responsible for Pex5p poly-ubiquitination? Deletion of Ubc4p, in combination with either Pex1p or Pex4p results in a reduction in the amount of ubiquitinated Pex5p, when compared to the *pex1Δ or pex6Δ* strain alone [28, 29]. This reduction becomes more severe when Ubc1p, a homolog of Ubc4p, was also deleted in an *ubc4Δ/pex4Δ* strain, suggesting that some redundancy exists between Ubc4p and Ubc1p [30]. Ubc4p is an E_2 enzyme involved in the degradation of short lived and abnormal proteins [37], and has already been shown to have a hand in peroxisome biogenesis, being involved in the ubiquitination and turnover of Pex18p [27, 37]. These observations, coupled with the fact that, at least in the *S. cerevisiae pex* deletion strains, ubiquitin chains rarely consist of more than three ubiquitin moieties, led Williams *et al.* [33] to refer to this form of ubiquitination as Ubc4p dependent ubiquitination, rather than poly-ubiquitination.

The current data suggest that poly-ubiquitination of Pex5p represents an attempt by the cell to degrade Pex5p. Assuming that, like many other Ubc4p substrates, Pex5p is non-functional under conditions that induce poly-ubiquitination of the protein, the question arises as to why Pex5p has to be removed? Poly-ubiquitination/degradation of Pex5p is observed in the absence of a number of peroxins implicated in Pex5p recycling. This, coupled with the fact that poly-ubiquitinated Pex5p predominantly associates with peroxisomes [28, 29], suggests that this modification may be a way to clear the peroxisomal membrane of unwanted Pex5p caused by inefficient recycling. Further supporting evidence is provided by the stability of Pex5p in certain double mutants. As already mentioned, Pex5p levels are severely reduced in a *P. pastoris pex4Δ* or *pex1Δ* strain [19, 38]. However, Pex5p becomes stable when elements of the docking complex are also deleted [19], indicating that docking at the peroxisomal membrane is a prerequisite for poly-ubiquitination.

The conjugation of ubiquitin usually occurs on to a lysine residue present in the substrate, although the α-NH_2 group can also be used [39, 40]. Mutation of the lysine residue present at position 21 (K_{21}) in *H. polymorpha* Pex5p stabilizes the protein in a *pex4Δ* strain, suggesting that this residue has an important role in the ubiquitin dependent turnover of the protein [35]. A role for lysine residues in the N-terminal domain of Pex5p was confirmed in *S. cerevisiae*. Replacing the lysines at positions 18 and 24 with arginines ($K_{18/24}R$) blocked Ubc4p dependent modification of Pex5p [21, 33]. This effect was specific for lysine 18 and 24, as mutation of the other 13 lysine residues present in the N-terminal region of Pex5p had no effect on poly-ubiquitination [33]. Sequence alignments of the N-terminal ~40 amino acids of a number of Pex5ps shows the presence of at least one lysine residue (Figure 41.2). Interestingly, a similar analysis of the N-terminal region of the Pex20p family of PTS2 co-receptors also shows the presence of at least one lysine residue (Figure 41.2). In the yeast *P. pastoris*, the poly-ubiquitination of Pex20p depends on lysine 19 ([31] and see below), confirming the role of N-terminal lysines in PTS (co-) receptor ubiquitination.

It appears likely that ubiquitin mediated degradation of Pex18p is required for its function (see above). Is this the same for the poly-ubiquitination of Pex5p, since poly-ubiquitinated Pex5p is only observed in mutants? Opinions are somewhat divided on this point. Platta and coworkers (2004) claimed that deletion of Ubc4p together with its homolog Ubc5p causes a growth defect on oleic acid media, suggesting that their presence is important for peroxisome function. However, two other groups reported the exact opposite result, that deletion of these E_2 enzymes has no effect of peroxisome function [28, 29]. It is also known that *ubc4Δubc5Δ* cells are temperature sensitive and grow quite slowly under most conditions [37], which may account for the observed growth defect on oleic acid. Other evidence against a vital role for Pex5p poly-ubiquitination in PTS import comes from work with Pex5p lysine mutants. In both *H. polymorpha* and *S. cerevisiae*, Pex5p mutants blocked in poly-ubiquitination through mutation of the target lysines can rescue the growth phenotype of a *pex5Δ* strain [33, 35]. Although these results are quite convincing, they do not completely rule out an important role for poly-ubiquitination in Pex5p function. It is conceivable that, due to the small amount of Pex5p that needs to be degraded at any one time, no effect on growth on oleic acid is observed when poly-ubiquitination is blocked. However, this mechanism may become important under stress conditions such as heat shock. Indeed, *pex5Δ* cells expressing Pex5p mutants blocked in poly-ubiquitination, exhibit growth retardation when grown on oleic acid for long periods of time (our unpublished results). The conserved nature of the N-terminal lysine residues support the notion that poly-ubiquitination is important at some stage of cellular life.

Pex5p is Mono-Ubiquitinated by Pex4p on a Cysteine Residue

Interestingly, the poly-ubiquitination of Pex5p is not the only ubiquitin related event associated with this protein. Kragt *et al.* [30] demonstrated that Pex5p is posttranslationally modified in wild-type cells. Immunoprecipitation analysis with cells overexpressing a myc tagged form of ubiquitin revealed the presence of a single, discreet, Pex5p band around 20 kDa heavier than unmodified Pex5p that specifically reacted with myc tag antibodies. A molecular weight increase of around 20 kDa is consistent with the attachment of two ubiquitin moieties (inclusive two myc tags), indicating that Pex5p is mono-ubiquitinated. Expression of a lysine-less form of myc tagged ubiquitin (Ub-K_0),

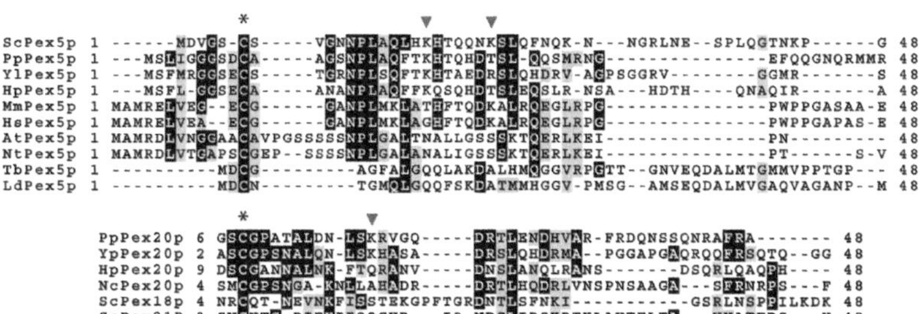

FIGURE 41.2 Sequence alignment showing the N-terminal 48 amino acids of a number of Pex5 (upper panel) and Pex18/20 (lower panel) proteins from different species.
*Indicates the conserved cysteine residue. Arrowheads indicate lysine residues shown to be involved in poly-ubiquitination (for Pex5p, lysines 18 and 24 in *S. cerevisiae* and 21 in *H. polymorpha*, and for Pex20p lysine 19 in *P. pastoris*). Sc, *Saccharomyces cerevisiae*; Pp, *Pichia pastoris*; Yl, *Yarrowia lipolytica*; Hp, *Hansenula polymorpha*; Mm, *Mus musculus*; Hs, *Homo sapiens*; At, *Arabidopsis thaliana*; Nt, *Nicotiana tabacum*; Tb, *Trypanosoma brucei*; Ld, *Leishmania donovani*; Nc, *Neurospora crassa*.

which in a similar way to the Ub-K_{48}R mutant cannot undergo chain elongation, did not reduce the levels of mono-ubiquitinated Pex5p nor result in the formation of a faster migrating band, leading the authors to suggest that Pex5p is mono-ubiquitinated at two different sites [30]. However, these data were obtained with cells where wild-type ubiquitin was still present, which if conjugated to Pex5p first, would be able to take part in chain elongation. In addition, the myc tag itself contains a lysine residue, which may also act as a conjugation site.

Experiments using mutants deficient in proteasomal and vacuolar degradation indicated that mono-ubiquitinated Pex5p is not a breakdown intermediate of either system. In addition, this modification does not require the E_2 enzymes Ubc1p or Ubc4p [30]. At the time, it was suggested that Pex4p is responsible for the mono-ubiquitination of Pex5p but, due to the observed Ubc4p dependent poly-ubiquitination in *pex4Δ* cells, this was difficult to prove [30]. However, the involvement of Pex4p in mono-ubiquitination was confirmed using lysine mutant versions of Pex5p, which can no longer be poly-ubiquitinated by Ubc4p. Such a mutant is still mono-ubiquitinated when introduced into a Pex5p wild-type or *pex6Δ* strain, but the mono-ubiquitination is lost in a *pex4Δ* strain [33]. In addition, mono-ubiquitination of Pex5p was seen in a cell free ubiquitination assay using Pex4p as E_2 [21].

As discussed above, poly-ubiquitination of Pex5p is Ubc4p dependent and targets conserved lysines present in the N-terminus [21, 33, 35]. These two lysine residues, however, are not involved in the mono-ubiquitination of Pex5p [21, 33], nor are other lysine residues present in the N-terminal half of Pex5p [33]. Furthermore, conjugation of ubiquitin to the α-NH2 group of Pex5p was also ruled out, as the α-NH_2 group is blocked by acetylation [33]. The ubiquitination of non-NH_2 groups by viral E_3 ligases, where cysteine, serine, and threonine residues are potential conjugation sites has recently become a hot topic [41, 42]. Sequence alignments of Pex5p show the presence of a well conserved cysteine residue in the N-terminal ~40 amino acids ([33] and Figure 41.2). Such a residue is also present in the N-terminal domain of the Pex20p family of proteins ([31] and Figure 41.2). Mutation of this cysteine residue renders Pex5p non-functional and causes the protein to be poly-ubiquitinated. Furthermore, when the cysteine mutant is combined with the poly-ubiquitination disturbing lysine mutations, Pex5p is no longer mono-ubiquitinated in wild-type cells. Further evidence for the cysteine as conjugation site came from experiments using reducing agents. Cysteine residues form thioester bonds with the C-terminus of ubiquitin, whereas lysine residues form amide (iso-peptide) bonds. Thioester bonds exhibit different chemical properties from amide bonds, one of which is their susceptibility to the reducing agent β-mercaptoethanol (β-me). Mono-ubiquitinated Pex5p is susceptible to β-me, whereas poly-ubiquitinated Pex5p is not [33]. Final proof of the role of the cysteine residue in mono-ubiquitination, such as mass spectrometry, so far is lacking. However, while this manuscript was in preparation, Carvalho *et al.* [43] reported that mammalian Pex5p is modified by ubiquitin in two distinct ways. One of these modifications was dependent on the conserved cysteine residue present in the N-terminal region of the protein, indicating that the cysteine dependent mono-ubiquitination of Pex5p is likely to occur in other organisms. In addition, the cysteine residue of *P. pastoris* Pex20p plays an important role in its recycling, again indicating the crucial nature of this residue in receptor function ([44] and see below).

THE UBIQUITINATION OF *P. PASTORIS* PEX20P: TWO INDEPENDENT UBIQUITINATION EVENTS?

As with the ubiquitination of Pex5p, many of the data concerning Pex20p ubiquitination comes from a single organism, in this case *P. pastoris*. Many similarities can be drawn between the behavior of Pex20p and that already mentioned for Pex5p from the same organism (see above). In the absence of Pex1p, Pex4p, or Pex6p, levels of Pex20p are severely reduced when cells are grown overnight on oleic acid medium [31]. Mutation of the lysine at position 19 in Pex20p to an arginine (K_{19}R) stabilizes the proteins in these deletion strains, analogous to the K_{21}R mutant in *H. polymorpha* Pex5p [35]. If cells lacking Pex1p, Pex4p, or Pex6p are exposed to oleic acid medium for a 6h period, Pex20p levels are not reduced. Instead, the formation of a ladder of higher molecular weight species is observed. Significantly, these higher molecular weight species are not observed in the K_{19}R mutant, suggesting poly-ubiquitination of the protein takes place via lysine 19 [31]. A conserved lysine residue is present in the N-terminal ~40 amino acids of all proteins from the Pex20p family, and, as already discussed, the N-terminal region of Pex5p ([31] and Figure 41.2). The same authors also expressed the Ub-K_{48}R mutant in *pex4Δ* cells and, again like the HpPex5p situation, observed a build up of ubiquitinated Pex20p. Poly-ubiquitinated Pex20p is also predominantly membrane associated, like its Pex5p counterparts. In addition, expression of the Pex20p K_{19}R mutant in one of the above mentioned deletion strains not only blocks degradation, it also results in a build up of the protein on the peroxisomal membrane, confirming the role already suggested for poly-ubiquitination, i.e., the removal of non-functional/unwanted PTS receptors from the peroxisomal membrane.

The story of PpPex20p does not end here. The sequence alignments of the Pex20p family of proteins show that besides the conserved lysine residue, also a conserved cysteine residue at position 8 (Figure 41.2). Although mono-ubiquitination of Pex20p has never been shown, a number of interesting observations concerning this cysteine residue in Pex20p are worth mentioning. Unlike the ScPex5p situation, mutation of this cysteine in Pex20p does not render the protein non-functional, although the protein is degraded over time. However, when this mutation is combined with the $K_{19}R$ mutant, cells can no longer grow on oleic acid medium. Co-expression of Pex20p C_8S with the Ub-$K_{48}R$ mutant, allowing ubiquitin conjugation to a substrate but interfering with ubiquitin chain formation, results in ubiquitinated Pex20p, suggesting that, in the absence of a cysteine residue, Pex20p is poly-ubiquitinated. Mutation of both the Cys_8 and Lys_{19} in Pex20p also results in a non-recycling phenotype [44].

Although direct proof for mono-ubiquitination is still lacking, it seems quite clear that the PTS2 co-receptor Pex20p, much like its PTS1 counterpart Pex5p, can undergo two distinct ubiquitination events, one involved in the recycling of the protein and the other in degradation of non-functional protein. An obvious difference between Pex20p and Pex5p is that mutation of the cysteine in Pex5p results in a growth defect on oleic acid, whereas the same mutation in Pex20p can partially rescue a *pex20Δ* strain. This may stem from the number of cargo proteins that are handled by the PTS2 co-receptors. So far, only one PTS2 protein has been identified in *P. pastoris*, thiolase. The number of PTS1 proteins, on the other hand, is considerable. Therefore, sufficient thiolase may be imported if Pex20p only undergoes one round of import and then, due to a lack of recycling, is degraded, the subsequent import being performed by newly synthesized Pex20p. It is conceivable that, due to the large number of PTS1 proteins, each Pex5p molecule must perform multiple rounds of import and that a block in recycling limits the amount of PTS1 proteins imported. On the other hand, a lack of data concerning the ubiquitination of *P. pastoris* Pex5p makes it difficult to predict if such a cysteine mutation would result a non-functional protein, as in *S. cerevisiae*.

INVOLVEMENT OF THE RING PROTEINS IN PTS (CO-) RECEPTOR UBIQUITINATION

The third and final step of the ubiquitination pathway is the attachment of ubiquitin to the substrate. This task is performed by an E_3 ligase enzyme. Currently, two main groups of E_3 ligases have been identified: the HECT (homologous to E6-AP C-terminus) E_3s and the RING (really interesting new gene) E_3s [45]. RING E_3s contain a zinc binding RING domain that acts as a bridge between the conjugating E_2 enzyme and the substrate, allowing transfer of ubiquitin from the E_2 to the substrate to occur [46, 47]. Three RING domain containing proteins are important in PTS import: Pex2p, Pex10p, and Pex12p [48–50]. However, zinc binding has only been shown for Pex10p and not for the other RING proteins [51]. In addition, Pex2p and Pex12p lack a complete set of cysteine/histidine residues necessary for zinc coordination. Analysis using SMART (http://smart.embl-heidelberg.de) predicts that both Pex2p and Pex12p contain a U-box, instead of a RING domain. U-box domain containing proteins represent a subgroup of the RING E_3 ligase family. Although they lack zinc coordinating residues, their overall fold is very similar to that of the RING domain [52].

The RING proteins are present as a complex at the peroxisomal membrane and RING domain of Pex10p can interact with the RING/U-box domains of both Pex2p and Pex12p [18]. In mammals, both Pex10p and Pex12p RING domains can interact with Pex5p [48, 49, 53]. Epistasis analysis places all three RING proteins downstream of the docking complex, suggesting that Pex5p is handed over from the docking complex to the RING complex [19, 48].

The presence of three potential E_3 ligase enzymes on the peroxisomal membrane raises many questions concerning their individual roles in the ubiquitination of the PTS (co-) receptors. Indeed, all three proteins are required for both the poly- and mono-ubiquitination of Pex5p [28–30]. The RING domain of Pex10p shares considerable homology with that of c-Cbl, a well characterized RING E_3 ligase [46] and can act as an E_3 ligase with UbcH5a, a homolog of *S. cerevisiae* Ubc4p [65]. Interestingly, Pex10p can interact with Pex4p in the split-ubiquitin system [54], which may suggest that Pex10p acts as the E_3 for Pex4p and Ubc4p. This theory does not explain the role of the other RING proteins in mono-ubiquitination, but does explain the need for Pex10p in Pex5p poly-ubiquitination [28, 29]. A number of RING E_3s cannot attach ubiquitin to a substrate alone and act in collaboration with other proteins. These complexes are known as multi-subunit E_3s and include the SCF (skip1-Cul-F-box) and the CBC/VCB (elongin C-elongin B-Cul2/Von Hippel-Lindau-elongin C/B) ligases [55]. A similar mechanism could be envisaged for the RING complex on the peroxisomal membrane, where Pex10p would function as E_3 ligase while the other two RING proteins, Pex2p and Pex12p, would have a role in binding of the substrate.

AAA PROTEIN MEDIATED (CO-) RECEPTOR RECYCLING: AN UBIQUITIN DEPENDENT EVENT?

The recycling of the receptors from the peroxisomal membrane requires ATP as an energy source [56]. So far, only two proteins essential for PTS import with the ability to bind and hydrolyse ATP have been identified: Pex1p and Pex6p. These AAA proteins are capable of forming a high molecular weight complex that can cycle between the cytosol and the peroxisomal membrane [57–59]. Membrane association is achieved through Pex6p's interaction with Pex15p in yeast [60] and Pex26p in mammals [61]. ATP plays a role in the Pex15p–Pex6p interaction, its hydrolysis being required for dissociation of the two proteins [60].

AAA-proteins are involved in a variety of cellular processes, in which they are often employed as protein complex dissociation factors. Indeed, the AAA protein Cdc48p (p97 or VCP in mammals) functions in, amongst other things, ER associated degradation (ERAD), where misfolded proteins are retrotranslocated from the ER into the cytosol for proteasomal degradation. Several reports suggest an important role for ubiquitin in this process, possibly acting as the signal for Cdc48p mediated removal [62–64]. Could Pex1p and Pex6p be involved in a similar process in peroxisomes? A role for the AAA proteins in recycling has been proposed on many occasions. This was largely based on genetic analysis, showing their involvement in a late step in the import process [19]. Recent data, however, gives us a clear indication that, at least for Pex5p, the AAA proteins perform such a function. *In vitro* export assays using membrane fractions have shown that the addition of a purified complex consisting of Pex1p and Pex6p is sufficient to remove Pex5p from the peroxisomal membrane [20]. These observations were expanded upon to include a role for ubiquitin in the recycling process. Results from Platta *et al.* [21] indicate that either mono- or poly-ubiquitination of Pex5p is required to remove the protein from the peroxisomal membrane. In the absence of one of these pathways, the other is capable of taking over. For example, AAA protein dependent Pex5p recycling is observed with the Pex5p $K_{18/24}R$ mutant and in a *pex4Δ* strain. Only when both modifications are blocked, by combining the Pex5p $K_{18/24}R$ mutant with the *pex4Δ* strain, does Pex5p fail to recycle [21]. Also, the addition of recombinant Pex4p, but not the active site cysteine mutant Pex4p $C_{115}S$, to the Pex5p $K_{18/24}R$/*pex4Δ* system can stimulate Pex5p recycling. Therefore, it seems likely that the AAA proteins play a major role in the ubiquitin dependent recycling of Pex5p. The mechanistic details of such a process, however, are not yet known. An interaction between ubiquitinated Pex5p and one or both of the AAA proteins may be expected. Further developments in this field are eagerly awaited.

CONCLUSIONS

It is clear that two independent ubiquitination events can occur in the Pex5p cycle. One, the mono-ubiquitination of Pex5p targets a cysteine residue, is dependent on the E_2 enzyme Pex4p and is likely to regulate the function of Pex5p, possibly the recycling of the protein to the cytosol. On the other hand, the poly-ubiquitination of Pex5p on lysine residues by the E_2 enzyme Ubc4p is implicated in quality control, resulting in degradation of non-functional protein stuck at the peroxisomal membrane. Although only one such event, ubiquitin mediated degradation, has been shown for the PTS2 co-receptors, the evidence heavily suggests the presence of the other. It is noteworthy that these two distinct events are both regulated by the same molecule, ubiquitin, and that such diversity can be achieved through the conjugation site as well as the action of different enzymes. Based on the results presented, we can draw up a hypothetical model concerning the role of ubiquitin in receptor function (Figure 41.3). After completion of the docking and PTS translocation steps, the membrane associated PTS (co-) receptor is mono-ubiquitinated by Pex4p, allowing recognition by the AAA proteins, Pex1p and Pex6p. The PTS (co-) receptor is then pulled out of the membrane and the ubiquitin is removed, possibly by the action of a de-ubiquitinating enzyme. The PTS (co-) receptor is then free to partake in another round of PTS protein import. In the situation that no efficient recycling is possible, due to the absence of one of the peroxins involved in recycling or in certain PTS (co-) receptor mutants, PTS (co-) receptor poly-ubiquitination, mediated by Ubc4p, is observed. This modified form is then removed from the membrane in an, as yet unknown, way and destroyed by the 26S proteasome, effectively removing the blockage.

Questions that still need answering concerning the role ubiquitin plays in PTS protein import include the identification of the true E_3 enzyme for poly- and mono-ubiquitination. The possibility that they are one and the same is very real and that the regulation comes from other factors, the E_2 enzyme for example. Alternatively, one RING protein

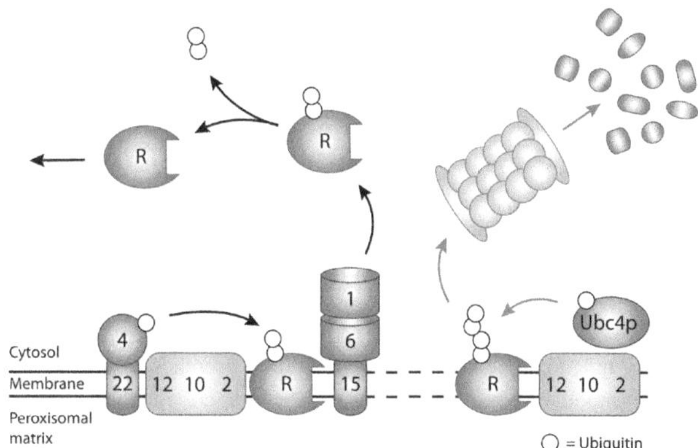

FIGURE 41.3 Hypothetical model for PTS (co-) receptor ubiquitination showing ubiquitin dependent recycling and degradation. Once the PTS (co-) receptor has released its cargo into the peroxisomal matrix, it is recycled to the cytosol. The E_2 enzyme Pex4p, with the aid of Pex2p, Pex10p, and Pex12p acting as RING E_3 ligases, mono-ubiquitinates the PTS (co-) receptor on the conserved cysteine residue. The modified PTS (co-) receptor then becomes a substrate for the AAA–protein complex consisting of Pex1p and Pex6p and is pulled out of the peroxisomal membrane, a step that requires ATP hydrolysis. Ubiquitin is then removed from the PTS (co-) receptor, allowing it to partake in another round of import. In the absence of functional recycling machinery, the PTS (co-) receptor is poly-ubiquitinated by the E_2 enzyme Ubc4p, with Pex10p possibly acting as an E_3 ligase. This modification targets the PTS (co-) receptor for degradation by the 26S proteasome. R represents the cycling (co-) receptors and numbers indicate specific peroxins. See text for details.

may be the E_3 for Pex4p and another for Ubc4p. Analysis of the different E_2 and E_3 enzymes in *in vitro* ubiquitination assays should resolve this issue. The confirmation that mono-ubiquitination on a cysteine residue also occurs in mammalian Pex5p adds further weight to the results obtained with *S. cerevisiae* Pex5p but we are still a long way from fully understanding the important role ubiquitin plays in the import of peroxisomal matrix proteins.

REFERENCES

1. Leon S, Goodman JM, Subramani S. Uniqueness of the mechanism of protein import into the peroxisome matrix: transport of folded, co-factor-bound and oligomeric proteins by shuttling receptors. *Biochim Biophys Acta* 2006;**1763**:1552–64.
2. Gould SJ, Keller GA, Hosken N, Wilkinson J, Subramani S. A conserved tripeptide sorts proteins to peroxisomes. *J Cell Biol* 1989;**108**:1657–64.
3. Gietl C, Faber KN, van der Klei IJ, Veenhuis M. Mutational analysis of the N-terminal topogenic signal of watermelon glyoxysomal malate dehydrogenase using the heterologous host *Hansenula polymorpha*. *Proc Natl Acad Sci USA* 1994;**91**:3151–5.
4. Schafer A, Kerssen D, Veenhuis M, Kunau WH, Schliebs W. Functional similarity between the peroxisomal PTS2 receptor binding protein Pex18p and the N-terminal half of the PTS1 receptor Pex5p. *Mol Cell Biol* 2004;**24**:8895–906.
5. Costa-Rodrigues J, Carvalho AF, Gouveia AM, Fransen M, Sa-Miranda C, Azevedo JE. The N terminus of the peroxisomal cycling receptor, Pex5p, is required for redirecting the peroxisome-associated peroxin back to the cytosol. *J Biol Chem* 2004;**279**:46,573–46,579.
6. Van der Leij I, Franse MM, Elgersma Y, Distel B, Tabak HF. PAS10 is a tetratricopeptide-repeat protein that is essential for the import of most matrix proteins into peroxisomes of *Saccharomyces cerevisiae*. *Proc Natl Acad Sci USA* 1993;**90**:11,782–11,786.
7. Klein AT, Barnett P, Bottger G, Konings D, Tabak HF, Distel B. Recognition of peroxisomal targeting signal type 1 by the import receptor Pex5p. *J Biol Chem* 2001;**276**:15,034–15,041.
8. Marzioch M, Erdmann R, Veenhuis M, Kunau WH. PAS7 encodes a novel yeast member of the WD-40 protein family essential for import of 3-oxoacyl-CoA thiolase, a PTS2-containing protein, into peroxisomes. *EMBO J* 1994;**13**:4908–18.
9. Rehling P, Marzioch M, Niesen F, Wittke E, Veenhuis M, Kunau WH. The import receptor for the peroxisomal targeting signal 2 (PTS2) in *Saccharomyces cerevisiae* is encoded by the PAS7 gene. *EMBO J* 1996;**15**:2901–13.
10. Purdue PE, Yang X, Lazarow PB. Pex18p and Pex21p, a novel pair of related peroxins essential for peroxisomal targeting by the PTS2 pathway. *J Cell Biol* 1998;**143**:1859–69.
11. Einwächter H, Sowinski S, Kunau WH, Schliebs W. *Yarrowia lipolytica* Pex20p, Saccharomyces cerevisiae Pex18p/Pex21p and mammalian Pex5pL fulfil a common function in the early steps of the peroxisomal PTS2 import pathway. *EMBO Rep* 2001;**2**:1035–9.
12. Purdue PE, Lazarow PB. Peroxisome biogenesis. *Annu Rev Cell Dev Biol* 2001;**17**:701–52.
13. Elgersma Y, Kwast L, Klein A, Voorn-Brouwer T, van den Berg M, Metzig B, America T, Tabak HF, Distel B. The SH3 domain of the *Saccharomyces cerevisiae* peroxisomal membrane protein Pex13p functions as a docking site for Pex5p, a mobile receptor for the import PTS1-containing proteins. *J Cell Biol* 1996;**135**:97–109.
14. Gould SJ, Kalish JE, Morrell JC, Bjorkman J, Urquhart AJ, Crane DI. Pex13p is an SH3 protein of the peroxisome membrane and a docking factor for the predominantly cytoplasmic PTS1 receptor. *J Cell Biol* 1996;**135**:85–95.
15. Erdmann R, Blobel G. Identification of Pex13p a peroxisomal membrane receptor for the PTS1 recognition factor. *J Cell Biol* 1996;**135**:111–21.
16. Albertini M, Rehling P, Erdmann R, Girzalsky W, Kiel JA, Veenhuis M, Kunau WH. Pex14p, a peroxisomal membrane protein binding both receptors of the two PTS-dependent import pathways. *Cell* 1997;**89**:83–92.
17. Huhse B, Rehling P, Albertini M, Blank L, Meller K, Kunau WH. Pex17p of *Saccharomyces cerevisiae* is a novel peroxin and component of the peroxisomal protein translocation machinery. *J Cell Biol* 1998;**140**:49–60.
18. Agne B, Meindl NM, Niederhoff K, Einwachter H, Rehling P, Sickmann A, Meyer HE, Girzalsky W, Kunau WH. Pex8p: an intraperoxisomal organizer of the peroxisomal import machinery. *Mol Cell* 2003;**11**:635–46.
19. Collins CS, Kalish JE, Morrell JC, McCaffery JM, Gould SJ. The peroxisome biogenesis factors pex4p, pex22p, pex1p, and pex6p act in the terminal steps of peroxisomal matrix protein import. *Mol Cell Biol* 2000;**20**:7516–26.
20. Platta HW, Grunau S, Rosenkranz K, Girzalsky W, Erdmann R. Functional role of the AAA peroxins in dislocation of the cycling PTS1 receptor back to the cytosol. *Nat Cell Biol* 2005;**7**:817–22.
21. Platta HW, El Magraoui F, Schlee D, Grunau S, Girzalsky W, Erdmann R. Ubiquitination of the peroxisomal import receptor Pex5p is required for its recycling. *J Cell Biol* 2007;**177**:197–204.
22. Imanaka T, Small GM, Lazarow PB. Translocation of acyl-CoA oxidase into peroxisomes requires ATP hydrolysis but not a membrane potential. *J Cell Biol* 1987;**105**:2915–22.
23. Mukhopadhyay D, Riezman H. Proteasome-independent functions of ubiquitin in endocytosis and signaling. *Science* 2007;**315**:201–5.
24. Thrower JS, Hoffman L, Rechsteiner M, Pickart CM. Recognition of the polyubiquitin proteolytic signal. *EMBO J* 2000;**19**:94–102.
25. Hicke L. Protein regulation by monoubiquitin. *Nat Rev Mol Cell Biol* 2001;**2**:195–201.
26. Wiebel FF, Kunau WH. The Pas2 protein essential for peroxisome biogenesis is related to ubiquitin-conjugating enzymes. *Nature* 1992;**359**:73–6.
27. Purdue PE, Lazarow PB. Pex18p is constitutively degraded during peroxisome biogenesis. *J Biol Chem* 2001;**276**:47,684–47,689.
28. Platta HW, Girzalsky W, Erdmann R. Ubiquitination of the peroxisomal import receptor Pex5p. *Biochem J* 2004;**384**:37–45.
29. Kiel JA, Emmrich K, Meyer HE, Kunau WH. Ubiquitination of the peroxisomal targeting signal type 1 receptor, Pex5p, suggests the presence of a quality control mechanism during peroxisomal matrix protein import. *J Biol Chem* 2005;**280**:1921–30.
30. Kragt A, Voorn-Brouwer T, van den Berg M, Distel B. The Saccharomyces cerevisiae peroxisomal import receptor Pex5p is monoubiquitinated in wild type cells. *J Biol Chem* 2005;**280**:7867–74.
31. Leon S, Zhang L, McDonald WH, Yates J, Cregg JM, Subramani S Dynamics of the peroxisomal import cycle of PpPex20p: ubiquitin-dependent localization and regulation. *J Cell Biol* 2006;**172**:67–78.
32. Lazarow PB. Peroxisome biogenesis: advances and conundrums. *Curr Opin Cell Biol* 2003;**15**:489–97.
33. Williams C, van den Berg M, Sprenger RR, Distel B. A conserved cysteine is essential for Pex4p-dependent ubiquitination of the peroxisomal import receptor Pex5p. *J Biol Chem* 2007;**282**:22,534–22,543.

34. Deffenbaugh AE, Scaglione KM, Zhang L, Moore JM, Buranda T, Sklar LA, Skowyra D. Release of ubiquitin-charged Cdc34-S-Ub from the RING domain is essential for ubiquitination of the SCF(Cdc4)-bound substrate Sic1. *Cell* 2003;**114**:611–22.
35. Kiel JA, Otzen M, Veenhuis M, van der Klei IJ. Obstruction of polyubiquitination affects PTS1 peroxisomal matrix protein import. *Biochim Biophys Acta* 2005;**1745**:176–86.
36. Kragt A, Benne R, Distel B. Ubiquitin: a new player in the peroxisome field. In: Mayer RJ, Ciechanover AJ, Rechsteiner M, editors. *Protein degradation, Vol 3, cell biology of the ubiquitin-proteasome system*. Weinheim, Germany: WILEY-VCH Verlag GmbH; 2006. p. 1–20.
37. Seufert W, Jentsch S. Ubiquitin-conjugating enzymes UBC4 and UBC5 mediate selective degradation of short-lived and abnormal proteins. *EMBO J* 1990;**9**:543–50.
38. Dodt G, Gould SJ. Multiple PEX genes are required for proper subcellular distribution and stability of Pex5p, the PTS1 receptor: evidence that PTS1 protein import is mediated by a cycling receptor. *J Cell Biol* 1996;**135**:1763–74.
39. Hershko A, Heller H, Eytan E, Kaklij G, Rose IA. Role of the alpha-amino group of protein in ubiquitin-mediated protein breakdown. *Proc Natl Acad Sci USA* 1984;**81**:7021–5.
40. Ciechanover A, Ben-Saadon R. N-terminal ubiquitination: more protein substrates join in. *Trends Cell Biol* 2004;**14**:103–6.
41. Cadwell K, Coscoy L. Ubiquitination on nonlysine residues by a viral E3 ubiquitin ligase. *Science* 2005;**309**:127–30.
42. Wang X, Herr RA, Chua WJ, Lybarger L, Wiertz EJ, Hansen TH. Ubiquitination of serine, threonine, or lysine residues on the cytoplasmic tail can induce ERAD of MHC-I by viral E3 ligase mK3. *J Cell Biol* 2007;**177**:613–24.
43. Carvalho AF, Pinto MP, Grou CP, Alencastre IS, Fransen M, Sá-Miranda C, Azevedo JE. Ubiquitination of mammalian Pex5p, the peroxisomal import receptor. *J Biol Chem* 2007;**282**:31,267–31,272.
44. Leon S, Subramani S. A conserved cysteine residue of *Pichia pastoris* Pex20p is essential for its recycling from the peroxisome to the cytosol. *J Biol Chem* 2007;**282**:7424–30.
45. Pickart CM. Mechanisms underlying ubiquitination. *Annu Rev Biochem* 2001;**70**:503–33.
46. Joazeiro CA, Wing SS, Huang H, Leverson JD, Hunter T, Liu YC. The tyrosine kinase negative regulator c-Cbl as a RING-type, E2-dependent ubiquitin-protein ligase. *Science* 1999;**286**:309–12.
47. Jackson PK, Eldridge AG, Freed E, Furstenthal L, Hsu JY, Kaiser BK, Reimann JD. The lore of the RINGs: substrate recognition and catalysis by ubiquitin ligases. *Trends Cell Biol* 2000;**10**:429–39.
48. Chang CC, Warren DS, Sacksteder KA, Gould SJ. PEX12 interacts with PEX5 and PEX10 and acts downstream of receptor docking in peroxisomal matrix protein import. *J Cell Biol* 1999;**147**:761–74.
49. Albertini M, Girzalsky W, Veenhuis M, Kunau WH. Pex12p of Saccharomyces cerevisiae is a component of a multi-protein complex essential for peroxisomal matrix protein import. *Eur J Cell Biol* 2001;**80**:257–70.
50. Fujiki Y, Okumoto K, Otera H, Tamura S. Peroxisome biogenesis and molecular defects in peroxisome assembly disorders. *Cell Biochem Biophys* 2000;**32**:155–64.
51. Kalish JE, Theda C, Morrell JC, Berg JM, Gould SJ. Formation of the peroxisome lumen is abolished by loss of *Pichia pastoris* Pas7p, a zinc-binding integral membrane protein of the peroxisome. *Mol Cell Biol* 1995;**15**:6406–19.
52. Hatakeyama S, Nakayama KI. U-box proteins as a new family of ubiquitin ligases. *Biochem Biophys Res Commun* 2003;**302**:635–45.
53. Okumoto K, Abe I, Fujiki Y. Molecular anatomy of the peroxin Pex12p: RING finger domain is essential for Pex12p function and interacts with the peroxisome-targeting signal type 1-receptor Pex5p and a ring peroxin, Pex10p. *J Biol Chem* 2000;**275**:25,700–25,710.
54. Eckert JH, Johnsson N. Pex10p links the ubiquitin conjugating enzyme Pex4p to the protein import machinery of the peroxisome. *J Cell Sci* 2003;**116**:3623–34.
55. Fang S, Weissman AM. A field guide to ubiquitylation. *Cell Mol Life Sci* 2004;**61**:1546–61.
56. Gouveia AM, Guimaraes CP, Oliveira ME, Reguenga C, Sa-Miranda C, Azevedo JE. Characterization of the peroxisomal cycling receptor, Pex5p, using a cell-free *in vitro* import system. *J Biol Chem* 2003;**278**:226–32.
57. Birschmann I, Rosenkranz K, Erdmann R, Kunau WH. Structural and functional analysis of the interaction of the AAA-peroxins Pex1p and Pex6p. *FEBS J* 2005;**272**:47–58.
58. Kiel JA, Hilbrands RE, van der Klei IJ, Rasmussen SW, Salomons FA, van der Heide M, Faber KN, Cregg JM, Veenhuis M. *Hansenula polymorpha* Pex1p and Pex6p are peroxisome-associated AAA proteins that functionally and physically interact. *Yeast* 1999;**15**:1059–78.
59. Faber KN, Heyman JA, Subramani S. Two AAA family peroxins, PpPex1p and PpPex6p, interact with each other in an ATP-dependent manner and are associated with different subcellular membranous structures distinct from peroxisomes. *Mol Cell Biol* 1998;**18**:936–43.
60. Birschmann I, Stroobants AK, van den Berg M, Schafer A, Rosenkranz K, Kunau WH, Tabak HF. Pex15p of Saccharomyces cerevisiae provides a molecular basis for recruitment of the AAA peroxin Pex6p to peroxisomal membranes. *Mol Biol Cell* 2003;**14**:2226–36.
61. Matsumoto N, Tamura S, Fujiki Y. The pathogenic peroxin Pex26p recruits the Pex1p-Pex6p AAA ATPase complexes to peroxisomes. *Nat Cell Biol* 2003;**5**:454–60.
62. Bays NW, Hampton RY. Cdc48-Ufd1-Npl4: stuck in the middle with Ub. *Curr Biol* 2002;**12**:366–71.
63. Tsai B, Ye Y, Rapoport TA. Retro-translocation of proteins from the endoplasmic reticulum into the cytosol. *Nat Rev Mol Cell Biol* 2002;**3**:246–55.
64. Richly H, Rape M, Braun S, Rumpf S, Hoege C, Jentsch S. A series of ubiquitin binding factors connects CDC48/p97 to substrate multiubiquitylation and proteasomal targeting. *Cell* 2005;**120**:73–84.
65. Williams C, van den Berg M, Geers E, Distel B. Pex10p functions as an E3 ligase for the Ubc4p dependent ubiquitination of Pex5p. *Biochem Biophys Res Commun* 2008;**374**:620–24.

Chapter 42

Mitochondrial Dynamics: Fusion and Division

Yasushi Tamura, Miho Iijima and Hiromi Sesaki
Department of Cell Biology, Johns Hopkins University School of Medicine, Baltimore, Maryland

INTRODUCTION

Mitochondria are highly dynamic organelles that continuously fuse and divide in highly regulated manners. These activities control number, distribution, and morphology of mitochondria in the cell, and therefore play important roles for diverse mitochondrial functions such as energy production, metabolism, intracellular signaling, and apoptosis [1–6]. Several neurodegenerative diseases are found to be defective in mitochondrial dynamics, illustrating the importance of these processes for human health [7–9]. Recent studies have identified key components required for mitochondrial fusion and division, including three dynamin related GTPases: mitofusin (Fzo1p in yeast) and OPA1 (Mgm1p in yeast) for fusion, and Drp1 (Dnm1p in yeast) for division [10, 11]. In contrast to classical dynamins, which promote membrane scission during endocytosis, Fzo1p/mitofusin and Mgm1p/OPA1 mediate the opposite reaction by facilitating membrane fusion (Figure 42.1). In this chapter, we describe the roles that these GTPases have in mitochondrial fusion and division.

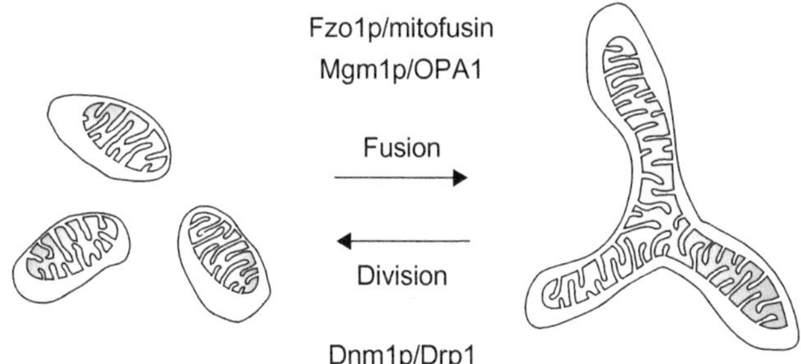

FIGURE 42.1 Mitochondrial fusion and division regulate mitochondrial number, morphology, and distribution.
Many small mitochondria would be produced by excessive mitochondrial division or reduced fusion. On the other hand, elongated, branched tubules are generated by stimulated mitochondrial fusion or compromised division. Mitochondrial fusion and division are mainly controlled by three dynamin related GTPases, Fzo1p/mitofusin, Mgm1p/OPA1, and Dnm1p/Drp1.

MITOCHONDRIAL FUSION

Two evolutionally conserved dynamin related GTPases, Fzo1p/mitofusin and Mgm1p/OPA1, are components of the mitochondrial fusion machinery, which differ from the components involved in secretory pathways such as NSF, SNAP, and SNARE [12–19]. Mitochondria consist of an inner and an outer membrane, and these two membranes carry their own fusion machineries, which are coupled to each other. Fzo1p/mitofusin spans the outer membrane twice with the N-terminal GTPase domain and C-terminus facing the cytosol [17], while Mgm1p/OPA is mainly associated with the inner membrane in the inter membrane space [12–15]. Cells lacking either Fzo1p/mitofusin or Mgm1p/OPA1 fail to fuse the mitochondrial outer membrane and the inner membrane, and display fragmented mitochondria. It is likely that the outer membrane fusion and inner membrane fusion are mechanically coupled to enable rapid mixing of inner membrane proteins and mitochondrial DNA immediately after outer membrane fusion. Consistent with this idea, *in vitro* reconstitution studies using isolated mitochondria have shown that Fzo1p is required for both inner and outer membrane fusion [20]. However, inner and outer membrane fusion can be biochemically separated during *in vitro* reactions and show differences in their dependence on GTP and a requirement for membrane potential [20, 21].

Fzo1p/mitofusins mediate tethering and fusion of the outer membrane via homotypic *trans* interactions on opposing

Handbook of Cell Signaling, Three-Volume Set 2 ed.

mitochondria [22]. In this homotypic interaction, the C-terminal coiled-coil domains of Fzo1p/mitofusin associate in an anti-parallel manner. Since hydrolysis of GTP, but not ATP, is required for mitochondrial fusion, GTP is the primary energy source to merge two lipid bi-layers [20]. This is in contrast to membrane fusion in exocytosis and endocytosis, in which both ATP and GTP are required. In mammals, two mitofusins, mitofusin 1 and 2, assemble into homo- and hetero-oligomers to regulate mitochondrial fusion activities [23, 24]. Furthermore, proapoptotic proteins Bax and Bak regulate assembly of mitofusin in non-apoptotic cells and control mitochondrial fusion [25].

Mgm1p/OPA1 also interacts with itself in trans in opposing inner membranes [21]. This mediates tethering of the inner membranes and merging of their lipid bi-layers. The GTPase and GTPase effector domains are thought to participate in this interaction. Interestingly, Mgm1/OPA1 are proteolytically cleaved in their transmembrane domain generating integral and peripheral membrane forms [26–32]. These two forms are required for full function of Mgm1p/OPA1 in the fusion and assembly into hetero-oligomers. Moreover, in mammals, mRNA of OPA1 undergoes alternative splicing and eight isoforms are produced prior to proteolytic processing [33]. These isoforms may contribute to cell-type specific regulation of OPA1. In addition to fusion, Mgm1p/OPA1 also regulates inner membrane cristae structures and is responsible for localizing cytochrome C inside cristae. Cells with defects in OPA1 spontaneously release cytochrome C from their mitochondria and become sensitive to different apoptotic stimuli [12, 13, 34].

Outer membrane fusion and inner membrane fusion might be coupled at contact sites where the outer membrane and inner membrane are structurally connected. Supporting this idea, Fzo1p, located in the outer membrane, is associated with contact sites via its inter membrane space loop and regulates the inner membrane [35]. A possible molecular link that couples the outer and inner membranes is Ugo1p, an outer membrane protein essential for mitochondrial fusion in yeast [36]. Ugo1p directly binds Fzo1p and Mgm1p and forms protein complexes containing both Fzo1p and Mgm1p [15, 37, 38]. Finding a mammalian Ugo1p would further our understanding of coordination between outer and inner membranes.

Membrane lipids are also important for mitochondrial fusion. Recent studies have shown that efficient mitochondrial fusion requires mitoPLD, a member of the phospholipase D family, which is located in the outer membrane and converts a mitochondria enriched lipid, cardiolipin, to phosphatidic acid [39]. Cardiolipin is located in the inner membrane. Thus, production of phosphatidic acid may occur and stimulate mitochondrial fusion at the contact site where cardiolipin could be transferred from the inner membrane to the outer membrane. Phosphatidic acid is also involved in exocytosis, which indicates that mechanisms underlying membrane fusion mediated by SNAREs and dynamin related GTPases may use fundamentally similar principles [40].

The physiological importance of mitochondrial fusion has been shown in many organisms. For example, mutations in Drosophila Fzo, a founding member of Fzo1p/mitofusin, cause male sterility due to mitochondrial fusion defects during spermatogenesis [19]. In *fzo* mutants, a number of small mitochondria fail to fuse into two giant structures called Nebenkern. This finding suggests that normal mitochondrial structure is critical for production of ATP, which is required for sperm to swim. In yeast, fusion defective mutants lacking Fzo1p, Mgm1p, or Ugo1p fragment otherwise tubular mitochondria and lose mitochondrial DNA leading to respiratory deficiency [4, 41]. Tubular mitochondrial shape might be required for efficient inheritance of mitochondrial DNA or its replication. Furthermore, mice carrying deletions of either mitofusin or OPA1 result in embryonic lethality, and Purkinje cell specific disruption of mitofusin 2 results in neurodegeneration [23, 42–44]. Neurodegeneration is thought to result from defects in mitochondrial distribution and in oxidative phosphorylation due to increased numbers of mitochondria lacking mitochondrial DNA [42]. With respect to the impact that these GTPases have on human health and disease, mutations of mitofusion 2 were found in Charcot-Marie-Tooth disease type 2A, which is manifested by degeneration of peripheral motor and sensory axons [45]. Similarly, OPA1 is mutated in autosomal dominant optic atrophy, and its polymorphisms are associated with glaucoma, both of which are the leading causes of blindness [46, 47].

MITOCHONDRIAL DIVISION

A key component for mitochondrial division is also a dynamin related GTPase, Dnm1p/Drp. Dnm1p/Drp1 is peripherally associated with the mitochondrial surface and shuttles between the cytosol and mitochondria [48–57]. The GTPase domain of Dnm1p regulates its assembly on to mitochondria and is essential for mitochondrial division. Like classical dynamin, purified Dnm1p/Drp1 has the ability to assemble into filaments, which can form ring-like structures on liposomes [58, 59]. The biochemical activity of Dnm1p/Drp1 suggests that they function in the deformation of membranes and possibly pinch off mitochondria as a mechano-chemical enzyme using GTP. Association of Dnm1p/Drp1 with mitochondria is highly regulated and is a crucial step in mitochondrial division. This association is mediated by several components including Fis1p, Mdv1p, Caf4p, and Num1p in yeast, and Fis1 in mammals [60–64]. In yeast, Fis1p is a transmembrane protein located in the outer membrane that binds two proteins containing WD40 repeats, Mdv1p and Caf4p. The Fis1p–Mdv1p–Caf4p complex facilitates the interaction of Dnm1p with mitochondria and stimulates mitochondrial division. In addition, Num1p functions as another anchor for Dnm1p on mitochondria independent of Fis1p [65]. Num1p was originally identified as a cortical anchor for dynein motors and is required for nuclear migration during mitosis [66, 67]. In addition to this well established role, Num1p also has an independent role in mitochondrial division. Cells lacking both Fis1p and Num1p show an increased amount of Dnm1p dissociated from mitochondria compared with cells lacking only Fis1p or Num1p [65]. In mammals, FIS1 is also suggested to recruit Drp1 to mitochondria [61]. Similar to yeast, there are probably additional mechanisms that anchor Drp1 to mitochondria, as RNAi knockdown of Fis1 does not completely remove Drp1 from the mitochondrial surface. Interestingly, Drp1 and Fis1 are also involved in peroxisomal division [68–70]. In contrast to mitochondria, peroxisomes do not fuse, but grow and divide to increase their volume and number in mammalian cells. Shared molecular machineries suggest a co-evolution of division machinery for these metabolism related organelles.

Regulation of mitochondrial division is also achieved by the posttranslational modification of Drp1. Ubiquitination of Drp1 by the outer membrane located E3 ubiquitin ligase MARCH-V stabilizes the association of Drp1 with mitochondria and promotes organelle division [71–73]. Cell cycle dependent phosphorylation of a serine residue in the GTPase effecter domain stimulates mitochondrial division and facilitates faithful segregation of mitochondria during mitosis, while PKA mediated phosphorylation of another serine residue in the same domain inhibits the GTPase activity [74–76]. Additionally, sumoylation of Drp1 stimulates mitochondrial division [50, 77, 78]. Integration of these regulations allows mitochondrial dynamics to be remarkably responsive to different intracellular and extracellular signals.

There may be separate mechanisms for outer membrane division and inner membrane division. Consistent with this idea, RNAi knockdown of DRP-1 in *C. elegans* blocks outer membrane division, but not inner membrane division [57]. Furthermore, an inner membrane protein, Mdm33p, in yeast is proposed to control division of the inner membrane. Understanding mechanisms and regulation of inner membrane division will be great interest for future studies [79].

Drp1 is also important for apoptosis as a proapoptotic factor [80]. Upon apoptotic stimulation, Drp1 is sumolylated in a Bax- and Bak dependent manner, becomes more tightly anchored to the mitochondrial surface, and fragment mitochondria. Inhibition of mitochondrial division slows down apoptosis [77, 81]. Mitochondrial fragmentation is suggested to stimulate the release of cytochrome C from mitochondria possibly by remodeling the inner membrane cristae structure [82, 83]. As described above, OPA1 acts as an antiapoptotic factor by regulating cristae structure. A balance between activities of Drp1 and OPA1 would be a key mechanism for regulation of apoptosis.

CONCLUSIONS

Mitochondria exhibit antagonistic activities of membrane fusion and division, and a balance between these two activities is a key mechanism in the regulation of mitochondrial structure and function. Several components involved in these dynamic processes have been identified. A complete understanding of mechanisms underlying mitochondrial dynamics as well as their physiological roles in different tissues where unique mitochondrial morphologies are taken awaits further studies.

ACKNOWLEDGEMENTS

This work was supported by American Heart Association to HS and by Uehara Memorial Foundation to YT.

REFERENCES

1. Osteryoung KW, Nunnari J. The division of endosymbiotic organelles. *Science* 2003;**302**:1698–704.
2. Cerveny KL, et al. Regulation of mitochondrial fusion and division. *Trends Cell Biol* 2007;**17**:563–9.
3. Karbowski M, Youle RJ. Dynamics of mitochondrial morphology in healthy cells and during apoptosis. *Cell Death Differ* 2003;**10**:870–80.
4. Okamoto K, Shaw JM. Mitochondrial morphology and dynamics in yeast and multicellular eukaryotes. *Annu Rev Genet* 2005;**39**:503–36.
5. Rube DA, van der Bliek AM. Mitochondrial morphology is dynamic and varied. *Mol Cell Biochem* 2004;**256–257**:331–9.
6. Scorrano L. Multiple functions of mitochondria-shaping proteins. *Novartis Found Symp* 2007;**287**:47–55.
7. Chan DC. Mitochondria: dynamic organelles in disease, aging, and development. *Cell* 2006;**125**:1241–52.
8. Santel A. Get the balance right: mitofusins roles in health and disease. *Biochim Biophys Acta* 2006;**1763**:490–9.
9. Olichon A, et al. Mitochondrial dynamics and disease, OPA1. *Biochim Biophys Acta* 2006;**1763**:500–9.
10. Hoppins S, et al. The machines that divide and fuse mitochondria. *Annu Rev Biochem* 2007;**76**:751–80.
11. McBride HM, et al. Mitochondria: more than just a powerhouse. *Curr Biol* 2006;**16**:R551–60.
12. Griparic L, et al. Loss of the intermembrane space protein Mgm1/OPA1 induces swelling and localized constrictions along the lengths of mitochondria. *J Biol Chem* 2004;**279**:18,792–18,798.
13. Olichon A, et al. Loss of OPA1 perturbates the mitochondrial inner membrane structure and integrity, leading to cytochrome c release and apoptosis. *J Biol Chem* 2003;**278**:7743–6.
14. Wong ED, et al. The dynamin-related GTPase, Mgm1p, is an intermembrane space protein required for maintenance of fusion competent mitochondria. *J Cell Biol* 2000;**151**:341–52.
15. Sesaki H, et al. Mgm1p, a dynamin-related GTPase, is essential for fusion of the mitochondrial outer membrane. *Mol Biol Cell* 2003;**14**:2342–56.
16. Santel A, Fuller MT. Control of mitochondrial morphology by a human mitofusin. *J Cell Sci* 2001;**114**:867–74.
17. Hermann GJ, et al. Mitochondrial fusion in yeast requires the transmembrane GTPase Fzo1p. *J Cell Biol* 1998;**143**:359–73.

18. Rapaport D, et al. Fzo1p is a mitochondrial outer membrane protein essential for the biogenesis of functional mitochondria in saccharomyces cerevisiae. *J Biol Chem* 1998;**273**:20,150–20,455.
19. Hales KG, Fuller MT. Developmentally regulated mitochondrial fusion mediated by a conserved, novel, predicted GTPase. *Cell* 1997;**90**:121–9.
20. Meeusen S, et al. Mitochondrial fusion intermediates revealed in vitro. *Science* 2004;**305**:1747–52.
21. Meeusen S, et al. Mitochondrial inner-membrane fusion and crista maintenance requires the dynamin-related GTPase Mgm1. *Cell* 2006;**127**:383–95.
22. Koshiba T, et al. Structural basis of mitochondrial tethering by mitofusin complexes. *Science* 2004;**305**:858–62.
23. Chen H, et al. Mitofusins Mfn1 and Mfn2 coordinately regulate mitochondrial fusion and are essential for embryonic development. *J Cell Biol* 2003;**160**:189–200.
24. Detmer SA, Chan DC. Complementation between mouse Mfn1 and Mfn2 protects mitochondrial fusion defects caused by CMT2A disease mutations. *J Cell Biol* 2007;**176**:405–14.
25. Karbowski M, et al. Role of bax and bak in mitochondrial morphogenesis. *Nature* 2006;**443**:658–62.
26. Sesaki H, et al. Cells lacking Pcp1p/Ugo2p, a rhomboid-like protease required for Mgm1p processing, lose mtDNA and mitochondrial structure in a Dnm1p-dependent manner, but remain competent for mitochondrial fusion. *Biochem Biophys Res Commun* 2003;**308**:276–83.
27. McQuibban GA, et al. Mitochondrial membrane remodelling regulated by a conserved rhomboid protease. *Nature* 2003;**423**:537–41.
28. Herlan M, et al. Processing of Mgm1 by the rhomboid-type protease Pcp1 is required for maintenance of mitochondrial morphology and of mitochondrial DNA. *J Biol Chem* 2003;**278**:27,781–27,788.
29. Cipolat S, et al. OPA1 requires mitofusin 1 to promote mitochondrial fusion. *Proc Natl Acad Sci USA* 2004;**101**:15,927–15,932.
30. Ishihara N, et al. Regulation of mitochondrial morphology through proteolytic cleavage of OPA1. *Embo J* 2006;**25**:2966–77.
31. Griparic L, et al. Regulation of the mitochondrial dynamin-like protein Opa1 by proteolytic cleavage. *J Cell Biol* 2007;**178**:757–64.
32. Song Z, et al. OPA1 processing controls mitochondrial fusion and is regulated by mRNA splicing, membrane potential, and Yme1L. *J Cell Biol* 2007;**178**:749–55.
33. Delettre C, et al. Mutation spectrum and splicing variants in the OPA1 gene. *Hum Genet* 2001;**109**:584–91.
34. Lee YJ, et al. Roles of the mammalian mitochondrial fission and fusion mediators Fis1, Drp1, and Opa1 in apoptosis. *Mol Biol Cell* 2004;**15**:5001–11.
35. Fritz S, et al. Connection of the mitochondrial outer and inner membranes by Fzo1 is critical for organellar fusion. *J Cell Biol* 2001;**152**:683–92.
36. Sesaki H, Jensen RE. UGO1 encodes an outer membrane protein required for mitochondrial fusion. *J Cell Biol* 2001;**152**:1123–34.
37. Sesaki H, Jensen RE. Ugo1p links the Fzo1p and Mgm1p GTPases for mitochondrial fusion. *J Biol Chem* 2004;**279**:28,298–28,303.
38. Wong ED, et al. The intramitochondrial dynamin-related GTPase, Mgm1p, is a component of a protein complex that mediates mitochondrial fusion. *J Cell Biol* 2003;**160**:303–11.
39. Choi SY, et al. A common lipid links Mfn-mediated mitochondrial fusion and SNARE-regulated exocytosis. *Nat Cell Biol* 2006;**8**:1255–62.
40. Chernomordik LV, et al. Membranes of the world unite! *J Cell Biol* 2006;**175**:201–7.
41. Jensen RE, et al. Yeast mitochondrial dynamics: fusion, division, segregation, and shape. *Microsc Res Techniq* 2000;**51**:573–83.
42. Chen H, et al. Mitochondrial fusion protects against neurodegeneration in the cerebellum. *Cell* 2007;**130**:548–62.
43. Davies VJ, et al. Opa1 deficiency in a mouse model of autosomal dominant optic atrophy impairs mitochondrial morphology, optic nerve structure and visual function. *Hum Mol Genet* 2007;**16**:1307–18.
44. Alavi MV, et al. A splice site mutation in the murine Opa1 gene features pathology of autosomal dominant optic atrophy. *Brain* 2007;**130**:1029–42.
45. Zuchner S, Vance JM. Molecular genetics of autosomal-dominant axonal Charcot-Marie-Tooth disease. *Neuromol Med* 2006;**8**:63–74.
46. Delettre C, et al. Nuclear gene OPA1, encoding a mitochondrial dynamin-related protein, is mutated in dominant optic atrophy. *Nat Genet* 2000;**26**:207–10.
47. Alexander C, et al. OPA1, encoding a dynamin-related GTPase, is mutated in autosomal dominant optic atrophy linked to chromosome 3q28. *Nat Genet* 2000;**26**:211–15.
48. Smirnova E, et al. A model for dynamin self-assembly based on binding between three different protein domains. *J Biol Chem* 1999;**274**:14,942–14,947.
49. Smirnova E, et al. Dynamin-related protein Drp1 is required for mitochondrial division in mammalian cells. *Mol Biol Cell* 2001;**12**:2245–56.
50. Harder Z, et al. Sumo1 conjugates mitochondrial substrates and participates in mitochondrial fission. *Curr Biol* 2004;**14**:340–5.
51. Taguchi N, et al. Mitotic phosphorylation of dynamin-related GTPase Drp1 participates in mitochondrial fission. *J Biol Chem* 2007;**282**: 11,521–11,529.
52. Sesaki H, Jensen RE. Division versus fusion: Dnm1p and Fzo1p antagonistically regulate mitochondrial shape. *J Cell Biol* 1999;**147**:699–706.
53. Bleazard W, et al. The dynamin-related GTPase Dnm1 regulates mitochondrial fission in yeast. *Nat Cell Biol* 1999;**1**:298–304.
54. Otsuga D, et al. The dynamin-related GTPase, Dnm1p, controls mitochondrial morphology in yeast. *J Cell Biol* 1998;**143**:333–49.
55. Fukushima NH, et al. The GTPase effector domain sequence of the Dnm1p GTPase regulates self-assembly and controls a rate-limiting step in mitochondrial fission. *Mol Biol Cell* 2001;**12**:2756–66.
56. Legesse-Miller A, et al. Constriction and Dnm1p recruitment are distinct processes in mitochondrial fission. *Mol Biol Cell* 2003;**14**:1953–63.
57. Labrousse AM, et al. C. elegans dynamin-related protein DRP-1 controls severing of the mitochondrial outer membrane. *Mol Cell* 1999;**4**:815–26.
58. Yoon Y, et al. Mammalian dynamin-like protein DLP1 tubulates membranes. *Mol Biol Cell* 2001;**12**:2894–905.
59. Ingerman E, et al. Dnm1 forms spirals that are structurally tailored to fit mitochondria. *J Cell Biol* 2005;**170**:1021–7.
60. Griffin EE, et al. The WD40 protein Caf4p is a component of the mitochondrial fission machinery and recruits Dnm1p to mitochondria. *J Cell Biol* 2005;**170**:237–48.
61. Yoon Y, et al. The mitochondrial protein hFis1 regulates mitochondrial fission in mammalian cells through an interaction with the dynamin-like protein DLP1. *Mol Cell Biol* 2003;**23**:5409–20.
62. Mozdy AD, et al. Dnm1p GTPase-mediated mitochondrial fission is a multi-step process requiring the novel integral membrane component Fis1p. *J Cell Biol* 2000;**151**:367–80.
63. Tieu Q, Nunnari J. Mdv1p is a WD repeat protein that interacts with the dynamin-related GTPase, Dnm1p, to trigger mitochondrial division. *J Cell Biol* 2000;**151**:353–66.

64. Cerveny KL, et al. Division of mitochondria requires a novel DNM1-interacting protein, Net2p. *Mol Biol Cell* 2001; **12**: 309–21.
65. Cerveny KL, et al. Yeast mitochondrial division and distribution require the cortical num1 protein. *Dev Cell* 2007;**12**:363–75.
66. Heil-Chapdelaine RA, et al. The cortical protein Num1p is essential for dynein-dependent interactions of microtubules with the cortex. *J Cell Biol* 2000;**151**:1337–44.
67. Farkasovsky M, Kuntzel H. Yeast Num1p associates with the mother cell cortex during S/G2 phase and affects microtubular functions. *J Cell Biol* 1995;**131**:1003–14.
68. Koch A, et al. A Role for Fis1 in both mitochondrial and peroxisomal fission in mammalian cells. *Mol Biol Cell* 2005;**16**:5077–86.
69. Schrader M. Shared components of mitochondrial and peroxisomal division. *Biochim Biophys Acta* 2006;**1763**:531–41.
70. Kobayashi S, et al. Fis1, DLP1, and Pex11p coordinately regulate peroxisome morphogenesis. *Exp Cell Res* 2007;**313**:1675–86.
71. Karbowski M, et al. The mitochondrial E3 ubiquitin ligase MARCH5 is required for Drp1 dependent mitochondrial division. *J Cell Biol* 2007;**178**:71–84.
72. Nakamura N, et al. MARCH-V is a novel mitofusin 2- and Drp1-binding protein able to change mitochondrial morphology. *EMBO Rep* 2006;**7**:1019–22.
73. Yonashiro R, et al. A novel mitochondrial ubiquitin ligase plays a critical role in mitochondrial dynamics. *Embo J* 2006;**25**:3618–26.
74. Cribbs JT, Strack S. Reversible phosphorylation of Drp1 by cyclic AMP-dependent protein kinase and calcineurin regulates mitochondrial fission and cell death. *EMBO Rep* 2007;**8**:939–44.
75. Chang CR, Blackstone C. Cyclic AMP-dependent protein kinase phosphorylation of Drp1 regulates its GTPase activity and mitochondrial morphology. *J Biol Chem* 2007;**282**:21,583–21,587.
76. Chang CR, Blackstone C. Drp1 phosphorylation and mitochondrial regulation. *EMBO Rep* 2007;**8**:1088–9. author reply 1089–1090.
77. Wasiak S, et al. Bax/Bak promote sumoylation of DRP1 and its stable association with mitochondria during apoptotic cell death. *J Cell Biol* 2007;**177**:439–50.
78. Zunino R, et al. The SUMO protease SENP5 is required to maintain mitochondrial morphology and function. *J Cell Sci* 2007;**120**:1178–88.
79. Messerschmitt M, et al. The inner membrane protein Mdm33 controls mitochondrial morphology in yeast. *J Cell Biol* 2003;**160**:553–64.
80. Frank S, et al. The role of dynamin-related protein 1, a mediator of mitochondrial fission, in apoptosis. *Dev Cell* 2001;**1**:515–25.
81. Karbowski M, et al. Spatial and temporal association of bax with mitochondrial fission sites, Drp1, and Mfn2 during apoptosis. *J Cell Biol* 2002;**159**:931–8.
82. Youle RJ, Karbowski M. Mitochondrial fission in apoptosis. *Nat Rev Mol Cell Biol* 2005;**6**:657–63.
83. Germain M, et al. Endoplasmic reticulum BIK initiates DRP1-regulated remodelling of mitochondrial cristae during apoptosis. *Embo J* 2005;**24**:1546–56.

Chapter 43

Signaling Pathways from Mitochondria to the Cytoplasm and Nucleus

Immo E. Scheffler
Division of Biology (Molecular Biology Section), University of California, San Diego, La Jolla, California

INTRODUCTION

Until about a decade ago mitochondria were generally viewed as an independent organelle that served as the "powerhouse of the cell." Much research was devoted to understanding their composition, the contribution of two genomes to their biogenesis, their role in compartmentalizing cellular metabolism, and a detailed elucidation of the mechanism of oxidative phosphorylation [1, 2]. An avalanche of studies in recent years has laid the foundation for the view that mitochondria are much more intricately integrated into a variety of cellular activities that include metabolism, cell proliferation and cell cycle control, programmed cell death (apoptosis), development and differentiation, control of the lifespan of an organism, and cellular responses to viruses and pathogens [3–5]. Thus, the challenge arises to understand the nature of the signals that are exchanged between the mitochondria, the cytosol, and the nucleus.

The number of papers on apoptosis and mitochondria exceeds 10,000, and they will be expertly summarized in the following chapter of this volume. A key finding was the discovery of the release of a mitochondrial protein, cytochrome c, from the intermembrane space (IMS), and its participation in triggering a caspase cascade. Subsequently the release of other proteins from the IMS was documented and their role in apoptosis was established. The present review will focus on signals originating from mitochondria that are primarily small molecules with targets in the cytosol and the nucleus. A discussion of some more general themes in the current literature will be followed by a few examples of specific signaling pathways, where our understanding has advanced significantly in recent years.

It should be stressed at the outset that there is no such object as a generic mitochondrion described in the textbooks, and the specific participation of mitochondria in cellular physiology must be considered in the context of the specialized function of differentiated cells and tissues. Electron transport and oxidative phosphorylation are a hallmark of all mitochondria, but recent proteomic studies and global assessments of transcripts for mitochondrial proteins in various tissues have established very significant differences between mitochondria from muscle, brain, liver, lymphocytes, etc. [6]. Although yeast as a model system has provided key insights into many aspects of mitochondrial biogenesis and function, this organism is likely to have developed some unique mechanisms for adapting to a broad spectrum of environmental conditions, and drawing parallels with mammalian cells should be done with caution [5]. Similarly, plant mitochondria in diverse tissues may have distinct properties that are beyond the scope of the present discussion.

In a discussion of signaling and various control mechanisms, a clear distinction should be made between mechanisms requiring a rapid or transient response (e.g., hypoxia or glucose deprivation), and mechanisms that are associated with long term adaptations, differentiation, and perhaps aging. In the former it is likely that allosteric mechanisms and reversible protein modifications can operate on a short time scale. Long term changes are more likely to be associated with changes in gene expression.

SMALL MOLECULES AS SIGNALS FROM MITOCHONDRIA

Protein traffic into mitochondria is a one way mechanism, with only scattered reports of protein export (other than during apoptosis). In contrast, there are many metabolites and ions that can enter and exit from mitochondria.

Handbook of Cell Signaling, Three-Volume Set 2 ed.

Such traffic is tightly controlled by a large number of mitochondrial transporters and ion channels. The transporters constitute a family of more than 50 proteins, localized in the inner membrane [7, 8]. Among them, the adenine nucleotide transporter (ANT) is the best characterized, and its high resolution structure [9, 10] provides clues for the similar structure/topology of many of the others [10]. Nevertheless, the primary sequence in combination with the general structure does not reveal the specificity of the transporters, and this specificity remains to be determined for more than half of the transporters identified by homology cloning in humans. Mitochondrial cation channels include a Ca^{+2} uniporter, a Ca^{+2} activated K^+ channel, an ATP sensitive K^+ channel, K^+/H^+ and Na^+/H^+ antiporters, and others [11]. The involvement of protons in the antiporters links ion transport/exchange directly to the proton pumps of the electron transport chain in the control of the membrane potential ($\Delta\Psi$), and indirectly to changes in mitochondrial volume and morphology. The present discussion will not dwell on the traffic of amino acids, keto acids and most of the other metabolic intermediates arising from the catabolic and anabolic reactions of the Krebs cycle, the urea cycle, and other metabolic pathways. The emphasis will be on several small molecules with a potential regulatory function in the cytosol and nucleus. These two compartments will not be distinguished with regard to the small molecules under consideration here.

SMALL CATIONS: NA^+, K^+, MG^{+2}, CA^{+2}, H^+

This is a broad subject that is difficult to summarize in a few paragraphs. The flux of these ions depends on the concentration gradients, on the membrane potential, and on gated channels. In general, mitochondria can act as buffers and take up Ca^{+2} when the cytosol is overloaded [12–16]. Excess Ca^{+2} can initiate a permeability transition and trigger apoptosis (excito-toxicity). Mitochondrial defects (from stroke or neurodegenerative diseases) can lead to failures in intracellular Ca^{+2} regulation [17]. The production of reactive oxygen species was invoked to "explain" such a behavior, but a recent reassessment suggests that "enhanced reactive oxygen species are a consequence rather than a cause of failed cytoplasmic calcium homeostasis" [18]. The outlook has changed dramatically in recent years, and instead of being viewed passive Ca^{+2} sinks, mitochondria are now known to contribute actively to the spatial and temporal regulation of intracellular Ca^{+2} concentrations [19, 20]. Ca^{+2} levels in the mitochondrial matrix control the activity of several enzymes in the Krebs cycle. In the cytosol the control by Ca^{+2} is exerted via its binding to regulatory proteins such as calmodulin and troponin.

Much emphasis has been placed on the study of ion channels in the heart, where relatively rapid changes in workload must be met with adaptations in the bioenergetic status of the cardiomyocytes even under normal conditions [21–24]. In pathological situations (ischemia, reperfusion, oxidative stress, and metabolite deprivation) a critical threshold may be crossed leading to apoptosis, necrosis, and cell death in cardiomyocytes and in other cells [11].

Uncoupling proteins (UCP1, 2, 3, etc.) are thought to transport protons, and this mechanism is well established for UCP1 in brown adipose tissue where uncoupling of respiration from phosphorylation leads to thermogenesis [25]. The specific role of the other isozymes in diverse tissues is still being debated [21, 26–30]. In particular, the control of their gating is dependent on purine nucleotides, but may also be influenced by free fatty acids and reactive oxygen species. They may act as a safety valve when the membrane potential is raised to above normal level. The expression of UCP2 in pancreatic β cells is repressed by SIRT1 by binding to the UCP2 promoter. Lowering UCP2 levels causes ATP levels to rise and hence enhances insulin secretion [31].

ATP, ADP, AND AMP

The adenine nucleotide transporter normally exports ATP and imports ADP by an antiport mechanism. In well coupled mitochondria the rate of ATP synthesis is coupled to ATP turnover by various types of biological work performed by the cell (metabolism, motility, ion pumps, proteasome activity, and more). From the point of view of signaling it is necessary to consider several parameters: (1) the total adenine nucleotide pool in the cell, (2) the concentrations of ATP, ADP, and AMP in the cytosol, and (3) their relative concentrations. ATP and AMP are common ligands acting in the allosteric control of key metabolic enzymes (e.g., glycogen metabolism, gluconeogenesis, and more). The reaction

$$2\ \text{ADP} \leftrightharpoons \text{ATP} + \text{AMP}$$

has an equilibrium constant near 1, and under physiological concentrations relatively small changes in [ATP] are associated with relatively large changes in [AMP]. An energy deficit (low [ATP], high [AMP]) may stimulate not only the flux through metabolic pathways (glycolysis), but an AMP activated protein kinase (AMPK) can initiate transcriptional activation of selected genes in combination with other activators, or it can directly phosphorylate and activate metabolic enzymes. The specific pathways are tissue specific. One pathway becoming defined in some detail proposes the following scenario [32]: a drop in ATP levels causes a rise in intracellular Ca^{+2}, followed by an increase in the level of the cAMP response element binding protein (CREB); a stimulation of PGC-1α expression in combination with PPARγ induces the transcription of many mitochondrial genes. PPARγ, the peroxisome-proliferator activated receptor-γ, was originally identified as the master regulator in adipogenesis [33], and it has now been found in other diverse tissues such as hepatocytes, where it regulates the expression of gluconeogenic genes. To determine the transcriptional specificity of PPARγ, a screen for proteins binding PPARγ led to the isolation of PGC-1α and other members of this group of coactivators. Many studies have confirmed that PGC-1α is a prominent regulator of mitochondrial biogenesis and oxidative phosphorylation (see also below). It has been described as a "docking platform" that, when bound to a transcription factor (e.g., NRF-1, nuclear respiratory factor 1), recruits histone modifying enzymes, bridges the factor to the basal transcriptional initiation complex, and takes part in the processing of the target gene mRNA [33]. Nuclear genes encoding subunits of complexes of the electron transport chain might be expected to be expressed coordinately, and a search for common promoter elements in these genes started long before whole mammalian genomes were sequenced [34–36]. Many but not all such promoters contain short sequence motifs for binding nuclear respiratory factors NRF-1 and/or NRF-2. These transcription factors also control the expression of genes for the mitochondrial protein import machinery and for the transcription of the mitochondrial genome.

NADH AND NAD$^+$

The nicotinamide adenine dinucleotides are found both in the cytosol and in the mitochondrial matrix, and at the elementary level it is taught that NADH "equivalents" are transported across the inner mitochondrial membrane by shuttle mechanisms involving other metabolites such as malate and oxaloacetate. Although it has been stated that mitochondrial membranes are impermeable to NAD+ and NADH [37], this cofactor has to be imported into the mitochondria in the first place, before the shuttles can operate. Physiological intracellular NADH concentrations are in the range of 1–10 μM, with the largest portion found in mitochondria. A recent review reports the cytosolic NAD$^+$/NADH ratio to be ~700 : 1, while the corresponding mitochondrial ratio is 7–8 : 1 [37]. These parameters were measured ~30 years ago for rat liver, and more up-to-date measurements for diverse tissues and cell types would be highly desirable for an interpretation of the postulated regulatory mechanisms involving this dinucleotide (see below). Mitochondria could be involved in controlling either the redox ratio or the absolute concentrations of these cofactor(s) in the cytosol (and nucleus). If total concentrations differ between cytosol and mitochondria, how quick is the equilibration? And, if the NADH/NAD$^+$ ratio changes in one compartment, how tightly is it coupled to the ratio in the other compartment? The NADH/NAD$^+$ ratio can in principle be measured spectrophotometrically, but it is still very difficult to determine it separately for the different compartments, because any fractionation procedure is likely to cause severe perturbations of this ratio.

In the mitochondrial matrix the NADH/NAD$^+$ ratio is expected to be dependent on the activity of the electron transport chain (respiration) and on the various metabolic redox reactions utilizing this cofactor in the Krebs cycle or fatty acid oxidation. When NADH levels rise due to a slow down of respiration, pyruvate dehydrogenase and α-ketoglutarate dehydrogenase are strongly inhibited, and therefore Krebs cycle activity and respiration are strongly coupled by a feedback mechanism. A closely related redox couple is represented by NADPH/NADP$^+$; NADH and NADPH are generally used in distinct reactions. The mitochondrial enzyme nicotinamde adenine transhydrogenase catalyzes the reaction.

$$\mathrm{NADH} + \mathrm{NADP}^+ + \mathrm{nH}^+_{\mathrm{out}} \leftrightharpoons \mathrm{NAD}^+ + \mathrm{NADPH} + \mathrm{nH}^+_{\mathrm{in}}$$

This reaction is coupled to proton translocation (n = ~1), and is therefore potentially sensitive to the mitochondrial membrane potential [38].

NAD$^+$ biosynthesis by a salvage pathway in mammalian cells requires niacin. At the same time there are several reactions leading to the degradation of the dinucleotide. The most notable among these are the reactions by the poly(ADP-ribose) polymerase 1 (PARP-1). PARP-1 is emerging as an important activator of caspase independent cell death [39, 40], and it may be the major NAD$^+$ consuming enzyme. How significant is this consumption in relation to the total available pool of NAD$^+$? A second NAD$^+$ consuming reaction is histone/protein deacetylation by the SIR2 family of proteins (sirtuins). Such a mechanism constitutes one of the most direct pathways that potentially links mitochondrial activity with histone deacetylation, chromatin remodeling, and changes in gene expression (see below). Different members of the sirtuin family have been found in different cellular compartments, where acetylated proteins other than histones are the substrates (see below).

GLUTATHIONE (GSH AND GSSG)

The activity and function of a significant number of proteins including transcription factors are dependent on the redox status of their sulfhydryl (-SH) groups [41]. The balance between oxidized and reduced forms of the protein is controlled by a combination of oxido-reductases, specifically thioredoxins and glutaredoxins and their cofactors. Isoforms of these enzymes are present both in the cytosol and in the mitochondrial matrix. One isozyme of thioredoxin is a critical component of the ribonucleotide reductase reaction providing deoxyribonucleotides for DNA synthesis. Glutaredoxin converts a disulfide bond of a protein into two thiol groups, while reduced glutathione (GSH) is converted to the oxidized glutathione (GSSG).

Therefore, the GSH/GSSG ratio influences the steady state distributions between active and inactive enzyme conformations, and overall enzyme activity thus becomes a function of the "redox state" of the cell (or compartment). In phototrophic organisms (plants) such reductions may be light driven, but in animal cells they are ultimately dependent on the $NADH/NAD^+$ ratio (or the $NADPH/NADP^+$ ratio), and hence metabolism and respiration. GSSG can be reduced to GSH by glutathione reductase, using NADPH as cofactor. GSH is also a key substrate for glutathione S-transferase, one of the scavenging enzymes for the removal of hydrogen peroxide (H_2O_2). The GSH/GSSG ratio is therefore dependent on normal metabolic, mitochondrial activity, but may also be perturbed, when the generation of reactive oxygen species becomes excessive. Redox sensitive green fluorescent proteins have recently been developed as probes to assess the thiol/disulfide status in different cellular compartments; mitochondria were found to be the most reduced compartment, followed by the nucleus and cytosol, confirming that redox signaling is compartmentalized [42].

REACTIVE OXYGEN SPECIES

Oxygen is relatively abundant in many cells. With its small size and zero charge it can gain access to many sites within proteins involved in redox reactions. In the cytosol many such reactions involve hydride transfers, but in mitochondria (and in some other redox enzymes in the cytosol) one electron transfers occur via the formation of organic free radicals such as the semiquinones of FAD, FMN, and ubiquinone. Single electrons from such semiquinones can escape to oxygen, if normal electron transport is impaired or perturbed at a downstream site. The resulting superoxide radical ($O_2^{-\cdot}$) is converted to H_2O_2 by superoxide dismutase (SOD), and a Fenton reaction (requiring Fe^{+3}) yields the highly reactive hydroxyl radical ($OH^\cdot$). The superoxide and the hydrogen peroxide have a low reactivity (long half-life) and they can diffuse readily away from their site of synthesis. The hydroxyl radical, by contrast, reacts rapidly with nucleic acids, proteins, and lipids. The multitude of potential targets makes it difficult to distinguish primary targets from which regulatory signals may originate from secondary widespread pathological modifications of lipids, proteins, and nucleic acids. These reactive oxygen species (ROS) are kept in check by scavenging enzymes in mitochondria and the cytosol (SOD, catalase, glutathione peroxidase). Knockouts of any of these enzymes in a variety of organisms lead to a reduced lifespan, emphasizing the importance of their protective function. Overexpression can increase the lifespan, and such findings have supported current theories of aging in their simplest form. The superoxide ($O_2^{-\cdot}$) can also react with nitric oxide (NO) to form a reactive nitrogen species, peroxynitrite. The latter reacts readily with reduced glutathione (GSH) or with thiol groups on proteins to produce nitrosothiols. It remains to be established whether such protein modifications are mainly harmful, or whether they may serve a regulatory function.

A PubMed search leads to the recovery of ~14,000 citations on ROS, of which ~2200 papers were published in 2006. ROS (and oxidative stress) can be blamed for a lot of problems, and they may have some positive functions as well [43, 44]. Oxygen sensing itself has been proposed to depend on mitochondria generated reactive oxygen species (ROS), independent of oxidative phosphorylation [45, 46]. Reactive oxygen species have been assigned a central role in the mechanisms of apoptosis, with arguments still unresolved about whether they are responsible for signaling, cell killing, or whether their increased rate of synthesis is a secondary consequence of mitochondrial degeneration.

The mitochondrial electron transport chain (complexes I, II, and III) is considered to be the main source of ROS, but there is still uncertainty about whether the ROS do their main damage to mitochondrial targets (mtDNA, cardiolipin, proteins containing [Fe-S] clusters such as aconitase), or whether they are exported, either placing cytosolic and nuclear targets at risk, or for serving a physiologically meaningful regulatory function (for a recent comprehensive review see [44]). Over the last 6 years, the existence of the NOX family of NADPH oxidases, comprising seven homologs including the phagocyte NADPH oxidase, has been recognized. Nox enzymes generate ROS in a variety of tissues as part of normal physiological functions, which include innate immunity, signal transduction, and biochemical reactions, e.g., to produce thyroid hormone [47]. The mechanisms are still being debated, and may be variable depending on the cell [48]. For example, they could play a role in controlling the levels of NO. NO is known to activate guanylate cyclase, and hence regulates cGMP levels. This signal induces the PPARγ coactivator 1α (PGC-1α) and PGC-1β, followed by expression of the nuclear respiratory factors NRF-1 and NRF-2 and other transcription factors. These transcription factors play crucial roles in the transcription of nuclear genes encoding subunits of the electron transfer chain complexes and the mitochondrial transcription factor A (Tfam alias mtTFA) [34, 49, 50]. Details on the mechanisms remain to be elucidated, but one very plausible signaling pathway now includes endothelial nitric acid synthase (eNOS) as a master regulator of mitochondrial biogenesis, with important implications for understanding physiological phenomena such as thermogenesis, obesity, diabetes, response to aerobic exercise, calorie restriction (CR), senescence, and more [51]. In this context the emphasis so far has been on eNOS, since in experiments with eNOS-negative (−/−) mice the anticipated responses to calorie restriction were not observed [52]. There is still controversy about whether a distinct mitochondrial nitric oxide synthase exists, producing NO with a regulatory function directly in the mitochondrial matrix.

IRON SULFUR CLUSTERS

One of the essential functions of mitochondria is the synthesis of [Fe-S] centers. Many proteins assembled in distinct complexes are involved [53, 54]. Redox reactions are required to synthesize the sulfide anion (or H_2S) from cysteine. The [Fe-S] clusters can be inserted into mitochondrial proteins (ETC, aconitase), but they can also be exported and subsequently inserted into cytosolic and even nuclear proteins. The cytosolic aconitase serves as an iron sensor and regulates the stability and translatability of several mRNAs encoding proteins associated with iron homeostasis and heme biosynthesis [55, 56].

SIRTUINS

Only a brief overview of this rapidly expanding subject can be given here. Sir2 was a gene first discovered in the study of gene silencing and lifespan expansion in yeast and *Drosophila* (see [31, 57] for authoritative reviews). The gene product was subsequently characterized as an NAD^+ dependent histone deacetylase. A homology search in mammalian organisms identified seven orthologs; all of them are NAD^+ dependent enzymes, but some catalyze deacetylation (Sirt1, Sirt3, etc.), while others (Sirt4) catalyze ADP-ribosylation of select target proteins. Sirt1, Sirt6, and Sirt7 are nuclear sirtuins, while Sirt3, Sirt4, and Sirt5 have been reported to be imported into the mitochondrial matrix. To give one specific example, Sirt1 associates with, deacetylates and stimulates PGC-1α, a coactivator of PPARγ. Other transcription factors (e.g., HNF4α) can also be stimulated by PGC-1α. In many tissues, members of the PGC-1 family are dominant regulators of mitochondrial functions [33]. Depending on the target tissue, other transcription factors and biological responses may be induced or repressed by Sirt1. These cell type specific actions of Sirt1 remain to be fully elucidated, but it is clear the Sirt1 is prominently involved in the regulation of metabolism, i.e., it "decodes the nutritional status" of tissues and cells such as hepatocytes, adipocytes, muscle, or pancreatic beta cells [57, 58]. Sirt3 has been shown to deacetylate the mitochondrial acetyl-CoA synthetase and hence it stimulates the synthesis of acetyl-CoA from acetate. The latter is released from the mammalian liver under ketogenic conditions (see [31] for a review and summary of recent work).

An expert review of the role of sirtuins in physiology has been written by Guarente [57] with the goal of integrating our understanding of two phenomena that the author considers to be opposites of a "metabolic spectrum": the "metabolic syndrome" and "calorie restriction." The metabolic syndrome is characterized by a combination of abnormal physiological conditions including obesity, high blood pressure, abnormal lipoprotein/cholesterol levels, dysfunctional glucose homeostasis, and a high risk of Type II diabetes. It appears to be clearly linked to our high calorie diet and sedentary lifestyle. The consequences of caloric excess constitute an enormous burden on our health care system. Calorie restriction (restricted food intake), by contrast, affects the above parameters in an opposite direction and leads to an extension of lifespan of up to 50 percent in a variety of organisms. The control mechanisms involving sirtuins and other factors have emerged as an overlapping set of regulatory mechanisms in response to overabundance or shortage of food [59].

MITOCHONDRIAL RETROGRADE SIGNALING

The power of genetics with yeast as a model organism has permitted a very detailed analysis of the adaptation of the yeast *Saccharomyces cerevisiae* to (1) different levels of nutrients and (2) to perturbations in mitochondrial functions due to mtDNA mutations. Defects or alterations in mitochondrial functions cause a "reconfiguration of metabolism" by altering the expression of select nuclear genes. The phenomenon has been termed "mitochondrial retrograde signaling" by Butow and colleagues, who have been primary contributors to this area and have written an up-to-date review and discussion of the various components of the RTG pathway [5, 60]. It is linked to the TOR signaling pathway, to mtDNA maintenance, and to the aging process in yeast. In much of the discussion of the retrograde pathway there is little emphasis on ATP synthesis, or the maintenance of the $NAD^+/NADH$ ratio. Instead, the focus is on the operation of a partial TCA cycle in mitochondria, β-oxidation and a glyoxylate cycle in peroxisomes (which proliferate under conditions of mitochondrial dysfunction). The net result is an adequate rate of synthesis of α-ketoglurate, which is then converted to glutamine for protein synthesis. Animal cells do not have a glyoxylate cycle. It was first shown by Scheffler's group that respiration deficient Chinese hamster cells in tissue culture have a high requirement for glutamine in the tissue culture medium and are auxotrophs for the (normally) non-essential amino acids aspartate and asparagine, because in tissues other than the liver these amino acids are derived from glutamine via the operation of the Krebs cycle and transamination reactions. The Krebs cycle was feedback inhibited by high levels of NADH resulting from a complex I deficiency [61]. In contrast to yeast, mammalian cells (fibroblasts) cannot reconfigure their metabolism under these conditions.

Mitochondrial diseases in humans have claimed much attention in the past decade. The subject is too vast to be adequately treated here. Typically, partial defects lead to an energy deficiency, but it should be noted that a severe reduction in the rate of ATP production alone will probably not explain the broad spectrum of symptoms, and the differential effects on specific tissues (blindness, deafness, neuropathies, myopathies).

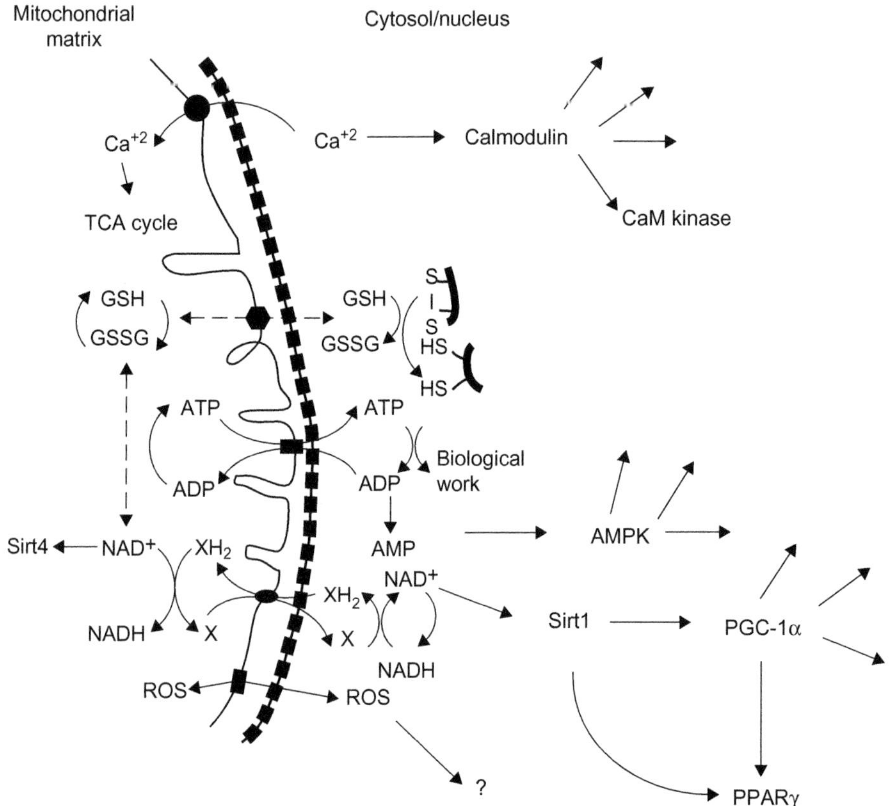

FIGURE 43.1 A schematic view of the major small molecules involved in signaling between mitochondria and the cytosol/nucleus.
X and XH_2 represent oxidized and reduced versions of a metabolite used in shuttling "reducing equivalents" in and out of mitochondria.

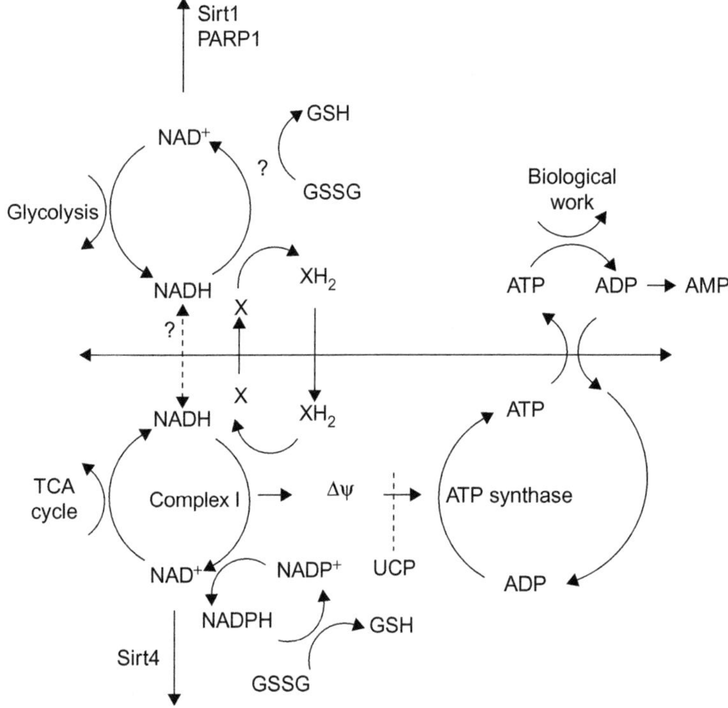

FIGURE 43.2 A schematic view of the cyclic interconversion of oxidized and reduced small molecules and the coupling of ATP synthesis to the cycling of NAD+/NADH as described by the chemiosmotic hypothesis.
Uncoupling proteins (UCP) can uncouple oxidative phosphorylation. The membrane potential (DY) may also control the flux of various cations (not shown). GSSG is reduced to GSH by NADPH.

CONCLUSION

The small molecules (ATP/ADP/AMP, NAD+/NADH, GSH/GSSG) that are continuously turned over in the cytosol and in the mitochondrial matrix are now recognized to be not only cofactors driving endergonic reactions, or donors/acceptors of electrons in redox reactions. In combination with Ca^{+2}, NO, cAMP, and cGMP they serve as allosteric effectors and in feedback mechanisms to control metabolic flux through numerous pathways, and they can also act as powerful elements in the modulation of gene expression. Major players involved in signaling pathways leading to transcriptional regulation have been identified: AMPK, PGC-1, PPARγ, the sirtuins. A detailed description of these pathways must focus on the distinct physiology of the tissues or organs being investigated. In the final analysis it must be recognized that metabolism is not a subject to be ignored as history, but metabolism and bioenergetics are indispensable aspects of the expression of genetic information, development and differentiation, and ultimately the phenomenon of senescence and death. Figures 43.1 and 43.2 represent schematic summaries. The first emphasizes the traffic of small molecules in and out of mitochondria and some of their potential targets in the two compartments. In the second figure the cyclic turnover of various small molecules is shown to highlight the interdependence of these cyclic interconversions. A challenge of the future is to measure accurate concentrations of each of these components in the various cellular compartments, to observe absolute and relative changes, and to incorporate all of these reactions into a computer model [62].

REFERENCES

1. Scheffler IE. A century of mitochondrial research: achievements and perspectives. *Mitochondrion* 2000;**1**:3–31.
2. Scheffler IE. *Mitochondria*. 2nd Edition Hoboken, NJ: John Wiley & Sons, Inc; 2008 1–462.
3. McBride HM, Neuspiel M, Wasiak S. Mitochondria: more than just a powerhouse. *Curr Biol* 2006;**16**:R551–60.
4. Ryan MT, Hoogenraad NJ. Mitochondrial-nuclear communications. *Annu Rev Biochem* 2007;**76**:701–22.
5. Liu Z, Butow RA. Mitochondrial retrograde signaling. *Annu Rev Genet* 2006;**40**:159–85.
6. Mootha VK, Bunkenborg J, Olsen JV, Hjerrild M, Wisniewski JR, Stahl E, Bolouri MS, Ray HN, Sihag S, Kamal M, Patterson N, Lander ES, Mann M. Integrated analysis of protein composition, tissue diversity, and gene regulation in mouse mitochondria. *Cell* 2003;**115**:629–40.
7. Wohlrab H. The human mitochondrial transport protein family: Identification and protein regions significant for transport function and substrate specificity. *Biochim Biophys Acta* 2005;**1709**:157–68.
8. Kunji ER. The role and structure of mitochondrial carriers. *FEBS Lett* 2004;**564**:239–44.
9. Dahout-Gonzalez C, Nury H, Trezeguet V, Lauquin GJ, Pebay-Peyroula E, Brandolin G. Molecular, functional, and pathological aspects of the mitochondrial ADP/ATP carrier. *Physiology (Bethesda)* 2006;**21**:242–9.
10. Nury H, Dahout-Gonzalez C, Trezeguet V, Lauquin GJM, Brandolin G, Pebay-Peyroula E. Relations between structure and function of the mitochondrial ADP/ATP carrier. *Annu Rev Biochem* 2006;**75**:713–41.
11. O'Rourke B, Cortassa S, Aon MA. Mitochondrial ion channels: gatekeepers of life and death. *Physiology (Bethesda)* 2005;**20**:303–15.
12. Szabadkai G, Simoni AM, Bianchi K, De Stefani D, Leo S, Wieckowski MR, Rizzuto R. Mitochondrial dynamics and Ca(2+) signaling. *Biochim Biophys Acta* 2006;**1763**:442–9.
13. Bianchi K, Rimessi A, Prandini A, Szabadkai G, Rizzuto R. Calcium and mitochondria: mechanisms and functions of a troubled relationship. *Biochim Biophys Acta* 2004;**1742**:119–31.
14. Szabadkai G, Simoni AM, Rizzuto R. Mitochondrial Ca2+ uptake requires sustained Ca2+ release from the endoplasmic reticulum. *J Biol Chem* 2003;**278**:15,153–15,161.
15. Rutter GA, Rizzuto R. Regulation of mitochondrial metabolism by RE Ca^{2+} release: an intimate connection. *Trends Biochem Sci* 2000;**25**:215–21.
16. Nicholls DG. Mitochondria and calcium signaling. *Cell Calcium* 2005;**38**:311–17.
17. Jacquard C, Trioulier Y, Cosker F, Escartin C, Bizat N, Hantraye P, Cancela JM, Bonvento G, Brouillet E. Brain mitochondrial defects amplify intracellular [Ca2+] rise and neurodegeneration but not Ca2+ entry during NMDA receptor activation. *Faseb J* 2006;**20**:1021–3.
18. Nicholls DG, Johnson-Cadwell L, Vesce S, Jekabsons M, Yadava N. Bioenergetics of mitochondria in cultured neurons and their role in glutamate excitotoxicity. *J Neurosci Res* 2007;**85**:3206–12.
19. Graier WF, Frieden M, Malli R. Mitochondria and Ca(2+) signaling: old guests, new functions. *Pflugers Arch* 2007;**455**:375–96.
20. Rizzuto R, Pozzan T. Microdomains of intracellular Ca2+: molecular determinants and functional consequences. *Physiol Rev* 2006;**86**:369–408.
21. Andrukhiv A, Costa AD, West IC, Garlid KD. Opening mitoKATP increases superoxide generation from Complex I of the electron transport chain. *Am J PhysiolHeart Circ Physiol* 2006;**291**:H2067–74.
22. Costa AD, Jakob R, Costa CL, Andrukhiv K, West IC, Garlid KD. The mechanism by which the mitochondrial ATP-sensitive K+ channel opening and H2O2 inhibit the mitochondrial permeability transition. *J Biol Chem* 2006;**281**:20,801–20,808.
23. Costa AD, Quinlan CL, Andrukhiv A, West IC, Jaburek M, Garlid KD. The direct physiological effects of mitoK(ATP) opening on heart mitochondria. *Am J Physiol Heart Circ Physiol* 2006;**290**:H406–15.
24. Garlid KD, Paucek P. Mitochondrial potassium transport: the K^+ cycle. *Biochim Biophys Acta* 2003;**1606**:23–41.
25. Mozo J, Emre Y, Bouillaud F, Ricquier D, Criscuolo F. Thermoregulation: what role for UCPs in mammals and birds?. *Biosci Rep* 2005;**25**:227–49.
26. Jaburek M, Garlid KD. Reconstitution of recombinant uncoupling proteins: UCP1, 2, and 3 have similar affinities for ATP and are unaffected by coenzyme Q10. *J Biol Chem* 2003;**278**:25,825–25,831,.
27. Garlid KD, Jaburek M, Jezek P, Varecha M. How do uncoupling proteins uncouple?. *Biochim Biophys Acta* 2000;**1459**:383–9.
28. Jezek P, Garlid KD. Mammalian mitochondrial uncoupling proteins. *Int J Biochem Cell Biol* 1998;**30**:1163–8.
29. Porter RK. A new look at UCP 1. *Biochim Biophys Acta* 2006;**1757**:446–8.
30. Brand MD. The efficiency and plasticity of mitochondrial energy transduction. *Biochem Soc Trans* 2005;**33**:897–904.

31. Schwer B, Verdin E. Conserved metabolic regulatory functions of sirtuins. *Cell Metabolism* 2008;**7**:104–12.
32. Rohas LM, St.-Pierre J, Uldry M, Jaeger S, Handschin C, Spiegelman BM. A fundamental system of cellular energy homeostasis regulated by PGC-1a. *Proc Natl Acad Sci U S A* 2007;**104**:7933–8.
33. Handschin C, Spiegelman BM. Peroxisome proliferator-activated receptor gamma coactivator 1 coactivators, energy homeostasis, and metabolism. *Endocr Rev* 2006;**27**:728–35.
34. Scarpulla RC. Nuclear control of respiratory gene expression in mammalian cells. *J Cell Biochem* 2006;**97**:673–83.
35. Scarpulla RC. Nuclear control of respiratory chain expression in mammalian cells. *J Bioenerg Biomembr* 1997;**29**:109–19.
36. Scarpulla RC. Nuclear respiratory factors and the pathways of nuclear-mitochondrial interaction. *Trends Cardiovasc Med* 1996;**6**:39–45.
37. Ying W. NAD+ and NADH in cellular functions and cell death. *Front Biosci* 2006;**11**:3129–48.
38. Hatefi Y, Yamaguchi M. Nicotinamide nucleotide transhydrogenase: a model for utilization of substrate binding energy for proton translocation. *FASEB J* 1996;**10**:444–52.
39. Hong SJ, Dawson TM, Dawson VL. Nuclear and mitochondrial conversations in cell death: PARP-1 and AIF signaling. *Trends Pharmacol Sci* 2004;**25**:259–64.
40. Dawson VL, Dawson TM. Deadly conversations: nuclear-mitochondrial cross-talk. *J Bioenerg Biomembr* 2004;**36**:287–94.
41. Yang Y, Song Y, Loscalzo J. Regulation of the protein disulfide proteome by mitochondria in mammalian cells. *Proc Natl Acad Sci U S A* 2007;**104**:10,813–10,817.
42. Hansen JM, Go YM, Jones DP. Nuclear and mitochondrial compartmentation of oxidative stress and redox signaling. *Annu Rev Pharmacol Toxicol* 2006;**46**:215–34.
43. Rhee SG. H2O2, a necessary evil for cell signalling. *Science* 2006;**312**:1882–3.
44. Valko M, Leibfritz D, Moncol J, Cronin MT, Mazur M, Telser J. Free radicals and antioxidants in normal physiological functions and human disease. *Int J Biochem Cell Biol* 2007;**39**:44–84.
45. Brunelle JK, Bell EL, Quesada NM, Vercauteren K, Tiranti V, Zeviani M, Scarpulla RC, Chandel NS. Oxygen sensing requires mitochondrial ROS but not oxidative phosphorylation. *Cell Metab* 2005;**1**:409–14.
46. Guzy RD, Hoyos B, Robin E, Chen H, Liu L, Mansfield KD, Simon MC, Hammerling U, Schumacker PT. Mitochondrial complex III is required for hypoxia-induced ROS production and cellular oxygen sensing. *Cell Metab* 2005;**1**:401–8.
47. Lambeth JD. Nox enzymes, ROS, and chronic disease: an example of antagonistic pleiotropy. *Free Radic Biol Med* 2007;**43**:332–47.
48. Genestra M. Oxyl radicals, redox-sensitive signalling cascades and antioxidants. *Cell Signal* 2007;**19**:1807–19.
49. Scarpulla RC. Nuclear activators and coactivators in mammalian mitochondrial biogenesis. *Biochim Biophys Acta* 2002;**1576**:1–14.
50. Scarpulla RC. Transcriptional activators and coactivators in the nuclear control of mitochondrial function in mammalian cells. *Gene* 2002;**286**:81–9.
51. Cadenas E. Mitochondrial free radical production and cell signaling. *Mol Aspects Med* 2004;**25**:17–26.
52. Nisoli E, Tonello C, Cardile A, Cozzi V, Bracale R, Tedesco L, Falcone S, Valerio A, Cantoni O, Clementi E, Moncada S, Carruba MO. Calorie restriction promotes mitochondrial biogenesis by inducing the expression of eNOS. *Science* 2005;**310**:314–17.
53. Lill R, Muhlenhoff U. Iron-sulfur protein biogenesis in eukaryotes: components and mechanisms. *Annu Rev Cell Dev Biol* 2006;**22**:457–86.
54. Lill R, Muhlenhoff U. Iron-sulfur-protein biogenesis in eukaryotes. *Trends Biochem Sci* 2005;**30**:133–41.
55. Klausner RD, Rouault TA, Harford JB. Regulating the fate of mRNA: The control of cellular iron metabolism. *Cell* 1993;**72**:19–28.
56. DeRusso PA, Philpott CC, Iwai K, Mostowski HS, Klausner RD, Rouault TA. Expression of a constitutive mutant of iron regulatory protein 1 abolishes iron homeostasis in mammalian cells. *J Biol Chem* 1995;**270**:15,451–15,454.
57. Guarente L. Sirtuins as potential targets for metabolic syndrome. *Nature* 2006;**444**:868–74.
58. Leibiger IB, Berggren PO. Sirt1: a metabolic master switch that modulates lifespan. *Nat Med* 2006;**12**:34–6.
59. Chen D, Guarente L. SIR2: a potential target for calorie restriction mimetics. *Trends Mol Med* 2007;**13**:64–71.
60. Butow Jr RA, Ferreira JR, Spírek M, Liu Z. Cross-talk between nucleus and organelles. *Gene* 2005;**354**:1–200.
61. Scheffler IE. Biochemical genetics of respiration -deficient mutants of animal cells. In: Morgan MJ, editor. *Carbohydrate metabolism in cultured cells*. 3rd ed London: Plenum Publishing Co; 1986. p. 77–109.
62. Wu F, Yang F, Vinnakota KC, Beard DA. Computer modeling of mitochondrial TCA cycle, oxidative phosphorylation, metabolite transport, and electrophysiology. *J Biol Chem* 2007;**282**:24,525–24,537.

Chapter 44

Quality Control and Quality Assurance in the Mitochondrion

Carolyn K. Suzuki

Department of Biochemistry and Molecular Biology, University of Medicine and Dentistry of New Jersey, Newark, New Jersey

INTRODUCTION

In 1989, Helenius and colleagues first applied the concept of quality control to describe mechanisms for selectively retaining and degrading misfolded and misassembled proteins within the endoplasmic reticulum (ER) [1, 2]. The editing of abnormal proteins within this entry point into the secretary pathway not only ensures that conformationally correct and functional proteins are transported out of the ER but also prevents the potentially deleterious effects of non-functional proteins. Such quality surveillance is governed principally by molecular chaperones that promote protein folding and prevent aggregation, as well as by proteases that degrade improperly folded and unassembled polypeptides [3–7]. Elimination of abnormal ER proteins requires the "dislocation" or "retrograde transport" of lumenal and membrane polypeptides into the cytosol followed by ubiquitin dependent degradation by the 26S proteasome [7]. In addition, certain abnormal protein complexes exit the ER, traverse the Golgi and are specifically targeted to lysosomes for degradation [8]. We now know that quality control mechanisms are not limited to the endoplasmic reticulum and secretory pathway, as systems for monitoring protein biogenesis and protein damage are present in nearly every intracellular compartment.

Within mitochondria, quality control is strictly applied to the operations carried out within each suborganellar compartment. Each mitochondrion is bordered by two phospholipid bilayers – the *outer membrane* (OM), which is freely permeable to small solutes, in contrast to the *inner membrane* (IM), which is impermeable or selectively permeable to small solutes. Between these two membrane systems resides the *intermembrane space* (IMS). The inner membrane surrounds the *matrix*, which constitutes the most voluminous space within mitochondria that is occupied by soluble enzymes and metabolic intermediates. In addition to ATP production, mitochondria provide essential metabolic processes that include heme and steroid hormone synthesis, iron-sulfur cluster assembly, and fatty acid oxidation. Mitochondria are also central to the regulation of apoptotic cell death [9]. To carry out these functions, mitochondria rely on a workforce that depends on two genomes located in the nucleus as well as in mitochondria, along with two distinct protein synthesis machineries in the cytosol and in mitochondria. The mitochondrial genome encodes a handful of mitochondrial proteins essential for generating ATP by the electron transport chain, as well as rRNAs and tRNAs required for their synthesis. By contrast, chromosomal DNA in the nucleus encodes the vast majority of mitochondrial proteins that are produced in the cytosol as precursor proteins, and imported to their respective work sites by amino terminal or internal targeting sequences that bind to import receptors at the outer and inner membranes. Some mitochondrial proteins are multisubunit complexes of the electron transport chain, which demand that the synthesis and import of cytosolically translated proteins are coordinated with that of mitochondrially made proteins in order to achieve properly assembled and functional end products. As in the world of commerce, the manufacture of products in mitochondria is subject to procedures of "quality assurance" aimed at ensuring that the products under development meet specified requirements. In mitochondria, such quality assurance is largely overseen by cytosolic and mitochondrial chaperones; however, a growing body of work demonstrates that import receptors at the outer and inner membranes, as well as the translocation channel itself function to prevent aggregation and promote unfolding (Tables 44.1 and 44.2). The final oversight step in the quality control of misfolded or improperly assembled proteins is overseen by ATP dependent proteases that are soluble in the matrix, or embedded in the inner membrane with their respective

Handbook of Cell Signaling, Three-Volume Set 2 ed.

TABLE 44.1 Quality assurance of protein import

Chaperone	Function	Co-chaperone
Cytsolic chaperones and co-chaperones in protein import		
Hsc70*	Maintain and deliver import competent preproteins to Tom70	DJA1, 2, 4*
Hsp90*	Maintain and deliver import competent preproteins to Tom70	p23, Aha1*
AIP*	Maintain and deliver import competent preproteins to Tom20	
Translocase of the outer membrane (TOM) components with chaperone-like function		
Tom40	Translocation pore prevents aggregation and actively unfolds preproteins	
Translocase of the inner membrane (TIM) components with chaperone-like function		
Tim9/10, Tim8/13	Transfer of hydrophobic preproteins to the TIM22 translocon	
Tim14 (Pam18)	Import motor associated with Tim44	
Tim16 (Pam16) mtHsp70 (Ssc1)	Import motor associated with Tim44	
Mge1	Import motor	

**Nomenclature for yeast proteins, except for some proteins described only in mammalian systems.*

TABLE 44.2 Quality Assurance of protein folding in the mitochondrial matrix (yeast proteins)

DnaK-like proteins	Function	Interacting partners
mtHsp70 (Ssc1)	Protein import	Tim44, Tim14, Tim16, Mge1p
mtHsp70 (Ssc1)	Protein folding	Mdj1, Mge1
Ssq1 (Ssc2)	Fe/S cluster biogenesis	Jac1, Mge1
Ecm10 (Ssc3)	Protein import and folding?	Mge1, Mdj1 (?)
Chaperones for DnaK-like proteins		
Zim17 (Hep1, Tim15)	Maintenance of Hsp70 solubility and function	mtHsp70, Ssq1
DnaJ-like proteins		
Mdj1	Protein folding	Ssc1
Mdj2	?	?
Jac1	Fe/S cluster biogenesis	Ssq1
Tim14 (Pam18)	Protein import	mtHsp70, Tim16, Tim44
Tim16 (Pam16)	Protein import	mtHsp70, Tim14, Tim44
GrpE-like proteins		
Mge1	Protein import, protein folding, Fe/S cluster biogenesis	Ssc1, Ssq1, Ecm10
GroEL/ES-like proteins		
Hsp60 (Cpn60)	Protein folding	Hsp10/Cpn10
Hsp10 (Cpn10)		Hsp60/Cpn60
Hsp100/Clp-like proteins		
Hsp78	Prevention of protein aggregation, refolding of denatured proteins	
Mcx1 (yeast)	?	
Cyclophilins		
Cpr3	Protein folding	
Translation elongation factor	Delivery of aminoacyl-tRNAs to mitochondrial ribosomes	
EF-Tumt*	Prevention of protein aggregation, refolding of denatured proteins	

**Nomenclature for yeast proteins, except for one protein described only in mammalian systems.*

proteolytic active sites located either in the IMS or matrix (Table 44.3) [10]. An important function of these energy powered proteases is to identify and eliminate abnormal proteins; however, recent work reveals that they may also function in protein biogenesis and regulatory processes.

QUALITY ASSURANCE MEDIATED BY CHAPERONES AND THE PROTEIN TRANSLOCATION COMPLEX

One-tenth of cellular proteins are localized to mitochondria, and most of these are synthesized in the cytosol as precursor polypeptides that are targeted, imported, and sorted to their final destinations [11–13]. Most mitochondrial precursor proteins are destined for the matrix and must traverse the double bilayer using distinct translocases.

Preprotein translocation across the outer membrane utilizes the TOM complex and must be coordinated with translocation across the inner membrane, which is mediated by the TIM23 complex (Figure 44.1). Proteins destined for insertion into the inner membrane generally use the TIM23 complex, with the exception of solute carriers and the hydrophobic subunits of the TIM complex itself, which use a dedicated translocation complex referred to as TIM22.

TABLE 44.3 Mitochondrial ATP dependent proteases

	Mammalian gene	Yeast gene	Chaperone assistance (in yeast)	Cofactors/ modulators (in yeast)
Matrix				
Lon	LON	PIM1	mtHsp70 system, Hsp78	
ClpXP	ClpX, ClpP	absent		
Inner membrane				
i-AAA	YME1L1	YME1	?	*i-AAA*
m-AAA	paraplegin	YTA10/ AFG3	mtHsp70 system	
	AFG3L1	YTA11/ RCA1		
	AFG3L2			

Translocation Across the Outer Membrane Assisted by Chaperones and Chaperone-Like Proteins

The principal function of quality assurance in the translocation process is to ensure that preproteins remain in an unfolded state for transport. The binding of the chaperones Hsp70 and Hsp90 in the cytosol to newly synthesized precursor proteins maintains these preproteins in an unfolded and import competent state while also preventing their degradation or aggregation. Most precursor proteins are targeted to the matrix by amino terminal targeting sequences, which are cleaved off by the mitochondrial processing peptidase (MPP) resulting in the mature functional protein. Other matrix proteins have targeting sequences that are uncleaved. The chaperone associated precursors are targeted to the TOM complex, which consists of two surface exposed receptors Tom70 and Tom20/22 that are associated with the translocation pore composed principally by Tom40. A detailed understanding of the translocation process is primarily derived from work using *Saccharomyces cerevisiae* and *Neurospora crassa*; however, the structure and function of the TOM complex in mammals are largely conserved [12, 13]. Tom70 has 11 tetratricopeptide repeat (TPR) domains, which in mammals interact with both Hsp70 and Hsp90 as well as with hydrophobic preproteins that carry an internal targeting signal [14, 15]. The docking of the chaperone–preprotein complex at Tom70 is required for the recognition of internal

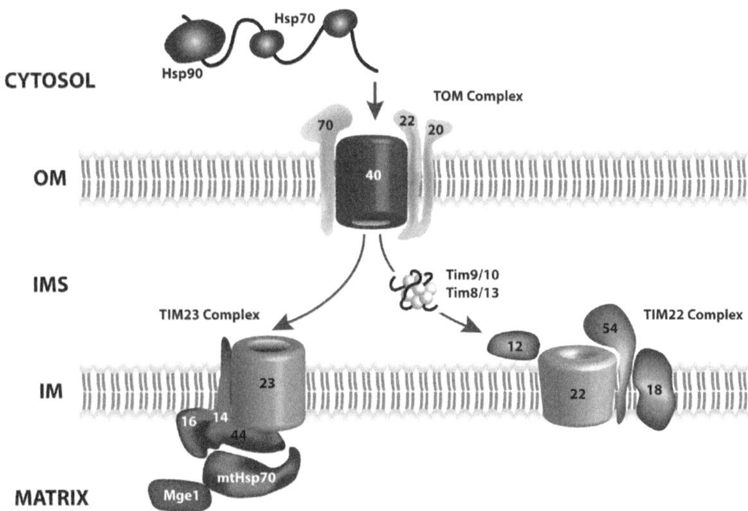

FIGURE 44.1 Quality assurance of protein import into the mitochondrion.
Most mitochondrial proteins are encoded by nuclear genes that are synthesized in the cytosol. The cytosolic chaperones Hsp70 and Hsp90 bind to many of these newly synthesized proteins and maintain them in an unfolded state for translocation across the outer membrane (OM) by the TOM complex. Proteins targeted to the matrix cross the inner membrane (IM) using the TIM23 complex. Components of the TOM and TIM23 complexes that exhibit chaperone-like functions are labeled in white. Carrier proteins inserted into the IM, are transported through the intermembrane space (IMS) by the chaperone-like proteins Tim9/10 and Tim8/13.

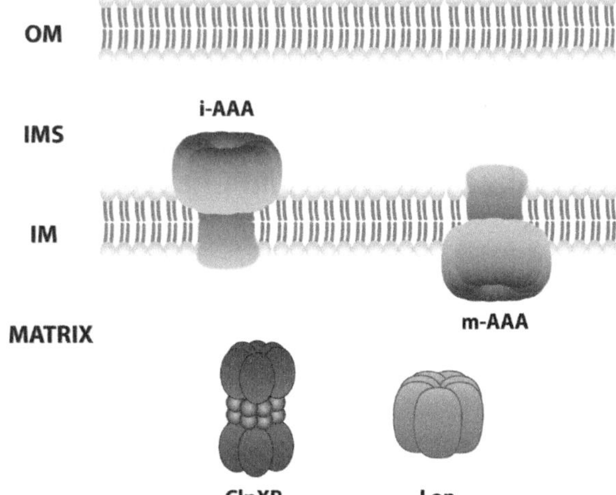

FIGURE 44.2 Quality control within the mitochondrion is mediated by ATP dependent proteases.
The soluble proteases within the matrix are Lon and the two-component ClpXP composed of the ClpP proteolytic complex and the ClpX ATPase complex. The inner membrane (IM) proteases are *i-AAA* with its catalytic sites for ATP hydrolysis and proteolysis located within the intermembrane space (IMS) and m-AAA with its catalytic sites located in the matrix.

targeting sequences by the receptor. It is postulated that ATP binding and hydrolysis by Hsp70 and Hsp90 promote the transfer of preproteins through the translocation pore. Other Hsp70 co-chaperones belonging to the Hsp40/DnaJ-like family of proteins also participate in Tom70 mediated protein import [16]. In addition, the Hsp90 co-chaperones, p23 and Aha1, modulate Hsp90–preprotein interactions [16]. By contrast, preprotein translocation mediated by Tom20/22 receptors appears to be less dependent on Hsp70 and Hsp90. Instead, the chaperone-like protein AIP (arylhydrocarbon receptor interacting protein) mediates import by binding to preproteins as well as to Tom20 [17]. The TOM complex itself prevents preprotein aggregation and unfolds precursor substrates. The IMS exposed or *trans* surface of the TOM complex has been shown to mediate the active unfolding and vectorial translocation of precursor proteins across the outer membrane [18, 19]. The inner wall and IMS surface of the Tom40 translocation channel exhibit an affinity for unfolded or loosely folded polypeptides [20]. This interaction with the Tom40 translocon accommodates preproteins until they are either (a) ready to fold properly and released as soluble proteins in the IMS, (b) transferred to the IMS chaperone-like proteins Tim9/10 or Tim8/13 that mediate insertion of TIM22 dependent proteins into the inner membrane, or (c) translocated by the TIM23 translocon into the matrix.

Translocation Across the Inner Membrane – Role of the mtHsp70 Chaperone

Quality assurance mechanisms are integrally associated with the protein import machinery across the inner membrane. Mitochondrial Hsp70 (mtHsp70) performs multiple functions as a key operator in protein translocation across the inner membrane, and also as a key mediator of protein folding in the matrix [21]. mtHsp70 is a structural and functional homolog of the bacterial DnaK ATPase, which cooperates with the co-chaperone DnaJ and the nucleotide exchange factor GrpE. Homologs of DnaJ and GrpE are also present within mitochondria [22]. DnaJ determines substrate specificity by binding and delivering proteins to ATP bound DnaK. Upon interaction with DnaJ, DnaK hydrolyzes ATP resulting in its ADP bound form, which associates with higher affinity to protein substrate and dissociates from DnaJ. The interaction of GrpE with ADP bound DnaK stimulates the exchange of ADP for ATP and the release of substrate. Non-native polypeptides are subject to cycles of reversible binding by the DnaK/DnaJ/GrpE system until the correct conformation required for protein function is achieved [22].

Results obtained from *S. cerevisiae* demonstrate that the reversible binding of mtHsp70 to protein substrates in the mitochondrial matrix is utilized in several ways: (a) to translocate preproteins across the inner membrane, (b) to stabilize newly imported polypeptides during folding, (c) to prevent protein aggregation, and (d) to assist in the efficient proteolysis of certain misfolded or unassembled substrates [21]. The TIM23 translocon mediates the transfer of the amino terminal targeting sequence of preproteins across the inner membrane; however, the complete translocation into the matrix requires an import motor that is powered by mtHsp70 (Figure 44.1). This motor is composed of mtHsp70 along with its nucleotide exchange factor Mge1, which together reversibly associate with structural subunits of the TIM23 complex – Tim44, Tim14 (Pam18), and Tim16 (Pam16) [11, 13]. Tim44 recruits ATP bound mtHsp70 to the translocon, and thereby promotes chaperone binding to the amino terminal targeting sequence of the preprotein as it emerges into the matrix. Atypical DnaJ-like proteins are associated with the mtHsp70 dependent import motor. Tim14 and Tim16 are membrane associated subunits of the TIM23 complex that each possess a signature J-domain characteristic of DnaJ-like proteins, which stimulates the ATPase activity of mtHsp70; Tim16, however, lacks this critical J-domain [11, 13]. Tim14 positively regulates mtHsp70 by stimulating ATP hydrolysis, thereby increasing the affinity of mtHsp70 for preprotein and decreasing its association with Tim44. By contrast, Tim16 negatively regulates this reaction. This dynamic cycling of reversible mtHsp70 binding to the preprotein mediates processive translocation across the inner membrane. The mitochondrial processing peptidase removes the amino terminal targeting sequence as the cleavage site emerges into the matrix. Translocation of the processed substrate proceeds to completion, whereupon protein folding and assembly ensue.

Protein Folding and Disaggregation Mediated by Chaperones in the Mitochondrial Matrix

Some proteins that are imported from the cytosol or synthesized within the matrix depend on chaperones for folding and complex assembly. mtHsp70 is required by some proteins (e.g., subunits of the F_1F_0 ATPase); whereas Hsp60 and Hsp10, which are mitochondrial homologs of the bacterial chaperones GroEL and GroES, respectively, are needed by others (e.g., bovine rhodanese) [23, 24]. In addition, there are proteins that fold spontaneously and independently of chaperones.

Mitochondrial Hsp70 Chaperones

In yeast there are three Hsp70s (mtHsp70/Ssc1, Ssc2/Ssq1, and Ssc3), whereas in mammals there is only mtHsp70. Studies in yeast have demonstrated that the role of mtHsp70 in protein folding is distinctly different from its role in protein translocation. This functional separation is demonstrated by yeast strains expressing different mutant forms of yeast mtHsp70 (Ssc1); some mutants exhibit defects in protein folding but not translocation (*ssc1-201* and *ssc1-202*), whereas others show defects in both processes (*ssc1-3* and *ssc1-3*) [25, 26]. The mtHsp70 system for protein folding shows mechanistic conservation with the bacterial DnaK/DnaJ/GrpE system as described above, and employs a mitochondrial DnaJ-like protein Mdj1 in addition to Mge1. In yeast there also exists a specialized chaperone system dedicated to the biogenesis of iron-sulfur (Fe/S) cluster containing proteins, which consists of the DnaK-like protein Ssq1, the DnaJ-like protein Jac1, in addition to Mge1, which also functions as an exchange factor in protein folding and import [25, 26].

The importance of Hsp70 dependent processes in mitochondria is highlighted by the existence of chaperones that function as specific chaperones for mtHsp70 and Ssq1. Data show that the nucleotide free form of mtHsp70 has a tendency to self-aggregate. Zim17 (also referred to as Hep1 or Tim15) is essential for maintaining the functions of Ssc1 and Ssq1 by preventing their self-aggregation [27–30]. Zim17 has two zinc finger motifs typically found in a subclass of DnaJ-like proteins, which are essential for activity. The maintenance of Hsp70 function in mitochondria also depends on Hsp78, which is a homolog of bacterial ClpB/Hsp100 chaperones that mediate the unfolding of proteins and disassembly of protein oligomers and aggregates [31]. By contrast to Zim17, Hsp78 functions as a disaggregase and resolubilizes aggregated Ssc1 and Ssq1 [31]. The other putative ClpB/Hsp100 in mitochondria, Mcx1p, has only limited ability to disaggregate Ssc1 [31] and its role in protein folding or disaggregation has not been well characterized.

Hsp60 and Hsp10

Assisted protein folding of mitochondrial matrix proteins is also mediated by the chaperonins Hsp60 and/or Hsp10. In contrast to mtHsp70, the GroEL related Hsp60 complex consists of two stacked rings of seven identical ATP hydrolyzing subunits, which form an inner chamber capable of accommodating proteins [32]. Non-native substrates entering this chamber achieve a folded state in isolation from the matrix milieu. The folding process involves cycles of substrate binding and release from the chamber wall, which is mediated by ATP dependent conformational changes in Hsp60. Some proteins that exceed the size limit of this inner cavity may also depend on Hsp60 to reach the folded state even though they are not encapsulated [33]. The GroES-like Hsp10 complex is a single ring of seven identical subunits that can cap one end of Hsp60 and coordinate cycles of ATP binding and hydrolysis [22]. Hsp60 and Hsp10 function as chaperones either alone or together [24].

The extent to which the mtHsp70 and Hsp60 systems cooperate in the folding of proteins once they are imported into the matrix has not been studied in detail. However, as mtHsp70 and Mge1 encounter precursor proteins during translocation into the matrix, a sequential order of chaperone assisted folding for some proteins is likely. In addition, there are auxiliary chaperones that accelerate protein folding. The yeast cyclophilin Cpr3 functions as a peptidyl-prolyl-isomerase within the matrix [34, 35]. By catalyzing *cis-trans* prolyl isomerizations, Cpr3 increases the rate at which a newly imported non-native polypeptide achieves its mature conformation. Cpr3 cooperates with mtHsp70 and Hsp60 and thereby increases the folding efficiency of these chaperones.

In mammalian mitochondria, Tid1 isoforms show significant sequence homology with DnaJ and with the *Drosophila melanogaster* tumor suppressor tumorous imaginal discs 56 (Tid56) [36, 37]. Tid1 exhibits a conserved mitochondrial DnaJ-like function by substituting for yeast Mdj1 [38]. Tid1 knockout mice die in early embryonic development [39], and mice that specifically lack Tid1 in the heart, develop dilated cardiomyopathy, respiratory chain deficiency, and reduced mitochondrial DNA (mtDNA) copy number [40]. Although Tid1 proteins localize overwhelmingly to mitochondria [37, 41], published data demonstrate principally non-mitochondrial functions in signaling pathways associated with stress responses [42–48]. Further work is required to determine Tid1 substrates and interactions with Hsp70-like chaperones inside and outside of the mitochondrion. In addition, the elongation factor of translation in mammalian mitochondria EF-Tumt has been shown to exhibit a chaperone-like activity *in vitro* by preventing protein aggregation and enhancing protein refolding [49]. Under stress conditions, EF-Tumt binds to unfolded proteins posttranslationally at the inner membrane and targets their degradation by mitochondrial proteases. Ef-Tumt may thus provide a mechanism for the quality control of mitochondrial protein synthesis within the matrix.

QUALITY CONTROL OF PROTEIN STRUCTURE AND FUNCTION MEDIATED BY ATP DEPENDENT PROTEASES

Mitochondrial ATP dependent proteases belong to the AAA^+ family of *A*TPases *A*ssociated with various cellular *A*ctivities, which include proteins that mediate proteolysis, membrane fusion, DNA replication, and transcription [50, 51]. Mitochondrial Lon and ClpXP proteases are soluble matrix enzymes that are named after their homologous bacterial ancestors, whereas *m*-AAA and *i*-AAA proteases are inner membrane embedded homologs of bacterial FtsH. These ATP dependent proteases oversee the final stage of protein quality control by ensuring that only correctly folded and assembled end products enter the mitochondrial work force (Figure 44.2).

A mechanistic understanding of ATP dependent proteolysis has emerged from studies principally with the 26S proteasome and the bacterial ClpP holoenzyme; these proteases generally degrade substrates tagged by peptide sequences such as ubiquitin and SsrA, respectively [52, 53]. Once a tagged protein substrate is recognized by the 26S proteasome or ClpP complex, the substrate is engaged at an unstructured or loosely folded region, after which it is unfolded and translocated to the proteolytic active site where it is degraded. Within mitochondria, discrete degradation sequences or tags are not used to mark proteins for degradation by ATP dependent proteases, which are generally unassembled, misfolded, or damaged proteins. Environmental changes such as thermal or oxidative stress, or genetic defects resulting in loss of protein subunit integrity and expression may cause abnormalities in the tertiary structure of proteins that signals degradation by mitochondrial ATP dependent proteases [54–56]. Certain substrates may also involve the participation of the mtHsp70 system and Hsp78 to assist in protein targeting and unfolding [57–60]. Recent work has demonstrated that the function of mitochondrial ATP dependent proteases extends beyond disposal management and is also directly linked to the biogenesis of proteins and mitochondria.

Lon

Lon is a soluble homo-oligomeric complex in which each subunit contains domains for ATP hydrolysis and proteolysis. Cryoelectron microscopy demonstrates that yeast Lon, also referred to as Pim1, is a ring-shaped complex with seven flexible subunits [61]. Absence of Lon/Pim1 in haploid yeast strains results in a lack of matrix ATP dependent proteolysis, accumulation of electron dense aggregates, and large mtDNA deletions [62, 63]. In yeast, Lon/Pim1 is the only ATP dependent protease within the matrix, in contrast to metazoans that have both Lon and ClpXP. The mtDNA deletion defect is uniquely associated with Lon-null yeast and is not observed in strains lacking the other membrane proteases *m*-AAA or *i*-AAA. Human Lon has been shown to bind mtDNA *in vitro* as well as in living cells, and preferentially associates with the control region for mtDNA replication and transcription [64–66]. mtDNA binding may function to recruit the protease to a site where it selectively degrades protein components of the replication, repair, and transcription machinery. However, the physiological importance of DNA binding by Lon remains unclear. Endogenous substrates of yeast Lon/Pim1 include subunits of the F_1/F_o ATPase, ribosomal proteins, and metabolic enzymes [54, 67–69]. Endogenous substrates of mammalian Lon include subunits of cytochrome c oxidase, which is an electron transport chain complex [67, 70], oxidized mitochondrial aconitase, which functions in the tricarboxylic acid cycle [96], and the steroidogenic acute regulatory protein StAR, which mediates cholesterol transfer from the cytosol to the mitochondrial inner membrane [74].

Some endogenous substrates of human Lon are degraded when they are in the folded state [55]. For example, the mitochondrial processing peptidase α subunit (MPPα) is degraded by Lon only when it is competent for assembly into active enzyme and trypsin resistant; by contrast, when MPPα is unfolded it is not degraded by Lon even though it is sensitive to limited trypsin digestion [55]. The initial cleavage sites within MPPα are hydrophobic residues located at the surface of the folded protein, which are surrounded by a highly charged environment. Such sites may function as recognition elements or patches that specify selective degradation by Lon. The finding that Lon degrades folded as well as abnormal proteins suggests that the protease may also have a role in regulating protein function.

Lon has also been shown to exhibit a chaperone-like activity that promotes protein complex assembly independent of its protease activity [70, 71]. In mammals, a stress response network between the endoplasmic reticulum (ER), nucleus, and mitochondria upregulates the expression of Lon and Yme1 (*i*-AAA), which provides distinctly different cytoprotective functions during ER stress or hypoxia [71]. Lon may promote protein complex assembly whereas Yme1/*i*-AAA may mediate selective proteolysis. ER stress is induced by the inhibition of ER functions such as protein glycosylation, Ca^{2+} uptake, or anterograde trafficking to the Golgi. Hypoxia can also potentially block these ER functions. Sustained inhibition leads to accumulation of abnormal proteins within the ER and activation of the unfolded protein response (erUPR) [72]. The erUPR stimulates the transcriptional upregulation of ER specific stress proteins. Another downstream effect of the erUPR is suppression of cytosolic protein synthesis, which alleviates the load of newly synthesized proteins within the ER. Partial suppression of protein translocation into mitochondria also occurs, reducing the levels of imported protein subunits that are needed to assemble electron transport chain complexes such as cytochrome c oxidase (COX). In the absence of proper COX assembly, there is a defect in respiration and a loss of mitochondrial membrane potential. Although ER stress and hypoxia lead to lower cytosolic synthesis of subunits Cox4 and 5, there is no significant effect on the mitochondrial synthesis of subunits Cox1 and 2 [71]. Instead, ER stress and oxygen deprivation lead to the rapid turnover of Cox2. The degradation of Cox2 is most likely mediated by the stress regulated increase of Yme1/*i*-AAA, in light of experiments in yeast showing that Yme1/*i*-AAA degrades Cox2 when it is unassembled [73]. By contrast, the upregulation of Lon upon ER stress or hypoxia has been proposed to increase Cox2 assembly into functional complexes and to partially restore mitochondrial membrane potential. Studies with yeast as well as mammalian cells show that defects in COX assembly can be alleviated upon overexpressing wild-type Lon or a protease inactive Lon mutant [70, 71]. Lon may thus have a chaperone-like function independent of its proteolytic activity.

New insight into the adaptive responses to hypoxia is demonstrated by the finding that Lon is involved in remodeling the COX holoenzyme [74]. During normoxia, an isoform of the Cox4 subunit, Cox4-1, is a stable component of the fully assembled enzyme. However, during oxygen deprivation, the hypoxia inducible transcription factor HIF-1α activates the expression of an alternate isoform, Cox4-2, as well as Lon. Increased levels of Lon are required for the degradation of Cox4-1, thereby permitting Cox4-2 assembly into the COX holoenzyme. It is hypothesized that subunit switching produces Cox4-1 or Cox4-2 containing complexes that are each optimized for transferring electrons under normoxic or hypoxic conditions, respectively. It is conceivable that Lon mediates not only the degradation of Cox4-1 but also the assembly of Cox4-2 into the holoenzyme complex. The upregulation of Lon by HIF-1α may modulate other processes in addition to COX function. HIF-1α is a master regulator of oxygen homeostasis and has essential roles in metazoan development, cellular metabolism, and disease pathogenesis (e.g., tumorogenesis) [75]. It will be interesting to determine the extent to which Lon has a role in HIF-1α induced processes.

ClpXP

In contrast to Lon, ClpXP is a two-component ATP dependent protease in which the catalytic sites for ATP hydrolysis and proteolysis are present on separate polypeptide chains. The protease component ClpP consists of two stacked homo-heptameric rings that are capped at each end by the ATPase component ClpX, which is a single homo-hexameric ring [76]. While ClpXP is present in mitochondria of metazoans, it is absent in yeast. Little is known about the quality control functions of ClpXP and its endogenous substrates. Human ClpXP shows little substrate overlap with human Lon [77, 78, C.K. Suzuki and M.R. Maurizi, unpublished observations], and does not degrade substrates recognized by bacterial ClpXP [79].

The accumulation of unfolded proteins in the mitochondrial matrix activates a signaling pathway referred to as the mitochondrial unfolded protein response (mtUPR), which upregulates the expression of stress related proteins within the matrix [80]. mtUPR is organelle specific and has not been shown to influence the levels of stress proteins within the ER. This is in contrast to erUPR, which increases stress proteins within mitochondria (see previous section). mtUPR upregulates ClpP and the *i*-AAA protease, as well as the matrix chaperones Hsp60, Hsp10, and the mitochondrial DnaJ-like protein Tid1 [80]. However, mtUPR has no effect on the expression of genes encoding Lon or the *m*-AAA subunits Afg3L2 and paraplegin. Although mtUPR and erUPR activate the transcription factor CHOP (*C*/*EBP ho*mology *p*rotein), each UPR pathway upregulates different mitochondrial and ER gene products. Recent work suggests that *cis*-acting elements and proteins modulate CHOP dependent transcription, which provide specificity to the mtUPR and erUPR pathways [81]. In *Caenorhabditis elegans*, ClpP may be important in mediating the mtUPR [82]. RNAi mediated knockdown of ClpP in worms blocked downstream events in the mtUPR signaling pathway. Further studies are required to elucidate the details of the mtUPR and the potential role of ClpP in initiating this stress response pathway between mitochondria and the nucleus.

m-AAA and i-AAA

Embedded within the inner membrane, *m*-AAA has its catalytic sites for ATP hydrolysis and proteolysis exposed to the matrix, whereas *i*-AAA has its active sites localized in the intermembrane space [10]. These large protease complexes are composed of individual subunits that are closely related to one another, and each polypeptide subunit carries an ATPase and protease domain, in addition to an HEXXH consensus motif for binding metal. In yeast, *i*-AAA is a homo-oligomeric complex whereas *m*-AAA is a hetero-oligomer composed of two different subunits. In mammals *i*-AAA is presumed to be a homo-oligomer, whereas *m*-AAA has five or possibly six isoforms that are homo-oligomers or hetero-oligomers [5]. The precise stoichiometry of these complexes is not known.

The functions and underlying mechanisms of *i*-AAA and *m*-AAA are understood in significant detail. Although *i*-AAA and *m*-AAA have their catalytic sites on opposite sides of the inner membrane they exhibit overlapping substrate specificity dependent on substrate topology and folding [83]. A model transmembrane substrate having unfolded regions on either side of the inner membrane is completely degraded by either *i*-AAA or *m*-AAA, thus demonstrating the dislocation of unfolded substrate polypeptides from the lipid bilayer in either direction during proteolysis [83]. The membrane spanning regions of *i*-AAA or *m*-AAA are likely to facilitate the extraction of substrate proteins across the membrane. The transmembrane regions of all *m*-AAA subunits are required for degradation of an integral membrane substrate but not for a peripheral membrane protein [84].

Modulators of substrate binding and degradation mediated by *i*-AAA and *m*-AAA have been identified. Cox20 functions as a chaperone in the processing, assembly, and *i*-AAA dependent turnover of the Cox2 subunit. Cox20 functions mediate the processing of Cox2 to the mature form and its assembly into the COX holoenzyme [85]. However, when the maturation of Cox2 is impaired, Cox20 facilitates its degradation by *i*-AAA [73]. Another factor Mgr1, also influences *i*-AAA proteolysis; however, its role is less clear. Mgr1 may function as an adaptor for substrate degradation by *i*-AAA and/or a factor in the assembly of the *i*-AAA complex [86]. The *m*-AAA associated prohibitin complex is unique as it appears to antagonize proteolysis [87]. The yeast prohibitin subunits Phb1 and Phb2 form a large inner membrane complex that may block substrate access to the active sites of *m*-AAA or alter the membrane environment of the protease thereby negatively regulating proteolysis [88].

m-AAA and *i*-AAA are not obligated to degrade protein substrates to completion. Substrates may be proteolytically cleaved but not fully degraded, or they may also be dislocated from the membrane in a process that is not coupled to protein degradation. *m*-AAA mediates the maturational processing of the ribosomal protein Mrp32L, which is required for its subsequent assembly into functional ribosomes that synthesize proteins in close proximity to the inner membrane [89]. *m*-AAA is also involved in the processing of OPA1, a dynamin GTPase that functions in mitochondrial fusion and inner membrane remodeling [90, 91]. Studies show that *i*-AAA also regulates the processing of OPA1 [92, 93]. Further studies are required to unravel the relationship between *m*- and *i*- AAA mediated cleavage of OPA1 and to determine the specificity of these proteases, as *m*-AAA as well as OPA1 have multiple isoforms. *m*-AAA also participates in the maturational processing of cytochrome c peroxidase Ccp1 by the rhomboid protease, which is an intramembrane cleaving peptidase [94]. Ccp1 processing depends only on the dislocase activity of m-AAA and not on its protease activity. In an ATP dependent reaction, *m*-AAA vectorially dislocates Ccp1 and thereby ensures its correct positioning within the inner membrane thereby facilitating its cleavage by the intramembrane rhomboid peptidase. This dislocation activity is also conserved in *i*-AAA, which is exploited in protein translocation rather than protein processing or degradation. *i*-AAA binds to the precursor of polynucleotide phosphorylase (PNPase) as it is being translocated across the double bilayer of the mitochondrion and functions to mediate the translocation of PNPase into the IMS [95]. These findings demonstrate that the roles of *m*-AAA and *i*-AAA in protein quality control include not only protein degradation but also protein and mitochondrial biogenesis.

PERSPECTIVE

From our current vantage point we can begin to chart out the diverse pathways by which the mediators of quality control determine cellular and organismal physiology. Inherited human diseases linked to defects in mitochondrial ATP dependent proteases and chaperones reveal how protein quality control extends beyond the boundaries of the mitochondrion. Mutations in the human gene encoding paraplegin, a subunit of the *m*-AAA protease, cause an autosomal recessive form of hereditary spastic paraplegia (HSP), resulting in neurodegeneration [96]. Whether HSP is caused by a general defect in paraplegin dependent proteolysis, or by defects in the turnover or processing of cell specific substrates remains unknown. In addition, loss of function mutations in the human gene encoding DDP1/TIMM8a, which are homologs of yeast Tim8–13 that chaperone the import of carrier proteins to the inner membrane [97, 98], are associated with Mohr-Tranebjaerg syndrome, leading to deafness, dystonia, mental deficiency, and blindness.

The potential functions of other quality control proteins in stress- and disease related processes await further elucidation. A role for Lon in the response to oxidative stress and damage of proteins and mtDNA [66, 99, 100] has direct relevance to aging. In addition, observations that Lon and Yme1/*i*-AAA are upregulated by ER stress- and HIF1 dependent pathways have implications for cancer biology and chemotherapy. In many solid tumors, these transcriptional pathways are activated to promote tumor survival and metastasis [101, 102]. Future work holds the promise of delineating the roles of mitochondrial ATP dependent proteases and chaperones not only in cellular metabolism and general physiology, but also in disease states such as neurodegeneration and cancer, or aging.

ACKNOWLEDGEMENTS

I am very grateful to Drs. D.M. Gordon, D. Pain and B.J. Wagner for their helpful discussions and critical review of this manuscript, and to Dr. F.J. Monsma, Jr. for artwork and design.

REFERENCES

1. de Silva AM, Balch WE, Helenius A. Quality control in the endoplasmic reticulum: folding and misfolding of vesicular stomatitis virus G protein in cells and *in vitro*. *J Cell Biol* 1990;**111**:857–66.

2. Hurtley SM, Bole DG, Hoover-Litty H, et al. Interactions of misfolded influenza virus hemagglutinin with binding protein (BiP). *J Cell Biol* 1989;**108**:2117–26.
3. Brodsky JL. The protective and destructive roles played by molecular chaperones during ERAD (endoplasmic-reticulum-associated degradation). *Biochem J* 2007;**404**:353–63.
4. Bukau B, Weissman J, Horwich A. Molecular chaperones and protein quality control. *Cell* 2006;**125**:443–51.
5. Koppen M, Metodiev MD, Casari G, et al. Variable and tissue-specific subunit composition of mitochondrial m-AAA protease complexes linked to hereditary spastic paraplegia. *Mol Cell Biol* 2007;**27**:758–67.
6. Leidhold C, Voos W. Chaperones and proteases: Guardians of protein integrity in eukaryotic organelles. *Ann N Y Acad Sci* 2007;**1113**:72–86.
7. Meusser B, Hirsch C, Jarosch E, Sommer T. ERAD: the long road to destruction. *Nat Cell Biol* 2005;**7**:766–72.
8. Klausner RD, Lippincott-Schwartz J, Bonifacino JS. The T cell antigen receptor: insights into organelle biology. *Annu Rev Cell Biol* 1990;**6**:403–31.
9. Green DR. Apoptotic pathways: ten minutes to dead. *Cell* 2005;**121**: 671–4.
10. Koppen M, Langer T. Protein degradation within mitochondria: versatile activities of AAA proteases and other peptidases. *Crit Rev Biochem Mol Biol* 2007;**42**:221–42.
11. Bohnert M, Pfanner N, van der Laan M. A dynamic machinery for import of mitochondrial precursor proteins. *FEBS Lett* 2007;**581**:2802–10.
12. Dolezal P, Likic V, Tachezy J, Lithgow T. Evolution of the molecular machines for protein import into mitochondria. *Science* 2006; **313**:314–18.
13. Neupert W, Herrmann JM. Translocation of proteins into mitochondria. *Annu Rev Biochem* 2007;**76**:723–49.
14. Fan AC, Bhangoo MK, Young JC. Hsp90 functions in the targeting and outer membrane translocation steps of Tom70-mediated mitochondrial import. *J Biol Chem* 2006;**281**:33,313–33,324.
15. Young JC, Hoogenraad NJ, Hartl FU. Molecular chaperones Hsp90 and Hsp70 deliver preproteins to the mitochondrial import receptor Tom70. *Cell* 2003;**112**:41–50.
16. Bhangoo MK, Tzankov S, Fan AC, et al. Multiple 40-kDa heat-shock protein chaperones function in Tom70-dependent mitochondrial import. *Mol Biol Cell* 2007;**18**:3414–28.
17. Yano M, Terada K, Mori M. Mitochondrial import receptors Tom20 and Tom22 have chaperone-like activity. *J Biol Chem* 2004; **279**:10,808–10,813.
18. Mayer A, Neupert W, Lill R. Mitochondrial protein import: reversible binding of the presequence at the trans side of the outer membrane drives partial translocation and unfolding. *Cell* 1995;**80**:127–37.
19. Rapaport D, Kunkele KP, Dembowski M, et al. Dynamics of the TOM complex of mitochondria during binding and translocation of preproteins. *Mol Cell Biol* 1998;**18**:5256–62.
20. Esaki M, Kanamori T, Nishikawa S, et al. Tom40 protein import channel binds to non-native proteins and prevents their aggregation. *Nat Struct Biol* 2003;**10**:988–94.
21. Voos W, Rottgers K. Molecular chaperones as essential mediators of mitochondrial biogenesis. *Biochim Biophys Acta* 2002;**1592**:51–62.
22. Bukau B, Horwich A. The Hsp70 and Hsp60 chaperone machines. *Cell* 1998;**92**:351–66.
23. Herrmann JM, Stuart RA, Craig EA, Neupert W. Mitochondrial heat shock protein 70, a molecular chaperone for proteins encoded by mitochondrial DNA. *J Cell Biol* 1994;**127**:893–902.
24. Rospert S, Looser R, Dubaquie Y, et al. Hsp60-independent protein folding in the matrix of yeast mitochondria. *EMBO J* 1996;**15**:764–74.
25. Kang PJ, Ostermann J, Shilling J, et al. Requirement for hsp70 in the mitochondrial matrix for translocation and folding of precursor proteins. *Nature* 1990;**348**:137–43.
26. Liu Q, Krzewska J, Liberek K, Craig EA. Mitochondrial Hsp70 Ssc1: role in protein folding. *J Biol Chem* 2001;**276**:6112–18.
27. Burri L, Vascotto K, Fredersdorf S, et al. Zim17, a novel zinc finger protein essential for protein import into mitochondria. *J Biol Chem* 2004;**279**:50,243–50,249.
28. Momose T, Ohshima C, Maeda M, Endo T. Structural basis of functional cooperation of Tim15/Zim17 with yeast mitochondrial Hsp70. *EMBO Rep* 2007;**8**:664–70.
29. Sanjuan Szklarz LK, Guiard B, Rissler M, et al. Inactivation of the mitochondrial heat shock protein zim17 leads to aggregation of matrix hsp70s followed by pleiotropic effects on morphology and protein biogenesis. *J Mol Biol* 2005;**351**:206–18.
30. Sichting M, Mokranjac D, Azem A, et al. Maintenance of structure and function of mitochondrial Hsp70 chaperones requires the chaperone Hep1. *Embo J* 2005;**24**:1046–56.
31. von Janowsky B, Major T, Knapp K, Voos W. The disaggregation activity of the mitochondrial ClpB homolog Hsp78 maintains Hsp70 function during heat stress. *J Mol Biol* 2006;**357**:793–807.
32. Sigler PB, Xu Z, Rye HS, et al. Structure and function in GroEL-mediated protein folding. *Ann Rev Biochem* 1998;**67**:581–608.
33. Chaudhuri TK, Farr GW, Fenton WA, et al. GroEL/GroES-mediated folding of a protein too large to be encapsulated. *Cell* 2001;**107**:235–46.
34. Rassow J, Mohrs K, Koidl S, et al. Cyclophilin 20 is involved in mitochondrial protein folding in cooperation with molecular chaperones Hsp70 and Hsp60. *Mol Cell Biol* 1995;**15**:2654–62.
35. Matouschek A, Rospert S, Schmid K, et al. Cyclophilin catalyzes protein folding in yeast mitochondria. *Proc Natl Acad Sci USA* 1995;**92**:6319–23.
36. Kurzik-Dumke U, Gundacker D, Renthrop M, Gateff E. Tumor suppression in *Drosophila* is causally related to the function of the lethal(2) tumorous imaginal discs gene, a dnaJ homolog. *Dev Genet* 1995;**16**:64–76.
37. Syken J, De-Medina T, Münger K. *TID1*, a human homolog of the *Drosophila* tumor suppressor *l(2) tid*, encodes two mitochondrial modulators of apoptosis with opposing functions. *Proc Natl Acad Sci USA* 1999;**96**:8499–504.
38. Lu B, Garrido N, Spelbrink JN, Suzuki CK. Tid1 isoforms are mitochondrial DnaJ-like proteins with unique carboxy-termini that determine cytosolic fate. *J Biol Chem* 2006;**281**:13,150–13,158.
39. Asai T, Takahashi T, Esaki M, et al. Reinvestigation of the requirement of cytosolic ATP for mitochondrial protein import. *J Biol Chem* 2004;**279**:19,464–19,470.
40. Hayashi M, Imanaka-Yoshida K, Yoshida T, et al. A crucial role of mitochondrial Hsp40 in preventing dilated cardiomyopathy. *Nat Med* 2006;**12**:128–32.
41. Kurzik-Dumke U, Debes A, Kaymer M, Dienes P. Mitochondrial localization and temporal expression of the *Drosophila* melanogaster DnaJ homologous tumor suppressor Tid50. *Cell Stress Chaperones* 1998;**3**:12–27.
42. Bae MK, Jeong JW, Kim SH, et al. Tid-1 interacts with the von Hippel-Lindau protein and modulates angiogenesis by destabilization of HIF-1alpha. *Cancer Res* 2005;**65**:2520–5.
43. Cheng H, Cenciarelli C, Nelkin G, et al. Molecular mechanism of hTid-1, the human homolog of Drosophila tumor suppressor l(2)Tid, in the regulation of NF-kappaB activity and suppression of tumor growth. *Mol Cell Biol* 2005;**25**:44–59.

44. Edwards KM, Munger K. Depletion of physiological levels of the human TID1 protein renders cancer cell lines resistant to apoptosis mediated by multiple exogenous stimuli. *Oncogene* 2004;**4**:8419–31.
45. Kim SW, Hayashi M, Lo JF, et al. Tid1 negatively regulates the migratory potential of cancer cells by inhibiting the production of interleukin-8. *Cancer Res* 2005;**65**:8784–91.
46. Sarkar S, Pollack BP, Lin K-T, et al. hTid-1, a human DnaJ protein, modulates the interferon signaling pathway. *J Biol Chem* 2001;**276**:49,034–49,042.
47. Sohn SY, Kim SB, Kim J, Ahn BY. Negative regulation of hepatitis B virus replication by cellular Hsp40/DnaJ proteins through destabilization of viral core and X proteins. *J Gen Virol* 2006;**87**:1883–91.
48. Trentin GA, He Y, Wu DC, et al. Identification of a hTid-1 mutation which sensitizes gliomas to apoptosis. *FEBS Lett* 2004;**578**:323–30.
49. Suzuki H, Ueda T, Taguchi H, Takeuchi N. Chaperone properties of mammalian mitochondrial translation elongation factor Tu. *J Biol Chem* 2007;**282**:4076–84.
50. Hanson PI, Whiteheart SW. AAA+ proteins: have engine, will work. *Nat Rev Mol Cell Biol* 2005;**6**:519–29.
51. Neuwald AF, Aravind L, Spouge JL, Koonin EV. AAA+: A class of chaperone-like ATPases associated with the assembly, operation, and disassembly of protein complexes. *Genome Res* 1999;**9**:27–43.
52. Baker TA, Sauer RT. ATP-dependent proteases of bacteria: recognition logic and operating principles. *Trends Biochem Sci* 2006;**31**:647–53.
53. Pickart CM, Cohen RE. Proteasomes and their kin: proteases in the machine age. *Nat Rev Mol Cell Biol* 2004;**5**:177–87.
54. Major T, von Janowsky B, Ruppert T, et al. Proteomic analysis of mitochondrial protein turnover: identification of novel substrate proteins of the matrix protease pim1. *Mol Cell Biol* 2006;**26**:762–76.
55. Ondrovicova G, Liu T, Singh K, et al. Cleavage Site Selection within a folded substrate by the ATP-dependent lon protease. *J Biol Chem* 2005;**280**:25,103–25,110.
56. von Janowsky B, Knapp K, Major T, et al. Structural properties of substrate proteins determine their proteolysis by the mitochondrial AAA+ protease Pim1. *Biol Chem* 2005;**386**:1307–17.
57. Bateman JM, Iacovino M, Perlman PS, Butow RA. Mitochondrial DNA instability mutants of the bifunctional protein Ilv5p have altered organization in mitochondria and are targeted for degradation by Hsp78 and the Pim1p protease. *J Biol Chem* 2002;**277**:47,946–47,953.
58. Leidhold C, von Janowsky B, Becker D, et al. Structure and function of Hsp78, the mitochondrial ClpB homolog. *J Struct Biol* 2006;**156**:149–64.
59. Savel'ev AS, Novikova LA, Kovaleva IE, et al. ATP-dependent proteolysis in mitochondria. m-AAA protease and PIM1 protease exert overlapping substrate specificities and cooperate with the mtHsp70 system. *J Biol Chem* 1998;**273**:20,596–20,602.
60. Wagner I, Arlt H, van Dyck L, et al. Molecular chaperones cooperate with PIM1 protease in the degradation of misfolded proteins in mitochondria. *EMBO J* 1994;**13**:5135–45.
61. Stahlberg H, Kutejova E, Suda K, et al. Mitochondrial Lon of *Saccharomyces cerevisiae* is a ring-shaped protease with seven flexible subunits. *Proc Natl Acad Sci USA* 1999;**96**:6787–90.
62. Suzuki CK, Suda K, Wang N, Schatz G. Requirement for the yeast gene LON in intramitochondrial proteolysis and maintenance of respiration. *Science* 1994;**264**:273–6. 891.
63. van Dyck L, Pearce DA, Sherman F. PIM1 encodes a mitochondrial ATP-dependent protease that is required for mitochondrial function in the yeast *Saccharomyces cerevisiae*. *J Biol Chem* 1994;**269**:238–42.
64. Liu T, Lu B, Lee I, et al. DNA and RNA binding by the mitochondrial Lon protease is regulated by nucleotide and protein substrate. *J Biol Chem* 2004;**279**:13,902–13,910.
65. Lu B, Liu T, Crosby JA, et al. The ATP-dependent Lon protease of Mus musculus is a DNA-binding protein that is functionally conserved between yeast and mammals. *Gene* 2003;**306**:45–55.
66. Lu B, Yadav S, Shah PG, et al. Roles for the human ATP-dependent Lon protease in mitochondrial DNA maintenance. *J Biol Chem* 2007;**282**:17,363–17,374.
67. Rep M, Grivell LA. The role of protein degradation in mitochondrial function and biogenesis. *Curr Genet* 1996;**30**:367–80.
68. Suzuki CK, Rep M, van Dijl JM, et al. ATP-dependent proteases that also chaperone protein biogenesis. *Trends Biochem Sci* 1997;**22**:118–23.
69. van Dyck L, Langer T. ATP-dependent proteases controlling mitochondrial function in the yeast *Saccharomyces cerevisiae*. *Cell Mol Life Sci* 1999;**56**:825–42.
70. Rep M, van Dijl JM, Suda K, et al. Promotion of mitochondrial membrane complex assembly by a proteolytically inactive yeast Lon. *Science* 1996;**274**:103–6.
71. Hori O, Icinoda F, Tamatani T, et al. Transmission of cell stress from endoplasmic reticulum to mitochondria: enhanced expression of Lon protease. *J Cell Biol* 2002;**157**:1151–60.
72. Ron D, Walter P. Signal integration in the endoplasmic reticulum unfolded protein response. *Nat Rev Mol Cell Biol* 2007;**8**:519–29.
73. Graef M, Seewald G, Langer T. Substrate recognition by AAA+ ATPases: distinct substrate binding modes in ATP-dependent protease Yme1 of the mitochondrial intermembrane space. *Mol Cell Biol* 2007;**27**:2476–85.
74. Fukuda R, Zhang H, Kim JW, et al. HIF-1 regulates cytochrome oxidase subunits to optimize efficiency of respiration in hypoxic cells. *Cell* 2007;**129**:111–22.
75. Hirota K, Semenza GL. Regulation of angiogenesis by hypoxia-inducible factor 1. *Crit Rev Oncol Hematol* 2006;**59**:15–26.
76. Kang SG, Dimitrova MN, Ortega J, et al. Human mitochondrial ClpP is a stable heptamer that assembles into a tetradecamer in the presence of ClpX. *J Biol Chem* 2005;**280**:35,424–35,432.
77. Granot Z, Kobiler O, Melamed-Book N, et al. Turnover of mitochondrial steroidogenic acute regulatory (StAR) protein by Lon protease: the unexpected effect of proteasome inhibitors. *Mol Endocrinol* 2007;**21**:2164–77.
78. Hansen J, Gregersen N, Bross P. Differential degradation of variant medium-chain acyl-CoA dehydrogenase by the protein quality control proteases Lon and ClpXP. *Biochem Biophys Res Commun* 2005;**333**:1160–70.
79. Kang SG, Ortega J, Singh SK, et al. Functional proteolytic complexes of the human mitochondrial ATP-dependent protease, hClpXP. *J Biol Chem* 2002;**277**:21,095–21,102.
80. Zhao Q, Wang J, Levichkin IV, et al. A mitochondrial specific stress response in mammalian cells. *EMBO J* 2002;**21**:4411–19.
81. Aldridge JE, Horibe T, Hoogenraad NJ. Discovery of genes activated by the mitochondrial unfolded protein response (mtUPR) and cognate promoter elements. *PLoS ONE* 2007;**2**:e874.
82. Haynes CM, Petrova K, Benedetti C, et al. ClpP mediates activation of a mitochondrial unfolded protein response in C. elegans. *Dev Cell* 2007;**13**:467–80.
83. Leonhard K, Guiard B, Pellecchia G, et al. Membrane protein degradation by AAA proteases in mitochondria: extraction of substrates from either membrane surface. *Mol Cell* 2000;**5**:629–38.
84. Korbel D, Wurth S, Käser M, Langer T. Membrane protein turnover by the m-AAA protease in mitochondria depends on the transmembrane domains of its subunits. *EMBO Rep* 2004;**5**:698–703.

85. Hell K, Tzagoloff A, Neupert W, Stuart RA. Identification of Cox20p, a novel protein involved in the maturation and assembly of cytochrome oxidase subunit 2. *J Biol Chem* 2000;**275**:4571–8.
86. Dunn CD, Lee MS, Spencer FA, Jensen RE. A genomewide screen for petite-negative yeast strains yields a new subunit of the i-AAA protease complex. *Mol Biol Cell* 2006;**17**:213–26.
87. Steglich G, Neupert W, Langer T. Prohibitins regulate membrane protein degradation by the m-AAA protease in mitochondria. *Mol Cell Biol* 1999;**19**:3435–42.
88. Tatsuta T, Model K, Langer T. Formation of membrane-bound ring complexes by prohibitins in mitochondria. *Mol Biol Cell* 2005;**16**:248–59.
89. Nolden M, Ehses S, Koppen M, et al. The m-AAA protease defective in hereditary spastic paraplegia controls ribosome assembly in mitochondria. *Cell* 2005;**123**:277–89.
90. Duvezin-Caubet S, Koppen M, Wagener J, et al. OPA1 processing reconstituted in yeast depends on the subunit composition of the m-AAA protease in mitochondria. *Mol Biol Cell* 2007;**18**:3582–90.
91. Ishihara N, Fujita Y, Oka T, Mihara K. Regulation of mitochondrial morphology through proteolytic cleavage of OPA1. *Embo J* 2006;**25**:2966–77.
92. Griparic L, Kanazawa T, van der Bliek AM. Regulation of the mitochondrial dynamin-like protein Opa1 by proteolytic cleavage. *J Cell Biol* 2007;**178**:757–64.
93. Song Z, Chen H, Fiket M, et al. OPA1 processing controls mitochondrial fusion and is regulated by mRNA splicing, membrane potential, and Yme1L. *J Cell Biol* 2007;**178**:749–55.
94. Tatsuta T, Augustin S, Nolden M, et al. m-AAA protease-driven membrane dislocation allows intramembrane cleavage by rhomboid in mitochondria. *Embo J* 2007;**26**:325–35.
95. Rainey RN, Glavin JD, Chen HW, et al. A new function in translocation for the mitochondrial i-AAA protease Yme1: import of polynucleotide phosphorylase into the intermembrane space. *Mol Cell Biol* 2006;**26**:8488–97.
96. Rugarli EI, Langer T. Translating m-AAA protease function in mitochondria to hereditary spastic paraplegia. *Trends Mol Med* 2006;**12**:262–9.
97. Hofmann S, Rothbauer U, Muhlenbein N, et al. The C66W mutation in the deafness dystonia peptide 1 (DDP1) affects the formation of functional DDP1.TIM13 complexes in the mitochondrial intermembrane space. *J Biol Chem* 2002;**277**:23,287–23,293.
98. Roesch K, Curran SP, Tranebjaerg L, Koehler CM. Human deafness dystonia syndrome is caused by a defect in assembly of the DDP1/TIMM8a-TIMM13 complex. *Hum Mol Genet* 2002;**11**:477–86.
99. Bota DA, Davies KJA. Lon protease preferentially degrades oxidized mitochondrial aconitase by an ATP-stimulated mechanism. *Nat Cell Biol* 2002;**4**:674–80.
100. Bulteau AL, Szweda LI, Friguet B. Mitochondrial protein oxidation and degradation in response to oxidative stress and aging. *Exp Gerontol* 2006;**41**:653–7.
101. Chan DA, Krieg AJ, Turcotte S, Giaccia AJ. HIF gene expression in cancer therapy. *Methods Enzymol* 2007;**435**:323–45.
102. Feldman DE, Chauhan V, Koong AC. The unfolded protein response: a novel component of the hypoxic stress response in tumors. *Mol Cancer Res* 2005;**3**:597–605.

Chapter 45

Mitochondria as Organizers of the Cellular Ca^{2+} Signaling Network

György Szabadkai and Michael R. Duchen

Department of Physiology and UCL Mitochondrial Biology Group, University College London, England, UK

INTRODUCTION

Over the past decade, mitochondria have emerged from relative obscurity to take center stage as remarkably autonomous and dynamic cellular organelles that are intimately involved in orchestrating a diverse range of cellular activities. Mitochondria are targeted by numerous cell signaling pathways, providing mechanisms to adapt to various stress conditions but also carrying the power to determine cell fate [1–4]. Thus, a major focus of current research in mitochondrial biology is to explore the signals generated by mitochondria to regulate gene expression and protein signaling networks. Ca^{2+} seems to be one of the most significant among these signals. Indeed, while the expression in the mitochondrial membrane of Ca^{2+} transporting mechanisms was established many years ago, the physiological significance of these pathways has only recently become apparent and is still being refined. There is now no question that mitochondria will take up and accumulate Ca^{2+} in all cells studied during the routine events of cellular $[Ca^{2+}]_c$ signaling, and that the pathway influences both mitochondrial function itself and the spatiotemporal and quantitative characteristics of the cellular $[Ca^{2+}]_c$ signal. As general issues relating to mitochondrial Ca^{2+} handling have recently been widely reviewed (see, for example, [5–10]), we propose in this chapter to highlight some of the more controversial and novel developments in the field over recent years, and some of the mechanistic, quantitative, and comparative questions that remain.

FUNDAMENTALS

The simplest model of mitochondrial Ca^{2+} handling is based on the cooperation of three processes: uptake of Ca^{2+} from the extramitochondrial space, Ca^{2+} buffering in the mitochondrial matrix, and extrusion into the surroundings. Curiously, and at variance with other Ca^{2+} handling systems (for example, that of the endoplasmic reticulum or the plasma membrane), movement of Ca^{2+} in both directions uses the same energy source; both are tightly coupled to respiration [11, 12]. Respiring mitochondria (supplied with oxygen and a carbon source) maintain a membrane potential, referred to as $\Delta\psi_m$, tightly coupled to a pH gradient, due to the activity of the electron transport chain (ETC). While Ca^{2+} accumulation depends on the electrochemical gradient for Ca^{2+} (defined by the $\Delta\psi_m$ and the low resting intramitochondrial Ca^{2+} concentration ($[Ca^{2+}]_m$)), Ca^{2+} extrusion through a xNa^+/Ca^{2+} exchanger is coupled to the H^+ gradient through Na^+/H^+ exchange. In this scenario, net Ca^{2+} accumulation in the mitochondrial matrix might occur only if the transport systems responsible for uptake and extrusion have different transport capacities. Indeed, as will be detailed below, above a certain $[Ca^{2+}]$ in the extramitochondrial space ($[Ca^{2+}]_e$), the rate of Ca^{2+} uptake exceeds its rate of extrusion, which is saturated at much lower Ca^{2+} levels, resulting in net Ca^{2+} accumulation in the matrix. The difference is excessive, but, as recently pointed out by Nicholls, it is matrix buffering, mainly by forming of insoluble insoluble xCa^{2+}–xPO_4^{x-}–xOH^- complexes, that rescues mitochondria from Ca^{2+} overload [11, 13]. Importantly, this latter process is also strictly pH dependent; thus, the high-capacity Ca^{2+} uptake, Ca^{2+} extrusion and the vast Ca^{2+} buffering systems are all controlled by the metabolic state (reflected in $\Delta\psi_m$ and ΔpH) of the mitochondria, providing a framework to control intracellular Ca^{2+} movements according to ever-changing cellular needs.

Handbook of Cell Signaling, Three-Volume Set 2 ed.

THE PLASTICITY OF THE MITOCHONDRIAL CA^{2+} HANDLING MACHINERY

The simple model described above becomes much more complex if we add even just some of the specific details accumulated during the past almost 60 years, counting from the first observations on mitochondrial Ca^{2+} accumulation [14, 15]. We can almost guarantee that there will be much more to discover, since virtually none of the molecules responsible for mitochondrial Ca^{2+} transport have yet been identified, precluding the use of the genetic studies that have proven so powerful in other systems [16].

The Ca^{2+} Uptake Pathway

Ca^{2+} has to traverse two membranes to enter into the mitochondrial matrix, but it has only recently been acknowledged that the outer membrane (OMM) can function as a regulated Ca^{2+} barrier as well as the inner (IMM). Overexpression of the voltage-dependent anion channel (VDAC) of the OMM increases matrix Ca^{2+} accumulation [17], and may function as a Ca^{2+}-activated Ca^{2+} channel [18, 19], contributing to the ruthenium red (RuR) sensitivity of the whole Ca^{2+} uptake machinery [20]. Accordingly, changes in conductivity or expression of VDAC might regulate Ca^{2+} accumulation under physiological conditions, and further studies on available genetic models could clarify this outstanding question. The insertion of truncated Bid into the OMM was also shown to increase net mitochondrial Ca^{2+} uptake, which might contribute to the Ca^{2+}-induced alteration of mitochondrial structure and function during apoptosis [21].

In contrast to the OMM, it has been recognized for many years that Ca^{2+} is taken up through the IMM by a uniporter. Remarkably, we do not know the molecular identity or even the precise nature of this pathway. Is it a channel or a carrier? Flux rates are equivalent to those measured for fast gated pores, but rather slower than those seen for channels (see [22] for review). The activity of the uniporter shows little sensitivity to changes in temperature, and it also shows a wide spectrum of cation selectivity, together suggesting that it is a channel rather than a carrier. Ca^{2+} uptake via the uniporter is inhibited by ruthenium red (RuR), a compound which inhibits a variety of cation channels, including L-type plasmalemmal Ca^{2+} channels [23], ryanodine-sensitive ER Ca^{2+} release channels [24], and vanilloid receptor operated channels [25], again suggesting that the uniporter may share channel properties. Moreover, an inwardly rectifying, highly Ca^{2+} selective, voltage-dependent Ca^{2+} channel (MiCa) was recently identified in patch-clamped mitoplasts (isolated IMM vesicles), with properties that match those predicted for the Ca^{2+} uniporter [26].

One of the most interesting features of the uniporter is an apparent gating by $[Ca^{2+}]_c$, identified primarily through studies of the Ca^{2+} sensitivity of RuR sensitive mitochondrial Ca^{2+} *efflux* in response to dissipation of $\Delta\psi_m$ [27, 28]. Montero and colleagues [28] showed that, while collapse of $\Delta\psi_m$ prevents mitochondrial Ca^{2+} uptake, collapse of $\Delta\psi_m$ *after* the accumulation of mitochondrial Ca^{2+} inhibited mitochondrial efflux – i.e., all mitochondrial efflux pathways were inhibited by depolarization. Addition of Ca^{2+} to the depolarized, Ca^{2+}-loaded mitochondria then promoted mitochondrial Ca^{2+} release sensitive to RuR, suggesting release through the uniporter. This is consistent with suggestions that the uniporter is allosterically gated by $[Ca^{2+}]_o$ [27], an observation that may also explain why local $[Ca^{2+}]_c$ needs to be higher than might be expected from the behavior of a conducting Ca^{2+} channel, in order to see significant increases in $[Ca^{2+}]_m$.

As mentioned above, the Ca^{2+} uniporter is surprisingly (and somewhat worryingly) resistant to attempts of identification. Some studies point to the existence of one or more glycoprotein complexes with regulated Ca^{2+} binding and channeling properties, but none of them have been shown unequivocally to fulfil this role (for reviews, see [6, 12]). Perhaps this uncertainty led to the recent proposal that other mitochondrial transporters might represent the long-sought IMM Ca^{2+} channel. A recent study from Graier's group provided genetic evidence that up- and downregulation of the uncoupling proteins (UCP-2 and 3) cause respective changes in mitochondrial Ca^{2+} uptake [29]. Despite the fascinating finding that knockdown of UCP2 completely blocks Ca^{2+} uptake in isolated liver mitochondria, the fact that the expression of the protein in yeast strains (which does not exhibit Ca^{2+} uniporter activity) does not give rise to a Ca^{2+} channel seems to show that these proteins cannot represent the uniporter itself. Indeed, the authors came to the conclusion the UCPs might reside in the same macromolecular complex with the molecular species responsible for IMM Ca^{2+} transfer.

An uptake pathway with properties distinct from those of the uniporter has also been described [30, 31], and dubbed the "rapid uptake mode" (RaM). This pathway has the capacity to transfer Ca^{2+} very rapidly into the mitochondria during the rising phase of a Ca^{2+} pulse. The properties of the pathway differ in different tissues [30], but in the heart the pathway saturates quickly and is slow to reset after activation. Again, the functional significance of the pathway remains to be established.

Ca^{2+} Extrusion Routes

The major route for Ca^{2+} efflux from mitochondria is a xNa^+/Ca^{2+} exchange. Identified about 20 years ago, it has a discrete pharmacology distinct from the plasmalemmal exchanger. Remarkably, the stoichiometry of the exchanger seems still to be controversial. Initially, it was thought to be an electroneutral $2Na^+/Ca^{2+}$ exchanger [32], but this has been questioned, as the exchanger can operate against a $[Ca^{2+}]$ gradient the energy of which is over twice that of the Na^+ gradient [33]. Jung and colleagues [33] suggested a stoichiometry of $3Na^+/Ca^{2+}$, in which case the operation of the exchanger will be dependent on $\Delta\psi_m$. The inhibition of mitochondrial Ca^{2+} efflux by mitochondrial depolarization (see above, [28], and also [34]) supports an electrogenic stoichiometry. An electrogenic stoichiometry also predicts that Ca^{2+} efflux should be associated with mitochondrial depolarization. The lack of evidence for this process might reflect the necessity of a Na^+ gradient for Ca^{2+} extrusion, built upon the H^+ gradient of polarized mitochondria through the Na^+/H^+ exchange process.

What role can we assign to the OMM in Ca^{2+} extrusion? VDAC appears to be part of the mitochondrial permeability transition pore (mPTP; [11, 35] and see below), itself regulated by $[Ca^{2+}]_m$. The mPTP provides a potential efflux pathway for Ca^{2+} [36], although the physiological relevance of this pathway is debated. Thus, even if it was never studied or even considered, we cannot rule out that the OMM might play a role as a significant permeability barrier not only to Ca^{2+} uptake but also to Ca^{2+} efflux.

Ca^{2+} Buffering in the Matrix, the "Set Point," and a Unified View of Mitochondrial Ca^{2+} Handling

Flux studies in isolated mitochondria revealed many years ago that net Ca^{2+} flux into the mitochondrial matrix might result in a drop of $\Delta\psi_m$ and an increase of ΔpH, since protons move out through the ETC to compensate for Ca^{2+} charge. This process would rapidly stop Ca^{2+} movement, as indeed occurs if charge compensating ions (e.g., P_i or acetate) are not transported together with Ca^{2+} [15]. In other words, in a physiological situation the presence of P_i (or a surrogate) is necessary for mitochondrial Ca^{2+} uptake. Interestingly, in the presence of P_i, mitochondria were able to accumulate much more Ca^{2+} than in the presence of acetate [37], pointing to an additional role of phosphate in Ca^{2+} handling. In a recent study, Chalmers and Nicholls elegantly demonstrated that the role of phosphate uptake accompanying Ca^{2+} is to maintain $[Ca^{2+}]_m$ at a constant value, independent of total Ca^{2+} load, over an amazingly wide range of concentrations [38]. The clue for this phenomenon is the formation of insoluble xCa^{2+}–xPO_4^{x-}–xOH^- complexes, "neutralizing" the physiologically and pathologically active calcium ions. The stoichiometry of these complexes is not defined unambiguously, but appears to be close to that of hydroxyapatite ($Ca_5(PO_4)_3OH$). Importantly, the solubility of this complex is extremely dependent on matrix pH, predicting a rapid, roughly 100-fold increase of $[Ca^{2+}]_m$ in the case of an acidification by a pH value of 1 – for example, following addition of a protonophore. This elevation of $[Ca^{2+}]$ has never been directly tested with mitochondrial Ca^{2+} sensors, but might explain a rapid release of Ca^{2+} into the cytoplasm, most probably through a reverse transport through the uniporter (see above).

As a result, the following picture emerges to depict the complex nature of mitochondrial Ca^{2+} transport between the extra- and intramitochondrial space (Figure 45.1). With small elevations of $[Ca^{2+}]_e$, the removal of Ca^{2+} from the matrix by the xNa^+/Ca^{2+} exchange keeps pace with the Ca^{2+} uptake rate, thus $[Ca^{2+}]_m$ equilibrates with $[Ca^{2+}]_e$. As $[Ca^{2+}]_e$ rises above ~4–500 nM, the Ca^{2+}-dependent activation of the Ca^{2+} uptake rate (the "uniporter" and maybe VDAC) exceeds the capacity of the exchanger, and mitochondria show net accumulation of Ca^{2+}. At this point, the P_i-dependent Ca^{2+} buffering mechanism serves to counterbalance excessive $[Ca^{2+}]_m$ increases (for a recent review, see [13]). Slow Ca^{2+} infusion in isolated mitochondria does not even lead to detectable $[Ca^{2+}]_m$ changes, due to the rapid formation of Ca^{2+}/P_i precipitates, while bolus application of Ca^{2+} (similarly to a physiological situation following Ca^{2+} mobilization following cell stimulation) causes a $[Ca^{2+}]_m$ signal which is readily detected by fluorescent or luminescent mitochondrial Ca^{2+} sensors. Thus, it is worth considering that Ca^{2+} flux into mitochondria is not necessarily synonymous with a net increase in $[Ca^{2+}]_m$. This is not purely semantic, as Ca^{2+} uptake by the uniporter is electrogenic and is therefore associated with small changes in $\Delta\psi_m$. Experimentally, changes in $\Delta\psi_m$ will reflect the rate of Ca^{2+} flux, and may therefore prove a more sensitive measurement of Ca^{2+} movement into mitochondria than measurement of $[Ca^{2+}]_m$.

The $[Ca^{2+}]_e$ value at which the rate of Ca^{2+} uptake exceeds that of the extrusion was termed the "set point;" however, this has another interpretation, maybe reflecting better the impact of mitochondria on cellular Ca^{2+} homeostasis. It also represents a $[Ca^{2+}]_e$ value which may be maintained by mitochondrial Ca^{2+} cycling. Indeed, isolated liver mitochondria could lower $[Ca^{2+}]_e$ to about 4–800{ts}nm (depending on the amount of total Ca^{2+} added to the medium), while if $[Ca^{2+}]_e$ was lowered below this value by the addition of a Ca^{2+} chelator there was a slow release of matrix Ca^{2+} until this same value was attained [39]. In a cellular context (see below), where several other mechanisms operate to maintain the equilibrium in the cytosol, this observation may have a restricted relevance, but it nicely illustrates the fundamental purpose of mitochondrial Ca^{2+} handling – i.e., to maintain and even dictate a Ca^{2+} level in its mother cell, through its Ca^{2+} uptake, buffering, and release capacities. Importantly, these capacities are the function of $\Delta\psi_m$ and ΔpH, thus we can interpret the set point as a signal of mitochondrial metabolism and integrity for the rest of cell, translated into a $[Ca^{2+}]$ value.

MITOCHONDRIAL CA^{2+} HANDLING IN THE CELLULAR CONTEXT

Early studies on the impact of Ca^{2+} accumulation on mitochondrial function led to the characterization of

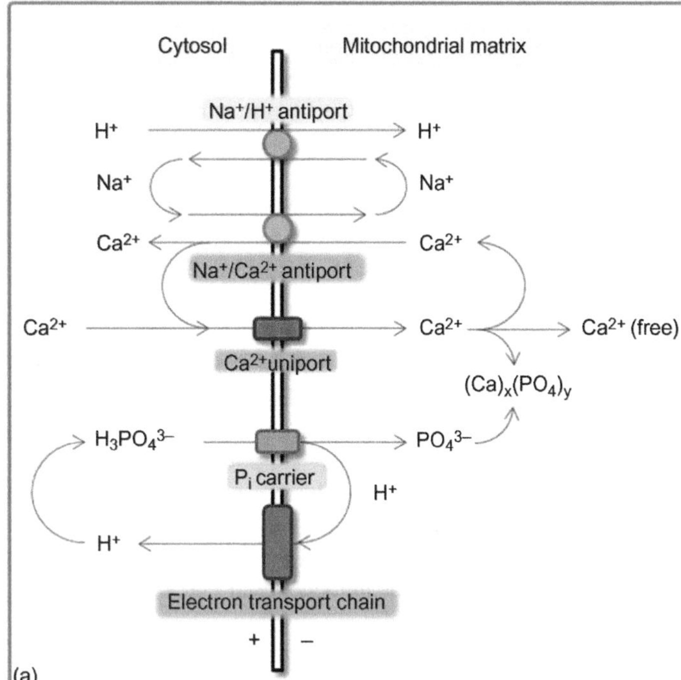

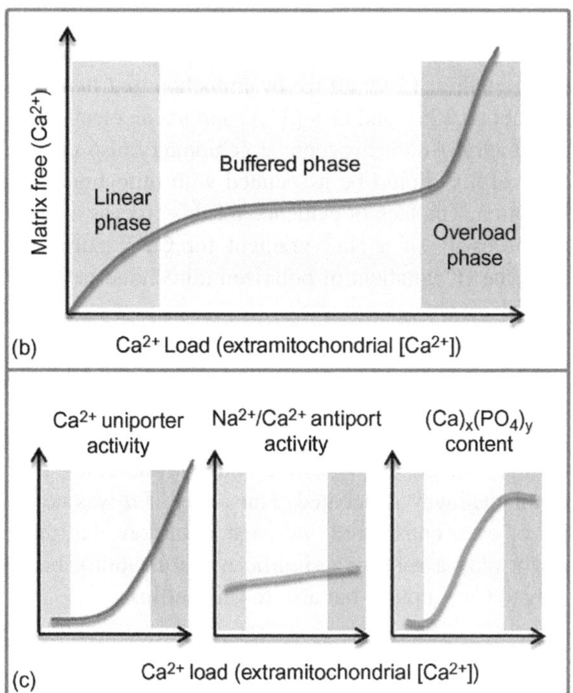

FIGURE 45.1 Fundamentals of mitochondrial Ca^{2+} handling.
(a) Schematic representation of Ca^{2+} cycling between the extramitochondrial (cytosolic) and mitochondrial matrix space. The uptake and extrusion of Ca^{2+} is tightly coupled to H^+ and Na^+ cycling, maintained by the electron transport chain in respiring mitochondria, while the reversible formation of the insoluble xCa^{2+}–xPO_4^{x-}–xOH^- precipitate depends on the accompanying P_i accumulation and the H^+ gradient. (b, c) Three-phase model of mitochondrial Ca^{2+} accumulation as the function of extramitochondrial $[Ca^{2+}]$. The kinetic parameters of the Ca^{2+} uniporter, the Na^+/Ca^{2+} exchanger and the formation of xCa^{2+}–xPO_4^{x-}–xOH^- precipitate (depicted in (c)), determines an initially linear relation between $[Ca^{2+}]$e and $[Ca^{2+}]_m$, followed by buffering without notable $[Ca^{2+}]_m$ elevation. When the Ca^{2+} load exceeds the buffering capacity, $[Ca^{2+}]_m$ again rises quickly, leading to mitochondrial permeability transition.

Ca^{2+}-dependent metabolic processes; this was further established in intact cells, through the development of imaging techniques to study mitochondrial function [40]. Further, the shift to more physiological systems also revealed that the interaction between mitochondria and cellular Ca^{2+} homeostasis is rigorously mutual, also leading to the acknowledgement that interaction between mitochondria and other cellular organelles has a large influence on cell function – doubtless even larger than merely regulating ATP production.

Impact of Ca^{2+} Uptake on Mitochondrial Function

In teleological terms, it seems that the major functional significance of mitochondrial Ca^{2+} uptake is in the regulation of mitochondrial metabolism. In the early 1990s, it was shown that the three major rate-limiting enzymes of the citric acid cycle are all upregulated by Ca^{2+} (for review, see [41]). The question that remained was the functional issue – is mitochondrial Ca^{2+} uptake during physiological signaling sufficient for this mechanism to provide a functional regulation of metabolism? First suggestions that such a system operates in intact cells came from measurements of changes in mitochondrial redox state, reflected as changes in mitochondrial NADH and flavoprotein autofluorescence, in response to changes in $[Ca^{2+}]_c$ [23, 42]. These observations showed clearly that (1) mitochondria must be taking up Ca^{2+} during $[Ca^{2+}]_c$ signals, and (2) this was sufficient to activate the TCA cycle, causing increased net reduction of the coenzymes. More recently, transfection of cells with firefly luciferase allowed a clear and unequivocal demonstration that mitochondrial Ca^{2+} uptake increases mitochondrial ATP production [43]. The relative importance of this mechanism in the regulation of mitochondrial oxidative phosphorylation over the more traditional model, in which the rate of ATP generation is regulated largely by the ATP/ADP ratio, is not clear. It is very attractive to suggest that the transfer of Ca^{2+} from the cytosol to mitochondria during $[Ca^{2+}]_c$ signals represents a major mechanism to couple ATP supply with demand, as, in almost all systems, increases in work are associated with increases in $[Ca^{2+}]_c$. Nevertheless, direct evidence for a significant role in intact systems is limited, and there are conflicting data (see, for example [44, 45]).

The time-course of the changes in $[Ca^{2+}]_m$ and in activation of the enzyme systems becomes crucial. $[Ca^{2+}]_c$ signals are typically brief, transient phenomena. Typically, it seems that the resultant mitochondrial activation is prolonged with respect to the change in $[Ca^{2+}]_c$ [23, 46, 47], and this in turn will be a function of the rate of mitochondrial Ca^{2+} efflux, and the half life of the activated states of the enzymes.

A further major question that is important in considering the impact of Ca^{2+} on mitochondrial function is: how high does $[Ca^{2+}]_m$ rise during these signals? The Ca^{2+} transport model depicted in the first part of the chapter gives some clues to answering this question (see also Figure 45.1). In isolated mitochondria, two phases of $[Ca^{2+}]_m$ changes were observed following elevation of $[Ca^{2+}]_e$. At submicromolar $[Ca^{2+}]_e$, $[Ca^{2+}]_m$ increases in a range (0.2–3μM) that allows the parallel activation of Ca^{2+}-dependent enzymes of the Krebs cycle, leading to increased supply of reducing equivalents ($NADH^+$/$NADPH^+$) [41, 42, 48]. This increase of $[Ca^{2+}]_m$ activates mitochondrial metabolism, being in the range where the activity of Ca^{2+}-dependent dehydrogenases is controlled by Ca^{2+}. At $[Ca^{2+}]_e$ above the μM level, the mitochondrial efflux mechanisms, assisted by the matrix Ca^{2+} buffering activity, keep $[Ca^{2+}]_m$ relatively stable, allowing mitochondria to accumulate as much as 700–1000 nmol $Ca^{2+}mg^{-1}$ mitochondrial protein, but one should not expect further activation of metabolism. In intact cells, a further increase of $[Ca^{2+}]_m$ was observed using low-affinity variants of targeted recombinant aequorin probes. This increase, as discussed below, seems to depend on interaction of mitochondria with other Ca^{2+}-handling organelles (ER and the plasma membrane), and most probably plays a role in the regulation of cell death processes by mitochondria.

The capacity to study mitochondria in intact cells that has accompanied the improvements in confocal microscopy has led to the (re)discovery of their dynamic morphological properties, adding a new layer to the plasticity of mitochondrial signaling. While it is out of the scope of this chapter to describe in detail the mechanisms controlling the continuous fusion, fission, and movement of the mitochondrial population (forming a dynamic interconnected network in several cell types; for a recent review see [49]), we would like to emphasize that shaping mitochondria seems to be closely interconnected with cellular and mitochondrial Ca^{2+} signals. The large mechanochemical GTPase enzyme, dynamin-like protein-1 (Drp-1), drives the mitochondrial division apparatus, translocating to the OMM following a cytoplasmic $[Ca^{2+}]$ rise in mammalian cells [50], and becomes activated by calcineurin-driven dephosphorylation [51]. Similarly, the intracellular distribution of mitochondria, as well as their fusion, is regulated by a newly discovered mitochondrial family of Rho GTPases (Miro1 & 2), bearing two functionally important Ca^{2+} binding EF-hand domains [52]. A recently emerging outstanding question is: how do mitochondrial dynamics (and their Ca^{2+} regulation) affect their metabolic properties and participation in the regulation of different cell death pathways?

In effect, the mitochondrial Ca^{2+} uptake may have profound consequences for mitochondrial function under pathological conditions. A combination of mitochondrial Ca^{2+} loading and oxidative stress and/or ATP depletion may promote opening of the mitochondrial permeability transition pore (mPTP). This appears to reflect a pathological conformation of a group of mitochondrial membrane proteins, notably the adenine nucleotide translocase (ANT) and VDAC, with the association of cyclophilin D, a regulatory protein that confers sensitivity of the complex to cyclosporin A, and a possible association of a number of other proteins, including the anti-apoptotic Bcl-2, and the benzodiazepine receptor (see [11, 35, 53] for reviews). Recent genetic evidence clearly identified cyclophilin D conferring Ca^{2+} sensitivity of the permeability transition [54–56], allowing too an initial molecular definition of a Ca^{2+}-dependent cell death pathway which is clearly distinct from apoptosis, but plays a fundamental role in reperfusion injury in the heart and in glutamate neurotoxicity in the CNS. Still, our knowledge of the potential role of mPTP in other cell death models (e.g., cancer cell death induced by antitumor agents, or degenerative diseases of the CNS) can now only be established on findings regarding cyclophilin D, since other proposed components (the ANT and VDAC) appeared to be dispensable for pore formation in recently developed genetic models [57, 58]. However, it is important to note that, similarly to cyclophilin $D^{-/-}$ cells, ANT1/2 knockouts also display less sensitivity to Ca^{2+} than their wild-type counterparts [57, 59].

The Reverse Signaling: Impact of Mitochondrial Ca^{2+} Uptake on Cellular Ca^{2+} Homeostasis

Many excitable cells respond to depolarization with a rise in $[Ca^{2+}]_c$, which rises rapidly and recovers with an initial rapid phase and a slower second phase that can even form a plateau [60–62]. It has been established in many studies that the slow recovery phase reflects the redistribution of mitochondrial Ca^{2+} through the activity of the Na^+/Ca^{2+} exchanger, reflecting the set point, typically initiated at a $[Ca^{2+}]_c$ of ~500 nM. The operation of this system has functional consequences at presynaptic terminals, where the $[Ca^{2+}]_c$ plateau that follows repetitive stimulation, maintained by the re-equilibration of mitochondrial Ca^{2+}, provides an elevated $[Ca^{2+}]_c$ baseline upon which subsequent stimulation initiates an enhanced synaptic response – the basis for post-tetanic potentiation of synaptic transmission [61, 63]. It is also intriguing that the post-stimulus plateau phase is not seen in non-excitable cells following the transmission of $[Ca^{2+}]_c$ signals from ER to mitochondria. Certainly in astrocytes, $[Ca^{2+}]_m$ remains high for a very prolonged period after stimulation [64], suggesting that mitochondrial Ca^{2+} efflux must be very slow and perhaps the activity of the exchanger differs between tissues or cell types.

There has been some debate about the quantitative relationships between ambient $[Ca^{2+}]$ and mitochondrial

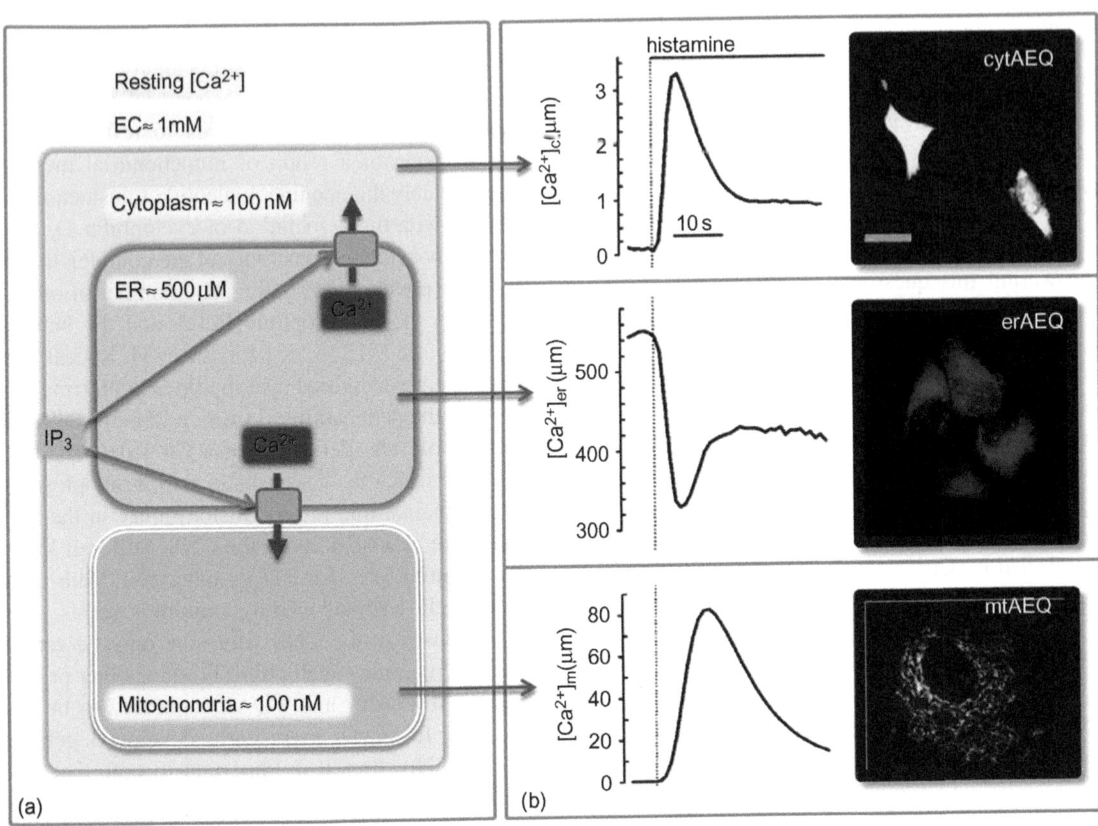

FIGURE 45.2 Compartmentalized cellular Ca^{2+} signaling.
A simplified three-compartment model of the cell (a) indicates the principal organelles participating in cellular Ca^{2+} handling and their resting $[Ca^{2+}]_c$. The endoplasmic reticulum (ER) represents the main intracellular Ca^{2+} store, having a basal Ca^{2+} level almost as high as in the extracellular space, roughly 10^4 times higher than the cytoplasm and mitochondria. The formation of IP_3 leads to Ca^{2+} release from the ER, triggering a reciprocal $[Ca^{2+}]$ increase in the cytosol and $[Ca^{2+}]$ drop in the ER (as measured by the luminescent recombinant aequorin Ca^{2+} probes; upper and middle panels on (b), respectively). Note the almost 100-fold greater response in mitochondria (lower panel in (b)) as compared to the cytosol, explained by the close physical association between the organelles.

uptake. In HeLa cells transfected with mitochondrially targeted aequorin and then permeabilized, net mitochondrial Ca^{2+} accumulation was only detectable if the added Ca^{2+} reached concentrations higher than 3μM, while $[Ca^{2+}]_c$ signals evoked by IP_3 mobilizing agonists were far more effective at raising $[Ca^{2+}]_m$ even though the mean $[Ca^{2+}]_c$ signal might rise to <1 μM [22]. This led to the recognition of a fundamental feature of the mitochondrial Ca^{2+} uptake machinery *in situ* that stems from its strategic localization at specific sources of the cellular Ca^{2+} signal (Figure 45.2). Indeed, intimate contacts between the PM and ER membranes and the OMM have been revealed by morphological and functional imaging studies [65–67], showing that immediate Ca^{2+} sequestration occurs preferentially at these sites into the organelle. Mitochondria, positioned at the cytoplasmic face of the ER Ca^{2+} release channels (inositol 1,4,5-trisphosphate receptors [IP_3R] and ryanodine receptors [RyRs]) as well as to different PM Ca^{2+} influx channels (capacitative Ca^{2+} entry [CCE] or ionotropic glutamate receptors), thus are exposed to Ca^{2+} concentrations well above those measured in the bulk cytosol, now generally termed as Ca^{2+} microdomains (Figure 45.3) [67–69].

The proximity of mitochondria to SR or ER Ca^{2+} release sites has been further emphasized through evidence that focal, non-propagating ER/SR Ca^{2+} release can cause a transient increase in $[Ca^{2+}]_m$ in mitochondria close to the release site. Thus, we found [23] that mitochondria in cardiomyocytes show spontaneous transient mitochondrial depolarizations that were dependent on local SR Ca^{2+} release and were blocked by inhibition of mitochondrial Ca^{2+} uptake. Hajnoczky and colleagues [24] have since shown that local $[Ca^{2+}]_c$ sparks may be associated with the direct transfer of Ca^{2+} to mitochondria visualized as transient increases in $[Ca^{2+}]_m$ – which the group termed Ca^{2+} "marks." Further data from cardiomyocytes [25] strongly suggest that, in cardiomyocytes, mitochondria and SR must show very close coupling, as the transfer of Ca^{2+} to mitochondria in response to SR Ca^{2+} release with caffeine in permeabilized cells was sustained despite Ca^{2+} buffering by BAPTA sufficient to suppress the cytosolic signal. The transfer of Ca^{2+} was prevented by disrupting the cytoskeleton, suggesting the maintained close apposition of mitochondria to SR was central to this signal. The significance of ER/SR–mitochondrial Ca^{2+} transfer has been recently confirmed by two different

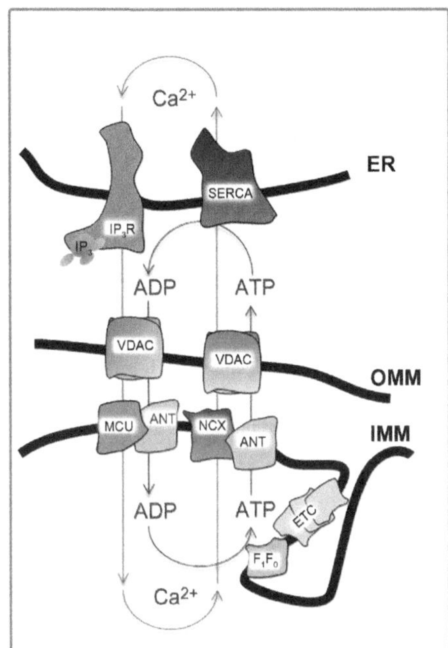

FIGURE 45.3 Metabolic and Ca^{2+} coupling between the endoplasmic reticulum and mitochondria.
The formation of Ca^{2+} and ADP/ATP microdomains at the ER-mitochondrial contact sites allows the direct control of ER Ca^{2+} content and direct Ca^{2+} channeling into the mitochondria. Ca^{2+} release from the ER occurs through the IP_3 receptor (IP_3R), which supplies Ca^{2+} to mitochondria through the OMM voltage-dependent anion channel (VDAC) and the IMM Ca^{2+} uniporter (MCU). Limited mitochondrial Ca^{2+} loads upregulate mitochondrial ATP production (by supplying reducing equivalents to the electron transport chain (ETC and F_1F_0 ATPase)), which is transported to the extramitochondrial space by the concerted activity of the ATP/ADP translocase (ANT) and VDAC. The re-accumulation of Ca^{2+} into the ER lumen through the sarco-endoplasmic Ca^{2+} ATPase (SERCA) is directly controlled by the ATP supplied by mitochondria. Current studies investigate how this framework of Ca^{2+} and ATP cycling will determine cellular function in diverse physiological and pathological conditions.

approaches. First, high-resolution fast kinetic imaging of $[Ca^{2+}]_m$, using fluorescent dyes or targeted recombinant probes, showed that intramitochondrial Ca^{2+} signals following IP_3-induced Ca^{2+} release propagate in the matrix from distinct foci (termed "hotspots") most probably representing ER-mitochondrial contacts [70, 71]. Moreover, a scaffolding chaperone (grp75) mediated coupling of ER and mitochondrial Ca^{2+} permeating channels (the IP_3R and VDAC, respectively) has been shown to mediate efficient Ca^{2+} transfer between the organelles [72].

Most importantly, the proximity of mitochondria to Ca^{2+} release sites has functional consequences for $[Ca^{2+}]_c$ signaling. Using $[Ca^{2+}]$ indicators in both mitochondria and ER in permeabilized cells, Csordas and colleagues [67] showed direct transfer of Ca^{2+} from ER to mitochondria, and suggested that the proximity must be ~10–20 nm. This work was extended to show that mitochondrial Ca^{2+} uptake enhances the release of Ca^{2+} from the ER in response to IP_3 by acting as a local buffer [73]. Thus, by removing Ca^{2+} from the microdomain close to the IP_3 Ca^{2+} release channel, mitochondria prevent the Ca^{2+}-dependent inactivation of the channel and facilitate ER Ca^{2+} release. This mechanism allows mitochondria to play a significant role in shaping the spatiotemporal patterning of $[Ca^{2+}]_c$ signals. In *Xenopus* oocytes, energization of mitochondria enhances the propagation and coordination of $[Ca^{2+}]_c$ waves [74], while in astrocytes, which express primarily IP_3 type 3 receptors, energized mitochondria served as a spatial buffer which limit the rate and extent of propagation of $[Ca^{2+}]_c$ waves [64]. Microdomains of $[Ca^{2+}]_c$ regulated by mitochondria also play a significant role in the regulation of capacitative Ca^{2+} influx ([75, 76]; for review see [77]), suggesting that the mitochondria must be positioned close to the plasma membrane. The principle is very much as outlined above for the IP_3 receptor, as the Ca^{2+} influx channel is desensitized by Ca^{2+}. By keeping $[Ca^{2+}]_c$ low in microdomains close to the channels, mitochondria keep the channels open and facilitate Ca^{2+} influx through them.

Intriguingly, studies concerning the dynamic positioning and morphological plasticity of mitochondrial structure provided important information on the role of mitochondria in shaping cellular Ca^{2+} signals. Thus, in pancreatic acinar cells, the mitochondria are concentrated into a band that isolates the secretory pole of these polarized cells, and they seem to act as a "firewall" that limits the spread of $[Ca^{2+}]_c$ signals from their initiation at the apical pole to the basal pole [78]. Furthermore, mitochondria localized close to the basal pole are more sensitive to local Ca^{2+} influx by capacitative entry, and so it seems that the positions of mitochondria within the cell may have a profound influence both on their interaction with cellular $[Ca^{2+}]_c$ signals and with the spatial regulation of the calcium signal itself [79].

Induction of mitochondrial biogenesis and consequent mitochondrial volume expansion was also shown to interact with Ca^{2+} uptake and intramitochondrial Ca^{2+} distribution [80]. The recent identification of the pathway that coordinates the concerted induction of nuclear and mitochondrial encoded mitochondrial proteins by the peroxisome proliferator-activated receptor-gamma (PPARγ) co-activator-1α (PGC-1α) served as a useful tool to assess this effect [81]. Importantly, the pathway induced by PGC-1α also increased mitochondrial volume, enhancing its capacity to accumulate and buffer Ca^{2+}, as evidenced by the increased distance of Ca^{2+} diffusion from the "hotspots." Overall, through this, and probably other mechanisms, PGC-1α-induced biogenesis leads to a reduced $[Ca^{2+}]_m$ signal.

Finally, further insight into the role of mitochondrial distribution on regulating global Ca^{2+} signals was obtained by analyzing changes of Ca^{2+} handling at the level of the plasma membrane in different models of forced mitochondrial division [82–84]. These studies, together with the use of inhibitors of mitochondrial function [85], shed light on the importance of the near-plasma membrane and perinuclear mitochondrial populations in endothelial and epithelial cell lines. As shown by a series of experiments from the group of

N. Demaurex, about 10 percent of MitoTracker Red labeled mitochondria are co-localized with plasma membrane-targeted protein tags, suggesting an important contribution of mitochondria to the Ca^{2+} transport between the intra- and extracellular space. Indeed, the redistribution of mitochondria following overexpression of hFis1 (a membrane anchor and downstream effector of Drp-1, see above) in HeLa cells led to changes in both the Ca^{2+} influx and extrusion pathways [82, 83]. First, they demonstrated that the fragmented perinuclear mitochondrial population only slowly accumulates Ca^{2+} from extracellular sources. These results, taken together with our previous observation that during long term cellular stimulation ER tunneling of Ca^{2+} from the EC space maintains sustained Ca^{2+} release, which is necessary for efficient mitochondrial Ca^{2+} uptake [86], imply two important deductions: (1) mitochondrial connectivity and the presence of mitochondria near to the plasmamembrane of endo/epithelial cells appears to be necessary to distribute Ca^{2+} not only from ER sources (see above: [70, 71]), but also from the EC space; and (2) Ca^{2+} tunneling through the ER, originally described in polarized pancreatic acinar cells [87], is also fundamental in these cell types, and only in the absence of the sub-plasma membrane mitochondrial population is it substituted by diffusion of Ca^{2+} in the cytoplasm. Indeed, confirming these postulations, further work depicted an elegant unifying scheme [82, 88], in which bidirectional Ca^{2+} channeling between the ER and mitochondria collaborate in the regulation of capacitative Ca^{2+} influx and Ca^{2+} extrusion through the plasma membrane Ca^{2+} ATPase (PMCA). The complete picture is detailed in the above citations; here we would like only to underline the importance of one of the findings: that the lack of the sub-plasma membrane mitochondrial population in hFis1 overexpressing HeLa cells led to increased Ca^{2+} cycling through the PMCA and capacitative influx channels, rendering the ER Ca^{2+} store very prone to Ca^{2+}depletion under low extracellular [Ca^{2+}] conditions. Since similar redistribution of the mitochondrial network (fragmentation and perinuclear clustering) is a major feature of several cell death processes, as well as the ER Ca^{2+} content being a key regulator of ER stress and apoptosis, undoubtedly the mechanisms discovered in these studies will find an important place in the regulation of cell fate.

CODA

Our perception of the mitochondrion has changed radically over the past decades, and continues to amaze us with newer aspects in each year. Mitochondrial function is critical for the viability of the cell, mitochondria play an integral role in shaping cell signaling, and the dysfunction of the pathways necessary for these functions may trigger cell death. These are not trivial and peripheral functions, but central to cell life and cell death. Thus, it seems more and more apparent that mitochondrial biology has undergone a paradigm shift; now, instead of studying how cells regulate the function of mitochondria according their needs, our studies are rather designed to explore how mitochondria determine cell function.

ACKNOWLEDGEMENTS

Work in our laboratories is supported by the Wellcome Trust, the Medical Research Council, and the Royal Society, whom we thank. We also thank Remi Dumollard, Jake Jacobson, and Laura Canevari for their invaluable discussion and comments on the manuscript. In so short an essay, it is not possible to describe all the fascinating activity in this field, and so we also apologize to those whose work is not cited here.

REFERENCES

1. Goldenthal MJ, Marin-Garcia J. Mitochondrial signaling pathways: a receiver/integrator organelle. *Mol Cell Biochem* 2004;**262**:1–16.
2. Ryan MT, Hoogenraad NJ. Mitochondrial–nuclear communications. *Annu Rev Biochem* 2007;**76**:701–22.
3. Green DR, Kroemer G. The pathophysiology of mitochondrial cell death. *Science* 2004;**305**:626–9.
4. Wallace DC. Why do we still have a maternally inherited mitochondrial DNA? Insights from evolutionary medicine. *Annu Rev Biochem* 2007;**76**:781–821.
5. Nicholls DG. Mitochondria and calcium signaling. *Cell Calcium* 2005;**38**:311–17.
6. Saris NEL, Carafoli E. A historical review of cellular calcium handling, with emphasis on mitochondria. *Biochemistry (Mosc)* 2005; **70**:187–94.
7. Bianchi K, Rimessi A, Prandini A, Szabadkai G, Rizzuto R. Calcium and mitochondria: mechanisms and functions of a troubled relationship. *Biochim Biophys Acta* 2004;**1742**:119–31.
8. Duchen MR. Mitochondria and calcium: from cell signalling to cell death. *J Physiol* 2000;**529**(Pt 1):57–68.
9. Rizzuto R, Bernardi P, Pozzan T. Mitochondria as all-round players of the calcium game. *J Physiol* 2000;**529**(Pt 1):37–47.
10. Gunter TE, Yule DI, Gunter KK, Eliseev RA, Salter JD. Calcium and mitochondria. *FEBS Letts* 2004;**567**:96–102.
11. Bernardi P. Mitochondrial transport of cations: channels, exchangers, and permeability transition. *Physiol Rev* 1999;**79**:1127–55.
12. O'Rourke B. Mitochondrial ion channels. *Annu Rev Physiol* 2007;**69**:19–49.
13. Nicholls DG, Chalmers S. The integration of mitochondrial calcium transport and storage. *J Bioenerg Biomembr* 2004;**36**:277–81.
14. Slater EC, Cleland KW. The effect of calcium on the respiratory and phosphorylative activities of heart-muscle sarcosomes. *Biochem J* 1953;**55**:566–90.
15. Rossi CS, Lehninger AL. Stoichiometry of respiratory stimulation, accumulation of Ca++ and phosphate, and oxidative phosphorylation in rat liver mitochondria. *J Biol Chem* 1964;**239**:3971–80.
16. Smyth JT, Dehaven WI, Jones BF, et al. Emerging perspectives in store-operated Ca(2+) entry: roles of Orai, Stim and TRP. *Biochim Biophys Acta* 2006;**1763**:1147–60.
17. Rapizzi E, Pinton P, Szabadkai G, et al. Recombinant expression of the voltage-dependent anion channel enhances the transfer of Ca^{2+} microdomains to mitochondria. *J Cell Biol* 2002;**159**:613–24.

18. Bathori G, Csordas G, Garcia-Perez C, Davies E, Hajnoczky G. Ca^{2+}-dependent control of the permeability properties of the mitochondrial outer membrane and voltage-dependent anion-selective channel (VDAC). *J Biol Chem* 2006;**281**:17,347–58.
19. Tan W, Colombini M. VDAC closure increases calcium ion flux. *Biochim Biophys Acta (BBA) – Biomembr* 2009;**6**. in press.
20. Zaid H, Abu-Hamad S, Israelson A, Nathan I, Shoshan-Barmatz V. The voltage-dependent anion channel-1 modulates apoptotic cell death. *Cell Death Differ* 2005;**12**:751–60.
21. Csordas G, Madesh M, Antonsson B, Hajnoczky G. tcBid promotes Ca(2+) signal propagation to the mitochondria: control of Ca(2+) permeation through the outer mitochondrial membrane. *EMBO J* 2002;**21**:2198–206.
22. Gunter TE, Buntinas L, Sparagna G, Eliseev R, Gunter K. Mitochondrial calcium transport: mechanisms and functions. *Cell Calcium* 2000;**28**:285–96.
23. Duchen MR. Ca(2+)-dependent changes in the mitochondrial energetics in single dissociated mouse sensory neurons. *Biochem J* 1992;**283**(Pt 1):41–50.
24. Lukyanenko V, Gyorke I, Subramanian S, Smirnov A, Wiesner TF, Gyorke S. Inhibition of Ca(2+) sparks by ruthenium red in permeabilized rat ventricular myocytes. *Biophys J* 2000;**79**:1273–84.
25. Wood JN, Winter J, James IF, Rang HP, Yeats J, Bevan S. Capsaicin-induced ion fluxes in dorsal root ganglion cells in culture. *J Neurosci* 1988;**8**:3208–20.
26. Kirichok Y, Krapivinsky G, Clapham DE. The mitochondrial calcium uniporter is a highly selective ion channel. *Nature* 2004;**427**:360–4.
27. Igbavboa U, Pfeiffer DR. EGTA inhibits reverse uniport-dependent Ca^{2+} release from uncoupled mitochondria. Possible regulation of the Ca^{2+} uniporter by a Ca^{2+} binding site on the cytoplasmic side of the inner membrane. *J Biol Chem* 1988;**263**:1405–12.
28. Montero M, Alonso MT, Albillos A, Garcia-Sancho J, Alvarez J. Mitochondrial Ca(2+)-induced Ca(2+) release mediated by the Ca(2+) uniporter. *Mol Biol Cell* 2001;**12**:63–71.
29. Trenker M, Malli R, Fertschai I, Levak-Frank S, Graier WF. Uncoupling proteins 2 and 3 are fundamental for mitochondrial Ca^{2+} uniport. *Nat Cell Biol* 2007;**9**:445–52.
30. Buntinas L, Gunter KK, Sparagna GC, Gunter TE. The rapid mode of calcium uptake into heart mitochondria (RaM): comparison to RaM in liver mitochondria. *Biochim Biophys Acta* 2001;**1504**:248–61.
31. Sparagna GC, Gunter KK, Sheu SS, Gunter TE. Mitochondrial calcium uptake from physiological-type pulses of calcium. A description of the rapid uptake mode. *J Biol Chem* 1995;**270**:27,510–15.
32. Brand MD. The stoichiometry of the exchange catalysed by the mitochondrial calcium/sodium antiporter. *Biochem J* 1985;**229**:161–6.
33. Jung DW, Baysal K, Brierley GP. The sodium–calcium antiport of heart mitochondria is not electroneutral. *J Biol Chem* 1995;**270**:672–8.
34. Bernardi P, Azzone GF. A membrane potential-modulated pathway for Ca^{2+} efflux in rat liver mitochondria. *FEBS Letts* 1982;**139**:13–16.
35. Crompton M, Barksby E, Johnson N, Capano M. Mitochondrial intermembrane junctional complexes and their involvement in cell death. *Biochimie* 2002;**84**:143–52.
36. Ichas F, Jouaville LS, Mazat JP. Mitochondria are excitable organelles capable of generating and conveying electrical and calcium signals. *Cell* 1997;**89**:1145–53.
37. Zoccarato F, Nicholls D. The role of phosphate in the regulation of the independent calcium-efflux pathway of liver mitochondria. *Eur J Biochem* 1982;**127**:333–8.
38. Chalmers S, Nicholls DG. The Relationship between Free and Total Calcium concentrations in the matrix of liver and brain mitochondria. *J Biol Chem* 2003;**278**:19,062–70.
39. Nicholls DG. The regulation of extramitochondrial free calcium ion concentration by rat liver mitochondria. *Biochem J* 1978;**176**:463–74.
40. Duchen MR, Surin A, Jacobson J. Imaging mitochondrial function in intact cells. *Methods Enzymol* 2003;**361**:353–89.
41. McCormack JG, Halestrap AP, Denton RM. Role of calcium ions in regulation of mammalian intramitochondrial metabolism. *Physiol Rev* 1990;**70**:391–425.
42. Pralong WF, Hunyady L, Varnai P, Wollheim CB, Spat A. Pyridine nucleotide redox state parallels production of aldosterone in potassium-stimulated adrenal glomerulosa cells. *Proc Natl Acad Sci USA* 1992;**89**:132–6.
43. Jouaville LS, Pinton P, Bastianutto C, Rutter GA, Rizzuto R. Regulation of mitochondrial ATP synthesis by calcium: evidence for a long-term metabolic priming. *Proc Natl Acad Sci USA* 1999;**96**:13,807–12.
44. Moravec CS, Desnoyer RW, Milovanovic M, Schluchter MD, Bond M. Mitochondrial calcium content in isolated perfused heart: effects of inotropic stimulation. *Am J Physiol* 1997;**273**:1432–9.
45. Horikawa Y, Goel A, Somlyo AP, Somlyo AV. Mitochondrial calcium in relaxed and tetanized myocardium. *Biophys J* 1998;**74**:1579–90.
46. Robb-Gaspers LD, Burnett P, Rutter GA, Denton RM, Rizzuto R, Thomas AP. Integrating cytosolic calcium signals into mitochondrial metabolic responses. *EMBO J* 1998;**17**:4987–5000.
47. Hajnoczky G, Robb-Gaspers LD, Seitz MB, Thomas AP. Decoding of cytosolic calcium oscillations in the mitochondria. *Cell* 1995;**82**:415–24.
48. Pitter JG, Maechler P, Wollheim CB, Spat A. Mitochondria respond to Ca^{2+} already in the submicromolar range: correlation with redox state. *Cell Calcium* 2002;**31**:97–104.
49. Hoppins S, Lackner L, Nunnari J. The machines that divide and fuse mitochondria. *Ann Rev Biochem* 2007;**76**:751–80.
50. Breckenridge DG, Stojanovic M, Marcellus RC, Shore GC. Caspase cleavage product of BAP31 induces mitochondrial fission through endoplasmic reticulum calcium signals, enhancing cytochrome c release to the cytosol. *J Cell Biol* 2003;**160**:1115–27.
51. Cribbs JT, Strack S. Reversible phosphorylation of Drp1 by cyclic AMP-dependent protein kinase and calcineurin regulates mitochondrial fission and cell death. *EMBO Rep* 2007;**8**:939–44.
52. Fransson S, Ruusala A, Aspenstrom P. The atypical Rho GTPases Miro-1 and Miro-2 have essential roles in mitochondrial trafficking. *Biochem Biophys Res Commun* 2006;**344**:500–10.
53. Jacobson J, Duchen MR. "What nourishes me, destroys me": towards a new mitochondrial biology. *Cell Death Differ* 2001;**8**:963–6.
54. Basso E, Fante L, Fowlkes J, Petronilli V, Forte MA, Bernardi P. Properties of the Permeability Transition Pore in Mitochondria Devoid of Cyclophilin D. *J Biol Chem* 2005;**280**:18,558–61.
55. Li Y, Johnson N, Capano M, Edwards M, Crompton M. Cyclophilin-D promotes the mitochondrial permeability transition but has opposite effects on apoptosis and necrosis. *Biochem J* 2004;**383**:101–9.
56. Nakagawa T, Shimizu S, Watanabe T, et al. Cyclophilin D-dependent mitochondrial permeability transition regulates some necrotic but not apoptotic cell death. *Nature* 2005;**434**:652–8.
57. Kokoszka JE, Waymire KG, Levy SE, et al. The ADP/ATP translocator is not essential for the mitochondrial permeability transition pore. *Nature* 2004;**427**:461–5.
58. Baines CP, Kaiser RA, Sheiko T, Craigen WJ, Molkentin JD. Voltage-dependent anion channels are dispensable for mitochondrial-dependent cell death. *Nat Cell Biol* 2007;**9**:550–5.

59. Zamzami N, Larochette N, Kroemer G. Mitochondrial permeability transition in apoptosis and necrosis. *Cell Death Differ* 2005,**12**(Suppl 2):1478–80.
60. Thayer SA, Miller RJ. Regulation of the intracellular free calcium concentration in single rat dorsal root ganglion neurones in vitro. *J Physiol* 1990;**425**:85–8115.
61. David G, Barrett JN, Barrett EF. Evidence that mitochondria buffer physiological Ca^{2+} loads in lizard motor nerve terminals. *J Physiol* 1998;**509**(Pt 1):59–65.
62. Babcock DF, Herrington J, Goodwin PC, Park YB, Hille B. Mitochondrial participation in the intracellular Ca2+ network. *J Cell Biol* 1997;**136**:833–44.
63. Tang Y, Zucker RS. Mitochondrial involvement in post-tetanic potentiation of synaptic transmission. *Neuron* 1997;**18**:483–91.
64. Boitier E, Rea R, Duchen MR. Mitochondria exert a negative feedback on the propagation of intracellular Ca^{2+} waves in rat cortical astrocytes. *J Cell Biol* 1999;**145**:795–808.
65. Rizzuto R, Pinton P, Carrington W, et al. Close contacts with the endoplasmic reticulum as determinants of mitochondrial Ca^{2+} responses. *Science* 1998;**280**:1763–6.
66. Lawrie AM, Rizzuto R, Pozzan T, Simpson AW. A role for calcium influx in the regulation of mitochondrial calcium in endothelial cells. *J Biol Chem* 1996;**271**:10,753–9.
67. Csordas G, Thomas AP, Hajnoczky G. Quasi-synaptic calcium signal transmission between endoplasmic reticulum and mitochondria. *EMBO J* 1999;**18**:96–108.
68. Rizzuto R, Brini M, Murgia M, Pozzan T. Microdomains with high Ca^{2+} close to IP3–sensitive channels that are sensed by neighboring mitochondria. *Science* 1993;**262**:744–7.
69. Montero M, Alonso MT, Carnicero E, et al. Chromaffin-cell stimulation triggers fast millimolar mitochondrial Ca2+ transients that modulate secretion. *Nat Cell Biol* 2000;**2**:57–61.
70. Gerencser AA, Adam-Vizi V. Mitochondrial Ca^{2+} dynamics reveals limited intramitochondrial Ca^{2+} diffusion. *Biophys J* 2005;**88**:698–714.
71. Szabadkai G, Simoni AM, Chami M, Wieckowski MR, Youle RJ, Rizzuto R. Drp-1 dependent division of the mitochondrial network blocks intraorganellar Ca^{2+} waves and protects against Ca^{2+} mediated apoptosis. *Mol Cell* 2004;**16**:59–68.
72. Szabadkai G, Bianchi K, Varnai P, et al. Chaperone-mediated coupling of endoplasmic reticulum and mitochondrial Ca^{2+} channels. *J Cell Biol* 2006;**175**:901–11.
73. Hajnoczky G, Hager R, Thomas AP. Mitochondria suppress local feedback activation of inositol 1,4, 5-trisphosphate receptors by Ca^{2+}. *J Biol Chem* 1999;**274**:14,157–62.
74. Jouaville LS, Ichas F, Holmuhamedov EL, Camacho P, Lechleiter JD. Synchronization of calcium waves by mitochondrial substrates in *Xenopus laevis* oocytes. *Nature* 1995;**377**:438–41.
75. Hoth M, Fanger CM, Lewis RS. Mitochondrial regulation of store-operated calcium signaling in T lymphocytes. *J Cell Biol* 1997,**137**. 633–48.
76. Gilabert JA, Parekh AB. Respiring mitochondria determine the pattern of activation and inactivation of the store-operated Ca(2+) current I(CRAC). *EMBO J* 2000;**19**:6401–7.
77. Parekh AB. Mitochondrial regulation of intracellular Ca^{2+} signaling: more than just simple Ca^{2+} buffers. *News Physiol Sci* 2003;**18**:252–6.
78. Tinel H, Cancela JM, Mogami H, et al. Active mitochondria surrounding the pancreatic acinar granule region prevent spreading of inositol trisphosphate-evoked local cytosolic Ca(2+) signals. *EMBO J* 1999;**18**:4999–5008.
79. Park MK, Ashby MC, Erdemli G, Petersen OH, Tepikin AV. Perinuclear, perigranular and sub-plasmalemmal mitochondria have distinct functions in the regulation of cellular calcium transport. *EMBO J* 2001;**20**:1863–74.
80. Bianchi K, Vandecasteele G, Carli C, Romagnoli A, Szabadkai G, Rizzuto R. Regulation of Ca(2+) signalling and Ca(2+)-mediated cell death by the transcriptional coactivator PGC-1α. *Cell Death Differ* 2005;**13**:586–96.
81. Puigserver P, Spiegelman BM. Peroxisome proliferator-activated receptor-gamma coactivator 1 α (PGC-1α): transcriptional coactivator and metabolic regulator. *Endocr Rev* 2003;**24**:78–90.
82. Frieden M, Arnaudeau S, Castelbou C, Demaurex N. Subplasmalemmal mitochondria modulate the activity of plasma membrane Ca^{2+}-ATPases. *J Biol Chem* 2005;**280**:43,198–208.
83. Frieden M, James D, Castelbou C, Danckaert A, Martinou JC, Demaurex N. Ca(2+) homeostasis during mitochondrial fragmentation and perinuclear clustering induced by hFis1. *J Biol Chem* 2004;**279**:22,704–14.
84. Varadi A, Johnson-Cadwell LI, Cirulli V, Yoon Y, Allan VJ, Rutter GA. Cytoplasmic dynein regulates the subcellular distribution of mitochondria by controlling the recruitment of the fission factor dynamin-related protein-1. *J Cell Sci* 2004;**117**:4389–400.
85. Malli R, Frieden M, Osibow K, et al. Sustained Ca^{2+} transfer across mitochondria is essential for mitochondrial Ca^{2+} buffering, store-operated Ca^{2+} entry, and Ca^{2+} store refilling. *J Biol Chem* 2003;**278**:44,769–79.
86. Szabadkai G, Simoni AM, Rizzuto R. Mitochondrial Ca2+ uptake requires sustained Ca^{2+} release from the endoplasmic reticulum. *J Biol Chem* 2003;**278**:15,153–61.
87. Mogami H, Nakano K, Tepikin AV, Petersen OH. Ca^{2+} flow via tunnels in polarized cells: recharging of apical Ca^{2+} stores by focal Ca^{2+} entry through basal membrane patch.. *Cell* 1997;**88**:45–9.
88. Malli R, Frieden M, Trenker M, Graier WF. The role of mitochondria for Ca^{2+} refilling of the endoplasmic reticulum. *J Biol Chem* 2005;**280**:12,114–22.

Chapter 46

Signaling during Organelle Division and Inheritance: Peroxisomes

Andrei D. Fagarasanu and Richard A. Rachubinski

Department of Cell Biology, University of Alberta, Edmonton, Alberta, Canada

INTRODUCTION TO PEROXISOMES

Peroxisomes are ubiquitous organelles involved in a wide spectrum of biochemical processes such as the β-oxidation of fatty acids and the metabolism of hydrogen peroxide [1, 2]. Peroxisomes are essential for normal human development, since failure to assemble functional peroxisomes results in a number of inherited lethal diseases collectively called the peroxisome biogenesis disorders [2].

Peroxisomes are highly dynamic, dividing constitutively in growing cell populations and expanding extensively in response to various nutrients that require peroxisome-housed enzymes for their metabolism. It is now well established that peroxisomes are organelles that derive from the endoplasmic reticulum (ER) [3–9]. Multiple studies have shown that several, if not all, peroxisomal membrane proteins (PMPs), initially target to the ER after their synthesis and are sequestered into discrete specialized regions of the ER membrane. These regions, which exclude resident ER proteins, are released from the ER donor membrane presumably as small preperoxisomal vesicles (for reviews, see [1, 10, 11]) (Figure 46.1). To form fully assembled, mature peroxisomes, preperoxisomes need to undergo a sequence of maturation steps that probably iolve protein and lipid import, as well

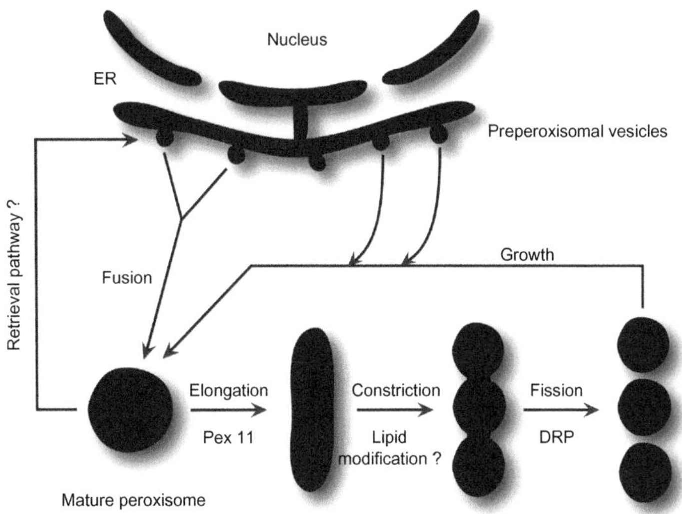

FIGURE 46.1 A model for peroxisome biogenesis and division.
Preperoxisomal vesicles originate at distinct specialized compartments of the ER. When *de novo* peroxisome formation is needed, these preperoxisomes mature through the import of membrane and matrix proteins as well as by homotypic membrane fusion. Alternatively, the ER derived preperoxisomes fuse with pre-existing peroxisomes, assisting in their growth. To divide, peroxisomes are first elongated by Pex11 proteins and constricted through modifications in their membrane lipid composition. DRPs are then recruited to execute the fission reaction. The growth of peroxisomes is probably mediated by both fusion with preperoxisomal vesicles and by the direct import of peroxisomal matrix and membrane proteins. A retrograde peroxisome-ER pathway ensures the retrieval of escaped ER proteins and the recycling of proteins involved in preperoxisome assembly at the ER.

Handbook of Cell Signaling, Three-Volume Set 2 ed.

as homotypic membrane fusion (Figure 46.1). To maintain organelle homeostasis, the ER-to-peroxisome pathway must be accompanied by retrograde, vesicle mediated transport for the retrieval both of escaped ER resident proteins and the PMPs that organize the initial steps of preperoxisome formation [1] (Figure 46.1).

Upon cell division, peroxisomes must be segregated equally between the two resulting cells to ensure their inheritance by future generations. This process also implies that the number of peroxisomes has to increase during the cell cycle to prevent their dilution with each round of cell division. Given that peroxisomes can form from the ER but also possess the ability to grow and divide, the question arises as to whether the doubling of peroxisome number during cell division is due solely to the fission of pre-existing peroxisomes or is also the result of the production of new peroxisomes from the ER. A recent study [5] showed that both processes, the division of peroxisomes and the synthesis of new ones, account for the formation of peroxisomes in constitutively dividing mammalian cells. However, Motley and Hettema recently provided evidence that peroxisomes multiply by growth and division and do not form *de novo* in wild-type yeast cells [12]. Only yeast cells lacking peroxisomes as a result of a segregation defect were observed to form peroxisomes *de novo* from the ER. Therefore, at least in yeast, *de novo* peroxisome formation may represent a rescue mechanism that becomes functional only in the eventuality that peroxisomes are lost [13]. It is tempting to speculate that during their growth, mature peroxisomes acquire their membrane constituents by fusion with ER derived preperoxisomes [1] (Figure 46.1). In this case, the ER-to-peroxisome pathway would function to supply existing peroxisomes with essential membrane components to sustain their growth and division. Such a scenario would also provide a means for those PMPs that constitutively transit through the ER to be incorporated into growing peroxisomes while maintaining their concentrations within the peroxisomal membrane during multiple rounds of peroxisomal growth and division. We will now focus on the molecular mechanisms and signaling pathways known to regulate and coordinate the division and inheritance of peroxisomes, paying particular attention to these processes as they occur in budding yeast.

PEROXISOME DIVISION

Peroxisome division must be coordinated with the cell cycle to maintain the number of peroxisomes in a cell during cell division. Morphological observations have shown that peroxisome division proceeds through three partially overlapping steps: elongation or tubulation of peroxisomes, constriction of the peroxisomal membrane, and fission of peroxisomes (Figure 46.1).

Pex11 proteins were the first PMPs to be implicated in peroxisome division [14, 15]. Based on sequence homology, all eukaryotic cells studied have been shown to contain multiple Pex11 isoforms. *S. cerevisiae* contains three proteins belonging to the Pex11 family, designated Pex11p, Pex25p, and Pex27p [14–18]. Cells lacking Pex11p contain fewer and larger peroxisomes as compared to wild-type cells, a phenotype indicative of impaired peroxisome division [14]. The same phenotype is observed in cells lacking either Pex25p or Pex27p or containing pairwise deletions of Pex11p, Pex25p, and Pex27p, suggestive of a partial functional redundancy among these proteins. Interestingly, overexpression of the *PEX11* gene leads to an increase in the number of peroxisomes but also in the appearance of elongated peroxisomal structures. A similar observation was made in mammalian cells, where the overproduction of one of the Pex11 isoforms, Pex11β, results in peroxisome tubulation followed by an increase in peroxisome number [19]. However, in the absence of a functional dynamin related protein (DRP) (see below), overproduction of Pex11β results in peroxisome tubulation without an increase in peroxisome number. This distinctive phenotype implicates Pex11p in the tubulation step of peroxisome division [15, 20] (Figure 46.1). Therefore, Pex11 proteins cannot constrict or divide peroxisomes themselves and most likely function upstream of DRPs by promoting peroxisome elongation through an as yet poorly understood mechanism [21–23].

Similar to other membrane fission events [24, 25], fission of the peroxisomal membrane must be preceded by destabilization of the membrane bilayer and strong membrane bending [26]. These energetically unfavorable processes usually require several protein complexes and a distinct set of membrane lipids, including phosphatidic acid and diacylglycerol [24, 26–29]. The molecular mechanism underlying the constriction of the peroxisomal membrane is largely unknown. One mechanism that allows the peroxisomal membrane to constrict only after peroxisomes have fully matured has been described in the budding yeast, *Yarrowia lipolytica*, and involves *Yl*Pex16p, a peripheral membrane protein that resides on the matrix side of the peroxisomal membrane [30, 31]. *Yl*Pex16p impedes peroxisome division by inhibiting a biosynthetic pathway that leads to the formation of phosphatidic acid and diacylglycerol, both of which are potent inducers of membrane curvature. The import of matrix proteins during peroxisome maturation promotes the gradual redistribution of acyl-CoA oxidase, an enzyme of peroxisomal fatty acid β-oxidation, from the peroxisomal matrix to the peroxisomal membrane. The interaction between membrane attached acyl-CoA oxidase and *Yl*Pex16p terminates the negative influence of *Yl*Pex16p on membrane bending, allowing mature peroxisomes to divide [26]. Peroxisome division in *Y. lipolytica* is therefore regulated by a distinctive mechanism that controls membrane fission in response to a signal emanating from within the peroxisome itself [26, 31]. Since a significant portion of acyl-CoA oxidase is relocated from the matrix to the inner face of the peroxisomal membrane only in mature peroxisomes, this mechanism prevents excessive proliferation of immature peroxisomal vesicles. It remains to be established whether the constriction of peroxisomal membranes in other cell types is regulated by similar intraperoxisomal signaling.

It is widely accepted that the final fission step of peroxisome division is catalyzed by DRPs. DRPs are small GTPases that probably function as mechanochemical enzymes that use GTP hydrolysis dependent conformational changes to drive fission [32]. Mammalian or yeast cells deficient in peroxisomal DRPs display elongated peroxisomes with segmented morphology that resemble beads on a string [21, 33] (Figure 46.1). Therefore, without DRPs, peroxisomes are able to elongate and constrict but unable to divide fully. These findings suggest that DRPs catalyze the last stage of peroxisome division by pinching off small peroxisomes from already constricted tubules [20]. The constriction of the peroxisomal tubule is probably required for the DRP ring to be efficiently assembled around the tubule and to promote fission of the peroxisome. There are other known instances in which membranes are first constricted by other factors and then subjected to scission by dynamins recruited to execute this final step. For example, in endocytosis, dynamin promotes membrane scission only after clathrin and other coat proteins have constricted the neck of an endocytic vesicle [34].

Vps1p and Dnm1p, two of the three DRPs in *S. cerevisiae*, were found to be required for peroxisome division [33, 35]. Vps1p is required for peroxisome fission under both peroxisome inducing and non-inducing conditions [33], whereas Dnm1p is required only under peroxisome inducing conditions [35]. Interestingly, the giant peroxisome present in a *vps1Δ/dnm1Δ* mutant cell is still able to undergo constitutive division and is correctly apportioned between mother and daughter cells upon cell division [35]. This finding points to the existence of a dynamin independent mode of peroxisome division. This type of division may result from the pulling forces exerted by the machinery that propels bud-directed movement of peroxisomes on the one hand and peroxisome retention mechanisms on the other, which act on the same giant peroxisome during peroxisome inheritance, eventually tearing it apart [35, 36] (see "Peroxisome inheritance," below).

Classical dynamins use their pleckstrin homology (PH) domains, which bind membrane lipids, to associate with membranes. However, because DRPs lack PH domains, other factors must recruit them to the membrane of their target organelles. Fis1 was shown to be the adaptor for DRPs on the peroxisomal membrane of mammalian cells. *S. cerevisiae* Fis1p recruits Dnm1p, but not Vps1p, to the peroxisomal membrane [35]. In line with this finding, Fis1p, like Dnm1p, plays a role in peroxisome division, especially under conditions of peroxisome induction. It remains to be established what proteins underlie the association of Vps1p with peroxisomes in *S. cerevisiae*, although Pex19p, the putative targeting receptor for PMPs, may be involved [37].

PEROXISOME INHERITANCE

At cell division, organelles have to be positioned at distinct locations at specific times to ensure their inheritance by daughter cells. *S. cerevisiae* has been used effectively to study the mechanisms of organelle inheritance. Since an *S. cerevisiae* cell divides by budding, organelle partitioning is achieved by the directional delivery of half of its organelles to the attached bud concomitant with the retention of the remaining organelles in the mother cell. This feature makes the detection and isolation of organelle inheritance mutants in *S. cerevisiae* easier than in cells that divide by median fission [38, 39].

S. cerevisiae cells undergo a repetitive pattern of growth and division [39–41]. After bud emergence in late G1, cell growth is restricted to the bud tip. In G2-M, a switch from apical to isotropic growth allows delivery of cell wall material over the entire bud surface. Later in the cell cycle, growth is directed to the mother-bud neck for assembly of a septum that will separate the mother cell from its bud [40, 41]. As is the case in most eukaryotes, the actin cytoskeleton provides the structural basis for polarized growth in *S. cerevisiae.* Actin cables, which direct the delivery of secretory vesicles, are subject to major rearrangements in response to cell cycle cues so that growth is polarized to discrete sites that are distinct for each stage of the cell cycle [42].

Actin cables are assembled by a conserved class of proteins called formins. Formins associate with the plus end of an actin filament and promote the incorporation of new actin monomers at this location while remaining attached to the growing actin filament [41] (Figure 46.2). Since formins are concentrated in the bud, actin cables will radiate from the bud deep into the mother cell [43] (Figure 46.2). In mother cells, Myo2p, a class V myosin, captures various cargoes, including secretory vesicles, and uses the actin cables as tracks toward the formin-rich regions, which thus become sites of growth. Changes in the localization of formins during the cell cycle result in alterations in the actin cytoskeleton that underlie the targeting of secretory vesicles to varying locations in the bud [39, 41]. In addition to polarizing secretion, Myo2p drives the bud-directed movement of many organelles, including peroxisomes, to ensure their faithful inheritance [33, 44].

Each *S. cerevisiae* cell contains on average about nine peroxisomes [33] that are localized at the cell periphery [33, 44, 45]. Peroxisomes display cell cycle coordinated movements that result in their correct segregation upon cell division [33, 44–46]. As soon as the bud emerges from the mother cell, peroxisomes detach one by one from their static cortical positions and travel toward the nascent bud. Recruitment of peroxisomes from the mother cell cortex to the bud continues until about half of the initial peroxisomal

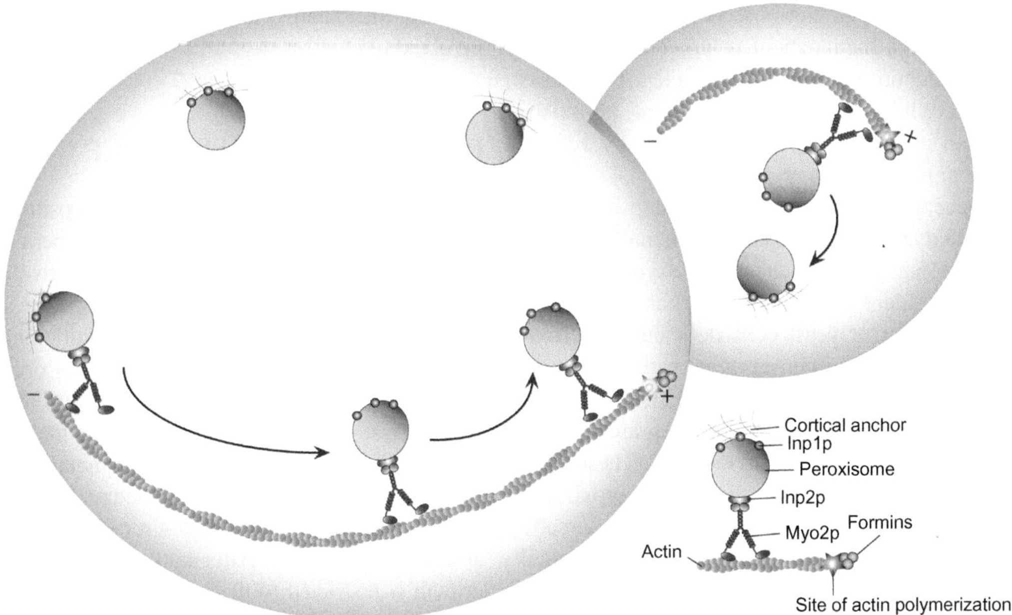

FIGURE 46.2 A model for peroxisome inheritance in *S. cerevisiae*.
Peroxisomes are immobilized at the mother cell cortex via an interaction between Inp1p and an unidentified cortical structure. At a specific point in the cell cycle, Inp2p is synthesized and preferentially associates with a subset of peroxisomes. The formation of multiple Inp2p-Myo2p transport complexes on the membrane of these peroxisomes results in a pulling force strong enough to dislodge them from their fixed cortical positions. Myo2p carries attached peroxisomes along actin cables to the bud tip. Later, peroxisomes spread throughout the bud cell periphery. The degradation of Inp2p later in the cell cycle results in the detachment of peroxisomes from the Myo2p motor, and only a few peroxisomes follow Myo2p to the mother-bud neck at cytokinesis. The release of peroxisomes from the grip of Myo2p is temporally coordinated with their Inp1p mediated capture at the cell cortex. The immobilization of peroxisomes at the bud periphery prepares the bud for the ensuing cell cycle, when, as a mother cell, it will have to retain half of its peroxisomes.

population is transferred to the daughter cell. The directional migration of peroxisomes is Myo2p driven and actin based, since a point mutation that affects Myo2p processivity compromises bud-directed peroxisome motility [33, 46] (Figure 46.2). Interestingly, inside the bud, peroxisomes follow the rearrangements of the actin cytoskeleton, being initially clustered at the bud tip and later distributed over the entire bud cortex. Before cytokinesis, a few peroxisomes in the bud and mother cell relocate to the mother-bud neck region, while the rest remain anchored at the bud and mother cell cortices. These findings indicate that the equitable distribution of peroxisomes between the mother and daughter cells is achieved through a tightly regulated interplay between peroxisome mobility and retention [1, 46]. Two PMPs called Inp1p and Inp2p have been identified in *S. cerevisiae* as having roles in the retention and motility of peroxisomes, respectively [44, 45].

Inp1p functions in immobilizing peroxisomes at the cell cortex [45]. In cells lacking Inp1p, all peroxisomes are very mobile, and the entire peroxisome population eventually gets transferred to the daughter cell. Consistent with a role for Inp1p in the retention of peroxisomes at the cell periphery, overproduction of Inp1p causes all peroxisomes in the mother cell to maintain fixed cortical positions, which prevents their normal transfer to the daughter cell [45, 46]. Interestingly, whereas Inp1p normally resides exclusively in peroxisomes, overproduced Inp1p decorates both peroxisomes and the cell cortex. This suggests that Inp1p has an intrinsic affinity for a structure that lines the cell periphery. Thus, Inp1p probably functions in peroxisome retention by linking these organelles to an as yet uncharacterized cortical anchor. Also, Inp1p probably also mediates the cortical retention of peroxisomes once they are transferred to the bud. The cortical immobilization of peroxisomes in the bud is necessary to prepare the bud for the next cell cycle, when, as a mother cell, it will have to retain about half of its peroxisomal population (Figure 46.2). Inp1p was the first protein identified to function in the anchoring of an organelle. Although it has long been proposed that cells possess devices for the site specific immobilization of their organelles, no component specifically involved in this process had previously been discovered [1, 46]. However, the molecular nature of the cortical structure to which Inp1p adheres remains undetermined.

Inp2p functions as the peroxisome specific receptor for Myo2p. Inp2p has been identified as an integral membrane protein of peroxisomes required for the distribution of peroxisomes to buds [44]. In cells lacking Inp2p, the migration of peroxisomes to buds is drastically compromised, whereas the segregation of other organelles is unaffected. Inp2p interacts directly with the Myo2p globular tail, and overproduction of Inp2p drives all peroxisomes to sites of growth at which Myo2p normally accumulates [39, 44, 46].

Inp2p levels oscillate during the cell cycle in a pattern that correlates with peroxisome dynamics [44]. It is believed that these fluctuations in Inp2p levels result in the assembly and then disassembly of Inp2p-Myo2p transport complexes at the peroxisomal membrane [39, 44, 46]. Inp2p starts to accumulate during early budding, and its levels reach a peak in medium sized budded cells when most vectorial movements of peroxisomes to daughter cells occur. Interestingly, once synthesized, Inp2p accumulates preferentially on a subset of peroxisomes. The peroxisomes enriched in Inp2p are those that are eventually delivered to the bud, indicating that Myo2p selectively carries those peroxisomes that contain high amounts of Inp2p. Once the number of Myo2p molecules on a given peroxisome reaches a critical level, the pulling force exerted on the peroxisomal membrane probably becomes great enough to detach that peroxisome from the Inp1p interacting cortical anchor. Notably, in cells lacking Inp1p where peroxisomes have lost their affinity for the cell cortex, the entire peroxisome population is eventually transferred to daughter cells, presumably in an Inp2p dependent and Myo2p dependent manner. This suggests that all peroxisomes contain enough Inp2p to promote their Myo2p driven movement in the absence of an opposing force [39, 46]. Therefore, peroxisomes that remain in mother cells in the wild-type strain are likely to still contain Inp2p but in quantities insufficient to overcome the adhesion forces between peroxisomes and cortical anchors. We have mentioned that *vps1Δ/dnm1Δ* mutant cells contain a single peroxisomal structure that extends between the mother and daughter cell and is split in two before cytokinesis (see "Peroxisome division," above). The molecular basis of this apparently dynamin independent fission event is unknown but could result from cooperation between retention mechanisms that anchor the peroxisome in the mother cell and the Inp2p-Myo2p translocation machinery that pulls it into the bud. Cytoskeletal tracks and motor proteins are known to exert tension on organelle membranes, thereby assisting organelle fission [47]. Indeed, it was recently shown that in cells of the triple deletion strain, *vps1Δ/dnm1Δ/inp2Δ*, peroxisomes failed to divide altogether [12], thus implicating Inp2p, and consequently Myo2p, in the fission of peroxisomes in *vps1Δ/dnm1Δ* cells at cell division. It remains to be established if Inp2p plays a role in peroxisome division in wild-type cells.

Inp2p is degraded late in the cell cycle. Before cytokinesis, the few peroxisomes that still contain detectable levels of Inp2p are moved to the mother-bud neck region by Myo2p, while the remaining peroxisomes have already assumed static positions at the bud cortex. This clearly suggests that the termination of peroxisome movement is caused by the disassembly of the Inp2p-Myo2p transport complex through proteolytic degradation of Inp2p. The termination of peroxisome motility seems to be coordinated with the capture of peroxisomes by an anchoring device at the bud cortex. This points to the existence of a regulated transfer of peroxisomes from Myo2p to a cortical anchor, a process that is exactly opposite to the one that initially recruited peroxisomes from the mother cell cortex. A tug of war between Inp1p and Inp2p, similar to the one envisaged for the mother cell, might determine whether a peroxisome becomes cortically anchored or remains attached to Myo2p within the bud. When Inp2p on a transferred peroxisome is degraded, Inp1p swings the balance of such a molecular "contest of strength" toward the establishment of peroxisome-cortex connections [1, 39, 46] (Figure 46.2).

The degradation machinery responsible for Inp2p turnover remains unknown. It also remains to be determined whether Inp2p degradation is intrinsically linked to cell cycle progression or is triggered by cellular surveillance mechanisms that monitor the extent of peroxisome transfer to buds [1, 39, 46].

CONCLUDING REMARKS

Even though the main molecular players involved in peroxisome division and inheritance appear to have been identified, the exact molecular mechanisms underlying their functions are poorly understood. Since the correct partitioning of peroxisomes involves both peroxisome division and peroxisome segregation, it is important to set aside the view of these two processes as being independent and self-contained. The challenge now is to unravel the signaling pathways that allow peroxisome division and inheritance to be temporally coordinated with each other and with cell cycle events to ensure the perpetuation of peroxisome populations during cell proliferation.

ACKNOWLEDGEMENTS

We thank members of the Rachubinski laboratory for insightful discussion and valuable comments on the manuscript. We apologize to those colleagues whose work was not cited owing to space constraints. Our work on peroxisome inheritance is supported by Grant 9208 from the Canadian Institutes of Health Research. RAR is an International Research Scholar of the Howard Hughes Medical Institute.

REFERENCES

1. Fagarasanu A, Fagarasanu M, Rachubinski RA. Maintaining peroxisome populations: a story of division and inheritance. *Annu Rev Cell Dev Biol* 2007;**23**:321–44.
2. Purdue PE, Lazarow PB. Peroxisome biogenesis. *Annu Rev Cell Dev Biol* 2001;**17**:701–52.
3. Geuze HJ, Murk JL, Stroobants AK, Griffith JM, Kleijmeer MJ, Koster AJ, Verkleij AJ, Distel B, Tabak HF. Involvement of the endoplasmic reticulum in peroxisome formation. *Mol Biol Cell* 2003;**14**:2900–7.
4. Hoepfner D, Schildknegt D, Braakman I, Philippsen P, Tabak HF. Contribution of the endoplasmic reticulum to peroxisome formation. *Cell* 2005;**122**:85–95.

5. Kim PK, Mullen RT, Schumann U, Lippincott-Schwartz J. The origin and maintenance of mammalian peroxisomes involves a de novo PEX16-dependent pathway from the ER. *J Cell Biol* 2006;**173**: 521–32.
6. Mullen RT, Lisenbee CS, Miernyk JA, Trelease RN. Peroxisomal membrane ascorbate peroxidase is sorted to a membranous network that resembles a subdomain of the endoplasmic reticulum. *Plant Cell* 1999;**11**:2167–85.
7. Tam YYC, Fagarasanu A, Fagarasanu M, Rachubinski RA. Pex3p initiates the formation of a preperoxisomal compartment from a subdomain of the endoplasmic reticulum in *Saccharomyces cerevisiae*. *J Biol Chem* 2005;**280**:34,933–9.
8. Titorenko VI, Ogrydziak DM, Rachubinski RA. Four distinct secretory pathways serve protein secretion, cell surface growth, and peroxisome biogenesis in the yeast *Yarrowia lipolytica*. *Mol Cell Biol* 1997;**17**:5210–26.
9. Titorenko VI, Rachubinski RA. The endoplasmic reticulum plays an essential role in peroxisome biogenesis. *Trends Biochem Sci* 1998;**23**:231–3.
10. Tabak HF, Hoepfner D, Zand A, Geuze HJ, Braakman I, Huynen MA. Formation of peroxisomes: present and past. *Biochim Biophys Acta* 2006;**1763**:1647–54.
11. Titorenko VI, Mullen RT. Peroxisome biogenesis: the peroxisomal endomembrane system and the role of the ER. *J Cell Biol* 2006;**174**:11–7.
12. Motley AM, Hettema EH. Yeast peroxisomes multiply by growth and division. *J Cell Biol* 2007;**178**:399–410.
13. Schrader M, Fahimi HD. The peroxisome: still a mysterious organelle. *Histochem Cell Biol* 2008;**129**:421–40.
14. Erdmann R, Blobel G. Giant peroxisomes in oleic acid-induced *Saccharomyces cerevisiae* lacking the peroxisomal membrane protein Pmp27p. *J Cell Biol* 1995;**128**:509–23.
15. Marshall PA, Krimkevich YI, Lark RH, Dyer JM, Veenhuis M, Goodman JM. Pmp27 promotes peroxisomal proliferation. *J Cell Biol* 1995;**129**:345–55.
16. Rottensteiner H, Stein K, Sonnenhol E, Erdmann R. Conserved function of Pex11p and the novel Pex25p and Pex27p in peroxisome biogenesis. *Mol Biol Cell* 2003;**14**:4316–28.
17. Smith JJ, Marelli M, Christmas RH, Vizeacoumar FJ, Dilworth DJ, Ideker T, Galitski T, Dimitrov K, Rachubinski RA, Aitchison JD. Transcriptome profiling to identify genes involved in peroxisome assembly and function. *J Cell Biol* 2002;**158**:259–71.
18. Tam YYC, Torres-Guzman JC, Vizeacoumar FJ, Smith JJ, Marelli M, Aitchison JD, Rachubinski RA. Pex11-related proteins in peroxisome dynamics: a role for the novel peroxin Pex27p in controlling peroxisome size and number in *Saccharomyces cerevisiae*. *Mol Biol Cell* 2003;**14**:4089–102.
19. Schrader M, Reuber BE, Morrell JC, Jimenez-Sanchez G, Obie C, Stroh TA, Valle D, Schroer TA, Gould SJ. Expression of PEX11β mediates peroxisome proliferation in the absence of extracellular stimuli. *J Biol Chem* 1998;**273**:29,607–14.
20. Yan M, Rayapuram N, Subramani S. The control of peroxisome number and size during division and proliferation. *Curr Opin Cell Biol* 2005;**17**:376–83.
21. Koch A, Thiemann M, Grabenbauer M, Yoon Y, McNiven MA, Schrader M. Dynamin-like protein 1 is involved in peroxisomal fission. *J Biol Chem* 2003;**278**:8597–605.
22. Koch A, Schneider G, Luers GH, Schrader M. Peroxisome elongation and constriction but not fission can occur independently of dynamin-like protein 1. *J Cell Sci* 2004;**117**:3995–4006.
23. Schrader M. Shared components of mitochondrial and peroxisomal division. *Biochim Biophys Acta* 2006;**1763**:531–41.
24. McMahon HT, Gallop JL. Membrane curvature and mechanisms of dynamic cell membrane remodelling. *Nature* 2005;**438**:590–6.
25. Zimmerberg J, Kozlov MM. How proteins produce cellular membrane curvature. *Nat Rev Mol Cell Biol* 2006;**7**:9–19.
26. Guo T, Gregg C, Boukh-Viner T, Kyryakov P, Goldberg A, Bourque S, Banu F, Haile S, Milijevic S, San KH, Solomon J, Wong V, Titorenko VI. A signal from inside the peroxisome initiates its division by promoting the remodeling of the peroxisomal membrane. *J Cell Biol* 2007;**177**:289–303.
27. Bankaitis VA. Slick recruitment to the Golgi. *Science* 2002;**295**: 290–1.
28. Behnia R, Munro S. Organelle identity and the signposts for membrane traffic. *Nature* 2005;**438**:597–604.
29. Farsad K, De CP. Mechanisms of membrane deformation. *Curr Opin Cell Biol* 2003;**15**:372–81.
30. Eitzen GA, Szilard RK, Rachubinski RA. Enlarged peroxisomes are present in oleic acid-grown *Yarrowia lipolytica* overexpressing the *PEX16* gene encoding an intraperoxisomal peripheral membrane peroxin. *J Cell Biol* 1997;**137**:1265–78.
31. Guo T, Kit YY, Nicaud J-M, Le Dall MT, Sears SK, Vali H, Chan H, Rachubinski RA, Titorenko VI. Peroxisome division in the yeast *Yarrowia lipolytica* is regulated by a signal from inside the peroxisome. *J Cell Biol* 2003;**162**:1255–66.
32. Praefcke GJ, McMahon HT. The dynamin superfamily: universal membrane tubulation and fission molecules? *Nat Rev Mol Cell Biol* 2004;**5**:133–47.
33. Hoepfner D, van den Berg M, Philippsen P, Tabak HF, Hettema EH. A role for Vps1p, actin, and the Myo2p motor in peroxisome abundance and inheritance in *Saccharomyces cerevisiae*. *J Cell Biol* 2001;**155**:979–90.
34. Osteryoung KW, Nunnari J. The division of endosymbiotic organelles. *Science* 2003;**302**:1698–704.
35. Kuravi K, Nagotu S, Krikken AM, Sjollema K, Deckers M, Erdmann R, Veenhuis M, van der Klei IJ. Dynamin-related proteins Vps1p and Dnm1p control peroxisome abundance in *Saccharomyces cerevisiae*. *J Cell Sci* 2006;**119**:3994–4001.
36. van der Zand A, Braakman I, Geuze HJ, Tabak HF. The return of the peroxisome. *J Cell Sci* 2006;**119**:989–94.
37. Vizeacoumar FJ, Vreden WN, Fagarasanu M, Eitzen GA, Aitchison JD, Rachubinski RA. The dynamin-like protein Vps1p of the yeast *Saccharomyces cerevisiae* associates with peroxisomes in a Pex19p-dependent manner. *J Biol Chem* 2006;**281**:12,817–23.
38. Chang J, Fagarasanu A, Rachubinski RA. Peroxisomal peripheral membrane protein YlInp1p is required for peroxisome inheritance and influences the dimorphic transition in the yeast *Yarrowia lipolytica*. *Eukaryot Cell* 2007;**6**:1528–37.
39. Fagarasanu A, Rachubinski RA. Orchestrating organelle inheritance in *Saccharomyces cerevisiae*. *Curr Opin Microbiol* 2007;**10**:528–38.
40. Bretscher A. Polarized growth and organelle segregation in yeast: the tracks, motors, and receptors. *J Cell Biol* 2003;**160**:811–6.
41. Pruyne D, Legesse-Miller A, Gao L, Dong Y, Bretscher A. Mechanisms of polarized growth and organelle segregation in yeast. *Annu Rev Cell Dev Biol* 2004;**20**:559–91.
42. Pruyne D, Gao L, Bi E, Bretscher A. Stable and dynamic axes of polarity use distinct formin isoforms in budding yeast. *Mol Biol Cell* 2004;**15**:4971–89.
43. Yang HC, Pon LA. Actin cable dynamics in budding yeast. *Proc Natl Acad Sci. U S A* 2002;**99**:751–6.

44. Fagarasanu A, Fagarasanu M, Eitzen GA, Aitchison JD, Rachubinski RA. The peroxisomal membrane protein Inp2p is the peroxisome-specific receptor for the myosin V motor Myo2p of *Saccharomyces cerevisiae*. *Dev Cell* 2006;**10**:587–600.
45. Fagarasanu M, Fagarasanu A, Tam YYC, Aitchison JD, Rachubinski RA. Inp1p is a peroxisomal membrane protein required for peroxisome inheritance in *Saccharomyces cerevisiae*. *J Cell Biol* 2005;**169**:765–75.
46. Fagarasanu M, Fagarasanu A, Rachubinski RA. Sharing the wealth: peroxisome inheritance in budding yeast. *Biochim Biophys Acta* 2006;**1763**:1669–77.
47. Schrader M, Fahimi HD. Growth and division of peroxisomes. *Int Rev Cytol* 2006;**255**:237–90.

Chapter 47

Bidirectional Crosstalk between Actin Dynamics and Endocytosis

Giorgio Scita[1,2] and Pier Paolo Di Fiore[1,2,3]
[1]*IFOM, Fondazione Istituto FIRC di Oncologia Molecolare, Milan, Italy*
[2]*Dipartimento di Medicina, Chirurgia ed Odontoiatria, Universita' degli Studi di Milano, Milan, Italy*
[3]*Dipartimento di Oncologia Sperimentale, Istituto Europeo di Oncologia, Milan, Italy*

INTRODUCTION

Endocytic trafficking of membranes/proteins and motility based on actin dynamics are intimately linked. However, our understanding of the molecular circuitry involved is still in its infancy. Results obtained in different species have established that endocytosis and trafficking events rely on propelling forces generated by actin polymerization and depolymerization. Accordingly, an increasing number of actin binding and regulatory proteins participate in a variety of internalization and trafficking processes, which ultimately control the signaling response of cells to extracellular stimuli. In addition, genetic and cellular biochemical evidence has revealed how cycles of endocytosis and recycling of plasma membranes and plasma membrane proteins are essential to promote the spatial restriction of signaling. This is achieved by confining and polarizing key actin regulatory molecules and generating self-controlled feedback loops that ensure the generation of uni-vectorial, actin based motility forces. These, in turn, regulate dynamic changes in cell shape and cell migration in response to motogenic stimuli. Thus, unraveling the interplay between trafficking and actin dynamics will bear important consequences for our understanding of the mechanisms of fundamental biological processes, such as the ability of cells to plastically adapt their shape and migration strategies in response to the various environmental conditions encountered in different tissues.

In this chapter we will initially cover the cell biology evidence that establishes the significance of actin dynamics for the correct execution of endocytosis; then we will survey the vast field of molecular connections between the endocytic and actin machineries, and finally we will describe results showing how endocytosis itself controls various forms of actin dynamics, especially those required for the spatial restriction of signaling events. Throughout this chapter, we will mainly focus on findings obtained in mammalian cells, while referring the reader to excellent recent reviews for work performed in yeast [1–3]

FROM ACTIN TO ENDOCYTOSIS

Actin Dynamics Promote Optimal Endocytosis

The dynamic elongation of actin filaments at the plasma membrane generates the forces that drive plasma membrane distortions, which often take the form of protrusions linked functionally to migratory behavior [4, 5]. Consistent with this, the fast growing or barbed ends of actin filaments are oriented toward the protruding plasma membrane, generating vectorial forces that are able to compensate and overcome membrane resilience. This enables net advancement of the cell edge in what has been recognized as the first step of cell migration. It has not yet been definitively established (at least in mammalian cells), whether similar actin based sources of force generation take part in the process that leads to inward invaginations of the plasma membrane, which then progressively deform into tubular bud-like ingressions allowing endocytosis; this subject is presently under intense scrutiny [6].

The first hints of the involvement of the actin cytoskeleton in endocytosis derived from early experiments with actin poisons that interfere with actin turnover [7, 8]. This allowed the establishment of actin dynamics as an absolute prerequisite for endocytic uptake in lower unicellular organisms, such as *S. cerevisiae* (see [1] for review).

Handbook of Cell Signaling, Three-Volume Set 2 ed.

In mammalian cells, however, the block in endocytosis caused by actin poisons seemed to be partial or limited to the apical surface of epithelial polarized cells [7–10]. Subsequent work exploited genetic approaches with actin mutants [11] and mutants of various actin regulators [2, 12] to establish that active filament assembly and disassembly is required for the first initial step of endocytosis, in *S. cerevisiae*. Furthermore, in mammalian cells, the ablation of the corresponding gene products resulted, at most, in partial defects [13–16]; the involvement of actin filament assembly and disassembly in mammalian endocytosis was unequivocally proven only after the development of sophisticated imaging techniques (see below).

Collectively, the above results suggest that actin dynamics were selected as the mechanism of choice for force generation during endocytosis in simple unicellular organisms. The evolution of multicellular organisms, however, required the development of additional means to invaginate plasma membranes; these were then optimized through actin based mechanisms. It is clear, however, that actin dynamics do play a role in endocytosis in mammals, as also witnessed by a large number of protein–protein interactions linking the two molecular machineries ([2, 17], see also below). A number of questions lie, therefore, at the heart of intense investigations: (i) how do actin dynamics contribute to endocytosis? (ii) are there cell-, or tissue-, or cargo specific differences? (iii) what are the biological end points that depend on actin dynamics via endocytosis (e.g., fate of internalized cargo, duration of signals, intensity and location of the signaling output)?

"Blinking lights" Over Actin in Endocytosis

Insights into some of these issues were gained from live-cell imaging and localization studies in *S. cerevisiae* and in mammalian cells. In both systems, there is an orderly recruitment of endocytic and actin regulatory proteins at sites of plasma membrane invagination [1]. Strikingly, both yeast and mammals utilize a common set of conserved molecules that assemble in a similarly sequential mode, somewhat paralleling the various steps of endocytic internalization. These phases include bending and invagination of the plasma membrane leading to pit formation, constriction of the pit to form a membrane-tethered vesicle, and, finally, scission that allows the vesicle to bud into the cytoplasm (Figure 47.1).

Electron microscopy (EM) of newly formed clathrin coats and of early invaginating pits containing the IgM receptor in human lymphoblastoid cells [18] first revealed that these structures are often associated with filamentous actin. Similarly, in neurons, F-actin co-localized with recycling vesicles at the synapse, and disruption of actin function interfered with synaptic vesicle recycling [19]. It was the advent of total internal reflection microscopy, however, that allowed researchers to zoom into the early and most dynamic events of internalization. This revealed that actin and a variety of actin regulators [6] are transiently recruited to motile clathrin spots, whose disappearance from the evanescent field coincided with the fission of clathrin coated vesicles and their internalization; this gives rise to a characteristic blinking behavior in real-time microscopy. Some 60–80 percent of the clathrin spots were shown to recruit

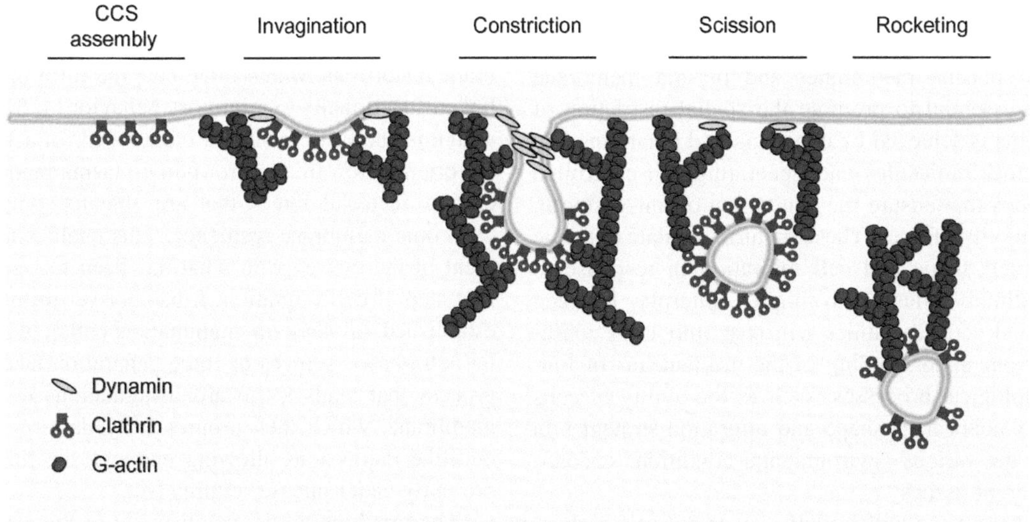

FIGURE 47.1 Possible roles of actin filament dynamics during the various phases of clathrin dependent internalization.
The various steps of internalization are depicted, from left to right: (i) assembly of clathrin coated structures (CCS) at the plasma membrane; (ii) invagination of the plasma membrane in correspondence to CCS, aided by phospholipid binding and bending proteins (not shown) in mammalian cells, and by actin assembly in yeast (note that the large GTPase dynamin localizes at the rim of forming pits); (iii) constriction and scission of the deeply invaginated pit – the neck of the pit is coated with Dynamin oligomers, which will constrict the pit upon GTP hydrolysis, and actin filament dynamics may generate additional forces, adding tension, thereby aiding scission of the vesicle; (iv) movement (rocketing) of the newly formed vesicle: short lived actin comet tails may propel the movement of the internalized vesicle.

actin or actin regulators transiently during constitutive internalization [13, 15, 16, 20], suggesting a role for actin dynamics in the early phases of endocytosis.

To define the endocytic phase in which actin recruitment takes place, Merrifield and colleagues [16] developed an elegant approach based on the use of pH sensitive GFP fused to the transferrin receptor (TfR), a cargo that is constitutively internalized in a clathrin dependent fashion. Cells were perfused with rapid fluxes of media at various pHs, which caused changes in the fluorescence of the GFP-TfR only when the receptor was exposed on the cell surface, but not when internalization had already occurred. This permitted a precise assessment of the timing of vesicle scission. Remarkably, a burst of polymerization, as assessed by the recruitment of the F-actin binding protein cortactin, reproducibly preceded the actual moment of vesicle scission [16]. These findings are compatible with a model in which actin polymerization may play a role along various steps of internalization (invagination, constriction, and scission), perhaps by generating a compressive or pushing force [21–23] that is coordinated with the final scission, thus facilitating the separation of a constricted pit from its site of formation [6, 17] (Figure 47.1).

It is to be noted, however, that not all fission events are associated with actin recruitment: actin poisons, or interference with the actin nucleator Arp2/3 complex, caused incomplete, albeit significant, inhibition of clathrin coated vesicle (CCV) internalization (about 80 percent, when measured microscopically) [16, 20], while an even smaller degree of inhibition (40–50 percent) was observed when the initial rate of clathrin dependent internalization of the EGFR or the TfR was measured biochemically [7, 14, 20]. The sum of these observations suggests the existence of additional, and concomitant, actin independent internalization mechanisms. This possibility is supported by EM coupled to live-cell imaging of cells treated with actin poisons (latrunculin B that sequesters actin monomers, or jasplakinolide that stabilizes actin filaments). These drugs, while causing a near complete cessation of all aspects of clathrin coated structure lateral motility at the plasma membrane, reduced, but did not abrogate TfR internalization; instead, they increased the formation of deep invaginations [20]. These latter findings indicate that invaginations can form even in the absence of actin polymerization, while constriction and scission may be more dependent on actin dynamics. Thus, actin may significantly contribute to various phases of clathrin dependent internalization, possibly including constriction and scission of vesicles, but has no obvious measurable role in the initial membrane bending and invagination phase. This is in contrast to findings in yeast, where the early phases of pit formation are, instead, dependent on actin [24] (Figure 47.1).

How do forces generated through actin dynamics aid the constriction and fission of vesicles? Hints in this direction came from the observation that the machinery of actin polymerization can propel vesicles *in vivo*, or functionalized beads or liposomes *in vitro* [21, 25]. This is achieved through Arp2/3 mediated branched elongation of actin filaments that self-organize into a propelling rocketing tail that pushes vesicles forward [26–28]. This mechanism might be relevant to endosome motility, although direct *in vivo* evidence in support of this model is still scarce [27]. Additionally, it is not immediately obvious how such a model could apply to the growth of a membrane invagination and to its budding. The major conceptual hurdle here is that actin filaments are normally oriented centrifugally, toward the plasma membrane, thereby generating forces that go in the opposite direction to the ones needed (which should be directed toward the center of the cell to allow budding). Despite these conceptual difficulties, it remains indisputable that, at least in yeast, the invaginating membranes advance at the same rate as actin elongation [1], suggesting tight coupling of the forces involved in the two processes.

A recently proposed model, which also takes into account the peculiar localization of some actin binding proteins in a newly forming pit, might help to reconcile these divergent observations [29]. According to this model, an endocytic coat invaginates the membrane and, at the same time, also recruits actin nucleators that remain confined at the rim of the pit (Figure 47.2). These actin nucleators activate the Arp2/3 complex and cause branched elongation of actin filaments. Since the barbed and growing ends of these filaments are oriented outwardly, toward the plasma membrane, the ensuing forces, generated by filament elongation, would push the membrane forward, in a fashion similar to that observed when migratory protrusions are formed. In yeasts, however, membrane tension is significantly higher than in mammalian cells, as consequence of the higher osmotic pressure of these cells that generates increased membrane resilience [29]. Under these conditions, when new actin monomers are added at the barbed ends, actin filament networks actually grow and elongate toward the center of cell. The sides of the actin filaments remain linked to the endocytic coat. As the actin network grows, forming a cone-like structure, it pulls the invading pit inwards, thereby facilitating internalization and generating forces at the rim of the pit that serve to promote its constriction [29] (Figure 47.2).

A set of intriguing, but circumstantial, experimental evidence supports this model at least in *S. cerevisiae*, where, for instance, clathrin binds to actin interacting proteins that potentially serve as a physical link between actin networks and vesicle coats. One such actin interacting protein is Sla2p, which binds simultaneously to clathrin and F-actin [30]. Sla2p localizes at sites of invaginating pits and, when bound to cortactin, modulates actin filament growth by blocking their barbed ends [22, 31]. The mammalian homolog of Sla2p, HIP1R (Huntingtin interacting protein-1 related) possesses similar properties [30, 32, 33]. Such a model also predicts that the integrity and stability of the actin meshwork, which is in part ensured by actin crosslinkers, is essential for endocytosis. Indeed, disruption of Sac6, the yeast homolog of the mammalian bundling

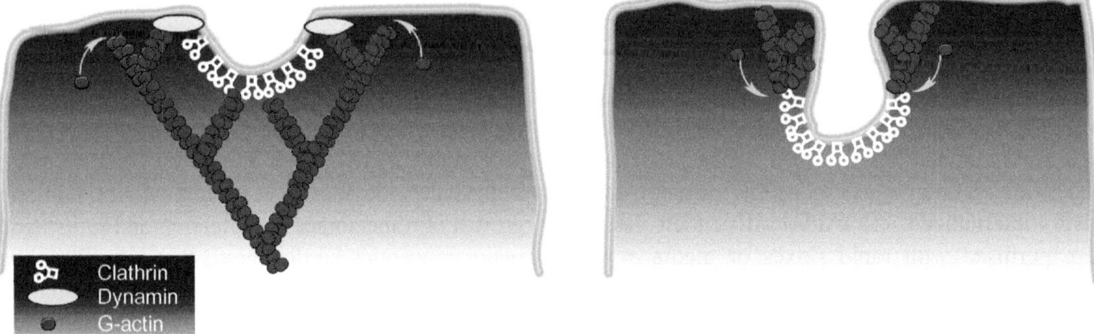

FIGURE 47.2 Polarized actin filament growth and invagination.
Left: actin branched nucleation leads to the generation of a dendritic array of actin filaments to which endocytic coats, via intermediate protein bridges, remain tightly connected. As new actin monomers are added at the barbed ends, actin filament networks grow and elongate toward the center of the cell generating a cone-like structure that pulls the clathrin coated structure inwardly. This model necessitates high membrane tension, such as the one typically observed in yeast cells, to provide sufficient resilience for the centripetal elongation of the actin meshwork, thus facilitating internalization. Right: an alternative model proposes that membranes first bend and start to invaginate deeply, in an actin independent manner; subsequently there could be a reorientation of the growing actin network centripetally, toward the center of the cells, in order to push the invagination away from the membrane.

protein Fimbrin, blocks internalization, without inhibiting actin polymerization [24].

A relevant question is whether a similar scenario is also at play in mammalian cells. The fact that membrane tension in mammalian cells is significantly lower than in *S. cerevisiae* creates some difficulties, in that it predicts that the forces exerted by the outwardly growing actin meshwork would push the membrane forward, thus generating evaginations or protrusions, rather than invaginations. Structures of this kind are observed during macropinocytosis, an exquisitely actin dependent, clathrin independent internalization process, but they have not been reported for clathrin mediated internalization. To accommodate an actin based mode of invagination in mammalian cells, alternative scenarios can be envisioned. For instance, membranes might first bend and start to invaginate, in an actin independent manner, followed, perhaps, by a reorientation of the growing actin network centripetally, toward the center of the cells, in order to push the invagination away from the membrane (Figure 47.2). In principle, this is feasible, since the initial phases of invagination (especially in mammals) apparently do not require actin dynamics, and are carried out by lipid binding proteins and clathrin coats [34]. However, direct experimental evidence in support of this mechanism is lacking [1, 12].

CONVERGING MOLECULAR MACHINERY IN ENDOCYTOSIS AND ACTIN DYNAMICS

Actin Binding Proteins in Endocytosis

The minimal requirement to support cycles of actin polymerization and depolymerization have been elegantly identified through *in vitro* reconstitution experiments; these showed that five different types of purified proteins are sufficient to generate forces for motility [5], including *de novo* actin nucleators and their regulators (Arp2/3 complex and N-WASP), actin severing/depolymerizing factors (ADF/Cofilin), barbed end cappers (Gelsolin, CP, Eps8), and an ATP-actin monomeric binding protein (profilin). The concerted activity of these proteins ensures rates of actin treadmilling that are similar to those observed *in vivo* and that are associated with actin based propulsion [5]. The expectation, therefore, is to find these proteins at endocytic sites and/or involved in internalization processes. Indeed, in *S. cerevisiae*, the Arp2/3 complex, a variety of its activators, profilin, and barbed end cappers are transiently recruited to endocytic sites, with slightly different timing, suggesting their participation in different stages of endocytosis [2, 12, 31]. The localization of cofilin, whose requirement for endocytosis was established by genetic or molecular genetic experiments in yeast and mammalian cells [35, 36], has not yet been reported.

In mammalian cells, critical components of the actin dynamics machinery – such as the Arp2/3 complex [15], N-WASP [13–15], and the capping protein Eps8 (GS, unpublished) – are also transiently recruited to clathrin-positive endocytic sites, and their ablation impairs clathrin mediated endocytosis to various extents [13–15]. In addition, a variety of binding and regulatory interactors of N-WASP, also localize to clathrin-rich spots. For instance, Abi-1, which activates N-WASP at nanomolar concentrations, co-localizes with N-WASP in clathrin- and EGFR containing sites [14] and dramatically enhances the lateral motility of clathrin structures that are thought to precede internalization. Other N-WASP interactors, which display less potent stimulatory activity, such as syndapins and endophilin [37, 38], in addition to binding and activating

N-WASP, might further contribute to the creation of a signaling platform, through their ability to oligomerize and to bind to several proteins. These include endocytic proteins (e.g., dynamin, synaptojanin, huntingtin), actin regulatory proteins (N-WASP, Filamin), and signaling molecules (Sos-1, Adam). Such platforms might, in turn, serve as integrators of signaling inputs that converge in on the actin and endocytic machinery [39]. A similar role might be played by the newly discovered N-WASP and dynamin interactor, Sortinexin-9 (SNX9), which can dimerize and is also capable of binding PIP2 [40 ,41], a lipid that is highly enriched at plasma membranes. These multiple interactions of SNX9 likely account for its function as a coordinator of actin assembly in several distinct endocytic processes.

Binding and bending of lipid bi-layers and the concomitant ability to associate with N-WASP and dynamin is, instead, a unique feature of the PCH/F-Bar family of proteins. These proteins are characterized by a F-BAR domain that can (i) dimerize by adopting a banana-like shape, which enables it to sense membrane curvature [42, 43], and (ii) multimerize, thus enveloping lipid tubules and further promoting tubulation [44]. Members of the family, including FBP17 and Toca-1, localize to clathrin sites [42, 43], and also possess an SH3 domain that can bind to dynamin and N-WASP. The activities of these two latter proteins appear to converge functionally, as they can both promote scission of FBP17- and Toca-1 generated lipid tubules; this suggests that the contribution of F-Bar proteins to endocytosis could be to favor membrane invagination in conjunction with clathrin coats, and to coordinate dynamin- and actin dependent constriction and scission events.

A more recent addition to the array of N-WASP regulators is Abp1 [45]. This protein can bind to both dynamin and F-actin [46]. In yeast, Abp1 binds to the Arp2/3 complex [47], whereas in mammals, it does so indirectly, through binding to N-WASP [45]. The ability of Abp1 to associate with F-actin and to the N-WASP/Arp2/3 complex may act in a positive synergic fashion to promote the formation of branched actin structures needed to sustain actin based motility in endocytosis. Intriguingly, mice devoid of Abp1 are immune compromised [48], as a consequence of defective T cell endocytosis [49], and display behavioral abnormalities, possibly due to reduced rates of endocytosis and synaptic recycling [50].

How all these N-WASP- and, often, dynamin binding proteins contribute to endocytosis, how they are regulated and temporally coordinated, and whether they control a subset of endocytic events (cargo specific? cell specific?) remains to be assessed.

Finally, the presence of actin filaments in close contact with clathrin coats opens up the possibility that myosin motors [51], which move cargo along filament tracks, might also participate in endocytosis. While this is well established in *S. cerevisiae* where class I myosins are essential for internalization [52], it is less studied in mammalian cells, with the exception of Myosin VI. Myosin VI moves clathrin coats toward the minus ends of actin filaments that are properly oriented toward the cell center ([53–55] and [56] for review). Adaptor proteins, such as DAB2, assist the motor in this process, serving as a molecular link to clathrin [57].

Endocytic Proteins in Actin Dynamics

Although actin binding proteins seem to be taking over the endocytic scene, a similar wealth of endocytic proteins has been implicated in actin dynamics; the principal player in this case is dynamin. Dynamin is a GTPase that causes the fission of vesicles from the plasma membrane [58]. The molecular mechanisms enabling this function have been the subject of heated debate. Although evidence suggests that both GTP hydrolysis and conformational changes are essential for dynamin function in endocytosis [59, 60], the question of whether the chemical energy from GTP hydrolysis is used to generate a "powerstroke" for mechanochemical work or to "flip a switch" to terminate signaling has not fully been resolved [58]. Recent live imaging analysis of *in vitro* dynamin coated lipid tubules revealed that the addition of GTP resulted in twisting and supercoiling of tubules, suggesting a rotatory movement of the helix turns of dynamin relative to each other during GTP hydrolysis. This GTP-hydrolysis induced dynamin twisting causes longitudinal tension along the tubules, which contributes, at least *in vitro*, to generate the forces leading to their scission [61, 62].

Dynamin, therefore, appears to act as a mechano-enzyme that wraps spirals around the neck of invaginating vesicles. The constricting activity of dynamin, following GTP hydrolysis, coupled with tension may then result in efficient scission, suggesting that cooperation of mechanisms that mediate constriction with mechanisms that create membrane tension must occur *in vivo*. Actin based mechanisms may provide the required tension (Figure 47.2). Within this scenario, proteins that link dynamin to the actin machinery may serve as molecular integrators of dynamin mediated constriction and actin dependent tension. This may thus be the ultimate and common function of proteins like Cortactin, intersectin, syndapin, amphyphysin, Abp1, Fbp17/Toca-1, Tuba (reviewed in [2] and mentioned above), which, to various extents and with distinct molecular mechanisms, connect or integrate dynamin with various regulators of the actin cytoskeleton, and of the endocytic and signaling machineries.

Another level of integration between endocytic proteins and actin dynamics is exerted by the scaffold protein family CIN85/CD2AP. CIN85 was identified as a protein interacting with the ubiquitin ligase Cbl, and with the endocytic protein endophilin, to form a ternary complex that regulates, via ubiquitination, the internalization of receptor tyrosine kinases [63–65]. CIN85 also interacts with synaptojanin,

HIP1, and HIP1R [66]; these proteins are all implicated in clathrin mediated endocytosis. The related protein CD2AP interacts with Rab4, a trafficking GTPase, cortactin, and AP2; CD2AP also regulates T cell receptor stability and endocytosis [67, 68]. Intriguingly both CIN85 and CD2AP have recently been reported to bind/bundle actin filaments, and to associate with capping protein, respectively [69, 70], a fact that reinforces the concept of a possible contribution of this family of proteins in orchestrating key events at the crossroads of endocytic and actin dynamics processes.

Molecular Switches in the Integration of Endocytosis and Actin Dynamics

Small GTPases function as molecular switches in a variety of molecular circuitries, guaranteeing precise outputs, in time and space, in response to stimuli. It is not surprising therefore that small GTPases are emerging as key regulators of the integrated circuitries built of proteins involved in endocytosis and actin dynamics. Rab5 and the Rho family of GTPases are the best characterized examples.

Rab5 and its Regulators

Rab5 is a master regulator of endocytosis and endosomal dynamics [71]. Rab5, however, is also involved in the control of actin dynamics [72]. This dual function is underscored by a complex system of regulation, in which several activators of Rab5 (including the GEFs RIN1, Rabex-5, and Alsin) are also involved in the control of actin dynamics [71]. RIN1 is an effector of Ras (itself a master regulator of signaling and actin dynamics); binding to Ras enhances the Rab5 specific GEF activity of RIN1 that, presumably, mediates EGFR mediated Rab5 activation via Ras *in vivo* [73]. Accordingly, interference with RIN1 expression or function impairs the clathrin mediated endocytosis of the EGFR [74, 75]. Thus, RIN1 bridges Ras to Rab5 signaling; this, in turn, regulates receptor endocytosis. Moreover RIN1 binds and activates the non-receptor ABL tyrosine kinase, thereby mediating actin remodeling associated with adhesion and migration of epithelial cells [76]. These findings together suggest integrated, yet ill defined, control of signaling, internalization, and actin remodeling.

A more direct way of achieving Rab5 activation by surface receptors is through Rabex-5. Rabex-5 possesses two ubiquitin binding domains [77, 78], which bind ubiquitinated proteins including EGFR. Ubiquitination has emerged as an EGF induced, internalization signal, so a simple scenario might be that, following EGF stimulation, EGFR receptors become ubiquitinated, thereby recruiting Rabex-5 and leading to the activation of Rab5. Finally, Alsin (also called ALS2), a causative gene for juvenile autosomal recessive amyotrophic lateral sclerosis 2 [79, 80], is a large protein containing GEF domains for Rab5 and, potentially, for Rac1 [80]. Indeed, Alsin activates Rab5 and is implicated in endosome dynamics in cultured cells and *in vivo* in null mouse models [81]. It is not clear yet whether Alsin acts as a GEF for Rac. Notably, a recent report provided evidence that Alsin might rather be an effector of Rac, which, once activated, recruits Alsin to membrane ruffles and subsequently to nascent macropinosomes. There, Alsin mediated Rab5 activity is required for promoting fusion with Rab5 positive endomembranes, thus suggesting a Rac to Rab5 signaling cascade [82].

Similar levels of coordination and integration apply to the deactivation phase of Rab5, which is controlled by GAPs, such as RN-tre [83]. The GAP activity of RN-tre is negatively modulated by EGF treatment and requires binding to Eps8, a signaling and actin binding molecule involved in Sos-1 mediated Rac activation; this represents yet another level of intersection of pathways that regulate actin remodeling (through Sos-1-Rac) and endocytosis (through RN-tre-Rab5) [83].

Rho Family GTPases

The Rho GTPase subfamily includes about 20 members, mostly involved in the control of different aspects of actin remodeling. In particular, (i) RhoA is a master regulator of stress fiber formation and focal adhesion, (ii) the three Rac isoforms Rac1-3 are essential for the extension of lamellipodia during cell migration through the control of Arp2/3 dependent actin branching, and (iii) Cdc42 promotes protrusions of filopodia, and controls cell polarization in directional migrating and epithelial cells [84]. Early studies implicated Rac and RhoA in the internalization of the TfR [85]; however, the definition of their exact role in endocytosis remains largely anecdotal. For instance, activated Rac associates with synaptojanin 2, a lipid phosphatase, and both activated Rac1 and a membrane targeted version of synaptojanin 2 inhibit EGFR mediated internalization; whether the two proteins play a direct role in clathrin mediated endocytosis remains unclear [86]. RhoA also appears to control EGFR-endocytosis indirectly, possibly by phosphorylating endophilin via its downstream effector, Rock (Rho associated kinase). This, in turn, inhibits binding between endophilin and CIN85, thus disrupting a complex previously implicated in endocytosis [87].

Cdc42, on the other hand, has been more firmly linked to endocytic routes, via its regulator/effector, Cool-1 (also called beta-Pix). Members of the Cool family of proteins contain tandem DH-PH domains [88] and have been assumed to function as GEFs [89]. EGFR induced phosphorylation of Cool-1 activates its GEF activity toward Cdc42 and also promotes the interaction between Cool-1 and the E3 ubiquitin ligase, Cbl. This, in turn, sequesters Cbl away from internalizing EGFR [90], something that ultimately regulates EGFR ubiquitination, internalization, and stability. Genetic studies in the nematode, *C. elegans*, have recently revealed an additional connection between Cdc42 pathways and endocytosis. The components of the signaling cascade Cdc42/Par-3/Par-6/PCK-ε, which are well established regulators of embryonic and epithelial cell polarity, were found to affect clathrin and non-clathrin dependent internalization of RME-2 receptor. This suggests that one way to maintain embryonic polarization during development may be accomplished through the regulation of membrane trafficking, thus revealing a novel and unexpected level of signal integration.

FROM ENDOCYTOSIS TO ACTIN DYNAMICS

It should come as no surprise, at this point – given the vast convergence of the endocytic and actin dynamics machineries – that not only does actin control endocytosis, but endocytosis also controls the correct execution of various forms of actin remodeling, with specific reference to the spatial restriction of actin dynamics.

Spatial restriction of signaling is critical for a number of fundamental cellular processes including directed cell migration, cell-fate decisions, epithelial-cell polarization, growth cone movement, and tissue morphogenesis during development [91–93]. One obvious way to achieve signal polarization is through localized redistribution of signaling molecules in response to extracellular cues. Consistent with this idea, endocytosis (which is a very effective way of redistributing molecules in the cell) has emerged as a fundamental process in the spatial restriction of signals. For instance, guided cell migration in *Drosophila* border cells strictly depends on the functional endocytosis of EGFR and PVR (the PDGF/VEGF Receptor) [94–96]. Moreover, disruption of typical endocytic regulators, such as Cbl, or the Rab5 activator Sprint (the homolog of mammalian Rin1), results in aberrant cell migration in response to stimulation by growth factors [96]. Analogously, disruption of Rab5 or the endocytic protein Syntaxin (Avalanche in *Drosophila*) causes hyperproliferation of wild-type imaginal tissues, likely by altering the cellular trafficking and distribution of the fate-decision receptor Notch, and the polarity determinant, Crumbs [97]. Collectively, these observations provided genetic evidence that one physiological role of endocytosis is to ensure a localized intracellular response to extracellular cues.

One instance in which the precise perception of extracellular cues is remarkably complex is when cells move toward a chemo-attractant, in a polarized fashion. Under these conditions, cells resolve external gradients of motogenic factors by generating a transient asymmetric localization of second messengers [98–100]. For example, RTKs act as guidance receptors by regulating, in space and in time, the dynamic assembly of actin filaments, in order to generate forces for the formation of different types of migratory protrusions, such as peripheral lamellipodia and dorsal surface circular ruffles (also called waves) [101]. Lamellipodia represent the first obligatory step in two-dimensional cell motility, whereas circular ruffles are sites of internalization events, such as fluid phase endocytosis [72, 102–105], and are required for migration in three-dimensional matrices [106]. Circular ruffles might therefore correspond to sites of integration of actin dynamics and endocytosis. Indeed, endocytic proteins, such as dynamin and Rab5, and actin regulators, such as cortactin, PI3K, and Rac, function coordinately in the generation of circular ruffles [105, 107, 108] (Figure 47.3).

The correct positioning of Rac, which is necessary for the extension of membrane protrusions, might also be regulated through endocytosis [109–112]. Upon activation of adhesion receptors, such as integrins, high affinity binding sites for Rac become available on the plasma membrane; this allows spatially restricted targeting of Rac to sites of adhesion. Such sites, probably containing membrane rafts, are internalized upon cell detachment (and termination of integrin signaling) in a dynamin- and caveolin dependent manner [111]. Notably, disruption of dynamin function also results in segregation of active Rac in aberrant membrane invaginations, away from the plasma membrane. This prevents the formation of regularly shaped lamellipodia, likely by blocking macropinocytic Rac internalization [112].

OUTLOOK

A remarkable change of perspective is obligatory in view of findings that endocytosis and actin dynamics are integrated processes that ultimately participate in the correct execution of cellular programs. Actin dynamics provide forces, and possibly also several levels of regulation, for endocytosis to occur. Endocytosis and recycling, in turn, provide routes and mechanisms to spatially restrict signaling for many actin dependent cellular processes, such as directed cell migration, epithelial polarization, and cellular morphogenesis. In a larger sense, both endocytosis and actin dynamics can be viewed as part of the signaling machinery that is needed to process information not only in a digital manner (on vs. off), but also in spatial and temporal manner. Consistent with this notion, recent mathematical models, which account for the dynamic redistribution of polarized cortical membrane proteins (including Cdc42) in budding yeast, revealed that cycles of endocytosis and recycling are sufficient to maintain dynamically polarized signaling through positive feedback with directed transport [113]. Thus, we should begin to view endocytosis as a critical mode through which biological systems process spatial information with a resulting polarized dynamic distribution of their molecular machineries. In this context, the bidirectional crosstalk between actin based forces that aid endocytosis, and trafficking of membrane/proteins to ensure a restricted signaling output of the actin machineries, provides an integrated

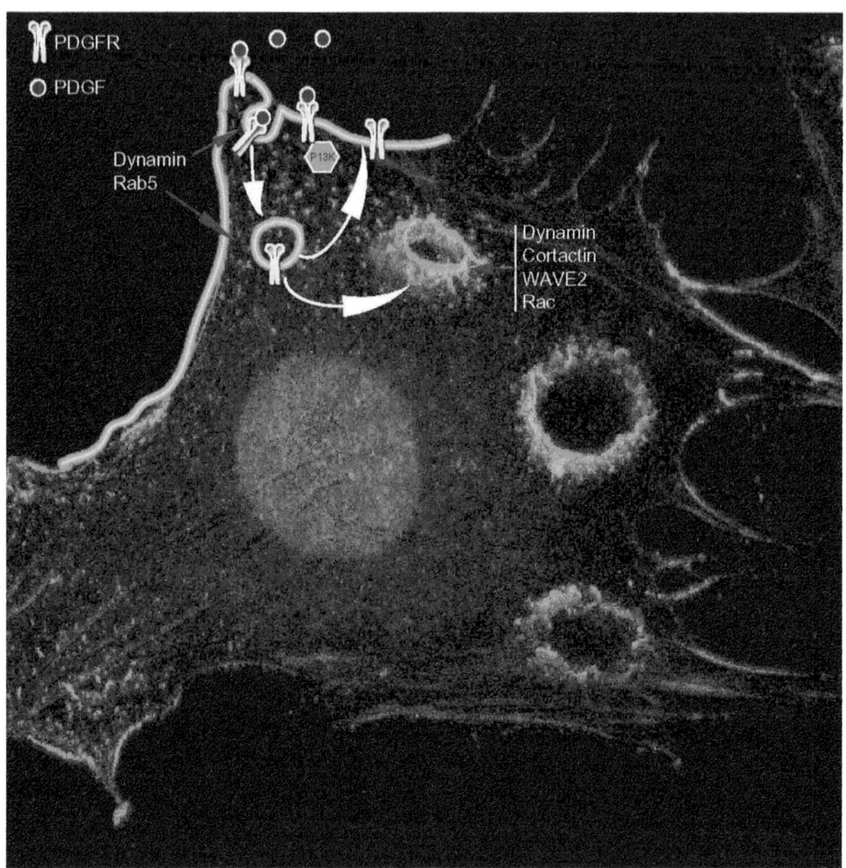

FIGURE 47.3 Endocytosis and spatial restriction of actin dynamics.
Stimulation with PDGF of mouse embryo fibroblasts causes the formation of protrusions containing actin (red) and the indicated actin bound and regulatory proteins (e.g., Eps8, green) on the dorsal surface of cells (also called circular ruffles). Endocytic internalization/recycling of plasma membrane receptors, such PDGFR, may contribute to confine spatially the signaling output emanating from various effectors such as PI3K, Rac, and Rab5 leading to the generation of these architecturally defined, actin based structures. A number of key molecules such dynamin and Rab5, may contribute to the process by regulating both endocytic and signaling events. The image of MEF stimulated with PDGF is courtesy of Emanuela Frittoli.

and highly coordinated mechanism to finely tune biological responses.

ACKNOWLEDGEMENTS

Work in the author's lab is supported by grants from AIRC (Italian Association for Cancer Research) (GS and PPDF); the European Community, VI Framework (GS and PPDF); the Cariplo Foundation (GS and PPDF); the Italian Ministries of Health and of Education and University, the Monzino Foundation and the Ferrari Foundation (PPDF). We thank Pascale Romano for critically editing the manuscript.

REFERENCES

1. Kaksonen M, Toret CP, Drubin DG. Harnessing actin dynamics for clathrin-mediated endocytosis. *Nat Rev Mol Cell Biol* 2006;**7**:404–14.
2. Smythe E, Ayscough KR. Actin regulation in endocytosis. *J Cell Sci* 2006;**119**:4589–98.
3. Ascough KR. Endocytosis: Actin in the driving seat. *Curr Biol* 2004;**14**:R124–6.
4. Pollard TD. The cytoskeleton, cellular motility and the reductionist agenda. *Nature* 2003;**422**:741–5.
5. Carlier MF, Pantaloni D. Control of actin assembly dynamics in cell motility. *J Biol Chem* 2007;**282**:23,005–9.
6. Merrifield CJ. Seeing is believing: imaging actin dynamics at single sites of endocytosis. *Trends Cell Biol* 2004;**14**:352–8.
7. Fujimoto LM, Roth R, Heuser JE, Schmid SL. Actin assembly plays a variable, but not obligatory role in receptor- mediated endocytosis in mammalian cells. *Traffic* 2000;**1**:161–71.
8. Lamaze C, Fujimoto LM, Yin HL, Schmid SL. The actin cytoskeleton is required for receptor-mediated endocytosis in mammalian cells. *J Biol Chem* 1997;**272**:20,332–5.
9. Gottlieb TA, Ivanov IE, Adesnik M, Sabatini DD. Actin microfilaments play a critical role in endocytosis at the apical but not the basolateral surface of polarized epithelial cells. *J Cell Biol* 1993;**120**:695–710.
10. Jackman MR, Shurety W, Ellis JA, Luzio JP. Inhibition of apical but not basolateral endocytosis of ricin and folate in Caco-2 cells by cytochalasin D. *J Cell Sci* 1994;**107**:2547–56.
11. Riezman H, Munn A, Geli MI, Hicke L. Actin-, myosin- and ubiquitin-dependent endocytosis. *Experientia* 1996;**52**:1033–41.
12. Toret CP, Drubin DG. The budding yeast endocytic pathway. *J Cell Sci* 2006;**119**:4585–7.

13. Benesch S, Polo S, Lai FP, et al. N-WASP deficiency impairs EGF internalization and actin assembly at clathrin-coated pits. *J Cell Sci* 2005;**118**:3103–15.
14. Innocenti M, Gerboth S, Rottner K, et al. Abi1 regulates the activity of N-WASP and WAVE in distinct actin-based processes. *Nat Cell Biol* 2005;**7**:969–76.
15. Merrifield CJ, Qualmann B, Kessels MM, Almers W. Neural Wiskott Aldrich Syndrome Protein (N-WASP) and the Arp2/3 complex are recruited to sites of clathrin-mediated endocytosis in cultured fibroblasts. *Eur J Cell Biol* 2004;**83**:13–8.
16. Merrifield CJ, Perrais D, Zenisek D. Coupling between clathrin-coated-pit invagination, cortactin recruitment, and membrane scission observed in live cells. *Cell* 2005;**121**:593–606.
17. Engqvist-Goldstein AE, Drubin DG. Actin assembly and endocytosis: from yeast to mammals. *Annu Rev Cell Dev Biol* 2003;**19**:287–332.
18. Salisbury JL, Condeelis JS, Satir P. Role of coated vesicles, microfilaments, and calmodulin in receptor-mediated endocytosis by cultured B lymphoblastoid cells. *J Cell Biol* 1980;**87**:132–41.
19. Shupliakov O, Bloom O, Gustafsson JS, et al. Impaired recycling of synaptic vesicles after acute perturbation of the presynaptic actin cytoskeleton. *Proc Natl Acad Sci U S A* 2002;**99**:14,476–81.
20. Yarar D, Waterman-Storer CM, Schmid SL. A dynamic actin cytoskeleton functions at multiple stages of clathrin-mediated endocytosis. *Mol Biol Cell* 2005;**16**:964–75.
21. Giardini PA, Fletcher DA, Theriot JA. Compression forces generated by actin comet tails on lipid vesicles. *Proc Natl Acad Sci USA* 2003;**100**:6493–8.
22. Kaksonen M, Sun Y, Drubin DG. A pathway for association of receptors, adaptors, and actin during endocytic internalization. *Cell* 2003;**115**:475–87.
23. Upadhyaya A, Chabot JR, Andreeva A, et al. Probing polymerization forces by using actin-propelled lipid vesicles. *Proc Natl Acad Sci USA* 2003;**100**:4521–6.
24. Kaksonen M, Toret CP, Drubin DG. A modular design for the clathrin- and actin-mediated endocytosis machinery. *Cell* 2005; **123**:305–20.
25. Wiesner S, Helfer E, Didry D, et al. A biomimetic motility assay provides insight into the mechanism of actin-based motility. *J Cell Biol* 2003;**160**:387–98.
26. Kaksonen M, Peng HB, Rauvala H. Association of cortactin with dynamic actin in lamellipodia and on endosomal vesicles. *J Cell Sci* 2000;**113**:4421–6.
27. Taunton J. Actin filament nucleation by endosomes, lysosomes and secretory vesicles. *Curr Opin Cell Biol* 2001;**13**:85–91.
28. Orth JD, Krueger EW, Cao H, McNiven MA. The large GTPase dynamin regulates actin comet formation and movement in living cells. *Proc Natl Acad Sci U S A* 2002;**99**:167–72.
29. Liu J, Kaksonen M, Drubin DG, Oster G. Endocytic vesicle scission by lipid phase boundary forces. *Proc Natl Acad Sci USA* 2006;**103**:10,277–82.
30. Engqvist-Goldstein AE, Warren RA, Kessels MM, et al. The actin-binding protein Hip1R associates with clathrin during early stages of endocytosis and promotes clathrin assembly in vitro. *J Cell Biol* 2001;**154**:1209–23.
31. Newpher TM, Smith RP, Lemmon V, Lemmon SK. In vivo dynamics of clathrin and its adaptor-dependent recruitment to the actin-based endocytic machinery in yeast. *Dev Cell* 2005;**9**:87–98.
32. Le Clainche C, Pauly BS, Zhang CX, et al. A Hip1R-cortactin complex negatively regulates actin assembly associated with endocytosis. *Embo J* 2007;**26**:1199–210.
33. Carreno S, Engqvist-Goldstein AE, Zhang CX, et al. Actin dynamics coupled to clathrin-coated vesicle formation at the trans-Golgi network. *J Cell Biol* 2004;**165**:781–8.
34. Itoh T, De Camilli P. BAR, F-BAR (EFC) and ENTH/ANTH domains in the regulation of membrane-cytosol interfaces and membrane curvature. *Biochim Biophys Acta* 2006;**1761**:897–912.
35. Lappalainen P, Drubin DG. Cofilin promotes rapid actin filament turnover in vivo. *Nature* 1997;**388**:78–82.
36. Nishimura Y, Yoshioka K, Bernard O, et al. A role of LIM kinase 1/cofilin pathway in regulating endocytic trafficking of EGF receptor in human breast cancer cells. *Histochem Cell Biol* 2006;**126**:627–38.
37. Kessels MM, Qualmann B. Syndapins integrate N-WASP in receptor-mediated endocytosis. *Embo J* 2002;**21**:6083–94.
38. Otsuki M, Itoh T, Takenawa T. Neural Wiskott-Aldrich syndrome protein is recruited to rafts and associates with endophilin A in response to epidermal growth factor. *J Biol Chem* 2003;**278**:6461–9.
39. Qualmann B, Kessels MM. Endocytosis and the cytoskeleton. *Int Rev Cytol* 2002;**220**:93–144.
40. Yarar D, Waterman-Storer CM, Schmid SL. SNX9 couples actin assembly to phosphoinositide signals and is required for membrane remodeling during endocytosis. *Dev Cell* 2007;**13**:43–56.
41. Shin N, Lee S, Ahn N, et al. Sorting nexin 9 interacts with dynamin 1 and N-WASP and coordinates synaptic vesicle endocytosis. *J Biol Chem* 2007;**282**:28,939–50.
42. Itoh T, Erdmann KS, Roux A, et al. Dynamin and the actin cytoskeleton cooperatively regulate plasma membrane invagination by BAR and F-BAR proteins. *Dev Cell* 2005;**9**:791–804.
43. Tsujita K, Suetsugu S, Sasaki N, et al. Coordination between the actin cytoskeleton and membrane deformation by a novel membrane tubulation domain of PCH proteins is involved in endocytosis. *J Cell Biol* 2006;**172**:269–79.
44. Shimada A, Niwa H, Tsujita K, et al. Curved EFC/F-BAR-domain dimers are joined end to end into a filament for membrane invagination in endocytosis. *Cell* 2007;**129**:761–72.
45. Pinyol R, Haeckel A, Ritter A, et al. Regulation of N-WASP and the Arp2/3 complex by Abp1 controls neuronal morphology. *PLoS ONE* 2007;**2**:e400.
46. Kessels MM, Engqvist-Goldstein AE, Drubin DG, Qualmann B. Mammalian Abp1, a signal-responsive F-actin-binding protein, links the actin cytoskeleton to endocytosis via the GTPase dynamin. *J Cell Biol* 2001;**153**:351–66.
47. Goode BL, Rodal AA, Barnes G, Drubin DG. Activation of the Arp2/3 complex by the actin filament binding protein Abp1p. *J Cell Biol* 2001;**153**:627–34.
48. Han J, Shui JW, Zhang X, et al. HIP-55 is important for T-cell proliferation, cytokine production, and immune responses. *Mol Cell Biol* 2005;**25**:6869–78.
49. Le Bras S, Foucault I, Foussat A, et al. Recruitment of the actin-binding protein HIP-55 to the immunological synapse regulates T cell receptor signaling and endocytosis. *J Biol Chem* 2004;**279**:15,550–60.
50. Connert S, Wienand S, Thiel C, et al. SH3P7/mAbp1 deficiency leads to tissue and behavioral abnormalities and impaired vesicle transport. *Embo J* 2006;**25**:1611–22.
51. Soldati T, Schliwa M. Powering membrane traffic in endocytosis and recycling. *Nat Rev Mol Cell Biol* 2006;**7**:897–908.
52. Jonsdottir GA, Li R. Dynamics of yeast Myosin I: evidence for a possible role in scission of endocytic vesicles. *Curr Biol* 2004;**14**:1604–9.
53. Rock RS, Rice SE, Wells AL, et al. Myosin VI is a processive motor with a large step size. *Proc Natl Acad Sci U S A* 2001; **98**:13,655–9.

54. Altman D, Goswami D, Hasson T, et al. Precise positioning of myosin VI on endocytic vesicles in vivo. *PLoS Biol* 2007;**5**:e210.
55. Park H, Li A, Chen LQ, et al. The unique insert at the end of the myosin VI motor is the sole determinant of directionality. *Proc Natl Acad Sci U S A* 2007;**104**:778–83.
56. Roberts R, Lister I, Schmitz S, et al. Myosin VI: cellular functions and motor properties. *Philos Trans R Soc Lond B Biol Sci* 2004;**359**:1931–44.
57. Spudich G, Chibalina MV, Au JS, et al. Myosin VI targeting to clathrin-coated structures and dimerization is mediated by binding to Disabled-2 and PtdIns(4,5)P2. *Nat Cell Biol* 2007;**9**:176–83.
58. Song BD, Schmid SL. A molecular motor or a regulator? Dynamin's in a class of its own. *Biochemistry* 2003;**42**:1369–76.
59. Marks B, Stowell MH, Vallis Y, et al. GTPase activity of dynamin and resulting conformation change are essential for endocytosis. *Nature* 2001;**410**:231–5.
60. Damke H, Baba T, Warnock DE, Schmid SL. Induction of mutant dynamin specifically blocks endocytic coated vesicle formation. *J Cell Biol* 1994;**127**:915–34.
61. Ferguson SM, Brasnjo G, Hayashi M, et al. A selective activity-dependent requirement for dynamin 1 in synaptic vesicle endocytosis. *Science* 2007;**316**:570–4.
62. Roux A, Uyhazi K, Frost A, De Camilli P. GTP-dependent twisting of dynamin implicates constriction and tension in membrane fission. *Nature* 2006;**441**:528–31.
63. Petrelli A, Gilestro GF, Lanzardo S, et al. The endophilin-CIN85-Cbl complex mediates ligand-dependent downregulation of c-Met. *Nature* 2002;**416**:187–90.
64. Soubeyran P, Kowanetz K, Szymkiewicz I, et al. Cbl-CIN85-endophilin complex mediates ligand-induced downregulation of EGF receptors. *Nature* 2002;**416**:183–7.
65. Take H, Watanabe S, Takeda K, et al. Cloning and characterization of a novel adaptor protein, CIN85, that interacts with c-Cbl. *Biochem Biophys Res Commun* 2000;**268**:321–8.
66. Kowanetz K, Husnjak K, Holler D, et al. CIN85 associates with multiple effectors controlling intracellular trafficking of epidermal growth factor receptors. *Mol Biol Cell* 2004;**15**:3155–66.
67. Cormont M, Meton I, Mari M, et al. CD2AP/CMS regulates endosome morphology and traffic to the degradative pathway through its interaction with Rab4 and c-Cbl.. *Traffic* 2003;**4**:97–112.
68. Hutchings NJ, Clarkson N, Chalkley R, et al. Linking the T cell surface protein CD2 to the actin-capping protein CAPZ via CMS and CIN85. *J Biol Chem* 2003;**278**:22,396–403.
69. Gaidos G, Soni S, Oswald DJ, et al. Structure and function analysis of the CMS/CIN85 protein family identifies actin-bundling properties and heterotypic-complex formation. *J Cell Sci* 2007;**120**:2366–77.
70. Bruck S, Huber TB, Ingham RJ, et al. Identification of a novel inhibitory actin-capping protein binding motif in CD2-associated protein. *J Biol Chem* 2006;**281**:19,196–203.
71. Zerial M, McBride H. Rab proteins as membrane organizers. *Nat Rev Mol Cell Biol* 2001;**2**:107–17.
72. Lanzetti L, Di Fiore PP, Scita G. Pathways linking endocytosis and actin cytoskeleton in mammalian cells. *Exp Cell Res* 2001;**271**:45–56.
73. Bliss JM, Venkatesh B, Colicelli J. The RIN Family of Ras Effectors. *Methods Enzymol* 2005;**407**:335–44.
74. Tall GG, Barbieri MA, Stahl PD, Horazdovsky BF. Ras-Activated Endocytosis Is Mediated by the Rab5 Guanine Nucleotide Exchange Activity of RIN1. *Dev Cell* 2001;**1**:73–82.
75. Barbieri MA, Fernandez-Pol S, Hunker C, et al. Role of rab5 in EGF receptor-mediated signal transduction. *Eur J Cell Biol* 2004;**83**:305–14.
76. Hu H, Bliss JM, Wang Y, Colicelli J. RIN1 is an ABL tyrosine kinase activator and a regulator of epithelial-cell adhesion and migration. *Curr Biol* 2005;**15**:815–23.
77. Penengo L, Mapelli M, Murachelli AG, et al. Crystal structure of the ubiquitin binding domains of rabex-5 reveals two modes of interaction with ubiquitin. *Cell* 2006;**124**:1183–95.
78. Lee S, Tsai YC, Mattera R, et al. Structural basis for ubiquitin recognition and autoubiquitination by Rabex-5. *Nat Struct Mol Biol* 2006;**13**:264–71.
79. Yang Y, Hentati A, Deng HX, et al. The gene encoding alsin, a protein with three guanine-nucleotide exchange factor domains, is mutated in a form of recessive amyotrophic lateral sclerosis. *Nat Genet* 2001;**29**:160–5.
80. Topp JD, Gray NW, Gerard RD, Horazdovsky BF. Alsin is a Rab5 and Rac1 guanine nucleotide exchange factor. *J Biol Chem* 2004;**279**:24,612–23.
81. Hadano S, Benn SC, Kakuta S, et al. Mice deficient in the Rab5 guanine nucleotide exchange factor ALS2/alsin exhibit age-dependent neurological deficits and altered endosome trafficking. *Hum Mol Genet* 2006;**15**:233–50.
82. Kunita R, Otomo A, Mizumura H, et al. The Rab5 activator ALS2/alsin acts as a novel Rac1 effector through Rac1-activated endocytosis. *J Biol Chem* 2007;**282**:16,599–611.
83. Lanzetti L, Rybin V, Malbarba MG, et al. The Eps8 protein coordinates EGF receptor signalling through Rac and trafficking through Rab5. *Nature* 2000;**408**:374–7.
84. Hall A. Rho GTPases and the control of cell behaviour. *Biochem Soc Trans* 2005;**33**:891–5.
85. Lamaze C, Chuang TH, Terlecky LJ, et al. Regulation of receptor-mediated endocytosis by Rho and Rac. *Nature* 1996;**382**:177–9.
86. Malecz N, McCabe PC, Spaargaren C, et al. Synaptojanin 2, a novel Rac1 effector that regulates clathrin-mediated endocytosis. *Curr Biol* 2000;**10**:1383–6.
87. Kaneko T, Maeda A, Takefuji M, et al. Rho mediates endocytosis of epidermal growth factor receptor through phosphorylation of endophilin A1 by Rho-kinase. *Genes Cells* 2005;**10**:973–87.
88. Whitehead IP, Campbell S, Rossman KL, Der CJ. Dbl family proteins. *Biochim Biophys Acta* 1997;**1332**:F1–F23.
89. Cerione RA, Zheng Y. The Dbl family of oncogenes. *Curr Opin Cell Biol* 1996;**8**:216–22.
90. Wu WJ, Tu S, Cerione RA. Activated Cdc42 sequesters c-Cbl and prevents EGF receptor degradation. *Cell* 2003;**114**:715–25.
91. Bretscher MS, Aguado-Velasco C. Membrane traffic during cell locomotion. *Curr Opin Cell Biol* 1998;**10**:537–41.
92. Le Roy C, Wrana JL. Signaling and endocytosis: a team effort for cell migration. *Dev Cell* 2005;**9**:167–8.
93. Polo S, Di Fiore PP. Endocytosis conducts the cell signaling orchestra. *Cell* 2006;**124**:897–900.
94. McDonald JA, Pinheiro EM, Montell DJ. PVF1, a PDGF/VEGF homolog, is sufficient to guide border cells and interacts genetically with Taiman. *Development* 2003;**130**:3469–78.
95. Wang X, Bo J, Bridges T, et al. Analysis of cell migration using whole-genome expression profiling of migratory cells in the Drosophila ovary. *Dev Cell* 2006;**10**:483–95.
96. Jekely G, Sung HH, Luque CM, Rorth P. Regulators of endocytosis maintain localized receptor tyrosine kinase signaling in guided migration. *Dev Cell* 2005;**9**:197–207.
97. Lu H, Bilder D. Endocytic control of epithelial polarity and proliferation in Drosophila. *Nat Cell Biol* 2005;**7**:1232–9.
98. Parent CA, Devreotes PN. A cell's sense of direction. *Science* 1999;**284**:765–70.

99. Loovers HM, Postma M, Keizer-Gunnink I, et al. Distinct roles of PI(3,4,5)P3 during chemoattractant signaling in Dictyostelium: a quantitative in vivo analysis by inhibition of PI3-kinase. *Mol Biol Cell* 2006;**17**:1503–13.
100. Andrew N, Insall RH. Chemotaxis in shallow gradients is mediated independently of PtdIns 3-kinase by biased choices between random protrusions. *Nat Cell Biol* 2007;**9**:193–200.
101. Buccione R, Orth JD, McNiven MA. Foot and mouth: podosomes, invadopodia and circular dorsal ruffles. *Nat Rev Mol Cell Biol* 2004;**5**:647–57.
102. Dowrick P, Kenworthy P, McCann B, Warn R. Circular ruffle formation and closure lead to macropinocytosis in hepatocyte growth factor/scatter factor-treated cells. *Eur J Cell Biol* 1993;**61**:44–53.
103. Warn R, Brown D, Dowrick P, et al. Cytoskeletal changes associated with cell motility. *Symp Soc Exp Biol* 1993;**47**:325–38.
104. Ballestrem C, Wehrle-Haller B, Hinz B, Imhof BA. Actin-dependent lamellipodia formation and microtubule-dependent tail retraction control-directed cell migration. *Mol Biol Cell* 2000; **11**:2999–3012.
105. Lanzetti L, Palamidessi A, Areces L, et al. Rab5 is a signalling GTPase involved in actin remodelling by receptor tyrosine kinases. *Nature* 2004;**429**:309–14.
106. Suetsugu S, Yamazaki D, Kurisu S, Takenawa T. Differential roles of WAVE1 and WAVE2 in dorsal and peripheral ruffle formation for fibroblast cell migration. *Dev Cell* 2003;**5**:595–609.
107. Krueger EW, Orth JD, Cao H, McNiven MA. A dynamin-cortactin-Arp2/3 complex mediates actin reorganization in growth factor-stimulated cells. *Mol Biol Cell* 2003;**14**:1085–96.
108. Orth JD, McNiven MA. Dynamin at the actin-membrane interface. *Curr Opin Cell Biol* 2003;**15**:31–9.
109. Radhakrishna H, Al-Awar O, Khachikian Z, Donaldson JG. ARF6 requirement for Rac ruffling suggests a role for membrane trafficking in cortical actin rearrangements. *J Cell Sci* 1999;**112**:855–66.
110. del Pozo MA, Alderson NB, Kiosses WB, et al. Integrins regulate Rac targeting by internalization of membrane domains. *Science* 2004;**303**:839–42.
111. del Pozo MA, Balasubramanian N, Alderson NB, et al. Phospho-caveolin-1 mediates integrin-regulated membrane domain internalization. *Nat Cell Biol* 2005;**7**:901–8.
112. Schlunck G, Damke H, Kiosses WB, et al. Modulation of Rac localization and function by dynamin. *Mol Biol Cell* 2004;**15**:256–67.
113. Marco E, Wedlich-Soldner R, Li R, et al. Endocytosis optimizes the dynamic localization of membrane proteins that regulate cortical polarity. *Cell* 2007;**129**:411–22.

Chapter 48

Signaling in Autophagy Related Pathways

Patrice Codogno[1] and Alfred J. Meijer[2]
[1]*INSERM U756, Faculté de Pharmacie, Université Paris-Sud 11, Châtenay-Malabry, France*
[2]*Department of Medical Biochemistry, Academic Medical Center, Amsterdam, The Netherlands*

INTRODUCTION

In the turnover of cell components, (macro)autophagy, which occurs in all eukaryotic cells, degrades long lived proteins and eliminates redundant or damaged organelles, e.g., peroxisomes, mitochondria, and the endoplasmic reticulum. During autophagy part of the cytoplasm is surrounded by a double membrane to form an autophagosome that acquires hydrolytic enzymes by fusing with lysosomes or endocytic compartments to form an autophagolysosome. This is followed by degradation of the sequestered material. The rate-limiting step in the entire process is the formation of the autophagosome, which starts with the expansion of a membrane core of unknown origin, the phagophore or isolation membrane: about 30 proteins known as Atg (autophagy related) proteins are involved in this process [1]. One of these proteins, Atg8 (LC3-I), when lipidated with phosphatidylethanolamine (LC3-II), can be used as a marker of autophagosomes [2]. It must be stressed, however, that the steady state level of autophagosomes does not necessarily reflect flux through the autophagic pathway [1, 3].

Autophagy not only produces amino acids for ATP formation under nutrient depleted conditions, e.g., during starvation or immediately after birth, but also protects against various adverse situations, e.g., cancer, neurodegeneration, invading pathogens, aging [1], and perhaps diabetes [4]. In addition, overactivation of autophagy can cause cell death by a mechanism that differs from apoptosis (reviewed in [5, 6]). However, autophagy can also act as a survival pathway by providing protection against apoptotic cell death, e.g., in the heart during reperfusion following a period of ischemia [3]. Autophagy is also essential for normal heart function [7].

SIGNALING CONTROL OF AUTOPHAGY

Autophagy is controlled by several signal transduction pathways (Figure 48.1). The insulin growth factor amino acid mTOR signaling pathway is the most important of these pathways, as the inhibition of autophagy by activation of this pathway occurs in all eukaryotic cells [5]. Other mechanisms, such as the regulation of autophagy by the Erk1,2- and p38-MAPK related signaling pathways, are less universal and do not operate in all cell types [5]. However, whether activation of these pathways inhibits (Erk1,2 [8]; p38 [9]) or stimulates autophagy (Erk1,2 [10], p38 [11]) is not clear. Uncertainty also exists with regard to the regulatory function of Gαi3 in autophagy regulation [12, 13]. Presumably, these differences may be cell-type dependent.

mTOR signaling not only inhibits autophagy, but also simultaneously stimulates protein synthesis [14]. This is

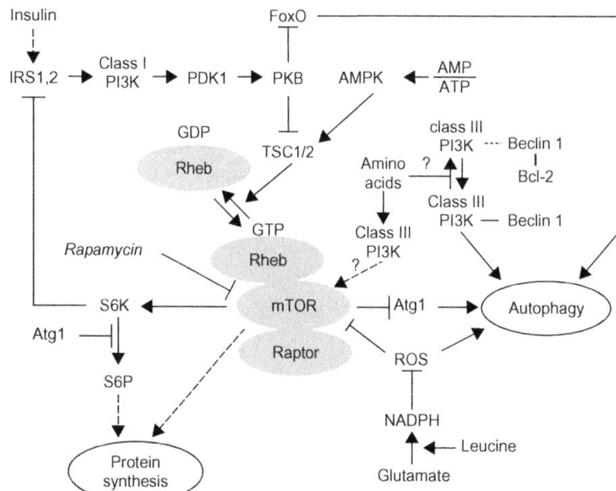

FIGURE 48.1 mTOR mediated signaling and the regulation of autophagy.
Abbreviations: IRS, insulin receptor substrate; PI3K, phosphatidylinositol 3-kinase; PDK1, phosphoinositide dependent kinase-1; PKB, protein kinase B; AMPK, AMP activated protein kinase; TSC, tuberous sclerosis complex; Rheb, Ras homolog enriched in brain; raptor, regulatory associated protein of mTOR; mTOR, mammalian target of rapamycin; S6K, 70kD S6 kinase; S6, ribosomal protein S6; ROS, reactive oxygen species. For aspects of signaling not indicated in the Figure, the reader is referred to the text and the relevant literature.

Handbook of Cell Signaling, Three-Volume Set 2 ed.

efficient from the point of view of metabolic regulation [5]. Activation of mTOR is strictly dependent on the presence of amino acids [14, 15]: high concentrations of amino acids alone (leucine in particular), or low concentrations of amino acids combined with insulin, fully activate mTOR. In the absence of amino acids, mTOR is inactive. The target of TOR in the autophagic machinery of yeast is the protein kinase Atg1, which is inactivated by phosphorylation; overexpression of this gene induces autophagy [16].

Phosphatidylinositol 3-phosphate (PI(3)P, the product of class III PI3K (or hVps34) is required for autophagosome formation. In contrast, PI(3,4,5)P_3, the product of class I PI3K, has an inhibitory effect (Figure 48.1). PI3K inhibitors, such as wortmannin, LY294002, and 3-methyladenine, cannot distinguish between these two lipid kinases and inhibition of class III PI3K by these compounds accounts for their ability to inhibit autophagy [5]. For autophagosome formation to occur, class III PI3K needs to be complexed with the tumor suppressor Beclin 1 (the mammalian homolog of the yeast Atg6). Formation of this complex is promoted by the UV irradiation resistance associated gene product (UVRAG) [17], and prevented by binding of the antiapoptotic protein, Bcl-2, to Beclin 1 [18]. An additional mechanism by which Bcl-2 regulates autophagy may be its binding to the endoplasmic reticulum, where it affects cellular Ca^{++} homeostasis [19, 20], Ca^{++} being required for autophagy [21]. Class III PI3K may stimulate autophagy by recruiting WIPI1-α (WD-repeat protein interacting with phosphoinositides (WIPI)-1α), the mammalian ortholog of yeast Atg18, to the autophagosome [22].

mTOR signaling network

Inactivation of mTOR by either amino acid depletion or the administration of rapamycin stimulates autophagy. Because amino acids sometimes inhibit autophagy by mechanisms that are unrelated to mTOR (e.g., by promoting the association of Beclin 1 and Bcl-2, cf. Figure 48.1), autophagy is not always fully restored by rapamycin in the presence of amino acids [4, 5].

mTOR can also be inhibited by the activation of AMPK, which acts as an energy sensor (Figure 48.1). In line with its function of stimulating catabolism, AMPK is essential for autophagy [19, 23, 24]. The p53 tumor suppressor activates AMPK, inhibits mTOR, and stimulates autophagy, of mitochondria in particular [25]. AMPK also underlies the stimulation of autophagy in cardiac [26] and cerebral [27] ischemia. Interestingly, not only does AMPK inhibit mTOR dependent signaling, but also the deletion of S6K (located downstream of mTOR; cf. Figure 48.1) in muscle cells stimulates AMPK activity and inhibits cell growth, which is restored when AMPK is blocked [28]. Although it is tempting to speculate that the inhibition of S6K by amino acid depletion would also result in the activation of AMPK, the evidence is against this, at least in hepatocytes [29].

It has been suggested that, apart from its ability to inhibit mTOR, AMPK may also activate autophagy in an mTOR independent manner by phosphorylating, and inhibiting, eEF2 (eukaryotic elongation factor-2) kinase [30].

The mechanism responsible for the activation of mTOR signaling by amino acids is still unknown. Several hypotheses have been proposed, but none of these has gained general support. It is most likely that the amino acid receptor is intra- rather than extracellular [5]. A recent hypothesis [4] is that the receptor is in fact glutamate dehydrogenase because this enzyme is specifically activated by leucine and by certain non-metabolizable leucine analogs that also activate mTOR signaling, and because it may be involved in the scavenging by NADPH of reactive oxygen species (ROS) that are involved in initiating autophagy and inhibit mTOR signaling [4, 31]. According to Nobukuni *et al.* [32], amino acids stimulate class III PI3K but do not affect the class III PI3K-Beclin 1 association. This seems to be inconsistent with the role of PI(3)P in autophagosome formation. The presence of class III PI3K in multiple protein complexes and PI(3)P compartmentalization may account for this apparent discrepancy [32]. This is consistent with the finding that a significant portion of class III PI3K is not bound to Beclin 1 [33]. It is possible that amino acids may stimulate the unbound class III PI3K.

Although TOR activation inhibits autophagy, paradoxically S6K appears to be required for autophagy in the *Drosophila* fat body [34]. This view, however, has recently been challenged by Lee *et al.* [35] who showed that overexpression of Atg1 in *Drosophila* stimulates autophagy, and inhibits cell growth, but at the same time inhibits S6K phosphorylation and activation. They also showed that Atg1 inhibited S6K in mammalian cells [35]. In addition, S6K deletion in muscle cells did not affect autophagy, at least as measured by LC3-II accumulation [36]. However, as indicated earlier, this does not provide any information on autophagic flux, and so no definite conclusions about the role of S6K in autophagy can yet be drawn. A possible way out of the controversy is that dephosphorylated S6K is structurally required for autophagy whereas phosphorylated S6K inhibits the process.

Unexpectedly, overactivation of mTOR signaling results in feedback inhibition of the insulin signaling pathway by the S6K dependent phosphorylation of IRS1, which reduces the activity of class I PI3K [32, 37] (Figure 48.1). Although autophagy was not measured in these studies [32, 37], this feedback mechanism could perhaps act as a safety mechanism to avoid the complete inhibition of autophagy because, even in the presence of excess nutrients, some autophagy is always needed to fulfill its housekeeping function, i.e., getting rid of redundant and/or damaged intracellular structures. Interestingly, Sch9 was recently identified as the yeast ortholog of mammalian S6 kinase 1 [38]. Sch9 inactivation triggers autophagy in the yeast *S. cerevisiae* in nutrient-rich medium, without inactivating

TORC1 (TOR complex 1), in an Atg1 dependent manner [39]. This would be consistent with the negative action of S6K on autophagy.

FoxO Proteins

A new development in the regulation of autophagy is the discovery that the expression of several autophagy genes is increased by the transcription factor FoxO (class O of forkhead box transcription factors) [40], the activity of which is inhibited by protein kinase B mediated phosphorylation [41] (Figure 48.1). The FoxO proteins mediate the fasting response, i.e., they inhibit progression of the cell cycle, suppress the expression of genes involved in glycolysis and lipogenesis, and stimulate the expression of the genes controlling gluconeogenesis and fatty acid oxidation. In addition, FoxO proteins are known to induce the expression of genes that provide protection against oxidative stress, e.g., superoxide dismutase and catalase, and therefore play an important role in combating aging and disease [41]. Controlling autophagy by FoxO, acting in concert with the mTOR signaling pathway, is therefore ideally suited to the role of autophagy as a survival pathway when nutrients become scarce.

Phosphoinositol Signaling

mTOR independent stimulation of autophagy has been observed in response to trehalose [42] and LiCl treatment [43]. After LiCl treatment, autophagy is induced via the inhibition of inositol monophosphatase independently of mTOR inhibition. The depletion of free inositol and reduced levels of myo-inositol-1,4,5 phosphate (IP3) stimulate autophagy. Interestingly, enhancing the IP3 level inhibits autophagy induced by nutrient depletion. These findings suggest that inositol and IP3 may regulate autophagy either via a signaling pathway parallel to mTOR, or impinge on the molecular machinery of autophagy downstream from mTOR. The induction of autophagy by inhibition of the endoplasmic reticulum IP3 receptor appears to be independent of changes in the level of Ca^{2+} in the endoplasmic reticulum [44]. However, the flux of Ca^{2+} into the cytosol from the endoplasmic reticulum has been shown to induce autophagy by activating Ca^{2+}/calmodulin dependent kinase kinase β, an AMPK kinase in mammalian cells, and AMPK [19].

eIF2α Kinases

eIF2α kinases belong to an evolutionarily conserved serine/threonine kinase family that regulates stress induced translational arrest. It has been demonstrated that the yeast eIF2α kinase GCN2 and the eIF2α regulated transcriptional transactivator GCN4 are essential for starvation induced autophagy [45]. It is worth noting that GCN2 is conserved in mammals, and so this protein may have a similar role in higher eukaryotes. Recently, endoplasmic reticulum stress induced by polyglutamine 72 repeat (polyQ72) aggregates has been shown to induce autophagy after activation of the eIF2α kinase PERK [46]. However, another study has reported that the autophagic process triggered in response to endoplasmic reticulum stress is PERK independent [47]. The mammalian interferon inducible eIF2α kinase PKR also plays an important role in triggering autophagy [45]. Overall, the eIF2α kinases regulate autophagy induced by various stress situations. However, further studies are required to elucidate the relationship between eIF2α kinases and other signaling pathways, and the molecular machinery of autophagy.

Specific Autophagy

Although starvation induced autophagy is non-specific in that cytoplasmic components are sequestered randomly, autophagy can also have very specific effects as redundant or damaged organelles can be recognized specifically and targeted for elimination. However, relatively little is known about the underlying signals that allow the autophagic system to recognize these structures.

In the case of mitochondria, the signals that could be involved include a low membrane potential, the mitochondrial permeability transition [48, 49], the Uth1 protein in the (yeast) mitochondrial outer membrane [50], the protein phosphatase Aup1p in the mitochondrial intermembrane space [51], and the proapoptotic mitochondrial protein BNIP3 in mammalian cells [52], but we do not know how these are linked to, for example, the mTOR pathway and the autophagic machinery. A low mitochondrial membrane potential may also result in a decrease in the local ATP concentration and activation of AMPK, and thus inhibit mTOR, as discussed above. Because damaged mitochondria are the source of ROS, and since ROS production contributes to aging, specific autophagic elimination of damaged mitochondria provides protection against aging [48, 53] and apoptosis [3].

Stress caused by the accumulation of misfolded proteins in the endoplasmic reticulum activates the unfolded protein response, which is accompanied by Atg1 kinase activation and stimulates autophagosome formation; it is not yet known whether TOR kinase activity is reduced under these conditions but this seems likely [54].

AUTOPHAGY AND CELL DEATH

As mentioned in the Introduction, autophagy can both protect against apoptotic cell death and assist in the execution of cell death. This is because on the one hand specific autophagic removal of damaged mitochondria and other intracellular structures can provide protection against apoptosis, whereas on the other hand overactivation of autophagy can result in autophagic cell death.

The death associated protein kinase (DAPK) family members and the related death associated related protein kinase-1 (DRP-1), induced by cell death stimuli like interferon-γ, activation of Fas receptors, TNFα and TGFβ [5, 6, 55], are all involved in the execution of autophagic cell death. DRP-1 is present in the lumen of autophagosomes, and its function may be the phosphorylation of one of the Atg proteins [5, 6]. The death receptor interacting protein RIP may cause autophagic cell death by activating JNK and c-Jun [5, 6].

A novel mechanism for the antiapoptotic function of NFκB was suggested by the observation that in sarcoma cells TNFα induced, ROS mediated autophagy was suppressed by activation of NFκB, a phenomenon that was accompanied by activation of mTOR. It has been suggested that stimulation of autophagy may provide a way of bypassing the resistance of cancer cells to anticancer agents that activate NFκB [56].

TNFα induced autophagy may, in fact, be mediated by ceramide (produced by spingomyelinase activation) because in cancer cells ceramide stimulates autophagy by upregulating Beclin1 and inhibiting protein kinase B phosphorylation [57]. Sphingosine 1-phosphate, another sphingolipid metabolite, also stimulates autophagy but it does so by a mechanism that is different from that of ceramide in that it inhibits mTOR in a protein kinase B independent manner, and not by changes in Beclin1; sphingosine 1-phosphate may play a role in starvation induced autophagy because the activity of sphingosine kinase-1 increases under these conditions [58]. It has been proposed that ceramide and sphingosine 1-phosphate constitute a rheostat system in which ceramide promotes cell death, and sphingosine 1-phosphate increases cell survival [58].

An interesting possible mechanism of the crosstalk between autophagy and apoptosis was recently revealed by the demonstration that Atg5 is not only involved in autophagosome formation but can also acquire apoptotic properties after being cleaved by calpain: the truncated Atg5 translocates from the cytosol to mitochondria, where it associates with the antiapoptotic molecule Bcl-xL and triggers cytochrome c release and caspase activation [59].

ACKNOWLEDGEMENTS

Work in P. Codogno's laboratory is supported by institutional funding from INSERM (Institut National de la Santé Et de la Recherche Médicale), University of Paris-Sud 11, and by grants from ARC (Association de Recherche sur le Cancer) and ANR (Agence Nationale pour la Recherche).

REFERENCES

1. Mizushima N, Klionsky DJ. Protein turnover via autophagy: implications for metabolism. *Annu Rev Nutr* 2007;**27**:19–40.
2. Mizushima N, Yoshimori T. How to Interpret LC3 Immunoblotting. *Autophagy* 2007;**3**:542–5.
3. Hamacher-Brady A, Brady NR, Gottlieb RA. The interplay between pro-death and pro-survival signaling pathways in myocardial ischemia/reperfusion injury: apoptosis meets autophagy. *Cardiovasc Drugs Ther* 2006;**20**:445–62.
4. Meijer AJ, Codogno P. Macroautophagy: protector in the diabetes drama? *Autophagy* 2007;**3**:523–6.
5. Codogno P, Meijer AJ. Autophagy and signaling: their role in cell survival and cell death. *Cell Death Differ* 2005;**12**:1509–18.
6. Gozuacik D, Kimchi A. Autophagy and cell death. *Curr Top Dev Biol* 2007;**78**:217–45.
7. Nakai A, Yamaguchi O, Takeda T, Higuchi Y, Hikoso S, Taniike M, Omiya S, Mizote I, Matsumura Y, Asahi M, Nishida K, Hori M, Mizushima N, Otsu K. The role of autophagy in cardiomyocytes in the basal state and in response to hemodynamic stress. *Nat Med* 2007;**13**:619–24.
8. Corcelle E, Nebout M, Bekri S, Gauthier N, Hofman P, Poujeol P, Fenichel P, Mograbi B. Disruption of autophagy at the maturation step by the carcinogen lindane is associated with the sustained mitogen-activated protein kinase/extracellular signal-regulated kinase activity. *Cancer Res* 2006;**66**:6861–70.
9. Häussinger D, Reinehr R, Schliess F. The hepatocyte integrin system and cell volume sensing. *Acta Physiol (Oxf)* 2006;**187**:249–55.
10. Pattingre S, Bauvy C, Codogno P. Amino acids interfere with the ERK1/2-dependent control of macroautophagy by controlling the activation of Raf-1 in human colon cancer HT-29 cells. *J Biol Chem* 2003;**278**:16,667–16,674.
11. Comes F, Matrone A, Lastella P, Nico B, Susca FC, Bagnulo R, Ingravallo G, Modica S, Lo Sasso G, Moschetta A, Guanti G, Simone C. A novel cell type-specific role of p38alpha in the control of autophagy and cell death in colorectal cancer cells. *Cell Death Differ* 2007;**14**:693–702.
12. Ogier-Denis E, Pattingre S, El Benna J, Codogno P. Erk1/2-dependent phosphorylation of Galpha-interacting protein stimulates its GTPase accelerating activity and autophagy in human colon cancer cells. *J Biol Chem* 2000;**275**:39,090–39,095.
13. Gohla A, Klement K, Piekorz RP, Pexa K, vom Dahl S, Spicher K, Dreval V, Häussinger D, Birnbaumer L, Nurnberg B. An obligatory requirement for the heterotrimeric G protein Gi3 in the antiautophagic action of insulin in the liver. *Proc Natl Acad Sci U S A* 2007;**104**:3003–8.
14. Dann SG, Thomas G. The amino acid sensitive TOR pathway from yeast to mammals. *FEBS Lett* 2006;**580**:2821–9.
15. van Sluijters DA, Dubbelhuis PF, Blommaart EF, Meijer AJ. Amino-acid-dependent signal transduction. *Biochem J* 2000;**351**:545–50.
16. Scott RC, Juhasz G, Neufeld TP. Direct induction of autophagy by Atg1 inhibits cell growth and induces apoptotic cell death. *Curr Biol* 2007;**17**:1–11.
17. Liang C, Feng P, Ku B, Dotan I, Canaani D, Oh BH, Jung JU. Autophagic and tumour suppressor activity of a novel Beclin1-binding protein UVRAG. *Nat Cell Biol* 2006;**8**:688–99.
18. Pattingre S, Tassa A, Qu X, Garuti R, Liang XH, Mizushima N, Packer M, Schneider MD, Levine B. Bcl-2 antiapoptotic proteins inhibit Beclin 1-dependent autophagy. *Cell* 2005;**122**:927–39.
19. Høyer-Hansen M, Bastholm L, Szyniarowski P, Campanella M, Szabadkai G, Farkas T, Bianchi K, Fehrenbacher N, Elling F, Rizzuto R, Mathiasen IS, Jäättelä M. Control of macroautophagy by calcium, calmodulin-dependent kinase kinase-beta, and Bcl-2. *Mol Cell* 2007;**25**:193–205.
20. Brady NR, Hamacher-Brady A, Yuan H, Gottlieb RA. The autophagic response to nutrient deprivation in the hl-1 cardiac myocyte is modulated by Bcl-2 and sarco/endoplasmic reticulum calcium stores. *FEBS J* 2007;**274**:3184–97.

21. Gordon PB, Holen I, Fosse M, Rotnes JS, Seglen PO. Dependence of hepatocytic autophagy on intracellularly sequestered calcium. *J Biol Chem* 1993;**268**:26,107–26,112.
22. Proikas-Cezanne T, Waddell S, Gaugel A, Frickey T, Lupas A, Nordheim A. WIPI-1alpha (WIPI49), a member of the novel 7-bladed WIPI protein family, is aberrantly expressed in human cancer and is linked to starvation-induced autophagy. *Oncogene* 2004;**23**:9314–25.
23. Meley D, Bauvy C, Houben-Weerts JH, Dubbelhuis PF, Helmond MT, Codogno P, Meijer AJ. AMP-activated protein kinase and the regulation of autophagic proteolysis. *J Biol Chem* 2006;**281**:34,870–34,879.
24. Liang J, Shao SH, Xu ZX, Hennessy B, Ding Z, Larrea M, Kondo S, Dumont DJ, Gutterman JU, Walker CL, Slingerland JM, Mills GB. The energy sensing LKB1-AMPK pathway regulates p27(kip1) phosphorylation mediating the decision to enter autophagy or apoptosis. *Nat Cell Biol* 2007;**9**:218–24.
25. Levine AJ, Feng Z, Mak TW, You H, Jin S. Coordination and communication between the p53 and IGF-1-AKT-TOR signal transduction pathways. *Genes Dev* 2006;**20**:267–75.
26. Matsui Y, Takagi H, Qu X, Abdellatif M, Sakoda H, Asano T, Levine B, Sadoshima J. Distinct roles of autophagy in the heart during ischemia and reperfusion: roles of AMP-activated protein kinase and Beclin 1 in mediating autophagy. *Circ Res* 2007;**100**:914–22.
27. Adhami F, Liao G, Morozov YM, Schloemer A, Schmithorst VJ, Lorenz JN, Dunn RS, Vorhees CV, Wills-Karp M, Degen JL, Davis RJ, Mizushima N, Rakic P, Dardzinski BJ, Holland SK, Sharp FR, Kuan CY. Cerebral ischemia-hypoxia induces intravascular coagulation and autophagy. *Am J Pathol* 2006;**169**:566–83.
28. Aguilar V, Alliouachene S, Sotiropoulos A, Sobering A, Athea Y, Djouadi F, Miraux S, Thiaudiere E, Foretz M, Viollet B, Diolez P, Bastin J, Benit P, Rustin P, Carling D, Sandri M, Ventura-Clapier R, Pende M. S6 Kinase Deletion Suppresses Muscle Growth Adaptations to Nutrient Availability by Activating AMP Kinase. *Cell Metab* 2007;**5**:476–87.
29. Krause U, Bertrand L, Hue L. Control of p70 ribosomal protein S6 kinase and acetyl-CoA carboxylase by AMP-activated protein kinase and protein phosphatases in isolated hepatocytes. *Eur J Biochem* 2002;**269**:3751–9.
30. Takagi H, Matsui Y, Sadoshima J. The Role of Autophagy in Mediating Cell Survival and Death During Ischemia and Reperfusion in the Heart. *Antioxid Redox Signal* 2007;**9**:1373–82.
31. Scherz-Shouval R, Shvets E, Fass E, Shorer H, Gil L, Elazar Z. Reactive oxygen species are essential for autophagy and specifically regulate the activity of Atg4. *EMBO J* 2007;**26**:1749–60.
32. Nobukuni T, Kozma SC, Thomas G. hvps34, an ancient player, enters a growing game: mTOR Complex1/S6K1 signaling. *Curr Opin Cell Biol* 2007;**19**:135–41.
33. Kihara A, Kabeya Y, Ohsumi Y, Yoshimori T. Beclin-phosphatidylinositol 3-kinase complex functions at the trans-Golgi network. *EMBO Rep* 2001;**2**:330–5.
34. Scott RC, Schuldiner O, Neufeld TP. Role and regulation of starvation-induced autophagy in the Drosophila fat body. *Dev Cell* 2004;**7**:167–78.
35. Lee SB, Kim S, Lee J, Park J, Lee G, Kim Y, Kim JM, Chung J. ATG1, an autophagy regulator, inhibits cell growth by negatively regulating S6 kinase. *EMBO Rep* 2007;**8**:360–5.
36. Mieulet V, Roceri M, Espeillac C, Sotiropoulos A, Ohanna M, Oorschot V, Klumperman J, Sandri M, Pende M. S6 kinase inactivation impairs growth and translational target phosphorylation in muscle cells maintaining proper regulation of protein turnover. *Am J Physiol Cell Physiol* 2007;**293**:C712–22.
37. Tremblay F, Lavigne C, Jacques H, Marette A. Role of Dietary Proteins and Amino Acids in the Pathogenesis of Insulin Resistance. *Annu Rev Nutr* 2007;**27**:293–310.
38. Urban J, Soulard A, Huber A, Lippman S, Mukhopadhyay D, Deloche O, Wanke V, Anrather D, Ammerer G, Riezman H, Broach JR, De Virgilio C, Hall MN, Loewith R. Sch9 is a major target of TORC1 in Saccharomyces cerevisiae. *Mol Cell* 2007;**26**:663–74.
39. Yorimitsu T, Zaman S, Broach JR, Klionsky DJ. Protein Kinase A and Sch9 Cooperatively Regulate Induction of Autophagy in Saccharomyces cerevisiae. *Mol Biol Cell* 2007;**18**:4180–9.
40. Juhasz G, Puskas LG, Komonyi O, Erdi B, Maroy P, Neufeld TP, Sass M. Gene expression profiling identifies FKBP39 as an inhibitor of autophagy in larval Drosophila fat body. *Cell Death Differ* 2007;**14**:1181–90.
41. van der Horst A, Burgering BM. Stressing the role of FoxO proteins in lifespan and disease. *Nat Rev Mol Cell Biol* 2007;**8**:440–50.
42. Sarkar S, Davies JE, Huang Z, Tunnacliffe A, Rubinsztein DC. Trehalose, a novel mTOR-independent autophagy enhancer, accelerates the clearance of mutant huntingtin and alpha-synuclein. *J Biol Chem* 2007;**282**:5641–52.
43. Sarkar S, Floto RA, Berger Z, Imarisio S, Cordenier A, Pasco M, Cook LJ, Rubinsztein DC. Lithium induces autophagy by inhibiting inositol monophosphatase. *J Cell Biol* 2005;**170**:1101–11.
44. Criollo A, Maiuri MC, Tasdemir E, Vitale I, Fiebig AA, Andrews D, Molgo J, Diaz J, Lavandero S, Harper F, Pierron G, di Stefano D, Rizzuto R, Szabadkai G, Kroemer G. Regulation of autophagy by the inositol trisphosphate receptor. *Cell Death Differ* 2007;**14**:1029–39.
45. Talloczy Z, Jiang W, Virgin HW, Leib DA, Scheuner D, Kaufman RJ, Eskelinen EL, Levine B. Regulation of starvation- and virus-induced autophagy by the eIF2alpha kinase signaling pathway. *Proc Natl Acad Sci U S A* 2002;**99**:190–5.
46. Kouroku Y, Fujita E, Tanida I, Ueno T, Isoai A, Kumagai H, Ogawa S, Kaufman RJ, Kominami E, Momoi T. ER stress (PERK/eIF2alpha phosphorylation) mediates the polyglutamine-induced LC3 conversion, an essential step for autophagy formation. *Cell Death Differ* 2007;**14**:230–9.
47. Ogata M, Hino S, Saito A, Morikawa K, Kondo S, Kanemoto S, Murakami T, Taniguchi M, Tanii I, Yoshinaga K, Shiosaka S, Hammarback JA, Urano F, Imaizumi K. Autophagy is activated for cell survival after endoplasmic reticulum stress. *Mol Cell Biol* 2006;**26**:9220–31.
48. Kim I, Rodriguez-Enriquez S, Lemasters JJ. Selective degradation of mitochondria by mitophagy. *Arch Biochem Biophys* 2007;**462**:245–53.
49. Mijaljica D, Prescott M, Devenish RJ. Different fates of mitochondria: alternative ways for degradation? *Autophagy* 2007;**3**:4–9.
50. Kissova I, Deffieu M, Manon S, Camougrand N. Uth1p is involved in the autophagic degradation of mitochondria. *J Biol Chem* 2004;**279**:39,068–39,074.
51. Tal R, Winter G, Ecker N, Klionsky DJ, Abeliovich H. Aup1p, a yeast mitochondrial protein phosphatase homolog, is required for efficient stationary phase mitophagy and cell survival. *J Biol Chem* 2007;**282**:5617–24.
52. Daido S, Kanzawa T, Yamamoto A, Takeuchi H, Kondo Y, Kondo S. Pivotal role of the cell death factor BNIP3 in ceramide-induced autophagic cell death in malignant glioma cells. *Cancer Res* 2004;**64**:4286–93.
53. Bonawitz ND, Shadel GS. Rethinking the mitochondrial theory of aging: the role of mitochondrial gene expression in lifespan determination. *Cell Cycle* 2007;**6**:1574–8.
54. Yorimitsu T, Klionsky DJ. Eating the endoplasmic reticulum: quality control by autophagy. *Trends Cell Biol* 2007;**17**:279–85.

55. Xiao G. Autophagy and NF-kappaB: Fight for fate. *Cytokine Growth Factor Rev* 2007;**18**:233–43
56. Djavaheri-Mergny M, Amelotti M, Mathieu J, Besancon F, Bauvy C, Souquere S, Pierron G, Codogno P. NF-kappaB activation represses tumor necrosis factor-alpha-induced autophagy. *J Biol Chem* 2006;**281**:30,373–30,382.
57. Scarlatti F, Bauvy C, Ventruti A, Sala G, Cluzeaud F, Vandewalle A, Ghidoni R, Codogno P. Ceramide-mediated macroautophagy involves inhibition of protein kinase B and up-regulation of beclin 1. *J Biol Chem* 2004;**279**:18,384–18,391.
58. Lavieu G, Scarlatti F, Sala G, Carpentier S, Levade T, Ghidoni R, Botti J, Codogno P. Regulation of autophagy by sphingosine kinase 1 and its role in cell survival during nutrient starvation. *J Biol Chem* 2006;**281**:8518–27.
59. Yousefi S, Perozzo R, Schmid I, Ziemiecki A, Schaffner T, Scapozza L, Brunner T, Simon HU. Calpain-mediated cleavage of Atg5 switches autophagy to apoptosis. *Nat Cell Biol.* 2006;**8**:1124–32.

Section D

Cell Cycle/ Cell Death Signaling

Chapter 49

Regulation of Cell Cycle Progression

Jennifer Scorah[1,3] and Clare H. McGowan[1,2,3]
[1]*Department of Molecular Biology*
[2]*Department of Cell Biology*
[3]*Scripps Research Institute, La Jolla, California*

INTRODUCTION

The fundamental purpose of the mitotic cell cycle is the generation of genetically identical daughter cells. Because growth, DNA replication, and organelle duplication are physically independent processes, successful division requires that the duplication and segregation of cellular components is tightly coordinated. Coordination is achieved by placing the execution of all cell cycle events under the control of a family of tightly regulated protein kinases. These kinases are known as cyclin dependent kinases (Cdks) and each phase of the cell cycle is defined by the activation and inactivation of distinct members of the family. The general principles by which the sequential activation and inactivation of Cdks are achieved are as follows: (1) processes that initiate Cdk activation also lead to downregulation or destruction, (2) the activity of one cyclin-Cdk sets up the conditions needed for the activation of the next, (3) the destruction of cyclins ensures a unidirectional cell cycle, (4) the inhibition of assembled cyclin-Cdk complexes, either by phosphorylation or by the binding of inhibitory proteins, allows the accumulation of inactive complexes, and (5) when cellular or environmental conditions are not favorable, checkpoint signals delay the activation of cyclin-Cdk complexes and delay the initiation or execution of the next step.

CYCLINS DEFINE CELL CYCLE PHASE

As the name implies, Cdks are protein kinases that need to bind a cyclin subunit to be active. Cyclins are synthesized and degraded in a highly coordinated process as cells progress through the cell cycle, thus, the most fundamental level of control exercised over the activity of Cdks is the periodic presence and absence of the cyclin subunit. Fluctuating levels of Cyclin-Cdk complexes help to regulate each phase of the cell cycle and ensure that the cell cycle progresses in one direction only.

In yeast, one Cdk (Cdc28 in *Saccharomyces cerevisiae* and Cdc2 in *Schizosaccharomyces pombe*) drives all stages of the cell cycle. Mammalian cells have at least four different Cdks (Cdk1, Cdk2, Cdk4, and Cdk6) and four cyclin families (Cyclin D1, D2, D3; cyclin E1, E2; cyclin A1, A2; and cyclin B1, B2, B3) that regulate different stages of the cell cycle. When resting (quiescent) cells are stimulated to enter the cell cycle, the first cyclin to be induced is cyclin D. Expression of cyclin D is dependent on sustained and coordinate signaling of growth factors, through receptor tyrosine kinases, and of extracellular matrix components through integrins (reviewed in [1]). Cyclin D abundance is also increased by increased translation and stabilization in a phosphoinositol-3-kinase dependent pathway (reviewed in [2]). D-type cyclins are growth factor sensors and Cyclin D transcription, assembly, nuclear transport, and turnover are mitogen dependent steps.

There are three D-type cyclins in human cells, with D1 being the most ubiquitously expressed [3]. The D-type cyclins assemble with their catalytic partners Cdk4 and Cdk6. Both Cdk4 and Cdk6 are constitutively expressed throughout the cell cycle. The monomeric kinase subunits are relatively unstable and thus the abundance of cyclin D is limiting for the stabilization and activation of Cdk4 and Cdk6. Cyclin D-Cdk complexes enter the nucleus where they must be phosphorylated by a Cdk activating kinase (CAK, see later section) to form an active kinase and be able to phosphorylate substrates. The same mitogenic signals that induce cyclin D expression also induce expression of a second cyclin, cyclin E, and of two inhibitors of cyclin dependent kinases, $p21^{Cip1}$ and $p27^{Kip1}$. The induction of Cdk inhibitory proteins as cells enter a new round of growth and division seems counterproductive. However, p21 and p27 are actually required for the efficient assembly and nuclear import of cyclin D-Cdk4 complexes. p21 and p27 bind to cyclin D-Cdk4 without inhibiting kinase activity [4, 5]. In contrast, they are effective inhibitors of cyclin

Handbook of Cell Signaling, Three-Volume Set 2 ed.

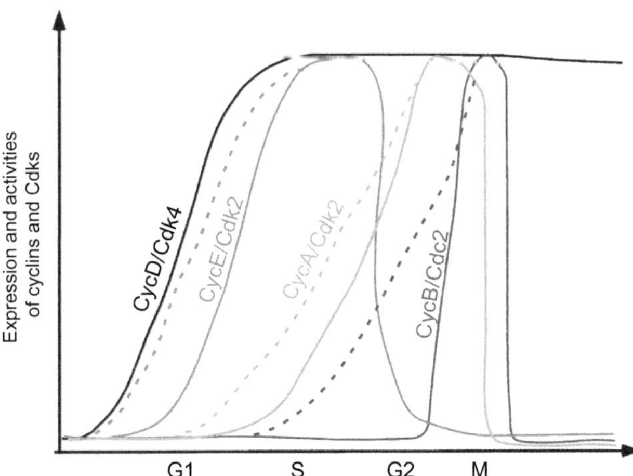

FIGURE 49.1 Expression pattern of cyclins and Cdk activities during the cell cycle.
Expression of the specified cyclin is indicated by the dashed lines; activity of the specified cyclin-Cdk is indicated by the solid line. Sustained expression of cyclin D is dependent on favorable extracellular conditions as explained in the text. Note that the delay between expression of cyclin E and activation of cyclin E-Cdk2 is primarily due to inhibition of kinase activity by p27. The delay between expression of cyclin B and activation of cyclin B-Cdc2 is primarily due to inhibitory phosphorylation. Note that cyclin D also binds to Cdk4 and that cyclin A also binds to Cdc2.

E-Cdk2 activity. Therefore, the presence of these proteins early in G1 promotes the formation of active cyclin D-Cdk4 complexes at the same time that it delays the activation of cyclin E-Cdk2 complexes. These processes drive two temporally distinct waves of Cdk activation (Figure 49.1).

The D and E-cyclin-Cdk complexes drive passage through the restriction point at the boundary of G_1 and S-phase. After the restriction point, further cell cycle progression will occur even if extracellular growth factors are removed. Internal or external events can delay, or prevent cell cycle progression. However, if no signal to stop is received, the initiated cycle of growth, replication, segregation, and division will continue without further extracellular input. If extracellular conditions remain favorable, mitogenic signaling maintains cyclin D expression throughout the cell cycle, and on completion of mitosis the new daughter cells remain ready to initiate a new round of cell division immediately. If, on the other hand, mitogenic stimuli are not maintained, cyclin D is phosphorylated in a glycogen synthase kinase 3 (GSK3) dependent process and is targeted for destruction by the proteosome (reviewed in [6]). Progression into another round of replication is thus dependent on sustained, favorable extracellular signals.

Cyclin D-Cdk4/6 plays two distinct roles in promoting cell cycle progression. One function stems from the ability of active Cyclin D-Cdk complexes to phosphorylate the retinoblastoma protein (Rb) and the related pocket proteins p107 and p130. Unphosphorylated Rb represses transcription from E2F dependent promoters by inhibiting the activation domain of the E2F family of transcription factors and by recruiting chromatin modifying enzymes to repress transcription. On phosphorylation of Rb, E2F is released and E2F dependent transcription of specific promoters including those of cyclin E, cyclin A, and the several proteins needed for S-phase (reviewed in [7]) is promoted. Induction of E2F dependent transcription of cyclin E leads to an increase of active Cdk2, which enforces Rb phosphorylation. Increased phosphorylation of Rb by Cyclin E-Cdk2 leads to full activation of E2F dependent transcription of the genes required for S-phase.

The second role cyclin D plays in promoting cell cycle progression is non-catalytic and depends on the ability of cyclin D-Cdk4 to bind $p27^{Kip1}$. As cyclin D expression increases in G_1, the increasing abundance of cyclin D-Cdk4 provides a binding site for $p27^{Kip1}$. $p27^{Kip1}$ is sequestered away from cyclin E-Cdk2, and the amount of cyclin E-Cdk2 relieved from the inhibitory effect of $p27^{Kip1}$ increases. This tips the balance toward active cyclin E-Cdk2 and contributes to driving G_1 progression. In addition, active cyclin E-Cdk2 phosphorylates its own inhibitor $p27^{Kip1}$ leading to degradation of $p27^{Kip1}$ and contributing to the positive feedback for cyclin E-Cdk2 activity (Figure 49.2).

Active cyclin E-Cdk2 phosphorylates a number of substrates including proteins involved in centrosome duplication [8], initiation of DNA synthesis [9], and induction of histone gene transcription [10]. Cyclin E-Cdk2 is therefore considered a master regulator for entering and executing S-phase and for coordinating DNA synthesis with other cellular processes at the G_1/S transition. Once S-phase has been initiated cyclin E-Cdk2 autophosphorylates on cyclin E, which contributes to its degradation and favors the switch to formation cyclin A containing complexes. Cyclin A is a rate limiting component required cell cycle progression and is expressed soon after cyclin E at the G_1/S boundary [11]. Both cyclin E-Cdk2 and cyclin A-Cdk2 activities are essential for the initiation and completion of DNA replication [12–14] and for ensuring that replication occurs only once in each cell cycle. In addition to its role in controlling replication, cyclin A-Cdk2 also promotes the efficient execution of S phase by increasing transcription of histone and other genes needed to accommodate replication [6].

In late mitosis and during G_1 replication origins are "licensed" for replication by loading of mini-chromosome maintenance (MCM)2–7 proteins to form the pre-replication complex (pre-RC) [9]. Licensing is the process that ensures that the entire genome is replicated only once in each cell cycle. This process begins with binding of the origin recognition complex (ORC), which is present throughout the cell cycle and is thought to bind specifically to replication origins [9]. ORC then recruits the initiation factors Cdc6 and Cdt1. These factors facilitate the loading of MCM2–7 on to chromatin. The assembly of pre-RCs occurs when Cdk activity is low and initiation of replication can only occur when Cdk activity increases [15]. The activation of replication is under the control of cyclin E-Cdk2 and the

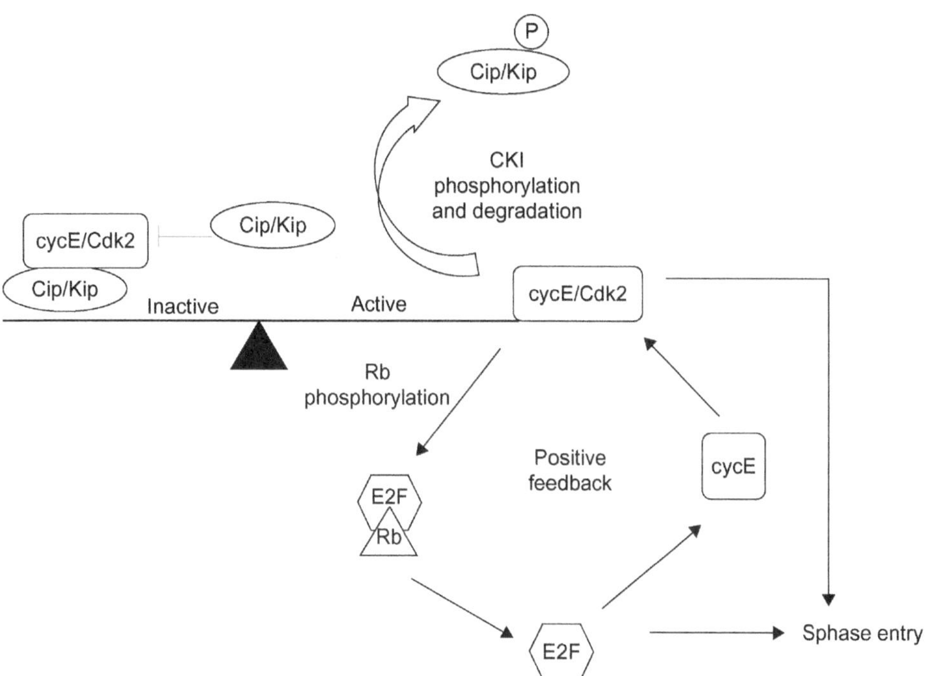

FIGURE 49.2 Activation of a small amount of cyclin E-Cdk2 leads to increased phosphorylation of Rb, which promotes more E2F dependent cyclin E transcription and assembly of cyclin E to cyclin-Cdk complexes. CKI(Cip/Kip) phosphorylation by cyclin E-Cdk2 contributes to destruction of the Cip/Kip proteins, which leads to further activation of cyclin E-Cdk2, reinforces the positive feedback loop and drives cells into S-phase.

S-phase activated kinase Cdc7 [16]. Our understanding of the mechanism by which cells ensure that DNA is replicated once, and only once, in each cycle is based on experiments in many species, and the exact details vary among organisms. Nevertheless, the principles by which it is achieved are at least partly conserved. Once replication has initiated, the pre-RC is disassembled to prevent a second round of replication [17]. Cdc6, Cdt1, and other essential replication cofactors, including the MCM proteins, are phosphorylated and inactivated (reviewed in [9, 18, 19]). Inactivating phosphorylation triggers the degradation of some proteins, and it prevents access to the nucleus or to chromatin of others. In addition to this, metazoan cells express Geminin, an inhibitor of origin firing in S-phase. Geminin binds and inhibits Cdt1 and this is one of the main mechanisms of inhibiting licensing during S and G_2 phases of the cell cycle [9, 17]. During G_2 and mitosis Geminin stabilizes Cdt1 allowing it to accumulate in its inactive form [9]. At the metaphase-to-anaphase transition Geminin is degraded by the anaphase promoting complex (APC) (reviewed in [15, 20] and see later section) releasing the inhibition on Cdt1 and allowing active Cdt1 to assemble into pre-RCs ready for initiation of replication in the next S-phase.

After growing and duplicating its contents, the cell is faced with the challenge of dividing itself into two viable daughter cells. This is achieved in mitosis, which begins at prophase, with the condensation of chromosomes and the formation of a mitotic spindle. In prometaphase, the spindle microtubules attach to the two sister chromatids via kinetochores, and the nuclear envelope breaks down. Once everything is correctly aligned, the spindle pulls the sister chromatids apart in anaphase. Following separation of the replicated genomes, the spindle is disassembled, the chromatids decondense, and the nuclear envelope begins to reform, in telophase. Finally, at cytokinesis, the formation of the two daughter cells is completed by the partitioning of the cytoplasm [21]. These dramatic morphological changes are under the control Cdc2 (Cdk1) in association with cyclin A and cyclin B. Expression of cyclin B lags behind that of cyclin A, rising late in S-phase and remaining high throughout G_2 and mitosis. The gradual increase in cyclin B protein is not reflected in gradually increased kinase activity (Figure 49.1) because the complex is largely maintained in a phosphorylated inactive form until the G_2/M boundary. Dephosphorylation and activation of cyclin B-Cdc2 closely correlate with the morphological changes that accompany mitosis. Nuclear lamins, nucleolar proteins, centrosomal proteins, and kinesin related motor proteins have all been described as cyclin B-Cdc2 substrates [6, 22]. Cdc2, in association with the second cyclin B subunit, cyclin B2, which localizes predominantly to the endoplasmic reticulum, may play a role in dispersing the Golgi apparatus so that cytokinesis will provide the two daughters with sufficient components to rebuild the secretory apparatus [23]. Cyclin B-Cdc2 also phosphorylates and activates at least two other kinases in mitosis, polo-like kinase (Plk) and the Aurora related kinases. Plk and Aurora have roles in centrosome separation, spindle assembly, and chromosome segregation [24, 25]. As an activator of Cdc25C, the phosphatase that activates cyclin B-Cdc2, Plk1 is thought to play a role in the feedback loop to ensure that cyclinB-Cdc2 is rapidly and fully activated in mitosis [25, 26]. Cyclin B-Cdc2 also stimulates its own destruction by phosphorylating and activating components of the APC (see later section).

SIGNALS TO SLOW PROGRESS: REGULATION OF CDKS BY INHIBITORY PROTEINS

An important mechanism for negatively regulating cyclin-Cdk is the interaction with small inhibitory proteins. Based on sequence homology and their specificity of action two distinct families of Cdk inhibitors (CKIs) have been described in mammals. The Ink4 family ($p15^{Ink4B}$, $p16^{Ink4A}$, $p18^{Ink4C}$, and $p19^{Ink4D}$), named for the ability to inhibit Cdk4/6, specifically inhibit the cyclin D containing Cdks and the Cip/Kip family ($p21^{Cip1}$, $p27^{Kip1}$, and $p57^{Kip2}$) strongly inhibits Cdk2 containing complexes [27].

The Ink4 family members act as brakes for G1 progression in response to mitogenic withdrawal, proliferation inhibition, differentiation signals, oncogenic stress, or senescence. Ink4 family members inhibit cyclin D by competing for binding to Cdk4 and/or Cdk6. The inhibitors bind Cdk4/6 inducing a conformational change that alters the cyclin binding site [28], therefore, high Ink4 expression blocks formation of the active cyclin D-Cdk4 complex. Because Ink4 prevents cyclin D-Cdk4 complex formation it also prevents binding of p27 to cyclin D-Cdk4. Increased Ink4 expression therefore increases the pool of p27 available to bind to and inhibit cyclin E-Cdk2. Hence, upregulation of Ink4 directly inhibits cyclin D-Cdk4 activity, and indirectly inhibits cyclin E-Cdk2 activity [27]. Ink4 family members play diverse and distinct cellular roles and they accumulate in anti-proliferative situations, e.g., senescence (p16), TGFβ treatment (p15) or specific differentiation processes [29]. The regulation of distinct Ink4 family members is complex and not fully elucidated. Nevertheless, the well characterized pathway by which the cytokine, TGFβ, induces Ink4 elegantly illustrates one mechanism by which signals from outside the cell delay progression if extracellular conditions are not favorable [30].

The Cip/Kip family of inhibitors binds to preformed cyclin-Cdk complexes blocking substrate access. The first of these inhibitors to be characterized, p21, is transcriptionally upregulated by p53 in response to DNA damage. The p21 inhibitor predominantly interacts with and inhibits cyclin E-Cdk2 and cyclin A-Cdk2, thereby delaying entry into S-phase. In addition, p21 binds and inhibits PCNA, a processivity factor for replicative DNA polymerases [31]. It has been suggested that these dual inhibitory actions might allow p21 to coordinately arrest the cell cycle by preventing E2F dependent transcription of S-phase genes and inhibiting PCNA dependent replication [28].

Another member of the Cip/Kip family of inhibitors, p27, plays a role in control of cell growth by inhibiting cyclin E-Cdk2. The activity of p27 is regulated by diverse mechanisms including changes in abundance, phosphorylation, subcellular localization, and regulated interactions with other cellular components [32].

CDKS ARE POSITIVELY AND NEGATIVELY REGULATED BY PHOSPHORYLATION

The association of kinase and cyclin subunits results in a partially activate kinase complex. Full activation requires phosphorylation of a threonine residue within the activation or T-loop region of the kinase. Crystallographic studies and biochemical analyses support a model in which cyclin binding to Cdks induces a conformational change in the kinase, exposing the T-loop threonine, and making it more accessible to the activating kinase [33]. Phosphorylation of the T-loop induces a second conformational shift that moves the T-loop away from the ATP binding region and thus facilitates catalysis. The kinases responsible for catalyzing the activating threonine phosphorylation are known as CAK, for Cdk activating kinases. In mammals, CAK is a nuclear holoenzyme comprising a Cdk subunit, Cdk7, and a cyclin subunit, cyclin H [34, 35]. The activity of cyclin H-Cdk7 is constitutive across the cell cycle and there is little evidence that modulation of CAK activity regulates cell cycle progression. Cyclin H-Cdk7 has a second function in transcription-coupled repair. A pool cyclin H-Cdk7 associates with the transcription machinery and phosphorylates the carboxyl terminal domain (CTD) of RNA polymerase II, promoting the transition from the initiation competent form to the elongation competent form of the complex [36]. The dual functions of cyclin H-Cdk7 in basal transcription and in cell cycle progression may help to coordinate these two processes.

Phosphorylation of Cdks also negatively regulates kinase activity. The activity of cyclin B-Cdc2 is perhaps the best characterized example of regulation by inhibitory phosphorylation; however, the activities of cyclin A-Cdk2, cyclin E-Cdk2, and Cylin D-Cdk4 are also, at least partially, controlled this way [37, 38]. As cyclin B-Cdc2 complexes accumulate, inhibitory phosphorylation occurs at sites corresponding to threonine 14 and tyrosine 15 in human Cdc2. Wee1 is a nuclear kinase that preferentially phosphorylates tyrosine 15, whereas Myt1, a cytoplasmic protein that associates with the membranes of the endoplasmic reticulum and Golgi, shows a preference for phosphorylating threonine 14. The distinct location of these two kinases suggests a mechanism by which Cdc2 can be negatively regulated in response to both cytoplasmic and nuclear conditions. Wee1 and Myt1 are active throughout interphase, and coincident with the activation of cyclin B-Cdc2 in mitosis, both kinases are inactivated [39–41]. Two kinases have been reported to inactivate Myt1 in maturing oocytes [42, 43] and the Wee1 ortholog is phosphorylated by three distinct kinases at different stages of the cell cycle. This cumulative phosphorylation leads to degradation and inactivation of Wee1 through ubiquitin mediated proteolysis [44, 45].

The inhibitory phosphates on Cdks are removed by the Cdc25 family of dual specificity phosphatases. Three forms of Cdc25 are expressed in mammalian cells. Cdc25A becomes active in mid to late G_1, expression of the predominantly cytoplasmic Cdc25B rises in S-phase and remains high in G_2 and although the abundance of Cdc25C is constant, its activity increases in mitosis [46]. On the basis of their expression pattern and/or activities throughout the cell cycle the different isoforms of Cdc25 were assigned functions in promoting progression at specific phases of the cell cycle [47]. Somatic cell studies using RNA interference and inhibitory antibodies to reduce the expression of each Cdc25 isoform demonstrate that Cdc25A and Cdc25B are both required for mitotic entry and that Cdc25C alone cannot induce mitosis [48]. However, given the significant sequence similarity between the three Cdc25 isoforms, it is perhaps not surprising that they have overlapping substrate specificity and they show functional redundancy. Mice lacking both Cdc25B and Cdc25C are viable and cells from these mice display normal cell cycle profiles, DNA damage responses, and normal Cdc25A regulation [49]. This suggests that, in the absence of Cdc25B and C, Cdc25A is sufficient to perform all Cdc25 functions required for normal cell division.

The Cdc25 proteins are themselves substrates of the Cdks: Cdc25A is phosphorylated and activated by cyclin E-Cdk2, and Cdc25C is activated by phosphorylation by cyclin B-Cdc2 [50, 51]. The activation of Cdc25 by the kinases that it activates led to the idea of a positive feedback loop that stimulates the rapid and full activation of Cdc2. In the case of the G_2/M transition, the feedback loop may require the activity of a non-Cdk kinase, Plk1, which phosphorylates and activates Cdc25 and is itself activated in mitosis [52, 53].

DEGRADATION: THE IMPORTANCE OF BEING ABSENT

Each cell cycle phase is defined by the form of the cyclin-Cdk complex that is present and active. It is therefore important that cyclin-Cdks be inactivated and destroyed in a timely manner. Ubiquitin mediated degradation is the main pathway for ridding cells of cell cycle proteins that have executed their function. The addition of ubiquitin residues to substrate proteins is the signal for degradation by the 26S proteasome, a multi-subunit cellular complex that specializes in the unfolding and proteolysis of ubiquitin tagged proteins. Traffic of cell cycle proteins into the 26S proteosome is controlled by the activity of two structurally and functionally similar E3 ubiquitin ligase complexes, the SCF (Skp-Cullin F box) and APC. The E3 ubiquitin ligase of SCF is active in G_1, through S-phase, and into G_2. The APC is also an E3 ubiquitin ligase; it is activated in mitosis and remains active in G_1. The specificity of these ubiquitin ligases is determined by substrate recognition proteins. In the case of SCF a large number of substrate recognition proteins, characterized by the presence of F-boxes, leucine-rich repeats, or WD-40 repeats, are known to exist [54]. In the case of the APC, two substrate specific interacting proteins, Cdc20 and Cdh1, are known, but more may exist [19]. Specificity and appropriate temporal degradation are achieved by a combination of the fact that Cdk catalyzed phosphorylation of target proteins is required for recognition by F-box proteins, and that components of APC are regulated by Cdk dependent phosphorylation.

As discussed earlier, the degradation of p27 and cyclin E is initiated when these proteins are phosphorylated by active cyclin E-Cdk2. In a similar manner, the transcriptional activator E2F is inactivated by cyclinA-Cdk2 dependent phosphorylation and degradation by the SCF dependent pathway [55]. In both cases degradation is initiated when the protein has been phosphorylated and thus degradation is under the direct control of Cdk activity. Compound mechanisms regulate the traffic of proteins through the APC. The rise in expression of the substrate recognition protein, Cdc20, in mitosis initiates degradation of APC-Cdc20 substrates. Phosphorylation of Cdc20 and core components of the APC by cyclinB-Cdc2 stimulates ubiquitin ligase activity.

A major role of the APC-Cdc20 complex is to promote the degradation of proteins that maintain sister chromatid cohesion and, thus, to allow the metaphase-to-anaphase transition [56]. The APC-Cdc20 complex ubiquitinates the anaphase inhibitor Securin, that blocks the onset of sister chromatid separation until all chromatids are aligned. Once Securin has been degraded its partner, Separase, becomes active, Cohesin is cleaved, and sister chromatids can separate [56, 57]. If all of the chromosomes are not correctly aligned, the presence of an unattached kinetochore generates a signal that inhibits APC-Cdc20 [58, 59]. By inhibiting the APC-Cdc20, the spindle checkpoint prevents sister chromatid separation until all chromatids are properly attached to the spindle microtubules. Expression of the second APC specificity factor, Cdh1, rises in S-phase and persists through G_1. However, the APC-Cdh1 complex is inhibited by high cyclinB-Cdc2 activity, thus degradation of APC-Cdh1 substrates cannot occur until after degradation of cyclin B. Once the APC-Cdh1 complex is activated, proteins that prevent anaphase, including remaining mitotic cyclins, are destroyed. The inhibition of Cdh1 by mitotic Cdks prevents the onset of anaphase if metaphase is delayed. APC-Cdh1 remains active into G_1 ensuring that components of the previous mitosis are destroyed before initiation of the next cycle. The APC also catalyzes the destruction of Cdc20, thereby preventing APC activity until Cdc20 is synthesized again late in the next cycle.

CHECKPOINT SIGNALING

Once it has passed the mitogen dependent stage, the cell cycle is an autonomous signaling system in which the completion of each step promotes progression through the next. Despite this constant drive forward, a problem in any of the subpathways of the cell cycle can generate a signal that stops progression in all other subpathways. The mechanisms that inhibit cell cycle progression in response to intracellular problems are known as *checkpoints*. Checkpoint responses may, in fact, be activated in every unperturbed cell cycle; however, they are most easily seen and best understood in the context of cells that have been subjected to unusual levels of damage or unanticipated nucleotide deprivation. The concept of checkpoint signaling and how they integrate with cell cycle regulation has been reviewed [60–62]. Upon encountering DNA damage, checkpoint pathways are activated to delay or arrest cell cycle progression until the damage can be repaired. As far as we know all DNA damage checkpoints function through two protein kinases, ATM and ATR [63]. DNA damage is sensed by specific proteins such as the Mre11-Rad50-Nbs1 (MRN) complex, which senses double-strand breaks [64] and ATR-ATRIP, which senses damage arising during replication [65]. The signal generated is amplified through mediators, transducers, and effector proteins. These signaling cascades ultimately inactivate Cdks to inhibit cell cycle progression. This goal is achieved through stimulation of CKI expression, dephosphorylation of Cdks, cyclin degradation, and/or cytoplasmic retention of Cdk-cyclin complexes.

Arrest in the G_1 phase of the cell cycle is mainly mediated by the p53 tumor suppressor, a transcription factor that induces cell cycle arrest, damage repair, and apoptosis [66]. The cell cycle effects of p53 activation include induction of transcription of several genes, including the Cdk inhibitor p21. The induction of p21 in response to DNA damage is the major mechanism enforcing arrest in G_1 in response to DNA damage or nucleotide pool perturbation [67, 68]. Recent reports have indicated that degradation of cyclin D and inactivation of Cdc25A also contribute to the G_1 checkpoint [69, 70].

The initial G_2 damage induced checkpoint works by maintaining the inhibitory phosphorylation of cyclin B-Cdc2. As discussed previously, Cdc2 dephosphorylation is catalyzed by Cdc25C. The checkpoint kinase Chk1 phosphorylates all three Cdc25 proteins and downregulates phosphatase activity by direct inhibition, by excluding the proteins from the nucleus and initiating their proteolytic degradation [71]. This prevents dephosphorylation and activation of Cdk, thereby delaying cell cycle progression.

An important factor in checkpoint controls is the timescale on which they operate. Checkpoint mechanisms that act through phosphorylation or degradation of key regulators might be expected to operate very rapidly in response to damage and be capable of delaying the onset of the next cell cycle step right up until the moment of initiation. On the other hand, the transcriptional induction of cell cycle inhibitor proteins is a process that functions over the course of hours rather than minutes. It is becoming increasingly clear that both quick acting and sustained checkpoint responses are both required to effectively inhibit the activity of the cyclin-Cdks that drive cell cycle progression (see Figure 49.3).

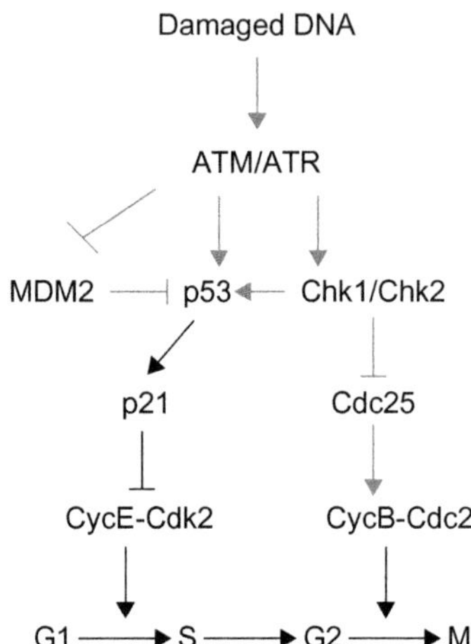

FIGURE 49.3 The DNA damage checkpoints inhibit Cdk activity. Phosphorylation and degradation events that are expected to function rapidly following DNA damage are indicated by gray lines and arrows. Processes that require transcription, and are required for the long term maintenance of the arrest are indicated in black.

LESSONS FROM MICE: CELL DIVISION WITHOUT CDKS

The generation of gene targeted mouse models has provided novel insight into the extent of functional redundancy between cyclin-Cdk complexes. These mouse models suggest that loss of specific cyclins or Cdks can be compensated for during development. This is perhaps not surprising given the sequence similarity of the different cyclin-Cdks complexes and has been explained partly by a concept known as embryonic plasticity. Embryonic plasticity is thought account for an increasing number of examples in which cell division and development is possible in the absence of gene products that are essential for proliferation in somatic cells [72]. As discussed earlier, cyclin D-Cdk and cyclin E-Cdk cooperatively control G_1 progression. This idea is both supported and expanded by genetic studies demonstrating that disruption of either Cdk4 or Cdk6 does not affect viability [11], whereas the double knockout leads to embryonic lethality [73].

Surprisingly mice lacking Cdk2 are viable and are the least affected of all cyclin- and Cdk deficient strains. However, Cdk2 probably does play crucial roles in cell proliferation and embryonic development but compensatory mechanisms may substitute for Cdk2 function [11, 73].

In support of this, the combined loss of Cdk2 and Cdk4 leads to severe defects in cell cycle progression and early embryonic lethality [11, 74]. The rounds of division seen in animals lacking Cdk2 and Cdk4, may be driven by Cdc2 (Cdk1); however, the reduction or misregulation of pRb phosphorylation limits proliferation to a few rounds.

As described above, the sequential activation of Cdk4/Cdk6 and Cdk2 is required to drive cells through interphase. and activation of Cdc2 (Cdk1) is required to proceed through mitosis. However, it is now clear that mouse cells can proliferate in the absence of two or even three Cdks [75]. Surprisingly, mice lacking all interphase Cdks (Cdk2, Cdk3, Cdk4, and Cdk6) undergo organogenesis and develop until mid-gestation [75]. In these embryos Cdc2 binds to all cyclins, phoshorylates pRb at residues normally phosphorylated by interphase Cdks, and expression of E2F dependent genes occurs in a timely manner [75]. Although Cdc2 can drive the cell cycle without the other Cdks, this is probably a specialized case where the cells have been reprogrammed early enough in development to allow division by reassigning all Cdk functions to Cdc2 (Cdk1). On the other hand, mice lacking Cdc2 are inviable [63].

KNOCKOUT MOUSE MODELS: CYCLING WITHOUT CYCLINS

Mice lacking one, two, or all three D-type cyclins have been generated [76, 77]. Studies of these cyclin D knockout mice show that the three family members are functionally interchangeable in most, if not all, cell types showing only subtle functional differences relating to temporal or spatial regulation. Triple knockout mice die late in embryonic development but cultured triple knockout fibroblasts have only minor defects in cell cycle progression *in vitro*, suggesting that cyclin D is not necessary for proliferation [11, 73]. In these cells pRb phosphorylation by Cdk2 appears to be temporally correct.

Disruption of the genes encoding the individual cyclin E1 and E2 genes does not affect viability. Cyclin E1 and E2 have significant amino acid homology and are expressed in an overlapping manner in almost all proliferating cells. Therefore, the fact that no significant phenotype was observed in single knockout mice suggests that E1 and E2 probably have overlapping functions [73]. In contrast, targeting both E-type cyclin genes leads to early embryonic lethality. The lethality in the double knockout is attributed to placental dysfunction since embryos lacking both E-type cyclins were rescued using donor placenta, thus the divisions needed for embryonic development can occur in a cyclin E independent manner [78]. Characterization of these knockout mice also highlighted the Cdk2 independent functions of E-type cyclins. Cdk2 null mouse embryonic fibroblasts are able to re-enter the cell cycle from quiescence, but cyclin E knockout cells are not. This is related to the fact that cyclin E, but not Cdk2, is essential for loading of MCMs on to replication origins to establish the pre-RC [79]. Curiously, it seems likely that the essential functions of cyclin E are independent of Cdk activity and phosphorylation [79, 80].

Disruption of cyclin A1 revealed that this cyclin also is not essential for cell cycle progression and mouse development as viable mice developmentally normal with only a male germ cell lineage defect were generated [73]. In stark contrast, knockout of cyclin A2, a ubiquitously expressed cyclin, caused early embryonic lethality, revealing a critical, non-redundant role for cyclin A2 in S- and M-phase progression [81]. Likewise, loss of cyclin B1 leads to embryonic lethality [82], thus cyclin A2 and B1 are the only cyclins shown to have an essential role in proliferation or early development.

The wealth of information obtained from gene targeted mouse models should be interpreted with caution. For example, in the absence of Cdk2, Cdk4, and Cdk6, it has been shown that Cdc2 (Cdk1) interacts with all cyclins; however, interactions only occur in the absence of the interphase Cdks, in normal circumstances when Cdk4/6 are present, Cdk2 does not interact with cyclin D [75]. Nevertheless, the lessons learned in these knockout studies may prove to have wider implications since evidence is emerging that cancer cells also differentially depend on specific cyclin-Cdks for successful progression through the cell cycle [11].

REFERENCES

1. Roovers K, Assoian RK. Integrating the MAP kinase signal into the G1 phase cell cycle machinery. *Bioessays* 2000;**22**(9):818–26.
2. Ekholm SV, Reed SI. Regulation of G(1) cyclin-dependent kinases in the mammalian cell cycle. *Curr Opin Cell Biol* 2000;**12**(6):676–84.
3. Sherr CJ. Mammalian G1 cyclins. *Cell* 1993;**73**(6):1059–65.
4. Cheng M, Olivier P, Diehl JA, Fero M, Roussel MF, Roberts JM, Sherr CJ. The p21(Cip1) and p27(Kip1) CDK "inhibitors" are essential activators of cyclin D-dependent kinases in murine fibroblasts. *Embo J* 1999;**18**(6):1571–83.
5. LaBaer J, Garrett MD, Stevenson LF, Slingerland JM, Sandhu C, Chou HS, Fattaey A, Harlow E. New functional activities for the p21 family of CDK inhibitors. *Genes Dev* 1997;**11**(7):847–62.
6. Obaya AJ, Sedivy JM. Regulation of cyclin-Cdk activity in mammalian cells. *Cell Mol Life Sci* 2002;**59**(1):126–42.
7. Harbour JW, Dean DC. The Rb/E2F pathway: expanding roles and emerging paradigms. *Genes Dev* 2000;**14**(19):2393–409.
8. Hinchcliffe EH, Sluder G. "It takes two to tango": understanding how centrosome duplication is regulated throughout the cell cycle. *Genes Dev* 2001;**15**(10):1167–81.
9. Blow JJ, Dutta A. Preventing re-replication of chromosomal DNA. *Nat Rev Mol Cell Biol* 2005;**6**(6):476–86.
10. Zhao J. Coordination of DNA synthesis and histone gene expression during normal cell cycle progression and after DNA damage. *Cell Cycle* 2004;**3**(6):695–7.
11. Berthet C, Kaldis P. Cell-specific responses to loss of cyclin-dependent kinases. *Oncogene* 2007;**26**(31):4469–77.

12. Girard F, Strausfeld U, Fernandez A, Lamb NJ. Cyclin A is required for the onset of DNA replication in mammalian fibroblasts. *Cell* 1991;**67**(6):1169–79.
13. Ohtsubo M, Theodoras AM, Schumacher J, Roberts JM, Pagano M. Human cyclin E, a nuclear protein essential for the G1-to-S phase transition. *Mol Cell Biol* 1995;**15**(5):2612–24.
14. Pagano M, Pepperkok R, Verde F, Ansorge W, Draetta G. Cyclin A is required at two points in the human cell cycle. *Embo J* 1992;**11**(3):961–71.
15. Diffley JF. Regulation of early events in chromosome replication. *Curr Biol* 2004;**14**(18):R778–86.
16. Zou L, Stillman B. Assembly of a complex containing Cdc45p, replication protein A, and Mcm2p at replication origins controlled by S-phase cyclin-dependent kinases and Cdc7p-Dbf4p kinase. *Mol Cell Biol* 2000;**20**(9):3086–96.
17. Arias EE, Walter JC. Strength in numbers: preventing rereplication via multiple mechanisms in eukaryotic cells. *Genes Dev* 2007;**21**(5):497–518.
18. Kelly TJ, Brown GW. Regulation of chromosome replication. *Annu Rev Biochem* 2000;**69**:829–80.
19. Zachariae W, Nasmyth K. Whose end is destruction: cell division and the anaphase-promoting complex. *Genes Dev* 1999;**13**(16):2039–58.
20. Diffley JF. DNA replication: building the perfect switch. *Curr Biol* 2001;**11**(9):R367–70.
21. Maller J, Gautier J, Langan TA, Lohka MJ, Shenoy S, Shalloway D, Nurse P. Maturation-promoting factor and the regulation of the cell cycle. *J Cell Sci Suppl* 1989;**12**:53–63.
22. Neef R, Klein UR, Kopajtich R, Barr FA. Cooperation between mitotic kinesins controls the late stages of cytokinesis. *Curr Biol* 2006;**16**(3):301–7.
23. Draviam VM, Orrechia S, Lowe M, Pardi R, Pines J. The localization of human cyclins B1 and B2 determines CDK1 substrate specificity and neither enzyme requires MEK to disassemble the Golgi apparatus. *J Cell Biol* 2001;**152**(5):945–58.
24. Bischoff JR, Plowman GD. The Aurora/Ipl1p kinase family: regulators of chromosome segregation and cytokinesis. *Trends Cell Biol* 1999;**9**(11):454–9.
25. Nigg EA. Polo-like kinases: positive regulators of cell division from start to finish. *Curr Opin Cell Biol* 1998;**10**(6):776–83.
26. Yoo HY, Kumagai A, Shevchenko A, Dunphy WG. Adaptation of a DNA replication checkpoint response depends upon inactivation of Claspin by the Polo-like kinase. *Cell* 2004;**117**(5):575–88.
27. Sherr CJ, Roberts JM. CDK inhibitors: positive and negative regulators of G1-phase progression. *Genes Dev* 1999;**13**(12):1501–12.
28. Pei XH, Xiong Y. Biochemical and cellular mechanisms of mammalian CDK inhibitors: a few unresolved issues. *Oncogene* 2005;**24**(17):2787–95.
29. Canepa ET, Scassa ME, Ceruti JM, Marazita MC, Carcagno AL, Sirkin PF, Ogara MF. INK4 proteins, a family of mammalian CDK inhibitors with novel biological functions. *IUBMB Life* 2007;**59**(7):419–26.
30. Massague J, Wotton D. Transcriptional control by the TGF-beta/Smad signaling system. *Embo J* 2000;**19**(8):1745–54.
31. Chen J, Jackson PK, Kirschner MW, Dutta A. Separate domains of p21 involved in the inhibition of Cdk kinase and PCNA. *Nature* 1995;**374**(6520):386–8.
32. Borriello A, Cucciolla V, Oliva A, Zappia V, Della Ragione F. p27Kip1 metabolism: a fascinating labyrinth. *Cell Cycle* 2007;**6**(9):1053–61.
33. Johnson LN, Noble ME, Owen DJ. Active and inactive protein kinases: structural basis for regulation. *Cell* 1996;**85**(2):149–58.
34. Bockstaele L, Coulonval K, Kooken H, Paternot S, Roger PP. Regulation of CDK4. *Cell Div* 2006;**1**:25.
35. Kaldis P. The cdk-activating kinase (CAK): from yeast to mammals. *Cell Mol Life Sci* 1999;**55**(2):284–96.
36. Nigg EA. Cyclin-dependent kinase 7: at the cross-roads of transcription, DNA repair and cell cycle control? *Curr Opin Cell Biol* 1996;**8**(3):312–17.
37. Dulic V, Lees E, Reed SI. Association of human cyclin E with a periodic G1-S phase protein kinase. *Science* 1992;**257**(5078):1958–61.
38. Terada Y, Tatsuka M, Jinno S, Okayama H. Requirement for tyrosine phosphorylation of Cdk4 in G1 arrest induced by ultraviolet irradiation. *Nature* 1995;**376**(6538):358–62.
39. Booher RN, Holman PS, Fattaey A. Human Myt1 is a cell cycle-regulated kinase that inhibits Cdc2 but not Cdk2 activity. *J Biol Chem* 1997;**272**(35):22,300–22,306.
40. McGowan CH, Russell P. Cell cycle regulation of human WEE1. *EMBO J* 1995;**14**(10):2166–75.
41. Watanabe N, Broome M, Hunter T. Regulation of the human WEE1Hu CDK tyrosine 15-kinase during the cell cycle. *EMBO J* 1995;**14**(9):1878–91.
42. Okumura E, Fukuhara T, Yoshida H, Hanada Si S, Kozutsumi R, Mori M, Tachibana K, Kishimoto T. Akt inhibits Myt1 in the signalling pathway that leads to meiotic G2/M-phase transition. *Nat Cell Biol* 2002;**4**(2):111–16.
43. Palmer A, Gavin AC, Nebreda AR. A link between MAP kinase and p34(cdc2)/cyclin B during oocyte maturation: p90(rsk) phosphorylates and inactivates the p34(cdc2) inhibitory kinase Myt1. *EMBO J* 1998;**17**(17):5037–47.
44. Lee KS, Asano S, Park JE, Sakchaisri K, Erikson RL. Monitoring the cell cycle by multi-kinase-dependent regulation of Swe1/Wee1 in budding yeast. *Cell Cycle* 2005;**4**(10):1346–9.
45. Watanabe N, Arai H, Iwasaki J, Shiina M, Ogata K, Hunter T, Osada H. Cyclin-dependent kinase (CDK) phosphorylation destabilizes somatic Wee1 via multiple pathways. *Proc Natl Acad Sci USA* 2005;**102**(33):11,663–11,668.
46. Nilsson I, Hoffmann I. Cell cycle regulation by the Cdc25 phosphatase family. *Prog Cell Cycle Res* 2004;**4**:107–14.
47. Takizawa CG, Morgan DO. Control of mitosis by changes in the subcellular location of cyclin-B1-Cdk1 and Cdc25C. *Curr Opin Cell Biol* 2000;**12**(6):658–65.
48. Lindqvist A, Kallstrom H, Lundgren A, Barsoum E, Rosenthal CK. Cdc25B cooperates with Cdc25A to induce mitosis but has a unique role in activating cyclin B1-Cdk1 at the centrosome. *J Cell Biol* 2005;**171**(1):35–45.
49. Ferguson AM, White LS, Donovan PJ, Piwnica-Worms H. Normal cell cycle and checkpoint responses in mice and cells lacking Cdc25B and Cdc25C protein phosphatases. *Mol Cell Biol* 2005;**25**(7):2853–60.
50. Hoffmann I, Clarke PR, Marcote MJ, Karsenti E, Draetta G. Phosphorylation and activation of human cdc25-C by cdc2: cyclin B and its involvement in the self-amplification of MPF at mitosis. *EMBO J* 1993;**12**(1):53–63.
51. Hoffmann I, Draetta G, Karsenti E. Activation of the phosphatase activity of human cdc25A by a cdk2-cyclin E dependent phosphorylation at the G1/S transition. *EMBO J* 1994;**13**(18):4302–10.
52. Kumagai A, Dunphy WG. Purification and molecular cloning of Plx1, a Cdc25-regulatory kinase from Xenopus egg extracts. *Science* 1996;**273**(5280):1377–80.
53. Qian YW, Erikson E, Taieb FE, Maller JL. The polo-like kinase Plx1 is required for activation of the phosphatase Cdc25C and cyclin B-Cdc2 in Xenopus oocytes. *Mol Biol Cell* 2001;**12**(6):1791–9.

54. Kipreos ET, Pagano M. The F-box protein family. *Genome Biol* 2000;**1**(5). REVIEWS3002.
55. Harper JW, Elledge SJ. Skipping into the E2F1-destruction pathway. *Nat Cell Biol* 1999;**1**(1):E5–7.
56. Yanagida M. Basic mechanism of eukaryotic chromosome segregation. *Philos Trans R Soc Lond B Biol Sci* 2005;**360**(1455):609–21.
57. Yu Q, Sicinski P. Mammalian cell cycles without cyclin E-CDK2. *Cell Cycle* 2004;**3**(3):292–5.
58. Gillett ES, Sorger PK. Tracing the pathway of spindle assembly checkpoint signaling. *Dev Cell* 2001;**1**(2):162–4.
59. Shah JV, Cleveland DW. Waiting for anaphase: Mad2 and the spindle assembly checkpoint. *Cell* 2000;**103**(7):997–1000.
60. Bartek J, Lukas J. Pathways governing G1/S transition and their response to DNA damage. *FEBS Lett* 2001;**490**(3):117–22.
61. Bartek J, Lukas J. Mammalian G1- and S-phase checkpoints in response to DNA damage. *Curr Opin Cell Biol* 2001;**13**(6):738–47.
62. Zhou BB, Elledge SJ. The DNA damage response: putting checkpoints in perspective. *Nature* 2000;**408**(6811):433–9.
63. Abraham RT. Cell cycle checkpoint signaling through the ATM and ATR kinases. *Genes Dev* 2001;**15**(17):2177–96.
64. Petrini JH, Stracker TH. The cellular response to DNA double-strand breaks: defining the sensors and mediators. *Trends Cell Biol* 2003;**13**(9):458–62.
65. Zou L, Elledge SJ. Sensing DNA damage through ATRIP recognition of RPA-ssDNA complexes. *Science* 2003;**300**(5625):1542–8.
66. Wahl GM, Linke SP, Paulson TG, Huang LC. Maintaining genetic stability through TP53 mediated checkpoint control. *Cancer Surv* 1997;**29**:183–219.
67. Brugarolas J, Chandrasekaran C, Gordon JI, Beach D, Jacks T, Hannon GJ. Radiation-induced cell cycle arrest compromised by p21 deficiency. *Nature* 1995;**377**(6549):552–7.
68. Deng C, Zhang P, Harper JW, Elledge SJ, Leder P. Mice lacking p21CIP1/WAF1 undergo normal development, but are defective in G1 checkpoint control. *Cell* 1995;**82**(4):675–84.
69. Agami R, Bernards R. Distinct initiation and maintenance mechanisms cooperate to induce G1 cell cycle arrest in response to DNA damage. *Cell* 2000;**102**(1):55–66.
70. Mailand N, Falck J, Lukas C, Syljuasen RG, Welcker M, Bartek J, Lukas J. Rapid destruction of human Cdc25A in response to DNA damage. *Science* 2000;**288**(5470):1425–9.
71. Niida H, Nakanishi M. DNA damage checkpoints in mammals. *Mutagenesis* 2006;**21**(1):3–9.
72. Sage J, Miller AL, Perez-Mancera PA, Wysocki JM, Jacks T. Acute mutation of retinoblastoma gene function is sufficient for cell cycle re-entry. *Nature* 2003;**424**(6945):223–8.
73. Santamaria D, Ortega S. Cyclins and CDKS in development and cancer: lessons from genetically modified mice. *Front Biosci* 2006;**11**:1164–88.
74. Hinds PW. A confederacy of kinases: Cdk2 and Cdk4 conspire to control embryonic cell proliferation. *Mol Cell* 2006;**22**(4):432–3.
75. Santamaria D, Barriere C, Cerqueira A, Hunt S, Tardy C, Newton K, Caceres JF, Dubus P, Malumbres M, Barbacid M. Cdk1 is sufficient to drive the mammalian cell cycle. *Nature* 2007;**448**(7155):811–15.
76. Fantl V, Stamp G, Andrews A, Rosewell I, Dickson C. Mice lacking cyclin D1 are small and show defects in eye and mammary gland development. *Genes Dev* 1995;**9**(19):2364–72.
77. Sicinski P, Donaher JL, Parker SB, Li T, Fazeli A, Gardner H, Haslam SZ, Bronson RT, Elledge SJ, Weinberg RA. Cyclin D1 provides a link between development and oncogenesis in the retina and breast. *Cell* 1995;**82**(4):621–30.
78. Geng Y, Yu Q, Sicinska E, Das M, Schneider JE, Bhattacharya S, Rideout WM, Bronson RT, Gardner H, Sicinski P. Cyclin E ablation in the mouse. *Cell* 2003;**114**(4):431–43.
79. Geng Y, Lee YM, Welcker M, Swanger J, Zagozdzon A, Winer JD, Roberts JM, Kaldis P, Clurman BE, Sicinski P. Kinase-independent function of cyclin E. *Mol Cell* 2007;**25**(1):127–39.
80. Zhang H. Life without kinase: cyclin E promotes DNA replication licensing and beyond. *Mol Cell* 2007;**25**(2):175–6.
81. Murphy M, Stinnakre MG, Senamaud-Beaufort C, Winston NJ, Sweeney C, Kubelka M, Carrington M, Brechot C, Sobczak-Thepot J. Delayed early embryonic lethality following disruption of the murine cyclin A2 gene. *Nat Genet* 1997;**15**(1):83–6.
82. Brandeis M, Rosewell I, Carrington M, Crompton T, Jacobs MA, Kirk J, Gannon J, Hunt T. Cyclin B2-null mice develop normally and are fertile whereas cyclin B1-null mice die in utero. *Proc Natl Acad Sci USA* 1998;**95**(8):4344–9.

Chapter 50

The Role of Rac and Rho in Cell Cycle Progression

Laura J. Taylor and Dafna Bar-Sagi

Department of Molecular Genetics and Microbiology, State University of New York at Stony Brook, Stony Brook, New York

INTRODUCTION

The ability of cells to progress through the cell cycle depends on the concerted action of mitogen- and anchorage-stimulated signal transduction pathways. The regulation of the cell cycle machinery is often disrupted in tumor cells, with the most common targets being proteins involved in G1 progression. Identifying the signaling pathways responsible for G1 progression is therefore important for increasing the understanding of control mechanisms that operate during normal cell proliferation and their subversion in tumor cells. One class of signaling proteins that has recently been recognized to play a significant role in G1 phase progression is the small GTP binding proteins of the Rho GTPase family. This chapter will focus on the role of two members of the Rho GTPase family, Rac and Rho, in G1 progression.

REGULATION OF G1 PROGRESSION

Cell cycle progression through G1 is a complex and tightly controlled process. It is regulated by stimulatory and inhibitory signals, both of which are targets of the Rho GTPases. Three activities are recognized to be important for progression through the G1 phase of the cell cycle: early-G1 transcriptional activation of immediate early genes, mid-G1 activation of cyclin D/cdk4/6, and late-G1 activation of cyclin E/cdk2 [1, 2]. Mitogenic stimulation results in the induction of many immediate early genes containing the serum response element (SRE) in their promoter [1]. The SRE is activated by binding to a ternary complex containing the transcription factors, serum response factor (SRF) and ternary complex factor (TCF) [1]. Although activation of immediate early genes is necessary for early G1 progression, it is not sufficient for progression to S phase, and progression through later phases of G1 requires the activity of cyclin-dependent kinases (cdks).

Cdks are a group of serine/threonine kinases that are activated by binding to their respective cyclin partners and by phosphorylation [2]. Two main cdk activities play a role in G1, cyclin D/cdk4/6 functions in mid G1 and cyclin E/cdk2 functions in late G1. The major substrate of the G1 kinase complexes is the retinoblastoma protein (Rb). In its unphosphorylated form, Rb functions as an inhibitor of E2F, a transcription factor that controls the expression of genes required for G1 progression [3]. The inhibitory effect of Rb on E2F transcriptional activity is exerted by two mechanisms, one involving the direct binding to E2F and the other involving the recruitment of histone deacetylase (HDAC) [4]. Both inhibitory effects are antagonized by the coordinated and sequential phosphorylation of Rb by cyclin D/cdk4/6 and cyclin E/cdk2 which in turn allows the ordered expression of E2Fdependent genes [4, 5]. Phosphorylation of Rb by cyclin D/cdk4/6 initially releases HDAC thereby alleviating transcriptional repression, and phosphorylation of Rb by cyclin E/cdk2 dissociates the Rb–E2F complex [6, 7].

An important mechanism for regulation of cyclin/cdk activity involves inhibition by cyclin-dependent kinase inhibitors. The main inhibitors of cyclin D/cdk4/6 complexes are $p16^{ink4a}$ and $p21^{Cip1}$, whereas inhibition of cyclin E/cdk2 occurs by $p21^{Cip1}$ and $p27^{Kip1}$[2]. The levels of these inhibitors are regulated by multiple mechanisms. $p21^{Cip1}$ is regulated predominantly at the level of transcription [8] and mRNA stability [9]. $p27^{Kip1}$ can be regulated at multiple levels including transcriptional [10], translational [11, 12], and posttranslational [13, 14] mechanisms dependent on the cell type and the extracellular signal. However, the predominant mechanism by which p27 ^{Kip1}

Handbook of Cell Signaling, Three-Volume Set 2 ed.

levels are controlled is through cyclin E/cdk2-dependent phosphorylation [15, 16], which targets p27[Kip1] for ubiquitination and proteolytic degradation [13].

THE FUNCTION OF RAC AND RHO IN CELL CYCLE PROGRESSION AND TRANSFORMATION

The Rho family of GTPases functions as molecular switches by oscillating between an active GTP-bound form and inactive GDP-bound form. Activation of the Rho GTPases can be induced by soluble growth factor stimulation and cell adhesion to the extracellular matrix (ECM). Their biological effects are exerted through the activation of multiple effector pathways that control transcription, cytoskeleton organization, and changes in the redox state [17].

The importance of the Rho GTPases in cell cycle progression was initially illustrated through studies demonstrating their involvement in both growth-factor-induced proliferation and oncogenic transformation. Rac1 and RhoA are each required for transformation by Ras and co-expression of a constitutively active form of Raf, a Ras effector, with either constitutively active RhoA or Rac1 synergistically enhances focus-forming activity [18, 19].

Rac and Rho have been shown to be necessary and sufficient for cell cycle progression. In Swiss 3T3 cells, introduction of dominant interfering forms of Rac1 and RhoA inhibits progression of growth-factor-induced cell cycle progression, while a dominant active form of each is sufficient to induce cell cycle progression [20]. However, the capacity of Rac1 and RhoA to promote cell cycle progression is cell-type specific. For example, in rat embryo fibroblasts, G1 to S transition requires the synergistic activities of Rac and Raf [21]. Using partial loss of function mutants of Rac it has been shown that the contribution of Rac to cell cycle progression is dependent on two distinct effector functions, cytoskeleton rearrangements and superoxide production [21, 22]. Rac-induced cytoskeleton rearrangements are mediated by the effector binding loop [21], a region spanning amino acids 26 to 40 that interacts with multiple downstream effector molecules. Rac dependent superoxide generation is controlled by the insert region, a sequence of 11 amino acids common to all of the RhoGTPase family members, but not found in the Ras GTPase family members [23]. Significantly, superoxide generation has been shown to be essential for Ras-induced proliferation [24], indicating that Rac-mediated superoxide production might be functionally relevant to Ras-induced proliferation. It is noteworthy that the insert region of Rho also plays an important role with regard to cell cycle progression through the activation of the Rho effector, Rho kinase, which cooperates with activated Raf to promote transformation [25, 26].

CELL CYCLE TARGETS OF RAC AND RHO

Increasing evidence indicates that Rac and Rho influence cell cycle progression by targeting multiple regulatory steps throughout G1. A well-documented mechanism by which Rac and Rho affects early-G1 involves the activation of genes controlled by the SRE. Rac, through its effector PAK, promotes the phosphorylation of both Raf and MEK, two components of the signaling cascade leading to ERK activation [27, 28]. Both PAK-mediated phosphorylation events act synergistically with the Ras pathway to promote full activation of ERK [27, 29]. Subsequently, activated ERK phosphor ylates and activates TCF thereby stimulating SRE-dependent transcription (Figure 50.1). Rac activity has also been demonstrated to potentiate SRF activity, but the signaling pathways regulating this response have not been identified [30, 31].

RhoA activity is essential for mitogen-induced activation of SRF [30]. The ability of Rho to activate SRF is linked to its effect on actin cytoskeleton dynamics. This is indicated by studies showing that the Rho effectors LIMK and mDia potentiate SRF activity independently of extracellular signals [32–34]. Although the relative contribution

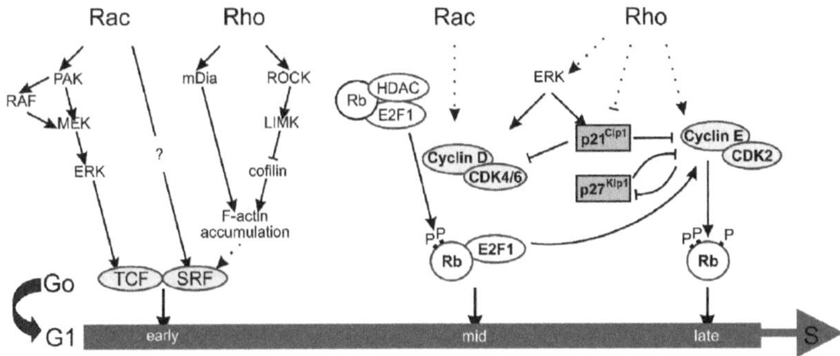

FIGURE 50.1 Integration of Rac and Rho signaling pathways with the cell cycle.
Rac and Rho target multiple regulatory events during the G1 stage of the cell cycle. The contribution of Rac and Rho to TCF/SRF and cyclin/cdk complexes promotes G1-phase progression.

of the Rho-dependent pathways to SRF activity seems to be cell-type dependent, both LIMK and mDia pathways contribute to F-actin accumulation, suggesting a causal role for F-actin levels in the activation of SRF (Figure 50.1).

Progression through mid-G1 of the cell cycle is dependent upon upregulation of cyclin D and formation of the cyclin D/cdk4/6 complex. Both Rac and Rho have been shown to contribute to the upregulation of cyclin D1 through ERK-dependent and -independent pathways (Figure 50.1). Rac-mediated induction of cyclin D1 occurs in part through the Rac effector PAK [31], and is also dependent on NFκB activation, as evident from the findings that an intact NFκB binding site in the cyclin D1 promoter is required for Rac-dependent cyclin D1 transcription [35].

Recent evidence by Welsh and colleagues demonstrates that Rho plays a central role in controlling adhesion- and mitogen-dependent cyclin D1 expression [36]. First, in early G1, Rho inhibits Rac-induced expression of cyclin D1 by antagonizing Rac through an unknown mechanism [36]. Second, in mid-G1 phase of the cell cycle, Rho promotes cyclin D1 induction by maintaining a sustained activation of ERK [36]. The mechanisms by which Rho might contribute to ERK activity are not well defined, but the Rho kinase pathway seems to be necessary for this effect [36]. Thus, Rho appears to have an important role in setting up the correct timing of cyclin D1 expression during cell cycle progression.

In addition to its role in the regulation of cyclin D1 expression, Rho regulates cyclin D/cdk4/6 activity by inhibiting the accumulation of the cdk inhibitor $p21^{Cip1}$ [37]. For example, in some cell types, high levels of Ras or Raf activities induce $p21^{Cip1}$ expression and cell cycle arrest [38–40], and this effect can be rescued by ectopic expression of activated Rho [37]. Furthermore, mouse embryo fibroblasts lacking $p21^{Cip1}$ do not require Rho for Ras-induced S-phase entry [37]. Rho can also be involved in the regulation of late-G1 progression by activating the cyclin E/cdk2 complex, which in turn promotes the degradation of the cdk inhibitor $p27^{Kip1}$ (Figure 50.1) [41]. Together, the effects of Rho on the levels of cdk inhibitors are likely to contribute to the ability of cells to undergo G1 to S progression in response to proliferative signals.

FUTURE PERSPECTIVES

Although, as outlined in this chapter, the involvement of Rac and Rho in regulating cell cycle progression is supported by many lines of evidence, the biochemical mechanisms that couple the signaling activities of these GTPases and the cell cycle machinery remain to be established. By virtue of their effects on the actin cytoskeleton, Rac and Rho play a key role in the regulation of cell shape changes that accompany adhesion and motility. It is well recognized that cell shape is an important determinant for the proliferative capacity of normal anchorage-dependent cells and loss of cell-shape-dependent growth control is a hallmark of oncogenically transformed cells. Thus, understanding the molecular basis for the involvement of Rac and Rho in cell cycle regulation should provide insights into the mechanisms by which alterations in cellular morphology can be sensed and converted to a growth response.

ACKNOWLEDGEMENT

This work was supported by a grant from the National Institutes of Health (CA55360).

REFERENCES

1. Treisman R. The SRE: a growth factor responsive transcriptional regulator. *Semin Cancer Biol* 1990;**1**:47–58.
2. Obaya AJ, Sedivy JM. Regulation of cyclin-Cdk activity in mammalian cells. *Cell Mol Life Sci* 2002;**59**:126–42.
3. Nevins JR. The Rb/E2F pathway and cancer. *Hum Mol Genet* 2001;**10**:699–703.
4. Harbour JW, Dean DC. The Rb/E2F pathway: expanding roles and emerging paradigms. *Genes Dev* 2000;**14**:2393–409.
5. Adams PD. Regulation of the retinoblastoma tumor suppressor protein by cyclin/cdks. *Biochim Biophys Acta* 2001;**1471**:M123–33.
6. Harbour JW, Luo RX, Dei Santi A, Postigo AA, Dean DC. Cdk phosphorylation triggers sequential intramolecular interactions that progressively block Rb functions as cells move through G1. *Cell* 1999;**98**:859–69.
7. Zhang HS, Gavin M, Dahiya A, Postigo AA, Ma D, Luo RX, Harbour JW, Dean DC. Exit from G1 and S phase of the cell cycle is regulated by repressor complexes containing HDAC-Rb-hSWI/SNF and Rb-hSWI/SNF. *Cell* 2000;**101**:79–89.
8. Gartel AL, Tyner AL. Transcriptional regulation of the p21(WAF1/CIP1) gene. *Exp Cell Res* 1999;**246**:280–9.
9. Macleod KF, Sherry N, Hannon G, Beach D, Tokino T, Kinzler K, Vogelstein B, Jacks T. p53-dependent and independent expression of p21 during cell growth, differentiation, and DNA damage. *Genes Dev* 1995;**9**:935–44.
10. Kolluri SK, Weiss C, Koff A, Gottlicher M. p27 (Kip1) induction and inhibition of proliferation by the intracellular Ah receptor in developing thymus and hepatoma cells. *Genes Dev* 1999;**13**:1742–53.
11. Hengst L, Reed SI. Translational control of p27Kip1 accumulation during the cell cycle. *Science* 1996;**271**:1861–4.
12. Millard SS, Yan JS, Nguyen H, Pagano M, Kiyokawa H, Koff A. Enhanced ribosomal association of p27(Kip1) mRNA is a mechanism contributing to accumulation during growth arrest. *J Biol Chem* 1997;**272**:7093–8.
13. Pagano M, Tam SW, Theodoras AM, Beer-Romero P, Del Sal G, Chau V, Yew PR, Draetta GF, Rolfe M. Role of the ubiquitin-proteasome pathway in regulating abundance of the cyclin-dependent kinase inhibitor p27. *Science* 1995;**269**:682–5.
14. Nguyen H, Gitig DM, Koff A. Cell-free degradation of p27(kip1), a G1 cyclin-dependent kinase inhibitor, is dependent on CDK2 activity and the proteasome. *Mol Cell Biol* 1999;**19**:1190–201.
15. Sheaff RJ, Groudine M, Gordon M, Roberts JM, Clurman BE. Cyclin E-CDK2 is a regulator of p27Kip1. *Genes Dev* 1997;**11**:1464–78.

16. Vlach J, Hennecke S, Amati B. Phosphorylation-dependent degradation of the cyclin-dependent kinase inhibitor p27. *EMBO J* 1997; **16**:5334–44.
17. Van Aelst L, D'Souza-Schorey C. Rho GTPases and signaling networks. *Genes Dev* 1997;**11**:2295–322.
18. Qiu RG, Chen J, Kirn D, McCormick F, Symons M. An essential role for Rac in Ras transformation. *Nature* 1995;**374**:457–9.
19. Qiu RG, Chen J, McCormick F, Symons M. A role for Rho in Ras transformation. *Proc Natl Acad Sci USA* 1995;**92**:11,781–11,785.
20. Olson MF, Ashworth A, Hall A. An essential role for Rho, Rac, and Cdc42 GTPases in cell cycle progression through G1. *Science* 1995;**269**:1270–2.
21. Joneson T, White MA, Wigler MH, Bar-Sagi D. Stimulation of membrane ruffling and MAP kinase activation by distinct effectors of RAS. *Science* 1996;**271**:810–12.
22. Joneson T, Bar-Sagi D. A Rac1 effector site controlling mitogenesis through superoxide production. *J Biol Chem* 1998;**273**:17,991–17,994.
23. Freeman JL, Abo A, Lambeth JD. Rac "insert region" is a novel effector region that is implicated in the activation of NADPH oxidase, but not PAK65. *J Biol Chem* 1996;**271**:19,794–19,801.
24. Irani K, Xia Y, Zweier JL, Sollott SJ, Der CJ, Fearon ER, Sundaresan M, Finkel T, Goldschmidt-Clermont PJ. Mitogenic signaling mediated by oxidants in Ras-transformed fibroblasts. *Science* 1997;**275**:1649–52.
25. Sahai E, Ishizaki T, Narumiya S, Treisman R. Transformation mediated by RhoA requires activity of ROCK kinases. *Curr Biol* 1999;**9**:136–45.
26. Zong H, Kaibuchi K, Quilliam LA. The insert region of RhoA is essential for Rho kinase activation and cellular transformation. *Mol Cell Biol* 2001;**21**:5287–98.
27. Frost JA, Steen H, Shapiro P, Lewis T, Ahn N, Shaw PE, Cobb MH. Cross-cascade activation of ERKs and ternary complex factors by Rho family proteins. *EMBO J* 1997;**16**:6426–38.
28. Sun H, King AJ, Diaz HB, Marshall MS. Regulation of the protein kinase Raf-1 by oncogenic Ras through phosphatidylinositol 3-kinase, Cdc42/Rac and Pak. *Curr Biol* 2000;**10**:281–4.
29. Chaudhary A, King WG, Mattaliano MD, Frost JA, Diaz B, Morrison DK, Cobb MH, Marshall MS, Brugge JS. Phosphatidylinositol 3-kinase regulates Raf1 through Pak phosphorylation of serine 338. *Curr Biol* 2000;**10**:551–4.
30. Hill CS, Wynne J, Treisman R. The Rho family GTPases RhoA, Rac1, and CDC42Hs regulate transcriptional activation by SRF. *Cell* 1995;**81**:1159–70.
31. Westwick JK, Lambert QT, Clark GJ, Symons M, Van Aelst L, Pestell RG, Der CJ. Rac regulation of transformation, gene expression, and actin organization by multiple, PAK-independent pathways. *Mol Cell Biol* 1997;**17**:1324–35.
32. Sotiropoulos A, Gineitis D, Copeland J, Treisman R. Signal-regulated activation of serum response factor is mediated by changes in actin dynamics. *Cell* 1999;**98**:159–69.
33. Tominaga T, Sahai E, Chardin P, McCormick F, Courtneidge SA, Alberts AS. Diaphanous-related formins bridge Rho GTPase and Src tyrosine kinase signaling. *Mol Cell* 2000;**5**:13–25.
34. Geneste O, Copeland JW, Treisman R. LIM kinase and Diaphanous cooperate to regulate serum response factor and actin dynamics. *J Cell Biol* 2002;**157**:831–8.
35. Joyce D, Bouzahzah B, Fu M, Albanese C, D'Amico M, Steer J, Klein JU, Lee RJ, Segall JE, Westwick JK, Der CJ, Pestell RG. Integration of Rac-dependent regulation of cyclin D1 transcription through a nuclear factor-kappaB-dependent pathway. *J Biol Chem* 1999;**274**:25,245–25,249.
36. Welsh CF, Roovers K, Villanueva J, Liu Y, Schwartz MA, Assoian RK. Timing of cyclin D1 expression within G1 phase is controlled by Rho. *Nature Cell Biol* 2001;**3**:950–7.
37. Olson MF, Paterson HF, Marshall CJ. Signals from Ras and Rho GTPases interact to regulate expression of p21Waf1/ Cip1. *Nature* 1998;**394**:295–9.
38. Lloyd AC, Obermuller F, Staddon S, Barth CF, McMahon M, Land H. Cooperating oncogenes converge to regulate cyclin/cdk complexes. *Genes Dev* 1997;**11**:663–77.
39. Pumiglia KM, Decker SJ. Cell cycle arrest mediated by the MEK/ mitogen-activated protein kinase pathway. *Proc Natl Acad Sci USA* 1997;**94**:448–52.
40. Sewing A, Wiseman B, Lloyd AC, Land H. High-intensity Raf signal causes cell cycle arrest mediated by p21Cip1. *Mol Cell Biol* 1997;**17**:5588–97.
41. Hu W, Bellone CJ, Baldassare JJ. RhoA stimulates p27(Kip) degradation through its regulation of cyclin E/CDK2 activity. *J Biol Chem* 1999;**274**:3396–401.

Chapter 51

The Role of Alternative Splicing During the Cell Cycle and Programmed Cell Death

Xialu Li[1] and James L. Manley[2]

[1]*National Institute of Biological Sciences, Beijing, China*

[2]*Department of Biological Sciences, Columbia University, New York*

INTRODUCTION

In metazoan organisms, most transcripts synthesized by RNA polymerase (RNAP) II contain non-coding intervening sequences, introns, that must be accurately and efficiently removed by the process of pre-mRNA splicing to form translatable mRNAs. Removing introns in different combinations, a phenomenon termed as alternative splicing, produces diverse mature mRNAs encoding structurally and functionally distinct protein isoforms from a single gene. Alternative splicing is widely involved in gene expression in higher eukaryotes, especially in vertebrates. Genome-wide analyses revealed that expression of more than 60 percent of human genes involves alternative splicing and more than 70 percent of these alternative splicing events alter the resulting protein products [1, 2]. These findings enhance the view that the process of alternative splicing not only constitutes an important mechanism to increase the complexity of the expressed proteome, but also provides an important extra layer to the control of gene regulation in higher eukaryotic cells.

Pre-mRNA splicing takes place within a large molecular complex, the spliceosome, that is composed of five small nuclear RNA (U1, U2, U4, U5, and U6) and a large number of protein factors [3]. Regulation of alternative splicing largely relies on a broad spectrum of interactions between sequence elements in the mRNA precursor and a complex repertoire of protein factors [4]. A growing body of evidence has shown that shifting the balance between alternatively spliced isoforms of a given pre-mRNA is important in modulating many biological processes. In this chapter, we summarize some of the progress made in the past years in understanding the regulation and the influence of splicing on two important cellular processes: programmed cell death (apoptosis) and cell cycle control.

APOPTOSIS AND SPLICING

Apoptosis, a process that removes deleterious or useless cells during animal development, is one of many cellular processes in which alternative splicing plays important regulatory roles [5–7]. This idea was originally suggested by the fact that a remarkable number of genes involved in apoptotic pathways utilize alternative splicing to generate mRNA isoforms encoding proteins with distinct or even opposite functions (see an excellent review in [6]). Furthermore, alterations in components of the splicing machinery, such as changes in phosphorylation of members of the serine/arginine-rich (SR) protein family, which has the potential to influence alternative splicing [8], have been observed upon apoptotic stimulation [9, 10]. Additionally, *in vitro* splicing and overexpression experiments have indeed indicated potential roles of splicing factors in modulating alternative splicing of pre-mRNAs encoding apoptotic regulators [11]. However, whether regulated splicing events can influence the progression of apoptosis *in vivo*, and whether and how individual splicing factor controls the process, are largely unclear.

Oligonucleosomal DNA fragmentation, which occurs in response to various apoptotic stimuli in a wide variety of cell types, is one of the hallmarks of apoptotic cell death [12, 13]. Regulated splicing has been shown to modulate apoptotic DNA fragmentation *in vivo* [14, 15]. It has been well documented that DNA fragmentation factor (DFF), a heterodimeric protein complex consisting of the caspase activated DNase (CAD) and inhibitor of caspase activated DNase (ICAD), is largely responsible for internucleosomal DNA fragmentation during apoptosis [16]. Genetic and biochemical studies showed that ICAD not only acts as an inhibitor of CAD, but is also required for the proper folding and nuclear accumulation of a catalytically competent CAD

Handbook of Cell Signaling, Three-Volume Set 2 ed.

[16, 17]. Interestingly, alternative splicing of intron 5 of *ICAD* pre-mRNA generates two mRNA isoforms, *Icad-s* and *Icad-l*, encoding proteins that exhibit similar inhibitory activity, but only one, ICAD-L, is capable of providing the positive chaperone effects [16, 17]. By employing a genetically engineered chicken DT-40 cell line, it has been shown that depletion of the SR protein ASF/SF2 resulted in a threefold increase at the ratio of ICAD mRNA small to large isoforms [14]. Remarkably, such a relatively modest shift in ICAD pre-mRNA splicing was found to be sufficient to block apoptotic DNA laddering *in vivo*. Importantly, the block can be partially recovered by stable expression of an appropriate level of exogenous ICAD-L. This study provides evidence illustrating that a subtle shift in pre-mRNA splicing can have a profound effect on fine-tuning a specific step in the programmed cell death pathway under physiological conditions.

The above described studies show that simply reducing the levels of a single SR protein can influence a specific step in apoptosis by modulating regulated splicing. In this view, a question of interest is whether alteration in concentrations and/or activity of splicing factors naturally occurs upon induction of apoptosis. As the activity of a number of splicing factors, including SR proteins, can be regulated by reversible phosphorylation, it is noteworthy that not only the phosphorylation status of SR proteins but also the activity of certain SR protein kinases as well as phosphatases vary during apoptosis induced by diverse stimuli. For example, SRPK1, a well characterized SR protein kinase [18], is initially activated but subsequently inactivated by caspase cleavage in response to multiple apoptotic stimuli, and this occurs concomitantly with a change in the phosphorylation status of SR proteins [19]. SR proteins have also been reported to be rapidly dephosphorylated through activation of protein phosphatase 1 (PP1) when apoptosis is induced by Fas ligand or exogenous ceramide in Jurkat cells [20]. More interestingly, alternative splicing of transcripts of two important apoptotic genes, *caspase-9* and *bcl-x*, is modulated in a PP1 dependent mechanism upon *de novo* ceramide generation [20]. Further studies identified two ceramide responsive RNA cis-elements (CRCEs) involved in regulating splice site selection of *bcl-x* pre-mRNA [20]. Splicing factor SAP155, a subunit of the U2 snRNP associated SF3b complex [21], was demonstrated to bind to the purine-rich CRCE-1 via the activation of PP1 in response to ceramide [20]. SAP155 is phosphorylated during splicing catalysis [22]. It is of note that PP1 is associated with SAP155 and regulates its phosphorylation status [23]. Together, these studies point to a possible link between stimulation of apoptosis, splicing factor activation, and alternative splicing during the apoptotic response.

But how mechanistically do splicing factors modulate alternative splicing of apoptotic transcripts? A well characterized example is provided by regulation of *caspase-2* (*casp2*) pre-mRNA splicing. Alternative inclusion/exclusion of a 61 bp exon leads to the formation of two *casp-2* mRNA isoforms encoding proteins with antagonistic activities in apoptosis: skipping of exon 9 generates proapoptotic Casp2L, whereas including of exon 9 produces antiapoptotic Casp2S [24]. By employing a murine caspase-2 minigene, an evolutionary conserved 100 nt intronic splicing regulatory element (In100) downstream of exon 9 was identified [25]. The In100 element contains a decoy 3' splice site (SS) upstream of a binding site for the well studied negative splicing regulator polypyrimidine tract binding protein (PTB) [26]. Binding of PTB to In100 promotes the non-productive interaction between the decoy 3'SS and the 5'SS of exon 9 and this facilitates the exon-skipping splicing event. Interestingly, In100-like elements are found in several human caspase pre-mRNAs near exons encoding the caspase active domain, and alternative splicing indeed generates isoforms predicted to affect their functional activity [27]. How, and whether, In100-like elements function in general to regulate the level of active caspases will be important questions for further study.

CELL CYCLE AND SPLICING

The cell cycle is a collection of highly ordered processes that lead to the duplication of a cell. A potential link between splicing and cell cycle progression was initially suggested by a number of genetic studies in yeast in which a group of genes was identified in independent screens for splicing factors and cell cycle regulators [28, 29]. Although the exact mechanisms underlying most of these cases are still unclear, impaired splicing of transcripts encoding proteins required for cell cycle transition is responsible for the cell cycle arrest phenotype observed in some mutants. For example, mutations in the *cdc40/prp17* gene, which encodes a splicing factor that functions in the second step of splicing [30], caused defects in G1/S and G2/M cell cycle transition [30–32]. Further studies indicated that inefficient splicing of the *anc1* transcript contributed to the cell cycle arrest phenotype observed in *cdc40/prp17* mutants [33]. CDC40/PRP17 is indeed required for the splicing of the *anc1* pre-mRNA, which is attributed to the unique sequence feature in the branchpoint and 3' splice site of the *anc1* intron [33]. ANC1 has been found to be associated with several protein complexes involving in transcription and chromatin remodeling [34–36]. Efficiency of *anc1* pre-mRNA splicing might influence the expression of ANC1 and thus the proper transcription of genes critical for cell cycle progression. Likewise, a G2/M arrest caused by mutations in the splicing factor CDC5/Cef1 can be suppressed by removing the intron from a single target gene *tub1*, encoding a-tubulin [37]. While the detailed mechanism is still unclear, *tub1* mRNA and a-tubulin protein levels decreased in *cdc5/cef1* mutant cells [37]. These findings indicate that specific splicing factors are required for cell cycle progression by modulating splicing of transcripts encoding cell cycle regulators. However, since true regulated alternative splicing is rare in yeast, it is unclear how these findings relate to the situation in metazoan organisms.

In mammals, a number of genes generate mRNA isoforms encoding proteins with different effects on cell cycle progression by alternative splicing. For example, the gene encoding Cyclin D1, the regulatory subunit of cyclin dependent kinases CDK4 and CDK6, generates two mRNA isoforms by alternative splicing. The proteins encoded by these alternatively spliced mRNAs differ in the last 55 amino acids at the carboxy-terminus, which is required for protein stability and subcellular localization. Ectopic expression studies demonstrated that the shorter isoform has no cell cycle regulatory properties, in contrast to the necessary role of the longer isoform on G1 to S phase transition [38].

Another example involves alternative splicing of the pre-mRNA encoding members of the p53 protein family. p53, together with two related genes p63 and p73, composes a family of transcription factors required for the control of cell cycle progress in response to stress and DNA damage [39, 40]. Interestingly, alternative splicing has been demonstrated to be a general mechanism for producing diverse isoforms of these proteins [41]. For example, intron 9 of human p53 can be alternatively spliced to produce three diverse C-terminal structures, where two of them, β and γ, lack the protein domain necessary for tetramerization [42]. As oligomerization facilitates binding of p53 to its target DNAs, it can be envisioned that alternative splicing contributes to the regulation of p53 transcription activity. Consistent with this idea, a recent study indeed indicated different accessibility of full length p53 and the p53β isoform to p53 responsive promoters [42]. Like p53, human p63 has at least three alternatively spliced C-terminal isoforms, whereas the human p73 gene expresses at least seven alternatively spliced C-terminal isoforms and at least four alternatively spliced N-terminal isoforms, which contain diverse variants of the transactivation domain [41]. Much remains to be learned concerning the different properties of these proteins, and how alternative splicing of their pre-mRNAs is regulated.

Several studies suggest that the roles of splicing factors in modulating the progress of the cell cycle extend beyond their function in splicing. A genome-wide RNA interference screen in HeLa cells for factors required for mitosis identified a number of genes encoding components of the splicing machinery, including SNRPA1, SNRPB, LSM6, SART1, DHX8, DDX5, and SNW1 [43]. SNRPA1, SNRPB, and LSM6 are Sm or like-Sm proteins, which are the important structural components of the splicing snRNPs [44–46]. SART1 has been reported to be essential for the recruitment of the U4/U6/U5 tri-snRNP to the prespliceosome in the spliceosome assembly pathway [47]. DHX8 and DDX5 are two DEAD box proteins with putative RNA helicase activity [48, 49]. SNW1 appears to play dual roles on transcription elongation and splicing [50, 51]. Cells treated with siRNAs specifically targeting these genes display impaired cell division phenotypes with spindle defects, resembling the phenotype observed upon depletion of TPX2, a protein required for chromosome induced microtubule assembly during spindle formation, by RNAi [52]. Interestingly, TPX2 was found to co-purify with the spliceosome [47], raising the possibility that splicing factors influence cell division by direct interactions with the spindle assembly machinery. Consistent with this possibility, a recent study in *Drosophila* showed that depletion of a group of spliceosome components that are predicted to function at various stages in the spliceosome assembly cycle consistently led to G2/M cell cycle arrest, indicative of a general correlation between impaired spliceosome activity and cell cycle transition [53]. But in both cases the detailed mechanisms remain largely unknown.

The influence of splicing on cell cycle progression likely involves additional complexities. Unexpectedly, a number of studies have revealed that RNA processing factors, including splicing factors, play important roles in the maintenance of genome stability, by preventing nascent transcripts from annealing to the template DNA strand during transcription. Genetic inactivation of a number of genes involved in RNA processing in yeast and at least one SR protein, ASF/SF2, in vertebrate cells resulted in the accumulation of RNA–DNA hybrids, or R loops, which in turn led to DNA damage and rearrangements [54, 55]. In the case of ASF/SF2 depletion, an increase in the cell population at G2-M phase occurred concomitantly with the appearance of transcriptional R loops and global double-strand DNA breaks [14, 54]. As delay in the cell cycle progress is a general response to DNA damage in eukaryotic cells, these studies reveal another mechanism underlying the role of splicing, or at least splicing factors, in regulating cell cycle progression. How widespread this mechanism is will be an interesting question for future research.

CONCLUSIONS

The split nature of gene makes splicing essential for gene expression in higher eukaryotic cells. The studies summarized here demonstrate that the efficiency, accuracy, and regulation of splicing are of paramount importance in the modulation of critical biological processes such as apoptosis and cell cycle control. However, in contrast to the large number of alternative splicing events that have been identified, our understanding of the physiological functions and molecular mechanism of splicing regulation are still limited. A challenge for the future will be to elucidate details of the mechanisms underlying the cell signaling pathways that regulate specific splicing events. Advances in global approaches to identify regulated splicing events, coupled with functional studies including analysis of protein – RNA and protein – protein interactions, will be helpful in solving these problems.

REFERENCES

1. Johnson JM, Castle J, Garrett-Engele P, Kan Z, Loerch PM, Armour CD, Santos R, Schadt EE, Stoughton R, Shoemaker DD. Genome-wide survey of human alternative pre-mRNA splicing with exon junction microarrays. *Science* 2003;**302**(5653):2141–4.
2. Brett D, Pospisil H, Valcarcel J, Reich J, Bork P. Alternative splicing and genome complexity. *Nat Genet* 2002;**30**(1):29–30.
3. Jurica MS, Moore MJ. Pre-mRNA splicing: awash in a sea of proteins. *Mol Cell* 2003;**12**(1):5–14.
4. Black DL. Mechanisms of alternative pre-messenger RNA splicing. *Annu Rev Biochem* 2003;**72**:291–336.
5. Adams JM. Ways of dying: multiple pathways to apoptosis. *Genes Dev* 2003;**17**(20):2481–95.
6. Schwerk C, Schulze-Osthoff K. Regulation of apoptosis by alternative pre-mRNA splicing. *Mol Cell* 2005;**19**(1):1–13.
7. Wu JY, Tang H, Havlioglu N. Alternative pre-mRNA splicing and regulation of programmed cell death. *Prog Mol Subcell Biol* 2003;**31**:153–85.
8. Xiao SH, Manley JL. Phosphorylation-dephosphorylation differentially affects activities of splicing factor ASF/SF2. *Embo J* 1998;**17**(21):6359–67.
9. Thiede B, Dimmler C, Siejak F, Rudel T. Predominant identification of RNA-binding proteins in Fas-induced apoptosis by proteome analysis. *J Biol Chem* 2001;**276**(28):26,044–26,050.
10. Utz PJ, Hottelet M, van Venrooij WJ, Anderson P. Association of phosphorylated serine/arginine (SR) splicing factors with the U1-small ribonucleoprotein (snRNP) autoantigen complex accompanies apoptotic cell death. *J Exp Med* 1998;**187**(4):547–60.
11. Jiang ZH, Zhang WJ, Rao Y, Wu JY. Regulation of Ich-1 pre-mRNA alternative splicing and apoptosis by mammalian splicing factors. *Proc Natl Acad Sci USA* 1998;**95**(16):9155–60.
12. Wyllie AH. Glucocorticoid-induced thymocyte apoptosis is associated with endogenous endonuclease activation. *Nature* 1980;**284**(5756):555–6.
13. Earnshaw WC. Nuclear changes in apoptosis. *Curr Opin Cell Biol* 1995;**7**(3):337–43.
14. Li X, Wang J, Manley JL. Loss of splicing factor ASF/SF2 induces G2 cell cycle arrest and apoptosis, but inhibits internucleosomal DNA fragmentation. *Genes Dev* 2005;**19**(22):2705–14.
15. Li X, Manley JL. Alternative splicing and control of apoptotic DNA fragmentation. *Cell Cycle* 2006;**5**(12):1286–8.
16. Sakahira H, Enari M, Ohsawa Y, Uchiyama Y, Nagata S. Apoptotic nuclear morphological change without DNA fragmentation. *Curr Biol* 1999;**9**(10):543–6.
17. Scholz SR, Korn C, Gimadutdinow O, Knoblauch M, Pingoud A, Meiss G. The effect of ICAD-S on the formation and intracellular distribution of a nucleolytically active caspase-activated DNase. *Nucleic Acids Res* 2002;**30**(14):3045–51.
18. Gui JF, Tronchere H, Chandler SD, Fu XD. Purification and characterization of a kinase specific for the serine- and arginine-rich pre-mRNA splicing factors. *Proc Natl Acad Sci USA* 1994;**91**(23):10,824–10,828.
19. Kamachi M, Le TM, Kim SJ, Geiger ME, Anderson P, Utz PJ. Human autoimmune sera as molecular probes for the identification of an autoantigen kinase signaling pathway. *J Exp Med* 2002;**196**(9):1213–25.
20. Massiello A, Roesser JR, Chalfant CE. SAP155 Binds to ceramide-responsive RNA cis-element 1 and regulates the alternative 5' splice site selection of Bcl-x pre-mRNA. *Faseb J* 2006;**20**(10):1680–2.
21. Das BK, Xia L, Palandjian L, Gozani O, Chyung Y, Reed R. Characterization of a protein complex containing spliceosomal proteins SAPs 49, 130, 145, and 155. *Mol Cell Biol* 1999;**19**(10):6796–802.
22. Wang C, Chua K, Seghezzi W, Lees E, Gozani O, Reed R. Phosphorylation of spliceosomal protein SAP 155 coupled with splicing catalysis. *Genes Dev* 1998;**12**(10):1409–14.
23. Boudrez A, Beullens M, Waelkens E, Stalmans W, Bollen M. Phosphorylation-dependent interaction between the splicing factors SAP155 and NIPP1. *J Biol Chem* 2002;**277**(35):31,834–31,841.
24. Wang L, Miura M, Bergeron L, Zhu H, Yuan J. Ich-1, an Ice/ced-3-related gene, encodes both positive and negative regulators of programmed cell death. *Cell* 1994;**78**(5):739–50.
25. Cote J, Dupuis S, Jiang Z, Wu JY. Caspase-2 pre-mRNA alternative splicing: Identification of an intronic element containing a decoy 3' acceptor site. *Proc Natl Acad Sci USA* 2001;**98**(3):938–43.
26. Cote J, Dupuis S, Wu JY. Polypyrimidine track-binding protein binding downstream of caspase-2 alternative exon 9 represses its inclusion. *J Biol Chem* 2001;**276**(11):8535–43.
27. Havlioglu N, Wang J, Fushimi K, Vibranovski MD, Kan Z, Gish W, Fedorov A, Long M, Wu JY. An intronic signal for alternative splicing in the human genome. *PLoS ONE* 2007;**2**(11):e1246.
28. Ben-Yehuda S, Dix I, Russell CS, McGarvey M, Beggs JD, Kupiec M. Genetic and physical interactions between factors involved in both cell cycle progression and pre-mRNA splicing in Saccharomyces cerevisiae. *Genetics* 2000;**156**(4):1503–17.
29. Russell CS, Ben-Yehuda S, Dix I, Kupiec M, Beggs JD. Functional analyses of interacting factors involved in both pre-mRNA splicing and cell cycle progression in Saccharomyces cerevisiae. *Rna* 2000;**6**(11):1565–72.
30. Vijayraghavan U, Company M, Abelson J. Isolation and characterization of pre-mRNA splicing mutants of Saccharomyces cerevisiae. *Genes Dev* 1989;**3**(8):1206–16.
31. Boger-Nadjar E, Vaisman N, Ben-Yehuda S, Kassir Y, Kupiec M. Efficient initiation of S-phase in yeast requires Cdc40p, a protein involved in pre-mRNA splicing. *Mol Gen Genet* 1998;**260**(2–3):232–41.
32. Chawla G, Sapra AK, Surana U, Vijayraghavan U. Dependence of pre-mRNA introns on PRP17, a non-essential splicing factor: implications for efficient progression through cell cycle transitions. *Nucleic Acids Res* 2003;**31**(9):2333–43.
33. Dahan O, Kupiec M. The Saccharomyces cerevisiae gene CDC40/PRP17 controls cell cycle progression through splicing of the ANC1 gene. *Nucleic Acids Res* 2004;**32**(8):2529–40.
34. Henry NL, Campbell AM, Feaver WJ, Poon D, Weil PA, Kornberg RD. TFIIF-TAF-RNA polymerase II connection. *Genes Dev* 1994;**8**(23):2868–78.
35. John S, Howe L, Tafrov ST, Grant PA, Sternglanz R, Workman JL. The something about silencing protein, Sas3, is the catalytic subunit of NuA3, a yTAF(II)30-containing HAT complex that interacts with the Spt16 subunit of the yeast CP (Cdc68/Pob3)-FACT complex. *Genes Dev* 2000;**14**(10):1196–208.
36. Poon D, Bai Y, Campbell AM, Bjorklund S, Kim YJ, Zhou S, Kornberg RD, Weil PA. Identification and characterization of a TFIID-like multiprotein complex from Saccharomyces cerevisiae. *Proc Natl Acad Sci U S A* 1995;**92**(18):8224–8.
37. Burns Jr. CG, Ohi R, Mehta S, O'Toole ET, Winey M, Clark TA, Sugnet CW, Ares M, Gould KL Removal of a single alpha-tubulin gene intron suppresses cell cycle arrest phenotypes of splicing factor mutations in Saccharomyces cerevisiae. *Mol Cell Biol* 2002;**22**(3):801–15.

38. Leveque C, Marsaud V, Renoir JM, Sola B. Alternative cyclin D1 forms a and b have different biological functions in the cell cycle of B lymphocytes. *Exp Cell Res* 2007;**313**(12):2719–29.
39. Levrero M, De Laurenzi V, Costanzo A, Gong J, Wang JY, Melino G. The p53/p63/p73 family of transcription factors: overlapping and distinct functions. *J Cell Sci* 2000;**113**(Pt 10):1661–70.
40. Schwartz D, Rotter V. p53-dependent cell cycle control: response to genotoxic stress. *Semin Cancer Biol* 1998;**8**(5):325–36.
41. Murray-Zmijewski F, Lane DP, Bourdon JC. p53/p63/p73 isoforms: an orchestra of isoforms to harmonise cell differentiation and response to stress. *Cell Death Differ* 2006;**13**(6):962–72.
42. Bourdon JC, Fernandes K, Murray-Zmijewski F, Liu G, Diot A, Xirodimas DP, Saville MK, Lane DP. p53 isoforms can regulate p53 transcriptional activity. *Genes Dev* 2005;**19**(18):2122–37.
43. Kittler R, Putz G, Pelletier L, Poser I, Heninger AK, Drechsel D, Fischer S, Konstantinova I, Habermann B, Grabner H, Yaspo ML, Himmelbauer H, Korn B, Neugebauer K, Pisabarro MT, Buchholz F. An endoribonuclease-prepared siRNA screen in human cells identifies genes essential for cell division. *Nature* 2004;**432**(7020):1036–40.
44. Achsel T, Brahms H, Kastner B, Bachi A, Wilm M, Luhrmann R. A doughnut-shaped heteromer of human Sm-like proteins binds to the 3'-end of U6 snRNA, thereby facilitating U4/U6 duplex formation in vitro. *Embo J* 1999;**18**(20):5789–802.
45. van Dam A, Winkel I, Zijlstra-Baalbergen J, Smeenk R, Cuypers HT. Cloned human snRNP proteins B and B' differ only in their carboxy-terminal part. *Embo J* 1989;**8**(12):3853–60.
46. Sillekens PT, Beijer RP, Habets WJ, van Verooij WJ. Molecular cloning of the cDNA for the human U2 snRNA-specific A' protein. *Nucleic Acids Res* 1989;**17**(5):1893–906.
47. Makarova OV, Makarov EM, Luhrmann R. The 65 and 110{ts}kDa SR-related proteins of the U4/U6.U5 tri-snRNP are essential for the assembly of mature spliceosomes. *Embo J* 2001;**20**(10):2553–63.
48. Ono Y, Ohno M, Shimura Y. Identification of a putative RNA helicase (HRH1), a human homolog of yeast Prp22. *Mol Cell Biol* 1994;**14**(11):7611–20.
49. Liu ZR. p68 RNA helicase is an essential human splicing factor that acts at the U1 snRNA-5' splice site duplex. *Mol Cell Biol* 2002;**22**(15):5443–50.
50. Bres V, Gomes N, Pickle L, Jones KA. A human splicing factor, SKIP, associates with P-TEFb and enhances transcription elongation by HIV-1 Tat. *Genes Dev* 2005;**19**(10):1211–26.
51. Figueroa JD, Hayman MJ. The human Ski-interacting protein functionally substitutes for the yeast PRP45 gene. *Biochem Biophys Res Commun* 2004;**319**(4):1105–9.
52. Gruss OJ, Wittmann M, Yokoyama H, Pepperkok R, Kufer T, Sillje H, Karsenti E, Mattaj IW, Vernos I. Chromosome-induced microtubule assembly mediated by TPX2 is required for spindle formation in HeLa cells. *Nat Cell Biol* 2002;**4**(11):871–9.
53. Andersen DS, Tapon N. Drosophila MFAP1 is required for pre-mRNA processing and G2/M progression. *J Biol Chem* 2008;**283**(45):31,256–31,267.
54. Li X, Manley JL. Inactivation of the SR protein splicing factor ASF/SF2 results in genomic instability. *Cell* 2005;**122**(3):365–78.
55. Huertas P, Aguilera A. Cotranscriptionally formed DNA : RNA hybrids mediate transcription elongation impairment and transcription-associated recombination. *Mol Cell* 2003;**12**(3):711–21.

Chapter 52

Cell-Cycle Functions and Regulation of Cdc14 Phosphatases

Harry Charbonneau
Department of Biochemistry, Purdue University, West Lafayette, Indiana

INTRODUCTION

The CDC14 gene of the budding yeast *Saccharomyces cerevisiae* encodes a protein phosphatase that is essential for cell-cycle progression [1] and serves as a prototype for a group of closely related enzymes within the protein tyrosine phosphatase (PTP) family. Orthologs of yeast Cdc14 have been identified in protists, fungi, flowering plants, and animals, suggesting that this phosphatase, like many other cell-cycle regulators, is conserved among all eukaryotes. Cdc14 from budding yeast is the founding member of this subgroup of protein phosphatases and has been most thoroughly studied.

THE CDC14 PHOSPHATASE SUBGROUP OF PTPs

The Cdc14 phosphatases [1, 2] utilize the Cys-dependent catalytic mechanism shared by all PTPs, but outside of a short segment surrounding their active sites they exhibit no sequence similarity to the classical tyrosine-specific enzymes of this family. Cdc14 phosphatases dephosphorylate Ser/Thr as well as Tyr residues in artificial substrates *in vitro* [1, 2], placing them among the dual-specificity phosphatases (DSPs), a distinct subgroup of the PTP family. The Cdc14 orthologs have little in common with other DSPs, many of which regulate MAP kinases. Cdc14 orthologs and these MAP kinase phosphatases differ in their domain organization, and the only sequence similarity is restricted to a 60-residue region flanking their active sites.

The basic structural organization of the prototypical budding yeast Cdc14 is shared by all orthologs identified to date (Figure 52.1). The 62-kDa yeast enzyme contains a conserved N-terminal catalytic domain (residues 1–374) and an Asn/Ser-rich, non-catalytic C-terminal segment that is not essential for its cell-cycle function [1]. The oligomerization of budding yeast Cdc14, observed both *in vitro* and *in vivo*, is mediated through an interaction requiring the catalytic domain [1, 3]. A non-catalytic domain is present at the C termini of all Cdc14 orthologs, but it varies in length and has diverged during speciation (Figure 52.1). Apart from a nuclear export sequence identified in human Cdc14A [4], no other functions have been assigned to the non-catalytic domain.

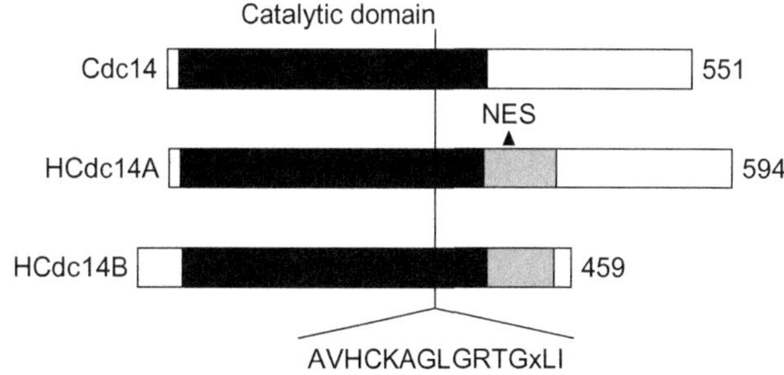

FIGURE 52.1 Schematic diagram illustrating the structural organization of budding yeast and human Cdc14 phosphatases. The yeast and human Cdc14A and B phosphatase sequences are depicted (accession numbers NP_116684, NP_003663, and NP_003662, respectively) with the total number of amino acid residues shown on the right. The solid black boxes delineate the position of the catalytic domain (≈330 residues) conserved among all Cdc14 orthologs, whereas the open boxes show divergent non-catalytic regions. The gray boxes depict additional sequences conserved only among the human enzymes and several other vertebrate orthologs. The vertical line denotes the position of the catalytic site; the active site sequence that is identical among all Cdc14 phosphatases is shown underneath (x indicates a variable position). The position of the nuclear export signal (NES) identified in human Cdc14A [4] is indicated by the triangle.

Handbook of Cell Signaling, Three-Volume Set 2 ed.

BUDDING YEAST CDC14 IS ESSENTIAL FOR EXIT FROM MITOSIS

Exit from Mitosis

Following their association with B-type cyclins, the activation of cyclin-dependent kinases (Cdk) triggers the onset of mitosis. At anaphase after sister chromatids have separated, mitotic Cdks must be inactivated in order for cells to exit from mitosis. During exit from mitosis, cells restore the nucleus to its premitotic state (e.g., disassemble the mitotic spindle) and prepare for cytokinesis (for review, see Morgan [5]). A prevailing mechanism for mitotic Cdk inactivation is the regulated destruction of mitotic cyclins.

The anaphase-promoting complex (APC) ubiquitinates cyclins and other mitotic regulators, triggering their recognition and proteolysis by the 26S proteosome [5]. In budding yeast, specificity factors known as Cdc20 and Cdh1/Hct1 interact with the APC to govern substrate selectivity and the order in which crucial regulators are ubiquitinated and destroyed during mitosis [5]. APC^{Cdc20} acts first to initiate anaphase by ubiquitinating the yeast securin Pds1. Upon its destruction, Pds1 liberates a protease necessary for sister chromatid separation. Subsequently, Cdh1 promotes the APC-mediated ubiquitination of mitotic cyclins and other targets that are destroyed during exit from mitosis. Cdh1 is expressed throughout the cell cycle but Cdk-mediated phosphorylation prevents its interaction with the APC during early mitosis [5].

Cdc14 Substrates

In budding yeast, Cdc14 dephosphorylates at least three substrates (Cdh1, Swi5, and Sic1) [6, 7] that ensure the inactivation of mitotic Cdk activity through two pathways: degradation of mitotic cyclins and expression of Sic1, a Cdk inhibitor (see Figure 52.2). Upon its dephosphorylation by Cdc14, Cdh1 activates the APC and directs the ubiquitination of mitotic cyclins and other protein targets [7]. Expression of the Cdk inhibitor Sic1 is dependent on the zinc finger transcription factor, Swi5. Prior to anaphase, Swi5 accumulates in the cytoplasm but is prevented from entering the nucleus because of Cdk-dependent phosphorylation at Ser residues adjacent to its nuclear localization signal (Figure 52.2). Cdc14 dephosphorylates Swi5, thus permitting it to enter the nucleus and activate Sic1 transcription [6]. Cdc14 also targets the Sic1 protein itself, preventing its destruction as a result of inopportune phosphorylation [6]. Cdh1, Swi5, and Sic1 undergo Cdk-dependent phosphorylation, and it is generally assumed that Cdc14 phosphatases prefer substrates phosphorylated by this group of kinases. Considerable evidence supports this notion, but it is premature to assume that Cdc14 opposes only Cdks as no sites dephosphorylated by this phosphatase, either *in vitro* or *in vivo*, have been directly mapped, and the substrate preference of Cdc14 has not yet been investigated.

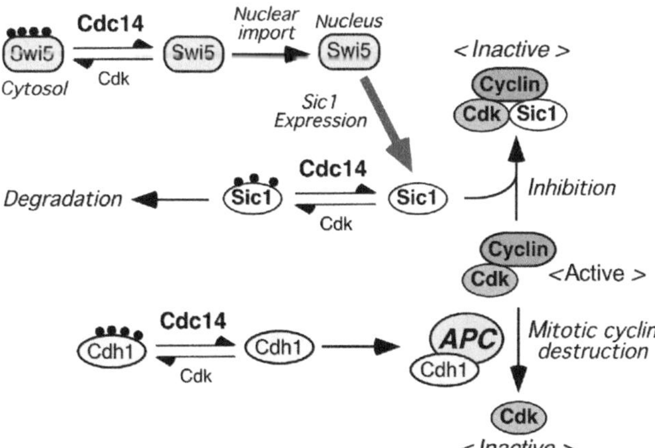

FIGURE 52.2 The role of budding yeast Cdc14 in promoting exit from mitosis.
The schematic diagram illustrates how Cdc14 dephosphorylates Swi5, Sic1, and Cdh1 to drive Cdk1 inactivation by two mechanisms: APC-mediated cyclin destruction and protein inhibition [6].

The Nucleolus and Cdc14 Regulation

Genetic and biochemical studies have begun to reveal how Cdc14, which is present at constant levels throughout the cell cycle, is held in check until its activity is required between anaphase and early G_1. Net1 (also known as Cfi1) [8, 9], a major player in the cell-cycle-dependent regulation of Cdc14, is a core subunit of the nucleolar RENT complex [8]. The RENT complex is also involved in maintenance of nucleolar integrity, repression of recombination among tandem rDNA repeats, recruitment of Pol I, and stimulation of rDNA transcription [10]. In interphase and early mitosis, most if not all Cdc14 is sequestered in the nucleolus by Net1 [8, 9], where its activity is fully inhibited [11] and its access to substrate is limited. Net1 is a highly specific and potent competitive inhibitor (K_i=3 nM) that contains a Cdc14-binding region (residues 1–341) at its N terminus [11].

Two distinct signaling pathways, known as the FEAR [12] and MEN [13] networks, control Cdc14 release from Net1 (Figure 52.3). For both pathways, it is not known how the protein kinases and other signaling components act on the RENT complex to induce the release of Cdc14, but it may involve phosphorylation of Net1 [14]. The FEAR pathway (Figure 52.3) is activated first at early anaphase when the securin Pds1 is degraded and active separase is released [12]. FEAR signaling triggers a transient release of Cdc14 into the nucleus that is not sufficient for exit from mitosis but ensures it occurs with proper timing [12]. The MEN pathway (Figure 52.3) is activated at late anaphase when the dividing nucleus spans the bud neck bringing the Tem1 G-protein into contact with its guanine nucleotide exchange factor Lte1 [15, 16]. Activation of MEN

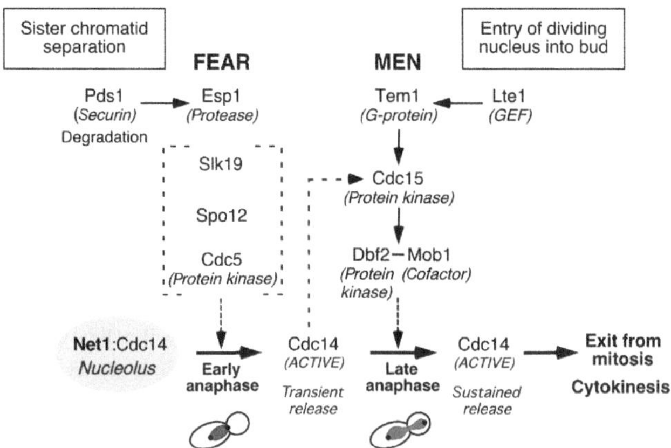

FIGURE 52.3 A model for the cell-cycle-dependent regulation of Cdc14 by the FEAR and MEN networks.
The signaling proteins involved in the FEAR [12] and MEN [26] pathways are depicted. Arrows are shown where the order of signaling within the pathways is known. The components (Slk19, Spo12, and Cdc5) enclosed in large brackets are necessary for FEAR signaling, but it is not clear how they interact or in what order they operate in the network. The dashed arrow depicts the potential role of Cdc14 in potentiating MEN signaling by targeting Cdc15 [12]. Although Cdc5 is thought to act in both pathways, it is shown here only in the FEAR network. As indicated by the dashed arrow, the exact mechanism triggering Cdc14 release from Net1 is not known for either pathway.

signaling is essential for exit from mitosis and produces a sustained release of Cdc14 into the nucleus and cytoplasm that promotes Cdk inactivation. Cdc14 released by FEAR signaling may act to potentiate subsequent signaling through the MEN pathway [12] by dephosphorylating the Cdc15 kinase that is known to be a Cdc14 substrate [17, 18]. It is likely that the FEAR network has additional roles during early anaphase. The dependency of Cdc14 release on the proteolysis of Pds1 [19–21] is not explained by the FEAR pathway alone, indicating there must be at least one other mechanism linking the two events. The FEAR and MEN pathways in conjunction with the requirement for Pds1 degradation ensure that mitotic exit does not occur unless sister chromatids are separated and the segregated chromosomes are correctly partitioned to mother and daughter cells (Figure 52.3). Like several other proteins of the MEN pathway, the role of Cdc14 may not be limited to mitotic exit but could also include functions required for cytokinesis [22].

FISSION YEAST CDC14 COORDINATES CYTOKINESIS WITH MITOSIS

An ortholog of *S. cerevisiae* Cdc14, named clp1 [23] or flp1 [24], has been identified in the fission yeast Schizosaccharomyces pombe. The role of fission yeast Cdc14 in cell-cycle progression differs considerably from that of the budding yeast enzyme. *S. pombe* Cdc14 is not an essential phosphatase and is not necessary for mitotic cyclin degradation or exit from mitosis [23, 24]. This is not completely surprising, as the fission yeast Cdc20 ortholog instead of Cdh1 appears to control the APC-dependent destruction of mitotic cyclins. Instead of exit from mitosis, *S. pombe* Cdc14 is involved in controlling the onset of mitosis [23, 24]. Through an undefined mechanism, Cdc14 suppresses Cdk activation at the G_2/M transition by opposing Tyr 15 dephosphorylation, a requirement for full mitotic kinase activity.

Recent analyses suggest that *S. pombe* Cdc14 is also involved in coordinating cytokinesis with the events of late mitosis [23]. In contrast to budding yeast, *S. pombe* divides by medial fission instead of budding [25]. During mitosis, *S. pombe* first assembles a medial ring containing actomyosin and then forms a septum at the middle of the cell. At the end of anaphase, a signaling pathway initiates septation, contraction of the medial actomyosin ring, and completion of cytokinesis [25]. Interestingly, most of the components of this signaling pathway, known as the septation initiation network (SIN), are orthologs of the MEN pathway of budding yeast, and the two pathways are thought to have the same organization and to propagate signals via similar mechanisms [25, 26]. Surprisingly, Cdc14 is not a major effector or target of the SIN pathway [24]. Instead, Cdc14 appears to potentiate the SIN pathway by suppressing Cdk activity that is known to antagonize SIN signaling and cytokinesis.

Like its budding yeast counterpart, *S. pombe* Cdc14 is localized to the nucleolus during interphase [23, 24]. Upon its release at early mitosis, Cdc14 diffuses throughout the nucleus and cytoplasm and accumulates at the spindle pole bodies, mitotic spindle, and medial ring [23, 24]. Fission yeasts have no homolog of budding yeast Net1 and it is not known whether Cdc14 is active within the nucleolus, but its sequestration could restrict access to substrates. The SIN network does not trigger Cdc14 release; instead, it is required to exclude the phosphatase from the nucleolus until cytokinesis is complete [23]. How Cdc14 is initially released is unknown. *S. pombe* Cdc14 is phosphorylated during mitosis, but how this modification might regulate the enzyme is not known [24]. Identification of substrates will be required to define how *S. pombe* Cdc14 modulates the G_2/M transition and coordinates cytokinesis with mitosis.

POTENTIAL CELL-CYCLE FUNCTIONS OF HUMAN CDC14A AND B

Two distinct Cdc14 phosphatases are expressed in humans [2] and several other vertebrates. Human Cdc14A and B exhibit 62 percent sequence identity over a 400-residue segment. Evidence suggests that Cdc14A is involved in regulating cell division, but so far there are few clues about the function of the B form.

Although many details differ, regulation of the APC during vertebrate mitosis is fundamentally the same as that observed in yeast. A Cdh1 ortholog must be dephosphorylated to direct the APC-dependent ubiquitination of mitotic cyclins that results in Cdk inactivation and exit from mitosis. A recent study [27] showed that human Cdc14A dephosphorylates Cdh1 *in vitro*, allowing it to activate APC-mediated cyclin ubiquitination. Moreover, human Cdc14A is found in a major fraction of Cdh1 phosphatase activity isolated from HeLa cell lysates [27]. Although this study [27] using *in vitro* reactions is not definitive, it provides evidence that human Cdc14A has the capacity to regulate the APC and to promote exit from mitosis. Thus, the function of budding yeast Cdc14 in promoting mitotic exit may have been conserved in humans.

Besides Cdh1, the only other potential substrate identified for human Cdc14 phosphatases is the tumor suppressor p53 [28]. Cdc14A and B associate with p53 *in vivo* and both dephosphorylate Ser 315 *in vitro* [28]. Ser 315 is targeted by Cdks, consistent with the notion that Cdc14 phosphatases oppose these kinases. Its binding to sequences in the N termini of the Cdc14 phosphatases [28] suggests that the interaction with p53 may be independent of its recognition as a phosphosubstrate and could permit the constitutive association of the two proteins. Thus far, evidence that Cdc14 controls the phosphorylation state of Ser 315 in cells is lacking, and there are conflicting reports regarding the role of this site in p53 regulation.

Several observations suggest that the regulation of human Cdc14A and B may differ from that observed in budding yeast. Both human phosphatases are insensitive to the yeast Net1 inhibitor, and no gene encoding a Net1 homolog can be identified in the human genome [11]. Targeting to specific organelles or subcellular compartments is at least partly responsible for human Cdc14 regulation. The majority of Cdc14A is localized to the centrosome, but some enzyme is also found in the cytosol [4]. During mitosis, most but not all of the Cdc14A leaves the centrosome and appears in the cytosol. A nuclear export signal (residues 352–367) (Figure 52.1) is necessary for the translocation of Cdc14A out of the nucleus and to prevent its sequestration in nucleoli, where Cdc14B is localized [4]. The nuclear export signal as well as N- and C-terminal sequences appear to be required for localization to the centrosome [4].

Recent findings have implicated human Cdc14A in centrosome duplication [4]. Like chromosomes, centrosomes must be duplicated exactly once in every round of cell division, and defects in this process lead to aberrant chromosome segregation and aneuploidy [29]. Overexpression or depletion of Cdc14A in human cells resulted in defective chromosome segregation that could be attributed to aberrations in the centrosome duplication cycle [4]. These data are fully consistent with the well-documented role of phosphorylation in regulating centrosome duplication. It will be important to identify substrates in order to define the role of Cdc14A in centrosome duplication. In this regard, it is intriguing that the potential Cdc14 substrate p53 has been linked to centrosome function [30, 31]. Cells lacking p53 accumulate multiple centrosomes, suggesting that they have defects in the duplication cycle [30]. The phosphorylation of Ser 315 is required for the binding of p53 to unduplicated centrosomes [31]. The possibility that Cdc14A could modulate centrosome duplication by controlling the phosphorylation state of Ser 315 in p53 certainly merits further study. Research on the human Cdc14 phosphatases is in its infancy; nevertheless, the clues we have obtained highlight the importance of investigating potential links between this group of enzymes and tumorigenesis.

REFERENCES

1. Taylor GS, Liu Y, Baskerville C, Charbonneau H. The activity of Cdc14p, an oligomeric dual specificity protein phosphatase from *Saccharomyces cerevisiae*, is required for cell cycle progression. *J Biol Chem* 1997;**272**:24,054–63.
2. Li L, Ernsting BR, Wishart MJ, Lohse DL, Dixon JE. A family of putative tumor suppressors is structurally and functionally conserved in humans and yeast. *J Biol Chem* 1997;**272**:29,403–6.
3. Grandin N, de Almeida A, Charbonneau M. The Cdc14 phosphatase is functionally associated with the Dbf2 protein kinase in *Saccharomyces cerevisiae*. *Mol Gen Genet* 1998;**258**:104–16.
4. Mailand N, Lukas C, Kaiser BK, Jackson PK, Bartek J, Lukas J. Deregulated human Cdc14A phosphatase disrupts centrosome separation and chromosome segregation. *Nat Cell Biol* 2002;**4**:318–22.
5. Morgan DO. Regulation of the APC and the exit from mitosis. *Nat Cell Biol* 1999;**1**:E47–53.
6. Visintin R, Craig K, Hwang ES, Prinz S, Tyers M, Amon A. The phosphatase Cdc14 triggers mitotic exit by reversal of Cdk-dependent phosphorylation. *Mol Cell* 1998;**2**:709–18.
7. Jaspersen SL, Charles JF, Morgan DO. Inhibitory phosphorylation of the APC regulator Hct1 is controlled by the kinase Cdc28 and the phosphatase Cdc14. *Curr Biol* 1999;**9**:227–36.
8. Shou W, Seol JH, Shevchenko A, Baskerville C, Moazed D, Chen ZW, Jang J, Charbonneau H, Deshaies RJ. Exit from mitosis is triggered by Tem1-dependent release of the protein phosphatase Cdc14 from nucleolar RENT complex. *Cell* 1999;**97**:233–44.
9. Visintin R, Hwang ES, Amon A. Cfi1 prevents premature exit from mitosis by anchoring Cdc14 phosphatase in the nucleolus. *Nature* 1999;**398**:818–23.
10. Shou W, Sakamoto KM, Keener J, Morimoto KW, Traverso EE, Azzam R, Hoppe GJ, Feldman RM, DeModena J, Moazed D, Charbonneau H, Nomura M, Deshaies RJ. Net1 stimulates RNA polymerase I transcription and regulates nucleolar structure independently of controlling mitotic exit. *Mol Cell* 2001;**8**:45–55.
11. Traverso EE, Baskerville C, Liu Y, Shou W, James P, Deshaies RJ, Charbonneau H. Characterization of the Net1 cell cycle-dependent regulator of the Cdc14 phosphatase from budding yeast. *J Biol Chem* 2001;**276**:21,924–31.
12. Stegmeier F, Visintin R, Amon A. Separase, polo kinase, the kinetochore protein Slk19, and Spo12 function in a network that controls Cdc14 localization during early anaphase. *Cell* 2002;**108**:207–20.

13. Jaspersen SL, Charles JF, Tinker-Kulberg RL, Morgan DO. A late mitotic regulatory network controlling cyclin destruction in *Saccharomyces cerevisiae*. *Mol Biol Cell* 1998;**9**:2803–17.
14. Shou W, Azzam R, Chen S, Huddleston M, Baskerville C, Charbonneau H, Annan R, Carr S, Deshaies R. Cdc5 influences phosphorylation of Net1 and disassembly of the RENT complex. *BMC Mol Biol* 2002;**3**:3.
15. Bardin AJ, Visintin R, Amon A. A mechanism for coupling exit from mitosis to partitioning of the nucleus. *Cell* 2000;**102**:21–31.
16. Pereira G, Hofken T, Grindlay J, Manson C, Schiebel E. The Bub2p spindle checkpoint links nuclear migration with mitotic exit. *Mol Cell* 2000;**6**:1–10.
17. Xu S, Huang HK, Kaiser P, Latterich M, Hunter T. Phosphorylation and spindle pole body localization of the Cdc15p mitotic regulatory protein kinase in budding yeast. *Curr Biol* 2000;**10**:329–32.
18. Jaspersen SL, Morgan DO. Cdc14 activates Cdc15 to promote mitotic exit in budding yeast. *Curr Biol* 2000;**10**:615–18.
19. Tinker-Kulberg RL, Morgan DO. Pds1 and Esp1 control both anaphase and mitotic exit in normal cells and after DNA damage. *Genes Dev* 1999;**13**:1936–49.
20. Cohen-Fix O, Koshland D. Pds1p of budding yeast has dual roles: inhibition of anaphase initiation and regulation of mitotic exit. *Genes Dev* 1999;**13**:1950–9.
21. Shirayama M, Toth A, Galova M, Nasmyth K. APC(Cdc20) promotes exit from mitosis by destroying the anaphase inhibitor Pds1 and cyclin Clb5. *Nature* 1999;**402**:203–7.
22. Tolliday N, Bouquin N, Li R. Assembly and regulation of the cytokinetic apparatus in budding yeast. *Curr Opin Microbiol* 2001;**4**:690–5.
23. Trautmann S, Wolfe BA, Jorgensen P, Tyers M, Gould KL, McCollum D. Fission yeast Clp1p phosphatase regulates G_2/M transition and coordination of cytokinesis with cell cycle progression. *Curr Biol* 2001;**11**:931–40.
24. Cueille N, Salimova E, Esteban V, Blanco M, Moreno S, Bueno A, Simanis V. Flp1, a fission yeast orthologue of the *S. cerevisiae* CDC14 gene, is not required for cyclin degradation or rum1p stabilisation at the end of mitosis. *J Cell Sci* 2001;**114**:2649–64.
25. McCollum D, Gould KL. Timing is everything: regulation of mitotic exit and cytokinesis by the MEN and SIN. *Trends Cell Biol* 2001;**11**: 89–95.
26. Bardin AJ, Amon A. Men and sin: what's the difference? *Nat Rev Mol Cell Biol* 2001;**2**:815–26.
27. Bembenek J, Yu H. Regulation of the anaphase-promoting complex by the dual specificity phosphatase human Cdc14a. *J Biol Chem* 2001;**276**:48,237–42.
28. Li L, Ljungman M, Dixon JE. The human Cdc14 phosphatases interact with and dephosphorylate the tumor suppressor protein p53. *J Biol Chem* 2000;**275**:2410–14.
29. Doxsey SJ. Centrosomes as command centres for cellular control. *Nat Cell Biol* 2001;**3**:E105–8.
30. Fukasawa K, Choi T, Kuriyama R, Rulong S, Vande Woude GF. Abnormal centrosome amplification in the absence of p53. *Science* 1996;**271**:1744–7.
31. Tarapore P, Tokuyama Y, Horn HF, Fukasawa K. Difference in the centrosome duplication regulatory activity among p53 "hot spot" mutants: potential role of Ser 315 phosphorylation-dependent centrosome binding of p53. *Oncogene* 2001;**20**:6851–63.

Chapter 53

Caspases: Cell Signaling by Proteolysis

Guy S. Salvesen

Program in Apoptosis and Cell Death Research, Burnham Institute, San Diego, California

PROTEASE SIGNALING

Proteolytic enzymes, including the cell's degrading machine the proteasome, calpains, and integral membrane proteases such as γ-secretase and rhomboid, participate in several intracellular signaling processes (Table 53.1). However, given that the human genome contains in excess of 500 genes that encode proteases, it seems odd that only the caspases constitute a formal multi-step pathway able to transmit intracellular signals by proteolysis. In contrast, extracellular multi-step signaling pathways frequently use the principle of proteolysis extensively for coagulation, fibrinolysis, complement activation in mammals, and gastrulation in flies. What is it about the caspases that makes them so suitable to transmit intracellular signals?

The consensus view of caspases places them in two main camps. First are the cytokine activators related to caspase 1, probably including mouse caspase 11 and its close homologs caspases 4 and 5 in humans. Their main role is to respond to bacterial infection by rapidly converting active cytokines (IL-1β, IL-18) from intracellular stores. Confirmation of the important roles of the caspases in the inflammatory cytokine response comes from gene ablation experiments in mice. Animals ablated in caspase 1 or 11 are deficient in cytokine processing [1, 2], but without any overt apoptotic phenotype. The second camp constitutes the apoptotic caspases that transduce and execute death signals. The phenotypes of these knockouts are very gross, evidently anti-apoptotic, and vary from early embryonic lethality (caspase 8) to perinatal lethality (caspases 3 and 9) [3–5], to relatively mild with defects in the process of normal oocyte ablation [6]. Techniques in biochemistry and cell biology have allowed us to place the apoptotic caspases in two converging pathways, such that some are activated by others (Figure 53.1). This core pathway probably represents a minimal apoptotic program, and certainly their simplicity is complicated by cell-specific additions that help to fine tune individual cell fates. Nevertheless, the basic order and at least some of the essential functions and, importantly, endogenous regulators of the caspases are now known.

TABLE 53.1 Proteases involved in intracellular signaling

Protease	Signaling function
Caspases	Apoptosis Pro-inflammatory cytokine activation
Proteasome	Cell cycle progression NFκB activation
Rhomboid	EGF signaling
γ-Secretase	Toll receptor signaling
SREB Site2 Protease	Upregulation of sterol synthesis genes
Separase	Anaphase
Calpains	Various signaling events.

While not an exhaustive survey, the table highlights the principle of proteolysis as a mechanism of signal transmission.

APOPTOSIS AND LIMITED PROTEOLYSIS

Apoptosis is a mechanism to regulate cell number, and is vital throughout the life of all metazoan animals. Though several different types of biochemical events have been recognized as important in apoptosis, perhaps the most fundamental is the participation of the caspases [7–9].

The name "caspase" is a contraction of *c*ysteine-dependent *asp*artate specific prote*ase* [10], thus their enzymatic properties are governed by a dominant specificity for protein substrates containing Asp, and by the use of a Cys side-chain for catalyzing peptide bond cleavage. The use of a Cys side-chain as a nucleophile during peptide bond hydrolysis

Handbook of Cell Signaling, Three-Volume Set 2 ed.

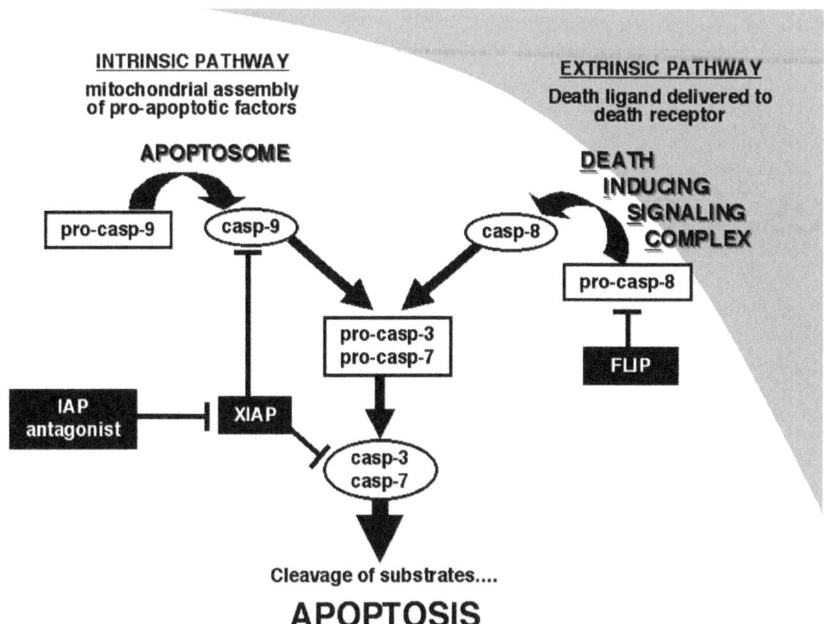

FIGURE 53.1 The framework of apoptosis.
Death may be signaled by direct ligand enforced clustering of receptors at the cell surface, which leads to the activation of initiator caspases 8 [49]. This caspase then directly activates the executioner caspases 3 and 7 (and possibly 6), which are predominantly responsible for the limited proteolysis that characterizes apoptotic dismantling of the cell. Alternatively, irreparable damage to the genome caused by mutagens, pharmaceuticals that inhibit DNA repair, or ionizing radiation–transmitted by a mechanism thought to involve the release of cytochrome c from mitochondria–engages the same executioner caspases [50]. The latter events progress through the initiator caspase 9 and its co-factor Apaf-1 [22]. Activation of the extrinsic pathway is regulated by FLIP, which serves to modulate the recruitment of caspase 8 to its adaptors [51]. The common execution phase is regulated through direct caspase inhibition by IAPs, some of which can also regulate the active form of caspase 9. In turn, the IAPs are under the influence of antagonist proteins that compete with caspases for IAPs [39]. Though other modulators may regulate the apoptotic pathway in a cell-specific manner, this framework is considered common to most mammalian cells.

is common to several protease families. However, the primary specificity for Asp turns out to be very rare among proteases throughout biotic kingdoms. Of all known mammalian proteases, only the caspase activator granzyme B, a serine protease, has the same primary specificity [11, 12]. Caspases cleave a number of cellular proteins [13], and the process is one of limited proteolysis where a small number of cuts, usually only one, are made. Sometimes cleavage results in activation of the protein, sometimes in inactivation [14], but never in degradation, since their substrate specificity distinguishes the caspases as among the most restricted of endopeptidases. This is an important distinction from the proteasome, which permits signaling by wholesale destruction of regulatory proteins such as IκB in NFκB signaling and PDS1 in anaphase promotion (see Chapter 255 of Handbook of Cell Signaling, Second Edition).

The most primitive organism with a *bona fide* caspase appears to be *Caenhorabditis elegans*. Indeed, the first apoptotic caspase, Ced3, was identified in this organism, galvanizing the apoptotic research field [15]. Initially it was thought that as the complexity of primitive cell death pathways developed, so apparently did the number of caspases. *Drosophila* have seven caspases [16] and humans have 11. However, this simple picture needs to be revised in light of the presence of multiple caspases in organisms more primitive than *C. elegans*, so that *C. elegans* and *Drosophila* may have lost geners encoding the more complex apoptotic network of primitive animals [17]. Mapping the inherent substrate specificity of caspases has allowed some broad consensuses to be recognized [18]. These consensuses also allow apoptotic caspases to be distinguished from pro-inflammatory caspases, since the latter have a rather distinct specificity that presumably allows them to carry out their job without threatening cell viability. Interestingly, there seems to have been a parallel evolution of apoptotic caspases along with their substrates. Consensus caspase targets in humans such as nuclear lamins and poly (ADP) ribose polymerase have easily recognizable caspase cleavage sites in *Drosophila*, but apparently not in organisms such as yeast and plants, which lack an apoptotic pathway.

CASPASE ACTIVATION

Induced Proximity

The seminal discovery that death receptor signaling required, in its most basic form, simply a transmembrane receptor, an adaptor molecule, and a caspase [19, 20] revealed a solution to the perplexing problem of how the first proteolytic signal was generated during apoptosis, since it implicated a caspase directly in the triggering event. Prior to this work, receptors were thought to signal by either altering the phosphorylation status of key signaling molecules, or by functioning as ion channels. Death receptors such as Fas signal by direct recruitment and activation of a protease (caspase 8). Concomitantly with this work, groundbreaking studies showed that the intrinsic pathway was activated by a co-factor known as apoptosis protease activating factor (Apaf-1) [21]. Subsequently, Apaf-1 was found to recruit and activate caspase 9, forming the "apoptosome" [22].

In common with most proteolytic enzymes, caspases reside as latent forms that are usually activated by limited proteolysis. It is relatively easy to imagine that the caspases operating at the bottom of the pathway are activated by ones above. Until recently, the question of how the first caspase in a pathway became activated, how the first death signal was generated, was a perplexing issue. Now several groups have focused on this issue, and a consensus has been arrived at to describe the intriguing operation of the initiation of the proteolytic pathways that execute apoptosis (reviewed in [23]). This mechanism is known as the "induced proximity hypothesis", and although it originally referred to death induced by caspase 8 (the extrinsic pathway), it has now been extended to death induced by caspase 9 (the intrinsic pathway)–see Figure 53.1. Notably, this mechanism does not apply to the executioner caspases.

How exactly does a recruited zymogen become active? To understand this, as a basis for formulating an adequate hypothesis, the unusual properties of caspase zymogens that set them apart from most other proteases must be understood. For, unlike most other proteases, simple expression of caspase zymogens in *E. coli* usually results in their activation by limited proteolysis within a "linker segment" that separated the large (~20-kDa) and small (~10-kDa) subunits of the catalytic domain. [24, 25]. This activation results from processing that is a consequence of intrinsic proteolytic activity residing in the caspase zymogens. It is not due to *E. coli* proteases, since catalytically disabled C285A (caspase 1 numbering convention) mutants fail to undergo processing.

Initiator Caspases

At the cytosolic concentration in human cells (<50 nM), pro-caspase 9 is a monomer [26] and requires dimerization within the apoptosome to become active–reviewed in [27]. Significantly, unlike the executioner caspases 3 and 7, pro-caspase 9 does not need to be cleaved in the linker region to become active [28, 29]. Not only is cleavage unnecessary, but also it is insufficient to produce an active enzyme. Instead, caspase 9 is activated by small-scale rearrangements of surface loops that define the substrate cleft and catalytic residues [26]. In the simplest model, this is achieved by dimerization of caspase 9 monomers within the apoptosome [30], with the dimer interface providing surfaces compatible with catalytic organization of the active site. A similar dimerization mechanism activates the caspase 8 zymogen to trigger the extrinsic pathway [31, 32].

Executioner Caspases

Once an in initiator caspase has become active, ensuing activation of the executioners is more straightforwardly explained. At cytosolic concentration in human cells, the caspase-3 and -7 zymogens are already dimers, but cleavage within their respective linker segments is required for activation [33, 34]. The same re-ordering of catalytic and substrate binding residues occurs in caspase 7 as seen in caspase 9, so the fundamental mechanism of zymogen activation is equivalent. Only the driving forces are distinct, since the linker segment of pro-caspase 7 blocks ordering of the active site, and upon cleavage the new N- and C-terminal sequences so generated aid in active site stabilization. The property that allows the distinct driving forces to converge on the same activation mechanism seems to be the unusual plasticity of the residues constituting the caspase active site, which, rather unusually for proteases, are predominantly placed on flexible loops and not ordered secondary structure.

REGULATION BY INHIBITORS

The first level of regulating proteolytic pathways is by zymogen activation, but an equally important level is achieved by the use of specific inhibitors that can govern the activity of the active components. The endogenous inhibitors of caspases, those present in mammalian cells, are members of the inhibitor of apoptosis (IAP) family. In addition to these endogenous regulators are the virally encoded cowpox virus CrmA and baculovirus p35 proteins that are produced early in infection to suppress caspase-mediated host responses. Each of the inhibitors has a characteristic specificity profile against human caspases, as determined *in vitro*, and these profiles, with few caveats [35], agree with the biologic function of the inhibitors (reviewed in [36, 37]). Though IAPs and CrmA would be expected to regulate mammalian caspases *in vivo*, p35 would never be present normally in mammals because it is expressed naturally by baculoviruses.

The best-characterized endogenous caspase inhibitor is the X-linked IAP (XIAP), a member of the IAP family. The

IAPs are broadly distributed and, as their name indicates, the founding members are capable of selectively blocking apoptosis, having initially been identified in baculoviruses (reviewed in [38]). Eight distinct IAPs have been identified in humans. XIAP (which is the human family paradigm) has been found by multiple research groups to be a potent but restricted inhibitor targeting caspases 3, 7, and 9 (reviewed in [39]. Despite earlier claims, other members of the IAP family probably do not directly inhibit caspases (reviewed in [40]), and have functions in addition to caspase inhibition because they have been found in organisms such as yeast, which neither contain caspases nor undergo apoptosis [41].

IAPs contain one, two, or three baculovirus IAP repeat (BIR) domains, which represent the defining characteristic of the family. The first BIR domain (BIR1) binds to TRAF2 and regulates interactions with TNF receptor complexes, [42, 43], the second BIR domain (BIR2) of XIAP specifically target caspases 3 and 7 (K_i~0.1–1 nM), and the third BIR domain (BIR3) specifically target caspase 9 (K_i~10 nM). This led to the general assumption that the BIR domain itself was important for caspase inhibition. Surprisingly, structures of BIR2 in complex with caspases 3 and 7, and BIR3 in complex with caspase 9, revealed the BIR domain to have almost no direct role in the inhibitory mechanism. Many of the important inhibitory contacts are made by the flexible region preceding the BIR domain [44–47]. Interestingly, the mechanism of inhibition of caspase 9 by the BIR3 domain requires cleavage in the inter subunit linker to generate the new sequence NH_2-ATPF [29]. In part this explains the cleavage of caspase 9 during apoptosis, which, as described above, is not required for its activation. Paradoxically, it seems required for its inactivation by XIAP.

Significantly, neither CrmA-like nor p35-like inhibitors, which operate by mechanism-based inactivation [36], have been chosen for endogenous caspase regulation; rather, IAPs have been adapted to regulate the executioner caspases. Although the reason for this is not certain, it seems likely that the IAP solution provides a degree of specificity that mechanism-based inhibitors cannot achieve. Thus, XIAP inhibition of caspases 3 and 7 requires a non-standard interaction with the extended 381 loop that is specific to these two caspases (reviewed in [36]). Possibly the 381 loop has evolved to achieve substrate specificity in the executioner caspases [48], but an equally likely possibility is that the 381 loop has been generated to enable the IAP scaffold to provide a unique control level over the execution phase of apoptosis. Adding to this level of sophistication, IAPs, but not CrmA nor p35-like proteins, are subject to negative regulation by IAP antagonists that go by the names of Hid, Grim, Reaper, and Sickle in *Drosophila*, and Smac/Diablo and HtrA2/Omi in mammals (reviewed in [38, 39]) (Figure 53.2).

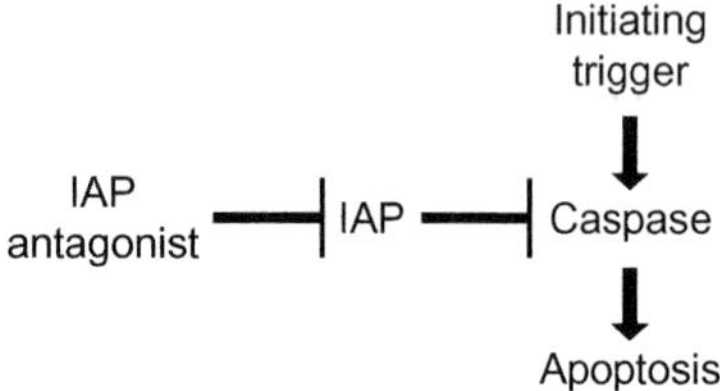

FIGURE 53.2 The basis of IAP antagonists as pro-apoptotic proteins.
Genetic screens in *Drosophila* initially revealed a genomic region encoding three proteins, head involution defective (Hid), Grim, and Reaper, that together seemed responsible for almost all developmental apoptosis in the fly [52]. These proteins, along with the more recently discovered Sickle, constitute the currently known IAP antagonists in flies. In flies there is evidence for a continuous low-level production of caspases, which are neutralized by *Drosophila* IAP-1 [53]. In this scenario simple upregulation of one or more fly IAP antagonists could send the system into apoptosis, and to this extent the system in transcriptionally regulated. Input from the left of the diagram may be the most important event. In contrast to flies, the currently known mammalian IAP antagonists are mitochondrial proteins and require translocation before they can influence the inhibitory activity of IAPs. This implies a more complex role for IAP antagonists in mammals, since both positive initiator caspase signaling and mitochondrial fluxes would presumably be required, and the system could only be transcriptionally regulated in an indirect manner. In mammals, input from the top of the diagram could be more important.

REFERENCES

1. Kuida K, Lippke JA, Ku G, Harding MW, Livingston DJ, Su MSS, Flavell RA. Altered cytokine export and apoptosis in mice deficient in interleukin-1-beta converting enzyme. *Science* 1995;**267**:2000–3.
2. Wang S, Miura M, Jung Y-K, Zhu H, Yuan J. Murine caspase-11, an ICE-interacting protease, is essential for the activation of ICE. *Cell* 1998;**92**:501–9.
3. Kuida K, Haydar TF, Kuan CY, Gu Y, Taya C, Karasuyama H, Su MS, Rakic P, Flavell RA. Reduced apoptosis and cytochrome c-mediated caspase activation in mice lacking caspase 9. *Cell* 1998;**94**:325–37.
4. Kuida K, Zheng TS, Na S, Kuan C-y, Yang D, Karasuyama H, Rakic P, Flavell RA. Decreased apoptosis in the brain and premature lethality in CPP32-deficient mice. *Nature* 1996;**384**:368–72.
5. Varfolomeev EE, Schuchmann M, Luria V, Chiannilkulchai N, Beckmann JS, Mett IL, Rebrikov D, Brodianski VM, Kemper OC, Kollet O, Lapidot T, Soffer D, Sobe T, Avraham KB, Goncharov T, Holtmann H, Lonai P, Wallach D. Targeted disruption of the mouse Caspase 8 gene ablates cell death induction by the TNF receptors, Fas/Apo1, and DR3 and is lethal prenatally. *Immunity* 1998;**9**:267–76.
6. Morita Y, Maravei DV, Bergeron L, Wang S, Perez GI, Tsutsumi O, Taketani Y, Asano M, Horai R, Korsmeyer SJ, Iwakura Y, Yuan J, Tilly JL. Caspase-2 deficiency prevents programmed germ cell death resulting from cytokine insufficiency but not meiotic defects caused by loss of ataxia telangiectasia-mutated (Atm) gene function. *Cell Death Differ* 2001;**8**:614–20.
7. Salvesen GS, Dixit VM. Caspases: intracellular signaling by proteolysis. *Cell* 1997;**91**:443–6.
8. Cohen GM. Caspases: the executioners of apoptosis. *Biochem J* 1997;**326**:1–16.
9. Thornberry NA, Lazebnik Y. Caspases: enemies within. *Science* 1998;**281**:1312–16.

10. Alnemri ES, Livingston DJ, Nicholson DW, Salvesen G, Thornberry NA, Wong WW, Yuan J. Human ICE/CED-3 protease nomenclature. *Cell* 1996;**87**:171.
11. Odake S, Kam CM, Narasimhan L, Poe M, Blake JT, Krahenbuhl O, Tschopp J, Powers JC. Human and murine cytotoxic T lymphocyte serine proteases: subsite mapping with peptide thioester substrates and inhibition of enzyme activity and cytolysis by isocoumarins. *Biochem USA* 1991;**30**:2217–27.
12. Harris JL, Backes BJ, Leonetti F, Mahrus S, Ellman JA, Craik CS. Rapid and general profiling of protease specificity by using combinatorial fluorogenic substrate libraries. *Proc Natl Acad Sci USA* 2000;**97**:7754–9.
13. Nicholson DW. Caspase structure, proteolytic substrates, and function during apoptotic cell death. *Cell Death Differ* 1999;**6**:1028–42.
14. Timmer JC, Salvesen GS. Caspase substrates. *Cell Death Differ* 2007;**14**:66–72.
15. Yuan J, Shaham S, Ledoux S, Ellis HM, Horvitz HM. The *C. elegans* cell death gene ced-3 encodes a protein similar to mammalian interleukin-1β-converting enzyme. *Cell* 1993;**75**:641–52.
16. Kumar S, Doumanis J. The fly caspases. *Cell Death Differ* 2000;**7**:1039–44.
17. Zmasek CM, Zhang Q, Ye Y, Godzik A. Surprising complexity of the ancestral apoptosis network. *Genome Biol* 2007;**8**:R226.
18. Thornberry NA, Rano TA, Peterson EP, Rasper DM, Timkey T, Garcia-Calvo M, Houtzager VM, Nordstrom PA, Roy S, Vaillancourt JP, Chapman KT, Nicholson DW. A combinatorial approach defines specificities of members of the caspase family and granzyme B. Functional relationships established for key mediators of apoptosis. *J Biol Chem* 1997;**272**:17,907–17,911.
19. Boldin MP, Goncharov TM, Goltsev YV, Wallach D. Involvement of MACH, a novel MORT1/FADD-interacting protease, in Fas/APO-1- and TNF receptor-induced cell death. *Cell* 1996;**85**:803–15.
20. Muzio M, Stockwell BR, Stennicke HR, Salvesen GS, Dixit VM. An induced proximity model for caspase-8 activation. *J Biol Chem* 1998;**273**:2926–30.
21. Zou H, Henzel WJ, Liu X, Lutschg A, Wang X. Apaf-1, a human protein homologous to *C. elegans* CED-4, participates in cytochrome c-dependent activation of caspase-3. *Cell* 1997;**90**:405–13.
22. Li P, Nijhawan D, Budihardjo I, Srinivasula SM, Ahmad M, Alnemri ES, Wang X. Cytochrome c and dATP-dependent formation of Apaf-1/caspase-9 complex initiates an apoptotic protease cascade. *Cell* 1997;**91**:479–89.
23. Fuentes-Prior P, Salvesen GS. The protein structures that shape caspase activity, specificity, activation and inhibition. *Biochem J* 2004;**384**:201–32.
24. Orth K, O'Rourke K, Salvesen GS, Dixit VM. Molecular ordering of apoptotic mammalian CED-3/ICE-like proteases. *J Biol Chem* 1996;**271**:20,977–20,980.
25. Stennicke HR, Salvesen GS. Biochemical characteristics of caspases-3, -6, -7, and -8. *J Biol Chem* 1997;**272**:25,719–25,723.
26. Renatus M, Stennicke HR, Scott FL, Liddington RC, Salvesen GS. Dimer formation drives the activation of the cell death protease caspase 9. *Proc Natl Acad Sci USA* 2001;**98**:14,250–14,255.
27. Riedl SJ, Salvesen GS. The apoptosome: signalling platform of cell death. *Nat Rev Mol Cell Biol* 2007;**5**:405–13.
28. Stennicke HR, Deveraux QL, Humke EW, Reed JC, Dixit VM, Salvesen GS. Caspase-9 can be activated without proteolytic processing. *J Biol Chem* 1999;**274**:8359–62.
29. Srinivasula SM, Hegde R, Saleh A, Datta P, Shiozaki E, Chai J, Lee RA, Robbins PD, Fernandes-Alnemri T, Shi Y, Alnemri ES. A conserved XIAP-interaction motif in caspase-9 and Smac/DIABLO regulates caspase activity and apoptosis. *Nature* 2001;**410**:112–16.
30. Pop C, Timmer J, Sperandio S, Salvesen GS. The apoptosome activates caspase-9 by dimerization. *Mol Cell* 2006;**22**:269–75.
31. Donepudi M, Mac Sweeney A, Briand C, Gruetter MG. Insights into the regulatory mechanism for caspase-8 activation. *Mol Cell* 2003;**11**:543–9.
32. Boatright KM, Renatus M, Scott FL, Sperandio S, Shin H, Pedersen I, Ricci J-E, Edris WA, Sutherlin DP, Green DR, Salvesen GS. A unified model for apical caspase activation. *Mol Cell* 2003;**11**:529–41.
33. Riedl SJ, Fuentes-Prior P, Renatus M, Kairies N, Krapp R, Huber R, Salvesen GS, Bode W. Structural basis for the activation of human procaspase-7. *Proc Natl Acad Sci USA* 2001;**98**:14,790–14,795.
34. Chai J, Wu Q, Shiozaki E, Srinivasula SM, Alnemri ES, Shi Y. Crystal structure of a procaspase-7 zymogen. Mechanisms of activation and substrate binding. *Cell* 2001;**107**:399–407.
35. Ryan CA, Stennicke HR, Nava VE, Lewis J, Hardwick JM, Salvesen GS. Inhibitor specificity of recombinant and endogenous caspase 9. *Biochem J* 2002;**366**:595–601.
36. Stennicke HR, Ryan CA, Salvesen GS. Reprieval from execution: the molecular basis of caspase inhibition. *Trends Biochem Sci* 2002;**27**:94–101.
37. Riedl SJ, Shi Y. Molecular mechanisms of caspase regulation during apoptosis. *Nat Rev Mol Cell Biol* 2004;**5**:897–907.
38. Callus BA, Vaux DL. Caspase inhibitors: viral, cellular and chemical. *Cell Death Differ* 2007;**14**:73–8.
39. Salvesen GS, Duckett CS. IAP proteins: blocking the road to death's door. *Nat Rev Mol Cell Biol* 2002;**3**:401–10.
40. Eckelman BP, Salvesen GS, Scott FL. Human inhibitor of apoptosis proteins: why XIAP is the black sheep of the family. *EMBO Rep* 2006;**7**:988–94.
41. Uren AG, Coulson EJ, Vaux DL. Conservation of baculovirus inhibitor of apoptosis repeat proteins (BIRPs) in viruses, nematodes, vertebrates and yeasts. *Trends Biochem Sci* 1998;**23**:159–62.
42. Samuel T, Welsh K, Lober T, Togo SH, Zapata JM, Reed JC. Distinct BIR domains of cIAP1 mediate binding to and ubiquitination of tumor necrosis factor receptor-associated factor 2 and second mitochondrial activator of caspases. *J Biol Chem* 2006;**281**:1080–90.
43. Varfolomeev E, Wayson SM, Dixit VM, Fairbrother WJ, Vucic D. The inhibitor of apoptosis protein fusion c-IAP2.MALT1 stimulates NF-kappaB activation independently of TRAF1 AND TRAF2. *J Biol Chem* 2006;**281**:29,022–29,029.
44. Chai J, Shiozaki E, Srinivasula SM, Wu Q, Dataa P, Alnemri ES, Yigong Shi Y. Structural basis of caspase-7 inhibition by XIAP. *Cell* 2001;**104**:769–80.
45. Huang Y, Park YC, Rich RL, Segal D, Myszka DG, Wu H. Structural basis of caspase inhibition by XIAP: differential roles of the linker versus the BIR domain. *Cell* 2001;**104**:781–90.
46. Riedl SJ, Renatus M, Schwarzenbacher R, Zhou Q, Sun S, Fesik SW, Liddington RC, Salvesen GS. Structural basis for the inhibition of caspase-3 by XIAP. *Cell* 2001;**104**:791–800.
47. Shiozaki EN, Chai J, Rigotti DJ, Riedl SJ, Li P, Srinivasula SM, Alnemri ES, Fairman R, Shi Y. Mechanism of XIAP-mediated inhibition of caspase-9. *Mol Cell* 2003;**11**:519–27.
48. Rotonda J, Nicholson DW, Fazil KM, Gallant M, Gareau Y, Labelle M, Peterson EP, Rasper DM, Tuel R, Vaillancourt JP, Thornberry NA, Becher JW. The three-dimensional structure of apopain/CPP32, a key mediator of apoptosis. *Nat Struct Biol* 1996;**3**:619–25.
49. Ashkenazi A, Dixit VM. Death receptors: signaling and modulation. *Science* 1998;**281**:1305–8.

50. Green DR, Reed JC. Mitochondria and apoptosis. *Science* 1998;**281**:1309–12.
51. Chang DW, Xing Z, Pan Y, Algeciras-Schimnich A, Barnhart BC, Yaish-Ohad S, Peter ME, Yang X. c-FLIP(L) is a dual function regulator for caspase-8 activation and CD95-mediated apoptosis. *EMBO J* 2002;**21**:3704–14.
52. White K, Grether ME, Abrams JM, Young L, Farrell K, Steller H. Genetic control of programmed cell death in *Drosophila. Science* 1994;**264**:677–83.
53. Rodriguez A, Chen P, Oliver H, Abrams JM. Unrestrained caspase-dependent cell death caused by loss of Diap1 function requires the Drosophila Apaf-1 homolog, Dark. *EMBO J* 2002;**21**:2189–97.

Chapter 54

Apoptosis Signaling: A Means to an End

Lisa J. Pagliari, Michael J. Pinkoski and Douglas R. Green
Division of Cellular Immunology, La Jolla Institute for Allergy and Immunology, San Diego, California

INTRODUCTION

Programmed cell death is a fundamental biological process of all multicellular organisms and plays important roles in tissue homeostasis, host defense, development, metamorphosis, and morphogenesis. In animals, programmed cell death occurs via apoptosis, a morphologically defined form of cell death that has a number of biochemical features. In addition to its physiological roles, apoptosis contributes to several pathological conditions, such as cancer, AIDS, aging, and cardiovascular, neurodegenerative, and autoimmune diseases. The mechanisms that govern a cellular decision to live or die are complex and tightly regulated by a plethora of molecules with distinct roles in the signaling process. Generally, cell death occurs after an initial apoptotic signal spurs a cascade of subsequent events from which a cell cannot recover.

Apoptosis is defined by its morphological features of membrane blebbing, cellular shrinkage, and chromosomal condensation. Endonuclease activation that results in a characteristic 200 bp nucleosomal DNA ladder is a common feature of apoptosis, although not definitive [1]. Intense research has indicated that the cellular events leading to apoptosis are complex and varied, often depending on the cell type and stimulus utilized. However, several aspects of these death pathways are common among various stimuli and cell types.

THE END OF THE ROAD

It is the cleavage of key cellular substrates and not a general proteolytic digestion that orchestrates the morphological and biochemical changes that characterize cell death by apoptosis. The degradative phase of apoptosis is mediated by a highly conserved family of cysteine proteases (caspases) that cleave specific proteins, including other caspases, at the C-terminal end of aspartic acid residues [2]. Figure 54.1 classifies caspases 1–10 and depicts several structural motifs. Most, if not all, caspases exist intracellularly as inactive zymogens. Following an apoptotic signal, the inactive zymogen is cleaved between what will be the

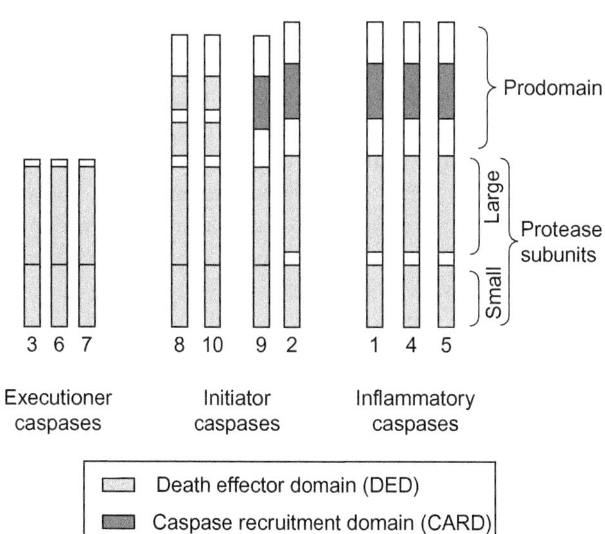

FIGURE 54.1 Caspase classification.
Caspases 1–10 can be divided into three categories based on function and structural composition. Executioner caspases-3, -6, and -7 contain the large and small protease subunits and a small prodomain. In contrast, initiator caspases, such as caspases-8, -9, and -10, possess large prodomains consisting of protein–protein interaction motifs, including death effector domains (DEDs) and caspase recruitment domains (CARDs). Caspases-1, -4, and -5 also contain CARDs within large prodomains, but are termed inflammatory caspases for their role in inflammation. Caspases are cleaved after aspartic acid residues that are located between the prodomain and the protease subunits and the large and small protease subunits.

Handbook of Cell Signaling, Three-Volume Set 2 ed.

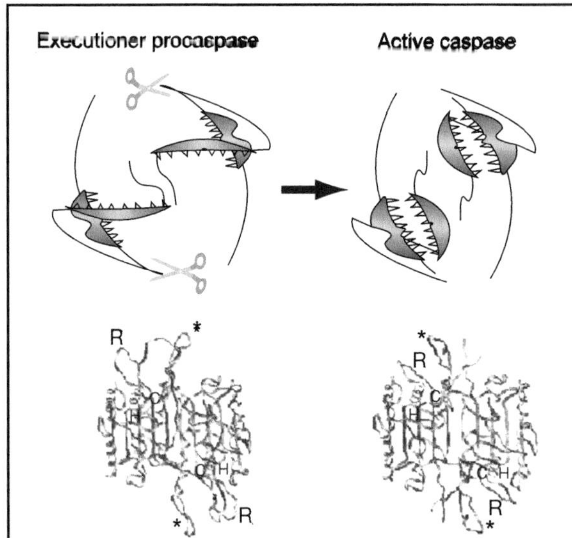

FIGURE 54.2 Activation of executioner caspases. Executioner caspases exist as inactive dimers that are activated following cleavage between the small and large subunits of each monomer. Although separated, the large and small subunits remain intimately associated and both contribute residues to the active site that are necessary for substrate binding and proteolysis. The drawing depicts where cleavage between the subunits occurs and the resultant conformational change. The crystal structure of an executioner caspase (caspase-7) shows the formation of the active sites (represented by asterisks) that occurs following cleavage. The amino acids critical for the active site (C, H) and for the aspartate specificity (R) of the protease are indicated.

large and small subunits of the mature enzyme to generate an active caspase [3, 4]. Caspases-3, -6, and -7 execute the degradative events and are, therefore, classified as effector caspases [2]. The inactive proforms of these appear to pre-exist as dimers, which can only be activated by proteolytic cleavage to create a mature executioner caspase with two active sites [5]. The activation of an executioner caspase by proteolytic cleavage is illustrated in Figure 54.2.

The active executioner caspases then cleave key substrates in the cell to promote apoptosis. For example, a complex of a nuclease and its chaperone/inhibitor is activated when caspase-3 cleaves the inhibitor of CAD (iCAD), releasing the nuclease CAD (caspase activated DNase), to cleave DNA [6]. Activated caspases also promote the blebbing of a dying cell through the cleavage and activation of several molecules, including gelsolin, p21 activated kinase, and ROCK-1 [7, 8]. Perhaps most importantly, caspases promote the exposure of phagocytosis markers on the surface of a dying cell, such as the externalization of phosphatidyl serine [9], which binds to specific receptors on phagocytic cells. However, exactly how caspases induce phosphatidylserine externalization remains unknown. Active caspases are therefore crucial for this specialized form of cell death that ensures that the dying cells will be packed and marked for clearance by professional phagocytes, including macrophages and surrounding epithelial cells, to avoid the induction of an inflammatory response by released intracellular molecules.

Although inhibition of caspase activity prevents many phenotypic characteristics of apoptosis, such as DNA fragmentation, cell death may still ensue in the absence of caspase activation [10–12]. Caspase independent cell death proceeds following the loss of mitochondrial function and may involve the release of alternative death inducing molecules, suggesting that mitochondrial damage may signify a "point of no return" for dying cells [10–12].

A family of naturally occurring inhibitors of apoptosis proteins (IAPs) blocks the activity of caspases and may target active caspases for degradation, thereby averting apoptosis at critical commitment steps of the pathway [13]. IAP orthologs have been identified in several species, including yeast, nematodes, and flies. The mammalian family members c-IAP1 and 2 function to inhibit the activation of caspase-8, an initiator caspase responsible for the cleavage and activation of downstream effector caspases. X-linked IAP (XIAP) inhibits both effector caspase-3 and caspase-9, another initiator caspase (Figure 54.3). Whereas certain family members appear to directly bind specific caspases (namely, caspases-3, -7, and -9), the mechanism by which IAPs inhibit caspase enzymatic activity has yet to be clarified. Some of the IAPs (such as yeast Bir-1 and possibly mammalian survivin) function in cell cycle regulation rather than by controlling caspases or apoptosis [14].

The initiator caspases bear long prodomains containing one or more protein interaction motifs (Figure 54.1) that solicit binding by adaptor molecules and subsequent activation by dimerization [2]. The process through which initiator caspases are activated often characterizes the pathways that lead to apoptosis. There are two general types of initiator caspase activation that are best understood: one is activation by receptor ligation and another occurs following the release of mitochondrial intermembrane proteins into the cytosol. The initiator caspases that are predominantly involved in these two pathways are caspase-8 and -9, respectively.

CASPASE-8 ACTIVATION VIA DEATH RECEPTORS

The death receptors are a subset of the tumor necrosis factor receptor (TNFR) superfamily and include Fas (CD95), TNFR1, TRAIL receptors-1 and -2, and death receptor-3 (DR3). These relay extracellular apoptotic signals through binding of specific ligands to trimeric receptors [15]. Death domains (DDs) found in the cytoplasmic portion of the death receptor interact with DDs on adaptor molecules, namely, Fas associated death domain (FADD) or TNFR associated death domain (TRADD), through homotypic interactions (Figure 54.3). FADD also associates with TRADD by DD interactions, providing a common scaffold for death receptor induced caspase activation. Engagement

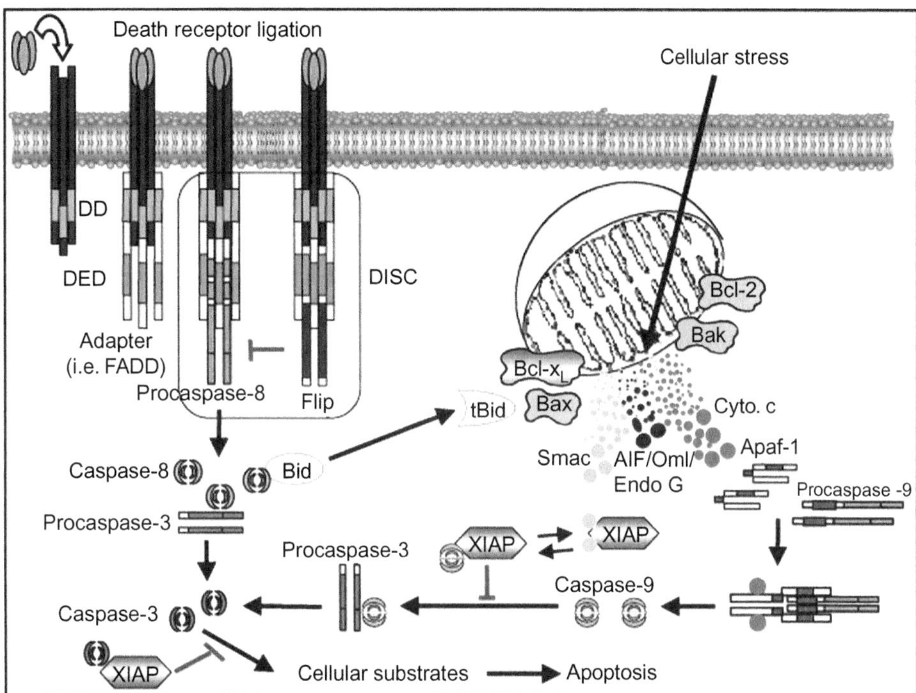

FIGURE 54.3 Apoptotic signaling from two distinct pathways.
The death receptor (i.e., Fas) pathway involves ligand binding, recruitment of adaptor proteins (i.e., FADD) through death domain (DD) interactions, binding of procaspase-8 through death effector domain (DED) interactions, and proteolytic activation of the initiator caspase-8. c-Flip expression blocks caspase-8 activation by inhibiting binding with FADD at the DISC. Caspase-8 may directly activate executioner caspase-3 or may cleave Bid to target the mitochondria. The mitochondria are regulated by the proapoptotic (Bid, Bax, Bak) and antiapoptotic (Bcl-2, Bcl-x_L) Bcl-2 family proteins, and disruption of the mitochondrial membrane results in the release of cytochrome c, Smac/Diablo, AIF, Omi/Htra2, and Endo G. The binding of cytochrome c to Apaf-1 and recruitment of procaspase-9 forms the apoptosome, which allows aggregation and activation of caspase-9. Caspase-9 then activates caspase-3 (and caspase-7) to cleave key cellular substrates and dismantle the dying cell. Proteolytic activity of caspase-9 or -3 may be inhibited by XIAP and cytosolic Smac reverses this inhibition upon binding XIAP.

of DDs on FADD exposes another domain termed the death effector domain (DED) that interacts with DEDs present in the prodomains of procaspases-8 and -10. Although distinct in sequence, the structures of the DD and DED are very similar, forming a so-called "death fold." The result is the formation of a DISC (death inducing signaling complex), which leads to rapid aggregation and activation of caspase-8 or -10 [16]. When two procaspase-8 molecules are brought together, each cleaves after the aspartic acid residues in the other's chain, separating the large and small subunits, removing the prodomain, and forming a mature and fully active caspase with two active sites [17]. It is likely that caspase-10, a similar caspase, is activated in this way as well [18]. The mature initiator caspases can now cleave and activate the executioner caspases to promote apoptosis.

The regulation of the DISC is complex and not fully understood. A naturally occurring inhibitor of DISC formation, termed c-Flip (also known as Casper, Clarp, Flame-1, I-Flice, Cash, Ursurpin, and Mrit), impedes the binding of procaspase-8 to FADD, thereby preventing apoptosis induced by death receptors [19] (Figure 54.3). Decreased Flip expression sensitizes cells to Fas and TNFα induced apoptosis; however, the precise mechanism by which this occurs remains to be elucidated [20–22]. Evidence suggests that Flip is a caspase regulator that may also function to promote caspase activation by facilitating caspase aggregation under some circumstances [23]. It appears that the level of Flip expression, relative to other members of the DISC, may determine its role in apoptosis.

Cells that are resistant to apoptosis induced by the death ligands TNFα, Fas ligand, or TRAIL can often be sensitized by inhibiting macromolecular synthesis (e.g., by addition of actinomycin D or cycloheximide), indicating that an active pathway of protection exists. In the case of TNFR1, the receptor induces both an apoptotic signal through TRADD and a survival signal through the activation of the transcription factor nuclear factor κB (NFκB) [24]. Although the inhibition of NFκB often sensitizes cells to death by TNFα, exactly how NFκB activation blocks TNFα induced apoptosis remains unclear. The ability of NFκB to block TNFR1 signaling may be in part through the expression of cIAP1 and cIAP2, which bind to TNFR1 adaptor molecules [25]. Moreover, NFκB has recently been implicated in the regulation of c-Jun N-terminal kinase (JNK) activity, which may play a role in TNFα induced apoptosis [26, 27]. Although NFκB can interfere with apoptosis in some cases, in others, such as Fas induced apoptosis [28], it does not. There are also instances in which NFκB activation actually enhances apoptosis, indicating that the response to NFκB activation may be stimulus specific [29]. The balance between signals for survival versus death appears to be crucial for determining a cell's ultimate fate.

Cells can be loosely categorized by the apoptotic pathway employed following death receptor ligation [30] (Figure 54.3). In short, ligation of a death receptor triggers apoptosis through activation of initiator caspases-8 or -10, which can cleave and activate executioner caspases that orchestrate the death of the cell. Alternatively, activated caspase-8 may cleave and activate the proapoptotic Bcl-2 family protein Bid [31, 32], which induces mitochondrial membrane permeabilization and the release of apoptotic factors from mitochondria [33] (discussed later). Certain cell types, termed type 1 cells, generate effective DISC formation and rapid, abundant caspase-8 cleavage that can result in direct activation of caspase-3. In contrast, type 2 cells form small amounts of DISC with only slight caspase-8 activation, necessitating an amplification of the apoptotic signal through Bid cleavage and disruption of the mitochondrial outer membrane [30].

MITOCHONDRIA AND THE ACTIVATION OF CASPASE-9

The activation of the second class of initiator caspase, caspase-9, represents a fundamentally different pathway from that involving caspase-8. Procaspase-9 also has a large prodomain containing a protein interaction domain called a CARD (caspase recruitment domain) (Figure 54.1) that structurally resembles DD and DED and also forms a death fold [34]. During its activation, the prodomain of caspase-9 is not removed. Unlike the other caspases discussed, cleavage between the large and small subunits is not necessary for activation of procaspase-9 [35]. Instead, procaspase-9 appears to be activated by binding to an adaptor, Apaf-1 (apoptotic protease activating factor-1) [36].

Inactive Apaf-1 is present as a monomer in the cytosol. Cytochrome c released from mitochondria binds the inert Apaf-1 and promotes Apaf-1 oligomerization and activation [37]. This is illustrated in Figure 54.4. In its active configuration, Apaf-1 forms a complex of seven Apaf-1 molecules with exposed CARDs in the central "hub." The CARDs of Apaf-1 bind the CARDs within the caspase-9 prodomains, allowing for aggregation and activation of the protease, which must remain associated to be active [38]. This complex has been termed the apoptosome and its formation leads to caspase-9 multimerization and a conformational change that permits only one of the two active sites in the caspase-9 tetramer to be active at a time [39]. Evidently, the conformational change necessary for the formation of a specificity determining groove pulls the other active site out of alignment (Figure 54.4). It is proposed that the binding of Apaf-1 to the prodomain of caspase-9 allows this conformational change to occur. Once caspase-9 is activated, it cleaves and activates executioner caspase-3 to induce the degradative events of apoptosis.

The release of cytochrome c from mitochondria is a critical event for apoptosome formation [36]. Cellular stresses, such as DNA damage, cytoskeletal damage, and metabolic disruption, induce proapoptotic Bcl-2 family molecules (discussed later) to compromise the barrier functions of the mitochondrial outer membrane, ultimately leading to the dissipation of the inner transmembrane potential and release of cytochrome c [40]. Normally found in the

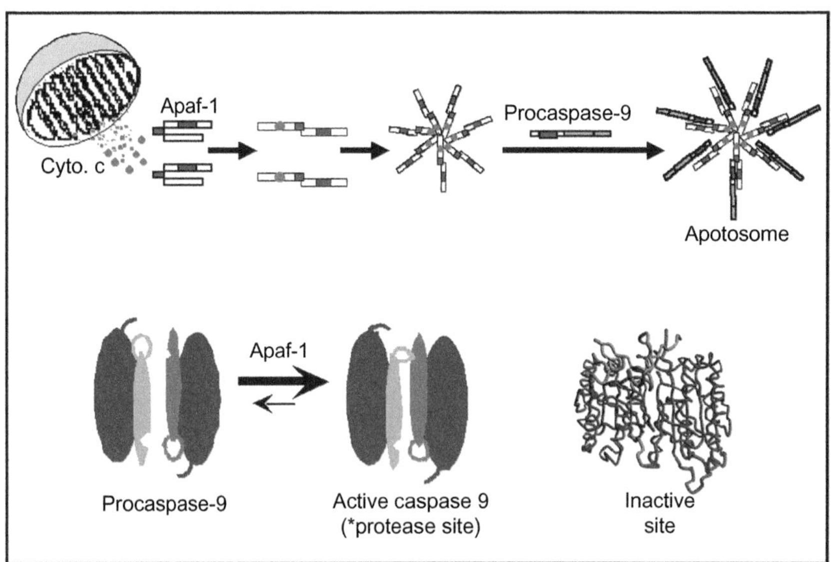

FIGURE 54.4 Apoptosome formation and activation of caspase-9.
Cytochrome c (cyto. c) released from mitochondria binds to inactive Apaf-1 and induces a conformational change and oligomerization of seven Apaf-1 molecules. This assembly exposes the caspase recruitment domain (CARD) of Apaf-1 and allows interaction of Apaf-1 CARD with the CARD of procaspase-9. The resultant formation is termed an apoptosome. Within the apoptosome, bound procaspase-9 is cleaved between the large and small subunits, which remain associated and form a proteolytic active site. Multimerization of caspase-9 induces a conformational change that allows for only one proteolytic site to be active at a time (identified by the asterisk).

intermembrane space of the mitochondria, cytochrome c must be released into the cytosol to induce apoptosome formation. Embryonic cells lacking cytochrome c did not undergo apoptosis in response to several inducers, demonstrating the importance for cytochrome c in the mitochondrial death pathway [41]. Cells from mice deficient in either Apaf-1 or caspase-9 display similar phenotypes (resistance to a variety of apoptotic stimuli) to cells lacking cytochrome c, suggesting that these three molecules serve a cooperative function in this pathway of apoptosis [42, 43].

In addition to the release of cytochrome c, other mitochondrial proteins, including Smac (second mitochondria derived activator of caspase)/Diablo (direct IAP binding protein with low pI), Endo G (endonuclease G), Omi/Htra2, and AIF (apoptosis inducing factor), are expelled into the cytosol following mitochondrial membrane disruption and potentially play a role in apoptosis. Because the function of cytochrome c in apoptosis appears to be fundamentally distinct from its role in electron transport, it is possible that several of the basic mitochondrial functions necessary for cell survival may also signal apoptosis when mitochondria are disrupted. Thus, various molecules associated with mitochondria may serve dual roles to maintain cellular function as well as to induce death, depending on cellular conditions. For instance, AIF, which possesses redox activity in the mitochondria, translocates to the nucleus and may promote DNA damage and cell death via a caspase independent mechanism [44, 45]. Endo G, an endonuclease potentially involved in mitochondrial replication, is released from mitochondria and has been suggested to mediate nuclear DNA damage in the absence of caspases [46]. Similarly, Omi/HtrA2, a mitochondrial serine protease, may promote caspase independent cell death via its protease activity (distinct from that of caspases) when released into the cytosol [47]. These proteins, along with cytochrome c, are conserved from mammals to yeast and apparently retain a potential to kill cells independent of their role in mitochondria function.

While some of the death inducing factors released from the mitochondria contribute to caspase independent cell death, others promote caspase activation to enhance apoptosis. Both Smac/Diablo [48, 49] and Omi/HtrA2 [47, 50–52] can block IAP mediated caspase inhibition by binding XIAP via an AVP(I/S) sequence at the N terminus and disrupting the association of caspase-9 with XIAP (Figure 54.3). Thus, these mitochondrial proteins function in the cytosol to negate IAPs' antiapoptotic function. Moreover, Smac/Diablo may serve an additional proapoptotic role that is independent from its ability to bind IAPs [53].

MITOCHONDRIAL OUTER MEMBRANE PERMEABILIZATION

The coincidental release of proteins from the mitochondrial intermembrane space during apoptosis suggests the occurrence of a mitochondrial outer membrane permeabilization. Although the onset of cytochrome c release may vary depending on the initial apoptotic stimulus utilized, once release occurs, it is rapid and complete [54]. In most instances, the molecules discussed earlier (e.g., Smac/Diablo, AIF) are also expelled into the cytosol with cytochrome c; however, the mechanism for this event remains unclear and it is possible that selective release can occur.

Essentially three models (reviewed in [55]) account for outer membrane permeabilization during apoptosis: (1) the pore forming model in which permeabilization is mediated by a change in the outer membrane, permitting protein release without involving the inner membrane [33]; (2) the permeability transition model, in which an opening of the adenine nucleotide transporter (ANT) in the inner membrane causes matrix swelling, leading to outer membrane disruption [56]; and (3) the voltage dependent anion channel (VDAC) closure model, in which metabolic signals trigger a closure of VDAC in the outer membrane, resulting in inner membrane perturbations that cause matrix swelling and outer membrane disruption [57]. Models 2 and 3 propose an active role for mitochondria in apoptosis, whereas model 1 defines this organelle as a repository of apoptogenic factors that are released by the actions of mediators on the outer membrane (discussed later). However, the function of the electron transport chain, which is essential for most mitochondrial functions including ATP generation, is lost following permeabilization, and this loss is greatly facilitated by caspase activation [54, 58]. Mitochondrial dysfunction may have a fundamental role in apoptotic cell death, but the extent to which the loss of mitochondrial function versus the activity of death promoting proteins accounts for the death of cells following mitochondrial outer membrane permeabilization is currently unresolved.

THE BCL-2 FAMILY

Although the precise nature of mitochondrial membrane permeabilization remains elusive, some of the factors controlling it are beginning to be understood. A group of related molecules known as the Bcl-2 family proteins can target the mitochondria and regulate membrane permeabilization [59]. Bcl-2 related proteins are categorized by their ability to either induce or inhibit apoptosis. Apoptotic inducers (Bax, Bad, Bak, Bik, Bok, Bim, Bip, Bid, Diva, Hrk, or Blk) and protectors (Bcl-x_L, Mcl-1, A1, and Bcl-w) that share homologous regions (BH domains) with antiapoptotic Bcl-2 have been identified (Figure 54.5) [60]. Antiapoptotic members Bcl-2 and Bcl-x_L prevent mitochondrial outer membrane permeabilization, whereas proapoptotic members Bax and Bak produce it [59]. Thus, the balance of antiapoptotic and proapoptotic Bcl-2 family members may be responsible for maintaining membrane integrity.

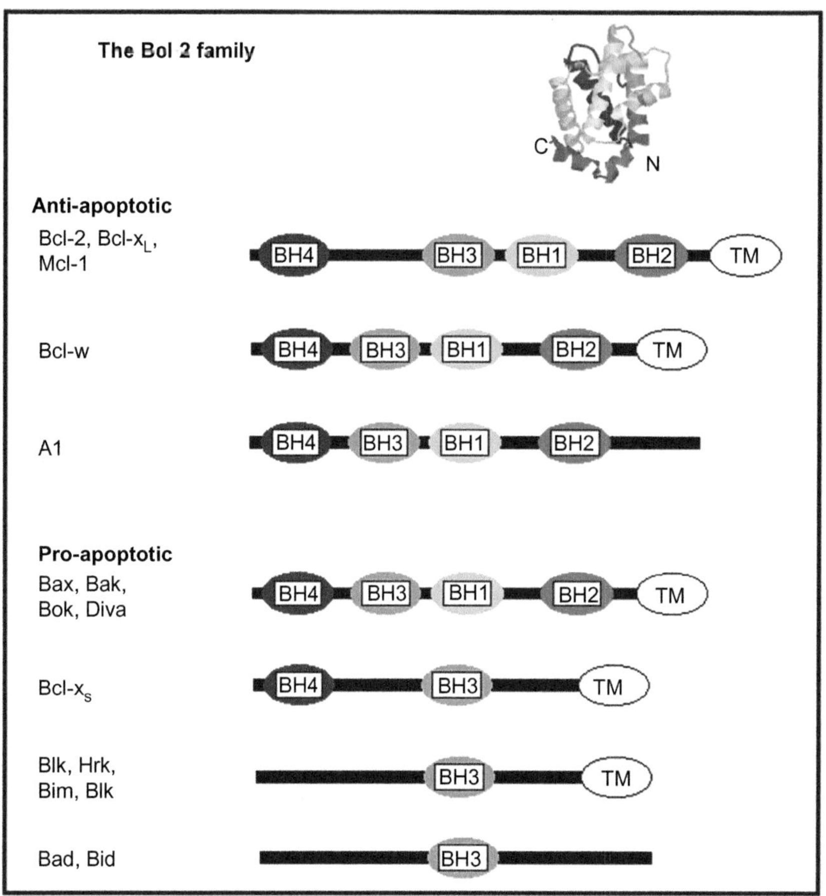

FIGURE 54.5 The Bcl-2 family.
The proteins belonging to the Bcl-2 family are related based on the presence of Bcl-2 homology (BH) domains, but are categorized by function. The family members are divided based on their ability to inhibit (antiapoptotic) or induce (proapoptotic) apoptosis. The Bcl-2 family is further sorted by the presence or absence of the four BH domains. For instance, the BH3-only proteins are a subdivision of the proapoptotic molecules that contain only the BH3 domain. Several of the family members also contain a hydrophobic transmembrane (TM) domain that allows for membrane insertion and localization to the mitochondria. The presence of common BH domains allows for oligomerization of family members, including binding between pro- and antiapoptotic proteins, which may determine the effect on mitochondrial integrity.

The proapoptotic Bcl-2 family members have been further divided into a subset termed BH3-only proteins (Bid, Bim, Bik, Blk, Bmf, Bad, Hrk, BNIP3, Puma, Noxa) that contain only the BH3 of the four BH domains and may function as "sensors" to assess the status of death or survival signals. For example, Bid "senses" protease activation and becomes activated itself through proteolytic cleavage by caspases [61], the cytotoxic lymphocyte protease granzyme B [62], and lysosomal proteases [63]. Bim [64] and Bmf [65] appear to "sense" cytoskeleton status, and Bad "senses" the status of growth factor receptor signaling. Noxa [66] and Puma [67, 68] "sense" the activation of the proapoptotic transcription factor p53 through a p53 dependent increase in their expression. Although they may not directly induce mitochondrial outer membrane effects, the BH3-only proteins interact with other proapoptotic Bcl-2 family members, namely, Bax and Bak, to induce mitochondrial membrane permeabilization [69]. Bax and Bak have three of the four BH domains and are called multidomain or BH-123 proteins. In this scenario, the BH3-only molecules sense a particular cellular apoptotic signal and direct Bax and Bak to disrupt the mitochondria. Isolated mitochondria from cells lacking both Bax and Bak (but not single knockouts) are strikingly resistant to the induction of mitochondrial permeabilization and this pathway of apoptosis [70], suggesting that these two proteins are intimately, perhaps directly, involved in the mechanism of membrane permeabilization.

The ability of Bax and Bak to induce permeabilization may depend on oligomerization. Bax and Bak, which are highly homologous, exist as inactive monomers that can be induced to oligomerize by BH3-only proteins [71, 72]. Oligomerization correlates with induction of membrane permeabilization; however, the precise role of oligomerization has not been directly demonstrated. It is possible that oligomerized proapoptotic Bcl-2 proteins form a pore or alter the lipids of the mitochondrial outer membrane to induce permeabilization. In support of this idea, Bcl-2, Bcl-x$_L$, Bax, and Bid contain similar structures that resemble the pore forming chain of some bacterial toxins and have been shown to have weak channel forming activity for small ions through lipid membranes [55]. It has yet to be determined if these molecules form pores on their own or through interactions with other proteins in the outer membrane, such as VDAC [55, 73]. Furthermore, the antiapoptotic Bcl-2 proteins may act by inhibiting oligomerization and possible pore formation by the proapoptotic Bcl-2 proteins [74]. However, whether or not Bcl-2 family members directly regulate apoptosis in this manner remains controversial.

CELL CYCLE VERSUS APOPTOSIS

To maintain homeostasis, cell death and proliferation must be precisely balanced, and communication between these two distinct signaling pathways is critical. Several proteins that engage the cell cycle, including c-Myc and E2F1, sensitize cells to apoptosis by increasing susceptibility to mitochondrial membrane permeabilization. In response to apoptotic stimuli, such as growth factor withdrawal or DNA damaging agents, c-Myc activation stimulates a caspase independent release of cytochrome c from mitochondria [75]. Furthermore, activation of c-Myc may provoke Fas mediated apoptosis, possibly through a c-Myc dependent upregulation of Fas ligand [76, 77] and/or sensitization to death receptor signaling [78].

Antiapoptotic Bcl-2 family members may function not only to protect mitochondria from membrane permeabilization, but also to repress cell cycle entry. The ability of Bcl-2 to prevent or delay entry into G_1 appears to be independent of its role in maintaining mitochondria integrity, as point mutations in Bcl-2 have been described that eliminate the cell cycle inhibitory activity without affecting the antiapoptotic activity [79]. Therefore, while entry into the cell cycle does not directly induce apoptosis, these two processes are linked and a variety of proteins have co-evolved to play distinct roles in each pathway. The connection between induction of the cell cycle and cell death, termed antagonistic pleiotropy [80], is a fundamental mechanism to prevent cancer and may be "hard wired" in animals in which cell proliferation is required for tissue homeostasis. Nevertheless, this does not mean that entry into the cell cycle is a requirement for apoptosis, nor that blockade of the cell cycle will necessarily promote (or inhibit) cell death via apoptosis. These are distinct but evolutionarily linked cellular processes.

CONCLUSIONS

When properly regulated, apoptosis is essential for maintaining homeostasis of several cellular systems, including development, tissue turnover, immune regulation, and control of oncogenesis. However, inappropriate cell death influences several disease states, because too much (i.e., Alzheimer's disease) or too little (i.e., cancer, autoimmunity) apoptosis creates an adverse imbalance. Considering the vast number of molecules involved in orchestrating an apoptotic response, defining the precise mechanisms and targets that may contribute to apoptosis in a specific setting is proving to be extremely challenging. In addition, many apoptotic signaling pathways are stimulus and cell-type specific, further complicating the process. Novel proteins and pathways are being discovered at a fervent pace and although tremendous gains have been made in elucidating the mechanisms of apoptosis, this is a highly complex process that remains only partially understood.

REFERENCES

1. Wyllie AH, Kerr JF, Currie AR. Cell death: the significance of apoptosis. *Int Rev Cytol* 1980;**68**:251–306.
2. Thornberry NA, Lazebnik Y. Caspases: enemies within. *Science* 1998;**281**:1312–16.
3. Wilson KP, Black JA, Thomson JA, Kim EE, Griffith JP, Navia MA, Murcko MA, Chambers SP, Aldape RA, Raybuck SA, et al. Structure and mechanism of interleukin-1 beta converting enzyme. *Nature* 1994;**370**:270–5.
4. Walker NP, Talanian RV, Brady KD, Dang LC, Bump NJ, Ferenz CR, Franklin S, Ghayur T, Hackett MC, Hammill LD, et al. Crystal structure of the cysteine protease interleukin-1 beta-converting enzyme: A (p20/p10)2 homodimer. *Cell* 1994;**78**:343–52.
5. Shi Y. Mechanisms of caspase activation and inhibition during apoptosis. *Mol Cell* 2002;**9**:459–70.
6. Enari M, Sakahira H, Yokoyama H, Okawa K, Iwamatsu A, Nagata S. A caspase-activated DNase that degrades DNA during apoptosis, and its inhibitor ICAD. *Nature* 1998;**391**:43–50.
7. Stroh C, Schulze-Osthoff K. Death by a thousand cuts: An ever increasing list of caspase substrates. *Cell Death Differ* 1998;**5**:997–1000.
8. Coleman ML, Sahai EA, Yeo M, Bosch M, Dewar A, Olson MF. Membrane blebbing during apoptosis results from caspase-mediated activation of ROCK I. *Nat Cell Biol* 2001;**3**:339–45.
9. Fadok VA, de Cathelineau A, Daleke DL, Henson PM, Bratton DL. Loss of phospholipid asymmetry and surface exposure of phosphatidylserine is required for phagocytosis of apoptotic cells by macrophages and fibroblasts. *J Biol Chem* 2001;**276**:1071–7.
10. Sun XM, MacFarlane M, Zhuang J, Wolf BB, Green DR, Cohen GM. Distinct caspase cascades are initiated in receptor-mediated and chemical-induced apoptosis. *J Biol Chem* 1999;**274**:5053–60.
11. Xiang J, Chao DT, Korsmeyer SJ. BAX-induced cell death may not require interleukin 1β-converting enzyme-like proteases. *Proc Natl Acad Sci USA* 1996;**93**:14,559–63.
12. McCarthy NJ, Whyte MK, Gilbert CS, Evan GI. Inhibition of Ced-3/ICE-related proteases does not prevent cell death induced by oncogenes, DNA damage, or the Bcl-2 homologue Bak. *J Cell Biol* 1997;**136**:215–27.
13. Salvesen GS, Duckett CS. IAP proteins: blocking the road to death's door. *Nat Rev Mol Cell Biol* 2002;**3**:401–10.
14. Silke J, Vaux DL. Two kinds of BIR-containing protein-inhibitors of apoptosis, or required for mitosis. *J Cell Sci* 2001;**114**:1821–7.
15. Locksley RM, Killeen N, Lenardo MJ. The TNF and TNF receptor superfamilies: Integrating mammalian biology. *Cell* 2001;**104**:487–501.
16. Green DR. Apoptotic pathways: the roads to ruin. *Cell* 1998; **94**:695–8.

17. Muzio M, Stockwell BR, Stennicke HR, Salvesen GS, Dixit VM. An induced proximity model for caspase-8 activation. *J Biol Chem* 1998;**273**:2926–30.
18. Wang J, Chun HJ, Wong W, Spencer DM, Lenardo MJ. Caspase-10 is an initiator caspase in death receptor signaling. *Proc Natl Acad Sci USA* 2001;**98**:13,884–8.
19. Irmler M, Thome M, Hahne M, Schneider P, Hofmann K, Steiner V, Bodmer J-L, Schroter M, Burns K, Mattmann C, et al. Inhibition of death receptor signals by cellular FLIP. *Nature* 1997;**388**:190–5.
20. Perlman H, Pagliari LJ, Georganas C, Mano T, Walsh K, Pope RM. FLICE-inhibitory protein expression during macrophage differentiation confers resistance to fas-mediated apoptosis. *J Exp Med* 1999;**190**:1679–88.
21. Yeh WC, Itie A, Elia AJ, Ng M, Shu HB, Wakeham A, Mirtsos C, Suzuki N, Bonnard M, Goeddel DV, Mak TW. Requirement for Casper (c-FLIP) in regulation of death receptor-induced apoptosis and embryonic development. *Immunity* 2000;**12**:633–42.
22. Scaffidi C, Schmitz I, Krammer PH, Peter ME. The role of c-FLIP in modulation of CD95-induced apoptosis. *J Biol Chem* 1999;**274**:1541–8.
23. Chang DW, Xing Z, Pan Y, Algeciras-Schimnich A, Barnhart BC, Yaish-Ohad S, Peter ME, Yang X. c-FLIP(L) is a dual function regulator for caspase-8 activation and CD95-mediated apoptosis. *EMBO J* 2002;**21**:3704–14.
24. Chen G, Goeddel DV. TNF-R1 signaling: a beautiful pathway. *Science* 2002;**296**:1634–5.
25. Wang C-Y, Mayo MW, Korneluk RG, Goeddel DV, Baldwin Jr. AS NF-κB antiapoptosis: Induction of TRAF1 and TRAF2 and c-IAP1 and c-IAP2 to suppress caspase-8 activation. *Science* 1998;**281**:1680–3.
26. Tang G, Minemoto Y, Dibling B, Purcell NH, Li Z, Karin M, Lin A. Inhibition of JNK activation through NF-kappaB target genes. *Nature* 2001;**414**:313–7.
27. De Smaele E, Zazzeroni F, Papa S, Nguyen DU, Jin R, Jones J, Cong R, Franzoso G. Induction of gadd45beta by NF-kappaB downregulates pro-apoptotic JNK signalling. *Nature* 2001;**414**:308–13.
28. Kasibhatla S, Brunner T, Genestier L, Echeverri F, Mahboubi A, Green DR. DNA damaging agents induce expression of Fas ligand and subsequent apoptosis in T lymphocytes via the activation of NF-kappa B and AP-1. *Mol Cell* 1998;**1**:543–51.
29. Ryan KM, Ernst MK, Rice NR, Vousden KH. Role of NF-kappaB in p53-mediated programmed cell death. *Nature* 2000;**404**:892–7.
30. Scaffidi C, Fulda S, Srinivasan A, Friesen C, Li F, Tomaselli KJ, Debatin KM, Krammer PH, Peter ME. Two CD95 (APO-1/Fas) signaling pathways. *EMBO J* 1998;**17**:1675–87.
31. Li H, Zhu H, Xu C, Yuan J. Cleavage of BID by caspase 8 mediates the mitochondrial damage in the Fas pathway of apoptosis. *Cell* 1998;**94**:491–501.
32. Luo X, Budihardjo I, Zou H, Slaughter C, Wang X. Bid, a Bcl-2 interacting protein mediates cytochrome c release from mitochondria in response to activation of cell surface death receptors. *Cell* 1998;**94**:481–90.
33. Wei MC, Zong WX, Cheng EH, Lindsten T, Panoutsakopoulou V, Ross AJ, Roth KA, MacGregor GR, Thompson CB, Korsmeyer SJ. Proapoptotic BAX and BAK: A requisite gateway to mitochondrial dysfunction and death. *Science* 2001;**292**:727–30.
34. Hofmann K, Bucher P, Tschopp J. The CARD domain: a new apoptotic signalling motif. *Trends Biochem Sci* 1997;**22**:155–6.
35. Stennicke HR, Deveraux QL, Humke EW, Reed JC, Dixit VM, Salvesen GS. Caspase-9 can be activated without proteolytic processing. *J Biol Chem* 1999;**274**:8359–62.
36. Li P, Nijhawan D, Budihardjo I, Srinivasula SM, Ahmad M, Alnemri ES, Wang X. Cytochrome c and dATP-dependent formation of Apaf-1/caspase-9 complex initiates an apoptotic protease cascade. *Cell* 1997;**91**:479–89.
37. Zou H, Henzel WJ, Liu X, Lutschg A, Wang X. Apaf-1, a human protein homologous to *C. elegans* CED-4, participates in cytochrome c-dependent activation of caspase-3. *Cell* 1997;**90**:405–13.
38. Shiozaki EN, Chai J, Shi Y. Oligomerization and activation of caspase-9, induced by Apaf-1 CARD. *Proc Natl Acad Sci USA* 2002;**99**:4197–202.
39. Renatus M, Stennicke HR, Scott FL, Liddington RC, Salvesen GS. Dimer formation drives the activation of the cell death protease caspase 9. *Proc Natl Acad Sci USA* 2001;**98**:14,250–5.
40. Wang X. The expanding role of mitochondria in apoptosis. *Genes Dev* 2001;**15**:2922–33.
41. Li K, Li Y, Shelton JM, Richardson JA, Spencer E, Chen ZJ, Wang X, Williams RS. Cytochrome c deficiency causes embryonic lethality and attenuates stress-induced apoptosis. *Cell* 2000;**101**:389–99.
42. Kuida K, Haydar TF, Kuan C, Gu Y, Taya C, Karasuyama H, Su MS, Rakic P, Flavell RA. Reduced apoptosis and cytochrome c-mediated caspase activation in mice lacking caspase 9. *Cell* 1998;**94**:325–37.
43. Yoshida H, Kong YY, Yoshida R, Elia AJ, Hakem A, Hakem R, Penninger JM, Mak TW. Apaf1 is required for mitochondrial pathways of apoptosis and brain development. *Cell* 1998;**94**:739–50.
44. Miramar MD, Costantini P, Ravagnan L, Saraiva LM, Haouzi D, Brothers G, Penninger JM, Peleato ML, Kroemer G, Susin SA. NADH oxidase activity of mitochondrial apoptosis-inducing factor. *J Biol Chem* 2001;**276**:16,391–8.
45. Susin SA, Lorenzo HK, Zamzami N, Marzo I, Brenner C, Larochette N, Prevost MC, Alzari PM, Kroemer G. Mitochondrial release of caspase-2 and -9 during the apoptotic process. *J Exp Med* 1999;**189**:381–93.
46. Li LY, Luo X, Wang X. Endonuclease G is an apoptotic DNase when released from mitochondria. *Nature* 2001;**412**:95–9.
47. Hegde R, Srinivasula SM, Zhang Z, Wassell R, Mukattash R, Cilenti L, DuBois G, Lazebnik Y, Zervos AS, Fernandes-Alnemri T, Alnemri ES. Identification of Omi/HtrA2 as a mitochondrial apoptotic serine protease that disrupts inhibitor of apoptosis protein-caspase interaction. *J Biol Chem* 2002;**277**:432–8.
48. Du C, Fang M, Li Y, Li L, Wang X. Smac, a mitochondrial protein that promotes cytochrome c-dependent caspase activation by eliminating IAP inhibition. *Cell* 2000;**102**:33–42.
49. Verhagen AM, Ekert PG, Pakusch M, Silke J, Connolly LM, Reid GE, Moritz RL, Simpson RJ, Vaux DL. Identification of DIABLO, a mammalian protein that promotes apoptosis by binding to and antagonizing IAP proteins. *Cell* 2000;**102**:43–53.
50. Suzuki Y, Imai Y, Nakayama H, Takahashi K, Takio K, Takahashi R. A serine protease, HtrA2, is released from the mitochondria and interacts with XIAP, inducing cell death. *Mol Cell* 2001;**8**:613–21.
51. Martins L, Iaccarino M, Tenev I, Gschmeissner T, Totty S, Lemoine NF, Savopoulos NR, Gray J, Creasy CW, Dingwall CL, Downward J. The serine protease Omi/HtrA2 regulates apoptosis by binding XIAP through a reaper-like motif. *J Biol Chem* 2002;**277**:439–44.
52. Verhagen AM, Silke J, Ekert PG, Pakusch M, Kaufmann H, Connolly LM, Day CL, Tikoo A, Burke R, Wrobel C, et al. HtrA2 promotes cell death through its serine protease activity and its ability to antagonize inhibitor of apoptosis proteins. *J Biol Chem* 2002;**277**:445–54.
53. Roberts DL, Merrison W, MacFarlane M, Cohen GM. The inhibitor of apoptosis protein-binding domain of Smac is not essential for its proapoptotic activity. *J Cell Biol* 2001;**153**:221–8.

54. Goldstein JC, Waterhouse NJ, Juin P, Evan GI, Green DR. The coordinate release of cytochrome c during apoptosis is rapid, complete and kinetically invariant. *Nat Cell Biol* 2000;**2**:156–62.
55. Martinou JC, Green DR. Breaking the mitochondrial barrier. *Nat Rev Mol Cell Biol* 2001;**2**:63–7.
56. Zamzami N, Kroemer G. The mitochondrion in apoptosis: how Pandora's box opens. *Nat Rev Mol Cell Biol* 2001;**2**:67–71.
57. Vander Heiden MG, Thompson CB. Bcl-2 proteins: regulators of apoptosis or of mitochondrial homeostasis? *Nat Cell Biol* 1999;**1**:E209–16.
58. Waterhouse NJ, Goldstein JC, von Ahsen O, Schuler M, Newmeyer DD, Green DR. Cytochrome c maintains mitochondrial transmembrane potential and ATP generation after outer mitochondrial membrane permeabilization during the apoptotic process. *J Cell Biol* 2001;**153**:319–28.
59. Harris MH, Thompson CB. The role of the Bcl-2 family in the regulation of outer mitochondrial membrane permeability. *Cell Death Differ* 2000;**7**:1182–91.
60. Adams JM, Cory S. The Bcl-2 protein family: arbiters of cell survival. *Science* 1998;**281**:1322–6.
61. Budihardjo I, Oliver H, Lutter M, Luo X, Wang X. Biochemical pathways of caspase activation during apoptosis. *Annu Rev Cell Dev Biol* 1999;**15**:269–90.
62. Pinkoski MJ, Waterhouse NJ, Heibein JA, Wolf BB, Kuwana T, Goldstein JC, Newmeyer DD, Bleackley RC, Green DR. Granzyme B-mediated apoptosis proceeds predominantly through a Bcl-2-inhibitable mitochondrial pathway. *J Biol Chem* 2001;**276**:12,060–7.
63. Stoka V, Turk B, Schendel SL, Kim TH, Cirman T, Snipas SJ, Ellerby LM, Bredesen D, Freeze H, Abrahamson M, et al. Lysosomal protease pathways to apoptosis. Cleavage of bid, not pro-caspases, is the most likely route. *J Biol Chem* 2001;**276**:3149–57.
64. Puthalakath H, Huang DC, O'Reilly LA, King SM, Strasser A. The proapoptotic activity of the Bcl-2 family member Bim is regulated by interaction with the dynein motor complex. *Mol Cell* 1999;**3**:287–96.
65. Puthalakath H, Villunger A, O'Reilly LA, Beaumont JG, Coultas L, Cheney RE, Huang DC, Strasser A. Bmf: A proapoptotic BH3-only protein regulated by interaction with the myosin V actin motor complex, activated by anoikis. *Science* 2001;**293**:1829–32.
66. Oda E, Ohki R, Murasawa H, Nemoto J, Shibue T, Yamashita T, Tokino T, Taniguchi T, Tanaka N. Noxa, a BH3-only member of the Bcl-2 family and candidate mediator of p53-induced apoptosis. *Science* 2000;**288**:1053–8.
67. Nakano K, Vousden KH. PUMA, a novel proapoptotic gene, is induced by p53. *Mol Cell* 2001;**7**:683–94.
68. Yu J, Zhang L, Hwang PM, Kinzler KW, Vogelstein B. PUMA induces the rapid apoptosis of colorectal cancer cells. *Mol Cell* 2001;**7**:673–82.
69. Lutz RJ. Role of the BH3 (Bcl-2 homology 3) domain in the regulation of apoptosis and Bcl-2-related proteins. *Biochem Soc Trans* 2000;**28**:51–6.
70. Lindsten T, Ross AJ, King A, Zong WX, Rathmell JC, Shiels HA, Ulrich E, Waymire KG, Mahar P, Frauwirth K, et al. The combined functions of proapoptotic Bcl-2 family members bak and bax are essential for normal development of multiple tissues. *Mol Cell* 2000;**6**:1389–99.
71. Eskes R, Desagher S, Antonsson B, Martinou JC. Bid induces the oligomerization and insertion of Bax into the outer mitochondrial membrane. *Mol Cell Biol* 2000;**20**:929–35.
72. Wei MC, Lindsten T, Mootha VK, Weiler S, Gross A, Ashiya M, Thompson CB, Korsmeyer SJ. tBID, a membrane-targeted death ligand, oligomerizes BAK to release cytochrome c. *Genes Dev* 2000;**14**:2060–71.
73. Shimizu S, Matsuoka Y, Shinohara Y, Yoneda Y, Tsujimoto Y. Essential role of voltage-dependent anion channel in various forms of apoptosis in mammalian cells. *J Cell Biol* 2001;**152**:237–50.
74. Cheng EH, Wei MC, Weiler S, Flavell RA, Mak TW, Lindsten T, Korsmeyer SJ. BCL-2, BCL-X(L) sequester BH3 domain-only molecules preventing BAX- and BAK-mediated mitochondrial apoptosis. *Mol Cell* 2001;**8**:705–11.
75. Juin P, Hueber AO, Littlewood T, Evan G. c-Myc-induced sensitization to apoptosis is mediated through cytochrome c release. *Genes Dev* 1999;**13**:1367–81.
76. Kasibhatla S, Beere HM, Brunner T, Echeverri F, Green DR. A "non-canonical" DNA-binding element mediates the response of the Fas-ligand promoter to c-Myc. *Curr Biol* 2000;**10**:1205–8.
77. Brunner T, Kasibhatla S, Pinkoski MJ, Frutschi C, Yoo NJ, Echeverri F, Mahboubi A, Green DR. Expression of Fas ligand in activated T cells is regulated by c-Myc. *J Biol Chem* 2000;**275**:9767–72.
78. Hueber AO, Zornig M, Lyon D, Suda T, Nagata S, Evan GI. Requirement for the CD95 receptor-ligand pathway in c-Myc-induced apoptosis. *Science* 1997;**278**:1305–9.
79. Huang DCS, O'Reilly LA, Strasser A, Cory S. The anti-apoptotic function of Bcl-2 can be genetically separated from its inhibitory effect on cell cycle entry. *EMBO J* 1997;**16**:4628–38.
80. Green DR, Evan GI. A matter of life and death. *Cancer Cell* 2002;**1**:19–30.

Chapter 55

The Role of Ceramide in Cell Regulation

Leah J. Siskind[1,2], Thomas D. Mullen[2] and Lina M. Obeid[1,2]

[1] *Ralph H. Johnson Veterans Administration Hospital, Charleston, South Carolina*

[2] *Department of Medicine, Division of General Internal Medicine and Geriatrics, Medical University of South Carolina, Charleston, South Carolina*

INTRODUCTION

Ceramide is a member of the sphingolipid family of lipids whose members all share a common 18 carbon sphingoid base backbone (Figure 55.1). Ceramide is comprised of a *N*-acylated (14–26 carbons) sphingosine (18 carbons); carbons 1–5 consist of hydroxyl groups at C1 and C3, a trans double bond across C4 and C5, and an amide group that serves as the fatty acyl linkage at C2 (Figure 55.1). Ceramides are classified by the length of the *N*-linked fatty acyl chain.

Ceramides and other sphingolipids are not merely structural components of membranes, but rather play important roles in coordinating cellular responses to extracellular stimuli and to stress [1]. Ceramide, at the heart of the sphingolipid family, is important to the regulation of several cellular processes, including apoptosis, autophagy, cell cycle arrest, and cellular senescence [1]. The purpose of this chapter is to give an overview of the role of ceramide in cellular regulation, with a focus on its role in apoptosis. In order to gain a better understanding of ceramide's role in cell regulation, it is important first to gain an appreciation of the complex process of ceramide metabolism, as it involves several sphingolipids with important and distinct cellular functions.

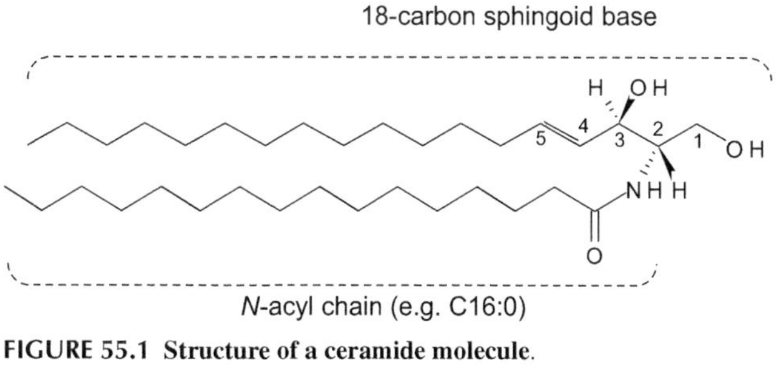

18-carbon sphingoid base

H OH

5 4 3 2 1 OH

NH H

O

N-acyl chain (e.g. C16:0)

FIGURE 55.1 Structure of a ceramide molecule.
Ceramide contains an *N*-acylated (14–26 carbons) sphingosine (18 carbons). Carbons 1–5 of the sphingosine backbone consist of hydroxyl groups at C1 and C3, a *trans* double bond across C4 and C5, and an amide group that serves as the fatty acyl linkage at C2. Ceramides are classified by the length of the *N*-linked fatty acyl chain, which varies from 14 to 26 carbons in length.

SPHINGOLIPID METABOLISM

Ceramide can be generated by multiple pathways in cells, namely sphingomyelin hydrolysis, *de novo* synthesis, the salvage pathway, or breakdown of more complex sphingolipids (Figure 55.2). Sphingomyelinases (SMases) catalyze the hydrolysis of sphingomyelin (SM) to form ceramide and phosphorylcholine, and are classified by their pH optima [2, 3]. Of these, the acid and Mg^{2+}-dependent neutral sphingomyelinases have been implicated in stress-induced ceramide generation [2, 3].

De novo ceramide synthesis (Figure 55.2) [4, 5] occurs in the ER [6]. The enzyme serine palmitoyl transferase (SPT) catalyzes the first and rate limiting step in *de novo* synthesis, namely the condensation of serine and palmitoyl-CoA to form 3-ketosphinganine. The enzyme 3-ketosphinganine reductase catalyzes the reduction of 3-ketosphinganine to form sphinganine. Ceramide synthases catalyze the *N*-acylation of sphinganine to form dihydroceramide (DH). A desaturase catalyzes the conversion of dihydroceramide to ceramide through the insertion of a *trans* double bond at the 4–5 position of the sphingoid base backbone.

Once generated, ceramide's *N*-linked fatty acyl chain can be removed to generate sphingosine as catalyzed by ceramidases (CDase). Several CDases have been identified in mammals, and are classified by their pH optima [7]. The sphingosine generated by the action of CDases can be used by ceramide synthases to regenerate ceramide in the recycling/ salvage pathway. Thus, ceramide synthases are thought to occupy a central postion in the sphingolipid metabolic pathway, as they catalyze the formation of ceramide in two distinct pathways: by *de novo* synthesis and the recycling/salvage pathway.

Handbook of Cell Signaling, Three-Volume Set 2 ed.

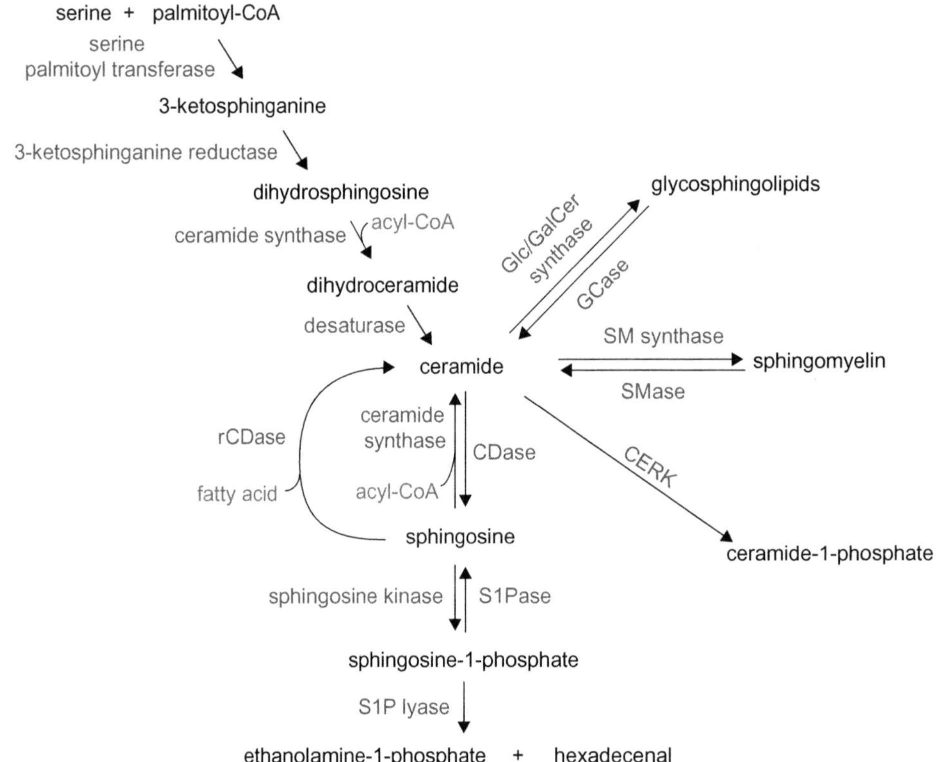

FIGURE 55.2 Overview of ceramide metabolism.
Ceramide is at the metabolic hub of the sphingolipid pathway such that many enzymes are responsible for its anabolism and catabolism. Many of the enzyme activities (e.g., sphingomyelinase) are represented by multiple gene products (e.g., neutral sphingomyelinase, acid sphingomyelinase, alkaline sphingomyelinase, etc.). Abbreviations: CoA, coenzyme A; CDase, ceramidase; rCDase, reverse ceramidase activity; S1Pase, sphingosine-1-phosphate phosphatase; S1P lyase, sphingosine-1-phosphate lyase; SM synthase, sphingomyelin synthase; SMase, sphingomyelinase; Glc/GalCer synthase, glucosylceramide synthase/galactosylceramide synthase; GCase, glucocerebrocidase.

In mammals, there are six ceramide synthase isoforms (CerS1–6) that have preferences for fatty acyl CoAs of particular chain lengths (Figure 55.3; [8–10]) to form the corresponding ceramides. Once generated, ceramide can serve as a metabolic precursor to complex sphingolipids, such as SM and glycosphingolipids. Of note, synthesis of complex sphingolipids can influence other lipid metabolic pathways; for example sphingomyelin synthase can catalyze the transfer of a phosphocholine headgroup from phosphatidylcholine to ceramide to generate SM and diacylglycerol in the Golgi apparatus. Ceramide can also be phosphorylated by ceramide kinase to generate ceramide-1-phosphate (C1P).

Ceramide is broken down when its *N*-linked fatty acyl chain is removed to liberate sphingosine (catalyzed by CDase), which in turn is phosphorylated to generate sphingosine-1-phosphase (S1P) as catalyzed by sphingosine kinases (SK). S1P is broken down by sphingosine lyase to yield hexadecenal and ethanolamine phosphate.

Two isoforms of SK have been cloned and characterized in mammals: SK1 and SK2. These enzymes are emerging as important and regulated enzymes that not only modulate the levels of S1P, but also those of sphingosine and ceramide [11, 12]. S1P is a pro-inflammatory, pro-proliferative, and anti-apoptotic sphingolipid [12]. The dynamic balance between the levels of S1P and ceramide is tightly regulated by the activities of the enzymes that catalyze their formation and breakdown. The relative cellular levels of ceramide and S1P are proposed to influence cellular fate because of their opposing effects and their ability to be interconverted.

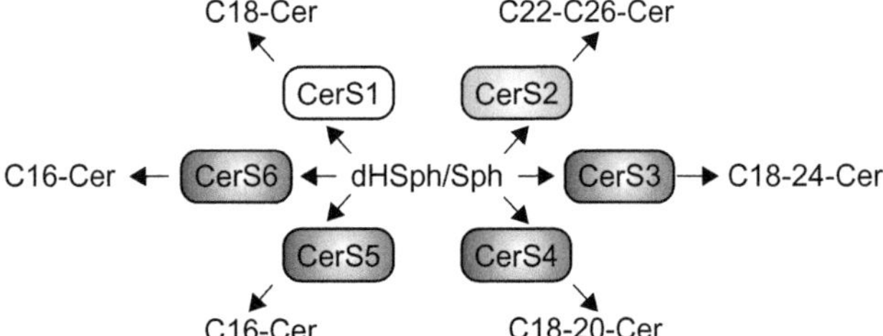

FIGURE 55.3 Ceramide synthase isoforms are responsible for generating particular types of ceramides. Six ceramide synthase isoforms have been identified in mammals. Each has preference for particular fatty acyl CoAs and thus are responsible for generating specific types of ceramides.

APOPTOSIS

Evidence for a Role of Ceramide in Apoptosis

There are a number of observations that support a pro-apoptotic role for ceramide. First, elevated cellular ceramide levels occur early in the apoptotic process [13, 14] in a variety of cell types and in response to a variety of apoptosis-inducing agents, including TNFα [15–18], UV irradiation [19, 20], ionizing radiation [21–25], serum withdrawal [26], etoposide [27], staurosporine [28], daunorubicin [29], taxol [30], and cisplatin [31–35]. Second, the effective doses of these agents required to induce ceramide generation closely matches the doses required to induce apoptosis [36]. Third, elevation in cellular ceramide in response to apoptotic agents has been shown to occur prior to the execution phase of apoptosis (i.e., upstream of caspases) [37–39]. Fourth, apoptosis can be inhibited upon blocking ceramide generation, and cells that are incapable of generating ceramide are often incapable of undergoing apoptosis [38, 40–42]. Fifth, exogenous addition of short-chain cell-permeable ceramide analogs induces apoptosis in a variety of cell lines [15, 43]. Finally, cancer cells can be induced to undergo apoptosis when endogenous ceramide levels are elevated via inhibitors of ceramide metabolism [24, 44], the addition of ceramide analogs [40, 42], or knock-down of enzymes of ceramide metabolism with small interfering RNA [45–49], highlighting the importance of ceramide in the initiation of apoptosis. While it is generally accepted that ceramide plays a role in apoptosis induction, the mechanism by which it does so is still highly debated.

Overview of Apoptosis

There are two main pathways for apoptosis, namely the extrinsic receptor-mediated pathway and an intrinsic mitochondrial pathway. The extrinsic pathway involves the binding of a death ligand (for example, tumor necrosis factor-α, TNFα) to its receptor (for example, TNF receptor-1) on the external surface of the plasma membrane, which leads to the formation of a death-inducing signaling complex (DISC). The DISC recruits multiple effectors, one of which is procaspase 8, which is cleaved and activated to caspase 8.

Ceramide has been proposed to play a role in the extrinsic form of cell death. There have been several pro-apoptotic receptors (for example, CD95 and DR5) that have been shown to activate acidic sphingomyelinase, resulting in its translocation to the cell surface and subsequent ceramide generation in the outer leaflet of the plasma membrane [50]. The formation of ceramide on the outer leaflet of the plasma membrane is thought to result in the formation of ceramide-enriched membrane platforms, which induce receptor clustering and amplification of the death signal [51, 52]. Disruption of these membrane rafts was shown to influence not only receptor clustering, but also DISC formation [53]. Other data indicate that stimulation of TNF receptors can induce activation of acidic sphingomyelinase and generation of ceramide in the endolysosomal compartment, which was shown to activate cathepsin D and mediate its translocation into the cytoplasm [54, 55]. In the cytoplasm, cathepsin D induces the cleavage of Bid to form t-Bid, which targets the intrinsic pathway of apoptosis through Bax and Bak activation (see below). There are also reports that ceramide activates caspase 8, which in turn cleaves Bid, resulting in its translocation to mitochondria [56].

The intrinsic pathway of apoptosis (Figure 55.4) involves intracellular stresses, including those that evoke DNA damage, endoplasmic reticulum (ER) stress, lysosomal stress, and mitochondrial dysfunction. These stresses engage several pro-death signaling pathways that converge on mitochondria [57], resulting in permeabilization of the mitochondrial outer membrane and release of proapoptotic proteins from the mitochondrial intermembrane space [58]. When released into the cytosol, these proteins ultimately lead to the activation of caspases 3, 6, and 7, which cleave proteins that are vital to cell function [59]. Once the caspase cascade is activated, cells display several features of apoptosis, including cleavage of PARP, externalization of plasma membrane phosphatidylserine, and chromatin condensation.

The Bcl-2 family of proteins regulates the initiation of apoptosis. There are both pro- and anti-apoptotic members in the Bcl-2 family that all share homology at one or more specific domains referred to as Bcl-2 homology domains (BH domains 1–4). The anti-apoptotic proteins include Bcl-2, Bcl-x_L, and Mcl-1, and their overexpression often rescues cells from cell death induced by several stimuli [60, 61]. The pro-apoptotic proteins are subdivided into two classes: the multi-domain proteins (for example, Bax and Bak) and BH3-only proteins (for example, Bid, Bad, Puma, Noxa, Bik). Cells lacking Bax and Bak are extremely resistant to death stimuli, highlighting their importance in cell death [62–65]. In healthy cells, Bax is a monomeric cytosolic and/ or nuclear protein; death signals, through unidentified mechanisms, result in its activation, translocation, and oligomerization (both with itself and other Bcl-2 proteins) at both mitochondrial and ER membranes. Bak is an ER and mitochondrial membrane protein in both its inactive and active forms. The activation of Bax and Bak is mediated in part by BH3-only pro-apoptotic Bcl-2 proteins, although the precise mechanism by which they do so is unclear. The initiation of apoptosis is governed by a critical balance between pro- and anti-apoptotic Bcl-2 family members, as well as by sphingolipids.

Ceramide and Mitochondria-Mediated Apoptosis

Many targets of ceramide have been identified *in vivo* that could indirectly alter mitochondrial function and lead to the initiation of apoptosis (Figure 55.5). In cells, elevated ceramide levels have been shown to alter the function of a number of proteins involved in the apoptotic process, including ceramide-activated protein kinase (CAPK [66]), ceramide-activated protein phosphatases PP1 and PP2A [67, 68], protein kinase C-ζ (PKCζ) [69,70], and the lysosomal protease cathepsin D [71]. Ceramide has been proposed to interact with PKCδ, resulting in its translocation to mitochondria and subsequent partial collapse in the

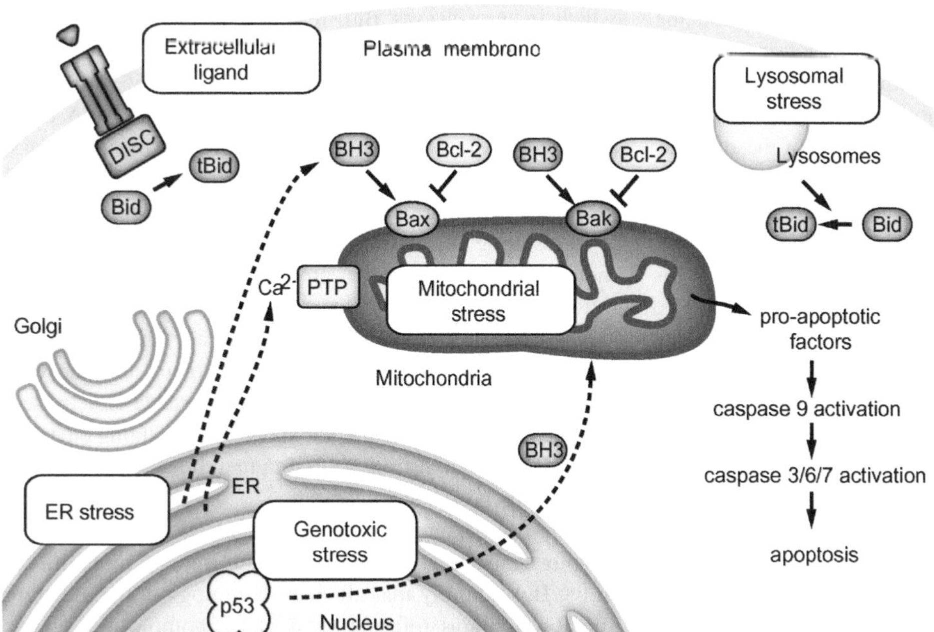

FIGURE 55.4 Several different types of cellular stresses result in activation of the mitochondrial cell death pathway.

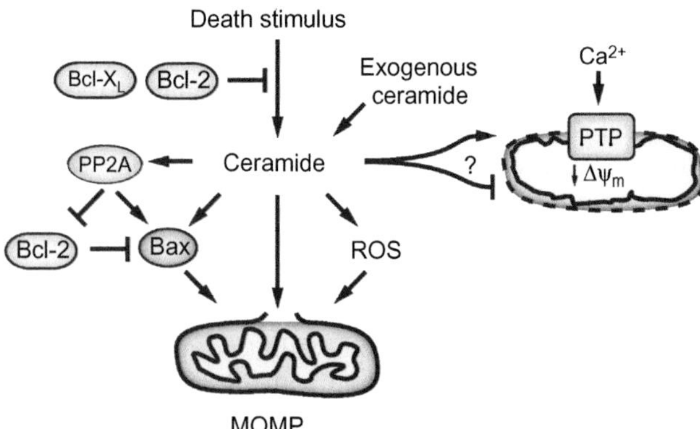

FIGURE 55.5 Ceramide interacts with Bcl-2 family members and mitochondria in a variety of ways that could explain its ability to promote mitochondrial outer membrane permeabilization (MOMP).

mitochondrial inner membrane potential [72, 73]. Multiple studies portray an intimate connection between ceramide signaling and mitochondrial function. Increases in cellular ceramide levels have been shown to occur prior to the mitochondrial phase of apoptosis [74–76], and are accompanied by the release of proteins from the mitochondrial intermembrane space, increased mitochondrial reactive oxygen species (ROS) generation, and collapse in the mitochondrial inner membrane potential [56, 77–79]. Interventions that suppress mitochondrial dysfunction also suppress ceramide-induced apoptosis, including bongkrekic acid [56,80] cyclosporine A [81], Bcl-2 [82, 83], and Bcl-x_L [84].

In vitro studies of the effect of ceramide on isolated mitochondria indicate what effects might be possible *in vivo*. Ceramides have been reported to have numerous effects on mitochondria (Figure 55.5), including enhanced ROS generation [17, 85, 86], alteration of Ca^{2+} homeostasis of mitochondria and ER [17, 87], ATP depletion [88], collapse in the inner mitochondrial membrane potential [85, 89], effects on various components of the mitochondrial electron transport chain [85, 90], inhibition of the permeability transition [91], release of intermembrane space proteins [89, 92, 93], and channel formation in the outer membrane [94]. In addition, ceramide-induced cytochrome c release and channel formation in the outer membrane are inhibited *in vitro* by Bcl-2 and Bcl-x_L [89, 94]. While these reductionist approaches are extremely valuable for determining *potential* mechanisms of action of ceramide, they do not identify *actual* mechanisms of action that occur *in vivo* unless verified *in vivo*.

Recent studies have questioned the existence of a mitochondrial pool of ceramide. When ceramide is generated specifically in mitochondria via targeting bacterial SMase (bSMase) to mitochondria, cytochrome c is released in a Bcl-2-inhibitable manner [95]. Alternatively, targeting bSMase to other subcellular compartments had no effect on apoptosis [95]. This study indicates a possible local action of ceramide on mitochondria in intact cells. Ceramide synthase and ceramidase activities have been detected in mitochondria enriched fractions [96–98]. In addition, ceramide is elevated in mitochondria enriched fractions during the initiation of apoptosis in response to PMA, CD95, TNFα, radiation, and UV-induced apoptosis [17, 20, 99–101]. Crude mitochondrial fractions contain not only mitochondria, but also contaminations from other fractions. Indeed, data indicate that ceramide can be generated directly in mitochondria *in vitro* [97], but it is unknown whether it is generated directly in mitochondria *in vivo*.

Ceramide and Bcl-2 Proteins

There are a number of studies that point to a complicated system of cross-talk between ceramide and the Bcl-2 family of proteins. Several theories exist for the underlying mechanisms of cross-talk between ceramide and Bcl-2 family proteins (Figure 55.5). One such theory is that anti-apoptotic Bcl-2 family proteins inhibit ceramide-mediated cell death. As stated above, ceramide-induced cytochrome c release is inhibited by the anti-apoptotic proteins Bcl-2 [82, 102] and Bcl-x_L [84]. However, there are conflicting reports in the literature as to whether this inhibition occurs via prevention of ceramide generation or inhibition of its downstream actions [82, 84, 95, 103]. Overexpression of Bcl-x_L in the B-lymphocyte cell line WEHI231 protected against ceramide-induced apoptosis, but not ceramide formation [84]. There have also been several reports that overexpression of Bcl-2 blocked ceramide-induced apoptosis without inhibiting ceramide generation, indicating that in some instances ceramide acts upstream of the anti-apoptotic Bcl-2 protein [82, 95, 104]. Alternatively, other studies *in vivo* have shown that Bcl-x_L inhibits ceramide-induced apoptosis by preventing ceramide accumulation [103]. Another theory for the mechanism of cross-talk between ceramide and Bcl-2 family proteins is that ceramide generation alters the expression patterns of Bcl-2 proteins and/or their splicing patterns. Ceramide elevations have been linked to an increase in the level of Bax [27, 105]. Ceramide elevations in response to gemcitabine has been shown to alter the splicing of Bcl-x to increase the pro-apoptotic variant Bcl-x_s [106]. Ceramide generation has also been shown to decrease protein and mRNA levels of Bcl-x_L [107]. The specific molecular mechanism responsible for these ceramide-induced changes in protein expression and/ or the splicing patterns of proteins is unknown. Due to its hydrophobic nature, they are most likely mediated by downstream targets of ceramide rather than by a direct effect of ceramide. Indeed, the ceramide-dependent decrease in protein and mRNA levels of Bcl-x_L was shown to be mediated by a ceramide-induced inhibition of protein kinase B activity [107], and the ceramide-induced increase in Bax protein levels was dependent on ERK and JNK activity [105]. An additional theory for mechanism(s) of cross-talk between ceramide and Bcl-2 family proteins is that generation of ceramide, either directly or indirectly, leads to the activation of pro-apoptotic Bcl-2 family members. Data indicate that elevations in ceramide levels precede and even influence translocation of Bax to the mitochondria and its consequent activation [49, 108–112]. Indeed, there was a report of ceramide-induced activation of Bax in *in vitro* studies as well, suggesting a potential direct interaction between ceramide and Bax [112]. For example, Kashkar and colleagues (2005) reported that *in vitro* C_{16}-ceramide addition to isolated mitochondria potentiated Bax insertion into mitochondrial membranes and release of intermembrane space proteins [112]. This group also reported that C_{16}-ceramide induced a conformational change in Bax in the presence of isolated mitochondria, but not with mitochondrial protein lysates or cytosolic fractions, suggesting the involvement of an additional unknown factor present in the mitochondrial membrane or membranes tightly associated with mitochondria thought to be of ER origin (MAMs) [112]. There are also several reports that ceramide influences the activation of pro-apoptotic Bcl-2 proteins through indirect mechanisms such as activation of the protein phosphatases PP1 and PP2A, which can lead to the dephosphorylation of Bcl-2, Bax, and Bad, and alter their function [113–115]. Thus, there is a variety of proposed mechanisms of cross-talk between ceramide and the Bcl-2 family of proteins that would all influence the induction of apoptosis.

The different mechanisms proposed for cross-talk between ceramide and proteins in the Bcl-2 family could be due to several factors. First, it may depend on the pathway of ceramide generation (i.e., SM hydrolysis, *de novo*, or recycling). Second, it could depend on the subcellular location of the ceramide, as ceramide generation has been observed in several subcellular compartments, including ER, mitochondria, lysosome, and plasma membrane. Likewise, Bcl-2 protein family members have been shown to act at sites other than just mitochondria, and therefore their site of action needs to be taken into account. Third, it could depend on the individual ceramide species. Indeed, ceramides of different chain lengths may contribute differently to the induction of apoptosis, and may interact differently with different Bcl-2 family members. Fourth, interactions between ceramide and any given Bcl-2 protein may be influenced by the expression levels of other Bcl-2 family members. A complex system of interactions between different members of the Bcl-2 family has been reported to influence the overall response of cells to apoptotic stimuli [116], and thus levels of other Bcl-2 family members need to be considered as well. Once we have a better understanding of the above issues, many of the varied reports of mechanisms of cross-talk between ceramide and Bcl-2 family members may provide a more coherent picture.

Ceramide and p53

The tumor suppressor p53 accumulates following genotoxic stress, and is responsible for maintaining the genome through initiation of a DNA repair mechanism, induction of cell cycle arrest, or induction of apoptosis [117–119]. In several systems, genotoxic stress has been shown to induce the accumulation of ceramide in a p53-dependent manner [120–122]. For example, in Molt-4 LXSN leukemia cells exposed to gamma irradiation and in EB-1 colon cancer cells treated with $ZnCl_2$, the p53-dependent ceramide generation was shown to be through *de novo* synthesis and with C_{16}-ceramide as the predominant ceramide species generated [122]. This group showed that the p53-dependent *de novo* synthesis was not accompanied with an increase in the activity of serine palmitoyl transferase, but rather ceramide synthase [122]. In Molt-4 LXSN cells, upregulation of p53 was shown to result in decreased activity of sphingosine kinase and its proteolytic cleavage [123]. As ceramide levels have been shown to be elevated following sphingosine kinase downregulation [49], p53 may regulate ceramide levels by more than one mechanism. However, others have reported p53-independent accumulations of ceramide following gamma irradiation in other cell lines [124]. In addition, exogenous addition of short-chain ceramide analogs to cells was reported to induce the accumulation of p53 [125]. The varied reports of the relationship between ceramide and p53 may be due to several factors. Ceramide generation following genotoxic stress may occur in both p53-dependent and -independent pathways, and this may be cell-type specific, or depend on the particular genotoxic stress employed and/or the degree of injury. In addition, the relationship between ceramide and p53 may depend on the pathway of ceramide accumulation and/or the subcellular compartment. Alternatively, ceramide may also function completely independently of p53 for some cell types, and in response to particular stress stimuli.

NON-APOPTOTIC CERAMIDE BIOLOGY

Autophagy is a means by which cells turn over cellular proteins and unwanted or damaged organelles. Autophagy is known to play an important role in cell maintenance, as well as in sustaining cells through periods of nutrient limitation. Autophagy has also been shown to play a role in a caspase-independent cell death. Recent work has shown that ceramide plays a role in induction of and regulation of autophagy [45, 126–129]. Exogenous addition of the short-chain ceramides and elevations in endogenous ceramides result in autophagy in a number of cell lines [45, 126–128, 130]. Several targets of ceramide are proposed to mediate its ability to induce autophagy, including nutrient transporter downregulation, PP2A, mTOR, beclin 1, Akt/PKB phosphorylation, and Bnip3 [45, 126–130].

A recent publication from the Edinger laboratory showed that ceramide-induced autophagy and eventual necrotic cell death was mediated by downregulation of nutrient transporters, which led to a bioenergetic catastrophe [129]. Ceramide-dependent toxicity was reversed when nutrient transporter downregulation was blocked or cells were supplied with cell-permeable nutrients [129]. Cells were sensitized to ceramide when autophagy was blocked, which indicates that the autophagy was a protective response of the cells to try and survive the downregulation of nutrient transporters. Cells were also sensitized to ceramide when extracellular nutrients were acutely limited or the AMP-activated protein kinase was deleted, indicating that the metabolic state of the cell influences the response to elevated ceramide levels [129].

Ceramide has been implicated in both cell cycle arrest and cellular senescence. Several studies have shown that ceramide generation is upstream of cell cycle regulators, and in some instances the generation of ceramide is necessary for the stimuli to induce the growth arrest. Ceramide is thought to induce cell cycle arrest through multiple targets, including dephosphorylation of the retinoblastoma gene product (Rb) [131], activation of the cyclin-dependent kinase inhibitor p21 [132], and inhibition of the cyclin-dependent kinase 2 (CDK2) [133]. Senescent cells have also been shown to have elevated ceramide levels as well as elevated neutral sphingomyelinase activity [134–136]. Ceramide may be involved in induction of cellular senescence through inhibition of phospholipase D/ protein kinase C (PLD/PKC), as this pathway is defective in senescent cells [137–140].

CONCLUSIONS

Ceramide has been suggested to play a number of regulatory roles in cells; however, it is not understood how one molecule can mediate so many diverse signaling events. Improved understanding of (1) how alterations of ceramide in particular subcellular locations impact downstream functions; (2) the functional significance of the individual ceramide species; and (3) how ceramide metabolism is regulated in response to individual stress stimuli, should permit more focused investigation into the molecular action of ceramide.

Much of the current literature on the role of ceramide in cell regulation involves the measurement of whole cell ceramides. However, ceramide can be generated in several subcellular compartments, and the location of its generation will determine its function. Ceramide is an extremely hydrophobic molecule, and does not spontaneously exit the hydrophobic membrane interior to enter the cytosol. Ceramide generated in one compartment can be transported to another compartment by way of vesicles or the transfer proteins (for example, CERT), or through metabolism to less hydrophobic metabolites (such as S1P or sphingosine) that are known to be more promiscuous than ceramide in partitioning between membrane and aqueous phases. These metabolites, once in another location, could be metabolized back into ceramide. However, aside from these few exceptions for transport of ceramide between intracellular locations, its site of generation will determine its site of action. Thus, future work aimed at determining the location of the ceramide generation for a particular stimulus needs to be taken into consideration.

Future work should also be aimed at understanding the functional significance of the six ceramide synthase isoforms and the individual ceramide species. It is not clear why mammals have six different CerS isoforms. However, it suggests that ceramides containing particular fatty acyl acids play important and distinct roles in cell regulation. This concept is just beginning to be explored with the development of more sensitive analytical techniques that can quantitatively measure ceramides of different chain lengths from cells and tissues. Once we gain an understanding of how the individual ceramide species contribute to cell regulation, then it will become clear how ceramides can mediate so many diverse signaling events.

A third area of much needed emphasis pertains to how the enzymes in the sphingolipid pathway are regulated in response to stress stimuli. Very little is known about what regulates these enzymes on the transcriptional, translational, and posttranslational levels. Understanding how ceramide metabolism is regulated will provide novel therapeutic targets for a variety of diseases for which ceramide and other members of the sphingolipid family have been shown to mediate, such as cancer, diabetes, heart disease, stroke, and neurodegenerative diseases.

REFERENCES

1. Hannun YA, Obeid LM. Principles of bioactive lipid signalling: lessons from sphingolipids. *Nat Rev Mol Cell Biol* 2008;**9**:139–50.
2. Clarke CJ, Snook CF, Tani M, Matmati N, Marchesini N, Hannun YA. The extended family of neutral sphingomyelinases. *Biochemistry* 2006;**45**:11,247–11,256.
3. Smith EL, Schuchman EH. The unexpected role of acid sphingomyelinase in cell death and the pathophysiology of common diseases. *FASEB J* 2008;**22**:3419–31.
4. Menaldino Jr DS, Bushnev A, Sun A, Liotta DC, Symolon H, Desai K, Dillehay DL, Peng Q, Wang E, Allegood J, Trotman-Pruett S, Sullards MC, Merrill AH. Sphingoid bases and de novo ceramide synthesis: enzymes involved, pharmacology and mechanisms of action. *Pharmacol Res* 2003;**47**:373–81.
5. Perry DK. Serine palmitoyltransferase: role in apoptotic *de novo* ceramide synthesis and other stress responses. *Biochim Biophys Acta* 2002;**1585**:146–52.
6. Mandon EC, Ehses I, Rother J, van Echten G, Sandhoff K. Subcellular localization and membrane topology of serine palmitoyltransferase, 3-dehydrosphinganine reductase, and sphinganine N-acyltransferase in mouse liver. *J Biol Chem* 1992;**267**:11,144–11,148.
7. Mao C, Obeid LM. Ceramidases: regulators of cellular responses mediated by ceramide, sphingosine, and sphingosine-1-phosphate. *Biochim Biophys Acta* 2008;**1781**:424–34.
8. Pewzner-Jung Y, Ben-Dor S, Futerman AH. When do Lasses (longevity assurance genes) become CerS (ceramide synthases)?: Insights into the regulation of ceramide synthesis. *J Biol Chem* 2006;**281**:25,001–25,005.
9. Riebeling C, Allegood JC, Wang E, Merrill Jr AH, Futerman AH. Two mammalian longevity assurance gene (LAG1) family members, trh1 and trh4, regulate dihydroceramide synthesis using different fatty acyl-CoA donors. *J Biol Chemistry* 2003;**278**:43,452–43,459,.
10. Mizutani Y, Kihara A, Igarashi Y. Mammalian Lass6 and its related family members regulate synthesis of specific ceramides. *Biochem J* 2005;**390**:263–71.
11. Taha TA, Mullen TD, Obeid LM. A house divided: ceramide, sphingosine, and sphingosine-1-phosphate in programmed cell death. *Biochim Biophys Acta* 2006;**1758**:2027–36.
12. Taha TA, Hannun YA, Obeid LM. Sphingosine kinase: biochemical and cellular regulation and role in disease. *J Biochem Mol Biol* 2006; **39**:113–31.
13. Hannun YA. Functions of ceramide in coordinating cellular responses to stress. *Science* 1996;**274**:1855–9.
14. Dbaibo GS, Perry DK, Gamard CJ, Platt R, Poirier GG, Obeid LM, Hannun YA. Cytokine response modifier A (CrmA) inhibits ceramide formation in response to tumor necrosis factor (TNF)-alpha: CrmA and Bcl-2 target distinct components in the apoptotic pathway. *J Exp Med* 1997;**185**:481–90.
15. Obeid LM, Linardic CM, Karolak LA, Hannun YA. Programmed cell death induced by ceramide. *Science NY* 1993;**259**:1769–71.
16. Modur V, Zimmerman GA, Prescott SM, McIntyre TM. Endothelial cell inflammatory responses to tumor necrosis factor alpha. Ceramide-dependent and -independent mitogen-activated protein kinase cascades. *J Biol Chem* 1996;**271**:13,094–13,102.
17. Garcia-Ruiz C, Colell A, Mari M, Morales A, Fernandez-Checa JC. Direct effect of ceramide on the mitochondrial electron transport chain leads to generation of reactive oxygen species. Role of mitochondrial glutathione. *J Biol Chem* 1997;**272**:11,369–11,377.
18. Geilen CC, Bektas M, Wieder T, Kodelja V, Goerdt S, Orfanos CE. 1alpha,25-dihydroxyvitamin D3 induces sphingomyelin hydrolysis in HaCaT cells via tumor necrosis factor alpha. *J Biol Chem* 1997;**272**:8997–9001.
19. Charruyer A, Jean C, Colomba A, Jaffrezou JP, Quillet-Mary A, Laurent G, Bezombes C. PKCzeta protects against UV-C induced apoptosis by inhibiting acid sphingomyelinase-dependent ceramide production. *Biochem J* 2007;**405**(1):77–83.
20. Dai Q, Liu J, Chen J, Durrant D, McIntyre TM, Lee RM. Mitochondrial ceramide increases in UV-irradiated HeLa cells and is mainly derived from hydrolysis of sphingomyelin. *Oncogene* 2004;**23**:3650–8.
21. Bruno AP, Laurent G, Averbeck D, Demur C, Bonnet J, Bettaieb A, Levade T, Jaffrezou JP. Lack of ceramide generation in TF-1 human myeloid leukemic cells resistant to ionizing radiation. *Cell Death Differ* 1998;**5**:172–82.
22. Chmura SJ, Nodzenski E, Kharbanda S, Pandey P, Quintans J, Kufe DW, Weichselbaum RR. Down-regulation of ceramide production abrogates ionizing radiation-induced cytochrome c release and apoptosis. *Mol Pharmacol* 2000;**57**:792–6.
23. Haimovitz-Friedman A, Kan CC, Ehleiter D, Persaud RS, McLoughlin M, Fuks Z, Kolesnick RN. Ionizing radiation acts on cellular membranes to generate ceramide and initiate apoptosis. *J Exp Med* 1994;**180**:525–35.
24. Rodriguez-Lafrasse C, Alphonse G, Aloy MT, Ardail D, Gerard JP, Louisot P, Rousson R. Increasing endogenous ceramide using inhibitors of sphingolipid metabolism maximizes ionizing radiation-induced mitochondrial injury and apoptotic cell killing. *Intl J Cancer* 2002;**101**:589–98.
25. Vit JP, Rosselli F. Role of the ceramide-signaling pathways in ionizing radiation-induced apoptosis. *Oncogene* 2003;**22**:8645–52.
26. Caricchio R, D'Adamio L, Cohen PL. Fas, ceramide and serum withdrawal induce apoptosis via a common pathway in a type II Jurkat cell line. *Cell Death Differ* 2002;**9**:574–80.
27. Sawada M, Nakashima S, Banno Y, Yamakawa H, Hayashi K, Takenaka K, Nishimura Y, Sakai N, Nozawa Y. Ordering of ceramide formation, caspase activation, and Bax/Bcl-2 expression during etoposide-induced apoptosis in C6 glioma cells. *Cell Death Differ* 2000;**7**:761–72.
28. Wiesner DA, Dawson G. Staurosporine induces programmed cell death in embryonic neurons and activation of the ceramide pathway. *J Neurochem* 1996;**66**:1418–25.
29. Come MG, Bettaieb A, Skladanowski A, Larsen AK, Laurent G. Alteration of the daunorubicin-triggered sphingomyelin-ceramide

pathway and apoptosis in MDR cells: influence of drug transport abnormalities. *Intl J Cancer* 1999,**81**.580–7.

30. Charles AG, Han TY, Liu YY, Hansen N, Giuliano AE, Cabot MC. Taxol-induced ceramide generation and apoptosis in human breast cancer cells. *Cancer Chemother Pharmacol* 2001;**47**:444–50.
31. Noda S, Yoshimura S, Sawada M, Naganawa T, Iwama T, Nakashima S, Sakai N. Role of ceramide during cisplatin-induced apoptosis in C6 glioma cells. *J Neurooncol* 2001;**52**:11–21.
32. Min J, Mesika A, Sivaguru M, Van Veldhoven PP, Alexander H, Futerman AH, Alexander S. (Dihydro)ceramide synthase 1 regulated sensitivity to cisplatin is associated with the activation of p38 mitogen-activated protein kinase and is abrogated by sphingosine kinase 1. *Mol Cancer Res* 2007;**5**:801–12.
33. Min J, Stegner AL, Alexander H, Alexander S. Overexpression of sphingosine-1-phosphate lyase or inhibition of sphingosine kinase in *Dictyostelium discoideum* results in a selective increase in sensitivity to platinum-based chemotherapy drugs. *Eukaryot Cell* 2004;**3**:795–805.
34. Saad AF, Meacham WD, Bai A, Anelli V, Elojeimy S, Mahdy AE, Turner LS, Cheng J, Bielawska A, Bielawski J, Keane TE, Obeid LM, Hannun YA, Norris JS, Liu X. The functional effects of acid ceramidase overexpression in prostate cancer progression and resistance to chemotherapy. *Cancer Biol Ther* 2007;**6**:1455–60.
35. Rebillard A, Tekpli X, Meurette O, Sergent O, LeMoigne-Muller G, Vernhet L, Gorria M, Chevanne M, Christmann M, Kaina B, Counillon L, Gulbins E, Lagadic-Gossmann D, Dimanche-Boitrel MT. Cisplatin-induced apoptosis involves membrane fluidification via inhibition of NHE1 in human colon cancer cells. *Cancer Res* 2007;**67**:7865–74.
36. Kolesnick RN, Kronke M. Regulation of ceramide production and apoptosis. *Annu Rev Physiol* 1998;**60**:643–65.
37. Kroesen BJ, Jacobs S, Pettus BJ, Sietsma H, Kok JW, Hannun YA, de Leij LF. BcR-induced apoptosis involves differential regulation of C16 and C24-ceramide formation and sphingolipid-dependent activation of the proteasome. *J Biol Chem* 2003;**278**:14,723–14,731.
38. Alphonse G, Aloy MT, Broquet P, Gerard JP, Louisot P, Rousson R, Rodriguez-Lafrasse C. Ceramide induces activation of the mitochondrial/caspases pathway in Jurkat and SCC61 cells sensitive to gamma-radiation but activation of this sequence is defective in radioresistant SQ20B cells. *Intl J Radiat Biol* 2002;**78**:821–35.
39. Basnakian AG, Ueda N, Hong X, Galitovsky VE, Yin X, Shah SV. Ceramide synthase is essential for endonuclease-mediated death of renal tubular epithelial cells induced by hypoxia-reoxygenation. *Am J Physiol Renal Physiol* 2005;**288**:F308–14.
40. Selzner M, Bielawska A, Morse MA, Rudiger HA, Sindram D, Hannun YA, Clavien PA. Induction of apoptotic cell death and prevention of tumor growth by ceramide analogues in metastatic human colon cancer. *Cancer Res* 2001;**61**:1233–40.
41. Chmura SJ, Nodzenski E, Beckett MA, Kufe DW, Quintans J, Weichselbaum RR. Loss of ceramide production confers resistance to radiation-induced apoptosis. *Cancer Res* 1997;**57**:1270–5.
42. Sautin Y, Takamura N, Shklyaev S, Nagayama Y, Ohtsuru A, Namba H, Yamashita S. Ceramide-induced apoptosis of human thyroid cancer cells resistant to apoptosis by irradiation. *Thyroid* 2000;**10**:733–40.
43. Cifone MG, De Maria R, Roncaioli P, Rippo MR, Azuma M, Lanier LL, Santoni A, Testi R. Apoptotic signaling through CD95 (Fas/Apo-1) activates an acidic sphingomyelinase. *J Exp Med* 1994;**180**:1547–52.
44. Alphonse G, Bionda C, Aloy MT, Ardail D, Rousson R, Rodriguez-Lafrasse C. Overcoming resistance to gamma-rays in squamous carcinoma cells by poly-drug elevation of ceramide levels. *Oncogene* 2004;**23**:2703–15.
45. Lavieu G, Scarlatti F, Sala G, Carpentier S, Levade T, Ghidoni R, Botti J, Codogno P. Regulation of autophagy by sphingosine kinase 1 and its role in cell survival during nutrient starvation. *J Biol Chem* 2006;**281**:8518–27.
46. Pchejetski D, Golzio M, Bonhoure E, Calvet C, Doumerc N, Garcia V, Mazerolles C, Rischmann P, Teissie J, Malavaud B, Cuvillier O. Sphingosine kinase-1 as a chemotherapy sensor in prostate adenocarcinoma cell and mouse models. *Cancer Res* 2005;**65**:11,667–11,675.
47. Liu YY, Han TY, Yu JY, Bitterman A, Le A, Giuliano AE, Cabot MC. Oligonucleotides blocking glucosylceramide synthase expression selectively reverse drug resistance in cancer cells. *J Lipid Res* 2004;**45**:933–40.
48. Bektas M, Jolly PS, Muller C, Eberle J, Spiegel S, Geilen CC. Sphingosine kinase activity counteracts ceramide-mediated cell death in human melanoma cells: role of Bcl-2 expression. *Oncogene* 2005;**24**:178–87.
49. Taha TA, Kitatani K, El-Alwani M, Bielawski J, Hannun YA, Obeid LM. Loss of sphingosine kinase-1 activates the intrinsic pathway of programmed cell death: modulation of sphingolipid levels and the induction of apoptosis. *FASEB J* 2006;**20**:482–4.
50. Grassme H, Schwarz H, Gulbins E. Molecular mechanisms of ceramide-mediated CD95 clustering. *Biochem Biophys Res Commun* 2001;**284**:1016–30.
51. Bollinger CR, Teichgraber V, Gulbins E. Ceramide-enriched membrane domains. *Biochim Biophys Acta* 2005;**1746**:284–94.
52. Gulbins E, Kolesnick R. Raft ceramide in molecular medicine. *Oncogene* 2003;**22**:7070–7.
53. Grassme H, Cremesti A, Kolesnick R, Gulbins E. Ceramide-mediated clustering is required for CD95-DISC formation. *Oncogene* 2003; **22**:5457–70.
54. Schneider-Brachert W, Tchikov V, Neumeyer J, Jakob M, Winoto-Morbach S, Held-Feindt J, Heinrich M, Merkel O, Ehrenschwender M, Adam D, Mentlein R, Kabelitz D, Schutze S. Compartmentalization of TNF receptor 1 signaling: internalized TNF receptosomes as death signaling vesicles. *Immunity* 2004;**21**:415–28.
55. Heinrich M, Wickel M, Winoto-Morbach S, Schneider-Brachert W, Weber T, Brunner J, Saftig P, Peters C, Kronke M, Schutze S. Ceramide as an activator lipid of cathepsin D. *Adv Exp Med Biol* 2000;**477**:305–15.
56. Lin CF, Chen CL, Chang WT, Jan MS, Hsu LJ, Wu RH, Tang MJ, Chang WC, Lin YS. Sequential caspase-2 and caspase-8 activation upstream of mitochondria during ceramideand etoposide-induced apoptosis. *J Biol Chem* 2004;**279**:40,755–40,761.
57. Yee KS, Vousden KH. Complicating the complexity of p53. *Carcinogenesis* 2005;**26**:1317–22.
58. Saelens X, Festjens N, Vande Walle L, van Gurp M, van Loo G, Vandenabeele P. Toxic proteins released from mitochondria in cell death. *Oncogene* 2004;**23**:2861–74.
59. Wolf BB, Green DR. Suicidal tendencies: apoptotic cell death by caspase family proteinases. *J Biol Chem* 1999;**274**:20,049–20,052.
60. Cory S, Huang DC, Adams JM. The Bcl-2 family: roles in cell survival and oncogenesis. *Oncogene* 2003;**22**:8590–607.
61. Kuwana T, Newmeyer DD. Bcl-2-family proteins and the role of mitochondria in apoptosis. *Curr Opin Cell Biol* 2003;**15**:691–9.
62. Wei MC, Zong WX, Cheng EH, Lindsten T, Panoutsakopoulou V, Ross AJ, Roth KA, MacGregor GR, Thompson CB, Korsmeyer SJ. Proapoptotic BAX and BAK: a requisite gateway to mitochondrial dysfunction and death. *Science* 2001;**292**:727–30.
63. Lindsten T, Thompson CB. Cell death in the absence of Bax and Bak. *Cell Death Differ* 2006;**13**:1272–6.

64. Ruiz-Vela A, Opferman JT, Cheng EH, Korsmeyer SJ. Proapoptotic BAX and BAK control multiple initiator caspases. *EMBO Rep* 2005;**6**:379–85.
65. Degenhardt K, Sundararajan R, Lindsten T, Thompson C, White E. Bax and Bak independently promote cytochrome C release from mitochondria. *J Biol Chem* 2002;**277**:14,127–14,134.
66. Basu S, Bayoumy S, Zhang Y, Lozano J, Kolesnick R. BAD enables ceramide to signal apoptosis via Ras and Raf-1. *J Biol Chem* 1998;**273**:30,419–30,426.
67. Chalfant CE, Kishikawa K, Mumby MC, Kamibayashi C, Bielawska A, Hannun YA. Long chain ceramides activate protein phosphatase-1 and protein phosphatase-2 A.Activation is stereospecific and regulated by phosphatidic acid. *J Biol Chem* 1999;**274**:20,313–20,317.
68. Dobrowsky RT, Hannun YA. Ceramide stimulates a cytosolic protein phosphatase. *J Biol Chem* 1992;**267**:5048–51.
69. Galve-Roperh I, Haro A, Diaz-Laviada I. Ceramide-induced translocation of protein kinase C zeta in primary cultures of astrocytes. *FEBS Letts* 1997;**415**:271–4.
70. Muller G, Ayoub M, Storz P, Rennecke J, Fabbro D, Pfizenmaier K. PKC zeta is a molecular switch in signal transduction of TNF-alpha, bifunctionally regulated by ceramide and arachidonic acid. *EMBO J* 1995;**14**:1961–9.
71. Heinrich M, Neumeyer J, Jakob M, Hallas C, Tchikov V, Winoto-Morbach S, Wickel M, Schneider-Brachert W, Trauzold A, Hethke A, Schutze S. Cathepsin D links TNF-induced acid sphingomyelinase to Bid-mediated caspase-9 and -3 activation. *Cell Death Differ* 2004; **11**:550–63.
72. Zeidan YH, Hannun YA. Activation of acid sphingomyelinase by protein kinase Cdelta-mediated phosphorylation. *J Biol Chem* 2007;**282**:11,549–11,561.
73. Grant S, Spiegel S. A chicken-or-egg conundrum in apoptosis: which comes first? Ceramide or PKCdelta? *J Clin Invest* 2002;**109**:717–19.
74. Rodriguez-Lafrasse C, Alphonse G, Broquet P, Aloy MT, Louisot P, Rousson R. Temporal relationships between ceramide production, caspase activation and mitochondrial dysfunction in cell lines with varying sensitivity to anti-Fas-induced apoptosis. *Biochem J* 2001;**357**:407–16.
75. Kroesen BJ, Pettus B, Luberto C, Busman M, Sietsma H, de Leij L, Hannun YA. Induction of apoptosis through B-cell receptor cross-linking occurs via de novo generated C16-ceramide and involves mitochondria. *J Biol Chem* 2001;**276**:13,606–13,614.
76. Thomas Jr RL, Matsko CM, Lotze MT, Amoscato AA. Mass spectrometric identification of increased C16 ceramide levels during apoptosis. *J Biol Chem* 1999;**274**:30,580–30,588.
77. Gentil B, Grimot F, Riva C. Commitment to apoptosis by ceramides depends on mitochondrial respiratory function, cytochrome c release and caspase-3 activation in Hep-G2 cells. *Mol Cell Biochem* 2003;**254**:203–10.
78. Hearps AC, Burrows J, Connor CE, Woods GM, Lowenthal RM, Ragg SJ. Mitochondrial cytochrome c release precedes transmembrane depolarisation and caspase-3 activation during ceramide-induced apoptosis of Jurkat T cells. *Apoptosis* 2002;**7**:387–94.
79. Andrieu-Abadie N, Gouaze V, Salvayre R, Levade T. Ceramide in apoptosis signaling: relationship with oxidative stress. *Free Radic Biol Med* 2001;**31**:717–28.
80. Stoica BA, Movsesyan VA, Lea PMt, Faden AI. Ceramide-induced neuronal apoptosis is associated with dephosphorylation of Akt, BAD, FKHR, GSK-3beta, and induction of the mitochondrial-dependent intrinsic caspase pathway. *Mol Cell Neurosci* 2003;**22**:365–82.
81. Pacher P, Hajnoczky G. Propagation of the apoptotic signal by mitochondrial waves. *EMBO J* 2001;**20**:4107–21.
82. Zhang J, Alter N, Reed JC, Borner C, Obeid LM, Hannun YA. Bcl-2 interrupts the ceramide-mediated pathway of cell death. *Proc Natl Acad Sci USA* 1996;**93**:5325–8.
83. Karasavvas N, Erukulla RK, Bittman R, Lockshin R, Hockenbery D, Zakeri Z. BCL-2 suppresses ceramide-induced cell killing. *Cell Death Differ* 1996;**3**:149–51.
84. Wiesner DA, Kilkus JP, Gottschalk AR, Quintans J, Dawson G. Anti-immunoglobulin-induced apoptosis in WEHI 231 cells involves the slow formation of ceramide from sphingomyelin and is blocked by bcl-XL. *J Biol Chem* 1997;**272**:9868–76.
85. Di Paola M, Cocco T, Lorusso M. Ceramide interaction with the respiratory chain of heart mitochondria. *Biochemistry* 2000;**39**: 6660–8.
86. Quillet-Mary A, Jaffrezou JP, Mansat V, Bordier C, Naval J, Laurent G. Implication of mitochondrial hydrogen peroxide generation in ceramide-induced apoptosis. *J Biol Chem* 1997;**272**:21,388–21,395.
87. Pinton P, Ferrari D, Rapizzi E, Di Virgilio F, Pozzan T, Rizzuto R. The Ca^{2+} concentration of the endoplasmic reticulum is a key determinant of ceramide-induced apoptosis: significance for the molecular mechanism of Bcl-2 action. *EMBO J* 2001;**20**:2690–701.
88. Arora AS, Jones BJ, Patel TC, Bronk SF, Gores GJ. Ceramide induces hepatocyte cell death through disruption of mitochondrial function in the rat. *Hepatology* 1997;**25**:958–63.
89. Ghafourifar P, Klein SD, Schucht O, Schenk U, Pruschy M, Rocha S, Richter C. Ceramide induces cytochrome c release from isolated mitochondria. Importance of mitochondrial redox state. *J Biol Chem* 1999;**274**:6080–4.
90. Gudz TI, Tserng KY, Hoppel CL. Direct inhibition of mitochondrial respiratory chain complex III by cell-permeable ceramide. *J Biol Chem* 1997;**272**:24,154–24,158.
91. Novgorodov SA, Gudz TI, Obeid LM. Long-chain ceramide is a potent inhibitor of the mitochondrial permeability transition pore. *J Biol Chem* 2008;**283**:24,707–24,717.
92. Di Paola M, Zaccagnino P, Montedoro G, Cocco T, Lorusso M. Ceramide induces release of pro-apoptotic proteins from mitochondria by either a Ca^{2+}-dependent or a Ca^{2+}-independent mechanism. *J Bioenerg Biomembr* 2004;**36**:165–70.
93. Siskind LJ, Kolesnick RN, Colombini M. Ceramide channels increase the permeability of the mitochondrial outer membrane to small proteins. *J Biol Chem* 2002;**277**:26,796–26,803.
94. Siskind LJ, Feinstein L, Yu T, Davis JS, Jones D, Choi J, Zuckerman JE, Tan W, Hill RB, Hardwick JM, Colombini M. Anti-apoptotic Bcl-2 family proteins disassemble ceramide channels. *J Biol Chem* 2008;**283**:6622–30.
95. Birbes H, El Bawab S, Hannun YA, Obeid LM. Selective hydrolysis of a mitochondrial pool of sphingomyelin induces apoptosis. *FASEB J* 2001;**15**:2669–79.
96. El Bawab S, Roddy P, Qian T, Bielawska A, Lemasters JJ, Hannun YA. Molecular cloning and characterization of a human mitochondrial ceramidase. *J Biol Chem* 2000;**275**:21,508–21,513.
97. Bionda C, Portoukalian J, Schmitt D, Rodriguez-Lafrasse C, Ardail D. Subcellular compartmentalization of ceramide metabolism: MAM (mitochondria-associated membrane) and/or mitochondria? *Biochem J* 2004;**382**:527–33.
98. Shimeno H, Soeda S, Sakamoto M, Kouchi T, Kowakame T, Kihara T. Partial purification and characterization of sphingosine N-acyltransferase (ceramide synthase) from bovine liver mitochondrion-rich fraction. *Lipids* 1998;**33**:601–5.
99. Vance JE. Phospholipid synthesis in a membrane fraction associated with mitochondria. *J Biol Chem* 1990;**265**:7248–56.

100. Matsko CM, Hunter OC, Rabinowich H, Lotze MT, Amoscato AA. Mitochondrial lipid alterations during Fas- and radiation-induced apoptosis. *Biochem Biophys Res Commun* 2001;**287**:1112–20.
101. Kitatani K, Idkowiak-Baldys J, Bielawski J, Taha TA, Jenkins RW, Senkal CE, Ogretmen B, Obeid LM, Hannun YA. Protein kinase C-induced activation of a ceramide/protein phosphatase 1 pathway leading to dephosphorylation of p38 MAPK. *J Biol Chem* 2006;**281**:36,793–36,802.
102. Geley S, Hartmann BL, Kofler R. Ceramides induce a form of apoptosis in human acute lymphoblastic leukemia cells that is inhibited by Bcl-2, but not by CrmA. *FEBS Letts* 1997;**400**:15–18.
103. El-Assaad W, El-Sabban M, Awaraji C, Abboushi N, Dbaibo GS. Distinct sites of action of Bcl-2 and Bcl-xL in the ceramide pathway of apoptosis. *Biochem J* 1998;**336**(3):735–41.
104. Allouche M, Bettaieb A, Vindis C, Rousse A, Grignon C, Laurent G. Influence of Bcl-2 overexpression on the ceramide pathway in daunorubicin-induced apoptosis of leukemic cells. *Oncogene* 1997;**14**:1837–45.
105. Oh HL, Seok JY, Kwon CH, Kang SK, Kim YK. Role of MAPK in ceramide-induced cell death in primary cultured astrocytes from mouse embryonic brain. *Neurotoxicology* 2006;**27**:31–8.
106. Chalfant CE, Rathman K, Pinkerman RL, Wood RE, Obeid LM, Ogretmen B, Hannun YA. De novo ceramide regulates the alternative splicing of caspase 9 and Bcl-x in A549 lung adenocarcinoma cells. Dependence on protein phosphatase-1. *J Biol Chem* 2002;**277**:12,587–12,595.
107. Hundal RS, Gomez-Munoz A, Kong JY, Salh BS, Marotta A, Duronio V, Steinbrecher UP. Oxidized low density lipoprotein inhibits macrophage apoptosis by blocking ceramide generation, thereby maintaining protein kinase B activation and Bcl-XL levels. *J Biol Chem* 2003;**278**:24,399–24,408.
108. Kim HJ, Mun JY, Chun YJ, Choi KH, Kim MY. Bax-dependent apoptosis induced by ceramide in HL-60 cells. *FEBS Letts* 2001;**505**:264–8.
109. Kim HJ, Oh JE, Kim SW, Chun YJ, Kim MY. Ceramide induces p38 MAPK-dependent apoptosis and Bax translocation via inhibition of Akt in HL-60 cells. *Cancer Letts* 2008;**260**:88–95.
110. Jin J, Hou Q, Mullen TD, Zeidan YH, Bielawski J, Kraveka JM, Bielawska A, Obeid LM, Hannun YA, Hsu YT. Ceramide generated by sphingomyelin hydrolysis and the salvage pathway is involved in hypoxia/reoxygenation-induced Bax redistribution to mitochondria in NT-2 cells. *J Biol Chem* 2008;**283**:26,509–26,517.
111. Birbes H, Luberto C, Hsu YT, El Bawab S, Hannun YA, Obeid LM. A mitochondrial pool of sphingomyelin is involved in TNFalpha-induced Bax translocation to mitochondria. *Biochem J* 2005;**386**:445–51.
112. Kashkar H, Wiegmann K, Yazdanpanah B, Haubert D, Kronke M. Acid sphingomyelinase is indispensable for UV light-induced Bax conformational change at the mitochondrial membrane. *J Biol Chem* 2005;**280**:20,804–20,813.
113. Lin SS, Bassik MC, Suh H, Nishino M, Arroyo JD, Hahn WC, Korsmeyer SJ, Roberts TM. PP2A regulates BCL-2 phosphorylation and proteasome-mediated degradation at the endoplasmic reticulum. *J Biol Chem* 2006;**281**:23,003–23,012.
114. Xin M, Deng X. Protein phosphatase 2 A enhances the proapoptotic function of Bax through dephosphorylation. *J Biol Chem* 2006;**281**:18,859–18,867.
115. Chiang CW, Harris G, Ellig C, Masters SC, Subramanian R, Shenolikar S, Wadzinski BE, Yang E. Protein phosphatase 2 A activates the proapoptotic function of BAD in interleukin-3-dependent lymphoid cells by a mechanism requiring 14–3–3 dissociation. *Blood* 2001;**97**:1289–97.
116. Adams JM, Cory S. Bcl-2-regulated apoptosis: mechanism and therapeutic potential. *Curr Opin Immunol* 2007;**19**:488–96.
117. Lowe SW, Ruley HE, Jacks T, Housman DE. p53-dependent apoptosis modulates the cytotoxicity of anticancer agents. *Cell* 1993;**74**:957–67.
118. Mummenbrauer T, Janus F, Muller B, Wiesmuller L, Deppert W, Grosse F. p53 Protein exhibits 3′-to-5′ exonuclease activity. *Cell* 1996;**85**:1089–99.
119. Vousden KH, Lu X. Live or let die: the cell's response to p53. *Nat Rev Cancer* 2002;**2**:594–604.
120. Dbaibo GS, Pushkareva MY, Rachid RA, Alter N, Smyth MJ, Obeid LM, Hannun YA. p53-dependent ceramide response to genotoxic stress. *J Clin Invest* 1998;**102**:329–39.
121. Sawada M, Nakashima S, Kiyono T, Nakagawa M, Yamada J, Yamakawa H, Banno Y, Shinoda J, Nishimura Y, Nozawa Y, Sakai N. p53 regulates ceramide formation by neutral sphingomyelinase through reactive oxygen species in human glioma cells. *Oncogene* 2001;**20**:1368–78.
122. Panjarian S, Kozhaya L, Arayssi S, Yehia M, Bielawski J, Bielawska A, Usta J, Hannun YA, Obeid LM, Dbaibo GS. De novo N-palmitoylsphingosine synthesis is the major biochemical mechanism of ceramide accumulation following p53 up-regulation. *Prostag Oth Lipid M* 2008;**86**:41–8.
123. Taha TA, Osta W, Kozhaya L, Bielawski J, Johnson KR, Gillanders WE, Dbaibo GS, Hannun YA, Obeid LM. Down-regulation of sphingosine kinase-1 by DNA damage: dependence on proteases and p53. *J Biol Chem* 2004;**279**:20,546–20,554.
124. Hara S, Nakashima S, Kiyono T, Sawada M, Yoshimura S, Iwama T, Banno Y, Shinoda J, Sakai N. p53-Independent ceramide formation in human glioma cells during gamma-radiation-induced apoptosis. *Cell Death Differ* 2004;**11**:853–61.
125. Kim SS, Chae HS, Bach JH, Lee MW, Kim KY, Lee WB, Jung YM, Bonventre JV, Suh YH. P53 mediates ceramide-induced apoptosis in SKN-SH cells. *Oncogene* 2002;**21**:2020–8.
126. Daido S, Kanzawa T, Yamamoto A, Takeuchi H, Kondo Y, Kondo S. Pivotal role of the cell death factor BNIP3 in ceramide-induced autophagic cell death in malignant glioma cells. *Cancer Res* 2004;**64**:4286–93.
127. Lavieu G, Scarlatti F, Sala G, Levade T, Ghidoni R, Botti J, Codogno P. Is autophagy the key mechanism by which the sphingolipid rheostat controls the cell fate decision?. *Autophagy* 2007;**3**:45–7.
128. Zeng X, Overmeyer JH, Maltese WA. Functional specificity of the mammalian Beclin-Vps34 PI3-kinase complex in macroautophagy versus endocytosis and lysosomal enzyme trafficking. *J Cell Sci* 2006;**119**:259–70.
129. Guenther GG, Peralta ER, Rosales KR, Wong SY, Siskind LJ, Edinger AL. Ceramide starves cells to death by downregulating nutrient transporter proteins. *Proc Natl Acad Sci USA* 2008;**105**:17,402–17,407.
130. Scarlatti F, Bauvy C, Ventruti A, Sala G, Cluzeaud F, Vandewalle A, Ghidoni R, Codogno P. Ceramide-mediated macroautophagy involves inhibition of protein kinase B and up-regulation of beclin 1. *J Biol Chem* 2004;**279**:18,384–18,391.
131. Lee JY, Leonhardt LG, Obeid LM. Cell-cycle-dependent changes in ceramide levels preceding retinoblastoma protein dephosphorylation in G2/M. *Biochem J* 1998;**334**(2):457–61.
132. Dbaibo GS, Pushkareva MY, Jayadev S, Schwarz JK, Horowitz JM, Obeid LM, Hannun YA. Retinoblastoma gene product as a downstream target for a ceramide-dependent pathway of growth arrest. *Proc Natl Acad Sci USA* 1995;**92**:1347–51.
133. Lee JY, Bielawska AE, Obeid LM. Regulation of cyclin-dependent kinase 2 activity by ceramide. *Exp Cell Res* 2000;**261**:303–11.

134. Mouton RE, Venable ME. Ceramide induces expression of the senescence histochemical marker, beta-galactosidase, in human fibroblasts. *Mech Ageing Dev* 2000;**113**:169–81.
135. Venable ME, Lee JY, Smyth MJ, Bielawska A, Obeid LM. Role of ceramide in cellular senescence. *J Biol Chem* 1995;**270**:30,701–30,708.
136. Venable ME, Webb-Froehlich LM, Sloan EF, Thomley JE. Shift in sphingolipid metabolism leads to an accumulation of ceramide in senescence. *Mech Ageing Dev* 2006;**127**:473–80.
137. Lee JY, Hannun YA, Obeid LM. Ceramide inactivates cellular protein kinase Calpha. *J Biol Chem* 1996;**271**:13,169–13,174.
138. Venable ME, Bielawska A, Obeid LM. Ceramide inhibits phospholipase D in a cell-free system. *J Biol Chem* 1996;**271**:24,800–24,805.
139. Venable ME, Blobe GC, Obeid LM. Identification of a defect in the phospholipase D/diacylglycerol pathway in cellular senescence. *J Biol Chem* 1994;**269**:26,040–26,044.
140. Venable ME, Obeid LM. Phospholipase D in cellular senescence. *Biochim Biophys Acta* 1999;**1439**:291–8.

Index

D

G

H

J

K

L

M

N

O

T

V

W

X

Y

Z

Printed and bound by CPI Group (UK) Ltd, Croydon, CR0 4YY
03/10/2024
01040318-0017